Biotransport: Principles and Applications

Robert J. Roselli · Kenneth R. Diller

Biotransport: Principles and Applications

Robert J. Roselli, Ph.D.
Vanderbilt University
Dept. Biomedical Engineering
Nashville, Tennessee
USA
robert.j.roselli@vanderbilt.edu

Kenneth R. Diller, Sc.D.
University of Texas, Austin
Dept. Biomedical Engineering
Austin, Texas
USA
kdiller@mail.utexas.edu

ISBN 978-1-4419-8118-9   e-ISBN 978-1-4419-8119-6
DOI 10.1007/978-1-4419-8119-6
Springer New York Dordrecht Heidelberg London

Library of Congress Control Number: 2011923229

© Springer Science+Business Media, LLC 2011
All rights reserved. This work may not be translated or copied in whole or in part without the written permission of the publisher (Springer Science+Business Media, LLC, 233 Spring Street, New York, NY 10013, USA), except for brief excerpts in connection with reviews or scholarly analysis. Use in connection with any form of information storage and retrieval, electronic adaptation, computer software, or by similar or dissimilar methodology now known or hereafter developed is forbidden.
The use in this publication of trade names, trademarks, service marks, and similar terms, even if they are not identified as such, is not to be taken as an expression of opinion as to whether or not they are subject to proprietary rights.

Printed on acid-free paper

Springer is part of Springer Science+Business Media (www.springer.com)

Foreword

The science of biotransport embraces the application of a large body of classical engineering knowledge of transport processes to the solution of problems in living systems covering a broad range of phenomena that are essential to homeostasis, and are encountered in routine experiences of human life, or in traumatic, diagnostic, or therapeutic contexts. Analyses of the transport of fluid, heat, and mass have been taught as fundamental components of engineering curricula for many decades, primarily with a focus on applications in industrial processes and design of various types of high-performance devices. The knowledge base that underpins this discipline derives from extensive high-quality, fundamental research conducted over the past century. Consequently, there have been hundreds of textbooks written for the instruction of undergraduates and graduates on the subject of transport processes.

In relatively recent times, a new arena of application for transport analysis has arisen dealing with processes in living systems. Although the fundamental physics of the governing transport phenomena remains unchanged, living systems tend to have constitutive properties that are quite distinct from those of typical inanimate systems, including anisotropy, complex geometries, composite materials, nonlinear dependence on state properties, and coupling across multiple energy domains. Therefore, it is important that bioengineering students be able to understand and appreciate the principles and subtleties of transport phenomena in the context of the types of problems that arise in their own field. The subject of biotransport is now widely accepted including the basics of fundamental transport science and the unique challenges that are encountered in dealing with biological cells, tissues, and organisms. A number of excellent texts have been published during the past decade that address various aspects of biotransport as a defined field of study.

The present text derives its genesis from a novel synthesis of cross-disciplinary intellect, that being the National Science Foundation sponsored Engineering Research Center (ERC) in Bioengineering Educational Technologies. This ERC was a multi-institutional consortium among Vanderbilt, Northwestern, Texas, and Harvard/MIT Universities (VaNTH) based on a collaboration among bioengineers, learning scientists, and learning technologists. It has the objective of developing state-of-the-art learning materials for students in bioengineering. One major outcome from this collaboration has been a refinement and application of the "How People Learn" (HPL) framework of student learning to higher education in the field of bioengineering. The HPL framework is explained in Part I of the text. The team of learning scientists and bioengineers has devised a set of materials designed to guide students into an experience of adaptive learning in which they gain expertise in both the core knowledge taxonomy of the subject material and the creative ability to innovatively apply the appropriate components of knowledge to solutions for problems presented in novel contexts.

This book is designed for use in either a traditional didactic style of course or in a learning environment achieved through a nontraditional course organization based on the HPL framework in which students are presented with a series of open-ended challenge problems. The suite of challenges is structured to drive learning through targeted components of the knowledge taxonomy while developing innovative problem definition and solving skills. HPL provides a learning context in which students can receive constructive formative feedback from instructors in both the knowledge and the innovation dimensions. It is important that the suite of challenge problems used for an entire course be constructed and choreographed to provide a logical progression through the knowledge taxonomy, although it is not at all necessary that the progression be linear in the traditional style of textbook organization.

The authors have accrued experience in teaching biotransport in the HPL format as a required core curriculum course in their home institutions over several years, and for an accrued total of more than 70 years. Extensive data have been gathered on student learning of the subject knowledge and on development of innovative analysis skills, and the data have been compared with control groups presented with the same subject material in a conventional didactic format. The results show the expected acquisition of knowledge along with a significant increase in innovative ability. Furthermore, surveys of student attitudes show that over the period of the course the students gained an understanding of the novel approach and an appreciation of how it helped them learn and prepare for their future careers, including lifelong learning. Further information on this research is presented in Chap. 1.

The text has been developed with dual objectives: to provide a coherent and concise pedagogical exposition of biotransport that includes the domains of fluid, heat and mass flows, and to present a guide for teaching and studying a core engineering subject in the HPL framework, with appropriate supporting materials for students and teachers. It is the authors' understanding that there is no other text that meets the latter objective.

The text is organized differently than standard transport textbooks. It is not designed to be a handbook of biotransport, where all aspects of a given topic are grouped together in sections or chapters. Instead, we have attempted to organize the text around principles for more effective learning. In Part I, we provide an extensive orientation for both instructors and students to the HPL framework. This provides a basis for understanding and appreciating the advantages of an active learning environment. The main portion of the text consists of an exposition of the taxonomy of knowledge in the field of biotransport. Part II presents enduring concepts and analogies that form the foundations of biotransport. In Parts III–V, these fundamentals are expanded in a progressive manner for momentum, heat, and mass transport. Each transport-specific section is further subdivided into four chapters.

The first chapter contains an expanded treatment of the fundamentals underlying the transport phenomena under consideration and treats topics unique to that transport mode (e.g., non-Newtonian fluids, radiation heat transfer, chemical reactions). The second chapter in each section deals with steady and unsteady-state transport in systems treated using a macroscopic approach, in which the focus of interest is on overall transport in a system, rather than on local property variations. Application of conservation principles in these problems leads to solutions involving algebraic equations or transient ordinary differential equations. We believe the first two chapters in each section will be of value not only to bioengineers, but also to those in the medical and life sciences. The third chapter deals with steady and unsteady-state transport in a single direction. For each problem, conservation principles are applied to a differential control volume. Steady-state solutions lead to ordinary differential equations or systems of ordinary differential equations. Unsteady-state applications lead to more complex partial differential equations that are first order in time and second order in position. The last chapter in each section of the text develops the general multidimensional microscopic transport equation(s) for that area. Our focus in these chapters is to identify situations when this more complex analysis is appropriate, how these general expressions can be simplified, how appropriate initial and boundary conditions can be specified, and how a limited number of important applications can be solved. These chapters also form the basis for more advanced studies in biotransport. In summary, the fundamental enduring concepts presented in Part II are reinforced in each transport-specific section, and are presented in an order that allows students to progressively analyze problems that are increasingly more difficult. Learning is further promoted by repeating this process for analogous aspects of momentum, heat, and mass transport.

A major objective of this text is to assist instructors in freeing up some of the time spent on classroom didactic lectures, thus allowing more teacher–student classroom contact opportunities for formative feedback, modeling problem definition and solution strategies, answering specific questions, and explaining difficult or subtle concepts. Therefore, we have endeavored to provide clear and complete explanations of the subject materials in the text along with numerous example problems with numerical solutions to help students learn effectively during self-study. We have included all intermediate steps in derivations to make it easier for students to follow along. Important equations have been highlighted to distinguish them from intermediate steps. We have avoided the use of tensors, which often are confusing for students who study introductory transport. The text includes extensive examples of various learning challenges that have been written by the authors for use in their own biotransport courses. The authors introduce physiological principles and data only to the extent that it is a requisite for learning the relevant biotransport principles. Likewise, they limit the coverage and depth of transport to the fundamentals necessary to achieve an integrated, working overview of the subject. There are numerous treatises that address both a broad physiological background and more comprehensive transport analysis.

The focus of this text is to cover the basics of biotransport sufficient for a stand-alone course on the subject in an undergraduate curriculum in the context of introducing and explaining an approach for students to learn the subject in the HPL framework. There is more material in this text that can be adequately covered in a single semester or quarter. Different institutions combine biotransport topics differently in their introductory courses. Some combine all three domains in a single course, others teach bioheat and biomass together with a separate course in biofluids, and others combine biofluids and biomass transfer. We have included enough material so that the text could be used for introductory semester courses in biofluid, biomass, and bioheat transfer. Some materials in the latter sections of the last two chapters of each section can be skipped in shorter courses. It is our hope that using this text would enable students to move more quickly and effectively along the pathway to becoming adaptive experts and productive practicing engineers. We expect that when students have completed a course using this text and learning method they will be able to demonstrate a breadth of knowledge across all three domains of biotransport and be able to sort and appropriately apply that knowledge to understand and solve problems in biotransport they have not encountered previously.

The text is also organized in a format that we hope will enable new adopters to move to the HPL framework with little required added investment of time beyond that associated with using any new teaching materials. We appreciate that there may be an upfront acculturation to understand HPL, but, based on our own experiences, this transition should lead to a teaching process that is no more demanding on a teacher's time than traditional pedagogical methods. The text should serve as a clear and effective resource for students to learn the basic components of the knowledge taxonomy for the subject so that a larger component of the faculty–student interaction can be focused on developing skills in adaptive thinking and solving open-ended problems.

The authors realize that many potential users of this text may not be ready to adopt the HPL framework for a complete course. In this context, we have tested many of the challenges as individual modules in both undergraduate and graduate courses. Colleagues at other institutions have done likewise. Our experience is that the challenges can be useful learning tools when used individually, and many faculty may find them to be quite helpful this way. Furthermore, this partial or progressive approach to adoption may provide a gradual pathway to using the HPL framework more fully. We have found that there is a considerable shift in the learning culture in which instructor and student mutually engage in this learning environment. In particular, for the HPL method to be effective, there needs to be an established level of trust and confidence of the students toward the teacher since the expectations for learning differ from the more traditional approach with which they are likely familiar that is more oriented toward memorization and repetition. Such a shift is not necessarily easy to effect in a step-wise manner. However, our experience is that the HPL framework can provide a much richer level of instructional interaction between faculty and students and that the level of enthusiasm exuded by the students in realizing a rapid learning curve toward adaptive expertise is rewarding for both student and teacher.

Developing a text that is compatible with learning in the HPL framework represents somewhat of a pioneering effort. We have tested and evolved the methodology with our own students and courses plus with some beta-phase adopters at sister institutions. It is certain that this approach will continue to be refined and improved; in that process, we hope that students will be enabled to learn with an enriched depth of understanding and perspective and that faculty will be stimulated to engage students in a community of learners and to acquire new and exciting dimensions in their careers as educators. We realize that the process of understanding the HPL methods and its implementation in higher education is an ongoing process requiring continuous improvement. Thus, we anticipate and request feedback on the structure and utility of this text.

We are most happy to acknowledge with a tremendous level of appreciation and gratitude our learning science colleagues, Professors Sean Brophy of Purdue University, Taylor Martin and Tony Petrosino of the University of Texas at Austin, and John Bransford of the University of Washington, for guiding us along the pathway of learning about the principles of HPL and applying these principles to our own teaching in biomedical engineering. We also thank Professor Jack Patzer at the University of Pittsburgh, Robby Sanders at Tennessee Tech University and Valerie Guenst at Vanderbilt University for their valuable assistance in reviewing the text, and Professor Todd Giorgio at Vanderbilt University for providing some end-of-chapter problems. Our colleague Thomas R. Harris of Vanderbilt University as leader of the VaNTH ERC has been a continual inspiration for us to engage in this endeavor. We have educated about 1,000 of our own students using early versions of this text and materials. The feedback and enthusiasm of these students has been highly motivational to us. Most importantly, we thank our wives Kathleen and JoAnn for their patient endurance, encouragement, and proofreading during the writing process over the past five years. And finally, we appreciate the continuing support of the editorial staff of the Springer Press.

March, 2011

Robert J. Roselli
Vanderbilt University

Kenneth R. Diller
The University of Texas at Austin

Contents

Part I Fundamentals of How People Learn (HPL)

1 Introduction to HPL Methodology 3
   1.1 Introduction 3
   1.2 Adaptive Expertise 4
   1.3 Learning for Adaptive Expertise 5
   1.4 Principles of Effective Learning 5
   1.5 Challenge-Based Instruction 6
   1.6 STAR.Legacy (SL) Cycle for Inquiry Learning 8
   1.7 Developing Innovation 9
       1.7.1 How to Use the Generate Ideas Model 9
       1.7.2 How to Use This Textbook to Develop Innovation 26
   1.8 Learning to Gain Understanding 27
   References 29

Part II Fundamental Concepts in Biotransport

2 Fundamental Concepts in Biotransport 33
   2.1 Introduction 33
   2.2 The System and Its Environment 34
   2.3 Transport Scales in Time and Space 35
       2.3.1 Continuum Concepts 37
   2.4 Conservation Principles 39
   2.5 Transport Mechanisms 40
       2.5.1 Molecular Transport Mechanisms 41
       2.5.2 Convective Transport Mechanisms 48
   2.6 Macroscopic Transport Coefficients 49
   2.7 Interphase Transport 52
   2.8 Transport in Biological Systems: Some Unique Aspects 56
   2.9 Summary of Key Concepts 58
   2.10 Questions 60
   2.11 Problems 61

    2.12 Challenges 64
    References 66

3 Modeling and Solving Biotransport Problems 67
    3.1 Introduction 67
    3.2 Theoretical Approach 68
        3.2.1 Geometric Considerations 69
        3.2.2 Governing Equations 70
        3.2.3 Solution Procedures 70
        3.2.4 Presentation of Results 72
        3.2.5 Scaling: Identification of Important Dimensionless
             Parameters 72
        3.2.6 Examples of the Theoretical Approach 75
    3.3 Empirical Approach 84
        3.3.1 The Buckingham Pi Theorem: Dimensional Analysis 84
    3.4 Summary of Key Concepts 94
    3.5 Questions 95
    3.6 Problems 96
    3.7 Challenges 103
    References 103

Part III Biofluid Transport

4 Rheology of Biological Fluids 107
    4.1 Introduction 107
    4.2 Solids and Fluids 107
    4.3 Flow Regimes: Laminar and Turbulent Flow 110
    4.4 Boundary Conditions 110
    4.5 Viscous Properties of Fluids 112
    4.6 Viscous Momentum Flux and Shear Stress 112
    4.7 Viscometers 115
    4.8 Newtonian and Non-Newtonian Fluid Models 119
        4.8.1 Newtonian Fluid Model 120
        4.8.2 Non-Newtonian Fluid Models 124
        4.8.3 Identification of Constitutive Model Equations .. 133
    4.9 Rheology of Biological Fluids 138
        4.9.1 Rheological Properties of Extravascular Body Fluids 139
        4.9.2 Blood Rheology 142
        4.9.3 Biorheology and Disease 154
    4.10 Summary of Key Concepts 158
    4.11 Questions 159
    4.12 Problems 161
    4.13 Challenges 165
    References 166

Contents

5 Macroscopic Approach for Biofluid Transport 169
5.1 Introduction 169
5.2 Conservation of Mass 169
5.3 Conservation of Momentum 180
5.4 Conservation of Energy 188
5.5 Engineering Bernoulli Equation 194
5.6 Friction Loss in Conduits 199
5.7 Friction Loss Factors, Flow Through Fittings 213
5.8 Laminar Flow and Flow Resistance in Noncircular Conduits . 223
5.9 Flow in Packed Beds 228
5.10 External Flow: Drag and Lift 230
5.11 Blood Flow in Microvessels 235
5.12 Steady Flow Through a Network of Rigid Conduits 237
5.13 Compliance and Resistance of Flexible Conduits 243
5.14 Flow in Collapsible Tubes 252
5.15 Fluid Inertia 261
5.16 Blood Flow in Organs 270
5.17 Osmotic Pressure and Flow 275
5.18 Summary of Key Concepts 289
5.19 Questions 291
5.20 Problems 294
5.21 Challenges 316
References 317

6 Shell Balance Approach for One-Dimensional Biofluid Transport .. 319
6.1 Introduction 319
6.2 General Approach 320
   6.2.1 Selecting an Appropriate Shell 321
   6.2.2 Fluid Mass Balance 322
   6.2.3 Fluid Momentum Balance 323
   6.2.4 Application of the Fluid Constitutive Relation
       to Find Fluid Velocity 328
   6.2.5 Examining and Applying Solutions for Shear Stress
       and Velocity 329
   6.2.6 Additional Shell Balances in Rectangular Coordinates . 332
6.3 One-Dimensional Shell Balances in Cylindrical Coordinates . 346
   6.3.1 Flow of a Newtonian Fluid Through a Circular Cylinder ... 346
   6.3.2 Flow of a Newtonian Fluid in an Annulus
       with Inner Wall Moving 356
   6.3.3 Flow Through an Inclined Tube or Annulus 359
   6.3.4 Flow of a Casson Fluid Through a Circular Cylinder . 362
   6.3.5 Osmotic Pressure and Flow in a Cylindrical Pore ... 366
6.4 Unsteady-State 1-D Shell Balances 373
6.5 Summary of Key Concepts 377

6.6	Questions	379
6.7	Problems	380
6.8	Challenges	387
References		388

7 General Microscopic Approach for Biofluid Transport 389
7.1 Introduction 389
7.2 Conservation of Mass 389
7.3 Conservation of Linear Momentum 391
7.4 Moment Equations 394
7.5 General Constitutive Relationship for a Newtonian Fluid 395
7.6 Substantial Derivative 398
7.7 Modified Pressure, $\wp$ 400
7.8 Equations of Motion for Newtonian Fluids 400
7.9 The Stream Function and Streamlines for Two-Dimensional Incompressible Flow 402
7.10 Use of Navier–Stokes Equations in Rectangular Coordinates 404
    7.10.1 Hydrostatics 404
    7.10.2 Reduction of the Equations of Motion 406
7.11 Navier–Stokes Equations in Cylindrical and Spherical Coordinate Systems 415
7.12 Use of Navier–Stokes Equations in Cylindrical and Spherical Coordinates 420
7.13 Scaling the Navier–Stokes Equation 436
7.14 General Momentum Equations for Use with Non-Newtonian Fluids 455
7.15 Constitutive Relationships for Non-Newtonian Fluids 457
    7.15.1 Power Law Fluid 459
    7.15.2 Bingham Fluid 461
    7.15.3 Casson Fluid 462
    7.15.4 Herschel–Bulkley Fluid 463
7.16 Setting Up and Solving Non-Newtonian Problems 464
7.17 Summary of Key Concepts 474
7.18 Questions 476
7.19 Problems 477
7.20 Challenges 484
References 485

Part IV Bioheat Transport

8 Heat Transfer Fundamentals 489
8.1 Introduction 489
8.2 Conduction 489
    8.2.1 Thermal Resistance in Conduction 492

8.3	Convection	493
	8.3.1 Four Principle Characteristics of Convective Processes	494
	8.3.2 Fundamentals of Convective Processes	495
	8.3.3 Forced Convection Analysis	503
	8.3.4 Free Convection Processes	514
	8.3.5 Thermal Resistance in Convection	522
	8.3.6 Biot Number	523
8.4	Thermal Radiation	524
	8.4.1 Three Governing Characteristics of Thermal Radiation Processes	524
	8.4.2 The Role of Surface Temperature in Thermal Radiation	524
	8.4.3 The Role of Surface Properties in Thermal Radiation	529
	8.4.4 The Role of Geometric Sizes, Shapes, Separation, and Orientation in Thermal Radiation	532
	8.4.5 Electrical Resistance Model for Radiation	539
8.5	Common Heat Transfer Boundary Conditions	547
8.6	Summary of Key Concepts	549
8.7	Questions	551
8.8	Problems	552
8.9	Challenges	554
References		556

9 Macroscopic Approach to Bioheat Transport ... 559

9.1	Introduction	559
9.2	General Macroscopic Energy Relation	559
9.3	Steady-State Applications of the Macroscopic Energy Balance	561
	9.3.1 Thermal Resistances	561
	9.3.2 Heat Transfer Coefficients	568
	9.3.3 Convective Heat Transport	574
	9.3.4 Biomedical Applications of Thermal Radiation	576
	9.3.5 Heat Transfer with Phase Change	587
9.4	Unsteady-State Macroscopic Heat Transfer Applications	588
	9.4.1 Lumped Parameter Analysis of Transient Diffusion with Convection	589
	9.4.2 Thermal Compartmental Analysis	595
9.5	Multiple System Interactions	598
	9.5.1 Convection: Multiple Well-Mixed Compartments	598
	9.5.2 Combined Conduction and Convection	601
	9.5.3 Radiation: Flame Burn Injury	602
	9.5.4 Human Thermoregulation	614
9.6	Summary of Key Concepts	618
9.7	Questions	619
9.8	Problems	620

	9.9	Challenges	624
	References		626

10 Shell Balance Approach for One-Dimensional Bioheat Transport ... 629
- 10.1 Introduction ... 629
- 10.2 General Approach ... 629
- 10.3 Steady-State Conduction with Heat Generation ... 630
  - 10.3.1 Steady-State Conduction with Heat Generation in a Slab ... 630
  - 10.3.2 Steady-State Conduction with Heat Generation in a Cylinder ... 633
  - 10.3.3 Steady-State Conduction with Heat Generation in a Sphere ... 639
- 10.4 Steady-State One-dimensional Problems Involving Convection ... 640
  - 10.4.1 Internal Flow Convection with a Constant Temperature Boundary Condition ... 642
  - 10.4.2 Internal Flow Convection with a Constant Heat Flux Boundary Condition ... 646
  - 10.4.3 Heat Exchangers ... 648
- 10.5 One-Dimensional Steady-State Heat Conduction ... 669
  - 10.5.1 Heat Conduction with Convection or Radiation at Extended Surfaces ... 669
  - 10.5.2 Heat Exchange in Tissue: Transient and Steady-State Pennes Equation ... 678
- 10.6 Transient Diffusion Processes with Internal Thermal Gradients ... 680
  - 10.6.1 Symmetric Geometries: Exact and Approximate Solutions for Negligible Heat Generation ... 682
  - 10.6.2 Semi-Infinite Geometry ... 692
  - 10.6.3 Graphical Methods ... 698
- 10.7 Summary of Key Concepts ... 709
- 10.8 Questions ... 711
- 10.9 Problems ... 712
- 10.10 Challenges ... 719
- References ... 720

11 General Microscopic Approach for Bioheat Transport ... 723
- 11.1 General Microscopic Formulation of Conservation of Energy ... 723
  - 11.1.1 Derivation of Conservation of Energy for Combined Conduction and Convection ... 723
  - 11.1.2 Simplifying the General Microscopic Energy Equation ... 726

Contents xvii

- 11.2 Numerical Methods for Transient Conduction: Finite Difference Analysis 729
  - 11.2.1 Forward Finite Difference Method 733
  - 11.2.2 Backward Finite Difference Method 747
- 11.3 Thermal Injury Mechanisms and Analysis 751
  - 11.3.1 Burn Injury 751
  - 11.3.2 Therapeutic Applications of Hyperthermia 761
- 11.4 Laser Irradiation of Tissue 764
  - 11.4.1 Distributed Energy Absorption 764
  - 11.4.2 Time Constant Analysis of the Transient Temperature Field 766
  - 11.4.3 Surface Cooling During Irradiation 769
- 11.5 Summary of Key Concepts 776
- 11.6 Questions 779
- 11.7 Problems 779
- 11.8 Challenges 783
- References 784

Part V Biological Mass Transport

12 Mass Transfer Fundamentals 789
- 12.1 Average and Local Mass and Molar Concentrations 789
- 12.2 Phase Equilibrium 795
  - 12.2.1 Liquid–Gas Equilibrium 795
  - 12.2.2 Liquid–Liquid, Gas–Solid, Liquid–Solid, Solid–Solid Equilibrium 803
- 12.3 Species Transport Between Phases 806
- 12.4 Species Transport Within a Single Phase 808
  - 12.4.1 Species Fluxes and Velocities 809
  - 12.4.2 Diffusion Fluxes and Velocities 810
  - 12.4.3 Convective and Diffusive Transport 811
  - 12.4.4 Total Mass and Molar Fluxes 812
  - 12.4.5 Molecular Diffusion and Fick's Law of Diffusion 817
  - 12.4.6 Mass Transfer Coefficients 829
  - 12.4.7 Experimental Approach to Determining Mass Transfer Coefficients 830
- 12.5 Relation Between Individual and Overall Mass Transfer Coefficients 840
- 12.6 Permeability of Nonporous Materials 842
  - 12.6.1 Membrane Permeability 842
  - 12.6.2 Vessel or Hollow Fiber Permeability 845
  - 12.6.3 Comparison of Internal and External Resistances to Mass Transfer 849
- 12.7 Transport of Electrically Charged Species 851

12.8 Chemical Reactions	855
12.8.1 Hemoglobin and Blood Oxygen Transport	858
12.8.2 Blood CO_2 Transport and pH	864
12.8.3 Enzyme Kinetics	866
12.8.4 Ligand–Receptor Binding Kinetics	872
12.9 Cellular Transport Mechanisms	876
12.9.1 Carrier-Mediated Transport	877
12.9.2 Active Transport	880
12.10 Mass Transfer Boundary Conditions	881
12.10.1 Mass or Molar Concentration Specified at a Boundary	881
12.10.2 Mass or Molar Flux Specified at a Boundary	882
12.10.3 No-Flux Boundary Condition	883
12.10.4 Concentration and Flux at an Interface	883
12.10.5 Heterogeneous Reaction at a Surface	883
12.11 Summary of Key Concepts	884
12.12 Questions	887
12.13 Problems	889
12.14 Challenges	895
References	896
13 Macroscopic Approach to Biomass Transport	**897**
13.1 Introduction	897
13.2 Species Conservation	897
13.3 Compartmental Analysis	901
13.3.1 Single Compartment	901
13.3.2 Two Compartments	910
13.3.3 Multiple Compartments	922
13.4 Indicator Dilution Methods	923
13.4.1 Stewart–Hamilton Relation for Measuring Flow Through a System	924
13.4.2 Volume Measurements	926
13.4.3 Permeability-Surface Area Measurements	927
13.5 Chemical Reactions and Bioreactors	934
13.5.1 Homogeneous Chemical Reactions	934
13.5.2 Heterogeneous Reactions	951
13.6 Pharmacokinetics	952
13.6.1 Renal Excretion	953
13.6.2 Drug Delivery to Tissue, Two Compartment Model	957
13.6.3 More Complex Pharmacokinetics Models	967
13.7 Mass Transfer Coefficient Applications	968
13.8 Solute Flow Through Pores in Capillary Walls	971
13.8.1 Small Solute Transport	972
13.8.2 Large Solute Transport Through Pores	973

	13.9	Summary of Key Concepts	980
	13.10	Questions	981
	13.11	Problems	983
	13.12	Challenges	1001
		References	1002
14	**Shell Balance Approach for One-Dimensional Biomass Transport**		1005
	14.1	Introduction	1005
	14.2	Microscopic Species Conservation	1005
	14.3	One-Dimensional Steady-State Diffusion Through a Membrane	1006
	14.4	1D Diffusion with Homogeneous Chemical Reaction	1014
		14.4.1 Zeroth Order Reaction	1014
		14.4.2 First-Order Reaction	1025
		14.4.3 Michaelis–Menten Kinetics	1031
		14.4.4 Diffusion and Reaction in a Porous Particle Containing Immobilized Enzymes	1032
	14.5	Convection and Diffusion	1041
		14.5.1 Conduits with Constant Wall Concentration	1042
		14.5.2 Hollow Fiber Devices	1045
		14.5.3 Capillary Exchange of Non-Reacting Solutes	1059
	14.6	Convection, Diffusion, and Chemical Reaction	1062
		14.6.1 Transcapillary Exchange of O_2 and CO_2	1062
		14.6.2 Tissue Solute Exchange, Krogh Cylinder	1072
		14.6.3 Bioreactors	1079
	14.7	One-Dimensional Unsteady-State Shell Balance Applications	1095
		14.7.1 Diffusion to Tissue	1095
		14.7.2 Unsteady-State 1D Convection and Diffusion	1116
	14.8	Summary of Key Concepts	1128
	14.9	Questions	1129
	14.10	Problems	1131
	14.11	Challenges	1147
		References	1148
15	**General Microscopic Approach for Biomass Transport**		1149
	15.1	Introduction	1149
	15.2	3-D, Unsteady-State Species Conservation	1149
		15.2.1 Comparison of the General Species Continuity Equation and the One-Dimensional Shell Balance Approach	1155
	15.3	Diffusion	1158
		15.3.1 Steady-State, Multidimensional Diffusion	1158
		15.3.2 Steady-State Diffusion and Superposition	1162

	15.3.3 Unsteady-State, Multidimensional Diffusion	1164
15.4	Diffusion and Chemical Reaction	1170
15.5	Convection and Diffusion	1175
	15.5.1 Steady-State, Multidimensional Convection and Diffusion	1179
15.6	Convection, Diffusion, and Chemical Reaction	1198
	15.6.1 Blood Oxygenation in a Hollow Fiber	1198
15.7	Summary of Key Concepts	1206
15.8	Questions	1207
15.9	Problems	1208
15.10	Challenges	1213
	References	1214

Appendix A Nomenclature 1217
Appendix B.1 Physical Constants 1236
Appendix B.2 Prefixes and Multipliers for SI units 1236
Appendix B.3 Conversion Factors 1237
Appendix C Transport Properties 1240
Appendix D Charts for Unsteady Conduction and Diffusion 1251

Index 1263

Part I
Fundamentals of How People Learn (HPL)

Chapter 1
Introduction to HPL Methodology

1.1 Introduction

The successful practice of engineering requires skills in both *technical expertise* and *innovation*. The new field of biomedical engineering is undergoing an exceptionally rapid intellectual evolution with new knowledge and applications being generated at an astounding pace. Thus, biomedical engineers need a solid understanding of fundamental principles that underpin the core knowledge in the discipline, such as in biotransport, and they also need to be able to be innovative in applying their knowledge as new opportunities and applications in this field arise.

The approach the authors have adopted in writing this textbook is designed explicitly to provide you with a clearly organized presentation of the principles of heat, mass, and momentum transport in living systems at an introductory level that you will hopefully find easy to access and apply in solving many types of biomedical engineering problems that you may encounter. At the same time, the textbook is designed to guide students to develop and hone adaptive skills while studying the subject of biotransport. The latter feature of this book may be unfamiliar to you because it requires that you practice being adaptive in your learning methods as well as in solving engineering problems. This approach is based on years of conducting cutting-edge research in engineering education in collaboration with some of the leading learning scientists in the world. We have labored very hard to devise and refine learning tools that work well for students to gain knowledge that is essential to the practice of biomedical engineering (in the case of this book, in biotransport) and to develop the ability to apply this knowledge adaptively to solve important and interesting new problems with which they may initially be unfamiliar. This two-dimensioned skill set will serve you exceptionally well in any direction you elect to pursue in your postgraduation career.

Our experience in teaching with this approach over the years is that some students are initially uncomfortable in a new learning environment and with a set of performance expectations that goes beyond the traditional approach of reading and studying textbook material and then reproducing it accurately on exams. In this textbook, we will be directing you to not only learn the knowledge associated with biotransport, but to also become adept in performing the initial thinking and reflection that are necessary to effectively define real-world problems so that they may be solved by applying appropriate subject knowledge materials. Our further experience is that undergraduate engineering students are able to master at least the initial phases of this rather daunting challenge and that they become enthusiastic about making significant progress toward becoming practicing engineers. Our educational objective for you is to develop technical expertise that can be adapted to solve real-world problems.

1.2 Adaptive Expertise

Learning scientists refer to the kind of skills and knowledge that employers value as *adaptive expertise* (Bransford et al. 2000; Hatano and Inagaki 1986; Schwartz et al. 2005). Being an adaptive expert means that you have sound knowledge skills and are innovative in your ability to apply them to solve new problems that you have not worked on previously. Knowledge and innovation are two different but complementary skills that are both important for defining and solving "real-world" types of problems. We believe your education should enable you to increase both your knowledge and innovation capabilities (Martin et al. 2006). Figure 1.1 shows a model for the process by which a person starts at a novice level of expertise and grows in knowledge and innovation to become an adaptive expert. Every student has a unique pathway for maturing toward adaptive expertise that depends on your own inherent capabilities, your prior educational background and life experiences, and the educational framework in which you study. It is in this latter aspect that we hope to provide a positive influence and resources for you in this textbook.

You are *efficient* in applying your *knowledge* base because you can use it quickly and appropriately to solve the problems, design the products, or approach the research questions needed in your job. You are *innovative* because you have the skills and

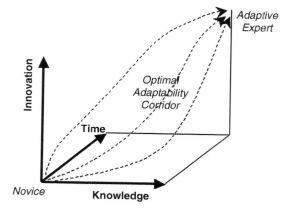

Fig. 1.1 Time trajectory for developing adaptive expertise based on skills in both knowledge and innovation (Martin et al. 2007b, with permission)

resources to discover what you need to know or be able to do to solve problems that are beyond your current knowledge and experience level. Learning scientists contrast adaptive expertise with *routine expertise*. Routine experts are as efficient as adaptive experts in the application of their knowledge. They can be accomplished in performing repetitive tasks that may actually be quite complex, but they are not innovative and tend to struggle in novel situations. Our objective is that you make good progress while studying biotransport toward being able to perform as an adaptive expert. That means that you need to add to both your knowledge and innovation capabilities.

1.3 Learning for Adaptive Expertise

Full adaptive expertise in any field takes much longer than an undergraduate career to develop. Many people estimate that it takes at least ten years of professional experience to become an expert (Anderson 1982). However, the way you learn in your undergraduate years can affect your ultimate effectiveness as the expert you become and how long it takes.

Traditional instruction in engineering is something you are very familiar with. It involves attending lectures, doing homework problem sets, and taking exams. In this approach, you are responsible for studying and learning the material presented, but not much else. This educational structure has plenty of benefits. Students receive a clear exposition of the information they need to learn, and they usually learn knowledge content well in a routine context. Teachers can be sure they have covered the required subject matter if they follow well-organized materials that are readily available in many textbooks. The expectations for student learning are very well defined for both the student and teacher.

However, this traditional approach has some disadvantages. Students sometimes learn material in a disconnected fashion and therefore have problems applying their knowledge in situations that are different from the classroom environment. They also tend to forget much of the material fairly quickly. Students can struggle to relate what they have learned to problems in the "real world" – at work or in graduate school. In other words, traditional instruction can help you develop routine knowledge efficiency, but it will not give you much practice with innovation. It frequently leads to routine expertise, but falls short in helping you to develop adaptive expertise.

1.4 Principles of Effective Learning

Several educators and learning scientists collaborated to write a report on learning for the National Research Council (The "How People Learn", or HPL, report; Bransford et al. 2000). This report recommends that effective learning environments follow four research-based principles.

1. These environments should be student centered, using the students' current capabilities as a starting point for learning.
2. They should be knowledge centered, focusing teaching on achieving mastery in the key content in the domain.
3. They should be assessment centered, building opportunities for students and teachers to get feedback on the students' progress throughout the learning process.
4. They should be community centered, appropriate to the discipline and the professional community context.

This textbook grows out of the efforts of a group of collaborating engineers and learning scientists who create instructional materials consistent with HPL principles and conduct research on how students learn using those materials. This group originally came together through the VaNTH (Vanderbilt, Northwestern, the University of Texas, and Harvard/MIT) National Science Foundation (NSF) Engineering Research Center (ERC) in Bioengineering Educational Technologies. We have developed a method of instruction called Challenge-Based Instruction, and we used this method to develop materials in several areas of bioengineering including bioimaging, biomechanics, biotransport, biotechnology, physiology, bio-optics, bioengineering ethics, and design. The results of more than five years of teaching in this format in all of the participating universities have been very positive (Cordray et al. 2009). Scientific assessments have documented that students learn core knowledge content as well as students in traditional educational environments, plus they make significant progress toward developing adaptive expertise. This textbook is the first to make use of HPL principles in the way and context in which subject knowledge is presented and applied to solve both routine and "real-world" problems.

1.5 Challenge-Based Instruction

This book is designed so that it can be adapted for many different styles of instruction. For example, the most common and traditional approach in engineering education is a didactic lecture format in which the teacher expounds all of the material that the students are expected to learn, either verbally and/or by writing on a board (or in more recent times via pre-prepared materials in digital format that can be projected and/or printed as handouts). This method has long been in use and has numerous demonstrated benefits (Schwartz and Bransford 1998). Students receive a clear exposition of the information they need to learn, teachers can be sure they have covered the content if they follow well-organized materials that are readily available, and students tend to learn content well as measured by performance on tests that replicate the content and context under which the material was presented. In this format, the teacher is clearly in the role of the expert and controls the flow of intellectual initiatives in the classroom. The responsibility of the students is to

1.5 Challenge-Based Instruction

receive and absorb the stream of knowledge from the teacher. However, there are drawbacks to the lecture approach. Students may learn the material in a disconnected fashion that makes it difficult for them to apply their knowledge out of context, and their long-term retention is often poor. Furthermore, students have difficulty in relating their accrued knowledge to problems in the "real world" – in the workplace or graduate school (Anderson 1982; Bransford et al. 2000).

An alternate teaching approach is to apply one of the several methods that can be grouped together under the moniker of Inquiry Learning. Problem- and Project-Based Learning, Authentic Inquiry, Challenge-Based Learning, and Discovery Learning are all examples of this approach (Prince and Felder 2006). There are many substantial differences among these methods, although all engage students in developing solutions to real-world problems that revolve around key concepts in the discipline. These approaches increase student motivation and awareness of the connections between their in-class experiences and their future work, lead to positive attitudes about learning for both students and teachers, and, when structured well, lead to significant increases in knowledge (Clough and Kaufmann 1999; de Jong 2006; Terezini et al. 2001). However, like traditional lecture, inquiry methods can have drawbacks as well. Students may have trouble structuring their approach to these open-ended problems if they have not also learned the fundamental principles for the subject and how to apply them with an effective analysis strategy (de Jong 2006). Thus, they may struggle with the processes needed such as hypothesis generation, defining appropriate systems for investigation, identifying the most relevant system variables and properties, and confining the breadth of their investigation to answer the question asked. Furthermore, if the application problems are not selected with adequate insight, the students may miss important concepts that they need to learn. Inquiry learning approaches that are not well structured may lead to students learning less than in traditional educational settings (Albanese and Mitchell 1993; Dochy et al. 2003).

We have worked extensively with the Challenge-Based Instruction method to provide a structured learning environment in which students do not miss key concepts and knowledge components that need to be gained from a particular course, and at the same time benefit from learning in the context of the challenge of solving open-ended real-world problems. We define a challenge as a real-world, open-ended problem, for which students do not have enough knowledge to solve when the challenge is first introduced. The desire to solve such a problem provides motivation for students to learn new material. This textbook is written to present a well-organized and easily readable treatise of knowledge concerning heat, mass and momentum transfer in living systems and biomedical devices and at the same time to present to students many opportunities for learning in a stimulating discovery-oriented environment. If the Challenge-Based Inquiry approach is adopted for an entire biotransport course, then it will be necessary to structure the course around a set of challenge problems that embrace the complete knowledge taxonomy that the students must learn. But, a strength of the Challenge-Based Instruction framework is that students are directed through a structured problem solving process that is scientifically documented to embody the crucial steps required for learning in the context of developing innovation. The key to this process is to adopt an approach for learning new material that ensures the challenge is knowledge centered, student centered assessment centered, and community centered. One way to accomplish this is to use an inductive learning method. We have adopted an inquiry sequence based on application of the STAR.Legacy (SL) Cycle as described in the following section.

1.6 STAR.Legacy (SL) Cycle for Inquiry Learning

Challenge-Based Instruction follows a cycle of inquiry called the SL Cycle (see Fig. 1.2: SL Cycle) (Schwartz et al. 1999). In the SL Cycle, instructors give students a realistic, complex problem (The Challenge). Students then generate ideas about what they already know and what they will need to learn to solve the challenge (Generate Ideas). Then students discover different views on important aspects of the challenge and key pieces of knowledge they will need to solve the challenge (Multiple Perspectives). Students might learn about multiple perspectives from

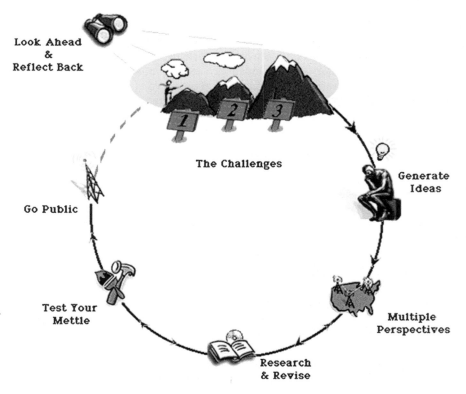

Fig. 1.2 Progressive steps of the STAR.Legacy Cycle for learning according to the How People Learn (HPL) educational framework (Martin et al. 2007b, with permission)

1.7 Developing Innovation

books, videos, a standard lecture, the internet, subject experts or other resources. Next, students revise their ideas, often in homework assignments (Research and Revise). Then students complete mid-challenge assessments with peers and/or the instructor (Test Your Mettle). These tests or assignments assess what students know and help students and instructors determine what students still need to learn. This information can inform students and instructors about how to revise prior to final assessments. Finally, students present their solutions to the challenge to the class (Go Public). At this stage, it is anticipated that the students will have the ability to transfer what they have learned to other situations. The process of solving a problem is often iterative such that at the end of one analysis cycle the understanding of the original challenge is refined so that a further analysis can be performed with a more accurate or complete perspective on addressing the key governing issues that are foundational to the problem.

Our research has shown positive outcomes for HPL-inspired Challenge-Based instruction. Students who study using this method learn core engineering knowledge just as well as students in traditional instruction courses, but they learn to innovate much better than students in traditional learning environments (Martin et al. 2005, 2007b; Pandy et al. 2004; Roselli and Brophy 2006). Challenge-Based instruction inspires students to solve complex problems (Pandy et al. 2004; Roselli and Brophy 2006), and they apply their knowledge more effectively in designing approaches to novel and difficult problems (Martin et al. 2007b). In Challenge-Based Instruction, students have the chance to practice innovation in a supporting context, while in traditional instruction students do not. We believe this difference accounts for the differences in outcomes found. Given this difference, it is worth considering how the SL Cycle helps students develop innovation. A key to the effectiveness of this approach is the Generate Ideas step of the SL Cycle (Martin et al. 2007a) as described in the next section.

1.7 Developing Innovation

1.7.1 *How to Use the* Generate Ideas Model

The *Generate Ideas Model* (GIM) is one of the most powerful and versatile tools available to guide you in solving engineering problems. We hope that you will learn to use the GIM in conjunction with studying this textbook. Moreover, we hope that the GIM will be become part of your enduring understanding of how to develop solutions for all types of engineering problems, and that it will serve you well in the coming years of your professional career. The authors consider that the GIM is one of the very few items you learn that is worth committing to long-term memory along with essentials such as the alphabet and multiplication tables. (This principle of becoming so familiar with a small number of critical pieces of information so that they may be recalled without effort is attributed to the renowned Greek scholar *Myosotis Palustris* and was reapplied to modern scholarship in a highly effective

Table 1.1 Steps in the Generate Ideas Method (GIM)

Component	Implementation	Description
Initial considerations	Formulate problem	Collect initial insights concerning the problem
Analysis	Define system	Ensure system is relevant to the process of interest
	Identify environmental interactions	Consider all exchanges between system and environment
	Identify conservation laws	Apply conservation of mass, momentum and energy, as appropriate
	Identify constitutive equations	Apply the appropriate empirical relationships for terms in the conservation equations
	Examine solution methods	Solve resulting equations, find properties, identify initial and boundary conditions
What to do next	Assess outcome	Evaluate what you still need to know about the solution and where to find this information

manner in the mid-twentieth century by the famous MIT professor Jacob Pieter Den Hartog[1] under whom one of the authors (KRD) had the privilege of studying.) Thus, the GIM should become part of your fundamental engineering toolbox that you can put to use without the need for reference to any external resources. It can provide a useful and effective perspective by which you can approach with confidence the solution of new problems that you may encounter.

We illustrate how to use the GIM via many example problems throughout the textbook. As an introduction, the following commentary is provided on how to follow the steps of the GIM, and what the outcome should be when the GIM is used effectively. Two examples are shown at the end of this section to illustrate its use. There are three major components of the GIM: *Initial Considerations*, *Analysis*, and *What To Do Next*, as shown in Table 1.1. The GIM is not a rigid set of rules that must be followed. Rather, it is a guide for a methodical approach to solving new problems. You will observe in the examples presented in this text that we will frequently depart from an exact replication of the GIM, depending on how a particular problem statement is presented and what the solution needs are. We do not mean to imply with the GIM that "one size fits all" for solving engineering problems. Nonetheless, the problem solving principles embodied therein will serve you well in addressing most engineering challenges that you may encounter.

A large majority of engineering problems involve a process that occurs while a system is changing over time, and/or there are internal gradients in important properties such as concentration, pressure, and temperature that cause transport to occur. As the process progresses, certain key features of a system of interest are altered. The features are termed the *properties of the system*. During *Initial*

---

[1]Den Hartog (1961) refers the reader to primary sources to verify the pedigree of Dr. Myosotis, whose principle he applied in his famous *Myosotis Method* of beam deflection analysis.

1.7 Developing Innovation

Considerations, the objective is to develop as accurate an understanding of the process as possible including what properties of the system are being affected. It is important to be very specific about as many of the details of the process as you can identify and to evaluate which aspects of the process are most influential on the direction and outcome of the process. The better the job you do initially in characterizing the process, the better your subsequent analysis of the problem will be. Alternatively, if you take a wrong direction in describing the process, later you will either need to go back and redo the analysis with a corrected direction in your analysis, or the results you obtain will be of limited relevance. Thus, it is an excellent investment of a relatively small amount of your time up front to engage in a clear consideration of exactly what are the key aspects of the problem you are attempting to solve. As you gain experience, the Initial Considerations step will become increasingly productive for you.

Analysis is the component of the GIM wherein you develop the equation or set of equations that you will use to describe the process of interest. The GIM will guide you through a logical series of steps to identify the governing equations for the process and how to solve them. The first two steps are to: (1) define a system and (2) identify how it interacts with its environment so that the process identified in *Initial Considerations* is integral to system/environment interaction. If the system and environmental interaction are defined without embedding the process of interest, then any subsequent analysis will not issue in relevant results. In the third step (3), the conservation laws that apply to the process are written with an emphasis on including all terms that contribute to the process. In the fourth step (4), each term is expressed in terms of its constitutive equation (the form of these equations is introduced in Chap. 2). The result is a differential or algebraic equation for each applied conservation law. The last step (5) is to define the initial and boundary conditions for the differential equation(s) and to determine the values for all the constitutive properties. At this point, it may be possible to solve the governing equation(s), but often you will not have all the information necessary to set up the equations, to determine all the needed data, or to identify the method for solving the equations. These issues are addressed in the next step. Nonetheless, you will have made substantial progress in establishing a framework for solving your problem.

What To Do Next may appear to be an unusual aspect of solving a problem, but it is a very important step in achieving an acceptable solution to a difficult open-ended challenge. The crux of this component is based on realizing what you do not know that you still need to learn in order to complete the solution of the problem. Learning scientists call the ability to engage in this step *metacognition,* meaning "knowing about knowing," or having a self-awareness about what you know (Wiggins and McTighe 2005). This capability is extremely useful to avoid "getting stuck" while working on a problem about which you have very little familiarity. It may initially seem uncomfortable for you to admit that you do not know everything necessary to solve a specific problem. However, an indicator of maturity and wisdom is the ability to recognize our limitations in a particular situation and to chart a course of action to move forward to reach a solution. To be able to provide substance and direction to our learning needs is one of the most empowering experiences we can have as a student (or professor). It is an ability that will serve you well over a lifetime of learning.

What To Do Next is based on having a "big picture" perspective on how to solve all kinds of engineering problems, and on realizing what pieces of the solution are missing from what you have accomplished by the completion of the *Analysis* component. As with the *Initial Considerations* step, *What To Do Next* requires that you give focused thought to the most fundamental and governing aspects of the problem. At first, you may not realize many aspects of what else there is to be learned about solving a problem. As you continue to work on a problem and understand more fully the nature of the process and the laws that govern it, more details for further analysis should become clear. Part of the *What To Do Next* step involves determining the sources for added information and resources. Many real-world problems involve iterative solution methods in which an initial analysis may be based on highly restrictive assumptions (which you will have made in the preceding steps), which you may go back to revisit and modify to obtain a solution that provides more relevant or complete information about the problem. The *What To Do Next* step is highly useful in guiding you through the iterative solution process.

As you hopefully realize, the GIM provides a rational and explicit framework for you to develop a solution for a problem you are addressing. For problems that are highly prescribed in great detail, GIM is of limited value because a prescriptive problem has much of the up-front thinking and assumption making already completed so that the solution may be simply a matter of "plug and chug" with an obvious equation. However, in general this is not the way the real world works. Many problems do not fit directly into obvious categories associated with explicit solutions. Therefore, the most important aspect of obtaining a solution that is meaningful and matches the need for information requires developing and mapping a strategy for solving the problem. Under these conditions, the GIM will be an effective tool for you in characterizing the nature of the problem you are solving, making appropriate assumptions, developing governing equations, and identifying added steps and further information that must be obtained to come to the final analysis required.

The steps for structuring your approach to solving nearly any biotransport problem (or for that matter, nearly any engineering problem) have been spelled out in terms of the GIM. Our expectation is that as you use this book in your course, you will develop a level of skill and confidence that will serve you well as you formulate strategies for solving all kinds of problems that you may encounter later in your education and in your professional career. Although it is not intuitively obvious, the most critical step in structuring a solution is the first one: defining the most appropriate system for the physical process you are considering. If your problem is specified to the extent that you understand what information about a particular behavior or feature you are required to determine with the solution, then you must define a system such that its interactions with its environment embody the processes relevant to that feature or behavior. It is all too easy to select a system such that the process of interest occurs totally on the interior so that it cannot be identified at a boundary. As a consequence, when the conservation equation(s) is

1.7 Developing Innovation

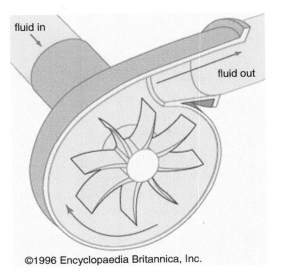

Fig. 1.3 Operation of a centrifugal pump with axial fluid flow in and radial fluid flow out as caused by a powered rotating impeller

©1996 Encyclopaedia Britannica, Inc.

written for the interaction between the system and environment the key process to be studied will not be included in the analysis, and it will miss the point of solving the problem. The following examples may help clarify this issue so that you can better define relevant systems on your own. They illustrate the GIM process for nonprescriptive problems that require significant thinking and assessment to determine an effective solution strategy.

Example 1.7.1.1. The Importance of Accurately Defining a System for Analysis.
Problem statement (for this example we also incorporate the Initial Considerations step): In this example, we will only discuss how to effectively define a system with its boundary in order to be able to analyze a specific process, illustrating that it is very important to initially give very careful consideration in choosing a system and boundary that match the actual problem to be solved.

A key component of a heart lung machine is the pump that supplies pressure and flow to the blood as it returns to the body. Over the years since development of the initial heart–lung machines, many different types of pumps have been tested and used. One of the most successful operational categories is the centrifugal pump. A centrifugal pump works on the principle shown in Fig. 1.3 in which fluid enters axially near the center and is accelerated by a spinning rotor to then exit at a higher velocity and pressure from the periphery. These pumps have several advantages in that they do not require valves, and their power consumption is relatively low. The flow and pressure are altered by adjusting the rate at which the rotor is spun. The pictures in Fig. 1.4 show a Medtronic Bio-Pump©. The Medtronic design is unique in that it does not use impeller blades to move the flow, which would create a lot of turbulence and could damage the blood cells by impacting them mechanically. Rather there is a set of three spinning nested cones with a large surface area that apply a viscous drag directly to the blood to cause a radial acceleration. You will learn in Chap. 4 that all fluids have a property called viscosity that causes a resistance to flow called the drag. Therefore, when the pump cone rotates the fluid adjacent to it tends to stick to the cone and be dragged along with it. It is by this mechanism that the rotating cone is able to cause the blood to move through the pump. The top two cones have openings in the center to allow for some of the inlet flow to move to one of the lower cones. This device can produce flows of 5–7 L/min with a pressure increase of 100 mm Hg at rotation rates of 1,400–3,000 rpm. The rotor is turned by a magnetic motor so that there is not the need for a mechanical shaft with seals. The ultimate goal of this problem could be to select the size and number of cones necessary to deliver a particular flow if inlet pressure, outlet pressure, and cone angular velocity are all known.

Solution. *System definition and environmental interactions:* This pump is a device for which performance can be analyzed using biotransport principles. An analysis strategy can be outlined using the GIM. Let us initially consider alternatives for how a system can be defined to analyze the performance of the pump.

When beginning students first evaluate an appropriate system to be analyzed, they most often choose the entire pump as shown in Fig. 1.5, with environmental interactions consisting of an axial blood flow in, a radial blood flow out, and work input to drive the rotor as it moves blood through the pump. Although this choice may seem logical, we need to evaluate what information we can learn about the pump function from this system representation. The key operating feature of this pump is that the rotor interacts with the blood to increase its pressure via the viscous drag on its considerable surface area as it rotates. The rotor and blood viscous interaction is essential to understanding the functioning of the pump. Any model of the pump function that is to be useful in designing the internal geometric features, rotational speed, shape of the rotors, and other characteristics that are essential to how it pumps blood must include this interaction between the rotating cone surface and the blood. A system with the boundary around the entire pump does not enable us to identify and analyze any of these pump features and interactions since they are all internal to the system boundary. Although a system with the boundary around the entire pump could be used to identify the relationship between the power input to the motor and the rate of blood flow being pumped, the actual mechanisms by which the pump operates could only be viewed as a black box, with no way to identify interactions internal to the boundary. None of the critical operating features of the pump occur across a boundary drawn around the entire pump. Cone–blood

1.7 Developing Innovation

Fig. 1.4 Medtronic centrifugal flow blood pump with an internal three part rotor that serves as the impeller (Kay and Munsch 2004, with permission)

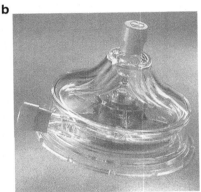

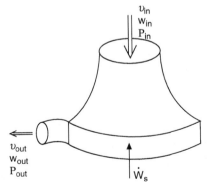

Fig. 1.5 System with boundary surrounding entire pump

interactions will not appear across a boundary of a system defined as the whole pump, and there will not be explicit terms in the governing conservation equations that describe these interactions. The result is that a system consisting of the entire pump will not lead us to an analysis that describes the essential elements of the operation, and it will be of very limited value for analysis and design of the pump.

If a system consisting of the entire pump is not the best choice, then there must be alternatives that will provide us with the type of information we want. Since the operating principle of the pump utilizes the viscous drag between the turning rotor and the blood, we should look for a system with a boundary that corresponds to the blood/rotor interface so that one of the environmental interactions is the drag force. There are two primary options that meet this criterion. One option is to define a closed system consisting of the rotor, Fig. 1.6. The environmental interactions are the drag force (F_{drag}) exerted by the viscous blood as it moves over the rotor surface (see Chap. 2) and the power input ($\dot{W}$) from the magnetic motor to turn the rotor at a speed ($\dot{\theta}$). The second option is to define an open system consisting of the internal volume of the pump through which the blood flows, Fig. 1.7. The environmental interactions for this system consist of the axial blood flow in, the radial blood flow out, both at identified pressures, velocities, and flow rates, and the viscous drag force (F_{drag}) applied to the blood by the turning rotor. We could further evaluate the flow in this system by looking at a microscopic shell boundary to determine the velocity profile of the blood between the rotating cones and then determining the pressure and velocity terms as functions of the rotor speed and geometry and the axial and radial position coordinates in the pump. The boundary interactions between the blood and rotor are equal and opposite, enabling us to couple the viscous flow pattern of blood through the pump with the power input requirements from the motor. Because the

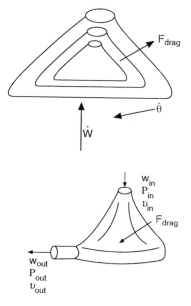

Fig. 1.6 System consisting of only the rotor

Fig. 1.7 System with boundary defined by the internal volume of the pump through which the blood flows

1.7 Developing Innovation

latter two systems both are defined to have a boundary at the viscous interface between the rotor and blood, they are much more useful than a model of the entire pump for deriving a model for the operation of the pump.

It may be that you start with a system that includes the entire pump. Later in the analysis when you are developing equations for the pump function you may have the realization that it is necessary to include the viscous interaction between the blood and the rotor. You might either write some equations that describe this viscous interaction, which would be inconsistent with the way your system is defined, or you will have to iterate back to the beginning of the GIM process and redefine your system to provide for addressing the viscous interaction.

When we give thoughtful consideration at the beginning of the GIM process about what we want to learn from the analysis and how we want to use the resulting model, it will help to avoid false starts and work that is not as productive as it should be or is outright wrong. With an effectively defined system along with an understanding of its interactions with the environment, we are now ready to move ahead with the GIM process and write the conservation equations that are relevant to the process of interest.

Example 1.7.1.2 Osmotic Lysis to Cells.

The process of lysis is used widely to rupture the cell plasma membrane to release the contents of the cytoplasm for analysis. A common method is via cytolysis in which cells are exposed to a hypotonic medium for which the concentration of solutes is less than that of the cytoplasm. This condition causes the cells to swell to the point where the membrane ruptures so that intracellular contents becomes mixed with the extracellular solution. This mixture may then be collected and analyzed. The objective of this problem is to build an analytical model of the cytolysis process. When such an analysis is completed and cell property data are available, it could be used to design protocols for the effective lysis of specific cell types.

This example will illustrate one approach to solving this problem. A different approach for analyzing transmembrane transport is shown in Example 5.17.2. In many real-world problems, there is not a single "correct" approach or answer. Multiple alternative approaches may be used, depending on what information is to be learned about a system undergoing a process, how detailed the analysis must be, and what resources (such as computing hardware and software or other expert collaborators with whom to work) are available, and how much time is allocated for obtaining a solution.

Solution. *Goal of the analysis:* First you may have noticed that the problem statement is presented in a nonprescriptive style. This was done intentionally in order to enable you to go through the all-important experience of evaluating the problem to determine what is the most significant and relevant information that pertains to the solution, to make initial assumptions, and to evaluate whether the analysis you perform actually addresses the problem at hand.

We will start with the *Initial Considerations* step. Let us consider what happens during osmotic cytolysis. A suspension of individual cells is exposed quickly to a hypotonic solution, either by being introduced into the solution or having the solution added to the suspension. As a consequence, the concentration of water will be larger outside the cells than inside. The cell membrane can be considered as a semipermeable transport barrier for the relatively short period of time over which the lysis occurs. For most cell types, the time to lysis is measured in seconds. A semipermeable barrier means that only certain chemical species move across it (in this case, only water traverses the membrane). Water will move from a region of higher concentration outside the cell to the cell interior where the concentration is lower. The rate of this movement can be anticipated to be governed by two primary factors: the magnitude of the water concentration difference between the cell exterior and interior, and the resistance of the membrane to the movement of water, as described by a property called *permeability* (to water). The membrane permeability varies greatly (by many orders of magnitude) among various cell species.

As water enters the cell at the membrane periphery, the concentration will be greater there than further to the interior until the water becomes redistributed inside the cell. The result is an internal water concentration gradient. As more water enters the cell, there is a proportional increase in the intracellular volume. The cell swells, and the elastic cell membrane is stretched. At some point enough added water enters the cell so that the membrane is stretched beyond its elastic limit, producing mechanical failure and causing the semipermeability function to be lost. As a consequence, the intracellular and extracellular contents are able to mix freely, and the cell is lysed.

When we evaluate the most important aspects of this process, it appears that cytolysis is governed by the transmembrane movement of water. Redistribution of water on the cell interior is less critical, although it does affect the concentration difference in the immediate vicinity of the membrane. The mechanical elastic properties of the membrane are another critical factor in the process. Thus, as we move forward to performing the *Analysis* we must bear in mind that the system should be defined in a manner that allows us to follow the movement of water across the membrane and the swelling of the cell that increases the stress in the membrane.

System definition and environmental interactions: Now we can start on the *Analysis* steps. (1) We consider how to *define a system* that will best embody the process of interest. A good option is to define the system as the cell, with the membrane as the boundary. Therefore, (2) the *interaction between the system and environment* will be the water flow across the membrane. The system is open since mass crosses the boundary. The system and boundary are defined as shown in Fig. 1.8. An important feature of this system is that it is open, meaning that mass is able to cross the boundary.

The state of the system is described by the properties: volume of the cell (V_{cell}), pressure (P), concentration of the water ($c_{w,in}$), tensile stress in the membrane (τ),

1.7 Developing Innovation

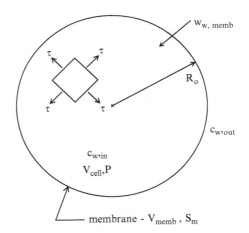

Fig. 1.8 System consisting of a cell that communicates with the environment via membrane transport of water. Relevant properties of the cell cytoplasm and membrane are denoted

and volume of the membrane (V_{memb}). The radius of the cell (R_o) is determined directly from the cell volume if the geometry is assumed to be spherical. The thickness of the membrane is given by δ_m. The membrane volume is small enough so that it has no appreciable contribution to mass storage in the system. But, the membrane has mechanical properties that are adequate to support the increasing pressure inside the cell as water enters until lysis occurs when the membrane strength is exceeded. The environment is characterized by a concentration of water ($c_{w,out}$) that can be assumed to be unchanged by interactions with the cell system. When the suspending solution is made hypotonic, the water concentration becomes greater outside than inside the cell resulting in a flow to the interior of the cell. The interaction of interest is the transmembrane flow of water ($w_{w,memb}$) into the cell that causes the osmotic cytolysis process.

Apprising the problem to identify governing equations: (3) The candidate relevant conservation equations are for mass, energy, and momentum. We will need to evaluate the system and process to see which of these are relevant to this problem. Because this is the introductory chapter of the text, it is expected that you will not yet be familiar with the details of the conservation and constitutive equations necessary to solve this problem. Thus, as needed we will both show these equations so you can gain an impression of their appearance and provide a verbal description of the meaning of the equation for the sake of your understanding in the context of this example.

As water crosses the membrane into the cell, the number of moles of water stored in the system is changed. This process is described by the conservation of mass equation (1.1).

(*Note*: The authors are very familiar with this textbook and all of the material in it, so we already know where to look for key knowledge resources to solve problems, such as (1.1). As you study biotransport and read this book, you also will become familiar with what the key knowledge concepts are and where and how to look them up when they are needed. You will observe that in working this example problem the authors pull in knowledge components from various locations (Chaps. 5 and 12, plus from the discipline of biomechanics that is outside of biotransport). This approach illustrates the fact that you do not need to learn and use the material in this book in the order of the chapters. The book makes available much information on an "as needed" basis, which is how you will make use of knowledge resources after you have completed your formal education. There is no reason to not get started in practicing this approach to problem solving right now. Also, since you have not yet studied the chapters from which the equations used in this solution are taken, you will not be familiar with the meaning of the symbols and notation. There is a comprehensive table of nomenclature in Appendix A to which you can refer. This table will be extremely important and useful to you as you go through the book. We absolutely do not expect you to try to memorize the meaning of symbols, although you may become familiar with some of them through repeated use. Your time is much better invested in developing your critical and creative thinking and problem-solving skills than in memorizing knowledge minutia.)

$$\frac{dm_w}{dt} = w_{w,memb} + w_{in} - w_{out}. \tag{1.1}$$

This equation describes the conservation of mass, which relates the change over time of the mass of water inside the cell to the net rate of exchange of water through the cell membrane.

The conservation of mass equation (1.1) states that the rate of change of mass of water (denoted by the subscript w) stored in the cell (the term on the left side of the equation) is equal to the net effects of all water transports across the membrane boundary (the terms on the right side of the equation). Mass fluxes are denoted by the symbol w. There are three options for mass flux: transport through the permeable membrane (first term on the right) and convective flow through defined conduits into and out of the system (last two terms). The net flow depends on the permeability of the membrane and the pressure difference across the membrane, both hydrostatic and osmotic. The pressure difference contributes to the net force on the membrane.

For this process, we can assume that there are no convective flows through conduits involved in the water transport. Rather, water moves across the membrane by a spatially uniform diffusion process. The conservation of mass equation reduces to

$$\frac{dm_w}{dt} = w_{w,memb}. \tag{1.2}$$

The result of the water flow into the cell is swelling to cause cytolysis. As the volume of the cell increases, the pressure inside the cell also rises, causing a stress in the membrane. When the stress reaches a critical state for which the strength of the membrane is exceeded, lysis occurs in conjunction with mechanical failure of

1.7 Developing Innovation

the membrane. This process is described by the conservation of momentum equation (1.3).

$$\frac{d\vec{p}}{dt} = \sum_{i=1}^{N_{\text{inlets}}} \left[K_2 \rho_i \langle v_i \rangle^2 + P_i \right] A_i \vec{e}_i - \sum_{j=1}^{N_{\text{outlets}}} \left[K_2 \rho_j \langle v_j \rangle^2 + P_j \right] A_j \vec{e}_j + m\vec{g} - \vec{R}. \quad (1.3)$$

This equation is derived for an open system with fluid flowing across specific ports on the boundary. The left side term denotes the rate of change of momentum stored within the system. On the right are summations for kinetic energy and pressure contributions associated with mass entering (positive) and leaving (negative) the system across the boundary. The final two terms on the right denote distributed internal forces, represented here by gravity, $m\vec{g}$, and boundary reaction forces, $\vec{R}$.

For the cytolysis process, there are no boundary momentum flows of significance, plus the internal momentum of the cell and gravity do not have important roles. Thus, all terms are zero except $\vec{R}$. Thus, the sum of forces operating on the system is zero.

$$\sum \vec{R} = 0. \quad (1.4)$$

The forces that are in balance are the internal pressure applied to the membrane and the elastic tension in the membrane shell. Under these conditions, the conservation of momentum equation reduces to

$$F_{\text{press}} = F_{\text{elast}}. \quad (1.5)$$

We can now explore whether the conservation of energy provides further independent understanding of the process. As the cell expands, there are two important energy changes. The increasing stress in the membrane results in a higher potential energy for the system. The expansion of the cell with an elevated internal pressure causes work to be done in displacing the environment that is at a lower pressure than the cell interior. These terms contribute to the conservation of energy equation (1.6).

$$\frac{dE}{dt} = -\sum_i \int_{A_i} \left(\hat{U}_i + \frac{v_i^2}{2} + \hat{\Phi}_i + \frac{P_i}{\rho_i} \right) \rho_i (\vec{v}_i \cdot \vec{n}_i) dA_i + \dot{Q}_S + \dot{Q}_{\text{gen}} - \dot{W}_S - \dot{W}_f. \quad (1.6)$$

All of the conservation equations are written in the same format in which the time rate of change in storage of the conserved property (energy in this case) within the system is equal to the sum of all interactions with the environment plus internal sources that can cause a change in that property. Thus, on the left side is an expression for the time rate of change of energy stored within the system. The first term on the right side is a summation of the contributions associated with each mass flow (i) crossing the boundary according to its internal energy ($\hat{U}_i$), kinetic energy ($v_i^2/2$), potential energy ($\hat{\Phi}_i$), and pressure (P_i/ρ). The next two terms account for the rate of heat transfer at the boundary ($\dot{Q}_s$) and the rate of internal energy generation ($\dot{Q}_{gen}$), such as via metabolism. The last two terms are for the rate work is done by the system on the environment ($\dot{W}_s$) and against frictional dissipative resistance ($\dot{W}_f$).

For our cell system and the cytolysis process, there are only two nonzero terms in the conservation of energy equation. There is a potential energy storage in the change in system energy term on the left side of the equation as the membrane is stretched and the internal elastic stress increases. The rate that displacement work is done on the environment as the cell expands is the second from last term on the right-hand side.

$$\frac{dE}{dt} = \frac{d}{dt} \int_V \hat{\Phi} \rho dV = \dot{W}_S = \frac{d}{dt} \int_V P dV. \tag{1.7}$$

In comparing the conservation of momentum and energy equations, we must be careful to note the subtle difference in notation between the expressions for momentum ($\vec{p}$), membrane permeability (P_A) and pressure (P). Although the symbols appear very similar, these are very different properties that should not be confused.

(4) After the three conservation laws are written to match the problem processes, each term of each equation can be described via an appropriate constitutive expression. The result will be the governing equations that can be solved to simulate the processes the system undergoes, and to which specific properties and boundary conditions can be applied to achieve a quantitative solution to the problem.

The left side term in (1.2) is already expressed in terms of the change in the mass contained within the cell. The right side term for the flow across the membrane can be rewritten with the constitutive equation for water flux as a function of the permeability and water concentration difference between the inside and outside of the cell according to (1.8)

$$N_A = \mathsf{P}_A [c_{A_1} - \Phi_{A_{12}} c_{A_2}], \tag{1.8}$$

where P_A is the membrane permeability to the transported species A, which in our case is water. The partition coefficient for water between the intracellular and extracellular solutions ($\Phi_{A_{12}}$) can be taken as 1.0, and water flow is the flux times the surface area S, so that the constitutive equation for the transmembrane water flux becomes

$$w_A = \mathsf{P}_A S [c_{w,out} - c_{w,in}]. \tag{1.9}$$

The concentration of water inside the cell is given by

$$c_{w,in} = \frac{m_w}{V_{cell}}. \tag{1.10}$$

1.7 Developing Innovation

By combining (1.2), (1.9), and (1.10), we obtain an expression for the rate of change of water contained within the cell while it is undergoing hypotonic cytolysis.

$$\frac{dm_w}{dt} = P_A S \left[c_{w,out} - \frac{m_w}{V_{cell}} \right]. \tag{1.11}$$

An important aspect of this process to note is that as the amount of water within the cell changes, the total cell volume is likewise proportionately altered. The simplest way to represent the contents of the cells is as a solution composed of solvent water and various solutes. The cell volume can then be expressed as

$$V_{cell} = \frac{m_w}{\rho_w} + \frac{m_{sol}}{\rho_{sol}}. \tag{1.12}$$

Finally,

$$\frac{dm_w}{dt} = P_A S \left[c_{w,out} - \frac{m_w}{\frac{m_w}{\rho_w} + \frac{m_{sol}}{\rho_{sol}}} \right] = P_A S \left[c_{w,out} - \frac{1}{\frac{1}{\rho_w} + \frac{m_{sol}}{m_w}\frac{1}{\rho_{sol}}} \right]. \tag{1.13}$$

The constitutive equation for the stress in the membrane is based on a classical mechanics analysis of tension in a thin spherical membrane shell of thickness δ_m. A simple force balance between the tension in the membrane and the pressure force applied to the inner surface of the membrane yields the relationship

$$\tau = \frac{PR_o}{2\delta_m}. \tag{1.14}$$

The outer radius of the cell, R_o, is related to the volume for simple spherical geometry.

$$V_{cell} = \frac{4}{3}\pi R_o^3. \tag{1.15}$$

It is likely that the total mass of the membrane material does not change substantially during the cytolysis process. Therefore, as the total cell volume increases, the membrane thickness will decrease proportionately.

$$V_{memb} = \delta_m S = \delta_m 4\pi R_o^2 \quad \text{or} \quad \delta_m = \frac{V_{memb}}{4\pi R_o^2} \tag{1.16}$$

so that the tensile stress in the membrane is related to the internal pressure and the cell geometry.

$$\tau = \frac{2\pi P R_o^3}{V_{memb}}. \tag{1.17}$$

Note that both the membrane tensile stress and the internal pressure vary with the cell radius. The single equation is not sufficient to determine the values of both dependent properties as the cell swells during cytolysis. An additional independent relationship is needed to solve for these properties. This relationship comes from the conservation of energy.

The energy stored in the system occurs in conjunction with the buildup of stress in the membrane as it is stretched; it is called strain energy. The formulation of the strain energy term involves advanced mathematical methods applied in the theory of elasticity (Naghdi and Rubin 1995; Nadler and Rubin 2009), and is beyond the scope of coverage in this text. Thus, to be able to pursue a full solution of this problem, it will be necessary to consult additional knowledge resources beyond those appropriate to discuss in a biotransport text. In order to develop an analysis of the cytolysis process, we are working on a true multidisciplinary problem. In solving real-world problems, this type of situation is encountered quite often. It is important to be able to identify conditions that demand that you do some further learning, and to be able to locate where you can find the appropriate reference materials. For now we will put off the strain energy aspect of the solution, and it can be addressed on a future iteration of the solution, although we anticipate that it may require considerable new learning to be able to incorporate that component of the analysis.

The right-hand term in the conservation of energy equation is the displacement work on the surroundings during cell swelling. The pressure inside the cell does PdV work on the environment as the cell expands. The work term is written as

$$\frac{d}{dt}\int_V P dV = 4\pi \frac{d}{dt}\int_{R_o} P R_o^2 dR_o. \qquad (1.18)$$

This expression may be equated to the strain energy term after it is identified to obtain a second equation for the relationship between the pressure and membrane tensile stress as the cell radius increases.

(5) The solution methods can be planned for the equations defined in the preceding section. This step involves determining the initial and boundary conditions for all differential equations, plus constitutive property values. For example, in the absence of other information, it may be assumed that the initial concentration of the solution in the cell is isotonic, and that the initial stress in the membrane is zero. The introduction of cells into a hypotonic solution usually occurs quite quickly so that the change in external concentration occurs in a step-wise manner. There are many property values to be determined including the membrane permeability to water, isotonic concentration and cell volume, membrane thickness that supports a tensile stress, the Young's modulus of elasticity, and the rupture strength of the membrane, plus how some of these properties may be altered as a function of state as the cell swells. Many of these properties are species specific.

(6) *What to do next*. We have made good progress in identifying a pathway to the solution of a rather challenging problem. However, there remain a number of

1.7 Developing Innovation

aspects of the problem to be refined before the solution can be completed. Some of these have been targeted in the previous analysis, including determining a constitutive equation for the strain energy in a membrane under tension and the values for a large number of constitutive properties for the cell species to be evaluated. It may be that some of these properties are not available in the literature, and if you have access to adequate experimental facilities, you will need to measure them yourself. If this is the case, the model and constitutive equations you have built will not be used for predictive simulation but for solving what is called an inverse problem. This analysis involves designing an experiment so that you can measure the behavior of your system in response to stimulation by explicit boundary conditions. The value of the property in your constitutive equation is then adjusted in an optimization process until you achieve an acceptably good match with the model for the behavior and the stimulating inputs. The value that gives the best match can be taken as the true constitutive property under the measurement conditions.

Examining and interpreting the results: When solving a challenge problem that requires us to perform many of the key thinking steps such as defining the system and its boundary interactions, what simplifying assumptions to apply in characterizing a real physical system undergoing a process, and what property values to apply, it is a good practice at the conclusion of your analysis to stand back and evaluate your work. You may address questions like: whether the work you performed actually solves the problem that you started with, do the results you obtained agree with your intuitive expectations for the outcome, how might changing some of your assumptions affect the results, are there other aspects of the problem that you have not yet addressed that should be investigated in order to cover significant omissions in the study, and have you expressed your analysis in a manner that can be shared in a clear and convincing manner with whomever you are responsible to for solving the problem. In many of the worked examples in this book, we will perform the *examining and interpreting the results* step since it is often an important component of performing a complete and good quality evaluation of a challenge problem.

For the current example problem, we can see that we have created a model that covers many of the most important physical phenomena that govern the cell lysis process, even including the effect of the mechanical properties of the membrane, a topic that is beyond the scope of the text but is very relevant to how a cell is damaged during lysis. However, we have not really arrived at a state in the solution wherein we can make calculations and evaluate actual lysis processes for specific cell types. This solution is already quite lengthy for presentation in the introductory chapter of a text. We could perform more detailed work to obtain and apply actual cell property values so as to simulate and design explicit lysis procedures. This would require many more pages in the text and probably result in the reader losing interest and focus. On the contrary, our primary objective in this example solution is to present an initial illustration of how one can use the GIM to generate a solution to a real-world problem that requires the student to perform much of the key thinking processes to move toward a useful solution. Hopefully, we have been successful in this objective.

1.7.2 How to Use This Textbook to Develop Innovation

You may not use the SL Cycle Challenge-Based method of instruction in your class. However, no matter how you use this textbook, you will be practicing the innovation you need to put you on a path toward becoming an adaptive expert because the approach to solving the example problems in this book reflects the important elements of the GI model.

The solution process in the GI model is similar to the way experts solve problems (Chi et al. 1988). When experts approach problems, they first think about the global perspective and identify the most important aspects of the problem. Once they have a good overall model of the problem, then they move toward developing specific equations or other solution methods.

The GI model is embedded in the structure of the worked solutions for all of the example problems. The problem-solving method follows the GI model pattern. In the *Initial Considerations* section, we start by helping you zero in on the main issues in the problem. This helps you start to think about the problem globally like experts do. Then, the *System Definition and Environmental Interactions* section follows the GIM by helping you get specific on the appropriate system and how it interacts with outside entities. Next, in the *Appraising the Problem to Identify Governing Equations* section, we show you how to use this global model for the problem to identify the assumptions about the system and the conditions for correctly applying conservation laws, constitutive equations, and boundary conditions for the resulting partial differential equation. In the *Analysis* section, we show you how to use the conditions and equations identified to develop a solution to the problem. In the next section, *Examining and Interpreting the Results*, we interpret the quantitative analysis developed in the previous section in terms of the problem context. Finally, in *Additional Comments*, we extend the analysis by considering how the solution would be different under different assumptions, boundary conditions, or environmental interactions.

Not every problem requires that you go through each step shown in the example solutions. Some problems prescribe several elements in the solution process for you. In these routine problem cases, you do not need to write out each step in the example solution process. However, it is helpful to recognize that these elements are included in some problems and that they are still useful to your solution.

We hope you find this textbook useful and that you can use the problem-solving approach in other courses and in your future work. We could say that other students have found the HPL approach helpful, but it is best to let students' comments speak for themselves. Here are some of the things our students have said on anonymous surveys:

- I had lost hope in the American educational system and thought the future was doomed to failure, but the direction you are heading with this course gives me hope.
- The first assignment was cool, because it was the first time I got to use my creative assets in an engineering class.

- Seriously, this class has changed my perception of the world as far as problem solving and epistemology is concerned. These skills will stick with me for the rest of my life. Thank you SO MUCH!
- I enjoyed the test. When I knew nothing right off, I went through the panic phase then, Bam! I thought hey, I can reason through to a start to the answer.

1.8 Learning to Gain Understanding

Much of the foregoing material presented in this chapter shares a common outlook with other educational innovators who may explain their insights and understandings concerning how to provide a framework for effective student learning with a slightly different vocabulary, but with a common foundational realization of the learning process. One particular exposition from which the authors have benefited significantly is the book *Understanding by Design* by Wiggins and McTighe (2005). The ideas and practices set forth in that text are in close alignment with the philosophy with which the present book was written, and they offer a very helpful and healthy perspective for both student and teacher on how to approach the learning process. We would like to summarize briefly our understanding of Wiggins and McTighe's approach (which we will designate as W&McT) to designing a learning environment with the expectation that it will be useful as you study from this particular textbook.

A recurring theme of W&McT is that students need to be driven continuously to a deeper level of thinking that goes beyond simply learning the objective knowledge content presented in a book or course. To this end, one of the features of the W&McT thesis is to design learning materials so that students grasp the *"big ideas"* worthy of *enduring understanding* in a particular course/topic. To reach this goal requires a balance between imparting knowledge tools and skills and pursing deeper thinking for a more comprehensive understanding concerning the foundational concepts that support specific knowledge components. *Understanding* goes beyond having *knowledge*.

To *understand* a topic or subject is to use knowledge and skill in sophisticated, flexible ways, to make conscious sense and apt use of the *knowledge* that is learned and the principles underlying it. *Understanding* involves the abstract and conceptual, not merely the concrete and discrete. *Understanding* also involves the ability to use knowledge and skill in context, as opposed to doing something routine and on cue. We want students to be able to use knowledge in authentic situations as well as to understand the background of that knowledge. It is very easy for students to *know* without having *understanding*.

Enduring understandings anchor a subject or course. *Enduring* refers to big ideas, the important understandings that students need to "get inside of" and retain after they have forgotten many of the details of a subject. In general, topics of enduring understanding are not retained by memorization but by gaining a big picture perspective of a subject in which the governing principles fit together to provide a cohesive impression of the subject.

Topics within a given course or subject area can be classified into three categories in terms of their priority for student learning. The topics of *enduring understanding* have the highest priority for learning and are used to define the design of a course and the assessment tools that are developed and employed. Of a lesser priority is *important knowledge*, including facts, concepts and principles, and skills, including processes, strategies, and methods that are essential to a course and without which student learning would be incomplete. The lowest priority is for knowledge that is *worth being familiar with* and that students should encounter at some point within a course or curriculum. An example of one type of this latter category is Appendix A containing the table of nomenclature definitions. You need to be very familiar with how and where to find the meaning of the various symbols in equations, but these symbol definitions are not foundational to understanding and applying the principles of biotransport. The objective of curriculum and course design is to identify the topics of enduring understanding and structure student learning and assessment around them. Many topics embody important knowledge that needs to be addressed, while other topics may be introduced to become familiar to the students as appropriate without a strong focus.

It is anticipated that there will be various opinions among instructors of biotransport as to which topics fall into the category of enduring understanding. After giving considerable thought to this matter, the authors have prepared the following list of biotransport topics that they think merit classification as requiring enduring understanding. You may notice, with justification, that there is a strong correspondence between the topics on this list and the process followed in the Generate Ideas Model. Other topics in a biotransport course may be treated as important knowledge or as worth being familiar with at the instructor's discretion.

Big Idea biotransport topics of enduring understanding

1. How to structure an analysis strategy for defining and solving an open-ended problem in biotransport presented in a context that is new to the solver.
2. The conservation principles for energy, mass, and momentum, what the individual terms represent, and how they apply to the interaction of a specific system with its environment. When is it appropriate to use a macroscopic approach vs. a microscopic approach?
3. The major constitutive equations that apply to biotransport (Fick, Fourier, Newton, internal storage of momentum, energy, mass, etc.) and how they are used with conservation equations to develop governing equations to describe biotransport processes.
4. The material derivative for the change in a property value over time based on the combination of time-wise alterations and motion through a spatial gradient.
5. How to define and identify the appropriate boundary conditions for a partial differential equation that describes a specific physical system.
6. How to perform dimensional analysis of a problem and to present analysis results in a dimensionless format applicable beyond a specific problem.
7. The domains of heat and mass transfer (conduction, convection, radiation, etc) and how they act cooperatively to constitute a complete biotransport phenomenon.

8. The domains of fluid flow (viscid/inviscid; boundary/free stream; internal/external; laminar/turbulent, incompressible/compressible, etc.) and how they combine to embody the movements of fluids external and internal to living systems.
9. Aspects of transport that are unique to living systems: complex geometries, nonlinear properties, cross-energy domain coupling, composite material structures, internal blood perfusion, etc.
10. Methods effective for solving biotransport problems in living systems that are more complex than many encountered in the inanimate world: approximation methods, scaling, finite difference, finite elements, etc.

There are many more nuggets of educational wisdom and very practical advice in W&McT, far beyond what we could ever attempt to review here. However, if you are interested in how to achieve superior educational outcomes from the perspective of either a student or a teacher, W&McT is an interesting, easy and stimulating read, and we heartily recommend it.

References

Albanese MA, Mitchell S (1993) Problem-based learning: a review of literature on its outcomes and implementation issues. Acad Med 68:52–81

Anderson JR (1982) Acquisition of a cognitive skill. Psychol Rev 89:369–406

Bransford JD, Brown AL, Cocking RR (eds) (2000) How people learn: mind, brain, experience, and school. National Academy Press, Washington, D.C

Chi MTH, Glaser R, Farr MJ (eds) (1988) The nature of expertise. Erlbaum, Hillsdale, NJ

Clough MP, Kaufmann KJ (1999) Improving engineering education: a research-based framework for teaching. J Eng Educ 88:527–534

Cordray DS, Harris TR, Klein S (2009) A research synthesis of the effectiveness, replicability, and generality of the VaNTH challenge-based instructional modules in bioengineering. J Eng Educ 98:335–348

de Jong T (2006) Computer simulations: technological advances in inquiry learning. Science 312:532–533

Den Hartog JP (1961) Strength of materials. Dover, New York, pp 85–91

Dochy F, Segersb M, Van den Bosscheb P, Gijbels D (2003) Effects of problem-based learning: a meta-analysis. Learn Instr 13:533–568

Hatano G, Inagaki K (1986) Two courses of expertise. In: Stevenson H, Azuma J, Hakuta K (eds) Child development and education in Japan. W. H. Freeman & Co, New York, NY, pp 262–272

Kay PH, Munsch CM (2004) Techniques in extracorporeal circulation, 4th ed., Edward Arnold, London

Martin T, Rayne K, Kemp NJ, Hart J, Diller KR (2005) Teaching for adaptive expertise in biomedical engineering ethics. Sci Eng Ethics 11(2):257–276

Martin T, Petrosino A, Rivale S, Diller KR (2006) The development of adaptive expertise in biotransport. New Dir Teach Learn 108:35–47

Martin T, Pierson J, Rivale S, Vye N, Bransford J, Diller K (2007a) The function of generating ideas in the legacy cycle. Innovations

Martin T, Rivale SR, Diller KR (2007b) Comparison of student learning in challenge-based and traditional instruction in biomedical engineering. Ann Biomed Eng 35(8):1313–1323

Nadler B, Rubin MB (2009) Analysis of constitutive assumptions for the strain energy of a generalized elastic membrane in a nonlinear contact problem. J Elast 97:77–95

Naghdi PM, Rubin MB (1995) Restrictions on nonlinear constitutive equations for elastic shells. J Elast 39:133–163

Pandy MG, Petrosino A, Austin BA, Barr RA (2004) Assessing adaptive expertise in undergraduate biomechanics. J Eng Educ 93(3):1–12

Prince MJ, Felder RM (2006) Inductive teaching and learning methods: definitions, comparisons, and research bases. J Eng Educ 95:123–138

Roselli RJ, Brophy SP (2006) Effectiveness of challenge-based instruction in biomechanics. J Eng Educ 95(4):311–324

Schwartz DL, Bransford JD (1998) A time for telling. Cogn Instr 16:475–522

Schwartz DL, Brophy S, Lin X, Bransford JD (1999) Software for managing complex learning: examples from an educational psychology course. Educ Technol Res Dev 47(2):39–59

Schwartz DL, Bransford JD, Sears D (2005) Innovation and efficiency in learning and transfer. In: Mestre J (ed) Transfer of learning from a modern multidisciplinary perspective. Erlbaum, Mahwah, NJ, pp 1–51

Terezini PT, Cabrera AF, Colbeck DL, Bjorklund SA (2001) Collaborative learning vs. lecture/discussion: students' reported learning gains. J Eng Educ 90:123–129

Wiggins G, McTighe J (2005) Understanding by design, expanded 2nd Edition. Pearson, Upper Saddle River, NJ

Part II
Fundamental Concepts in Biotransport

Chapter 2
Fundamental Concepts in Biotransport

2.1 Introduction

Biotransport is concerned with understanding the movement of mass, momentum, energy, and electrical charge in living systems and devices with biological or medical applications. It is often subdivided into four disciplines: biofluid mechanics, bioheat transfer, biomass transfer, and bioelectricity. These topics are often taught together because of the great similarities in the principles that govern the transport of mass, heat, and momentum of charged and uncharged species.

Many different types of problems may be encountered in the study of biotransport processes. However, we believe that a standard approach, as introduced in Chap. 1, can be exercised to formulate a solution strategy for all biotransport problems. The formalism of the approach is straightforward, but the details will vary depending on the nature of the problem of interest and the extent to which the problem description leads directly to limiting assumptions and identification of constraints on how system and process are understood and modeled.

In this text at the introductory level, we will take a standard approach in defining systems for evaluation and in developing solution methods. Thus, the knowledge organization and presentation aspect of much of this text will resemble the very large number of preceding texts that have been written for transport. In some cases, we will introduce more advanced topics describing methods for handling the unique features of biosystems. The reader should beware that in many practical applications it will be necessary to address these features, and that special and more difficult modeling and solution methods will be required.

This chapter provides a brief introduction to a unified understanding of biotransport processes and how they can be modeled for analysis. First we will discuss some of the physical mechanisms that give rise to transport processes in a single material. Next we will address transport properties that can provide quantitative measures of a material's ability to participate in specific types of transport. Finally, we will consider transport across the interface between two different materials.

2.2 The System and Its Environment

The starting point for analyzing transport processes and problems is to define and understand the system of interest. In the most general sense, a system is identified as that portion of the universe that is involved directly in a particular process. The remainder of the universe is called the environment. The system interacts with the environment across its boundary. These interactions are directly responsible for changes that occur to the state of the system. The boundary surface provides a locus at which interactions can be identified (Fig. 2.1). Knowledge of these interactions can be used to predict resulting changes that will occur to the system.

There are two different approaches to identifying the boundary of a system. In one case the system is determined by a fixed mass. The system includes this specified mass and nothing else. As this mass moves or changes its shape, so does the boundary. Although the system may change over time in many different ways, a key feature is that the mass remains constant. Thus, this type of system is called a *closed system* since no mass can be added or removed (Fig. 2.2).

Alternatively, a system may be identified in terms of a boundary surface specified in three-dimensional space. This type of system is called an *open system* since mass may be exchanged across the boundary with the environment (Fig. 2.3).

The state of a system is described in terms of an independent set of measurable characteristics called properties. These properties can be either extensive (extrinsic) or intensive (intrinsic). *Intensive properties* are independent of the size of the system and include familiar properties, such as pressure, temperature, and density.

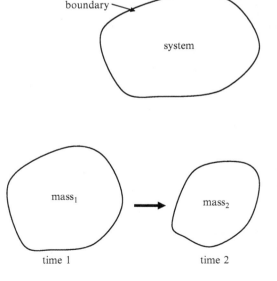

Fig. 2.1 A system is separated from the environment by a surface called the boundary. Interactions between a system and its environment are identified as they occur across the boundary

Fig. 2.2 A *closed* system is defined by a fixed mass which may change in position and shape as well as other properties. However, there is no movement of mass across the boundary. Thus, $mass_1$ is the same as $mass_2$

2.3 Transport Scales in Time and Space

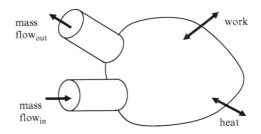

Fig. 2.3 An *open* system is defined by a boundary in space across which a system and the environment interact. The interactions may include mass exchange as well as work and heat

In addition, transport properties such as viscosity, diffusivity, and thermal conductivity are intensive properties. *Extensive properties* depend on the size of the system and include the transport properties mass, volume, heat, momentum, and electrical charge. Intensive properties can vary spatially or temporally within a system, but they do not flow into or out of the system. Extensive properties, however, can move across the system boundaries, and prediction of their movement is one of the primary objectives of this text.

2.3 Transport Scales in Time and Space

We can think of the transport of momentum, energy, mass, and charge to occur at three fundamental levels or scales, as illustrated by blood flow through the left ventricle in Fig. 2.4. Random molecular interactions can be associated with the transfer of all transport variables as shown on the right. As molecules collide, mass, energy, momentum, and electrical charge can be transferred from one molecule to another. If one considers a nanoscale open system consisting of a spherical volume with diameter of perhaps ten solvent molecules, the mass of the system will be proportional to the number of molecules in the sphere. As molecules enter and leave the boundaries of the spherical volume, the mass, energy, momentum, and charge within the system change. Changes can also occur as different species react, transfer electrons, or dissociate within the system. If molecules are treated as particles, one can write conservation equations for energy, mass, momentum, and charge for each species and add (integrate) the contributions from all molecules to produce the total for each transport variable within the system. Because of the large number of molecules per unit volume, this approach is only practical when we are dealing with a nanoscale system.

A more practical approach for larger systems is to neglect the particulate nature of matter and treat the system as if it consists of material that is continuously distributed in space, indicated by the middle panel of Fig. 2.4. We will evaluate the validity of that assumption shortly. This *microscopic* approach uses the conservation of mass, charge, momentum, and energy to the microsystem. Empirical relationships are used to relate fluxes of heat, mass, momentum, and electrical

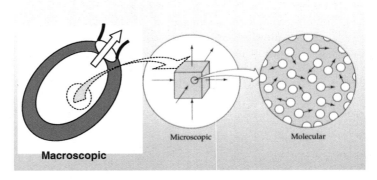

Fig. 2.4 Transport scales

charge (i.e., extrinsic or "through variables") through the microsystem to gradients of driving "forces" such as temperature, concentration, pressure, or electrical potential (intrinsic or "across variables"). The resulting relations allow us to determine how transport variables vary with position (x, y, z) and time (t).

At the *macroscopic* scale, we are interested in how the average momentum, heat, mass, and electrical charge vary with time inside the system as a whole. Consider the system in the left panel of Fig. 2.4 which consists of the blood within the left ventricle. If the blood is well mixed, then variations in the transport variables change only with time and do not change with position within the system. In many situations, even if the system is not well mixed, we are only interested in how the average temperature, mass, or concentration varies in the system, so a macroscopic approach is appropriate.

Transport processes in living systems are manifested across length scales extending from physiological to molecular. Until recently, most analysis has been focused on processes that can be measured and analyzed at the macroscopic and microscopic levels. Advances in adjuvant sciences such as molecular biology have demonstrated that heat transfer can be used to manipulate the genetic expression of specific molecules for purposes of prophylaxis and therapy for targeted medical disease states. An illustrative example is the application of a spatially and temporally varying macroscopic scale thermal stress to control the pattern of genetic expression of specific proteins within cells of a tissue.

Common transport processes and their effects have been identified across a broad range of length scales. The greatest length is on the order of the size of the human body (1 m) and is typically encountered in environmental thermal interactions at the surface of the skin. At the opposite extreme is the profound effect of temperature on the genetic expression of individual protein molecules. In many instances, there is a direct coupling of the transport processes across disparate length scales. For example, transport originating at the physiologic scale can have its most important manifestation within individual cells.

There also exists a wide range of time scales for physiological transport, from near instantaneous to days, weeks, and longer. Here again coupling across time scales is significant. The feedback control systems that regulate all aspects of life are among the most complex encountered in nature. Typically there exist many options for parallel pathways and for counterbalancing effects. There can be an interaction among transport processes having very different scales of length and time that is not apparent by superficial inspection. These differences in scales can provide a major challenge to modeling the integrated behavior of a physiological system. When encountering new arenas of application, it may be important to ensure that multiscale effects are accounted for.

Improved understanding of the constitutive behavior of living systems across the full range of scales has enabled meaningful application of biotransport modeling techniques which were not previously possible. It has been a continual challenge to develop mechanistically accurate models of biotransport processes since these are highly coupled and generally of a more complex nature than are processes in inanimate systems. The recent acceleration in learning about life at the cellular and molecular scales will lead to the development of more accurate and comprehensive biotransport models.

Complex geometric and nonlinear properties of living systems must be accounted for in building realistic models of living systems. This requirement remains one of the major challenges in the field of biotransport. Recent dramatic improvements in medical imaging techniques enable acquisition of more complete and accurate geometric and property data that can be used for developing patient specific models. This area of analysis holds great potential for future exploitation with applications such as computer-controlled surgical procedures using energy-intensive sources.

A primary conclusion of these observations is that currently there is a great potential for defining and solving new and important problems in biomedical transport (Schmid-Schönbein and Diller 2005). It is anticipated that there will be forthcoming significant advances in both theory and applications of biotransport in the near future.

In this text, we will be concerned primarily with treatments at the macroscopic and microscopic scales. In addition, the bulk of the text deals with the transport of uncharged species.

2.3.1 Continuum Concepts

How do we define the density at a specific point (x_0, y_0, z_0) in a system? The classic mathematical definition of density in a truly continuous system (i.e., a continuum) would be to measure the mass per unit volume as the volume approaches zero. However, because of the molecular nature of the material, the density can oscillate wildly as molecules jump into and out of a molecular sized control volume, ΔV (Fig. 2.5). If our point (x_0, y_0, z_0) is centered on one of the molecules, the final density would approach that of a nucleon. But an instant later the molecule might move away and the density would be zero. We are not interested in whether or not

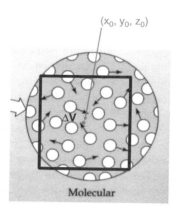

Fig. 2.5 Volume of molecular proportions

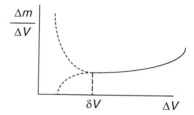

Fig. 2.6 Local density versus volume

molecules are present at the point, but rather in the average local density in the neighborhood of the point.

At what volume can we safely ignore the molecular nature of matter, but still compute a representative local density? As we reduce the local volume surrounding the point (x_0, y_0, z_0), the mass per unit volume will change. At large volumes this reflects true differences in density caused by spatial variations. However, as the volume gets smaller, the computed density will eventually become independent of the size of the control volume, representing the true local density in the vicinity of the point (x_0, y_0, z_0). As the volume drops below a critical value, δV, the density will oscillate in time between the two dotted lines in Fig. 2.6 as molecules move in and out of the volume. A practical definition for the *local density* would be:

$$\rho(x_0, y_0, z_0) = \lim_{\Delta V \to \delta V} \left\{ \frac{\Delta m}{\Delta V} \right\}. \tag{2.1}$$

How big is δV? Let us postulate that the critical volume δV is reached when the point (x_0, y_0, z_0), is surrounded by 1,000 molecules. Consider first the volume occupied by 1,000 molecules of an ideal gas. We can use Avogadro's number (number of molecules per mole) and the ideal gas law. Assuming a pressure of

1 atmosphere and a temperature of 298 K, we find this volume to be equivalent to a cube with each side having a length of 0.0344 μm. By comparison, the smallest structural unit in the lung, an alveolus, has a diameter of about 300 μm. For all practical purposes, air can be considered a continuum.

The value of δV will be even smaller in a liquid. To calculate the volume occupied by 1,000 molecules of water, we use Avogadro's number and the molar density of water at a pressure of 1 atmosphere and a temperature of 298 K. We find this volume to be equivalent to a cube with each side having a length of 0.0031 μm (3.1 nm). By comparison, a red blood cell has a volume that is more than a billion times larger than the critical volume. Again, for all practical purposes, problems involving biological materials of interest can be treated using the *continuum approach*.

2.4 Conservation Principles

In its simplest form, biotransport can be considered as the study of the movement of extensive properties across the boundaries of a biological or biomedical system. The first step in formulating a biotransport problem is to identify the system and its boundaries. The next step is to apply the appropriate conservation principles governing the movement of an extensive property, such as mass or energy. A general *conservation statement* for any extensive quantity can be expressed in words as:

$$\left\{ \begin{array}{c} \text{Rate of} \\ \text{accumulation} \\ \text{of the quantity} \\ \text{within a system} \end{array} \right\} = \left\{ \begin{array}{c} \text{Net rate the} \\ \text{quantity is produced} \\ \text{within the system} \end{array} \right\} + \left\{ \begin{array}{c} \text{Net rate the} \\ \text{quantity enters} \\ \text{through the} \\ \text{system boundary} \end{array} \right\}. \quad (2.2)$$

Let us refer to an extensive property of interest as "X". The rate of accumulation of X within the system refers to the change in X per unit time. The accumulation of X in the system between times t and $t + \Delta t$ is $X(t + \Delta t) - X(t)$, and the rate of accumulation can be found by dividing this difference by the time increment Δt as it becomes very small:

$$\left\{ \begin{array}{c} \text{Rate of} \\ \text{accumulation of } X \\ \text{within a system} \end{array} \right\} = \lim_{\Delta t \to 0} \left\{ \frac{X(t + \Delta t) - X(t)}{\Delta t} \right\} = \frac{\partial X}{\partial t}. \quad (2.3)$$

If the rate of accumulation of X is positive, then X will increase with time, and if it is negative, X will decrease within the system as time increases.

The net rate of production of the quantity X in (2.2) refers to the rate at which X is produced or generated within the system minus the rate at which X is being consumed or depleted within the system. The net rate of production of quantities that are truly conserved such as mass, net electrical charge, and total energy is zero. Those quantities can only change if there is a net movement through the boundary. For this reason, some scientists prefer to call (2.2) an "accounting equation", rather than a conservation equation. Quantities that are not conserved, such as the mass of cations or anions in a system, can change because of chemical dissociation or reaction without any cations or anions entering the system. Similarly, molecular species can be produced or depleted in the system by chemical reaction, irrespective of whether the species traverse the system boundary. Momentum in a system can be changed without momentum entering or leaving through the system boundaries. According to Newton's second law, momentum will be altered if a net force is applied to the system. Heat can be produced in a system by viscous dissipation or chemical reaction; so the production term in the conservation relationship must be included if those sources of heat are present.

2.5 Transport Mechanisms

The final step necessary in the formulation of a biotransport problem is to identify appropriate expressions for the last term in (2.2) that accounts for the movement of extensive properties across the boundary. To answer this question, it is useful to first take a look at equilibrium situations. A system in *equilibrium* with its surroundings has no net exchange of any extensive property, such as mass or energy, with its surroundings. Thus, the net mass flow of each individual species between system and surroundings is zero. Consequently, there will be no current flow or total mass flow into or out of the system. Finally, there cannot be any net heat gain or loss from a system in equilibrium.

In addition, if we were to measure the temperature at all positions within a system that is in equilibrium with its surroundings, we would find no spatial variations. Similarly, we would find no spatial variations in pressure or in the concentrations of any of the molecular species within a system that is in equilibrium with its surroundings. The temperature and pressure within the system would be the same as the temperature and pressure of the surroundings. However, the concentration of each species within the system may be different than the concentration of the same species in the surroundings. This is because the solubility of a species in the system can be different than the solubility of the species in the surroundings. For example, if the system is a pane of glass immersed in the ocean, the solubility coefficients for various salts in glass are generally much lower than they are in water. Consequently, under equilibrium conditions, even though there is no net movement of any salt between system and surroundings, the salt concentrations in the system will generally be different than the concentrations of the same salts in the surroundings.

2.5 Transport Mechanisms

Now, consider two systems in contact that are not in equilibrium. If the temperatures of the two systems are different, then heat will flow from the system having the greater temperature to the system having the lower temperature. Consequently, the temperature difference between two systems that are not in equilibrium is an appropriate driving force for inducing heat transfer between the systems. However, a simple concentration difference cannot be considered as an appropriate driving force for mass transfer between systems because differences in species concentration can occur under equilibrium conditions, where no mass transport can occur. Instead, the appropriate driving force would be the concentration in the first system minus the concentration in the second system that would be in equilibrium with the concentration in the first system.

What causes heat, momentum, and mass to flow under nonequilibrium conditions? There are two basic transport mechanisms: random molecular motion and bulk fluid motion. Heat, momentum, mass, and electrical charge can be transported by both of these mechanisms. Other important transport mechanisms also exist, including radiation, evaporation, condensation, and freezing.

2.5.1 Molecular Transport Mechanisms

Let us begin with a description of transport by molecular motion. If we open a bottle containing an odiferous gas at the center of a large room containing stagnant air, we will smell the gas several feet away within a short time. The transport of the gas is by random molecular motion, known as diffusion. The more molecules of gas present at the release site, the greater will be the movement of gas away from the release site. Consequently, the higher the concentration gradient in a particular direction, the greater will be the movement of gas in the opposite direction.

This and other transport processes can be described in terms of a constitutive equation. A *constitutive relationship* for molecular transport mechanisms is an empirical equation relating the motion of an extensive transport property to the negative gradient of an intensive transport property. A unique constitutive equation is associated with each transport process, and many of these equations have been known for more than a century based on the observation of naturally occurring transport phenomena. The constitutive equations are usually written in terms of the transport flux in a particular coordinate direction n and the precipitating potential gradient in that direction. This can be expressed as:

$$\text{Flux}_n = (\text{constitutive property}) \cdot \left[-\frac{\partial(\text{potential})}{\partial n} \right]. \qquad (2.4)$$

Let us formally define flux and gradient. The *flux* of a quantity X (e.g., species, mass, momentum in the n-direction, heat, charge) at a point (x_0, y_0, z_0) is a vector representing the rate at which X passes through a unit area A that is perpendicular to the n-direction per unit time (Fig. 2.7):

$$(\text{Flux of } X)_{x_0,y_0,z_0} = \frac{1}{A}\left(\frac{\partial X}{\partial t}\right)_{x_0,y_0,z_0}. \tag{2.5}$$

Let us define a potential Ψ, which is an intensive property responsible for inducing the flux of X. In the case of heat transfer, Ψ would be temperature and X would be heat. The *gradient* of the potential Ψ in the n-direction at the point (x_0, y_0, z_0) is simply the rate at which the potential varies in the n-direction at that point:

$$(\text{Gradient of } \Psi)_{x_0,y_0,z_0} = \left(\frac{\partial \Psi}{\partial n}\right)_{x_0,y_0,z_0}. \tag{2.6}$$

In general, the potential is a scalar property, and the flux is a vector expressed as the flow per unit area normal to the direction of the applied potential gradient.

The *constitutive property* is a measure of the ability of the system material to facilitate the transport process. It is dependent on the chemical composition of the system material and the state of the system. For example, changes in temperature and pressure can often cause significant alterations in the flux of extensive properties, and the magnitude of the effect will depend on the composition of the system.

Because of the random nature of molecular collisions, regions of space with an initially high population of molecules possessing a particular transport characteristic will lose some of those molecules to surrounding regions with time. Energy and momentum can also be exchanged to surrounding regions via molecular collisions. A completely random process cannot concentrate mass, charge, momentum, or energy. This would violate the second law of thermodynamics. In a random process, each of these quantities must move from regions of high potential to low potential. A positive potential gradient is one in which the potential increases with n. Therefore, the flux of transport quantities such as mass, momentum, charge, and heat must be in the opposite direction as the potential gradient. Consequently, the flux in the constitutive equation is proportional to the negative of the potential gradient. To generate a transport flow with a positive vector, it is necessary to apply

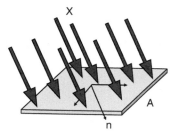

Fig. 2.7 Flux in n-direction through a surface with area A that is perpendicular to n

Fig. 2.8 One-dimensional flux of X generated in a positive direction by application of a negative gradient in driving potential along the axis of flow

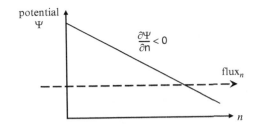

a negative potential gradient, i.e., with a magnitude decreasing along the coordinate axis (Fig. 2.8).

2.5.1.1 Basic Constitutive Equations for Molecular Transport

There are characteristic constitutive equations that are applied uniquely to describe all observed transport processes. However, processes based on random molecular motion all have analogous constitutive equations even though they have been derived for processes in different domains. The diffusion of chemical species in dilute solutions is represented by a constitutive equation known as *Fick's law of diffusion*, which states that the transfer of species A in a binary mixture of A and B is expressed for a one-dimensional Cartesian coordinate as:

$$J_{Ax} = -D_{AB} \frac{dc_A}{dx} \quad \text{(diffusion)}, \quad (2.7)$$

where J_{Ax} (mol/(s m^2)) is the molar flux of species A in the x-direction, c_A (mol/m^3) is the molar concentration of species A in solution, and D_{AB} (m^2/s) is the binary diffusion coefficient of species A in the mixture. Fick's law is also used in mixtures that are not binary mixtures if species A is dilute. For multidimensional applications, (2.7) can be written in vector form as:

$$\vec{J}_A = -D_{AB} \vec{\nabla} c_A = -D_{AB} \left(\vec{i} \frac{\partial c_A}{\partial x} + \vec{j} \frac{\partial c_A}{\partial y} + \vec{k} \frac{\partial c_A}{\partial z} \right). \quad (2.8)$$

For the transport of heat, the equivalent constitutive equation is known as *Fourier's law of conduction* or thermal diffusion. One-dimensional heat flux in Cartesian coordinates is given by:

$$q_x = \frac{\dot{Q}_x}{A} = -k \frac{dT}{dx} \quad \text{(conduction)}, \quad (2.9)$$

where x is the direction of heat flow, q_x is the heat flux in that direction (W/m^2), $\dot{Q}_x$ is heat flow (W), k (W/(m K)) is the thermal conductivity, T (K) is the local temperature within the system, and A (m^2) is the area normal to the temperature gradient. If heat also flows in the y and z directions, then a more general expression for the heat flux is written as a vector quantity, $\vec{q}$:

$$\vec{q} = -k\vec{\nabla}T = -k\left(\vec{i}\frac{\partial T}{\partial x} + \vec{j}\frac{\partial T}{\partial y} + \vec{k}\frac{\partial T}{\partial z}\right). \tag{2.10}$$

For the transport of electrical charge, Q (Coulombs), *Ohm's law* can be written on a molecular scale as:

$$i_x = \frac{I_x}{A} = \frac{1}{A}\frac{\partial Q}{\partial t} = -\frac{1}{\Re_e'}\frac{dE}{dx} = -k_e\frac{dE}{dx}, \tag{2.11}$$

where i_x is the flux of electrical charge or the current density, I_x is the current, E is the electrical potential, $\Re_e'$ is the electrical resistivity, k_e is its inverse, or electrical conductivity, and A is the area normal to the electrical potential gradient. In more general terms, the current density can be written as a vector:

$$\vec{i} = -k_e\vec{\nabla}E. \tag{2.12}$$

If you move your hand in the x-direction through a fluid, like water, fluid layers above and below your hand will also move in the x-direction. The fluid closest to your hand will move the fastest, while fluid far from your hand will be nearly stationary. Because of the viscous or frictional nature of the fluid, x-momentum induced in the fluid layer next to your hand will be transmitted in the y-direction, perpendicular to the direction of movement. *Newtonian fluids* are those in which the flux of x-momentum in the y-direction is proportional to the negative of the velocity gradient. *Newton's law of viscosity* can be written:

$$\tau_{yx} = -\mu\dot{\gamma} = -\mu\frac{dv_x}{dy}, \tag{2.13}$$

where $\dot{\gamma}$ is the shear rate in the fluid, which is the velocity gradient orthogonal to the flow vector, in units of (1/s). The constant of proportionality μ is called the viscosity in units of (N s/m^2) or (Pa s). The momentum flux can also be interpreted as the shear stress τ_{yx} that acts on a plane normal to the y axis in the direction of the x axis in units of (N/m^2). These interpretations will be discussed in Chap. 4.

Other transport mechanisms are known to exist. A large temperature gradient can induce a mass flux (*Soret effect*) and a large concentration gradient can induce an energy flux (*Dufour effect*). However, in biomedical applications, these effects are expected to be negligible. The basic constitutive relations for one-dimensional molecular transport are summarized in Table 2.1. There are clear analogies between these four relations, showing why these transport mechanisms are often studied together.

2.5 Transport Mechanisms

Table 2.1 Constitutive relationships for 1D molecular transport [$x = n$ in (2.4)]

	Species	Momentum	Heat	Electrical charge
Extensive variable (flux)	Molar flux J_{Ax}	x-Momentum flux in y-direction τ_{yx}	Heat flux q_x	Current density i_x
Intensive variable (potential)	Molar concentration c_A	Velocity in x-direction v_x	Temperature T	Electrical potential E
Transport coefficient or property	Diffusivity of A in B D_{AB}	Fluid viscosity (Newtonian) μ	Thermal conductivity k	Electrical conductivity k_e
Constitutive relation	$J_{Ax} = -D_{AB}\dfrac{dc_A}{dx}$	$\tau_{yx} = -\mu\dfrac{dv_x}{dy}$	$q_x = -k\dfrac{dT}{dx}$	$i_x = -k_e\dfrac{dE}{dx}$

2.5.1.2 Molecular Transport Properties

The ability of a material to participate in a particular type of transport is related to its molecular structure. Consequently, solids, liquids, and gases will vary in their ability to enable different types of transport processes. This ability can be quantified and measured in terms of a unique property value associated with each mechanism of transport. These features of a material are called constitutive properties, and they are used in constitutive equations, as described in the foregoing section, to calculate the magnitude of a transport flow in a particular type of material subjected to a driving potential. The value of a property gives a measure of the rate at which a transport flow will occur in response to a given driving potential.

Values for many transport properties are available in tables, some of which can be found in Appendix C of this text. These values are obtained by monitoring transport processes in specific materials under very tightly controlled experimental conditions. Fluid viscosity of Newtonian fluids, for instance, is computed from direct measurements of wall force and shear rate in a viscometer having a known wall area. When these measured values are inserted in the appropriate constitutive equation (2.13), the only unknown is the fluid viscosity. Similar experiments can be designed to provide direct measures of other constitutive properties, such as thermal conductivity and diffusion coefficients. This approach to analysis is termed the "inverse solution method", since it contrasts with the common application where a constitutive equation is used to calculate a transport flow from known values of the applied driving potential and the transport property.

Transport properties generally depend on the state conditions of the material under investigation, particularly temperature and pressure. Therefore, it is important that state variables be controlled in experiments designed to measure transport properties. Likewise, when attempting to identify transport properties for a particular application, it is important to ensure that the transport properties selected are valid under the applicable state conditions.

In some cases, large variations in state variables may exist in particular transport processes. This situation may arise frequently for heat transfer applications where large temperature gradients are applied to a system. An analogous situation may arise for fluid flow resulting from a steep gradient in applied pressure. In such cases, transport properties might vary significantly with position. If significant property gradients are present, it will be necessary to either use an average approximation for the magnitude of the property throughout the system or to account for local property variations when applying the constitutive relation.

Sometimes, it is useful to combine constitutive properties with other properties to form a new property that has particular physical significance. Examples include the *kinematic viscosity*, v, and the *thermal diffusivity*, α, both of which have units of (m²/s):

$$v = \frac{\mu}{\rho}. \tag{2.14}$$

$$\alpha = \frac{k}{\rho c_p}. \tag{2.15}$$

The quantity c_p is the *heat capacity per unit mass at constant pressure* (or simply *specific heat*) of the material. It is not accidental that the properties v and α from fluid and heat transport match the units for the diffusion coefficient, D_{AB}, for mass transport. We will discuss the analogies among the different transport domains in the next section.

2.5.1.3 1D Molecular Transport Analogies

An extensive variable can be converted to an intensive variable by dividing by the volume of the system. For example, the number of moles of species A in a system N_A, an extensive variable, can be converted to the average molar concentration of species A, c_A, an intensive variable, by dividing by the system volume. The flux of an extensive property X can be written as the product of a transport coefficient γ_X and the negative gradient of the same extensive property per unit volume, $\tilde{X}$, where the tilde above the symbol can be interpreted as the quantity "per unit volume." For one-dimensional transport in the y-direction:

$$\text{Molecular flux of } X \text{ in } y\text{-direction} = -\gamma_X \frac{\partial \tilde{X}}{\partial y}. \tag{2.16}$$

A consequence of writing the flux equation in this form is that the dimensions of the transport coefficient γ_X will always be the same. The flux of X will have dimensions of X per unit time per unit length squared and the dimensions of $\partial \tilde{X}/\partial y$ will be the dimensions of X per unit length to the fourth power. Consequently, γ_X will have dimensions of length²/time.

2.5 Transport Mechanisms

Returning to our previous example, if the extensive property X is the number of moles of species A, then:

$X = N_A$, number of moles
Molecular flux of species A in y-direction $= J_{Ay}$ (molar flux by diffusion)
$\tilde{X} = \tilde{N}_A = c_A$ (moles per unit volume or concentration)

Therefore, according to (2.16):

$$J_{Ay} = -\gamma_{N_A} \frac{\partial c_A}{\partial y}. \tag{2.17}$$

Comparing this with Fick's law of diffusion (2.7), we found that the transport coefficient γ_{N_A} is equal to the diffusion coefficient D_{AB}, which has the appropriate dimensions of length^2/time.

If the extensive variable is the internal energy, U, then:

$X = U = mc_p(T - T_R)$, where T_R is a reference temperature.
Molecular flux of energy in y-direction $= q_{Ay}$ (conduction heat flux)
$\tilde{X} = \tilde{U} = \rho c_p(T - T_R)$, internal energy per unit volume

Therefore, according to (2.16):

$$q_y = -\gamma_U \frac{\partial \tilde{U}}{\partial y} = -\gamma_U \frac{\partial}{\partial y}(\rho c_p T). \tag{2.18}$$

Comparing this with Fourier's law of conduction (2.9), we find $\gamma_U = k/\rho c_p = \alpha$, the thermal diffusivity, from (2.15). Indeed, this has the same dimensions as the diffusion coefficient.

Finally, let us take x-momentum as our extensive variable:

$X = mv_x$
Molecular flux of x-momentum in the y-direction $= \tau_{yx}$
$\tilde{X} = \rho v_x$.

The flux equation becomes:

$$\tau_{yx} = -\gamma_{mv_x} \frac{\partial(\rho v_x)}{\partial y}. \tag{2.19}$$

Comparing this with Newton's law of viscosity, (2.13), the transport variable γ_{mv_x} must be μ/ρ, which is the kinematic viscosity ν defined previously in (2.14). Again, the momentum transport variable will have dimensions of length^2/time.

Equations (2.17)–(2.19) all have similar forms and have transport coefficients with the same dimensions: D_{AB}, α, and ν. The similarities in the relations between the fluxes and gradients for 1D transport are known as the *transport analogies*.

2.5.2 Convective Transport Mechanisms

The second major transport mechanism is the transport of materials, energy, momentum, and charge by bulk fluid motion. This is known as convective transport. Let us take as an example the flow of a fluid into a control volume as shown in Fig. 2.9. Recall that the flux of a quantity X is the rate of change of X per unit area per unit time. Let us apply that definition to an element of mass Δm that enters a system in the x-direction in a time interval Δt. The bolus passes through a cross-sectional area A_c in time Δt by bulk fluid motion. The mass that enters is equal to the fluid density times the fluid volume that enters the control volume. The volume of fluid entering the system in time Δt is equal to the cross-sectional area times the length of the bolus ΔL. But the length of the bolus relative to the time it enters is simply equal to the fluid velocity in the x-direction. Therefore, the convective mass flux in the x-direction is found to be ρv_x:

$$\text{Convective mass flux} = \frac{1}{A_c}\frac{\Delta m}{\Delta t} = \frac{1}{A_c}\frac{\rho A_c \Delta L}{\Delta t} = \rho\frac{\Delta L}{\Delta t} = \rho v_x. \quad (2.20)$$

Note that this is also equal to the x-momentum of the entering fluid per unit volume. In a similar manner, if species A also enters the control volume, the convective molar flux of A will be equal to $c_A v_x$:

$$\text{Convective molar flux of } A = \frac{1}{A_c}\frac{c_A A_c \Delta L}{\Delta t} = c_A\frac{\Delta L}{\Delta t} = c_A v_x. \quad (2.21)$$

Using the same procedure, the convective momentum flux is found to be ρv_x^2:

$$\text{Convective } x\text{-momentum flux} = \frac{1}{A_c}\frac{\Delta(m v_x)}{\Delta t} = \rho v_x^2. \quad (2.22)$$

The convective heat flux arises primarily from the transport of internal energy U into the system by bulk fluid movement. This is equal to product of the mass flux,

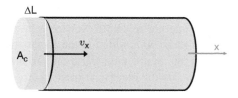

Fig. 2.9 Convection of volume $A_c \Delta L$ into control volume

2.6 Macroscopic Transport Coefficients

the specific heat capacity at constant pressure, and the temperature relative to some standard reference temperature T_R:

$$\text{Convective heat flux} = \frac{1}{A_c}\frac{\Delta U}{\Delta t} = \frac{1}{A_c}\frac{\Delta[mc_p(T-T_R)]}{\Delta t} = \rho v_x c_p(T-T_R). \quad (2.23)$$

In the case of heat transfer, we also need to consider a third mechanism, radiation heat flux, which will be introduced in Chap. 8. The total flux in 1D flow situations can be found by adding the flux contributions from all mechanisms. Neglecting all but molecular and convective fluxes, the total fluxes are:

$$\text{Total mass flux} = \rho v_x. \quad (2.24)$$

$$\text{Total flux of } x\text{-momentum} = \rho v_x^2 + \tau_{yx} = \rho v_x^2 - \mu\frac{dv_x}{dy}. \quad (2.25)$$

$$\text{Total molar flux of species } A = N_{Ax} = c_A v_x + J_{Ax} = c_A v_x - D_{AB}\frac{dc_A}{dx}. \quad (2.26)$$

$$\text{Total heat flux} = \rho v_x c_p(T-T_R) + q_x = \rho v_x c_p(T-T_R) - k\frac{dT}{dx}. \quad (2.27)$$

The student should be aware that some texts refer to transport by bulk fluid motion as *advection* and the combined transport by molecular motion and bulk fluid motion as *convection*. Most engineering texts, including this one, associate convection with transport by bulk fluid motion.

2.6 Macroscopic Transport Coefficients

In many situations, we are interested only in transport at the macroscopic scale where convection and molecular motion may both play a role. Rather than applying the expressions for molecular and convective fluxes derived in previous sections, a more empirical approach involving the use of dimensional analysis has proven to be quite fruitful. Details of the general approach will be presented in Chap. 3 and applications will be presented in Chaps. 5, 9, and 13. For macroscopic transport in a single medium, we can express the flow of an extrinsic transport quantity as the product of a transport coefficient and a driving force. To use a familiar example, the flow of charge in an electrical wire is expressed as:

$$I = \frac{\Delta E}{\Re_e}, \quad (2.28)$$

where I is the electric current in the wire, ΔE is the electrical potential difference across the wire, and $\Re_e$ is the electrical resistance. In some instances, the electrical resistance is nearly constant and in others it may depend on the potential difference. For a wire with length L and cross-sectional area A_c, the average flux of electrical charge or *current density* $\langle i \rangle$ can be found by dividing (2.28) by A_c:

$$\text{Average charge flux} = \langle i \rangle = \frac{I}{A_c} = \frac{\Delta E}{A_c \Re_e} = \left(\frac{L}{A_c \Re_e}\right) \frac{\Delta E}{L}. \qquad (2.29)$$

A quantity in braces $\langle \rangle$ is defined in this text as the *average value* of that quantity over the cross-section of the system. Since ΔE is the inlet potential minus the outlet potential, then the average gradient of electrical potential in the wire over its length L is $-\Delta E/L$. Therefore, (2.29) can be interpreted as a macroscopic version of (2.11), a constitutive relationship between the average current density and the average negative gradient of electrical potential. In this case, the electrical resistance is directly proportional to the electrical resistivity of the material, $\Re_e = \Re_e' L/A_c$, or $\Re_e = L/k_e A_c$, where k_e is the electrical conductivity of the material.

An empirical expression analogous to (2.28) can be written for the *volumetric flow rate* Q_V of fluid through a horizontal conduit:

$$Q_V = \frac{\Delta P}{\Re_f}. \qquad (2.30)$$

ΔP is the hydrostatic pressure difference between inlet and outlet of the conduit and $\Re_f$ is the resistance to flow. The resistance will depend on conduit geometry, fluid properties, and flow rate. If the cross-sectional area of the conduit is constant and flow is laminar, then resistance is constant. However, if flow is turbulent, the resistance will depend on the pressure difference and empirical methods relating the flow resistance to a friction factor must be used. The technique is described in Sect. 5.6. If the conduit is not horizontal, then (2.30) must be modified to include gravitational effects, as described in Sect. 5.5.

Total *mass flow* w through a conduit is equal to the volumetric flow multiplied by the fluid density:

$$w = \rho Q_V = \frac{\rho \Delta P}{\Re_f}. \qquad (2.31)$$

The area perpendicular to the direction of flow in a conduit is the cross-sectional area of the conduit, A_c. If A_c is constant for the entire length L of the conduit, the average mass flux can be found by dividing the mass flow by A_c:

$$\text{Average mass flux} = \frac{w}{A_c} = \rho \frac{Q_V}{A_c} = \rho \langle v \rangle = \frac{\rho \Delta P}{\Re_f A_c} = \frac{\rho L}{\Re_f A_c}\left(\frac{\Delta P}{L}\right). \qquad (2.32)$$

The quantity $\langle v \rangle$ *is defined as the average velocity* in the conduit. Since ΔP is the upstream pressure minus the downstream pressure, then $-\Delta P/L$ is the average

2.6 Macroscopic Transport Coefficients

pressure gradient in the conduit. Therefore, (2.32) can be interpreted as a macroscopic version of a constitutive relationship between the average mass flux and the average negative pressure gradient, with the transport coefficient equal to $\rho L/(\Re_F A_c)$.

Since heat flows from regions of high temperature to low temperature, an empirical expression for the flow of heat through a material can be written as:

$$\dot{Q} = \frac{\Delta T}{\Re_T}. \tag{2.33}$$

where $\dot{Q}$ is the *flow of thermal energy or heat flow*, ΔT is the temperature difference across the media, and $\Re_T$ is the thermal resistance. The thermal resistance in a solid will depend on geometry and material properties such as thermal conductivity and density. Thermal resistance in fluids will also depend on how the velocity is distributed in the fluid, which in turn depends on other fluid properties, such as viscosity.

If the area perpendicular to the direction of heat flow is constant and equal to A, and the material has length L, then the average heat flux $\langle q \rangle$ will be:

$$\text{Average heat flux} = \frac{\dot{Q}}{A} = \langle q \rangle = \frac{\Delta T}{\Re_T A} = \frac{L}{\Re_T A}\left(\frac{\Delta T}{L}\right). \tag{2.34}$$

The average temperature gradient in a solid material under these conditions is $-\Delta T/L$. Consequently, (2.34) can be thought of as a macroscopic version of the constitutive relation for 1D conduction, (2.9). In that case, the thermal resistance can be shown to be inversely proportional to the thermal conductivity of the material ($\Re_T = L/kA$). However, if the material is a flowing fluid, the thermal resistance will also depend on other fluid properties, on the velocity distribution, and on whether the flow is laminar or turbulent. Thus, the macroscopic thermal resistance must account for both convective and molecular heat transfer.

The macroscopic expression for molar flow W_A of species A through a solid material B is:

$$W_A = \frac{\Delta c_A}{\Re_{AB}}, \tag{2.35}$$

where $\Re_{AB}$ is the resistance to the transport of species A in material B and Δc_A is the difference in concentration of species A between the two ends of the material. For a solid material or stationary fluid with length L and with area A perpendicular to the direction of flow, the average molar flux $\langle N_A \rangle$ is:

$$\langle N_A \rangle = \frac{W_A}{A} = \frac{L}{\Re_{AB} A}\left(\frac{\Delta c_A}{L}\right). \tag{2.36}$$

This is a macroscopic version of (2.7), with $\Re_{AB} = L/(D_{AB} A)$ for a solid. However, as in the case of heat transfer, if the material is a flowing fluid, then convective transport mechanisms will contribute to the mass transfer resistance.

Table 2.2 Relationships for 1D macroscopic transport

	Species	Mass	Heat	Electrical charge
Extensive variable (flow)	Molar flow W_A	Mass flow w	Heat flow $\dot{Q}$	Current flow I
Intensive variable (difference)	Molar concentration Δc_A	Pressure ΔP	Temperature ΔT	Electrical potential ΔE
Empirical relation	$W_A = \dfrac{\Delta c_A}{\Re_{AB}}$	$w = \dfrac{\rho \Delta P}{\Re_f}$	$\dot{Q} = \dfrac{\Delta T}{\Re_T}$	$I = \dfrac{\Delta E}{\Re_e}$
Constitutive relation	$\langle N_A \rangle = \dfrac{L}{\Re_{AB}A}\left(\dfrac{\Delta c_A}{L}\right)$	$\rho \langle v \rangle = \dfrac{\rho L}{\Re_f A_c}\left(\dfrac{\Delta P}{L}\right)$	$\langle q \rangle = \dfrac{L}{\Re_T A}\left(\dfrac{\Delta T}{L}\right)$	$\langle i \rangle = \dfrac{L}{\Re_e A_c}\left(\dfrac{\Delta E}{L}\right)$
Transport coefficient	$L/(\Re_{AB}A)$	$\rho L/(\Re_f A_c)$	$L/(\Re_T A)$	$L/(\Re_e A_c)$

The macroscopic transport relationships are summarized in Table 2.2.

The analogous nature of these expressions provides motivation for studying these transport phenomena together.

2.7 Interphase Transport

By interphase transport, we refer to transport across the interface separating two media. For example, we might be interested in the transport of energy, momentum, or species between a swimmer and the water or between blood and the wall of a blood vessel.

Let us consider first the transport of momentum. Since fluid exhibits the phenomenon of "no slip", the motion of a swimmer's hand will impart momentum to the fluid in contact with the hand, and because water is viscous, momentum will be transferred from the water in contact with the hand to the water far away from the hand. The flow of momentum has the dimensions of momentum flux multiplied by area. Since momentum flux ρv^2 has the same dimensions as a force per unit area, the force F_s exerted by the hand on the fluid would be expected to be proportional to the product of area and the change in momentum flux:

$$F_s = \frac{f}{2} A \left[\rho v_s^2 - \rho v_\infty^2 \right], \tag{2.37}$$

where v_s is the velocity of the solid (hand), v_∞ is the fluid velocity, f is defined as a *friction factor*, and A is a characteristic area, defined differently for internal and external flow situations. The factor of 2 arises from the fact that f was originally defined in terms of the kinetic energy per unit volume ($\frac{1}{2}\rho v^2$). In many instances, we are interested in the force exerted by the fluid on a stationary object ($v_s = 0$), such as the force of flowing blood on a vessel wall or the force on a fisherman standing in a flowing river. This force by the fluid on the solid is commonly referred to as the *drag force* ($F_D = -F_s$). In this case, the velocity will vary from zero at the

2.7 Interphase Transport

wall to v_∞ in a thin fluid layer near the solid, referred to as a *boundary layer*. The friction factor will depend on the thickness of the boundary layer, which in turn depends on fluid properties and the average fluid velocity.

For external flow past a stationary body, the drag force is:

$$F_D = fA_f \left(\frac{1}{2}\rho v_\infty^2\right) \quad \text{(external flow)}. \quad (2.38)$$

The area A_f used to define the friction factor for external flow is the frontal area of the object or the area that the object projects onto a plane that is perpendicular to the flow. For internal flow in a conduit, the wetted surface area of the conduit S_w and the average fluid velocity $\langle v \rangle$ in the conduit are used to define the friction factor f. For steady flow through the conduit, the drag force on the conduit wall is equal to the pressure difference across the ends of the conduit ΔP multiplied by the cross-sectional area of the conduit A_c:

$$F_D = \Delta P A_c = fS_w \left(\frac{1}{2}\rho \langle v \rangle^2\right) \quad \text{(internal flow)}. \quad (2.39)$$

This can be used to find the pressure difference across a tube for a given geometry and flow rate. The friction factors for both internal and external flows depend on the geometry of the object and properties of the fluid.

A *thermal boundary layer* will also exist in the fluid region near the interface. This is defined as the region where the temperature varies from the interfacial temperature to a value that is very near the fluid temperature far from the interface. For external flow past a solid object, heat transfer across the interface is governed by *Newton's Law of cooling*. This states that the heat flux across the interface from solid to fluid is proportional to the difference in temperature between the solid surface T_S at the interface and the fluid temperature far from the interface T_∞:

$$q = \frac{\dot{Q}_s}{S} = h[T_S - T_\infty] \quad \text{(external flow)}. \quad (2.40)$$

A positive value of the heat flux q or heat flow $\dot{Q}_s$ indicates heat loss from the surface with area S (cooling). The proportionality factor h is known as the *convective heat transfer coefficient*. For internal flow, the heat transfer coefficient is defined slightly differently:

$$q = h[T_w - T_m] \quad \text{(internal flow)}. \quad (2.41)$$

T_m is the mixing cup temperature or the bulk fluid temperature in the fluid flowing through the conduit and T_w is the wall temperature (i.e., interface temperature). T_m is also known as the *flow averaged temperature*, which is the temperature that would be measured if the conduit were cut at the axial location of interest, the fluid collected in a cup, and the temperature of the well-mixed fluid in the cup measured. The heat flux is positive when heat flows from the wall to the fluid. The interfacial heat transfer coefficient h is a transport coefficient that depends on the thermal boundary layer and is normally determined experimentally.

The flow of species A across a fluid–solid interface is slightly more complex. Within the fluid phase, we can rewrite (2.36) in a manner analogous to Newton's law of cooling for external and internal flow situations:

$$N_A = k_{Af}\left([c_{Af}]_S - [c_{Af}]_\infty\right) \text{ (external flow).} \tag{2.42}$$

$$N_A = k_{Af}\left([c_{Af}]_w - [c_{Af}]_m\right) \text{ (internal flow).} \tag{2.43}$$

The subscripts are analogous to those used for heat transfer, with $[c_{Af}]_m$ representing the flow averaged mixing cup or bulk concentration of species A in fluid f. These expressions serve as definitions for k_{Af}, the *convective mass transfer coefficient for species A in fluid f*. The mass transfer coefficient depends on geometry, fluid properties, and the distribution of velocity in the boundary layer. The factor k_{Af} is the inverse of $\Re_{AB}A$ in (2.36). Concentrations in (2.42) and (2.43) reflect concentrations of species A *in the fluid phase*, as indicated by the subscripts "Af". Similar expressions could also be written for the transport of species A through the solid. Since species A may have a different solubility in the fluid than in the solid, a concentration difference will normally exist between the fluid and solid at the interface, even when the flux of species A is zero. Let us define a *partition coefficient* Φ_{A12} as the equilibrium ratio of concentration of species A in material 1 $(c_{A1})_{eq}$ relative to material 2 $(c_{A2})_{eq}$:

$$\Phi_{A12} = \left[\frac{c_{A1}}{c_{A2}}\right]_{eq}. \tag{2.44}$$

In this case, material 1 is the fluid f and material 2 is the solid s. Φ_{Afs} should be interpreted as the partition of species A in the fluid relative to the solid under equilibrium conditions. The order of the last two subscripts is important. Alternatively, we could define a partition coefficient Φ_{Asf} for species A between the solid relative to the liquid under equilibrium conditions:

$$\Phi_{Asf} = \left[\frac{c_{As}}{c_{Af}}\right]_{eq} = \frac{1}{\Phi_{Afs}}. \tag{2.45}$$

The two partition coefficients are inversely related to each other. If we assume that local equilibrium occurs at the interface, then:

$$[c_{Af}]_{\text{interface}} = \Phi_{Afs}[c_{As}]_{\text{interface}}. \tag{2.46}$$

Although the fluid far from the interface is not in equilibrium with the solid, a key assumption made in interfacial transport is that species A in the fluid and solid are in local equilibrium *at the interface*. For the internal flow situation, the interface

2.7 Interphase Transport

represents the conduit wall. Substituting (2.45) and (2.46) into (2.43) provides an expression for the mass flux in terms of the concentration of A in the solid at the interface and the bulk concentration of A in the fluid:

$$N_A = k_{Af}\left(\Phi_{Afs}[c_{As}]_w - [c_{Af}]_m\right) \text{ (internal flow).} \qquad (2.47)$$

Therefore, the appropriate driving force for the movement of species A across the interface is not $[c_{As}]_w - [c_{Af}]_m$, but rather $\Phi_{Afs}[c_{As}]_w - [c_{Af}]_m$. The convective mass transfer coefficient k_{Af} accounts for both convective and diffusive fluxes.

For external flow situations, the bulk concentration in the liquid is replaced by the concentration far from the wall, $[c_{Af}]_\infty$, and the concentration in the solid at the interface is $[c_{Af}]_s$. The flux of species A from wall to fluid, N_A, can be computed from the empirical relation:

$$N_A = k_{Af}\left[\Phi_{Afs}[c_{As}]_s - [c_{Af}]_\infty\right] \text{ (external flow).} \qquad (2.48)$$

Unfortunately, we rarely know the interfacial concentration in either the fluid or the solid at a solid–fluid interface. For an internal flow application, where the solid represents the conduit wall, it is more common to know the concentration at the outside surface of the conduit wall, $[c_{As}]_o$, rather than the concentration at the inside surface. Applying an analog of (2.43) to the flux of species A through the conduit wall:

$$N_A = k_{As}\left([c_{As}]_o - [c_{As}]_w\right). \qquad (2.49)$$

The factor k_{As} is a mass transfer coefficient governing the transport of species A in the conduit wall. It is directly proportional to the diffusion coefficient D_{As} for species A in the wall material. Assuming steady-state transport, we can eliminate $[c_{As}]_w$ from (2.47) to (2.49), to provide the flux in terms of the outside wall concentration and the bulk fluid concentration:

$$N_A = P_A\left(\Phi_{Afs}[c_{As}]_o - [c_{Af}]_m\right). \qquad (2.50)$$

P_A is defined as the *permeability of species A* and is related in this case to the mass transfer coefficients in the fluid and solid as follows:

$$\frac{1}{P_A} = \frac{1}{k_{Af}} + \frac{\Phi_{Afs}}{k_{As}}. \qquad (2.51)$$

In biotransport applications, we are often interested in finding the flux of solutes from one fluid region to another across a barrier, such as a cell membrane or a capillary wall. This actually involves transport across three barriers in series: two fluid–solid interfaces and a barrier material. Although the concentrations at either fluid–barrier interface are not generally known, an analysis that includes all three

Table 2.3 Interphase transport relationships

	Species	Momentum	Heat
Through variable (flux)	Molar flux N_A	Momentum flux F_{Dx}/A	Heat flux q
Across variable (external)	Molar concentration $[c_{Af}]_s - [c_{Af}]_\infty$ or $\Phi_{Afs}[c_{As}]_s - [c_{Af}]_\infty$	KE per unit volume $\frac{1}{2}\rho v_\infty^2$	Temperature $T_S - T_\infty$
Across variable (internal)	Molar concentration $[c_{Af}]_w - [c_{Af}]_m$ or $\Phi_{Afs}[c_{As}]_w - [c_{Af}]_m$ (interface) or $C_{A2} - \Phi_{A21}C_{A1}$ (across barrier)	KE per unit volume $\frac{1}{2}\rho\langle v\rangle^2$	Temperature $T_w - T_m$
Transport coefficient	Mass transfer coefficient k_{Af} or P_A	Friction factor f	Heat transfer coefficient h

barriers (see Chap. 12) allows us to express the flux from side 2 to side 1 in terms of bulk concentrations on the two sides of the barrier, $[c_{A1}]_m$ and $[c_{A2}]_m$:

$$N_A = \mathsf{P}_A\left([c_{A2}]_m - \Phi_{A21}[c_{A1}]_m\right) = \mathsf{P}_A(C_{A2} - \Phi_{A21}C_{A1}). \tag{2.52}$$

We use an uppercase "C" to refer to mixing cup or bulk concentrations in the two fluids. The *permeability* P_A *of the membrane to species A* is an overall mass transfer coefficient that includes effects of the boundary layers on each side of the barrier and conductance through the barrier itself.

A summary of empirical relationships used to describe interfacial transport of momentum, species, and heat is shown in Table 2.3.

2.8 Transport in Biological Systems: Some Unique Aspects

Transport phenomena are studied by students in many disciplines of engineering and science, and they are analyzed and used in design by professionals in a very wide range of occupations. Although the spectrum of applications in transport is large, there is a common thread of principles and methods that runs through all of these different domains. However, it is our observation over decades of research and teaching that biotransport tends to fall at an extreme end of this spectrum owing to unique features of the systems that are frequently encountered.

In multiple ways, biological and biomedical systems tend to give rise to more complex transport processes than do inanimate systems. It is helpful to be aware of some of the more commonly encountered features unique to living systems to plan

2.8 Transport in Biological Systems: Some Unique Aspects

the best problem-solving strategies. The following list is not exhaustive, but it presents a flavor of some of the characteristics and behaviors that can make dealing with biological systems challenging.

- Biotransport often occurs in systems having irregular-shaped geometries. Organs, limbs, muscles, tissues, vasculature and respiratory networks, bones, and many other biological structures have shapes that are asymmetric, curvilinear, and three-dimensional. These geometries are typically not compatible with standard coordinate systems, requiring simplifying assumptions, approximations to closed form mathematical solutions, or numerical solution methods such as finite difference and finite element that can accommodate specific geometries via meshing techniques.
- Many biotransport processes are nonlinear in which the behavior is not a simple function of the inputs that cause a process to occur. As a consequence, the constitutive relations that describe these behaviors can introduce mathematical functions that are difficult to incorporate into analytical functions. An example is that blood flow can be altered drastically as a function of many different tissue properties such as temperature, pressure, and mechanical stress, as well as systemic functions such as heart rate and muscle work level. As a consequence, the behavior of a system can be altered due to internally driven changes causing nonlinear function. Again, numerical and approximation methods can be applied to obtain a solution.
- Nearly all biological materials are composite, consisting of areas that have differing properties. Since the spatial domain is not homogeneous, it is difficult to apply a simple analytical solution for the entire process. At material interfaces it is necessary to ensure that the boundary conditions embody both continuity of potential (such as concentration and temperature) and flow (such as molecules and heat).
- Many biological materials have anisotropic properties, meaning that their behavior is directionally dependent. This is another feature that makes the use of closed form analytical solutions problematic.
- Some biological systems may change their properties and/or phase state during a process in response to imposed stresses. Examples are the denaturation of molecules at elevated temperatures and the freezing of tissues at subzero temperatures. The result is that there can be dramatic changes in properties over both time and position during a process.

Since all of these unique features of biological systems make the most simple closed form analytical solutions problematic, numerical simulations find very widespread use in biomedicine. Fortunately, there now exists commercial software that is powerful and user friendly so that the difficult process of self-generating code is no longer an impediment in many cases. The theory behind the use of these advanced techniques is largely beyond the scope of this introductory text. Many excellent sources are available that present directions and details for these methods.

All of these phenomena require that special experimental methods be applied to characterize the phenomena and to measure the relevant transport properties. The availability of accurate constitutive property values is important in successfully modeling and solving biotransport problems. Despite these complexities, the fundamental principles and methods of analysis presented in this text will provide a good starting point for approaching all biotransport problems and will often provide good first-order approximations to their solutions.

2.9 Summary of Key Concepts

In Chap. 1, we presented the Generate Ideas Model, which provides a systematic way of thinking about a biotransport problem before attempting to solve the problem. In this chapter, we consider the essential components of the problem that should be resolved before a formal analysis is attempted.

Fundamental Concepts. We introduce the fundamental concepts that define the field of biotransport, and will refer back to these over and over again throughout the text. In Sect. 2.2 we discuss how to define the system of interest and all of the interactions that the system has with the environment at the system boundaries. We define the extensive transport properties that cross the boundaries and the intensive properties that tend to cause the motion. In Sect. 2.3, we discuss how to identify the appropriate scale of interest. The two scales of interest in most biotransport applications are the macroscopic scale, in which average properties are of interest, and the microscopic scale, in which spatial variations in properties are of interest.

Conservation Relationships. Once a system and the appropriate transport properties have been identified, and a scale selected, the appropriate conservation relationship for each extensive property must be selected. General conservation principles applied to mass, momentum, energy, species, and electrical charge can be stated simply as in (2.2):

$$\left\{ \begin{array}{c} \text{Rate of} \\ \text{accumulation} \\ \text{of the quantity} \\ \text{within a system} \end{array} \right\} = \left\{ \begin{array}{c} \text{Net rate the} \\ \text{quantity is produced} \\ \text{within the system} \end{array} \right\} + \left\{ \begin{array}{c} \text{Net rate the} \\ \text{quantity enters} \\ \text{through the} \\ \text{system boundary} \end{array} \right\}.$$

Constitutive Relationships. The final components that must be assembled before a biotransport problem can be analyzed are the constitutive relationships that govern the transport of extensive properties across the system boundaries. For microscopic systems, the molecular flux of an extensive property is proportional to the negative gradient of an appropriate intensive property, given by (2.4):

$$\text{Flux}_n = (\text{constitutive property}) \cdot \left[-\frac{\partial (\text{potential})}{\partial n} \right].$$

2.9 Summary of Key Concepts

The flux of an extensive property is defined as the rate the property crosses a boundary per unit time per unit area of the boundary. Specific examples that will be used often in this text are Newton's law of viscosity (2.13), Fourier's law of thermal conduction (2.9), Fick's law of diffusion (2.7), and Ohm's law of electrical conduction (2.11). The analogies between these molecular transport mechanisms are summarized in Table 2.1 and in Sect. 2.5.1.3, providing motivation for studying these concepts together. If the process involves fluid motion, then we must consider convection of the properties of interest in addition to molecular motion (Sect. 2.5.2).

Macroscopic Relationships. In macroscopic applications, the constitutive relationship often relates flow of an extensive property to a difference in the potential across the system:

$$\text{Flow} = \frac{\Delta(\text{potential})}{\Re},$$

where $\Re$ is the resistance to flow. These relationships are summarized in Table 2.2.

Transport Coefficients (Internal Flow). When fluid flows past an object, the empirical relationships that govern transport from the fluid in contact with the boundary and fluid far from the boundary can be summarized as:

$$\text{Flux} = (\text{transport coefficient}) \cdot (\Psi_s - \Psi_\infty) \text{ (external flow)}.$$

For heat transfer applications, this is known as Newton's law of cooling, with the flux equal to the heat flux q away from the surface, the transport coefficient equal to the heat transfer coefficient h, Ψ_s equal to the surface temperature, and Ψ_∞ equal to the temperature in the fluid far away from the system surface. For mass transfer applications, the flux represents the flux of species A, N_A, away from the surface, the transport coefficient is called the mass transfer coefficient k_A, and the potential difference is the difference in concentration between the fluid in contact with the system surface and fluid far from the surface. In fluid momentum applications, the flux represents the momentum flux or shear stress at the wall, the transport coefficient is the friction factor, and the potential difference is the kinetic energy per unit mass of the fluid flowing past the object.

Transport Coefficients (External Flow). If the fluid flows inside a conduit, the flux at the surface is given by the following expression:

$$\text{Flux} = (\text{transport coefficient}) \cdot (\Psi_w - \Psi_m) \text{ (internal flow)}.$$

The potential difference in this case represents the difference between the potential at the wall and the mean (flow averaged or mixing cup) potential in the fluid flowing through the conduit. The analogous expressions are given in Table 2.3.

Interphase Transport. The analogy between heat and mass transport breaks down when dealing with macroscopic transport from one phase to another or across a barrier connecting two materials. In general we can write:

$$\text{Flux}_{1\to 2} = (\text{transport coefficient}) \cdot (\Psi_{1m} - \Phi \Psi_{2m}) \text{ (across barrier)}.$$

The subscripts "m" refer to mean properties in the materials. The factor Φ equals unity for the case of heat transfer and equals the equilibrium partition coefficient between materials 1 and 2 (2.44) for the case of mass transfer.

2.10 Questions

2.10.1. Why do we study momentum, heat, and mass transfer together?
2.10.2. Distinguish between a closed system and an open system.
2.10.3. What separates a system from the environment?
2.10.4. Distinguish between intensive (intrinsic) and extensive (extrinsic) properties and give examples of each.
2.10.5. Distinguish between macroscopic and microscopic approaches to transport problems. Which independent variables will not appear in macroscopic problems?
2.10.6. Should a microscopic or a macroscopic species mass balance be used to determine blood concentration as a function of position from the inlet of an alveolar capillary sheet? Justify your answer.
2.10.7. Under what circumstances can matter be treated as a continuum?
2.10.8. Give an example where the assumption that matter can be treated as a continuum is probably invalid.
2.10.9. Write the general expression for the conservation of a quantity "X" in a system.
2.10.10. Which conservation statement describes the conservation of oxygen inside a red blood cell?
2.10.11. Can a system in equilibrium with its surroundings exchange mass, energy, or momentum with the surroundings?
2.10.12. Two systems are in equilibrium. Can the temperatures of the two systems be different? Can the concentrations of species "A" in each system be different?
2.10.13. Can a system that is in equilibrium with its surroundings have a temperature gradient at the system boundary?
2.10.14. Explain the negative sign in the molecular flux vs. gradient expressions for momentum, heat, and species transport.
2.10.15. How do molecular and convective transport mechanisms differ?
2.10.16. Distinguish between flux, diffusive flux, and convective flux for mass transfer of species "A."
2.10.17. What are Newton's Law of viscosity, Fick's Law of diffusion, and Fourier's Law of conduction? What quantities in these equations are analogous? Define any symbols used.
2.10.18. What transport mechanisms are involved in lung CO_2 exchange?
2.10.19. Which *gradients* influence the total passive flux of Cl^- ions through a channel in a cell membrane? (Assume that there is a net flow of water through the same channel.)

2.10.20. In an electrical system, current is the "flow variable" while a voltage difference provides the driving force. A fluid mechanics system may be modeled similarly if we treat volumetric flow as the flow variable and which variable as the driving force?

2.10.21. Distinguish between resistance and conductance.

2.10.22. How are heat transfer and mass transfer coefficients defined for internal and external flow situations?

2.10.23. What area is used when converting fluxes to flows?

2.10.24. How is a heat transfer coefficient related to thermal resistance?

2.10.25. What factors influence heat and mass transfer coefficients for internal and external flow situations?

2.10.26. Two materials in contact are assumed to be in "local equilibrium" at the interface. What does this mean in terms of the temperature of each material at the interface? What does this mean in terms of the concentrations of nonreacting chemical species in each material at the interface?

2.10.27. The concentration of O_2 in a cell membrane is higher than the concentration of O_2 in the intracellular fluid. O_2 will: (a) be transported into the cell, (b) be transported out of the cell, (c) not be transported into or out of the cell, or (d) not enough information is given. Justify your answer.

2.10.28. The concentration of CO_2 in the cell membrane near the intracellular side is higher than the concentration of CO_2 in the membrane near the extracellular side. CO_2 will: (a) be transported into the cell, (b) be transported out of the cell, (c) not be transported into or out of the cell, or (d) not enough information is given. Justify your answer.

2.10.29. Species A can diffuse from material B to material C even though the concentration of A in C is higher than the concentration of A in B if: (a) the solubility of A in C is higher than the solubility of A in B; (b) the solubility of A in B is lower than the solubility of A in C, (c) the solubility of A in B is equal to the solubility of A in C, or (d) species A cannot move from B to C under these circumstances

2.10.30. What is a partition coefficient? Why is it important? Explain the subscript convention used for Φ_{ABC}.

2.10.31. Why doesn't the partition coefficient appear in equations (2.42) and (2.43)? Why does it appear in (2.47) and (2.48)?

2.10.32. What factors affect membrane permeability?

2.11 Problems

2.11.1 Transport Examples

Identify one specific example each of heat, mass, and momentum transport in a mammalian system. Briefly describe each example and identify the driving force and the resistance. In each example, identify a parameter that can change in value, how that parameter is likely to change, and the effects of that change on heat, mass, or momentum transfer in your example.

2.11.2 Convective and Diffusive Fluxes

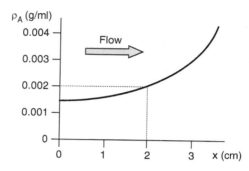

A solution containing species A in water moves to the right with a mass average velocity of 0.06 cm/min. Species A obeys Fick's law with a diffusion coefficient of 10^{-5} cm^2/s. The mass concentration of species A vs. position is shown in the figure. Find the diffusion flux and convective flux of species A at the position $x = 2$ cm. Is the total flux of A greater or less than the convective flux of A? Justify your answer.

2.11.3 Relation Between Macroscopic and Microscopic Transport Coefficients, Heat Transfer

Show that the resistance to heat transfer is inversely proportional to the thermal conductivity, $\Re_T = L/kA$, for steady-state, 1D conduction through a slab of material with thickness L and cross-sectional area A.

2.11.4 Relation Between Macroscopic and Microscopic Transport Coefficients, Mass Transfer

Show that the resistance to mass transfer is inversely proportional to the diffusion coefficient, $\Re_{AB} = L/D_{AB}A$, for steady-state, 1D diffusion of species A through a slab of material B with thickness L and cross-sectional area A.

2.11 Problems

2.11.5 Conservation Principles

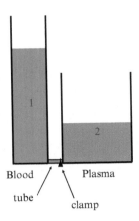

Blood in reservoir 1 is separated from plasma in reservoir 2 by a rubber tube, which is initially clamped. The height of blood in reservoir 1 is greater than the height of plasma in reservoir 2, and the initial blood temperature is different than the initial plasma temperature. The hematocrit value (i.e., ratio of cell volume to total volume) of the blood in reservoir 1 is 0.45. After the clamp is released, blood from reservoir 1 flows into reservoir 2. We wish to predict how the following factors vary with time: height in each reservoir, pressure difference across the tube, blood flowrate from reservoir 1 to reservoir 2, hematocrit value in each reservoir, and temperature in each reservoir. How would you go about setting up this problem? What principles will you apply? What assumptions will you make? What additional information do you need?

2.11.6 Conservation and Chemical Reaction

Species A and species B are added to a well-mixed volume of water V, where they combine chemically $A + 2B \rightarrow C$. Write molar conservation equations for each species if the rate of production of species C (mol/s) is proportional to the product of the molar concentration of A and the molar concentration of B squared. The proportionality factor is k. What additional information do we need to know before we can predict the concentrations of A, B, and C as functions of time? (Do not solve).

2.11.7 Measurement of Lung Gas Volumes

A pneumotachograph is a device used to measure airflow at the mouth. Air flow rate is computed from the pressure drop measured across a fixed resistance such as a fine screen or honeycomb structure during breathing. Integration of the flow provides the volume of air that passes through the mouth. The amount of air left in the lung after a maximal expiration is known as the residual volume (RV). The amount of air in the lung at the end of a maximal inspiration is known as the total lung capacity (TLC). The amount of air left in the lung at the end of a normal expiration is known as the functional residual capacity (FRC). FRC, TLC, and RV cannot be detected with a pneumotachograph since all of the air in the lung does not flow through the device. However, several important lung volumes can be measured with this device. Design a procedure that uses a pneumotachograph to measure the following lung volumes:

(a) Expiratory reserve volume (ERV) = FRC − RV
(b) Inspiratory capacity (IC) = TLC − FRC
(c) Tidal volume (TV) = volume at the end of a normal inspiration minus FRC
(d) Inspiratory reserve volume (IRV) = IC − TV
(e) Vital capacity (VC) = TLC − RV

2.11.8 Measurement of FRC or RV

FRC or RV cannot be measured using a pneumotachograph, since, by definition, RV cannot pass through the device. These volumes can be detected by breathing in a gas mixture containing a small amount of a tracer gas that does not penetrate the alveolar wall. Apply conservation of tracer before and after inspiring a gas mixture with known volume and tracer concentration to estimate the total volume of gas in the lung. Describe a procedure that involves the pneumotachograph and inspiration of the gas mixture to compute FRC or RV. Neglect the volume of the conducting airways in relation to alveolar volume.

2.11.9 Continuum Concepts

The volume of air that contains 1,000 molecules at standard temperature and pressure is approximately 4×10^{-7} cm^3. This is small enough that we can consider it to be a "point" in space. Outer space consists primarily of hydrogen atoms with a density of 1.66×10^{-24} g/cm^3. What volume of outer space contains 1,000 molecules? Compare that with the volume of air above.

2.12 Challenges

2.12.1 Preliminary Research, Design of a Heart–Lung Machine

Background: New surgical devices and procedures have helped to improve cardiac care of many patients. Open heart surgery requires opening the chest cavity which collapses the lungs, and the heart is usually stopped so that a surgeon can perform

2.12 Challenges

techniques such as a bypass or valve replacement. The functions of the heart and lung must be maintained, not to mention preservation of the heart tissue once it has stopped beating. A young biotechnology company produces various medical devices such as IV units and blood filtration systems. They believe that they could use some of their technology to design an advanced heart lung machine. The company hires you, a biomedical engineer, to conduct a feasibility study. *Challenge:* You are to determine which major components the system should contain, and you must consider the potential design challenges these systems may encounter while interfacing with the human body. *Generate Ideas:* What are the important system variables that you should consider in your design? What system components would you recommend? If these components were placed in series, in what order would you place them? Why?

2.12.2 Alternate Applications for a Device

The company that hired you as a consultant on a heart–lung machine project is concerned that an open heart surgery application for the proposed device is too specialized. *Challenge:* Are there other applications that you might suggest that will expand the market for this device? *Generate ideas:* Use the transport principles presented in this chapter to provide a list of applications that might convince the company executives to build the device.

2.12.3 Design of a Heart–Lung Machine

The company decides to move ahead with the project and assigns you to lead the team that designs and builds the machine. *Challenge:* Suggest preliminary designs for each subsystem. *Generate Ideas:* Estimate specific input values for each subsystem. What are the desired output variables of each subsystem? Suggest a specific design for each subsystem that can provide the desired outputs. What conservation relationships are relevant for each subsystem? What constitutive relationships are relevant for each subsystem? What additional information will you need before you can actually design a heart–lung machine that provides the desired outlet variables?

2.12.4 Heat Loss by Conduction

The temperature difference across the skin layer of a resting individual is measured to be 1°C. *Challenge:* Assuming all steady-state heat loss from the body is by conduction through the skin, estimate basal metabolic rate (BMR) of the body. *Generate Ideas:* What general principle governs the conduction of heat through the skin? Should we use a macroscopic or microscopic approach to determine heat flow through the skin? What system should be analyzed to determine BMR? What conservation principle applies? What additional information is required if we are to solve this problem? Search the literature to find this information and estimate BMR. Compare this with values found in the literature and provide explanations if the values are different.

2.12.5 Respiratory Heat Loss

Tidal volume for an average human is of 500 ml of air and normal breathing rate is about 12 breaths/min. *Challenge:* What is the rate that heat is lost via respiration if inspired air at 20°C is heated to body temperature by the time it is expired? *Generate Ideas:* What system should be analyzed? What assumptions will you make? Is a macroscopic or microscopic analysis appropriate? What additional information is needed? Perform a literature search to find any missing information and estimate the respiratory heat loss. What fraction of normal heat generated by the body does this represent?

References

Schmid-Schönbein GW, Diller K (2005) Transport processes in biomedical systems: a roadmap for future research directions. Ann Biomed Eng 33:1136–1141

Chapter 3
Modeling and Solving Biotransport Problems

3.1 Introduction

The objective of building a model of a transport process is to predict the behavior of a system under a specified set of operating conditions or to understand the interplay among various factors that contribute to a process of interest. Models can be either mathematical models or physical models. Consequently, there are two general approaches used to solve problems in biotransport: the theoretical approach and the empirical approach.

The theoretical approach is based on the application of conservation principles and relevant constitutive relationships to the system under analysis. Application of a systematic method of analysis, which we have termed the "Generate Ideas Model" (GIM), results in a mathematical model of the physical system. Once the model is developed and a general solution is obtained, model parameters can be adjusted until the computed output is equal to the desired output.

The empirical approach is based upon experimental measurements made either on the system of interest itself or a physical model of the system. This approach is often used to find the dependency of friction factors, heat transfer coefficients, mass transfer coefficients, thermal resistance, fluid resistance or mass transfer resistance on system geometry, material properties, flow rates and other relevant parameters. Dimensional analysis, described in detail later in this chapter, can be used to reduce the number of important parameters by combining them to form dimensionless numbers. This allows results based on measurements made under one set of experimental conditions to predict system behavior under a completely different set of experimental conditions.

As a simple example, consider the problem of selecting a 100 cm length of tubing that links the femoral vein of a patient to a blood oxygenator. The tubing must be capable of delivering a flow rate of, say, 3 L/min when a pressure difference of 2 cmH$_2$O is applied across the ends of the tubing. The theoretical approach would involve the application of conservation of mass, conservation of momentum, the constitutive relationship for blood, and other auxiliary conditions and assumptions. The resulting expression would relate the flow rate to the pressure difference, fluid viscosity, tubing length, and tubing diameter. This expression could then be used to select an appropriate tubing diameter that will deliver the appropriate flow for the specified pressure difference and tubing length. In the experimental approach, system components are systematically adjusted until the measured output equals the desired output. In this case, 100 cm lengths of tubing with different diameters would be tested until one was found that provided the correct flow rate for the applied pressure difference. An alternate approach would be to rely on relevant published experimental data, which in this case will involve the determination of an appropriate friction factor for flow in a tube.

There are advantages and disadvantages to each approach. The empirical approach is accurate, but purchasing materials for a trial-and-error procedure could be very costly. The apparatus and instruments used in experiments can be expensive and experimental set up times and measurement times can be quite lengthy. The theoretical approach is generally less expensive and the predicted result can often be computed in a relatively short period of time. However, the theoretical prediction is based on idealized geometry and assumptions that can lead to inaccurate results. The empirical approach, although accurate for the design conditions, cannot often be used to reliably predict behavior under different operating conditions. Often, the most practical method is to first use the theoretical approach to select a prototype, then make adjustments with the empirical approach until the degree of accuracy is achieved.

3.2 Theoretical Approach

The primary goal of this text is to provide students with a general approach for identifying, formulating, and solving biotransport problems. We have defined an explicit theoretical approach by which any new problem can be addressed in order to develop a set of equations that describe the governing elements of a process. The method is described as the Generate Ideas Model, which is an essential step of the Legacy Cycle described in Chap. 1. The method involves the following steps:

1. Clearly identify the goal of the problem analysis. Knowing the desired output will help you identify the appropriate principles and procedures that should be applied.
2. Define the system as it pertains to the process of interest. How does the system interact with the environment to cause the process of interest? What geometric factors characterize the system? Can the system geometry be simplified for the purposes of analysis without introducing large errors?
3. Determine which principles need to be included in the analysis procedure. In particular, identify appropriate:
   (a) Conservation principles
   (b) Constitutive laws
   (c) Other relevant empirical relationships
4. Develop a mathematical model of the transport processes based on the appropriate conservation principles and constitutive relationships.
5. Simplify the model if appropriate. List all simplifying assumptions and provide justifications for making them.

3.2 Theoretical Approach

6. Develop a procedure for solving the resulting equation or set of equations and identify auxiliary relationships necessary for their solution, such as boundary and initial conditions.
7. Find appropriate property values and transport coefficients from the literature, and compute the final numerical results. Provide a graphical display of the results as appropriate to enhance interpretation and understanding of the solution.
8. Examine the meaning of the solution. Is it realistic? Does it agree with published data? If possible, relax one or more of the assumptions made to see how it affects the final result. What are the limitations of the solution? Remember, this is a mathematical model of a real system that depends on the original formulation. Maybe something important was left out of the formulation. Are there conditions where the model predicts unreasonable results, such as negative concentration? These need to be identified and the limitations specified.
9. A very important aspect of solving a problem is to realize what important aspects about obtaining a solution that you do not yet know. Ask yourself "What additional important information is needed to develop a good solution for this problem?" Understanding what you do not know about solving a particular problem is one of the most helpful aspects of being able to make progress on solving a challenging problem.

You will find that in this textbook we have placed a strong emphasis on learning and following the above strategy for solving problems. We present many examples in which we describe in detail how to follow the solution method. We believe learning effective problem-solving methods that will serve you well throughout a lifetime career involves much more than just being able to identify an equation to apply to a specific problem. Real-world problem solving requires creative and innovative thinking to be able to translate the characteristics of a physical process into an appropriate working system definition, a set of assumptions that match the process and physical aspects of the system, and the governing equations, initial and boundary conditions, and property values to be used in simulating the process. We have tried to emphasize materials and exercises that will guide students in developing this type of problem-solving ability, in addition to providing a well-organized and clear introductory exposition of the knowledge and principles that undergird the discipline of biotransport. Our hope is that the readers would not only learn the subject of biotransport but also become versatile and adept problem solvers. We view this combination as the true essence of being a good engineer.

3.2.1 *Geometric Considerations*

Details of system geometry are often ignored as a first approximation. Tapered or elliptically shaped blood vessels are modeled initially as circular cylinders. The head is modeled as a sphere for the sake of heat and mass transfer analyses. Curvature effects are neglected when analyzing the diffusion of solutes through the skin. These idealized geometries will not provide exact results, but will often be accurate enough to provide reasonable predictions of actual transport.

When defining a system, it is necessary to select a coordinate system. If the system has one or more axes of symmetry, it will generally simplify the problem if coordinate axes are aligned with the planes of symmetry or axes of symmetry. For instance, when dealing with fluid flow between parallel plates, we would select a rectangular coordinate system with the origin in the center of the fluid, between the plates. This is better than selecting the origin at either wall, since we would expect velocity to be symmetrical about the central plane. Similarly, for heat transfer from a circular cylinder, we would select a cylindrical coordinate system with origin at the center of the cylinder. Diffusion of a solute in the intracellular space of a spherical cell calls for a spherical coordinate system with origin at the center of the cell. Because symmetry is assumed in these problems, an appropriate boundary condition at the plane, the axis or the point of symmetry is $(\partial \phi/\partial r)_{r=0} = 0$, where ϕ represents velocity, temperature, or concentration, and r is the distance from the plane, axis, or point of symmetry.

3.2.2 Governing Equations

The governing equations for transport problems are formed by substituting appropriate constitutive equations into the relevant mass, species, momentum, and/or energy conservation relationships. The exact form of the governing equation or equations for a particular transport problem will depend on the specifics of the problem, but certain generalities can be identified, depending on whether the macroscopic or microscopic approach is used.

If the system can be analyzed using a steady-state macroscopic approach, then the governing expression(s) will consist of one or more algebraic equations. If the system is modeled using a "lumped" or time-dependent macroscopic approach, one or more ordinary differential equations (odes) will result. The number of odes will depend upon the number of species involved and whether or not energy or mass exchange also occurs simultaneously.

If the system is modeled with the microscopic approach, one or more partial differential equations will generally result. The most complex analyses will involve systems in which spatial and temporal variations in velocity, species concentrations, and temperature occur. In this text, we will normally confine our treatment to simpler microscopic systems in which only one or two transport variables depend on time and/or position.

3.2.3 Solution Procedures

Algebraic equations resulting from steady-state macroscopic balances can usually be solved analytically. In some instances, when transport coefficients depend on the

3.2 Theoretical Approach

transported quantity, it may be necessary to use a root finding algorithm, such as Newton's method (e.g., Matlab function fzero), to find an appropriate solution. Since there are no spatial or temporal variations of transported quantities within the system, initial conditions and boundary conditions are not applicable to such systems.

Ordinary differential equations that arise from macroscopic balances are first-order equations that can often be solved by separating variables and integrating over time. These are known as initial value problems. An initial condition for each transport variable is required before a solution can be found. If the equations are coupled, as in the case of chemical reactions, analytic solutions may be complex, or even impossible to obtain. In such cases, a first-order ordinary differential equation (ode) solver, such as Matlab's ode45, can be used to find the numerical solution for even large systems of coupled ordinary differential equations.

Although it is generally not difficult to specify the governing equations and auxiliary conditions that apply to complex microscopic systems, it is usually much more difficult to solve the resulting set of coupled partial differential equations (pde). We give a few relatively simple examples of analytic solutions in the text, but we do not expect the student to have had a course in partial differential equations. Numerical pde solvers, such as Matlab's pdetool, are now available in several computer packages. To use these solvers, the student must be able to specify the correct pde(s) and identify the appropriate initial and/or boundary conditions necessary to obtain a solution. The governing equations for unsteady-state transport problems are all first order in time, so if the problem has time dependency, a single initial condition for each transport variable must be specified. Initial conditions may be constant, or they may have known spatial distributions. If the transport variables within the system depend on position, then boundary conditions are needed before a solution can be obtained. The number of boundary conditions will depend on the order of the governing equation. If the governing equation is second order in y and first order in z, then two boundary conditions must be specified in y and one in z if a unique solution is to be found. Common boundary conditions used in fluid mechanics, heat transfer, and mass transfer applications are discussed in Sects. 4.4, 8.5 and 12.10.

Simplifying assumptions are often made to reduce the complexity of the problem. In addition to idealizations concerning the shape of the system, transport in specific directions is often ignored. For instance, heat transfer is ignored along the axis of a long bar immersed in a temperature bath relative to heat transferred from the sides of the bar. Formal scaling methods will be introduced in Sect. 3.2.5 and discussed in greater detail in Chap. 7. Scaling can assist in identifying terms in the governing equations that are not exactly zero, but are negligible relative to other important terms.

Many important transport problems analyzed with the microscopic approach can be reduced to situations where transport is steady state and the transport variable depends on a single dimension. The governing equation is a second-order ordinary differential equation, known as a boundary value problem. This can often be solved by separating variables and integrating twice. The constants of integration can then be determined by applying the two boundary conditions appropriate for that system. Other methods for obtaining analytic solutions for various common boundary conditions are presented in the text.

3.2.4 Presentation of Results

How should we present the solution to a transport problem in the most meaningful way? Too often students approach a problem with a single application in mind, and substitute specific values into the governing equations before generating the solution. Substituting a number for fluid density, rather than a symbol, will almost inevitably lead to problems with units in the solution and will limit the result to fluids with that particular density. Students who use this approach will completely miss the generality of their solution. Always use symbols for variables in the development of your solution. Do not substitute in values for system dimensions or material properties until the form of the solution is found. Only then should you try to apply the general solution to the problem at hand. The same solution can then be used for systems with different dimensions and for materials with different properties.

Analytic solutions to transport problems are often complex and system behavior is not easily visualized by simply presenting the solution in the form of an equation. Graphical presentations of results are often more useful. However, to gain the most generality from graphical solutions, they should be presented in dimensionless form, where a dimensionless transport parameter is plotted as a function of a dimensionless time and/or dimensionless position. The appropriate normalizing factors depend on the problem at hand, and arise from the solution to the problem. Excellent examples are the charts found in Appendix D, which are used to predict the unsteady-state temperature or concentration profiles in solid objects of various shapes after immersion in an environment with a different temperature or concentration. The same chart can be used to predict the caffeine concentration in a spherical coffee bean during a decaffeination process that can be used to predict the temperature variation in an orange during a cold snap.

3.2.5 Scaling: Identification of Important Dimensionless Parameters

In most instances, it is not possible to obtain analytic solutions to the governing partial differential transport equations. However, it is often possible to neglect certain terms in the full equations relative to other terms. In nearly stationary fluids, convective transport mechanisms may be unimportant relative to molecular mechanisms. An order of magnitude estimate of the ratio of convective to molecular

3.2 Theoretical Approach

transport can be made using a technique known as scaling. With this method, we first identify or estimate the maximum values that each dependent variable and each independent variable can attain for the problem at hand. We then divide the actual variables by the maximum value to construct dimensionless dependent and independent variables that are of order unity. Finally, we substitute these into the convective to molecular flux ratio. A dimensionless group will then be formed that provides an estimate of the magnitude of the convective flux relative to the molecular flux.

The procedure is best illustrated with an example. Consider the flow of fluid between two parallel plates that have a length L and gap H (Fig. 3.1). The inlet temperature is T_0 and inlet concentration of species A is c_{A_0}. The average velocity in the fluid between the plates is $\langle v \rangle$. The inside surface temperature and concentration are kept at T_1 and c_{A_1}, respectively. We can define the following dimensionless variables that range from 0 to about 1:

$$x^* = \frac{x}{L} \tag{3.1}$$

$$y^* = \frac{y}{H} \tag{3.2}$$

$$v_x^* = \frac{v_x}{\langle v \rangle} \tag{3.3}$$

$$T^* = \frac{T - T_0}{T_1 - T_0} \tag{3.4}$$

$$C^* = \frac{c_A - c_{A_0}}{c_{A_1} - c_{A_0}}. \tag{3.5}$$

If the maximum velocity is known, then this could be used in (3.3) in place of the average velocity. Substituting (3.1)–(3.3) into the ratio of convective to molecular momentum transport:

$$\frac{\text{Convective transport of } x\text{-momentum}}{\text{Molecular transport of } x\text{-momentum}} = \frac{(\rho v_x^2)}{\left(-\mu \dfrac{\partial v_x}{\partial y}\right)} = \frac{\rho \langle v \rangle H}{\mu} \left\{ \frac{(v_x^*)^2}{\left(-\dfrac{\partial v_x^*}{\partial y^*}\right)} \right\}. \tag{3.6}$$

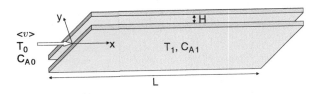

Fig. 3.1 Flow between parallel plates

Because of the scaling procedure, the term in brackets should be of order unity. Consequently, the ratio of momentum transfer by convection relative to molecular transport is approximately:

$$\frac{\text{Convective transport of } x\text{ - momentum}}{\text{Molecular transport of } x\text{ - momentum}} \approx \frac{\rho \langle v \rangle H}{\mu} = Re_H, \quad (3.7)$$

where Re_H is a dimensionless number known as the Reynolds number based on the distance between the plates H. This can be interpreted as the ratio of inertial forces to viscous forces in the fluid. At high Reynolds numbers, inertial terms dominate and the flow may tend to become turbulent or chaotic. At low Reynolds numbers, viscous effects will dominate and flow will be laminar.

Let us compare the conduction of heat in the x-direction with the conduction of heat in the y-direction:

$$\frac{\text{Heat conduction in } x\text{ - direction}}{\text{Heat conduction in } y\text{ - direction}} = \frac{-k\frac{\partial T}{\partial x}}{-k\frac{\partial T}{\partial y}} = \frac{H}{L}\left\{\frac{\left(\frac{\partial T^*}{\partial x^*}\right)}{\left(\frac{\partial T^*}{\partial y^*}\right)}\right\} \approx \frac{H}{L}. \quad (3.8)$$

If the distance between plates is small with respect to the length of the plates, conduction in the x-direction can be safely neglected with respect to conduction in the y-direction. The same is true if we compare diffusion in the x-direction to diffusion in the y-direction.

Taking the ratio of convective heat transfer in the x-direction to conduction heat transfer in the y-direction:

$$\frac{\text{Convective heat flux}}{\text{Conduction heat flux}} = \frac{\rho v_x c_P (T - T_R)}{-k\frac{\partial T}{\partial y}} = \frac{\rho \langle v \rangle c_P H}{k} \left\{ \frac{v_x^* T^*}{\left(-\frac{\partial T^*}{\partial y^*}\right)} \right\}$$

$$\approx \left(\frac{\rho \langle v \rangle H}{\mu}\right)\left(\frac{\mu c_P}{k}\right). \quad (3.9)$$

Therefore, the ratio of convection to conduction is found to be the product of two dimensionless numbers. The first is the Reynolds number, Re_H. The second is known as the Prandtl number Pr. The Prandtl number can be rewritten as the ratio of kinematic viscosity (2.14) to thermal diffusivity (2.15):

$$Pr = \left(\frac{\mu c_P}{k}\right) = \left(\frac{\mu/\rho}{k/(\rho c_P)}\right) = \frac{\nu}{\alpha}. \quad (3.10)$$

Finally, comparing the relative magnitudes of diffusion in the y-direction to convection of solute A in the x direction:

3.2 Theoretical Approach

$$\frac{\text{Species A convection in } x\text{-direction}}{\text{Species A diffusion in } y\text{-direction}} = \frac{C_A v_x}{-D_{AB}\frac{\partial C_A}{\partial y}} = \frac{\langle v \rangle H}{D_{AB}} \left\{ \frac{v_x^* C_A^*}{\left(-\frac{\partial C_A^*}{\partial y^*}\right)} \right\} \quad (3.11)$$

$$\approx \frac{\langle v \rangle H}{D_{AB}}.$$

The dimensionless group $\langle v \rangle H / D_{AB}$ is known as the *Peclet number* based on the dimension H, Pe_H. The Peclet number can also be written in terms of the Reynolds number:

$$Pe_H = \frac{\langle v \rangle H}{D_{AB}} = \left(\frac{\rho \langle v \rangle H}{\mu}\right)\left(\frac{\mu}{\rho D_{AB}}\right) = Re_H \left(\frac{\nu}{D_{AB}}\right) = Re_H Sc. \quad (3.12)$$

The ratio of kinematic viscosity to diffusion coefficient is known as the *Schmidt number*, Sc. If the Peclet number is large, then convection dominates, but if the Peclet number is small, diffusion dominates. Since diffusion coefficients for solutes in liquids are quite small, convection often dominates in flowing liquids, even at very low velocities. If mass transfer by diffusion is an objective, then it is necessary to make H and $\langle v \rangle$ as small as is practically possible.

3.2.6 Examples of the Theoretical Approach

We are now ready to apply the theoretical approach outlined in the previous sections to solve actual problems in biotransport. We provide two example problems. The first illustrates a problem that can be solved using the macroscopic approach and the second is a problem that is solved using the microscopic approach. In each case, we begin by identifying the system of interest. We then apply conservation principles to the system and identify the relevant transport mechanisms that occur at the system boundaries. The macroscopic approach for momentum, heat and mass transport is discussed in greater detail in Chaps. 5, 9, and 13, respectively. The microscopic approaches for one-dimensional momentum, heat and mass transport are discussed in more detail in Chaps. 6, 10 and 14. These two examples will serve as an introduction to the two different approaches.

> **Example 3.2.6.1 Species Conservation in a Bioreactor.**
> Consider a well-mixed bioreactor system with constant volume V as shown in Fig. 3.2. Species A is produced at a constant rate per unit volume R_A by cells suspended in the system. The walls of the system are impermeable to fluid, but are permeable to species A with a permeability surface area product of $P_A S$. The concentration of species A is maintained at zero in the fluid surrounding the system. Our task is to find how the concentration varies in the reactor with time after the cells are introduced.

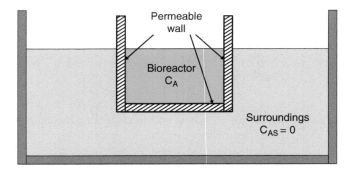

Fig. 3.2 Mass flow in a permeable bioreactor

Solution. *Initial considerations*: Since the fluid in the reactor is well mixed, the concentration of species A will vary with time, but not with position within the reactor. Consequently, a macroscopic approach is appropriate for analysis of this problem.

System definition and environmental interactions: Potential systems for analysis include the reactor fluid, the reactor wall, or a combination of the reactor fluid and wall. Analysis of the reactor wall can be used to predict the flux of species A through the wall if the concentrations of species A on each side of the wall are known. However, we only know the concentration on one side of the wall. Therefore, we must include the reactor fluid in the analysis, since this is where cells are producing species A. We could analyze the reactor fluid and reactor wall separately, then combine the solutions by realizing that the flow of species A through the wall is equal to the flow of species A out of the reactor fluid. Alternately, we can treat the reactor fluid and wall as a single system, "the reactor." This is the approach we will take.

Apprising the problem to identify governing equations: The reactor fluid and the cells within the reactor are prevented from moving to the surroundings by the presence of the reactor walls. Only species A can move from the reactor fluid into the surrounding fluid. Consequently, an appropriate governing principle is the conservation of species A in the system.

Analysis: We begin by applying conservation of species A to the bioreactor:

$$\left\{ \begin{array}{c} \text{Rate of} \\ \text{accumulation} \\ \text{of species } A \\ \text{within bioreactor} \end{array} \right\} = \left\{ \begin{array}{c} \text{Net rate species A} \\ \text{is produced} \\ \text{within bioreactor} \end{array} \right\} + \left\{ \begin{array}{c} \text{Net rate species A} \\ \text{enters through the} \\ \text{bioreactor boundary} \end{array} \right\}.$$

The rate of accumulation of species A on a molar basis is:

3.2 Theoretical Approach

$$\left\{\begin{array}{c} \text{Rate of} \\ \text{accumulation} \\ \text{of species A} \\ \text{within bioreactor} \end{array}\right\} = \frac{dN_A}{dt} = \frac{d}{dt}[C_A V] = V\frac{dC_A}{dt}.$$

N_A is the number of moles of species A present in the reactor fluid, V is the volume of the reactor, and C_A is the concentration of species A in the reactor fluid. The rate of production is equal to the rate of production per unit volume multiplied by the bioreactor volume:

$$\left\{\begin{array}{c} \text{Net rate species A} \\ \text{is produced} \\ \text{within bioreactor} \end{array}\right\} = R_A V.$$

Finally, the net rate at which species A enters the bioreactor is equal to minus the rate at which species A leaves the bioreactor through the walls. The outward flow can be found using the overall mass transfer coefficient from (2.52), $P_A S$ and the partition coefficient Φ_{ARS} between the reactor R and surrounding fluid S:

$$\left\{\begin{array}{c} \text{Net rate species A} \\ \text{enters through the} \\ \text{bioreactor boundary} \end{array}\right\} = -P_A S[C_A - \Phi_{ARS} C_{AS}].$$

Combining the three terms and recognizing that the concentration in the reservoir surrounding the bioreactor, C_{AS}, is zero, species conservation reduces to:

$$V\frac{dC_A}{dt} = R_A V - P_A S[C_A].$$

The solution to this differential equation for an initial condition $C_A(0) = 0$ is:

$$C_A = \frac{R_A V}{P_A S}\left[1 - \exp\left(-\frac{P_A S}{V}t\right)\right].$$

Examining and interpreting the results: The concentration will continue to rise in the reactor with time until it levels off at $C_{A\infty} = (R_A V)/(P_A S)$. The ultimate concentration in the reactor can be raised by adding more cells, which will increase R_A, by decreasing the permeability of the reactor walls, or by decreasing the surface area of the reactor walls. Increasing the volume of the reactor without increasing the number of cells will have no effect on the final concentration, since the product of R_A and V will remain constant. However, raising V will increase the total number of moles of species A that is ultimately contained in the system, and will affect the time constant of the system, $\tau = V/(\mathsf{P_A}S)$. The concentration can be written in dimensionless form:

$$\frac{C_A}{C_{A\infty}} = 1 - \exp\left(-\frac{t}{\tau}\right).$$

A plot of dimensionless reactor concentration vs. time relative to the time constant is shown in Fig. 3.3. The concentration will be within 99% of the final concentration after a period equal to five time constants. Note that the same dimensionless graph can be used for any combination of parameters of V, $\mathsf{P_A}$, S, or R_A. This is far more useful than a plot of C_A vs t for a single set of parameters.

Additional comments: We might ask what will happen when the permeability of the reactor wall approaches zero. Our analytic solution indicates that the concentration becomes undefined (i.e., equals 0/0). Using L'Hopital's rule, we can take the derivate of the numerator relative to $\mathsf{P_A}S$ and divide by the derivative of the denominator to find that the concentration in the reactor increases linearly with time as expected, having a slope equal to R_A, since $C_A(t)|_{\mathsf{P_A}S=0} = R_A t$. The same result could be obtained by setting $\mathsf{P_A}S = 0$ in the differential form of the conservation statement before integrating.

> **Example 3.2.6.2 Mass Flow and Heat Transfer in a Tapered Bronchiole.**
> Room temperature air is normally heated to body temperature by the time respiratory gases enter alveoli in the lung. Our goal is to determine the outlet air temperature T_L from a tapered bronchiole, given the inlet temperature T_0 and the wall temperature, which is assumed to be maintained at body temperature T_b. The cross-sectional area $A(x)$ is known as a function of axial position. Let us confine our analysis to mid-inspiration, where the flow is relatively constant. A schematic is shown in Fig. 3.4.

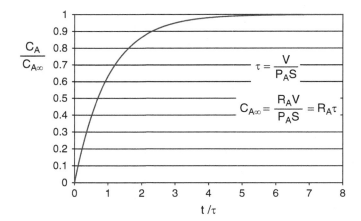

Fig. 3.3 Dimensionless reactor concentration vs. dimensionless time

3.2 Theoretical Approach

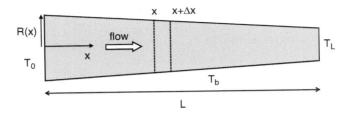

Fig. 3.4 Flow and heat transfer in a tapered bronchiole

Solution. *Initial considerations*: Since flow is steady and the bronchiole is tapered, velocity will increase with axial position, x. The mean temperature of air moving through the bronchiole will also increase as it moves downstream, since it is being heated by the bronchiolar wall. Since temperature and velocity must change with position, a microscopic approach should be used to solve this problem.

System definition and environmental interactions: Since the microscopic approach must be used, we will select a system that can shrink to an infinitesimally small volume. Let us assume that the air in a bronchiole at any axial position has uniform temperature and velocity, but that these both vary with axial position. For 1-D transport, our system can consist of the mass confined to the volume between position x and position $x + \Delta x$ within the bronchiole. The volume of the microscopic system is $A(x)\Delta x$, where A is the area perpendicular to the direction of air flow.

Apprising the problem to identify governing equations: The solution to this problem will involve application of the conservation of mass to determine how the velocity varies with position and conservation of energy to estimate how the temperature varies with position.

Analysis: Since mass can neither be created nor be destroyed, conservation of mass within the microscopic volume can be written:

$$\left\{ \begin{array}{c} \text{Rate of} \\ \text{accumulation} \\ \text{of mass} \\ \text{within system} \end{array} \right\} = \left\{ \begin{array}{c} \text{Net rate mass} \\ \text{enters through the} \\ \text{system boundary} \end{array} \right\}.$$

Since the flow is steady, the rate at which mass accumulates in the system volume is zero. The rate at which mass enters the system is the product of the inlet mass flux and the area at axial position x. The mass rate out is the same product at $x + \Delta x$. The net rate at which mass enters the volume $A\Delta x$ is:

$$\left\{ \begin{array}{c} \text{Net rate mass} \\ \text{enters through the} \\ \text{system boundary} \end{array} \right\} = (\rho v_x A)|_{x=x} - (\rho v_x A)|_{x=x+\Delta x}.$$

Substituting these expressions into the conservation statement:

$$0 = (\rho v_x A)|_{x=x} - (\rho v_x A)|_{x=x+\Delta x}.$$

Dividing this expression by Δx and letting Δx approach zero, we find:

$$-\frac{\partial}{\partial x}(\rho v_x A) = 0.$$

The density of air under these circumstances is nearly constant, v_x and A are assumed to vary only with axial position x, and their product is the volumetric flowrate, Q_V. Therefore:

$$\frac{d}{dx}(v_x A) = \frac{dQ_V}{dx} = 0.$$

Now, let us turn our attention to the rate of change of energy. Changes in kinetic energy and potential energy will be small in comparison with changes in internal energy. Neglecting these contributions, the conservation statement becomes:

$$\left\{\begin{array}{c}\text{Rate of}\\ \text{accumulation}\\ \text{of internal energy}\end{array}\right\} = \left\{\begin{array}{c}\text{Net rate of production}\\ \text{of thermal energy}\\ \text{in system}\end{array}\right\}$$
$$+ \left\{\begin{array}{c}\text{Net rate thermal energy}\\ \text{enters through the}\\ \text{system boundary}\end{array}\right\}.$$

During steady flow, the rate of accumulation of internal energy is zero. We will neglect any production of energy by viscous dissipation within the system. According to (2.23), energy enters the volume at x by convection and conduction. It will also leave at $x + \Delta x$ by the same mechanisms. In addition, heat is added to the air by conduction through the walls of the bronchiole, which is described by an interphase heat flux with internal heat transfer coefficient h (2.41) and a wall temperature T_w equal to body temperature T_b. The net flow of thermal energy across the system boundary is:

$$\left\{\begin{array}{c}\text{Net rate thermal energy}\\ \text{enters through the}\\ \text{system boundary}\end{array}\right\} = 0 = \{A[\rho v_x c_p(T-T_R) + q_x]\}_x$$
$$-\{A[\rho v_x c_p(T-T_R) + q_x]\}_{x+\Delta x}$$
$$+ h(x)(2\pi R(x)\Delta x)[T_b - T].$$

3.2 Theoretical Approach

The heat transfer coefficient is written as a function of axial position since it is known to depend on the Reynolds number, which increases as we move downstream. The cross-section of the bronchiole is assumed to be circular and the wall surface area of the small element of volume is $2\pi R \Delta x$. Dividing by Δx, and taking the limit as Δx approaches zero:

$$0 = -\rho c_p \frac{d}{dx}[v_x A(T - T_R)] - \frac{d}{dx}[A q_x] + 2\pi h(x) R(x)[T_b - T].$$

Comparison of the first two terms using scaling methods introduced in Sect. 3.2.5 shows the convection term relative to the conduction term to be $Re_L Pr$, where Re_L is the Reynolds number based on the length of the bronchiole and Pr is the Prandtl number. This product is large, so the axial conduction term can be neglected with respect to the axial convection term. With these simplifications, and after introducing the definition of the flow rate $v_x A = Q_V$ (which is independent of x}, we find:

$$\rho c_p Q_V \frac{dT}{dx} = 2\pi h(x) R(x)[T_b - T].$$

Our physical problem has now been reduced to a mathematical expression. Separating variables and integrating:

$$\int_{T_b - T_0}^{T_b - T_L} \frac{d[T_b - T]}{[T_b - T]} = -\frac{2\pi}{\rho C_p Q_V} \int_{x=0}^{x=L} h(x) R(x) dx.$$

T_0 is the temperature of air at the inlet of the bronchiole and T_L is the temperature at the outlet of the bronchiole ($x = L$). The final expression for air temperature at the exit of the bronchiole becomes:

$$T_L = T_b - [T_b - T_0] \exp\left[-\frac{\bar{h} S}{\rho c_p Q_V}\right],$$

where we have defined an average overall heat transfer coefficient as:

$$\bar{h} = \frac{\int_{x=0}^{x=L} 2\pi R(x) h(x) dx}{\int_{x=0}^{x=L} 2\pi R(x) dx} = \frac{1}{S} \int_{x=0}^{x=L} 2\pi R(x) h(x) dx.$$

S is equal to the inside surface area of the bronchiole. Both experimental and theoretical studies have shown that for laminar flow, the heat transfer coefficient for heat exchange in tubes with a constant wall temperature is given by the following expression (see Chap. 9):

$$h = 3.66 \, (k/d) = 1.83 \, (k/R),$$

where k is the thermal conductivity of air and d is the diameter of the tube. Assuming that we can apply this to a bronchiole with a slight taper, the product $h(x)R(x)$ is constant and equal to 1.83 k. This makes evaluation of the integral trivial:

$$\bar{h} = \frac{3.66\pi kL}{S} = 3.66\frac{k}{\langle d \rangle},$$

where $\langle d \rangle$ is the average diameter of the tapered bronchiole, equal to $S/\pi L$. If the taper is linear from inlet to outlet, then $\langle d \rangle$ will be the diameter at the mid point of the bronchiole. Substituting this expression for $\bar{h}$ above, we obtain a final expression for the dimensionless temperature at the bronchiole outlet:

$$\frac{T_L - T_0}{T_b - T_0} = 1 - \exp\left[-3.66\pi \frac{\alpha L}{Q_v}\right].$$

The thermal diffusivity of air α has been introduced in place of $k/(\rho c_p)$ (2.15).

Examining and interpreting the results: Note that if the negative exponent in the above expression is greater than five, the temperature at the outlet will be very nearly body temperature. To estimate actual air temperature at the outlet of a bronchiole, we need to provide estimates of the inlet temperature, the thermal diffusivity of air, the volumetric flow rate through the bronchiole and the length of the bronchiole. The heat transfer coefficient that we have selected is conditional on flow being laminar in the bronchiole. For flow in a tube, this is true if the Reynolds number based on the tube diameter is less than 2,200:

$$Re_{\langle d \rangle} = \frac{v_x \langle d \rangle}{v} = \frac{4Q_v}{\pi v \langle d \rangle}.$$

We will also need to know the mean diameter of the bronchiole and the kinematic viscosity of air. The kinematic viscosity of air at 37°C is 1.65×10^{-5} m^2 s^{-1} and the thermal diffusivity of air is 2.35×10^{-5} m^2 s^{-1}. In Table 3.1, we provide mean dimensions of airways all the way from the trachea (generation 0) to the bronchi (generations 1–4) to the conducting bronchioles (generations 5–16), and finally to the respiratory bronchioles (generations 17–19). For an average inspiratory flow rate of 500 ml s^{-1}, we have also computed the average volumetric flow through airways in each generation, the Reynolds number, the dimensionless exponent $3.66\pi \frac{\alpha L}{Q_v}$, and the dimensionless temperature. For a given environmental temperature, we could then compute the outlet temperatures for each airway generation.

Additional comments: For the inspiration flow rate selected, the flow is laminar for all generations and the expression used to compute h should apply. Figure 3.5 shows the temperature of air as we move through the respiratory tree for two different environmental temperatures: $-25°C$ and $+25°C$. Note that the air is predicted to be heated to 37°C by the tenth generation in both cases. The exponent

3.2 Theoretical Approach

Table 3.1 Geometric factors and transport parameters for heating of air in the lung

Order	Average diameter (mm)	Average length (mm)	Number of airways	Flow per airway (ml/s)	$Re_{(d)}$	$3.66\pi \dfrac{\alpha L}{Q_V}$	$\dfrac{T_L - T_0}{T_b - T_0}$
0	18	120	1	500	2136.49	0.06	0.063
1	12.2	47.6	2	250	1576.10	0.05	0.050
2	8.3	19	4	125	1158.34	0.04	0.040
3	5.6	7.6	8	62.5	858.41	0.03	0.032
4	4.5	12.7	16	31.25	534.12	0.11	0.104
5	3.5	10.7	32	15.625	343.36	0.19	0.169
6	2.8	9	64	7.8125	214.60	0.31	0.267
7	2.3	7.6	128	3.90625	130.63	0.53	0.409
8	1.86	6.4	256	1.953125	80.76	0.89	0.587
9	1.54	5.4	512	0.976563	48.77	1.49	0.776
10	1.3	4.6	1,024	0.488281	28.89	2.55	0.922
11	1.09	3.9	2,048	0.244141	17.23	4.32	0.987
12	0.95	3.3	4,096	0.12207	9.88	7.30	0.999
13	0.82	2.7	8,192	0.061035	5.72	11.95	1.000
14	0.74	2.3	16,384	0.030518	3.17	20.36	1.000
15	0.66	2	32,768	0.015259	1.78	35.42	1.000
16	0.6	1.65	65,536	0.007629	0.98	58.44	1.000
17	0.54	1.41	131,072	0.003815	0.54	99.88	1.000
18	0.5	1.17	262,144	0.001907	0.29	165.75	1.000

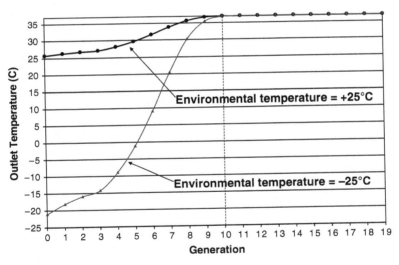

Fig. 3.5 Predicted outlet temperature as a function of airway generation for two different environmental temperatures

$3.66\pi\ \alpha L/Q_V$ becomes very large for generations 11 and above. Experiments have shown that air inspired at temperatures as low as $-80°C$ is heated to body temperature before it reaches the alveoli.

3.3 Empirical Approach

In many situations, the theoretical approach may prove to be too difficult to provide useful results. Perhaps, a theoretical relationship is lacking, the theory is extremely complex, or it is simply impractical to analyze the problem theoretically. This is particularly true in complex situations such as turbulent flow or boiling heat transfer. We can identify the important physical variables that influence a transport problem, but we do not know their interdependence. An empirical approach is necessary, but we would like to limit the number of experiments that need to be performed to determine the relevant empirical relationships.

3.3.1 The Buckingham Pi Theorem: Dimensional Analysis

The Buckingham Pi theorem provides a formalized method for reducing the number of variables that characterize a problem (Buckingham 1914, 1915; Langhaar 1951; Taylor 1974). It is based on the fact that the relationship between a dependent variable and the independent variables must be consistent in all of the fundamental dimensions. Fundamental dimensions are unique and cannot be expressed in terms of other fundamental dimensions.

In biotransport, we ordinarily deal with the set of fundamental dimensions shown in Table 3.2. A force, for instance, has fundamental dimensions of MLT^{-2}, while velocity has fundamental dimensions of LT^{-1}. The fundamental dimensions for the important variables in biotransport are shown in the Nomenclature section (Appendix A).

Application of the Buckingham Pi Theorem is best introduced through the use of an example. Consider blood flow through an elliptically shaped heart valve. Experimental evidence shows that the average blood velocity through the valve $\langle v \rangle$ is influenced by the valve geometry (major axis d_1, minor axis d_2), the pressure difference across the valve (ΔP) and the fluid density (ρ), but is not greatly influenced by the blood apparent viscosity (η). We are interested in finding how the average velocity varies with valve geometry. We can write a functional relationship between velocity and the other variables:

Table 3.2 Fundamental dimensions

Dimension	Symbol
Mass	M
Length	L
Time	T
Moles	N
Temperature	Θ
Current	I

3.3 Empirical Approach

$$\langle v \rangle = f(d_1, d_2, \rho, \Delta P). \tag{3.13}$$

This equation must be dimensionally homogeneous, so the function f must have fundamental dimensions of LT^{-1}. Therefore, the variables d_1, d_2, ρ, and ΔP must be combined in such a way that the dimensions of the function are LT^{-1}. The first three variables in the list do not contain the fundamental dimension of time T, so no combination of these variables will produce a variable with dimensions of velocity. In addition, since ΔP contains the dimension of mass M, and d_1 and d_2 do not, these three variables cannot be combined to produce a variable with dimensions of velocity. The variables ΔP and ρ can be combined such that the square root of the ratio $\Delta P/\rho$ has units of velocity. A second variable with these same units could be found by multiplying $(\Delta P/\rho)^{1/2}$ by d_2/d_1 or an appropriate dimensionless function of d_2/d_1. With these simplifications, based on dimensional homogeneity, we can rewrite (3.13):

$$\langle v \rangle = \left[\sqrt{\frac{\Delta P}{\rho}} \right] \phi \left(\frac{d_2}{d_1} \right), \tag{3.14}$$

where ϕ is an unknown function of d_2/d_1. According to the functional relationship in (3.13), we might first conduct an experiment where we alter the pressure difference across a valve with fixed d_1 and d_2, while keeping the fluid density the same. This single graph can probably be characterized by measuring the velocity at 10 different values of ΔP. To determine the influence of fluid density on velocity, we might repeat the series of experiments using ten fluids having different densities. A family of 10 curves, based on 100 measurements could be displayed on a single graph. To explore the effect of altering d_2, these same measurements should be repeated for 10 values of the minor diameter of the valve. These results could be displayed on 10 pages, reflecting 1,000 data points. Finally, all of these measurements would need to be repeated for ten different values of d_1, resulting in a total of 10,000 data points and 100 pages to display all of the data. This is a tremendous amount of data to collect, although accurate, interpolation between imposed values of the four independent variables might prove to be difficult.

Now, consider the simplification in experiments posed by (3.14) over (3.13). Only the ratios $\Delta P/\rho$ and d_2/d_1 influence the velocity. We can hold d_1 and ρ constant and measure velocity vs. pressure difference at 10 different values of d_2. It is not necessary to vary ρ and ΔP or d_1 and d_2 independently, because they only appear as ratios. Consequently, data representing $\langle v \rangle$ vs. $(\Delta P/\rho)^{1/2}$ with d_2/d_1 as a parameter can all fit on a single page. What a tremendous simplification to go from 10,000 measurements to 100 measurements! The experiments can all be performed with the same fluid, using 10 different valves rather than with 10 different fluids and 100 valves.

However, this problem can be simplified even further. Multiplying (3.14) by $(\rho/\Delta P)^{1/2}$ and squaring both sides leads to the following dimensionless equation:

$$\frac{\rho \langle v \rangle^2}{\Delta P} = \phi^2 \left(\frac{d_2}{d_1} \right). \tag{3.15}$$

This equation shows that a single dimensionless variable involving $\langle v \rangle$ is a function of another dimensionless variable, the ratio of minor to major valve axes. This relationship can be plotted on a single graph. Therefore, through the process of dimensional analysis, we have reduced this problem from one in which the velocity depends on four distinct parameters ($d_1, d_2, \Delta P$, and ρ) to one where a dimensionless velocity depends on a single dimensionless geometric variable. Not only does this procedure reduce the number of experiments required to establish an empirical relationship, but it also allows us to make theoretical predictions based on experimental measurements made by others. It is not necessary to conduct additional experiments if the functional relationship between these two dimensionless groups has been established and is published.

We now present a formal method for identifying important dimensionless parameters for a given problem. The Buckingham Pi theorem states that a dimensionally homogeneous equation involving a number of variables can be restated in terms of a smaller number of interdependent dimensionless variables. Let N_v represent the number of variables (independent and dependent) that influence a transport problem:

$$f(v_1, v_2, ..., v_{N_v}) = 0. \tag{3.16}$$

If N_D is the number of fundamental dimensions contained within those N_v variables, then the number of dimensionless variables N_Π that characterize the problem can ordinarily be reduced by N_D:

$$N_\Pi = N_v - N_D. \tag{3.17}$$

There are exceptions to this "rule", so N_Π in (3.17) actually represents the maximum number of dimensionless groups that can be expected to be formed from N_v variables. We will consider these exceptions as they arise in the text.

In the case above, we have $N_v = 5$ dimensional variables (i.e., $\langle v \rangle, d_1, d_2, \rho, \Delta P$) and $N_D = 3$ fundamental dimensions (i.e., M, L, T). We would expect to reduce the number of relevant dimensionless groups to $N_\Pi = 5-3 = 2$, which is indeed what is found in (3.15).

The following procedure can be followed to construct appropriate dimensionless groups from the original variables.

1. Identify the variables ($v_1, v_2, ..., v_{N_v}$) that influence the transport problem under investigation. Be careful not to select variables that can be calculated from other variables in the set. For instance, the surface area S of a cylinder depends on its radius R and length L, so only two of these three variables should be considered, since the third can be computed from the other two. In the example above, the variables are $\langle v \rangle, d_1, d_2, \rho$, and ΔP. The valve area is not included since it can be

3.3 Empirical Approach

computed from d_1 and d_2. Flow through the valve is not included since it can be computed from $\langle v \rangle$, d_1 and d_2. The number of variables N_v for this example is 5.

2. Inspect the fundamental dimensions associated with each variable to determine the total number of fundamental dimensions represented in the set of N_v variables. This is defined as N_D. In the example above, the fundamental dimensions of M, L, and T appear in our set of variables. Temperature, molar quantity, and electrical charge do not appear, so $N_D = 3$.

3. Divide the N_v variables into two groups: $N_v - N_D$ dependent variables (or excluded variables) and N_D independent variables (or core variables). Core variables can potentially appear in every dimensionless group while excluded variables will appear only in one dimensionless group. If you are interested in how a dimensionless parameter involving variable v_1 depends on a dimensionless group involving v_2, then you will want to exclude both of those variables from the core. There are certain guidelines that should be followed when selecting core variables:

   (a) The set of core variables must contain all of the fundamental dimensions associated with the entire set of variables. To assist with this process, it is useful to construct a table showing the fundamental dimensions for all the variables. This is shown in Table 3.3 for our sample problem. In our example, between them, the core variables must contain three fundamental dimensions: M, L, and T. Thus, the set ($\langle v \rangle$, d_1, d_2) cannot be selected as a core set because the fundamental dimension of mass M is not represented in the core. Similarly, the group (d_1, d_2, ρ) cannot be selected for the core because the fundamental dimension of time T would not be present in the core.

   (b) No two core variables should have the same fundamental dimensions. Pressure drop and shear stress have the same set of fundamental dimensions ($ML^{-1}T^{-2}$), so if both appear in a problem, only one of these variables can be selected as a core variable. If cylinder length and radius are important parameters in a problem, they cannot both appear in the core because they both have the same fundamental dimensions. In our example, d_1 and d_2 each have fundamental dimensions of L, so both variables cannot be present in the core set of variables.

   (c) The dimensions of a core variable must be independent of the dimensions of the other core variables. In other words, the set of fundamental dimensions associated with one core variable cannot be a multiple of the set of dimensions of another core variable. This is a more general guideline, which includes

Table 3.3 Dimensions of variables in example problem

Variable	M	L	T
$\langle v \rangle$	0	1	−1
d_1	0	1	0
d_2	0	1	0
ρ	1	−3	0
ΔP	1	−1	−2

guideline b above as a special case. For example, tissue volume $V_T(L^3)$, capillary surface area $S_c(L^2)$, and red cell radius $R_{rbc}(L)$ may be important parameters in a problem. However, since their dimensions are linearly related, only one of these variables can be placed in the core. In our example, the fundamental dimensions of variables d_1, and d_2 are linearly dependent.

(d) If the fundamental dimensions of two or more variables are not independent, then at least one must be placed in the core. They cannot all be treated as excluded variables. So, again, if tissue volume $V_T(L^3)$, capillary surface area $S_c(L^2)$, and red cell radius $R_{rbc}(L)$ are important parameters in a problem, one of these variables must be selected as a core variable. In our example, d_1 and d_2 cannot both be treated as excluded variables. One of them must be a core variable.

(e) Dimensionless variables must be excluded from the core set of variables. Since the objective of this method is to produce a set of dimensionless variables, it is not necessary to apply the method to variables that are already dimensionless. Thus, quantities such as hematocrit value and oxyhemoglobin saturation values are treated as excluded variables. No such quantities appear in our example.

(f) Variables that you wish to appear in a single dimensionless group should not be selected as a core variable. In our example, the mean velocity is of particular interest and should be excluded from the core.

Based on these guidelines, we select a set of core variables and a set of excluded variables:

$$\begin{array}{ll} v_{c_1}, v_{c_2}, ..., v_{cN_D} & N_D \text{ core variables} \\ v_{e_1}, v_{e_2}, ..., v_{eN_\Pi} & N_\Pi \text{ excluded variables.} \end{array} \quad (3.18)$$

The relationship between dimensional variables (3.17) is reduced by N_D variables:

$$\Pi_{v_{e1}} = f(\Pi_{v_{e2}}, ..., \Pi_{v_{eN_\Pi}}). \quad (3.19)$$

For our example problem, we should exclude the mean velocity $\langle v \rangle$ and must exclude either the major axis d_1 or the minor axis d_2. Let us select the core and excluded variables as follows:

core variables: $\quad d_1, \rho, \Delta P$
excluded variables: $\langle v \rangle, d_2$

Therefore, one dimensionless group involving $\langle v \rangle$ will depend only on one other group involving d_2:

$$\Pi_{\langle v \rangle} = f(\Pi_{d_2}).$$

3.3 Empirical Approach

4. Once the core variables and excluded variables have been selected, we are ready to construct N_Π dimensionless groups. Each dimensionless group Π will contain one excluded variable and all of the core variables. Each excluded variable will be raised to the power 1 and each core variable will be raised to an unknown power n_{ec}.

$$\Pi_{v_{e1}} = v_{e1} v_{c1}^{n_{11}} v_{c2}^{n_{12}} \ldots v_{cN_D}^{n_{1N_D}}$$
$$\Pi_{v_{e2}} = v_{e2} v_{c1}^{n_{21}} v_{c2}^{n_{22}} \ldots v_{cN_D}^{n_{2N_D}}$$
$$\vdots$$
$$\Pi_{v_{eN_\Pi}} = v_{eN_\Pi} v_{c1}^{n_{N_\Pi 1}} v_{c2}^{n_{N_\Pi 2}} \ldots v_{cN_D}^{n_{N_\Pi N_D}}. \tag{3.20}$$

For our example of blood flow through a heart valve:

$$\Pi_{\langle v \rangle} = \langle v \rangle d_1^a \rho^b \Delta P^c,$$
$$\Pi_{d_2} = d_2 d_1^d \rho^e \Delta P^f,$$

where the exponents a, b, c, d, e and f are to be determined such that $\Pi_{\langle v \rangle}$ and Π_{d_2} are dimensionless.

5. The next step is to find appropriate values for each exponent that will ensure that each product in (3.20) is dimensionless. Molar concentrations can be converted to mass concentrations by multiplying by molecular weights of the relevant species. Therefore, a fairly general case would be a problem involving the five fundamental dimensions M, L, T, Θ, and I. In that case, $N_D = 5$. Since (3.20) must be dimensionally homogeneous, we can restate (3.20) in terms of its fundamental dimensions:

$$\Pi_{v_{e1}}[M^0 L^0 T^0 \Theta^0 I^0] [=] v_{e1}[M^{M_{e1}} L^{L_{e1}} T^{T_{e1}} \Theta^{\theta_{e1}} I^{I_{e1}}] \times v_{c1}[M^{M_{c1}} L^{L_{c1}} T^{T_{c1}} \Theta^{\theta_{c1}} I^{I_{c1}}]^{n_{11}}$$
$$\times v_{c2}[M^{M_{c2}} L^{L_{c2}} T^{T_{c2}} \Theta^{\theta_{c2}} I^{I_{c2}}]^{n_{12}} \times v_{c3}[M^{M_{c3}} L^{L_{c3}} T^{T_{c3}} \Theta^{\theta_{c3}} I^{I_{c3}}]^{n_{13}}$$
$$\times v_{c4}[M^{M_{c4}} L^{L_{c4}} T^{T_{c4}} \Theta^{\theta_{c4}} I^{I_{c4}}]^{n_{14}} \times v_{c5}[M^{M_{c5}} L^{L_{c5}} T^{T_{c5}} \Theta^{\theta_{c5}} I^{I_{c5}}]^{n_{15}}$$

$$\Pi_{v_{e2}}[M^0 L^0 T^0 \Theta^0 I^0] [=] v_{e2}[M^{M_{e2}} L^{L_{e2}} T^{T_{e2}} \Theta^{\theta_{e2}} I^{I_{e2}}] \times v_{c1}[M^{M_{c1}} L^{L_{c1}} T^{T_{c1}} \Theta^{\theta_{c1}} I^{I_{c1}}]^{n_{21}}$$
$$\times v_{c2}[M^{M_{c2}} L^{L_{c2}} T^{T_{c2}} \Theta^{\theta_{c2}} I^{I_{c2}}]^{n_{22}} \times v_{c3}[M^{M_{c3}} L^{L_{c3}} T^{T_{c3}} \Theta^{\theta_{c3}} I^{I_{c3}}]^{n_{23}}$$
$$\times v_{c4}[M^{M_{c4}} L^{L_{c4}} T^{T_{c4}} \Theta^{\theta_{c4}} I^{I_{c4}}]^{n_{24}} \times v_{c5}[M^{M_{c5}} L^{L_{c5}} T^{T_{c5}} \Theta^{\theta_{c5}} I^{I_{c5}}]^{n_{25}}$$

$$\vdots$$

$$\Pi_{v_{eN_\Pi}}[M^0 L^0 T^0 \Theta^0 I^0] [=] v_{eN_\Pi}[M^{M_{eN_\Pi}} L^{L_{eN_\Pi}} T^{T_{eN_\Pi}} \Theta^{\theta_{eN_\Pi}} I^{I_{eN_\Pi}}]$$
$$\times v_{c1}[M^{M_{c1}} L^{L_{c1}} T^{T_{c1}} \Theta^{\theta_{c1}} I^{I_{c1}}]^{n_{N_\Pi 1}} \times v_{c2}[M^{M_{c2}} L^{L_{c2}} T^{T_{c2}} \Theta^{\theta_{c2}} I^{I_{c2}}]^{n_{N_\Pi 2}}$$
$$\times v_{c3}[M^{M_{c3}} L^{L_{c3}} T^{T_{c3}} \Theta^{\theta_{c3}} I^{I_{c3}}]^{n_{N_\Pi 3}} \times v_{c4}[M^{M_{c4}} L^{L_{c4}} T^{T_{c4}} \Theta^{\theta_{c4}} I^{I_{c4}}]^{n_{N_\Pi 4}}$$
$$\times v_{c5}[M^{M_{c5}} L^{L_{c5}} T^{T_{c5}} \Theta^{\theta_{c5}} I^{I_{c5}}]^{n_{N_\Pi 5}}. \tag{3.21}$$

Here, the symbol [=] is interpreted as "has the fundamental dimensions of". The capital letters in (3.21) represent the fundamental dimensions raised to the appropriate powers for each of the core and excluded variables. The sum of exponents on the right-hand side of (3.21) must equal zero for each fundamental dimension in each dimensionless group. Starting with the first dimensionless group, this leads to a linear set of five algebraic equations in five unknown exponents: n_{11}, n_{12}, n_{13}, n_{14}, and n_{15}.

$$\begin{aligned} M &: 0 = M_{e_1} + n_{11}M_{c_1} + n_{12}M_{c_2} + n_{13}M_{c_3} + n_{14}M_{c_4} + n_{15}M_{c_5} \\ L &: 0 = L_{e_1} + n_{11}L_{c_1} + n_{12}L_{c_2} + n_{13}L_{c_3} + n_{14}L_{c_4} + n_{15}L_{c_5} \\ T &: 0 = T_{e_1} + n_{11}T_{c_1} + n_{12}T_{c_2} + n_{13}T_{c_3} + n_{14}T_{c_4} + n_{15}T_{c_5} \\ \Theta &: 0 = \theta_{e_1} + n_{11}\theta_{c_1} + n_{12}\theta_{c_2} + n_{13}\theta_{c_3} + n_{14}\theta_{c_4} + n_{15}\theta_{c_5} \\ I &: 0 = I_{e_1} + n_{11}I_{c_1} + n_{12}I_{c_2} + n_{13}I_{c_3} + n_{14}I_{c_4} + n_{15}I_{c_5}. \end{aligned} \quad (3.22)$$

Similar sets of equations can be constructed for each of the other excluded variables. Once all of the exponents have been computed, they are inserted back into (3.20) to identify terms in each of the dimensionless groups.

Now, let us apply these methods to our example. Referring to Table 3.3, (3.21) can be written:

$$\Pi_{\langle v \rangle}[M^0L^0T^0][=]\langle v \rangle[M^0L^1T^{-1}]d_1[M^0L^1T^0]^a \rho[M^1L^{-3}T^0]^b \Delta P[M^1L^{-1}T^{-2}]^c$$
$$\Pi_{d_2}[M^0L^0T^0][=]d_2[M^0L^1T^0]d_1[M^0L^1T^0]^d \rho[M^1L^{-3}T^0]^e \Delta P[M^1L^{-1}T^{-2}]^f.$$

Equation (3.22) for $\Pi_{\langle v \rangle}$ becomes

$$\begin{aligned} M &: 0 = 0 + a[0] + b[1] + c[1] \\ L &: 0 = 1 + a[1] + b[-3] + c[-1] \\ T &: 0 = -1 + a[0] + b[0] + c[-2]. \end{aligned}$$

The solution to this set of equations is:

$$a = 0, b = \tfrac{1}{2}, c = -\tfrac{1}{2}$$

Similarly, (3.22) for Π_{d_2} is:

$$\begin{aligned} M &: 0 = 0 + d[0] + e[1] + f[1] \\ L &: 0 = 1 + d[1] + e[-3] + f[-1] \\ T &: 0 = 0 + d[0] + e[0] + f[-2]. \end{aligned}$$

The solution (which could easily be found from inspection) is:

$$d = -1, e = 0, f = 0.$$

Substituting the exponents back into the general expressions for $\Pi_{\langle v \rangle}$ and Π_{d_2}:

3.3 Empirical Approach

$$\Pi_{\langle v \rangle} = \langle v \rangle d_1^0 \rho^{\frac{1}{2}} \Delta P^{-\frac{1}{2}} = \langle v \rangle \sqrt{\frac{\rho}{\Delta P}}$$

$$\Pi_{d_2} = d_2 d_1^{-1} \rho^0 \Delta P^0 = \frac{d_2}{d_1}.$$

Sets of dimensionless groups identified with the Buckingham Pi method are not necessarily unique. Groups can be inverted or raised to any power and still remain dimensionless. The product or quotient of two dimensionless groups is also dimensionless. So if Π_1, Π_2, and Π_3 are three dimensionless groups identified with the Buckingham Pi method, the relationship between them can be written in many different ways, including:

$$\Pi_1 = f_1(\Pi_2, \Pi_3)$$
$$\Pi_1 = f_2(\Pi_2, \Pi_3/\Pi_2)$$
$$\Pi_1 = f_3(\Pi_1 \Pi_2, \Pi_3 / \left[\sqrt{\Pi_1 \Pi_2}\right])$$
$$\Pi_1^2 = f_4(\sqrt{\Pi_2}, \Pi_3)$$
etc.

For our sample problem, we can write either of the following relationships:

$$\langle v \rangle \sqrt{\frac{\rho}{\Delta P}} = \phi\left(\frac{d_2}{d_1}\right)$$

$$\text{or,} \quad \frac{\rho \langle v \rangle^2}{\Delta P} = \phi^2\left(\frac{d_2}{d_1}\right).$$

The last expression corresponds with that found from dimensional analysis (3.15). Although the number of such combinations can be large, the number of *independent* dimensionless groups is properly identified with the Buckingham Pi Method.

6. The final step is to perform experiments to determine the relationships that exist between the dimensionless groups:

$$\Pi_{v_{e1}} = \Pi_{v_{e1}}(\Pi_{v_{e2}}, ..., \Pi_{v_{eN_\Pi}}).$$

Often, it is assumed that one of the groups is proportional to the products of the other groups raised to different powers:

$$\Pi_1 = A \Pi_2^a \Pi_3^b ... \Pi_{N_\Pi}^n,$$

where $A, a, b, ..., n$ are constants determined by experiments. In this way, a set of experiments performed using a particular fluid in a particular device can be used to predict the behavior of a different fluid flowing through a different, but geometrically similar, device.

The Buckingham Pi Theorem is used extensively in many areas of biotransport, particularly in finding functional relationships for friction factors, heat transfer coefficients, and mass transfer coefficients. These applications will be discussed more thoroughly when those topics are covered in the text.

> **Example 3.3.1. Permeability of a Porous Membrane.**
> A membrane consists of a solid plate permeated by a number of small pores. The permeability of the membrane to helium P is found to be a function of the following parameters:
>
> 1. *The thickness of the membrane, δ.*
> 2. *The porosity of the membrane (pore volume relative to total membrane volume), ε.*
> 3. *The diffusion coefficient for helium in the membrane, $D_{He,m}$.*
> 4. *The viscosity of the fluid that passes through the pores, μ.*
> 5. *The density of the fluid that passes through the pores, ρ.*
> 6. *The mean velocity of fluid in the membrane pores, $\langle v \rangle$.*
>
> We are interested in finding a relationship between a dimensionless variable containing the permeability and a dimensionless variable containing the mean velocity in the membrane pores.

Solution. *Initial considerations*: The first thing we need to do is to construct a table of each of the variables and their dimensions:

Variable	M	L	T
P	0	1	−1
δ	0	1	0
ε	0	0	0
$D_{He,m}$	0	2	−1
μ	1	−1	−1
ρ	1	−3	0
$\langle v \rangle$	0	1	−1

There are seven important variables and three fundamental dimensions. Therefore, we should be able to form $7 - 3 = 4$ dimensionless groups.

System definition and environmental interactions: The system in this case is the membrane, including the porous and non porous regions.

Apprising the problem to identify governing relationships: We will use the Buckingham Pi Theorem to identify an appropriate set of dimensionless groups.

Analysis: The number of excluded variables is equal to the number of dimensionless groups. Therefore, four excluded variables must be selected. Two of these must be the variables of interest, i.e., P and $\langle v \rangle$. The core must include three variables, since this is the number of fundamental dimensions in this problem. Between them, the variables must include each fundamental dimension. The core cannot include ε, since a dimensionless number must be excluded from the core.

3.3 Empirical Approach

Let us select ρ, μ, and δ for the core variables. The four dimensionless groups will be:

$$\Pi_P = P\rho^a \mu^b \delta^c$$
$$\Pi_{\langle v \rangle} = \langle v \rangle \rho^d \mu^e \delta^f$$
$$\Pi_{D_{He}} = D_{He,m} \rho^g \mu^h \delta^i$$
$$\Pi_\varepsilon = \varepsilon.$$

Starting with the dimensionless permeability, we write the equation in terms of fundamental dimensions on each side of the equation:

$$M^0 L^0 T^0 = \left[L^1 T^{-1}\right] \left[M^1 L^{-3}\right]^a \left[M^1 L^{-1} T^{-1}\right]^b \left[L^1\right]^c.$$

This reduces to three equations, one for each dimension:

$$M : 0 = a + b$$
$$L : 0 = 1 - 3a - b + c$$
$$T : 0 = -1 - b.$$

The solutions are $b = -1$, $a = 1$, and $c = 1$
Therefore, the dimensionless permeability is:

$$\Pi_P = \frac{P\rho\delta}{\mu} = \frac{P\delta}{\nu},$$

where ν is the kinematic viscosity defined by (2.14). Since $\langle v \rangle$ and P have the same dimensions, then $d = a = 1$, $e = b = -1$, and $f = c = 1$. Consequently, the dimensionless velocity is:

$$\Pi_{\langle v \rangle} = \frac{\langle v \rangle \rho \delta}{\mu} = \frac{\langle v \rangle \delta}{\nu} = Re_\delta,$$

where Re_δ is the Reynolds number based on the membrane thickness. The exponents for the dimensionless diffusion coefficient is found from:

$$M^0 L^0 T^0 = \left[L^2 T^{-1}\right] \left[M^1 L^{-3}\right]^g \left[M^1 L^{-1} T^{-1}\right]^h \left[L^1\right]^i$$

or

$$M : 0 = g + h$$
$$L : 0 = 2 - 3g - h + i$$
$$T : 0 = -1 - h.$$

The solution is $h = -1$, $g = 1$, $i = 0$. The dimensionless diffusion coefficient is:

$$\Pi_{D_{\text{He}}} = \frac{D_{\text{He,m}}\rho}{\mu} = \frac{D_{\text{He}}}{\nu} = \frac{1}{Sc}.$$

The ratio of kinematic viscosity to diffusion coefficient is defined as the Schmidt number, Sc.

Examining and interpreting the results: We have gone from the following relationship between dimensional parameters

$$\mathsf{P} = f(\langle v \rangle, \delta, \varepsilon, \rho, \mu, D_{\text{He,m}})$$

to the following dimensionless relationship:

$$\frac{\mathsf{P}\delta}{\nu} = F(Re_\delta, Sc, \varepsilon),$$

where f and F represent functional relationships. Since the Schmidt number is constant for helium transport through the membrane, a single graph of $\mathsf{P}\delta/\nu$ versus Re_δ with families of curves for different values of the porosity is sufficient to characterize the permeability of the membrane. This is a vast improvement over the direct dependency of permeability on six different parameters.

Additional Comments: This type of analysis also guides the types of experiments that need to be done to characterize the permeability of the membrane. Experiments need to be performed in which permeability is measured over a range of velocities, and the experiments repeated at different porosities.

3.4 Summary of Key Concepts

Two major approaches to solving transport problems are presented in this chapter: the theoretical approach and the empirical or experimental approach. These approaches are not mutually exclusive. In reality, constitutive relationships originally determined with the empirical approach are almost always applied while solving problems using the theoretical approach.

Theoretical Approach. The theoretical approach forms the basis for the bulk of this textbook, and so a detailed general procedure for solving biotransport problems is provided in Sect. 3.2. We recommend that students follow this procedure for every biotransport problem they attempt until the procedure becomes second nature. The procedure builds on the Generate Ideas Model presented in Chap. 1 and employs the transport fundamentals introduced in Chap. 2. More specifics relating to assumptions made on the basis of the system geometry, selection of the constitutive relationships, and governing conservation equations are provided in this chapter. Auxiliary conditions, such as initial conditions and boundary conditions are also discussed, as are general solution methods. Many examples of setting up and solving biotransport problems will be given throughout the text, so only an overview is presented in this chapter.

In addition to presenting methods that allow us to arrive at a solution, we also discuss how to present solutions in a meaningful way. Analytic solutions to many transport problems contain complex mathematical functions, so the behavior of the solution is difficult to ascertain by simply examining the final form of the solution. Graphical solutions are often much easier to understand. In addition, appropriate selection of nondimensional dependent and independent variables can make a single diagram applicable to many more situations than just plotting the solution in dimensional form for a single case. Additional information is given in Sect. 3.2.4 on presentation of results and in Sect. 3.2.5 on the use of scaling to identify important dimensionless variables and to find terms that can be ignored in the analysis. We conclude our discussion on the use of the theoretical approach with two examples: one that emphasizes the macroscopic approach and the other that emphasizes the microscopic approach to biotransport.

Empirical Approach. The empirical approach is examined in Sect. 3.3. We introduce the Buckingham Pi Theorem, which is a powerful method for reducing the number of variables in a problem. This forms the basis of identifying dimensionless heat transfer and mass transfer coefficients, as well as friction factors. These dimensionless coefficients allow results based on measurements made under one set of experimental conditions to predict system behavior under a completely different set of experimental conditions. Students should understand how to choose core variables and excluded variables from a set of variables and their fundamental dimensions. Examples are provided.

3.5 Questions

3.5.1. Compare the experimental and theoretical approaches to solving transport problems in biomedical engineering. Discuss issues of cost, accuracy, and time.

3.5.2. What steps are involved in developing a mathematical model of a physical system?

3.5.3. Is neglecting the curvature of the skin an assumption that is likely to cause major errors in transport computations?

3.5.4. What is the most appropriate coordinate system to use for oxygen exchange in a capillary? Where would you place the origin of the coordinate system?

3.5.5. Distinguish between conservation equations and constitutive equations. Give examples of each. Which are based on theory and which are based on experimental measurements?

3.5.6. Transport in a microscopic system is generally described by one or more: (a) algebraic equations, (b) ordinary differential equations, or (c) partial differential equations.

3.5.7. Steady-state transport in a macroscopic system is generally described by one or more: (a) algebraic equations, (b) ordinary differential equations, or (c) partial differential equations.

3.5.8 What is meant by an Initial condition? What types of problems require an initial condition for a solution?

3.5.9. What is a boundary condition? What types of problems require boundary conditions for their solution? How many boundary conditions are needed?

3.5.10. The solution of macroscopic transport problems will usually require: (a) one or more boundary conditions, (b) one or more initial conditions, (c) both initial and boundary conditions, or (d) neither initial nor boundary conditions.

3.5.11. Why is it preferable to express a solution in terms of symbols for each variable, before substituting values in for the variables?

3.5.12. What is the value of plotting solutions in terms of dimensionless variables, such as the graph shown in Fig. 3.3, rather than plotting dimensional variables?

3.5.13. What is the purpose of scaling governing equations?

3.5.14. What is the Reynolds number and what is its physical significance?

3.5.15. What is the physical significance of the Peclet number, Prandtl number, and Schmidt number?

3.5.16. What is the Buckingham Pi theorem and why is it useful?

3.5.17. What is meant by a core variable? How many of these should be specified for a given problem?

3.5.18. What is an excluded variable? How many of these will exist for a particular problem?

3.5.19. If you ultimately wish to find how a dimensionless group involving parameter 1 varies with changes in a second group involving parameter 2, would you select these parameters as core variables or excluded variables? Explain.

3.5.20. What restrictions are made on selecting core variables?

3.5.21. How do you go about constructing the dimensionless groups? How do you solve for the exponents in each dimensionless group?

3.5.22. What is the maximum number of dimensionless groups that can be formed from a group of n parameters with N fundamental dimensions?

3.6 Problems

3.6.1 Setting Up a Transport Problem

Blood in reservoir 1 is separated from plasma in reservoir 2 by a rubber tube, which is initially clamped. The height h_1 of blood in reservoir 1 is greater than the height h_2 of plasma in reservoir 2, and the initial blood temperature T_{10} is different than the initial plasma temperature T_{20}. The hematocrit value (i.e., cell volume to total volume) of the blood in reservoir 1 is H_1. After the clamp is released, blood from reservoir 1 flows into reservoir 2. We wish to predict how the following factors vary as a function of time: (1) fluid volume in each reservoir, (2) pressure difference

3.6 Problems

across the tube, (3) blood flowrate from reservoirs 1 to 2, (4) hematocrit value in each reservoir, and (5) temperature in each reservoir. Assume the reservoirs are well mixed and thermally insulated at all times. Follow the procedure outlined in Sect. 3.2 to set up each of these problems. Begin by describing the applicable conservation statements and constitutive relationships in words. Then insert the appropriate relationships from Chap. 2. List all of your assumptions and provide justification for each of them. Check to see whether the number of equations are sufficient to solve for the number of unknowns. What additional information do you need to actually solve the set of equations? (Do not solve).

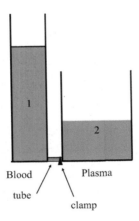

3.6.2 Flow Between Reservoirs

Neglect the difference in density between blood and plasma in Problem 3.6.1, and assume that the pressure at the bottom of each reservoir is equal to the hydrostatic pressure, $\rho g h$. The tube between the reservoirs has a constant flow resistance $\Re$. The initial reservoir heights are h_{10} and h_{20}. Show that after the clamp is released, the flow rate decreases exponentially with time.

3.6.3 Hematocrit in Reservoir 2

If the flow between reservoirs in Problem 3.6.1 is $Q_V = Q_{V_0} \exp(-t/\tau)$, where Q_{V_0} and τ are known, use conservation principles to find the hematocrit in reservoir 2 as a function of time. The initial reservoir height is h_{20}.

3.6.4 Temperature in Reservoir 2

If the flow between reservoirs in Problem 3.6.1 is $Q_V = Q_{V_0} \exp(-t/\tau)$, where Q_{V_0} and τ are known, use conservation principles to find the temperature in reservoir 2 as a function of time. Assume equal densities and specific heats for blood and plasma. The initial reservoir height is h_{20}. Assume no heat is lost to the environment.

3.6.5 Transport Principles

Identify a transport problem involving a biological or a physiological system, a medical device or a component of a medical device. Clearly formulate the problem and provide a solution procedure. In particular:

(a) Clearly state the goal of the problem analysis.
(b) Define the system as it pertains to the process of interest; how does the system interact with the environment to cause the process of interest?
(c) List all assumptions and provide justification for making them.
(d) Discuss the analysis procedure: include all details; conservation principles, constitutive laws, development and solution of equations, identification of boundary and initial conditions, include all steps of the solution process. Find appropriate geometric and property values in the literature so you can obtain a final numerical answer. Show graphical output as appropriate to enhance interpretation and understanding of the solution.
(e) Comments: discuss the meaning of the solution; interpretations of what the results mean; discuss further points of analysis that may be relevant; limitations of the solution.

3.6.6 Tapered Tube

Use the Buckingham Pi method to identify appropriate dimensionless groups for describing flow of a Newtonian fluid in a region of a smooth tapered blood vessel that tapers linearly from R_0 to R_L in a length of tube L. Describe a series of experiments that could be used to identify the relationship between dimensionless flow and other relevant dimensionless groups.

3.6.7 External Flow

Use dimensional analysis to reduce the number of parameters used in finding the *pressure drop* ΔP across a rectangular shaped object that is lodged in a blood vessel.

3.6 Problems

The important parameters and their dimensions in parentheses are: $l(L)$, $h(L)$, $w(L)$, $v(L\ T^{-1})$, $\mu(M\ LT^{-1})$, $\rho(M\ L^{-3})$, and $\Delta P(MT^{-2}L^{-1})$. How many independent dimensionless groups can be formed? What variables would you select for the core? Find a set of appropriate dimensionless groups.

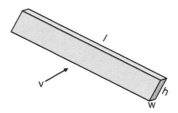

3.6.8 Dimensional Analysis, Lung Alveoli

The flow of blood in the alveolar wall of the lung has been likened to the flow of traffic through one level of a parking garage. Red cells must go around the segments of tissue that connect the sheet-like endothelial layers together in the same way that cars must go around the pillars that separate the ceiling from the floor of a parking garage. The presence of red cells and the tissue segments increase the overall flow resistance. We wish to construct dimensionless groups based on the following parameters:

Pressure difference, ΔP.
Plasma volumetric flow, Q_V.
Alveolar channel height, h.
Alveolar channel width, w.
Alveolar channel length, L.
Red cell diameter, D_c.
Tissue segment diameter, D_t.
Distance between tissue segments, d.
Blood hematocrit value, H.
Plasma physical properties, ρ, μ.

(a) How many dimensionless groups can be formed from these variables? Why?
(b) We would like to have Q_V and ΔP appear in separate dimensionless groups. Which variables would you select for the core? Why?
(c) Using the core variables in (b) above, find the dimensionless group that involves ΔP.

3.6.9 Dimensional Analysis, Parallel Plates

We wish to find dimensionless groups that describe flow through horizontal, parallel plates. The flow Q is believed to be a function of the following factors:

$$Q = Q(L, h, w, \rho, v, \Delta P).$$

(a) How many dimensionless groups can be formed from these 7 variables? Why?
(b) If one dimensionless group is to involve Q and another is to involve ΔP, which variables would you select for the core? Why?
(c) Using the core variables in (b) above, find the dimensionless group that involves Q.

3.6.10 Dimensional Analysis, Cylindrical Tube

The pressure drop ΔP ($ML^{-1}T^{-2}$) across a cylindrical tube depends on the following variables:
  Fluid density ρ (ML^{-3}).
  Fluid viscosity ($ML^{-1}T^{-1}$).
  Tube diameter D (L).
  Tube roughness k (L).
  Tube Length L (L).
  Fluid average velocity $\langle v \rangle$ (LT^{-1}).

We wish to see how a dimensionless group involving ΔP depends on a dimensionless group involving $\langle v \rangle$ and perhaps other dimensionless groups.

(a) How many dimensionless groups can be formed from this set of variables?
(b) Select an appropriate set of core variables for this problem
(c) Find the dimensionless group involving ΔP

3.6.11 Dimensional Analysis, Cylindrical Cell

A cylindrically shaped cell is placed in an isotonic saline solution, where it begins to slowly sink. The terminal velocity v (LT^{-1}) is thought to depend on the following variables:
  Cell density ρ_c (ML^{-3}).
  Cell radius R_c (L).
  Cell length L (L).
  Saline density ρ_s (ML^{-3}).
  Saline viscosity μ ($ML^{-1}T^{-1}$).
  Gravitational acceleration g (LT^{-2}).

(a) How many independent dimensionless groups can be constructed in this problem?

3.6 Problems

(b) Our primary interest is in predicting how a dimensionless variable involving cell velocity depends on a dimensionless group involving the saline viscosity. Select an appropriate group of core variables.

(c) Find an appropriate dimensionless group that can be constructed from the cell velocity and your core variables.

3.6.12 Flow Measurement

The fluid velocity (v) of a flow-measuring nozzle is presumed to be a function of the pipe diameter (d), fluid density (ρ), nozzle diameter (a), fluid viscosity (μ), and pressure drop (Δp). Find an appropriate set of dimensionless parameters that describe this system using the Buckingham Pi method.

3.6.13 Dimensional Analysis: Flow Through an Elliptical Blood Vessel

Blood flows through a blood vessel with an elliptical cross section. The length of the vessel is L, major axis is D_1, and minor axis is D_2. The blood has a hematocrit value of H (dimensionless), density ρ (ML^{-3}), and viscosity μ ($ML^{-1}T^{-1}$). We are interested in determining how a dimensionless variable involving the velocity through the vessel (LT^{-1}) depends on a dimensionless variable involving the pressure difference between the two ends of the vessel, ΔP ($ML^{-1}T^{-2}$).

(a) How many total variables are important in this problem?
(b) How many dimensionless groups can be constructed from this group of variables?
(c) By inspection of the variables and their dimensions, list three independent dimensionless groups.
(d) How many core variables are needed?
(e) Which variables must be excluded from the core?
(f) Select an appropriate set of core variables.
(g) Use this set to find the dimensionless group involving the velocity.

3.6.14 Forced Convection from a Circular Cylinder

The heat transfer coefficient (h) governing forced convection of heat from the outside surface of a cylinder to the surrounding fluid is found to be a function of the following variables:
  Fluid viscosity (μ).
  Fluid density (ρ).

Cylinder diameter (d).
Cylinder length (L).
Fluid average velocity ($\langle v \rangle$).
Fluid thermal conductivity (k).
Fluid specific heat (c_P).

We are particularly interested in finding how a dimensionless group involving h depends on a dimensionless group involving $\langle v \rangle$.

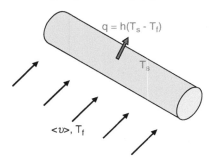

(a) What variables will you select as core variables and as excluded variables. Explain.
(b) Find an appropriate set of dimensionless variables.

3.6.15 Natural Convection from a Circular Cylinder

The heat transfer coefficient (h) governing natural convection of heat from the outside surface of a cylinder to the surrounding fluid is found to be a function of the following variables:

fluid viscosity (μ)
Fluid density (ρ).
Tube diameter (d).
Tube length (L).
Buoyancy factor ($g\beta (T_s - T_\infty)$).
Thermal conductivity (k).
Specific heat (c_P).

We are interested in finding how a dimensionless group involving h depends on a dimensionless group involving $g\beta (T_s - T_\infty)$.

The buoyancy factor is the product of the thermal coefficient of expansion β (dimensions of 1/temperature), the acceleration due to gravity g, and the temperature difference between the cylinder surface and the fluid far from the surface.

(a) What variables will you select as core variables and as excluded variables. Explain.
(b) Find an appropriate set of dimensionless variables.

3.7 Challenges

3.7.1 Heat Loss to the Environment

In Problem 3.6.1, blood at one temperature is added to plasma at a lower temperature over a period equal to t^* seconds. We wish to use this system to raise the temperature of the fluid in the second reservoir by an amount ΔT. Our initial calculations neglected heat loss to the environment during the mixing process. We suspect this is not minimal and want to add a heater to reservoir 1. *Challenge*: What size heater must we buy to maintain the temperature in reservoir 1 constant during the mixing process? *Generate ideas*: What are the mechanisms of heat loss to the environment at the top, bottom, and sides of the reservoir? Can we apply the general problem solution method outlined in Sect. 3.2 to estimate heat exchange between the fluid and air at the top of the reservoir, and between fluid and the reservoir walls? Can we apply a steady-state analysis for heat flux to the air and walls? How do we estimate the necessary heater output from calculations of rates of heat loss from the top of the reservoir and losses through the walls?

3.7.2 Application of Buckingham Pi Theorem to Heart–Lung Machine

You are responsible for designing the heart–lung machine described in Challenge 2.12.1–2.12.3. *Challenge*: Use the Buckingham Pi Theorem to identify important dimensionless parameters for each subsystem of your design. *Generate ideas*: What are the important parameters in each system, including geometry, fluid properties, transport properties, flow properties, etc.? How many fundamental dimensions are contained in these parameters? How many dimensionless groups can be found? What are appropriate core and excluded variables? Find the dimensionless groups. Are some of these less likely to be important than others?

References

Buckingham E (1914) On physically similar systems; illustrations of the use of dimensional equations. Phys Rev 4:345–376
Buckingham E (1915) The principle of similitude. Nature 96:396–397
Langhaar HL (1951) Dimensional analysis and theory of models. Wiley, New York
Taylor ES (1974) Dimensional analysis for engineers. Clarendon Press, Oxford

Part III
Biofluid Transport

Chapter 4
Rheology of Biological Fluids

4.1 Introduction

Rheology is a branch of mechanics that studies the deformation and flow of materials. The field of biorheology is primarily concerned with understanding the relationship between the application of forces and the deformation and flow of biomaterials, including bone, soft tissue, synovial fluid, mucus, blood, and other biomaterials. However, biorheology is also concerned with the interaction between nonbiological materials and biological systems. The study of cell motility in the ocean, fluid movement in the esophagus, or the deformation of an implanted artificial heart valve are all examples from the field of biorheology.

Fluid mechanics deals with the rheology of fluids. Biofluid transport is a subclassification in which the fluid or the interacting system is biological in nature. The fluid can be a biological fluid in a biological system, such as synovial fluid in a knee joint, or a biological fluid in a nonbiological system, such as blood flow in a heart-lung machine, or a non biological fluid flowing in a biological system, such as the flow of air in the respiratory system.

In this chapter, we will introduce some fundamental concepts in fluid mechanics and discuss the rheological models and properties of fluids encountered by bioengineers and biomedical engineers.

4.2 Solids and Fluids

Ideal solids subjected to shear (Fig. 4.1a) will deform rapidly and remain in the same deformed state until the shear is changed (Fig. 4.1b).

Many engineering materials, such as steel, wood, and hard plastics, behave like ideal solids over large ranges of shear. The deformation, depicted in Fig. 4.1b, is the shear strain $\Delta l/\Delta y$. For small deformations, the shear strain is equal to the angle γ:

$$\frac{\Delta l}{\Delta y} = \tan \gamma \approx \gamma. \tag{4.1}$$

R.J. Roselli and K.R. Diller, *Biotransport: Principles and Applications*,
DOI 10.1007/978-1-4419-8119-6_4, © Springer Science+Business Media, LLC 2011

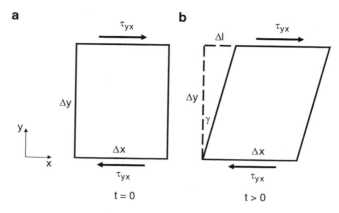

Fig. 4.1 Deformation of a solid exposed to a shear stress τ_{yx}

Hookean elastic solids will exhibit a linear relationship between applied shear stress τ and angular deformation.

$$\tau = G\gamma. \qquad (4.2)$$

The factor G is known as the *shear modulus*. The material will eventually fail if high enough shear stresses are applied. Brittle materials fail immediately when exposed to a critical shear, but ductile materials exhibit plastic behavior just before failure, where adjacent material layers begin to slide relative to each other.

Soft tissues such as skin, and even harder tissues such as bone, will continue to deform for some time after shear is applied. But eventually, the deformation will cease (Fig. 4.2a). Such materials, for which application of a constant shear results in a noticeably transient change in shape, are called viscoelastic materials. From a molecular viewpoint, there is some transient slippage of molecules between adjacent planes in the material that occurs in the direction of the applied shear. This slippage of adjacent layers produces internal friction and accounts for the viscous properties of the viscoelastic material.

Although viscoelastic materials will ultimately cease all motion when a shear stress is applied (Fig. 4.2a), a fluid, by definition, will continue to deform when subjected to a constant shear stress (Fig. 4.2b). For steady parallel flow, the distance traveled by a fluid particle at a position y is equal to the velocity at y multiplied by the time the shear stress has been applied (Fig. 4.3b). The deformation occurs because the layers at y and $y + \Delta y$ move at different velocities. The distance moved at y in time Δt is $\Delta l(y)$ and the distance moved at position $y + \Delta y$ in time Δt is $\Delta l(y + \Delta y)$. If the time increment Δt is small, the deformation will be small and we can write:

$$\gamma \approx \tan \gamma = \frac{\Delta l(y + \Delta y) - \Delta l(y)}{\Delta y} = \frac{v_x(y + \Delta y)\Delta t - v_x(y)\Delta t}{\Delta y}. \qquad (4.3)$$

4.2 Solids and Fluids

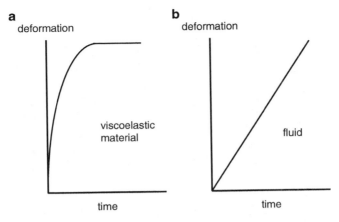

Fig. 4.2 Deformation vs. time of (**a**) a viscoelastic material and (**b**) a fluid when a constant shear stress is applied to the material

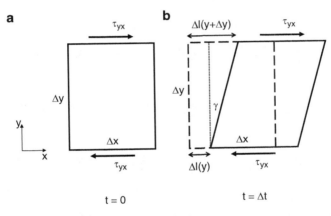

Fig. 4.3 Fluid element (**a**) before exposure to a shear stress and (**b**) deformation at a time Δt after exposure to a shear stress

The instantaneous shear rate is defined as:

$$\dot{\gamma} = \lim_{\Delta t \to 0} \left\{ \frac{\gamma(t + \Delta t) - \gamma(t)}{\Delta t} \right\} = \lim_{\Delta y \to 0} \left\{ \frac{v_x(y + \Delta y) - v_x(y)}{\Delta y} \right\} = \frac{dv_x}{dy}. \quad (4.4)$$

Thus, the rate of deformation (or shear rate) is the same as the velocity gradient. For fluids, we would expect to find a relationship between the applied shear stress and the *rate of deformation*, rather than the deformation itself. However, different fluids behave differently when subjected to the same shear stress. Our main objective in this chapter is to examine the constitutive relationships between shear stress and shear rate in biological fluids and to introduce useful models that describe them.

4.3 Flow Regimes: Laminar and Turbulent Flow

In our everyday experiences with flowing fluids, we have come to recognize two distinct flow regimes. These are known as laminar flow and turbulent flow. Laminar flow is characterized by the smooth, orderly movement of one fluid layer past another, while turbulent flow is characterized by chaotic fluid motion. Water flowing in the same creek can be as smooth as glass upstream of a restriction, but completely chaotic as it moves through a narrow channel. The smoke of a cigarette rising in a still room is characterized by a very orderly region near the ash in which all of the smoke particles rise in parallel streams. All smoke particles that leave the cigarette at a specific site will pass through the same point in space at a later time τ. Flow in this region is laminar. However, after a short distance, the layer becomes unstable and one or more waves begin to develop downstream of the laminar region. Ultimately, the smoke plume becomes completely disorganized. A smoke particle that leaves the laminar region at a specific place will rarely pass through the same location as a different smoke particle that left from the same place at a different time. Flow in this region is turbulent.

There is little hope of predicting the exact path that a fluid particle will take in turbulent flow. Computational schemes for turbulence often focus on predicting the time-averaged velocity of all fluid particles that pass through each point occupied by the flowing fluid. In this text, we will not be concerned with the details of the velocity distribution in turbulent flow. This does not mean that turbulence is unimportant in biofluid applications. Turbulence is desirable in some medical devices where mixing is important, and is purposely avoided in others. Turbulence is known to occur in the upper airways and in large arteries. However, we are normally not so much concerned with the velocity distribution as we are with pressure-flow relationships in such applications. In Chap. 5, we will make use of dimensional analysis, empirical measurements and macroscopic mass and energy balances to account for frictional losses in turbulent flow. A key result from that analysis is that we can predict whether a flow regime is laminar or turbulent on the basis of the value of the Reynolds number introduced in Chap. 3, which compares the relative magnitudes of inertial forces to viscous friction forces.

4.4 Boundary Conditions

The boundary of a fluid system refers to the interface between external solids or fluids and the fluid system being analyzed. Boundary conditions refer to conditions that apply at the boundaries of the fluid. Four boundary conditions are often used in fluid mechanics:

4.4 Boundary Conditions

1. *Boundary condition at a solid–liquid interface*

Fluid in the thin layer right next to a solid interface adheres to the solid surface. This is known as the *no-slip boundary condition*. If the solid is moving, then the fluid at the solid boundary moves with the velocity of the solid. If the solid is stationary, like a container wall, the fluid at the interface is also stationary. If v_i is the component of velocity in the i direction, then for each component, the fluid velocity $(v_i)_f$ must equal the solid velocity $(v_i)_s$:

$$(v_i)_f = (v_i)_s. \qquad (4.5)$$

Thus, for blood moving through a capillary, the blood plasma in contact with the capillary wall is stationary, while plasma in contact with flowing blood elements, such as red cells, white cells or platelets, has the same velocity as the formed element surfaces.

2. *Boundary conditions at the interface between immiscible liquids*

When the surfaces of two immiscible fluids (f_1 and f_2) are in contact along a plane with unit normal n, the boundary conditions require that both the velocity v_i and shear stress τ_{ni} be continuous at the interface:

$$(v_i)_{f1} = (v_i)_{f2}, \qquad (4.6)$$

$$(\tau_{ni})_{f1} = (\tau_{ni})_{f2}, \qquad (4.7)$$

3. *Boundary conditions at a liquid–gas interface*

The shear stress exerted by a gas on a liquid at a gas–liquid interface is often neglected, so:

$$\tau_{ni} \approx 0. \qquad (4.8)$$

4. *Symmetry boundary condition*

In many applications, we will deal with flow between parallel plates or flow in a conduit where the velocity is symmetrical about the centerline. Taking $y = 0$ at the center, with velocity in the x-direction, this can be written:

$$\left.\frac{\partial v_x}{\partial y}\right|_{y=0} = 0. \qquad (4.9)$$

With regards to conditions (2) and (3), it should also be noted that pressure must also be continuous across the interface. If the pressure in the gas phase of a liquid-gas system is maintained constant, then the pressure in the liquid at the interface is also constant.

We will see shortly how these conditions can help us find variations in fluid velocity as a function of position.

4.5 Viscous Properties of Fluids

Real fluids exhibit internal friction. Friction exists between sliding fluid layers, and this tends to slow the faster moving fluid layers when they are in contact with slower moving fluid layers. The friction force at the interface between the layers will also tend to speed up slower moving layers. This frictional force, which occurs at the surface of a fluid element, is called the *viscous force*.

Consider a thin lubricating fluid layer that is placed between a block and a flat surface. This is analogous to placing synovial fluid between two cartilage surfaces in a joint. What happens in the fluid film of Fig. 4.4 when a force is applied to the block? The block will accelerate in the direction of the applied force until the viscous force exerted by the fluid on the bottom of the block is equal in magnitude and opposite in direction to the applied force. As the block accelerates, the no-slip boundary condition ensures that the fluid at the top of the film moves at the same velocity as the block. The fluid in contact with the surface at the bottom of the film remains stationary for all times, again because of the no-slip boundary condition. At very short times after the force is applied to the block, the fluid velocity will be nearly zero throughout the fluid, except in a thin region near the top of the film. As time progresses, friction forces exerted by the upper fluid layers on layers below will begin to drag additional fluid in the direction that the block is moving. Ultimately, a time-independent distribution of fluid velocities will develop from the bottom of the film, where the velocity is zero, to the top of the film, where the velocity is a maximum. This spatial variation in velocity is called the *velocity profile*. Velocity profiles at progressively increasing times are shown in Fig. 4.5. The shape of the velocity profile depends on the viscous nature of the fluid, as will be discussed later in this chapter. The ultimate velocity of the block will depend on the thickness of the fluid layer and on a property of the fluid known as the *fluid viscosity*. Consider next, factors that govern the rate at which these viscous effects propagate through the fluid.

4.6 Viscous Momentum Flux and Shear Stress

Let the direction of the applied force in the film problem above be defined as the x-direction and direction normal to flow be defined as the n-direction. Flow in the film is one dimensional, so the net velocity in the direction normal to flow (n-direction)

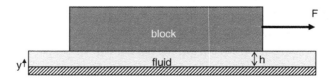

Fig. 4.4 Movement of a block over a lubricating fluid layer

4.6 Viscous Momentum Flux and Shear Stress

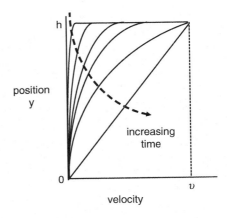

Fig. 4.5 Velocity profiles in fluid layer as time progresses

is zero. If we were to examine the fluid on a molecular scale, molecules on each side of a plane of constant n would take random excursions in the n-direction. All molecules which cross the plane from below bring with them x-momentum in the positive n-direction and all molecules which cross the plane from above carry x-momentum in the negative n-direction. Since there is no net movement of mass in the n-direction, the number of molecules per unit time that move across the plane in the positive n-direction will be the same as the number that move in the negative direction. If the velocities of the two layers just above and below the plane are the same, then there will be no net x-momentum crossing the plane. However, if the layers are moving at different velocities, then a velocity gradient dv_x/dn will exist across the plane and x-momentum will be transported across the plane. Thus, x-momentum will be transported in the n-direction from regions of high velocity to regions of low velocity (i.e., down a velocity gradient). The flux of x-momentum in the n-direction, $p_x|_n$, will generally be some function f of the negative velocity gradient:

$$p_x|_n = f\left(-\frac{dv_x}{dn}\right) = f(-\dot{\gamma}_{nx}). \qquad (4.10)$$

We showed in Sect. 4.2 that the velocity gradient is also known as the shear rate $\dot{\gamma}_{nx}$. The first subscript denotes the plane on which the shear is imposed and the second subscript denotes the direction of shear.

$$\dot{\gamma}_{nx} = \frac{dv_x}{dn}. \qquad (4.11)$$

The functional relationship between momentum flux and shear rate is often written as the product of an *apparent viscosity*, η and the negative shear rate:

$$p_x|_n = -\eta\dot{\gamma}_{nx}. \qquad (4.12)$$

The apparent viscosity is also known as the *effective viscosity*. This equation for molecular momentum flux has the same form as the constitutive relations for heat flux and mass flux. The apparent viscosity may itself depend on the shear rate.

While undergoing these random excursions, molecules with mass m_i and velocity $v_{xi}(t)$ will collide with neighboring molecules with different velocities in adjacent fluid layers, and their velocity will change in a time interval Δt to $v_{xi}(t + \Delta t)$. Friction forces will increase the x-momentum of molecules that jump into fluid regions having a higher velocity, and will decrease the x-momentum of molecules that move into regions of lower velocity. The net shear force acting on a plane of constant n in the x-direction is equal to the sum of all the frictional forces F_x resulting from molecular interactions on that plane. Applying Newton's second law and dividing by the area of the plane A_n:

$$\frac{1}{A_n \Delta t} \sum_i \left[m_i v_{xi}(t + \Delta t) - m_i v_{xi}(t) \right] = \frac{1}{A_n} \sum_i F_{xi}. \quad (4.13)$$

But the term on the left is simply the definition of the flux of x-momentum in the n-direction, $p_x|_n$ and the term on the right is equal to the net shear stress τ_{nx}. The first subscript refers to the plane on which the stress acts and the second subscript refers to the direction of the applied stress. Therefore, (4.13) suggests that the shear stress τ_{nx} can be interpreted as a flux of x-momentum in the n-direction:

$$|\tau_{nx}| = |p_x|_n|. \quad (4.14)$$

We write this in terms of absolute values because, although τ_{nx} and $p_x|_n$ have the same magnitude and same dimensions, the sign of τ_{nx} depends on the sign convention adopted for shear stress. If we take a cubic element of fluid, there are two ways that we can apply shear on a face of constant n, as shown in Fig. 4.6. Since arrows point in both directions when shear is applied, which direction is considered positive?

There are two possible sign conventions, and both are used in engineering applications (Fig. 4.6). The sign convention often adopted in the field of engineering mechanics, and sometimes in biomedical engineering (e.g., Truskey et al. 2004) is that the shear stress τ_{nx} is positive when the stress on the face of greater n is in the positive x-direction or the stress on the face of lesser n is in the negative x-direction. The opposite sign convention has been applied in most chemical engineering (e.g., Bird et al. 2002) and biomedical engineering applications. In that case, shear stress τ_{nx} is considered positive when the stress on the face of lower n is in the positive x-direction.

Since we are at liberty to adopt either sign convention, it seems natural to adopt the second approach, because shear stress is then positive when viscous momentum

4.7 Viscometers

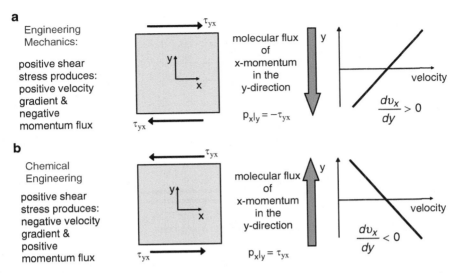

Fig. 4.6 Sign conventions for shear stress (**a**) Engineering Mechanics, (**b**) Chemical Engineering

flux is positive. In that case, shear stress can be directly interpreted as a molecular momentum flux resulting from a velocity gradient in the fluid, and the general expression relating τ_{nx} and shear rate is:

$$\tau_{nx} = p_x|_n = -\eta \frac{dv_x}{dn}. \tag{4.15}$$

The relationship between shear stress (or momentum flux) and shear rate for one-dimensional flow is known as the *constitutive relationship* for the fluid.

4.7 Viscometers

Viscometers are devices used to measure the effective viscosity of fluids. Some of the more common types are shown in Fig. 4.7.

The falling ball viscometer shown in Fig. 4.7a is a very simple device in which a ball is dropped into a cylindrical tube containing the fluid with unknown effective viscosity. The ball will accelerate for a short time, but eventually the gravitational force on the ball will be balanced by the upward viscous resistance force. The terminal velocity of the ball is determined by measuring the time the ball takes to drop a known distance. Empirical data collected from fluids with known viscosity

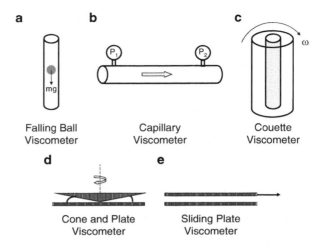

Fig. 4.7 Different types of viscometers

or mathematical models can be applied to estimate the apparent viscosity of the fluid. This is fine if the apparent viscosity is constant, where only a single measurement is necessary to compute fluid viscosity. However, if the apparent viscosity is not constant, different shear rates need to be imposed, so experiments must be performed with several different sized balls and/or tubes.

The capillary viscometer shown in Fig. 4.7b is another commonly used viscometer. Fluid is forced at steady flow rate through a cylindrical tube with known radius R and length L. The pressure drop ΔP across the tube is measured. The net pressure force applied to the fluid must be balanced by the shear force applied by the viscometer wall:

$$\Delta P \left(\pi R^2 \right) = 2\pi R L \tau_w. \tag{4.16}$$

This allows us to compute the shear stress exerted by the fluid on the tube wall, τ_w:

$$\tau_w = \frac{\Delta P R}{2L} = -\eta \frac{dv_x}{dr}\bigg|_{r=R}. \tag{4.17}$$

If we were able to measure the velocity gradient at the tube wall, we could compute η directly from measurements. Unfortunately, the capillary viscometer does not allow direct measurement of shear rate at the wall, but does provide a direct measure of fluid flowrate. In Chaps. 5 and 6, we will show how we can relate wall shear rate to fluid flow for various types of fluids.

A Couette viscometer (Fig. 4.7c) is one in which an unknown fluid is placed between two concentric cylinders. One cylinder is rotated while the other is held stationary. The torque required to rotate the moving cylinder at a known angular

4.7 Viscometers

velocity is measured and the wall shear stress is computed. The shear rate at the wall must be known to compute the apparent viscosity, but, as with the capillary viscometer, the shear rate varies with radial position and is not measured directly. If the cylinder radii are large and the gap thin, then the shear rate would be nearly constant in the fluid. However, such a device might require a large sample of fluid. Using methods developed in Chap. 7, we can relate the known angular velocity of the rotating cylinder to the shear rate at the wall for various fluids, allowing us to compute apparent viscosity from measured quantities.

The most accurate viscometers are capable of controlling or directly measuring both the shear stress applied to the fluid and the shear rate induced in the fluid by the device. If these two quantities are known, apparent viscosity can be computed directly from their ratio. A cone and plate viscometer (Fig. 4.7d) is a device in which a drop of unknown fluid is placed between a flat surface and a cone with an obtuse angle. The cone is rotated with a known angular velocity and the torque necessary to do so is measured. This is similar in principle to operation of the Couette viscometer. However, the distinguishing feature of the cone and plate viscometer is that the shear rate is very nearly constant throughout the entire fluid, independent of radial position (see Example 7.16.3). The shear rate can be computed directly as the ratio of the cone angular velocity to the angle between the cone and plate. Therefore, the device can provide a direct measurement of the apparent viscosity without the need to apply mathematical models.

The simplest viscometer, at least in concept, is the sliding plate viscometer (Fig. 4.7e). The unknown fluid is placed between two parallel plates. The force necessary to move one plate at a constant velocity is measured, while the second plate remains stationary. If a force balance is applied to fluid contained between the moving plate and a fluid surface at an arbitrary position y, the friction force f applied by the adjacent fluid layer will balance the applied force F, as shown in Fig. 4.8.

$$F = f. \tag{4.18}$$

The applied force can be measured. The friction force equals the shear stress at y multiplied by the known contact area A between the moving plate and the fluid at y. According to the sign convention we have adopted, shear stress is considered positive when the stress on the face of lower y is in the positive x-direction.

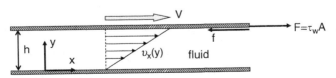

Fig. 4.8 Sliding plate viscometer. The fluid lies between the stationary *bottom plate* and the *upper plate*, which moves at velocity V

In Fig. 4.8, the friction force is directed in the negative x-direction, and therefore the shear stress at y is negative:

$$\tau_{yx} = -\frac{f}{A}. \qquad (4.19)$$

Combining these two equations with (4.15) shows the shear stress is constant throughout the entire fluid film:

$$\tau_{yx} = -\frac{F}{A} = -\tau_w. \qquad (4.20)$$

The wall shear stress τ_w is defined as the applied force per unit contact surface. Combining this with the definition of the apparent viscosity, (4.15), we find the following relationship between shear rate and wall shear stress:

$$\frac{dv_x}{dy} = \frac{\tau_w}{\eta}. \qquad (4.21)$$

Consider two fluid elements at different y-positions in the viscometer. Both of these experience the same shear stress and their rates of deformation must be the same. Therefore, the shear rate and apparent viscosity will be constant throughout the film. Since the plate at $y = h$ moves at velocity V and the plate at $y = 0$ is stationary, the shear rate at each point in the film will be

$$\frac{dv_x}{dy} = \frac{V}{h}. \qquad (4.22)$$

Therefore, the effective viscosity can be computed directly from the measured or imposed values F, A, h, and V:

$$\eta = \frac{Fh}{AV}. \qquad (4.23)$$

The shear rate dependence can be easily examined with a sliding plate viscometer by increasing the force applied to the plate.

In practice, there are a number of restrictions we must place on a simple sliding plate viscometer. We have neglected "end effects" near the edges of the moving plate, where the velocity profile may be quite different from the linear velocity profile computed above. These effects can be minimized by making the length and width of the plate much larger than the film thickness. An alternate approach is to locate a stress transducer centrally on the plate surface, so it is exposed only to the uniform shear stress. In addition, the film thickness must be kept thin. Otherwise,

Fig. 4.9 Vertical sliding plate viscometer

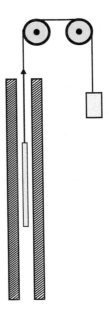

the block velocity may be high enough to induce turbulence in the film. We will see later that turbulence can be induced at high Reynolds numbers. For this device, the Reynolds Number is $Re = \rho V h/\eta$, where ρ is the fluid density. The smaller we keep h, the smaller will be V, and so their product, and thus Re, will also be small. The final restriction is that, in the configuration shown in Fig. 4.8 the moving plate must be nearly weightless, otherwise it would squeeze some of the fluid out from the film during an experiment and h would not remain constant. This can be achieved by suspending the plate, perhaps magnetically, during the experiment. A more practical solution might be to pull the plate vertically through the center of a fluid-filled channel, so there is uniform shearing on both sides of the plate (Fig. 4.9), thus doubling the accuracy of the viscometer.

The sliding plate viscometer will be used to characterize Newtonian and non-Newtonian fluids in the sections to follow.

4.8 Newtonian and Non-Newtonian Fluid Models

If the constitutive relationship relating shear stress to strain rate is nonlinear, the fluid is classified as a *non-Newtonian fluid*. If the relationship is linear, then the effective viscosity is constant and the fluid is said to be a *Newtonian fluid*. All gases and most common liquids, including water, are Newtonian fluids. However, many biological fluids including blood and mucus are non-Newtonian.

4.8.1 Newtonian Fluid Model

The momentum flux or shear stress of a Newtonian fluid is directly proportional to the negative of the velocity gradient (or shear rate $\dot{\gamma}$). The proportionality factor is called the *fluid viscosity,* rather than the effective viscosity η, and is given its own defining symbol, μ:

$$\tau_{nx} = -\mu \frac{dv_x}{dn} = -\mu \dot{\gamma}. \tag{4.24}$$

This constitutive relationship is known as *Newton's law of viscosity* and is analogous to Fourier's Law of heat conduction and Fick's law of diffusion. Fluid viscosity μ has dimensions of $ML^{-1}T^{-1}$. The SI unit of viscosity is the Pascal-second (1 Pa s = 1 kg m^{-1} s^{-1}). Units traditionally used in the literature include the poise (1 poise = 1 g cm^{-1} s^{-1} = 0.1 Pa s) and centipoise (1 cp = 1 mPa s). Other conversion factors for viscosity are provided in Appendix B. In the English system, viscosity is sometimes expressed in units of lb s ft^{-2} (1 lb s ft^{-2} = 478.8 mPa s).

The viscosity of a pure Newtonian fluid is a fluid property and is constant for a given temperature and pressure. The viscosity of common fluids is given in Appendix C. The viscosity of gases increases with increasing temperature while the viscosity of liquids decreases with increasing temperature. The temperature dependence of some common gases and liquids is shown in Fig. 4.10. Note that the viscosity of gases is considerably lower than the viscosity of liquids.

In many applications, the ratio of viscosity μ to fluid density ρ will consistently appear as a single transport parameter known as the *kinematic viscosity,* ν:

$$\nu \equiv \frac{\mu}{\rho}. \tag{4.25}$$

The dimensions of kinematic viscosity are L^2T^{-1}. SI units for ν are m^2/s, but units of cm^2/s are in common use. In the English system, units of ft^2/s are used. The temperature dependence of kinematic viscosity is shown in Fig. 4.11. Note that since gases have low densities, the kinematic viscosity of gases is generally higher than liquids. Values of ν for fluids of interest are also listed in Appendix C.

Most fluids of interest in bioengineering applications are not "pure" fluids, but instead are mixtures of various solutes in water. The viscosity of the mixture will depend on the composition and concentration of the various solutes in the mixture. For example, aqueous solutions of polyethylene glycol (PEG) exhibit Newtonian behavior. However, the kinematic viscosity depends on the average molecular weight of the PEG and on the concentration of PEG in water. The relationship is nonlinear and the viscosity of a 10,000 molecular weight PEG solution is much higher than the viscosity of a 1,000 molecular weight solution of PEG with the same mass fraction (Fig. 4.12). Thus, one needs to know not only the temperature of the solution to predict its viscosity, but also the concentration and molecular weight of the PEG as well.

4.8 Newtonian and Non-Newtonian Fluid Models

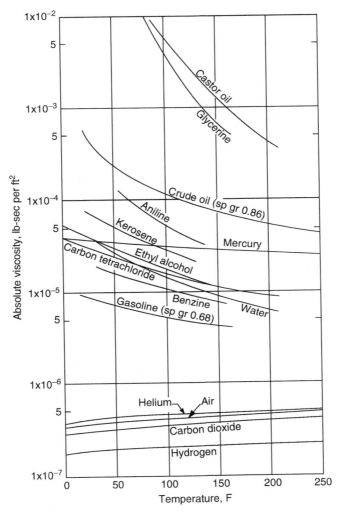

Fig. 4.10 Viscosity vs. temperature for common fluids, from Streeter et al. (1998) with permission

Example 4.8.1.1 Measurement of the Viscosity of Water.

The sliding plate viscometer shown in Fig. 4.8 is used to measure the viscosity of water, a Newtonian fluid, at 20°C. The force applied to the viscometer plate is 1.0 mN and the velocity of the wall is measured to be 1 cm/s. The viscometer plate area is 0.1 m² and the distance between the plate and stationary wall is 1 mm. Show that the velocity profile for a Newtonian fluid in a sliding plate viscometer is linear and use the data to find the viscosity of water at 20°C.

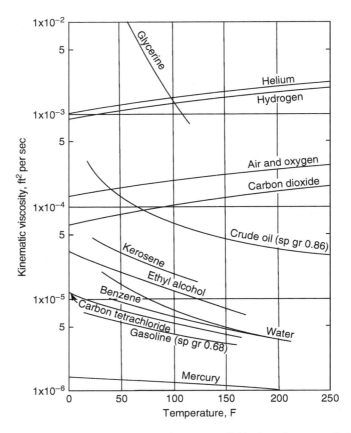

Fig. 4.11 Kinematic viscosity vs. temperature for common fluids, from Streeter et al. (1998) with permission

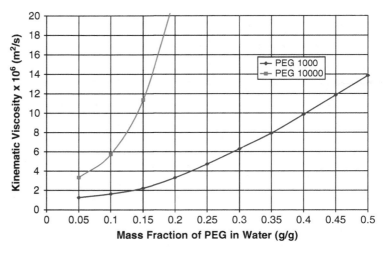

Fig. 4.12 Kinematic viscosity vs. mass fraction of Polyethelyne Glycol (PEG) for two different molecular weights

4.8 Newtonian and Non-Newtonian Fluid Models

Solution. *Initial considerations*: There are two goals. The first goal is to show that if a constant force is applied to the plate, the velocity in the fluid will vary linearly from a minimum value at the stationary wall ($y = 0$) to a maximum of 1 cm/s at the surface of the moving plate ($y = h = 1$ mm). The second goal is to compute the viscosity of water from the applied force, the measured plate velocity, and the viscometer geometry.

System definition and environmental interactions: The system being analyzed is the fluid within the viscometer (Fig. 4.8) between $y = 0$ and $y = h$. Since the wall at $y = 0$ is stationary, the fluid in contact with the stationary wall will also have zero velocity according to the no-slip boundary condition. Likewise, the fluid in contact with the moving plate will have a velocity of $V = 1$ cm/s.

Apprising the problem to identify governing equations: We have examined the operation of a sliding plate viscometer in Sect. 4.7. Equations (4.18)–(4.23) are all valid for any type of fluid in a sliding plate viscometer. We need to modify those results for a constitutive relationship for a Newtonian fluid with $\eta = \mu$.

Analysis: From (4.20), we know that fluid shear stress is constant and equal to $-\tau_w$ in the sliding plate viscometer. Inserting this into the constitutive relation for a Newtonian fluid:

$$-\mu \frac{dv_x}{dy} = -\tau_w.$$

Integrating with respect to y:

$$v_x = \frac{\tau_w y}{\mu} + C.$$

C is a constant of integration, which can be found by applying the no-slip boundary condition at the bottom of the fluid film. Since $v_x = 0$ at $y = 0$, then C must be zero, and the velocity increases linearly with y for a Newtonian fluid:

$$v_x(y) = \left(\frac{\tau_w}{\mu}\right) y.$$

Thus, our first goal is accomplished. At the top of the film ($y = h$), the film velocity is equal to the plate velocity V:

$$V = v_x(h) = \frac{\tau_w h}{\mu}.$$

Rearranging this to solve for the fluid viscosity, and letting $\tau_w = F/A$:

$$\mu = \frac{Fh}{AV}.$$

This could also have been obtained by setting $\mu = \eta$ in (4.23). Introducing the parameters for this experimental measurement of the viscosity of water, we find:

$$\mu = \frac{(10^{-3}\text{N})(10^{-3}\text{m})}{(10^{-1}\text{m}^2)(10^{-2}\text{ms}^{-1})} = 1 \times 10^{-3}\text{Pa s}.$$

Examining and interpreting the results: This value for the viscosity of water compares favorably with the value of 1.002×10^{-3} Pa s listed in Appendix C for water at 20°C.

Additional comments: A difference of 2 parts in 1,000 between this measured value and the tabulated value is well within experimental error encountered in such measurements. Nonlinear velocity profiles will actually exist near the ends of the moving plate, so a more accurate method might be to use a sensitive strain gauge to measure the wall shear stress on either the stationary or the moving plate, far from the ends. Alternately, plate velocity measurements at several values of applied force can be made. The analysis above indicates that the ratio of wall force to plate velocity should be constant for a Newtonian fluid. End effects should be smaller at lower velocities, so an asymptotic value of F/V at low velocities, assuming accurate force and velocity measurements, should give the most accurate measurement of viscosity.

It takes time for the viscous effects to be felt at the stationary wall after the top plate is set in motion. Consequently, the plate will accelerate until the applied force is balanced by the viscous force applied by the fluid on the plate. The thicker the fluid film between the two plates, the longer it will take for friction to induce motion in the fluid near the bottom of the film. The plate will then have a longer time to accelerate, and will reach a higher maximum velocity. Consequently, it will take a longer time before a steady-state velocity is attained, resulting in a shorter measurement time. Therefore, it is desirable to keep the fluid film as thin as is practically possible in a sliding plate viscometer.

4.8.2 Non-Newtonian Fluid Models

Many biological liquids exhibit non-Newtonian behavior over some range of shear rates. If the apparent viscosity decreases with increasing shear rate, the fluid is labeled a *pseudoplastic* fluid. If η increases with increasing shear rate, the fluid is called a *dilatant* fluid.

Mathematical models have been developed that characterize the shear stress vs. shear rate relationship for many different non-Newtonian fluids. Most non-Newtonian liquids of biological interest can be characterized by the constitutive relationships for the four models we will consider in this section: the power law model, the Bingham model, the Casson model, or the Herschel–Bulkley model. The reader should be aware that many other models for more complex non-Newtonian

4.8 Newtonian and Non-Newtonian Fluid Models

fluids and viscoelastic biological fluids can be found in the literature; however, these will not be treated in this text.

4.8.2.1 Power Law Model

A power law fluid model is one in which fluid shear stress is proportional to the shear rate raised to a power n, known as the *behavior index* or *power law index*. The proportionality factor K is known as *the flow consistency index*:

$$\tau_{yx} = -K \left|\frac{dv_x}{dy}\right|^{n-1} \frac{dv_x}{dy}. \tag{4.26}$$

The apparent viscosity for a power law fluid is:

$$\eta = K \left|\frac{dv_x}{dy}\right|^{n-1} = K|\dot{\gamma}|^{n-1}. \tag{4.27}$$

The absolute value symbol assures that the apparent viscosity is positive and also assures that momentum travels from regions of high velocity to low velocity when τ_{yx} is interpreted as a viscous momentum flux. Note that if $n = 1$, the fluid is Newtonian with $K = \mu$, the fluid viscosity. For pure fluids, the value of n will depend on temperature and pressure. Dimensions of K will depend on the value of n (i.e., K [=] $ML^{-1}T^{n-2}$). If $0 < n < 1$, the fluid is a pseudoplastic fluid and its shear stress vs. shear rate relationship will exhibit concave upward behavior. If $n > 1$, the fluid is a dilatant fluid, and its shear stress vs. shear rate relationship will exhibit concave downward behavior.

> **Example 4.8.2.1.1 Power Law Fluid in a Sliding Plate Viscometer.**
> Find the velocity distribution for a power law fluid with $n = 2$ in the sliding plate viscometer shown in Fig. 4.8. If a force F is applied to the plate, how can we predict the steady-state plate velocity?

Solution. *Initial considerations*: Our goals are to determine the velocity distribution between the stationary and sliding plates in the viscometer and to compute the velocity of the sliding plate from measurements of applied force and the viscometer geometry. Since the relation between shear rate and shear stress is nonlinear in this case, we might expect the velocity profile to be nonlinear.

System definition and environmental interactions: The system being analyzed is the fluid within the viscometer (Fig. 4.8) between $y = 0$ and $y = h$. Since the wall at $y = 0$ is stationary, the fluid in contact with the stationary wall will also have zero velocity according to the no-slip boundary condition. Likewise, the fluid in contact with the moving plate will have a velocity V, which is to be determined by the analysis.

Apprising the problem to identify governing equations: Equations (4.18)–(4.23), applicable to any type of fluid in a sliding plate viscometer, are our starting point. These are to be modified for the constitutive relationship for a power law fluid as described by (4.26) and (4.27).

Analysis: Rearranging (4.21):

$$-\eta \frac{dv_x}{dy} + \tau_w = 0.$$

Inserting the power law expression (4.26) for η with $n = 2$:

$$-\left(K \left|\frac{dv_x}{dy}\right|^1\right) \frac{dv_x}{dy} + \tau_w = 0.$$

For the coordinate system used in Fig. 4.8, $dv_x/dy > 0$ everywhere in the film, so this can be rearranged as follows:

$$K \left(\frac{dv_x}{dy}\right)^2 = \tau_w.$$

Dividing by K, and taking the square root of both sides:

$$\frac{dv_x}{dy} = \pm\sqrt{\frac{\tau_w}{K}}.$$

But, since $dv_x/dy > 0$, $K > 0$ and $\tau_w > 0$, the plus sign must be selected, so

$$\frac{dv_x}{dy} = \sqrt{\frac{\tau_w}{K}}.$$

Integrating with respect to y and applying the no-slip boundary condition at the bottom of the film:

$$v_x = \left(\sqrt{\frac{\tau_w}{K}}\right) y.$$

This provides an answer to our first goal. In contrast to our original expectations, the velocity is linearly distributed with position y, similar to the velocity profile of a Newtonian fluid. Our second goal, finding the plate velocity, can be attained by setting $v_x = V$ at $y = h$:

$$V = \left(\sqrt{\frac{\tau_w}{K}}\right) h.$$

Examining and interpreting the results: As in the Newtonian fluid case, the plate velocity is directly proportional to the thickness of the fluid layer, h. However, in contrast to the Newtonian case, the plate velocity is not directly proportional to the

4.8 Newtonian and Non-Newtonian Fluid Models

wall shear stress, but instead is proportional to the square root of the shear stress. Increasing the force on the plate by a factor of 4 will only double the plate velocity for the power law case with $n = 2$, but will quadruple the plate velocity if the fluid is Newtonian.

Additional comments: The results from this example can be extended to any arbitrary power law with exponent n:

$$v_x = \left(\frac{\tau_w}{K}\right)^{\frac{1}{n}} y$$

and the steady-state velocity of the plate will be:

$$V = \left(\frac{\tau_w}{K}\right)^{\frac{1}{n}} h. \quad (4.27a)$$

Comparing this expression to the one derived for a Newtonian fluid, we find that the plate velocity still depends linearly on the film thickness, but it increases as $\tau_w^{1/n}$ rather than as τ_w. The larger is n, the greater will be the shear stress necessary to move the plate at the same velocity.

If we wish to use a sliding plate viscometer to estimate the power law parameters K and n, we must make measurements of wall stress at two or more plate velocities, as will be discussed in Sect. 4.8.3.

> **Example 4.8.2.1.2 Influence of the Coordinate System.**
> Repeat Example 4.8.2.1.1 for the case where a different coordinate system is used in the fluid film, as shown in Fig. 4.13

Solution. *Initial considerations*: The velocity distribution will be physically independent of the placement of the coordinate system. Fluid in contact with either plate moves with the same velocity as the plate, and the velocity decreases linearly from the moving plate to the stationary plate. However, the form of the mathematical solution must depend on the location and orientation of the coordinate system. Our goal is to see how the solution to Example 4.8.2.1.1 is influenced by the selection of the coordinate direction, y.

System definition and environmental interactions: The system of interest is still the fluid between the stationary and sliding plates, but now the sliding plate is

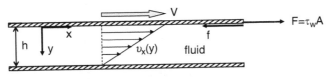

Fig. 4.13 Sliding plate viscometer with a different coordinate system

located at $y = 0$, and the stationary plate is at $y = h$. Consequently, the boundary conditions change to $v_x(0) = V$ and $v_x(h) = 0$.

Apprising the problem to identify governing equations: The governing equations listed in Example 4.8.2.1.1 still apply in this situation.

Analysis: The velocity gradient dv_x/dy will be negative if a coordinate system is adopted in which $y = 0$ at the top of the film and $y = h$ at the bottom. In that case, the flux of x-momentum in the y-direction is positive ($\tau_{yx} = +\tau_w$) and (4.26) becomes:

$$K\left(-\frac{dv_x}{dy}\right)^2 = \tau_w.$$

Taking the square root of both sides:

$$-\frac{dv_x}{dy} = \sqrt{\frac{\tau_w}{K}}.$$

The positive sign was selected for the right-hand side because $-dv_x/dy > 0$. Integrating this with respect to y:

$$v_x = -\left(\sqrt{\frac{\tau_w}{K}}\right)y + C,$$

where C is a constant of integration. Applying the no-slip boundary condition at $y = h$, we find:

$$C = \left(\sqrt{\frac{\tau_w}{K}}\right)h.$$

Substituting this into the expression above:

$$v_x = \left(\sqrt{\frac{\tau_w}{K}}\right)(h - y).$$

Examining and interpreting the results: As in the previous example, the velocity profile is linear, going from zero at the bottom to a maximum at the top. The plate velocity at $y = 0$ is the same as in the previous example for $y = h$. If we made a change of variable in the original problem, such that $y = h - y'$, this would have the same effect as moving the origin from the bottom plate to the top plate and reorienting the positive direction so it points downward, as in this problem.

Additional comments: These two examples illustrate the need for the absolute value of the velocity gradient in the definition of the power law fluid model. The sign of the velocity gradient and shear stress both depend on the orientation of the coordinate system selected.

Fig. 4.14 Constitutive Relation for a Bingham fluid

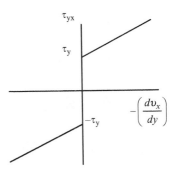

4.8.2.2 Bingham Fluid Model

A Bingham fluid behaves like an elastic solid (not like a fluid) until the applied shear stress exceeds a value termed the *yield stress*, τ_y. For applied shear stresses greater than τ_y, the fluid behaves like a Newtonian fluid with viscosity μ_0. If a shear stress greater than τ_y is applied, the fluid will shear such that the shear stress is linearly proportional to the shear rate. Constitutive equations for the model are shown below. A graph of the relationship is shown in Fig. 4.14.

$$\tau_{yx} = \tau_y + \mu_0 \left(-\frac{dv_x}{dy}\right); \quad \frac{dv_x}{dy} < 0, \qquad (4.28)$$

$$\tau_{yx} = -\tau_y - \mu_0 \left(\frac{dv_x}{dy}\right); \quad \frac{dv_x}{dy} > 0, \qquad (4.29)$$

$$-\tau_y \leqslant \tau_{yx} \leqslant \tau_y; \quad \frac{dv_x}{dy} = 0. \qquad (4.30)$$

Example 4.8.2.2.1 Bingham Fluid in a Sliding Plate Viscometer.
Using the original coordinate system in Fig. 4.8, show the velocity profile for a Bingham fluid in a sliding plate viscometer is linear, and find the plate velocity when the plate is pulled with a force F.

Solution. *Initial considerations*: As in the previous problems, our goals are to determine the velocity profile for a Bingham fluid in a sliding plate viscometer and to determine the velocity of the sliding plate when a force F is applied. In particular, we are interested in the effect of the yield stress on the upper plate velocity and the velocity profile.

System definition and environmental interactions: The system is a Bingham fluid confined between a stationary plate at $y = 0$ and a sliding plate at $y = h$, as shown in Fig. 4.8.

Apprising the problem to identify governing equations: The general equations applicable to a sliding plate viscometer (4.18)–(4.23), along with the constitutive relationships for a Bingham fluid (4.28)–(4.30) are the appropriate governing equations.

Analysis: If the force is less than the product of the yield stress and fluid-plate contact area, then the fluid behaves like a solid (4.30) and the plate velocity must be constant. Since the plate is initially stationary, then it remains stationary until the yield stress is exceeded at the wall:

$$V = 0, \quad \tau_w \leq \tau_y.$$

For $\tau_w > \tau_y$, the Bingham fluid constitutive relationship in the film with constant shear stress becomes:

$$-\tau_w = -\tau_y + \mu_0\left(-\frac{dv_x}{dy}\right), \quad \tau_w > \tau_y.$$

Solving for velocity profile:

$$v_x = \frac{1}{\mu_0}\left[\tau_w - \tau_y\right]y, \quad \tau_w > \tau_y.$$

We have made use of the no-slip boundary condition at $y = 0$. Thus, the velocity profile is linear from the bottom of the film to the top, with the maximum velocity V attained in this case at $y = h$:

$$V = \frac{1}{\mu_0}\left[\tau_w - \tau_y\right]h, \quad \tau_w > \tau_y.$$

Examining and interpreting the results: Note that for the same wall shear stress, the same film thickness and for $\mu_0 = \mu$, the plate velocity will be slower for a Bingham fluid than a Newtonian fluid by an amount equal to $\tau_y h/\mu_0$. Otherwise, it will appear to behave as a Newtonian fluid for $\tau_w > \tau_y$.

Additional comments: If the yield stress is zero, then a Bingham fluid reduces to a Newtonian fluid. The key feature that distinguishes it from a Newtonian fluid is the fact that the plate will not move until the wall shear stress exceeds the yield stress.

4.8.2.3 Casson Fluid Model

Like a Bingham fluid model, the constitutive relation for a Casson model exhibits a yield stress τ_y (Fig. 4.15). If a shear stress less than τ_y is applied to the fluid, it behaves like a solid. If a shear stress greater than τ_y is applied, the fluid will shear such that the square root of the shear stress is linearly related to the square root of

4.8 Newtonian and Non-Newtonian Fluid Models

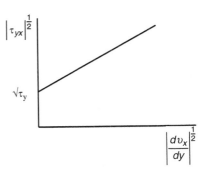

Fig. 4.15 Constitutive relationship for a Casson fluid

the shear rate. The slope is a material property, S, having dimensions the same as the square root of viscosity:

$$\sqrt{\tau_{yx}} = S\sqrt{-\frac{dv_x}{dy}} + \sqrt{\tau_y}; \quad \frac{dv_x}{dy} < 0, \quad (4.31)$$

$$\sqrt{-\tau_{yx}} = S\sqrt{\frac{dv_x}{dy}} + \sqrt{\tau_y}; \quad \frac{dv_x}{dy} > 0, \quad (4.32)$$

$$|\tau_{yx}| < \tau_y; \quad \frac{dv_x}{dy} = 0. \quad (4.33)$$

Blood is often modeled with the Casson fluid model at low shear rates, as we will see in Sect. 4.9.

Example 4.8.2.3.1 Blood in a Sliding Plate Viscometer.
Blood, modeled as a Casson fluid, is placed in a sliding plate viscometer. Confirm that the velocity profile is linear and find the plate velocity when the plate is pulled with a force F.

Solution. *Initial considerations*: Like the power law fluid, the relationship between shear stress and velocity gradient is not linear for a Casson fluid. Despite the nonlinear relationship, we found the velocity profile to be linear for a power law fluid. Our goals are to find the velocity profile for a Casson fluid in a sliding plate viscometer and to determine how the plate velocity depends on the yield stress and the material property S.

System definition and environmental interactions: The system is blood, modeled as a Casson fluid, confined between a stationary plate at $y = 0$ and a sliding plate at $y = h$, as shown in Fig. 4.8.

Appraising the problem to identify governing equations: The general equations applicable to a sliding plate viscometer (4.18)–(4.23), along with the constitutive relationships for a Casson fluid (4.31)–(4.33) are the appropriate governing equations.

Analysis: As with the Bingham fluid, the plate will not move until the wall shear stress exceeds the yield stress:

$$V = 0, \tau_w \leq \tau_y.$$

For larger shear stresses, select the constitutive equation for $dv_x/dy > 0$ and $\tau_{yx} = -\tau_w$:

$$\sqrt{\tau_w} = S\sqrt{\frac{dv_x}{dy}} + \sqrt{\tau_y}, \quad \tau_w > \tau_y.$$

Rearranging and solving for the velocity profile:

$$v_x = \left[\frac{\left(\sqrt{\tau_w} - \sqrt{\tau_y}\right)}{S}\right]^2 y, \quad \tau_w > \tau_y.$$

Thus, despite the nonlinear relationship between shear rate and shear stress, the velocity profile is linear. Our second goal is achieved by solving for the velocity at $y = h$:

$$V = \left[\frac{\left(\sqrt{\tau_w} - \sqrt{\tau_y}\right)}{S}\right]^2 h, \quad \tau_w > \tau_y.$$

Examining and interpreting the results: If the applied stress is less than the yield stress, the plate will not move. If it is greater than the yield stress, the plate velocity is proportional to the thickness of the fluid film, but, unless the yield stress is zero, it is not proportional to the applied wall stress. The relation between plate velocity and wall shear stress will be nonlinear, and V will be lower than a Newtonian fluid with $\mu = S^2$.

Additional comments: If the yield stress is zero, a Casson fluid reduces to a Newtonian fluid with $\mu = S^2$.

4.8.2.4 Herschel–Bulkley Fluid Model

The Herschel–Bulkley non-Newtonian fluid model is a 3 parameter model that combines the nonlinear behavior of a power law fluid with a yield stress, characteristic of Casson or Bingham fluids:

$$\tau_{yx} = \tau_y + K\left(-\frac{dv_x}{dy}\right)^n; \quad \frac{dv_x}{dy} < 0, \tag{4.34}$$

4.8 Newtonian and Non-Newtonian Fluid Models

$$\tau_{yx} = -\tau_y - K\left(\frac{dv_x}{dy}\right)^n ; \quad \frac{dv_x}{dy} > 0, \quad (4.35)$$

$$-\tau_y \leq \tau_{yx} \leq \tau_y ; \quad \frac{dv_x}{dy} = 0. \quad (4.36)$$

The Newtonian, power law, and Bingham models are all special cases of the Herschel–Bulkley model, but the Casson model is fundamentally different. We leave it as an exercise (Problem 4.12.2) to show that the plate velocity for a sliding plate viscometer filled with a Herschel–Bulkley fluid will be zero for $\tau_w < \tau_y$ and for $\tau_w > \tau_y$, V will be:

$$V = \left(\frac{\tau_w - \tau_y}{K}\right)^{\frac{1}{n}} h. \quad (4.37)$$

Once again, the velocity profile is linear and the plate velocity is proportional to the thickness of the fluid layer between the plates.

4.8.3 Identification of Constitutive Model Equations

In the previous sections, we have characterized how Newtonian and non-Newtonian fluids behave in a sliding plate viscometer. Now, consider how we might use the viscometer to characterize a biological fluid with unknown viscous behavior. In particular, we would like to: (1) identify the appropriate constitutive relationship, and (2) estimate the viscous properties.

Experiments are performed by placing the unknown fluid between the two plates and a series of forces are applied to the sliding plate. For each applied force, we measure the final steady-state plate velocity. From these measured quantities, we compute the wall shear stress ($-\tau_{yx} = \tau_w = F/A_s$) and the shear rate ($dv_x/dy = V/h$) at each value of the applied force, then plot τ_w vs. V/h. The local slope will be the apparent viscosity, η.

If this is a linear relationship, and it passes through zero, the fluid is Newtonian, and the slope is the fluid viscosity. If the relationship is nonlinear, but passes through zero, it may be a power law fluid. If the relationship is linear, with a positive intercept, the fluid is a Bingham fluid with a slope μ_0 and an intercept τ_y. If the relationship is nonlinear but has an intercept, it may be either a Casson fluid or a Herschel–Bulkley fluid with yield stress equal to the intercept.

To test for a Casson fluid, plot $(\tau_w)^{\frac{1}{2}}$ vs. $(V/h)^{\frac{1}{2}}$. If this is linear, the intercept is $(\tau_y)^{\frac{1}{2}}$ and the slope is S.

To test for a power law fluid, we can plot $\log(\tau_w)$ vs. $\log(V/h)$. What will this tell us? If we take the logarithm of both sides of (4.27a), we get:

$$\log(\tau_w) = \log(K) + n\log\left(\frac{V}{h}\right). \quad (4.38)$$

Therefore, if this relationship is linear, the fluid obeys a power law model with slope n and intercept log (K).

Finally, to test for a Herschel–Bulkley fluid, we plot log $(\tau_w - \tau_y)$ vs. log (V/h), where τ_y is obtained from the initial plot of τ_w vs. V/h. The parameters K and n are determined from (4.37) in the same manner as for the power law fluid, since:

$$\log(\tau_w - \tau_y) = \log(K) + n\log\left(\frac{V}{h}\right). \tag{4.39}$$

Therefore, data from this simple viscometer can be used to characterize the viscous properties of the Newtonian and non-Newtonian fluids considered in this text.

Example 4.8.3.1 Characterization of Constitutive Relation of an Unknown Fluid Sample.

τ_w (dynes/cm^2)	V (mm/s)
40.0	0.01
81.6	0.05
120.0	0.1
328.5	0.5
521.2	1
834.2	2

The data at right were collected when a biological fluid was tested in a sliding plate viscometer with film thickness $h = 0.1$ mm. Identify the fluid model and its viscous properties.

Solution. *Initial considerations*: Our goal is to determine the appropriate constitutive relationship for the fluid and to determine the viscous properties. Our first step should be to determine whether the fluid is Newtonian or non-Newtonian by plotting shear stress as a function of shear rate. If it is non-Newtonian, we will need to distinguish between a Bingham, Casson, Power Law, and Herschel–Bulkley fluid by using the methods outlined in the section above.

System definition and environmental interactions: The system consists of the unknown fluid confined within the viscometer in Fig. 4.8.

Apprising the problem to identify governing equations: The appropriate governing relationships are the equations for a fluid in a sliding plate viscometer (4.18)–(4.23) and the constitutive relationships for Newtonian and non-Newtonian fluids (4.24) and (4.26)–(4.36).

Analysis: First, we divide the plate velocity by film thickness to compute the shear rate, then plot shear stress vs. shear rate (Fig. 4.16). The plot is nonlinear with a nonzero intercept, so the fluid is non-Newtonian, with a yield stress of 20 dynes/cm^2.

A graph of the square root of shear stress vs. the square root of shear rate is also nonlinear, so the fluid is not a Casson fluid. Our final test is to see whether the

4.8 Newtonian and Non-Newtonian Fluid Models

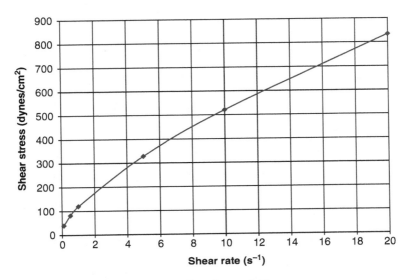

Fig. 4.16 Plot of shear stress vs. shear rate for unknown fluid

Herschel–Bulkley (H–B) model is appropriate for this unknown non-Newtonian fluid. To avoid possible confusion with units, it is useful to make the H–B constitutive relation dimensionless before making a log–log plot. First, we divide both sides of the H–B equation by the yield stress to make the equation dimensionless. Next we multiply and divide the right side of the equation by a constant $\dot{\gamma}_0^n = 1 \mathrm{s}^{-n}$, allowing us to normalize shear rate. The normalized B–H relation for $dv_x/dy < 0$ (i.e., (4.37)) becomes:

$$\frac{\tau_w - \tau_y}{\tau_y} = \frac{K \dot{\gamma}_0^n}{\tau_y} \left(\frac{\dot{\gamma}^n}{\dot{\gamma}_0^n} \right).$$

Taking the natural logarithm of both sides:

$$\ln \left(\frac{\tau_w - \tau_y}{\tau_y} \right) = \ln \left(\frac{K \dot{\gamma}_0^n}{\tau_y} \right) + n \cdot \ln \left(\frac{\dot{\gamma}}{\dot{\gamma}_0} \right). \quad (4.39a)$$

A plot of $\ln ((\tau_w - \tau_y)/\tau_y)$ vs. $\ln(\dot{\gamma}/\dot{\gamma}_0)$ with $\dot{\gamma}_0 = 1 \mathrm{s}^{-1}$ is shown in Fig. 4.17. This is a straight line, so the H–B model fits the data quite well. The slope of the line is n, the behavior index. Taking the smallest and largest points to compute the slope, we find:

$$n = \frac{3.7 - 0}{3 - (-2.3)} = 0.7.$$

The flow consistency index K can be computed from the zero shear rate intercept, which from Fig. 4.17 equals 1.6:

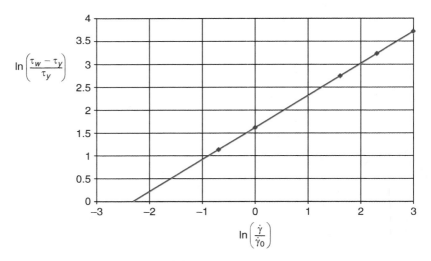

Fig. 4.17 Experimental data plotted according to (4.39a)

$$\text{intercept} = 1.6 = \ln\left(\frac{K\dot{\gamma}_0^n}{\tau_y}\right),$$

$$K = \frac{\tau_y}{\dot{\gamma}_0^n} e^{1.6} = \frac{20 \text{ dynes/cm}^2}{(1\text{s}^{-1})^{0.7}} (5.0) = 100 \text{ dynes cm}^{-2} \text{s}^{0.7}.$$

Converting from dynes/cm² to Pa (1 Pa = 10 dynes/cm²), we find $K = 10$ Pa s$^{0.7}$.

Examining and interpreting the results: This biological fluid is well characterized by the Herschel–Bulkley constitutive model with $\tau_y = 2$ Pa, $n = 0.7$ and $K = 10$ Pa s$^{0.7}$.

Example 4.8.3.2 Measuring Rheological Coefficients.
Apparent viscosity vs. shear rate data are measured with a cone and plate viscometer for a fluid suspected of obeying the Casson constitutive equation. The resulting data are shown in the table below. Confirm that the data are in agreement with the Casson relationship and determine the coefficients S and τ_y.

$\dot{\gamma}(\text{s}^{-1})$	η (mPa s)
0.1	38.58
0.5	13.98
1	9.94
5	5.63
10	4.78
50	3.76
100	3.54
500	3.25
1,000	3.18

4.8 Newtonian and Non-Newtonian Fluid Models

Solution. *Initial considerations*: To confirm that the data are consistent with behavior of a Casson fluid, we need to show that a linear relationship exists between the square root of the shear stress and the square root of the shear rate. The Casson viscous parameters S and τ_y can be estimated from the slope and intercept of this relationship.

System definition and environmental interactions: The system is the fluid contained in a cone and plate viscometer. The cone was rotated at various angular velocities, providing nearly uniform shear rates at all locations in the fluid.

Apprising the problem to identify governing equations: If the fluid is a Casson fluid, then it should obey the constitutive relationships given by (4.31)–(4.33).

Analysis: The constitutive relationship for a Casson fluid with positive shear can be found by substituting the definition of apparent viscosity into (4.32):

$$\sqrt{-\tau_{yx}} = \sqrt{\eta \dot{\gamma}} = S\sqrt{\dot{\gamma}} + \sqrt{\tau_y}.$$

Therefore, if the fluid is a Casson fluid, a plot of $\sqrt{\eta \dot{\gamma}}$ vs. $\sqrt{\dot{\gamma}}$ will produce a straight line with slope S and intercept $\sqrt{\tau_y}$. This plot is indeed linear for the data in the table, as shown in Fig. 4.18. The intercept from the inset in the figure is 0.045 Pa$^{1/2}$. Squaring this, we find the yield stress to be 0.002 Pa or 2 mPa. The slope, based on the first and last points, is equal to S:

$$S = \frac{(1.78 \text{ Pa}^{1/2} - 0.062 \text{ Pa}^{1/2})}{(31.62 \text{ s}^{-1/2} - 0.316 \text{ s}^{-1/2})} = 0.055 (\text{Pa s})^{1/2}.$$

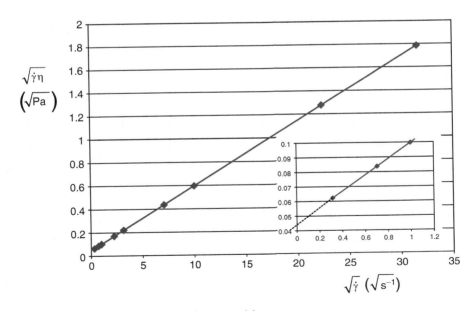

Fig. 4.18 Data are consistent with a Casson model

Examining and interpreting the results: The data are consistent with a Casson fluid having parameters $S^2 = 3.025 \times 10^{-3}$ Pa s and $\tau_y = 2 \times 10^{-3}$ Pa.

Additional comments: Similar approaches can be taken when apparent viscosity vs. shear rate data are provided for suspected power law fluids, Bingham fluids or Herschel–Bulkley fluids. Substituting the definition of apparent viscosity into the constitutive relation for a power law fluid yields:

$$\eta\dot{\gamma} = K\dot{\gamma}^n \text{ (power law fluid)}.$$

Taking the logarithm of both sides provides a straight line relationship from which K and n can be found:

$$\log(\eta\dot{\gamma}) = \log(K) + n(\log(\dot{\gamma})) \text{ (power law fluid)}.$$

A similar procedure can be used to find a linear relationship for a Bingham fluid:

$$\eta\dot{\gamma} = \tau_y + \mu_0\dot{\gamma} \text{ (Bingham fluid)}.$$

Graphical determination of the rheological parameters for a Herschel–Bulkley fluid is more complicated because there are three parameters to be determined. This can usually be accomplished in two steps. The constitutive relationship for a Herschel–Bulkley fluid can be written:

$$\eta\dot{\gamma} = \tau_y + K\dot{\gamma}^n \text{ (Herschel–Bulkley fluid)}.$$

The first step is to estimate the yield stress by plotting $\eta\dot{\gamma}$ vs. $\dot{\gamma}$. Although this will not produce a straight line, the intercept as $\dot{\gamma} \to 0$ will equal the yield stress. Once we have a reasonable estimate of the yield stress, we can estimate the other parameters by taking the logarithm of the constitutive equation and plotting $\log(\eta\dot{\gamma} - \tau_y)$ vs. $\log\dot{\gamma}$:

$$\log(\eta\dot{\gamma} - \tau_y) = \log K + n\log\dot{\gamma} \text{ (Herschel–Bulkley fluid)}.$$

The parameter K can be found from the intercept and n from the slope.

4.9 Rheology of Biological Fluids

The human body is composed primarily of water. Body fluids can be classified as vascular (i.e., contained within the cardiovascular system) and extravascular. Extravascular and intravascular body fluids can each be further subdivided into cellular (67%) and extracellular (33%) components. Blood is a suspension of cellular components (about 2.2 L) in blood plasma (3.8 L), and occupies the intravascular space. There are many different extravascular, extracelluar body fluids. Interstitial fluid (11.2 L) is a filtrate of blood plasma that bathes cells in all body tissues and is returned to the cardiovascular system via the lymphatic system. Glomerular fluid is another plasma filtrate that has its composition modified as it travels through the nephrons in the kidney, ultimately stored in the bladder as urine. Pericardial fluid provides lubrication between the outside surface of the beating heart and surrounding structures. Similar functions are performed by pleural fluid (lung) and peritoneal fluid (intestines). Cerebrospinal fluid (CSF) cushions the brain and is formed in the choroid plexuses in the ventricles of the brain, flows between all of the meningeal layers over the entire surface of the brain, and is ultimately reabsorbed into the venous system. Mucus is secreted by glands in the respiratory, gastrointestinal, and reproductive tracts, and serves many functions, including the protection of underlying tissues and protection of the body from infection. Synovial fluid is expressed in joints to lubricate the articulating cartilage surfaces. Bile is produced by the liver and released into the gastrointestinal tract to assist with the digestion of food. Chyme is the suspension of solids in fluid that is released by the stomach into the small intestines. Many other body fluids perform important functions, including sweat, intraocular fluid, saliva, and semen.

4.9.1 *Rheological Properties of Extravascular Body Fluids*

Cytoplasm. The cytoplasm of most intact cells is gel-like and is probably best modeled as a viscoelastic fluid. The cytoplasm of endothelial cells and chondrocytes, behave as a solid when subjected to shear (Jones et al. 1999; Sato et al. 1990). However, the cytoplasm of other cells exhibit non-Newtonian fluid properties. For instance, the axoplasm from the giant squid axon behaves like a classic Bingham fluid with a yield stress of 109 ± 46 dynes/cm^2 and an apparent viscosity μ_o of 14.6 Pa s (Rubinson and Baker 1979), while the cytoplasm of neutrophils has been modeled as a power law fluid with $n = 0.48$ and $K = 130$ Pa s$^{0.48}$ (Tsai et al. 1993).

Interstitial Fluid (ISF). Interstitial fluid is a Newtonian fluid with a viscosity that lies between the viscosity of water and the viscosity of blood plasma. The actual viscosity depends on the concentration of total protein and the relative fractions of large to small proteins.

Cerebrospinal fluid (CSF). CSF has been shown to be a Newtonian fluid for strain rates between 25 and 460 s^{-1} with a viscosity at 37°C between 0.7 and 1.0 mPa s (Bloomfield et al. 1998). The viscosity is not greatly affected by moderate changes in protein or cellular composition.

Synovial Fluid. Synovial fluid is viscoelastic and exhibits shear thinning at shear rates between 10 and 250 s^{-1} (Lumsden et al. 1996). The rheological properties depend primarily on the concentration of hyaluronic acid, and proteoglycan. Synovial fluid has been modeled as a Bingham fluid (Tandon et al. 1994), but more recent measurements show the apparent viscosity follows a power law model, decreasing by a factor of 10–12 as shear rates are raised from 1 s^{-1} (900 mPa s) to 250 s^{-1} (70 mPa s) (Conrad 2001). It is considerably more viscous than water, even at high shear rates.

Urine. Urine is a Newtonian fluid, at least for shear rates above 300 s^{-1} (Roitman et al. 1995) with a viscosity of 0.85 ± 0.07 mPa s (Kienlen et al. 1990). It should be noted that certain contrast agents, such as iotrolan, can raise urine viscosity by a factor of three or more (Ueda et al. 1998)]. This can raise tubular hydrostatic pressure and decrease glomerular filtration. Thus caution should be exercised when selecting contrast agents to make sure that they do not alter the physiological system during an experimental investigation.

Saliva. Saliva coats and lubricates the oral surfaces and ingested food. It contains proteins, which initiate the digestion of food. It exhibits some shear thinning, but for the most part can be considered to be a Newtonian fluid with viscosity of 0.95–1.1 mPa s at 25°C (Waterman et al. 1988).

Mucus. Mucus is a non-Newtonian biological fluid composed of mucin molecules of high molecular weight. Mucus exhibits a yield stress and its apparent viscosity decreases with increasing shear rate. Mucus with different compositions performs important functions in different organs.

Gastric mucus protects the stomach from high levels of hydrochloric acid in the lumen. The apparent viscosity of gastric mucus depends on at least five factors, including pH, ionic strength, mucin concentration, shear rate, and temperature as shown in Fig. 4.19 (List et al. 1978). Mucus viscosity increases by a factor of 100 as the pH drops from 7 to 2 at low ionic strength (Bhaskar et al. 1991). At low pH, mucin molecules aggregate to form a gel layer 50–500 μm thick that covers the gastric epithelium and protects it from auto-digestion. Mucin aggregation is impaired as the ionic strength of mucus is increased, and apparent viscosity drops. Characteristic rheological parameters vary greatly in the literature with species as well as pH, mucin concentration, ionic strength, and shear rate. The yield stress of gastric mucus and of duodenal mucus have been reported as 24.9 ± 8.5 Pa and 12.9 ± 3 Pa, respectively (Zahm et al. 1989). Apparent viscosity of human gastric mucus varied from 6,000 mPa·s at 1.15 s^{-1} to less than 500 mPa s at 23 s^{-1} (Markesich et al. 1995).

The respiratory tract is lined by an epithelial cell layer that contains beating cilia. An aqueous periciliary layer is found between the epithelial cell surface and the tips of the cilia. A layer of mucus gel is deposited between the cilia and the airway lumen. The mucus layer is moved by the beating cilia up the respiratory tree toward the mouth. The function of this layer is to protect the body from airborne particulates and bacteria by trapping them in the mucus layer, transporting them to the throat, where they are swallowed and destroyed in the highly acidic digestive juices in the stomach. Respiratory mucus is a gel-like substance that exhibits a yield stress and an apparent viscosity that decreases with increasing shear rate, obeying a Herschel–Buckley rheological model (Low et al. 1997). Yield stress of airway mucus is in the range of 4–7 dynes/cm^2 (Hsu et al. 1996), apparent viscosity at a shear rate of 1 s^{-1} ranges from 1,000 to 2,320 mPa·s (Hsu et al. 1996), and the power law exponent is in the range 0.68–0.72 (McCullagh et al. 1995).

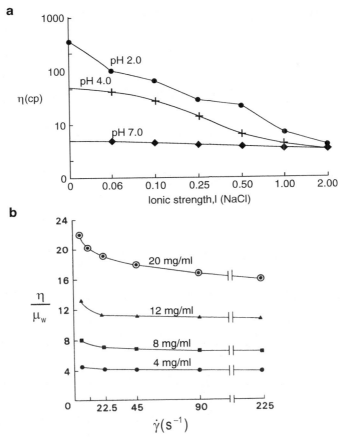

Fig. 4.19 Factors which affect the viscosity of gastric mucus, in addition to temperature, include (**a**) pH and ionic strength, from Bhaskar et al. 1991, (**b**) mucin concentration and shear rate, from List et al. 1978

Vaginal fluid and cervical mucus have both been shown to be non-Newtonian fluids with properties that vary during different phases of the menstrual cycle (Rutlant et al. 2002). Fertility studies have shown that changes in these properties can lead to lower or higher resistance to sperm transport in the female reproductive tract. Human semen also changes its properties with time, but on a much more rapid time scale. The yield stress of semen drops from 3,000 to 60 mPa within 5 min of ejaculation and its apparent viscosity also decreases with time and with shear rate (Shi et al. 2004).

Other Extravascular Biofluids. Amniotic fluid, which surrounds the fetus during development, has been shown to be non-Newtonian, following the power law model (Dasari et al. 1995). The rheology of intestinal contents is described by the Bingham model (Takahashi and Sakata 2002).

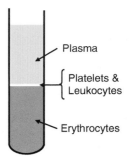

Fig. 4.20 Centrifuged blood sample showing separation of plasma, red cells, and white cells

4.9.2 Blood Rheology

Blood has some unusual rheological characteristics because of its particulate nature. Blood is a mixture of plasma and three primary cellular components: red blood cells (erythrocytes), white blood cells, and platelets. The density of the cellular components is slightly higher than blood plasma, so they settle to the bottom of a test tube when blood, collected in an anticoagulant, is allowed to stand. The volume percentage of red blood cells in blood is known as the hematocrit value. This can be measured by spinning a blood sample in a centrifuge and measuring the volume of red cells relative to the total sample volume Fig. 4.20). The hematocrit value H ordinarily ranges from 40 to 50%. The reader should be aware that H is sometimes expressed as a fraction between 0 and 1, rather than as a percentage. Platelets and white blood cells have densities between blood plasma and erythrocytes, and will accumulate in a thin layer at the interface when blood is centrifuged.

4.9.2.1 Blood Plasma

Blood plasma ordinarily contains about 7% proteins by weight. Plasma is normally a Newtonian fluid with a viscosity at 37°C of about 1.24 mPa s. Plasma viscosity increases with increasing protein concentration. Species differences in plasma viscosity exist primarily because of differences in total protein concentration and the relative amounts of large proteins to smaller proteins. Cattle, for instance, have a higher plasma viscosity (1.74 mPa s) than humans because bovine plasma has a higher total protein concentration and a higher concentration of fibrinogen. The influence of changing the concentrations of albumin (MW = 66.5 kDa) and large immunoglobulins (MW = 166 kDa) on the viscosity of aqueous protein solutions is shown in Fig. 4.21. The concentration of immunoglobulins is elevated during disease states, and this will have a greater effect on raising plasma viscosity than a similar increase in albumin concentration.

4.9.2.2 Erythrocytes

Red cells are by far the major cellular component of blood. Erythrocytes have the shape of a biconcave disc with diameter of 8 µm. The thickness near the outer edge is about 2 µm, tapering to about 1 µm at the center. They are normally quite flexible, and can change shape readily when exposed to shear stress. Erythrocytes contain very high concentrations of the protein hemoglobin. This protein combines with oxygen in lung capillaries and transports oxygen via the cardiovascular system to oxygen-consuming tissues throughout the body. About one third of the weight of the red cell cytoplasm is due to the presence of hemoglobin (34 g/dl). Despite this high concentration, intracellular hemoglobin solutions up to concentrations of at least 45 g/dl exhibit Newtonian behavior for shear rates between 1 and 200 s^{-1}. Intracellular viscosity of red cells has been shown to be the same as the viscosity of hemoglobin solutions with the same concentration. The relationship between viscosity and concentration is shown in Fig. 4.21 for a hemoglobin solution at 25C.

Hemoglobin has a molecular weight of 68 kDa, which is slightly larger than the molecular weight of albumin. However, when comparing Figs. 4.21 and 4.22, it is clear that the viscosity of a hemoglobin solution is lower than the viscosity of an albumin solution with the same concentration. This is a consequence of the more compact, spherical shape of the hemoglobin molecule. Thus, molecular size and shape, and not just molecular weight are important in determining the viscosity of protein solutions. As a consequence, the intracellular viscosity of red cells is rather low, normally around 5 mPa s.

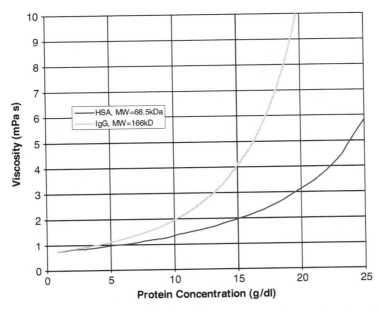

Fig. 4.21 Effect of protein concentration and molecular weight on plasma viscosity at 35°C

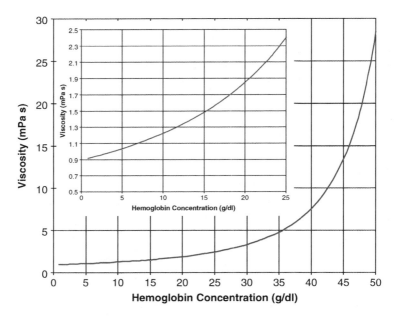

Fig. 4.22 Effect of concentration on hemoglobin solution viscosity

It was once thought that the benefit of confining hemoglobin to red cells was that the blood viscosity of a suspension of red cells in plasma was less than the viscosity of plasma with the same amount of hemoglobin. However, viscosity measurements have shown that to be incorrect. If all the hemoglobin in erythrocytes were deposited directly in plasma, the hemoglobin concentration would be about 15 g/dl. Measurements show that plasma viscosity increases by only 50–60%. Total blood viscosity would drop in half because of the loss of the suspended cells. This reduction in blood viscosity would appear to be beneficial from the standpoint of the energy necessary to pump blood through the cardiovascular system. However, this benefit would be blunted considerably by the loss of the Fahraeus–Lindqvist effect, which we shall discuss in the next section. In addition, if all the hemoglobin were dissolved in the plasma, the colloid osmotic pressure of blood plasma would be tripled, causing water retention and an elevation in blood pressure.

Liposome encapsulated hemoglobin solutions have been proposed as a blood substitute. These suspensions exhibit non-Newtonian behavior, with the apparent viscosity increasing with hemoglobin concentration and decreasing with shear rate.

4.9.2.3 Leukocytes

Leukocytes, or white blood cells (WBC), are circulating cells associated with the body's immune and defense system. White cells are larger than red blood cells,

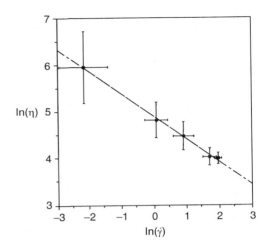

Fig. 4.23 Viscosity of neutrophil cytoplasm, from Tsai et al. (1996) with permission

typically with diameters around 20 μm, and red cells outnumber them by a ratio of about 700:1. Polymorphonuclear granulocytes, such as neutrophils, contain multiple nuclei and many small granules in the cytoplasm. They are relatively short lived, spending a day or so in the vascular system, before migrating between endothelial cells, entering tissue, and removing unwanted materials and bacteria via the process of phagocytosis. The ease with which a WBC migrates, a process known as extravasation, is dependent on the rheological properties of the cell. Recent measurements have shown that the cytoplasm of neutrophils is a non-Newtonian fluid corresponding to a power law model (Fig. 4.23), with apparent viscosities ranging from 50 to 500 Pa s, depending on shear rate. This is 4 to 5 orders of magnitude greater than the cytoplasmic viscosity of erythrocytes. Similar non-Newtonian behavior has been measured in pulmonary macrophages. The cytoplasm of red cells and white cells behave as fluids, albeit with different rheological characteristics. Both are quite different from the cytoplasm of endothelial cells, which behaves like an elastic solid.

4.9.2.4 Whole Blood

Experimental measurements made on whole blood indicate that the apparent viscosity of blood:

- Increases with increasing hematocrit.
- Decreases with increasing shear rate.
- Decreases as the radius of the conduit through which it flows decreases (Fahraeus–Lindqvist effect).

Let us examine each of these observations in more detail.

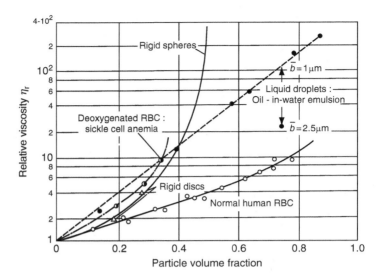

Fig. 4.24 Relative viscosity vs. particle volume fraction for various suspensions, from Goldsmith and Mason (1967) with permission

Hematocrit value. The dependence of relative viscosity on hematocrit value (expressed as a volume fraction) is shown in Fig. 4.24. Relative apparent viscosity of blood η_r is defined as the apparent viscosity of blood η relative to plasma viscosity μ_p:

$$\eta_r = \frac{\eta}{\mu_p}. \tag{4.40}$$

Figure 4.24 is a semilog plot, so the relationship is nonlinear. Anemic patients, with normal cell rigidity and low hematocrit values, have lower blood viscosity than normal, while individuals with polycythemia (an excess of red cells) have significantly higher viscosity than normal. The viscosity of particulate solutions increases with increasing concentration of particles in a nonlinear fashion. The viscosity of normal blood is less than the viscosity of other particulate mixtures shown in Fig. 4.24, including rigid discs and spheres, and a suspension of oil droplets in water. This is largely because of the ability of normal red cells to deform when sheared. Deoxygenated sickle cells behave more like rigid spheres than normal erythrocytes. We will examine this in more detail later.

Shear rate dependence. The shear rate dependence of apparent blood viscosity is shown in Fig. 4.25, where the shear rate is plotted on a log scale for blood with a normal hematocrit value. Blood is a pseudoplastic fluid in which the apparent viscosity decreases with increasing shear rate for shear-rates less than about $50\ \text{s}^{-1}$.

At high shear rates, cells deform into parachute-like shapes as they pass though capillaries (Fig. 4.26, left). However, at low shear rates, cells tend to form stacked aggregates called rouleaux (Fig. 4.26, right).

4.9 Rheology of Biological Fluids

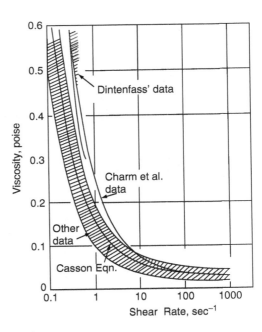

Fig. 4.25 Shear rate dependence of apparent blood viscosity, adapted from Whitmore (1968)

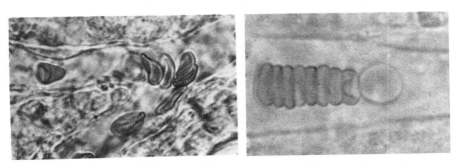

Fig. 4.26 Erythrocytes moving through capillaries (*left*), and stagnant in capillary (*right*), from Per-ingvar (1971) with permission

The relative effects of cell deformation and aggregation/disaggregation are shown in Fig. 4.27, where the apparent viscosity is measured as a function of shear rate for red blood cells (RBC) suspended in plasma, RBC suspended in an albumin-Ringers solution (no aggregation) and hardened (undeformable) RBC suspended in albumin-Ringers. Aggregation raises blood viscosity at low shear rates and deformation lowers relative viscosity at high shear rates. Note the range of relative viscosities from 200 at very low shear rates to 4–5 at high shear rates.

Figure 4.28 shows a plot of the square root of shear stress as a function of the square root of shear rate for blood with normal hematocrit.

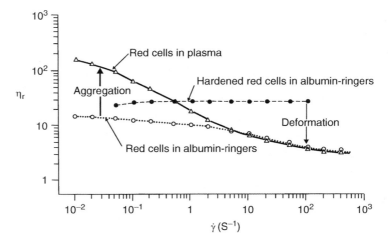

Fig. 4.27 Relative viscosity vs. shear rate for red cells suspended in plasma, red cells in an albumin-ringers solution, and hardened red cells in albumin-ringers, from Chen (1970) with permission

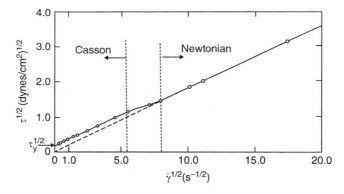

Fig. 4.28 Casson plot for blood with $H = 40\%$, $T = 37°C$, and fibrinogen concentration $= 0.27$ g/100 ml, from Merrill and Pelletier (1967) with permission

This relationship is linear over two different ranges, indicating that normal blood behaves as (1) a Casson fluid for shear rates less than about 30 s^{-1} and (2) a Newtonian fluid for shear rates above 50 s^{-1}. The slopes in each region are virtually identical, indicating an apparent viscosity of about 3 mPa s. For other hematocrit values, the Casson parameter S (mPa·s$^{1/2}$) can be computed from:

$$S = \frac{\mu_p^{1/2}}{1 - H}, \quad (4.41)$$

where μ_p is plasma viscosity (mPa s) and H is hematocrit expressed as a volume fraction. The yield stress (dynes/cm^2) is known to depend on the hematocrit value

4.9 Rheology of Biological Fluids

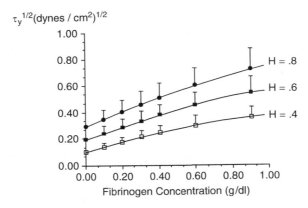

Fig. 4.29 Influence of plasma fibrinogen concentration on yield stress, from Morris et al. (1989) with permission

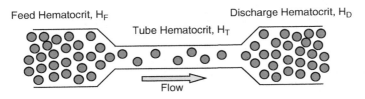

Fig. 4.30 Hematocrit value in small tubes is lower than in large feed or discharge vessels

and fibrinogen concentration ρ_F (g/dl), as shown in the empirical equation below, based on measurements made by Morris et al. (1989), shown in Fig. 4.29.

$$\tau_y^{1/2} = 0.47H + 0.48H\rho_F + 0.22\rho_F - 0.14\rho_F^2 - 0.091. \quad (4.42)$$

Fibrinogen is a protein responsible for clot formation in wounds and contributes to the formation of rouleaux. Thus, the yield stress of blood increases with increasing fibrinogen concentration.

Fahraeus–Lindqvist effect. The dependence of blood apparent viscosity on vessel radius is known as the Fahraeus–Lindqvist effect. This has been examined extensively in glass tubes of various radii using an apparatus similar to that sketched in Fig. 4.30. In steady flow, the discharge hematocrit, H_D must equal the feed hematocrit H_F.

Experimental data compiled by Pries et al. (1992) for a feed or discharge hematocrit of 45% are summarized in Fig. 4.31.

The relative apparent viscosity is independent of tube size if the tube diameter is greater than 500 μm. However, as the tube diameter gets smaller than 500 μm, the apparent viscosity decreases until it reaches a minimum at a tube diameter of about 7 μm. This is very similar to the diameter of an undeformed erythrocyte (8 μm). The apparent blood viscosity increases dramatically when the tube diameter drops

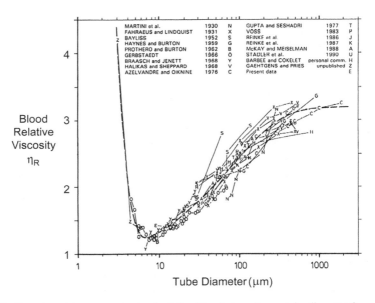

Fig. 4.31 Experimental measurements of relative blood viscosity vs. tube diameter for a feed or discharge hematocrit of 45%, from Pries et al. (1992) with permission. See original for data citations

below 7 μm, with an asymptote at about 3 μm. The minimum blood viscosity at 7 μm is only 20–25% greater than the viscosity of the plasma itself. Pries et al. (1992) have fitted a model to these data which provides a prediction of the relative apparent viscosity at $H_D = 45\%$ as a function of tube diameter D (μm):

$$\eta_r(D, 45\%) = 220 \cdot e^{-1.3D} + 3.2 - 2.44 \cdot e^{-0.06 D^{0.645}}. \tag{4.43}$$

Pries et al. have expanded their empirical description of blood viscosity in tubes to include the effects of varying discharge hematocrit:

$$\eta_r(D, H_D) = 1 + (\eta_r(D, 45\%) - 1) \cdot \left[\frac{\left(1 - \frac{H_D}{100}\right)^\alpha - 1}{\left(1 - \frac{45}{100}\right)^\alpha - 1} \right], \tag{4.44}$$

where the exponent α is given by the following function of tube diameter D (μm):

$$\alpha = (0.8 + e^{-0.075D}) \cdot \left(-1 + \frac{1}{1 + 10^{-11} \cdot D^{12}}\right) + \frac{1}{1 + 10^{-11} \cdot D^{12}}. \tag{4.45}$$

The effects of changing the hematocrit value between 15% and 65%, based on (4.44), are shown in Fig. 4.32. What can be responsible for the shape of these graphs? To answer this, we consider red cell and plasma flow in a narrow bore tube like that shown in Fig. 4.33.

4.9 Rheology of Biological Fluids

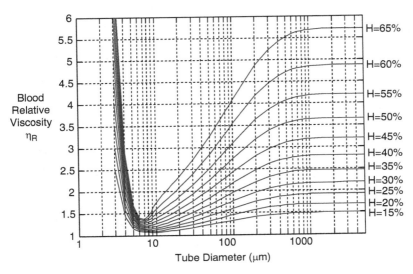

Fig. 4.32 Effect of hematocrit value on the relationship between blood relative viscosity and vessel dimeter for small tubes (4.44)

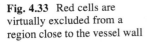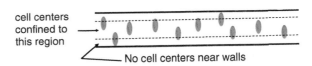

Fig. 4.33 Red cells are virtually excluded from a region close to the vessel wall

Even the largest molecules in plasma are much smaller than red cells. Plasma in the tube can be distributed evenly between the center of the tube and the tube wall. However, since all portions of an erythrocyte move as a unit and because red cells have a finite size, the centers of erythrocytes are restricted to the central portion of the tube, between the dotted lines in Fig. 4.33. Although plasma and red cells move at the same speed in the region between the dotted lines, plasma near the wall moves much slower than the slowest red cell. As a consequence, erythrocytes move through the tube with a higher average velocity than plasma. The tube hematocrit is defined as the volume percent of erythrocytes inside the capillary tube. Although the feed hematocrit must equal the discharge hematocrit, the tube hematocrit H_T must be lower than the discharge hematocrit H_D because the red cells move through the tube at a higher velocity. This ratio H_T/H_D will decrease as the tube diameter decreases. This is known as the Fahraeus effect and is shown for a discharge hematocrit of 45% in Fig. 4.34.

The tube hematocrit is a volume-averaged hematocrit while the discharge hematocrit is a flow averaged hematocrit. Stated in mathematical terms for a cylindrical vessel with volume V and cross-sectional area A_c:

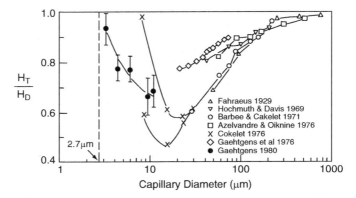

Fig. 4.34 Ratio of tube hematocrit to discharge hematocrit in narrow tubes as a function of tube radius, from Gaehtgens (1980) with permission. See original for data citations

$$H_T = \frac{\int_V H dV}{\int_V dV} = \frac{\int_0^L \int_0^R H(2\pi r dr) dz}{\pi R^2 L} = \frac{2}{R^2} \int_0^R Hr dr, \quad (4.46)$$

$$H_D = \frac{\int_{A_c} Hv dA_c}{\int_{A_c} v dA_c} = \frac{\int_0^R Hv(2\pi r dr)}{\pi R^2 \langle v \rangle}, \quad (4.47)$$

where v is the local velocity, $\langle v \rangle$ is the average velocity in the capillary, and R is the capillary radius. To compute the tube hematocrit, we must know how the hematocrit varies with radial position in the vessel. A reasonable assumption based on Fig. 4.33 might be that the local hematocrit value is uniform in the core region ($H = H_0$), but is zero in the region near the wall, where red cell centers can come no closer to the wall than the radius of the cell. If the red cell radius is R_c, then we can write:

$$\begin{aligned} H &= H_0, r \leq R - R_c, \\ H &= 0, R - R_c < r \leq R. \end{aligned} \quad (4.48)$$

Substituting (4.48) into (4.46) and integrating, we obtain the following expression for the tube hematocrit:

$$H_T = H_0 \left(1 - \frac{R_c}{R}\right)^2. \quad (4.49)$$

Computation of the discharge hematocrit requires that we also know how the velocity varies from the wall to the center of the vessel. Determination of the

4.9 Rheology of Biological Fluids

velocity profile under various situations is treated extensively in Chaps. 6 and 7. If blood can be considered a Newtonian fluid as it flows through the vessel, then the velocity profile will be parabolic with the velocity in the center of the tube being twice the average velocity:

$$v(r) = 2 <v> \left[1 - \left(\frac{r}{R}\right)^2\right]. \quad (4.50)$$

Details of the derivation of (4.50) can be found in Sect. 6.3. Substituting (4.50) and (4.48) into (4.47) yields the following prediction for the discharge hematocrit:

$$H_D = H_0 \left(1 - \frac{R_c}{R}\right)^2 \left[2 - \left(1 - \frac{R_c}{R}\right)^2\right]. \quad (4.51)$$

Taking the ratio of predicted discharge hematocrit to tube hematocrit:

$$\frac{H_T}{H_D} = \frac{1}{2 - (1 - R_c/R)^2}. \quad (4.52)$$

This relationship depends only on the ratio of erythrocyte radius to vessel radius and is plotted in Fig. 4.35 for a red cell radius or 3.5 μm. Despite the oversimplifications represented by (4.48) and (4.50), the agreement with experimental data in Fig. 4.34 is remarkable, and implies that exclusion of red cells from the wall region of blood vessels can explain the Fahraeus effect.

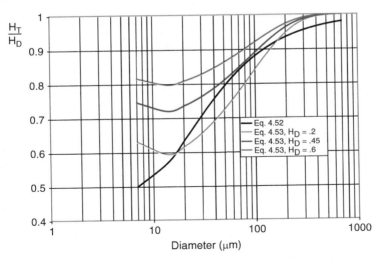

Fig. 4.35 Theoretical prediction of tube hematocrit relative to discharge hematocrit vs. tube diameter for a red cell radius of 3.5 μm (4.52) and empirical relationships from Pries et al. 1992 (4.53) for three different discharge hematocrit values

An empirical relationship for the dependence of H_T/H_D on D(μm) based on the available experimental data in glass tubes has been developed by Pries et al. 1992:

$$\frac{H_T}{H_D} = H_D + (1 - H_D) \cdot (1 + 1.7e^{-0.415D} - 0.6e^{-0.011D}). \tag{4.53}$$

This relationship is also shown in Fig. 4.35 for three different discharge hematocrit values. The Fahraeus effect appears to be more pronounced at lower hematocrit values. In vivo experimental data support, the existence of the Fahraeus effect, as shown in Fig. 4.36, where microvascular hematocrit clearly decreases as vessel size decreases.

Although the data presented for glass tubes are relevant to the design of devices through which blood will flow, Pries et al. (1994) have found that the in vivo resistance to blood flow appears to be greater than the in vitro resistance. This is demonstrated in Fig. 4.37, where relative apparent viscosity measured in vivo is compared to the in vitro value. The differences are substantial and are probably due in large part to the presence of the endothelial surface layer (ESL), which includes the glycocalyx and an adsorbed layer of macromolecules, which protrude into the lumen of the capillary, reducing the available diameter of the microvessels (Weinbaum et al. 2007). Pries and Secomb (2005) showed that the difference between in vivo and in vitro flow resistance for blood in small vessels could be explained by an ESL of 1 μm or less.

4.9.3 Biorheology and Disease

Koenig et al. (2000) found a strong positive correlation between plasma viscosity and mortality of all causes in middle-aged men, and suggested that plasma viscosity may have considerable potential to predict death from all causes. There are many disease states associated with abnormal rheology of biological fluids. We mention here just a few.

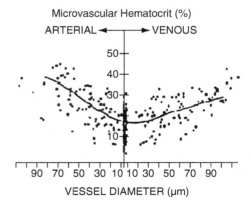

Fig. 4.36 Average hematocrit observed in microvessels, from House and Lipowski (1987) with permission

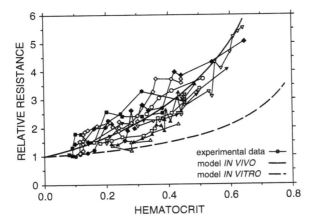

Fig. 4.37 Comparison of in vivo experimental data with in vitro prediction of relative viscosity, from Pries et al. (1994) with permission

4.9.3.1 Polycythemia

Primary polycythemia is a condition which results from an excessive production of red blood cells, leading to an increased hematocrit and an elevated blood viscosity. In severe cases, the hematocrit value may exceed 70% and blood viscosity can increase more than three fold. The elevated blood viscosity puts an additional strain on the heart, leading to high blood pressure and possibly microvascular damage. Studies have shown that there is a positive correlation between elevated blood viscosity and the severity of coronary artery disease (Lowe et al. 1980). Hypoxia, resulting from living at high altitudes where the partial pressure of oxygen is low, induces the bone marrow to produce more erythrocytes. The resulting elevation in hematocrit is known as secondary polycythemia. Although blood viscosity is raised slightly, the body is able to compensate by an overall reduction in vascular tone.

Blood hematocrit can be artificially increased by a transfusion of packed cells. Some athletes have sought to gain an unfair advantage over their opponents by increasing the number of red cells in their body just before an athletic event. This is known as blood doping, and is banned not only for its unethical consequences, but also for the potential physiological risks inherent in transfusions and the accompanying elevation in blood viscosity.

Hematocrit can be elevated by mechanisms which do not raise the number of red cells, but instead by decreases in the plasma volume. This is known as relative polycythemia. An increase in capillary permeability will cause a greater amount of fluid to cross the capillary wall and pool in the interstitial space, resulting in a decrease in plasma volume and an increase in blood viscosity. Plasma volume will also be lost if capillary pressure is increased, either through an increase in arterial blood pressure, or through an increase in the postcapillary to precapillary resistance ratio.

4.9.3.2 Cancer

Plasma viscosity has been observed to be non-Newtonian in oral cancer patients having poor prognosis and to be much higher than normal at low shear rates (Ranade et al. 1995). Plasma viscosity was found to be an independent prognostic marker for survival of women with breast cancer, probably because of the increased breakdown products associated with tumor cell dissemination and high rates of fibrinogen/fibrin turnover (von Tempelhoff et al. 2002). A significant rise in plasma viscosity is observed when metastasis becomes clinically apparent in gynecologic cancer (von Tempelhoff et al. 2003), and this is useful in monitoring patients on follow-up visits. They also found plasma viscosity to be a significant preoperative risk factor for survival in ovarian cancer.

4.9.3.3 Sickle Cell Anemia

We mentioned previously the abnormally high viscosity of deoxygenated sickle cells shown in Fig. 4.24. These cells contain hemoglobin S (HbSS), as opposed to normal hemoglobin A. HbSS tends to polymerize when exposed to oxygen tensions below 80 mmHg, increasing the apparent viscosity of sickle blood (Fig. 4.38).

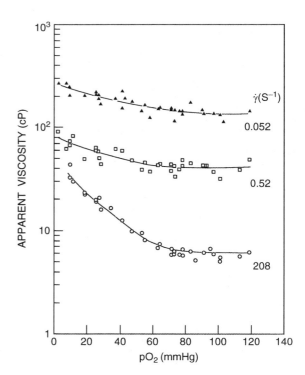

Fig. 4.38 Apparent viscosity of sickle blood as a function of partial pressure of oxygen at three different shear rates, from Usami et al. (1975) with permission

4.9 Rheology of Biological Fluids

The effect is most pronounced at high shear rates. Such oxygen tensions are certain to exist in capillaries of oxygen-consuming tissues. The polymerized HbSS increases cell rigidity and deforms the cell into a characteristically sickle shape. Both of these features tend to increase the viscosity dramatically. Sickled cells are more fragile than normal cells, and therefore they hemolyze more readily. Consequently, individuals with nearly all hemoglobin in the form of HbSS have hematocrit values near 20%. The low hematocrit is the origin of the "anemia" in sickle cell anemia. Because of this low hematocrit and high cell fragility, people with sickle cell anemia tend to have relatively normal blood viscosity in large vessels (Fig. 4.24), but have abnormally high viscosity in the microvascular system. Sickle blood also obeys a Casson relationship for various hematocrit values and for oxygenated and deoxygenated blood as shown in Fig. 4.39.

4.9.3.4 Cystic Fibrosis

Inhaled contaminants and bacteria are ordinarily trapped in the airway surface liquid and propelled out of the lung before they can do damage. The airway surface liquid is composed of two separate regions: a low viscosity periciliary fluid and

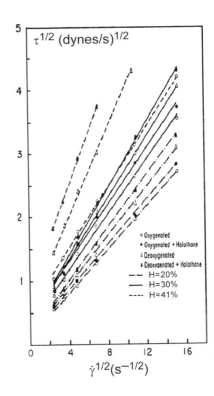

Fig. 4.39 Casson plots of oxygenated and deoxygenated sickle blood at various hematocrit values, adapted from Laasberg and Hedley-Whyte (1973)

a higher viscosity mucus layer. The periciliary layer is in contact with the epithelial cells of the bronchial wall and surrounds the cilia. The periciliary fluid layer is normally slightly thicker than the length of the cilia. The cilia, by virtue of rapid forward strokes (power strokes) and slow reverse strokes, impose a velocity of about 45 μm/s at the top of the layer toward the trachea.

The mucus layer is ordinarily about 25 μm thick and is located between the airway lumen and the periciliary fluid. Mucin molecules have many side chains and can readily attach to bacteria and contaminants when they come close to the wall of the airway. Mucus also has a relatively high yield stress, and so it can ride as a plug on top of the periciliary layer. This mucus "conveyor belt" ordinarily moves fast enough so that bacteria are swept out of the lung before they can penetrate the layer.

In cystic fibrosis, both the mucus and the periciliary fluid layers become abnormally thin. This is caused by the presence of an abnormal gene that is responsible for low secretion rates by submucosal glands of higher viscosity secretions into the airway surface liquid. This is most critical in the periciliary fluid because cilia do not have adequate room to rotate properly, and consequently the velocity at the top of the periciliary fluid layer is reduced almost tenfold. Mucociliary clearance of bacteria is severely reduced and some airways become completely occluded. As a result, bacteria have time to cross the mucus barrier, causing infection, inflammation, and shortening life. Next to sickle cell anemia, cystic fibrosis is the second most common life-shortening disease (Grosse et al. 2004). Average survival of patients with cystic fibrosis is about 33 years.

4.10 Summary of Key Concepts

A fluid, by definition, will continue to deform for as long as a shear stress is applied. All fluids are viscous, meaning that they exhibit internal friction. If the appropriate sign convention is adopted, a shear stress applied in the x-direction on a plane of constant y can also be interpreted as the molecular flux of x-momentum in the y-direction.

Constitutive Relationships. A fluid in which the shear rate or negative velocity gradient is directly proportional to the applied shear stress is said to be a Newtonian fluid. The proportionality factor is known as the fluid viscosity, μ. The viscosity generally increases with temperature for gases and decreases with temperature for liquids. Kinematic viscosity ν is defined as the viscosity divided by the density of the fluid. All other fluids are classified as non-Newtonian fluids, but can be described by a constitutive relationship similar to Newton's law of viscosity (4.15):

$$\tau_{nx} = -\eta \frac{\mathrm{d}v_x}{\mathrm{d}n}.$$

Here, η is defined as the apparent viscosity of the fluid and is a function of the shear rate. Several types of non-Newtonian fluids are of interest in biotransport applications, including the power law fluid (4.26), the Bingham fluid (4.28)–(4.30), the Casson fluid (4.31)–(4.33), and the Herschel–Bulkley fluid (4.34)–(4.36). The Bingham, Casson, and Herschel–Bulkley fluids behave as solids until a yield stress is exceeded. Viscometers can be used to measure the apparent viscosity of unknown types of fluids under various conditions. The cone and plate viscometer is often used to characterize non-Newtonian fluids, since it can expose the entire fluid being tested to a uniform shear rate.

Blood Rheology. Blood is a mixture of cells (primarily red blood cells) and plasma. The ratio of red cell volume to total volume is known as the hematocrit value. Plasma is a Newtonian fluid with viscosity dependent on protein concentration. Blood behaves like a Newtonian fluid at high shear rates and as a Casson fluid at low shear rates (Fig. 4.28). The relative viscosity of blood η_r is defined as the apparent viscosity of blood relative to the viscosity of plasma. The relative viscosity increases with increasing hematocrit value and decreases with increasing shear rate. The yield stress of blood increases with hematocrit value and with plasma fibrinogen concentration. The non-Newtonian nature of blood is attributed to the tendency of red cells to aggregate into rouleaux formations at low shear rates and to dissociate and deform at high shear rates. The apparent viscosity of blood decreases as vessel diameter decreases, increasing again as the vessel size is smaller than the size of the cells passing through it. This is known as the Fahraeus–Lindqvist effect and is caused by the fact that cells cannot come any closer to the vessel wall than the cell radius, while plasma can come much closer. Plasma moves slower on average than red cells through small vessels, so the small vessel hematocrit is lower than large vessel hematocrit. A number of disabling diseases, such as sickle cell anemia and cystic fibrosis, can be directly attributed to abnormal rheology of a biological fluid.

4.11 Questions

4.11.1. Distinguish between a solid and a fluid.
4.11.2. What is the "no-slip" boundary condition?
4.11.3. What boundary conditions would you apply at the interface between two immiscible fluids in motion?
4.11.4. What approximate boundary condition is normally applied at a gas–liquid interface?
4.11.5. Distinguish between laminar and turbulent flow.
4.11.6. Discuss how shear stress can be interpreted as either a flux of momentum or a force per unit area.
4.11.7. What is the meaning of the subscripts y and x used to define the shear stress, τ_{yx}?
4.11.8. Explain the sign convention adopted for shear stress in this text.

4.11.9. What is the advantage of a cone and plate viscometer over the other viscometers shown in Fig. 4.7?

4.11.10. What is a Newtonian fluid?

4.11.11. What is a non-Newtonian fluid?

4.11.12. How does temperature affect the viscosity of gases and liquids?

4.11.13. What factors might affect the viscosity of blood plasma?

4.11.14. What features of Bingham, Herschel–Bulkley and Casson fluids are similar? Different? How do they differ from a Newtonian fluid?

4.11.15. What is yield stress? What causes it?

4.11.16. How would you analyze data collected at several plate velocities with a parallel plate viscometer to distinguish between Newtonian, power law, Bingham, Herschel–Bulkley and Casson fluids?

4.11.17. What is hematocrit value?

4.11.18. What is the difference between apparent viscosity and relative viscosity of blood?

4.11.19. Is blood a Newtonian fluid? Do large errors result if we assume blood to be Newtonian?

4.11.20. Describe three rheological characteristics of blood that distinguish it from plasma or water? Explain the relationship between these properties and the presence of red cells in blood.

4.11.21. Draw a graph of apparent viscosity of blood as a function of the logarithm of shear rate.

4.11.22. What causes the shear rate dependence of the apparent viscosity of blood?

4.11.23. Draw a graph of the relative viscosity of blood as a function of the logarithm of blood vessel diameter. Discuss reasons for the shape of the curve.

4.11.24. What is the Fahraeus–Lindqvist effect? What causes this? If you could measure the hematocrit in microvessels, how would you go about estimating the viscosity?

4.11.25. What is the qualitative relationship between blood viscosity and hematocrit value?

4.11.26. What influences the yield stress of blood?

4.11.27. Why is microvessel hematocrit different from large vessel hematocrit? How would you estimate microvessel viscosity from hematrocrit measurements and plasma viscosity?

4.11.28. What is the difference between volume averaged and flow averaged quantities, such as hematocrit, concentration, or temperature?

4.11.29. Is resistance to blood flow in vivo greater or less than the in vitro resistance of a tube with the same diameter and length

4.11.30. Explain how sickle cell crisis is initiated in terms of transport properties of gas exchange and blood rheology.

4.11.31. How does the loss of fluid from plasma in glomerular capillaries affect viscosity?

4.11.32. Describe the rheological ramifications of blood doping via transfusion of packed red cells.

4.12 Problems

4.12.1 Cytoplasm Viscosity

Tsai et al. (1996) used the following formula to describe the apparent viscosity η of a power law fluid as a function of shear rate, $\dot{\gamma}$:

$$\eta = \eta_c (\dot{\gamma}/\dot{\gamma}_c)^{-b},$$

where η_c is a characteristic apparent viscosity measured at a characteristic shear rate, $\dot{\gamma}_c = 1\text{s}^{-1}$ and b is a material constant. They found $\eta_c = 276$ Pa s and $b = 0.51$ for the cytoplasm of HL-60 cells measured during the S phase. (a) Find the coefficients K and n in (4.26) for this fluid. Assume $dv_x/dy = \dot{\gamma} > 0$. (b) What is the apparent viscosity of cytoplasm at a shear rate of 10 s^{-1}? (c) Compare this with the viscosity of water at 25°C.

4.12.2 Herschel–Bulkley Fluid

Show that the steady-state plate velocity in a sliding plate viscometer is given by (4.37) for a Herschel–Bulkley fluid for $\tau_w > \tau_y$.

4.12.3 Yield Stress

Blood is placed in a sliding plate viscometer with a fluid thickness $h = 0.1$ cm and plate area $A = 50$ cm^2. A force of 2 dynes must be applied to the wall of the viscometer before the plate begins to move. What is the yield stress of the blood sample?

4.12.4 Constitutive Relationships

Five biological fluid samples were tested using a sliding plate viscometer with fluid thickness h. The plate was pulled at several specific velocities V and the wall stress τ_w was measured for each plate velocity. Results are presented in the table below, where the shear rate $\dot{\gamma} = V/h$. For each sample in the table, identify the type of fluid and the appropriate rheological parameters.

$\dot{\gamma}=V/h(\text{s}^{-1})$	τ_w (mPa)				
	Sample A	Sample B	Sample C	Sample D	Sample E
0.01	96.1	347.4	18.9	0.04	4
0.02	98.7	379.8	28.7	0.07	5
0.04	102.4	434.2	43.5	0.14	7
0.08	107.8	525.6	65.9	0.28	11
0.16	115.6	679.5	99.9	0.56	19
0.32	127.1	938.2	151.4	1.12	35
0.64	144.4	1,373.3	229.5	2.24	67
1.32	172.1	2,147.2	354.4	4.62	135
2.64	213.9	3,406.7	537.1	9.24	267
5.12	277.0	5,405.6	799.3	17.92	515

4.12.5 Apparent Viscosity

The apparent viscosities of several biological fluids are examined with a cone and plate viscometer as a function of shear rate. Results are shown in the table below. Determine the rheological classification for each fluid sample and provide values for the rheological properties.

$\dot{\gamma}(\text{s}^{-1})$	η (mPa·s)				
	Sample 1	Sample 2	Sample 3	Sample 4	Sample 5
0.1	72.9	2.05	804.48	1,505.0	354.8
0.5	25.4	1.98	162.55	305.0	146.4
1	17.8	2.02	82.00	155.0	100.0
5	9.7	1.97	17.14	35.0	41.3
10	8.1	1.96	8.89	20.0	28.2
50	6.2	2.07	2.11	8.0	11.6
100	5.8	2.01	1.20	6.5	7.9
500	5.3	1.99	0.39	5.3	3.3
1,000	5.2	2.02	0.26	5.2	2.2

4.12.6 Blood Yield Stress

Compare the yield stress of blood from a normal individual with a hematocrit of 45% and fibrinogen concentration of 0.2 g/dl with blood from a polycythemia patient with hematocrit of 65% and fibrinogen concentration of 0.6 g/dl. Estimate the apparent viscosity of each blood sample if plasma viscosity is 1.1 mPa s for the normal sample and 1.2 mPa s for the polycythemia sample.

4.12.7 Casson Fluid Parameters

Estimate the Casson rheological parameters S and τ_y for blood with a hematocrit of 40% and fibrinogen content of 0.27 g/dl from the data in Fig. 4.28. Is the yield stress consistent with the value predicted from (4.42)? Using the rheological parameters found for S and τ_y and data from Fig. 4.28, compute the apparent viscosity of blood at shear rates of 1 s^{-1} and 36 s^{-1}. Compare these with the apparent viscosity of blood at high shear rates.

4.12.8 Fahraeus–Lindqvist Effect

Blood with a hematocrit of 45% flows through a 10 μm diameter tube. Flow is directly proportional to the pressure difference across the tube and inversely proportional to the apparent viscosity of blood. Compare the actual pressure difference across the tube with the pressure difference necessary to provide the same flow if the apparent viscosity in the tube remained the same as the apparent viscosity in a very large tube. Discuss implications of the Fahraeus–Lindqvist effect on the work of the heart.

4.12.9 Apparent Viscosity of Blood

Compare the apparent viscosity of blood with a hematocrit value of 60% flowing in an 8 μm diameter tube with the viscosity of a hemoglobin solution with the same overall hemoglobin concentration. Does this argue for or against the packaging of hemoglobin inside red cells?

4.12.10 Apparent Viscosity of Blood

Compare the apparent viscosity of blood with a hematocrit value of 20% flowing in a 100 μm diameter tube with the viscosity of a hemoglobin solution with the same overall hemoglobin concentration. Does this argue for or against the packaging of hemoglobin inside red cells?

4.12.11 Tube Hematocrit

Estimate the tube hematocrit in a 50 μm diameter tube if discharge hematocrit is (a) 25% and (b) 50%.

4.12.12 Casson Fluid

The experimental data shown in the accompanying table was collected in a cone-and-plate viscometer filled with whole human blood. The blood hematocrit was 45% and the measurements were conducted at a constant temperature of 37°C.

Shear rate, s^{-1}	Shear stress, Pa
1.03E-02	1.77E-02
5.99E-02	1.98E-02
1.21E-01	2.10E-02
5.27E-01	2.08E-02
1.17E+00	3.50E-02
5.99E+00	6.71E-02
1.14E+01	9.83E-02
5.81E+01	3.16E-01
1.18E+02	5.44E-01
2.30E+02	9.29E-01

Find the yield stress (τ_y) in Pa and the viscosity parameter S^2 in Pa s for whole human blood described by the Casson model using the experimental data.

4.12.13 Determination of Fluid Properties

Two parallel plates are separated by a thin gap of thickness h containing an unknown fluid. The following experiments are performed to determine rheological properties of the fluid. A known force per unit area F/A is applied to the upper plate and the steady-state velocity of the plate is measured to be v_1. Next, the force is (a) increased to 2F and then (b) increased to 4F, the velocity v of the plate is measured in each case. Results for three fluids show:

1. (a) $v = 3\ v_1$ and (b) $v = 7\ v_1$.
2. (a) $v = 2\ v_1$ and (b) $v = 4\ v_1$.
3. (a) $v = 4\ v_1$ and (b) $v = 16\ v_1$.

Determine whether these velocities reflect Newtonian, power law or Bingham fluid behavior, and describe how you would use the data to estimate viscosity for a Newtonian fluid, n and K for a power law fluid and μ_0 and τ_y for a Bingham fluid.

4.12.14 Fahraeus Lindqvist Effect

Estimate blood viscosity in 30 and 50 μm diameter vessels with feed $H = 60\%$ and plasma viscosity $= 1.2$ cp.

4.13 Challenges

4.12.15 Blood Rheology in a Small Tube

Blood flows through a tube with a radius of 25 μm and a length of 1,000 μm. The hematocrit in large tubes (feed hematocrit or discharge hematocrit H_D) is 45%. Plasma viscosity is 1.05 cp.

(a) Estimate the tube hematocrit value H_T from experimental data in Fig. 4.34 or from (4.46).
(b) Use Fig. 4.24 to estimate the apparent viscosity of blood in large vessels. Use this to estimate resistance of the 25 μm tube.
(c) Use the same graph to estimate the apparent viscosity of blood in the 25 μm tube, based on the tube hematocrit. Compare this with the value estimated using Fig. 4.31 or (4.43). How much is the tube resistance overestimated when large vessel hematocrit is used rather than tube hematocrit?

4.12.16 Reynolds Numbers for Blood Vessels

Typical diameters and mean velocities of various blood vessels are listed in the table below. For each vessel, estimate the apparent blood viscosity and Reynolds number Re_D at 37°C if large vessel hematocrit is a) 20% and b) 60%. Assume ESL thickness is 1 μm.

Vessel	Diameter (cm)	Average velocity (cm/s)
Aorta	2.5	40
Artery	0.5	15
Arteriole	0.005	0.6
Capillary	0.0008	0.08
Venule	0.007	0.3
Vein	0.6	10
Vena cava	3.0	28

4.13 Challenges

4.13.1 Respiratory Clearance

We inhale contaminants and bacteria every day. If these were allowed to travel through the respiratory airways to the alveolar epithelium, toxic materials might diffuse into the blood stream and inhaled bacteria could produce infection. Instead, these are ordinarily trapped in a mucus layer that lines many airways, propelling contaminants toward the trachea. *Challenge*: When standing, why does mucus not flow down the airways toward the alveoli instead of up toward the trachea?

Generate Ideas: Which important factors affect the clearance of contaminants from the airways? Is mucus a Newtonian fluid? What boundary condition is applied at the interface between the periciliary and mucus layers? What property of mucus is most important in ensuring that the mucus-air surface is directed away from alveoli?

4.13.2 Blood Rheology

When blood flows through an in vitro microcirculatory system, the resistance is found to be less than half the resistance of another fluid with the same viscosity as blood. However, the resistance appears to be very similar to the in vivo system it was meant to mimic. *Challenge*: Can these phenomena be explained on the basis of unique properties of blood and blood vessels? *Generate Ideas*: What is the composition of blood? Does blood behave like a Newtonian fluid? Does blood exhibit a yield stress? Is the apparent viscosity of blood independent of the size of the blood vessel? What unique features of blood might explain its lower resistance in microvessels? What unique features of the microcirculation might explain an elevated resistance relative to an in vitro model?

4.13.3 Blood Doping

Many types of performance enhancers have been developed that improve the ability of athletes to perform. Nearly all of these enhancers are artificial and are banned from organized sport competitions. Blood doping, consisting of increasing the percentage of red cells in the circulation, has been the subject of extensive ethical and scientific debate. *Challenge*: As a biomedical engineer, you are commissioned to prepare a report on the advantages and dangers to an athlete of increasing his/her hematocrit above the normal levels from the perspective of altering the affected biotransport processes in the body. *Generate Ideas*: What are the advantages and disadvantages of blood doping? What important cardiovascular variables might be altered when blood hematocrit is changed? Sketch plots that provide qualitative relationships between each of these variables and blood hematocrit value during maximal exercise. Perform a literature search to provide quantitative evidence for your report.

References

Bhaskar KR, Gong D, Bansil R, Pajevic S, Hamilton JA, Turner BS, Lamont JT (1991) Profound increase in viscosity and aggregation of pig gastric mucin at low PH. Am J Physiol 261 (Gastrointest. Liver Physiol. 24):G827–G832

Bird RB, Stewart WE, Lightfoot EN (2002) Transport phenomena, 2nd edn. Wiley, New York

References

Bloomfield IG, Johnson IH, Bilston LE (1998) Effects of proteins, blood cells and glucose on the viscosity of cerebrospinal fluid. Ped Neurosurg 28:246–251

Chen S (1970) Shear dependence of effective cell volume as a determinant of blood viscosity. Science 168:977–978

Conrad BP (2001) The effects of glucosamine and chondroitin on the viscosity of synovial fluid in patients with osteoarthritis. MS Thesis, University of Florida

Dasari G, Prince I, Hearn MT (1995) Investigations into the rheological characteristics of bovine amniotic fluid. J Biochem Biophys Methods 30:217–225

Gaehtgens P (1980) Flow of blood through narrow capillaries: rheological mechanisms determining capillary hematocrit and appaent viscosity. Biorheology 17:183–189

Goldsmith HL, Mason SG (1967) The microrheology of dispersions. In: Eirich FR (ed) Rheology: theory and applications, vol 4. Academic, New York, pp 85–250, Chap. 2

Grosse SD, Boyle CA, Botkin JR, Comeau AM, Kharrazi M, Rosenfeld M, Wilfond BS (2004) Newborn screening for cystic fibrosis: evaluation of benefits and risks and recommendations for state newborn screening programs. MMWR Recomm Rep 53(RR-13):1–36

House SD, Lipowski HH (1987) Microvascular hematocrit and red cell flux in rat cremaster muscle. Am J Physiol 252:H211–H222

Hsu SH, Strohl KP, Haxhiu MA, Jamieson AM (1996) Role of viscoelasticity in tube model of airway reopening. II. Non-Newtonian gels and airway simulatuion. J Appl Physiol 80:1649–1659

Jones WR, Ting-Beall HP, Lee GM, Kelley SS, Hochmuth RM, Guilak F (1999) Alterations in the Young's modulus and volumetric properties of chondrocytes isolated from normal and osteoarthritic human cartilage. J Biomech 32:119–127

Kienlen J, Mathieudaude JC, Passeron D, Dellord A, Dathis F, Dacailar J (1990) Pharmacokinetics of Dextran 60000 and physical effects on blood and urine. Ann Fran Anesth Reanimation 9:495–500

Koenig W, Sund M, Loè Wel H, Doè Ring A, Ernst E (2000) Association between plasma viscosity and all-cause mortality: results from the MONICA-Augsburg Cohort Study 1984–92. Br J Haematol 109:453–458

Laasberg LH, Hedley-Whyte J (1973) Viscosity of sickle cell disease and sickle trait blood: changes with anesthesia. J Appl Physiol 35:837–843

List SJ, Findlay BP, Forstner GG, Forstner JF (1978) Enhancement of the viscosity of mucin by serum albumin. Biochem J 175:565–571

Low HT, Chew YT, Zhou CW (1997) Pulmonary airway reopening: effects of non-Newtonian fluid viscosity. J Biomech Eng – Trans ASME 119:298–308

Lowe GD, Drummond MM, Lorimer AR, Hutton I, Forbes CD, Prentice CR, Barbenel JC (1980) Relation between extent of coronary artery disease and blood viscosity. Br Med J 280 (6215):673–674

Lumsden JM, Caron JP, Steffe JF, Briggs JL, Arnoczky SP (1996) Apparent viscosity of the synovial fluid from mid-carpal tibiotarsal, and distal interphalangeal joints of horses. Am J Vet Res 57:879–883

Markesich DC, Anand BS, Lew GM, Graham DY (1995) Helicobacter pylori infection does not reduce the viscosity of human gastric mucus gel. Gut 36:327–329

McCullagh CM, Jamieson AM, Blackwell J, Gupta R (1995) Viscoelastic properties of human tracheobronchial mucin in aqueous solution. Biopolymers 35:149–159

Merrill EW, Pelletier GA (1967) Viscosity of human blood: transition from Newtonian to non-Newtonian. J App Physiol 23:178–182

Morris CL, Rucknagle DL, Shukla R, Gruppo RA, Smith CM II, Blackshear P Jr (1989) Evaluation of the yield stress of normal blood as a function of fibrinogen concentration and hematocrit. Microvas Res 37:323–338

Per-ingvar B (1971) Intravsacular anatomy of blood cells in man. S. Karger, Basel

Pries AR, Neuhaus D, Gaehtgens P (1992) Blood viscosity in tube flow: dependence on diameter and hematocrit. Am J Physiol 263 (Heart Circ Physiol 32): H1770–H1778

Pries AR, Secomb TW (2005) Microvascular blood viscosity in vivo and the endothelial surface layer. Am J Physiol 289 (Heart Circ Physiol) H2657–H2664

Pries AR, Secomb TW, Gessner T, Sperandio MB, Gross JF, Gaehtgens P (1994) Resistance to blood flow in microvessels in vivo. Circ Res 75:904–915

Ranade GG, Puniyani RR, Huilgol NG (1995) Plasma viscosity: indicator of severity of disease in cancer patients. Biorheology 32:347–348

Roitman EV, Dement'eva II, Kolpakov PE (1995) Urine viscosity in the evaluation of homeostasis in heart surgery patients in the early postoperative period. Klin Lab Diagn Jul-Aug;(4):29–31

Rubinson KA, Baker PF (1979) The flow properties of axoplasm in a defined chemical environment: influence of anions and calcium. Proc Roy Soc Lond B Biol Sci 205:323–345

Rutlant J, Lopez-Bejar M, Santolaria P, Yaniz J, Lopez-Gatius F (2002) Rheological and ultrastructural properties of bovine vaginal fluid obtained at oestrus. J Anat 201:553–60

Sato M, Theret DP, Wheeler LT, Ohshima N, Nerem RM (1990) Application of the micropipette technique to the measurement of cultured porcine aortic endothelial cell viscoelastic properties. J Biomech Eng 112:263–268

Shi Y, Pan L, Yang F, Wang S (2004) A preliminary study on the rheological properties of human ejaculate and changes during liquefaction. Asian J Androl 6:299–304

Streeter VL, Wylie EB, Bedford KW (1998) Fluid mechanics, 9th edn. McGraw-Hill, New York

Takahashi T, Sakata T (2002) Large particles increase viscosity and yield stress of pig cecal contents without changing basic viscoelastic properties. J Nutr 132:1026–1030

Tandon PN, Bong NH, Kushwaha K (1994) A new model for synovial joint lubrication. Int J Biomed Comp 35:125–140

Tsai MA, Frank RS, waugh RE (1993) Passive mechanical behavior of human neutrophils: power law fluid. Biophys J 65:2078–2088

Tsai MA, Waugh RE, Keng PC (1996) Cell cycle-dependence of HL-60 cell deformability. Biophys J 70:2023–2029

Truskey GA, Yuan F, Katz DF (2004) Transport phenomena in biological systems. Prentice Hall, Upper Saddle, NJ

Ueda J, Nygren A, Sjoquist M, Jacobsson E, Ulfendahl HR, Araki Y (1998) Iodine concentrations in the rat kidney measured by X-ray microanalysis. Comparison of concentrations and viscosities in the proximal tubules and renal pelvis after intravenous injections of contrast media. Acta Radiol 39:90–95

Usami S, Chien S, Scholtz PM, Bartles JF (1975) Effect of deoxygenation on blood rheology in sickle cell disease. Microvasc Res 9:324–334

von Tempelhoff G, Schönmann N, Heilmann L, Pollow K, Hommel G (2002) Prognostic role of plasmaviscosity in breast cancer. Clin Hemorheol Microcirc 26:55–61

von Tempelhoff G, Heilmann L, Hommel G, Pollow K (2003) Impact of rheological variables in cancer. Semin Thromb Hemost 29(5):499–513

Waterman HA, Blom C, Holterman HJ, 's-Gravenmade EJ, Mellema J (1988) Rheological properties of human saliva. Arch Oral Biol 33:589–596

Weinbaum S, Tarbell JM, Damiano ER (2007) The structure and function of the endothelial glycocalyx layer. Annu Rev Biomed Eng 9:121–167

Whitmore RL (1968) Rheology of the circulation. Pergamon Press, New York

Zahm JM, Pierrot D, Fuchey C, Levrier J, Duval D, Lloyd KG, Puchelle E (1989) Comparative rheological profile of rat gastric and duodenal gel mucus. Biorheology 26:813–822

Chapter 5
Macroscopic Approach for Biofluid Transport

5.1 Introduction

The macroscopic approach to biofluid transport is used to analyze fluid flow in the system as a whole. This is in contrast to the microscopic approach that we will describe in Chap. 6, which is used to predict spatial variations of shear stress, pressure, and velocity within the fluid. The macroscopic approach is appropriate when we are not interested in the spatial variations within the system, but instead are interested in average transient values or output values. Common objectives of the macroscopic approach are to find the rate of accumulation of fluid in the system, the fluid flow into or out of the system, or the forces exerted by the fluid on the system. These objectives are accomplished by applying general conservation principles to the fluid in the system.

5.2 Conservation of Mass

We begin by applying conservation of mass to fluid inside a system. Since mass can be neither produced nor destroyed in the system, a conservation of mass statement is simple:

$$\left\{\begin{array}{c} \text{Rate of} \\ \text{accumulation} \\ \text{of mass} \\ \text{within a system} \end{array}\right\} = \left\{\begin{array}{c} \text{Rate mass enters} \\ \text{the system} \end{array}\right\} - \left\{\begin{array}{c} \text{Rate mass leaves} \\ \text{the system} \end{array}\right\}. \quad (5.1)$$

Consider the system shown in Fig. 5.1. Let the symbol w represent a mass flow rate (e.g., kg/s). Mass can enter or leave the system by two mechanisms: (1) convection in or out through the conduits at the inlet w_{in} and outlet w_{out}, and (2) seepage of fluid into the system through the boundary or walls of the system w_{wall}:

$$\frac{dm}{dt} = w_{wall} + w_{in} - w_{out}. \quad (5.2)$$

R.J. Roselli and K.R. Diller, *Biotransport: Principles and Applications*,
DOI 10.1007/978-1-4419-8119-6_5, © Springer Science+Business Media, LLC 2011

Fig. 5.1 Fluid system with an inlet and outlet conduit and a leaky wall

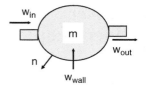

If there is a net loss of mass through the system walls, then w_{wall} is negative. Let us examine each term in (5.2). The term on the left side, the rate of change of mass in the system, is:

$$\frac{dm}{dt} = \frac{d}{dt}\int_V \rho dV, \tag{5.3}$$

where the volume integral represents the sum of all elements of mass ρdV in the system. The mass flow through the walls of the system is equal to the mass flux perpendicular to the walls integrated over the entire surface of the system:

$$w_{\text{wall}} = -\int_S (\rho \vec{v}) \cdot (\vec{n} dS). \tag{5.4}$$

The quantity dS represents a small element of system wall surface and $\vec{n}$ is an outward-directed unit normal relative to the system surface. The total mass flux is $\rho \vec{v}$, and the mass flow out through the surface dS is the dot product $(\rho \vec{v}) \cdot (\vec{n} dS)$. Inclusion of the negative sign in (5.4) indicates that w_{wall} represents the total inward mass flow. The surface S in (5.4) is the entire wall surface, but does not include the system surface associated with the inlet and outlet conduits, A_{in} and A_{out}. Mass flow into the system through the cross-sectional area A_{in} is:

$$w_{\text{in}} = \int_{A_{\text{in}}} \rho_{\text{in}} v_{\text{in}} dA, \tag{5.5}$$

and mass flow out of the system through the cross-sectional area A_{out} is:

$$w_{\text{out}} = \int_{A_{\text{out}}} \rho_{\text{out}} v_{\text{out}} dA. \tag{5.6}$$

If density is invariant over the inlet and outlet cross sections, then density can be brought outside the integration signs in (5.5) and (5.6). The definition of *volumetric flow* Q_V is:

$$Q_V = \int_{A_c} v dA = \langle v \rangle A_c, \tag{5.7}$$

5.2 Conservation of Mass

where $\langle v \rangle$ is the average velocity in the conduit, defined by (5.7), and A_c is the cross-sectional area of the conduit. The convective mass flow terms for uniform density at the inlet and outlet can be written as:

$$w_{in} = \rho_{in} Q_{V,in} = \rho_{in} \langle v_{in} \rangle A_{in}, \tag{5.8}$$

$$w_{out} = \rho_{out} Q_{V,out} = \rho_{out} \langle v_{out} \rangle A_{out}. \tag{5.9}$$

Similarly, if the system itself is well mixed, then the density in the system is uniform, and (5.3) reduces to:

$$\frac{dm}{dt} = \frac{d}{dt}(\rho V) = \rho \frac{dV}{dt} + V \frac{d\rho}{dt} \quad \text{(well mixed)}. \tag{5.10}$$

Thus, with the restrictions that density is uniform in the system, $\rho_{out} = \rho$, the general macroscopic conservation expression, (5.2), becomes:

$$\frac{d}{dt}(\rho V) = w_{wall} + \rho_{in} Q_{V,in} - \rho Q_{V,out} \quad \text{(well mixed)}. \tag{5.11}$$

If the fluid is incompressible, then conservation of volume requires

$$\frac{dV}{dt} = Q_{wall} + Q_{V,in} - Q_{V,out} \quad \text{(incompressible fluid)}, \tag{5.12}$$

where Q_{wall} is the inward fluid flow through the wall, w_{wall}/ρ.

In steady-state flow, the left-hand side of (5.2) vanishes, and we have:

$$0 = w_{wall} + w_{in} - w_{out} \quad \text{(steady flow)}. \tag{5.13}$$

For steady flow with uniform densities at the inlet and outlet, (5.13) reduces to:

$$0 = w_{wall} + \rho_{in} Q_{V,in} - \rho_{out} Q_{V,out}. \tag{5.14}$$

For steady flow of an incompressible fluid into a system with a single inlet and a single outlet, with no loss of fluid in the system ($w_{wall} = 0$) conservation of mass reduces to:

$$Q_{V,in} = Q_{V,out} = \langle v_{in} \rangle A_{in} = \langle v_{out} \rangle A_{out}. \tag{5.15}$$

In the unsteady-state case where the system has N_{inlets} inlets and $N_{outlets}$ outlets, the general conservation of mass expression can be written as:

$$\frac{dm}{dt} = w_{\text{wall}} + \sum_{i=1}^{N_{\text{inlets}}} w_{\text{in},i} - \sum_{j=1}^{N_{\text{outlets}}} w_{\text{out},j}. \tag{5.16}$$

If the fluid in the system is incompressible, then conservation of volume requires:

$$\frac{dV}{dt} = Q_{\text{wall}} + \sum_{i=1}^{N_{\text{inlets}}} Q_{V,\text{in},i} - \sum_{j=1}^{N_{\text{outlets}}} Q_{V,\text{out},j} \quad (\text{incompressible}). \tag{5.17}$$

Example 5.2.1 Flushing a Catheter.
A 1 ml catheter is initially filled with blood having density ρ_b. At time $t = 0$, the catheter is flushed at a constant rate of 1 ml/s with saline having a density ρ_s. How long will it take to flush out 99% of the blood? What volume of saline is needed?

Solution. *Initial considerations:* Our goal is to determine the volume of saline that must be infused and the time needed to flush out 99% of the blood originally contained in the catheter. Syringes come in various standard sizes, and so we need to select a syringe large enough to accomplish our goal. We can take three different approaches to solving this problem, depending on how the blood and saline interact in the catheter. The simplest approach is to assume that the velocity profile is flat and saline and blood do not mix at the interface. In that case, the saline will simply push the blood ahead of it until the saline front emerges from the other end of the tube. This should provide a minimum estimate of the syringe size. An alternate approach is to assume that the blood and saline are well mixed in the catheter; so the hematocrit value at any point in the catheter is uniform at any given time. Another less restrictive approach is to assume that the fluid is viscous; so fluid velocity depends on radial position in the tube. In addition, it is assumed that the blood and saline do not mix at all; so the red blood cells near the catheter wall will take a longer than average time to move through the catheter, while cells near the center will emerge before cells moving at the average transit time through the catheter.

System definition and environmental interactions: Irrespective of our approach, the system of interest is the fluid (blood plus saline) contained within the catheter. This fluid is bounded by the catheter wall (assumed to be rigid and impermeable to blood or saline) at $r = R$, the inlet end at $x = 0$, which is attached to the syringe, and the outlet end at $x = L$.

Apprising the problem to identify governing equations: Conservation of mass is clearly the governing principle that is to be applied in this case. However, which of the several equations derived in the previous section apply in this case? The answer depends on the assumptions we make. It is best to start with the most general case, (5.2), as a starting point, then apply simplifications based on what we consider to be reasonable assumptions. We will assume that the catheter wall is impermeable to fluid, so $w_{wall} = 0$. Next we assume that saline, blood, and a mixture of saline and blood are all incompressible fluids. Since the inflow is constant and the tube volume is constant, conservation of system volume requires the volumetric outflow and inflow to match:

$$Q_{V,in} = Q_{V,out} = Q_V.$$

Only saline with density ρ_s enters the catheter; so the inlet mass flow at all times from (5.8) is:

$$w_{in} = \rho_s Q_V.$$

The assumptions made thus far seem quite reasonable, and (5.2) based on these assumptions reduces to:

$$\frac{dm}{dt} = Q_V(\rho_s - \rho_{out}).$$

Our task now is to determine the nature of the fluid density at the outlet, ρ_{out}. We know that this will be a weighted average of blood density ρ_b and saline density ρ_s, but what determines the appropriate weight factor as a function of time? This depends on which of the assumptions discussed above is appropriate. Let us take two different approaches to estimate the density at the outlet: (1) the saline and blood mix rapidly as more saline is introduced at the inlet, and (2) the saline and blood do not mix at all, but instead, the saline pushes the blood ahead of it as it travels through the catheter. We analyze each case in the following sections.

Analysis (well-mixed case): If the fluid in the catheter is well mixed, then the outlet density ρ_{out} is the same as the intracatheter fluid density ρ and the mass of fluid in the catheter is ρV. Since the catheter volume V is constant, the conservation equation above reduces to:

$$V\frac{d\rho}{dt} = Q_V(\rho_s - \rho).$$

Since ρ_s is constant, this can also be written as:

$$\frac{d}{dt}(\rho - \rho_s) + \frac{Q_V}{V}(\rho - \rho_s) = 0.$$

Initially, only blood is in the catheter, so $\rho(t=0) = \rho_b$. The solution with this initial condition is:

$$\frac{\rho - \rho_s}{\rho_b - \rho_s} = e^{-\frac{Q_V t}{V}}.$$

If t^* is the time when 99% of the blood in the catheter is replaced by saline, then the mass m of blood and saline in the catheter at time t^* is:

$$m(t^*) = 0.01\rho_b V + 0.99\rho_s V.$$

Subtracting $\rho_s V$ from both sides and rearranging, we have:

$$\frac{\rho(t^*) - \rho_s}{\rho_b - \rho_s} = 0.01 = e^{-\frac{Q_V t^*}{V}}.$$

Solving for the time t^* required for 99% of the blood to be flushed from the catheter:

$$t^* = -\frac{V}{Q_V} \ln(0.01) = \frac{1\,\text{ml}}{1\,\text{ml/s}} (4.6) = 4.6\,\text{s}.$$

The volume of saline necessary to flush 99% of the blood out of the catheter is 4.6 s × 1 ml/s = 4.6 ml. This is 4.6 times the volume of the catheter. A simple rule of thumb for flushing out a well-mixed volume is that it takes a period equivalent to five system time constants before a new steady state is reached. The system time constant in this case is $V/Q_V = 1$ s. After a time equivalent to five time constants (5 s), 5 ml of fluid will have emerged from the catheter. This is in reasonable agreement with the computed value of 4.6 ml.

Analysis (unmixed case): In this section, we consider the case where blood and saline do not mix, so that a sharp interface exists between saline and blood in the catheter. The location of the saline–blood interface will depend on time and on the velocity profile in the catheter. If the velocity profile is flat (plug flow), all of the blood will be pushed out of the catheter after 1 ml of saline is introduced (i.e., 1.0 s). Nearly flat velocity profiles with significant radial mixing can be attained in turbulent flow situations. If, however, the flow is laminar and fully developed, the velocity will depend on radial position r in a parabolic manner (see Sect. 6.3.1):

$$v(r) = 2\langle v \rangle \left[1 - \left(\frac{r}{R}\right)^2\right],$$

where R is the tube radius and $\langle v \rangle$ is the average velocity. Saline moves more rapidly through the central portion of the catheter than it does near the catheter walls. In laminar flow, a fluid particle that enters the catheter at a radial position r will remain at that same radial position until it leaves the catheter. The time required for saline, initially at a radial position R^*, to reach the exit of a catheter with length L is:

$$t = \frac{L}{v(R^*)}.$$

5.2 Conservation of Mass

The interface between saline and blood can be found from the two expressions above, and is shown for various dimensionless times after saline is introduced in Fig. 5.2.

Saline first appears at the catheter exit in the center of the tube at a time $t_0 = L/(2\langle v \rangle) = V/(2Q_V)$. Only blood emerges from the exit of the tube at times less than or equal to t_0:

$$\rho_{out} = \rho_b, \quad t \leq \frac{1}{2}\left(\frac{V}{Q_V}\right).$$

At the catheter outlet ($x = L$) for times greater than t_0, saline will occupy a region near the center of the catheter ($0 \leq r \leq R^*$), and blood will occupy a region near the wall ($R^* \leq r \leq R$). The location of R^*/R at the outlet is shown in Fig. 5.2 for a time equal to V/Q_V. To find how R^* varies with time at the exit of the catheter, we substitute the parabolic profile for $v(R^*)$:

$$\left(\frac{R^*}{R}\right)^2 = 1 - \frac{L}{2\langle v \rangle t}, \quad t \geq \frac{1}{2}\left(\frac{V}{Q_V}\right).$$

For the case $t \geq t_0$, the outlet mass flow is given by adding the contributions from saline in the central region and blood near the wall. Applying (5.6) for a parabolic velocity profile with $dA = 2\pi r dr$:

$$\rho_{out} Q_V = \int_0^{R^*} \rho_s \left\{ 2\langle v \rangle \left[1 - \left(\frac{r}{R}\right)^2 \right] \right\} [2\pi r dr]$$
$$+ \int_{R^*}^{R} \rho_b \left\{ 2\langle v \rangle \left[1 - \left(\frac{r}{R}\right)^2 \right] \right\} [2\pi r dr], \quad t \geq \frac{1}{2}\left(\frac{V}{Q_V}\right).$$

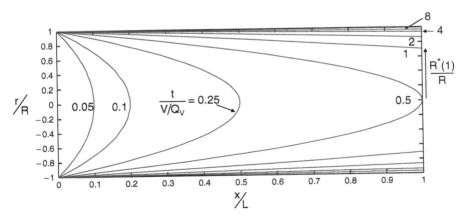

Fig. 5.2 Interface between saline and blood at various dimensionless times after saline begins flowing through the catheter

Performing the integrations and converting R^* to t using the relationship above, we find:

$$\rho_{out} = \rho_s + (\rho_b - \rho_s)\frac{1}{4}\left(\frac{V}{Q_v}\right)^2\left(\frac{1}{t^2}\right), \quad t \geq \frac{1}{2}\left(\frac{V}{Q_v}\right).$$

Thus, for times less than or equal to t_0, the conservation of mass equation is:

$$\frac{dm}{dt} = (\rho_s - \rho_b)Q_v, \quad t \leq \frac{1}{2}\frac{V}{Q_v}.$$

Integrating between $t = 0$ and $t = t$:

$$m(t) - m(0) = (\rho_s - \rho_b)Q_v t, \quad t \leq \frac{1}{2}\frac{V}{Q_v}.$$

Since $m(0) = \rho_b V$ and $t_0 = V/(2Q_v)$, $m(t_0)$ is:

$$m(t_0) = \rho_b V + (\rho_0 - \rho_b)Q_v\left(\frac{1}{2}\frac{V}{Q_v}\right) = \frac{(\rho_0 + \rho_b)V}{2}.$$

Therefore, when $t = t_0 = 0.5$ s, half of the blood initially in the catheter has been replaced with saline. Next, we look at the loss of blood for $t > t_0$. The same conservation equation applies, but the expression for w_{out} when $t > t_0$ must be used:

$$\frac{dm}{dt} = \rho_0 Q_v - \left[\rho_s Q_v + (\rho_b - \rho_s)\left(\frac{V^2}{4Q_v}\right)\left(\frac{1}{t^2}\right)\right], \quad t \geq \frac{1}{2}\frac{V}{Q_v}.$$

Integrating this expression from $t = t_0$ to t, the mass of blood and saline in the catheter is:

$$m(t) = \rho_s V + (\rho_b - \rho_s)\left(\frac{V^2}{4Q_v}\right)\frac{1}{t}, \quad t \geq \frac{1}{2}\frac{V}{Q_v}.$$

We are now in a position to find the time required for 99% of the blood to be washed out of the catheter. When $t = t^*$ the mass of fluid inside the catheter will be composed of 99% saline and 1% blood:

$$m(t^*) = \rho_s(0.99\,V) + \rho_b(0.01\,V).$$

Substituting this in for $m(t)$ above and solving for t^*:

$$t^* = 25\left(\frac{V}{Q_v}\right).$$

5.2 Conservation of Mass

The volume of saline required to flush out 99% of the blood is:

$$V_{\text{flush}} = Q_V t^* = 25V.$$

Examining and interpreting the results: In summary, if the blood and saline are well mixed in the catheter, flushing the catheter with a saline volume equal to 4.6 times the catheter volume is sufficient to remove 99% of the blood initially in the catheter. If the saline does not mix with the blood in the catheter, the flush volume will depend on the velocity profile. If the velocity profile is flat, a saline volume equal to the catheter volume will flush all of the blood out of the catheter. However, if flow is laminar and no radial mixing occurs, a saline volume equal to 25 times the catheter volume is required to force 99% of the blood from the catheter. In this case, a 25 ml syringe would be required to flush blood out of a 1 ml catheter.

The well-mixed and unmixed laminar flow predictions are compared in Fig. 5.3, which plots dimensionless mass in the catheter vs. either dimensionless flush time or flush volume. The unmixed model prediction is more efficient than the well-mixed prediction in removing blood until about 85% of the blood is flushed from the catheter. This corresponds to a flush volume of about twice the catheter volume. Beyond this, the well-mixed model is much more efficient. This reflects the difference between an exponential loss, in the well-mixed case, and a loss in the unmixed case that is inversely proportional to time.

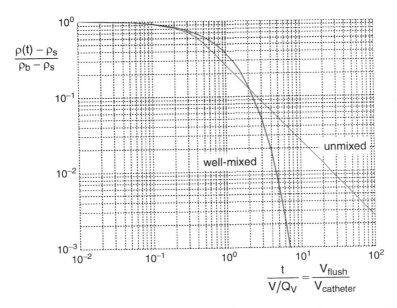

Fig. 5.3 Comparison of dimensionless blood mass remaining in catheter as a function of dimensionless saline flush time or volume. *Blue line* is based on the assumption that saline and blood are well mixed; *red line* is based on a parabolic velocity profile with no radial or axial mixing

Additional comments: So, which model is more accurate? Only a comparison with experimental results can tell us for sure. However, most catheters are much longer than their inside radius. Axial mixing, even with turbulent flow, is highly unlikely under these circumstances. Consequently, the well-mixed assumption would seem faulty in this case. Whether or not flow is laminar is determined by the Reynolds number, as we will learn in Sect. 5.6. For a catheter with a diameter of 1 mm and flow of 1 ml/s, the Reynolds number will be low enough for flow to be laminar. Pushing the saline through at a higher rate could change the flow regime to turbulent, reducing the saline volume needed to flush the blood out.

Example 5.2.2 Steady-State Mixing.
Blood from two capillaries merge to form a small venule as shown in Fig. 5.4. If the flow rate and density in each capillary are constant, find the flow rate and density of fluid emerging from the venule.

Solution. *Initial considerations:* Our goal is to predict the mean density and volumetric flow rate at the outlet of the venule. We will assume that the walls of the venule are impermeable to all species.

System definition and environmental interactions: The system of interest is the blood inside the venule. Since this is a steady-state problem, the volume of the venule will remain constant.

Apprising the problem to identify governing equations: We can apply a macroscopic mass balance to the venule with the following assumptions: (1) the walls of the venule are impermeable, so $w_{wall} = 0$; (2) blood densities at the venule inlets and at the venule outlet are uniform; (3) the volume of fluid in the venule remains constant; and (4) the system has reached a steady state.

Analysis: Starting with conservation of volume for an incompressible fluid in a rigid system (5.17):

$$\frac{dV}{dt} = Q_{V,1} + Q_{V,2} - Q_{V,out} = 0.$$

Applying the steady-state assumption and solving for flow out of the venule:

$$Q_{V,out} = Q_{V,1} + Q_{V,2}.$$

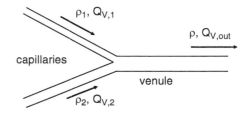

Fig. 5.4 Merging of two blood streams from different capillaries to provide flow to a venule

5.2 Conservation of Mass

Next, we apply the conservation of mass statement for multiple inlets and outlets, (5.16). Introducing assumptions (1) and (4):

$$\frac{dm}{dt} = 0 = w_{in,1} + w_{in,2} - w_{out}.$$

Applying assumption (2):

$$w_{out} = \rho_{out} Q_{V,out} = \rho_1 Q_{V,1} + \rho_2 Q_{V,2}.$$

Therefore, the density of the blood emerging from the venule is:

$$\rho_{out} = \frac{\rho_1 Q_{V,1} + \rho_2 Q_{V,2}}{Q_{V,1} + Q_{V,2}}.$$

Examining and interpreting the results: The capillary with the highest mass flow rate contributes the most to the outlet density of the venule and the capillary with the highest volumetric flow rate contributes the greatest to the venule flow rate. The outlet density will always lie between the densities of the two inlet streams.

Additional comments: Note that we did not assume the blood in the venule was well mixed over its length. The densities ρ_1, ρ_2, and ρ_{out} represent the flow-averaged mean densities over the cross sections of the inlet and outlet streams ($\rho_{mean} = w/Q_V$).

Example 5.2.3 Height Change in a Tank.

Blood in an isolated organ perfusion system flows out of the bottom of a vertical holding tank with cross-sectional area A_c. The flow rate is proportional to the height of fluid in the tank. How long will it take for half of the fluid to leave the tank (Fig. 5.5)?

Solution. *Initial considerations:* Our key assumptions are that (1) blood is incompressible, (2) there is no flow into the tank, and (3) blood does not seep through the walls of the tank.

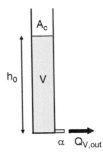

Fig. 5.5 Flow from holding tank

System definition and environmental interactions: The system of interest in this case is the blood in the tank at any time t.

Apprising the problem to identify governing equations: Conservation of mass is the only relationship necessary to solve this problem.

Analysis: Starting with (5.12) and applying assumptions (2) and (3):

$$\frac{dV}{dt} = -Q_{V,\text{out}}.$$

But the volume of blood V in the tank is equal to the product of the cross-sectional area of the tank A_c and the fluid level h in the tank: $V = hA_c$. Also, from the problem statement, $Q_{V,\text{out}} = \alpha h$, where α is a constant. Substituting these relations into the expression above, we obtain an ordinary differential equation for h, the height of blood in the tank:

$$\frac{dh}{dt} + \frac{\alpha}{A_c} h = 0.$$

The solution for $h(t=0) = h_0$ is:

$$h = h_0 e^{-\frac{\alpha t}{A_c}}.$$

The time t^* for h to fall to $h_0/2$ can be found by rearranging the above expression:

$$t^* = -\frac{A_c}{\alpha} \ln\left(\frac{1}{2}\right) = 0.693 \frac{A_c}{\alpha}.$$

Examining and interpreting the results: Note that the time for half the blood to leave the tank is independent of the initial height of fluid in the tank, h_0. In addition, the larger the cross section of the tank, the longer it will take to drain. Finally, the larger the value of α, the faster the tank will drain.

Additional comments: Assuming the tank drains through a tube to atmospheric pressure, the hydrostatic pressure difference across the outlet tube will be $\rho g h$. Referring back to the definition of flow resistance, (2.30), the proportionality factor α is inversely proportional to the flow resistance of the outlet tube. The higher the outlet resistance, the longer it will take to drain the tank to any given level.

5.3 Conservation of Momentum

The macroscopic momentum equation is most often used to estimate the force exerted by the fluid in the system on the walls of the system. Consider the system in Fig. 5.6. Fluid enters and leaves the system through a number of conduits.

5.3 Conservation of Momentum

Fig. 5.6 Macroscopic system

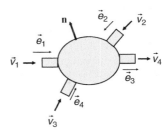

We neglect momentum added to the system by flow through the walls of the system. The unit vector $\vec{e}_i$ associated with each conduit points in the same direction as the velocity vector in the conduit. The general conservation of momentum equation is a vector equation stating that the rate of change of momentum in the system is equal to the net addition of momentum contributed by the fluid streams and the production of momentum resulting from the application of external forces on the system (fluid):

$$\left\{\begin{array}{c} \text{Rate of} \\ \text{accumulation} \\ \text{of momentum} \\ \text{within a system} \end{array}\right\} = \left\{\begin{array}{c} \text{Rate momentum} \\ \text{enters the system} \\ \text{by convection} \end{array}\right\} - \left\{\begin{array}{c} \text{Rate momentum} \\ \text{leaves the system} \\ \text{by convection} \end{array}\right\} + \left\{\begin{array}{c} \text{Rate of production} \\ \text{of momentum} \\ \text{within the system} \end{array}\right\}. \quad (5.18)$$

Let us consider each of these terms individually. The total momentum of the system is given the symbol p. The rate that momentum accumulates in the system is given by integrating the momentum of all fluid particles within the system:

$$\left\{\begin{array}{c} \text{Rate of} \\ \text{accumulation} \\ \text{of momentum} \\ \text{within a system} \end{array}\right\} = \frac{d}{dt}\int_V \rho \vec{v} \, dV = \frac{d\vec{p}}{dt}. \quad (5.19)$$

Now, let us consider the net rate at which momentum enters the system via convection through the various inlet and outlet conduits. The mass flux into the system via a single conduit i is $\rho \vec{v}$. The net momentum brought into the system through a single conduit is:

$$\left\{\begin{array}{c} \text{Momentum added} \\ \text{to system via} \\ \text{a single conduit } i \end{array}\right\} = -\int_{A_i} (\rho_i \vec{v}_i)(\vec{v}_i \cdot \vec{n}_i dA_i) = -\int_{A_i} (\rho_i v_i^2) dA_i (\vec{e}_i \cdot \vec{n}_i)\vec{e}_i. \quad (5.20)$$

Recall that $\vec{e}_i$ is a unit vector in the direction of the velocity vector $\vec{v}_i$ and $\vec{n}_i$ is an outward-directed unit normal at the surface of conduit i. Therefore, the rate at which momentum enters the system via a single conduit is the product of conduit mass flux and velocity integrated over the conduit inlet cross-sectional area. If the velocity $\vec{v}_i$ and the outward unit normal $\vec{n}_i$ are in opposite directions, then the dot product in (5.20) will be negative. However, positive momentum will be added to the system in such a case, so the minus sign is necessary. Thus, momentum will be added to the system at all inlets where $\vec{e}_i \cdot \vec{n}_i$ is negative and will be removed from the system at all outlets where $\vec{e}_i \cdot \vec{n}_i$ is positive. The net rate at which momentum is added to the system from all inlets and outlets is:

$$\left\{ \begin{array}{c} \text{Rate momentum} \\ \text{enters the system} \\ \text{by convection} \end{array} \right\} - \left\{ \begin{array}{c} \text{Rate momentum} \\ \text{leaves the system} \\ \text{by convection} \end{array} \right\}$$

$$= \sum_{i=1}^{N_{\text{inlets}}} \left[\int_{A_i} \left(\rho_i v_i^2 \right) dA_i \right] \vec{e}_i - \sum_{j=1}^{N_{\text{outlets}}} \left[\int_{A_j} \left(\rho_j v_j^2 \right) dA_j \right] \vec{e}_j. \quad (5.21)$$

If the density is assumed to be uniform over the cross section, then:

$$\int_{A_i} \left(\rho_i v_i^2 \right) dA_i = \rho_i \int_{A_i} \left(v_i^2 \right) dA_i = \rho_i A_i \langle v_i^2 \rangle = \rho_i A_i K_{2i} \langle v_i \rangle^2, \quad (5.22)$$

where the bracketed term $\langle v^k \rangle$ represents the average of v^k over the cross section, and K_{ki} is defined as:

$$K_{ki} \equiv \frac{\langle v_i^k \rangle}{\langle v_i \rangle^k} = \frac{1}{\langle v_i \rangle^k} \left[\frac{1}{A_i} \int_{A_i} \left(v_i^k \right) dA_i \right]. \quad (5.23)$$

The value of K_{ki} depends on the exponent k, the type of conduit, and whether the flow is laminar or turbulent. If $k = 1$, then (5.23) reduces to (5.7), the definition of $\langle v_i \rangle$. Consequently, $K_{1i} = 1$. The velocity profile is relatively flat for turbulent flow, and so for any value of k, $K_{ki} \approx 1$. However, for laminar flow it can be shown that $K_{2i} = 6/5$ for a conduit composed of parallel plates and $K_{2i} = 4/3$ for a conduit with a circular cross section. In general, for a parabolic velocity profile in a circular tube with radius R, $v_i = 2\langle v_i \rangle \left(1 - (r/R)^2\right)$ and $dA = 2\pi r dr$. Substituting these relations into (5.23), we find:

$$K_{ki} = \frac{2^k}{k+1} \quad \text{(parabolic velocity, circular tube)}. \quad (5.24)$$

5.3 Conservation of Momentum

Introducing (5.22) into (5.21):

$$\left\{ \begin{array}{c} \text{Rate momentum} \\ \text{enters the system} \\ \text{by convection} \end{array} \right\} - \left\{ \begin{array}{c} \text{Rate momentum} \\ \text{leaves the system} \\ \text{by convection} \end{array} \right\}$$

$$= \sum_{i=1}^{N_{inlets}} \rho_i A_i K_{2i} \langle v_i \rangle^2 \vec{e}_i - \sum_{j=1}^{N_{outlets}} \rho_j A_j K_{2j} \langle v_j \rangle^2 \vec{e}_j. \quad (5.25)$$

If momentum enters the system through the walls, this must be added to (5.25). Let us turn our attention now to the production term in (5.18). Momentum is generated when external forces are applied to the fluid. The forces acting on a system with four conduits are shown in Fig. 5.7. These include the pressure forces at the stream inlets and outlets, the weight of the fluid $m\vec{g}$, and the force by the system wall on the fluid, $-\vec{R}$ (where $\vec{R}$ = force by fluid on system wall).

$$\left\{ \begin{array}{c} \text{Rate of production} \\ \text{of momentum} \\ \text{within the system} \end{array} \right\} = \sum_i (\vec{F}_i)_{\text{on system}}$$

$$= m\vec{g} - \vec{R} + \sum_{i=1}^{N_{inlets}} \left[\int_{A_i} P_i dA_i \right] \vec{e}_i - \sum_{j=1}^{N_{outlets}} \left[\int_{A_j} P_j dA_j \right] \vec{e}_j. \quad (5.26)$$

If we assume that the pressure is uniform across the cross section of each conduit, then:

$$\left\{ \begin{array}{c} \text{Rate of production} \\ \text{of momentum} \\ \text{within the system} \end{array} \right\} = \sum_i (\vec{F}_i)_{\text{on system}} = m\vec{g} - \vec{R} + \sum_{i=1}^{N_{inlets}} P_i A_i \vec{e}_i - \sum_{j=1}^{N_{outlets}} P_j A_j \vec{e}_j. \quad (5.27)$$

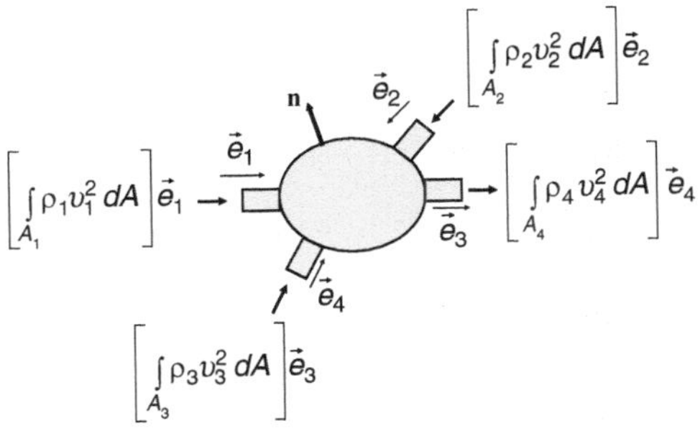

Fig. 5.7 Forces on a macroscopic system with four conduits

Finally, substitution of (5.19), (5.25), and (5.27) into (5.18) yields:

$$\frac{d\vec{p}}{dt} = \sum_{i=1}^{N_{\text{inlets}}} \left[K_{2i}\rho_i \langle v_i \rangle^2 + P_i\right] A_i \vec{e}_i - \sum_{j=1}^{N_{\text{outlets}}} \left[K_{2j}\rho_j \langle v_j \rangle^2 + P_j\right] A_j \vec{e}_j + m\vec{g} - \vec{R}. \quad (5.28)$$

This represents the general unsteady-state macroscopic momentum equation which is valid for multiple inlets and outlets. This is a vector equation with contributions from each of the inlets and outlets, in addition to the external forces applied to the fluid. In the steady state, this is often used to compute the force exerted by the fluid on the walls of the system, $\vec{R}$:

$$\vec{R} = \sum_{i=1}^{N_{\text{inlets}}} \left[K_{2i}\rho_i \langle v_i \rangle^2 + P_i\right] A_i \vec{e}_i - \sum_{j=1}^{N_{\text{outlets}}} \left[K_{2j}\rho_j \langle v_j \rangle^2 + P_j\right] A_j \vec{e}_j + m\vec{g} \quad \text{(steady state)}. \quad (5.29)$$

Next, consider the case where there is a single inlet and a single outlet, and the unit vectors at the inlet and outlet are both in the positive x-direction. Then (5.29) reduces to a scalar equation:

$$R_x = \left[K_{2\text{in}}\rho_{\text{in}} \langle v_{\text{in}} \rangle^2 + P_{\text{in}}\right] A_{\text{in}} - \left[K_{2\text{out}}\rho_{\text{out}} \langle v_{\text{out}} \rangle^2 + P_{\text{out}}\right] A_{\text{out}} + mg_x. \quad (5.30)$$

Finally, if the inlet and outlet have the same cross-sectional areas A_c, and fluid density is constant, then the net force by the fluid on the system will be:

$$R_x = [P_{\text{in}} - P_{\text{out}}] A_c + mg_x. \quad (5.31)$$

Example 5.3.1 Left Ventricular Force.
Determine the force that the left ventricle must exert on blood during the ejection phase of the cardiac cycle. Assume that blood is ejected through the aortic valve at an angle of 50° with the horizontal (Fig. 5.8). Blood density is 1.03 g/ml, radius of the aortic valve is 1 cm, and ventricular volume at the start of ejection is 150 ml. Traces for aortic flow and aortic pressure are given in Fig. 5.9, and these can be approximated as:

$$Q_V(\text{ml/s}) = 283.3 \sin(\omega t),$$
$$P(\text{mmHg}) = 80 + 35 \cos(\omega t) + 20t,$$

in the range $0 \le t \le 0.5$ s with $\omega = 2\pi s^{-1}$.

5.3 Conservation of Momentum

Fig. 5.8 Fluid in left ventricle

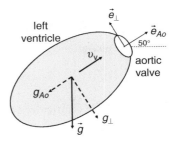

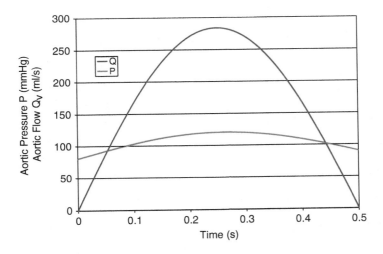

Fig. 5.9 Aortic pressure and flow vs. time

Solution. *Initial considerations:* During systole, the mitral valve is closed, and hence no blood enters the left ventricle. Blood leaves through the aortic valve because contraction of the ventricular wall pushes on the blood in the ventricle, increasing ventricular pressure. We will assume that the wall is impermeable to fluid and all blood that leaves the ventricle passes through the aortic valve.

System definition and environmental interactions: The system of interest in this problem is the blood within the left ventricle. The volume of this system will change with time as the force of the ventricular wall changes.

Apprising the problem to identify governing equations: Since our ultimate goal is to estimate the force applied to the blood in the ventricle by the ventricular wall, we must ultimately apply the momentum equation. In addition, a change in volume of the left ventricle produces a flow through the aortic valve; so we will also need to apply conservation of mass.

Analysis: We begin with a mass balance on the ventricle. Equation (5.2) with constant density reduces to:

$$\frac{dV_V}{dt} = -Q_V.$$

Solving for ventricular volume with $Q_{vo} = 283.3$ ml/s:

$$V_V = V_{V0} - \int_0^t Q_V dt = V_{V0} - Q_{V0}\int_0^t \sin(\omega t)dt = V_{V0} - \frac{Q_{V0}}{\omega}[1 - \cos(\omega t)].$$

This is shown in Fig. 5.10.

Next, we introduce the unsteady-state momentum equation for no inlets and a single outlet:

$$\frac{d\vec{p}}{dt} = -\left[K_{2,Ao}\rho\langle v_{Ao}\rangle^2 + P_{Ao}\right]A_{Ao}\vec{e}_{Ao} + m\vec{g} - \vec{R}.$$

The subscript Ao indicates values at the aortic valve. We can resolve this into two scalar equations: one parallel with the axis of the aorta (in the direction $\vec{e}_{Ao}$) and one perpendicular to that direction $\vec{e}_\perp$. We assume that the rate of change of momentum in the ventricle is along the axis of the ventricle. The components of force by the ventricle on the fluid are:

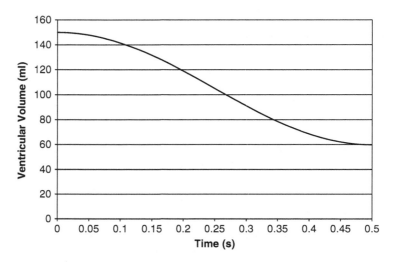

Fig. 5.10 Ventricular volume vs. time

5.3 Conservation of Momentum

$$F_\perp = -\vec{R} \cdot \vec{e}_\perp = -m\vec{g} \cdot \vec{e}_\perp = \rho g V_V \cos(50°),$$

$$F_{Ao} = -\vec{R} \cdot \vec{e}_{Ao} = \frac{d}{dt}(\rho V_V \langle v_V \rangle) + \left[K_{2,Ao}\rho\langle v_{Ao}\rangle^2 + P_{Ao}\right]A_{Ao} + \rho g V_V \sin(50°).$$

The velocity profile is quite flat at the aortic valve, so $K_{2,Ao} = 1$. The quantity $\langle v_v \rangle$ is the average velocity of blood in the ventricle. This is equal to the flow out of the ventricle divided by the cross-sectional area A_V of the ventricle. Since this is not given, we must estimate how this changes with time when the ventricle contracts. Let us assume that the ventricle behaves as a cylinder of constant length $L = 10$ cm. Then, the cross-sectional area can be computed from the ventricular volume

$$A_V = \frac{V_V}{L},$$

and the average velocity of blood in the ventricle can be computed as:

$$\langle v_v \rangle = \frac{Q_V}{A_V} = \frac{LQ_V}{V_V}.$$

Also, we can write the velocity at the aortic valve in terms of the flow and area:

$$\langle v_{Ao} \rangle = \frac{Q_V}{A_{Ao}}.$$

We are now in a position to compute the force by the ventricle on the fluid in the direction of flow:

$$F_{Ao} = \rho L \frac{dQ_V}{dt} + \frac{\rho Q_V^2}{A_{Ao}} + P_{Ao}A_{Ao} + \rho g V_V \sin(50°). \quad (5.32)$$

Examining and interpreting the results: The force F_{Ao} and the relative contributions of each of the terms in this equation are shown in the upper panel of Fig. 5.11, and the total force and its components are shown in the lower panel. Note that the momentum of the blood in the ventricle and the momentum of fluid leaving through the aorta contribute very little to the overall force. The major contribution is the force needed to overcome the pressure in the aorta. The next most important component is the force caused by the weight of the fluid in the ventricle. The maximum force by the ventricular wall on the fluid is about 6.2 N, directed 7° above the axis of the ventricle, or 57° with the horizontal.

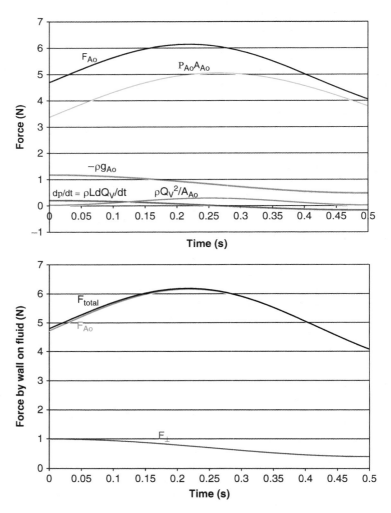

Fig. 5.11 *Upper*: Force F_{Ao} and the relative contributions of each of the terms in (5.32); *lower*: total force and its components

5.4 Conservation of Energy

An open system, such as that shown in Fig. 5.6, can exchange energy with its surroundings. However, the total energy of the system plus its surroundings is conserved. A general statement of conservation of energy can be expressed as follows:

5.4 Conservation of Energy

$$\left\{\begin{array}{c}\text{Rate of}\\\text{accumulation}\\\text{of energy}\\\text{within a system}\end{array}\right\} = \left\{\begin{array}{c}\text{Rate energy}\\\text{enters}\\\text{the system}\end{array}\right\} - \left\{\begin{array}{c}\text{Rate energy}\\\text{leaves}\\\text{the system}\end{array}\right\}$$
$$+ \left\{\begin{array}{c}\text{Rate of energy}\\\text{production}\\\text{within the system}\end{array}\right\}. \quad (5.33)$$

The "system" is defined as the bounded fluid in Fig. 5.6, and the energy within the system includes kinetic energy, potential energy, and internal energy. Energy can enter and leave the system across the walls of the system and via the inlet and outlet channels. Energy can be produced within the system through friction or chemical reaction, or if work is performed by the surroundings on the system. Energy can be lost (i.e., negative energy production) when the system does work on the surroundings or energy is lost to chemical reaction.

The accumulation of energy E within the system is the energy per unit mass integrated over the entire mass of the system:

$$\left\{\begin{array}{c}\text{Rate of}\\\text{accumulation}\\\text{of energy}\\\text{within a system}\end{array}\right\} = \frac{dE}{dt} = \frac{d}{dt}\int_m \hat{E}dm = \frac{d}{dt}\int_V (\hat{U} + \hat{K} + \hat{\Phi})\rho dV, \quad (5.34)$$

where

$\hat{E}$ = Total energy per unit mass = $\hat{U} + \hat{K} + \hat{\Phi}$;
$\hat{U}$ = Internal energy per unit mass = $c_p[T - T_R]$;
$\hat{K}$ = Kinetic energy per unit mass = $v^2/2$; and
$\hat{\Phi}$ = Potential energy per unit mass = gh.

The internal energy is measured relative to a reference temperature T_R, c_p is the heat capacity per unit mass, T is the absolute temperature, v is the velocity, h is the height relative to a datum height, and g is the acceleration due to gravity. Energy can enter or leave the system through the solid walls of the system and the fluid channels. If the walls are permeable to the fluid, then some kinetic energy can be transferred to or from the system by convection. We will neglect this in comparison with energy exchange across the system boundaries by other mechanisms. In particular, we will account for heat added through the system boundaries, $\dot{Q}_S$ (excluding convection at the inlets and outlets), the work done by the system on the surroundings, and the net energy added by fluid entering and leaving through the system conduits. The net rate at which energy is transferred across the system boundaries is:

$$\left\{\begin{array}{c}\text{Rate energy}\\\text{enters}\\\text{the system}\end{array}\right\} - \left\{\begin{array}{c}\text{Rate energy}\\\text{leaves}\\\text{the system}\end{array}\right\} = \dot{Q}_S - \sum_i \int_{A_i} (\hat{E}_i)(\rho_i \vec{v}_i) \cdot (\vec{n}_i dA_i). \quad (5.35)$$

The first term represents the heat added to the system through the surface, as well as the heat added at the inlets and outlets by conduction and radiation. The second term represents the net energy added to the system by convection at the inlets and outlets. The negative sign ensures that energy is added to the system at the inlets, where $\vec{v}_i$ and $\vec{n}_i$ point in opposite directions, and that energy leaves the system at the outlets, where $\vec{v}_i$ and $\vec{n}_i$ point in the same direction. The net energy production in the system is caused by two phenomena: the rate at which work is done by the surroundings on the system and the rate at which heat is generated within the system.

$$\left\{\begin{array}{c} \text{Rate energy} \\ \text{is produced} \\ \text{in system} \end{array}\right\} = \dot{Q}_{\text{gen}} - \dot{W}. \tag{5.36}$$

Heat can be generated in the system by several mechanisms, including radioactive decay, metabolic heat generation from chemical reactions, viscous dissipation in the system, and electrical heating. $\dot{W}$ is the rate at which work is performed *by the system on the surroundings* [which explains the negative sign in (5.36)], and is equal to the surface integral of the product of fluid velocity $\vec{v}$ and force per unit surface area $\vec{f}_s$ applied by the fluid to the surroundings:

$$\dot{W} = \int_S \vec{f}_s \cdot \vec{v} \, dS. \tag{5.37}$$

If the fluid causes part of the surroundings to move, such as the rotation of turbine blades or the leaflets of a heart valve, then we can divide the fluid surface of the system into three regions: (1) the moving boundary, (2) the solid, stationary boundary, and (3) the fluid boundaries at the inlets and outlets of the system. Since the velocity is zero at the stationary surface, the rate at which work is done to the stationary surface is zero. The rate at which work is done by the fluid to the moving boundary is referred to as the shaft work rate, $\dot{W}_s$. The rate at which fluid does work on the surroundings is therefore equal to the rate at which work is done to the moving solid boundary and the rate at which work is done by the fluid at each of the i inlets and outlets:

$$\dot{W} = \dot{W}_s + \sum_i \int_{A_i} \vec{v}_i \cdot (\vec{f}_{si} dA_i). \tag{5.38}$$

Several different sources can contribute to the force per unit area at the surface of each inlet and outlet. The two most important forces are pressure forces and frictional forces. However, in special cases, we may need to account for other forces, such as electrical or magnetic forces, as well. Accounting for viscous and pressure forces, (5.38) becomes:

$$\dot{W} = \dot{W}_s + \dot{W}_f + \sum_i \int_{A_i} \vec{v}_i \cdot (P_i \vec{n}_i dA_i). \tag{5.39}$$

5.4 Conservation of Energy

Substituting (5.34)–(5.36) and (5.39) into (5.34):

$$\frac{dE}{dt} = -\sum_i \int_{A_i} \left(\hat{U}_i + \frac{v_i^2}{2} + \hat{\Phi}_i + \frac{P_i}{\rho_i}\right)\rho_i(\vec{v}_i \cdot \vec{n}_i)dA_i + \dot{Q}_S + \dot{Q}_{gen} - \dot{W}_s - \dot{W}_f. \tag{5.40}$$

The summation in (5.40) can be separated into summations over inlet and outlet conduits:

$$-\sum_k \int_{A_k} \left(\hat{U}_k + \frac{v_k^2}{2} + \hat{\Phi}_k + \frac{P_k}{\rho_k}\right)\rho_k(\vec{v}_k \cdot \vec{n}_k)dA_k$$

$$= \sum_{i=1}^{N_{inlets}} \int_{A_i} \left(\rho_i v_i \hat{U}_i + \frac{\rho_i v_i^3}{2} + \rho_i v_i \hat{\Phi}_i + P_i v_i\right)dA_i$$

$$- \sum_{j=1}^{N_{outlets}} \int_{A_j} \left(\rho_j v_j \hat{U}_j + \frac{\rho_j v_j^3}{2} + \rho_j v_j \hat{\Phi}_j + P_j v_j\right)dA_j. \tag{5.41}$$

Finally, if the density, internal energy, potential energy, and pressure are assumed to be uniform over the cross section of each inlet and outlet conduit, (5.40) can be simplified as:

$$\frac{dE}{dt} = \dot{Q}_S + \dot{Q}_{gen} - \dot{W}_s - \dot{W}_f + \sum_{i=1}^{N_{inlets}} w_i\left(\hat{U}_i + \frac{1}{2}\frac{\langle v_i^3\rangle}{\langle v_i\rangle} + \hat{\Phi}_i + \frac{P_i}{\rho_i}\right)$$

$$- \sum_{j=1}^{N_{outlets}} w_j\left(\hat{U}_j + \frac{1}{2}\frac{\langle v_j^3\rangle}{\langle v_j\rangle} + \hat{\Phi}_j + \frac{P_j}{\rho_j}\right), \tag{5.42}$$

where $\langle v^n\rangle$ is defined by (5.23), and w_i and w_j are the mass flow rates in the inlet and outlet conduits [(5.5) and (5.6)]. The steady-state energy balance for a system with multiple inlets and outlets (Fig. 5.12) is:

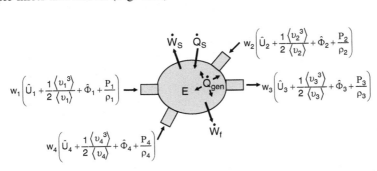

Fig. 5.12 Macroscopic energy balance on system with three inlet conduits (1, 2, and 4) and one outlet conduit (3)

$$\dot{Q}_S + \dot{Q}_{gen} - \dot{W}_s - \dot{W}_f = \sum_{j=1}^{N_{outlets}} w_j \left(\hat{U}_j + \frac{1}{2}K_{3j}\langle v_j\rangle^2 + \hat{\Phi}_j + \frac{P_j}{\rho_j} \right)$$

$$- \sum_{i=1}^{N_{inlets}} w_i \left(\hat{U}_i + \frac{1}{2}K_{3i}\langle v_i\rangle^2 + \hat{\Phi}_i + \frac{P_i}{\rho_i} \right). \tag{5.43}$$

Friction at the fluid inlets and outlets is usually small and is often neglected. If a system is in steady state with only a single inlet and outlet, with the outlet at a different height than the inlet, then $w_{in} = w_{out} = w$, $\hat{U} = c_p \Delta T$, and the energy equation with negligible friction at the inlet and outlet becomes:

$$\dot{Q}_S + \dot{Q}_{gen} - \dot{W}_s = w\left(c_P[T_{out} - T_{in}] + \frac{1}{2}\left[K_{3out}\langle v_{out}\rangle^2 - K_{3in}\langle v_{in}\rangle^2 \right] \right.$$

$$\left. + g[h_{out} - h_{in}] + \frac{1}{\rho}[P_{out} - P_{in}] \right). \tag{5.44}$$

Many important applications of the energy equation involve situations in which fluid flows through horizontal conduits having constant cross-sectional areas under conditions in which no work is performed by the fluid on the surroundings and heat generation is negligible. Under these circumstances, $\langle v_{out}\rangle = \langle v_{in}\rangle$, $h_{out} = h_{in}$, $\dot{W}_s = 0$, $\dot{Q}_{gen} = 0$, and (5.44) simplifies to:

$$\dot{Q}_S = w\left(c_P[T_{out} - T_{in}] + \frac{1}{\rho}[P_{out} - P_{in}] \right) = w(\hat{H}_{out} - \hat{H}_{in}), \tag{5.45}$$

where $\hat{H}$ is defined as the *enthalpy per unit mass* relative to the enthalpy per unit mass for the fluid at a particular reference state:

$$\hat{H} = \hat{U} + \frac{P - P_R}{\rho} = c_P(T - T_R) + \frac{P - P_R}{\rho}. \tag{5.46}$$

P_R and T_R are pressures and temperatures at the reference state, respectively, which can be chosen arbitrarily. The pressure term in (5.45) is usually negligible in comparison with the temperature term in bioheat transfer applications, as is shown in the following example. Note that if a system in steady state is isothermal, so that $T_{out} = T_{in}$, and the inlet and outlet pressures are different, heat will flow between the system and the boundary. The rate that heat crosses the boundary will be the product of the pressure difference and the flow rate. If, on the other hand, the system is adiabatic (i.e., insulated, so no heat passes through the boundary), then the fluid enthalpy will remain constant in a horizontal tube; so a positive pressure difference will give rise to a negative temperature difference. Friction in the tube, which produces a positive pressure difference, will cause the outlet temperature to be greater than the inlet temperature. The change in internal energy per unit volume

5.4 Conservation of Energy

between the outlet and the inlet fluid will equal the difference in pressure between the inlet and the outlet. Consequently, the rate at which heat is generated by friction within the tube is the same as in the isothermal case: $Q_V \Delta P$.

> **Example 5.4.1 Contribution of Pressure to Enthalpy Change Per Unit Mass.**
> The pressure drop across an arteriole is 50 mmHg. Heat is transferred through the walls of the blood vessel at a rate sufficient to raise the outlet temperature by 1°C. Compare the contributions of internal energy and pressure to the change in enthalpy.

Solution. *Initial considerations:* We will assume that flow through the arteriole is constant and that there is no fluid loss through the walls of the vessel.

System definition and environmental interactions: The system of interest is the blood within the arteriole. Energy in the form of heat is added to the blood by the warmer walls as it flows through the arteriole. The pressure difference across the ends of arteriole provides the energy necessary to cause blood to move through the vessel.

Apprising the problem to identify governing equations: We are interested in the relative contributions of pressure and internal energy to the change in enthalpy. Therefore, the relationship that applies in this case is the definition of enthalpy, (5.45).

Analysis: We can write (5.45) as follows:

$$\dot{Q}_S = wc_P[T_{out} - T_{in}]\left(1 + \frac{1}{\rho c_P}\frac{[P_{out} - P_{in}]}{[T_{out} - T_{in}]}\right).$$

Examining and interpreting the results: The second term in brackets indicates the departure from assuming that the rate of change in enthalpy is equal to the rate of change of internal energy. Inserting the appropriate values for blood flow in an arteriole:

$$\frac{1}{\rho c_P}\frac{[P_{out} - P_{in}]}{[T_{out} - T_{in}]} = \frac{(-50\,\text{mmHg})}{(1{,}050\,\text{kg m}^{-3})(3{,}740\,\text{J kg}^{-1})(+1°\text{K})}$$
$$\times \left[\frac{133.32\,\text{kg m}^{-1}\text{s}^{-2}\text{mmHg}^{-1}}{1\,\text{kg m}^2\text{s}^{-2}\text{J}^{-1}}\right] = -0.0017.$$

Therefore, the heat gained in this case is:

$$\dot{Q}_S = 0.9983\,wc_P[T_{out} - T_{in}].$$

Neglecting the pressure term in (5.45) results in less than 0.2% error in computing the heat gained.

Additional comments: We conclude that for heat exchanger applications, where the temperature change is normally greater than 1°C, neglecting the contribution of the pressure change in the energy equation is justified.

5.5 Engineering Bernoulli Equation

According to the second law of thermodynamics, some mechanical energy in the system is irreversibly converted into heat. The starting point for such an analysis must be the differential form of the momentum equation, which we will derive later in Chap. 7. This is a vector equation with dimensions of a force per unit volume. Taking the dot product between the local velocity vector and the momentum equation, then integrating the resulting expression over the volume of the system produces what has been termed the macroscopic mechanical energy equation. The resulting expression for the rate of change of mechanical energy (i.e., potential energy plus kinetic energy) is also known as the Engineering Bernoulli equation. For an incompressible, isothermal system with multiple inlets and outlets, this expression is similar in form to (5.42) (see Bird et al. 2002 for details):

$$\frac{d}{dt}\int_V (\hat{K}+\hat{\Phi})\rho dV = -\dot{W}_s - E_v + \sum_{i=1}^{N_{inlets}} w_i \left(\frac{1}{2}\frac{\langle v_i^3\rangle}{\langle v_i\rangle} + \hat{\Phi}_i + \frac{P_i}{\rho}\right) \\ - \sum_{j=1}^{N_{outlets}} w_j \left(\frac{1}{2}\frac{\langle v_j^3\rangle}{\langle v_j\rangle} + \hat{\Phi}_j + \frac{P_j}{\rho}\right). \tag{5.47}$$

All the variables in (5.47) were defined in Sect. 5.4, with the exception of E_V, which is the rate at which mechanical energy within the system is irreversibly converted into heat by viscous dissipation. This is also known as the friction loss and is positive for Newtonian fluids. If the mechanical energy is not changing with time, and a system has only a single inlet and a single outlet ($w_{in} = w_{out} = w$), then the Engineering Bernoulli equation reduces to:

$$\frac{\dot{W}_s}{w} + \hat{E}_v = \left(\frac{1}{2}\left[\frac{\langle v_{in}^3\rangle}{\langle v_{in}\rangle} - \frac{\langle v_{out}^3\rangle}{\langle v_{out}\rangle}\right] + g[h_{in} - h_{out}] + \frac{1}{\rho}[P_{in} - P_{out}]\right), \tag{5.48}$$

where $\hat{E}_v$ is defined as E_v divided by the mass flow rate:

$$\hat{E}_v \equiv \frac{E_v}{w}. \tag{5.49}$$

If there is no work performed ($\dot{W}_s = 0$) and friction is negligible ($\hat{E}_v = 0$, $\langle v^n\rangle/\langle v\rangle = v^n$), (5.48) reduces to a form similar to the Bernoulli equation encountered

5.5 Engineering Bernoulli Equation

in physics, which is valid for steady, frictionless flow along a streamline: $\frac{1}{2}\rho v^2 + \rho g h + P =$ constant. Variations in hydrostatic pressure can be accounted for automatically by defining a modified pressure $\wp$:

$$\wp \equiv P - \rho g_x x - \rho g_y y - \rho g_z z, \tag{5.50}$$

where g_x, g_y, and g_z are the components of the gravitational vector. In this case $g_x = g_z = 0$ and, since y is positive upward, $g_y = -g$. Therefore, $\wp_{in} = P_{in} + \rho g h_{in}$ and $\wp_{out} = P_{out} + \rho g h_{out}$. Substituting these into (5.48) and using the definition of K_{ki} in (5.23), we can simplify the Engineering Bernoulli equation still further:

$$\frac{\dot{W}_s}{w} + \hat{E}_v = \frac{1}{2}\left[K_{3,in}\langle v_{in}\rangle^2 - K_{3,out}\langle v_{out}\rangle^2\right] + \frac{1}{\rho}[\wp_{in} - \wp_{out}]. \tag{5.51}$$

This expression applies to steady-state flow through a system with a single inlet and a single outlet. In a round tube, K_{3i} is 2 for laminar flow (5.24) and is approximately 1 for turbulent flow. Equation (5.51) is an extremely useful relation for finding pressure–flow relationships in conduits and for finding power requirements for moving fluids through those conduits.

> **Example 5.5.1 Work to Overcome Friction in a Closed System.**
> What is the relationship between the rate at which the heart performs work and the total friction loss in the cardiovascular system?

Solution. *Initial considerations:* Although flow through the cardiovascular system is periodic, we will simplify the problem by assuming flow to be steady. This should provide an estimate of the relationship between friction loss and average work of the heart.

System definition and environmental interactions: The circulatory system is normally divided between the systemic circulation, supplied by the left heart, and the pulmonary circulation, supplied by the right heart. When joined together in series, the two systems form a single system, the cardiovascular system. Blood inside the cardiovascular system is the system of interest. Environmental interactions occur between the blood and active contractions of cardiac muscle, which propel blood through the system, and passive viscous interactions with blood vessel walls, which tend to resist motion.

Apprising the problem to identify governing equations: We seek an expression that relates the rate at which work is performed to frictional losses. The Engineering Bernoulli equation contains terms for the rate work is performed and the rate that heat is lost by viscous dissipation.

Analysis: Assuming steady flow, we can apply (5.48) to the systemic circulation which is depicted in Fig. 5.13 as the system between the inlet at the ascending aorta

Fig. 5.13 System Inlet and outlet

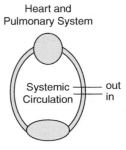

and the outlet, also in the ascending aorta. The distance between inlet and outlet is allowed to shrink to zero, so that the entire systemic circulation is contained between the inlet and the outlet. In the limit, we find $h_{in} = h_{out}$, $v_{in} = v_{out}$, and $P_{in} = P_{out}$. Consequently, the Engineering Bernoulli equation tells us that:

$$\dot{W}_s = -w\hat{E}_v = -E_v.$$

Examining and interpreting the results: Recall that $\dot{W}_s$ is the rate at which fluid does work on the surroundings and that E_v is the viscous dissipation, which is always positive. Therefore, $\dot{W}_s$ is negative and represents the rate that work is done by the heart on the blood in the systemic circulation. This relation tells us that the rate at which the left heart performs work is just enough to overcome frictional losses in the systemic circulation.

Example 5.5.2 Work of the Left Ventricle.
Estimate the steady-state rate that work is performed by the left ventricle to maintain a cardiac output of 5 L/min. The density of blood is 1.05 g/ml, the cross-sectional area of the mitral valve is 7.6 cm^2, and the cross-sectional area of the aortic valve is 3.1 cm^2. Neglect friction losses in the heart. Mean left atrial pressure is 10 Torr and mean aortic pressure is 100 Torr. The mitral and aortic valves are at approximately the same elevation.

Solution. *Initial considerations:* Although this is actually an unsteady-state problem that takes place in two stages (ventricular filling or diastole, and ventricular emptying or systole), we will treat it as though flow is steady. In addition, we will assume that blood velocity is uniform across each valve so that $\langle v^3 \rangle / \langle v \rangle$ is approximately $\langle v \rangle^2 = (Q_V/A_c)^2$, where Q_V is the cardiac output and A_c is the cross-sectional area of the valve. We will neglect any difference in height between the inlet and outlet valves.

System definition and environmental interactions: Our system consists of the blood within the left ventricle between the inlet (mitral valve) and the outlet (aortic valve). Work is done by the ventricular wall on the blood to move it into the aorta.

5.5 Engineering Bernoulli Equation

Apprising the problem to identify governing equations: Inlet and outlet pressures are known and average inlet and outlet velocities can be computed from the given valve areas and cardiac output. As in the previous example, the Engineering Bernoulli equation can be used to compute the rate that the left ventricle performs work.

Analysis:
Equation (5.48) becomes:

$$\dot{W}_s = \rho Q_V \left(\frac{1}{2} \left[\left(\frac{Q_V}{A_{in}} \right)^2 - \left(\frac{Q_V}{A_{out}} \right)^2 \right] + \frac{1}{\rho} [P_{in} - P_{out}] \right).$$

Inlet quantities are evaluated at the mitral valve, and outlet quantities are evaluated at the aortic valve. All quantities on the right-hand side are known and both terms provide negative values, indicating that the rate the heart does work on the blood is positive.

Examining and interpreting the results: Substituting the values from above:

$$\dot{W}_s = \frac{\left(1.05 \frac{g}{ml}\right)\left(83.3 \frac{ml}{s}\right)}{10^7 \, g \, cm^2/Ws^3}$$

$$\times \left(\frac{1}{2} \left[\left(\frac{83.3 \, cm^3/s}{7.6 \, cm^2} \right)^2 - \left(\frac{83.3 \, cm^3/s}{3.1 \, cm^2} \right)^2 \right] + \frac{1}{1.05 \frac{g}{ml}} \frac{[10 - 100 \, mmHg]}{7.5 \times 10^{-4} \frac{mmHg}{g/(cm \, s^2)}} \right).$$

This provides an estimate of $\dot{W}_s$. For a heart rate of 72 beats/min, the work per beat is 0.833 J/beat. This is in reasonable agreement with measurements by Kameyama et al. (1992), who found an average of 0.7 J/beat.

Additional comments: Note that the net contributions from the kinetic energy terms are very small in comparison with the contribution from the pressure difference.

Example 5.5.3 Power Rating of a Pump.
What power rating is required of a pump if it has inlet and outlet ports of equal diameters and heights, and is to deliver water at a flow of 10 L/min when the pressure difference added by the pump is 300 mmHg? The pump efficiency is 85%.

Solution. *Initial considerations:* The efficiency of the pump refers to the output power that can be delivered by the pump relative to the energy supplied to the pump. Pump frictional losses are accounted for by the pump efficiency, η. Water is an incompressible fluid and we will assume that the pump operates in a steady state.

System definition and environmental interactions: Water inside the pump between the inlet and outlet ports of the pump comprises the system. Motion is imparted to the water by the pump in some manner, such as a rotating set of impellers, rollers, a moving piston, or another mechanism. It is not necessary to specify the exact interaction between the system and the environment to perform the analysis.

Apprising the problem to identify governing equations: Once again, since the necessary pump power is sought, the Engineering Bernoulli equation is ideally suited to the analysis of this problem.

Analysis: Since the efficiency is given, the power delivered to the pump is equal to the rate at which shaft work is performed divided by the efficiency:

$$\left(\dot{W}\right)_{\text{pump}} = \dot{W}_s/\eta.$$

Since the E_V term is included in the pump efficiency, we can drop the E_v term in (5.48) to compute the rate at which shaft work is performed. For inlet and outlet at the same height and same diameter:

$$\dot{W}_s = \frac{w}{\rho}[P_{\text{in}} - P_{\text{out}}] = Q_V[P_{\text{in}} - P_{\text{out}}].$$

The minimum power rating of the pump will be:

$$\left(\dot{W}\right)_{\text{pump}} = \frac{w}{\rho\eta}[P_{\text{in}} - P_{\text{out}}] = \frac{Q_V}{\eta}[P_{\text{in}} - P_{\text{out}}].$$

Examining and interpreting the results: For the case at hand, the minimum power required to pump the fluid is:

$$\left(\dot{W}\right)_{\text{pump}} = \frac{\left(10\,\frac{\text{L}}{\text{min}}\right)(300\,\text{mmHg})}{0.85\quad 10^7\,\frac{\text{g cm}^2}{\text{W s}^3}}\left(1{,}333.2\,\frac{\text{g}}{\text{mmHg cm s}^2}\right)\left(1{,}000\,\frac{\text{cm}^3}{\text{L}}\right)\left(\frac{1\,\text{min}}{60\,\text{s}}\right)$$

$$= 7.85\,\text{W}.$$

Since the pump has an efficiency of 85%, we need a pump rated at 7.85 W. If the pump were 100% efficient, we could select a pump with a minimum rating of 6.67 W, which is the actual power needed to move a frictionless fluid through a pressure rise of 300 mmHg at a rate of 10 L/min.

Additional comments: The power that must be delivered by a pump to cause fluid to flow at a rate Q_V from low-pressure to high-pressure is $Q_V\Delta P$. This is analogous to the power $I\Delta V$ that must be delivered by a battery that provides a current flow I from low voltage to high voltage ΔV.

5.6 Friction Loss in Conduits

We now develop methods for estimating the friction loss which occurs as fluids move through straight conduits with uniform cross section. Consider a straight section of a conduit with length L and two dimensions which characterize its cross section, d and B. This could represent a duct with a rectangular cross section $d \times B$, a triangular cross section (base × height), an elliptical cross section (major and minor axes) or other cross sections. We will use dimensional analysis, as discussed in Chap. 3, to help us determine which experiments need to be performed to characterize the friction loss in these conduits. The quantity $\hat{E}_v$ in (5.49) has dimensions of L^2T^{-2} and would be expected to depend on the following variables:

Fluid properties	Density ρ (ML^{-3}) and viscosity μ (ML^{-1}T^{-1})
Geometry	Length L (L), dimension 1 d (L), dimension 2 B (L)
Flow property	$\langle v \rangle$ (LT^{-1})

The number of fundamental dimensions $N_D = 3$
The number of variables $N_v = 7$ (including $\hat{E}_v$)
The number of dimensionless groups $= N_D - N_v = 4$

Following the rules set forth in Sect. 3.3 for selecting the core variables, we must include one of the geometric parameters in the core, say d, and we must include $\langle v \rangle$ in the core, since $\langle v \rangle$ and $\hat{E}_v$ do not have linearly independent dimensions. Consequently, we select the following core and excluded variables:

Core variables: $\mu, d, \langle v \rangle$
Excluded variables: $\hat{E}_v, \rho, L, B$
From this we can write:

$$\Pi_{\hat{E}_v} = \Pi_{\hat{E}_v}(\Pi_\rho, \Pi_L, \Pi_B), \tag{5.52}$$

where the dimensionless groups can be found easily using the Buckingham Pi Theorem:

$$\Pi_{\hat{E}_v} = \frac{\hat{E}_v}{\langle v \rangle^2}, \quad \Pi_\rho = \frac{\rho \langle v \rangle d}{\mu}, \quad \Pi_L = \frac{L}{d}, \quad \Pi_B = \frac{B}{d}. \tag{5.53}$$

The *Fanning friction factor* f is used to account for friction losses in straight conduits. In its most general form, a friction factor can be defined for flow past a stationary body or flow through a stationary conduit, as discussed in Chap. 2. In either case, the fluid will exert a frictional force on the stationary object. The friction factor is defined by the relation:

$$F_k = fKA, \tag{5.54}$$

where F_k is the kinetic force applied to the conduit or solid object. This is the force exerted on the conduit or solid object because of the kinetic nature of the fluid, over and above the force experienced by the object when the fluid is stationary. The factor K in (5.54) is the mean kinetic energy of the fluid per unit volume. A is a characteristic area. In the case of flow through a conduit, A is the *wetted surface area of the conduit* S_w, which is the product of the *wetted perimeter* P_w and the length of the conduit L. Thus, for a fluid in a conduit, the friction factor can be computed as:

$$f = \frac{F_k}{S_w\left(\frac{1}{2}\rho\langle v\rangle^2\right)} = \frac{F_k}{P_w L\left(\frac{1}{2}\rho\langle v\rangle^2\right)}. \quad (5.55)$$

The kinetic force can be found from a macroscopic force balance on fluid within the conduit (5.30):

$$F_k = [P_{in} - P_{out}]A_c + \rho A_c L g_x, \quad (5.56)$$

where x is the axial direction and A_c refers to the cross-sectional area of the fluid within the conduit. From geometry (see Fig. 5.14), the component of g in the axial direction can be related to the height difference between the inlet and outlet of the conduit:

$$\sin\theta = \frac{g_x}{g} = \frac{h_{in} - h_{out}}{L}. \quad (5.57)$$

So, (5.56) can be rewritten as:

$$F_k = A_c\{[P_{in} - P_{out}] + \rho g[h_{in} - h_{out}]\}. \quad (5.58)$$

The Engineering Bernoulli equation for a straight conduit, with no shaft work is:

$$\hat{E}_v = \frac{[P_{in} - P_{out}]}{\rho} + g[h_{in} - h_{out}] = \frac{\wp_{in} - \wp_{out}}{\rho}, \quad (5.59)$$

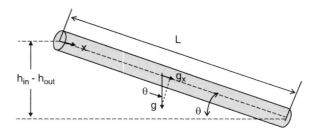

Fig. 5.14 Geometric relationships for (5.57)

5.6 Friction Loss in Conduits

where $\wp$ is the modified pressure defined in (5.50). Substituting this into (5.58) provides a relation between $\hat{E}_v$ and F_k:

$$\hat{E}_v = \frac{F_k}{\rho A_c}. \tag{5.60}$$

Replacing F_k with (5.55), an expression involving the friction factor, we find a relation between $\hat{E}_v$ and f:

$$\hat{E}_v = f \frac{S_w}{A_c} \left(\frac{1}{2} \langle v \rangle^2 \right). \tag{5.61}$$

The *hydraulic diameter* D_h of a conduit is equal to four times the cross-sectional area of the fluid in the conduit divided by the wetted perimeter:

$$D_h = \frac{4A_c}{P_w} = \frac{4A_c L}{S_w}. \tag{5.62}$$

Substitution of the definition of the hydraulic diameter into (5.61) provides an expression that relates $\hat{E}_v$ to the Fanning friction factor f:

$$\hat{E}_v = 4f \frac{L}{D_h} \left(\frac{1}{2} \langle v \rangle^2 \right). \tag{5.63}$$

If the friction factor can be found from experimental measurements, then the pressure difference across the ends of the conduit can be computed by substituting (5.63) back into (5.59):

$$\wp_{in} - \wp_{out} = \Delta \wp = 4f \frac{L}{D_h} \left(\frac{1}{2} \rho \langle v \rangle^2 \right). \tag{5.64}$$

In cases where the pressure difference and height difference are specified, (5.64) can be rearranged to solve for the average fluid velocity:

$$\langle v \rangle = \sqrt{\frac{D_h}{L} \frac{\Delta \wp}{2f \rho}}. \tag{5.65}$$

Returning to our dimensional analysis, (5.53), we can now determine the relationship between the friction factor and our dimensionless viscous dissipation $\Pi_{\hat{E}_v}$:

$$2f \left(\frac{D_h}{L} \right) = \Pi_{\hat{E}_v} \left(Re_d, \frac{L}{d}, \frac{B}{d} \right). \tag{5.66}$$

The dependence on L/d is caused by changes in the velocity distribution over the cross section as the fluid moves downstream. The velocity is relatively uniform at the inlet of the conduit, but because of the presence of the wall, fluid decelerates near the wall and accelerates in the central portion of the conduit for some distance from the inlet. This is known as the *hydrodynamic entry length*. Beyond this length, no additional variations occur in the velocity profile and the flow is said to be *fully developed*. In turbulent flow, the entry length is extremely short (approximately 20 diameters) and dependence of friction factor on L/d is minor if the entrance is rounded (Olson and Sparrow 1963). Furthermore, if the Reynolds number is based on the hydraulic diameter, the friction factor in turbulent flow is nearly independent of the shape of the conduit cross section. However, the turbulent friction factor depends on the roughness of the conduit wall k relative to the hydraulic diameter of the conduit. Thus, for turbulent flow:

$$f = f\left(Re_{D_h}, \frac{k}{D_h}\right) \quad \text{(turbulent)}. \tag{5.67}$$

In laminar flow, the friction factor is independent of the tube roughness, but is highly dependent on the shape of the conduit:

$$f = f\left(Re_d, \frac{L}{d}, \frac{B}{d}\right) \quad \text{(laminar)}. \tag{5.68}$$

For fully developed laminar flow in a circular tube, the Fanning friction factor is inversely related to the Reynolds number:

$$f = 16/Re_d \quad \text{(laminar)}. \tag{5.69}$$

The pressure gradient in the laminar hydrodynamic entry length is higher than the pressure gradient in the fully developed region. The dimensionless pressure drop in the entry region of a circular tube is shown in Fig. 5.15 for laminar flow. Shown for comparison is the dimensionless pressure drop which would occur if entry effects are neglected. The additional pressure drop caused by entry effects can be significant for short tubes.

The *hydrodynamic entry length* L_e is defined as the distance between the tube entrance and the point where the centerline velocity has come to within 1% of its final velocity. Theoretical and experimental results show that the dimensionless entry length for laminar flow depends on the Reynolds number (Durst et al. 2005):

$$\frac{L_e}{d} = \left[(0.619)^{1.6} + (0.0567 Re_d)^{1.6}\right]^{1/1.6}. \tag{5.70}$$

At low Reynolds numbers, the entry length is approximately 0.619 diameters in length. At high Reynolds numbers (i.e. $Re_d > 50$), the laminar entry length is approximately $0.0567\, Re_d$.

5.6 Friction Loss in Conduits

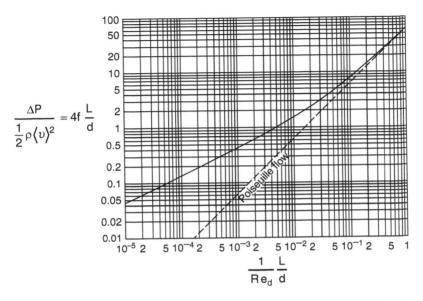

Fig. 5.15 Pressure losses at entrance to a tube, laminar flow (adapted from Perry and Chilton, 1973)

If the tube is much longer than L_e, then the additional pressure drop in the entry region is negligible. In fact, from Fig. 5.15, corrections for pressure changes in the laminar flow entry region are not needed if:

$$\frac{L}{d} > 0.2 Re_d. \quad (5.71)$$

This criterion is met in very small blood vessels, where the Reynolds number is less than unity. However, there are many larger blood vessels and conducting airways in the human body in which entry effects cannot be neglected (see Example 5.6.1).

A set of graphs relating the friction factor (actually $4f$) to Reynolds number based on hydraulic diameter for various ratios of tube roughness to hydraulic diameter is shown in Fig. 5.16. This is known as the *Moody diagram*, and can be used to find the friction factor in laminar or turbulent flow in circular tubes and for turbulent flow in tubes of noncircular, but uniform cross section. Laminar flow in noncircular conduits is treated in Sect. 5.8.

The transition from laminar to turbulent flow in conduits occurs at Reynolds numbers based on hydraulic diameter of approximately 2,200. For $Re_{D_h} < 2,200$, flow is always laminar, and for $Re_{D_h} > 2,200$, flow is usually turbulent. However, if disturbances are minimized at the inlet to the conduit, laminar flow can persist at much higher Reynolds numbers. Special applications of the Moody diagram include:

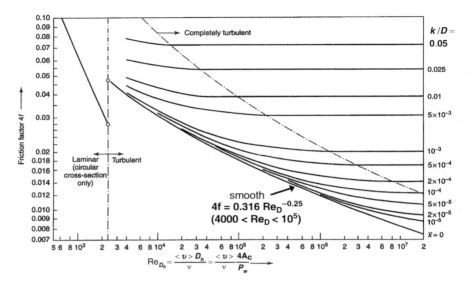

Fig. 5.16 Moody diagram. Friction factor vs. Reynolds number (modified from Beek et al. 1999 with permission)

- Laminar flow in circular tube: $f = 16/Re_D$ or $4f = 64/Re_D$ ($Re_D < 2{,}200$)
- Laminar flow in noncircular ducts (see Sect. 5.8)
- Turbulent flow in *smooth* ducts: $4f = 0.316\, Re_{D_h}^{-0.25}$ ($4{,}000 < Re_{D_h} < 10^5$)

There are four typical applications of the Moody diagram:

- Given $\langle v \rangle$, L, D, ρ, ν, k; Find $\Delta \wp$
- Given $\Delta \wp$, L, D, ν, ρ, k; Find $\langle v \rangle$ or Q_V
- Given $\langle v \rangle$, $\Delta \wp$, L, ν, ρ, k; Find D
- Given Q_V, $\Delta \wp$, L, ν, ρ, k; Find D

We will examine each of these by way of example.

Example 5.6.1 Pressure Drop Across Large Bronchi in the Respiratory System.

Bronchi that are located two generations down the respiratory tree from the trachea have a mean diameter of 9.4 mm and a length of 30.3 mm. If air flow in the lung during inspiration is 1 L/s (0.25 L/s per bronchus), estimate the pressure difference across these bronchi.

Solution. *Initial considerations:* We will assume steady flow through the bronchi during inspiration. Since these bronchi are relatively short, the assumption of a uniform velocity profile along the entire length of a bronchus may be invalid. Consequently, it may be inappropriate to neglect entrance effects in this case. The exact nature of the velocity profile at the inlets to these bronchi is not specified,

5.6 Friction Loss in Conduits

but it seems reasonable to assume that they are relatively flat. The temperature is not specified, so from Example 3.2.6.2 and Fig. 3.5, we will estimate the temperature of air in second generation bronchi to be 27°C.

System definition and environmental interactions: If all four of the bronchi have identical shapes and the flow through each is the same, then the pressure difference across each will also be the same. Therefore, it is only necessary to analyze one of the bronchi. The system of interest is the air in a single bronchus. Interactions with the environment include the pressure forces at the two ends of the bronchus and the viscous force at the bronchus wall. The viscous force is highest in the region near the inlet to the bronchus where the velocity gradient at the wall is greatest.

Apprising the problem to identify governing relationships: If entrance effects are ignored, the Moody diagram (Fig. 5.16) can be used to find the friction factor for flow through a bronchus and the pressure difference can then be computed using (5.64). Hydrostatic pressure differences in pressure caused by a difference in height between inlet and outlet can be ignored in air, so $\Delta \wp = \Delta P$. This approach would lead to an underestimate in the pressure difference. If entrance effects are included, Fig. 5.15 can be used if flow is laminar to find the pressure difference. This assumes that the velocity profile at the inlet to the bronchus is flat. It would be expected to provide an overestimate to the actual pressure difference.

Analysis: The average velocity in a second generation bronchus is:

$$\langle v \rangle = \frac{4 Q_v}{\pi d^2} = \frac{4(0.25\,\text{l/s})(1{,}000\,\text{ml/l})}{\pi (0.94\,\text{cm})^2} = 360\,\frac{\text{cm}}{\text{s}}.$$

The Reynolds number in each of these four bronchi is:

$$Re_d = \frac{\langle v \rangle d}{v_\text{air}} = \frac{(360\,\text{cm/s})(0.94\,\text{cm})}{(1.57 \times 10^{-5}\,\text{m}^2/\text{s})(10^4\,\text{cm}^2/\text{m}^2)} = 2{,}155.$$

The Fanning friction factor from the Moody diagram, based on fully developed flow, is $f = 16/2{,}155 = 0.0074$. Since flow is laminar (barely), the friction factor that includes entry effects can be computed using Fig. 5.15. The value of $L/(d\,Re_d)$ is 0.0015. From the graph, $4fL/d = 0.5$, so $f = 0.0388$, which is about five times higher than the friction factor estimated if entry effects are not included. The pressure difference across one of the bronchi is calculated from (5.64):

$$\Delta P = 4f \frac{L}{D_h}\left(\frac{1}{2}\rho \langle v \rangle^2\right)$$

$$= \frac{4(0.0388)}{\left(\frac{980\,\text{g}}{\text{cm}\,\text{s}^2\text{cmH}_2\text{O}}\right)}\left(\frac{3.03\,\text{cm}}{0.94\,\text{cm}}\right)\left[\frac{1}{2}\left(1.1 \times 10^{-3}\,\frac{\text{g}}{\text{cm}^3}\right)\left(360\,\frac{\text{cm}}{\text{s}}\right)^2\right]$$

$$= 0.036\,\text{cmH}_2\text{O}.$$

Examining and interpreting the results: Failure to account for entry effects in these airways would lead to considerable underestimates of the actual pressure difference.

Additional comments: The entry length computed using (5.70) is greater than the actual airway length for nearly every airway generation. However, as air moves further into the respiratory system, the velocity at the entrance to each daughter vessel begins to approach fully developed flow (Olson et al. 1970). Consequently, the use of Fig. 5.15, which is based on a flat inlet velocity profile at the inlet, would be expected to lead to an overcorrection for entry effects on airways which are far down the respiratory tree. Nevertheless, Olson et al., after correcting for developing velocity profiles at the inlet to each airway, still found that entry effects accounted for 75% of the pressure drop in the respiratory tree.

Example 5.6.2 Computing the Pressure Drop Across a Square Duct.
Find the pressure drop across a square horizontal duct with 2 mm sides and a length of 20 cm when water flows through the duct at a rate of 10 ml/s. The duct walls have a roughness of 0.002 mm.

Solution. *Initial considerations:* Since this is a square duct, we can only use the Moody diagram to estimate the friction factor if flow is turbulent. If flow is laminar, we will need to use methods discussed in Sect. 5.8 for laminar flow though noncircular ducts.

System definition and environmental interactions: The system to be analyzed is the water inside the square duct. The walls of the duct apply frictional forces to the fluid in the direction opposite to flow. These forces are balanced by the difference in pressure forces between the inlet and outlet of the duct.

Apprising the problem to identify governing relationships: Assuming turbulent flow, we can find the pressure drop using (5.64). The friction factor can be found from the Moody diagram. To use the Moody diagram, we need to compute the Reynolds number based on the hydraulic diameter and the roughness factor.

Analysis: The hydraulic diameter, from (5.62), is:

$$D_h = \frac{4A_c}{P_w} = \frac{4(2\,\text{mm})(2\,\text{mm})}{(2\,\text{mm} + 2\,\text{mm} + 2\,\text{mm} + 2\,\text{mm})} = 2\,\text{mm}.$$

The average velocity of water in the duct is:

$$\langle v \rangle = \frac{Q_v}{A_c} = \frac{(10\,\text{cm}^3/\text{s})}{0.04\,\text{cm}^2} = 250\,\text{cm/s}.$$

The Reynolds number for water flow based on D_h is:

$$Re_{D_h} = \frac{\langle v \rangle D_h}{\nu} = \frac{(250\,\text{cm/s})(0.2\,\text{cm})}{0.01\,\text{cm}^2/\text{s}} = 5 \times 10^3.$$

5.6 Friction Loss in Conduits

Therefore, the flow is turbulent. The roughness factor k/D_h is:

$$\frac{k}{D_h} = \frac{0.002 \text{ mm}}{2 \text{ mm}} = 1 \times 10^{-3}.$$

From the Moody diagram, we find $4f$ at the intersection of $k/D_h = 10^{-3}$ and $Re_{D_h} = 5 \times 10^3$ to be:

$$4f = 0.04.$$

Inserting this value into (5.64) (with $h_{in} = h_{out}$):

$$P_{in} - P_{out} = 4f \frac{L}{D_h} \left(\frac{1}{2}\rho\langle v\rangle^2\right) = 0.04 \left(\frac{200 \text{ mm}}{2 \text{ mm}}\right) \left(\frac{1}{2}(1 \text{ g cm}^{-3})(250 \text{ cm s}^{-1})^2\right)$$
$$\times (7.5 \times 10^{-4} \text{ mmHg cm s}^2\text{g}^{-1}) = 93.8 \text{ mmHg}.$$

Examining and interpreting the results: Therefore, the pressure difference that must be applied across the duct to provide a flow of 10 ml/s is 93.8 mmHg. Since flow is turbulent and the duct length is much greater than 20 hydraulic diameters, any additional pressure drop caused by entry effects can be neglected.

Example 5.6.3 Finding Velocity and Flow for a Given Pressure Drop.
Find the velocity and flow rate of blood flowing through a circular tube with diameter 2 cm, roughness 0.02 cm, and length 25 cm when a pressure difference of 4 cmH$_2$O is applied across the tube.

Solution. *Initial considerations:* The properties of blood are not given; hence, we will assume that blood behaves as a Newtonian fluid with a kinematic viscosity of 0.04 cm^2/s and density of 1.04 g/cm^3. We will also assume that the tube is horizontal and so gravitational forces can be neglected.

System definition and environmental interactions: Blood inside the tube is the system of interest. The net viscous force by the walls and the pressure force at the outlet oppose the flow of blood, while the pressure force at the inlet tends to push the blood through the tube.

Apprising the problem to identify governing relationships: Since this problem involves flow through a cylindrical tube, the Moody diagram can be used. However, since we do not know the velocity, we cannot compute the Reynolds number explicitly. We can get around this difficulty by finding a relationship between $4f$ and Re in terms of parameters that we do know. If we plot this relationship on the Moody diagram, it will intersect the line for roughness $k/D = 0.01$, and the velocity can be computed from the Reynolds number found at the intersection.

Let us start first by solving (5.64) for the unknown velocity in terms of the friction factor. For a horizontal tube (with $h_{in} = h_{out}$):

$$\langle v \rangle = \frac{1}{\sqrt{4f}} \left[\sqrt{\frac{2\Delta P}{\rho} \frac{D}{L}} \right].$$

Rearranging the definition of Re to solve for velocity:

$$\langle v \rangle = \frac{\nu Re}{D}.$$

Setting the velocities in the above expressions equal to each other and rearranging:

$$Re\sqrt{4f} = \frac{D}{\nu} \left[\sqrt{\frac{2\Delta P}{\rho} \frac{D}{L}} \right].$$

Analysis: All quantities on the right side of the above expression are known:

$$\frac{D}{\nu} \sqrt{\frac{2\Delta P}{\rho} \frac{D}{L}} = \frac{(2\,\text{cm})}{0.04 \frac{\text{cm}^2}{\text{s}}} \sqrt{\frac{2\left(\frac{980\,\text{g}}{\text{cms}^2\text{cm H}_2\text{O}}\right)(4\,\text{cmH}_2\text{O})}{1.04 \frac{\text{g}}{\text{cm}^3}} \frac{2\,\text{cm}}{25\,\text{cm}}} = 1,228.$$

Therefore, for this particular problem, the following relationship is valid:

$$Re\sqrt{4f} = 1,228.$$

This relationship will produce a straight line on the Moody diagram. To define a straight line, we need to select two points. Let us select $4f = 0.02$ as our first point. The Reynolds number corresponding to this point is $1,228/(0.02)^5 = 8,682$. Selecting our second point with $4f = 0.06$ yields an Re of $1,228/(0.06)^5 = 5,014$. These two points (P_1 and P_2) are plotted on the Moody diagram in Fig. 5.17 and a straight line drawn through them.

Where this line intersects the $k/D = 0.01$ line is the point consistent with the Reynolds number and friction factor for $k/D = 0.01$. At the point of intersection: $4f = 0.046$. Substituting this into the equation above, we find $Re = 5,726$ (turbulent). The velocity can now be found from the definition of the Reynolds number:

$$\langle v \rangle = \frac{\nu Re}{D} = \frac{(0.04\,\text{cm}^2/\text{s})(5,726)}{2\,\text{cm}} = 115\,\text{cm/s}.$$

5.6 Friction Loss in Conduits

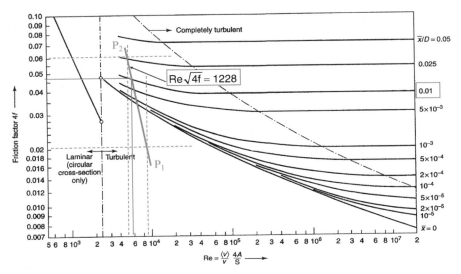

Fig. 5.17 Finding friction factor and Reynolds number for $Re\sqrt{4f} = 1,228$ and $k/D = 0.01$

The blood flow rate can be computed from:

$$Q_V = \frac{\pi D^2}{4} \langle v \rangle = \frac{\pi}{4}(2\,\text{cm})^2(115\,\text{cm/s}) = 361\,\text{ml/s}.$$

Examining and interpreting the results: These results can be checked using the solution to recompute the pressure drop. For an $Re = 5{,}726$ and $k/D = 0.01$, we found $4f = 0.046$. The pressure difference for this case would be:

$$P_{\text{in}} - P_{\text{out}} = 4f \frac{L}{D_h}\left(\frac{1}{2}\rho\langle v\rangle^2\right)$$

$$= \frac{0.046}{\left(\frac{980\,\text{g}}{\text{cm s}^2\text{cmH}_2\text{O}}\right)}\left(\frac{25}{2}\right)\left[\frac{1}{2}\left(1.04\,\frac{\text{g}}{\text{cm}^3}\right)\left(115\,\frac{\text{cm}}{\text{s}}\right)^2\right] = 4\,\text{cmH}_2\text{O}.$$

This is the correct pressure drop, as stated in the problem statement.

Example 5.6.4 Finding the Correct Conduit Size to Provide a Desired Velocity.
We wish to fabricate a smooth square duct with sides a and length 200 cm that will provide an average velocity for water of 200 cm/s when a pressure difference of 500 cmH$_2$O is applied across the duct. What is an appropriate length for the sides of the conduit?

Solution. *Initial considerations:* We will assume that the duct will be in the horizontal position for the average velocity and pressure drop specified. We will also assume initially that the flow is turbulent; hence, the Moody diagram can be used. This must be checked after the length of a side is determined.

System definition and environmental interactions: Although our goal is to determine the length of the sides for the duct, the system to be analyzed is the water inside the duct.

Apprising the problem to identify governing relationships: Assuming flow is turbulent, the Moody diagram will once again be the starting point for this analysis. As in the previous example, we are unable to compute the Reynolds number. In this case we do not know the hydraulic diameter. The procedure is to eliminate the unknown hydraulic diameter by rearranging (5.64) and the definition of the Reynolds number, as follows:

$$D_h = 4f \frac{L}{\Delta P}\left(\frac{1}{2}\rho \langle v \rangle^2\right) = \frac{v Re_{D_h}}{\langle v \rangle}.$$

Rearranging this, so the unknown quantities are grouped on the left side and known quantities on the right side, yields:

$$(4f)Re_{D_h}^{-1} = \frac{2v}{\rho \langle v \rangle^3}\frac{\Delta P}{L}.$$

Analysis: Substituting in the known values on the right side:

$$(4f)Re_{D_h}^{-1} = \frac{2(0.01\,\text{cm}^2/\text{s})}{(1\,\text{g cm}^{-3})\langle 200\,\text{cm/s}\rangle^3}\frac{(500\,\text{cm H}_2\text{O})\left(980\,\text{g cm}^{-1}\text{s}^{-2}\text{cm H}_2\text{O}^{-1}\right)}{200\,\text{cm}}$$
$$= 6.125 \times 10^{-6}.$$

We could plot this on the Moody diagram, as in the previous example. However, since the tube is smooth, and assuming the flow is turbulent, the relationship between $4f$ and Re is:

$$4f = 0.316\,Re_{D_h}^{-0.25}\,(4,000 < Re_{D_h} < 10^5).$$

Substituting this into the previous expression, we find:

$$\left(0.316 Re_{D_h}^{-0.25}\right)Re_{D_h}^{-1} = 6.125 \times 10^{-6}.$$

Solving for the Reynolds number:

$$Re_{D_h} = 5,889.$$

5.6 Friction Loss in Conduits

The flow is turbulent and the Reynolds number is within the acceptable range for the use of the analytic expression $(4,000 < Re_{D_h} < 10^5)$. Solving for the hydraulic diameter, which is equal to the channel dimension a:

$$D_h = \frac{4A_c}{P_w} = \frac{4a^2}{4a} = a = \frac{\nu Re_{D_h}}{\langle v \rangle} = \frac{(0.01\text{ cm}^2\text{s}^{-1})(5,889)}{200\text{ cm s}^{-1}} = 0.295\text{ cm}.$$

Examining and interpreting the results: If we fabricate the square duct with sides of 0.3 cm instead of 0.295 cm, the pressure drop necessary to produce the same average velocity of 200 cm/s would be 490 cmH$_2$O, slightly less than the value specified in the problem statement (500 cmH$_2$O).

Example 5.6.5 Finding the Correct Conduit Size to Provide a Desired Flow Rate.
We wish to design a rectangular conduit with height a and width $2a$ that will allow water to flow at a rate of 18 ml/s when a pressure difference of 400 cmH$_2$O is applied across a 200 cm long conduit.

Solution. Initial considerations: We will assume that the duct is smooth and that flow is turbulent.

System definition and environmental interactions: Again, although our goal is to determine the dimensions of the conduit, the system to be analyzed is the water within the conduit, not the conduit itself.

Apprising the problem to identify governing equations: In this case, we do not know the hydraulic diameter of the conduit and we do not know the fluid velocity, so are unable to explicitly compute the Reynolds number. Our approach will be to write the velocity in terms of the known flow rate and the channel height a, substitute this into the definitions of the friction factor and Reynolds number, rearrange each to solve for the unknown channel height, and set the two expressions equal to each other.

Analysis: First, writing the velocity in terms of the known flow rate Q_V and the unknown hydraulic diameter:

$$\langle v \rangle = \frac{Q_V}{A_c} = \frac{Q_V}{2a^2}.$$

The hydraulic diameter is:

$$D_h = \frac{4A_c}{P_w} = \frac{4(2a^2)}{6a} = \frac{4a}{3}.$$

The Reynolds number is:

$$Re_{D_h} = \frac{\left(\frac{Q_V}{2a^2}\right)\left(\frac{4a}{3}\right)}{v} = \frac{2}{3}\frac{Q_V}{va}.$$

The friction factor equation can be expressed in terms of Q_V and a:

$$\Delta P = 4f\frac{L}{D_h}\left(\frac{1}{2}\rho\langle v\rangle^2\right) = 4f\left(\frac{L}{4a/3}\right)\left[\frac{1}{2}\rho\left(\frac{Q_V}{2a^2}\right)^2\right] = 4f\left[\frac{3\rho L Q_V^2}{32}\right]a^{-5}.$$

Substituting the above expression for Reynolds number and rearranging to solve for a:

$$a = \left\{4f\left[\frac{3\rho L Q_V^2}{32\Delta P}\right]\right\}^{\frac{1}{5}} = \left(\frac{2}{3}\frac{Q_V}{v}\right)Re_{D_h}^{-1}.$$

Grouping the unknowns on the left and the known values on the right:

$$(4f)^{\frac{1}{5}}Re_{D_h} = \left(\frac{2}{3}\frac{Q_V}{v}\right)\left\{\left[\frac{3\rho L Q_V^2}{32\Delta P}\right]\right\}^{-\frac{1}{5}}.$$

Inserting values for all of the known quantities on the right:

$$\left(\frac{2}{3}\frac{Q_V}{v}\right)\left\{\left[\frac{3\rho L Q_V^2}{32\Delta P}\right]\right\}^{-\frac{1}{5}} = \left(\left(\frac{2}{3}\right)\frac{18\,\text{cm}^3\text{s}^{-1}}{0.01\,\text{cm}^2\text{s}^{-1}}\right)$$

$$\times \left\{\frac{3(1\,\text{gcm}^{-3})(200\,\text{cm})(18\,\text{cm}^3\text{s}^{-1})^2}{32(400\,\text{cmH}_2\text{O})(980\,\text{gcm}^{-1}\text{s}^{-2}\text{cmH}_2\text{O})}\right\}^{-\frac{1}{5}} = 2,761.$$

And so the relationship between Re and $4f$ which satisfies the definitions of the Reynolds number and friction factor is:

$$(4f)^{\frac{1}{5}}Re_{D_h} = 2,761.$$

Combining this with the expression in Fig. 5.16 for the friction factor in a smooth tube:

$$\left(0.316 Re_{D_h}^{-0.25}\right)^{\frac{1}{5}} Re_{D_h} = 2,761.$$

From this, we compute a Reynolds number of 5,340. Therefore, flow is turbulent and this value is consistent with the restrictions placed on the analytic expression for finding friction factor in terms of the Reynolds number. The length of side a can be computed from:

$$a = \frac{2}{3}\frac{Q_V}{\nu Re_{D_h}} = \frac{2}{3}\frac{(18\,\text{cm}^3\text{s}^{-1})}{(0.01\,\text{cm}^2\text{s}^{-1})(5,340)} = 0.225\,\text{cm}.$$

Examining and interpreting the results: The rectangular conduit should have sides of length 0.225 by 0.45 cm if water is to flow at 18 ml/s for an applied pressure of 400 cmH$_2$O. This result can be checked by computing the hydraulic diameter (0.3 cm), average velocity (177.8 cm/s), and Reynolds number (5,333), which is close to that found above. The friction factor for a smooth tube is $4f = 0.037$, and the computed pressure difference is 398 cmH$_2$O, which is very close to the imposed pressure difference of 400 cmH$_2$O.

5.7 Friction Loss Factors, Flow Through Fittings

By analogy to (5.53), $\Pi_{\hat{E}_v}$ for fittings, expansion sections, contraction sections of tubing, etc. can be shown to be a function of the Reynolds number and appropriate geometric ratios. A *friction loss factor* K_w is used extensively in the literature to account for friction losses in fittings or sudden changes in cross-sectional area. This is a simple multiple of the $\Pi_{\hat{E}_v}$ defined in (5.53):

$$K_w \equiv \frac{\hat{E}_v}{\frac{1}{2}\langle v \rangle^2} = 2\Pi_{\hat{E}_v}. \tag{5.72}$$

Figure 5.18 provides values of K_w for a limited set of fittings, sudden expansions, etc. It is important to note that the average velocity used in the definition of K_w is the *downstream velocity*, unless otherwise noted. More complete information about friction loss factors can be found in engineering handbooks.

If a tubing section contains N_f fittings (including sections with cross-sectional area changes) and N_c sections of straight conduit in series, then we can sum the contributions caused by friction losses as follows:

$$\hat{E}_v = \sum_{i=1}^{N_f} K_{w,i}\left(\frac{1}{2}\langle v_{\text{out},i}\rangle^2\right) + \sum_{j=1}^{N_c} 4f_j \frac{L_j}{D_{h,j}}\left(\frac{1}{2}\langle v_j\rangle^2\right). \tag{5.73}$$

The term $\langle v_{\text{out},i}\rangle$ refers to the average velocity at the outlet of fitting i or change in cross-sectional area i, and the term $\langle v_j\rangle$ refers to the velocity in section j of a straight conduit. This expression for $\hat{E}_v$ can then be substituted back into the general Engineering Bernoulli equation for a system with a single inlet located at station 1, a single outlet located at section 2, and N_f fittings and N_c sections of straight conduit joined together between the inlet and outlet:

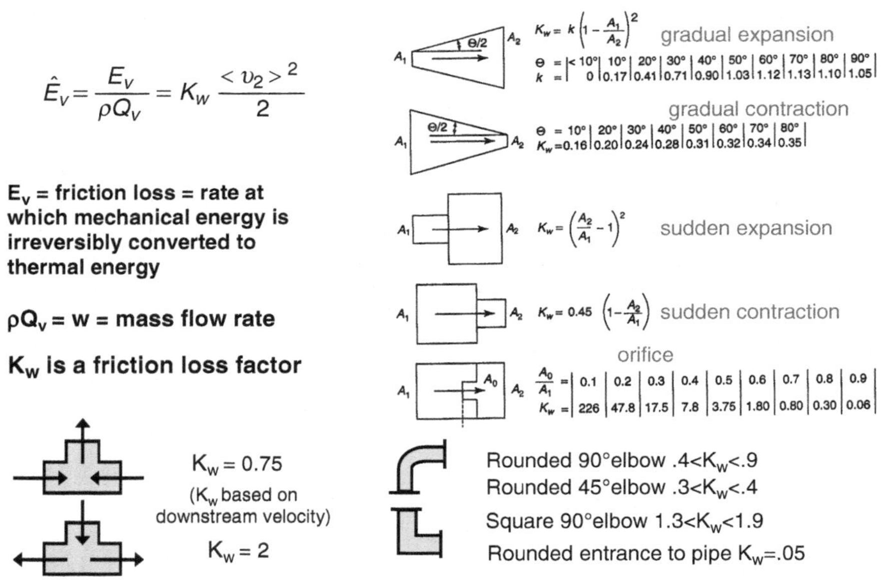

$$\hat{E}_v = \frac{E_v}{\rho Q_v} = K_w \frac{\langle v_2 \rangle^2}{2}$$

E_v = friction loss = rate at which mechanical energy is irreversibly converted to thermal energy

ρQ_v = w = mass flow rate

K_w is a friction loss factor

$K_w = 0.75$
(K_w based on downstream velocity)

$K_w = 2$

$K_w = k\left(1 - \frac{A_1}{A_2}\right)^2$ gradual expansion

θ =	<10°	10°	20°	30°	40°	50°	60°	70°	80°	90°
k =	0	0.17	0.41	0.71	0.90	1.03	1.12	1.13	1.10	1.05

gradual contraction

θ =	10°	20°	30°	40°	50°	60°	70°	80°
K_w =	0.16	0.20	0.24	0.28	0.31	0.32	0.34	0.35

$K_w = \left(\frac{A_2}{A_1} - 1\right)^2$ sudden expansion

$K_w = 0.45\left(1 - \frac{A_2}{A_1}\right)$ sudden contraction

orifice

$\frac{A_0}{A_1}$ =	0.1	0.2	0.3	0.4	0.5	0.6	0.7	0.8	0.9
K_w =	226	47.8	17.5	7.8	3.75	1.80	0.80	0.30	0.06

Rounded 90° elbow $.4 < K_w < .9$
Rounded 45° elbow $.3 < K_w < .4$
Square 90° elbow $1.3 < K_w < 1.9$
Rounded entrance to pipe $K_w = .05$

Fig. 5.18 Friction loss factors for various fittings (modified from Beek et al. 1999 with permission). If $A_2 \gg A_1$ for a sudden expansion, then $K_w = 1$ and $\langle v \rangle$ is based on the upstream velocity. These are strictly valid for turbulent flow, with $Re > 10^5$

$$\frac{\dot{W}_s}{w} + \sum_{i=1}^{N_f} K_{w,i}\left(\frac{1}{2}\langle v_{\text{out},i}\rangle^2\right) + \sum_{j=1}^{N_c} 4f_j \frac{L_j}{D_{h,j}}\left(\frac{1}{2}\langle v_j\rangle^2\right)$$
$$= \frac{1}{2}\left[\frac{\langle v_1^3\rangle}{\langle v_1\rangle} - \frac{\langle v_2^3\rangle}{\langle v_2\rangle}\right] + g[h_1 - h_2] + \frac{1}{\rho}[P_1 - P_2]. \quad (5.74)$$

Applying conservation of mass at the inlet, the outlet, at each fitting, and at each segment of tubing between the inlet and outlet:

$$w = \rho\langle v_1\rangle A_1 = \rho\langle v_2\rangle A_2 = \rho\langle v_{\text{out},i}\rangle A_{\text{out},i} = \rho\langle v_j\rangle A_j. \quad (5.75)$$

The ratio $\langle v_i^3\rangle/\langle v_i\rangle$ at the inlet ($i = 1$) or outlet ($i = 2$) can be written in terms of K_{3i} and K_{1i} defined by (5.23):

$$\frac{\langle v_i^3\rangle}{\langle v_i\rangle} = \frac{K_{3i}\langle v_i\rangle^3}{\langle v_i\rangle} = K_{3i}\langle v_i\rangle^2. \quad (5.76)$$

If the velocity profile is flat over the entire cross section, then $K_{3i} = 1$ and if the velocity profile is parabolic, (5.24) can be used to show $K_{3i} = 2$. In other situations, K_{3i} must be determined by integrating (5.23) (see Example 5.7.1). Substituting (5.75) and (5.76) into (5.74):

5.7 Friction Loss Factors, Flow Through Fittings

$$P_1 - P_2 = \rho \left\{ \frac{\dot{W}_s}{w} + \sum_{i=1}^{N_f} K_{w,i} \left(\frac{1}{2}\langle v_{\text{out},i}\rangle^2\right) + \sum_{j=1}^{N_c} 4f_j \frac{L_j}{D_{h,j}} \left(\frac{1}{2}\langle v_j\rangle^2\right) \right.$$
$$\left. + \frac{1}{2}\left[K_{32}\langle v_2\rangle^2 - K_{31}\langle v_1\rangle^2\right] - g[h_1 - h_2] \right\}. \tag{5.77}$$

This is a form of the Engineering Bernoulli equation that can be used to compute the pressure drop in a system composed of several segments in series with the inlet and outlet at different heights. The system may be passive between the inlet and outlet $(\dot{W}_s = 0)$, it may produce power $(\dot{W}_s > 0)$, or power might be applied to the system $(\dot{W}_s < 0)$.

Consider now a system composed of a short segment of horizontal conduit in which a single fitting or change in cross section occurs. Assuming the wall friction is small, and no work is done on the fluid, (5.77) reduces to the following expression relating outlet velocity to the pressure drop across the segment:

$$P_1 - P_2 = \frac{\rho \langle v_2 \rangle^2}{2} \left\{ K_w + K_{32} - K_{31} \left(\frac{A_2}{A_1}\right)^2 \right\}. \tag{5.78}$$

Let us define γ as the factor in brackets:

$$\gamma = K_w + K_{32} - K_{31} \left(\frac{A_2}{A_1}\right)^2. \tag{5.79}$$

This will be a constant for a particular combination of velocity profiles and geometry.

The relationship between flow and pressure drop resulting from a change in cross section or the presence of a single fitting is:

$$Q_V = A_2 \sqrt{\frac{2}{\rho \gamma}[P_1 - P_2]}. \tag{5.80}$$

Note that the pressure–flow relationship across a fitting or change in area is not linear. If the velocity profiles at the inlet and outlet are nearly flat and the cross-sectional areas are the same, then $\gamma = K_w/A_2^2$, and the pressure drop across the fitting is given by:

$$P_1 - P_2 = \frac{1}{2} K_w \rho \langle v_2 \rangle^2. \tag{5.81}$$

In some cases, such as flow across a sharp orifice, the friction loss factor inherently accounts for variations in the velocity profile, and (5.81) should be used directly. For other situations, (5.79) should be used to compute γ, followed by (5.80) to determine the pressure–flow relation across a single fitting.

Example 5.7.1 Friction Loss Factor for a Sudden Expansion.
Fluid passes through a conduit with a sudden expansion, as shown in Fig. 5.19. Over what region of the expansion does the friction loss factor given in Fig. 5.18 apply? Compare the friction loss factors computed on the basis of pressure differences measured between stations 1 and 2 and stations 1 and 3.

Solution. *Initial considerations:* The flow pattern downstream of a sudden expansion is complex. Except under very low Reynolds number conditions, fluid entering the downstream expansion continues forward under its own momentum rather than making the turn around the sharp bend. This is called flow separation, and the fluid entering the expansion continues as a jet for a short distance (station 2 in the figure). A vortex is formed in the corner region of the expansion, where flow near the wall is slightly negative. The flow field eventually returns to its normal radial profile at some distance downstream (station 3). We will assume that the fluid is incompressible and the flow is turbulent.

System definition and environmental interactions: There are two systems of interest in this problem. The first is the fluid in the conduit between stations 1 and 2, and the second is the fluid in the conduit between stations 1 and 3.

Apprising the problem to identify governing equations: To estimate the friction loss factor, we need to apply conservation of mass, conservation of momentum, and the Engineering Bernoulli equation to each system. For steady flow of an incompressible fluid through a system with a single inlet and outlet, conservation of mass (5.15) and conservation of momentum (5.30) for a horizontal system are:

$$\langle v_{in} \rangle A_{in} = \langle v_{out} \rangle A_{out},$$

$$R_x = \left[K_{2in} \rho_{in} \langle v_{in} \rangle^2 + P_{in} \right] A_{in} - \left[K_{2out} \rho_{out} \langle v_{out} \rangle^2 + P_{out} \right] A_{out}.$$

R_x is the force by the fluid on the wall. Combining these, we have:

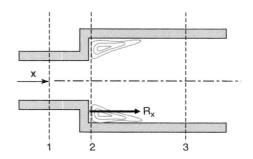

Fig. 5.19 Sudden expansion (modified from Beek et al. 1999 with permission)

5.7 Friction Loss Factors, Flow Through Fittings

$$R_x = \rho A_{out} \langle v_{out} \rangle^2 \left[K_{2in} \left(\frac{A_{out}}{A_{in}} \right) - K_{2out} \right] + P_{in} A_{in} - P_{out} A_{out}.$$

This has two components: friction force in the x-direction by the fluid on the wall and pressure force by the fluid on the wall in the x-direction. Since the expansion segment of the system is short, we will neglect wall friction. Let us apply this expression to the system between sections 1 and 2. If the pressure just downstream of the expansion region can be assumed to be the same as the pressure at the inlet, then the force by the fluid on the wall of the system at the inlet to the expansion section is:

$$R_x = -P_1(A_2 - A_1).$$

The negative sign indicates that this force is in the negative x-direction. Substituting this into the momentum equation with $P_{in} = P_1$ and $\langle v_{in} \rangle = \langle v_1 \rangle$, we find:

$$P_{out} - P_1 = \rho \langle v_{out} \rangle^2 \left[K_{21} \left(\frac{A_{out}}{A_1} \right) - K_{2out} \right].$$

This expression is valid for both outlet section 2 and outlet section 3. We now introduce the Engineering Bernoulli Equation (5.78) with a friction loss factor K_w for a sudden expansion based on the downstream average velocity. The downstream subscript "out" can represent either section 2 or section 3:

$$\langle v_{out} \rangle^2 \left\{ K_w + K_{3out} - K_{31} \left(\frac{A_{out}}{A_1} \right)^2 \right\} = \frac{2}{\rho} [P_1 - P_{out}].$$

Substituting for $P_1 - P_{out}$ from the momentum equation, recognizing that $A_{out} = A_2 = A_3$, and solving for the friction loss factor, K_w:

$$K_w = K_{31} \left(\frac{A_2}{A_1} \right)^2 - K_{3out} - 2K_{21} \left(\frac{A_2}{A_1} \right) + 2K_{2out}.$$

This expression can be used to determine the friction loss factor at any position downstream from the sudden contraction. The only factors that vary with position are K_{2out} and K_{3out}.

Analysis: Consider first the case of turbulent flow through the expansion with the inlet region located at section 1 and the outlet region located at section 3. The velocity profile is relatively flat at both the inlet and the outlet, so $K_{21} = K_{31} = K_{23} = K_{33} = 1$. For this case:

$$K_w = \left(\frac{A_2}{A_1} \right)^2 - 1 - 2 \left(\frac{A_2}{A_1} \right) + 2 = \left(\frac{A_2}{A_1} - 1 \right)^2.$$

This agrees with the expression for K_w for a sudden expansion shown in Fig. 5.18.

Next, let us consider the turbulent flow case where the outlet section (section 2) is just downstream of the expansion. Conditions at the inlet section are the same, but the velocity profile at the outlet is not flat. Let us assume that the radius of the jet in the neighborhood of section 2 is the same as the radius of the inlet section, R_1. Because of the presence of the jet, the velocity from $r=0$ to $r=R_1$ is approximately equal to the inlet velocity, $v_2 = \langle v_1 \rangle = \langle v_2 \rangle (A_2/A_1)$, and for $r > R_1$ the velocity is approximately zero. To find K_{22} and K_{32} for this nonuniform cross section, we must integrate (5.23) over the cross section:

$$K_{22} = \frac{1}{A_2 \langle v_2 \rangle^2} \int_{A_2} v_2^2 \, dA = \frac{1}{A_2 \langle v_2 \rangle^2} \int_{A_1} \left(\frac{A_2}{A_1} \langle v_2 \rangle \right)^2 dA = \frac{A_2}{A_1},$$

$$K_{32} = \frac{1}{A_2 \langle v_2 \rangle^3} \int_{A_2} v_2^3 \, dA = \frac{1}{A_2 \langle v_2 \rangle^3} \int_{A_1} \left(\frac{A_2}{A_1} \langle v_2 \rangle \right)^3 dA = \left(\frac{A_2}{A_1}\right)^2.$$

Substituting these values back into our expression for K_w:

$$K_w = (1)\left(\frac{A_2}{A_1}\right)^2 - \left(\frac{A_2}{A_1}\right)^2 - 2(1)\left(\frac{A_2}{A_1}\right) + 2\left(\frac{A_2}{A_1}\right) = 0.$$

Examining and interpreting the results: Our analysis suggests that the friction loss occurs between sections 2 and 3, rather than right at the sudden expansion between 1 and 2. Therefore, the expression for K_w for a sudden expansion in Fig. 5.18 governs the section between the sudden expansion and the downstream location where the velocity profile has recovered from the effects of the expansion.

Additional comments: If friction losses were ignored, the pressure difference between the inlet and the outlet would be:

$$P_1 - P_2 = \frac{\rho \langle v_2 \rangle^2}{2} \left\{ K_{32} - K_{31}\left(\frac{A_2}{A_1}\right)^2 \right\} = \frac{\rho Q_V^2}{2 A_2^2} \left\{ 1 - \left(\frac{A_2}{A_1}\right)^2 \right\} \quad \text{(no friction loss)}.$$

Since $A_2 > A_1$ for a sudden expansion, the right side of this expression is negative. Therefore, the downstream pressure is greater than the upstream pressure, even though the flow is in the positive direction from station 1 to station 3. The pressure difference when the friction loss term is included is:

$$P_1 - P_2 = \frac{\rho \langle v_2 \rangle^2}{2} \left\{ K_w + K_{33} - K_{31}\left(\frac{A_2}{A_1}\right)^2 \right\} = \frac{\rho Q_V^2}{A_2^2} \left\{ 1 - \left(\frac{A_2}{A_1}\right) \right\}$$

(with friction loss).

5.7 Friction Loss Factors, Flow Through Fittings

This pressure difference is also less than zero. The ratio of the pressure rise with friction relative to the pressure rise without friction can be shown to be less than unity for a sudden expansion:

$$\frac{[P_2 - P_1]_{viscous}}{[P_2 - P_1]_{inviscid}} = \frac{2\left(\left(\frac{A_2}{A_1}\right) - 1\right)}{\left(\left(\frac{A_2}{A_1}\right)^2 - 1\right)} < 1, A_2 > A_1.$$

Therefore, the effect of friction is to reduce the magnitude of the pressure rise.

> **Example 5.7.2 Pressure Difference Across the Mitral Valve.**
> Find the pressure–flow relationship across the mitral valve in the heart. Estimate the pressure at the valve opening and the pressure difference between left atrium and left ventricle during diastolic filling. Assume a diastolic flow of 100 ml/s, a mitral valve area of 2.5 cm^2, and area of the left atrium of 5 cm^2.

Solution. *Initial considerations:* A schematic of flow through the valve is shown in Fig. 5.20. Although this is not generally a steady-state problem, we will assume flow to be relatively constant for the purpose of this analysis. We will also assume the velocity profile in blood to be flat as it flows through the left atrium and the flow through the valve to be flow in a sudden expansion.

System definition and environmental interactions: We will consider two systems in our analysis. The blood in the heart between stations 1 and 0, and the blood between stations 1 and 2. Section 2 in the left ventricle is assumed to have the same cross-sectional area as section 0 in the left atrium. We will begin by estimating the pressure difference between station 1 (left atrium) and station 0 (mitral valve).

Apprising the problem to identify governing equations: The pressure difference across the valve will arise from friction and from changes in the cross-sectional

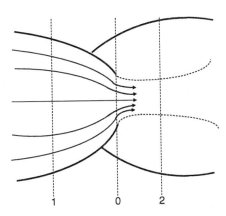

Fig. 5.20 Flow through mitral valve

area. The Engineering Bernoulli equation can be used to estimate the pressure difference. We will take two approaches. In the first, we will analyze the problem as a contraction section followed by an expansion section. In the second approach, we will treat the valve as an orifice.

Analysis: Our starting point is (5.80) with γ given by (5.79). Since we assume that the velocity profiles at stations 1 and 0 are flat, then $K_{31} = K_{30} = 1$. With these assumptions, (5.79) becomes:

$$\gamma = K_w + 1 - \left(\frac{A_0}{A_1}\right)^2.$$

The ratio $A_0/A_1 = 0.5$. Assuming a sudden contraction between atrium and valve, we can compute K_w from the equation given in Fig. 5.18:

$$K_w = 0.45\left(1 - \frac{A_0}{A_1}\right) = 0.45(0.5) = 0.225.$$

Substituting this above and computing γ:

$$\gamma = 0.225 + 1 - (0.5)^2 = 0.975.$$

The pressure difference between left atrium and the plane of the mitral valve during ventricular filling for a diastolic flow of 100 ml/s is:

$$P_1 - P_0 = \frac{\rho\gamma Q_V^2}{2A_0^2} = \frac{(1.05 \text{ gcm}^{-3})(0.975)(100\,\text{cm}^3\,\text{s}^{-1})^2}{2(2.5\,\text{cm}^2)^2 980\,\text{g cm}^{-1}\,\text{s}^{-2}\,\text{cmH}_2\text{O}^{-1}} = 0.836\,\text{cmH}_2\text{O}.$$

Therefore the pressure at the valve plane will be slightly lower than left atrial pressure. The pressure difference from the valve plane to the ventricle can be estimated assuming a sudden expansion, as in the previous example. For this case, $K_w = (2-1)^2 = 1$ and:

$$\gamma = 1 + 1 - (2)^2 = -2.$$

The negative value indicates that the pressure will rise in the expansion region beyond the valve. The pressure difference will be:

$$P_0 - P_2 = \frac{\rho\gamma Q_V^2}{2A_2^2} = \frac{(1.05\,\text{gcm}^{-3})(-2)(100\,\text{cm}^3\,\text{s}^{-1})^2}{2(5\text{cm}^2)^2 980\,\text{g cm}^{-1}\,\text{s}^{-2}\,\text{cmH}_2\text{O}^{-1}} = -0.429\,\text{cmH}_2\text{O}.$$

Adding the effect of the contraction and the expansion, the total pressure difference across the valve, $P_1 - P_0$, is $P_1 - P_0 + P_0 - P_2 = 0.407$ cmH$_2$O.

Another way to estimate the pressure difference across the valve is to assume that it behaves like a sharp orifice. For our case, $A_0/A_1 = 0.5$, and from the orifice

5.7 Friction Loss Factors, Flow Through Fittings

table in Fig. 5.18, $K_w = 3.75$. Since we assume $A_1 = A_2$, we find that $\gamma = K_w = 3.75$, and the pressure difference is estimated to be:

$$P_1 - P_2 = \frac{\rho K_w Q_V^2}{2A_2^2} = \frac{(1.05 \text{ gcm}^{-3})(3.75)(100 \text{ cm}^3 \text{ s}^{-1})^2}{2(5 \text{ cm}^2)^2 980 \text{ g cm}^{-1} \text{ s}^{-2} \text{cmH}_2\text{O}^{-1}} = 0.804 \text{ cmH}_2\text{O}.$$

Examining and interpreting the results: The value computed from the sharp orifice equation is closer to the pressure drop we computed between the left atrium and the valve plane. The reason for this is that the pressure taps used to compute the pressure difference across an orifice are usually placed within one diameter on either side of the orifice. Pressures used to compute K_w values for a sudden expansion are generally measured farther downstream.

Additional comments: Peak flow through the valve is usually about 200 ml/s. Under peak flow conditions, we might expect pressure at the valve plane to be 3.34 cmH$_2$O below left atrial pressure.

Example 5.7.3 Pressure Drop in a Syringe–Needle System During a Rapid Bolus Injection.

A syringe with a diameter of 1 cm is connected to a needle with a diameter of 1 mm and length of 10 cm. What force must be applied to the syringe plunger to deliver a flow of 9 ml/s (rapid bolus injection) through the needle into a large blood vessel? The fluid has a viscosity of 1 cp and density of 1 g/ml. The pressure in the blood vessel is 1 cmH$_2$O.

Solution. *Initial considerations:* The force applied to the syringe plunger must overcome the pressure force in the fluid within the syringe. We will neglect friction between the plunger and syringe wall. Pressure losses in the fluid between the end of the plunger and the syringe outlet will be small relative to pressure losses in the small diameter needle. Therefore, pressure in the fluid within the syringe will be assumed to be constant. The force delivered to the end of the plunger to keep it moving at a constant velocity will equal the pressure in the syringe multiplied by the cross-sectional area of the syringe.

System definition and environmental interactions: The fluid between the syringe outlet and the blood vessel slightly downstream of the needle outlet will constitute the system to be analyzed. There are four major sources of pressure losses in this system: (1) the sudden contraction at the inlet to the needle, (2) the friction loss in the needle, (3) the sudden expansion where the needle is in contact with blood in the blood vessel, and (4) the difference in velocity between fluid in the syringe and fluid in the blood vessel. We will assume that the major component of velocity in the blood vessel during the injection is from fluid passing through the needle. We will neglect any losses along the length of the syringe in comparison with the four sources listed above.

Apprising the problem to identify governing equations: The Engineering Bernoulli equation can be used to analyze pressure losses in this system. Friction loss factors for a sudden contraction and sudden expansion from Fig. 5.18 must be applied, and the fanning friction factor for flow in the needle must be found. The overall pressure difference is found using (5.77).

Analysis: Computing the friction loss factor associated with the sudden contraction between the outlet of the syringe and the needle inlet from Fig. 5.18:

$$K_w = 0.45\left(1 - \frac{A_{needle}}{A_{syringe}}\right) = 0.45\left(1 - \frac{\pi R^2_{needle}}{\pi R^2_{syringe}}\right) = 0.45\left(1 - \frac{\pi(0.05\,\text{cm})^2}{\pi(0.5\,\text{cm})^2}\right) = 0.446.$$

The average velocity in the syringe is:

$$\langle v \rangle_{syringe} = \frac{Q_V}{\pi R^2_{syringe}} = \frac{9\,\text{cm}^3\,\text{s}^{-1}}{\pi(0.5\,\text{cm})^2} = 11.5\,\text{cm/s}.$$

The average velocity in the needle is:

$$\langle v \rangle_{needle} = \frac{Q_V}{\pi R^2_{needle}} = \frac{9\,\text{cm}^3\,\text{s}^{-1}}{\pi(0.05\,\text{cm})^2} = 1,146\,\text{cm/s}.$$

The pressure drop across the sudden contraction is:

$$\Delta P_{contraction} = \frac{K_w}{2}\rho\langle v \rangle^2_{needle} = \frac{0.446}{2}\left(\frac{1\,\text{gcm}^{-3}}{980\,\text{g cm}^{-1}\,\text{s}^{-2}\text{cmH}_2\text{O}^{-1}}\right)(1,146\,\text{cm/s})^2$$
$$= 299\,\text{cmH}_2\text{O}.$$

At the needle outlet, the friction loss, based on the upstream velocity, for a sudden expansion from a small area to a large area, with $K_w = 1$ is:

$$\Delta P_{expansion} = \frac{1}{2}\rho\langle v \rangle^2_{needle} = \frac{1}{2}\left(\frac{1\,\text{gcm}^{-3}}{980\,\text{g cm}^{-1}\,\text{s}^{-2}\,\text{cmH}_2\text{O}^{-1}}\right)(1,146\,\text{cm/s})^2$$
$$= 670\,\text{cmH}_2\text{O}.$$

The pressure drop across the needle can be determined once the Reynolds number is known:

$$Re = \frac{\langle v \rangle D}{\nu} = \frac{(1,146\,\text{cm/s})(0.1\,\text{cm})}{0.01\,\text{cm}^2/\text{s}} = 11,460.$$

Therefore, flow is turbulent. Assuming the needle to be smooth, the friction factor is found from Fig. 5.16:

$4f = 0.316\,(11,460)^{-0.25} = 0.09665,$

and the pressure drop across the needle is:

$$\Delta P_{\text{needle}} = 4f \frac{L}{D_h} \left(\frac{1}{2} \rho \langle v \rangle^2 \right) = (0.0967) \left(\frac{10 \text{ cm}}{0.1 \text{ cm}} \right) (670 \text{ cmH}_2\text{O}) = 6{,}476 \text{ cmH}_2\text{O}.$$

Finally, the pressure difference resulting from the different average velocities at the inlet and outlet of the system must be computed:

$$\Delta P_{\text{area difference}} = \frac{\rho}{2} \left[K_{3\text{vessel}} \langle v_{\text{vessel}} \rangle^2 - K_{3\text{syringe}} \langle v_{\text{syringe}} \rangle^2 \right].$$

Since the area in the vessel is assumed large, the average velocity downstream of the needle outlet will be assumed to be zero. The Reynolds number in the syringe is:

$$Re = \frac{\langle v \rangle D}{\nu} = \frac{(11.5 \text{ cm/s})(1 \text{ cm})}{0.01 \text{ cm}^2/\text{s}} = 115.$$

Therefore, flow in the syringe is laminar and $K_{3\text{syringe}} = 2$. The pressure difference caused by the area difference between inlet and outlet is:

$$\Delta P_{\text{area difference}} = \frac{1}{2} \left(\frac{1 \text{ gcm}^{-3}}{980 \text{ g cm}^{-1} \text{ s}^{-2} \text{ cmH}_2\text{O}^{-1}} \right) \left[0 - 2(11.5 \text{ cm/s})^2 \right]$$

$$= -0.27 \text{ cmH}_2\text{O}.$$

This is negative because the area increases, but it is negligible in comparison with the other terms. The total pressure drop necessary to push the fluid through the syringe at a rate of 9 ml/s would be 7,445 cmH$_2$O plus the pressure in the blood vessel (1 cmH$_2$O). The pressure in the syringe is 7.2 atmospheres, which is very large.

The force needed to push the syringe plunger so that this pressure can be developed is:

$$F = \Delta P A_{\text{syringe}} = (7{,}446 \text{ cm H}_2\text{O})(\pi)(0.5 \text{ cm})^2 (980 \text{ dynes cm}^{-2} \text{ cm H}_2\text{O}^{-1})$$

$$= 5.73 \times 10^6 \text{ dynes} = 12.88 \text{ lb}.$$

Examining and interpreting the results: Although the force is not excessive, the pressure inside the syringe is large. The syringe must be able to withstand this pressure without leaking and without breaking.

Additional comments: The force required to push fluid through the needle represents 87% of the total force, while the force needed to overcome friction at the contraction is 4% of the total and the force required to overcome friction at the outlet is 9% of the total.

5.8 Laminar Flow and Flow Resistance in Noncircular Conduits

The Moody diagram cannot be used to evaluate the friction factor when fluid flow is laminar and the conduit does not have a circular cross section. If flow is fully developed through a horizontal conduit with a cross section characterized by two dimensions d and B, then the flow through the conduit is directly proportional to the pressure drop across the conduit:

$$Q_V = \frac{\Delta P}{\Re_f}, \tag{5.82}$$

where $\Delta P = P_{in} - P_{out}$ and $\Re_f$ is the resistance to flow through the conduit and is a function of tube geometry and fluid viscosity:

$$\Re_f = \frac{12\mu L}{Bd^3 M_0} \quad \text{(laminar)}. \tag{5.83}$$

If the flow is in non-horizontal tubes, then ΔP would be replaced by $\Delta \wp$. Since flow is directly proportional to the pressure drop in laminar flow, one can compute the flow if given the pressure drop or compute the pressure drop if given the flow. This is an easier procedure than in turbulent flow, where the relationship between pressure drop and flow is nonlinear [i.e., (5.65)].

The dimensionless coefficient M_0 is a function of B/d and is shown in Fig. 5.21 for conduits with different geometries. If the tube has a circular cross section (i.e., ellipse with $B = d$), then M_0 is a constant and equals $3\pi/32$, or 0.295, which is consistent with the value for a circle shown in Fig. 5.21. Thus, for laminar flow in a circular tube with $R = d = B$, the resistance to flow is:

$$\Re_f = \frac{8\mu L}{\pi R^4}. \tag{5.84}$$

The viscous dissipation term for laminar flow in a horizontal conduit can be written as:

$$\hat{E}_v = \frac{12\mu L A_c \langle v \rangle}{\rho B d^3 M_0}. \tag{5.85}$$

In dimensionless form:

$$\Pi_{\hat{E}_v} = \frac{\hat{E}_v}{\langle v \rangle^2} = \frac{12\mu L A_c}{\rho \langle v \rangle B d^3 M_0} = \frac{12}{M_0} \frac{A_c}{Bd} \frac{L}{d} \frac{1}{Re_d}. \tag{5.86}$$

It can be shown that for all of the shapes in Fig. 5.21, the dimensionless quantity $A_c/(Bd)$ is simply a function of B/d, as is M_0. Therefore, (5.86) is consistent with our original result from dimensional analysis in (5.53):

$$\Pi_{\hat{E}_v} = \Pi_{\hat{E}_v}\left(Re_d, \frac{L}{d}, \frac{B}{d}\right). \tag{5.87}$$

5.8 Laminar Flow and Flow Resistance in Noncircular Conduits

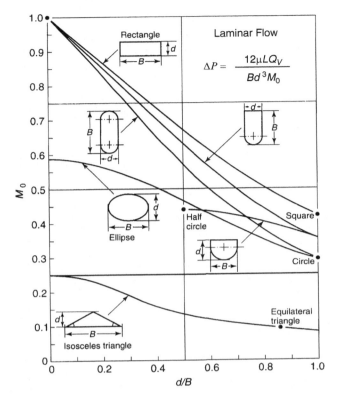

Fig. 5.21 Values of M_0 for laminar flow in several different types of conduits with uniform cross-sectional area (modified from Beek et al. 1999 with permission)

Since different relationships are used to compute pressure drops across conduits for laminar and turbulent flow, it is essential that the Reynolds number be computed in each application to ensure that the correct relationship is used. We use the same criteria as discussed in Sect. 5.6, where the Reynolds number is based on the hydraulic diameter:

$$\text{if } \text{Re}_{D_h} \leq 2200, \text{ flow is laminar, use Fig. 5.21,} \qquad (5.88)$$
$$\text{if } \text{Re}_{D_h} > 2200, \text{ flow is turbulent, use Fig. 5.16.}$$

Thus, if you are given a pressure difference across a conduit and you do not know the velocity, then you cannot compute the Reynolds number to determine which relationship to use. You must choose to use either the laminar flow relationship [(5.82) and (5.83)] based on a value of M_0, or the turbulent relationship (5.65) based on a value of the friction factor, f. Once the velocity is computed, it is essential that the Reynolds number be computed to verify that your original assumption of laminar or turbulent flow was correct. If not, then you must recompute the velocity based on the correct pressure–flow relationship.

Example 5.8.1 Flow Through a Vein with an Elliptical Cross Section.
Find blood flow through an elliptical tube with major axis $B = 1$ cm and minor axis $d = 0.5$ cm. Compare this with the flow through a round tube with the same cross-sectional area. The pressure difference across the tube is 1 cmH$_2$O, blood density is 1.04 g/cm^3, blood viscosity is 4 cp, and the vessel length is 20 cm.

Solution. *Initial considerations:* Since we do not know the average velocity in the vein, we cannot compute the Reynolds number. Therefore, we must make an assumption about the nature of the flow through the vein. If we assume flow to be turbulent, we can use the Moody diagram as described in Sect. 5.6. After finding the velocity, we must check to make sure the Reynolds number is greater than 2,200. An alternate approach is to assume that the flow is laminar and use the methods described in this section. Again, when the velocity is found, we must check to make sure that the flow is laminar. If we find the Reynolds number to be greater than 2,200, our result would be inconsistent with our original assumption and we would need to use the methods found in Sect. 5.6. We will adopt this approach.

System definition and environmental interactions: The system to be analyzed in this problem is the fluid within the elliptically shaped vein.

Apprising the problem to identify governing equations: Assuming the flow through the vessel is laminar, (5.82) and (5.83) are the appropriate relationships for estimating flow. We must first use Fig. 5.21 to estimate the factor M_0.

Analysis: The line for the ellipse in Fig. 5.21 intersects $d/B = (0.5\ \text{cm})/(1\ \text{cm}) = 0.5$ at $M_0 = 0.47$. Flow through the elliptically shaped vessel will be:

$$Q_V = \frac{\Delta P}{12 \mu L} B d^3 M_0 = \frac{(1\,\text{cmH}_2\text{O})\,(980\,\text{g cm}^{-1}\,\text{s}^{-2}\,\text{cmH}_2\text{O}^{-1})}{12(0.04\,\text{g cm}^{-1}\,\text{s}^{-1})\,(20\,\text{cm})}(1\,\text{cm})\,(0.5\,\text{cm})^3 (0.47)$$
$$= 6\,\text{ml/s}.$$

Examining and interpreting the results: Before we can accept this flow estimate, we must confirm our initial assumption that the flow is laminar. Therefore, we need to compute the Reynolds number based on the hydraulic diameter of the vessel which is four times the cross-sectional area divided by the wetted perimeter. The cross-sectional area of our ellipse is:

$$A_c = \pi \left(\frac{B}{2}\right)\left(\frac{d}{2}\right) = \pi (0.5\,\text{cm})\,(0.25\,\text{cm}) = 0.393\,\text{cm}^2.$$

The exact solution for the perimeter of an ellipse can be found in terms of an infinite series. An excellent approximation was developed by the mathematician Srinivasa Ramanujan, 1913–1914:

5.8 Laminar Flow and Flow Resistance in Noncircular Conduits

$$P_w = \frac{\pi}{2}\left[3(B+d) - \sqrt{(3d+B)(3B+d)}\right] = \frac{\pi}{2}\left[4.5\,\text{cm} - \sqrt{(2.5\,\text{cm})(3.5\,\text{cm})}\right]$$
$$= 2.42\,\text{cm}.$$

Note that this formula is exact in the case of a circular tube. The hydraulic diameter of our ellipse is:

$$D_h = \frac{4A_c}{P_w} = \frac{4(0.393\,\text{cm}^2)}{2.42\,\text{cm}} = 0.649\,\text{cm}.$$

The average velocity in the elliptical tube is:

$$\langle v \rangle = \frac{Q_V}{A} = \frac{6\,\text{ml/s}}{0.393\,\text{cm}^2} = 15.28\,\text{cm/s}.$$

The Reynolds number is:

$$Re_{D_h} = \frac{\rho \langle v \rangle D_h}{\mu} = \frac{(1.04\,\text{gcm}^{-3})(15.28\,\text{cm/s})(0.649\,\text{cm})}{0.04\,\text{g cm}^{-1}\,\text{s}^{-1}} = 258.$$

Since this is less than 2,200, flow will be laminar and the procedure used above is justified.

A circular tube with the same cross-sectional area will have a diameter D that must conform to the following equation:

$$\pi \left(\frac{D}{2}\right)^2 = \pi \left(\frac{B}{2}\right)\left(\frac{d}{2}\right).$$

The diameter of the circular tube with the same cross-sectional area as the ellipse will be:

$$D = \sqrt{Bd} = \sqrt{(1\,\text{cm})(0.5\,\text{cm})} = 0.707\,\text{cm}.$$

M_0 for a circular tube is 0.295. Therefore, blood flow through a tube with a diameter of 0.707 cm will be:

$$Q_V = \frac{\Delta P}{12\mu L}D^4 M_0 = \frac{(1\,\text{cmH}_2\text{O})(980\,\text{g cm}^{-1}\,\text{s}^{-2}\,\text{cmH}_2\text{O}^{-1})}{12(0.04\,\text{g cm}^{-1}\,\text{s}^{-1})(20\,\text{cm})}(0.707\,\text{cm})^4(0.295)$$
$$= 7.52\,\text{ml/s}.$$

Comparing this with the flow through an elliptical tube with the same cross-sectional area under the same pressure gradient, we find flow to be increased by 27.5%. A circular tube has the smallest surface area to volume ratio of any duct. Consequently, it will have the smallest resistance and the largest flow when compared with other vessels with the same cross-sectional area.

5.9 Flow in Packed Beds

There are many applications in bioengineering where certain chemical species are removed from the suspending fluid by passing the solution through a reactive packed bead bed. We will study the mass transfer that takes place in packed columns in a later chapter. In this section, we are interested in the pressure–flow relationship for the bead bed.

We begin by assuming that the column is uniformly packed; hence, there are no preferential flow channels in the column. Fluid particles that pass through the bed follow a tortuous path through the interstices between each bead. We can think of this as flow through an irregularly shaped tube. A friction factor can be defined in a manner exactly the same as f in straight conduits:

$$\Delta P = 4f \frac{L}{D_h} \left(\frac{1}{2} \rho \langle v \rangle^2 \right). \tag{5.89}$$

The hydraulic diameter D_h is considered to be the average diameter of an equivalent tube formed by the interstices of the beads as the fluid flows in the packed bed. At any cross section in the bed, the cross-sectional area available to flow is A^* and the wetted perimeter P_w is the perimeter of all of the beads at that cross section. Since the bead bed is uniform, $A^* = V^*/L$, where V^* is the *void volume* of the bed (i.e., the fluid volume contained between the beads). In addition, because the bed is uniform, the wetted perimeter will equal the surface area of the beads per unit length of column, $P_w = S/L$. Consequently, the hydraulic diameter is:

$$D_h = \frac{4A^*}{P_w} = \frac{4V^*/L}{S/L} = \frac{4V^*}{S} = 4\left(\frac{V}{S}\right)\left(\frac{V^*}{V}\right) = 4\varepsilon \frac{V}{S}. \tag{5.90}$$

The *void fraction* ε is defined as the ratio of fluid volume in the bed V^* to total bed volume V. The ratio V/S can be rewritten as:

$$\frac{V}{S} = \left(\frac{V - V^*}{S}\right)\left(\frac{V}{V - V^*}\right) = \left(\frac{V - V^*}{S}\right)\left(\frac{1}{1 - \varepsilon}\right). \tag{5.91}$$

The quantity $V - V^*$ is the bead volume; so $(V - V^*)/S$ is the volume to surface ratio of the beads. If the beads are spherical particles with diameter D_p:

$$\left(\frac{V - V^*}{S}\right) = \frac{\left(4\pi (D_p/2)^2\right)}{\left(\frac{4}{3}\pi (D_p/2)^3\right)} = \frac{D_p}{6}. \tag{5.92}$$

If the particles are not spherical, then (5.92) can be used to define an equivalent particle diameter D_p from the volume to surface ratio. Substituting (5.92) and (5.91)

5.9 Flow in Packed Beds

into (5.90) provides the following relationship that allows us to compute the hydraulic diameter from the particle diameter and the void fraction of the bed:

$$D_h = \frac{2}{3}D_p\left(\frac{\varepsilon}{1-\varepsilon}\right). \tag{5.93}$$

If the flow rate through the packed bed is Q_v, then the average velocity in the fluid between beads is Q_v/A^*. Let the cross-sectional area of the device without packing material be A_c. We can define a *superficial velocity* $v_0 = Q_v/A_c$. The average velocity in the interstices can then be written in terms of the superficial velocity:

$$\langle v \rangle = \frac{Q_v}{A^*} = \frac{Q_v}{A}\frac{A}{A^*} = \frac{Q_v}{A}\frac{V}{V^*} = v_0/\varepsilon. \tag{5.94}$$

Substituting (5.93) and (5.94) into the definition of the friction factor (5.89):

$$f = \frac{1}{3}\frac{D_p}{L}\left(\frac{\varepsilon^3}{1-\varepsilon}\right)\left(\frac{\Delta P}{\rho v_0^2}\right). \tag{5.95}$$

The friction factor will depend on the Reynolds number based on the hydraulic diameter. Again, using (5.93) and (5.94) in the definition of the Reynolds number:

$$Re_{D_h} = \frac{\langle v \rangle D_h}{\nu} = \frac{2}{3}\left(\frac{1}{1-\varepsilon}\right)\left(\frac{v_0 D_p}{\nu}\right). \tag{5.96}$$

In laminar flow situations, we would expect the friction factor to vary inversely with Reynolds number, much like it does in a straight tube with constant cross section. Experiments have shown that for laminar flow:

$$f = \frac{33.3}{Re_{D_h}} = \frac{50}{\left(\frac{1}{1-\varepsilon}\right)\left(\frac{D_p v_0}{\nu}\right)}, \quad \left(\frac{1}{1-\varepsilon}\right)\left(\frac{D_p v_0}{\nu}\right) < 10. \tag{5.97}$$

Substituting this back into (5.89), we find that for laminar flow in a packed bed:

$$\Delta P = 66.7\left(\frac{\mu \langle v \rangle L}{D_h^2}\right) = 150\left(\frac{(1-\varepsilon)^2}{\varepsilon^3}\right)\left(\frac{\mu L v_0}{D_p^2}\right) \quad \text{(laminar)}. \tag{5.98}$$

This is known as the *Blake–Kozeny* equation. This can be compared with the pressure drop for laminar flow in a straight circular tube with length l and diameter d. From (5.69), $f = 16/Re_d$ for a straight tube and the pressure drop across the tube would be:

$$\Delta P = 32\left(\frac{\mu \langle v \rangle l}{d^2}\right). \tag{5.99}$$

Therefore, the actual pressure drop across a packed bead bed would be about 2.08 times greater than the pressure drop across a tube with $d = D_h$ and $l = L$. This is not surprising, since the actual effective diameter of the interstices is sometimes less than the average hydraulic diameter and the tortuous path taken by a fluid particle through a bead bed is certainly longer than the length of the bed.

For fully turbulent flow, the friction factor is found experimentally to remain constant:

$$f = 0.583, \quad \left(\frac{1}{1-\varepsilon}\right)\left(\frac{D_p v_0}{\nu}\right) > 1,000. \tag{5.100}$$

The pressure drop across the bed from (5.89) will be:

$$\Delta P = 1.167 \frac{L}{D_h}(\rho <v>^2) = 1.75\left(\frac{1-\varepsilon}{\varepsilon^3}\right)\left(\frac{L}{D_p}\right)(\rho v_0^2) \quad \text{(turbulent)}. \tag{5.101}$$

This is known as the *Burke–Plummer* equation. In the intermediate region of Reynolds numbers, Ergun (1952) showed that the actual pressure drop can be accurately estimated by adding the pressure drops computed for the laminar and turbulent regimes:

$$\Delta P = \left(\frac{1-\varepsilon}{\varepsilon^3}\right)\left(\frac{L}{D_p}\right)(\rho v_0^2)\left[\frac{150}{\left(\frac{1}{1-\varepsilon}\right)\left(\frac{v_0 D_p}{\nu}\right)} + 1.75\right], \tag{5.102}$$

$$10 \leq \left(\frac{1}{1-\varepsilon}\right)\left(\frac{v_0 D_p}{\nu}\right) \leq 1,000.$$

This is known as the *Ergun* equation. Comparing this with (5.95), we obtain an expression involving the friction factor in this region:

$$3f = \frac{150}{\left(\frac{1}{1-\varepsilon}\right)\left(\frac{v_0 D_p}{\nu}\right)} + 1.75, \quad 10 \leq \left(\frac{1}{1-\varepsilon}\right)\left(\frac{v_0 D_p}{\nu}\right) \leq 1,000. \tag{5.103}$$

The three relationships are plotted in Fig. 5.22 for comparison.

5.10 External Flow: Drag and Lift

Drag forces caused by fluid movement past stationary bodies or bodies moving through stationary fluid have two components: frictional drag caused by the shear stress exerted by the fluid on the wall of the object, and form drag, which is the force on the object caused by a pressure difference that might exist from the upstream to

5.10 External Flow: Drag and Lift

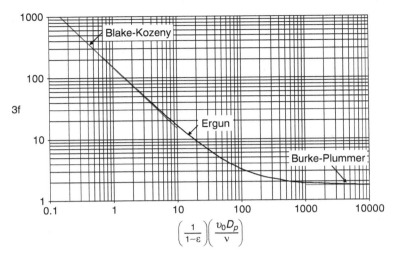

Fig. 5.22 Log–log plot of $3f$ vs. $3Re_{D_h}/2$ for flow through a packed bed

the downstream sides of the object. There can also be a net lift force exerted on the object, which is caused primarily by the net difference in pressure forces exerted on the underside of a body relative to the forces exerted on the top surface of the body. Birds, bees, and airplanes make use of lift forces that are greater than the weight of the body to enable them to fly.

When we introduced the friction factor in Sect. 5.6, we indicated that the kinetic energy per unit volume and the area were defined differently for external flow vs. internal flow. The definition remains the same:

$$F_k = fKA. \qquad (5.104)$$

But for external flows we define $K = \frac{1}{2}\rho v_\infty^2$, where v_∞ is the velocity far from the wall of the object (or the velocity of the object if moving in stationary fluid), and A is the frontal area of the body as seen when looking directly downstream at the body. Unlike the case for internal flow, the friction factor for external flow includes both friction and form drag. For flow past a sphere, the friction factor is a function of the sphere Reynolds number based on the sphere diameter D_s, $Re_{D_s} = v_\infty D_s/\nu$. Experimental data relating the friction factor to the Reynolds number is shown in Fig. 5.23. The force on the sphere can be computed from:

$$F_k = f\left(\frac{1}{2}\rho v_\infty^2\right)\left(\frac{\pi D_s^2}{4}\right). \qquad (5.105)$$

The friction factor is relatively constant in the range $10^3 \leq Re_{D_s} \leq 10^5$. However, f decreases for Reynolds numbers between 10^5 and 10^6, where a turbulent

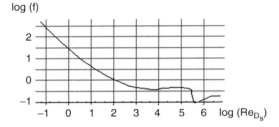

Fig. 5.23 Friction factor vs. Reynolds number for flow past a sphere

boundary layer forms in front of the sphere and a narrow wake forms behind the sphere. For low Reynolds numbers ($Re_{D_s} < 0.5$), the friction factor is inversely related to the Reynolds number, similar to laminar flow in a round tube:

$$f = 24/Re_{D_s}. \quad (5.106)$$

The friction factor is also referred to as the drag coefficient, C_D. The friction and drag force (5.54) exerted on a sphere at low Reynolds numbers is:

$$F_k = \left(\frac{24}{Re_{D_s}}\right)\left(\frac{1}{2}\rho v_\infty^2\right)\left(\frac{\pi D_s^2}{4}\right). \quad (5.107)$$

Expanding the Reynolds number and simplifying:

$$F_k = 3\pi\mu D_s v_\infty = 6\pi\mu R_s v_\infty \quad \text{(laminar)}. \quad (5.108)$$

This relationship between the drag force on a stationary sphere and the fluid velocity, fluid viscosity and sphere radius R_s is known as *Stokes' Law*. If the sphere moves in the same direction with velocity v_s as an unbounded fluid moving at v_∞, then Stokes' Law can be written as:

$$F_k = -6\pi\mu R_s (v_s - v_\infty). \quad (5.109)$$

Assuming $v_\infty > v_s$, this represents the force by the fluid on the moving particle in the direction of flow. In cases of biological interest, particles are often swept along by fluid in a confined environment, where the particles are nearly as large as the vessels they travel through. Examples include red cells moving through a capillary or large proteins moving through the pores in capillary walls. If the particles can be approximated by spheres and the vessel by a round tube, then the drag force is:

$$F_k = -6\pi\mu R_s (K_1 v_s - K_2 v). \quad (5.110)$$

The factors K_1 and K_2 are drag coefficients that depend on the radius of the particle relative to the radius of the tube and on the position of the particle center relative to the axis of the tube (Paine and Scherr 1975).

5.10 External Flow: Drag and Lift

Relationships for friction factors relating the kinetic force on other objects, such as cylinders and disks, can be found in the literature and in fluid mechanics handbooks.

Example 5.10.1 Terminal Velocity of a Cell.

Fluid and a few lymphocytes from a lymph duct are placed in a test tube. The fluid has a density of 1.03 g/dl and kinematic viscosity of 0.01 cm²/s. Lymphocytes have a density of 1.075 g/ml and volume of 250 µm³. How long will it take a cell located 1 cm from the bottom of the test tube to reach the bottom under the influence of gravity? Assume the cell has a spherical shape.

Solution. *Initial considerations:* We will neglect the initial acceleration of the cell and assume that the net force on the cell is zero.

System definition and environmental interactions: The system of interest is a lymphocyte initially located 1 cm from the bottom of the test tube. There are three forces acting on the cell: the weight of the cell (downward), the buoyant force by the fluid on the cell (upward), and the drag force by the fluid on the cell (upward).

Apprising the problem to identify governing equations: Assuming laminar flow, the drag force is given by (5.107) with $Re_{D_s} = v_\infty D_s / \nu$.

Analysis: Since there is no net force on the cell:

$$\sum F = 0 = W - F_B - F_k = \rho_c V_c g - \rho_f V_c g - \left(\frac{24\nu}{v_\infty D_s}\right)\left(\frac{1}{2}\rho_f v_\infty^2\right)\left(\frac{\pi D_s^2}{4}\right).$$

The diameter of the cell is $2(3V_c/4\pi)^{1/3} = 7.82$ µm. Solving for the cell velocity:

$$v_\infty = \frac{1}{3\pi} \frac{(\rho_c - \rho_f) V_c g}{\rho_f \nu D_s} = \frac{1}{3\pi} \frac{(0.045 \text{ g/ml})(250 \text{ µm}^3)(980 \text{ cm/s}^2)}{(1.03 \text{ g/ml})(0.01 \text{ cm}^2/\text{s})(7.82 \text{ µm})} (10^{-4} \text{ cm/µm})^2$$

$$= 1.45 \times 10^{-4} \text{ cm/s}.$$

The time for the cell to travel 1 cm will be 6.9×10^3 s, or almost 2 h.

Examining and interpreting the results: Needless to say, this is not a very effective way to separate cells from the suspending fluid. Before we accept this result, we must confirm that the flow is laminar. The Reynolds number is $(1.45 \times 10^{-4} \text{ cm/s})(7.82 \times 10^{-4} \text{ cm})/(0.01 \text{ cm}^2/\text{s}) = 1.13 \times 10^{-5}$, which is consistent with laminar flow.

Example 5.10.2 Cell Velocity in a Centrifuge.

A better way to separate cells from the suspending fluid would be to put the test tube into a centrifuge with a radius $R_{cent} = 10$ cm and spin it at 500 rpm. Estimate the time required for the same cell in Example 5.10.1 to travel 1 cm in the centrifuge. Assume that the test tube remains horizontal while it is spun and neglect startup time of the centrifuge.

Solution. *Initial considerations:* Once the centrifuge reaches its final angular velocity, the direction of travel of the lymphocyte will be horizontal. The weight and buoyancy forces do not influence cell velocity toward the end of the tube.

System definition and environmental interactions: The system of interest is a lymphocyte initially located 1 cm from the bottom of the test tube. There will be two forces on the cell: the centrifugal force tending to move it away from the center of rotation and the drag force opposing the motion. After a short period of acceleration, these two forces will balance and the cell will reach a constant velocity.

Apprising the problem to identify governing equations: Assuming laminar flow, the drag force is given by (5.107) with $Re_{D_s} = v_\infty D_s/\nu$. The centrifugal force on the cell is:

$$F_{cent} = \rho_c V_c \frac{(\Omega R_{cent})^2}{R_{cent}}$$

where Ω is the angular velocity of the centrifuge and R_{cent} is the radius of the centrifuge. We will assume that the distance from the center of rotation of the centrifuge to the lymphocyte remains constant and is equal to R_{cent}.

Analysis: The angular velocity Ω (radians/s) of the centrifuge is related to the number of revolutions per minute as follows:

$$\Omega = \left(500\frac{\text{rev}}{\text{min}}\right)\left(2\pi\frac{\text{rad}}{\text{rev}}\right)\left(\frac{1\,\text{min}}{60\,\text{s}}\right) = 52.4\frac{\text{rad}}{\text{s}}.$$

Assuming the drag force balances the centrifugal force:

$$\sum F = 0 = F_{cent} - F_k = \rho_c V_c \frac{(\Omega R_{cent})^2}{R_{cent}} - \left(\frac{24\nu}{v_\infty D_s}\right)\left(\frac{1}{2}\rho v_\infty^2\right)\left(\frac{\pi D_s^2}{4}\right).$$

Solving for cell velocity:

$$v_\infty = \frac{1}{3\pi}\left(\frac{\rho_c V_c \Omega^2 R_{cent}}{\rho_f \nu D_s}\right) = \left(\frac{1}{3\pi}\right)\frac{(1.075\,\text{g/ml})\,(250\,\mu\text{m}^3)\,(52.4\,\text{s}^{-1})^2(10\,\text{cm})}{(1.03\,\text{g/ml})\,(0.01\,\text{cm}^2/\text{s})\left(7.82\,\mu\text{m}\left(10^8\frac{\mu\text{m}^2}{\text{cm}^2}\right)\right)}$$

$$= 0.097\,\text{cm/s}.$$

Therefore, once the centrifuge reaches 500 rpm, the cell will travel 1 cm in 10.3 s.

Examining and interpreting the results: Rotating the cells in a centrifuge is much more efficient than gravitational separation, since it causes them to move considerably faster than if falling under their own weight. However, care must be taken in selecting the angular velocity and radius of a centrifuge, since spinning cells at very high rates can result in large drag forces, producing cell damage and lysis.

5.11 Blood Flow in Microvessels

Blood flow in small tubes is difficult to predict from a straightforward application of theory. Blood can only be treated using a continuum approach when the ratio of red cell radius to tube radius is small. The rheology of blood was discussed in Sect. 4.9. The key results are that blood behaves as a non-Newtonian pseudoplastic fluid for shear rates below 36 s^{-1}, that blood exhibits a yield stress, and that the apparent viscosity of blood depends on hematocrit value. In addition, the apparent viscosity decreases with vessel size (Fahraeus Lindqvist Effect), and this is particularly applicable in the microcirculation. An empirical approach is often used to predict the flow of blood in small tubes. Since the Reynolds number is usually very small, the flow Q_V is laminar and, according to the analysis in Sect. 5.8, is proportional to the pressure drop across the tube, ΔP:

$$Q_V = \frac{\Delta P}{\Re_f}. \tag{5.111}$$

The factor $\Re_f$ is the resistance to flow, which depends on fluid properties and tube geometry. In Sect. 6.3, we use a microscopic balance to derive an expression for the fluid resistance when a Newtonian fluid is forced through a cylindrical tube. The same expression for $\Re_f$ can be used to characterize the resistance of a non-Newtonian fluid in a cylindrical tube with radius R and length L if we replace the fluid viscosity μ in (6.53) with the apparent viscosity η:

$$\Re_f = \frac{8\eta L}{\pi R^4}. \tag{5.112}$$

The flow rate obtained when (5.111) is substituted into (5.112) is known as the Hagen-Poiseuille equation.

> **Example 5.11.1 Blood Flow in a Small Hollow Fiber.**
> Blood having a hematocrit value of 40% flows through an extracorporeal device consisting of many hollow fibers arranged in parallel. Each fiber has an inside diameter of 75 μm and length of 10 cm. Plasma viscosity is 1.05 cp. A pressure difference of 50 mmHg is applied across the fibers. What will be the blood flow through a single fiber? Compare this to the flow computed if the Fahraeus-Lindqvist effect is neglected.

Solution. *Initial considerations:* Our objective is to compute blood flow in a single fiber under the conditions stated. Since the vessel size is small, we must realize that the apparent viscosity of blood will be lower than the apparent viscosity of blood flowing through a large blood vessel (Fahraeus-Lindqvist effect). We will assume flow to be laminar, but will check this assumption after we compute the velocity in the hollow fiber.

System definition and environmental interactions: The system of interest is the blood contained within the hollow fiber. The pressure force is balanced by the viscous force at the fiber walls.

Apprising the problem to identify governing relationships: Since we must account for blood rheology in this problem, we must first find the apparent relative viscosity of blood in the 75 μm tube, multiply by the plasma viscosity to obtain the apparent viscosity of blood, and find the blood flow through the fiber from the Hagen–Poiseuille equation. We must then compute the Reynolds number to confirm that flow is laminar. If not, use of the Hagen–Poiseuille equation cannot be justified and methods in Sect. 5.6 must be used.

Analysis: First, we find the apparent viscosity of blood in the 75 μm tube. Using Fig. 4.32, we must interpolate η_r for a hematocrit H_D of 0.4 and $D = 75$ μm. The relative hematocrit is approximately 2.1. A more accurate approach would be to use (4.43)–(4.45) to compute $\eta_r = 2.05$. Multiplying this by the viscosity of plasma (given as 1.05 cp) yields an apparent viscosity of blood of 2.15 cp in the 75 μm tube.

Next, we use (5.112) to compute resistance to blood flow in a single fiber. If resistance is to be expressed as mmHg/(ml/min), then:

$$\Re_f = \frac{8\eta L}{\pi R^4} = \left[\frac{8}{\pi}\right]\left[\frac{(2.15\,\text{cp})(10\,\text{cm})}{(37.5\times 10^{-4}\,\text{cm})^4}\right]\left[\frac{1\,\text{mmHgs}}{1.333\times 10^5\,\text{cp}}\right]\left[\frac{1\,\text{min}}{60\,\text{s}}\right] = 3.46\times 10^4 \frac{\text{mmHg}}{\text{ml/min}}.$$

The flow in a single hollow fiber can be computed using (5.111):

$$Q_V = \frac{\Delta P}{\Re_f} = \frac{50\,\text{mmHg}}{3.46\times 10^4\,\text{mmHg}\,\text{ml}^{-1}\,\text{min}} = 1.44\times 10^{-3}\frac{\text{ml}}{\text{min}}.$$

If the Fahraeus-Lindqvist effect is not included, the relative viscosity of blood in large vessels at $H = 40\%$ can be found from Fig. 4.24. The relative viscosity is about 2.85 and the apparent viscosity is about 3 cp. The flow per fiber would be less than computed above if blood viscosity in small tubes was equal to blood viscosity in large tubes. The resistance would be increased by a factor equal to the ratio of the apparent viscosities, which is (3 cp)/(2.15 cp). Consequently the flow per fiber will be decreased by a factor of 2.15/3, or $Q_v = 1.032\times 10^{-3}$ ml/min.

Examining and interpreting the results: Flow computed on the basis of apparent viscosity in large tubes is only 46% of the actual blood flow through a fiber. Failure to account for the Fahraeus–Lindqvist effect can lead to significant under-predictions of

5.12 Steady Flow Through a Network of Rigid Conduits

flow in small vessels. Before we can accept these results, we must confirm that flow is laminar. The Reynolds number is:

$$Re_d = \frac{4\rho Q_V}{\pi d \eta} = \frac{4\left(1.05 \frac{g}{ml}\right)\left(1.44 \times 10^{-3} \frac{ml}{min}\right)\left(\frac{1 \, min}{60 \, s}\right)}{\pi (75 \times 10^{-4} \, cm)(2.15 \, cp)\left(0.01 \frac{g}{cm \, s \, cp}\right)} = 0.20.$$

Therefore, the flow is laminar and analysis based on the Hagen–Poiseuille equation is justified.

5.12 Steady Flow Through a Network of Rigid Conduits

Fluid flow through a network of vessels can be computed using (5.111), where the resistance represents the overall resistance of the network. If entry lengths are short, relative to vessel lengths, and energy losses are negligible at bifurcations, then the overall resistance can be computed using an electrical circuit analog. The resistance of two vessels in series (Fig. 5.24) can be computed from the resistance of each vessel by applying (5.111) twice:

$$P_{in} - P_1 = Q_V \Re_1, \qquad (5.113)$$

$$P_1 - P_{out} = Q_V \Re_2. \qquad (5.114)$$

Adding these two expressions:

$$P_{in} - P_{out} = Q_V(\Re_1 + \Re_2). \qquad (5.115)$$

Comparing (5.111) and (5.115), the total resistance for two resistors in series is:

$$\Re_f = \Re_1 + \Re_2. \qquad (5.116)$$

If more than two resistors are in series, then an additional equation similar to (5.113) or (5.114) can be written for each additional resistor. The total resistance for n resistors in series is:

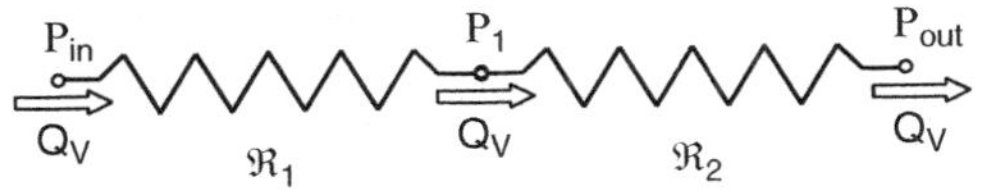

Fig. 5.24 Series resistance to flow

Fig. 5.25 Resistors in parallel

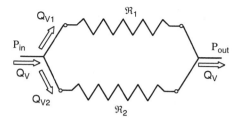

$$\Re_f = \sum_{i=1}^{n} \Re_i \quad \text{(series)}. \quad (5.117)$$

For two resistors in parallel (Fig. 5.25):

$$Q_{V1} = \frac{P_{in} - P_{out}}{\Re_1}, \quad (5.118)$$

$$Q_{V2} = \frac{P_{in} - P_{out}}{\Re_2}, \quad (5.119)$$

$$Q_V = Q_{V1} + Q_{V2} = (P_{in} - P_{out})\left(\frac{1}{\Re_1} + \frac{1}{\Re_2}\right). \quad (5.120)$$

Comparing (5.120) with (5.111), the total resistance is related to the resistance of the parallel components as follows:

$$\frac{1}{\Re_f} = \frac{1}{\Re_1} + \frac{1}{\Re_2}. \quad (5.121)$$

If n resistors are placed in parallel, then the total resistance will be:

$$\frac{1}{\Re_f} = \sum_{i=1}^{n} \frac{1}{\Re_i} \quad \text{(parallel)}. \quad (5.122)$$

Equations (5.117) and (5.122) can be used to reduce a complex network of resistors into a single equivalent resistor. Examples that illustrate this are as follows.

Example 5.12.1 Blood Flow in a Hollow Fiber Device.
The hollow fiber device described in Example 5.11.1 is used to oxygenate blood with a hematocrit value of 40%. How many fibers are needed to allow a flow rate of 100 ml/min through the device when the pressure difference

5.12 Steady Flow Through a Network of Rigid Conduits

across the fibers is 50 mmHg? Compare this with the number of fibers computed when the Fahraeus effect is neglected.

Solution. *Initial considerations:* In Example 5.11.1, we found the resistance of each fiber to be $\Re_F = 3.46 \times 10^4$ mmHg/(ml/min) when the Fahraeus effect is included and $\Re_F = 5.06 \times 10^4$ mmHg/(ml/min) if the Fahraeus effect is neglected.

System definition and environmental interactions: Our system consists of the blood in all of the fibers in the oxygenator.

Apprising the problem to identify governing equations: Since the fibers are in parallel and each fiber has the same resistance, (5.122) can be used to determine the resistance of the device from the resistance of a single fiber.

Analysis: For n fibers in parallel, (5.122) becomes:

$$\frac{1}{\Re_f} = \sum_{i=1}^{n} \frac{1}{\Re_F} = \frac{n}{\Re_F},$$

and (5.111) becomes:

$$Q_V = \frac{n\Delta P}{\Re_F}.$$

Solving for the number of fibers, n when the Fahraeus–Lindqvist effect is included:

$$n = \frac{\Re_F Q_V}{\Delta P} = \frac{\left[3.46 \times 10^4 \, \frac{\text{mmHg}}{\text{ml/min}}\right]\left[100 \, \frac{\text{ml}}{\text{min}}\right]}{50 \, \text{mmHg}} = 69,200.$$

If the Fahraeus-Lindqvist effect is neglected, then $\Re_F = 5.06 \times 10^4$ mmHg/(ml/min), and the number of fibers needed to produce a flow of 100 ml/min would increase to 101,000.

Examining and interpreting the results: Ignoring the Fahraeus-Lindqvist effect would result in an oxygenator with 46% more fibers than are actually needed to attain the desired flow. This would add significantly to the cost of the device.

Example 5.12.2 Resistance of an In Vitro Network of Microtubules.
The pressure drop across a network of microtubules with dichotomous branching shown in Table 5.1 is 35 mmHg. The hematocrit in large vessels is 40% and the viscosity of plasma is 1.05 cp.

(a) Estimate flow through the entire in vitro network, the pressure drop across each segment, and the "tube hematocrit" in each segment.
(b) Compare the flow in part (a) to the flow computed on the basis of blood having a constant viscosity equal to large vessel viscosity.

(c) Estimate blood flow through the same network in vivo, assuming that the presence of a rigid glycocalyx on the inside surface of each vessel reduces the effective diameter of each microvessel by 1.5 μm.

Solution. *Initial considerations:* We will neglect losses at the branches and assume that flow is fully developed in each branch. Construction of a spreadsheet will prove most useful in solving this problem.

System definition and environmental interactions: The system consists of the blood in the network shown in Fig. 5.26. Each branch leads to two daughter vessels of equal dimensions until generation 5 is reached. The branches then converge until a single outlet vessel is formed.

Apprising the problem to identify governing relationships: For part (a), we need to account for the Fahraeus–Lindqvist effect in each of the microtubules. The

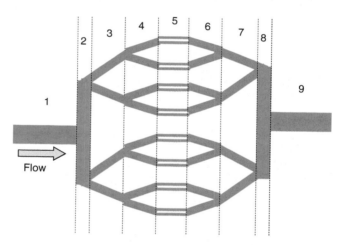

Fig. 5.26 Network with dichotomous branching

Table 5.1 Vessel dimensions in a hypothetical network

Generation	Number of vessels	Diameter (μm)	Length (μm)
1	1	75	2,000
2	2	45	1,600
3	4	25	1,200
4	8	15	800
5	16	8	600
6	8	18	800
7	4	30	1,200
8	2	55	1,600
9	1	90	2,000

5.12 Steady Flow Through a Network of Rigid Conduits

apparent viscosity will be different for each tube size. Once the viscosity is known, the flow resistance for each vessel can be determined and the combined resistance of parallel vessels in each generation can be found. The series resistances of each generation can then be added to find the total resistance.

Analysis: The detailed procedure is shown below and the results are entered for each step in Table 5.2:

1. Use (4.43) to compute the relative viscosity $\eta_r(D,45\%)$ of blood at a hematocrit of 45% for each of the tube diameters in Table 5.1.
2. Compute the exponent α for each tube diameter using (4.45).
3. Correct the relative viscosity $\eta_r(D,H_D)$ in each tube for the correct discharge hematocrit $H_D = 40\%$ using (4.44).
4. Compute the apparent viscosity by multiplying $\eta_r(D,H_D)$ by the plasma viscosity of 1.05 cp.
5. Compute the resistance of a single tube $\Re_F$ for each generation using (5.112).
6. Compute the resistance of all parallel vessels in each generation, $\Re_{gen}$, using (5.122).
7. Compute the resistance of all generations in series using (5.117). This is equivalent to adding the resistances from each generation, $\Re_{total} = \Sigma \Re_{gen} = 5{,}734 \; [\text{mmHg}/(\mu\text{l/s})]$.
8. Compute the flow through the entire network using (5.111)

$$Q_V = \frac{\Delta P}{\Re_{total}} = \frac{35 \text{ mmHg}}{5{,}734 \text{ mmHg} \, \mu l^{-1} \, s} = 0.0061 \; \mu\text{l/s}.$$

9. Compute the pressure difference across each generation using (5.111), and subtract this from the upstream pressure to obtain the downstream pressure for that generation:

$$P_{down} = P_{up} - Q_V \Re_{gen}.$$

Since we assume no losses at the branches, this is also the upstream pressure for the next generation.

Table 5.2 Computations for each network generation accounting for the Fahraeus-Lindqvist effect

Gen	η_r (D,45)	α	η_r (D,40)	η_a (cp)	$\Re_F$ [mmHg/(μl/s)]	$\Re_{gen}$ [mmHg/(μl/s)]	P_{up} (mmHg)	P_{down} (mmHg)	ΔP (mmHg)	H_T/H_D	H_T
1	2.277	−0.804	2.051	2.153	41.6	41.6	35.000	34.746	0.254	0.842	0.337
2	1.987	−0.834	1.811	1.902	226.8	113.4	34.746	34.054	0.692	0.781	0.312
3	1.688	−0.953	1.562	1.640	1,539.8	384.9	34.054	31.704	2.350	0.727	0.291
4	1.470	−1.123	1.381	1.450	7,003.3	875.4	31.704	26.361	5.343	0.697	0.279
5	1.267	0.043	1.229	1.290	57,754.6	3,609.7	26.361	4.329	22.032	0.707	0.283
6	1.543	−1.059	1.441	1.513	3,524.9	440.6	4.329	1.640	2.689	0.705	0.282
7	1.775	−0.905	1.635	1.717	777.3	194.3	1.640	0.454	1.186	0.741	0.296
8	2.099	−0.816	1.904	1.999	106.8	53.4	0.454	0.128	0.326	0.803	0.321
9	2.382	−0.801	2.138	2.244	20.9	20.9	0.128	0.000	0.128	0.866	0.346

10. Compute relative tube hematocrit for each microvessel using (4.53).
11. Compute the tube hematocrit by multiplying by the discharge hematocrit in large vessels (0.4).

Part (b): If blood viscosity is constant, we find from (4.43) that $\eta_r(D,45\%) = 3.2$ for large vessels and from (4.45), $\alpha = -0.8$. Therefore, from (4.44) $\eta_r(D,40\%) = 2.81$ and multiplying by the plasma viscosity, $\eta_a = 2.95$ cp. The resistances and pressures are computed as before and the results are given in Table 5.3.

Table 5.3 Computations for each network generation assuming blood viscosity is independent of tube diameter

Gen	$\Re_F$ [mmHg/(μl/s)]	$\Re_{gen}$ [mmHg/(μl/s)]	P_{up}(mmHg)	P_{down} (mmHg)	ΔP (mmHg)
1	57.0	57.0	35.000	34.837	0.163
2	352.0	176.0	34.837	34.335	0.502
3	2,771.3	692.8	34.335	32.358	1.977
4	14,255.5	1,781.9	32.358	27.274	5.084
5	132,144.6	8,259.0	27.274	3.709	23.565
6	6,874.8	859.3	3.709	1.257	2.452
7	1,336.5	334.1	1.257	0.303	0.953
8	157.7	78.9	0.303	0.078	0.225
9	27.5	27.5	0.078	0.000	0.078

The total resistance in this case is 12,266 [mmHg/(μl/s)] and the total flow is 0.00285 μl/s. This is less than half the actual flow when the Fahraeus–Lindqvist effect is included.

Part (c): Subtracting 1.5 μm from the diameters in Table 5.1 and following the procedure from part (a), we obtain the results in Table 5.4. The total resistance in this case is 11,128 [mmHg/(μl/s)] and the total flow is 0.00315 μl/s. Thus the flow is reduced to almost half of the flow in an in vitro system without a glycocalyx.

Table 5.4 Results with a 1 μm glycocalyx in each vessel

Gen	η_r (D,45)	α	η_r (D,40)	η_a (cp)	$\Re_F$ [mmHg/(μl/s)]	$\Re_{gen}$ [mmHg/(μl/s)]	P_{up} (mmHg)	P_{down} (mmHg)	ΔP (mmHg)	H_T/H_D	H_T
1	2.265	−0.804	2.041	2.143	44.9	44.9	35.000	34.859	0.141	0.840	0.336
2	1.969	−0.838	1.796	1.886	257.5	128.8	34.859	34.454	0.405	0.777	0.311
3	1.659	−0.972	1.538	1.615	1,942.1	485.5	34.454	32.927	1.527	0.722	0.289
4	1.431	−1.157	1.348	1.416	10,421.8	1,302.7	32.927	28.829	4.097	0.693	0.277
5	1.251	0.870	1.222	1.283	131,800.1	8,237.5	28.829	2.920	25.909	0.734	0.293
6	1.508	−1.090	1.412	1.482	4,890.5	611.3	2.920	0.997	1.923	0.701	0.280
7	1.750	−0.918	1.614	1.695	942.1	235.5	0.997	0.256	0.741	0.737	0.295
8	2.083	−0.818	1.891	1.985	118.5	59.3	0.256	0.070	0.186	0.800	0.320
9	2.372	−0.801	2.130	2.236	22.3	22.3	0.070	0.000	0.070	0.864	0.346

Examining and interpreting the results: The primary pressure drop occurs in the generation with the smallest diameter vessels. Accounting for both the presence of a glycocalyx and the Fahraeus effect yields a flow that is similar to the flow obtained by ignoring both the presence of the glycocalyx and the Fahraeus effect. Except for the smallest vessels, the in vivo tube hematocrit values, based on the reduced diameters, are very similar to the in vitro tube hematocrit values.

5.13 Compliance and Resistance of Flexible Conduits

Most blood vessels, airways, lymph vessels, and other physiological conduits are not rigid, but change their shape when subjected to a *transmural pressure*. A transmural pressure P_{tm} refers to the difference between the internal (P) and external (P_e) pressures applied to the vessel wall:

$$P_{tm} = P - P_e. \tag{5.123}$$

It is important not to confuse this with the pressure drop inside the tube ΔP, which refers to the upstream minus downstream pressure inside the tube. If transmural pressure is positive, the lumen of the conduit will remain open; but if the transmural pressure becomes sufficiently negative, the vessel wall may collapse. We will consider conditions where transmural pressure is positive in this section and cases when it is negative in the next section.

If the conduit wall is flexible, then an increase in transmural pressure will generally increase the cross-sectional area within the conduit, while a decrease in transmural pressure will decrease the cross-sectional area of the conduit. For example, contraction of the diaphragm decreases pleural pressure in the lung, causing transmural pressure across the alveolar membrane to increase. Since the alveolar membrane is flexible, the positive transmural pressure causes the alveolar volume to expand, which in turn draws fresh air into the alveolus. When the diaphragm relaxes, the transmural pressure decreases, and gas is flushed out of the alveolus. The rate at which the volume changes with respect to a change in transmural pressure is defined as the *compliance C* of the conduit:

$$C = \frac{dV}{dP_{tm}}. \tag{5.124}$$

A highly compliant conduit or compartment is one where large changes in volume occur for relatively small changes in transmural pressure. Conduits with stiff walls have a low compliance. The compliance is often relatively constant over some range of transvascular pressures, but will generally decrease (become stiffer) as the transmural pressure increases. If compliance of a conduit is constant over a range of transmural pressures between P_{tm0} and P_{tm}, then the volume at P_{tm} can be predicted by integrating (5.124):

$$V - V_0 = C[(P - P_e) - (P_0 - P_{e0})]. \tag{5.125}$$

If the conduit has a circular cross section and the length L of the vessel remains constant as the transmural pressure changes, then the volume and diameter are related as follows:

$$V - V_0 = \frac{\pi L}{4}[D^2 - D_0^2]. \tag{5.126}$$

The volumes can be eliminated from (5.125) and (5.126) to provide an expression relating tube diameter, tube compliance, and transmural pressure difference:

$$D^2 = D_0^2 + \frac{4C}{\pi L}[(P - P_e) - (P_0 - P_{e0})]. \tag{5.127}$$

The resistance to flow for laminar flow through a circular tube with diameter D is:

$$\Re_f = \frac{128 \mu L}{\pi D^4}. \tag{5.128}$$

Although the volume increases linearly with transmural pressure, the increase in tube diameter and the decrease in flow resistance are nonlinear functions of transmural pressure. The ratio of the resistance at a new transmural pressure to the resistance at the original transmural pressure is:

$$\frac{\Re_f}{\Re_{f0}} = \frac{1}{\left[1 + \frac{4C}{\pi D_0^2 L}[(P - P_e) - (P_0 - P_{e0})]\right]^2}. \tag{5.129}$$

The rate at which mass accumulates in a compartment when a positive transmural pressure is applied across the conduit wall for the case when fluid density is constant is:

$$\frac{dm}{dt} = \rho \frac{dV}{dt} = \rho \left(\frac{dV}{dP_{tm}}\right) \frac{dP_{tm}}{dt} = \rho C \frac{d(P - P_e)}{dt}. \tag{5.130}$$

If the compliant compartment has N_i inlets and N_j outlets, then conservation of mass becomes:

$$C\left(\frac{dP}{dt} - \frac{dP_e}{dt}\right) = \sum_{i=1}^{N_i} Q_{Vi,in} - \sum_{j=1}^{N_j} Q_{Vj,out}, \tag{5.131}$$

where $Q_{Vi,in}$ is the flow rate in the ith inlet and $Q_{Vj,out}$ is the flow rate in the jth outlet. Applications of (5.131) are shown in the following examples.

Example 5.13.1 Simple Respiratory Flow Model.
A simple model of the respiratory system is shown in Fig. 5.27. The figure at the left represents alveolar volume at the end of a normal expiration (subscript 1), and the figure at the right represents the alveolar volume at the end of a normal inspiration (subscript 2). Pleural pressure surrounding the alveolar compartment is generated by contraction and relaxation of the diaphragm. Find the compliance of the lung if alveolar volume changes under static conditions from 2 L when pleural pressure is -2 cmH$_2$O to

5.13 Compliance and Resistance of Flexible Conduits

2.6 L when pleural pressure is held at −5 cmH$_2$O. Find alveolar pressure, volume, and flow as a function of time if pleural pressure is sinusoidal with an end-expiratory pleural pressure (relative to atmospheric pressure) of −2 cmH$_2$O and an end-inspiratory pleural pressure of −8 cmH$_2$O. How are these values modified by altering the normal breathing rate of 10 breaths/min? Airway resistance is elevated in asthma. Examine how the volume delivered to the alveolar compartment changes as airway resistance is elevated above the normal value of 2 cmH$_2$O/(L/s)?

Solution. *Initial considerations:* Although air is a compressible fluid, we will ignore any changes in density during normal breathing. The conducting airways are assumed to be rigid and all compliance is attributed to the alveoli. The compliance of the alveoli can be computed from the pressure–volume information given in the problem statement.

System definition and environmental interactions: The system of interest in this respiratory flow model is the air inside the lung. This can be conveniently divided into two systems in series: the respiratory airways, modeled as a single rigid tube with constant resistance, and a compliant alveolar volume. The flow out of the conducting airways is equal to the flow into the alveoli, and the pressure at the outlet of the conducting airways is equal to the pressure in the alveoli.

Apprising the problem to identify governing relationships: We must apply conservation of mass in the airways and in the alveoli, and must apply the pressure–volume relationship to the compliant alveolar compartment and the pressure–flow relationship to the conducting airways.

Analysis: If the resistance $\Re$ of the conducting airways is constant, then flow into the alveolar compartment is given by:

$$Q_{V,in} = \frac{P_{atm} - P_A}{\Re},$$

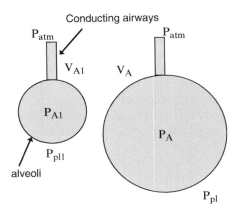

Fig. 5.27 One-compartment model of the respiratory system. *Left*: end expiration, *right*: end inspiration

where P_A is the internal pressure in the alveolar compartment and P_{atm} is the pressure at the mouth (which equals atmospheric pressure). Alveolar compliance C can be estimated by integrating (5.124) between the inspiration and expiration limits:

$$C = \frac{V_{A2} - V_{A1}}{(P_{A2} - P_{pl2}) - (P_{A1} - P_{pl1})} = \frac{V_{A2} - V_{A1}}{P_{pl1} - P_{pl2}}.$$

The external pressure P_e in this case is equal to the pleural pressure, P_{pl}. Since the measurements were made under static conditions, there is no flow into or out of the alveolar compartment. Because flow is zero, alveolar pressure during the measurement was equal to atmospheric pressure (i.e., $P_{A_1} = P_{A_2} = P_{atm}$), and the denominator becomes the pleural pressure difference. The compliance for our example is computed as $C = (2.6\ L - 2.0\ L)/((-2\ cmH_2O) - (-5\ cmH_2O)) = 0.2\ L/cmH_2O$.

Now, let us turn our attention to the dynamic case. Arbitrarily selecting $t = 0$ as the time when alveolar pressure is at its average value during the cycle, a periodic pleural pressure can be written as:

$$P_{pl} = \langle P_{pl} \rangle + \Delta P_{pl} \sin(\omega t),$$

where $\langle P_{pl} \rangle$ is the mean pleural pressure and ΔP_{pl} is half the maximum variation in pleural pressure between inspiration and expiration. In terms of end-inspiratory and end-expiratory pressures:

$$\langle P_{pl} \rangle = \frac{P_{pl1} + P_{pl2}}{2}, \quad \Delta P_{pl} = \frac{P_{pl2} - P_{pl2}}{2}.$$

Substituting the expressions for flow and pleural pressure into (5.131), we obtain the following ODE for pressure in the alveolar compartment:

$$\frac{dP_A}{dt} - \omega \Delta P_{pl} \cos(\omega t) = \frac{P_{atm} - P_A}{\Re C}.$$

Since atmospheric pressure is constant, this can be rewritten as follows:

$$\frac{d}{dt}(P_A - P_{atm}) + \frac{(P_A - P_{atm})}{\Re C} = \omega \Delta P_{pl} \cos(\omega t).$$

The solution to this equation can be divided into two parts: the transient solution and the forced solution. The transient solution is simply the solution to the homogeneous expression, where the right-hand side is set equal to zero. This will be a simple exponential with a time constant equal to $\Re C$. After waiting several time constants, the transient solution will be negligible and the solution will reduce to the forced solution. The forced solution will have a form:

$$P_A - P_{atm} = a \sin(\omega t) + b \cos(\omega t) + c,$$

where a, b, and c are constants. Substituting this back into the original ODE and requiring that the coefficients of the sine, cosine, and constant terms on each side of the equation are equal, we find:

5.13 Compliance and Resistance of Flexible Conduits

$$a = \frac{\Delta P_{pl}(\omega \Re C)^2}{1+(\omega \Re C)^2}, \quad b = \frac{\Delta P_{pl}(\omega \Re C)}{1+(\omega \Re C)^2}, \quad c = 0.$$

The flow into the alveolar compartment is:

$$Q_V = \frac{P_{atm} - P_A}{\Re} = \frac{-\Delta P_{pl}(\omega C)}{\left[1+(\omega \Re C)^2\right]} \{\omega \Re C \sin(\omega t) + \cos(\omega t)\}.$$

Alveolar volume can be found by integrating the flow. The volume relative to the volume at $t = 0$ is:

$$\Delta V(t) = \int_0^t Q_V(t) dt = \frac{-\Delta P_{pl} C}{\left[1+(\omega \Re C)^2\right]} \{\omega \Re C[1 - \cos(\omega t)] + \sin(\omega t)\}.$$

The dynamic response for pressures, flow, and volume is shown in Fig. 5.28.

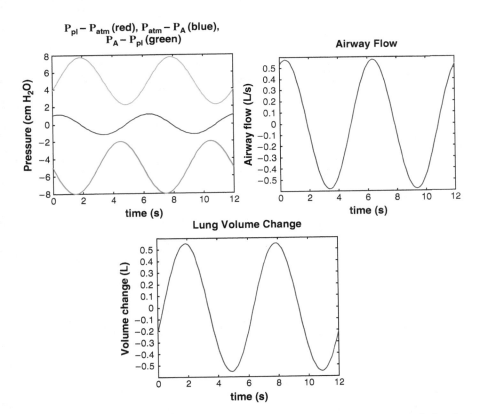

Fig. 5.28 Simulated dynamic respiratory pressures (*left*), alveolar flow (*right*), and alveolar volume changes (*lower*)

Examining and interpreting the results: Note the lag between peak pressure and peak flow. The effect of altering respiratory frequency on alveolar volume is shown in Fig. 5.29. A dramatic decrease in volume occurs as frequency is increased.

The volume of air inspired over a 1 min period (minute volume) is shown in Fig. 5.30. Shown for comparison is the minute volume that would result if the volume per breath remained the same. The minute volume is reduced significantly at high frequencies.

Additional comments: The effect of simulating asthma is shown in Fig. 5.31. Airway resistance is increased from 2 to 5 $cmH_2O/(L/s)$. Comparing the tidal volume with the normal tidal volume in Fig. 5.28 shows a decrease from about 1,100 to 800 ml. To make up for the reduced volume, an asthmatic must increase his or her respiratory rate, which will further reduce the tidal volume.

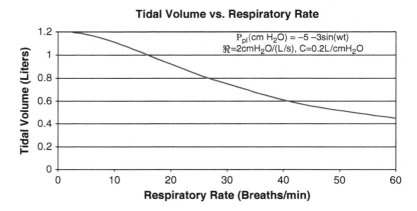

Fig. 5.29 Effect of respiratory rate on tidal volume

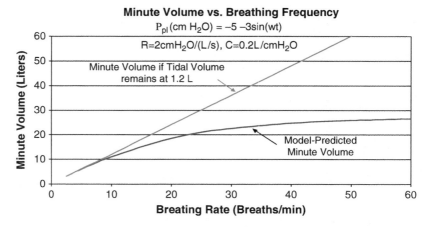

Fig. 5.30 Influence of breathing frequency on minute volume

5.13 Compliance and Resistance of Flexible Conduits

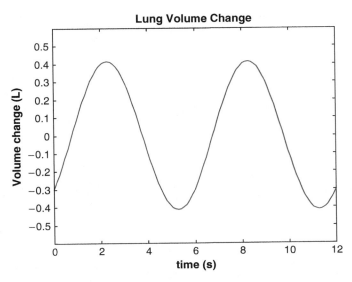

Fig. 5.31 Simulation of changes in alveolar volume for an asthmatic breathing with an airway resistance of 5 mmHg/(L/s)

Example 5.13.2 Windkessel Model of Arterial Pressure.
The Windkessel model (Fig. 5.32) is a simple model that predicts blood pressure vs. time in the aorta. All of the resistance in the systemic circulatory system is assumed to be located downstream of the large arteries (lumped here as the aorta), while all of the compliance in the system is assumed to be located in the large arteries. Input to the model is the cardiac output $Q_{V,in}(t)$, compliance of the aorta, systemic resistance, and downstream pressure. Model output is the pressure in the aorta and large arteries. Find the pressure in the aorta as a function of time and flow out of the aorta under conditions simulating normal cardiac output.

Solution. *Initial considerations:* To estimate systolic pressure in the aorta, we need to know the resistance of the systemic circulation $\Re$, the pressure downstream of the systemic circulation (right atrial pressure) P_v, and cardiac output $Q_{V,in}$. Normal systemic vascular resistance is about 20 mmHg/(L/min), and right atrial pressure is about 1 mmHg. Let us assume that the average cardiac output is 4 L/min, the cardiac frequency is 1 beat/s, and each cardiac cycle is assumed to consist of a systolic phase (0.4 s) where outflow is 10 L/min and a diastolic phase (0.6 s) where outflow is 0 (Fig. 5.33).

System definition and environmental interactions: The system of interest is blood in the large arteries, which we assume forms a single compliant compartment, the aorta. Flow into the aorta is determined by the cardiac output and flow out of the

Fig. 5.32 Windkessel Model

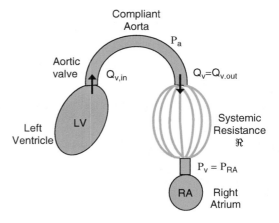

Fig. 5.33 Idealized cardiac output

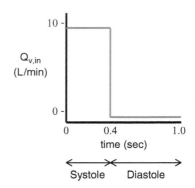

aorta is determined by the resistance of the systemic circulation and the downstream pressure, both which we will assume to be constants.

Apprising the problem to identify governing relationships: The relationships needed to predict the pressure in the aorta and flow in the systemic circulation include a mass balance on the aorta compartment, cardiac output as a function of time, the definition of large artery compliance, and the definition of systemic vascular resistance.

Analysis: The equation for flow out of the aorta is:

$$Q_{V,\text{out}} = \frac{P_A - P_V}{\Re},$$

where $\Re$ is the systemic resistance, or total peripheral resistance. A mass balance on blood (assumed incompressible) in the aorta is:

$$\frac{dV_A}{dt} = Q_{V,\text{in}} - Q_{V,\text{out}}.$$

5.13 Compliance and Resistance of Flexible Conduits

Inserting the expression for the outflow into the mass balance, and expressing the rate of volume change in terms of the compliance of the aorta:

$$C\frac{dP_A}{dt} = Q_{V,\text{in}} - \frac{P_A - P_V}{\Re}.$$

The solution for aortic pressure is:

$$P_A(t) = P_{A0}e^{-\frac{t}{\Re C}} + (\Re Q_{V,\text{in}} + P_V)\left(1 - e^{-\frac{t}{\Re C}}\right), \quad 0 \le t \le T_s \text{ (systole)},$$

$$P_A(t) = P_V + (P_A(T_s) - P_V)e^{-\left(\frac{t - T_s}{\Re C}\right)}, \quad T_s \le t \le T \text{ (diastole)}.$$

The value computed for aortic pressure at the end of systole $P_A(T_s)$ is used as the initial value for diastole. We are now in position to account for the effect of vascular compliance of the aorta in the Windkessel model. Pressure–volume experiments allow us to measure the compliance, $C = 0.8$ ml/mmHg. From Fig. 5.33, we use 10 L/min for cardiac output during systole and 0 during diastole. To solve the ODE, we need to know an initial condition, say $P_A(0) = P_{A0} = 80$ mmHg, normal diastolic pressure. The solutions for pressure and outflow are shown in Fig. 5.34.

Examining and interpreting the results: After about five cycles, the model predicts consistent values for systolic and diastolic pressure and flow. Peak pressure is below 120 mmHg, and after five cycles the pulse pressure is less than 50 mmHg. The flow does not go to zero, as it would if aortic compliance was neglected. We used the solution at the end of each cycle as the initial condition for the next cycle [replacing the Pa(0) = 80 mmHg]. Aortic compliance causes peak flow to drop to less than 6 L/min and aortic outflow remains above 2.5 L/min throughout the entire cycle.

Additional comments: If the aorta behaved as a rigid vessel, the flow out would equal the flow in during systole and the flow out would be zero during diastole. In that case, the pulse pressure would be 200 mmHg, a huge pressure swing with each beat of the heart. The normal pulse pressure is about 40 mmHg. A simple rigid

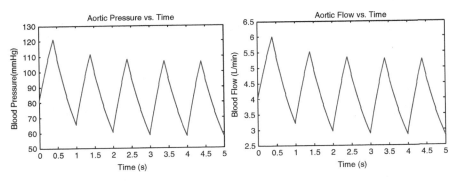

Fig. 5.34 Response to Windkessel model over five cardiac cycles. *Left*: Aortic pressure (mmHg) vs. time (s). *Right*: Aortic outflow (L/min) vs. time (s)

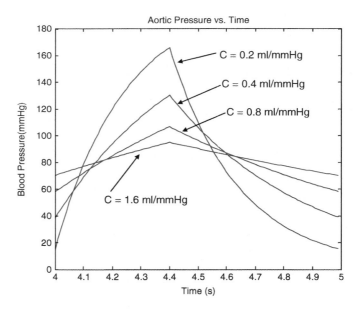

Fig. 5.35 Response to Windkessel model over a cardiac cycle for four different values of aortic compliance

model of the aorta is not realistic. Flow to the systemic circulation is nonzero, even during diastole when the aortic valve is closed. The aorta and other large arteries are actually compliant vessels. Consequently, some of the blood that enters the arteries during systole expands the vessels rather than flowing directly to the systemic circulation. The same vessels discharge the excess fluid during diastole, thus delivering nonzero flow to the systemic circulation during the entire cardiac cycle.

More realistic aortic pressure traces would result if actual ventricular outflow traces were used as input to the model, rather than rectangular wave functions. However, this example illustrates the important effects of aortic compliance in protecting the vascular system from high pressures. Normal aging and certain diseases, such as atherosclerosis, can stiffen the vessels and widen the pulse pressure without affecting mean pressure. This is simulated in Fig. 5.35 by changing the compliance. Increasing the systemic vascular resistance or venous pressure will elevate the mean aortic pressure, but will not significantly affect pulse pressure.

5.14 Flow in Collapsible Tubes

There are many occasions when transmural pressure differences across conduit walls in the human body are negative. If the conduit is relatively rigid, like the trachea, this will not influence flow through the tube. However, flow through

5.14 Flow in Collapsible Tubes

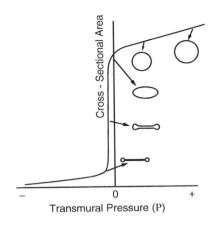

Fig. 5.36 Cross-sectional area vs. transmural pressure for a collapsible tube, from Fung 1996 with permission

compliant vessels will generally decrease as the transmural pressure difference becomes more negative. This can occur in veins in the upper body, in small airways during forced expiration, in lung and heart capillaries, and in initial lymphatic vessels.

The pressure–volume relationships for compliant vessels change dramatically as the transmural pressure difference becomes negative, particularly for thin-walled vessels. Thin-walled vessels begin to collapse when small negative transmural pressures are applied (Fig. 5.36). The cross-sectional area changes from a circular shape under positive transmural pressures to an elliptical shape at slight negative transmural pressures. As the transmural pressure becomes more negative, the opposite walls of the vessel will touch and two channels with very small cross-sectional area will form along the sides of the collapsed vessel.

Consider fluid movement through a horizontal compliant tube. The fluid pressure $P(x)$ will decrease from its value of P_{in} at the inlet to P_{out} at the outlet of the conduit. Since the tube is compliant, its local cross section $A(x)$ will be a function of the local transmural pressure $P(x) - P_e$. Ernest Starling recognized three possible flow regimes for a compliant tube: (1) the situation when the inlet and outlet fluid pressures are both greater than the external pressure ($P_{in} > P_{out} > P_e$); (2) the case when the external pressure is greater than the inlet pressure ($P_e > P_{in} > P_{out}$); and (3) the situation shown in Fig. 5.37, where the external pressure lies between the inlet and outlet pressures ($P_{in} > P_e > P_{out}$). Let us consider an ideal resistor, known as a *Starling resistor*, in which the local tube resistance per unit length assumes one of two values: $\mathcal{R}'$ in the open state under positive transmural pressures and $\mathcal{R}'_c$ in the collapsed state caused by negative transmural pressures.

When the transmural pressure is positive along the entire length of the tube, the flow through the tube is:

$$Q_V = \frac{P_{in} - P_{out}}{\mathcal{R}'L} = \frac{P_{in} - P_{out}}{\mathcal{R}}, \quad P_{in} > P_{out} > P_e. \quad (5.132)$$

Fig. 5.37 Pressure vs. position in a partially collapsed vessel

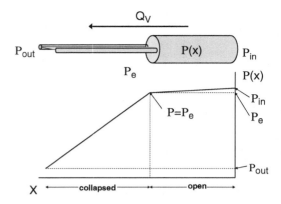

This is the situation we have dealt with to this point in the text. If the external pressure exceeds the inlet pressure, there will be no flow through an ideal Starling resistor:

$$Q_V = 0, \ P_e > P_{in} > P_{out}. \qquad (5.133)$$

Finally, if the external pressure lies between the inlet and outlet pressures, Starling postulated that the flow was independent of downstream pressure, and that the inlet minus external pressure was the driving force for flow through the tube:

$$Q_V = \frac{P_{in} - P_e}{\Re}, \ P_{in} > P_e > P_{out}. \qquad (5.134)$$

Although it is not intuitive, (5.134) has been verified by many sets of experiments over the years. The effect has been likened to a waterfall in which the flow over the cliff is independent of the height of the cliff. Experimental evidence of flow limitation in thin-walled surgical tubing with a flow resistance of 2 cmH$_2$O/(ml/s) is shown in Fig. 5.38. External pressure is held constant at atmospheric pressure and inlet pressure is held at 10 cmH$_2$O above atmospheric pressure. Outlet pressure is adjusted for each flow measurement. As long as outlet pressure is greater than external pressure, flow increases with increasing $P_{in} - P_{out}$. However, when external pressure is greater than outlet pressure, further increases in $P_{in} - P_{out}$ do not produce an increase in flow. The maximum flow occurs when $P_{out} = P_e = 0$. When the outlet pressure is dropped below the external pressure, the flow does not change, and the appropriate driving force becomes $P_{in} - P_e$.

Example 5.14.1 Derivation of (5.134).
Show that (5.134) is a direct consequence of the resistance per unit length of the collapsed section being much greater than the resistance per unit length of the uncollapsed segment of tube.

5.14 Flow in Collapsible Tubes

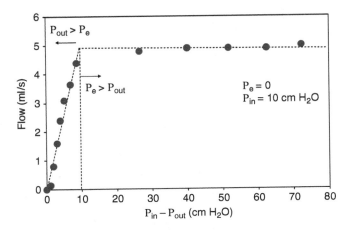

Fig. 5.38 Flow vs. negative transmural pressure in a collapsible tube

Solution. *Initial considerations:* The equation for resistance of a Starling resistor is not exactly correct. It results from an assumption that we can ignore the resistance of the uncollapsed portion of the tube relative to the collapsed portion.

System definition and environmental interactions: The system of interest in a partially collapsed tube consists of fluid in both the collapsed and uncollapsed portions of the tube.

Appraising the problem to identify governing relationships: The fluids in the uncollapsed and collapsed regions of the tube constitute two resistors in series. In the steady state, the flow through each resistor will be the same and the pressure at the outlet of the uncollapsed region will equal the pressure at the inlet of the collapsed region. If the resistor is an ideal Starling resistor, then the pressure at the interface between these two resistors will be equal to the surrounding pressure.

Analysis: Let the length of the collapsed segment be L_c. Then its resistance will equal $\Re'_c L_c$. The length of the uncollapsed section of tubing will be $L - L_c$ and its resistance will be $\Re'(L - L_c)$. The pressure at the downstream end of the uncollapsed region of a Starling resistor is the external pressure P_e. The flow through the collapsed and uncollapsed segments must be equal:

$$Q_V = \frac{P_{in} - P_e}{\Re'(L - L_c)} = \frac{P_e - P_{out}}{\Re'_c L_c}.$$

Solving for the length of the collapsed segment:

$$L_c = L \left\{ \frac{\Re'(P_e - P_{out})}{\Re'_c(P_{in} - P_e) + \Re'(P_e - P_{out})} \right\}.$$

Substituting this back into the expression for flow in the collapsed segment:

$$Q_V = \frac{P_e - P_{out}}{\mathfrak{R}'_c L_c} = \frac{P_{in} - P_e}{\mathfrak{R}'L} + \frac{\mathfrak{R}'}{\mathfrak{R}'_c}\left(\frac{P_e - P_{out}}{\mathfrak{R}'L}\right).$$

But the quantity $\mathfrak{R}'L$ is equal to $\mathfrak{R}$, the resistance of the uncollapsed tube. Since the ratio of uncollapsed to collapsed resistance per unit length is very small and $(P_e - P_{out})/\mathfrak{R}$ is of the same order of magnitude as $(P_{in} - P_e)/\mathfrak{R}$, then the second term can be neglected with respect to the first. Hence, we are left with:

$$Q_V \approx \frac{P_{in} - P_e}{\mathfrak{R}'L} = \frac{P_{in} - P_e}{\mathfrak{R}}.$$

Examining and interpreting the results: The expression derived above is the same as (5.134). Note that the resistance used in (5.134) is the resistance of the tube in its uncollapsed state. It is not the resistance of the uncollapsed portion of the tube. If 90% of the tube is collapsed, the denominator in (5.134) is the resistance of the entire tube in its uncollapsed state.

Example 5.14.2 Forced Expiration in the Respiratory System.
Inspiration is an active process normally resulting from contraction of the diaphragm. This can be aided during periods of activity by contraction of the external intercostal muscles, which expands the thorax. Contraction of these muscles reduces pleural pressure, causing air to flow into the lungs. Expiration is normally passive, resulting from elastic recoil of expanded alveoli and compliant airways. However, during exercise, expiration can be assisted by contraction of the internal intercostal muscles, which increases pleural pressure and forces air out of the lungs. This is known as forced expiration. Experimental results (Pride et al. 1967) show that forced expiratory flow increases with expiratory effort up to a point, but beyond that point, increasing alveolar pressure does not cause an increase in flow. Furthermore, the maximum expiratory flow increases with the initial volume of air in the lungs at the beginning of the forced expiration, as shown in Fig 5.39. Finally, the alveolar pressure at which flow limitation is reached increases with increasing lung volume. How can these observations be explained on the basis of the anatomy of the respiratory system and the principles presented in this section?

Solution. *Initial considerations:* At the end of a normal inspiration, alveolar pressure will be zero (relative to atmospheric pressure) and pleural pressure will be negative. The difference between alveolar pressure and pleural pressure at zero flow represents the elastic recoil of the lung and can be defined as the recoil pressure. The magnitude of the pleural pressure or recoil pressure will depend on the lung volume. The larger the lung volume, the greater will be the recoil pressure.

5.14 Flow in Collapsible Tubes

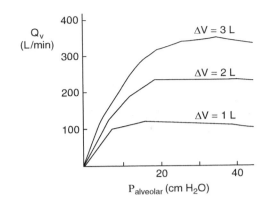

Fig. 5.39 Forced expiration as a function of alveolar pressure and initial lung volume relative to functional residual capacity (FRC)

Experiments demonstrating these effects can be performed by connecting the mouth to a valve which is initially closed at a specified lung volume. The subject then raises pleural pressure by contracting the expiratory muscles. This pressure is also transmitted to the air within the lung where it is measured. The valve is then opened causing air to flow from alveoli toward the mouth. The initial flow rate is measured.

System definition and environmental interactions: The system under analysis consists of the air in the conducting airways in the lung. Some of these are easily collapsible when exposed to negative pressures and others are not.

Apprising the problem to identify governing relationships: For low values of pleural pressures, the transmural pressures at all points in the respiratory system will remain positive. However, since outlet pressure at the mouth is constant, transmural pressure at the mouth will become negative when pleural pressure is raised above atmospheric pressure. Because the oral cavity and trachea are relatively rigid, negative transmural pressures in these portions of the respiratory system will not cause the walls to collapse. However, the large extrapulmonary bronchi have thin, compliant walls which can collapse when exposed to negative transmural pressure differences, behaving as ideal Starling resistors. A diagram of the idealized system is shown in Fig. 5.40.

Analysis: Pressure from alveoli to mouth in the respiratory system are sketched in Fig. 5.40 for each increase in pleural pressure (lines a–h). Lines (a) to (e) reflect linear pressure distributions in the tube (lung), and flow increases linearly with alveolar pressure. Under conditions (b) to (d), the transmural pressure in the downstream portion of the rigid section is negative, but the transmural pressure in the collapsible section is positive, so the tube does not collapse. Under condition (e), where pleural pressure is P^*, the transmural pressure at the downstream end of the collapsible section is zero and the tube is on the verge of collapsing. As pleural pressure increases further, condition (f), the transmural pressure at the collapsible section is negative and the tube collapses. Flow remains constant for additional increases in pleural pressure. For all other increases in pleural pressure, the difference between alveolar pressure and pleural pressure will remain constant and will equal $P^*_{alv} - P^*$.

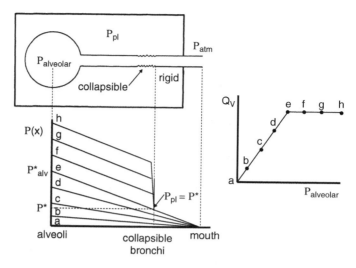

Fig. 5.40 Idealized respiratory system, airway pressure vs. position for multiple increases in pleural pressure, and flow vs. alveolar pressure for each pleural pressure

Analysis of the data in Fig. 5.39 shows that a nearly constant flow of about 110 L/min is reached when alveolar pressure is raised beyond about 10 cmH$_2$O at $\Delta V = 1$ L. The data suggest that for $\Delta V = 1$ L the system begins to collapse at an alveolar pressure of about 10 cmH$_2$O. This corresponds to P^*_{alv} in Fig. 5.39. Resistance of the uncollapsed respiratory tree is estimated to be 10 cmH$_2$O/110 L/min = 0.091 cmH$_2$O/(L/min). If we knew P^*, we could separate the total resistance into a component upstream of the collapsible section and one downstream of the collapsible segment.

The data at $\Delta V = 2$ L suggest P^*_{alv} to be about 20 cmH$_2$O, a forced expiratory flow of about 230 L/min, and an uncollapsed resistance of the respiratory system of 0.087 cmH$_2$O/(L/min). Finally, using data at $\Delta V = 3$ L, we estimate P^*_{alv} to be about 35 cmH$_2$O, maximum flow to be about 350 L/min, and resistance to be about 0.1 cmH$_2$O/(L/min).

Examining and interpreting the results: The estimates of respiratory resistance from the three graphs are consistent. The differences in P^*_{alv} reflect differences in elastic recoil pressure at different lung gas volumes.

Example 5.14.3 Respiratory Flow During Exercise.
Under exercise conditions, contraction of the external intercostal muscles during expiration may increase pleural pressure above airway pressure at the location of the extrapulmonary collapsible bronchi. Develop a model of respiration during exercise that includes the collapsible bronchi and compare flow predictions to those when the bronchi do not collapse during exercise. Make the following assumptions: (1) total resistance of the respiratory system

5.14 Flow in Collapsible Tubes

$\mathfrak{R}$ is 2 cmH$_2$O/(L/s); (2) alveolar compliance C is 0.2 L/cmH$_2$O; (3) the collapsible bronchi are situated such that 75% of the respiratory resistance lies between them and the alveoli and 2% between them and the mouth; (4) pressure at the mouth is atmospheric; (5) respiratory frequency is 20 breaths/min during exercise (i.e., $\omega = \pi$ radians/s); (6) pleural pressure can be approximated as $P_{\text{pl}}(t) = -5 - 30$ cmH$_2$O sin (ωt).

Solution. *Initial considerations:* The solution for the case where no collapse occurs is given in Example 5.13.1. Numerical values will be different in this example since the breathing frequency is higher and the pleural oscillations are greater.

System definition and environmental interactions: As in Example 5.13.1, the system of interest in this respiratory flow model is the air inside the lung. However, in this case we include the fact that airways can collapse when exposed to negative transmural pressures. Rather than modeling the respiratory airways as a rigid tube, we will model it as an ideal Starling resistor.

Apprising the problem to identify governing relationships: The governing relationships are (1) a mass balance on the alveolar compartment, (2) treatment of conducting airways as an ideal Starling Resistor, and (3) compliance relationship for the alveolar compartment.

Analysis: During inspiration, a mass balance on the collapsible respiratory system will be the same as for the rigid respiratory system, since the transmural pressure is positive during inspiration:

$$C \frac{d(P_A - P_{\text{pl}})}{dt} = Q_V = \frac{P_{\text{atm}} - P_A}{\mathfrak{R}}.$$

The nomenclature is the same as was used in Example 5.13.1. During expiration, the expression for flow depends on the transmural pressure at the collapsible section. Let P^* equal the airway pressure in the collapsible bronchi (see Fig. 5.40). During expiration, the flow from the alveoli to the collapsible segment must equal the flow from the collapsible segment to the mouth:

$$-Q_V = \frac{P_A - P^*}{\mathfrak{R}_u} = \frac{P^* - P_{\text{atm}}}{\mathfrak{R}_d},$$

where $\mathfrak{R}_u$ is the resistance upstream of the collapsible bronchi during expiration and $\mathfrak{R}_d$ is the downstream resistance. Solving for P^* with the pressure at the mouth $P_{\text{atm}} = 0$:

$$P^* = \left(\frac{\mathfrak{R}_d}{\mathfrak{R}_d + \mathfrak{R}_u}\right) P_A = \left(\frac{\mathfrak{R}_d}{\mathfrak{R}}\right) P_A.$$

Expiratory flow will depend on the transmural pressure at the collapsible section:

$$-Q_V = \frac{P_A(t) - P^*(t)}{\Re_u}, \quad P^* > P_{pl}(t),$$

$$-Q_V = \frac{P_A(t) - P_{pl}(t)}{\Re_u}, \quad P^* \leq P_{pl}(t).$$

The method of solution during inspiration is the same as outlined in Example 5.13.1. During expiration ($P_A > P_{atm}$), a numerical approach using Matlab is applied in which we first compute pleural pressure and P^*, compare them, and use the appropriate expression for Q_V, depending on the transmural pressure difference.

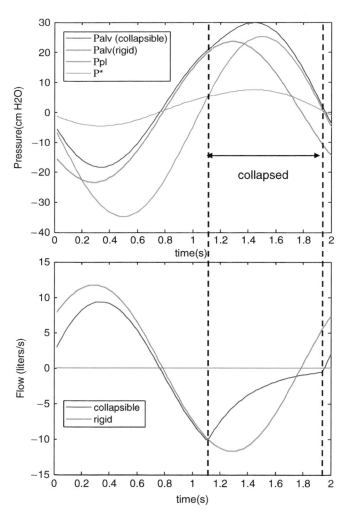

Fig. 5.41 Simulated pressures and flows during exercise, with and without airway collapse. $\Re_d/\Re = 0.25$

5.15 Fluid Inertia

Results for a single respiratory cycle are shown in Fig. 5.41. The upper panel shows the pleural pressure trace, alveolar pressures for the collapsible and rigid airways, and the pressure at the collapsible section, $P^*(t)$.

Examining and interpreting the results: The points where P^* and P_{pl} intersect are shown by the dotted lines. The airways in the collapsible section are collapsed for nearly the entire expiratory portion of the respiratory cycle. This additional resistance prolongs expiration and reduces the maximal and average inspiratory and expiratory flow rates achievable under the same exercise conditions.

5.15 Fluid Inertia

We know from Newton's first law that objects in motion tend to stay in motion and objects at rest tend to stay at rest. Fluid particles have inertia as well. When a force imbalance is applied to a fluid particle, the particle will accelerate until the forces become balanced once again, and the particle attains a new velocity. Because of fluid inertia, the new velocity is not attained instantaneously. Let us consider a fluid with mass m initially at rest in a tube with length ℓ and cross-sectional area A_c, as shown in Fig. 5.42.

A pressure difference $\Delta P = P_1 - P_2$ is suddenly applied between the ends of the tube. Ignoring the effects of friction, Newton's second law applied to the fluid is:

$$\Delta P A_c = \frac{d}{dt}(mv_x). \qquad (5.135)$$

The fluid mass m is equal to the product of the fluid density and tube volume, $\rho A_c \ell$, and the fluid velocity in the axial direction v_x is the ratio of fluid flow Q_V to tube cross-sectional area. Substituting these relationships into (5.135), we obtain:

$$\Delta P = I\left(\frac{dQ_V}{dt}\right), \qquad (5.136)$$

where I is defined as the fluid inertance:

$$I = \frac{\rho \ell}{A_c}. \qquad (5.137)$$

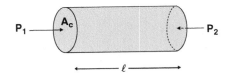

Fig. 5.42 Fluid in tube exposed to pressure difference

It should be noted that fluid inertance depends not only on the fluid density, ρ, but also on the ratio of the tube length to its cross-sectional area, a geometric factor. Equation (5.136) is analogous to the expression defining inductance L in an electrical circuit:

$$\Delta E = L\left(\frac{dI}{dt}\right). \tag{5.138}$$

Comparing (5.136) and (5.138), the potential difference across an inductive element is analogous to the pressure difference across the fluid in a tube and the current through the element is analogous to the flow through the tube.

> **Example 5.15.1 Startup Flow in a Rigid Tube.**
> Estimate the time it takes for blood, originally at rest, to come to a new steady-state flow after a step increase in pressure is applied at the inlet to an artery or arteriole.

Solution. *Preliminary considerations:* A new steady state will be attained after a time equal to about five time constants has elapsed. Therefore, our goal is to determine the time constant of the system when a step change in pressure is applied.

System definition and environmental interactions: The fluid has both resistance and inertia. A model of the fluid system is represented by series elements reflecting fluid resistance and fluid inertance, as shown in Fig. 5.43.

Apprising the problem to identify governing equations: The pressure drop across the vessel is equal to the pressure drop across the resistor plus the pressure drop across the inertance element. These pressure differences can be expressed in terms of the flow rate through the vessel:

$$\Delta P = (P_1 - P_2) + (P_2 - P_3) = Q_V \Re + I \frac{dQ_V}{dt}. \tag{5.139}$$

Analysis: Solving for Q_V with the initial condition that $Q_V = 0$, we find:

$$Q_V = \frac{\Delta P}{\Re}\left(1 - \exp\left(-\frac{\Re}{I}t\right)\right).$$

Therefore, the flow starts at zero and after a long period of time levels off at $\Delta P/\Re$, as expected. The time constant governing the transient is:

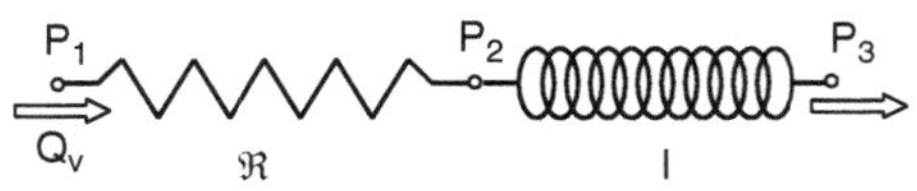

Fig. 5.43 Circuit analogy for fluid inertance and resistance elements

5.15 Fluid Inertia

$$\tau = \frac{I}{\mathfrak{R}} = \frac{\rho \ell}{\mathfrak{R} A_c}.$$

The resistance is given by (5.112) for laminar flow of a Newtonian fluid in a round, rigid tube. The cross-sectional area is πR^2, so the time constant can be computed from the following relation:

$$\tau = \frac{R^2}{8\nu}.$$

Examining and interpreting the results: Note that the time constant depends only on the tube radius and kinematic viscosity of the fluid. It is independent of the pressure difference applied across the tube, independent of the tube length, and independent of other fluid properties. A tube with a radius twice as large as another will have a time constant four times greater than the smaller tube. The time constant reflects the ratio of inertial effects to viscous effects. Inertial effects can be safely ignored in very small blood vessels, but not in large arteries. An artery with $R = 0.5$ cm and $\nu = 0.04$ cm^2/s will have a time constant of 0.78 s. Inertial effects would be virtually absent after a period of five time constants, or 3.9 s. An arteriole with radius of 50 μm will have a time constant of 7.8×10^{-5} s; so the response is virtually instantaneous.

Additional comments: It should be noted that the velocity profile in a tube is not parabolic during the early phase of startup flow; so the intravascular resistance actually increases with time. The times estimated above, based on a constant resistance, will underestimate the actual times necessary to reach a steady state. A more exact solution (Schlichting 1968) shows that flow stabilizes when $\nu t/R^2 = 1$. The corresponding time is 6.25 s, rather than 3.9 s for the artery and 6.25×10^{-4} s for the arteriole.

Example 5.15.2 Periodic Flow in a Rigid Artery.
Periodic pressure variations generated by the heart induces periodic flows in large arteries. We are interested in predicting flow through a rigid artery when exposed to a periodic pressure at the inlet $P_1 = P_{10} + P_{11}\sin(\omega t)$ and a constant pressure at the outlet, P_3.

Solution. *Preliminary considerations:* Since blood has inertia, we would expect a lag time to exist between the time when the maximum pressure is applied and the time when the maximum flow is attained in the vessel. In addition, the oscillations in flow for a given pressure pulse would be expected to vary with the inertance, and hence to vary with vessel size. Our goal is to examine how the magnitude and phase of the flow relative to pressure vary with vessel size.

System definition and environmental interactions: Our model of flow through the blood vessel remains the same as in the previous example, as shown in Fig. 5.43.

Apprising the problem to identify governing equations: Applying (5.139) for flow through a vessel with inertance and resistance, and summing the pressure differences across the vessel in Fig. 5.43, we arrive at the appropriate governing equation:

$$P_{10} + P_{11}\sin(\omega t) - P_3 = Q_V \Re + I\frac{dQ_V}{dt}.$$

Analysis: The general solution to this expression is:

$$Q_V = a\sin(\omega t) + b\cos(\omega t) + c + d\exp\left(-\frac{\Re t}{I}\right).$$

The constants a, b, c, and d can be determined by substituting this expression for Q_V back into the governing expression and applying the initial condition. However, we are not interested in the transient response, which we know from the previous example will be completed within a few seconds. Thus, for $t \gg I/\Re$, the last term will be negligible and the forced response will be:

$$Q_V(t) = \frac{P_{11}}{\Re}\left(\frac{1}{1+\left(\frac{\omega I}{\Re}\right)^2}\right)\sin(\omega t) - \frac{P_{11}}{\Re}\left(\frac{\frac{\omega I}{\Re}}{1+\left(\frac{\omega I}{\Re}\right)^2}\right)\cos(\omega t) + \frac{P_{10} - P_3}{\Re}.$$

Examining and interpreting the results: The mean flow $\bar{Q}_V$ is given by the last term in the above expression, which is what we would expect if a constant pressure difference of $P_{10} - P_3$ was applied. The flow lags the pressure oscillation and oscillates about the mean with periodic components given by the first two terms. Dividing the solution by the mean flow:

$$\frac{Q_V(t)}{\bar{Q}_V} = 1 + \left[\frac{P_{11}}{P_{10} - P_3}\right]\left[\frac{1}{1+\left(\frac{\omega I}{\Re}\right)^2}\right]\left[\sin(\omega t) - \left(\frac{\omega I}{\Re}\right)\cos(\omega t)\right].$$

Two dimensionless parameters determine how the periodic flow varies relative to the average flow. The first is the ratio of the periodic pressure component to the mean pressure difference across the tube. If pressure oscillations are small, then the flow will be nearly constant. The second dimensionless group is $\omega I/\Re$. The flow ratio is plotted as a function of ωt for a single cycle in Fig. 5.44. The oscillating pressure relative to mean pressure difference is 0.5 and plots are shown for several values of $\omega I/\Re$.

Additional comments: Substituting the values of I and $\Re$ for a tube, the ratio $\omega I/\Re$ can be rewritten:

$$\frac{\omega I}{\Re} = \frac{1}{8}\left(\frac{\omega R^2}{\nu}\right).$$

5.15 Fluid Inertia

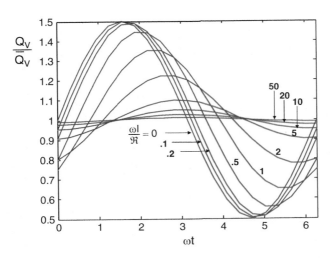

Fig. 5.44 Effect of $\omega l/\Re$ on flow through a vessel relative to mean flow as a function of ωt for a sinusoidal inlet pressure with $P_{11}/(P_{10} - P_3) = 0.5$

The dimensionless group in parentheses is related to a well-known dimensionless group which arises in periodic flow problems, known as the *Womersley number* α:

$$\alpha \equiv R\sqrt{\frac{\omega}{\nu}} = \sqrt{8\left(\frac{\omega l}{\Re}\right)}. \tag{5.140}$$

If this group is very small, then the cosine term in our solution above can be neglected and the flow will be in phase with the pressure.

$$\frac{Q_V(t)}{\bar{Q}_V} = 1 + \left[\frac{P_{11}}{P_{10} - P_3}\right]\sin(\omega t), \quad \frac{\alpha^2}{8} \ll 1.$$

On the other hand, if the Womersley number is much larger than unity, then the sine term will be small and:

$$\frac{Q_V(t)}{\bar{Q}_V} = 1 - \left[\frac{\left(\frac{P_{11}}{P_{10} - P_3}\right)}{\left(\frac{\alpha^2}{8}\right)}\right]\cos(\omega t), \quad \frac{\alpha^2}{8} \gg 1.$$

Therefore, for high values of the Womersley number, the oscillating component of flow is completely out of phase with the pressure, and the magnitude of the oscillating component becomes smaller as α becomes larger. For very large values of α, the oscillating component is negligible and although the inlet pressure oscillates with time, the flow through the tube remains relatively constant.

A typical value of the angular frequency for blood flow in arteries is $\omega = 2\pi$ radians/s, corresponding to one heart beat per second. The Womersley number for an artery with radius of 0.5 cm is about 6.3 and $\alpha^2/8$ is 4.9. Therefore, from Fig. 5.44 with $\omega I/\Re = 5$, we might expect the oscillating component of flow in an artery this size or larger to be nearly out of phase with the oscillating component of pressure. For an arteriole, however, the Womersley number is 6.3×10^{-3} and the flow will oscillate in phase with the pressure. In reality, the oscillating component of blood pressure is damped as the vessels become smaller; so flow through small blood vessels like arterioles is nearly constant. A more detailed analysis of periodic flow in rigid arteries is given in Example 7.12.3.

> **Example 5.15.3 Periodic Flow in a Compliant Artery.**
> Repeat the analysis in the previous example for flow in a brachial artery, accounting for resistance, compliance, and inertia. Provide a relationship between periodic flow through the vessel and average flow through the vessel.

Solution. *Preliminary considerations:* We anticipate two time constants in this case, one associated with inertia, $\tau_{\Re I} = I/\Re$, and another associated with compliance, $\tau_{\Re C} = \Re C$. Our goal is to determine how the magnitude and phase lag for blood outflow is affected by dimensionless time constants $\omega I/\Re$ and $\omega \Re C$, particularly for values appropriate for the brachial artery.

System definition and environmental interactions: There are several possible models of blood flow in a compliant vessel that accounts for inertia and resistance. One model is shown in Fig. 5.45, where a compliant element is placed centrally in the vessel, so that half the resistance and half the inertance lie on each side of the compliant section of the vessel. In reality, the compliance is distributed throughout the vessel, but we can understand the fundamental principles by examining the behavior of the symmetric model depicted in Fig. 5.45. Inertia will cause a lag between pressure and flow through the vessel. In addition, some of the blood entering the vessel goes to expanding the vessel volume. Consequently, the outflow Q_{23} will generally not equal the inflow, Q_{12}, causing an additional phase lag between pressure and flow.

Apprising the problem to identify governing equations: Our previous analysis of the pressure–flow relationship through an $\Re I$ element (5.139) provides appropriate expressions for Q_{12} and Q_{23}:

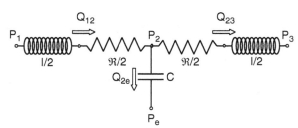

Fig. 5.45 Model of a compliant blood vessel with inertance

5.15 Fluid Inertia

$$P_1 - P_2 = \left(\frac{\Re}{2}\right)Q_{12} + \left(\frac{I}{2}\right)\frac{dQ_{12}}{dt},$$

$$P_2 - P_3 = \left(\frac{\Re}{2}\right)Q_{23} + \left(\frac{I}{2}\right)\frac{dQ_{23}}{dt}.$$

Subtracting the second expression from the first:

$$P_1 - 2P_2 - P_3 = \left(\frac{\Re}{2}\right)(Q_{12} - Q_{23}) + \left(\frac{I}{2}\right)\frac{d}{dt}(Q_{12} - Q_{23}).$$

The flow to the compliant portion of the tube is:

$$Q_{2e} = Q_{12} - Q_{23} = \frac{dV}{dt} = C\frac{d(P_2 - P_e)}{dt} = C\frac{dP_2}{dt},$$

where V is the vessel volume and C is the compliance of the vessel. This expression can be substituted into the one above to eliminate $Q_{12} - Q_{23}$. If the downstream pressure P_3 is constant and the upstream pressure P_1 is periodic, then the only unknown is P_2. The governing differential equation is:

$$\frac{d^2 P_2}{dt^2} + \frac{\Re}{I}\left(\frac{dP_2}{dt}\right) + \frac{4}{CI}P_2 = \frac{2}{CI}[P_{10} + P_{11}\sin(\omega t) + P_3].$$

Once the pressure at node 2 is found by solving this equation, we can compute Q_{12}, Q_{23}, and Q_{2e} from the expressions above. Our goal is to compute the outlet flow relative to the mean flow, as in the previous example; so it will only be necessary to compute P_2 and Q_{23}.

Analysis: We are only interested in the forced responses for P_2; so the general solutions to the governing differential equation for P_2 will be:

$$P_2(t) = a\sin(\omega t) + b\cos(\omega t) + c,$$

where a, b, and c are constants to be determined by substituting these expressions back into the differential equations for P_2. Solving the governing equation for P_2, we find the coefficients to be:

$$a = \frac{2P_{11}\left(4 - \omega\Re C\left(\frac{\omega I}{\Re}\right)\right)}{\left[4 - \omega\Re C\left(\frac{\omega I}{\Re}\right)\right]^2 + (\omega\Re C)^2}, \quad b = \frac{-2P_{11}(\omega\Re C)}{\left[4 - \omega\Re C\left(\frac{\omega I}{\Re}\right)\right]^2 + (\omega\Re C)^2},$$

$$c = \frac{P_{10} + P_3}{2}.$$

Substituting the solution for P_2 into the differential equation for Q_{23}:

$$\frac{dQ_{23}}{dt} + \left(\frac{\Re}{I}\right) Q_{23} = \left(\frac{2}{I}\right) [a \sin(\omega t) + b \cos(\omega t) + c - P_3].$$

The forced response for outflow is given by:

$$Q_{23}(t) = A \sin(\omega t) + B \cos(\omega t) + C^*,$$

where A, B, and C^* are constants (note: we use C^* to distinguish this constant from the compliance, C). Solving for the new coefficients in terms of a, b, and c, and dividing by the average inlet flow, $(P_{10} - P_3)/\Re$, we obtain the final solution:

$$\frac{Q_{23}}{\bar{Q}_V} = 1 + \frac{\left(\frac{2a}{P_{10} - P_3}\right)}{\left(1 + \left(\frac{\omega I}{\Re}\right)^2\right)} \left[\left(1 + \frac{b}{a}\left(\frac{\omega I}{\Re}\right)\right) \sin(\omega t) + \left(\frac{b}{a} - \left(\frac{\omega I}{\Re}\right)\right) \cos(\omega t) \right].$$

The factor $a/(P_{10} - P_3)$ is proportional to the ratio of the oscillating pressure component relative to the mean pressure difference across the vessel. Recall that the time constant for a section with resistance and inertance is $\tau_{\Re I} = R^2/8v$. The brachial artery has a radius of about 0.23 cm. For blood, with $v = 0.04$ cm^2/s, and for an angular frequency $\omega = 2\pi$ s^{-1}, this corresponds to $\tau_{\Re I} = 0.159$ s and $\omega I/\Re = \omega \tau_\Re = 1.0$. We plot the outflow relative to the average inflow vs. ωt over one cycle for $P_{11}(P_{10} - P_3) = 0.5$, $\omega I/\Re = 1.0$ and for various values of $\omega \Re C$ in Fig. 5.46.

Examining and interpreting the results: Note that when $\omega \Re C = 0$, the coefficients a and b are $P_{11}/2$ and 0, respectively. In that case, the expression above reduces to the expression found for a rigid tube in the previous example.

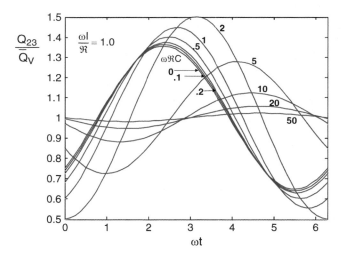

Fig. 5.46 Outflow from a compliant vessel relative to the average flow in the vessel. The oscillating component of pressure is half the mean pressure difference across the vessel and $\omega I/\Re = 1.0$

5.15 Fluid Inertia

This corresponds to the graph for $\omega I/\Re = 1$ in Fig. 5.44. Note that as vascular compliance increases, the curves shift further to the right, indicating a larger phase lag relative to the input pressure. This is caused by fluid which expands the vessel during a positive pulse, rather than flowing through it. As the pressure drops in the tube, the vessel volume becomes smaller, pushing the stored fluid out. This produces a delay in the flow peak relative to the pressure peak. The peak of the pressure pulse occurs at $\omega t = \pi/2$ (i.e., $\omega t = 1.571$ in Fig. 5.46). The peak outflow initially increases with increasing compliance, but after reaching a maximum at $\omega \Re C = 2$, peak outflow begins to fall off with increasing compliance. The phase lag between peak pressure and peak flow increases with increasing values of $\omega \Re C$. The flow peak lags the pressure peak at $\omega \Re C = 5$ by about 2.6 radians, or nearly 150°.

Compliance measurements reported in the literature range over several orders of magnitude. Most estimates are below 0.01 ml/mmHg. Using this as an upper limit, we can compute the time constant $\tau_{\Re C} = \Re C$ from the following:

$$\tau_{\Re C} = \Re C = \left(\frac{8\mu\ell}{\pi R^4}\right) C = \left(\frac{8\nu}{R^2}\right)\left(\frac{\rho\ell}{A_c}\right) C = \left(\frac{8\nu}{R^2}\right) IC = \frac{IC}{\tau_{\Re I}}.$$

For blood with $\rho = 1.0$ g/ml in a 10 cm long segment of artery, the inertance $I = 60.2$ g. The time constant $\tau_{\Re C}$ is:

$$\tau_{\Re C} = \frac{IC}{\tau_{\Re I}} = \frac{(60.2\text{ g})(0.01\text{ cm}^3\text{mmHg}^{-1})}{0.159\text{ s}}\left(\frac{1\text{ mmHg}}{1333.2\text{ g cm}^{-1}\text{s}^{-2}}\right) = 2.84 \times 10^{-3}\text{ s}.$$

Therefore, $\omega \Re C = 0.018$ and we might expect the outflow to be similar to the plot for $\omega \Re C = 0$ in Fig. 5.46. In this case, the outflow peak is only slightly damped and lags the pressure peak by about 50°.

Additional comments: The results presented in Fig. 5.46 are for a single value of $\omega I/\Re = 1.0$. This is representative of small arteries with diameters of about 5 mm, such as the brachial artery or carotid artery. Predictions for other combinations of $\omega \Re C$ and $\omega I/\Re$ are shown in Fig. 5.47. The peak outflow relative to the mean flow is

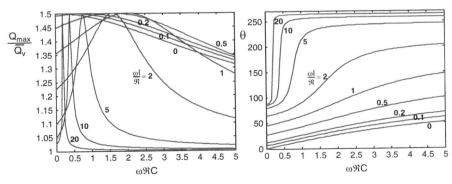

Fig. 5.47 Influence of the dimensionless time constants $\omega \Re C$ and $\omega I/\Re$ on the peak outflow relative to average flow (*left*) and phase lag of the outflow in degrees relative to the pressure (*right*)

shown in the left panel as a function of $\omega \Re C$ for several values of $\omega l/\Re$. The right panel shows the corresponding phase lag between the pressure peak and the flow peak. All curves show the same maximum value for peak outflow of 1.5 times mean flow, but the location of the maximum depends on the combination of dimensionless time constants.

5.16 Blood Flow in Organs

When presenting the Windkessel model of arterial flow in Sect. 5.13, we treated systemic resistance as though it were a constant resistance. This is a reasonable assumption for small variations in cardiac output or mean arterial pressure, but is not reasonable for larger variations. Blood flow through any organ is generally not a linear function of the pressure difference applied across the organ. There are many factors that contribute to the overall resistance. Neural and humeral factors can cause active alterations to vascular resistance, primarily by causing arteriolar vasoconstriction or vasodilation.

Passive mechanisms also contribute to changes in vascular resistance. As the pressure difference across small diameter vessels increases, red cells deform into more streamlined shapes (see Fig. 4.26), resulting in a decrease in vascular resistance. Conversely, when capillary pressure decreases, the lubrication layer between red cells and the vessel wall in small capillaries becomes smaller, and can even break down entirely, causing cells to seize up and halt flow through that capillary. This is known as capillary derecruitment, which can reduce the surface area available for solute exchange. Increasing vascular pressure can recruit flow through those same vessels by providing enough force to overcome friction between erythrocytes and the capillary wall.

Vascular distension is another passive mechanism that influences vascular resistance. Since blood vessels are compliant, their resistance decreases as transmural pressure increases. An elevation in intravascular pressure tends to reduce vascular resistance, while an increase in perivascular pressure tends to elevate vascular resistance. Negative transmural pressure differences can collapse vessels, also causing derecruitment. This is known to occur in the lung when alveolar capillary pressure drops below alveolar gas pressure. It also occurs in the coronary circulation during ventricular contraction, where extravascular compression briefly occludes the left coronary artery.

Vascular resistance in the lung is unique because it depends on the orientation of the lung relative to gravity. When an individual stands or sits, the uppermost portions of the lung are above the level of the heart and the bottom of the lung lies below the level of the heart. If we can neglect pressure losses in the large vessels, then pressures at the arterial and venous ends of a lung capillary bed relative to their height y above the level of the left atrium are:

5.16 Blood Flow in Organs

$$P_a = P_{PA} - \rho g y, \tag{5.141}$$

$$P_v = P_{LA} - \rho g y, \tag{5.142}$$

where ρ is the density of blood, g is the gravitational acceleration, P_{PA} is the pulmonary artery pressure, and P_{LA} is the left atrial pressure. Note that the pressure difference across any capillary bed $P_a - P_v$ is independent of the position of the bed relative to the left atrium. The difference between arterial pressure and venous pressure is independent of position for other organs besides the lung, so why is vascular resistance in the lung unique? Because, in other organs, the pressure surrounding the capillary bed also has a gravitational component, the transmural pressure difference is also independent of the vertical position of the capillary bed. This is not true in alveolar capillaries, where the surrounding pressure is atmospheric pressure at all vertical positions. Since the alveolar capillary wall is collapsible, the capillary bed behaves like a Starling Resistor. If we let P_A be alveolar pressure, then blood flow Q_B through a given capillary bed with resistance $\Re_{cap}$ will obey the Starling Resistor equations:

$$\begin{aligned} Q_B &= 0 \quad \text{if } P_A > P_a, \\ Q_B &= \frac{P_a - P_A}{\Re_{cap}} \quad \text{if } P_a > P_A > P_v, \\ Q_B &= \frac{P_a - P_v}{\Re_{cap}} \quad \text{if } P_a > P_v > P_A. \end{aligned} \tag{5.143}$$

Consequently, we can identify three potential regions in the lung, one for each of the conditions expressed by these equations. These are known as the three zones of the lung, as shown in Fig. 5.48.

Let the position of the boundary between zones 1 and 2 equal h_{12} and the position of the boundary between zones 2 and 3 equal h_{23}. When $y = h_{12}$, $P_a = P_A$ and when $y = h_{23}$, $P_v = P_A$. Substituting these values into (5.141) and (5.142), we find:

$$h_{12} = \frac{P_{PA} - P_A}{\rho g}, \tag{5.144}$$

$$h_{23} = \frac{P_{LA} - P_A}{\rho g}. \tag{5.145}$$

For example, if $\rho = 1.1$ g/ml, $P_{PA} = 13$ cmH$_2$O, $P_{LA} = -1$ cmH$_2$O, and $P_A = 1$ cmH$_2$O, then $h_{12} = 10.9$ cm and $h_{23} = -1.8$ cm. Any portion of the lung 1.8 cm below the level of the left atrium will be in zone 3 (fully open). Lung capillaries that lie anywhere from 1.8 cm below to 10.9 cm above the level of the left atrium will be in zone 2, where the driving pressure difference is $P_a - P_A$. If any portion of the lung is located more than 10.9 cm above the left atrium, then

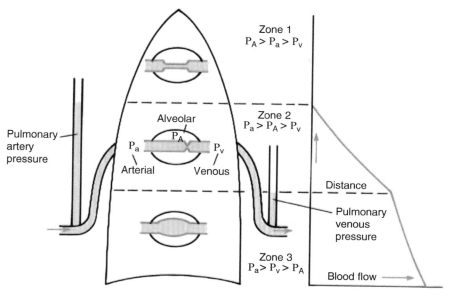

Fig. 5.48 Lung zones adapted from West et al. 1964

there will be no flow in those capillaries. To find the total flow through the lung, we need to know the conductance per unit volume L' and the distribution of volume from the bottom to the top of the lung. The flow per unit volume is:

$$\frac{dQ_B}{dV} = L'\Delta P. \tag{5.146}$$

L' is assumed constant, and $dV = A(y)dy$, where $A(y)$ is the lung cross-sectional area perpendicular to the vertical direction y. Integrating from the bottom of the lung $y = -h_b$ to the top, $y = h_t$ and substituting (5.141) and (5.142) for P_a and P_v:

$$Q_B = L'\left\{ (P_{PA} - P_{LA}) \int_{h_b}^{h_{23}} A(y)dy + \int_{h_{23}}^{h_{12}} (P_{PA} - \rho g y - P_A)A(y)dy \right\}. \tag{5.147}$$

This assumes that the top of the lung h_t is above h_{12} and the bottom of the lung h_b is below h_{23}, so that all three zones exist in the lung. If $h_{12} > h_t$, then there is no zone 1, and zone 2 will extend from h_{23} to h_t. In that case, the upper limit of the second term in (5.147) should be replaced with h_t. If $h_{23} > h_t$, then only zone 3 will exist in the lung. In that case the second term would be dropped altogether and the upper limit on the first term would be h_t. This might occur when the individual is in the horizontal position, rather than in the vertical position.

5.16 Blood Flow in Organs

Example 5.16.1 Pulmonary Resistance.
Assume that the right lung can be modeled as a cylinder with radius 3 cm and length of 22 cm. When placed in the vertical position, the top of the lung lies 11 cm above the level of the left atrium. Left atrial pressure is 4 cm, alveolar pressure is 0 cmH$_2$O, and pulmonary artery pressure is 15 cmH$_2$O. Blood density is 1.1 g/dl and the flow conductance per unit volume L' is 0.365 (liter blood)/(liter tissue)/min/cmH$_2$O. Find blood flow and nominal flow resistance when the lung is oriented in the vertical and horizontal positions.

Solution. *Initial considerations:* We will assume that the resistances of the arterial and venous conducting vessels are negligible in comparison with the resistance of the pulmonary capillary bed. Therefore, the inlet pressure for an alveolar capillary is the pressure in the pulmonary artery plus the hydrostatic pressure in the capillary relative to the level of the left atrium. Outlet pressure is the pressure in the left atrium plus the hydrostatic pressure in the capillary. Since capillaries are very short, we can neglect any hydrostatic variations in the capillary from arterial to venous ends. We will also assume that the capillary walls are very thin and behave as ideal Starling resistors.

System definition and environmental interactions: The system to be analyzed is the blood contained inside the pulmonary capillaries. This can be further divided into two subsystems: blood flowing through partially collapsed (zone 2) capillaries and blood flowing through completely uncollapsed capillaries (zone 3). Total blood flow through the lung depends on the fraction of the lung in each zone.

Apprising the problem to identify governing relationships: The boundaries of zones 1–2 and 2–3 can be found using (5.144) and (5.145). The expressions for flow through a Starling resistor can be used to find blood flow through capillaries at a particular height in the lung. Integration of these from the bottom of the lung to the top of the lung, as shown in (5.147), will allow us to calculate blood flow through the entire lung.

Analysis: We begin by computing the boundaries of zones 1–2 and 2–3 using (5.144) and (5.145):

$$h_{12} = \frac{15\,\text{cmH}_2\text{O} - 0\,\text{cmH}_2\text{O}}{(1.1\,\text{g/ml})(980\,\text{cm s}^{-2})}\left(980\,\text{g cm}^{-1}\,\text{s}^{-2}\,\text{cmH}_2\text{O}^{-1}\right) = 13.64\,\text{cm},$$

$$h_{23} = \frac{4\,\text{cmH}_2\text{O} - 0\,\text{cmH}_2\text{O}}{(1.1\,\text{g/ml})(980\,\text{cm s}^{-2})}\left(980\,\text{g cm}^{-1}\,\text{s}^{-2}\,\text{cmH}_2\text{O}^{-1}\right) = 3.64\,\text{cm}.$$

Therefore the region between -11 and $+3.64$ cm is in zone 3 and the region between 3.64 and 11 cm is in zone 2. Using (5.147) to compute the flow for constant cross-sectional area A, and using the top of the lung as the upper limit for zone 2:

$$Q_B = L'A \left\{ (P_{PA} - P_{LA})(h_{23} - h_b) + (P_{PA} - P_A)(h_t - h_{23}) - \rho g \frac{(h_t^2 - h_{23}^2)}{2} \right\}.$$

All values on the right-hand side are known; so we can compute the flow:

$$Q_B = \left(\frac{0.365 \, \text{Lmin}^{-1} \, \text{cmH}_2\text{O}}{1,000 \, \text{ml}} \right) (28.27 \, \text{cm}^2) \times \left\{ (11 \, \text{cmH}_2\text{O}) \, (3.64 \, \text{cm} - (-11 \, \text{cm})) \right.$$

$$\left. + (15 \, \text{cmH}_2\text{O}) \, (11 - 3.64 \, \text{cm}) - 1.1 \frac{\text{cmH}_2\text{O}}{\text{cm}} \frac{\left((11 \, \text{cm})^2 - (3.64 \, \text{cm})^2 \right)}{2} \right\}.$$

Or, finally:
$Q_B = 2.19$ L/min (vertically oriented lung)
The nominal resistance in the vertical position is:

$$\Re = \frac{P_{PA} - P_{LA}}{Q_B} = \frac{11 \, \text{cmH}_2\text{O}}{2.19 \, \text{L min}^{-1}} = 5 \, \text{cmH}_2\text{O} \, \text{L}^{-1} \text{min}.$$

For the horizontal case, the top of the lung is at +3 cm and the bottom of the lung is at −3 cm. Consequently, since $h_{23} > h_t$, the entire lung is in zone 3. In this case, the resistance can be computed from the conductance of the entire lung:

$$\Re = \frac{1}{L'A(h_t - h_b)} = \frac{1}{\left(\frac{0.365 \, \text{Lmin}^{-1} \, \text{cmH}_2\text{O}}{1,000 \, \text{ml}} \right) (28.27 \, \text{cm}^2) \, (22 \, \text{cm})}$$

$$= 4.4 \, \text{cmH}_2\text{O} \, \text{L}^{-1} \, \text{min},$$

and the blood flow in the horizontal case is:

$$Q_B = \frac{P_{PA} - P_{LA}}{\Re} = \frac{11 \, \text{cmH}_2\text{O}}{4.4 \, \text{cmH}_2\text{O} \, \text{L}^{-1} \, \text{min}} = 2.5 \, \text{L/min}.$$

Examining and interpreting the results: When perfused in the vertical orientation, 33% of the lung is in zone 2 and 67% in zone 1. We predict the flow to be about 14% higher when the lung is perfused in the horizontal position than when perfused in the vertical position with the same pulmonary artery and left atrial pressures. However, a cylindrical model of the lung tends to overestimate the effects of lung zones. The lung is tapered from apex to base; so the actual volume of capillaries in zone 2 is less and in zone 3 is more than the corresponding volumes in a cylindrical model.

5.17 Osmotic Pressure and Flow

There are many instances in physiology and biology in which fluids are separated by a barrier that allows water, and perhaps some solutes to pass through it. Cell membranes and the blood–tissue barrier are primary examples. The barriers are often idealized as a solid material permeated by several channels that are continuous with the fluids on each side of the membrane. The volumetric flow from side 1 to side 2 through a single channel i separating fluids 1 and 2 (Fig. 5.49) is:

$$Q_{Vi} = \frac{P'_1 - P'_2}{\Re_i} = L_i A_{ci}(P'_1 - P'_2), \qquad (5.148)$$

where P'_1 is the pressure inside the channel at the interface with fluid 1; P'_2 is the pressure inside the channel at the interface with fluid 2; $\Re_i = 1/(L_i A_{ci})$ is the resistance of a single channel; A_{ci} is the cross-sectional area of the channel (assumed constant); L_i is defined as the hydraulic conductivity or hydraulic permeability of the channel.

If the channel is a circular pore with constant radius R_i, then $A_{ci} = \pi R_i^2$ and from (5.84), $L_i = R_i^2/8 \, \mu l$. Note that the pressure difference driving flow through the channel is the pressure difference *inside the channel*, signified using the prime symbols. Since the concentrations of some solutes in the bulk fluid are different from the concentrations just inside the channel, the pressure inside the channel P' will be different from the pressure P in the bulk fluid. We would like to write (5.148) in terms of the bulk fluid pressures at each end of the channel, rather than in terms of the internal pressures at each end of the channel. If we assume that water is in thermodynamic equilibrium at the interface between bulk fluid and fluid inside the channel, then the chemical potential of water will be equal in the two fluids:

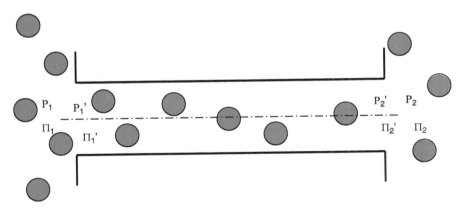

Fig. 5.49 Channel through barrier separating fluids 1 and 2. Solute concentration is different inside channel at each end than outside the channel

$$P - P' = -\frac{RT}{V_w} \ln\left(\frac{a_w}{a'_w}\right), \qquad (5.149)$$

where R is the universal gas constant; T is absolute temperature; V_w is the partial molar volume of water; a_w is the activity of water in the bulk fluid at the interface; and a'_w is the activity of water in the channel fluid at the interface.

If the channel excludes all solutes except for water, the activity in the channel will be unity and the activity in the bulk fluid will be less than one. We see from (5.148) that the equilibrium pressure in the bulk fluid will be greater than the pressure in the channel.

Let us define the osmotic pressure Π as the pressure that must be added to the bulk solution to keep it in equilibrium with pure water:

$$\Pi \equiv -\frac{RT}{V_w} \ln(a_w). \qquad (5.150)$$

If the channel admits solutes, then it too will require an additional pressure to keep it in equilibrium with pure water:

$$\Pi' \equiv -\frac{RT}{V_w} \ln(a'_w). \qquad (5.151)$$

Applying (5.150) and (5.151) at each end of the channel and substituting the resulting expressions into (5.148) yields:

$$Q_{Vi} = L_i A_{ci}[P_1 - P_2 - \sigma_{di}(\Pi_1 - \Pi_2)], \qquad (5.152)$$

where σ_{di} is known as the *osmotic reflection coefficient* for the channel and is defined as:

$$\sigma_{di} = 1 - \left[\frac{\Pi'_{1i} - \Pi'_{2i}}{\Pi_1 - \Pi_2}\right]. \qquad (5.153)$$

If all solutes are excluded at both ends of the channel, then $\Pi_{1i}' = \Pi_{2i}'$ and $\sigma_{di} = 1$. This would apply to channels in a semipermeable membrane. On the other hand, if the channel does not exclude solutes at either end, then $\Pi_{1i} = \Pi_{2i}$ and $\sigma_{di} = 0$, and the driving pressure for flow is simply the pressure difference between the two bulk fluids, $P_1 - P_2$. Between these two extremes, the reflection coefficient can be interpreted as the fraction of total bulk fluid osmotic pressure that is exerted across a channel that allows at least one solute to pass through it.

Equation (5.152) describes volumetric flow through a single channel connecting the two fluids. A barrier consisting of N_i identical channels in a biological barrier will have a flow equal to N_i times the flow in a single channel:

5.17 Osmotic Pressure and Flow

$$Q_v = \sum_{i=1}^{N_i} Q_{vi} = K_f[P_1 - P_2 - \sigma_d(\Pi_1 - \Pi_2)], \tag{5.154}$$

where $\sigma_d = \sigma_{di}$ and K_f is a fluid conductance known as the filtration coefficient of the barrier and is equal to $N_i L_i A_{ci}$. Flow is positive when directed from side 1 of the barrier to side 2. Equation (5.154) is known as the *Starling Equation* and is used to describe flow across both heterogeneous and homogeneous barriers. If a barrier consists of i different types of channels and the number of each type of channels is n_i, then the overall coefficients are related to the individual channel coefficients as follows:

$$K_f = \sum_i n_i L_i A_{ci}, \tag{5.155}$$

$$\sigma_d = \frac{\sum_i n_i L_i A_{ci} \sigma_{di}}{K_f}. \tag{5.156}$$

The distribution of channels through the barrier is usually unknown, and the overall coefficients in the Starling Equation are determined experimentally.

Channels in biological barriers can vary in both number and size. Aquaporin is a water-specific channel and is the primary channel for water transport across many cell membranes. The number of channels that span the cell membrane are under cellular control, and so K_f can change in response to cellular needs. However, the osmotic reflection coefficient for all aquaporin pathways is independent of the number of channels, since the reflection coefficient of each individual channel is unity. In capillaries, intracellular junctions between endothelial cells form channels that connect plasma to interstitial fluid. These junctions can vary in size and consequently will have different reflection coefficients. Endothelial cells themselves contain additional channels on the lumenal and ablumenal surfaces, which allow fluid to move directly through endothelial cells. Consequently, the blood–tissue barrier is heterogeneous and both LS and σ_d will change as the distribution of pathways changes.

Let us now turn our attention to methods for estimating the osmotic pressure terms in the Starling Equation. The activity of water is equal to the product of the activity coefficient for water and the mole fraction of water, x_w. For dilute solutions, the activity coefficient of water is very nearly unity and so (5.150) can be written:

$$\Pi \approx -\frac{RT}{V_w} \ln(x_w). \tag{5.157}$$

The mole fraction of water is one minus the mole fractions of all solutes present:

$$x_w = 1 - \sum_s x_s. \tag{5.158}$$

For a dilute solution, the sum of all the mole fractions of the solutes will be small, and the logarithm of one plus a small number can be expressed using the following series expansion:

$$\ln(1+\phi) = \phi - \frac{1}{2}\phi^2 + \frac{1}{3}\phi^3 + \cdots, \quad \phi = -\sum_s x_s. \tag{5.159}$$

Substituting (5.158) and (5.159) into (5.157) and noting that by definition, the mole fraction $x_s = C_s/c$, we obtain an approximate expression for osmotic pressure in terms of bulk solute concentrations:

$$\Pi = -\frac{RT}{V_w c}\left[-\sum C_s - \frac{1}{2c}\left(\sum C_s\right)^2 - \frac{1}{3c^2}\left(\sum C_s\right)^3 + \cdots\right]. \tag{5.160}$$

However, the product of total molar concentration and the partial molar volume of water is nearly unity for a dilute solution. Therefore, osmotic pressure is related to total molar concentration of all solutes other than water as follows:

$$\Pi = RT\left[C_T + AC_T^2 + BC_T^3 + \cdots\right], \quad C_T \equiv \sum C_s. \tag{5.161}$$

The coefficients $A, B, \ldots$ are known as the virial coefficients, and since they are only applied in situations when concentrations are not dilute, they are determined experimentally. For dilute solutions, the higher order terms are quite small, and the osmotic pressure is approximated well by *van't Hoff's Law*:

$$\Pi = RT\sum_s C_s = RTC_T \quad \text{(dilute)}. \tag{5.162}$$

At 37°C, the osmotic pressure difference in mmHg can be expressed in terms of the concentration difference in mM:

$$\Delta\Pi(\text{mmHg}) = 19.3\,\Delta C\,(\text{mM}). \tag{5.163}$$

The molar concentration of a solute in solution is proportional to the number of molecules present in solution. A solute like sodium chloride, when dissolved in an aqueous solution, dissociates into sodium and chloride ions. Consequently, NaCl must be treated as two separate solutes when considering the osmotic pressure of the solution. In general, if a compound s dissociates into i_s ions in solution, then the osmotic pressure difference, in terms of the concentration of the original undissociated compound, is:

$$\Delta\Pi_s = RT[i_s \Delta C_s], \tag{5.164}$$

where $\Delta\Pi_s$ is the contribution to total osmotic pressure made by solute s, which dissociates into i_s ions. The product $i_s \Delta C_s$ is known as the osmolarity difference for

5.17 Osmotic Pressure and Flow

solute s and is expressed in milliosmoles/L or milliosmolar units (mOsM). The number of dissociating ions i can also be interpreted as the ratio of mOsM per mM of solute. The cytoplasm of normal cells has an osmolarity of approximately 300 mOsM, which translates to an uncorrected osmotic pressure of 5,790 mmHg. This is equivalent to the pressure at the bottom of a column of water 258 ft tall!

Ordinarily there is no difference in hydrostatic pressure across mammalian cell membranes. Thus, if there is no net flow of water across the cell membrane, the osmolarity of the cytoplasm will equal the osmolarity of the extracellular fluid. The extracellular fluid is said to be *isotonic* in this case. If the osmolarity of the extracellular fluid is lower than that of the intracellular fluid, the extracellular fluid is said to be *hypotonic*. In that case, water will tend to go down its concentration gradient from the hypotonic extracellular fluid into the cytoplasm, where the water concentration is lower. Consequently, the cell will swell. This is consistent with the Starling Equation prediction. Likewise, if the extracellular fluid osmolarity is higher than intracellular osmolarity, fluid will flow out of the cell. The extracellular fluid in this case is said to be *hypertonic*. Intracellular concentrations of macromolecules are ordinarily higher than extracellular concentrations. These macromolecules are charged and they attract extracellular ions of the opposite charge. This effect, known as the *Gibbs–Donnan effect*, tends to increase intracellular osmolarity relative to extracellular osmolarity, and by itself would continually cause water to flow into the cell. The rigid walls of plant cells allow intracellular pressure to increase, ultimately halting the flow of water. Mammalian cell walls are not rigid and cannot support significant hydrostatic pressure differences. Fortunately, cells have active mechanisms that help regulate intracellular osmolarity in response to changes in extracellular osmolarity, but these are relatively slow. Rapid exposure of cells to hypotonic media can cause the cells to swell beyond their capacity to compensate and the cells will burst. This is known as *cell lysis*.

Not all intracellular or extracellular water is osmotically active. Water that is bound to structural elements within the cell or bound to other intracellular molecules cannot be considered as bulk water in the osmotic pressure calculation. This osmotically unresponsive water can make up a substantial fraction of the total water in some cells.

The channels in the capillary barrier separating blood plasma and interstitial fluid are often large enough so that small solutes can freely pass through them. Consequently, the difference in concentrations of macromolecules across the capillary barrier is primarily responsible for the osmotic pressure difference. This is referred to as the *oncotic pressure difference* or the colloid oncotic pressure difference. The oncotic pressure in plasma is normally about 25 mmHg, which is 232 times smaller than the osmotic pressure in either the intracellular or extracellular fluid.

Although (5.161) indicates that oncotic pressures should be related to molar concentrations of proteins, the physiological literature typically reports oncotic pressures in terms of mass concentrations of proteins ρ_s(g/dl) rather than molar concentrations. For example, Landis and Pappenheimer (1963) found that the oncotic pressure of a solution of human albumin at 37°C was related to mass concentration of albumin as follows:

$$\Pi_{\text{alb}} = 2.8\rho_{\text{alb}} + 0.18\rho_{\text{alb}}^2 + 0.012\rho_{\text{alb}}^3, \quad (5.165)$$

where Π is given in units of mmHg and ρ_{alb} is in g/dl. The molecular weight of albumin can be estimated by letting the concentration become very small and setting (5.165) equal to van't Hoff's law, with $\rho_{\text{alb}} = M_{\text{alb}} C_{\text{alb}}$.

$$\Pi_{\text{alb}}(\text{mmHg}) = 19.3 C_{\text{alb}}(\text{mM})$$
$$= 2.8 \frac{\rho_{\text{alb}}(\text{g/dl})}{M_{\text{alb}}(\text{g/mol})} \left(\frac{10\,\text{dl}}{\text{liter}}\right) \left(1,000 \frac{\text{mmol}}{\text{mol}}\right). \quad (5.166)$$

Solving for M_{alb} provides an estimate of 68,930 for the molecular weight of albumin, which is slightly higher than the accepted molecular weight for human albumin. Landis and Pappenheimer (1963) also measured the relationship between plasma oncotic pressure and total plasma protein concentration under normal conditions at 37°C:

$$\Pi_{\text{plasma}} = 2.1\rho_{\text{plasma}} + 0.16\rho_{\text{plasma}}^2 + 0.009\rho_{\text{plasma}}^3. \quad (5.167)$$

There are many other empirical relationships found in the literature, primarily because oncotic pressure depends on the distribution of individual plasma proteins, and not just on the total mass concentration (see Example 5.17.4). Albumin is normally the most abundant of the plasma proteins and consequently contributes significantly to total plasma oncotic pressure. If the distribution of proteins is uniform, and only the overall concentration varies, then relationships like (5.167) are appropriate. If not, the relative fraction of albumin to globulin will affect the virial coefficients when oncotic pressure is expressed in terms of protein mass concentration. The effect should be less when the osmotic pressure is expressed in terms of molar concentration [e.g., (5.161)].

Example 5.17.1 Flow Across a Cell Membrane.
A cell with radius of 5 μm is initially in osmotic equilibrium with its surroundings at 300 mOsM and 25°C. The cell is placed in a 155 mM solution of NaCl. Which way will water flow through the membrane and what is the initial flow rate if the filtration coefficient of the cell is 4×10^{-4} μm^3/s/mmHg? Estimate the final cell volume.

Solution. *Initial considerations:* We will assume that the solute concentrations are small enough to allow use of the van't Hoff equation. Also, we will assume that sodium chloride dissociates completely into two ions. We will assume that the hydrostatic pressure difference across the membrane is initially zero and the osmotic reflection coefficient is unity.

5.17 Osmotic Pressure and Flow

System definition and environmental interactions: The system of interest is the cytoplasm of the cell. Water can pass through the membrane, driven by osmotic and hydrostatic pressure differences, but solutes are prevented from moving through the membrane.

Apprising the problem to identify governing relationships: Flow through the membrane is governed by the Starling equation, (5.154).

Analysis: Since NaCl dissociates completely into Na$^+$ and Cl$^-$ ions, the osmolarity of the NaCl solution is $2 \times 155 = 310$ mOsM, which is greater than the intracellular osmolarity. Therefore, water concentration is higher in the cytoplasm than in the surroundings. Water will flow down its concentration gradient from cytoplasm to the surroundings. From our assumption that the osmotic pressure in cytoplasm (c) and surrounding solution (s) can be estimated using the van't Hoff equation, we find:

$$\Pi_c = 19.3(300\,\text{mOsM}) \left(\frac{273\,°\text{K} + 25\,°\text{K}}{273\,°\text{K} + 37\,°\text{K}} \right) = 5,566\,\text{mmHg},$$

$$\Pi_s = 19.3(2)(155\,\text{mM}) \left(\frac{273\,°\text{K} + 25°\text{K}}{273\,°\text{K} + 37°\text{K}} \right) = 5,751\,\text{mmHg}.$$

The initial flow from cell to surroundings through the cell membrane will be:

$$Q_V = K_f \Delta\Pi = \left(4 \times 10^{-4}\,\mu\text{m}^3\,\text{s}^{-1}\,\text{mmHg}^{-1} \right)(186\,\text{mmHg}) = 0.074\,\mu\text{m}^3\,\text{s}^{-1}.$$

Examining and interpreting the results: If the cell is initially spherical in shape, its initial volume is:

$$V_0 = \frac{4\pi}{3}(5\,\mu\text{m})^3 = 78\,\mu\text{m}^3.$$

If the cell were to lose fluid continuously at a rate of 0.074 µm³/s, it would take about 17.5 min for the cell to lose all of its water. However, the flow decreases with time because the osmotic pressure on the inside of the cell increases as water leaves the cell. The final cell volume will be reached when the osmotic pressure on the inside of the cell balances the osmotic pressure on the outside. The number of moles of intracellular solutes remains the same while fluid passes through the membrane, but the osmolarity increases because the cell volume is decreasing. The final volume V_∞ can be found by applying a simple molar balance on intracellular solutes:

$$310\,\text{mOsM}(V_\infty) = 300\,\text{mOsM}(V_0)$$

$$V_\infty = \frac{300\,\text{mOsM}(78\,\mu\text{m}^3)}{310\,\text{mOsM}} = 75.48\,\mu\text{m}^3.$$

Therefore, the cell ultimately loses 3.2% of its initial volume. The rate of fluid loss with time will be examined in the next example.

Example 5.17.2 Dynamics of Passive Cell Water Loss.
Consider the same cell as in Example 5.17.1, placed in the same NaCl solution. How long will it take for 99% of the cell volume change to occur?

Solution. *Initial considerations:* We will make the same assumptions as in the previous example. The system is still the cytoplasm and its interactions with the environment are limited to the flow of water through the semipermeable membrane.

Apprising the problem to identify governing relationships: In addition to van't Hoff's law and the Starling equation, which governs fluid exchange at the cell boundary, we must also perform a mass balance on the cytoplasm.

Analysis: A balance on cell volume yields:

$$\frac{dV}{dt} = -Q_V = -K_f[P_c - P_s - \sigma_d(\Pi_c - \Pi_s)].$$

The flow rate Q_V represents flow from the cytoplasm to the NaCl solution. This is positive and consequently dV/dt is negative. Osmotic pressure in the cytoplasm will depend on cell volume. If N represents the number of moles of intracellular solutes, the osmotic pressure of the cytoplasm can be expressed as follows:

$$\Pi_c = RT\left(\frac{N}{V}\right).$$

Since N is constant, it can be expressed in terms of the initial osmotic pressure in the cytoplasm Π_{c0} and the initial volume V_0.

$$N = \frac{\Pi_{c0}}{RT} V_0.$$

We will make the same assumptions about hydrostatic pressures and osmotic reflection coefficient as in the previous example. The resulting differential equation is:

$$\frac{dV}{dt} = K_f\left(\Pi_{c0}\left(\frac{V_0}{V}\right) - \Pi_s\right).$$

From the previous example, the final volume relative to the initial volume is:

$$\frac{V_\infty}{V_0} = \frac{\Pi_{c0}}{\Pi_s}.$$

Therefore, the differential equation can be written as:

5.17 Osmotic Pressure and Flow

$$\frac{dV}{dt} = K_f \Pi_s \left(\left(\frac{V_\infty}{V}\right) - 1 \right).$$

Since the solution volume is assumed to be much larger than the cell volume, osmotic pressure in the NaCl solution will remain essentially constant while fluid exits from the cell. Separating variables and integrating:

$$\int_{V_0}^{V} \frac{V dV}{(V - V_\infty)} = -K_f \Pi_s \int_0^t dt.$$

Consulting a table of integrals for the left side of the equation:

$$\int \frac{x dx}{a + bx} = \frac{x}{b} - \frac{a}{b^2} \ln(a + bx).$$

For our case $x = V$, $a = -V_\infty$, and $b = 1$. Thus, the final solution is an implicit solution for the cell volume:

$$V - V_0 + V_\infty \ln \left[\frac{V - V_\infty}{V_0 - V_\infty} \right] = -K_f \Pi_s t.$$

The cell volume change is 99% complete when:

$$V = V_0 + 0.99(V_\infty - V_0).$$

So the specified volume is 75.5 µm³. Solving for the time required to attain this volume:

$$t = \frac{1}{K_f \Pi_s} \{0.99(V_0 - V_\infty) - V_\infty \ln[0.01]\}.$$

Inserting the appropriate values:

$$t = \frac{1}{(4 \times 10^{-4} \text{ µm}^3 \text{ s}^{-1} \text{ mmHg}^{-1})(5{,}751 \text{ mmHg})}$$
$$\times \{0.99(78 - 75.48 \text{ µm}^3) - 75.48 \text{ µm}^3 \ln[0.01]\}.$$

Examining and interpreting the results: The time required to reach the specified volume is 152 s. The volume vs. time relationship is shown in Fig. 5.50. Note that as time becomes very large, dV/dt approaches zero, and the final steady-state cell volume becomes 75.48 µm³. This is consistent with the value computed in the previous example.

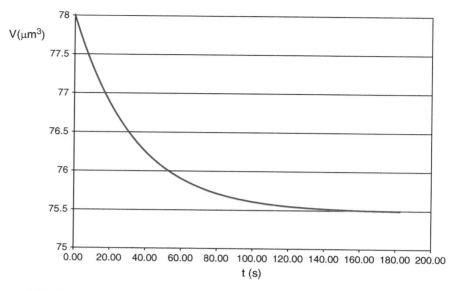

Fig. 5.50 Cell volume vs. time after immersion in hypertonic NaCl solution

Example 5.17.3 Glomerular Filtration.

The functional unit of the kidney is the nephron. Fluid enters a nephron through the walls of glomerular capillaries into a structure called Bowman's capsule. The capillary walls allow ions and small solutes to pass from plasma to the filtrate in Bowman's capsule, but proteins are ordinarily excluded. Estimate the combined filtration coefficient of all glomerular capillaries if total glomerular filtration rate (GFR) is measured to be 125 ml/min, average glomerular capillary pressure is 60 mmHg, pressure in Bowman's capsule is 20 mmHg, and plasma protein concentration is 7 g/dl. If glomerular capillary pressure was lowered to 50 mmHg, what would be the expected GFR?

Solution. *Initial considerations:* We shall assume that the glomerular membrane is permeable to small solutes, but impermeable to proteins. Consequently, the osmotic reflection coefficient is 1 and the protein oncotic pressure in Bowman's capsule is zero. We will also assume that the relationship between plasma oncotic pressure and plasma protein concentration is given by (5.167).

System definition and environmental interactions: The system in this case is the fluid in Bowman's capsule. We are interested in the steady-state flow of fluid from the glomerular capillary into Bowman's capsule through the wall of the capillary.

Apprising the problem to identify governing relationships: The Starling equation can be used to predict flow through the membrane under a combined hydrostatic and oncotic pressure difference.

5.17 Osmotic Pressure and Flow

Analysis: For a plasma concentration equal to 7 g/dl, the plasma protein oncotic pressure from (5.167) is 25.6 mmHg. Solving for the overall filtration coefficient, using (5.154):

$$K_f = \frac{\text{GFR}}{P_{cap} - P_{BC} - \Pi_{cap}} = \frac{125\,\text{ml/min}}{(60 - 20 - 25.6)\,\text{mmHg}} = 8.68\,\text{ml min}^{-1}\,\text{mmHg}^{-1}.$$

If the capillary pressure was lowered to 50 mmHg, and all other factors remained the same, the predicted GFR would be:

$$\text{GFR} = K_f \left[P_{cap} - P_{BC} - \Pi_{cap} \right]$$
$$= \left(8.68\,\text{ml min}^{-1}\,\text{mmHg}^{-1}\right)(50 - 20 - 25.6)\,\text{mmHg} = 38.2\,\text{ml/min}.$$

Examining and interpreting the results: The analysis predicts a decrease in GFR to 31% of its normal value for a 10 mmHg drop in glomerular capillary pressure. To maintain normal GFR, the kidney generally compensates by increasing resistance of the efferent arterioles (downstream of the glomerular capillaries) to raise glomerular capillary pressure.

Example 5.17.4 Oncotic Pressure of Protein Mixtures.

Express (5.165) in terms of molar concentration, rather than mass concentration. Then use the virial coefficients from albumin to compute oncotic pressures of mixtures of albumin and globulin. Show that a single set of virial coefficients expressed in molar terms results in different sets of virial coefficients expressed in mass concentration terms, depending on the composition of albumin and globulin.

Solution. *Initial considerations:* We might expect an oncotic pressure relationship based on molar concentration to be more accurate, since the derivation of (5.160) was based on molar concentrations, not mass concentrations.

System definition and environmental interactions: The system of interest here is a mixture of proteins in plasma. In particular, we will focus on a mixture of albumin and larger globulin proteins.

Apprising the problem to identify governing relationships: Our starting point will be the polynomial expression for oncotic pressure in terms of total protein mass concentration reported by Landis and Pappenheimer. We will use the definitions of mass and molar concentration and the assumption that the mixture contains only two protein species in addition to the solvent.

Analysis: The relationship between mass concentration and molar concentration of albumin is:

$$\rho_{alb} = M_{alb} C_{alb}.$$

Substituting this into (5.165):

$$\Pi_{\text{alb}} = 2.8 M_{\text{alb}} C_{\text{alb}} + 0.18 (M_{\text{alb}} C_{\text{alb}})^2 + 0.012 (M_{\text{alb}} C_{\text{alb}})^3.$$

Comparing this term with (5.161):

$$RT = 2.8 M_{\text{alb}},$$
$$RTA = 0.18 M_{\text{alb}}^2,$$
$$RTB = 0.012 M_{\text{alb}}^3.$$

The first equation provides an estimate of the molecular weight of albumin, as shown in (5.166). If molar concentrations are expressed in units of mM, mass concentrations in terms of g/dl, and oncotic pressures in terms of mmHg, the virial coefficients can be expressed as:

$$A = \frac{0.18 \frac{\text{mmHg}}{(\text{g/dl})^2}}{2.8 \frac{\text{mmHg}}{\text{g/dl}}} \left(68{,}930 \frac{\text{g}}{\text{mol}}\right) \left(\frac{1 \text{ mol}}{1{,}000 \text{ mmol}}\right) \left(\frac{1 \text{ l}}{10 \text{ dl}}\right) = 0.443 \text{ mM}^{-1},$$

$$B = \frac{0.012 \frac{\text{mmHg}}{(\text{g/dl})^2}}{2.8 \frac{\text{mmHg}}{\text{g/dl}}} \left(68{,}930 \frac{\text{g}}{\text{mol}}\right)^2 \left(\frac{1 \text{ mol}}{1{,}000 \text{ mmol}}\right)^2 \left(\frac{1 \text{ l}}{10 \text{ dl}}\right)^2 = 0.204 \text{ mM}^{-2}.$$

Equation (5.161) is based on the assumption that osmotic pressure depends only on the number of nonsolvent molecules present, not on the mass of the solutes or the type of solutes. If albumin (alb) and globulin (g) proteins interact in the same manner as albumin molecules alone, the relationship between oncotic pressure of a protein mixture of albumin and globulin expressed in mmHg and total molar protein concentration C_P expressed in mM is:

$$\Pi = RT \left[C_P + 0.443 C_P^2 + 0.204 C_P^3 \right],$$

where $C_P = C_{\text{alb}} + C_g$. Let us now consider a protein solution in which the fraction of albumin relative to the mass of total protein present is f_{alb}.

$$f_{\text{alb}} = \frac{\rho_{\text{alb}}}{\rho_{\text{alb}} + \rho_g} = \frac{\rho_{\text{alb}}}{\rho_P}.$$

We would like to express the above relationship in terms of total protein mass concentration ρ_P, rather than molar concentration C_P. In this case:

5.17 Osmotic Pressure and Flow

$$\rho_{alb} = f_{alb}\rho_P,$$
$$\rho_g = (1 - f_{alb})\rho_P,$$
$$C_P(mM) = \beta\left(\frac{\rho_{alb}(g/dl)}{M_{alb}} + \frac{\rho_g(g/dl)}{M_g}\right) = \beta\left(\frac{f_{alb}}{M_{alb}} + \frac{1-f_{alb}}{M_g}\right)\rho_P(g/dl),$$

where $\beta = 10^4 \dfrac{dl\ mM}{mol}$.

Writing the oncotic pressure–concentration relationship in terms of total protein mass concentration:

$$\Pi = A^*\rho_P + B^*\rho_P^2 + C^*\rho_P^3,$$

where:

$$A^* = RT\left[\beta\left(\frac{f_{alb}}{M_{alb}} + \frac{1-f_{alb}}{M_g}\right)\right],$$

$$B^* = RTA\left[\beta\left(\frac{f_{alb}}{M_{alb}} + \frac{1-f_{alb}}{M_g}\right)\right]^2,$$

$$C^* = RTB\left[\beta\left(\frac{f_{alb}}{M_{alb}} + \frac{1-f_{alb}}{M_g}\right)\right]^3.$$

Examining and interpreting the results: Although the coefficients A and B are independent of composition, the coefficients A*, B*, and C* are all strongly dependent on the relative mass fraction of albumin, f_{alb}. If the molecular weight of the globulin protein is 120,000, the coefficients for $f_{alb} = 0, 0.2, 0.4, 0.6, 0.8$, and 1.0 are shown in Table 5.5.

Graphs of oncotic pressure vs. mass concentration are shown in Fig. 5.51 for various values of albumin protein mass fraction. The Landis–Pappenheimer relationship for plasma proteins (5.167) is also plotted on the graph for comparison. This corresponds to an albumin mass fraction of about 0.67, which agrees with measurements on human plasma (Navar and Navar 1977). Note that all of these curves are generated from a single relationship between oncotic pressure and total molar protein concentration shown in Fig. 5.52. The assumption that protein–protein interactions are the same for proteins of all sizes is generally not valid, and so the coefficients A and B will vary somewhat with composition. However, the errors in predicted osmotic pressure resulting from small changes in protein composition would be

Table 5.5 Virial coefficients for various albumin fractions

Virial coefficients	$f_{alb} = 1$	$f_{alb} = 0.8$	$f_{alb} = 0.6$	$f_{alb} = 0.4$	$f_{alb} = 0.2$	$f_{alb} = 0$
A*(mmHg/gdl)	2.800	2.562	2.323	2.085	1.847	1.608
B*(mmHg/g²dl²)	0.180	0.151	0.124	0.100	0.078	0.059
C*(mmHg/g³dl³)	0.012	0.009	0.007	0.005	0.003	0.002
β (dl mM/mol)	0.145	0.133	0.120	0.108	0.096	0.083

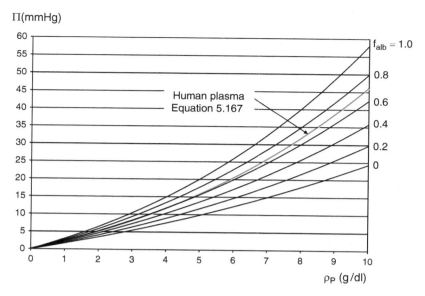

Fig. 5.51 Oncotic pressure vs. total protein mass concentration for mixtures with different protein mass fractions of albumin

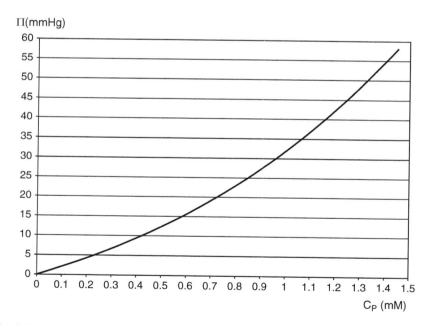

Fig. 5.52 Oncotic pressure vs. total protein molar concentration

expected to be smaller if the predictions were based on the molar concentration virial coefficients (A and B) than the mass concentration coefficients (A*, B*, and C*).

5.18 Summary of Key Concepts

Macroscopic Approach. Macroscopic fluid mechanics is generally concerned with estimation of pressure–flow relationships in conduits of various shapes or entire internal flow systems, with forces applied by flowing fluids internally on conduit walls or externally on solid objects, and with energy requirements necessary to move fluids through various types of systems. Local property variations within the system are ignored and only average properties are considered in macroscopic analyses.

General Macroscopic Equations. The general unsteady-state relationships governing macroscopic fluid flow include (1) conservation of mass (5.2), (2) conservation of momentum (5.28), (3) conservation of energy (5.42), and (4) the Engineering Bernoulli equation (5.47). It is important to know the meaning of each term in these equations, but there is no reason to commit any of the equations to memory. It is much more important to understand whether or not these macroscopic equations are appropriate for a particular situation, and if so, to learn how to simplify them for the problem under investigation.

Steady Flow Equations. A relatively general set of equations for the steady-state flow of an incompressible fluid through a system consisting of many conduits and fittings in series is given below. There are N total sections, each with a single inlet and outlet. The inlet to the system as a whole is designated with a subscript 1, and the outlet to the system as a whole is denoted by subscript 2. Inlets to individual segments are designated with subscript i and outlets with subscript j:

Conservation of mass (5.15):

$$Q_V = \langle v_1 \rangle A_1 = \langle v_2 \rangle A_2 = \{\langle v_i \rangle A_i\}_{\text{inlets}} = \{\langle v_j \rangle A_j\}_{\text{outlets}}.$$

Conservation of momentum (5.29):

$$\vec{R} = \sum_{i=1}^{N} \left\{ \left[K_{2i}\rho_i \langle v_i \rangle^2 + P_i \right] A_i \vec{e}_i \right\}_{\text{inlets}} - \sum_{j=1}^{N} \left\{ \left[K_{2j}\rho_j \langle v_j \rangle^2 + P_j \right] A_j \vec{e}_j \right\}_{\text{outlets}} + \sum_{i=1}^{N} m_i \vec{g}.$$

Conservation of energy (5.44):

$$\dot{Q}_S + \dot{Q}_{\text{gen}} - \dot{W}_s = w \left(c_P [T_2 - T_1] + \frac{1}{2} \left[K_{32} \langle v_2 \rangle^2 - K_{31} \langle v_1 \rangle^2 \right] + g[h_2 - h_1] + \frac{1}{\rho}[P_2 - P_1] \right).$$

Engineering Bernoulli equation (5.48):

$$\frac{\dot{W}_s}{w} + \hat{E}_v = \left(\frac{1}{2}\left[K_{31}\langle v_1\rangle^2 - K_{32}\langle v_2\rangle^2\right] + g[h_1 - h_2] + \frac{1}{\rho}[P_1 - P_2]\right),$$

where $\langle v_i \rangle \equiv \dfrac{1}{A_i} \displaystyle\int_{A_i} v_i \, dA_i$ (5.7) and $K_{ki} \equiv \dfrac{\langle v_i^k \rangle}{\langle v_i \rangle^k} = \dfrac{1}{\langle v_i \rangle^k}\left[\dfrac{1}{A_i}\displaystyle\int_{A_i}(v_i^k)\,dA_i\right]$ (5.23).

The rate at which energy is lost by friction per unit mass flow rate, $\hat{E}_v$, must generally be determined by experiment.

Constitutive Relationships. We must introduce constitutive relationships that relate fluid motion to forces applied to the fluid by the surroundings. The Buckingham Pi Theorem, introduced in Chap. 3, can be used to convert experimental data collected under controlled situations to dimensionless parameters that apply under entirely different experimental conditions.

Friction Factor and the Moody Diagram. The Fanning friction factor f is an important dimensionless number that relates the pressure difference applied between the ends of a conduit to the average fluid velocity $\langle v \rangle$ inside the conduit. The Fanning friction factor for flow in a cylindrical tube is found to be a function of another important dimensionless number, the Reynolds number $Re = \langle v\rangle d/\nu$. The relationship is plotted in Fig. 5.16, which is known as the Moody diagram. If flow is not fully developed, the friction factor will be higher in the entry region of the conduit where velocity gradients at the conduit walls are steeper (Fig. 5.15). When flow is turbulent ($Re > 2,200$), the friction factor depends on a second dimensionless factor, the roughness of the tube wall relative to the tube diameter. For turbulent flow, the Moody diagram can be used to estimate the friction factor in tubes with noncircular cross sections using the hydraulic diameter D_h (5.62) in place of the tube diameter in the definition of the Reynolds number.

Laminar flow in noncircular ducts. The Moody diagram cannot be used for noncircular tubes when flow is laminar. A separate dimensionless parameter, M_0, obtained from Fig. 5.21, governs the laminar pressure–flow relationship in noncircular ducts.

Friction loss in fittings and with changes in flow area. Empirical relationships for friction losses across fittings and cross-sectional area changes are governed by friction loss factors K_w (5.72), which are summarized in Fig. 5.18 for turbulent flow. The friction term in the Engineering Bernoulli equation can be found by considering the friction loss in all fittings and changes in cross-sectional area and the friction loss in each segment of straight tubing (5.73):

$$\hat{E}_v = \sum_{i=1}^{N_f} K_{w,i}\left(\frac{1}{2}\langle v_{\text{out},i}\rangle^2\right) + \sum_{j=1}^{N_c} 4f_j \frac{L_j}{D_{h,j}}\left(\frac{1}{2}\langle v_j\rangle^2\right).$$

Friction factors for flow through a packed bed (Sect. 5.9) and flow around a sphere (Sect. 5.10) are also relevant for many biomedical applications.

Flow through porous structures. Another important constitutive relationship in biofluid transport is the Starling equation (5.154), which describes fluid flow through porous structures:

$$Q_v = K_f[P_1 - P_2 - \sigma_d(\Pi_1 - \Pi_2)].$$

This is characterized by two coefficients, a filtration coefficient K_f and an osmotic reflection coefficient σ_d, and two driving forces, the hydrostatic pressure difference ΔP and the osmotic pressure difference $\Delta \Pi$ across the structure. The osmotic pressure difference depends on the difference in the number of solute molecules (excluding water) on the two sides of the membrane. For dilute solutions, osmotic pressure can be estimated using van't Hoff's law (5.164). The Starling equation has important applications in finding fluid flow across cell membranes and fluid flow through capillary walls.

Flow in microvascular networks. Since in vivo blood flow is laminar in most blood vessels, we can use the Hagen–Poiseuille equation to describe the pressure–flow relationship as long as we account for the dependence of apparent viscosity of blood on hematocrit value, shear rate, and vessel size, as discussed in Chap. 4. Flow resistance of vessels in parallel (5.122) and vessels in series (5.117) can be combined to determine the pressure–flow relationship in a complex network of blood vessels.

Compliance, Inertia, and Collapsible Tubes. Not all blood vessels can be considered to have rigid walls. Blood in a compliant vessel increases in volume when a positive transmural pressure is applied to the wall of the vessel. The transmural pressure P_{tm} is defined as the pressure inside a vessel minus the pressure surrounding the vessel. A linearly compliant vessel is one in which the change in volume is directly proportional to the change in pressure, $dV/dP_{tm} = C$, where C is the compliance of the vessel (Sect. 5.13). Vascular compliance influences resistance in a nonlinear manner. It is most important when dealing with unsteady flow situations where pressure and flow can vary with time. Fluid inertia is unimportant in steady flow applications, but causes a lag in flow relative to an imposed pressure gradient in unsteady-state situations. The importance of this effect is dependent on the Womersley number, defined by (5.14). When the transmural pressure is negative, vessels with thin walls will have a tendency to collapse. An ideal Starling resistor is one which collapses at all locations where the transmural pressure is negative. Flow through such vessels is zero if the surrounding pressure is greater than the inlet pressure, but is proportional to the difference between the inlet pressure and the surrounding pressure if the surrounding pressure is greater than the outlet pressure. This has practical effects for flow through microvessels in the heart and lung (Sect. 5.16), where surrounding pressure varies periodically during cardiac and respiratory cycles.

5.19 Questions

5.19.1. Write the general unsteady-state macroscopic mass balance in words and write it as an equation.

5.19.2. What is the relationship between steady-state flow of an incompressible fluid through a duct and the average velocity of the fluid in the duct?

5.19.3. What procedure would you use to apply a macroscopic mass balance to flow in physiological systems such as air flow in the lungs or blood flow in vascular beds?

5.19.4. An *incompressible fluid* flows at *steady state* through a *rigid pipe* flow properties in the list (you will not necessarily use all terms). Use subscript/flow properties in the list (you will not necessarily use all terms). Use subscript 1 or 2 to denote inlet or outlet. Additionally, compare inlet and outlet conditions to determine whether inlet is greater than (>), less than (<), or the same (=) as the outlet condition (circle correct answer).

Variable List: ρ, μ, α, τ, A (cross-sectional area), σ, $\langle v \rangle$ (average velocity in direction of flow)

	Inlet	Outlet	Inlet vs. outlet comparison
Velocity			< = >
Mass flow rate			< = >
Mass flux			< = >
Volumetric flow rate			< = >

5.19.5. Define flow resistance. What are the key factors that influence vascular resistance to blood flow?

5.19.6. Define hydrostatic pressure and be able to derive an expression for hydrostatic pressure in terms of fluid density, gravitational constant, and fluid height.

5.19.7. What can cause the average momentum in a fluid system with multiple inlets and outlets to change with time?

5.19.8. What is the relationship between the rate of production of momentum in a fluid system and the sum of forces applied to the fluid by the surroundings?

5.19.9. What factors influence the value of the coefficient K_{ki} defined by (5.23)?

5.19.10. What is the primary application of the macroscopic momentum equation?

5.19.11. What can cause the average energy in a fluid system with multiple inlets and outlets to change with time?

5.19.12. Provide examples where the source term $\dot{Q}_{gen}$ is not zero.

5.19.13. How can work be performed by the fluid on the surroundings?

5.19.14. What are typical applications of the Engineering Bernoulli equation?

5.19.15. What is meant by entry length of a conduit? How would you account for the additional friction loss in laminar flow, relative to Poiseuille flow, in the entry region of tubes?

5.19.16. What is meant by fully developed flow in a conduit?

5.19 Questions

5.19.17. Define hydraulic diameter and Reynolds number for tube flow based on hydraulic diameter. What is the significance of the Reynolds number in characterizing flow in tubes?

5.19.18. What is the difference between laminar and turbulent flow? How can we estimate whether flow is laminar or turbulent in a tube? How can you determine experimentally if flow is laminar or turbulent?

5.19.19. Would you expect the pressure differences across a tube with turbulent flow to be larger or smaller than the pressure drop computed using the Hagen–Poiseuille equation?

5.19.20. Given the definition of the friction factor, how would you use the Moody diagram to find pressure difference, flow or tube hydraulic diameter?

5.19.21. What is a friction loss factor and how are friction losses in fittings incorporated into the Engineering Bernoulli equation?

5.19.22. How would you use Fig. 5.21 to estimate laminar flow through tubes with noncircular cross sections?

5.19.23. What is meant by superficial velocity and void fraction in relation to flow through packed beds?

5.19.24. What restrictions apply to the use of Stokes' law to find the drag force on a sphere?

5.19.25. How would you determine the overall resistance of a vascular network composed of vessels with known resistance arranged in series and/or in parallel?

5.19.26. What determines how flow is distributed to the two daughter vessels at a branch in a network? Why might flow resistance depend on flow rate? What equations apply at each node and vessel segment?

5.19.27. What is transmural pressure?

5.19.28. Define compliance. How does vascular compliance influence vascular resistance?

5.19.29. What are some of the benefits of blood vessel compliance?

5.19.30. Would you expect airway resistance to be lower during inspiration than it is in expiration? Explain.

5.19.31. What effect does increasing the frequency of pressure oscillations have on flow through a compliant system?

5.19.32. What is a Starling Resistor? How does it differ from a standard flow resistor? What is the relationship between flow and inlet pressure, outlet pressure, and surrounding pressure for a Starling Resistor?

5.19.33. What is fluid inertance and when is it important?

5.19.34. What is the significance of the Womersley number?

5.19.35. How does hydrostatic pressure affect flow through the lung? What are the zones of the lung? How are the zone boundaries identified? What assumption is made in this calculation of zone boundaries? Will zones be found in organs other than the lung? Discuss.

5.19.36. What is van't Hoff's Law? What is a filtration coefficient? How would you use van't Hoff's law to compute flow through a membrane, given filtration coefficient, hydrostatic pressures, and dilute solute concentrations on each side of the membrane?

5.19.37. What is the difference between osmotic pressure and oncotic pressure?

5.19.38. The osmotic pressure inside a cell is higher than the osmotic pressure outside the cell and there is no difference in hydrostatic pressure. If the cell membrane behaves like a semipermeable membrane, water will: (a) be transported into the cell, (b) be transported out of the cell, (c) not be transported into or out of the cell, or (d) not enough information is given. Justify your answer.

5.18.39. The osmotic pressure inside a cell is lower than the osmotic pressure outside the cell and there is no difference in hydrostatic pressure. If the cell membrane behaves like a semipermeable membrane, glucose will: (a) be transported into the cell, (b) be transported out of the cell, (c) not be transported into or out of the cell, or (d) not enough information is given. Justify your answer.

5.19.40. What is hemolysis? Why does it occur?

5.20 Problems

5.20.1 Conservation of Mass

Two capillaries merge at the entrance to a rigid venule with volume V. Initially, the density of fluid in the venule is ρ_0. At time $t = 0$, blood from each capillary begins to flow into the venule. Blood from capillary 1 has a constant density ρ_1 and constant flow Q_{V1}, while fluid from capillary 2 has a constant density ρ_2 and constant flow Q_{V2}. Assume blood in the venule is well mixed. Find: (a) the mass flow rate into the venule; (b) the volumetric flow rate into the venule; (c) an expression for venule fluid density vs. time $\rho(t)$; (d) the final steady-state density in the venule after a long time.

5.20.2 Organ Reperfusion

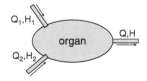

An organ has two arterial inlets and a single venous outlet. Both arteries are clamped while surgery is performed on the organ. Blood in the organ is initially stagnant with a hematocrit value of H_0. At time $t = 0$, the two arteries are unclamped and blood flows through the organ. The flow in each artery is constant (Q_1 and Q_2) and the two arteries have different hematocrit values (H_1 and H_2). The organ blood volume V remains constant and can be assumed to be well mixed. We are interested in how the

5.20 Problems

hematocrit value H varies in the organ as a function of time after releasing the clamps. Neglect the dependency of blood density on hematocrit value.

(a) Provide an expression for blood flow Q at the venous exit. Which conservation principle is applied?
(b) Provide the appropriate conservation equation governing hematocrit in the organ. Use words to describe each term in the equation.
(c) Convert your word equation in part (b) to a differential equation that can be solved for organ hematocrit. Describe how you would solve the resulting equation to find H. What is the initial condition?
(d) What is the *steady-state* hematocrit value in the organ after a very long period of reperfusion?
(e) Solve the ODE derived in part (c) and show that it reduces to the solution in part (d) after a long time has passed.
(f) How long should we wait for the time interval in part (e) to qualify as a "long time"?

5.20.3 Flow Out of a Tank

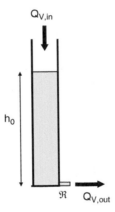

A tank with cross-sectional area A initially contains plasma with density ρ to a height of h_0. At time $t = 0$, additional plasma is introduced at a constant inflow rate of $Q_{v,in}$. At the same time, the drain tube at the outlet of the tank, having resistance $\Re$, is opened. Find an expression for the outlet flow through the drain as a function of time.

5.20.4 Flow Between Reservoirs

A short tubing segment separates two tanks. The relationship between flow through the tube and the pressure drop across the tube is shown in the graph below. Initially the tubing is clamped so that no flow occurs between the tanks. The fluid level in tank 1 is $h_1 = 3$ cm and the fluid level in tank 2 is $h_2 = 23$ cm. The cross-sectional area of each reservoir is 30 cm^2 and the fluid density is 1.05 g/cm^3. At time $t = 0$, the clamp is removed and fluid begins to flow through the tube. Assume that the pressure at either end of the tubing segment is equal to the hydrostatic pressure at the bottom of the reservoir.

(a) Find the resistance of the tubing segment.
(b) Use conservation of mass in the two tanks to show that the height difference decreases exponentially with time.
(c) Find the height difference 60 s after the clamp is released.

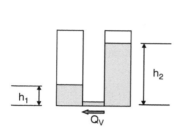
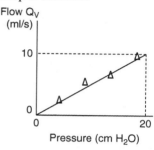

5.20.5 *Mass and Momentum Balance*

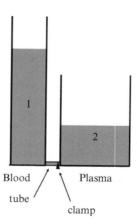

Two tanks are separated by a tube with a flow resistance of 30 cmH$_2$O/(ml/s). The tank on the left has a cross-sectional area of 5 cm^2 and initially contains 100 ml of blood with a density of 1.05 g/ml. The tank on the right has a cross-sectional area of 10 cm^2 and initially contains 50 ml of plasma with a density of 1.01 g/ml. The flow through the tube connecting the tanks is found to decrease exponentially with time, $Q_V = Q_{V0}\exp(-\alpha t)$, where Q_{V0} and α are constants.

(a) Provide values for Q_{vo} and α, and derive expressions for the height in each tank as a function of time after the clamp is released.

(b) Estimate the rate of change in momentum of the fluid in tank 1 during the first second after the clamp is removed. You can approximate this as:

[$(m(1s)v(1s) - m(0)v(0))]/1s$, where $v(0) = Q_{vo}/A_1 =$ velocity of fluid surface in tank 1 immediately after the clamp is removed. Compare this with the average weight of the column during that 1 s period. Are we justified in neglecting the rate of change of momentum in the conservation of momentum expression (i.e., assuming the pressure at the bottom of the tank is the hydrostatic pressure, $\rho g h$)? By how much would the pressure change if we included the rate of change of momentum?

5.20.6 Isolated Perfused Organ

The system used to perfuse an artificial organ is shown in the figure. Perfusion fluid is pumped at a constant rate from the collecting tank to the head tank. The combined flow resistance $\Re$ of the tubing and organ is assumed to be constant and equal to 50 cmH$_2$O/(ml/s). Fluid density is 1 g/cm^3. The head tank cross-sectional area A is 5 cm^2 and the initial level of fluid in the head tank h is 30 cm.

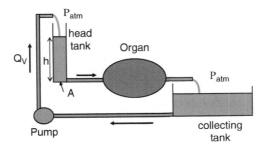

(a) Find the initial pump flow rate.
(b) If the pump flow is suddenly cut in half, how long will it take the tank height to come to within 99% of its final value?

5.20.7 Flow Through a Curved, Tapered Vessel

Blood flows in a horizontal plane through the vessel shown at a rate of 100 ml/min. Downstream pressure is 10 mmHg.

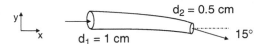

(a) Find the average blood velocity at the upstream and downstream locations.
(b) Find the upstream pressure (neglect friction loss in the vessel).
(c) Compute the net force exerted by the blood on the walls of the vessel (neglect weight of the blood).

5.20.8 *Laminar Flow Through an Aortic Aneurysm*

An aortic aneurysm is a ballooning of the aorta caused by a weakened vessel wall.

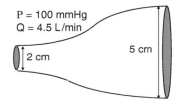

(a) Use a macroscopic mass balance to find the velocity at the outlet of an aortic aneurysm. Assume a steady horizontal flow of 4.5 L/min.
(b) Use a mechanical energy balance to estimate the pressure in the aneurysm. The inlet pressure is 100 mmHg. Does this tend to increase or decrease the diameter of the aneurysm? Explain.
(c) Use a momentum balance to compute the force by the blood on the vessel wall.

5.20.9 *Steady Flow in a Blood Oxygenator*

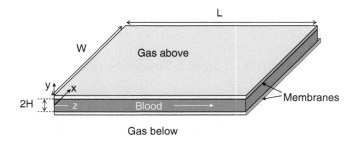

A blood oxygenator is composed of N identical parallel compartments. Blood in each compartment flows between two stiff oxygen-permeable membranes. The width W of each membrane is 10 cm and length L (in direction of flow) is 5 cm. Total blood flow through the device is 22.7 ml/s, blood kinematic viscosity is 0.04 cm^2/s, blood density is 1 g/ml, and total blood volume in the device is 500 ml. For parallel plates, the hydrodynamic entry length relative to the half height of the channel is $Le/H = 0.176\ Re_H$. What is the minimum number of compartments N necessary to ensure that the hydrodynamic entry length in each compartment is no more than 5% of the device length? What is the half height H for each compartment?

5.20.10 Pressure and Flow in a Vein

(a) Find blood flow through a vein with a circular cross section. The vein has a diameter of 0.5 cm and is 15 cm long. Blood has a density of 1.05 g/ml and a viscosity of 5 cp. The inlet pressure is 5 cmH$_2$O greater than the outlet pressure.
(b) The same vein is partially compressed so that it takes on an elliptical shape with the same perimeter, but with a major axis twice the minor axis. Find the pressure difference needed to maintain the same flow as in part a. Use the Ramanujan formula given in Example 5.8.1 to estimate the perimeter of an ellipse.

5.20.11 Blood Flow in a Microvessel

Blood flows through a small arteriole with length 1,000 μm and diameter 20 μm. The hematocrit is 45% and the pressure difference across the vessel is 6 mmHg. Plasma viscosity is 1.05 cp. Compute flow through the vessel using the Hagen–Poiseuille Equation, if: (a) the Fahraeus–Lindqvist effect is not included and (b) the Fahraeus–Lindqvist effect is included. Discuss the physiological implications.

5.20.12 Pressure Drop Across a Blood Oxygenator

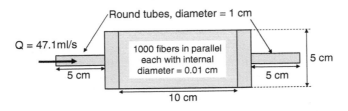

Blood ($\rho = 1.05$ g/ml) with a hematocrit value of 45% flows through a blood oxygenator at a flow rate of 47.1 ml/s. (a) Estimate the pressure difference across the device. Neglect energy losses at contractions and expansions. All tubes are smooth with circular cross sections. Plasma has a viscosity of 1.03 cp and density of 1.02 g/ml. Compare your result for blood with the pressure difference that would occur if the Fahraeus–Lindqvist effect was not present.

(b) Would you expect the blood flow rate through a similar device constructed of 16,000 fibers, each with a diameter of 50 µm, to be greater than, equal to, or less than the flow in part (a) under the same circumstances? Provide computational evidence for your answer.

5.20.13 Friction in a Vascular Network

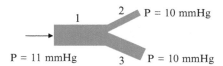

Vessel	Vascular resistance [mmHg/(ml/s)]
1	12,230
2	117,470
3	17,940

(a) The Hagen–Poiseuille resistance for blood flow ($\mu = 4$ cp) in each branch of a network is given in the table. If the diameter of vessel 2 is 0.005 cm, what is the length of that vessel?
(b) What is the pressure at the branch point and the flow through vessel 2?
(c) Discuss the validity of your computations. What factors might influence this calculation?

5.20.14 Friction Losses in Branching Networks

Calculate the flow of air through the airway branch below based on (a) an assumption of Poiseuille flow and (b) accounting for entry length effects in laminar flow. Parent vessel: diameter = 2 cm, length = 10 cm; daughter vessels: diameter = 1.2 cm, length = 8 cm. Upstream pressure = 0.025 cmH$_2$O, downstream pressure = 0.

5.20 Problems

5.20.15 Laminar Flow of Blood Through a Vein

.042 cm = 2a
.1 cm = 2b
Area of an ellipse = πab
Perimeter of an ellipse $\approx \pi(a+b)$

A vein is 10 cm long with an elliptical cross section as shown. What is the pressure drop (in mmHg) across the vein if the blood flow rate is 0.2 ml/min? Assume that blood is a Newtonian fluid with a relative viscosity of 4, and that plasma physical properties can be found in Appendix C at 37°C. Explain your procedure and confirm that the flow is laminar.

5.20.16 Flow Resistance in Laminar and Turbulent Flow

(a) By setting the pressure difference in the definition of the friction factor equal to the pressure difference in the definition of flow resistance for fluid flow in a smooth circular tube, show that the flow resistance $\Re$ is related to the friction factor f as follows:

$$\Re = \left(\frac{\rho L Q_V}{\pi^2 R^5}\right) f.$$

(b) Use information relating the friction factor to Reynolds number in the Moody diagram to show that the resistance is constant when flow is laminar:

$$\Re_{\text{laminar}} = \frac{8\mu L}{\pi R^4}.$$

(c) When the flow is turbulent, show that the resistance increases with the flowrate Q_V as follows:

$$\frac{\Re_{\text{turbulent}}}{\Re_{\text{laminar}}} = 0.006 \left(\frac{Q_V}{D\nu}\right)^{\frac{3}{4}},$$

where D is the tube diameter and ν is the fluid kinematic viscosity.

(d) Find the ratio of turbulent to laminar resistance at Reynolds numbers of 4,000 and 10^5.

(e) Show that the pressure drop for turbulent flow in a smooth tube obeys the relationship:

$$Q_V = B(\Delta P)^{\frac{4}{7}},$$

where B depends on the fluid properties μ and ρ, and on the tube geometry D and L. Find the expression for B.

(f) Plot flow vs. pressure drop for water flowing through a 1 m long smooth tube having a diameter of 1 cm. Cover the range of Reynolds numbers from 10^0 to 10^5. Discuss your results.

5.20.17 Friction Factor in a Rough Tube

Find the tube diameter D(cm) necessary to deliver a velocity of 200 cm/s when a pressure drop of 3 cmH$_2$O is applied across a 20-cm long pipe with roughness $k = 0.01$ cm. Fluid properties are $\rho = 1.04$ g/ml and $\nu = 0.04$ cm^2/s.

5.20.18 Pressure Drop, Circular Tube

Cerebrospinal fluid (CSF) at 37°C is transported through a circular pipe with diameter of 0.5 mm and length of 15 mm. What pressure drop (Pa) applied across the length of the pipe is required to achieve an average CSF velocity of 0.85 cm/s? Would you consider your estimate to be greater or less than the actual measured pressure difference at this velocity? Explain.

5.20.19 Flow Rate, Circular Tube

CSF at 37°C is transported through a circular pipe with a diameter of 18 μm and axial length of 60 μm. A pressure drop of $1.33 \times 10^{+4}$ Pa is applied across the length of the pipe. Estimate the volumetric flow rate and Reynolds number in this system. Comment on the likely validity of your calculation.

5.20.20 Pressure and Flow

CSF at 37°C is transported through a circular pipe with a diameter of 0.5 mm and length of 15 mm. You find some incomplete data from a previous experiment that suggests Re_D for this system was 1.5. What was the volumetric flow rate $\langle$cm^3/h$\rangle$ of

CSF in this system? Estimate the pressure drop applied across the length of pipe in this application.

5.20.21 Blood Flow in the Portal Vein

The portal vein transports approximately 75% of the blood flow through the adult human liver, resulting in a volumetric flow rate through the portal vein of 1,050 cm^3/min. The typical adult human portal vein diameter in the absence of cardiovascular disease is 10 mm. Consider blood to be a Casson fluid with $S = 0.0597$ (Pa s)$^{1/2}$ and $\tau_y = 0.0072$ Pa. The roughness factor (k) of the portal vein may be presumed to be 1.0×10^{-6} m.

(a) Is blood flow through the portal vein laminar, transition, or turbulent? State any assumptions you make. Explicitly identify the fluid viscosity used in your calculation.
(b) Estimate the pressure drop per unit length ($-\Delta P/L$ in Pa/m) across the portal vein.

A second patient with cardiovascular disease has a reduced portal vein diameter of 3.16 mm over an extended length without reduction of the volumetric flow rate.

(c) Is blood flow through the portal vein laminar, transition, or turbulent in this patient? State any assumptions you make. Explicitly identify the fluid viscosity used in your calculation.
(d) Estimate the pressure drop per unit length in the portal vein in this patient.

5.20.22 Flow Through a Semicircular Duct

A duct is constructed by machining a semicircular grove down a 25-cm long sheet of acrylic plastic and then gluing a flat piece of acrylic above the grove. The radius of the semicircular grove is 1 mm. Water at 25°C is to be forced through the duct. What is the maximum pressure difference that can be imposed across the duct if flow through the duct is to remain laminar?

5.20.23 Flow Through an Annulus

Water at 30°C is forced at a rate of 90 cm^3/s through a 25-cm long annulus with inside diameter of 0.4 cm and outside diameter of 0.5 cm. Both walls of the annulus are smooth. What pressure difference between the inlet and outlet of the annulus is necessary to produce this flow rate?

5.20.24 Pressure Drop in a Fluid System

Two pressure vessels are connected by a rectangular duct with dimensions 1 mm × 4 mm × 100 cm. We need to adjust the pressure in the upstream tank P_1 so that a Newtonian fluid with kinematic viscosity $v = 0.016 \text{ cm}^2/\text{s}$ and $\rho = 1$ g/ml can flow at 40 ml/s through the duct. The duct has a wall roughness of 0.016 mm. Provide a step-by-step procedure, complete with calculations, for estimating the total pressure drop that is necessary to achieve this velocity. Do not neglect the sudden changes in cross-sectional area at the ends of the duct.

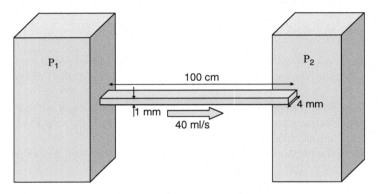

5.20.25 Flow in a Channel

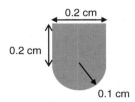

A circular saw is used to make a smooth groove in a piece of plastic. When covered by a flat piece of plastic, a channel is formed like that shown. Water at 20°C is forced through the channel with a pressure difference of 334 mmHg.

(a) What will be the mean velocity and flow rate in the channel if the channel length is 10 cm? Explain your method of solution. Verify that the flow is laminar or turbulent.
(b) Find the flow rate if the pressure difference is reduced to 15 mmHg.

5.20 Problems

5.20.26 Design of a Device to Study Chemical Reactions

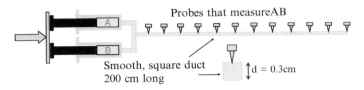

Two fluid streams containing chemicals A and B, respectively, are mixed at the entrance to a smooth square duct having the dimensions shown. We wish to study the reaction A + B → AB as the species travel through the duct. Probes that detect the concentration of AB are equally spaced along the length of the duct. The reaction is thought to be virtually complete in τ seconds. (a) What flow rate through the square duct would you suggest for this experiment if the outlet flows from each syringe are identical? (b) What minimum syringe volume do you need for an experiment that lasts 5τ? (c) Would laminar or turbulent flow be preferable? Explain.

5.20.27 Syringe Selection

You are asked to purchase syringes for the experiment described in Problem 5.20.26. Since high-pressure syringes are expensive, you need to determine the expected pressure in each syringe. Neglect pressure drops in the fittings between the syringes and the duct. Find the minimum pressure rating required for the syringes if the mean transit time through the square duct is (a) 4 s and (b) 1 s. Fluid density = 1 g/ml and kinematic viscosity = 0.01 cm²/s for both fluids.

5.20.28 Pressure Drop in Fittings

Repeat Problem 5.20.27, but account for pressure drops in the fittings and connecting tubing segments between the syringes and the duct. Dimensions of the fittings are given on the figure below. Is it safe to neglect the effect of fittings?

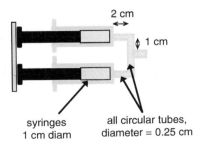

5.20.29 Pressure Drop Across a Blood Oxygenator

Sterile *water* at 20°C is used to test the blood oxygenator in Problem 5.20.12 for leaks. (a) Estimate the pressure drop across the device for a flow rate of 47.1 ml/s. Neglect energy losses at contractions and expansions. All tubes are smooth with circular cross sections. (b) Estimate the additional pressure difference caused by the presence of expansions and contractions.

5.20.30 Pressure Drop in an Extracorporeal Device

Blood plasma at 25°C flows horizontally through the device below at a steady flow rate of 4 ml/s from right to left. The device is composed of a smooth square channel connected to a smooth round tube. What is the total pressure drop across the device?

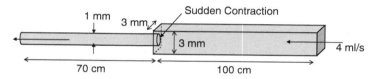

5.20.31 Blood Flow Through an In Vitro Network

The pressure drop across a simulated microvascular network with dichotomous branching is 40 mmHg. Dimensions of each vessel are given in the table below. The hematocrit in large vessels is 50% and the viscosity of plasma is 1.05 cp.

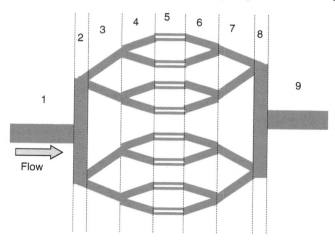

5.20 Problems

(a) Estimate flow through the entire in vitro network, the pressure drop across each segment, and the "Tube hematocrit" in each segment.
(b) Compare the flow in part (a) to the flow computed on the basis of blood having a constant viscosity equal to large vessel viscosity.
(c) Estimate blood flow through the same network in vivo, assuming that the presence of a rigid glycocalyx on the inside surface of each vessel reduces the effective diameter of each microvessel by 1.6 μm.

Generation	Number of vessels	Diameter (μm)	Length (μm)
1	1	75	2,000
2	2	45	1,600
3	4	25	1,200
4	8	15	800
5	16	8	600
6	8	18	800
7	4	30	1,200
8	2	55	1,600
9	1	90	2,000

5.20.32 Vascular Compliance

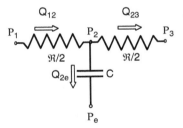

A compliant blood vessel is modeled as shown in the figure. Upstream pressure is periodic: $P_1 = P_{10} + P_{11}\sin(\omega t)$, and downstream pressure P_3 is constant. The pressure surrounding the vessel, P_e, is also constant. Plot the outflow from the vessel relative to the mean flow through the vessel for various values of $\omega \Re C$. Show that the solution is the same as the solution for $\omega/I\Re = 0$ in Example 5.15.3.

5.20.33 Vascular Compliance

Rework Problem 5.20.32 for the case where the resistances on each side of node 2 are different, but the total resistance is $\Re$.

5.20.34 Left Atrial Filling

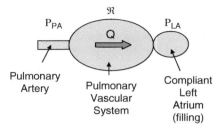

The left atrium is considered to be a compliant vessel with compliance $C_{LA} = 15.0$ ml/mmHg. Left atrial pressure is measured to be -5.0 mmHg immediately after the mitral valve closes ($t = 0$). The pulmonary circulation offers a fixed resistance to flow of $\Re = 2$ mmHg/(L/min). Pressure in the pulmonary artery is constant, $P_{PA} = 15$ mmHg. Derive expressions for left atrial pressure P_{LA} and blood flow Q through the pulmonary circulatory system as a function of time. Compute flow 0.5 s after the mitral valve closes.

5.20.35 Simple Ventilation Model

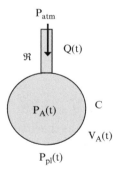

Derive an expression for alveolar pressure in the simple model shown if airway resistance $= 2$ cmH$_2$O/(L/s), alveolar compliance $= 0.2$ L/cmH$_2$O, and pleural pressure (relative to P_{atm}) suddenly jumps from -2 to -8 cmH$_2$O. Plot pleural pressure, alveolar pressure, the change in alveolar volume, and respiratory flow as a function of time for 12 s after pleural pressure is changed.

5.20.36 Air Flow to Compliant Lungs with Different Airway Resistances

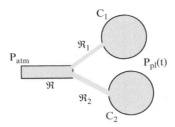

Air is distributed to each lung differently because of differences in airway resistance and compliance. Plot the pressure in each lung, the change in alveolar volume of each lung, and the flow to each lung as a function of time. Pleural pressure $P_{pl}(t) = 755 - 3\sin(\omega t)$ [mmHg], $\omega = 0.4\pi$ radians/s, $\Re = 0.5$ mmHg/(L/s), $\Re_1 = 2$ mmHg/(L/s), $\Re_2 = 2$ mmHg/(L/s), $C_1 = 0.1$ L/mmHg, $C_2 = 0.2$ L/mmHg, and $P_{A1}(0) = P_{A2}(0) = 760$ mmHg.

5.20.37 Response of a Fluid-Filled Compliant Pressure Transducer

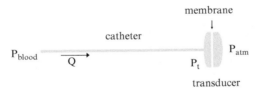

A fluid-filled catheter system for measuring blood pressure consists of a catheter filled with saline and a pressure transducer. One end of the catheter is inserted into an artery and the other end is attached to the transducer. As blood pressure increases, a small amount of fluid enters the catheter and deflects the compliant membrane in the transducer. The transducer is calibrated under static conditions and is assumed to remain accurate in dynamic situations. Assume the catheter is relatively rigid with resistance 2 mmHg/(L/s) and the transducer has compliance 0.05 L/mmHg. Compare the response of the transducer $P_t(t)$ to the stimulus $P_{blood}(t) = 100 + 20$ mmHg*$\sin(\omega t)$ for cardiac frequencies of 1 Hz (man) and 20 Hz (shrew) ($\omega = 2\pi f$).

5.20.38 Aortic Pressure and Flow in a Compliant System

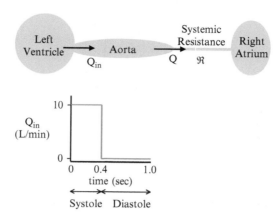

Blood flows from the left ventricle of the heart into the aorta (Q_{in} vs. time is shown in the figure below). Blood leaves the aorta through the systemic circulation and flows into the right atrium as shown in the figure below. The aorta can be considered to be a compliant region (0.8 ml/mmHg) and resistance of the systemic circulation is 20 mmHg/(L/min). The pressure in the right atrium is maintained constant at 1 mmHg, initial pressure in the aorta is 80 mmHg, and pressure surrounding the aorta is constant.

(a) Perform an unsteady-state macroscopic mass balance on the aorta during systole.
(b) Use the pressure–volume relationship to replace volume in part (a) with an expression involving aortic pressure.
(c) Provide an expression for flow out of the aorta.
(d) Substitute your result from part (c) into your result from part (b).
(e) Solve the resulting ODE in part (d) (What is the initial condition?).
(f) Find the pressure in the aorta at the end of systole.
(g) Find the flow out of the aorta at the end of systole.
(h) Repeat the process above for diastole.

5.20.39 Venous Occlusion Experiment

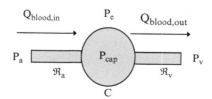

5.20 Problems

Capillary pressure in an isolated, perfused, organ can be estimated by holding inlet flow constant and suddenly occluding the outlet. Venous pressure, upstream from the occlusion, is measured during the experiment. Assume that virtually all of the vascular compliance is associated with the capillary bed. Model this system. How would you estimate capillary pressure and capillary compliance using the venous pressure measurement?

5.20.40 Flow in a Collapsed Venule

Consider blood flow through a very small collapsible venule, where the Reynolds Number is very low. The fluid pressure outside the vessel P_e is lower than the inlet pressure P_i but higher than the outlet pressure P_o. The vessel wall is assumed to collapse from its fully open state when $P > P_e$ to its fully collapsed state for $P < P_e$. When fully open, the venule resistance per unit length is $\mathcal{R}'$ and when fully closed the resistance per unit length is $\mathcal{R}_c'$. If $\mathcal{R}_c' = 100\, \mathcal{R}'$, how much error is introduced by estimating the flow using the expression for a Starling Resistor, $Q_V = (P_i - P_e)/\mathcal{R}$, where $\mathcal{R} = \mathcal{R}'L$? Assume that $P_i = 20$ Torr, $P_e = 10$ Torr, and $P_o = 5$ Torr. Is the actual flow greater or less than the flow predicted from the Starling equation?

5.20.41 Lung Zones

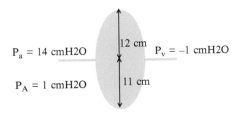

Blood having a density of 1.05 g/cm³ flows through a lung with horizontal dimensions as shown. Alveolar pressure P_A is 1 cmH$_2$O, microvascular arterial pressure P_a is 14 cmH$_2$O, and microvascular venous pressure P_v is −1 cmH$_2$O. Assume that large arteries and veins offer negligible resistance. (a) Find the location of the borders between zone I to zone II and zone II to zone III. (b) What fractions of lung height are in zone I, zone II, and zone III? (c) How are the locations of each of these two boundaries influenced by the fact that arteries and veins between the central and peripheral portions of the lung have finite resistance? Will it move each boundary upwards or downwards? Explain.

5.20.42 Lung Zones

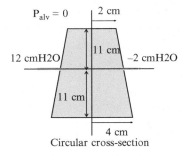

Circular cross-section

Consider a lung that has a circular cross section with a radius that changes linearly from 4 cm at the bottom to 2 cm at the top. The top of the lung is 11 cm above the left atrium and the bottom is 11 cm below the left atrium. The flow conductance per unit tissue volume is 0.4 $L_{blood}/(L_{tissue})$/min/cmH$_2$O. Left atrial pressure is -2 cmH$_2$O, pulmonary artery pressure is 12 cm H$_2$O, and alveolar pressure is 0 cmH$_2$O. Blood density is 1.1 g/ml. Compute flow and nominal resistance through the lung when the individual stands erect, so the lung is perfused in the vertical position.

5.20.43 Osmotic Pressure

Find the osmotic pressure of (a) a 5 mM glucose solution at 37°C and (b) a 2 mM sodium chloride solution at 25°C.

5.20.44 Osmotic Pressure

What is the osmotic pressure of the cytoplasm of a red blood cell with Osmolarity = 281.3 mOsM at 37 and 20°C? What mass of NaCl must be added to 1 L of sterile water to provide the same osmotic pressure?

5.20.45 Osmotic Flow

A semipermeable membrane with filtration coefficient K_f of 1 μl/s/mmHg separates water from a 2 mM NaCl solution at 23°C. The hydrostatic pressure on the water side is 75 mmHg higher than on the NaCl side. (a) What is the water flow through the membrane (μl/s) assuming van't Hoff's law is valid? (b) What is the molar concentration of water (moles/L) if mass density of water is 1 g/ml? (c) What is the molar concentration of the NaCl solution?

5.20 Problems

5.20.46 Osmotic flow

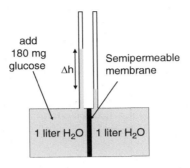

One hundred and eighty milligrams of glucose ($M = 180$ g/mol) is added to 1 L of water on one side of a semipermeable membrane. One liter of pure water is placed on the other side. Initially, the fluid heights in each column are the same. Find the difference in height Δh between the two columns as a function of time after the glucose is added. The filtration coefficient of the membrane is 0.001 ml/s/mmHg, the cross-sectional area of each column is 0.01 cm^2, and the temperature is 37°C. Assume that the tube volumes are very small and relative to the reservoir volumes.

5.20.47 Cell Fluid Loss

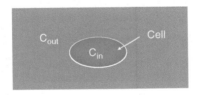

A cell with initial volume $V_0 = 10^{-6}$ ml, $T = 37°C$, membrane filtration coefficient $K_f = 8.6 \times 10^{-6}$ μl/(mmHg s), and an intracellular protein concentration $C_{in} = 3$ mmol/L is placed in a large beaker containing a similar protein solution with a protein concentration $C_{out} = 5$ mmol/L. There is no difference in hydrostatic pressure across the cell membrane. Protein cannot move through the membrane.

(a) Derive a differential equation for cell volume vs. time.
(b) Use the resulting differential equation to determine the final cell volume as $t \to \infty$.
(c) How much fluid volume does the cell ultimately lose?
(d) Separate variables and integrate to solve for V(t).
(e) How long does it take for the cell to lose 95% of the final fluid loss?

5.20.48 Concentrating Proteins

Proteins in solution can be concentrated using a semipermeable membrane. In one such separation, pure saline is contained on one side (compartment A) of the membrane and saline containing 80 g/L of a valuable, nondissociating protein is placed on the other side (compartment B) of the membrane. The membrane has a hydraulic conductance of 7.39 cm^3/(h-m^2-mmHg). Fluid volumes on each side of the membrane are 10 L. The hydrostatic pressure is equal in both fluid compartments, the temperature is 37°C, the reflection coefficient of the protein is unity, and the reflection coefficient of the NaCl and the water is 0. The molecular weight of the protein is 69,600 g/mol.

(a) Under these circumstances, what will happen to protein concentration in compartment B?
(b) The hydrostatic pressure in compartment B is increased to 76 mmHg greater than the hydrostatic pressure in compartment A. Which way will water flow in this case?
(c) What membrane area (m^2) is required to achieve an initial water filtration rate of 500 cm^3/h when the hydrostatic pressure in part (b) is applied? Assume that van't Hoff's law applies.
(d) How long will it take before the protein concentration is within 99% of its final concentration? What is the ultimate protein concentration?

5.20.49 Renal Vascular Resistance and GFR

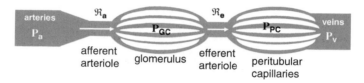

A model of renal blood flow containing two capillary beds in series is shown in the figure above. P_a represents arterial pressure, P_{GC} represents pressure in the glomerular capillaries, P_{PC} represents pressure in the peritubular capillaries, and P_v is venous pressure. Hematocrit value in the afferent arteriole is $H_a = 45\%$. The A drug influences blood pressure and renal vascular resistance as shown in the table below. Pressures before and after use of the drug are also provided in the table.

	Arterial pressure P_a (mmHg)	Venous pressure P_v (mmHg)	Afferent resistance $\mathcal{R}_a$ (mmHg/(ml/min))	Efferent resistance $\mathcal{R}_e$ (mmHg/(ml/min))
Before drug	155	5	0.082	0.055
After drug	125	5	0.055	0.027

5.20 Problems

(a) Find the following quantities both predrug and postdrug:
   1. Renal blood flow
   2. Plasma flow in afferent arteriole
   3. Glomerular capillary pressure
(b) Find glomerular filtration rate (GFR) under predrug and postdrug conditions. Assume that $T = 37°C$ and GFR obeys Starling's law with filtration coefficient $K_f = 15$ ml/min/mmHg between plasma and Bowman's capsule (B). Pressure in Bowman's capsule surrounding the glomerular capillaries is $P_B = 15$ mmHg. Assume proteins are not present in Bowman's capsule and that van't Hoff's law can be used to compute plasma protein concentration. Assume that plasma concentration in afferent arteriole is $C_a = 1.1$ mmol/L.
(c) Compute the following predrug and postdrug quantities at the distal end of the efferent arteriole:
   1. Plasma flow
   2. Protein concentration
   3. Hematocrit value

5.20.50 Osmotic Water Flow

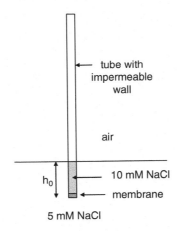

A semipermeable membrane with negligible thickness is attached to the bottom of a tube with cross-sectional area A. The tube is filled to a height of h_0 with a 10 mM solution of NaCl. At time $t = 0$, the tube is immersed to a depth of h_0 in a large reservoir filled with 5 mM NaCl. The concentration of NaCl in the reservoir can be assumed to remain constant.

(a) Use species conservation to provide an expression relating NaCl concentration in the tube to the height of fluid in the tube.
(b) Derive an ordinary differential equation for the height of fluid in the tube as a function of time. All variables in your resulting equation except height and time should be known quantities.

(c) Provide an expression for the height of fluid in the tube a very long time after the tube was placed in the reservoir.

(d) Solve the ODE derived in part (b) and compare the final height to the value computed in part (c).

5.21 Challenges

5.21.1 Rapid Reaction Experiment

The device described in Problem 5.20.26 is used to study the rate of chemical reaction between two species. Estimates from the literature suggest that when the two species are mixed, the reaction is virtually complete within τ_d seconds. The time constant of the probes which measure concentration of the product is τ_p. You have been asked to purchase syringes for the device that will allow an accurate estimate of the reaction rate constant. *Challenge:* Estimate the minimum volume and pressure rating that each syringe should have to provide an accurate study of the reaction rate. *Generate ideas:* What would you suggest for the mean transit time for fluid to flow through the device? How does this translate to mean velocity and mean flow through the device? How long should the experiment last? What minimum syringe volume is necessary? From the standpoint of radial mixing, would you prefer the flow to be laminar or turbulent? What is the relationship between the mean transit time and the Reynolds number? How can you find the pressure drop across a square duct? What additional information do you need to estimate the pressure in the syringes during an experiment? Would the solution procedure be the same if flow was laminar or turbulent?

5.21.2 Vascular Resistance and GFR

Drug "A" is effective in reducing blood pressure in the kidney. Because arterial pressure is reduced, we might expect the drug to reduce GFR. However, measurements of urinary output indicate that the drug does not affect GFR. *Challenge:* Can we use transport principles to explain this apparent mystery? *Generate Ideas:* What mechanisms are primarily responsible for regulating blood pressure and flow? Why might we expect the drug to lower GFR? Describe the vascular system of the kidney. What are afferent and efferent arterioles? What factors influence arteriolar resistance? If renal blood flow remains relatively constant when blood pressure is raised, what happens to renal vascular resistance? If GFR is also unchanged when arterial pressure is raised, is this regulated by a change in afferent resistance, efferent resistance, or both?

5.21.3 Osmotic Shock

Red blood cells are observed to shrink, swell, or maintain their volume, depending on the nature of the fluid they are suspended in. Cell volume affects cell function. *Challenge:* Can we predict cell volume changes when a red cell is placed in a hypertonic solution? *Generate ideas:* What is a hypertonic solution? What factors influence red cell volume? If the cell membrane is a lipid bilayer, how can fluid or hydrophilic solutes pass in or out? What is the driving force for fluid passage through the cell membrane? What constitutive relationship describes flow through the membrane? What conservation equation is appropriate for cell fluid balance? How can we estimate osmotic pressure?

References

Beek WJ, Muttzall KMK, Van Heuven JW (1999) Transport phenomena, 2nd edn. Wiley, New York

Bird RB, Stewart WE, Lightfoot EN (2002) Transport phenomena, 2nd edn. Wiley, New York

Durst F, Ray S, Unsal B and Bayoumi OA (2005) The development lengths of laminar pipe and channel flows. Trans ASME 127:1154–1160

Ergun S (1952) Fluid flow through packed columns. Chemical Engineering Progress 48:89–94

Fung YC (1996) Biomechanics: circulation, 2nd edn. Springer, New York

Kameyama T, Asanoi H, Ishizaka S, Yamanishi K, Fujita M and Sasayama S (1992) Energy conversion efficiency in human left ventricle. Circulation 85:988–996

Landis EM, Pappenheimer JR (1963) Exchange of substances through the capillary walls. In: Hamilton WF, Dow P (eds) Handbook of physiology, section 2, Circulation, part II. American Physiological Society, Washington, DC, pp 961–1034

Navar PD, Navar LG (1977) Relationship between colloid osmotic pressure and plasma protein concentration in the dog. Am J Physiol 233:H295–H298

Olson DE, Dart GA, Filley GF (1970). Pressure drop and fluid flow regime of air inspired into the human lung. J Appl Physiol 28:483–494

Olson RM, Sparrow EM (1963). Development in tubes and annuli with square or rounded entrances. AIChE J 9:766–770

Paine PL, Scherr P (1975) Drag coefficients for the movement of rigid spheres through liquid-filled cylindrical pores. Biophys J 15:1087–1091

Perry RH, Chilton CH (1973) Chemical engineer's handbook, 5th edn. McGraw-Hill, New York

Pride NB, Permut S, Riley RL, Bromberger-Barnea B (1967) Determinants of maximal expiratory flow from the lungs. J Appl Physiol 23:646–662

Ramanujan S (1913–1914) Modular Equations and Approximations to Π. Quart J Pure Appl Math 45:350–372

Schlichting, H., 1968, Boundary Layer Theory, McGraw-Hill, NY

West JB, Dollery CT, Naimark A (1964) Distribution of blood flow in isolated lung; relation to vascular and alveolar pressures. J Appl Physiol 19:713–724

Chapter 6
Shell Balance Approach for One-Dimensional Biofluid Transport

6.1 Introduction

In the previous chapter, we took a macroscopic view for solving transport problems. The macroscopic approach is sufficient if you are interested in how the average transport variables change with time in a system. However, if you also want to know how transport properties such as mass, concentration, charge, velocity, or temperature vary with position (x, y, z), then it is necessary to use a microscopic approach. This allows the system of interest to shrink to an infinitesimally small volume surrounding an arbitrary point within the larger system.

In this chapter, we will apply the microscopic shell balance approach for solving one-dimensional (1-D) flow problems. Multi-dimensional problems will be treated in Chap. 7. In 1-D flow problems, fluid flows in a single coordinate direction, but some fluid properties, such as velocity, will generally vary in the direction perpendicular to the direction of flow. This is known as parallel flow. We are generally interested in how the velocity varies with position in a parallel flow device, the forces exerted by the fluid on the walls of the device, the total flow through the device, and the distribution of pressure in the fluid.

Many important applications can be considered to be parallel flow problems. The capillary and sliding plate viscometers discussed in Chap. 4 are examples of 1-D flow applications. There are many physiological, medical, and biological applications in which fluids are forced through circular tubes or between parallel plates. Physiological examples include flow in blood vessels, airways, lymph vessels, nephron tubules, and other body conduits. Medical devices include intravenous tubing for drug infusion, and hollow fiber and parallel membrane devices used for blood oxygenation and blood dialysis. Tubing and parallel flow chambers are used in the biotechnology industry to transport and separate cells or proteins in solution, and for many other applications.

6.2 General Approach

The solution to parallel flow problems can be found by applying what is known as the "shell balance" approach. The steps involved are:

1. Select an appropriate fluid shell.
2. Apply conservation of mass to fluid in the shell.
3. Apply conservation of momentum to fluid in the shell.
4. Apply the relevant fluid constitutive relation.
5. Compute the velocity profile using appropriate boundary conditions.
6. Compute flow rate and wall shear stress.

Details for each step are provided in the following sections. We will confine our initial discussion to the solution of steady-state problems in rectangular coordinates. Later in the chapter, we will consider problems in cylindrical coordinates, and apply the same technique to an unsteady-state parallel flow problem.

To introduce the shell balance approach, we will consider the problem of blood flow in a film oxygenator. In this example, blood is pumped at a constant rate over several flat plates, each with width W and length L. The plates are surrounded with an oxygen-rich gas at constant pressure and temperature. We will analyze flow over one plate, which is inclined at an angle α with the horizontal (see Fig. 6.1).

The force of gravity will initially accelerate the blood as it flows down the plate. Since the fluid is incompressible, conservation of mass dictates that the film thickness will decrease as the velocity increases. We also know from the viscous properties of fluids that, as the velocity gradient at the wall increases, the shear stress exerted by the wall on the fluid will also increase. At some distance down the plate, indicated by l' in Fig. 6.1, the gravitational force is balanced by the wall force and the blood will no longer accelerate as it moves down the plane. We take this point as the origin of our coordinate system, as shown in Fig. 6.1. The film thickness will be constant (equal to h in Fig. 6.1) and blood flow will be parallel until a

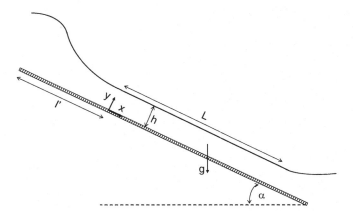

Fig. 6.1 Blood flow down an inclined plane

6.2 General Approach

disturbance is encountered at $x = L$, where the fluid begins to flow off of the plate. Therefore, the problem can be analyzed using the 1-D shell balance approach in the region $0 \leq x \leq L$.

To simplify the analysis, we will assume the blood to be Newtonian, but will investigate the effect of that assumption after completion of the Newtonian analysis. Our goals are to find:

1. The distribution of shear stress in the blood film.
2. The velocity profile in the blood film.
3. The total blood flow rate down the plate as a function of film thickness, h.

6.2.1 Selecting an Appropriate Shell

Since our first two goals involve finding the local shear stress and local velocity, we cannot use a macroscopic approach. Determination of local flow properties requires a microscopic approach, in which the system to be analyzed includes only a small fluid shell that surrounds the point of interest. In fact, we will ultimately allow the volume of the shell to shrink to zero so the analysis will apply to a single point within the fluid.

In general, a microscopic shell is constructed about an arbitrary point $P(x, y)$ within the fluid. The shape of the shell depends on two primary factors: the coordinate system adopted and the number of independent spatial variables to be considered in the problem. In a parallel flow problem in Cartesian coordinates, there are two spatial directions of interest: the flow direction (x), and the direction perpendicular to the flow direction (y). The material within the shell is confined to the volume between planes of constant x and $x + \Delta x$ and planes of constant y and $y + \Delta y$. The lengths Δx and Δy are normally taken to be very small, so that as Δx and Δy approach zero, the volume will represent fluid only in the neighborhood of the point $P(x, y)$. The width of the shell in the z-direction is W. The shell is drawn in Fig. 6.2. The shell volume is $W \Delta x \Delta y$. Ultimately, we will develop equations that apply at any arbitrary point $P(x, y)$ in the system by allowing $\Delta y \to 0$ and $\Delta x \to 0$.

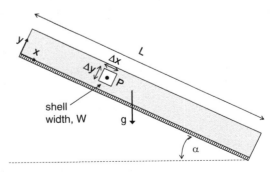

Fig. 6.2 Location of the shell surrounding the point $P(x, y)$ in the fluid film

6.2.2 Fluid Mass Balance

A general 1-D mass balance is performed on the fluid within the shell in Fig. 6.3. This can be expressed in words as follows:

$$\left\{\begin{array}{c}\text{Rate of}\\ \text{accumulation of}\\ \text{mass in shell}\end{array}\right\} = \left\{\begin{array}{c}\text{Rate that}\\ \text{mass}\\ \text{enters shell}\end{array}\right\} - \left\{\begin{array}{c}\text{Rate that}\\ \text{mass}\\ \text{leaves shell}\end{array}\right\} + \left\{\begin{array}{c}\text{Rate of}\\ \text{production of}\\ \text{mass in shell}\end{array}\right\}. \quad (6.1)$$

We are only interested in the steady-state solution, so the rate at which mass accumulates in the shell must be zero. Also, mass cannot be created or destroyed in the shell, so the rate of production of mass in the shell must also equal zero. The rate that fluid enters is equal to the mass flux into the shell at x $(\rho v_x)|_x$ multiplied by the shell inlet flow area at x ($W\Delta y$), where W is the width of the shell. Similarly, the rate that fluid leaves the shell is equal to the mass flux out of the shell at $x = x + \Delta x$, $(\rho v_x)|_{x=x+\Delta x}$, multiplied by the shell outlet flow area at $x = x + \Delta x$ ($W\Delta y$), so the mass balance reduces to:

$$0 = \left\{\begin{array}{c}\text{Rate that}\\ \text{mass}\\ \text{enters shell}\end{array}\right\} - \left\{\begin{array}{c}\text{Rate that}\\ \text{mass}\\ \text{leaves shell}\end{array}\right\} = \left[(\rho v_x)|_x - (\rho v_x)|_{x+\Delta x}\right] W\Delta y. \quad (6.2)$$

The next step in the shell balance procedure is to divide the conservation equation by the shell volume ($W \Delta x \Delta y$) and let the volume shrink to zero:

$$0 = \lim_{\substack{\Delta x \to 0 \\ \Delta y \to 0}} \left\{ \frac{\left[(\rho v_x)|_x - (\rho v_x)|_{x+\Delta x}\right] W\Delta y}{W\Delta x \Delta y} \right\} = -\frac{\partial(\rho v_x)}{\partial x}. \quad (6.3)$$

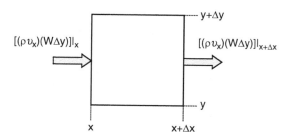

Fig. 6.3 Rate at which fluid enters and leaves the shell

6.2 General Approach

But the density of blood ρ is constant, so this reduces to:

$$\frac{\partial v_x}{\partial x} = 0. \tag{6.4}$$

This is known as the continuity equation for parallel flow, and its interpretation is that v_x is independent of x, or $v_x = v_x(y)$ only. We conclude that a mass balance confirms our initial assumption that the velocity does not depend on the axial position x in parallel flow. Since the velocity does not change with x, it would have been acceptable to have initially selected a shell that spanned the entire length of the plate, rather than a length of Δx.

6.2.3 Fluid Momentum Balance

The next step in the fluid shell balance approach is to apply conservation of momentum to the fluid shell. Momentum is vector quantity. Since velocities in the y and z directions are zero in parallel flow, we will consider the conservation of x-momentum. In words:

$$\begin{Bmatrix} \text{Rate of} \\ \text{accumulation of} \\ x\text{ - momentum in shell} \end{Bmatrix} = \begin{Bmatrix} \text{Rate that} \\ x\text{ - momentum} \\ \text{enters shell} \end{Bmatrix} - \begin{Bmatrix} \text{Rate that} \\ x\text{ - momentum} \\ \text{leaves shell} \end{Bmatrix} \\ + \begin{Bmatrix} \text{Rate of production} \\ \text{of } x\text{ - momentum} \\ \text{in shell} \end{Bmatrix}. \tag{6.5}$$

Let us look first at the left-hand side of this expression. As fluid accelerates in its passage down the plate, this term is positive and nonzero, decreasing as it moves down the plate. However, our shell is only located in the region, where the steady-state velocity is constant. Thus, the rate of accumulation of momentum in this region must be zero.

How is x-momentum produced in the fluid inside the shell? According to Newton's second law, x-momentum is produced when forces are applied to the element in the x-direction, so:

$$\begin{Bmatrix} \text{Rate of production} \\ \text{of } x\text{ - momentum} \\ \text{in fluid element} \end{Bmatrix} = \begin{Bmatrix} \text{Sum of forces} \\ \text{applied in } x\text{ - direction} \\ \text{to fluid element} \end{Bmatrix}. \tag{6.6}$$

What forces are applied in the x-direction to the fluid element in the shell? Potential forces are illustrated in Fig. 6.4.

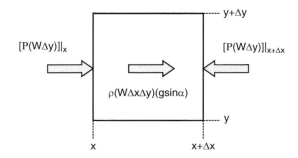

Fig. 6.4 Forces on the fluid element within the shell in the x-direction

Certainly, there is a component of the gravitational force in the x-direction:

$$\left\{ \begin{array}{c} \text{gravitational force} \\ \text{in } x \text{ - direction} \end{array} \right\} = mg_x = \rho[W\Delta x \Delta y]g_x. \tag{6.7}$$

Here, we have made use of the fact that the mass of blood m in the shell is equal to the product of blood density ρ and shell volume. The gravitational component in the x-direction is:

$$g_x = g \sin \alpha. \tag{6.8}$$

If fluid pressure were to vary in the x-direction, then there would be a net force on the shell caused by the pressure differences between the faces at x and $x + \Delta x$:

$$\left\{ \begin{array}{c} \text{net pressure force} \\ \text{in } x \text{ - direction} \end{array} \right\} = [P|_x - P|_{x+\Delta x}][W\Delta y]. \tag{6.9}$$

We know that gravity will cause pressure to vary with y, but will it also vary with x? To help answer this question, it is useful to perform a momentum balance on our fluid shell in the y-direction. Since the fluid does not move in the y-direction, the y-momentum equation reduces to a force balance in the y-direction (Fig. 6.5):

$$0 = \left[P|_y - P|_{y+\Delta y}\right][W\Delta x] - \rho g \cos \alpha [W\Delta x \Delta y]. \tag{6.10}$$

Dividing by the shell volume and taking the limit as the volume shrinks to zero:

$$0 = -\frac{\partial P}{\partial y} - \rho g \cos \alpha. \tag{6.11}$$

Integrating the y-momentum equation with respect to y:

$$P = -(\rho g \cos \alpha)y + f(x), \tag{6.12}$$

6.2 General Approach

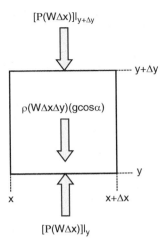

Fig. 6.5 Forces acting on the fluid within the shell in the y-direction

where $f(x)$ is an arbitrary function of x. Now consider any point along the free surface $(x, y = h)$. The pressure must equal atmospheric pressure P_{atm} at all points on the free surface:

$$f(x) = P_{atm} + \rho g h \cos \alpha = \text{constant}. \tag{6.13}$$

Therefore, the pressure distribution in the film is:

$$P = P_{atm} + \rho g \cos \alpha \cdot (h - y). \tag{6.14}$$

Because of the presence of the free surface at the top of the blood film, the pressure is independent of x everywhere within the film. Consequently, the pressures at the two ends of the shell, $P|_x$ and $P|_{x+\Delta x}$, are equal and the net pressure force on the shell in the x-direction is zero for this example.

What other forces might be responsible for changing the momentum of the fluid in the shell? Recall that in Chap. 4 we stated that shear stress τ_{yx} can be interpreted either as a friction force per unit area (viscous stress) or as the molecular flux of x-momentum in the y-direction. In the current application, we will treat viscous stress as a momentum flux. Therefore, the gravitational force is the only force applied to the shell and the production term in the momentum equation becomes:

$$\left\{ \begin{array}{c} \text{Rate of production} \\ \text{of } x \text{ - momentum} \\ \text{in fluid element} \end{array} \right\} = \rho g \sin \alpha [W \Delta x \Delta y]. \tag{6.15}$$

Let us now return to the first term on the right-hand side of the x-momentum equation, (6.5). There are two ways that x-momentum can enter the shell:

1. x-momentum enters the shell by molecular transport across the boundary located at y:

$$\left\{\begin{array}{c} \text{Rate that} \\ x\text{ - momentum} \\ \text{enters shell} \\ \text{by molecular} \\ \text{transport at } y \end{array}\right\} = \left[(\tau_{yx})(W\Delta x)\right]\big|_y. \tag{6.16}$$

2. x-momentum enters the shell as fluid flows into it via bulk fluid motion at the boundary located at x:

$$\left\{\begin{array}{c} \text{Rate that} \\ x\text{ - momentum} \\ \text{enters shell} \\ \text{by bulk fluid} \\ \text{transport at } x \end{array}\right\} = \left[(\rho v_x)(W\Delta y)(v_x)\right]\big|_x. \tag{6.17}$$

The product of the first two terms (mass flux × area) is the rate at which mass enters the shell by bulk fluid motion. Multiplying mass rate by fluid velocity provides the rate at which x-momentum enters the shell by bulk fluid motion. The rates at which x-momentum leaves the shell by these two mechanisms can be written as follows:

$$\left\{\begin{array}{c} \text{Rate that} \\ x\text{ - momentum} \\ \text{leaves shell} \\ \text{by molecular} \\ \text{transport at } y+\Delta y \end{array}\right\} = \left[(\tau_{yx})(W\Delta x)\right]\big|_{y+\Delta y}, \tag{6.18}$$

$$\left\{\begin{array}{c} \text{Rate that} \\ x\text{ - momentum} \\ \text{leaves shell} \\ \text{by bulk fluid} \\ \text{transport at } x+\Delta x \end{array}\right\} = \left[(\rho v_x)(W\Delta y)(v_x)\right]\big|_{x+\Delta x}. \tag{6.19}$$

The rate at which x-momentum passes through each shell boundary is shown in Fig. 6.6.

Combining all terms in the original x-momentum balance:

$$0 = \left[(\tau_{yx})_y - (\tau_{yx})_{y+\Delta y}\right](W\Delta x) + \left[(v_x)_x^2 - (v_x)_{x+\Delta x}^2\right](\rho W\Delta y) + \rho g \sin\alpha [W\Delta x\Delta y]. \tag{6.20}$$

6.2 General Approach

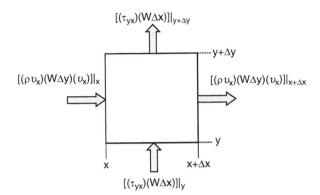

Fig. 6.6 Momentum entering and leaving the shell by convection at x and $x + \Delta x$ and by viscous molecular motion at y and $y + \Delta y$

But the mass balance performed in the previous section shows $v_x|_x = v_x|_{x+\Delta x}$, so there is no net change in x-momentum caused by bulk fluid motion, and the second term vanishes. Dividing by the shell volume $W\Delta x \Delta y$, and letting the volume approach zero, we obtain:

$$\lim_{\substack{\Delta x \to 0 \\ \Delta y \to 0}} \left\{ \frac{\left[(\tau_{yx})_y - (\tau_{yx})_{y+\Delta y}\right]}{\Delta y} \right\} + \rho g \sin \alpha = 0. \qquad (6.21)$$

The left-hand side of this equation is simply the negative of the derivative of shear stress with respect to y. Therefore, the x-momentum balance results in the following ordinary differential equation:

$$\frac{d\tau_{yx}}{dy} = \rho g \sin \alpha. \qquad (6.22)$$

The right hand side of this equation is a constant, so it can be easily integrated:

$$\tau_{yx} = (\rho g \sin \alpha) y + C_1. \qquad (6.23)$$

Here, C is a constant of integration that can be determined by applying the boundary condition at the gas-fluid interface ($y = h$). We assume from (4.8) that the shear stress exerted by the gas on the liquid at the interface is negligible, so:

$$\tau_{yx}(h) \approx 0 = (\rho g \sin \alpha) h + C_1. \qquad (6.24)$$

So the constant $C_1 = -\rho g h \cdot \sin(\alpha)$ and the general expression for the distribution of shear stress in the blood film is:

$$\tau_{yx} = \rho g (y - h) \sin \alpha. \qquad (6.25)$$

Fig. 6.7 Distribution of shear stress in the film

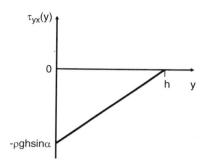

Thus, our first goal has been accomplished. The shear stress varies linearly from zero at the top to $-\rho gh \cdot \sin \alpha$ at the wall (Fig. 6.7). The momentum flux at the wall is negative, which is consistent with a positive velocity gradient. In addition, the viscous momentum flux (or shear stress) is independent of the width and length of the film. Finally, it is important to note that we have managed to derive this relationship without specifying anything about the type of fluid in the film. Therefore, the same expression is valid for Newtonian and non-Newtonian fluids.

6.2.4 Application of the Fluid Constitutive Relation to Find Fluid Velocity

Computation of the velocity profile requires that we select the appropriate constitutive relationship for the fluid at hand. For this case, we have assumed the shear rate is high enough so blood can be approximated as a Newtonian fluid. We will examine this assumption later. Combining the expression derived from a momentum balance with the constitutive relationship for a Newtonian fluid:

$$\tau_{yx} = -\mu \frac{dv_x}{dy} = \rho g (y - h) \sin \alpha. \tag{6.26}$$

Dividing by $-\mu$ and integrating with respect to y:

$$v_x = -\frac{\rho g \sin \alpha}{\mu} \left(\frac{y^2}{2} - hy \right) + C_2. \tag{6.27}$$

The constant of integration C_2 can be found by applying the no-slip boundary condition (4.5) at the wall of the device (i.e., $y = 0$). Therefore, $C_2 = 0$, and the velocity as a function of position is given by:

$$v_x = \frac{\rho g h^2 \sin \alpha}{\mu} \left(\frac{y}{h} - \frac{1}{2} \frac{y^2}{h^2} \right). \tag{6.28}$$

6.2 General Approach

Fig. 6.8 Velocity profile in fluid film

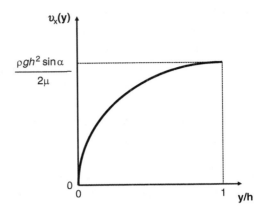

Thus, we have accomplished our second goal. The velocity profile for a Newtonian fluid flowing down an inclined plane has a parabolic shape (Fig. 6.8) with zero velocity at the wall and a maximum velocity at the top of the film.

6.2.5 Examining and Applying Solutions for Shear Stress and Velocity

Let us now examine the solution we have obtained for a Newtonian fluid flowing down an inclined plane. The maximum velocity v_{max} at $y = h$ is equal to

$$v_{max} = \frac{\rho g h^2 \sin \alpha}{2\mu}. \tag{6.29}$$

The average velocity $\langle v_x \rangle$ in the film is:

$$\langle v_x \rangle = \frac{1}{h}\int_0^h v_x dy = \frac{\rho g \sin \alpha}{\mu}\int_0^h \left(y - \frac{y^2}{2h}\right) dy = \frac{\rho g h^2 \sin \alpha}{3\mu}. \tag{6.30}$$

Thus, the average velocity is two thirds the maximum velocity for flow down an inclined plane. The volumetric flow rate Q_V at which blood flows down an inclined plane is given as the product of the average velocity $\langle v_x \rangle$ and the cross-sectional area $W h$ of the film:

$$Q_V = \langle v_x \rangle W h = \frac{\rho g W h^3 \sin \alpha}{3\mu}. \tag{6.31}$$

We have thus satisfied our third goal. Since the flowrate is related to the film height cubed, small changes in h can lead to large variations in flow. If the flowrate were unknown, we could measure the film height and compute the flowrate. This could be used to compute flow in a river, for instance, if all the flow were diverted over an inclined plane with known width W and angle of inclination, α. In a blood film oxygenator, the flowrate will generally be known, so (6.31) can be used to predict the steady-state film height, h for a plate of width W:

$$h = \left(\frac{3\mu Q_V}{\rho g W \sin \alpha}\right)^{\frac{1}{3}}. \tag{6.32}$$

Example 6.2.5.1.
A blood oxygenator with a total flow rate of 1 L/min consists of one hundred 10 cm wide by 10 cm long plates arranged in parallel. Each plate makes an angle of 45° with the horizontal and blood at the film surface is exposed to an oxygen-rich gas. The degree to which blood is oxygenated will depend in part on the thickness of the blood film on each plate. Find the film thickness if blood is considered to be a Newtonian fluid with a density of 1.04 g/cm^3 and viscosity of 3 mPa s. Comment on the validity of modeling blood as a Newtonian fluid in this situation.

Solution. *Initial considerations*: Our goal is to find the thickness of the film of flowing blood formed on each of the plates comprising the blood oxygenator. We will neglect the short region near the upstream end of the plates where fluid accelerates and the film thickness decreases in the downstream direction. We will also assume that total flow in the oxygenator is evenly distributed at the inlet to each of the 100 plates.

System definition and environmental interactions: The assumption that each plate receives the same flow allows us to confine the analysis to film flow on a single plate. Therefore, our system consists of the blood flowing within the film above the surface of a single plate.

Apprising the problem to identify governing equations: Two fundamental relations govern flow in this situation: (1) conservation of linear momentum, and (2) the constitutive relationship for the fluid. Flow is steady (i.e., not changing with time) in this problem. If we further assume that flow in the film is parallel, then conservation of linear momentum leads to a linear distribution of shear stress in the film (6.25). Blood behaves as a Newtonian fluid at high shear rates. Use of the Newtonian constitutive relationship leads to the method of analysis presented in Sect. 6.2.4, ultimately leading to (6.32) as an expression for the film thickness.

Analysis: Flow over a single plate will be .01 L/min = 10 ml/min. Substituting into (6.32):

$$h = \left(\frac{3(0.003\,\text{Pa}\cdot\text{s})(10\,\text{gcm}^{-1}\text{s}^{-2}\text{Pa}^{-1})(10\,\text{cm}^3/\text{min})(1\,\text{min}/60\,\text{s})}{(1.04\,\text{g/cm}^3)(980\,\text{cms}^{-2})(10\,\text{cm})(0.707)}\right)^{\frac{1}{3}} = 0.013\,\text{cm}.$$

So the predicted film thickness is 0.13 mm.

6.2 General Approach

Examining and interpreting the results: Let us look now at the assumption that blood can be considered to be Newtonian under these circumstances. Rewriting (6.26):

$$\dot{\gamma} = \frac{dv_x}{dy} = \frac{\rho g}{\mu}(h-y)\sin\alpha.$$

The shear rate is not constant, but instead varies linearly with y. At the bottom of the film, the maximum shear rate is

$$\dot{\gamma}_{max} = \frac{\rho g h \sin\alpha}{\mu} = \frac{(1.04\,\text{gcm}^{-3})(980\,\text{cms}^{-2})(0.013\,\text{cm})(0.707)}{(0.003\,\text{Pas})(10\,\text{gcm}^{-1}\,\text{s}^{-2}\,\text{Pa}^{-1})} = 312\,\text{s}^{-1}.$$

Experimental measurements discussed in Chap. 4 indicate that blood can be considered a Newtonian fluid when the shear rate is greater than 50 s^{-1}. Therefore, blood can certainly be considered a Newtonian fluid in the vicinity of the plate wall. We can estimate the film thickness for which blood in the device can be considered Newtonian by solving (6.26) for the position y^* at which the shear rate is 50 s^{-1}:

$$y^* = h - \frac{\dot{\gamma}_{50}\mu}{\rho g \sin\alpha}.$$

Fluid in the top 16% of the film is exposed to shear rates that are too low for blood to be considered a Newtonian fluid. Because blood exhibits a yield stress, a thin layer at the top of the film will move as a plug. We will examine non-Newtonian behavior later in this chapter. Although the use of the Newtonian relationship provides a good estimate of film thickness, a more accurate measure would involve the use of the Casson constitutive equation, rather than the Newtonian relation. We leave the more detailed analysis as an exercise for the student.

Additional comments: The average film velocity from (6.30) is:

$$\langle v_x \rangle = \frac{(1.04\,\text{gcm}^{-3})(980\,\text{cms}^{-2})(0.013\,\text{cm})^2(0.707)}{3(0.003\,\text{Pas})(10\,\text{gcm}^{-1}\,\text{s}^{-2}\,\text{Pa}^{-1})} = 1.317\,\text{cm/s}.$$

The average film-contact time will be 10 cm/1.317 cm/s = 7.6 s. It remains to be seen if this is sufficient time for significant uptake of oxygen. We will examine this in more detail in Chap. 13. it is useful to check that total flow over all 100 plates is equal to the flow to the blood oxygenator:

$$Q_V = 100\,Wh\,\langle v_x \rangle = 100(10\,\text{cm})(0.013\,\text{cm})(1.317\,\text{cm/s})(60\,\text{s/min})$$
$$= 1{,}000\,\text{ml/min}.$$

This agrees with the imposed flow of 1 L/min. It is also useful to check to see whether the volume of the device is reasonable. Neglecting the effects of fluid acceleration near the leading edge, total blood film volume V in a device with 100 plates is

$$V = 100\,WLh = 100\,(10\,\text{cm})(10\,\text{cm})(0.013\,\text{cm}) = 130\,\text{ml}.$$

This is a small fraction of the total blood volume for normal individuals (about 5,000 ml).

6.2.6 Additional Shell Balances in Rectangular Coordinates

In this section, we will apply the shell balance technique outlined in the previous sections to set up and solve other parallel flow problems in Cartesian coordinates. First, we will consider the problem of two immiscible Newtonian fluids flowing down an inclined plane. Then we will consider an internal flow problem, where a Newtonian fluid is confined between two walls. Finally, we will consider two examples of non-Newtonian fluids in parallel flow situations. Our goals are to reinforce the shell balance procedure, to illustrate the most common boundary conditions used in fluid mechanics problems, and to provide useful solutions for some important parallel flow problems.

> **Example 6.2.6.1 Marginal Zone Theory Applied to Blood Flowing Down an Inclined Plane.**
> It was shown in Sect. 4.9 that red blood cells travel faster, on average, than plasma in small tubes. This is caused by the inability of red cell centers to come any closer to vessel walls than the radius of the cell, while slower moving plasma can come much closer to the wall. Therefore, a thin region near the wall of the vessel, known as the marginal zone, is composed primarily of plasma. This is sometimes modeled as though a cell-free layer of plasma exists between the wall and the whole blood. In this example, we examine the flow rate of blood flowing down an inclined plane, with blood treated as a two-fluid system, and compare it with the flow computed by assuming blood is a homogeneous, Newtonian fluid. Neglect differences between plasma and blood density.

Solution. *Initial considerations*: Our goal is to compute the volumetric flow rate of blood as it moves down an inclined plane. What makes this problem different from the one treated in Sects. 6.2.1–6.2.5 is that blood is not considered to be a single, homogeneous Newtonian fluid, but consists, instead, of two distinct fluids in the regions shown in Fig. 6.9: region "p" near the wall that consists of plasma alone and region "b" that consists of whole blood. We assume parallel flow in both regions.

System definition and environmental interactions: There are two systems that must be analyzed in this problem: the fluids in regions "p" and in region "b." Because red cells are excluded from the plasma region, the two fluids are assumed to be

6.2 General Approach

Fig. 6.9 Shells for blood and plasma layers

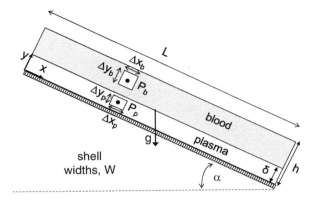

immiscible. Plasma and whole blood are both assumed to be Newtonian fluids with density ρ and viscosities μ_p and μ_b, respectively. The viscosity of blood will depend on the hematocrit value H, as discussed in Chap. 4, and will be three to four times the viscosity of plasma for $H = 0.45$. The thickness of the plasma layer is designated as δ, which will also depend on the hematocrit value. In addition, interactions between the two fluids at the interface must be specified. According to the boundary conditions discussed in Sect. 4.4, the velocities of the plasma and the blood at the interface ($y = \delta$) are equal, and the shear stress applied by the plasma on the blood is also equal to the shear stress applied by the blood on the plasma at the interface.

Apprising the problem to identify governing equations: The governing equations for this problem are derived by applying conservation of mass, conservation of momentum, and the constitutive relation for a Newtonian fluid to each region.

Analysis: We begin by selecting two shells, one in each fluid, as shown in Fig. 6.9. Application of a mass balance to each shell will tell us that the velocity is independent of x in both regions. Following the procedure in Sect. 6.2.3 for conservation of momentum, we derive an equation for shear stress in each fluid, similar to (6.26):

$$(\tau_{yx})_b = (\rho g \sin \alpha) y + C_b, \delta \leq y \leq h,$$
$$(\tau_{yx})_p = (\rho g \sin \alpha) y + C_p, 0 \leq y \leq \delta.$$

The unknown constants C_p and C_b are found by applying appropriate boundary conditions. First, in the blood phase, the shear stress at the top surface ($y = h$) is assumed to be negligible, so $\tau_{yx}(h) = 0$ and therefore $C_b = -\rho g h \cdot \sin \alpha$. At the interface between two fluids ($y = \delta$), the shear stresses must be identical in the two fluids.

$$(\tau_{yx}(\delta))_b = (\tau_{yx}(\delta))_p$$
$$(\rho g \sin \alpha)(\delta - h) = (\rho g \sin \alpha)\delta + C_p.$$
$$C_p = C_b = -\rho g h \sin \alpha$$

Therefore, the expression for shear stress is identical in both fluids, and is the same as that derived in Sect. 6.2.3:

$$\tau_{yx} = \rho g(y-h)\sin\alpha.$$

The next step is to substitute the Newtonian constitutive relationship into the momentum equation for each fluid, as in Sect. 6.2.4:

$$-\mu_b \frac{dv_x}{dy} = \rho g(y-h)\sin\alpha, \; \delta \leq y \leq h,$$

$$-\mu_p \frac{dv_x}{dy} = \rho g(y-h)\sin\alpha, \; 0 \leq y \leq \delta.$$

Integrating:

$$v_x = -\frac{\rho g \sin\alpha}{\mu_b}\left(\frac{y^2}{2} - hy\right) + C_{b_2}, \; \delta \leq y \leq h,$$

$$v_x = -\frac{\rho g \sin\alpha}{\mu_p}\left(\frac{y^2}{2} - hy\right) + C_{p_2}, \; 0 \leq y \leq \delta.$$

The constants of integration C_{b2} and C_{p2} can be found by applying appropriate boundary conditions. First, we can apply the no-slip boundary condition in the plasma layer at the wall ($y = 0$) to show that $C_{p2} = 0$. Next, we apply the no-slip boundary condition at the interface between the blood and plasma ($y = \delta$) to solve for the unknown constant C_{b2}:

$$(v_x(\delta))_b = (v_x(\delta))_p,$$

$$-\frac{\rho g \sin\alpha}{\mu_b}\left(\frac{\delta^2}{2} - h\delta\right) + C_{b_2} = -\frac{\rho g \sin\alpha}{\mu_p}\left(\frac{\delta^2}{2} - h\delta\right),$$

$$C_{b_2} = \frac{\rho g h^2 \sin\alpha}{\mu_b}\left(\frac{\mu_b}{\mu_p} - 1\right)\left(\frac{\delta}{h} - \frac{1}{2}\frac{\delta^2}{h^2}\right).$$

Substitution of C_{b1} and C_{b2} back into the original expressions provides expressions for velocity in the blood and plasma regions:

$$v_x = -\frac{\rho g h^2 \sin\alpha}{\mu_b}\left[\frac{1}{2}\frac{y^2}{h^2} - \frac{y}{h} + \left(\frac{\mu_b}{\mu_p} - 1\right)\left(\frac{1}{2}\frac{\delta^2}{h^2} - \frac{\delta}{h}\right)\right], \; \delta \leq y \leq h,$$

$$v_x = -\frac{\rho g h^2 \sin\alpha}{\mu_p}\left(\frac{1}{2}\frac{y^2}{h^2} - \frac{y}{h}\right), \; 0 \leq y \leq \delta.$$

As a check, this reduces to the expression for a homogeneous film if $\mu_p = \mu_b$. Also, it is easy to show that the plasma and blood velocities are equal at $y = \delta$. However,

6.2 General Approach

the velocity gradients are discontinuous at the interface because the shear stresses must be equal but the viscosities of the plasma and blood regions are different:

$$\left(\frac{dv_x}{dy}\right)_b = \frac{\mu_p}{\mu_b}\left(\frac{dv_x}{dy}\right)_p \quad \text{at } y=\delta.$$

Our goal is to compare the flow of this bi-layer film down the inclined plane to the flow computed on the basis of a homogeneous fluid with the viscosity of whole blood. The total volumetric flowrate for the bi-layer is found by adding the contributions of each layer:

$$Q_V = \int_0^\delta (v_x(y))_p W dy + \int_\delta^h (v_x(y))_b W dy.$$

Substituting for plasma and blood velocities and integrating, the total flow down the inclined plane is:

$$Q_V = \frac{\rho g W h^3 \sin\alpha}{3\mu_b}\left\{1 + 3\left(\frac{\mu_b}{\mu_p} - 1\right)\left(\frac{\delta}{h}\right)\left[1 - \frac{\delta}{h} + \frac{1}{3}\left(\frac{\delta}{h}\right)^2\right]\right\}.$$

This expression reduces to (6.31) for flow of whole blood down the plane when blood viscosity approaches plasma viscosity (i.e., low hematocrit value), or when the total film thickness is much greater than the plasma layer thickness (i.e., $\delta/h \to 0$). The influence of plasma layer thickness and total film thickness on the ratio of bi-layer flow to flow of a homogeneous blood layer is shown in Fig. 6.10. The thickness of the plasma layer is assumed to lie between half the radius of a red cell (2 μm, bottom line)

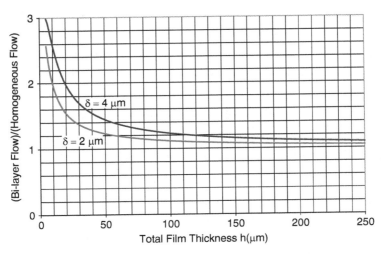

Fig. 6.10 Flow of blood with plasma layers of 4 μm and 2 μm, relative to flow of blood without plasma layer, as a function of total layer thickness, h

and one red cell radius (4 μm, top line). The viscosity of blood is assumed to be three times the viscosity of plasma ($\mu_b/\mu_p = 3$).

Examining and interpreting the results: Note that the presence of a plasma layer can significantly increase flow for very thin layers, and 5–10% increases are predicted for film layers as thick as 250 μm. This "plasma skimming" phenomena is responsible for reducing the apparent viscosity of whole blood in thin layers and narrow tubes.

Additional comments: The student should realize that this is a model of how blood flows down an inclined plane, and as such is subject to the assumptions made in developing the model. The assumption that plasma and blood are two distinct fluids separated at a constant distance δ from the wall is certainly an oversimplification. The assumption that blood is a Newtonian fluid is probably a good one near the interface, but will not be valid near the top of the blood layer, particularly if the yield stress is high.

Example 6.2.6.2 Flow of a Newtonian Fluid Through an Inclined Channel with a Pressure Gradient.

There are many devices used in biology and medicine in which fluid is forced through the narrow gap between parallel plates. Parallel plate membrane blood oxygenators or dialysis units are prime examples. In this example, we replace the free surface in the falling film problem with a second stationary wall. In addition, we impose a pressure P_0 at the inlet ($x = 0, y = 0$) and a pressure P_L at the outlet ($x = L, y = 0$) of the channel. Our objective is to predict the flow of a Newtonian fluid through a channel of height h, width W and length L under the influence of a known pressure difference $P_0 - P_L$.

Solution. *Initial considerations*: Our goal is to derive an expression that can be used to predict the flow of a Newtonian fluid through a channel of known geometry under an imposed pressure difference. We will neglect entrance effects, so flow can be assumed to be 1-D. The pressures at the ends of the channel are constant, so flow will be steady.

System definition and environmental interactions: The system under investigation is the fluid contained inside a channel in which the longitudinal axis makes an angle α with the horizontal. The fluid is an incompressible Newtonian fluid. The pressure is known at two points in the fluid, ($x = 0, y = 0$) and ($x = L, y = 0$).

Apprising the problem to identify governing equations: Since our ultimate goal is the prediction of fluid flow, the appropriate governing principle is the conservation of linear momentum. The primary difference between this application and that from previous examples is the presence of pressure forces in the momentum balance.

Analysis: The shell selection and mass balance steps will be identical to those used in Sects. 6.2.1 and 6.2.2. Conservation of x-momentum, after canceling x-momentum carried into and out of the shell by bulk fluid motion, becomes:

$$0 = \left[\left(\tau_{yx}\right)_y - \left(\tau_{yx}\right)_{y+\Delta y}\right](W\Delta x) + \left[P|_x - P|_{x+\Delta x}\right](W\Delta y) + \rho g \sin\alpha [W\Delta x \Delta y].$$

6.2 General Approach

The pressure terms do not cancel in this case because a free surface is not present in this problem. In fact, a pressure gradient is explicitly imposed by application of the external pressures P_0 and P_L. Dividing by the shell volume, and letting the volume approach zero:

$$0 = \lim_{\Delta y \to 0} \left\{ \frac{\left[(\tau_{yx})_y - (\tau_{yx})_{y+\Delta y}\right]}{\Delta y} \right\} + \lim_{\Delta x \to 0} \left\{ \frac{[P|_x - P|_{x+\Delta x}]}{\Delta x} \right\} + \rho g \sin \alpha.$$

We obtain the following partial differential equation:

$$0 = -\frac{\partial \tau_{yx}}{\partial y} - \frac{\partial P}{\partial x} + \rho g \sin \alpha.$$

We have used partial derivatives, because those are the correct interpretations of the limiting processes, and we know that pressure will in general vary both with x and with y. But Newton's Law can be written as a total derivative since v_x only depends on y in parallel flow:

$$\tau_{yx} = -\mu \frac{dv_x}{dy}.$$

Since τ_{yx} does not depend on x, the partial derivative associated with shear stress reduces to an ordinary derivative, and the x-momentum equation becomes:

$$\frac{d\tau_{yx}}{dy} = -\frac{\partial P}{\partial x} + \rho g \sin \alpha.$$

Another expression involving pressure can be obtained by applying a momentum balance in the y-direction as shown in Sect. 6.2.3:

$$P = -(\rho g \cos \alpha) y + f(x).$$

Here, $f(x)$ is an arbitrary function of x to be determined by applying boundary conditions. Taking the derivative of this expression with respect to x, yields:

$$\frac{\partial P}{\partial x} = \frac{df}{dx}.$$

Substituting this into the x-momentum equation:

$$\frac{d\tau_{yx}}{dy} = -\frac{df}{dx} + \rho g \sin \alpha.$$

The left-hand side of this expression cannot be a function of x, and the right-hand side cannot be a function of y. The only way the relationship can be valid is if each side of the equation is equal a constant, say C_1:

$$\frac{d\tau_{yx}}{dy} = C_1,$$

$$-\frac{df}{dx} + \rho g \sin \alpha = C_1.$$

Integrating the expression and solving for f:

$$f(x) = \lceil \rho g \sin \alpha - C_1 \rceil x + C_2.$$

C_2 is another constant of integration. Substituting this into the y-momentum expression for $f(x)$:

$$P(x, y) = -(\rho g \cos \alpha) y + \lceil \rho g \sin \alpha - C_1 \rceil x + C_2.$$

We can now use the boundary conditions for known pressures at the inlet and outlet to find C_1 and C_2:

$$P(0, 0) = P_0 = C_2,$$
$$P(L, 0) = P_L = [\rho g \sin \alpha - C_1] L + P_0.$$

Solving for C_1:

$$C_1 = \frac{\Delta P}{L} + \rho g \sin \alpha,$$

where ΔP is defined as the inlet minus outlet pressure difference:

$$\Delta P \equiv P_0 - P_L.$$

The solution for the pressure distribution in the film is:

$$P(x, y) - P_0 = -(\rho g \cos \alpha) y - \left\lceil \frac{\Delta P}{L} \right\rceil x.$$

For any given value of y, the pressure varies linearly with x. Therefore, the pressure gradient in the x-direction is independent of y. At any given x, the pressure variation in the y-direction is the hydrostatic pressure caused by the weight of the column of fluid above it. Substituting for C_1 into the momentum equation:

$$\frac{d\tau_{yx}}{dy} = \frac{\Delta P}{L} + \rho g \sin \alpha.$$

6.2 General Approach

Integrating and applying Newton's law of viscosity:

$$\tau_{yx} = -\mu \frac{dv_x}{dy} = \left(\frac{\Delta P}{L} + \rho g \sin \alpha\right) y + C_3.$$

Integrating again:

$$v_x = -\frac{1}{\mu}\left(\frac{\Delta P}{L} + \rho g \sin \alpha\right)\frac{y^2}{2} - \frac{C_3}{\mu} y + C_4.$$

The constants of integration C_3 and C_4 can be found by applying the no-slip boundary conditions at the walls at $y = 0$ and $y = h$:

$$v_x(0) = 0 = C_4,$$

$$v_x(h) = 0 = -\frac{1}{\mu}\left(\frac{\Delta P}{L} + \rho g \sin \alpha\right)\frac{h^2}{2} - \frac{C_3}{\mu} h.$$

Substituting the solutions for C_3 and C_4 back into the expression for v_x, we obtain the final solution for the velocity profile between the two walls:

$$v_x = \frac{h^2}{2\mu}\left(\frac{\Delta P}{L} + \rho g \sin \alpha\right)\left(\frac{y}{h} - \frac{y^2}{h^2}\right).$$

The velocity profile is symmetric about the center with a parabolic shape. The larger the pressure difference $P_0 - P_L$, the greater will be the velocity for any position y. If the pressure P_L is raised above P_0, then the velocity will be reduced, eventually becoming zero everywhere when

$$\Delta P = -\rho g L \sin \alpha.$$

If P_L is raised further, the fluid will begin to flow in the reverse direction, up the inclined plane. The velocity will be a maximum at the center of the tube, where $y = h/2$:

$$v_{\max} = \frac{h^2}{8\mu}\left(\frac{\Delta P}{L} + \rho g \sin \alpha\right). \tag{6.32a}$$

A graph of the parabolic relation between $v_x/v_{\max}$ and y/h is shown in Fig. 6.11. Flow between the plates is found to be:

$$Q_V = \int_0^h v_x W dy = \frac{Wh^3}{2\mu}\left(\frac{\Delta P}{L} + \rho g \sin \alpha\right)\int_0^1 \left(\frac{y}{h} - \frac{y^2}{h^2}\right) d\left(\frac{y}{h}\right)$$

$$= \frac{Wh^3}{12\,\mu L}(\Delta P + \rho g L \sin \alpha).$$

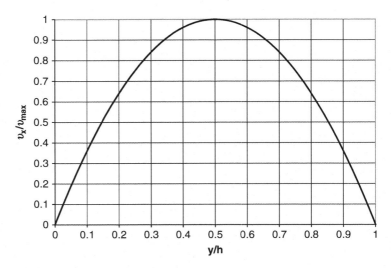

Fig. 6.11 Parabolic velocity profile between parallel plates

Examining and interpreting the results: Total flow is directly proportional to the length and width of the channel, and inversely proportional to the fluid viscosity. In contrast, flow is proportional to the third power of the height of the channel. Thus, small changes in channel height can produce substantially larger changes in flow. There will be no flow if the channel is horizontal ($\alpha = 0$) and a pressure difference is not imposed ($\Delta P = 0$). Flow will be positive if $\Delta P > \rho g L \sin(\alpha)$ and negative if $\Delta P < \rho g L \sin(\alpha)$. The flow resistance $\Re_f$ for this situation is:

$$\Re_f = \frac{\text{Driving Force per Unit Area}}{\text{Flowrate}} = \frac{12\,\mu L}{W h^3}.$$

Note that flow resistance depends on both geometric properties of the channel and fluid properties. Note also that the velocity and shear stress depend only on y, while the pressure depends on both x and y. However, the pressure gradient with respect to the x-direction is a constant, independent of x and y. Therefore, every fluid element in the channel is exposed to the same pressure gradient.

Additional comments: Note that when a pressure drop is not imposed across an inclined channel ($\Delta P = 0$), the flow through a channel of height H is twice the flow of the same fluid flowing down an inclined plane with fluid height $h = H/2$ (6.31). This follows from the symmetry of the channel problem when $\Delta P = 0$. The shear stress at the center of the channel must be zero, just as it is at the free surface in the inclined plane problem. The velocity distribution in the channel from the wall to the center is identical to the velocity distribution in the film from the wall to the free surface [(6.28) with $h = H/2$].

Example 6.2.6.3 Flow of a Power Law Fluid Down an Inclined Plane.
Repeat the falling film analysis for a fluid that obeys a power-law fluid.

6.2 General Approach

Solution. *Initial considerations*: Our goal is to determine distributions of shear stress and velocity in a Power law fluid flowing down an inclined plane. Ultimately, we wish to develop a relationship for computing total flow of a power law fluid down the plane as a function of fluid properties and the angle of inclination. We will assume flow to be steady and parallel to the inclined plane.

System definition and environmental interactions: The system is identical to that treated in Sects. 6.2.1–6.2.5, namely the flowing fluid in the film above the inclined plane.

Apprising the problem to identify governing equations: Procedures for selecting the shell, performing the mass balance, and performing the momentum balance are all identical to the analysis presented in Sects. 6.2.1–6.2.3. The only difference is that the constitutive equation for a power law fluid should be substituted into the momentum equation.

Analysis: Equation (6.26) becomes:

$$\tau_{yx} = -K\left|\frac{dv_x}{dy}\right|^{n-1}\frac{dv_x}{dy} = \rho g(y-h)\sin\alpha.$$

Since $dv_x/dy > 0$ throughout the film, this can be rewritten:

$$-K\left(\frac{dv_x}{dy}\right)^n = \rho g(y-h)\sin\alpha.$$

Rearranging this and separating variables:

$$\int dv_x = \int \left[\frac{\rho g \sin\alpha}{K}(h-y)\right]^{\frac{1}{n}}dy.$$

Introducing a change of variable: $z = h - y$, then $dy = -dz$:

$$\int dv_x = -\left(\frac{\rho g \sin\alpha}{K}\right)^{\frac{1}{n}}\int z^{\frac{1}{n}}dz.$$

Integrating and replacing z with $h - y$:

$$v_x = -\left(\frac{n}{n+1}\right)\left(\frac{\rho g \sin\alpha}{K}\right)^{\frac{1}{n}}(h-y)^{\frac{n+1}{n}} + C.$$

C is a constant of integration. Applying the no-slip condition at the solid boundary at $y = 0$, we find

$$C = \frac{n}{n+1}\left(\frac{\rho g \sin\alpha}{K}\right)^{\frac{1}{n}}h^{\frac{n+1}{n}}.$$

Therefore, the power law velocity profile becomes:

$$v_x = \left(\frac{n}{n+1}\right)\left(\frac{\rho g h^{n+1} \sin\alpha}{K}\right)^{\frac{1}{n}}\left[1 - \left(1 - \frac{y}{h}\right)^{\frac{n+1}{n}}\right].$$

The maximum velocity is found ay $y = h$:

$$v_{\max} = \left(\frac{n}{n+1}\right)\left(\frac{\rho g h^{n+1} \sin\alpha}{K}\right)^{\frac{1}{n}}.$$

Therefore, $v_x/v_{\max}$ is

$$\frac{v_x}{v_{\max}} = 1 - \left(1 - \frac{y}{h}\right)^{\frac{n+1}{n}}.$$

This relation is plotted for various values of n in Fig. 6.12. The solid line ($n = 1$) represents the velocity profile for a Newtonian fluid. The lines above this are for pseudoplastic or shear-thinning fluids ($n < 1$), while the lines below the Newtonian line are for dilatant fluids ($n > 1$). The physical interpretation is straightforward if we recall that the shear stress increases linearly from $-\rho g h \sin\alpha$ at the wall ($y/h = 0$) to zero at the liquid–gas interface ($y/h = 1$). Rearranging the constitutive relationship for a power law fluid to solve for the velocity gradient at the wall:

$$\frac{dv_x}{dy} = \left(-\frac{\tau_w}{K}\right)^{\frac{1}{n}} = \left(\frac{\rho g \sin\alpha}{K}\right)^{\frac{1}{n}}.$$

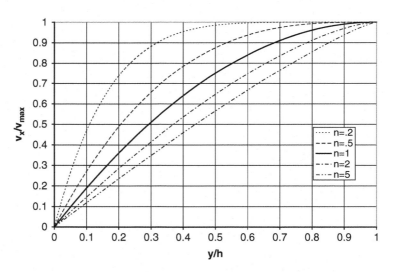

Fig. 6.12 Velocity profile for a power law fluid with different values of the exponent n flowing down an inclined plane

6.2 General Approach

If the numerical value of K is independent of n, then we see that the smaller the value for n, the larger will be the velocity gradient at the wall, and the velocity profile will be blunter for pseudoplastic fluids than for Newtonian or dilatant fluids. We would expect the flow of a pseudoplastic fluid to be higher than a Newtonian fluid with a viscosity having the same numerical value as K. The flowrate for a power law fluid down the inclined plane is:

$$Q_V = W \int_0^h v_x dy = \left(\frac{n}{2n+1}\right)\left(\frac{\rho g h \sin \alpha}{K}\right)^{\frac{1}{n}} W h^2.$$

Examining and interpreting the results: Using values from Example 6.2.5.1 for h and α, the flowrate of a power law fluid is plotted in Fig. 6.13 as a function of K for three similar values of n. The data point on the graph represents the flowrate for the Newtonian fluid in Example 6.2.5.1, which has a viscosity of 3 mPa s. If the magnitude of K remains relatively constant as n changes by $\pm$ 10%, Fig. 6.13 shows that disproportionately large changes in flow will occur. Flow could more than double if n is decreased by 10%, and it could be cut in half if n increases by 10%.

Additional comments: The above analysis assumes, of course, that the film thickness is maintained constant by adjusting the flow upward or downward as n is changed. If the flow is not altered, then the thickness of the fluid film will change.

> **Example 6.2.6.4 Falling Film Analysis for a Bingham Fluid.**
> Repeat the falling film analysis for a fluid that obeys the constitutive relation for a Bingham fluid.

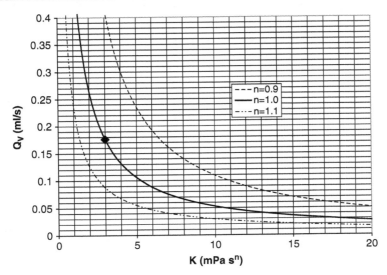

Fig. 6.13 Flowrate of a power law fluid down an inclined plane with similar values of n as a function of K. Parameters used are discussed in the text

Solution. *Initial considerations*: Our goal is to determine the velocity profile and the total flow rate for a Bingham fluid flowing down an inclined surface. Flow is assumed to be steady and the direction of flow is parallel to the inclined plane.

System definition and environmental interactions: The system is identical to that treated in Sects. 6.2.1–6.2.5, namely the material in the film above the inclined plane. However, because we are dealing with a Bingham fluid, we will need to divide the film into two regions: the solid portion, where the shear stress is less than the yield stress, and the fluid portion, where the shear stress is greater than the yield stress of the film material.

Apprising the problem to identify governing equations: Procedures for selecting the shell, performing the mass balance, and performing the momentum balance are all identical to the analysis presented in Sects. 6.2.1–6.2.3. In each of the preceding falling film examples, we found that the shear stress distribution in the film is independent of the type of fluid flowing down the inclined plane. This will be true in this case as well, since the solution for shear stress is independent of the constitutive relationship for the fluid. The appropriate starting point will be the momentum equation for a falling film, (6.25). However, to determine the velocity distribution, we need to split the fluid film into two regions, applying the appropriate constitutive relationship in each.

Analysis: Reference to Fig. 6.7 shows that the shear stress is negative throughout the film. When the shear stress in the Bingham fluid lies in the range $0 < \tau_{yx} < -\tau_y$, the constitutive relation is $dv_x/dy = 0$. Therefore, the velocity is constant, or the fluid moves as a solid plug until the yield stress is exceeded. Let y_y represent the location, measured from the bottom of the film, where the shear stress is equal to $-\tau_y$. The momentum equation (6.25) at y_y can be written:

$$-\tau_y = \rho g (y_y - h) \sin \alpha.$$

Solving for y_y in terms of known quantities:

$$y_y = h - \frac{\tau_y}{\rho g \sin \alpha}.$$

This defines the boundary between the solid or "plug" region and the fluid region. As the yield stress increases or the angle α decreases, the solid–fluid interface migrates from a position near the top of the fluid film toward the wall. When $y_y = 0$, the plug region comes in contact with the wall, and since the wall does not move, the plug cannot move. Therefore, if $y_y \leq 0$, then the film behaves as a stationary solid:

If $\dfrac{\tau_y}{\rho g \sin \alpha} \geq h$, then $v_x = 0$ everywhere in the film.

If $y_y > 0$, then shearing will occur in a region between the wall and the position y_y. For $y < y_y$, $dv_x/dy > 0$ the Bingham relationship is:

$$\tau_{yx} = -\tau_y - \mu_0 \frac{dv_x}{dy} = \rho g (y - h) \sin \alpha.$$

6.2 General Approach

Separating variables:

$$\int dv_x = \int \left[\left(\frac{\rho g h \sin \alpha - \tau_y}{\mu_0} \right) - \left(\frac{\rho g \sin \alpha}{\mu_0} \right) y \right] dy.$$

Integrating and applying the no-slip Boundary condition at $y = 0$:

$$v_x = \left(\frac{\rho g h \sin \alpha - \tau_y}{\mu_0} \right) y - \left(\frac{\rho g \sin \alpha}{\mu_0} \right) \frac{y^2}{2}.$$

Substituting for τ_y:

$$v_x = \left(\frac{\rho g \sin \alpha}{\mu_0} \right) \left[y_y y - \frac{y^2}{2} \right], \quad y \leq y_y.$$

This expression is valid for $y \leq y_y$. The velocity at the plug interface, $y = y_y$ is also the velocity for $y \geq y_y$:

$$v_x(y_y) = v_x(y) = \left(\frac{\rho g \sin \alpha}{2\mu_0} \right) y_y^2, \quad y \geq y_y.$$

Thus, our goal of specifying the velocity profile is now determined by the two expressions above. Our second goal, computing the total flow rate, can be accomplished by integrating the velocity over the two regions and using the definition of y_y:

$$Q_V = \int_0^{y_y} v_x W dy + v_x(y_y) W (h - y_y) = W \left[\frac{\rho g h \sin \alpha}{3\mu_o} - \frac{\tau_y}{2\mu_o} \right] \left[h - \frac{\tau_y}{\rho g \sin \alpha} \right]^2.$$

Examining and interpreting the results: A graphical display of the velocity profile in a Bingham fluid is shown in Fig. 6.14. Two distinct regions are shown. In the region closest to the wall, where the shear stresses are the highest, the material behaves like a fluid, providing a parabolic profile. Since dv_x/dy is zero for $y > y_y$, the fluid travels as a solid plug with velocity $v_x(y_y)$ in the upper portion of the film. The thickness of the plug is a function of fluid properties and the angle of inclination, and can be found from:

$$h - y_y = \frac{\tau_y}{\rho g \sin \alpha}.$$

Additional comments: As a check, the velocity profile reduces to the parabolic profile for a Newtonian fluid when $y_y \rightarrow h$ (i.e., $\tau_y \rightarrow 0$). The volumetric flow rate also reduces to the Newtonian expression for film flow when $\tau_y \rightarrow 0$.

Fig. 6.14 Velocity profile for a Bingham fluid flowing down an inclined plane

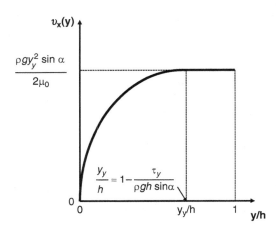

6.3 One-Dimensional Shell Balances in Cylindrical Coordinates

The shell balance method is difficult to apply when the fluid moves in a curved trajectory. Couette flow between rotating cylinders is a good example. We will treat such problems in cylindrical and spherical coordinate systems using methods developed in Chap. 7. However, there are many parallel flow situations that can best be analyzed using a cylindrical coordinate system. The most important involve fluid flow through tubes and annuli. In the following sections, we will consider examples of each of these for Newtonian and non-Newtonian fluids.

6.3.1 Flow of a Newtonian Fluid Through a Circular Cylinder

The human body contains millions of tubes distributed between the circulatory system, respiratory system, digestive system, urinary system, and reproductive system. Many of these can be approximated with a hollow circular cylinder. Circular tubes of all sizes are also used in medical devices and in the biotechnology industry. It is important for engineers to understand how to predict the pressure-flow relationship in tubes, and a simple shell balance can be used to do this.

In this section, we will focus on the flow of a Newtonian fluid through a horizontal circular tube. We will neglect pressure variations caused by gravitational effects in the radial and tangential directions. Methods for including these will be presented in Sect. 6.3.3. It is assumed that flow is laminar and parallel. This excludes fluid in what is known as the "entry region," where the velocity can vary with axial position near the inlet of the tube.

The first step in our analysis is to select an appropriate fluid shell. Since flow is in the z-direction, and z-momentum travels in the r-direction by molecular transport,

6.3 One-Dimensional Shell Balances in Cylindrical Coordinates

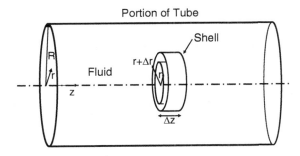

Fig. 6.15 Fluid shell used for parallel flow problems in cylindrical coordinates

an appropriate shell is one that is constrained between two planes of constant z (z and $z + \Delta z$) and two planes of constant r (r and $r + \Delta r$). The shell is shown in Fig. 6.15. A steady-state mass balance for fluid entering and leaving the shell gives:

$$(\rho v_z)|_z (2\pi r \Delta r) - (\rho v_z)|_{z+\Delta z}(2\pi r \Delta r) = 0. \tag{6.33}$$

The volume of the shell is $[\pi(r + \Delta r)^2 - \pi r^2]\Delta z$. Expanding this, and neglecting the $\pi \Delta r^2 \Delta z$ term, the shell volume is $2\pi r \Delta r \Delta z$. Dividing (6.33) by the shell volume and letting the volume approach zero, we obtain the following for a fluid with constant density:

$$\frac{\partial v_z}{\partial z} = 0. \tag{6.34}$$

Therefore, the velocity v_z is only a function of radial position, r.

Next, we introduce a z-momentum balance on fluid in the shell. Momentum enters by bulk fluid motion at z and leaves by the same mechanism at $z + \Delta z$. Momentum entering and leaving the shell by convection is equal to the product of mass flux, velocity, and axial shell surface area. Momentum also enters by molecular flux at r and leaves by molecular transport at $r + \Delta r$. This is the product of shear stress and shell radial surface area. The change in momentum is caused by an imbalance in forces. Since the tube is horizontal, the gravitational force in the z-direction is zero. The only force acting on the fluid in the z-direction is the net pressure force. Thus, conservation of z-momentum can be written:

$$0 = [(\rho v_z)(v_z)(2\pi r \Delta r)]|_z - [(\rho v_z)(v_z)(2\pi r \Delta r)]|_{z+\Delta z} + [(\tau_{rz})(2\pi r \Delta z)]|_r \\ - [(\tau_{rz})(2\pi r \Delta z)]|_{r+\Delta r} + [(P)(2\pi r \Delta r)]|_z - [(P)(2\pi r \Delta r)]|_{z+\Delta z}. \tag{6.35}$$

The first two terms cancel, since the velocity does not change in the z-direction. Dividing the remainder of the equation by the shell volume, we find:

$$0 = \frac{2\pi \Delta z \{[(r\tau_{rz})]|_r - [(r\tau_{rz})]|_{r+\Delta r}\}}{2\pi \Delta z r \Delta r} + \frac{2\pi r \Delta r \{P|_z - P_{z+\Delta z}\}}{2\pi \Delta z r \Delta r}. \tag{6.36}$$

Note that in the first term we were able to pull the common factor $2\pi\Delta z$ out of the brackets because it is independent of radial position. In the second term, we were able to pull $2\pi r \Delta r$ out of the brackets because it does not depend on z. Now, letting the volume approach zero, we find:

$$0 = \lim_{\Delta r \to 0} \left[\frac{\{[(r\tau_{rz})]|_r - [(r\tau_{rz})]|_{r+\Delta r}\}}{r\Delta r} \right] + \lim_{\Delta z \to 0} \left[\frac{\{P|_z - P_{z+\Delta z}\}}{\Delta z} \right]$$

$$0 = -\frac{1}{r}\frac{\partial(r\tau_{rz})}{\partial r} - \frac{\partial P}{\partial z}. \tag{6.37}$$

Since we are neglecting hydrostatic effects, the pressure will only depend on axial position. Therefore, the second term becomes an ordinary derivative. In parallel flow, the shear stress will only vary in a direction perpendicular to the flow. Therefore, the z-momentum equation reduces to:

$$\frac{1}{r}\frac{d(r\tau_{rz})}{dr} = -\frac{dP}{dz}. \tag{6.38}$$

This equation states that a function of r is equal to a function of z. This can only be true if both functions equal a constant, say C_1.

$$\frac{1}{r}\frac{d(r\tau_{rz})}{dr} = C_1$$

$$-\frac{dP}{dz} = C_1. \tag{6.39}$$

Integrating the second equation, we see that pressure depends linearly on axial position:

$$P = -C_1 z + C_2. \tag{6.40}$$

Applying the boundary conditions at the two ends of the fluid:

$$P(0) = P_0 = -C_1(0) + C_2,$$
$$P(L) = P_L = -C_1(L) + C_2. \tag{6.41}$$

So the constants are:

$$C_2 = P_0$$
$$C_1 = \frac{P_0 - P_L}{L}. \tag{6.42}$$

Substituting C_1 back into (6.39) and integrating:

$$\tau_{rz} = \left(\frac{P_0 - P_L}{L}\right)\frac{r}{2} + \frac{C_3}{r}, \tag{6.43}$$

6.3 One-Dimensional Shell Balances in Cylindrical Coordinates

where C_3 is a constant of integration. This expression indicates that as the radial position becomes smaller and smaller, the shear stress gets larger and larger, until, at $r = 0$, the shear stress becomes infinitely large. But we know that shear stress is a flux of momentum caused by molecular transfer. Fluid layers converging at the tube center will have identical velocities from symmetry, so the net momentum flux τ_{rz} at the center of the tube will be zero. Thus from physical arguments we can set the integration constant $C_3 = 0$. This argument can only be made if fluid is present at the center of the tube. If we were dealing with flow in an annulus, where a solid object occupies the central portion of the tube, the constant would not be zero (see Sect. 6.3.2).

Our solution shows that shear stress is a linear function of radial position

$$\tau_{rz} = \left(\frac{P_0 - P_L}{2L}\right) r. \qquad (6.44)$$

The shear stress varies from zero in the center to a maximum τ_w at the wall ($r = R$):

$$\tau_w = \tau_{rz}|_{r=R} = \left(\frac{P_0 - P_L}{2L}\right) R. \qquad (6.45)$$

Blood vessels are lined with endothelial cells and they can be damaged when wall shear stress becomes high. For a given pressure gradient, the wall shear stress decreases as the vessel radius decreases. However, as shown in Table 6.1, the pressure gradient increases dramatically in the circulatory system as vessel size is reduced. The largest wall shear stresses are actually present in the arterioles, where shear stresses typically exceed 1,000 dynes/cm^2. Although the highest pressure gradients are found in the capillaries, the wall shear stress is lower than in arterioles because capillary radii are much smaller than arteriole radii.

Note that the analysis thus far has been independent of the type of fluid that is flowing in the tube. Therefore, the expressions for shear stress distribution and wall shear stress are valid for Newtonian and non-Newtonian fluids. Let us turn our attention now to determining the velocity profile in the tube for a Newtonian fluid. Substituting the Newtonian constitutive relation into the expression for τ_{rz}:

$$\tau_{rz} = -\mu \frac{dv_z}{dr} = \left(\frac{P_0 - P_L}{2L}\right) r. \qquad (6.46)$$

Table 6.1 Average wall shear stress for blood vessels with different radii

	Radius R (cm)	Length L (cm)	ΔP (mmHg)	τ_w (dynes/cm^2)
Aorta	1.25	40	1	42
Arteries	0.3	15	5	67
Arterioles	0.008	0.2	50	1333
Capillaries	0.0004	0.05	25	133
Venules	0.009	0.2	8	240
Veins	0.4	18	3	44
Vena Cava	1.5	40	1	25

Dividing by $-\mu$ and integrating:

$$v_z = -\left(\frac{P_0 - P_L}{4\mu L}\right)r^2 + C_4, \tag{6.47}$$

where C_4 is another constant of integration. Applying the no-slip boundary condition at $r = R$, we find:

$$C_4 = \left(\frac{P_0 - P_L}{4\mu L}\right)R^2 = \frac{\Delta P R^2}{4\mu L}. \tag{6.48}$$

Here, we have defined ΔP as the upstream minus downstream pressure difference, $P_0 - P_L$. The final velocity profile for a Newtonian fluid in a circular tube is:

$$v_z = \frac{\Delta P R^2}{4\mu L}\left(1 - \frac{r^2}{R^2}\right). \tag{6.49}$$

The velocity profile is parabolic with zero velocity at the walls and a maximum velocity v_{max} in the center:

$$v_{max} = v_x(r = 0) = \frac{\Delta P R^2}{4\mu L}. \tag{6.50}$$

The average velocity is:

$$\langle v \rangle = \frac{1}{A}\int_A v_z dA = \frac{1}{\pi R^2}\int_0^R \frac{\Delta P R^2}{4\mu L}\left(1 - \frac{r^2}{R^2}\right)(2\pi r dr)$$

$$\langle v \rangle = \frac{\Delta P R^2}{8\mu L} = \frac{v_{max}}{2}. \tag{6.51}$$

Total flow rate through the tube is:

$$Q_V = \langle v_z \rangle (\pi R^2) = \frac{\pi \Delta P R^4}{8\mu L}. \tag{6.52}$$

This is known as the *Hagen-Poiseuille* equation for flow of a Newtonian fluid through a cylindrical tube. Flow resistance $\Re$ is defined as the pressure difference divided by the flow rate:

$$\Re = \frac{\Delta P}{Q_V} = \frac{8\mu L}{\pi R^4}. \tag{6.53}$$

The relationship between flow resistance and tube radius is highly nonlinear. Doubling the tube radius does not double the flow. Instead, it increases flow 16-fold! Slight changes in vessel radius can have significant effects on blood flow. Arterioles,

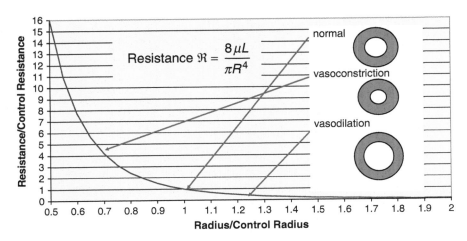

Fig. 6.16 Flow resistance relative to control as a function of tube radius relative to control

with their muscular walls, are the principal sites in the body where this type of control occurs, as illustrated in Fig. 6.16.

> **Example 6.3.1.1 Flow in a Cylindrical Vessel with Leakage Through the Wall.**
> Use a shell balance to derive an expression for fluid flow and pressure as a function of axial position in a horizontal vessel with a leaky wall. The pressure surrounding the wall, P_S, is assumed constant. Assume further that the fluid is Newtonian, the filtration coefficient of the wall per unit length is constant and the wall freely admits all solutes.

Solution. *Initial considerations*: Since some fluid flows in the radial direction through the conduit walls and some flows in the axial direction through the conduit lumen, this is not truly a 1-D problem. However, if the leakage is small, we would expect the axial velocity to retain its parabolic shape, but the magnitude would be reduced as fluid moves downstream.

System definition and environmental interactions: Since flow through the conduit varies with axial position, we will select a system consisting of the fluid in the conduit between an arbitrary axial position z and a position $z + \Delta z$, as shown in Fig. 6.17. The system extends in the radial direction all the way to the wall of the conduit. Environmental interactions occur with fluid in the lumen at the two ends, where the pressures are known, and with the conduit wall.

Apprising the problem to identify governing relationships: Flow through the wall is assumed to be governed by the Starling equation (5.154) for flow through a membrane. Since solutes can freely move through the wall, the osmotic reflection coefficient is zero. Flow through the lumen is governed by conservation of mass and conservation of momentum.

Fig. 6.17 Shell for flow through a vessel with a leaky wall

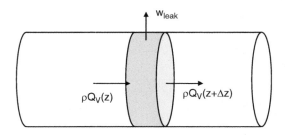

Analysis: We begin by relating flow at an axial position to the pressure gradient. Conservation of momentum in the axial direction for flow in a circular cylinder led to the Hagen–Poiseuille equation, (6.52). Assuming this relationship can be applied to a portion of a leaky tube with length Δz:

$$Q_V = \frac{\pi R^4}{8\mu}\left(\frac{P(z) - P(z+\Delta z)}{\Delta z}\right).$$

Letting the tube length shrink to a point at z, this becomes:

$$Q_V = -\left(\frac{\pi R^4}{8\mu}\right)\frac{dP}{dz}.$$

This could also have been obtained by inserting (6.39) and (6.42) into (6.52). Applying a steady-state mass balance to the shell in Fig. 6.17 bounded by $z, z + \Delta z$ and the leaky wall:

$$(\rho Q_V)|_z - (\rho Q_V)|_{z+\Delta z} - w_{\text{leak}} = 0.$$

The leakage flow between z and $z + \Delta z$ is determined using Starling's law, assuming a constant filtration coefficient per unit length, K_f/L:

$$w_{\text{leak}} = \rho\frac{K_f}{L}\Delta z\left\{\left(P|_{z+\frac{\Delta z}{2}} - P_S\right) - \sigma_d\left(\Pi|_{z+\frac{\Delta z}{2}} - \Pi_S\right)\right\}.$$

In this case, solutes can easily pass through the wall, so σ_d is zero. Assuming the fluid density to be constant, conservation of mass becomes:

$$\frac{(Q_V)|_z - (Q_V)|_{z+\Delta z}}{\Delta z} = \frac{K_f}{L}\left(P|_{z+\frac{\Delta z}{2}} - P_S\right).$$

Taking the limit as Δz goes to zero:

$$-\frac{dQ_V}{dz} = \frac{K_f}{L}(P(z) - P_S).$$

6.3 One-Dimensional Shell Balances in Cylindrical Coordinates

Substituting the expression for flow in terms of the pressure gradient:

$$-\frac{d}{dz}\left[-\left(\frac{\pi R^4}{8\mu}\right)\frac{dP}{dz}\right] = \frac{K_f}{L}(P(z) - P_S).$$

Since P_S, R, and μ are constant, this can be written:

$$\frac{d^2(P - P_S)}{dz^2} = \frac{8\mu K_f}{\pi R^4 L}(P - P_S) = \frac{\Re K_f}{L^2}(P - P_S),$$

where the tube resistance $\Re$ is given by (6.53). The general solution is:

$$P - P_S = A \sinh\left(\frac{z}{L}\sqrt{\Re K_f}\right) + B \cosh\left(\frac{z}{L}\sqrt{\Re K_f}\right)$$

The boundary conditions are (1) $P = P_0$ at $z = 0$, and (2) $P = P_L$ at $z = L$. Applying these, we can solve for the constants A and B to obtain the solution for pressure as a function of axial position:

$$\frac{P - P_S}{P_0 - P_S} = \left\{\frac{P_L - P_S}{P_0 - P_S} - \cosh\left(\sqrt{\Re K_f}\right)\right\} \frac{\sinh\left(\frac{z}{L}\sqrt{\Re K_f}\right)}{\sinh\left(\sqrt{\Re K_f}\right)} + \cosh\left(\frac{z}{L}\sqrt{\Re K_f}\right).$$

Dividing the negative pressure gradient by the resistance per unit length provides an expression that can be used to determine the flow rate as a function of axial position:

$$Q_V\left(\frac{z}{L}\right) = -(P_0 - P_S)\frac{\sqrt{\Re K_f}}{\Re}$$
$$\times \left\{\left[\frac{P_L - P_S}{P_0 - P_S} - \cosh\left(\sqrt{\Re K_f}\right)\right] \frac{\cosh\left(\frac{z}{L}\sqrt{\Re K_f}\right)}{\sinh\left(\sqrt{\Re K_f}\right)} + \sinh\left(\frac{z}{L}\sqrt{\Re K_f}\right)\right\}.$$

Examining and interpreting the results: Flow relative to flow through a tube with impermeable walls, $Q_{V,\text{Poiseuille}}$, is:

$$\frac{Q_V}{Q_{V,\text{Poiseuille}}} = \left\{\frac{\sqrt{\Re K_f}}{\sinh\left(\sqrt{\Re K_f}\right)}\right\}\left[P_0^* \cosh\left(\sqrt{\Re K_f}\right) - P_L^*\right]\cosh\left(\frac{z}{L}\sqrt{\Re K_f}\right)$$
$$- P_0^* \sqrt{\Re K_f} \sinh\left(\frac{z}{L}\sqrt{\Re K_f}\right),$$

where the dimensionless pressures are defined as:

$$P_0^* = \left(\frac{P_0 - P_S}{P_0 - P_L}\right),$$

$$P_L^* = \left(\frac{P_L - P_S}{P_0 - P_L}\right).$$

We write the dimensionless flow this way to show that as K_f approaches zero, the flow rate is equal to the Hagen–Poiseuille flow rate, since the term in braces { } approaches unity as K_f approaches zero. This also shows that there are three dimensionless parameters that influence the flow rate in a leaky vessel: $\Re K_f$, P_0^* and P_L^*. The pressure equation can also be rearranged in terms of the same three dimensionless groups:

$$\frac{P-P_L}{P_0-P_L} = \left[P_L^* - P_0^* \cosh\left(\sqrt{\Re K_f}\right)\right] \left\{ \frac{\sinh\left(\frac{z}{L}\sqrt{\Re K_f}\right)}{\sinh\left(\sqrt{\Re K_f}\right)} \right\} + P_0^* \cosh\left(\frac{z}{L}\sqrt{\Re K_f}\right) - P_L^*.$$

As K_f approaches zero, the term in braces approaches z/L and the pressure is independent of P_S. In this case, pressure decreases linearly from P_0 at $z = 0$ to P_L at $z = L$, as predicted for a tube with impermeable walls.

As is often the case with analytic solutions, it is difficult to visualize how the pressure and flow vary with position in the vessel by simply examining the equations for pressure and flow. A graphical presentation is nearly always more instructive. In Fig. 6.18, we examine the effect of varying $\Re K_f$ while keeping the upstream pressure constant and setting the downstream and surrounding pressures at zero. This is similar to the situation encountered in a leaky systemic capillary bed. The pressure in the vessel becomes more nonlinear as the vessel wall becomes more porous. However, the nonlinearity is barely noticeable, even when the wall filtration coefficient is equal to the conductance of fluid through the tube (i.e., $1/\Re$). When the wall is impermeable to fluid, the flow is the same at all axial positions and is equal to the Hagen–Poiseuille flow, as expected. If the wall is porous, some fluid flows out of the tube through the wall, and as we might expect, the outlet flow will be less than the inlet flow. However, it is interesting to note that the inlet flow into a permeable tube is greater than the corresponding Hagen–Poiseuille (H–P) flow through the same tube with impermeable walls. Since the surrounding pressure is less than the fluid pressure in the tube, suction is applied to the outside surface of the tube over the

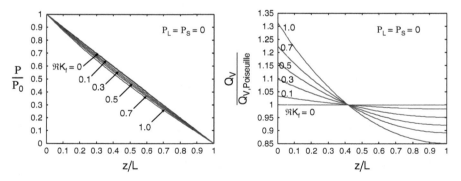

Fig. 6.18 Dimensionless pressure and flow in a leaky tube as functions of axial position for various products of tube resistance and wall filtration coefficient. Surrounding pressure and outlet pressure are maintained at zero

6.3 One-Dimensional Shell Balances in Cylindrical Coordinates

entire length of the tube wall, increasing the inlet flow. The greater the filtration coefficient, the larger will be the inlet flow relative to the H–P flow. Irrespective of the filtration coefficient, the flow in the tube will equal the H–P flow at a position that is about 41% of the way along the axis of the tube. The difference between the inlet flow and the outlet flow is equal to the leakage flow from the tube to the surroundings. The leakage flow relative to the Hagen–Poiseuille flow is shown to increase almost linearly with an increase in wall filtration coefficient in Fig. 6.19.

Additional comments: Surrounding pressure was held constant in the previous simulation. Let us now look at a series of simulations where the inlet pressure, outlet pressure, and $\Re K_f$ are held constant, but the surrounding pressure is varied. Results are shown in Fig. 6.20 for $-1 \leq P_S/P_0 \leq +1$. As surrounding pressure is increased above outlet pressure, the pressure distribution in the lumen rises above the corresponding pressures for flow through an impermeable wall, inlet flow is reduced, and outlet flow is elevated. When $P_S/P_0 = 0.5$, the inlet flow and outlet flow are the same, so the volume of fluid lost in the first half of the tube is gained in the last half. Even then,

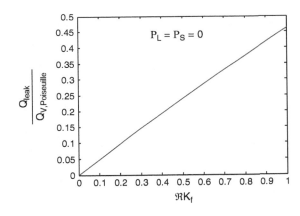

Fig. 6.19 Leakage flow through tube wall to surroundings relative to the Hagen–Poiseuille flow as a function of $\Re K_f$

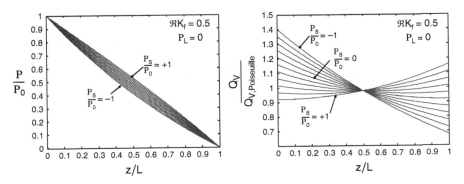

Fig. 6.20 Dimensionless pressure and flow in a leaky tube as functions of axial position. Surrounding pressure is changed from $-P_0$ to $+P_0$. Upstream and downstream pressures are constant ($P_L = 0$) and $K_f \Re = 0.5$

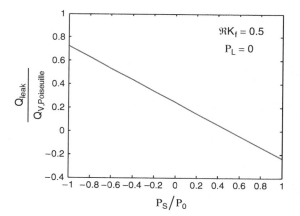

Fig. 6.21 Leakage flow relative to Hagen–Poiseuille flow as surrounding pressure is changed from minus the inlet pressure to the inlet pressure. Positive leakage is from lumen to surroundings

the inlet flow is slightly higher than the Hagen–Poiseuille flow and the pressure distribution is nonlinear. Flow in the tube at the midpoint is the same in all cases, but is slightly lower than the Hagen–Poiseuille flow. Figure 6.20 shows that fluid is lost from the tube for $P_S/P_0 < 0.5$ and fluid is gained for $P_S/P_0 > 0.5$. Leakage flow is shown in Fig. 6.21 with positive leakage directed from lumen to surroundings.

There are a number of potential applications for this type of analysis. Blood vessels in the microcirculation generally allow fluid to pass through the walls. Continuous capillaries allow fluid to pass through pores formed by intercellular junctions between adjoining endothelial cells. However, these vessels also tend to restrict the passage of macromolecules, so osmotic pressure effects cannot be ignored. An approach for treating transvascular flow through these pore-like structures is introduced in Sect. 6.3.5. Discontinuous capillaries, found in organs such as the liver or spleen, have much larger gaps in the walls and the approach outlined above is likely to be more applicable to these types of leaky blood vessels. The terminal ends of lymph vessels, or lymph capillaries, are composed of endothelial cells that partially overlap, forming gaps. When interstitial fluid pressure is higher than the pressure within the lymph vessel, the cells are forced apart, opening the gap and allowing fluid and solutes to freely enter the vessel. The cells are forced together when lymph pressure rises above surrounding pressure, closing the gap and preventing leakage out of the vessel. The analysis presented above would be appropriate for the inward flow into terminal lymphatics with $P_0 = P_S$.

6.3.2 Flow of a Newtonian Fluid in an Annulus with Inner Wall Moving

When a catheter is inserted into a blood vessel, the resistance is increased and flow will be reduced. In this example, we will analyze flow of fluid in an annulus with an

6.3 One-Dimensional Shell Balances in Cylindrical Coordinates

inner wall at $r = R_i$ and an outer wall at $r = R_o$. Our goals are to determine the velocity profile and fluid resistance when the catheter is stationary, and when it is moving.

The analysis is identical to that in the previous example, up through the derivation of (6.43):

$$\tau_{rz} = \frac{\Delta P}{2L} r + \frac{C_3}{r}. \tag{6.54}$$

We cannot make the argument that C_3 is zero in this case, because fluid is not present at $r = 0$, so no boundary condition can be specified at that position. Instead, we simply insert the constitutive equation for a Newtonian fluid and integrate this equation to get an expression for velocity in terms of two unknown constants:

$$v_z = -\left(\frac{\Delta P}{4\mu L}\right) r^2 - \frac{C_3}{\mu} \ln(r) + C_4. \tag{6.55}$$

The fluid is bounded by two solid walls and we can apply the no-slip boundary condition to each. If the catheter wall is moving with velocity V, and the vessel wall is stationary, we can write the following two equations for the unknowns, C_3 and C_4:

$$\begin{aligned} v_z(R_i) = V &= -\left(\frac{\Delta P}{4\mu L}\right) R_i^2 - \frac{C_3}{\mu} \ln(R_i) + C_4, \\ v_z(R_o) = 0 &= -\left(\frac{\Delta P}{4\mu L}\right) R_o^2 - \frac{C_3}{\mu} \ln(R_o) + C_4. \end{aligned} \tag{6.56}$$

Solving for the constants, and defining $\kappa = R_i/R_o$ we find:

$$\begin{aligned} C_3 &= -\frac{\mu V}{\ln(\kappa)} + \frac{\Delta P R_o^2}{4L}\left[\frac{(1-\kappa^2)}{\ln(\kappa)}\right], \\ C_4 &= \frac{\Delta P R_o^2}{4\mu L}\left[1 + \frac{(1-\kappa^2)}{\ln(\kappa)} \ln(R_o)\right] - \frac{V \ln(R_o)}{\ln(\kappa)}. \end{aligned} \tag{6.57}$$

Substituting these back into the expressions for shear stress and velocity:

$$\tau_{rz} = \frac{\Delta P}{2L}\left[r + \frac{(1-\kappa^2)}{2\ln(\kappa)} \frac{R_o^2}{r}\right] - \frac{\mu V}{\ln(\kappa)}\left(\frac{1}{r}\right). \tag{6.58}$$

$$v_z = \frac{\Delta P R_o^2}{4\mu L}\left(1 - \frac{r^2}{R_o^2} - (1-\kappa^2)\frac{\ln(r/R_o)}{\ln(\kappa)}\right) + V \frac{\ln(r/R_o)}{\ln(\kappa)}. \tag{6.59}$$

As a check, we see that the velocity is zero at $r = R_o$ and is V at $r = R_i$. In addition, this reduces to the parabolic velocity profile in a tube if $\kappa \to 0$. The flow rate can be found by integrating the velocity over the cross section of the annulus:

$$Q_V = \int_{R_i}^{R_o} v_z(r) \cdot 2\pi r \, dr = \frac{\pi \Delta P R_o^4}{8\mu L}\left[1 - \kappa^4 + \frac{(1-\kappa^2)^2}{\ln(\kappa)}\right] - \pi R_o^2 V \left[\frac{1-\kappa^2}{2\ln(\kappa)} + \kappa^2\right]. \tag{6.60}$$

Finally, the flow in an annulus, relative to the Hagen-Poiseuille flow in a hollow tube with the same outer radius is:

$$\frac{Q_{V,\text{annulus}}}{Q_{V,\text{tube}}} = \left[1 - \kappa^4 + \frac{(1-\kappa^2)^2}{\ln(\kappa)}\right] - \frac{V}{\langle v \rangle}\left[\frac{1-\kappa^2}{2\ln(\kappa)} + \kappa^2\right]. \tag{6.61}$$

where $\langle v \rangle$ is the average velocity in the tube, given by (6.51). The first term shows the influence of the position of the inner wall of the annulus, and the second term shows the effect of moving the wall with velocity V. The influences of varying the relative velocity and κ for the catheter are shown in Fig. 6.22. Consider first the influence of geometry on flow with a stationary catheter (bold line in Fig. 6.22). Even if the catheter radius is only 1% of the vessel radius, the flow is reduced to about 78% of the flow that would occur if the catheter was not present. This is caused by the no-slip condition along the catheter surface. If the catheter radius is half the vessel radius, the flow will be reduced to 27% of the flow that would result if the catheter was not present. Thus, flow measurements based on a device placed at the tip of a catheter may be accurate, but may not represent the normal flow in the vessel. This would be a classic case of the instrument altering the measurement.

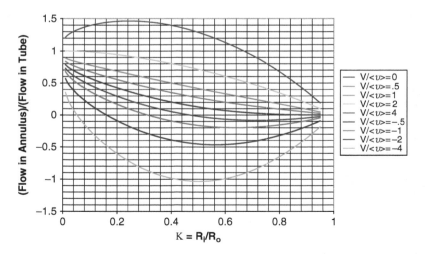

Fig. 6.22 Flow of fluid in an annulus relative to flow in a tube vs. ratio of inner to outer radius. Different curves represent different velocities of the inner wall (i.e., catheter movement)

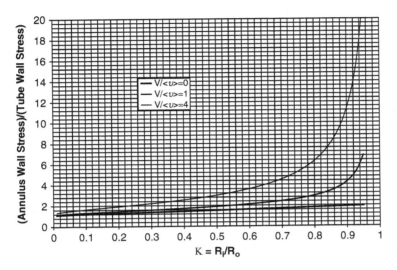

Fig. 6.23 Ratio of shear stress on vessel wall relative to shear stress in a tube with the same radius and average velocity. Curves are for a stationary catheter or a catheter moving at velocities equal to one or four times the average flow velocity

Moving the catheter back and forth can cause the flow to go higher than normal or can cause flow to go in the opposite direction, depending on the size and the velocity of the catheter (Fig. 6.22).

The shear stress on the vessel wall, relative to that for the same vessel without the catheter in place is shown in Fig. 6.23. The elevated shear stresses are not large for a stationary catheter, but large shear stresses can be applied to vascular endothelial cells if the catheter has a radius approaching that of the vessel, and the catheter is moved rapidly through the vessel.

6.3.3 Flow Through an Inclined Tube or Annulus

Consider the case where the tube or annulus in the above sections is oriented at an angle β relative to the vertical, as shown in Fig. 6.24. Momentum balances on the shells for the previous sections in the z, r, and θ directions provide the following three relations:

$$-\frac{1}{r}\frac{\partial}{\partial r}(r\tau_{rz}) = \frac{\partial P}{\partial z} - \rho g_z (z \text{ - momentum}), \tag{6.62}$$

$$0 = \frac{\partial P}{\partial r} - \rho g_r (r \text{ - momentum}), \tag{6.63}$$

$$0 = \frac{1}{r}\frac{\partial P}{\partial \theta} - \rho g_\theta (\theta \text{ - momentum}), \tag{6.64}$$

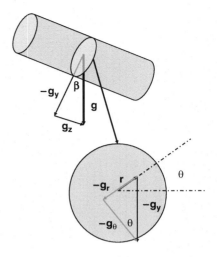

Fig. 6.24 Geometry for a nonhorizontal tube

where, from Fig. 6.24, the components of the gravitational acceleration are:

$$\begin{aligned} g_x &= 0, \\ g_z &= g\sin\beta, \\ g_y &= -g\cos\beta, \\ g_r &= -|g_y|\sin\theta = -g\cos\beta\sin\theta, \\ g_\theta &= -|g_y|\cos\theta = -g\cos\beta\cos\theta. \end{aligned} \qquad (6.65)$$

Integrating the r-momentum equation:

$$P = -\rho g r \cos\beta \sin\theta + f(z, \theta). \qquad (6.66)$$

Taking the derivative of P with respect to θ and setting it equal to $\rho g_\theta r$ from the θ-momentum equation:

$$\frac{\partial P}{\partial \theta} = -\rho g r \cos\beta \cos\theta + \frac{\partial f}{\partial \theta} = -\rho g r \cos\beta \cos\theta. \qquad (6.67)$$

Therefore, $\partial f/\partial \theta = 0$, and f is at most a function of z, $f(z)$. Now, taking the derivative of P with respect to z:

$$\frac{\partial P}{\partial z} = \frac{df}{dz} = f_1(z). \qquad (6.68)$$

6.3 One-Dimensional Shell Balances in Cylindrical Coordinates

Substituting this into the z-momentum equation:

$$\frac{\partial P}{\partial z} = f_1(z) + \rho g \sin \beta = \frac{1}{r}\frac{d}{dr}(r\tau_{rz}). \tag{6.69}$$

Therefore, $f_1(z)$ must be a constant, say C_1, and the pressure gradient in the z-direction is independent of z. Integrating (6.68) with $f_1 = C_1$:

$$f = C_1 z + C_2. \tag{6.70}$$

Substituting this back into (6.66):

$$P = -\rho g r \cos \beta \sin \theta + C_1 z + C_2. \tag{6.71}$$

To find the constants C_1 and C_2 we use the boundary conditions for pressure at the ends of the cylinder:

$$\begin{aligned} P(z=0, r=0) &= P_0 = C_2, \\ P(z=L, r=0) &= P_L = C_1 L + C_2. \end{aligned} \tag{6.72}$$

The final solution for pressure is:

$$P(r, \theta, z) = -\rho g r \cos \beta \sin \theta + \left(\frac{P_L - P_0}{L}\right) z + P_0. \tag{6.73}$$

Taking the derivative of P with respect to z, we see that $\partial P / \partial z = dP/dz = (P_L - P_0)/L$. Letting $\Delta P = P_0 - P_L$, and substituting back into the z-momentum equation:

$$\frac{\Delta P}{L} + \rho g \sin \beta = \frac{1}{r}\frac{d}{dr}(r\tau_{rz}). \tag{6.74}$$

This has the same form as (6.39) for a horizontal tube except for the addition of the gravitational term, $\rho g \sin(\beta)$. Thus, the solution for nonhorizontal flow through a tube is the same as the solution for horizontal flow in a tube or annulus if we replace ΔP with $\Delta P + \rho g L \sin \beta$ in the expressions derived for flow in a horizontal tube. In Sect. 5.5, we introduced the *modified pressure*, $\wp$,:

$$\wp \equiv P - \rho g_x x - \rho g_y y - \rho g_z z. \tag{6.75}$$

Substituting the values from (6.65) for g_x, g_y, and g_z into (5.50), along with the conversion from rectangular to cylindrical coordinate ($x = r \cos(\theta)$, $y = r \sin(\theta)$), this can be written:

$$\wp = P - \rho g z \sin \beta + \rho g r \cos \beta \sin \theta. \tag{6.76}$$

Taking the derivative of $\wp$ with respect to z, r, and θ:

$$\frac{\partial \wp}{\partial z} = \frac{\partial P}{\partial z} - \rho g \sin \beta = -\left[\frac{\Delta P}{L} + \rho g \sin \beta\right], \quad (6.77)$$

$$\frac{\partial \wp}{\partial r} = \frac{\partial P}{\partial r} + \rho g \cos \beta \sin \theta = \frac{\partial P}{\partial r} - \rho g_r = 0, \quad (6.78)$$

$$\frac{1}{r}\frac{\partial \wp}{\partial \theta} = \frac{1}{r}\left[\frac{\partial P}{\partial \theta} + \rho g r \cos \beta \cos \theta\right] = \frac{1}{r}\frac{\partial P}{\partial \theta} - \rho g_\theta = 0. \quad (6.79)$$

The last two equations show that $\partial \wp/\partial r = 0$ satisfies the r-momentum equation, and $\partial \wp/\partial \theta = 0$ satisfies the θ-momentum equation. Therefore, the modified pressure $\wp$ is a function of z alone and $-\mathrm{d}\wp/\mathrm{d}z$ is the same constant that appears in the z-momentum equation:

$$-\frac{\mathrm{d}\wp}{\mathrm{d}z} = \frac{\Delta P}{L} + \rho g \sin \beta = \frac{1}{r}\frac{\mathrm{d}}{\mathrm{d}r}(r\tau_{rz}). \quad (6.80)$$

It is common practice to neglect the force of gravity in deriving expressions for shear stress, velocity and flow, and then to correct for gravity by replacing P with $\wp$ in the final expressions. Thus for flow in a tube at arbitrary angle β, (6.49)–(6.52) can be replaced with:

$$v_z = \frac{\Delta \wp R^2}{4\mu L}\left(1 - \frac{r^2}{R^2}\right), \quad (6.81)$$

$$v_{\max} = \frac{\Delta \wp R^2}{4\mu L}, \quad (6.82)$$

$$\langle v \rangle = \frac{\Delta \wp R^2}{8\mu L}, \quad (6.83)$$

$$Q_V = \frac{\pi \Delta \wp R^4}{8\mu L}. \quad (6.84)$$

Flow in an annulus at arbitrary angle can also be corrected in a similar manner. The examples presented in Sect. 6.2, based on rectangular coordinates, can also be easily expressed in terms of the modified pressure.

6.3.4 Flow of a Casson Fluid Through a Circular Cylinder

In Sect. 4.9.2, we showed experimental data suggesting that blood can be modeled as a Casson fluid at low shear rates. In this section, we will apply an analysis similar

6.3 One-Dimensional Shell Balances in Cylindrical Coordinates

to that in Sect. 6.3.1, but for flow of a Casson fluid in a circular tube. We will remove the restriction that the tube be horizontal and use the coordinate system and the modified pressure applied in Sect. 6.3.3. Our goal is to estimate the error introduced in various anatomical blood vessels by assuming blood to be a Newtonian fluid, rather than a Casson fluid.

We begin the analysis with results of the z-momentum balance, which was independent of the type of fluid:

$$\tau_{rz} = \left(\frac{\Delta\wp}{2L}\right)r. \tag{6.85}$$

At this point, we introduce the Casson model for blood:

$$\frac{dv_z}{dr} = 0; \ |\tau_{rz}| < \tau_y,$$

$$\frac{dv_z}{dr} < 0; \ \sqrt{\tau_{rz}} = S\sqrt{-\frac{dv_z}{dr}} + \sqrt{\tau_y},$$

$$\frac{dv_z}{dr} > 0; \ \sqrt{-\tau_{rz}} = S\sqrt{\frac{dv_z}{dr}} + \sqrt{\tau_y}. \tag{6.86}$$

The blood will behave as a solid until $\tau_{rz} > \tau_y$. Let us define a yield radius R_y as the radial position when $\tau_{rz} = \tau_y$. Substituting R_y into (6.85):

$$R_y \equiv \frac{2L\tau_y}{\Delta\wp}. \tag{6.87}$$

Thus if $R < R_y$, then $dv_z/dr = 0$ everywhere in the tube and the velocity is constant. Since there is no slip at the wall:

$$v_z = 0 \text{ for all } r \text{ if } R < R_y. \tag{6.88}$$

Consider the case where $R > R_y$. Since shear stress varies linearly with radial position, there will be a "core" region in which $0 \leq r \leq R_y$ and a shearing region where $r > R_y$ (Fig. 6.25). Applying the appropriate constitutive relationship in the core region, $dv_z/dr = 0$, so:

$$v_z = C_1, \ r \leqslant R_y, \tag{6.89}$$

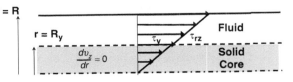

Fig. 6.25 Shear stress distribution in Casson fluid

where C_1 is a constant to be determined later. Therefore, blood in the entire region between the center of the tube and the yield radius moves as a solid plug with nonzero velocity. Let us turn our attention now to flow in the shearing region ($R_y < r \leq R$). Assuming flow is in the positive z-direction, then dv_z/dr will be less than zero in this region and the appropriate constitutive relationship will be:

$$\sqrt{\tau_{rz}} = S\sqrt{-\frac{dv_z}{dr}} + \sqrt{\tau_y}. \quad (6.90)$$

The shear stress can be written in terms of the yield radius:

$$\tau_{rz} = \tau_y \frac{r}{R_y}. \quad (6.91)$$

Substituting this into the constitutive relationship, rearranging, and squaring both sides:

$$-\frac{dv_z}{dr} = \frac{\tau_y}{S^2}\left[\frac{r}{R_y} - 2\sqrt{\frac{r}{R_y}} + 1\right]. \quad (6.92)$$

Integrating from $r = r$ to $r = R$ in the fluid region, and applying the no slip boundary condition at $r = R$:

$$v_z(r) = \frac{\tau_y R}{S^2}\left\{\frac{R}{2R_y}\left[1 - \left(\frac{r}{R}\right)^2\right] - \frac{4}{3}\sqrt{\frac{R}{R_y}}\left[1 - \left(\frac{r}{R}\right)^{3/2}\right]\right.$$
$$\left. + \left[1 - \left(\frac{r}{R}\right)\right]\right\}, R_y \leq r \leq R. \quad (6.93)$$

Finally, equating this to the constant C_1 at $r = R_y$:

$$v_z(R_y) = C_1 = \frac{\tau_y R}{S^2}\left\{\frac{R}{2R_y}\left[1 - \left(\frac{R_y}{R}\right)^2\right] - \frac{4}{3}\sqrt{\frac{R}{R_y}}\left[1 - \left(\frac{R_y}{R}\right)^{3/2}\right]\right.$$
$$\left. + \left[1 - \left(\frac{R_y}{R}\right)\right]\right\}, 0 \leq r \leq R_y. \quad (6.94)$$

The velocity profile is shown in Fig. 6.26. Integrating over the core and wall regions, we find the flow rate:

$$Q_V = \frac{\pi \Delta \wp R^4}{8S^2 L}\left\{1 - \frac{16}{7}\left(\frac{R_y}{R}\right)^{1/2} + \frac{4}{3}\left(\frac{R_y}{R}\right) - \frac{1}{21}\left(\frac{R_y}{R}\right)^4\right\}. \quad (6.95)$$

Fig. 6.26 Velocity profile for Casson fluid in a tube; R_y is the yield radius

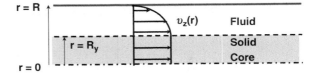

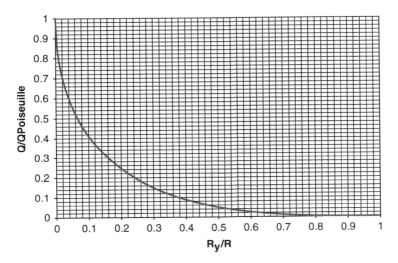

Fig. 6.27 Flow of a Casson fluid relative to a Newtonian fluid in a circular tube as a function of the ratio of yield radius to tube radius

Note that the term in front of the brackets in (6.95) is the Hagen–Poiseuille flow for a Newtonian fluid if $\mu = S^2$. Recall that:

$$\frac{R_y}{R} \equiv \frac{2L\tau_y}{R\Delta\wp} = \frac{\tau_y}{\tau_w}. \tag{6.96}$$

Therefore, the relationship between flow and modified pressure gradient is highly nonlinear. The ratio of Casson flow to Newtonian flow with $\mu = S^2$ is shown in Fig. 6.27 as a function of R_y/R.

Note that the flow of the Casson fluid is always lower than the flow of a Newtonian fluid having $\mu = S^2$. In addition, significant decreases in flow occur when the core occupies a small fraction of the total cross-sectional area. For instance, a core region with R_y/R of 0.1 (i.e., 1% of the total cross sectional area) slows flow to 41% of the Newtonian value. It would appear that approximating blood as a Newtonian fluid might lead to serious errors. However, the ratio of yield radius to blood vessel radius is generally quite small. Assuming the yield stress for blood is 0.04 dynes/cm^2, and using vessel wall stresses from Table 6.1, the

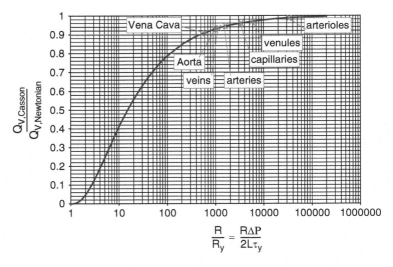

Fig. 6.28 Location of various size blood vessels on the relative flow vs. R/R_y graph, assuming a yield stress of .04 dynes/cm^2. Inclusion of capillary flow on this plot is inappropriate, because of the particulate nature of flow in these vessels

flow ratio for various blood vessels is shown in Fig. 6.28 as a function of the dimensionless pressure drop across the tube. The worst estimates of blood flow using a Newtonian model should occur in the larger blood vessels, where overestimates of 6–10% might occur. However, use of the Newtonian model for venules and arterioles leads to very little error. Although the position of capillaries is shown on the graph, the Fahraeus Lindqvist effect makes application of the Casson model to flow in capillaries inappropriate.

6.3.5 Osmotic Pressure and Flow in a Cylindrical Pore

Consider a pore that restricts admittance of a solute through the wall of a capillary, as shown in Fig. 6.29. We idealize the pore as a cylindrical tube and the solute as having a spherical shape. The pore radius is R_p and the solute radius is R_s. We are interested in predicting fluid flow through the pore, given hydrostatic and osmotic pressures on each side of the pore. Since the solute centers can come no closer to the wall of the pore than the radius of the solute, we idealize the pore as consisting of two regions: a region that contains solute in the central portion of the pore ($0 \leq r < R_p - R_s$) and a peripheral region that excludes solute ($R_p - R_s \leq r \leq R_p$). This model provides only an approximation to reality.

Flow is in the z-direction and the radial pressure gradient $\partial P/\partial r$ is assumed to be zero in both the core and peripheral regions. However, there will be a discontinuity in hydrostatic pressure at the interface between these regions. At the interface, we

6.3 One-Dimensional Shell Balances in Cylindrical Coordinates

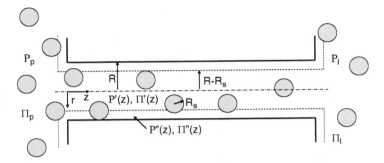

Fig. 6.29 Flow through a capillary pore

assume the two regions are in equilibrium. Assuming the fluids in each region are Newtonian, a shell balance in each region will yield (6.38):

$$\frac{1}{r}\frac{d}{dr}\left(r\frac{dv'_z}{dr}\right) = -\frac{\Delta P'}{\mu'\Delta z}. \tag{6.97}$$

$$\frac{1}{r}\frac{d}{dr}\left(r\frac{dv''_z}{dr}\right) = -\frac{\Delta P''}{\mu''\Delta z}, \tag{6.98}$$

where Δz is the length of the pore, $\Delta P'$ is the hydrostatic pressure difference in the core region between the plasma and interstitial ends, $\Delta P''$ is the hydrostatic pressure difference in the peripheral region between the two ends, and μ' and μ'' are the fluid viscosities in the two regions. For dilute solutions, the fluid viscosities in the core and periphery will be nearly the same, so we let $\mu' = \mu'' = \mu$. These two second-order equations can be solved subject to the following four boundary conditions:

1. Radial symmetry in the core region: $dv'_z/dr(r=0) = 0$.
2. Equal velocities at $r = R_p - R_s$: $v'_z = v''_z$.
3. Equal shear stress at $r = R_p - R_s$: $-\mu dv'_z/dr = -\mu dv''_z/dr$.
4. No slip at the pore wall: $v''(R_p) = 0$.

In addition, since there are no differences in solute concentration between the bulk fluid regions and the core region at $z = 0$ or $z = \Delta z$, the hydrostatic pressures inside the pore at the interface are equal to the hydrostatic pressures in the bulk fluid, so $P'(0) = P_p$ and $P'(\Delta z) = P_I$. However, since solute is excluded from the peripheral region, equilibrium at the interface requires that $P''(0) = P_p - \Pi_p$ and $P''(\Delta z) = P_I - \Pi_I$, where Π_p and Π_I are the osmotic pressures in the plasma and interstitial fluids. Therefore, the pressure differences inside the pore can be written in terms of pressure differences in the bulk fluids:

$$\Delta P' = P_p - P_I = \Delta P, \tag{6.99}$$

$$\Delta P'' = (P_p - \Pi_p) - (P_I - \Pi_I) = \Delta P - \Delta \Pi. \tag{6.100}$$

The solutions for the velocity in the core and peripheral regions are:

core region ($0 \leq r \leq R_p - R_s$)

$$v'_z(r) = \frac{R_p^2}{4\mu\Delta z}\left\{\left[1-\left(\frac{r}{R_p}\right)^2\right]\Delta P - \left[1-\left(1-\left(\frac{R_s}{R_p}\right)\right)^2 \right.\right.$$
$$\left.\left. \times\left(1 - 2\ln\left(1-\left(\frac{R_s}{R_p}\right)\right)\right)\right]\Delta\Pi\right\}, \quad (6.101)$$

solute - free peripheral region ($R_p - R_s < r \leq R_p$)

$$v''_z(r) = \frac{R_p^2}{4\mu\Delta z}\left\{\left[1-\left(\frac{r}{R_p}\right)^2\right]\Delta P \right.$$
$$\left. -\left[1-\left(\frac{r}{R_p}\right)^2 + 2\left(1-\left(\frac{R_s}{R_p}\right)\right)^2\ln\left(\frac{r}{R_p}\right)\right]\Delta\Pi\right\}. \quad (6.102)$$

Note that unless $\Delta\Pi$ is zero, the velocity profile will generally not be parabolic. In fact, if ΔP is zero, all of the variations in velocity will take place in the peripheral region and the fluid in the core will move as a plug (Anderson and Malone 1974), similar to the flow of a Bingham fluid. The average velocity can be found by integrating the velocity over the entire cross section and dividing by the pore area:

$$\langle v_z \rangle = \frac{2}{R_p^2}\int_0^{R_p-R_s} v'_z r\,dr + \frac{2}{R_p^2}\int_{R_p-R_s}^{R_p} v''_z r\,dr. \quad (6.103)$$

The mean concentration of solute just inside the pore at either end of the pore is different than the concentration in the bulk fluid just outside the pore. The ratio of concentrations inside the pore relative to the bulk fluid concentration is known as the steric partition coefficient for solute s, Φ_s. Assuming the concentration in the core region c'_s to be uniform and equal to the bulk concentration c_s, while the concentration c''_s near the wall is zero, the partition coefficient between pore and bulk fluid is given as

$$\Phi_s = \frac{c_{s,\text{pore}}}{c_{s,\text{bulk}}} = \frac{\frac{2}{R_p^2}\left[\int_0^{R_p-R_s} c'_s r\,dr\right]}{c_s} = \left(1 - \frac{R_s}{R_p}\right)^2. \quad (6.104)$$

After substituting (6.101) and (6.102) for v'_z and v''_z and integrating, we obtain an amazingly simple relationship for the average velocity:

$$\langle v_z \rangle = \frac{Q_{\text{Vp}}}{\pi R_p^2} = L\{\Delta P - \sigma_d \Delta\Pi\}, \quad (6.105)$$

where σ_d is the osmotic reflection coefficient and L is the hydraulic conductivity of the pore:

$$L = \frac{R_p^2}{8\mu\Delta z}, \quad (6.106)$$

6.3 One-Dimensional Shell Balances in Cylindrical Coordinates

$$\sigma_d = \left[1 - \left(1 - \left(\frac{R_s}{R_p}\right)^2\right)^2\right] = [1 - \Phi_s]^2. \tag{6.107}$$

The flow though a single pore can be found by multiplying the average velocity in (6.105) by the cross-sectional area of the pore, πR_p^2. The resulting expression is identical to the Starling equation for flow through a single pore (5.154).

> **Example 6.3.5.1 Velocity Profile in a Restrictive Pore.**
> Equivalent pores spanning the capillary membrane with radii of about 20 nm have been reported to exist in the microvasculature of many organs. Albumin is the predominant macromolecule that passes through capillary pores. Let us consider the flow through a pore that is 100 nm long with an osmotic pressure difference across the pore of 10 mmHg. If the only solute in the pore is albumin ($R_s = 3.6$ nm), find the velocity profile as the hydrostatic pressure difference across the pore is increased from zero to 20 mmHg.

Solution. *Initial considerations*: Since the solute size is not negligible in comparison with the pore size, we need to account for the effects of protein osmotic pressure in addition to hydrostatic pressure. We will assume that the fluid in the pore is Newtonian, that albumin has a spherical shape, and the shape of the equivalent pore is a circular cylinder.

System definition and environmental interactions: The system of interest is the mixture of fluid and albumin inside an equivalent pore. Hydrostatic and osmotic pressure forces are exerted at the two ends of the system, and these are balanced by friction forces between the system and the pore wall.

Apprising the problem to identify governing relationships: The idealized relationships based on conservation of mass and momentum, derived in Sect. 6.3.5, are applicable to this situation. Flow through the pore is divided two regions: a core region that contains albumin and fluid and a peripheral region that is assumed to be devoid of albumin.

Analysis: For albumin $R_s/R_p = 0.18$ and the steric partition coefficient for albumin between the pore interior and bulk fluid is $\Phi_{\text{albumin}} = 0.672$ and the osmotic reflection coefficient is $\sigma_d = 0.107$. Equations (6.101) and (6.102) can be used to compute the velocity profiles at various values of the hydrostatic pressure difference.

Examining and interpreting the results: Results are shown in Fig. 6.30. The various profiles might represent conditions in pores located in capillary walls as one moves from the arterial end of a capillary (higher capillary pressure) to the venous end (lower capillary pressure). The velocity profile goes from nearly parabolic at the higher hydrostatic pressure difference to a much flatter profile as the hydrostatic pressure drops. When $\Delta P = 0$, the velocity in the central region of the tube ($r/R_p < 1 - R_s/R_p$) is constant and negative, and flow is directed from the interstitial fluid into the capillary. These profiles are only as valid as the assumptions

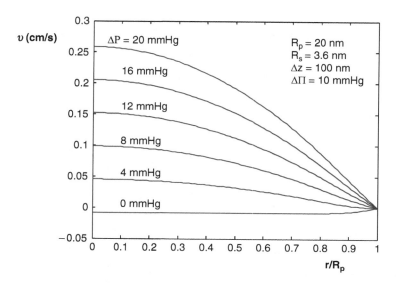

Fig. 6.30 Velocity profiles in pores for a fixed $\Delta\Pi$ and diminishing ΔP

used in the analysis, particularly the assumption that a protein-free fluid layer exists near the pore wall and that there is thermodynamic equilibrium at the interface between the layers. More complete analyses that include other effects, such as electrical charge, can be found in the literature.

6.3.5.1 Accounting for Multiple Solutes

In reality, there are many other proteins present in plasma and interstitial fluid, and most of these are larger than albumin. Each of these proteins will contribute to the overall osmotic reflection coefficient, so σ_d will be higher than the value computed for albumin. Interactions between proteins can be complex, so the best estimate of σ_d would be to compute it from experimental measurements of $\Delta\Pi$, ΔP and transvascular flow using the Starling equation. An estimate can be made if the osmotic effects from each protein are additive. In that case, the overall osmotic reflection coefficient σ_d can be estimated from:

$$\sigma_d = \frac{\sum_s \sigma_{ds}\Delta\Pi_s}{\Delta\Pi} = \frac{\sum_s \sigma_{ds}\Delta\Pi_s}{\sum_s \Delta\Pi_s}, \qquad (6.108)$$

where $\Delta\Pi_s$ is the partial osmotic pressure difference across the pore contributed by protein s, and σ_{ds} is the osmotic reflection coefficient for protein s. We can estimate the partial osmotic pressure contributed by protein s as follows:

$$\Pi_s = RTC_s\left[1 + AC_T + BC_T^2 + \ldots\right], C_T = \sum_s C_s, \qquad (6.109)$$

6.3 One-Dimensional Shell Balances in Cylindrical Coordinates

where the Virial coefficients A, B, etc. are assumed to be the same for all proteins. When the contributions of all protein are added together, we obtain the correct expression for the total osmotic pressure (5.161):

$$\Pi = \sum_s \Pi_s = RT\left[C_T + AC_T^2 + BC_T^3 + ...\right]. \tag{6.110}$$

Equation (6.107) can be used to estimate σ_{ds} for each protein. Proteins larger than the pore radius would have an osmotic reflection coefficient of unity. It should be emphasized that this approach becomes less accurate if the concentrations of proteins are not dilute and if the Virial coefficients are significantly different for the proteins in the mixture.

Example 6.3.5.2 Steady Flow Through an Equivalent Pore.
Estimate the flow rate through an equivalent pore that spans a junction between endothelial cells in a capillary. The equivalent pore radius is 20 nm and its length Δz is 100 nm. The pressure on the plasma side of the pore is 20 mmHg higher than the pressure on the interstitial side. The principal plasma proteins, which contribute to the oncotic pressure, are albumin (alb) and globulin (g). The concentrations of these proteins on the plasma (p) and interstitial (i) sides of the pore are: $C_{\text{alb,p}} = 0.682$ mM, $C_{\text{g,p}} = 0.192$ mM, $C_{\text{alb,i}} = 0.435$ mM, and $C_{\text{g,i}} = 0.083$ mM. The radii of albumin and globulin molecules are approximately 3.6 nm and 4.5 nm, respectively.

Solution. *Initial considerations*: Our goal is to find the total flow through the equivalent pore, not to find the velocity profile. We will assume the solutes are spherical and the fluid is Newtonian. Flow through the pore is assumed to be steady.

System definition and environmental interactions: The system of interest is the fluid and proteins inside an equivalent pore. Viscous forces at the pore wall are balanced by hydrostatic and osmotic pressures at the capillary and interstitial fluid ends of the pore. Each protein contributes to the total osmotic pressure and to the osmotic reflection coefficient.

Apprising the problem to identify governing relationships: The Starling equation is assumed to govern flow through the pore. Before it can be used, we must estimate the filtration coefficient and the osmotic reflection coefficient for this case.

Analysis: We begin with the Starling equation for flow through the equivalent pore:

$$Q_V = K_f\{\Delta P - \sigma_d \Delta \Pi\},$$

where the filtration coefficient K_f for a circular pore is:

$$K_f = \frac{\pi R_p^4}{8\mu \Delta z} = \frac{\pi (20 \text{ nm})^4 \left(1.333 \times 10^5 \frac{\text{cp}}{\text{mmHg s}}\right)\left(10^{-7} \frac{\text{cm}}{\text{nm}}\right)^3}{8(1 \text{ cp})(100 \text{ nm})}$$

$$= 8.38 \times 10^{-14} \frac{\text{ml}}{\text{mmHg s}}.$$

Next, we compute the partition coefficients for albumin and globulin between the pore and bulk fluids using (6.104):

$$\Phi_{\text{alb}} = \left(1 - \frac{R_{\text{alb}}}{R_{\text{p}}}\right)^2 = \left(1 - \frac{3.6 \text{ nm}}{20 \text{ nm}}\right)^2 = 0.672,$$

$$\Phi_{\text{g}} = \left(1 - \frac{R_{\text{g}}}{R_{\text{p}}}\right)^2 = \left(1 - \frac{4.5 \text{nm}}{20 \text{ nm}}\right)^2 = 0.601.$$

The osmotic reflection coefficients for albumin and globulin can be estimated from (6.107):

$$\sigma_{\text{d,alb}} = [1 - \Phi_s]^2 = [1 - 0.672]^2 = 0.107,$$

$$\sigma_{\text{d,g}} = [1 - \Phi_g]^2 = [1 - 0.601]^2 = 0.160.$$

Partial protein osmotic pressures for albumin and globulin can be estimated from (6.109) using the Virial coefficients A and B found in Example 5.17.4

$$\Pi_s(\text{mmHg}) = RTC_s \left[1 + 0.443\, C_P + .204\, C_P^2 \right],$$

where the concentrations are expressed in units of mM and C_p is the total protein concentration:

$$C_P = C_{\text{alb}} + C_{\text{g}}.$$

Total protein concentration in the plasma is:

$$C_{P,p} = 0.682 \text{ mM} + 0.192 \text{ mM} = 0.874 \text{ mM}.$$

Total protein concentration in the interstitial fluid is:

$$C_{P,i} = 0.435 \text{ mM} + 0.083 \text{ mM} = 0.518 \text{ mM}.$$

Estimated partial protein osmotic pressures at 37°C for albumin and globulin in plasma and interstitial fluid are:

$$\Pi_{\text{alb},p} = (19.3 \text{ mmHg})(0.682)\left[1 + 0.443(.874) + .204(.874)^2\right] = 20.3 \text{ mmHg},$$

$$\Pi_{\text{g},p} = (19.3 \text{ mmHg})(0.192)\left[1 + 0.443(.874) + .204(.874)^2\right] = 5.72 \text{ mmHg},$$

$$\Pi_{\text{alb},i} = (19.3 \text{ mmHg})(0.435)\left[1 + 0.443(.518) + .204(.518)^2\right] = 10.8 \text{ mmHg},$$

6.4 Unsteady-State 1-D Shell Balances

$$\Pi_{g,i} = (19.3\,\text{mmHg})(0.083)\left[1 + 0.443(.518) + .204(.518)^2\right] = 2.06\,\text{mmHg},$$

$$\Delta\Pi_{\text{alb}} = \Pi_{\text{alb,p}} - \Pi_{\text{alb,i}} = 20.3\,\text{mmHg} - 10.8\,\text{mmHg} = 9.5\,\text{mmHg},$$

$$\Delta\Pi_g = \Pi_{g,p} - \Pi_{g,i} = 5.72\,\text{mmHg} - 2.06\,\text{mmHg} = 3.66\,\text{mmHg},$$

$$\Delta\Pi = \Delta\Pi_{\text{alb}} + \Delta\Pi_g = 9.5\,\text{mmHg} + 3.66\,\text{mmHg} = 13.16\,\text{mmHg}.$$

We can now estimate the osmotic reflection coefficient using (6.108):

$$\sigma_d = \frac{\sigma_{d,\text{alb}}\Delta\Pi_{\text{alb}} + \sigma_{d,g}\Delta\Pi_g}{\Delta\Pi} = \frac{(0.107)(9.5\,\text{mmHg}) + (0.160)(3.66\,\text{mmHg})}{13.16\,\text{mmHg}} = 0.122.$$

Finally, we can substitute the values computed above into the Starling equation to compute the flow through the pore

$$Q_V = K_f\{\Delta P - \sigma_d\Delta\Pi\} = \left(8.38 \times 10^{-14}\,\frac{\text{ml}}{\text{mmHg s}}\right)$$
$$\times \{20\,\text{mmHg} - (0.122)(13.16\,\text{mmHg})\},$$

$$Q_V = 1.54 \times 10^{-12}\,\frac{\text{ml}}{\text{s}}.$$

Examining and interpreting the results: Note that since $\sigma_d = 0.122$, only 12.2% of the actual protein osmotic pressure difference across the pore influences flow through the pore. If the pore radius was smaller than albumin, then the reflection coefficient would be unity and the full osmotic pressure difference would be exerted across the pore. In that case, the net driving force would be reduced significantly from 18.7 mmHg to 6.84 mmHg. Consequently, the effective pressure difference, $\Delta P - \sigma_d\Delta\Pi$, can be quite different for different size pores exposed to the same ΔP and $\Delta\Pi$. Remember, the computation of the osmotic reflection coefficient is only an estimate based on the assumptions introduced by (6.108) and (6.110). Other models can be found in the literature.

6.4 Unsteady-State 1-D Shell Balances

In this section, we will expand the shell balance approach to illustrate how it can be used to analyze unsteady-state problems in one dimension. The rate of accumulation of x-momentum within the shell will not be zero in unsteady-state momentum balances. The rate of accumulation of x-momentum in the shell is:

$$\frac{(m_{\text{shell}}v_x)|_{t+\Delta t} - (m_{\text{shell}}v_x)|_t}{\Delta t} = m_{\text{shell}}\frac{\partial v_x}{\partial t}, \tag{6.111}$$

where m_{shell} is the shell mass. Both the shell volume and the fluid density are assumed to be independent of time, so m_{shell} does not depend on time.

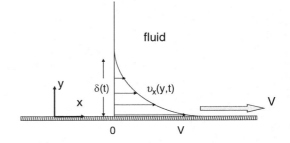

Fig. 6.31 Velocity profile $v_x(y,t)$ and penetration thickness $\delta(t)$ in fluid after wall is moved horizontally at a constant velocity V, starting at $t = 0$

Consider a problem in which a stationary fluid sits on top of a stationary wall. At time $t = 0$, the wall is moved in the x-direction with constant velocity V (Fig. 6.31).

Fluid far from the wall will remain stationary, while fluid close to the wall will gain x-momentum by molecular interaction, and this will propagate into the fluid in the y-direction. Our goal is to determine how quickly the disturbance is felt in the fluid at some distance δ away from the wall.

We will assume that the component of gravitational acceleration in the x-direction is zero and v_y is zero. Furthermore, the pressure is assumed to be independent of x. Since shell mass remains constant, a mass balance provides the same equation as did the mass balance for the steady-state case:

$$\frac{\partial v_x}{\partial x} = 0. \tag{6.112}$$

Therefore, $v_x = v_x(y,t)$, and there can be no net contribution of convection to x-momentum in the shell. The x-momentum equation is:

$$(\rho W \Delta x \Delta y)\frac{\partial v_x}{\partial t} = (W \Delta x)\left[\tau_{yx}\big|_y - \tau_{yx}\big|_{y+\Delta y}\right]. \tag{6.113}$$

Dividing by the shell volume and letting the volume shrink to zero:

$$\rho \frac{\partial v_x}{\partial t} = -\frac{\partial \tau_{yx}}{\partial y}. \tag{6.114}$$

Applying Newton's law of viscosity and introducing the definition of kinematic viscosity, $\nu = \mu/\rho$:

$$\frac{\partial v_x}{\partial t} = \nu \frac{\partial^2 v_x}{\partial y^2}. \tag{6.115}$$

This is a partial differential equation that is first order in t and second order in y. We need to specify auxiliary conditions (two in y, one in t) to solve for the velocity profile. The fluid is initially stationary at $t = 0$. The no-slip boundary condition requires that the fluid at the wall moves with the velocity of the wall, V at all times.

6.4 Unsteady-State 1-D Shell Balances

A second boundary condition is that far away from the wall, the fluid remains stationary. In mathematical terms, the auxiliary conditions are:

$$v_x(y, 0) = 0,$$
$$v_x(0, t) = V, \qquad (6.116)$$
$$v_x(\infty, t) = 0.$$

Solving partial differential equations is generally outside the scope of an introductory text, such as this. However, this problem can be solved using Laplace transforms. Multiplying both sides of (6.115) by e^{-st} and integrating from 0 to ∞:

$$\int_0^\infty \frac{\partial v_x}{\partial t} e^{-st} dt = \nu \int_0^\infty \frac{\partial^2 v_x}{\partial y^2} e^{-st} dt = \nu \frac{d^2}{dy^2} \int_0^\infty v_x e^{-st} dt. \qquad (6.117)$$

Defining the Laplace transform $\bar{v}_x$:

$$\bar{v}_x = \int_0^\infty v_x e^{-st} dt. \qquad (6.118)$$

Integrating the first term by parts, the transformed equation becomes:

$$-v_x(y, 0) + s\bar{v}_x = \nu \frac{d^2 \bar{v}_x}{dy^2}. \qquad (6.119)$$

The first term is zero because the velocity is initially zero everywhere. The general solution for the transformed variable is:

$$\bar{v}_x = C_1 e^{-\sqrt{\frac{s}{\nu}} y} + C_2 e^{+\sqrt{\frac{s}{\nu}} y}, \qquad (6.120)$$

where C_1 and C_2 are constants. Transforming the boundary conditions:

$$\bar{v}_x(0) = \frac{V}{s} \qquad (6.121)$$
$$\bar{v}_x(\infty) = 0.$$

Applying the boundary condition at $y \to \infty$, we find $C_2 = 0$. In addition, C_1 must equal V/s to satisfy the boundary condition at $y = 0$. The final solution for the transformed variable is:

$$\bar{v}_x = V \left(\frac{e^{-\sqrt{\frac{s}{\nu}} y}}{s} \right). \qquad (6.122)$$

Consulting a table of Laplace transforms (Abramowitz and Stegun 1972), we find:

$$v_x(t,y) = V\left\{1 - \frac{2}{\sqrt{\pi}} \int_0^{\frac{y}{2\sqrt{vt}}} e^{-t^2} dt\right\} = V \cdot \text{erfc}\left(\frac{y}{2\sqrt{vt}}\right). \quad (6.123)$$

The function on the right side of (6.123) is known as the complementary error function, and values are tabulated in standard mathematics handbooks (e.g., Abramowitz and Stegun 1972). Plots of the velocity profile at various times in water ($v = 0.01$ cm^2/s) is shown in Fig. 6.32.

We can define a boundary layer thickness δ as the distance from the wall to the point, where the velocity is some fraction of the wall velocity. This is often taken to be $v_x/V = 0.01$. Fluid velocity is 1% of the wall velocity when the argument of the complementary error function equals 1.82, so the boundary layer thickness is given by:

$$\delta = 3.64\sqrt{vt}. \quad (6.124)$$

Fluid outside of the boundary layer is nearly stationary. The boundary layer thickness increases as the square root of time after the wall is set in motion. The growth of the boundary layer with time is shown in Fig. 6.33 for water. The boundary layer penetrates to 5 cm in just over three minutes, and the rate of penetration slows with time.

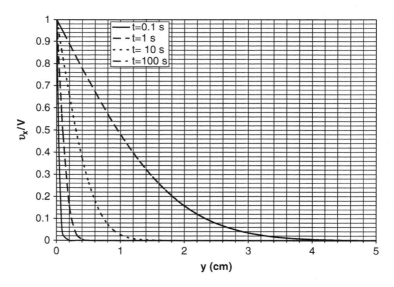

Fig. 6.32 Velocity profile vs. time after plate is set in motion at constant velocity, V

6.5 Summary of Key Concepts

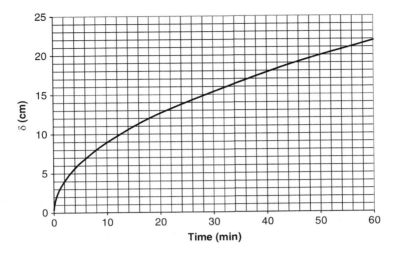

Fig. 6.33 Boundary layer thickness (6.124) vs. time after plate is set in motion in water

Since the fluid velocity in the neighborhood of the wall increases with time, the flux of x-momentum in the y-direction must decrease with time. Computing the shear stress from Newton's law of viscosity:

$$\tau_{yx} = -\mu \frac{\partial v_x}{\partial y} = -\mu \frac{\partial}{\partial y}\left[V\left\{1 - \frac{2}{\sqrt{\pi}}\int_0^{\frac{y}{2\sqrt{vt}}} e^{-t^2}\,dt\right\}\right] = V\sqrt{\frac{\rho\mu}{\pi t}}e^{-\frac{y^2}{4vt}}. \quad (6.125)$$

Shear stress at the plate wall ($y = 0$) is:

$$\tau_w = V\sqrt{\frac{\rho\mu}{\pi t}}. \quad (6.126)$$

Thus, the momentum flux from the wall to the fluid decreases inversely with the square root of time. Recall that an alternate interpretation of τ_w is the force per unit area that must be applied to the plate to move it at velocity V. This force gets smaller and smaller with time. However, at times near zero, very large forces would be required to move the plate at constant velocity, and, at least theoretically, an infinitely large force would be required to jerk the plate into motion at $t = 0$.

6.5 Summary of Key Concepts

Our goal in this chapter is to analyze how shear stress and velocity vary perpendicular to the direction of flow in parallel flow situations. Once these distributions are known, they can be used to find wall forces and flow rates in conduits for Newtonian and non-Newtonian fluids.

Shell Balance Approach. A general procedure for analysis of 1-D fluid transport is provided in Sect. 6.2. These are summarized below.

1. A subsystem within the general system of interest, known as a shell, is selected. Assuming no leakage of fluid through the system walls, the shell can be as long as the system in the direction of flow, and should have a cross-section bounded by r and $r + \Delta r$, where r is the coordinate direction perpendicular to the direction of flow.
2. Conservation of mass is applied to fluid within the shell. In parallel flow situations, this will always lead to $\partial v_x/\partial x = 0$, confirming that the velocity is independent of axial location.
3. Conservation of momentum, a vector equation, is applied to the shell. In directions perpendicular to flow, this results in a simple force balance between pressure and gravitational forces. In the axial direction, x-momentum is transferred in the r-direction by molecular interactions between adjacent fluid layers. This change in x-momentum is balanced by pressure and gravitational forces in the x-direction. This results in an ordinary differential equation that governs the distribution of shear stress on a plane of constant r in the x-direction. If the fluid is in contact with a free surface, there will be no net pressure force in the x-direction. The results generated up to this point are independent of the constitutive relationship for the fluid.
4. The appropriate constitutive relationship for the fluid in the system is introduced into the x-momentum equation. This generally leads to a second-order ordinary differential equation.
5. Two boundary conditions must be applied to solve the equation generated with step 4. Common conditions are: (a) For symmetric flows, the shear stress or velocity gradient are zero at $r = 0$; (b) At a gas–liquid interface, the shear stress by the gas on a liquid surface is negligible; (c) The velocity of fluid in contact with a solid wall or another fluid is the same as the wall or contacting fluid (no slip); and (d) The shear stresses are the same at an interface between two immiscible fluids.
6. Once the velocity profile is known, the distribution of shear stress can be found from the fluid constitutive relationship. Flow rate can be calculated by integrating the velocity over the flow area, $Q_V = \int_A v dA$. A friction factor can be computed for internal flows using (5.64) and fluid resistance can be found from $\Re_f = \Delta\wp/Q_V$. Net force and torque applied to the wall can also be computed from the shear stress distribution.

The above principles were applied to Newtonian and non-Newtonian fluids throughout this chapter.

Laminar flow in circular tubes. Fluid flow through circular tubes is fundamental to the study of bioengineering. Key relationships for flow of a Newtonian fluid through a circular tube from Sect. 6.3.1 are given below:

Shear stress distribution: $\tau_{rz} = \left(\dfrac{\Delta P}{2L}\right) r$.

Velocity distribution: $v_z = \dfrac{\Delta P R^2}{4\mu L}\left(1 - \dfrac{r^2}{R^2}\right)$.

Flow rate (Hagen–Poiseuille equation): $Q_V = \dfrac{\pi \Delta P R^4}{8\mu L}$.

Fluid resistance: $\Re_f = \dfrac{8\mu L}{\pi R^4}$.

There is no need for students to memorize these expressions. However, students should be able to derive the velocity distribution from the shear stress distribution and Newton's law of viscosity, and should be able to compute the average velocity and flow rate from the velocity distribution.

Modified Pressure. For flow in tubes that are not horizontal, the modified pressure $\wp$ should be used in place of the hydrodynamic pressure, P:

Modified pressure: $\wp \equiv P - \rho g_x x - \rho g_y y - \rho g_z z$.

Blood. Although blood is considered to be a Casson fluid, use of the Hagen–Poiseuille equation is accurate in describing the pressure-flow relationship in vessels the size of arterioles and to within 10% for larger vessels.

Flow through small pores. Osmotic pressure effects must be included in very small pores, where the radii of the macromolecules passing through pores are of the same order of magnitude as the radius of the pore. The velocity profile in these pores is not generally parabolic. Flow through a single small pore is governed by the Starling equation:

$$Q_{Vp} = K_{fp}\{\Delta P - \sigma_d \Delta \Pi\},$$

where

$$K_{fp} = \frac{\pi R_p^4}{8\mu\Delta z} \quad \text{and} \quad \sigma_d = \left[1 - \left(1 - \left(\frac{R_s}{R_p}\right)^2\right)^2\right]^2 = [1 - \Phi_s]^2.$$

This reduces to the Hagen–Poiseuille equation if all the macromolecules are small relative to the pore radius.

6.6 Questions

6.6.1. Under what circumstances is it necessary to apply the shell balance approach?

6.6.2. Why do we start with a thin shell when applying a microscopic balance? Why do we let the thickness of the shell shrink to zero?

6.6.3 How do we determine the shape of the shell?

6.6.4. What is responsible for changing fluid momentum within a shell?

6.6.5. What is meant by 1-D flow?

6.6.6. What is meant by a "velocity profile?" If you are given a velocity profile in a duct, how would you compute fluid flow through the duct?

6.6.7. What procedures would you follow to find the shear stress distribution, velocity profile, and flow rate in a tube or between plates?

6.6.8 What fluid shells are appropriate for applying microscopic mass and momentum balances to flow in circular tubes? Identify a shell that would not be appropriate for analyzing flow through an annulus

6.6.9. What boundary conditions must be applied to find velocity profiles of fluid in a tube? An annulus? Between parallel plates? Flow down an inclined plane?

6.6.10. How was the Hagen–Poiseuille equation derived? Under what conditions does this equation apply?

6.6.11. Why do moderate levels of bronchoconstriction cause significant increases in respiratory resistance?

6.6.12. At an air–liquid interface, we often use the boundary condition that the shear stress at the air–liquid interface is zero. This is an approximation. What are the actual boundary conditions that apply at that interface?

6.6.13. For a given pressure difference in the same tube, will a Casson fluid or a Newtonian fluid (with $\mu = S^2$) provide a larger flow rate? What magnitude of error would you expect by treating blood as a Newtonian fluid rather than a Casson fluid for flow in vessels under physiological conditions? Newtonian analysis of blood flow in which vessels would produce the largest errors?

6.6.14. Why do arterioles have so much influence on vascular resistance?

6.6.15. Why does water tend to flow down the respiratory tract, while mucus flows up?

6.6.16. Given a distribution of shear stress in the direction perpendicular to flow, how would you determine the location of the boundary between the sheared and plug regions (where $dv_x/dy = 0$) if the fluid exhibits a yield stress?

6.6.17. For a given pressure difference across the ends of a horizontal tube, is the pressure gradient in the direction of flow the same at all radial positions in the tube?

6.6.18. For a given pressure difference across the ends of a tube, is the pressure gradient in the direction of flow the same for an inclined tube as it is in a horizontal tube?

6.6.19. How does the presence of macromolecules in small pores influence the velocity profile and flow rate? How do the partition coefficient, osmotic reflection coefficient and the filtration coefficient of the pore depend on pore radius and solute radius?

6.6.20. What is a fluid boundary layer? How is the thickness of the boundary layer estimated? If an object is set in motion in a fluid, how does the thickness of the boundary layer vary with time and fluid properties?

6.7 Problems

6.7.1 Marginal Zone Theory

Rework example 6.2.6.1 for the application of blood flowing down an inclined plane using the marginal zone theory with the assumption that plasma is a Newtonian fluid and blood is a Casson fluid. Compare your results to the Newtonian case.

6.7.2 Power Law Fluid

A polymer solution behaves like a power law fluid. Derive an expression for the velocity profile, mean velocity, and flow rate when a thickness h of this fluid flows down an inclined plane.

6.7 Problems

6.7.3 Respiratory Mucus Transport

A large airway is lined with a mucus layer of thickness h, which lies above a periciliary fluid layer. Beating cilia maintain the velocity at the top of the periciliary layer at v_0, directed at an angle θ from the horizontal toward the mouth. Assume airway mucus can be modeled as a Newtonian fluid. Neglect curvature of the vessel and assume rectangular geometry with layer width w and length L.

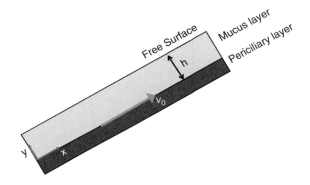

(a) Use the shell balance approach to determine the velocity profile in the mucus layer, the velocity at the lumenal surface of the mucus layer, and the flow rate of mucus in the airway. Neglect pressure gradients in the airway and the shear stress exerted by the air on the mucus surface.

(b) Derive an equation for the maximum mucus layer thickness that ensures that the velocity in all portions of the mucus layer is directed in the positive x-direction. Use the resulting equation to compute, this maximum mucus thickness for the following conditions: $v = 0.05$ cm^2/s, $g = 980$ cm/s^2, $v_0 = 45$ μm/s, $\theta = 90°$. Compare this to measured values of about 25 μm.

(c) What is the net flow rate (μl/s) for the following conditions: $h = 10$ μm, $W = 1$ cm, $v = 0.05$ cm^2/s, $g = 980$ cm/s^2, $v_0 = 45$ μm/s, $\theta = 30°$. Is the net flow in the positive x-direction? Will particles at the surface of this layer be swept in the positive x-direction?

6.7.4 Boundary Conditions at Air–Mucus Interface

We wish to examine the assumption in problem 6.7.3 that the shear stress at the air–mucus interface is approximately zero. Air is a Newtonian fluid with a viscosity of 0.019 cp at body temperature (1 cp = 10^{-2} dyne cm^{-2} s). The maximum velocity gradient in the air at the air–mucus interface is 4.5×10^{-3} s^{-1}. The density of mucus is 1 g/cm^3, the thickness of the mucus layer is 10 μm, and the layer makes an angle θ with the horizontal. Show that a momentum balance performed on the mucus layer provides the following expression for shear stress in the mucus layer:

$$\tau_{yx} = -[\rho g \sin \theta] y + C,$$

where C is a constant of integration. Find C when (a) shear stress is assumed to be zero at $y = h$ and (b) shear stress in the fluid is set equal to shear stress in the mucus at $y = h$. (c) Comment on the validity of the assumption that shear stress is negligible at the interface.

6.7.5 Mucus Plug

An airway mucus layer behaves as a Bingham fluid with yield stress of 1.6 dynes/cm^2 and viscosity μ_0 of 5 cp. Plot the thickness of the plug flow portion of the mucus layer and the velocity of the mucus at the top of the layer as a function of the angle made with the horizontal from zero to 90°. The density of mucus is 1 g/cm^3, the thickness of the mucus layer is 25 µm, and the velocity at $y = 0$ is 45 µm/s. Based on this prediction, what recommendations would you make to a patient with these measured parameters?

6.7.6 Power Law Fluid

Repeat Problem 6.7.3 when mucus behaves as a Power Law fluid. Use the same boundary conditions made in class to derive an expression for mucus velocity as a function of position y.

6.7.7 Thin Film Blood Oxygenator

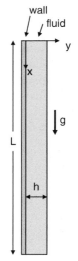

A blood oxygenator consists of several gas permeable rectangular plates having length L and width W. A thin film of blood with uniform thickness h flows at a constant rate vertically down the wall of each plate. The film surface is exposed to air at atmospheric pressure.

(a) Use an x-momentum balance to find the distribution of shear stress in the film.
(b) Find the shear stress τ_w exerted by the blood on the oxygenator wall.
(c) If blood is a non-Newtonian fluid with a yield stress $\tau_y = \tau_w/2$, provide a qualitative sketch of the velocity profile $v_x(y)$ in the film from $y = 0$ to $y = h$.
(d) Derive an expression for the velocity as a function of y in the blood film if blood is modeled as a Newtonian fluid.

6.7.8 Flow Between Inclined Parallel Plates

A Newtonian fluid with viscosity μ flows upward at a steady rate between two parallel plates that make an angle θ with the horizontal. The fluid thickness h is much smaller than the width of the channel W. The pressures at each end, $P(0)$ and $P(L)$, are known and the pressure variations in the y-direction are small. Assume that $v_y = v_z = 0$ and v_x is a function of y alone. Use the shell balance approach to find the following quantities:

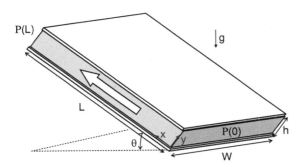

(a) Shear stress, $\tau_{yx}(y)$
(b) Wall shear stress, $\tau_w = \tau_{yx}(y = 0)$
(c) Force that needs to be applied to each plate to keep them from moving
(d) Fluid velocity profile, $v_x(y)$
(e) Total flowrate, Q_V
(f) Average fluid velocity, $\langle v \rangle$
(g) Minimum pressure difference $P(0) - P(L)$ necessary to move fluid in the positive x-direction
(h) Explain why the pressure gradient in the x direction is constant (i.e., $(P_L - P_0)/L$
(i) What aspects of your analysis also apply to non-Newtonian fluids flowing in the same system with the same pressure gradient?

6.7.9 Flow in the Lung Microcirculation

Blood flow in the alveolar microvasculature has been likened to blood flow between parallel plates as shown in Problem 6.7.8. Blood exhibits a yield stress equal to τ_y. What minimum pressure difference $P_0 - P_L$ must be imposed before the blood in the capillary will flow? If the pressure difference is greater than this, what will be the boundaries of the plug flow region?

6.7.10 Flow of a Casson Fluid Between Two Plates

A Casson fluid with yield stress τ_y flows between two plates a distance $2b$ apart. The length of the plates is L and width is W, both which are much greater than b. The pressure difference applied over the length of the plates is ΔP. Show that for $\Delta P > \tau_y L/b$, the thickness of the plug region is $2h$, where $h = \tau_y L/\Delta P$, and the flow relative to the flow of a Newtonian fluid with $\mu = S^2$ is:

$$\frac{Q_V}{Q_{V,\text{Newtonian}}} = 1 - \frac{1}{10}\left(\frac{h}{b}\right)^3 - \frac{12}{5}\left(\frac{h}{b}\right)^{\frac{1}{2}} + \frac{3}{2}\left(\frac{h}{b}\right),$$

where:

$$Q_{V,\text{Newtonian}} = \frac{2}{3}\left(\frac{\Delta P W b^3}{\mu L}\right), \mu = S^2.$$

6.7.11 Laminar Flow in a Tube

(a) Show that for steady parabolic flow in a circular tube $\langle v^3 \rangle / \langle v \rangle = 2\langle v \rangle^2$.
(b) Find $\langle v^2 \rangle$ in terms of $\langle v \rangle$ for laminar flow in a tube.

6.7.12 Newtonian Film Oxygenator

A blood film oxygenator consists of a number of parallel tubes oriented vertically. Blood flows from below through the lumen of each tube with outside radius R, out the end of the tube, and then forms a film of thickness h as it falls down over the outside surface of the tube. Use the shell balance approach to find the velocity profile and total flow rate of blood as it falls down the outside of the tube. Assume blood is Newtonian.

6.7.13 Non-Newtonian Film Oxygenator

Repeat problem 6.7.12 assuming blood is a Bingham fluid.

6.7.14 Alternate Shell Approach

Derive an expression for shear stress vs. radial position in a horizontal tube using a shell with radius r and length Δz. The tube has radius R, pressure difference ΔP, and length L.

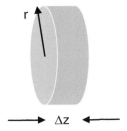

6.7.15 Marginal Zone Theory

Apply the marginal zone theory of Example 6.2.6.1 to derive an equation for blood flow in a circular tube, assuming the two fluids are Newtonian with different viscosities.

6.7.16 Marginal Zone Theory

Blood flows through a tube with a radius of 25 μm and a length of 1,000 μm. The pressure drop across the tube is 30 mmHg. The hematocrit in large vessels (feed hematocrit or discharge hematocrit H_D) is 45%. Plasma viscosity is 1.03 cp. Assume that blood in the tube can be modeled as two immiscible layers: a plasma layer near the wall of thickness 4 μm and a blood core region of radius 21 μm.

(a) Compute the hematocrit of blood in the core region of the tube from tube geometry and the tube hematocrit.
(b) Estimate the viscosity of blood in the core region.
(c) Specify the boundary conditions necessary to find the four unknown coefficients that arise in the plasma and core velocity equations.
(d) Use conservation of momentum to find the shear stress and velocity distributions in the plasma region and in the blood region.
(e) Plot the velocity vs. radial position in the plasma and in the core.
(f) Derive an expression for flow through the tube, and compute the flow.

6.7.17 Yield Stress

A non-Newtonian fluid will not flow through a circular tube of radius 0.1 mm and length 10 cm until a pressure difference of 10 cmH$_2$O is applied across the tube. What is the yield stress of the fluid?

6.7.18 Power Law Fluid in a Cylindrical Tube

Derive expressions for the velocity profile and flow rate of a power law fluid that flows through a cylindrical tube of length L and radius R.

6.7.19 Bingham Fluid in a Cylindrical Tube

Derive expressions for the velocity profile and flow rate of a Bingham fluid that flows in a cylindrical tube with length L and radius R. What is the relationship between the radius of the plug R_y and the yield stress for a Bingham fluid? Derive a relationship between Bingham Flow/Poiseuille Flow vs. R_y/R, similar to the relationship derived in Sect. 6.3.4 for a Casson fluid. Plot this over a physiological range of R_y/R and comment on the effect of yield stress on flow.

6.7.20 Bingham Fluid in an Annulus

Derive an expression for the flow of a Bingham fluid in an annulus where $\tau_{rz} > \tau_y$ in the annulus.

6.7.21 Flow of Air and Mucus Through a Small Bronchiole

A mucus layer is in contact with the periciliary surface at $r = R_2$ and with air at $r = R_1$. The surface at R_2 is propelled along the bronchiole at a velocity V. The bronchiole is horizontal and the same pressure difference exists in both the air and the mucus.

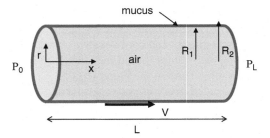

(a) How does shear stress vary with radial position in the air and in the mucus?
(b) If the mucus behaves as a Bingham fluid, what is the minimum yield stress needed to ensure that the entire mucus layer moves as a plug?
(c) Derive an expression for the velocity vs. radial position *in the air*, assuming air is a Newtonian fluid and that the entire mucus layer moves at velocity V.

6.7.22 Respiratory Flow of Non-Newtonian Fluid

Repeat Problem 6.7.3 for flow of mucus modeled as a Bingham fluid with density of 1 g/ml, viscosity of 5 cp and yield stress of 2.0 dynes/cm^2.

6.7.23 Restricted Fluid Flow Through a Narrow Slit

Consider the flow of a spherical solute with radius R_s through the space between two endothelial cells. The slit has height $2H$, length L and width w. The width is much greater than H, so this is equivalent to the flow between two parallel plates. The hydrostatic pressures on the plasma and interstitial sides of the slit are P_p and P_i, respectively. The osmotic pressures on the two sides of the slit are Π_p and Π_i. The solute center can come no closer to the wall of the slit than its radius. (a) Apply the methods introduced in Sect. 6.3.5 to find appropriate expressions for the velocity profile in the slit. Assume fluid viscosities in the core and peripheral regions are the same. (b) If $R_s = 3.6$ nm, $H = 15$ nm, $L = 100$ nm, $w = 200$ nm, $\mu = 1$ cp, and $\Delta\Pi = 5$ Torr, plot velocity vs. y for $\Delta P = 5, 3, 1, 0.5, 0.2, 0.1$ and 0 Torr.

6.7.24 Starling's Law for Flow Through a Narrow Slit

Integrate the velocity profile found in Problem 6.6.23 to show: $Q_V = K_f[\Delta P - \sigma_d \Delta\Pi]$. Note: there is a lot of algebra here – be careful and check this over a few times. Provide an appropriate expression for K_f in terms of geometric factors and fluid viscosity. As a check, this should be the same as what would be computed for flow between parallel plates. Provide an expression for σ_d in terms of $\lambda_s = R_s/H$.

6.8 Challenges

6.8.1 Blood Flow in Vessels

In Chap. 4, we found that blood is best modeled as a Casson Fluid. *Challenge*: Under physiological conditions, how much error is introduced if blood is assumed to behave as a Newtonian fluid? *Generate Ideas*: What are the differences between a Newtonian Fluid and a Casson Fluid? Sketch velocity vs. radial position in a blood vessel for Newtonian and Casson fluids. Sketch Flow vs. Pressure for each fluid. Would you expect flow to be lower or higher for a Casson Fluid than a Newtonian fluid with the same viscosity and pressure drop? Using the shell balance approach, what is the difference between analysis for blood modeled as a Newtonian fluid and a Casson fluid? What is the yield stress of blood?

6.8.2 Modeling Respiratory Clearance

We wish to develop a model of respiratory clearance that predicts the essential features of mucus flow up the respiratory tree. *Challenge*: How would you model the transport of mucus in the large airways? *Generate Ideas*: What is the system of interest? What assumptions about geometry might be appropriate? Is a macroscopic or microscopic approach called for in this problem? What boundary conditions might you impose at the two surfaces of the mucus layer? If mucus behaved as a Newtonian fluid, what minimum viscosity would be required to ensure that the mucus–air surface always moves away from alveoli? Is this realistic? What non-Newtonian property of mucus can ensure that the layer moves away from the alveoli, even if the mucus viscosity is much lower than that needed for a Newtonian fluid?

References

Abramowitz M, Stegun IA (1972) Handbook of mathematical functions with formulas, graphs, and mathematical tables. U.S. Department of Commerce, National Bureau of Standards, Applied Mathematics Series, 55

Anderson JL, Malone DM (1974) Mechanism of osmotic flow in porous membranes. Biophys J 14:957–982

Chapter 7
General Microscopic Approach for Biofluid Transport

7.1 Introduction

Many problems in biofluid mechanics can be treated with the one-dimensional shell balance approach described in Chap. 6. However, the solution of multidimensional problems requires a more general approach. The result of this approach is a set of partial differential equations that can be applied to a variety of different problems, including many of the one-dimensional problems we have already solved in Chap. 6. It is not our intention to provide extensive methods for solving partial differential equations in this chapter. We will solve some multidimensional problems with relevant biofluids applications, but our main goals are to show how the general equations are derived, what restrictions are placed on them, how to simplify them by applying scaling concepts and reasonable assumptions, and how to specify auxiliary conditions, such as initial conditions and boundary conditions, that will provide us with a well-posed problem.

7.2 Conservation of Mass

We begin by deriving a general statement of the conservation of mass. For convenience, we will select a rectangular coordinate system, and a cube-shaped shell in the fluid with volume of $\Delta x \Delta y \Delta z$, as shown in Fig. 7.1.

The rate of accumulation of fluid in the shell between time t and time $t + \Delta t$ is:

$$\left\{ \begin{array}{c} \text{Rate of accumulation} \\ \text{of mass in shell} \end{array} \right\} = \frac{m(t + \Delta t) - m(t)}{\Delta t} = \left[\frac{\rho(t + \Delta t) - \rho(t)}{\Delta t} \right] \Delta x \Delta y \Delta z. \tag{7.1}$$

Since mass cannot be spontaneously created or destroyed, the only way that mass can enter or leave the shell is through one or more of the six faces of the cube, as shown in Fig. 7.2.

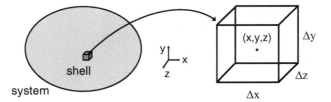

Fig. 7.1 General three-dimensional fluid shell in rectangular coordinates

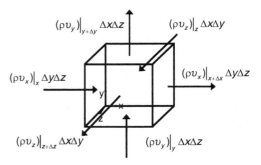

Fig. 7.2 Rate at which mass enters shell via bulk fluid motion (convection)

The rate at which mass enters the shell across any single face of the cube is equal to the mass flux through that face multiplied by the surface area of that face. The net rate at which mass enters through all six faces is:

$$\left\{\begin{array}{l}\text{Net rate mass}\\ \text{flows into shell}\end{array}\right\} = (\rho v_x \Delta y \Delta z)_x - (\rho v_x \Delta y \Delta z)_{x+\Delta x} + (\rho v_y \Delta x \Delta z)_y$$
$$- (\rho v_y \Delta x \Delta z)_{y+\Delta y} + (\rho v_z \Delta x \Delta y)_z - (\rho v_z \Delta x \Delta y)_{z+\Delta z}. \quad (7.2)$$

Conservation of mass requires that (7.1) and (7.2) balance. Setting the rate of accumulation equal to the net mass flow into the shell, dividing by the shell volume, and letting the volume and Δt approach zero:

$$\lim_{\Delta t \to 0}\left\{\frac{\rho|_{t+\Delta t}-\rho|_t}{\Delta t}\right\}$$
$$= \lim_{\substack{\Delta x \to 0 \\ \Delta y \to 0 \\ \Delta z \to 0}}\left\{\frac{(\rho v_x|_x - \rho v_x|_{x+\Delta x})}{\Delta x} + \frac{(\rho v_y|_y - \rho v_y|_{y+\Delta y})}{\Delta y} + \frac{(\rho v_z|_z - \rho v_z|_{z+\Delta z})}{\Delta z}\right\}.$$

$$(7.3)$$

7.3 Conservation of Linear Momentum

In the limit, conservation of mass in rectangular coordinates reduces to:

$$\frac{\partial \rho}{\partial t} + \frac{\partial}{\partial x}(\rho v_x) + \frac{\partial}{\partial y}(\rho v_y) + \frac{\partial}{\partial z}(\rho v_z) = 0. \quad (7.4)$$

This is known as the continuity equation for a compressible fluid. Appropriate expressions in cylindrical and spherical coordinates will be presented in Sect. 7.11.

In most biofluids applications, the density is constant. Thus, for incompressible flow the continuity equation reduces to:

$$\frac{\partial v_x}{\partial x} + \frac{\partial v_y}{\partial y} + \frac{\partial v_z}{\partial z} = 0. \quad (7.5)$$

7.3 Conservation of Linear Momentum

The rate at which x-momentum accumulates in the shell shown in Fig. 7.1 is:

$$\left\{\begin{array}{c} \text{Rate of accumulation} \\ \text{of } x\text{-momentum in shell} \end{array}\right\} = \frac{(mv_x)|_{t+\Delta t} - (mv_x)|_t}{\Delta t} = \left[\frac{(\rho v_x)|_{t+\Delta t} - (\rho v_x)|_t}{\Delta t}\right] \Delta x \Delta y \Delta z. \quad (7.6)$$

The x-momentum in the shell can change if: (1) there is an imbalance in x-momentum transported by bulk fluid motion through the surfaces of the shell, (2) there is an imbalance in x-momentum by molecular momentum transport across the six surfaces of the shell, (3) there is generation of x-momentum in the shell caused by an imbalance of forces applied to the shell, or (4) some combination of the above.

Consider the flux of x-momentum into the shell by bulk fluid motion at the bottom surface of the shell. The mass flux through that surface is $\rho v_y|_y$. The flux of x-momentum by bulk fluid motion through the bottom surface is the mass flux multiplied by the velocity in the x-direction, $\rho v_y v_x|_y$. The rate that x-momentum is gained in the shell equals the momentum flux multiplied by the bottom surface area, $(\rho v_y v_x|_y)(\Delta x \Delta z)$. The same reasoning can be used to account for the rate x-momentum crosses all six surfaces by bulk fluid motion, as shown in Fig. 7.3. The net rate that x-momentum is added by bulk fluid motion is:

$$\left\{\begin{array}{c} \text{Rate } x\text{-momentum} \\ \text{is added by} \\ \text{bulk fluid motion} \end{array}\right\}$$
$$= \left(\rho v_x v_x|_x - \rho v_x v_x|_{x+\Delta x}\right) \Delta y \Delta z + \left(\rho v_y v_x|_y - \rho v_y v_x|_{y+\Delta y}\right) \Delta x \Delta z$$
$$+ \left(\rho v_z v_x|_z - \rho v_z v_x|_{z+\Delta z}\right) \Delta x \Delta y. \quad (7.7)$$

Fig. 7.3 x-Momentum entering and leaving the shell by bulk movement

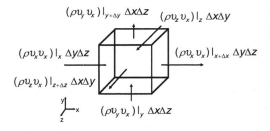

Fig. 7.4 x-Momentum entering and leaving shell by viscous molecular transport

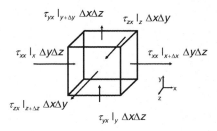

Fig. 7.5 Forces applied to shell by surroundings

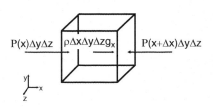

Next, consider the flux of x-momentum into the shell at z by molecular transport down a velocity gradient. This is simply equal to $\tau_{zx}|_z$. The rate that x-momentum enters the shell through this surface is the product of the viscous momentum flux and surface area, $\left(\tau_{zx}|_z\right)(\Delta x \Delta y)$. The rate that x-momentum crosses each of the surfaces of the shell by viscous transport is shown in Fig. 7.4. The net contribution to the rate of accumulation of x-momentum by diffusion is:

$$\left\{\begin{array}{c}\text{Rate } x\text{-momentum}\\ \text{is added by}\\ \text{molecular motion}\end{array}\right\} = \left(\tau_{xx}|_x - \tau_{xx}|_{x+\Delta x}\right)\Delta y \Delta z + \left(\tau_{yx}|_y - \tau_{yx}|_{y+\Delta y}\right)\Delta x \Delta z$$

$$+ \left(\tau_{zx}|_z - \tau_{zx}|_{z+\Delta z}\right)\Delta x \Delta y. \tag{7.8}$$

Finally, consider the forces applied to the shell which will cause the generation of x-momentum in the shell. The pressure and gravitational forces applied to the shell in the x-direction are shown in Fig. 7.5. The generation term is:

7.3 Conservation of Linear Momentum

$$\left\{\begin{array}{c}\text{Rate } x\text{-momentum}\\ \text{is generated by}\\ \text{forces in } x\text{-direction}\end{array}\right\} = (P|_x - P|_{x+\Delta x})\Delta y \Delta z + \rho g_x \Delta x \Delta y \Delta z. \quad (7.9)$$

Setting the accumulation term equal to the sum of the three sources of momentum, and dividing by the shell volume:

$$\frac{(\rho v_x)|_{t+\Delta t} - (\rho v_x)|_t}{\Delta t} = \frac{(P|_x - P|_{x+\Delta x})}{\Delta x} + \rho g_x + \frac{(\rho v_x v_x|_x - \rho v_x v_x|_{x+\Delta x})}{\Delta x}$$

$$+ \frac{(\rho v_y v_x|_y - \rho v_y v_x|_{y+\Delta y})}{\Delta y} + \frac{(\rho v_z v_x|_z - \rho v_z v_x|_{z+\Delta z})}{\Delta z}$$

$$+ \frac{(\tau_{xx}|_x - \tau_{xx}|_{x+\Delta x})}{\Delta x} + \frac{(\tau_{yx}|_y - \tau_{yx}|_{y+\Delta y})}{\Delta y} + \frac{(\tau_{zx}|_z - \tau_{zx}|_{z+\Delta z})}{\Delta z}$$

(7.10)

Taking the limit as $\Delta t \to 0$, $\Delta x \to 0$, $\Delta y \to 0$, and $\Delta z \to 0$, we find:

$$\frac{\partial(\rho v_x)}{\partial t} = -\frac{\partial P}{\partial x} + \rho g_x - \frac{\partial(\rho v_x v_x)}{\partial x} - \frac{\partial(\rho v_y v_x)}{\partial y} - \frac{\partial(\rho v_z v_x)}{\partial z} - \frac{\partial \tau_{xx}}{\partial x} - \frac{\partial \tau_{yx}}{\partial y} - \frac{\partial \tau_{zx}}{\partial z}.$$

(7.11)

Expanding the bulk momentum terms and using the continuity equation, we finally obtain the x-momentum equation in rectangular coordinates:

$$\rho \frac{\partial v_x}{\partial t} = -\frac{\partial P}{\partial x} + \rho g_x - \rho v_x \frac{\partial v_x}{\partial x} - \rho v_y \frac{\partial v_x}{\partial y} - \rho v_z \frac{\partial v_x}{\partial z} - \frac{\partial \tau_{xx}}{\partial x} - \frac{\partial \tau_{yx}}{\partial y} - \frac{\partial \tau_{zx}}{\partial z}. \quad (7.12)$$

Repeating the momentum balance for y-momentum and z-momentum, we arrive at the following set of three scalar equations, known as the momentum equations:

x-momentum:
$$\rho \frac{\partial v_x}{\partial t} = -\frac{\partial P}{\partial x} + \rho g_x - \rho v_x \frac{\partial v_x}{\partial x} - \rho v_y \frac{\partial v_x}{\partial y} - \rho v_z \frac{\partial v_x}{\partial z} - \frac{\partial \tau_{xx}}{\partial x} - \frac{\partial \tau_{yx}}{\partial y} - \frac{\partial \tau_{zx}}{\partial z},$$

y-momentum:
$$\rho \frac{\partial v_y}{\partial t} = -\frac{\partial P}{\partial y} + \rho g_y - \rho v_x \frac{\partial v_y}{\partial x} - \rho v_y \frac{\partial v_y}{\partial y} - \rho v_z \frac{\partial v_y}{\partial z} - \frac{\partial \tau_{xy}}{\partial x} - \frac{\partial \tau_{yy}}{\partial y} - \frac{\partial \tau_{zy}}{\partial z}, \quad (7.13)$$

z-momentum:
$$\rho \frac{\partial v_z}{\partial t} = -\frac{\partial P}{\partial z} + \rho g_z - \rho v_x \frac{\partial v_z}{\partial x} - \rho v_y \frac{\partial v_z}{\partial y} - \rho v_z \frac{\partial v_z}{\partial z} - \frac{\partial \tau_{xz}}{\partial x} - \frac{\partial \tau_{yz}}{\partial y} - \frac{\partial \tau_{zz}}{\partial z}.$$

The combinations of these momentum equations and the continuity equation are known as the *equations of motion*.

7.4 Moment Equations

Thus far, we have derived four scalar equations to describe fluid motion. Assuming the fluid is incompressible, we have 13 unknown quantities: three velocity components, pressure, and nine stress components. As it stands, this is a highly underdetermined system of equations. More information is needed. To predict the motion of a rigid body in mechanics, we must not only apply Newton's second law in each direction, but must also apply the moment equations about each axis. We shall apply the moment equation about each axis of our rectangular cube-shaped fluid element. Shear stress will be interpreted as a force per unit area in this analysis. The sum of all moments acting about the z-axis on a fluid element such as that shown in Fig. 7.1 is equal to the product of the centroidal mass moment of inertia $\bar{I}_z$ and the angular acceleration of the fluid element α_z.

$$\sum M_z = \bar{I}_z \alpha_z = \left[\frac{m}{12}(\Delta x^2 + \Delta y^2)\right]\alpha_z = \frac{1}{12}(\rho \Delta x \Delta y \Delta z)(\Delta x^2 + \Delta y^2)\alpha_z. \quad (7.14)$$

Forces caused by the normal stresses, pressure, and gravity all pass through the center of the fluid element. Therefore, they do not contribute to the angular acceleration of the element. The shear forces are the only moments acting about the centroid:

$$\sum M_z = \left[\tau_{xy}(x + \Delta x, y, z) + \tau_{xy}(x, y, z)\right][\Delta y \Delta z]\frac{\Delta x}{2}$$
$$- \left[\tau_{yx}(x, y + \Delta y, z) + \tau_{yx}(x, y, z)\right][\Delta x \Delta z]\frac{\Delta y}{2}. \quad (7.15)$$

Equating these two expressions for the sum of moments and dividing by the fluid element volume:

$$\frac{\rho}{6}(\Delta x^2 + \Delta y^2)\alpha_z = \left[\tau_{xy}(x + \Delta x, y, z) + \tau_{xy}(x, y, z)\right] - \left[\tau_{yx}(x, y + \Delta y, z) + \tau_{yx}(x, y, z)\right]. \quad (7.16)$$

In the limit as the dimensions of the fluid element shrink to the point (x, y, z), we find:

$$\tau_{xy}(x, y, z) = \tau_{yx}(x, y, z). \quad (7.17)$$

Applications of the moment equation about the y and the x-axes provide similar relations:

$$\tau_{xz}(x, y, z) = \tau_{zx}(x, y, z), \quad (7.18)$$

$$\tau_{yz}(x, y, z) = \tau_{zy}(x, y, z). \quad (7.19)$$

Therefore, the order of the subscripts is interchangeable for shear stress. Consequently, the shear stress in the x direction on a surface of constant y is the same as the stress in the y direction on a surface of constant x. These three relationships reduce the number of unknowns from 13 to 10, but we still only have four independent equations.

7.5 General Constitutive Relationship for a Newtonian Fluid

Thus far, we have not introduced a constitutive relationship for the fluid. Therefore, all of the equations derived to this point are independent of specific properties of the fluid.

The constitutive relationships presented to this point describe the relationship between shear stress and shear rate in parallel flow situations, where only a single component of velocity is present. In addition, the magnitude of the velocity was assumed to vary in a direction perpendicular to the direction of flow. The rate of deformation in such a case is equal to the velocity gradient, as we showed in Sect. 4.2. Although this covers an important classification of problems, there are many situations where the velocity field is complex, with velocity gradients in more than one direction. In such cases, the rate of deformation cannot be described by (4.11). Deformation will occur in all three directions. Let us consider the deformation that occurs in a time interval Δt about the z-axis. The shear strain can be defined as the average of the two angles γ_1 and γ_2 in Fig. 7.6. Velocity variations in the z-direction will not influence the deformation about the z-axis. Assuming small time intervals so the deformation is small, the shear strain caused by a variation of the y-component of velocity in the x-direction is:

$$\gamma_1 \simeq \tan(\gamma_1) = \frac{v_y(x+\Delta x, y)\Delta t - v_y(x,y)\Delta t}{\Delta x}. \quad (7.20)$$

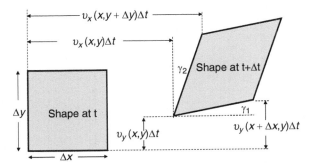

Fig. 7.6 Deformation of a fluid element about the z-axis when subjected to shear

Similarly, the shear strain that results from a variation of the x-component of velocity in the y-direction is:

$$\gamma_2 \simeq \tan(\gamma_2) = \frac{v_x(x, y + \Delta y)\Delta t - v_x(x, y)\Delta t}{\Delta y}. \tag{7.21}$$

The average *rate of deformation* in the x–y plane D_{xy} about the z-axis can be found by dividing the mean of the shear strains by the time increment Δt and letting the volume of the fluid element shrink to zero:

$$D_{xy} = D_{yx} = \frac{1}{2}(\dot{\gamma}_1 + \dot{\gamma}_2) = \frac{1}{2}\left(\frac{\partial v_y}{\partial x} + \frac{\partial v_x}{\partial y}\right). \tag{7.22}$$

This rate of deformation is associated with the shear stress τ_{yx} (or τ_{xy}), which tends to rotate fluid about the z-axis. Therefore if shear stress is just a function of strain rate, then the constitutive relationship for τ_{yx} must take the functional form:

$$\tau_{yx} = \tau_{xy} = f(D_{xy}). \tag{7.23}$$

Here, $f()$ represents a functional relationship. The same reasoning can be applied to deformations about the x and y axes:

$$\begin{aligned}\tau_{yz} = \tau_{zy} = f(D_{yz}), D_{zy} = D_{yz} = \frac{1}{2}\left(\frac{\partial v_y}{\partial z} + \frac{\partial v_z}{\partial y}\right), \\ \tau_{xz} = \tau_{zx} = f(D_{xz}), D_{xz} = D_{zx} = \frac{1}{2}\left(\frac{\partial v_x}{\partial z} + \frac{\partial v_z}{\partial x}\right).\end{aligned} \tag{7.24}$$

Viscous shear stresses cause the fluid element to rotate, but viscous normal stresses induce deformation of the fluid element. Consider deformation of an incompressible fluid subjected to a positive normal viscous stress, τ_{xx}. As shown in Fig. 7.7, this will induce elongation of a microscopic fluid element in the x-direction, resulting in a net

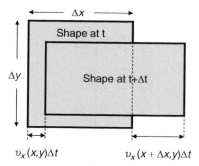

Fig. 7.7 Deformation of a fluid element subjected to a normal stress τ_{xx}

7.5 General Constitutive Relationship for a Newtonian Fluid

strain in the x-direction. The strain ε_x is defined as the change in length divided by the original length:

$$\varepsilon_x = \frac{[v_x(x+\Delta x, y)\Delta t + \Delta x - v_x(x,y)\Delta t] - \Delta x}{\Delta x}. \tag{7.25}$$

Letting Δx approach zero in (7.25), and dividing by Δt, we find the strain rate in the x-direction (or rate of deformation in the x-direction) to be equal to the velocity gradient:

$$\dot{\varepsilon}_x = D_{xx} = \frac{\partial v_x}{\partial x}. \tag{7.26}$$

Similarly, rates of deformation in the y and z directions can be written:

$$D_{yy} = \frac{\partial v_y}{\partial y} \quad D_{zz} = \frac{\partial v_z}{\partial z}. \tag{7.27}$$

Rates of deformation in the x-, y-, and z-directions are induced by normal stresses in those directions:

$$\tau_{xx} = f(D_{xx}), \quad \tau_{yy} = f(D_{yy}), \quad \tau_{zz} = f(D_{zz}). \tag{7.28}$$

Equations (7.22), (7.24), (7.26), and (7.27) can be generalized as follows:

$$D_{ij} = D_{ji} = \frac{1}{2}\left(\frac{\partial v_j}{\partial x_i} + \frac{\partial v_i}{\partial x_j}\right), \tag{7.29}$$

where i and j can take on the values x, y, or z. The general functional relationship between stress and rate of deformation for an isotropic, incompressible fluid is:

$$\tau_{ij} = f(D_{ij}) = -2\eta D_{ij} = -\eta\left(\frac{\partial v_j}{\partial x_i} + \frac{\partial v_i}{\partial x_j}\right), \tag{7.30}$$

where η is the *apparent viscosity* of the fluid, which can be a function of the rate of deformation. We will discuss non-Newtonian constitutive relationships in Sect. 7.15. The simplest possible constitutive relationship for multidimensional flow occurs when $\eta = \mu$, a constant. The resulting constitutive expression reflects a linear relationship between stress components and strain rate components:

$$\tau_{ij} = -\mu\left(\frac{\partial v_i}{\partial x_j} + \frac{\partial v_j}{\partial x_i}\right). \tag{7.31}$$

This is the constitutive relationship for a Newtonian fluid in three dimensional flow. The constant μ is the coefficient of viscosity, a positive value. The negative sign ensures that a negative velocity gradient produces a positive momentum flux. The six viscous stress terms in rectangular coordinates can be written:

$$\tau_{xy} = \tau_{yx} = -\mu\left(\frac{\partial v_x}{\partial y} + \frac{\partial v_y}{\partial x}\right), \quad \tau_{xz} = \tau_{zx} = -\mu\left(\frac{\partial v_x}{\partial z} + \frac{\partial v_z}{\partial x}\right),$$

$$\tau_{yz} = \tau_{zy} = -\mu\left(\frac{\partial v_y}{\partial z} + \frac{\partial v_z}{\partial y}\right), \quad \tau_{xx} = -2\mu\left(\frac{\partial v_x}{\partial x}\right), \quad \tau_{yy} = -2\mu\left(\frac{\partial v_y}{\partial y}\right),$$

$$\tau_{zz} = -2\mu\left(\frac{\partial v_z}{\partial z}\right). \tag{7.32}$$

After substituting these values into the x-, y-, and z-momentum equations we obtain what are known as the Navier–Stokes equations for an incompressible Newtonian fluid:

x-component

$$\rho\frac{\partial v_x}{\partial t} + \rho\left[v_x\frac{\partial v_x}{\partial x} + v_y\frac{\partial v_x}{\partial y} + v_z\frac{\partial v_x}{\partial z}\right] = \mu\left[\frac{\partial^2 v_x}{\partial x^2} + \frac{\partial^2 v_x}{\partial y^2} + \frac{\partial^2 v_x}{\partial z^2}\right] - \frac{\partial P}{\partial x} + \rho g_x,$$

y-component

$$\rho\frac{\partial v_y}{\partial t} + \rho\left[v_x\frac{\partial v_y}{\partial x} + v_y\frac{\partial v_y}{\partial y} + v_z\frac{\partial v_y}{\partial z}\right] = \mu\left[\frac{\partial^2 v_y}{\partial x^2} + \frac{\partial^2 v_y}{\partial y^2} + \frac{\partial^2 v_y}{\partial z^2}\right] - \frac{\partial P}{\partial y} + \rho g_y, \tag{7.33}$$

z-component

$$\rho\frac{\partial v_z}{\partial t} + \rho\left[v_x\frac{\partial v_z}{\partial x} + v_y\frac{\partial v_z}{\partial y} + v_z\frac{\partial v_z}{\partial z}\right] = \mu\left[\frac{\partial^2 v_z}{\partial x^2} + \frac{\partial^2 v_z}{\partial y^2} + \frac{\partial^2 v_z}{\partial z^2}\right] - \frac{\partial P}{\partial z} + \rho g_z.$$

The first term on the left side of each equation is the acceleration per unit volume of the fluid at the point (x, y, z) in the appropriate coordinate direction. The other terms on the left represent the convective acceleration per unit volume, also referred to as the inertial terms. These terms arise from momentum transfer by bulk fluid motion, and more will be discussed about these terms in the next section. The first term on the right results from viscous momentum transfer, and the last two are momentum production caused by pressure and gravitational forces per unit volume.

7.6 Substantial Derivative

There are several time derivates that are used in biotransport, and we will attempt to distinguish between them in this section. Consider a scalar quantity $\Phi(x, y, z, t)$, where Φ represents a variable such as air temperature in a large room, which can

7.6 Substantial Derivative

vary both spatially and temporally. Applying the chain rule for differentiation, the total time derivative of Φ is:

$$\frac{d\Phi}{dt} = \frac{\partial \Phi}{\partial t} + \frac{\partial \Phi}{\partial x}\frac{dx}{dt} + \frac{\partial \Phi}{\partial y}\frac{dy}{dt} + \frac{\partial \Phi}{\partial z}\frac{dz}{dt}. \tag{7.34}$$

What is the meaning of each of these terms? Imagine that you are standing in the room holding a thermometer and Φ in (7.34) represents temperature. The first term on the right side represents the variation of temperature with time at a given point (x, y, z). The spatial partial derivatives represent temperature gradients that may exist in the room. Thus, if you were standing at a different point in the room, perhaps closer to the air duct, you might measure a completely different rate of change of temperature. Consider now the case where the temperature in the room depends on position, but is independent of time. If you stand in one place, you will measure one temperature, and if you stand in another place, you will measure a different temperature, but neither of them will change with time ($d\Phi/dt = \partial \Phi/\partial t = 0$). Now, begin to walk at a velocity q_x in the x-direction, while holding the thermometer. The temperature you measure will change with time, even though the temperature does not change with time at any of the points you pass. Your velocity in this case is $q_x = dx/dt$, and so the second term in the equation above arises from the velocity of the observer. In general, if the observer also moves in the y and z-directions, with $q_y = dy/dt$ and $q_z = dz/dt$:

$$\frac{d\Phi}{dt} = \frac{\partial \Phi}{\partial t} + q_x\frac{\partial \Phi}{\partial x} + q_y\frac{\partial \Phi}{\partial y} + q_z\frac{\partial \Phi}{\partial z}. \tag{7.35}$$

The last three terms can be written as the dot product of the velocity and the gradient of temperature:

$$\frac{d\Phi}{dt} = \frac{\partial \Phi}{\partial t} + \vec{q} \cdot \vec{\nabla}\Phi, \tag{7.36}$$

where:

$$\vec{q} = q_x\vec{i} + q_y\vec{j} + q_z\vec{k}, \tag{7.37}$$

and $\vec{\nabla}$ is the del operator:

$$\vec{\nabla} = \vec{i}\frac{\partial}{\partial x} + \vec{j}\frac{\partial}{\partial y} + \vec{k}\frac{\partial}{\partial z}. \tag{7.38}$$

If an observer moves in a direction perpendicular to the temperature gradient, no variation in temperature will be associated with the movement. The greatest variations will be detected when the observer's velocity is in the same direction as the temperature gradient.

Consider now the case where the observer is moving with the local fluid velocity, $\vec{q} = \vec{v}$. This is a special case of the time derivative, known as the *substantial derivative*, and is denoted with an upper case D, DΦ/Dt:

$$\frac{D\Phi}{Dt} = \frac{\partial \Phi}{\partial t} + \vec{v} \cdot \vec{\nabla} \Phi. \tag{7.39}$$

The interpretation of the substantial derivate is that it is the total derivative of Φ at a point in the fluid as the point moves with the local fluid velocity.

7.7 Modified Pressure, $\wp$

In Chap. 5, we introduced the concept of the modified pressure, which accounts for the effects of both pressure and gravitational forces per unit volume. In rectangular coordinates, this was defined (5.50):

$$\wp \equiv P - \rho g_x x - \rho g_y y - \rho g_z z. \tag{7.40}$$

Consequently, the gradient of the modified pressure can be written:

$$\vec{\nabla} \wp = \vec{\nabla} P - \rho \vec{g}. \tag{7.41}$$

The three components of this vector equation are:

$$\frac{\partial \wp}{\partial x} = \frac{\partial P}{\partial x} - \rho g_x, \quad \frac{\partial \wp}{\partial y} = \frac{\partial P}{\partial y} - \rho g_y, \quad \frac{\partial \wp}{\partial z} = \frac{\partial P}{\partial z} - \rho g_z. \tag{7.42}$$

These are the same as the pressure and gravitational terms found in (7.33). Therefore, the effects of pressure and gravitational forces can be combined in the momentum equation by introducing the modified pressure.

7.8 Equations of Motion for Newtonian Fluids

Using the definition of the substantial derivative and the modified pressure, the components of the momentum equation for a Newtonian fluid simplify to:

$$x\text{-component: } \rho \frac{Dv_x}{Dt} = \mu \nabla^2 v_x - \frac{\partial \wp}{\partial x}, \tag{7.43}$$

$$y\text{-component: } \rho \frac{Dv_y}{Dt} = \mu \nabla^2 v_y - \frac{\partial \wp}{\partial y}, \tag{7.44}$$

7.8 Equations of Motion for Newtonian Fluids

$$z\text{-component:}\quad \rho\frac{\mathrm{D}v_z}{\mathrm{D}t} = \mu\nabla^2 v_z - \frac{\partial\wp}{\partial z}, \tag{7.45}$$

where ∇^2 is the Laplacian operator:

$$\nabla^2 = \vec{\nabla}\cdot\vec{\nabla} = \frac{\partial^2}{\partial x^2} + \frac{\partial^2}{\partial y^2} + \frac{\partial^2}{\partial z^2}. \tag{7.46}$$

The three components of the Navier–Stokes equation can be written as a single vector equation, which is independent of the coordinate system used:

$$\rho\frac{\mathrm{D}\vec{v}}{\mathrm{D}t} = \mu\nabla^2\vec{v} - \vec{\nabla}\wp. \tag{7.47}$$

This is a simple statement of Newton's second law of mechanics for a fluid. The mass times acceleration per unit volume of a fluid element is equal to the sum of forces per unit volume acting on the fluid element. The forces per unit volume arise from viscous stresses, pressure, and gravity.

The equation of continuity can also be written in a more compact form with the help of the substantial derivative:

$$\frac{\mathrm{D}\rho}{\mathrm{D}t} + \rho(\vec{\nabla}\cdot\vec{v}) = 0. \tag{7.48}$$

This form of the continuity equation is independent of the coordinate system adopted for a particular problem. If the fluid is incompressible, the continuity equation reduces to:

$$\vec{\nabla}\cdot\vec{v} = 0\quad\text{(incompressible).} \tag{7.49}$$

Equations (7.47) and (7.49) can be used to find the velocity and pressure distribution in incompressible Newtonian fluids. The continuity equation for compressible fluids (7.48) and an additional equation of state, such as the ideal gas law, would need to accompany the Navier–Stokes equation if problems involving compressible fluids in isothermal systems are to be solved. For nonisothermal systems, the energy equation must also be introduced. In most physiological applications, the assumption of an isothermal fluid is appropriate, and compressibility effects are generally quite small. Therefore, (7.47) and (7.49) are often used to describe biological and physiological problems that involve incompressible, Newtonian fluids. Examples are provided in the next sections. Equation (7.49) is applicable to incompressible, non-Newtonian biological fluids. However, (7.47) cannot be used to describe such fluids. Methods for approaching the flow of non-Newtonian fluids are discussed near the end of this chapter.

7.9 The Stream Function and Streamlines for Two-Dimensional Incompressible Flow

In two-dimensional, incompressible flow with $v_z = 0$, a *stream function* $\Psi(x, y)$ can be defined, which automatically satisfies the continuity equation:

$$v_x = \frac{\partial \Psi}{\partial y}, \qquad (7.50)$$

$$v_y = -\frac{\partial \Psi}{\partial x}. \qquad (7.51)$$

This can be checked by substituting (7.50) and (7.51) back into (7.49):

$$\frac{\partial v_x}{\partial x} + \frac{\partial v_y}{\partial y} = \frac{\partial}{\partial x}\left(\frac{\partial \Psi}{\partial y}\right) + \frac{\partial}{\partial y}\left(-\frac{\partial \Psi}{\partial x}\right) = 0. \qquad (7.52)$$

How should we interpret the stream function? Let us begin by writing the differential of Ψ, and inserting (7.50) and (7.51):

$$d\Psi = \frac{\partial \Psi}{\partial x}dx + \frac{\partial \Psi}{\partial y}dy = -v_y dx + v_x dy. \qquad (7.53)$$

Now consider a line where the stream function is constant, with $\Psi = \Psi_0$. Along this line $d\Psi = 0$, or:

$$0 = -v_y dx + v_x dy, \qquad (7.54)$$

rearranging this expression:

$$\left(\frac{dy}{dx}\right)\bigg|_{\Psi=\Psi_0} = \frac{v_y}{v_x}. \qquad (7.55)$$

But the ratio v_y/v_x is simply the tangent of the angle between the x-axis and the velocity vector. Therefore, the slope of a line with constant Ψ at a point has the same slope as the velocity vector at that point. Such a line is called a *streamline*. The tangent of the streamline at any point provides the direction of the velocity vector at that point (Fig. 7.8). Simply looking at a contour plot of equidistant streamlines provides a spatial view of the velocity field. The distance between streamlines has physical significance. Since a streamline is parallel to the local velocity vector, fluid will not flow across a streamline. Therefore, the flow rate between any two streamlines is the same at all axial locations. When two streamlines

7.9 The Stream Function and Streamlines for Two-Dimensional Incompressible Flow

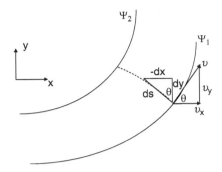

Fig. 7.8 Geometric relationship between ds and v

get closer together, the fluid velocity increases, and when the streamlines diverge the velocity decreases.

The dimensions of the stream function represent an area per unit time, or a volumetric flow per unit channel width. The flow per unit width Q'_V between two streamlines Ψ_1 and Ψ_2 can be found by integrating the velocity along any path s that is everywhere perpendicular to the velocity vector, as shown in Fig. 7.8:

$$Q'_V = \int_{\Psi_1}^{\Psi_2} v \, ds. \tag{7.56}$$

Figure 7.8 shows the following geometric relationships between the path s and velocity v:

$$\sin\theta = -\frac{dx}{ds} = \frac{v_y}{v}, \tag{7.57}$$

$$\cos\theta = \frac{dy}{ds} = \frac{v_x}{v}. \tag{7.58}$$

Dividing (7.53) by ds and introducing (7.57) and (7.58):

$$\frac{d\Psi}{ds} = -v_y \frac{dx}{ds} + v_x \frac{dy}{ds} = -v_y \left(\frac{-v_y}{v}\right) + v_x \left(\frac{v_x}{v}\right) = \frac{v_x^2 + v_y^2}{v} = v. \tag{7.59}$$

Therefore, the flow per unit width between two streamlines (7.56) is equal to the difference in the stream function associated with each streamline:

$$Q'_V = \int_{\Psi_1}^{\Psi_2} v \, ds = \int_{\Psi_1}^{\Psi_2} d\Psi = \Psi_2 - \Psi_1. \tag{7.60}$$

7.10 Use of Navier–Stokes Equations in Rectangular Coordinates

In this section, we will consider a few examples of how to reduce the general Navier–Stokes equations for fluid motion, (7.47) and (7.48), under specific sets of conditions and with certain assumptions.

7.10.1 Hydrostatics

Consider the case where the fluid is an incompressible liquid and there is no fluid motion. Interstitial fluid in most tissues, for instance, moves very slowly and can be considered to be nearly stationary for fairly long periods of time. In such a case, all velocity components and their derivatives are zero. Every term in the continuity equation is identically zero, and the equations of motion reduce to:

$$\frac{\partial \wp}{\partial x} = \frac{\partial \wp}{\partial y} = \frac{\partial \wp}{\partial z} = 0, \qquad (7.61)$$

where $\wp$ is the modified pressure discussed in Sect. 7.7. The general solution to this equation is $\wp =$ constant. Thus, if we know the pressure P^* at some point (x^*, y^*, z^*), then the general solution for pressure P at any point (x, y, z) is:

$$P - P^* = \rho g_x(x - x^*) + \rho g_y(y - y^*) + \rho g_z(z - z^*). \qquad (7.62)$$

This can be used to estimate interstitial pressure at any point in the tissue, knowing the pressure at another point.

> **Example 7.10.1 Calibration of a Pressure Transducer.**
> A practical application of hydrostatics is the use of a manometer to calibrate a pressure transducer. A typical setup is shown in Fig. 7.9. The transducer is used to monitor arterial pressure when valve 1 is connected to a catheter that is inserted into an artery. The catheter and the connection between the two valves are filled with saline. The transducer can be calibrated by turning valve 1 so the transducer is connected to a monometer at a known pressure. Air is introduced into the manometer via valve 2 at the top of a U-tube partially filled with water or mercury. The imposed pressure forces the fluid column downward on the right and upward on the left until equilibrium occurs. The transducer can be calibrated by turning valve 2 until the manometer with a known pressure is connected to the pressure transducer. How is the pressure in the manometer related to the height of fluid in the manometer columns?

7.10 Use of Navier–Stokes Equations in Rectangular Coordinates

Fig. 7.9 Calibration system for pressure transducer

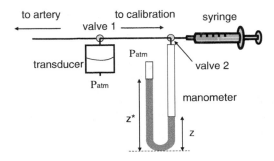

Solution. *Initial considerations:* Our goal is to impose a known pressure to the transducer so the output voltage can be associated with the pressure. Calibration of a transducer requires imposition of at least two known pressures. The procedure we provide above must be repeated for each measurement. When valve 2 is opened to the manometer, there could be a flow of air from the manometer toward the transducer (or vice-versa, depending on the initial pressure in the transducer). Consequently, with the loss (or gain) of air, the final heights of the two columns may not be the same before valve 2 is turned as they are after the valve is turned. The heights of the two columns after the valve is turned should be used in the pressure calculation.

System definition and environmental interactions: Since fluid is stationary during calibration and the connection between the two valves is horizontal, the pressure at the transducer is the same as the pressure in the air on the right side of the manometer. This pressure can be found by analyzing the fluid in the manometer, which is our system of interest.

Apprising the problem to identify governing relationships: Equation (7.62) is the general expression for determining pressure in a static fluid. We are interested in the pressure that must be applied to the right column by the air in contact with fluid in the column.

Analysis: If we take the z-axis as positive in the upward direction, with origin at the bottom of the manometer, then $g_x = g_y = 0$, and $g_z = -g$. The pressure at the surface of the column at the left ($z = z^*$) is atmospheric pressure, P_{atm}. The pressure at the surface of the column at the right ($z = z$) is the pressure imposed by the air, P. Thus, from (7.62), the gauge pressure imposed on the transducer is:

$$P - P_{atm} = \rho g(z^* - z) = \rho g \Delta z.$$

Examining and interpreting the results: Once calibrated, valve 1 is turned so the transducer is connected to an arterial catheter to measure the blood pressure. The use of manometers to measure the pressure led to early adoption of units such as "mm Hg" when the fluid is mercury and Δz is measured in millimeters, or "cm H_2O" when the fluid is water and Δz is measured in centimeters.

7.10.2 Reduction of the Equations of Motion

The continuity equation and the Navier–Stokes equation can be used to find velocity and pressure as a function of position and time for the isothermal, laminar flow of an incompressible, Newtonian fluid. In addition to the governing equations of motion, we must also provide sufficient initial conditions and boundary conditions before the pressure and velocity distributions can be found. Even when the problem is well posed, it is often not possible to find an analytic solution, so numerical computations might need to be employed. However, it is always beneficial to remove terms in the Navier–Stokes equations that are identically zero or negligible before attempting to find a solution. We provide examples in this section that illustrate the process of reducing the number of terms in the Navier–Stokes equations in rectangular coordinates on the basis of assumptions made about the nature of the flow and the requirement that the continuity equation be satisfied. Identifying additional terms that are small can be found with the process of scaling, which will be presented in Sect. 7.13.

> **Example 7.10.2 Entry Flow of a Newtonian Fluid Between Parallel Plates.**
> Many extracorporeal devices require that blood pass between one or more parallel plates in the device. Reduce the continuity equation and Navier–Stokes equations for the steady flow of an incompressible Newtonian fluid in the entry region between two parallel plates. Assume fluid velocity at $x = 0$ is uniform ($v_x = V$, $v_y = 0$), and far downstream the velocity profile is parabolic.

Solution. *Initial considerations:* We begin by making a list of what we know about the problem along with all of the assumptions we will make. It is useful to number our list for easy reference later. Let us postulate the following:

1. Flow is steady-state: $\partial()/\partial t = 0$, where () refers to any component of velocity.
2. The plates are very wide, so there are no changes in the z-direction: $\partial()/\partial z = 0$
3. There is no flow in z-direction: $v_z = 0$.
4. The fluid is incompressible and Newtonian

System definition and environmental interactions: The system consists of the fluid bounded by two parallel walls, one at the top ($y = h/2$) and the other at bottom ($y = -h/2$). Because of the no-slip condition, both components of the velocity are zero at $y = \pm h/2$ and all values of x. The velocity profile is assumed to be flat at $x = 0$ and all values of y.

Apprising the problem to identify governing relationships: The continuity equation and the three components of the Navier–Stokes equation are the governing relationships for developing flow between parallel plates.

7.10 Use of Navier–Stokes Equations in Rectangular Coordinates

Analysis: Since the fluid is incompressible (condition 4), the incompressible continuity equation will be our starting point. Conditions (2) and (3) reduce (7.49) to:

$$\frac{\partial v_x}{\partial x} + \frac{\partial v_y}{\partial y} = 0,$$

where v_x and v_y are both functions of x and y. Applying the conditions above to the Navier–Stokes equations, we can eliminate many of the terms in the general equations:

X – momentum

$$\rho \cancel{\frac{\partial v_x}{\partial t}}^1 = -\rho\left[v_x \frac{\partial v_x}{\partial x} + v_y \frac{\partial v_x}{\partial y} + \cancel{v_z \frac{\partial v_x}{\partial z}}^{3,2}\right] + \mu\left[\frac{\partial^2 v_x}{\partial x^2} + \frac{\partial^2 v_x}{\partial y^2} + \cancel{\frac{\partial^2 v_x}{\partial z^2}}^2\right] - \frac{\partial \wp}{\partial x}$$

Y – momentum

$$\rho \cancel{\frac{\partial v_y}{\partial t}}^1 = -\rho\left[v_x \frac{\partial v_y}{\partial x} + v_y \frac{\partial v_y}{\partial y} + \cancel{v_z \frac{\partial v_y}{\partial z}}^{3,2}\right] + \mu\left[\frac{\partial^2 v_y}{\partial x^2} + \frac{\partial^2 v_y}{\partial y^2} + \cancel{\frac{\partial^2 v_y}{\partial z^2}}^2\right] - \frac{\partial \wp}{\partial y}$$

Z – momentum

$$\rho \cancel{\frac{\partial v_z}{\partial t}}^{1,2} = -\rho\left[\cancel{v_x \frac{\partial v_z}{\partial x}}^2 + \cancel{v_y \frac{\partial v_z}{\partial y}}^2 + \cancel{v_z \frac{\partial v_z}{\partial z}}^{2,3}\right] + \mu\left[\cancel{\frac{\partial^2 v_z}{\partial x^2}}^2 + \cancel{\frac{\partial^2 v_z}{\partial y^2}}^2 + \cancel{\frac{\partial^2 v_z}{\partial z^2}}^{2,3}\right] - \frac{\partial \wp}{\partial z}$$

Therefore, the final set of equations to be solved is:

$$\frac{\partial v_x}{\partial x} + \frac{\partial v_y}{\partial y} = 0,$$

$$\rho\left[v_x \frac{\partial v_x}{\partial x} + v_y \frac{\partial v_x}{\partial y}\right] = \mu\left[\frac{\partial^2 v_x}{\partial x^2} + \frac{\partial^2 v_x}{\partial y^2}\right] - \frac{\partial \wp}{\partial x},$$

$$\rho\left[v_x \frac{\partial v_y}{\partial x} + v_y \frac{\partial v_y}{\partial y}\right] = \mu\left[\frac{\partial^2 v_y}{\partial x^2} + \frac{\partial^2 v_y}{\partial y^2}\right] - \frac{\partial \wp}{\partial y},$$

$$0 = \frac{\partial \wp}{\partial z}.$$

The x and y-components of the Navier–Stokes equations are nonlinear because of the presence of the convective acceleration terms on the left-hand side of the equations. The z-component indicates that the modified pressure depends at most on x and y.

Specification of the governing equations meets only half of our objectives. If we are to find a unique solution to this set of equations, we must specify all of the appropriate boundary conditions. Before reading on, pause to see whether you can specify the number of boundary conditions that are necessary to solve this set of equations, and why that number is necessary. Then, write out each boundary condition. Refrain from reading on until you have finished this exercise.

Okay, let us see how you did. The velocities v_x and v_y are both second order with respect to x and y. We must specify two boundary conditions in x and two boundary conditions in y for each. In addition the modified pressure is first order with respect to x and y, so we must specify one boundary condition in x and one boundary condition in y for pressure. That corresponds to ten total conditions which must be specified. Consider first the boundary conditions that are valid at all values of x, but particular values of y:

1. v_x: no slip at $y = +h/2$,	$v_x(x, +h/2) = 0$
2. v_x: no slip at $y = -h/2$,	$v_x(x, -h/2) = 0$
3. v_y: no slip at $y = +h/2$,	$v_y(x, +h/2) = 0$
4. v_y: no slip at $y = -h/2$,	$v_y(x, -h/2) = 0$
5. $\wp$: by symmetry about $y = 0$,	$(\partial \wp / \partial y)\|_{y=0} = 0$

Next, consider the boundary conditions that are valid at all values of y for particular values of x:

6. v_x: constant inlet velocity,	$v_x(0, y) = V$
7. v_x: fully developed,	$(\partial v_x / \partial x)\|_{x \to \infty} = 0$
8. v_y: zero at inlet,	$v_y(0, y) = 0$
9. v_y: zero far downstream,	$v_y(\infty, y) = 0$
10. $\wp$: pressure known at inlet,	$\wp(0, y) = \wp_0$

Examining and interpreting the results: The problem is now uniquely defined. This problem can be simplified further through the use of scaling, as we will see later in this chapter. Although our objective was to set the problem up, and not to show details of its solution, we will provide a brief explanation of its solution. One approach to solving problems, where v_x and v_y are each functions of x and y is to introduce a stream function $\Psi(x, y)$, defined by (7.50) and (7.51). Use of the stream function automatically satisfies the continuity equation. Pressure can be eliminated mathematically by taking the derivative of the x-component of the Navier–Stokes equation with respect to y and the derivative of the y-component with respect to x, then equating the resulting $\partial^2 \wp / \partial x \partial y$ terms, which will appear in each equation. This produces a single partial differential equation for $\Psi(x, y)$ which is fourth order in x and fourth order in y. The eight boundary conditions necessary for solving the resulting PDE can be found by transforming the boundary conditions for v_x and v_y to boundary conditions in Ψ. A numerical method, such as the finite difference method or finite element method, can be used to solve for the stream function, and equations (7.50) and (7.51) used to find $v_x(x, y)$ and $v_y(x, y)$. Finally, the pressure distribution can be found by numerical integration of the Navier–Stokes equations. The solution will depend on the values of V, ρ, μ, and h, but in fact depends only on the combination of these variables $\rho V h / \mu$, which is the Reynolds number.

Solutions for $v_x(x, y)$ for various Reynolds numbers from Darbandi and Hosseinizadeh (2004) are shown in Fig. 7.10. Note that a region near the wall initially accelerates more rapidly than at the center, but ultimately, as $x^+ = x/h$ becomes

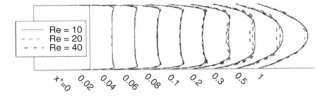

Fig. 7.10 Axial velocity profiles vs. $x^* = x/h$ in entrance region between parallel plates for three different Reynolds numbers, from Darbandi et al. 2004 with permission

Fig. 7.11 Centerline pressure in entrance region between parallel plates ($Re = 20$), from Mokheimer 2002 with permission

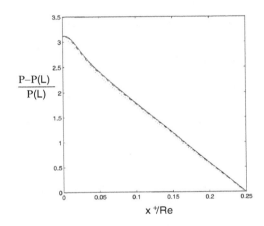

large, the velocity profiles at all Reynolds numbers reduce to the same parabolic profile. The entry length L_e is longer for higher Reynolds numbers:

$$L_e = 0.04 h Re_h$$

The centerline pressure for $Re = 20$ is shown in Fig 7.11. This is steeper in the entry region than in the fully developed region of the duct because of the higher velocity gradients at the walls.

> **Example 7.10.3 Flow of a Newtonian Fluid in a Vessel with Elliptical Cross Section.**
> Blood vessels can be compressed by surrounding soft tissue, thus altering the cross section and resistance to flow. Simplify the equations of motion for steady-state, fully developed, laminar flow of an incompressible Newtonian fluid through a horizontal vessel having length L, and an elliptical cross-section with major axis $2c$ and minor axis $2b$ as shown in Fig. 7.12. Derive expressions for the velocity profile and flow rate in the vessel, and compare the flow to Poiseuille flow through a circular vessel with the same perimeter.

Fig. 7.12 Vessel with elliptical cross section

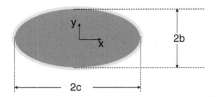

Derive a relationship for the factor M_0 for an ellipse as a function of the ratio of minor to major axes in (5.83) and show it is consistent with the line for an ellipse in Fig. 5.21. The vessel geometry (b, c, L), fluid properties (ρ, μ), and upstream and downstream pressures (P_0, P_L) are all assumed to be known quantities.

Solution. *Initial considerations:* We begin by making a list of appropriate assumptions:

1. Fluid is Newtonian and incompressible, so $\rho = $ constant.
2. Flow is steady, so $\partial()/\partial t = 0$.
3. Fully developed flow, so $v_x = 0$.
4. Fully developed flow, so $v_y = 0$.
5. Neglect effects of gravity (or use $\wp$ instead of P): $g_x = g_y = g_z = 0$.

System definition and environmental interactions: The system consists of blood inside the elliptically shaped vessel. The velocity of blood at the vessel wall is zero.

Apprising the problem to identify governing relationships: Since blood is assumed to be Newtonian and incompressible, the Navier–Stokes equation is applicable. We must also make use of the continuity relationship.

Analysis: Simplifying the continuity equation under these conditions:

$$\cancel{\frac{\partial \rho}{\partial t}}^{1,2} + \rho\left[\cancel{\frac{\partial v_x}{\partial x}}^{3} + \cancel{\frac{\partial v_y}{\partial y}}^{4} + \frac{\partial v_z}{\partial z}\right] + \left[\cancel{v_x \frac{\partial \rho}{\partial x}}^{1,3} + \cancel{v_y \frac{\partial \rho}{\partial y}}^{1,4} + \cancel{v_z \frac{\partial \rho}{\partial z}}^{1}\right] = 0$$

This leads to a sixth condition:

6. $\partial v_z/\partial z = 0$

Coupling this with conditions (2)–(4) above, the only nonzero component of velocity is v_z, and since v_z cannot depend on time or axial position, then $v_z = v_z(x,y)$.

Turning next to the x-component of the Navier–Stokes equation:

$$\cancel{\rho\frac{\partial v_x}{\partial t}}^{2,3} + \rho\left[\cancel{v_x\frac{\partial v_x}{\partial x}}^{3} + \cancel{v_y\frac{\partial v_x}{\partial y}}^{4,3} + \cancel{v_z\frac{\partial v_x}{\partial z}}^{3}\right] = \mu\left[\cancel{\frac{\partial^2 v_x}{\partial x^2}}^{3} + \cancel{\frac{\partial^2 v_x}{\partial y^2}}^{3} + \cancel{\frac{\partial^2 v_x}{\partial z^2}}^{3}\right] - \frac{\partial P}{\partial x} + \cancel{\rho g_x}^{5}$$

7.10 Use of Navier–Stokes Equations in Rectangular Coordinates

This provides a seventh condition:

7. $\partial P/\partial x = 0$

Application of our assumptions to the y-component of the Navier–Stokes equation:

$$\cancel{\rho \frac{\partial v_y}{\partial t}}^{2,4} + \rho \left[\cancel{v_x \frac{\partial v_y}{\partial x}}^{3\;4} + \cancel{v_y \frac{\partial v_y}{\partial y}}^{4} + \cancel{v_z \frac{\partial v_y}{\partial z}}^{4} \right] = \mu \left[\cancel{\frac{\partial^2 v_y}{\partial x^2}}^{4} + \cancel{\frac{\partial^2 v_y}{\partial y^2}}^{4} + \cancel{\frac{\partial^2 v_y}{\partial z^2}}^{4} \right] - \frac{\partial P}{\partial y} + \cancel{\rho g_y}^{5}$$

which simplifies to:

8. $\partial P/\partial y = 0$.

Finally, the z-component of the Navier–Stokes equation simplifies as follows:

$$\cancel{\rho \frac{\partial v_z}{\partial t}}^{2} + \rho \left[\cancel{v_x \frac{\partial v_z}{\partial x}}^{3} + \cancel{v_y \frac{\partial v_z}{\partial y}}^{4} + \cancel{v_z \frac{\partial v_z}{\partial z}}^{6} \right] = \mu \left[\frac{\partial^2 v_z}{\partial x^2} + \frac{\partial^2 v_z}{\partial y^2} + \cancel{\frac{\partial^2 v_z}{\partial z^2}}^{6} \right] - \frac{\partial P}{\partial z} + \cancel{\rho g_z}^{5}$$

Conditions (2), (7) and (8), taken together indicate that the pressure can depend only on z, so $P = P(z)$. If gravity cannot be neglected, we can simply replace $P(z)$ with the modified pressure, $\wp(z)$. The z-component reduces to:

$$\frac{\partial^2 v_z}{\partial x^2} + \frac{\partial^2 v_z}{\partial y^2} = \frac{1}{\mu} \frac{dP}{dz}.$$

The left-hand side of this expression is a function of x and y, while the right-hand side is a function of z. This equality can only be valid if both sides are constant. The right-hand side of the expression becomes:

$$\frac{1}{\mu} \frac{dP}{dz} = C.$$

Integration from $z = 0$ to $z = L$, we find the constant C in terms of the difference between the upstream and downstream pressure, ΔP:

$$C = \frac{P_L - P_0}{\mu L} = -\frac{\Delta P}{\mu L}.$$

Substituting this back into the z-component of the Navier–Stokes equation:

$$\frac{\partial^2 v_z}{\partial x^2} + \frac{\partial^2 v_z}{\partial y^2} = -\frac{\Delta P}{\mu L}.$$

There are many different functions that can satisfy this partial differential equation. One of these, which can be confirmed by inspection, is shown below:

$$v_z(x,y) = A_0 + A_1 x + A_2 x^2 + A_3 y + A_4 y^2 + A_5 xy,$$

where the six coefficients A_0 to A_5 are constants. Although this expression satisfies the differential equation, we seek a solution that also satisfies the appropriate boundary conditions. Since the velocity is symmetrical with respect to x and y at the center of the elliptical vessel:

$$\left.\frac{\partial v_z}{\partial x}\right|_{x=0} = 0 = A_1 + A_5 y,$$

$$\left.\frac{\partial v_z}{\partial y}\right|_{y=0} = 0 = A_3 + A_5 x.$$

Consequently, $A_1 = A_3 = A_5 = 0$, and the expression for velocity becomes:

$$v_z(x,y) = A_0 + A_2 x^2 + A_4 y^2.$$

The no-slip boundary condition states that $v_z = 0$ on the vessel boundary, which is governed by the equation of an ellipse:

$$\frac{x^2}{c^2} + \frac{y^2}{b^2} = 1.$$

Setting $v_z = 0$ and dividing by A_0:

$$0 = 1 + \frac{A_2}{A_0} x^2 + \frac{A_4}{A_0} y^2.$$

The last two expressions will be identical when:

$$A_2 = -\frac{A_0}{c^2},$$

$$A_4 = -\frac{A_0}{b^2}.$$

The velocity can now be expressed in terms of a single constant A_0:

$$v_z(x,y) = A_0 \left(1 - \frac{x^2}{c^2} - \frac{y^2}{b^2}\right).$$

Taking second derivatives with respect to x and y, and substituting them back into the z-component of the Navier–Stokes equation, we find A_0

7.10 Use of Navier–Stokes Equations in Rectangular Coordinates

$$A_0 = \frac{\Delta P}{2\mu L}\left(\frac{b^2 c^2}{b^2 + c^2}\right).$$

Therefore, the final expression for the velocity profile in the elliptical tube is:

$$v_z(x,y) = \frac{\Delta P}{2\mu L}\left(\frac{b^2 c^2}{b^2 + c^2}\right)\left(1 - \frac{x^2}{c^2} - \frac{y^2}{b^2}\right).$$

Examining and interpreting the results: Note that the expression for velocity satisfies the condition that the velocity must be zero at the boundary and also satisfies the symmetry boundary conditions. A graphical representation of the velocity profile is shown in Fig. 7.13.

The flow through the elliptically shaped vessel can be computed from:

$$Q_V = \int_A v_z(x,y)\,dA = \frac{\Delta P}{2\mu L}\left(\frac{b^2 c^2}{b^2 + c^2}\right)\int_{-b}^{+b}\left\{\int_{-c\sqrt{1-(y/b)^2}}^{+c\sqrt{1-(y/b)^2}}\left(1 - \frac{x^2}{c^2} - \frac{y^2}{b^2}\right)dx\right\}dy.$$

After performing the double integration, we find:

$$Q_V = \frac{\pi}{8}\left(\frac{\Delta P}{\mu L}\right)\left(\frac{2b^3 c^3}{b^2 + c^2}\right).$$

Note that this reduces to the Hagen–Poiseuille equation for flow in a round tube of radius R when $a = b = R$. The perimeter of an ellipse p is approximately related to the minor and major axes as follows:

$$p \approx 2\pi\sqrt{\frac{b^2 + c^2}{2}}.$$

Therefore, a circular tube with the same perimeter as an elliptical tube will have a radius R:

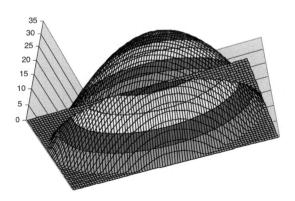

Fig. 7.13 Velocity profile for flow in an elliptical tube with $c = 2b$

$$R = \sqrt{\frac{b^2 + c^2}{2}}.$$

Flow Q_{VP} through a circular tube with the same perimeter as the elliptical tube will obey the Hagen–Poiseuille equation:

$$Q_{VP} = \frac{\pi}{8}\left(\frac{\Delta P}{\mu L}\right)\left(\frac{b^2 + c^2}{2}\right)^2.$$

Flow through an elliptical tube relative to flow through a circular tube with the same perimeter is:

$$\frac{Q_V}{Q_{VP}} = 8\left(\frac{b^3 c^3}{(b^2 + c^2)^3}\right) = \frac{8(c/b)^3}{\left[1 + (c/b)^2\right]^3}.$$

A semilog plot of Q_V/Q_{VP} relative to the ratio of major to minor axes of the ellipse is shown in Fig. 7.14. Note that the maximum flow occurs when the cross-section is circular and the flow drops off dramatically as c/b increases or decreases. If $c/b = 2$, the flow drops to 51% of the flow through a circular tube and if $c/b = 4$, the flow is reduced to only 10% of the flow through a circular tube. Therefore, the resistance of a distensible tube with constant perimeter can be greatly increased if the vessel is squeezed into an elliptical shape by surrounding tissue.

The resistance to flow in an elliptical vessel can be written:

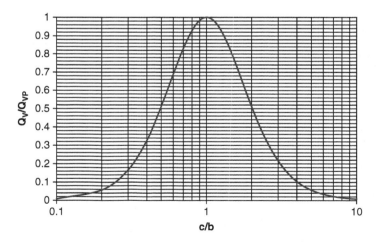

Fig. 7.14 Flow through an elliptical tube relative to flow through a circular tube as a function of the ratio of major to minor axis of the ellipse

$$\Re = \frac{\Delta P}{Q_V} = \frac{4\mu L}{\pi}\left(\frac{b^2 + c^2}{b^3 c^3}\right).$$

This can be compared to the general result presented in Chap. 5 for laminar resistance in a tube of elliptical cross section with $B = 2c$ and $d = 2b$:

$$\Re = \frac{12\mu L}{(2c)(2b)^3 M_0} = \frac{3\mu L}{4cb^3 M_0}.$$

From which we derive the following expression for M_0:

$$M_0 = \frac{3\pi}{16}\left(\frac{1}{1 + \left(\frac{b}{c}\right)^2}\right) \quad \text{(ellipse)}.$$

It can be easily confirmed that the line for the ellipse in Fig. 5.21 is described by the above expression, with limits of $3\pi/16$ for $b/c = d/B = 0$ and $3\pi/32$ for $b/c = d/B = 1$.

7.11 Navier–Stokes Equations in Cylindrical and Spherical Coordinate Systems

One advantage to writing the equations of motion in the forms shown in Sect. 7.8 is that the equations are independent of the coordinate system selected. One simply needs to know the expressions for D/Dt, $\vec{\nabla}$ and ∇^2 in the coordinate system of interest. These are shown in Table 7.1 for cylindrical and spherical coordinate systems, as defined in Fig. 7.15. Note that the definitions of r and θ are different in cylindrical and spherical coordinates.

Unit vectors in cylindrical coordinates $\vec{e}_r, \vec{e}_\theta, \vec{e}_z$ are related to $\vec{e}_x, \vec{e}_y, \vec{e}_z$ (or $\vec{i}, \vec{j}, \vec{k}$) in rectangular coordinates as follows:

$$\begin{aligned}\vec{e}_r &= \cos\theta\vec{e}_x + \sin\theta\vec{e}_y \\ \vec{e}_\theta &= -\sin\theta\vec{e}_x + \cos\theta\vec{e}_y \\ \vec{e}_z &= \vec{e}_z.\end{aligned} \quad (7.63)$$

Since $\vec{e}_x, \vec{e}_y, \vec{e}_z$ are all independent of r and z, and $\vec{e}_z$ is also independent of θ, we can write:

$$\frac{\partial \vec{e}_r}{\partial z} = \frac{\partial \vec{e}_\theta}{\partial z} = \frac{\partial \vec{e}_z}{\partial z} = \frac{\partial \vec{e}_r}{\partial r} = \frac{\partial \vec{e}_\theta}{\partial r} = \frac{\partial \vec{e}_z}{\partial r} = \frac{\partial \vec{e}_z}{\partial \theta} = 0. \quad (7.64)$$

It is important to realize that $\vec{e}_r$ and $\vec{e}_\theta$ are both functions of θ. Unlike the other spatial derivates, derivates of these two unit vectors with respect to θ are not zero:

Table 7.1 Definitions of the del operator, the substantial derivative, and the Laplacian operator in cylindrical and spherical coordinate systems

Cylindrical	Spherical
$\vec{\nabla} = \vec{e}_r \dfrac{\partial}{\partial r} + \vec{e}_\theta \dfrac{1}{r}\dfrac{\partial}{\partial \theta} + \vec{e}_z \dfrac{\partial}{\partial z}$	$\vec{\nabla} = \vec{e}_r \dfrac{\partial}{\partial r} + \vec{e}_\theta \dfrac{1}{r}\dfrac{\partial}{\partial \theta} + \vec{e}_\phi \dfrac{1}{r\sin\theta}\dfrac{\partial}{\partial \phi}$
$\dfrac{D}{Dt} = \dfrac{\partial}{\partial t} + v_r \dfrac{\partial}{\partial r} + \dfrac{v_\theta}{r}\dfrac{\partial}{\partial \theta} + v_z \dfrac{\partial}{\partial z}$	$\dfrac{D}{Dt} = \dfrac{\partial}{\partial t} + v_r \dfrac{\partial}{\partial r} + \dfrac{v_\theta}{r}\dfrac{\partial}{\partial \theta} + \dfrac{v_\phi}{r\sin\theta}\dfrac{\partial}{\partial \phi}$
$\nabla^2 = \dfrac{1}{r}\dfrac{\partial}{\partial r}\left(r\dfrac{\partial}{\partial r}\right) + \dfrac{1}{r^2}\dfrac{\partial^2}{\partial \theta^2} + \dfrac{\partial^2}{\partial z^2}$	$\nabla^2 = \dfrac{1}{r^2}\dfrac{\partial}{\partial r}\left(r^2 \dfrac{\partial}{\partial r}\right) + \dfrac{1}{r^2 \sin\theta}\dfrac{\partial}{\partial \theta}\left(\sin\theta \dfrac{\partial}{\partial \theta}\right)$
	$+ \dfrac{1}{r^2 \sin^2\theta}\dfrac{\partial^2}{\partial \phi^2}$

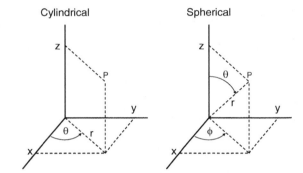

Fig. 7.15 Relationship between cylindrical and rectangular coordinate systems (*left*) and spherical and rectangular coordinate systems (*right*). Note that some texts reverse the definitions of φ and θ

$$\frac{\partial \vec{e}_r}{\partial \theta} = -\sin\theta \vec{e}_x + \cos\theta \vec{e}_y = +\vec{e}_\theta$$
$$\frac{\partial \vec{e}_\theta}{\partial \theta} = -\cos\theta \vec{e}_x - \sin\theta \vec{e}_y = -\vec{e}_r. \quad (7.65)$$

We can use these expressions to write the incompressible continuity equation in cylindrical coordinates:

$$\vec{\nabla} \cdot \vec{v} = \left[\vec{e}_r \frac{\partial}{\partial r} + \vec{e}_\theta \frac{1}{r}\frac{\partial}{\partial \theta} + \vec{e}_z \frac{\partial}{\partial z}\right] \cdot [v_r \vec{e}_r + v_\theta \vec{e}_\theta + v_z \vec{e}_z] = 0. \quad (7.66)$$

Expanding and using the chain rule for the derivative of a product:

$$\vec{\nabla} \cdot \vec{v} = [\vec{e}_r] \cdot \left[\frac{\partial v_r}{\partial r}\vec{e}_r + v_r \frac{\partial \vec{e}_r}{\partial r} + \frac{\partial v_\theta}{\partial r}\vec{e}_\theta + v_\theta \frac{\partial \vec{e}_\theta}{\partial r} + \frac{\partial v_z}{\partial r}\vec{e}_z + v_z \frac{\partial \vec{e}_z}{\partial r}\right]$$
$$+ [\vec{e}_\theta] \cdot \left[\frac{1}{r}\frac{\partial v_r}{\partial \theta}\vec{e}_r + \frac{v_r}{r}\frac{\partial \vec{e}_r}{\partial \theta} + \frac{1}{r}\frac{\partial v_\theta}{\partial \theta}\vec{e}_\theta + \frac{v_\theta}{r}\frac{\partial \vec{e}_\theta}{\partial \theta} + \frac{1}{r}\frac{\partial v_z}{\partial \theta}\vec{e}_z + \frac{v_z}{r}\frac{\partial \vec{e}_z}{\partial \theta}\right]$$
$$+ [\vec{e}_z] \cdot \left[\frac{\partial v_r}{\partial z}\vec{e}_r + v_r \frac{\partial \vec{e}_r}{\partial z} + \frac{\partial v_\theta}{\partial z}\vec{e}_\theta + v_\theta \frac{\partial \vec{e}_\theta}{\partial z} + \frac{\partial v_z}{\partial z}\vec{e}_z + v_z \frac{\partial \vec{e}_z}{\partial z}\right] = 0. \quad (7.67)$$

7.11 Navier–Stokes Equations in Cylindrical and Spherical Coordinate Systems

Taking the dot products and substituting for the spatial derivatives of unit vectors from (7.64) to (7.65):

$$\vec{\nabla} \cdot \vec{v} = \frac{\partial v_r}{\partial r} + \frac{v_r}{r} + \frac{1}{r}\frac{\partial v_\theta}{\partial \theta} + \frac{\partial v_z}{\partial z} = 0. \quad (7.68)$$

The same procedure can be used to convert the continuity equation for compressible fluids or the components of the Navier–Stokes equation into cylindrical or spherical coordinates. An alternate method is to take the expanded equations in rectangular coordinates and transform them term by term into the coordinate system of interest. To do this, you must know how the spatial derivates and velocity components between the coordinate system of interest are related to the rectangular coordinate system. In addition, if the fluid is non-Newtonian, we need to be able to express the stress components in the coordinate system of interest in terms of the stress components in rectangular coordinates. Methods for transforming stress components from Cartesian to cylindrical or spherical coordinates can be found in Richardson (1989). Relationships necessary for this approach are shown for cylindrical and spherical coordinates using the definitions in Fig. 7.15 and Table 7.2.

Table 7.2 Expressions in Cartesian coordinates in terms of cylindrical or spherical components

Cartesian	Cylindrical	Spherical
x	$r\cos\theta$	$r\sin\theta\cos\phi$
y	$r\sin\theta$	$r\sin\theta\sin\phi$
z	z	$r\cos\theta$
$\dfrac{\partial}{\partial x}$	$(\cos\theta)\dfrac{\partial}{\partial r} - \left(\dfrac{\sin\theta}{r}\right)\dfrac{\partial}{\partial \theta}$	$(\sin\theta\cos\phi)\dfrac{\partial}{\partial r} + \left(\dfrac{\cos\theta\cos\phi}{r}\right)\dfrac{\partial}{\partial \theta} - \left(\dfrac{\sin\phi}{r\sin\theta}\right)\dfrac{\partial}{\partial \phi}$
$\dfrac{\partial}{\partial y}$	$(\sin\theta)\dfrac{\partial}{\partial r} + \left(\dfrac{\cos\theta}{r}\right)\dfrac{\partial}{\partial \theta}$	$(\sin\theta\sin\phi)\dfrac{\partial}{\partial r} + \left(\dfrac{\cos\theta\sin\phi}{r}\right)\dfrac{\partial}{\partial \theta} + \left(\dfrac{\cos\phi}{r\sin\theta}\right)\dfrac{\partial}{\partial \phi}$
$\dfrac{\partial}{\partial z}$	$\dfrac{\partial}{\partial z}$	$(\cos\theta)\dfrac{\partial}{\partial r} - \left(\dfrac{\sin\theta}{r}\right)\dfrac{\partial}{\partial \theta}$
v_x	$v_r\cos\theta - v_\theta\sin\theta$	$v_r\sin\theta\cos\phi + v_\theta\cos\theta\cos\phi - v_\phi\sin\phi$
v_y	$v_r\sin\theta + v_\theta\cos\theta$	$v_r\sin\theta\sin\phi + v_\theta\cos\theta\sin\phi + v_\phi\cos\phi$
v_z	v_z	$v_r\cos\theta - v_\theta\sin\theta$
τ_{xx}	$\tau_{rr}\cos^2\theta + \tau_{\theta\theta}\sin^2\theta - 2\tau_{r\theta}\sin\theta\cos\theta$	$[\tau_{rr}\sin^2\theta + \tau_{\theta\theta}\cos^2\theta + 2\tau_{r\theta}\sin\theta\cos\theta]\cos^2\phi$ $-2\sin\phi\cos\phi[\tau_{r\phi}\sin\theta + \tau_{\theta\phi}\cos\theta] + \tau_{\phi\phi}\sin^2\phi$
τ_{yy}	$\tau_{rr}\sin^2\theta + \tau_{\theta\theta}\cos^2\theta + 2\tau_{r\theta}\sin\theta\cos\theta$	$[\tau_{rr}\sin^2\theta + \tau_{\theta\theta}\cos^2\theta + 2\tau_{r\theta}\sin\theta\cos\theta]\sin^2\phi$ $+2\cos\phi\sin\phi[\tau_{r\phi}\sin\theta + \tau_{\theta\phi}\cos\theta] + \tau_{\phi\phi}\cos^2\phi$
τ_{zz}	τ_{zz}	$\tau_{rr}\cos^2\theta + \tau_{\theta\theta}\sin^2\theta - 2\tau_{r\theta}\sin\theta\cos\theta$
τ_{xy}	$\tau_{r\theta}(\cos^2\theta - \sin^2\theta)$ $+ (\tau_{rr} - \tau_{\theta\theta})\sin\theta\cos\theta$	$[\tau_{rr}\sin^2\theta + \tau_{\theta\theta}\cos^2\theta + 2\tau_{r\theta}\sin\theta\cos\theta - \tau_{\phi\phi}]\cos\phi\sin\phi$ $+ (\cos^2\phi - \sin^2\phi)[\tau_{r\phi}\sin\theta + \tau_{\theta\phi}\cos\theta]$
τ_{xz}	$\tau_{rz}\cos\theta - \tau_{\theta z}\sin\theta$	$[\tau_{rr}\cos\phi - \tau_{\theta\theta}\sin\phi]\sin\theta\cos\theta$ $+ \sin\phi[\tau_{\theta\phi}\sin\theta - \tau_{r\phi}\cos\theta]$ $+ (\cos^2\theta - \sin^2\theta)\tau_{r\theta}\cos\phi$
τ_{yz}	$\tau_{rz}\sin\theta + \tau_{\theta z}\cos\theta$	$\sin\phi[(\tau_{rr} - \tau_{\theta\theta})\sin\theta\cos\theta + \tau_{r\theta}(\cos^2\theta - \sin^2\theta)]$ $+ \cos\phi[\tau_{r\phi}\cos\theta - \tau_{\theta\phi}\sin\theta]$

The general continuity equations in cylindrical and spherical coordinates are:

Cylindrical: $\dfrac{\partial \rho}{\partial t} + \dfrac{1}{r}\dfrac{\partial (\rho r v_r)}{\partial r} + \dfrac{1}{r}\dfrac{\partial (\rho v_\theta)}{\partial \theta} + \dfrac{\partial (\rho v_z)}{\partial z} = 0,$ (7.69)

Spherical: $\dfrac{\partial \rho}{\partial t} + \dfrac{1}{r^2}\dfrac{\partial (\rho r^2 v_r)}{\partial r} + \dfrac{1}{r \sin \theta}\dfrac{\partial (\rho v_\theta \sin \theta)}{\partial \theta} + \dfrac{1}{r \sin \theta}\dfrac{\partial (\rho v_\phi)}{\partial \phi} = 0.$ (7.70)

For incompressible fluids, these become:

Cylindrical: $\dfrac{1}{r}\dfrac{\partial (r v_r)}{\partial r} + \dfrac{1}{r}\dfrac{\partial v_\theta}{\partial \theta} + \dfrac{\partial v_z}{\partial z} = 0,$ (7.71)

Spherical: $\dfrac{1}{r^2}\dfrac{\partial (r^2 v_r)}{\partial r} + \dfrac{1}{r \sin \theta}\dfrac{\partial (v_\theta \sin \theta)}{\partial \theta} + \dfrac{1}{r \sin \theta}\dfrac{\partial v_\phi}{\partial \phi} = 0.$ (7.72)

Newton's Law of viscosity in cylindrical coordinates for incompressible fluids is:

$$\tau_{r\theta} = \tau_{\theta r} = -\mu\left(\dfrac{1}{r}\dfrac{\partial v_r}{\partial \theta} + r\dfrac{\partial}{\partial r}\left(\dfrac{v_\theta}{r}\right)\right), \quad \tau_{\theta z} = \tau_{z\theta} = -\mu\left(\dfrac{\partial v_\theta}{\partial z} + \dfrac{1}{r}\dfrac{\partial v_z}{\partial \theta}\right),$$

$$\tau_{rz} = \tau_{zr} = -\mu\left(\dfrac{\partial v_r}{\partial z} + \dfrac{\partial v_z}{\partial r}\right), \quad \tau_{rr} = -2\mu\left(\dfrac{\partial v_r}{\partial r}\right), \quad \tau_{\theta\theta} = -2\mu\left(\dfrac{1}{r}\dfrac{\partial v_\theta}{\partial \theta} + \dfrac{v_r}{r}\right),$$

$$\tau_{zz} = -2\mu\left(\dfrac{\partial v_z}{\partial z}\right).$$

(7.73)

In spherical coordinates, the relation between stress and velocity gradients for an incompressible Newtonian fluid is:

$$\tau_{r\theta} = \tau_{\theta r} = -\mu\left(\dfrac{1}{r}\dfrac{\partial v_r}{\partial \theta} + r\dfrac{\partial}{\partial r}\left(\dfrac{v_\theta}{r}\right)\right), \quad \tau_{\theta\phi} = \tau_{\phi\theta} = -\mu\left(\dfrac{1}{r \sin \theta}\dfrac{\partial v_\theta}{\partial \phi} + \dfrac{\sin \theta}{r}\dfrac{\partial}{\partial \theta}\left(\dfrac{v_\phi}{\sin \theta}\right)\right),$$

$$\tau_{r\phi} = \tau_{\phi r} = -\mu\left(\dfrac{1}{r \sin \theta}\dfrac{\partial v_r}{\partial \phi} + r\dfrac{\partial}{\partial r}\left(\dfrac{v_\phi}{r}\right)\right), \quad \tau_{rr} = -2\mu\left(\dfrac{\partial v_r}{\partial r}\right),$$

$$\tau_{\theta\theta} = -2\mu\left(\dfrac{1}{r}\dfrac{\partial v_\theta}{\partial \theta} + \dfrac{v_r}{r}\right), \quad \tau_{\phi\phi} = -2\mu\left(\dfrac{v_r}{r} + \dfrac{v_\theta \cot \theta}{r} + \dfrac{1}{r \sin \theta}\dfrac{\partial v_\phi}{\partial \phi}\right).$$

(7.74)

For incompressible Newtonian fluids in cylindrical coordinates: the Navier–Stokes equations become:

7.11 Navier–Stokes Equations in Cylindrical and Spherical Coordinate Systems

r-momentum

$$\frac{\partial v_r}{\partial t} = -\left\{ v_r \frac{\partial v_r}{\partial r} + \frac{v_\theta}{r} \frac{\partial v_r}{\partial \theta} - \frac{v_\theta^2}{r} + v_z \frac{\partial v_r}{\partial z} \right\}$$
$$+ \nu \left\{ \frac{\partial}{\partial r}\left(\frac{1}{r}\frac{\partial}{\partial r}(r v_r)\right) + \frac{1}{r^2}\frac{\partial^2 v_r}{\partial \theta^2} - \frac{2}{r^2}\frac{\partial v_\theta}{\partial \theta} + \frac{\partial^2 v_r}{\partial z^2} \right\} - \frac{1}{\rho}\frac{\partial P}{\partial r} + g_r$$

θ-momentum

$$\frac{\partial v_\theta}{\partial t} = -\left\{ v_r \frac{\partial v_\theta}{\partial r} + \frac{v_\theta}{r}\frac{\partial v_\theta}{\partial \theta} - \frac{v_r v_\theta}{r} + v_z \frac{\partial v_\theta}{\partial z} \right\}$$
$$+ \nu \left\{ \frac{\partial}{\partial r}\left(\frac{1}{r}\frac{\partial}{\partial r}(r v_\theta)\right) + \frac{1}{r^2}\frac{\partial^2 v_\theta}{\partial \theta^2} + \frac{2}{r^2}\frac{\partial v_r}{\partial \theta} + \frac{\partial^2 v_\theta}{\partial z^2} \right\} - \frac{1}{\rho r}\frac{\partial P}{\partial \theta} + g_\theta$$

z-momentum

$$\frac{\partial v_z}{\partial t} = -\left\{ v_r \frac{\partial v_z}{\partial r} + \frac{v_\theta}{r}\frac{\partial v_z}{\partial \theta} + v_z \frac{\partial v_z}{\partial z} \right\} + \nu \left\{ \frac{1}{r}\frac{\partial}{\partial r}\left(r \frac{\partial v_z}{\partial r}\right) + \frac{1}{r^2}\frac{\partial^2 v_z}{\partial \theta^2} + \frac{\partial^2 v_z}{\partial z^2} \right\} - \frac{1}{\rho}\frac{\partial P}{\partial z} + g_z.$$

(7.75)

For Newtonian fluids in spherical coordinates, the Navier–Stokes equations become:

r-momentum

$$\frac{\partial v_r}{\partial t} = -\left\{ v_r \frac{\partial v_r}{\partial r} + \frac{v_\theta}{r}\left(\frac{\partial v_r}{\partial \theta}\right) + \frac{v_\phi}{r \sin\theta}\left(\frac{\partial v_r}{\partial \phi}\right) - \left(\frac{v_\phi^2 + v_\theta^2}{r}\right) \right\}$$
$$+ \nu \left\{ \nabla^2 v_r - \frac{2}{r^2}\left(v_r + \frac{\partial v_\theta}{\partial \theta} + v_\theta \cot\theta + \frac{1}{\sin\theta}\frac{\partial v_\phi}{\partial \phi} \right) \right\} - \frac{1}{\rho}\frac{\partial P}{\partial r} + g_r$$

θ-momentum

$$\frac{\partial v_\theta}{\partial t} = -\left\{ v_r \frac{\partial v_\theta}{\partial r} + \frac{v_\theta}{r}\left(v_r + \frac{\partial v_\theta}{\partial \theta} \right) + \frac{v_\phi}{r \sin\theta}\left(\frac{\partial v_\theta}{\partial \phi} - v_\phi \cos\theta \right) \right\}$$
$$+ \nu \left\{ \nabla^2 v_\theta + \frac{2}{r^2}\frac{\partial v_r}{\partial \theta} - \frac{1}{r^2 \sin^2\theta}\left(v_\theta + 2\cos\theta \frac{\partial v_\phi}{\partial \phi} \right) \right\} - \frac{1}{\rho r}\frac{\partial P}{\partial \theta} + g_\theta$$

ϕ-momentum

$$\frac{\partial v_\phi}{\partial t} = -\left\{ v_r \frac{\partial v_\phi}{\partial r} + \frac{v_\theta}{r}\frac{\partial v_\phi}{\partial \theta} + \frac{v_\phi}{r \sin\theta}\left(\frac{\partial v_\phi}{\partial \phi} + v_r \sin\theta + v_\theta \cos\theta \right) \right\}$$
$$+ \nu \left\{ \nabla^2 v_\phi - \frac{1}{r^2 \sin^2\theta}\left(v_\phi - 2\sin\theta \frac{\partial v_r}{\partial \phi} - 2\cos\theta \frac{\partial v_\theta}{\partial \phi} \right) \right\} - \frac{1}{\rho r \sin\theta}\frac{\partial P}{\partial \phi} + g_\phi.$$

(7.76)

Stream functions can be defined which satisfy the continuity equation for two-dimensional and axisymmetric flow fields in cylindrical and spherical coordinate systems. For axisymmetric flow in cylindrical coordinates with $v_\theta = 0$:

$$v_z = \frac{1}{r}\frac{\partial \Psi}{\partial r}, \quad v_r = -\frac{1}{r}\frac{\partial \Psi}{\partial z}. \tag{7.77}$$

For axisymmetric flow about the z-axis in spherical coordinates with $v_\phi = 0$ and no velocity dependence on the angle ϕ, the continuity equation is satisfied if:

$$v_\theta = \frac{1}{r \sin \theta}\frac{\partial \Psi}{\partial r}, \quad v_r = -\frac{1}{r^2 \sin \theta}\frac{\partial \Psi}{\partial \theta}. \tag{7.78}$$

7.12 Use of Navier–Stokes Equations in Cylindrical and Spherical Coordinates

There are many important applications in bioengineering where Newtonian fluids flow with radial symmetry. Some examples where the use of cylindrical or spherical coordinates simplify the analysis are provided in this section.

Example 7.12.1 Couette Flow of a Newtonian Fluid.
As mentioned in Chap. 4, Couette viscometers are often used to measure fluid viscosity. Fluid is placed in the annulus between an inner cylinder and an outer cylinder, and the torque required to rotate one cylinder relative to the other is measured. Fluid viscosity can be computed from the imposed angular velocity of the cylinder and geometry of the system.

Solution. *Initial considerations:* A Newtonian fluid is placed in the viscometer, and our goal is to compute its viscosity. Since the viscometer has a cylindrical shape, this problem is best analyzed in cylindrical coordinates. The velocity is directed in the tangential direction, θ. We postulate the following:

1. There is no velocity in the r-direction, $v_r = 0$.
2. There is no velocity in the z-direction, $v_z = 0$.
3. The modified pressure is independent of θ, $\partial \wp/\partial \theta = 0$.
4. Velocity components are independent of z, $\partial v_i/\partial z = 0$.
5. Incompressible fluid, $\rho = $ constant.
6. Steady-state, nothing changes with time, $\partial ()/\partial t = 0$.

System definition and environmental interactions: The system is shown in Fig. 4.7c. Fluid lies between the two cylinders at $r = R_i$ and $R = R_o$. The inner cylinder rotates at angular velocity ω and the torque required to keep the outer cylinder stationary is measured. The top surface is in contact with air, so the pressure at the top surface is equal to atmospheric pressure.
Apprising the problem to identify governing relationships: Since the fluid is assumed to be Newtonian and incompressible, the Navier–Stokes equation is

7.12 Use of Navier–Stokes Equations in Cylindrical and Spherical Coordinates

the appropriate governing relationship. In addition, we will apply the continuity equation.

Analysis: Applying the above conditions to the continuity equation:

$$\cancel{\frac{\partial \rho}{\partial t}}^{5,6} + \frac{1}{r}\cancel{\frac{\partial(\rho r v_r)}{\partial r}}^{1} + \frac{1}{r}\frac{\partial(\rho v_\theta)}{\partial \theta} + \cancel{\frac{\partial(\rho v_z)}{\partial z}}^{2,4} = 0$$

Thus, from continuity and condition (5), we find:

7. $\partial v_\theta / \partial \theta = 0$.

Therefore, from (4), (6), and (7), we can postulate that v_θ is a function only of r. Turning to the Navier–Stokes equation:

r - momentum

$$\cancel{\frac{\partial v_r}{\partial t}}^{1,6} = -\left\{ \cancel{v_r \frac{\partial v_r}{\partial r}}^{1} + \cancel{\frac{v_\theta}{r}\frac{\partial v_r}{\partial \theta}}^{1} - \frac{v_\theta^2}{r} + \cancel{v_z \frac{\partial v_r}{\partial z}}^{1,2,4} \right\}$$

$$+ \nu \left\{ \cancel{\frac{\partial}{\partial r}\left(\frac{1}{r}\frac{\partial}{\partial r}(rv_r)\right)}^{1} + \frac{1}{r^2}\cancel{\frac{\partial^2 v_r}{\partial \theta^2}}^{1} - \frac{2}{r^2}\cancel{\frac{\partial v_\theta}{\partial \theta}}^{7} + \cancel{\frac{\partial v_r}{\partial z^2}}^{1,4} \right\} - \frac{1}{\rho}\frac{\partial \wp}{\partial r}$$

θ - momentum

$$\cancel{\frac{\partial v_\theta}{\partial t}}^{2,6} = -\left\{ \cancel{v_r \frac{\partial v_\theta}{\partial r}}^{1,2} + \cancel{\frac{v_\theta}{r}\frac{\partial v_\theta}{\partial \theta}}^{7} - \cancel{\frac{v_r v_\theta}{r}}^{1} + \cancel{v_z \frac{\partial v_\theta}{\partial z}}^{1,4} \right\}$$

$$+ \nu \left\{ \frac{\partial}{\partial r}\left(\frac{1}{r}\frac{\partial}{\partial r}(rv_\theta)\right) + \frac{1}{r^2}\cancel{\frac{\partial^2 v_\theta}{\partial \theta^2}}^{7} + \frac{2}{r^2}\cancel{\frac{\partial v_r}{\partial \theta}}^{1} + \cancel{\frac{\partial^2 v_\theta}{\partial z^2}}^{4} \right\} - \frac{1}{\rho r}\cancel{\frac{\partial \wp}{\partial \theta}}^{3}$$

z - momentum

$$\cancel{\frac{\partial v_z}{\partial t}}^{6} = -\left\{ \cancel{v_r \frac{\partial v_z}{\partial r}}^{1,2} + \cancel{\frac{v_\theta}{r}\frac{\partial v_z}{\partial \theta}}^{2} + \cancel{v_z \frac{\partial v_z}{\partial z}}^{2,4} \right\} + \nu \left\{ \cancel{\frac{1}{r}\frac{\partial}{\partial r}\left(r\frac{\partial v_z}{\partial r}\right)}^{2} + \frac{1}{r^2}\cancel{\frac{\partial^2 v_z}{\partial \theta^2}}^{2} + \cancel{\frac{\partial^2 v_z}{\partial z^2}}^{2,4} \right\} - \frac{1}{\rho}\frac{\partial \wp}{\partial z}$$

The z-component reduces to:

$$\frac{\partial \wp}{\partial z} = 0.$$

Therefore, $\wp = \wp(r)$ alone. The r and θ components reduce to the following ODEs:

$$\frac{d\wp}{dr} = \frac{\rho v_\theta^2}{r},$$

$$\frac{d}{dr}\left[\frac{1}{r}\frac{d}{dr}(rv_\theta)\right] = 0.$$

The appropriate boundary conditions are:

1. $v_\theta(r = R_i) = \omega R_i$.
2. $v_\theta(r = R_o) = 0$.
3. $\wp(r = R_o) = \wp_o$.

The solution to the θ-component of the Navier–Stokes equation, with boundary conditions (1) and (2) is:

$$v_\theta = \frac{\omega \left[\dfrac{R_o^2}{r} - r \right]}{\left(\dfrac{R_o}{R_i}\right)^2 - 1}.$$

The viscous stress on the outer cylinder $\tau_{r\theta}$ can be found by applying Newton's law of viscosity (7.73):

$$\tau_{r\theta} = -\mu \left(\frac{1}{r} \frac{\partial v_r}{\partial \theta} + r \frac{\partial}{\partial r}\left(\frac{v_\theta}{r}\right) \right) = 0 + \left[\frac{2\mu\omega \left(\dfrac{R_o}{R_i}\right)^2}{\left(\dfrac{R_o}{R_i}\right)^2 - 1} \right] \frac{R_o^2}{r^2}.$$

The torque $\mathfrak{I}$ applied to the stationary wall at $r = R_o$ is equal to the product of the stress at the wall, the wetted area, and the radius R_o.

$$\mathfrak{I} = \left\{ \left[\frac{2\mu\omega \left(\dfrac{R_o}{R_i}\right)^2}{\left(\dfrac{R_o}{R_i}\right)^2 - 1} \right] \right\} [2\pi R_o z_o][R_o] = 4\pi\mu\omega R_o^2 z_o \left[\frac{\left(\dfrac{R_o}{R_i}\right)^2}{\left(\dfrac{R_o}{R_i}\right)^2 - 1} \right].$$

Examining and interpreting the results: If the torque on the outer cylinder and the angular velocity of the inner cylinder are measured, then the above expression can be used to compute the fluid viscosity from experimental measurements:

$$\mu = \frac{\mathfrak{I}}{4\pi\omega z_o}\left[\frac{R_o^2 - R_i^2}{R_o^4}\right].$$

Integration of the r-component of the Navier–Stokes equation with boundary condition (3), we find the modified pressure in the fluid:

$$\wp - \wp_o = \rho \left(\frac{\omega R_i}{1 - \left(\dfrac{R_o}{R_i}\right)^2} \right)^2 \left[\frac{r^2}{2R_i^2} - 2\left(\frac{R_o}{R_i}\right)^2 \ln\left(\frac{r}{R_o}\right) - \frac{R_o^4}{2R_i^2 r^2} \right].$$

Since $g_r = g_\theta = 0$ and $g_z = -g$, then $\wp = P + \rho g z$. At a particular point z_o, $\wp_o = P_o + \rho g z_o$. If we let z_o be the location of the free surface at $r = R_o$, then P_o is atmospheric pressure. Setting $P = P_o$, the expression can be used to find the location of the free surface of the fluid, z_f as a function of radial position:

$$z_f - z_o = \frac{1}{g}\left(\frac{\omega R_i}{1 - \left(\frac{R_o}{R_i}\right)^2}\right)^2 \left[\frac{r^2}{2R_i^2} - 2\left(\frac{R_o}{R_i}\right)^2 \ln\left(\frac{r}{R_o}\right) - \frac{R_o^4}{2R_i^2 r^2}\right].$$

Fluid near the outside radius of the viscometer will be higher than the fluid near the inner radius.

Example 7.12.2 Creeping Flow Around a Spherical Cell, Stokes Law.

A spherical cell with density $\rho_s = 1.05$ g/cm³ radius $R = 10$ μm falls under its own weight in a stationary Newtonian fluid having density $\rho_f = 1.01$ g/cm³ and viscosity $\mu = 1$ cp. Find the terminal velocity v_0 of the cell, the drag force on the cell and the velocity and pressure distributions in the fluid.

Solution. *Initial considerations:* If we adopt a coordinate system with its origin at the center of the cell, we can also interpret this problem as finding the force necessary to hold the cell stationary as fluid with approach velocity v_0 flows past the cell. We will orient the spherical coordinate system so the positive z-axis is aligned with the direction of approaching fluid far from the sphere, and opposite to the direction of gravity. From symmetry arguments, we can make the following assumptions;

1. $\partial()/\partial\phi = 0$.
2. $v_\phi = 0$.

System definition and environmental interactions: The system to be analyzed is not the solid sphere, but rather the fluid that flows past the sphere in Fig. 7.16. The fluid is bounded by the radius of the sphere and is assumed to extend many diameters from the sphere surface. Fluid far from the sphere has a vertical velocity v_0, and fluid in contact with the sphere surface has zero velocity.

Apprising the problem to identify governing relationships: Since the fluid is incompressible and Newtonian, the appropriate governing equations are the incompressible continuity equation and the Navier–Stokes equation.

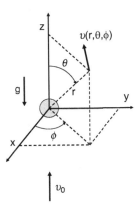

Fig. 7.16 Flow around a sphere

Analysis: The continuity equation for spherical coordinates (7.72) with the assumptions above reduces to:

$$\frac{1}{r^2}\frac{\partial(r^2 v_r)}{\partial r} + \frac{1}{r\sin\theta}\frac{\partial(v_\theta \sin\theta)}{\partial \theta} = 0.$$

This is satisfied by the stream function defined by (7.77):

$$v_\theta = \frac{1}{r\sin\theta}\frac{\partial \Psi}{\partial r}, \quad v_r = -\frac{1}{r^2 \sin\theta}\frac{\partial \Psi}{\partial \theta}.$$

These expressions for velocity can be substituted into the Navier–Stokes equations. However, before doing so, let us stipulate that the fluid velocity is independent of time and the velocity is very slow. This is known as the *creeping flow assumption* and allows us to neglect the convective acceleration terms relative to the viscous terms. Applying these restrictions, the resulting Navier–Stokes equations, (7.75), become:

r - momentum

$$\frac{\partial \wp}{\partial r} = \mu\left\{\nabla^2 v_r - \frac{2}{r^2}\left(\frac{\partial v_\theta}{\partial \theta} + v_r + v_\theta \cot\theta\right)\right\}$$

θ - momentum

$$\frac{\partial \wp}{\partial \theta} = \mu r \left\{\nabla^2 v_\theta + \frac{2}{r^2}\frac{\partial v_r}{\partial \theta} - \frac{v_\theta}{r^2 \sin^2\theta}\right\}$$

ϕ - momentum

$$\frac{\partial \wp}{\partial \phi} = 0.$$

The Laplacian operator given in Table 7.1 reduces to:

$$\nabla^2 = \frac{1}{r^2}\frac{\partial}{\partial r}\left(r^2 \frac{\partial}{\partial r}\right) + \frac{1}{r^2 \sin\theta}\frac{\partial}{\partial \theta}\left(\sin\theta \frac{\partial}{\partial \theta}\right).$$

The pressure can be eliminated from the problem by taking the derivative of the r-momentum equation with respect to θ and setting it equal to the derivative of the θ-momentum equation with respect to r. Finally, the equations above can be used to replace the velocities v_θ and v_r with the stream function. After much manipulation, the resulting differential equation can be expressed as:

$$\left[\frac{\sin\theta}{r^2}\frac{\partial}{\partial \theta}\left(\frac{1}{\sin\theta}\frac{\partial}{\partial \theta}\right) + \frac{\partial^2}{\partial r^2}\right]\left[\frac{\sin\theta}{r^2}\frac{\partial}{\partial \theta}\left(\frac{1}{\sin\theta}\frac{\partial \Psi}{\partial \theta}\right) + \frac{\partial^2 \Psi}{\partial r^2}\right] = 0.$$

7.12 Use of Navier–Stokes Equations in Cylindrical and Spherical Coordinates

The no-slip boundary conditions for v_θ and v_r at $r = R$ in terms of the stream function become:

$$0 = \frac{1}{R \sin \theta} \frac{\partial \Psi}{\partial r}\bigg|_{r=R}, \quad 0 = -\frac{1}{R^2 \sin \theta} \frac{\partial \Psi}{\partial \theta}\bigg|_{r=R}.$$

As r becomes very large, the velocity is v_0 in the positive z direction. Therefore, $-v_\theta(r \to \infty) = v_0 \sin \theta$ and $v_r(r \to \infty) = v_0 \cos \theta$. Substituting the stream function into the expressions for v_θ or v_r and integrating either expression, we find:

$$\Psi(r \to \infty) = -\left(\frac{v_0}{2}\right) r^2 \sin^2 \theta.$$

This form of the solution far from the sphere suggests that the solution at other values of radial position might behave like:

$$\Psi(r) = F(r) \sin^2 \theta.$$

The function $F(r)$ is just a function of radial position that must be determined by satisfying the momentum equation and the boundary conditions. Substituting the assumed form of the solution into the expression for the stream function leads to the following ordinary differential equation:

$$\frac{d^4 F}{dr^4} - \frac{4}{r^2} \frac{d^2 F}{dr^2} + \frac{8}{r^3} \frac{dF}{dr} - \frac{8F}{r^4} = 0.$$

Letting $F = ar^n$, we obtain the following equation for n:

$$0 = n(n-1)(n-2)(n-3) - 4n(n-1) + 8n - 8.$$

This can be factored as follows:

$$(n-1)(n+1)(n-2)(n-4) = 0.$$

Therefore, n has four solutions, $n = +1, -1, +2$ and $+4$. Consequently, the general form for the stream function is:

$$\Psi(r) = \left[\frac{a_1}{r} + a_2 r + a_3 r^2 + a_4 r^4\right] \sin^2 \theta.$$

The coefficients a_1, a_2, a_3, and a_4 need to be determined from boundary conditions. As r approaches infinity, the r^4 term would dominate. However, we know that the stream function is proportional to r^2 as r becomes large. Therefore, a_4 must be zero and $a_3 = -v_0/2$. The coefficients a_1 and a_2 can be determined from the no-slip boundary conditions. Substituting the expression for ψ into the no-slip expressions, we obtain the following two algebraic equations:

$$0 = -\frac{a_1}{R^2} + a_2 + 2\left(-\frac{v_0}{2}\right)R,$$

$$0 = \frac{a_1}{R} + a_2 R - \frac{v_0}{2}R^2.$$

Solving these equations, we find $a_1 = -v_0 R^3/4$ and $a_2 = 3v_0 R/4$. Therefore, the final solutions for the velocity components are:

$$v_r = v_0 \cos\theta \left[1 - \frac{3}{2}\left(\frac{R}{r}\right) + \frac{1}{2}\left(\frac{R}{r}\right)^3\right],$$

$$v_\theta = -v_0 \sin\theta \left[1 - \frac{3}{4}\left(\frac{R}{r}\right) - \frac{1}{4}\left(\frac{R}{r}\right)^3\right].$$

Substituting these back into the Navier–Stokes equations, we find:

$$\frac{\partial \wp}{\partial r} = \frac{3\mu R v_0 \cos\theta}{r^3},$$

$$\frac{\partial \wp}{\partial \theta} = \frac{3\mu R v_0 \sin\theta}{2r^2}.$$

Integrating the first equation, we find

$$\wp = -\frac{3\mu R v_0 \cos\theta}{2r^2} + h(\theta),$$

where $h(\theta)$ is an arbitrary function of θ. Taking the derivative of $\wp$ with respect to θ, we find:

$$\frac{\partial \wp}{\partial \theta} = \frac{3\mu R v_0 \sin\theta}{2r^2} + \frac{dh}{d\theta}.$$

Comparing this with the expression above for $\partial \wp/\partial \theta$, we see that $dh/d\theta = 0$, so h is a constant. If we let $\wp_0$ be the modified pressure as $r \to \infty$, then the final expression for the modified pressure is:

$$\wp - \wp_0 = -\frac{3\mu R v_0 \cos\theta}{2r^2}.$$

In terms of actual pressure in the fluid, we can use (7.40) with $g_z = -g$, $g_x = g_y = 0$, and P_0 equal to the pressure far from the sphere in the plane at $z = 0$:

7.12 Use of Navier–Stokes Equations in Cylindrical and Spherical Coordinates

$$P - P_0 = -\rho_f g z - \frac{3\mu R v_0 \cos\theta}{2r^2},$$

where ρ_f is the density of the fluid.

The force exerted by the fluid on the sphere in the z-direction is caused by two phenomena. First, there is a force resulting from the pressure integrated over the surface of the sphere. The component of the pressure in the z-direction on the surface of the sphere is $P(r = R)\cos\theta$. If z_0 is the distance from the origin of the coordinate system to the center of the sphere, then the axial position of the sphere surface is $z = z_0 + R\cos\theta$. The element of surface area on the sphere is $R^2 \sin\theta \, d\theta d\phi$, so the total pressure force F_P on the sphere surface is given by:

$$F_P = \int_0^{2\pi}\int_0^{\pi} \left\{\left[-P_0 + \rho_f g(z_0 + R\cos\theta) + \frac{3\mu v_0}{2R}\cos\theta\right]\cos\theta\right\}\{R^2 \sin\theta d\theta d\phi\}.$$

Integrating, we find the net pressure force to be:

$$F_P = \frac{4}{3}\pi R^3 \rho_f g + 2\pi \mu R v_0.$$

The first term represents a buoyant effect while the second term is known as form drag. No net contribution is made by the constant terms P_0 or $\rho_f g z_0$.

The second force in the z-direction is caused by shear stress at the surface of the sphere. The stress on a surface of constant r in the θ direction is given by (7.74):

$$\tau_{r\theta} = \mu\left(\frac{1}{r}\frac{\partial v_r}{\partial \theta} + r\frac{\partial}{\partial r}\left(\frac{v_\theta}{r}\right)\right).$$

Evaluating this at the surface of the sphere, where $r = R$:

$$\tau_{r\theta}|_{r=R} = \frac{3\mu v_0 \sin\theta}{2R}.$$

Recall that $\tau_{r\theta}$ is the stress exerted by the surroundings on the fluid, so the stress exerted by the fluid on the sphere will be $-\tau_{r\theta}(r = R)$. The component of shear stress in the positive z-direction is $-(-\tau_{r\theta}(R))(\sin\theta) = \tau_{r\theta}(R)\sin\theta$. Integrating this over the surface provides the friction force, F_f:

$$F_f = 2\pi \int_0^{\pi} \left\{\left[\frac{3\mu v_0 \sin\theta}{2R}\right]\sin\theta\right\}[R^2 \sin\theta d\theta] = 4\pi \mu R v_0.$$

The net force opposing motion of the sphere is the sum of the pressure and friction forces, F_P and F_f.

$$F_P + F_f = \frac{4}{3}\pi R^3 \rho_f g + 6\pi \mu R v_0.$$

The first term would vanish if the fluid flows horizontally past the sphere, since the z-axis would be perpendicular to the gravity vector. In that case, the net drag force on the sphere would be:

$$F_D = 6\pi\mu R v_0. \tag{7.79}$$

This relationship is known as *Stokes law*. Recall the definition of the friction factor or drag coefficient for external flow past a sphere given by rearranging (5.105):

$$f = \frac{2}{\pi}\left(\frac{F_D}{\rho_f v_0^2 R^2}\right).$$

Substituting Stokes law for the drag force on the sphere:

$$f = \frac{12\mu}{\rho_f v_0 R} = \frac{24}{Re_D}, \tag{7.80}$$

where Re_D is the Reynolds number based on the diameter of the sphere. This is the origin of (5.106) and is valid for $Re_D < 0.5$. Figure 5.23 should be used for larger Reynolds numbers.

In the vertical flow situation, the forces F_P and F_f are balanced by the weight of the sphere when the sphere falls at constant velocity. The weight W of a sphere with mass m_s and density ρ_s is:

$$W = m_s g = \rho_s \left(\frac{4}{3}\pi R^3\right) g = F_P + F_f.$$

Substituting the values found above for F_P and F_f, we can solve for the terminal velocity of the sphere:

$$v_0 = \frac{2}{9}\left(\frac{R^2 g (\rho_s - \rho_f)}{\mu}\right). \tag{7.81}$$

Examining and interpreting the results: Turning our attention to the spherical cell specified in the problem statement:

$$v_0 = \frac{2}{9}\left(\frac{(10 \times 10^{-4}\,\text{cm})^2 (980\,\text{cm s}^{-2})(1.05 - 1.01\,\text{g cm}^{-3})}{(1\,\text{cp})(10^{-2}\,\text{g cm}^{-1}\,\text{s}^{-1}\,\text{cp}^{-1})}\right) = 8.7 \times 10^{-4}\,\text{cm/s}.$$

The Reynolds number is:

$$Re_D = \frac{\rho_f v_0 D}{\mu} = \frac{(1.01\,\text{g cm}^{-3})(8.7 \times 10^{-4}\,\text{cm s}^{-1})(20 \times 10^{-4}\,\text{cm})}{10^{-2}\,\text{g cm}^{-1}\,\text{s}^{-1}} = 1.76 \times 10^{-4}.$$

7.12 Use of Navier–Stokes Equations in Cylindrical and Spherical Coordinates

This confirms that the creeping flow assumption is justified. The drag force on the sphere is:

$$F_D = 6\pi\mu R v_0 = 6\pi(10^{-2}\,\text{g cm}^{-1}\,\text{s}^{-1})(10 \times 10^{-4}\,\text{cm})(8.7 \times 10^{-4}\,\text{cm s}^{-1})$$
$$= 1.64 \times 10^{-7}\,\text{dynes}.$$

The maximum deviation in pressure caused by the presence of the moving sphere is at the sphere surface, where $\theta = 0$:

$$\Delta P_{\max} = -\frac{3\mu v_0}{2R} = -\frac{3(10^{-2}\,\text{g cm}^{-1}\,\text{s}^{-1})(8.7 \times 10^{-4}\,\text{cm s}^{-1})}{2(10 \times 10^{-4}\,\text{cm})} = -0.0131\,\text{dyne cm}^{-2}.$$

Example 7.12.3 Periodic Flow in a Tube.
Blood flow in arteries is periodic, with the fundamental frequency equal to the heart rate. The axial pressure gradient in an artery consists of a steady component A_0 and an oscillating component. The oscillating pressure gradient can be represented as a Fourier series consisting of N terms:

$$-\frac{\partial P}{\partial z} = A_0 + \sum_{p=1}^{N} A_p \cos(p\omega_0 t + \phi_p).$$

The angular frequency ω_0 represents the fundamental frequency of the heart ($\omega_0 = f_0/2\pi$, where f_0 is the heart rate in Hz). Each value of p is an integer, so each term in the series contributes a higher harmonic to the pressure gradient. A_p is the magnitude of the pth component of the pressure gradient and ϕ_p represents a phase lag for the pth component relative to the beginning of the cardiac cycle ($t = 0$). Womersley (1955) used Fourier analysis to decompose the periodic portion of the pressure gradient from a canine femoral artery into six harmonics. The artery had a radius $a = 0.15$ cm and the heart rate was $f_0 = 3$ Hz. Values of A_p and ϕ_p are shown in Table 7.3.

Assume that the steady component of the pressure gradient is $A_0 = 0.16$ mmHg/cm. Blood flowing through the artery has a viscosity equal to 0.04 poise and density of 1.05 g/ml. Pressure gradients with and without the steady component are shown in Fig. 7.17. Note that the magnitude of the oscillating component is generally much larger than the steady component, suggesting that the oscillating components of the flow and wall shear stress might be much higher in arteries than their average values.

Reduce the continuity and Navier–Stokes equations for this case. Find the velocity profile in the artery at various times during the cardiac cycle. Integrate the velocity expression to determine how flow rate varies with time during the cardiac cycle. Find the maximum wall stress in the vessel. When in the cycle does this occur? Discuss the effects of varying the fundamental frequency from 0.5 to 20 Hz.

Table 7.3 Magnitude and phase angle for the first six harmonics of the pressure gradient

p	A_p (mmHg/cm)	ϕ_p (degrees)
1	1.1050	+40.23
2	1.5316	−69.28
3	−0.9668	+34.73
4	−0.2857	−33.78
5	0.2821	+87.52
6	−0.1924	−4.97

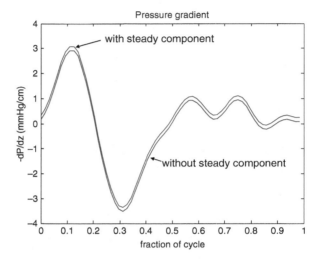

Fig. 7.17 Pressure gradient in femoral artery of a dog

Solution. *Initial considerations:* We begin by assuming that the femoral artery can be modeled as a rigid cylinder with constant radius. In addition, we will assume that measurements are made far enough downstream from the inlet of the artery that flow can be considered to be fully developed (independent of axial location). In addition, blood is assumed to be an incompressible, Newtonian fluid. We will also neglect gravitational effects, which can be added later using the methods introduced in Sect. 7.7. We will adopt a cylindrical coordinate system with $r = 0$ at the center of the artery. The velocity is not expected to vary with the angle θ. Our list of conditions becomes:

1. Blood is incompressible, $\rho = $ constant.
2. Fully developed flow: $v_r = 0$.
3. Fully developed flow: $v_\theta = 0$.
4. $\partial v_z / \partial \theta = 0$.
5. Blood is Newtonian.

System definition and environmental interactions: Blood in the femoral artery is the system of interest. Pressure forces on the two ends of the vessel and the viscous force by the wall are the environmental interactions.

7.12 Use of Navier–Stokes Equations in Cylindrical and Spherical Coordinates

Apprising the problem to identify governing relationships: Since blood is assumed to be Newtonian in this vessel, the Navier–Stokes equation is the appropriate governing equation, along with the continuity equation.

Analysis: Applying assumptions (1)–(3), the continuity equation reduces to:

6. $\partial v_z / \partial z = 0$.

Taken together, conditions (4) and (6) suggest that $v_z = v_z(r,t)$. Since blood is assumed to be a Newtonian fluid, we can use the Navier–Stokes equations. The r-component of the Navier–Stokes equation reduces to:

7. $\partial P / \partial r = 0$.

The θ-component of the Navier–Stokes equation reduces to:

8. $\partial P / \partial \theta = 0$.

Conditions (7) and (8) indicate that pressure depends only on axial position and time, $P(z,t)$. After using conditions (1)–(6), the z-component of the Navier–Stokes equation reduces to:

$$\rho \frac{\partial v_z}{\partial t} - \mu \frac{1}{r} \frac{\partial}{\partial r}\left(r \frac{\partial v_z}{\partial r}\right) = -\frac{\partial P}{\partial z}.$$

Note that the left side of this equation depends on r and t and the right side depends on z and t. The only way this can occur is if each side depends on a function of t alone. This function is simply the pressure gradient defined in the problem statement. Consequently, the equation can be written:

$$\rho \frac{\partial v_z}{\partial t} = \mu \frac{1}{r} \frac{\partial}{\partial r}\left(r \frac{\partial v_z}{\partial r}\right) + A_0 + \sum_{p=1}^{\infty} A_p \cos(p\omega_0 t + \phi_p).$$

We are only interested in finding the periodic solution (forced response), so it is not necessary to specify an initial condition. The boundary conditions are:

$$v_z(r = a, t) = 0,$$

$$\frac{\partial v_z(r = 0, t)}{\partial r} = 0.$$

The PDE is linear, so we might try breaking the solution into $p + 1$ components, as follows:

$$v_z(r, t) = v_{z0}(r) + \sum_{p=1}^{N} v_{zp}(r, t).$$

Substituting this into the PDE and rearranging, we find:

$$\left\{\rho\frac{\partial v_{z0}}{\partial t} - \mu\frac{1}{r}\frac{\partial}{\partial r}\left(r\frac{\partial v_{z0}}{\partial r}\right) - A_0\right\} + \left\{\rho\frac{\partial v_{z1}}{\partial t} - \mu\frac{1}{r}\frac{\partial}{\partial r}\left(r\frac{\partial v_{z1}}{\partial r}\right) - A_1\cos(\omega_0 t + \phi_1)\right\}$$
$$+ \cdots + \left\{\rho\frac{\partial v_{zp}}{\partial t} - \mu\frac{1}{r}\frac{\partial}{\partial r}\left(r\frac{\partial v_{zp}}{\partial r}\right) - A_p\cos(p\omega_0 t + \phi_p)\right\} + \cdots = 0.$$

Note that superposition of the equations in brackets satisfies the original PDE. The original boundary condition becomes:

$$v_z(a,t) = 0 = v_{z0}(a) + v_{z1}(a,t) + \cdots + v_{zp}(a,t) + \cdots v_{zN}(a,t),$$
$$\frac{\partial v_z(0,t)}{\partial r} = 0 = \frac{\partial v_{z0}(0)}{\partial r} + \frac{\partial v_{z1}(0,t)}{\partial r} + \cdots + \frac{\partial v_{zp}(0,t)}{\partial r} + \cdots + \frac{\partial v_{zN}(0,t)}{\partial r}.$$

Since the boundary conditions are homogeneous, the original PDE and boundary conditions can be constructed by superposition of the following PDEs:

$$0 = \mu\frac{1}{r}\frac{d}{dr}\left(r\frac{dv_{z0}}{dr}\right) + A_0, \quad v_{z0}(a) = 0, \quad \left.\frac{dv_{z0}}{dr}\right|_{r=0} = 0,$$
$$\rho\frac{\partial v_{z1}}{\partial t} = \mu\frac{1}{r}\frac{\partial}{\partial r}\left(r\frac{\partial v_{z1}}{\partial r}\right) + A_1\cos(\omega_0 t + \phi_1), \quad v_{z1}(a,t) = 0, \quad \left.\frac{\partial v_{z1}}{\partial r}\right|_{r=0} = 0,$$
$$\rho\frac{\partial v_{zp}}{\partial t} = \mu\frac{1}{r}\frac{\partial}{\partial r}\left(r\frac{\partial v_{zp}}{\partial r}\right) + A_p\cos(p\omega_0 t + \phi_p), \quad v_{zp}(a,t) = 0, \quad \left.\frac{\partial v_{zp}}{\partial r}\right|_{r=0} = 0.$$

A total of N equations can be written in the form of the last two PDEs with $p = 1, 2, \ldots, N$. The PDE constructed by adding all of the component differential equations is equal to the original PDE. Likewise, the sum of all of the component boundary conditions at $r = 0$ is identical to the original boundary condition at $r = 0$. The same can be said for the boundary condition at $r = a$. Consequently, the sum of the solutions for the component equations will equal the solution for the velocity profile. The solution for the v_{z0} component is the same as that computed previously for steady flow in a tube:

$$v_{z0}(r) = \frac{A_0 a^2}{4\mu}\left(1 - \left(\frac{r}{a}\right)^2\right).$$

We can write the form for each of the remaining PDEs as follows:

$$\rho\frac{\partial v_{zp}}{\partial t} = \mu\frac{1}{r}\frac{\partial}{\partial r}\left(r\frac{\partial v_{zp}}{\partial r}\right) + A_p\mathrm{Re}\left(e^{i(p\omega_0 t + \phi_p)}\right),$$

where Re() refers to the real part of the quantity in parentheses and $i = \sqrt{-1}$. Let us postulate that $v_z(r,t)$ is a product of a function of r and a function of t, as follows:

$$v_{zp}(r,t) = \mathrm{Re}\left(F_p(r)e^{i(p\omega_0 t + \phi_p)}\right).$$

7.12 Use of Navier–Stokes Equations in Cylindrical and Spherical Coordinates

Substituting this into the PDE, and dropping the explicit reference to the real part of the result:

$$\rho \frac{\partial}{\partial t}\left(F_p(r)e^{i(p\omega_0 t+\phi_p)}\right) = \mu \frac{1}{r}\frac{\partial}{\partial r}\left(r\frac{\partial}{\partial r}\left(F_p(r)e^{i(p\omega_0 t+\phi_p)}\right)\right) + A_p\left(e^{i(p\omega_0 t+\phi_p)}\right).$$

This substitution converts the PDE to an ODE:

$$i p \omega_0 \rho F_p(r) = \mu \left[\frac{d^2 F_p(r)}{dr^2} + \frac{1}{r}\frac{dF_p(r)}{dr}\right] + A_p.$$

This equation can be converted to an equation with a known solution by defining the following dimensionless variables:

$$y = \frac{r}{a} \quad \alpha = a\sqrt{\frac{\omega_0}{\nu}} \quad \alpha_p = a\sqrt{\frac{p\omega_0}{\nu}} = \alpha\sqrt{p} \quad x = i^{3/2}\alpha_p y,$$

where $\nu = \mu/\rho$, the kinematic viscosity and α is known as the Womersley number. After making these substitutions, we obtain the following ODE:

$$x^2 \frac{d^2 F_p}{dx^2} + x\frac{dF_p}{dx} + x^2 F_p = -x^2\left[\frac{iA_p a^2}{\mu \alpha_p^2}\right].$$

The solution for F can be separated into a particular solution $F_{p,p}$ and a homogeneous solution $F_{p,h}$. The particular solution can be found by letting F_p equal a constant:

$$F_{p,p} = -\frac{iA_p a^2}{\mu \alpha_p^2}.$$

The homogeneous portion of the solution is the solution to Bessel's equation of zeroth order:

$$x^2 \frac{d^2 F_{p,h}}{dx^2} + x\frac{dF_{p,h}}{dx} + x^2 F_{p,h} = 0.$$

This has a known solution:

$$F_{p,h}(x) = C_1 J_0(x) + C_2 Y_0(x),$$

where C_1 and C_2 are constants and J_0 and Y_0 are Bessel functions of the first kind and second kind of zeroth order. The total solution is the sum of the particular solution and the homogeneous solution. Returning to the original variables, the solution for F is:

$$F_p = C_1 J_0\left(i^{3/2}\alpha_p \frac{r}{a}\right) + C_2 Y_0\left(i^{3/2}\alpha_p \frac{r}{a}\right) - \frac{iA_p a^2}{\mu \alpha_p^2}.$$

The function Y_0 approaches negative infinity as r approaches zero. Consequently, the coefficient C_2 must be zero. The other constant can be found by applying the boundary condition that $F(a) = 0$. Therefore, the solution for F_p is:

$$F_p = \frac{iA_p a^2}{\mu \alpha_p^2} \left[\frac{J_0\left(i^{3/2}\alpha_p \frac{r}{a}\right)}{J_0\left(i^{3/2}\alpha_p\right)} - 1\right].$$

The solution for the pth oscillating component of velocity is

$$v_{zp}(r,t) = \frac{A_p a^2}{\mu \alpha_p^2} \text{Re}\left(\left[\frac{J_0\left(i^{3/2}\alpha_p \frac{r}{a}\right)}{J_0\left(i^{3/2}\alpha_p\right)} - 1\right] i e^{i(p\omega_0 t + \phi_p)}\right).$$

Finally, the solution for $v_z(r, t)$ is found by summing the steady term and the N oscillating terms:

$$v_z(r,t) = \frac{A_0 a^2}{4\mu}\left(1 - \left(\frac{r}{a}\right)^2\right) + \sum_{p=1}^{N} \frac{A_p a^2}{\mu \alpha_p^2} \text{Re}\left(\left[\frac{J_0\left(i^{3/2}\alpha_p \frac{r}{a}\right)}{J_0\left(i^{3/2}\alpha_p\right)} - 1\right] i e^{i(p\omega_0 t + \phi_p)}\right).$$

Examining and interpreting the results: The velocity profiles, computed using Matlab at 22.5 degree increments of the cardiac cycle, are shown in Fig. 7.18. A single cycle represents one third of a second, corresponding to a canine heart rate of 180 beats/min. Each profile is numbered in sequence, beginning with profile 1 at $t = 0$, profile 2 at $t = 0.0208$ s, etc., and ending again with profile 1 at $t = 0.333$ s.

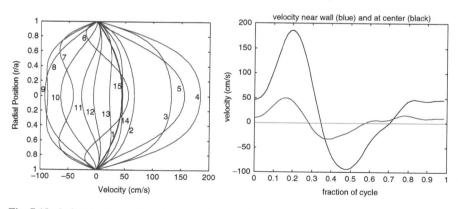

Fig. 7.18 *Left*: velocity profiles at consecutive time intervals of 0.0208 s in the cardiac cycle and *Right*: velocity vs. fraction of cardiac cycle for dimensionless radial positions $r/a = 0$ (*black*) and 0.9 (*blue*)

7.12 Use of Navier–Stokes Equations in Cylindrical and Spherical Coordinates

The figure at the right shows time traces during a single cycle for fluid at the center of the tube and near the wall ($r/a = 0.9$). Fluid in the central portion of the tube accelerates rapidly over the first 20% of the cycle. Fluid near the wall also accelerates initially, but begins to decelerate earlier than fluid in the center. The slower moving fluid near the wall is able to respond more rapidly to the pressure gradient than the fluid in the central portion of the tube, which has a greater momentum.

The instantaneous flow rate can be found by integrating the velocity over the tube cross section from the center of the artery to the wall:

$$Q(t) = 2\pi \int_0^a v_z(r,t) r \, dr = \frac{A_0 \pi a^4}{8\mu} + \sum_{p=1}^{N} \frac{A_p \pi a^4}{\mu \alpha_p^2} Re\left(\left[1 - \frac{2J_1(i^{3/2}\alpha_p)}{i^{3/2}\alpha_p J_0(i^{3/2}\alpha_p)}\right] \frac{e^{i(p\omega_0 t + \phi_p)}}{i}\right).$$

The Bessel function J_1 of the first kind and first order arises from the integration of J_0. Note that since the oscillating components of flow are periodic, the integral over a complete cardiac cycle for each term in the summation will be zero. Therefore, the net flow through the artery is given by the Poiseuille equation, where A_0 is the steady component of $-\partial P/\partial z$. However, the magnitudes of the maximum and minimum flow rates can be much larger than the average flow. This is illustrated in Fig. 7.19. The steady component of the pressure gradient provides an average flow in the vessel of 1.06 ml/s. The maximum flow is more than seven times the average flow, while the minimum flow is about -4 times the average flow. In addition, Fig. 7.19 shows that the flow lags behind the pressure gradient by a time equal to about one tenth of the cardiac cycle. This effect becomes even more pronounced at higher cardiac frequencies.

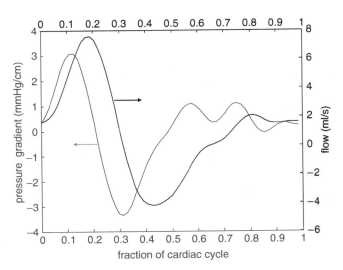

Fig. 7.19 Comparison of predicted flow rate in femoral artery to pressure gradient for one cardiac cycle

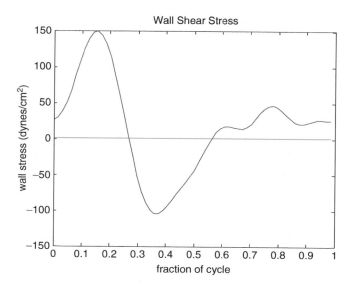

Fig. 7.20 Predicted femoral artery wall shear stress as a function of cardiac cycle

The shear stress can be computed from the velocity at any point in the fluid:

$$\tau_{rz}(r,t) = -\mu \frac{\partial v_z}{\partial r} = \frac{A_0 r}{2} - \sum_{p=1}^{N} \frac{A_p a}{\alpha_p} Re \left\{ \left[\frac{J_1(\alpha_p i^{3/2} \frac{r}{a})}{J_0(\alpha_p i^{3/2})} \right] \sqrt{i} e^{i(p\omega_0 t + \phi_p)} \right\}.$$

Shear stress at the wall of the vessel ($r = a$) is given by:

$$\tau_w(t) = \tau_{rz}(a,t) = \frac{aA_0}{2} - \sum_{p=1}^{N} \frac{A_p a}{\alpha_p} Re \left\{ \left[\frac{J_1(\alpha_p i^{3/2})}{J_0(\alpha_p i^{3/2})} \right] \sqrt{i} e^{i(p\omega_0 t + \phi_p)} \right\}.$$

This relationship is shown in Fig. 7.20. The shear stress caused by the steady component of the pressure gradient is $A_0 a/2 = 16$ dynes/cm^2. However, endothelial cells lining the vessel wall are exposed to significantly higher shear stresses because of the oscillating pressure components. The maximum shear stress of 150 dynes/cm^2 is nearly ten times larger than the average, and the magnitude of the minimum wall stress is more than six times the average wall stress. The zero cross-over points correlate well with the velocity gradients shown in Fig. 7.18.

7.13 Scaling the Navier–Stokes Equation

There are many occasions when specific terms in the Navier Stokes equations are not identically zero, but are small enough to be neglected when compared with other terms. The process of scaling the equations of motion is a systematic approach

7.13 Scaling the Navier–Stokes Equation

that can be used to identify such terms. The goal of scaling is to estimate the order of magnitude of each term in the equation relative to a term considered to be important. If these terms are of the same order of magnitude, they should be retained, but if they are much smaller, they can be neglected. Terms with the same order of magnitude are designated as having order unity, O(1), meaning that the ratio of the two terms is roughly one. By "roughly" we mean the ratio lies between about one third and three. If the ratio is greater than 10, the term is considered to have a higher order of magnitude, and if the ratio is less than 0.1, the term is considered to have a lower order of magnitude. These boundaries are all somewhat arbitrary. The validity of neglecting terms on the basis of scaling is improved the farther their ratio is from O(1).

The first step in scaling the equations of motion is to select appropriate length scales for each coordinate direction. For example, consider fully developed flow in a rectangular shaped conduit with length L, width $2w$, and height $2h$. If the origin of the coordinate system is located at the upstream end in the center of the conduit, then $0 \leq z \leq L$, $0 \leq x \leq w$, and $0 \leq y \leq h$. We can then define dimensionless variables that vary from zero to one:

$$z^* = z/L, \tag{7.82}$$

$$x^* = x/w, \tag{7.83}$$

$$y^* = y/h. \tag{7.84}$$

In addition, the maximum velocity is of the same order of magnitude as the average velocity $\langle v_z \rangle$, so a dimensionless velocity can be defined as:

$$v_z^* = v_z/\langle v_z \rangle. \tag{7.85}$$

If the velocity varies gradually between $x = 0$ and $x = w$, then we might expect the first derivative of $\langle v_z \rangle$ with respect to x to be of order $\langle v_z \rangle/w$, since:

$$\frac{\partial v_z}{\partial x} = \frac{\partial (\langle v_z \rangle v_z^*)}{\partial (wx^*)} = \frac{\langle v_z \rangle}{w} \left\{ \frac{\partial v_z^*}{\partial x^*} \right\} = \frac{\langle v_z \rangle}{w} \{O(1)\}. \tag{7.86}$$

Similarly, the second derivative will be of order $\langle v_z \rangle/w^2$:

$$\frac{\partial^2 v_z}{\partial x^2} = \frac{\partial^2 (\langle v_z \rangle v_z^*)}{\partial (wx^*)^2} = \frac{\langle v_z \rangle}{w^2} \left\{ \frac{\partial^2 v_z^*}{\partial x^{*2}} \right\} = \frac{\langle v_z \rangle}{w^2} \{O(1)\}. \tag{7.87}$$

The quantities in brackets in (7.86) and (7.87) are assumed to be of O(1). The second derivative of v_z with respect to y is:

$$\frac{\partial^2 v_z}{\partial y^2} = \frac{\partial^2 (\langle v_z \rangle v_z^*)}{\partial (hy^*)^2} = \frac{\langle v_z \rangle}{h^2} \left\{ \frac{\partial^2 v_z^*}{\partial y^{*2}} \right\} = \frac{\langle v_z \rangle}{h^2} \{O(1)\}. \tag{7.88}$$

The next step is to substitute these scaled variables into the Navier–Stokes equation.

Since the flow is fully developed, $v_x = v_y = 0$ and the continuity equation reduces to $\partial v_z/\partial z = 0$. If body forces are neglected, the x-component and y-component of the Navier–Stokes equation reduces to $\partial P/\partial x = 0$ and $\partial P/\partial y = 0$, respectively. Consequently, $P = P(z)$. The z-component for fully developed flow in a rectangular conduit is:

$$\frac{\partial^2 v_z}{\partial x^2} + \frac{\partial^2 v_z}{\partial y^2} = \frac{1}{\mu} \frac{dP}{dz}. \tag{7.89}$$

After introducing the scaled variables, this becomes:

$$\left(\frac{h}{w}\right)^2 \frac{\partial^2 v_z^*}{\partial x^{*2}} + \frac{\partial^2 v_z^*}{\partial y^{*2}} = \left(\frac{h^2}{\mu \langle v_z \rangle L}\right) \frac{dP}{dz^*}.$$

Comparison of the first two terms shows that the second term is of order unity and the first term is of order $(h/w)^2$. Thus, if the height is much smaller than the width of the conduit, the first term can be neglected. Furthermore, this provides a scale for the pressure, so a dimensionless pressure P^* could be defined so that the second and third terms are of the same order of magnitude.

$$P^* = \frac{P}{\left(\frac{\mu \langle v_z \rangle L}{h^2}\right)}. \tag{7.90}$$

In Example 6.2.6.2, we examined the flow of a Newtonian fluid between two parallel plates a distance h apart. We can compare that solution to the current situation if we replace h in (6.32a) with $2h$ (height in the current problem) and assume the plates are horizontal:

$$v_{\max} = \frac{3}{2} \langle v_z \rangle = \frac{(2h)^2}{8\mu} \frac{\Delta P}{L}.$$

Comparing this with (7.90), we estimate P^* to be of order 3 for the upstream pressure. If we had used $v_{\max}$ as the scale factor for v_z, rather than $\langle v_z \rangle$, then P^* would be of order 2 for the upstream pressure and of order unity for the average pressure in the channel.

7.13 Scaling the Navier–Stokes Equation

Example 7.13.1 Flow in a Slightly Tapered Blood Vessel.
The radius of a compliant blood vessel will change in response to applied transmural pressure. Consider a compliant vessel which has a length of 1 cm and constant radius $a_0 = 50$ μm when blood flow is zero. Blood is allowed to flow through the vessel with an upstream pressure of 100 mmHg, a downstream pressure of 25 mmHg, and an external pressure (P_e) of zero. Assume flow is laminar with a Reynolds number based on a_0 much smaller than the ratio of tube length to tube radius, and the tube radius is small relative to tube length. Consider blood to be a Newtonian fluid with density of 1.04 g/ml and viscosity of 4 cp. The relationship between vessel radius and transmural pressure is linear:

$$a - a_0 = \frac{\alpha}{2}(P - P_e), \tag{7.91}$$

where the compliance factor $\alpha = 0.2667$ μm/mmHg. Find the following relationships for flow through the tapered tube:

(a) Flow rate through the tapered tube relative to Poiseuille flow through a tube with radius a_0.
(b) Vessel radius vs. distance from the inlet.
(c) Intravascular pressure variation with axial position.
(d) Average velocity vs. axial position in the tube.
(e) Axial velocity vs. radial position in the tube at the inlet, center and outlet axial locations.
(f) Radial velocity vs. radial position in the tube at the inlet, center and outlet axial locations.

Solution. *Initial considerations:* We begin by adopting a cylindrical coordinate system and make the following assumptions:

1. Velocity in the circumferential direction v_θ is zero (no swirl).
2. Velocity in the r and z directions are independent of θ.
3. Flow is steady.
4. Blood is Newtonian and incompressible.
5. The body force is negligible.
6. $\langle v \rangle a_0 / \nu \ll L/a_0$.
7. $a_0 \ll L$.

System definition and environmental interactions: The system of interest is the blood in the tapered vessel. The radius of the vessel varies with axial position because the transmural pressure varies with axial location.

Apprising the problem to identify governing relationships: The appropriate governing equations of motion are the continuity and Navier–Stokes equations since the fluid is Newtonian and incompressible.

Analysis: Applying these conditions to the continuity (7.69) and Navier–Stokes equations (7.75) yields:

Continuity:

$$\frac{1}{r}\frac{\partial(rv_r)}{\partial r} + \frac{\partial(v_z)}{\partial z} = 0.$$

r-Momentum:

$$0 = -\left\{v_r\frac{\partial v_r}{\partial r} + v_z\frac{\partial v_r}{\partial z}\right\} + \nu\left\{\frac{\partial}{\partial r}\left(\frac{1}{r}\frac{\partial}{\partial r}(rv_r)\right) + \frac{\partial^2 v_r}{\partial z^2}\right\} - \frac{1}{\rho}\frac{\partial P}{\partial r}.$$

θ-Momentum:

$$0 = -\frac{1}{\rho r}\frac{\partial P}{\partial \theta}.$$

z-Momentum:

$$0 = -\left\{v_r\frac{\partial v_z}{\partial r} + v_z\frac{\partial v_z}{\partial z}\right\} + \nu\left\{\frac{1}{r}\frac{\partial}{\partial r}\left(r\frac{\partial v_z}{\partial r}\right) + \frac{\partial^2 v_z}{\partial z^2}\right\} - \frac{1}{\rho}\frac{\partial P}{\partial z}.$$

From the θ-momentum equation, we learn that the pressure is not a function of θ and at most will be a function of radial location and axial position, $P(r,z)$. Although we have eliminated all of the terms from the continuity and Navier–Stokes equations that are identically zero, there are still terms that may be small in relation to other terms in each expression. Let us start by scaling the continuity equation to estimate the relative magnitude of the velocity in the r direction. Let us define the following dimensionless values based on the vessel geometry and average velocity, $\langle v \rangle$:

$$r^* = r/a_0,$$

$$z^* = z/L,$$

$$v_z^* = v_z/\langle v \rangle.$$

Each of these values will be of order unity. Substituting these into the continuity equation:

$$\frac{1}{a_0}\left(\frac{1}{r^*}\frac{\partial(r^* v_r)}{\partial r^*}\right) + \frac{\langle v \rangle}{L}\frac{\partial(v_z^*)}{\partial z^*} = 0.$$

Let us define a dimensionless radial velocity so that both terms are of the same order of magnitude:

7.13 Scaling the Navier–Stokes Equation

$$v_r^* = \frac{L}{a_0} \frac{v_r}{\langle v \rangle}.$$

Substituting the dimensionless values into the r-momentum equation, we find:

$$0 = -\frac{a_0 \langle v \rangle^2}{L^2} \left\{ v_r^* \frac{\partial v_r^*}{\partial r^*} + v_z^* \frac{\partial v_r^*}{\partial z^*} \right\} + \frac{\nu \langle v \rangle}{a_0 L} \left\{ \frac{\partial}{\partial r^*} \left(\frac{1}{r^*} \frac{\partial}{\partial r^*} (r^* v_r^*) \right) \right\}$$
$$+ \frac{\nu \langle v \rangle a_0}{L^3} \left\{ \frac{\partial^2 v_r^*}{\partial z^{*2}} \right\} - \frac{1}{\rho a_0} \frac{\partial P}{\partial r^*}.$$

Multiplying the entire equation by the inverse of the coefficient of the second term, we obtain:

$$0 = -\frac{a_0}{L} \left(\frac{\langle v \rangle a_0}{\nu} \right) \left\{ v_r^* \frac{\partial v_r^*}{\partial r^*} + v_z^* \frac{\partial v_r^*}{\partial z^*} \right\} + \left\{ \frac{\partial}{\partial r^*} \left(\frac{1}{r^*} \frac{\partial}{\partial r^*} (r^* v_r^*) \right) \right\}$$
$$+ \frac{a_0^2}{L^2} \left\{ \frac{\partial^2 v_r^*}{\partial z^{*2}} \right\} - \frac{L}{\mu \langle v \rangle} \frac{\partial P}{\partial r^*}.$$

By design, the terms in brackets { } are all of order unity. Since a_0/L is small (assumption 7), the third term is small in relation to the second term. If the Reynolds number based on the tube radius is much less than L/a_0 (assumption 6), then the first term is also small in relation to the second term. Although these terms are not identically zero, their contribution can be neglected. As a consequence, the dimensionless r-component of the momentum equation can be approximated as:

$$\frac{\partial}{\partial r^*} \left(\frac{1}{r^*} \frac{\partial}{\partial r^*} (r^* v_r^*) \right) = \frac{L}{\mu \langle v \rangle} \frac{\partial P}{\partial r^*}.$$

Introducing the dimensionless values into the z-momentum equation leads to

$$0 = -\frac{a_0}{L} \left(\frac{\langle v \rangle a_0}{\nu} \right) \left\{ v_r^* \frac{\partial v_z^*}{\partial r^*} + v_z^* \frac{\partial v_z^*}{\partial z^*} \right\} + \left\{ \frac{1}{r^*} \frac{\partial}{\partial r^*} \left(r^* \frac{\partial v_z^*}{\partial r^*} \right) \right\} + \frac{a_0^2}{L^2} \left\{ \frac{\partial^2 v_z^*}{\partial z^{*2}} \right\} - \frac{a_0^2}{\mu \langle v \rangle L} \frac{\partial P}{\partial z^*}.$$

Applying the assumptions (6) and (7) to this expression allows us to drop the first and third terms:

$$\frac{1}{r^*} \frac{\partial}{\partial r^*} \left(r^* \frac{\partial v_z^*}{\partial r^*} \right) = \frac{a_0^2}{\mu \langle v \rangle L} \frac{\partial P}{\partial z^*}.$$

Taking the ratio of the r-momentum equation to the z-momentum equation, we find:

$$\frac{\dfrac{\partial P}{\partial r^*}}{\dfrac{\partial P}{\partial z^*}} = \left(\frac{a_0^2}{L^2}\right) \left\{ \frac{\dfrac{\partial}{\partial r^*}\left(\dfrac{1}{r^*}\dfrac{\partial}{\partial r^*}(r^* v_r^*)\right)}{\dfrac{1}{r^*}\dfrac{\partial}{\partial r^*}\left(r^* \dfrac{\partial v_z^*}{\partial r^*}\right)} \right\} = \left(\frac{a_0^2}{L^2}\right)\{O(1)\}.$$

If the problem has been scaled properly, the symbol O(1) indicates that the quantity in brackets { } is of order unity. Therefore, since $a_0 \ll L$, the pressure gradient in the r-direction is much smaller than the pressure gradient in the z-direction, and to a first approximation the radial dependence of pressure can be neglected altogether. Consequently, we can assume that $P = P(z)$. Returning to the dimensional form of the z-momentum equation, we can now write:

$$\frac{1}{r}\frac{\partial}{\partial r}\left(r\frac{\partial v_z}{\partial r}\right) = \frac{1}{\mu}\frac{dP}{dz} = f(z).$$

Multiplying this by r and integrating with respect to r yields:

$$\frac{\partial v_z}{\partial r} = \frac{f(z)r}{2} + \frac{g(z)}{r},$$

where $g(z)$ is an arbitrary function of z. Applying the symmetry boundary condition at the tube center, we find that $g(z)$ must equal zero. Integrating again, we find:

$$v_z = \frac{f(z)r^2}{4} + h(z),$$

where $h(z)$ is determined by the application of the no-slip boundary condition ($v_z = 0$ at $r = a(z)$) to be $-a^2 f(z)/4$. The expression for velocity now becomes:

$$v_z = \frac{f(z)}{4}(r^2 - a^2) = \frac{1}{4\mu}\frac{dP}{dz}(r^2 - a^2).$$

The flow rate Q_V at any position z in the tapered tube can be found by integrating the velocity over the tube cross section:

$$Q_V = 2\pi \int_0^a v_z r\, dr = \frac{\pi}{2\mu}\frac{dP}{dz}\int_0^a (r^2 - a^2) r\, dr = -\frac{\pi a^4}{8\mu}\frac{dP}{dz}.$$

Since the vessel radius varies linearly with transmural pressure, we can write:

$$\frac{dP}{dz} = \left(\frac{dP}{da}\right)\left(\frac{da}{dz}\right) = \frac{2}{\alpha}\frac{da}{dz}.$$

7.13 Scaling the Navier–Stokes Equation

Substituting this back in for dP/dz in the flow equation:

$$Q_V = -\frac{\pi a^4}{4\mu\alpha}\frac{da}{dz}.$$

Separating variables and integrating from the inlet ($z=0$) to an arbitrary position z in the tapered vessel:

$$Q_V z = \frac{\pi}{20\mu\alpha}\left[a(0)^5 - a(z)^5\right].$$

Solving for the variation of vessel radius with axial location, $a(z)$:

$$a(z) = \left\{a^5(0) - \left(\frac{20\mu\alpha Q_V}{\pi}\right)z\right\}^{\frac{1}{5}}. \quad (7.92)$$

Note that unless the transmural pressure at $z=0$ is zero (i.e., $P(0)=P_e$), a_0 will generally not be the same as $a(0)$. If we let $z=L$ in the flow expression, flow can be determined in terms of the upstream and downstream pressures:

$$Q_V = -\frac{\pi}{20\mu\alpha L}\left[\left\{a_0 + \frac{\alpha}{2}(P(0)-P_e)\right\}^5 - \left\{a_0 + \frac{\alpha}{2}(P(L)-P_e)\right\}^5\right]. \quad (7.93)$$

Now, turning our attention to the axial velocity profile, $v_z(r,z)$:

$$v_z = \frac{1}{4\mu}\frac{dP}{dz}(r^2 - a^2) = \frac{1}{2\mu\alpha}\frac{da}{dz}(r^2 - a^2) = \frac{1}{2\mu\alpha}\left(\frac{-4\mu\alpha Q_V}{\pi a^4}\right)(r^2 - a^2).$$

But since $a(z)$ is now known, this can be written:

$$v_z(r,z) = \frac{2Q_V}{\pi a^2}\left(1 - \left(\frac{r}{a}\right)^2\right). \quad (7.94)$$

Note that this is parabolic, as it is in a straight tube, and if $r=0$, the velocity at the center of the tube is twice the average velocity.

Velocity in the radial direction can be determined by integrating the continuity equation:

$$rv_r = -\int r\frac{\partial v_z}{\partial z}dr = -\frac{\partial}{\partial z}\int\frac{f(z)}{4}(r^3 - ra^2)dr = -\frac{1}{4}\frac{df}{dz}\int(r^3 - ra^2)dr$$
$$+ \frac{f}{4}\left(2a\frac{da}{dz}\right)\int r\,dr + k(z),$$

where $k(z)$ is an arbitrary function of z. Solving for v_r, and recognizing that $k(z)$ must be zero since v_r must be finite at $r = 0$:

$$v_r(r,z) = -\frac{1}{8}\frac{df}{dz}\left(\frac{r^3}{2} - ra^2\right) + \frac{fa}{4}\left(\frac{da}{dz}\right)r,$$

The no-slip boundary condition at $r = a$ indicates that

$$\frac{1}{f}\frac{df}{dz} = -\frac{4}{a}\frac{da}{dz},$$

and this can be confirmed by expanding the definition of f and the expression for $a(z)$. Using this to eliminate df/dz from the expression for v_r:

$$v_r(r,z) = \frac{1}{4}(f)\left(\frac{da}{dz}\right)\left\{\left(\frac{r}{a}\right)(a^2 - r^2)\right\} = \frac{1}{4}\left(-2\frac{Q_V}{\pi a^4}\right)\left(-\frac{4\mu\alpha Q_V}{\pi a^4}\right)\left\{\left(\frac{r}{a}\right)(a^2 - r^2)\right\}$$

or, finally, velocity in the radial direction is:

$$v_r(r,z) = 2\left(\frac{Q_V}{\pi a^2}\right)^2\left(\frac{\mu\alpha}{a^2}\right)\left(\frac{r}{a}\right)\left(1 - \frac{r^2}{a^2}\right). \quad (7.95)$$

The radial position where the radial velocity is maximum is found by setting the derivative of v_r with respect to r equal to zero. This shows the radial velocity is a maximum when $r = a/\sqrt{3}$.

Examining and interpreting the results: We have now derived all of the relationships needed to answer the questions posed in the problem statement. The radii at the vessel inlet and outlet are 63.3 and 53.3 μm, respectively. The ratio $a(0)/L = 6.33 \times 10^{-3}$, so the assumption that this ratio is small is valid. The flow through the tube is 0.0012 ml/s, which is 1.88 times higher than the Poiseuille flow through a tube with the same pressure gradient and a constant radius of 50 μm. The Reynolds numbers at the tube inlet and outlet are 1.5 and 1.79, respectively, so the assumption that the product of Reynolds number and $a/L \ll 1$ is valid. Graphical solutions for $a(z)$, $P(z)$ and average axial velocity vs. axial position are shown in Fig. 7.21 for the conditions given in the problem statement. The radius and pressure vary axially in the same nonlinear fashion. The average velocity increases from 9.15 cm/s at the inlet to 12.91 cm/s at the outlet. The axial and radial velocity profiles are shown in Fig. 7.21 for the inlet, center and outlet of the vessel. The separate graphs for axial velocity collapse onto a single curve when $v_z(r, z)$ is normalized by $v_z(0, z)$:

$$\frac{v_z(r,z)}{v_z(0,z)} = 1 - \left(\frac{r}{a}\right)^2$$

7.13 Scaling the Navier–Stokes Equation

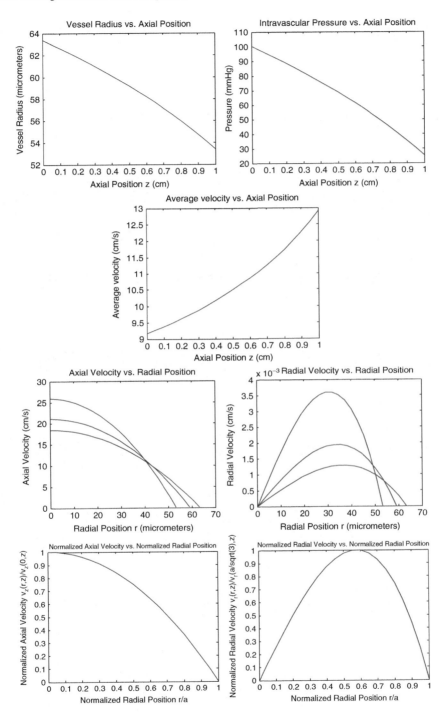

Fig. 7.21 Tapered tube predictions

and the three graphs for $v_r(r, z)$ will collapse onto a single curve when $v_r(r, z)$ is normalized by $v_r(r/\sqrt{3}, z)$ and r is normalized by a:

$$\frac{v_r(r,z)}{v_r\left(\frac{a}{\sqrt{3}}, z\right)} = \frac{3\sqrt{3}}{2}\left(\frac{r}{a}\right)\left[1 - \left(\frac{r}{a}\right)^2\right].$$

These velocity profiles are known as self-similar profiles.

Example 7.13.2 Scaling: Womersley Number in Periodic Flow

In Example 7.12.3, we found that when a periodic pressure gradient is applied to fluid in a cylindrical tube, the velocity of fluid near the wall slightly lags the pressure gradient, while fluid in the center of the tube shows a more pronounced phase lag. Comparisons between the pressure gradient in Fig. 7.17 and the velocities in Fig. 7.18 illustrate this for a Womersley number of 3.33. If a simple periodic pressure gradient $-\partial P/\partial z = A\cos \omega t$ is applied at various frequencies, the velocity computed at the center of the tube is shown in Fig. 7.22.

Thus, at low values of α, the magnitude of the velocity oscillations are the same and the velocity is found to be in phase with the pressure gradient. However, at high values of α, the magnitude is attenuated and the velocity lags the pressure gradient. Use scaling to show that when the Womersley number α is high, the velocity will be 90° out of phase with the pressure gradient; however, at very low values of α the two will be in phase.

Solution. *Initial considerations:* We will use the same set of assumptions, same system, and the same governing equations that were used in the original analysis, Example 7.12.3.

Analysis: The Navier–Stokes equation reduces to:

$$\frac{\partial v_z}{\partial t} = \nu \frac{1}{r}\frac{\partial}{\partial r}\left(r\frac{\partial v_z}{\partial r}\right) - \frac{1}{\rho}\frac{\partial P}{\partial z}.$$

Appropriate scales for the velocity and radial position would be the maximum velocity v_{max} and the vessel radius a, respectively:

$$v_z^* = \frac{v_z}{v_{max}},$$
$$r^* = \frac{r}{a}.$$

If the pressure gradient is periodic with angular frequency ω, then an appropriate dimensionless scale would be:

7.13 Scaling the Navier–Stokes Equation

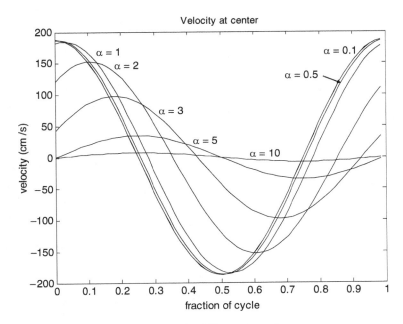

Fig. 7.22 Velocity at center of tube for various Womersly numbers

$$t^* = \frac{t}{1/\omega} = \omega t.$$

Substituting these into the z-component of the Navier–Stokes equation yields the following equation:

$$\omega v_{max} \frac{\partial v_z^*}{\partial t^*} = \frac{\nu v_{max}}{a^2} \frac{1}{r^*} \frac{\partial}{\partial r^*}\left(r^* \frac{\partial v_z^*}{\partial r^*}\right) - \frac{1}{\rho} \frac{\partial P}{\partial z}.$$

The ratio of the periodic inertial term to the viscous term is:

$$\frac{\text{inertial term}}{\text{viscous term}} = \frac{\omega a^2}{\nu} \left\{ \frac{\frac{\partial v_z^*}{\partial t^*}}{\frac{1}{r^*} \frac{\partial}{\partial r^*}\left(r^* \frac{\partial v_z^*}{\partial r^*}\right)} \right\} \approx \frac{\omega a^2}{\nu} = \alpha^2.$$

Since the term in brackets should be of order unity, the Womersley number squared represents the ratio of the magnitude of the periodic inertial term to the viscous term. If α is very small, the inertial term can be ignored relative to the viscous term. If it is very large, the viscous term can be neglected while the inertial term is retained.

Let us consider the following form for the periodic pressure gradient:

$$-\frac{\partial P}{\partial z} = A\cos(\omega t).$$

If the Womersley number is very small, we can neglect the inertial term, and the Navier–Stokes equation becomes:

$$\frac{1}{r}\frac{\partial}{\partial r}\left(r\frac{\partial v_z}{\partial r}\right) = -\frac{A}{\mu}\cos(\omega t).$$

This can be solved if we let the velocity be the product of a function of time, $G(t)$ and a function of radial position, $U(r)$:

$$v_z(r,t) = U(r)G(t).$$

Substituting this into the Navier–Stokes equation:

$$G(t)\left\{\frac{1}{r}\frac{\mathrm{d}}{\mathrm{d}r}\left(r\frac{\mathrm{d}U(r)}{\mathrm{d}r}\right)\right\} = -\frac{A}{\mu}\cos(\omega t).$$

Since the right-hand side of this equation is only a function of time, then the term in brackets must be equal to a constant, say k.

$$\frac{1}{r}\frac{\mathrm{d}}{\mathrm{d}r}\left(r\frac{\mathrm{d}U(r)}{\mathrm{d}r}\right) = k.$$

The solution for $U(r)$ is a parabolic velocity profile. Therefore, the velocity for the case of periodic flow in a tube with small Womersley number can be written:

$$v_z(r,t) = -\frac{A}{\mu k}U(r)\cos(\omega t). \tag{7.96}$$

Examining and interpreting the results: Note that the velocity for small α^2 is predicted to be in phase with the pressure gradient and will have a magnitude directly proportional to the magnitude of the pressure gradient. Thus, the velocity profile at low Womersley number will be parabolic at all times, with the magnitude of the average velocity being proportional to the oscillating pressure gradient. This is consistent with the analytic solution shown at various times over a single cycle in Fig. 7.23 for $\alpha = 0.1$.

7.13 Scaling the Navier–Stokes Equation

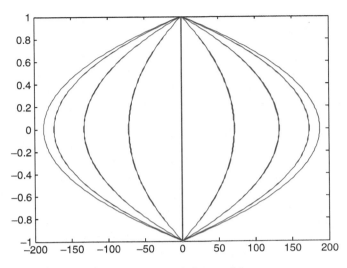

Fig. 7.23 Velocity profiles at equal cycle intervals for $\alpha = 0.1$

If the Womersley number is high, the Navier–Stokes equation for oscillating flow in a tube reduces to:

$$\frac{\partial v_z}{\partial t} = \frac{A}{\rho} \cos(\omega t).$$

The right-hand side of this equation depends only on time. Integrating, we find:

$$v_z(r,t) = v_z(r,0) - \frac{A}{\rho \omega} \sin(\omega t). \tag{7.97}$$

Consequently, at high values of α^2, all fluid particles will oscillate back and forth 90° out of phase with the pressure gradient, and the relative radial positions of all fluid particles will remain the same throughout the cycle. Finally, the magnitude of the velocity will be proportional to the pressure gradient and inversely proportional to the angular frequency of oscillations. This is the type of behavior we observe in the central portion of the tube in Example 7.12.3 at high frequencies. The analytic solution for $\alpha = 20$ is shown in Fig. 7.24. This is in general agreement with our scaled simplification. The velocity is flat as predicted, except in a very thin layer near the wall, known as a hydrodynamic boundary layer. Viscous effects cannot be neglected in this region, which is more closely in phase with the pressure gradient. More on hydrodynamic boundary layers can be found in the following example.

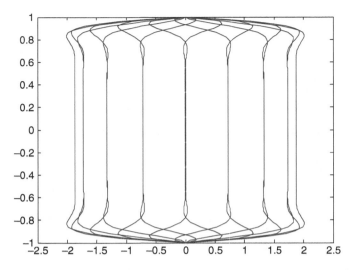

Fig. 7.24 Velocity profiles for periodic flow with $\alpha = 20$

Example 7.13.3 Hydrodynamic Boundary Layers.
Viscous effects are often confined to a thin region near the surface of solid objects, known as the hydrodynamic boundary layer. Estimate the thickness of this layer and the drag force on a solid flat plate for the steady flow of blood and air. The plate is 10 cm long and 10 cm wide. The steady-state velocity far from the solid surface is $v_0 = 100$ cm/s, parallel to the surface.

Solution. *Initial considerations:* Let the thickness of the boundary layer be δ. The thickness will grow with distance from the leading edge x, but we will assume that $\delta(x)$ is always much smaller than the length of the solid surface, L. In addition to a velocity v_x parallel to the solid surface, there is a velocity v_y directed away from the surface. However, the velocity in the z-direction is zero and nothing changes in the z-direction, so $\partial(\ )/\partial z = 0$. Flow is steady, so $\partial(\ )/\partial t = 0$.

System definition and environmental interactions: The system of interest is the fluid flowing past a stationary wall. The wall exerts a viscous drag force on the fluid, which changes with downstream position.

Apprising the problem to identify governing relationships: We assume that the fluid flowing past the surface is an incompressible, Newtonian fluid, so the continuity equation and Navier–Stokes equation are the governing relationships.

Analysis: Eliminating terms from the Navier–Stokes equations that are identically zero, we find:

7.13 Scaling the Navier–Stokes Equation

x-momentum

$$0 = -\rho\left[v_x\frac{\partial v_x}{\partial x} + v_y\frac{\partial v_x}{\partial y}\right] + \mu\left[\frac{\partial^2 v_x}{\partial x^2} + \frac{\partial^2 v_x}{\partial y^2}\right] - \frac{\partial \wp}{\partial x},$$

y-momentum

$$0 = -\rho\left[v_x\frac{\partial v_y}{\partial x} + v_y\frac{\partial v_y}{\partial y}\right] + \mu\left[\frac{\partial^2 v_y}{\partial x^2} + \frac{\partial^2 v_y}{\partial y^2}\right] - \frac{\partial \wp}{\partial y},$$

z-momentum

$$0 = \frac{\partial \wp}{\partial z}.$$

The continuity equation is:

$$\frac{\partial v_x}{\partial x} + \frac{\partial v_y}{\partial y} = 0.$$

We now introduce the dimensionless variables:

$$v_x^* = v_x/v_0,$$
$$x^* = x/L,$$
$$y^* = y/\delta.$$

Substituting these into the continuity equation, we find the following dimensionless scale for v_y:

$$v_y^* = \left(\frac{v_y}{v_0}\right)\left(\frac{L}{\delta}\right).$$

Substituting these dimensionless values into the momentum equations, we find:

x-momentum

$$\frac{\partial \wp}{\partial x^*} = \rho v_0^2 \left\{ -\left[v_x^* \frac{\partial v_x^*}{\partial x^*} + v_y^* \frac{\partial v_x^*}{\partial y^*}\right] + \frac{\mu L}{\rho v_0 \delta^2}\left[\left(\frac{\delta}{L}\right)^2 \frac{\partial^2 v_x^*}{\partial x^{*2}} + \frac{\partial^2 v_x^*}{\partial y^{*2}}\right]\right\},$$

y-momentum

$$\frac{\partial \wp}{\partial y^*} = \rho v_0^2 \left(\frac{\delta}{L}\right)^2 \left\{ -\left[v_x^* \frac{\partial v_y^*}{\partial x^*} + v_y^* \frac{\partial v_y^*}{\partial y^*}\right] + \frac{\mu L}{\rho v_0 \delta^2}\left[\left(\frac{\delta}{L}\right)^2 \frac{\partial^2 v_y^*}{\partial x^{*2}} + \frac{\partial^2 v_y^*}{\partial y^{*2}}\right]\right\}.$$

Therefore the pressure gradient in the *y*-direction is small relative to the pressure gradient in the *x*-direction, and to a first approximation can be neglected altogether. Consequently, the modified pressure can at most be a function of x. Far from the flat solid surface, the pressure is constant, so $dP/dx = 0$. In addition, scaling tells us that $\partial^2 v_x/\partial x^2 \ll \partial^2 v_y/\partial y^2$, and the former can be neglected. Under these assumptions, the *x*-momentum equation reduces to:

$$\left[v_x \frac{\partial v_x}{\partial x} + v_y \frac{\partial v_x}{\partial y}\right] = \nu \frac{\partial^2 v_x}{\partial y^2}.$$

Introducing a stream function from (7.50) and (7.51), we obtain the following third-order, nonlinear partial differential equation:

$$\frac{\partial \psi}{\partial y}\left(\frac{\partial^2 \psi}{\partial x \partial y}\right) - \frac{\partial \psi}{\partial x}\left(\frac{\partial^2 \psi}{\partial y^2}\right) = \nu\left(\frac{\partial^3 \psi}{\partial y^3}\right). \tag{7.98}$$

The boundary conditions that must be satisfied are:

1. At $y = 0$, $v_x = 0$, or $\left.\dfrac{\partial \psi}{\partial y}\right|_{y=0} = 0$.
2. At $y = 0$, $v_y = 0$, or $\left.\dfrac{\partial \psi}{\partial x}\right|_{y=0} = 0$.
3. As $y \to \infty$, $v_x \to v_0$, or $\left.\dfrac{\partial \psi}{\partial y}\right|_{y=\infty} = v_0$.

We can postulate that the solution takes the form:

$$v_x = \frac{\partial \psi}{\partial y} = v_0 F(\eta). \tag{7.99}$$

The variable η is a combination of the variables x and y:

$$\eta = ayx^n, \tag{7.100}$$

where a and n are constants to be determined from the analysis. Integrating (7.99):

$$\psi = \int v_0 F(\eta) \mathrm{d}y = v_0 \int \frac{F(\eta)}{\left(\dfrac{\partial \eta}{\partial y}\right)} \mathrm{d}\eta = \frac{v_0}{ax^n} \int F(\eta) \mathrm{d}\eta = \frac{v_0}{ax^n} f(\eta),$$

the function $f(\eta)$ is defined as $\int F(\eta) \mathrm{d}\eta$. The terms in (7.98) become:

$$\frac{\partial \psi}{\partial y} = v_0 \frac{\mathrm{d}f}{\mathrm{d}\eta},$$

$$\frac{\partial \psi}{\partial x} = \frac{nyv_0}{x}\frac{\mathrm{d}f}{\mathrm{d}\eta} - \frac{nv_0}{ax^{n+1}}f,$$

$$\frac{\partial^2 \psi}{\partial x \partial y} = nayx^{n-1} v_0 \frac{\mathrm{d}^2 f}{\mathrm{d}\eta^2},$$

$$\frac{\partial^2 \psi}{\partial y^2} = ax^n v_0 \frac{\mathrm{d}^2 f}{\mathrm{d}\eta^2},$$

$$\frac{\partial^3 \psi}{\partial y^3} = a^2 x^{2n} v_0 \frac{\mathrm{d}^3 f}{\mathrm{d}\eta^3}.$$

7.13 Scaling the Navier–Stokes Equation

Substituting these back into (7.98), we obtain:

$$\frac{d^3 f}{d\eta^3} = \left(\frac{n v_0}{a^2 \nu x^{2n+1}}\right)\left[f \frac{d^2 f}{d\eta^2}\right].$$

If this is to be an ordinary differential equation, the exponent of x, which is $2n + 1$, must equal zero. Consequently, $n = -1/2$. In addition, for the equation to be dimensionless, the factor a must be some multiple of $(v_0/\nu)^{1/2}$. If we let $a = (v_0/\nu)^{1/2}$, then the ordinary differential equation becomes:

$$f \frac{d^2 f}{d\eta^2} + 2 \frac{d^3 f}{d\eta^3} = 0, \qquad (7.101)$$

where:

$$\eta = y \sqrt{\frac{v_0}{\nu x}}. \qquad (7.102)$$

Equation (7.101) needs to be solved subject to the following boundary conditions:

1. $\eta = 0, f = 0$.
2. $\eta = 0, df/d\eta = 0$.
3. $\eta \to \infty, df/d\eta \to 1$.

An analytic solution to this nonlinear equation does not exist. Blasius (1908) solved the equation numerically. Matlab code for solving this problem is given in Sect. 15.5, where the concentration profile is also found. The solution for f and the first two derivatives are shown in Fig. 7.25. From this solution, we can find the components of the velocity in the x and y directions and the shear stress as a function of position:

$$v_x = \frac{\partial \Psi}{\partial y} = v_0 \frac{df}{d\eta}, \qquad (7.103)$$

$$v_y = -\frac{\partial \Psi}{\partial x} = \frac{1}{2}\sqrt{\frac{v_0 \nu}{x}}\left[\eta \frac{df}{d\eta} - f\right], \qquad (7.104)$$

$$\tau_{yx} = -\mu\left(\frac{\partial^2 \psi}{\partial y^2} - \frac{\partial^2 \psi}{\partial x^2}\right) = -\mu\left[\sqrt{\frac{v_0}{\nu x}}\left(v_0 + \eta \sqrt{\frac{v_0 \nu}{x}}\right)\frac{d^2 f}{d\eta^2} - \frac{1}{2x}\sqrt{\frac{v_0 \nu}{x}}\left(\eta \frac{df}{d\eta} - f\right)\right]. \qquad (7.105)$$

Examining and interpreting the results: It is customary to define the boundary layer thickness δ as the distance from the wall to the point in the flow, where the velocity is 99% of the free stream velocity. From Fig. 7.25, we see that $v_x/v_0 = 0.99$ when $\eta = 5$, or:

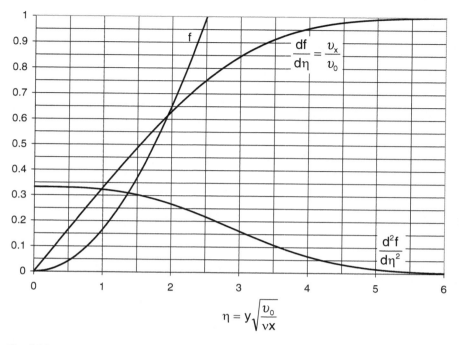

Fig. 7.25 Blasius solution for flow over a flat plate

$$\delta = 5\sqrt{\frac{vx}{v_0}}. \tag{7.106}$$

The thickness of the boundary layer grows as the square root of the distance from the leading edge of the solid surface. As the free stream velocity increases, the boundary layer thickness decreases. The higher the kinematic viscosity, the thicker will be the boundary layer. The shear stress at the solid surface, where $\eta = 0$ is:

$$\tau_w = \tau_{yx}|_{\eta=0} = -\mu\left[\sqrt{\frac{v_0}{vx}}v_0\frac{d^2f}{d\eta^2}\bigg|_{\eta=0} + \frac{1}{2x}\sqrt{\frac{vv_0}{x}}\left(f|_{\eta=0}\right)\right]. \tag{7.107}$$

The Blasius solution (Fig. 7.25) shows that $f(0) = 0$ and $d^2f/d\eta^2$ at $\eta = 0$ is 0.332. Therefore,

$$\tau_w = -0.332\mu v_0\sqrt{\frac{v_0}{vx}}. \tag{7.108}$$

The total force by the fluid on the top and bottom surfaces of a flat plate of width w and length L is:

$$F_D = -2w \int_0^L \tau_w dx = 0.664 \mu w v_0 \sqrt{\frac{v_0}{\nu}} \int_0^L x^{-1/2} dx = 1.328 \rho w v_0 \sqrt{v_0 L \nu}. \quad (7.109)$$

The friction factor for flow past a flat plate is defined by (5.54) with $F_k/A = -\tau_w$, and K equal to the kinetic energy per unit volume of the approaching fluid:

$$f = \frac{F_k}{KA} = \frac{-\tau_w}{\left(\frac{1}{2}\rho v_0^2\right)}. \quad (7.110)$$

Inserting (7.108) into (7.110), we find:

$$\frac{f}{2} = 0.332 \sqrt{\frac{\nu}{x v_0}} = 0.332 Re_x^{-\frac{1}{2}}. \quad (7.111)$$

The thickness of the boundary layer at the end of the plate is:

$$\delta(L) = 5 \sqrt{\frac{\nu L}{v_0}}. \quad (7.112)$$

Additional comments: These results are valid for laminar flow over a flat plate, which is generally appropriate when the Reynolds number based on the axial position is less than 300,000:

$$Re_x = \frac{v_0 x}{\nu} \leq 3 \times 10^5. \quad (7.113)$$

For blood with $\nu = 0.04$ cm^2/s, approach velocity of 100 cm/s and length of 10 cm, the boundary layer thickness at the end of the plate is 0.32 cm. For air ($\nu = 0.17$ cm^2/s) flowing past the same plate at the same velocity, the boundary layer thickness at the end of the plate is 0.65 cm. The Reynolds number at the end of the plate is 5,882 for air and 25,000 for blood. The boundary layer is laminar for both fluids. In both cases, the ratio δ/L is much smaller than 1, in agreement with our original assumption in simplifying the Navier–Stokes equation. The drag force on a 10 cm wide plate for blood ($\rho = 1.05$ g/cm^3) is 8,819 dynes, and for air ($\rho = 1.29 \times 10^{-3}$ g/cm^3) is 22.3 dynes.

7.14 General Momentum Equations for Use with Non-Newtonian Fluids

The continuity equation derived in Sect. 7.2 is independent of properties of the fluid. The identity of stress components ($\tau_{ij} = \tau_{ji}$) found in Sect. 7.4, was also derived independently of any properties. However, the Navier–Stokes equations were derived under the assumption that the fluid obeyed Newton's Law of Viscosity. Therefore, the Navier–Stokes equations cannot be used to analyze the flow of non-Newtonian fluids. The starting point for the analysis of non-Newtonian fluids, as presented in Sect. 7.5, is based on the conservation of linear momentum. In rectangular coordinates, the three components of the momentum equation are:

x-momentum:

$$\rho \frac{\partial v_x}{\partial t} = -\frac{\partial \wp}{\partial x} - \rho \left[v_x \frac{\partial v_x}{\partial x} + v_y \frac{\partial v_x}{\partial y} + v_z \frac{\partial v_x}{\partial z} \right] - \left[\frac{\partial \tau_{xx}}{\partial x} + \frac{\partial \tau_{yx}}{\partial y} + \frac{\partial \tau_{zx}}{\partial z} \right],$$

y-momentum:

$$\rho \frac{\partial v_y}{\partial t} = -\frac{\partial \wp}{\partial y} - \rho \left[v_x \frac{\partial v_y}{\partial x} + v_y \frac{\partial v_y}{\partial y} + v_z \frac{\partial v_y}{\partial z} \right] - \left[\frac{\partial \tau_{xy}}{\partial x} + \frac{\partial \tau_{yy}}{\partial y} + \frac{\partial \tau_{zy}}{\partial z} \right],$$

z-momentum:

$$\rho \frac{\partial v_z}{\partial t} = -\frac{\partial \wp}{\partial z} - \rho \left[v_x \frac{\partial v_z}{\partial x} + v_y \frac{\partial v_z}{\partial y} + v_z \frac{\partial v_z}{\partial z} \right] - \left[\frac{\partial \tau_{xz}}{\partial x} + \frac{\partial \tau_{yz}}{\partial y} + \frac{\partial \tau_{zz}}{\partial z} \right]. \quad (7.114)$$

The general momentum equations in cylindrical coordinates are:

r-component

$$\frac{\partial v_r}{\partial t} = -\left\{ v_r \frac{\partial v_r}{\partial r} + \frac{v_\theta}{r} \frac{\partial v_r}{\partial \theta} - \frac{v_\theta^2}{r} + v_z \frac{\partial v_r}{\partial z} \right\} - \frac{1}{\rho} \left\{ \frac{1}{r} \frac{\partial}{\partial r}((r\tau_{rr})) + \frac{1}{r} \frac{\partial \tau_{r\theta}}{\partial \theta} - \frac{\tau_{\theta\theta}}{r} + \frac{\partial \tau_{rz}}{\partial z} \right\}$$
$$- \frac{1}{\rho} \frac{\partial P}{\partial r} + g_r,$$

θ-component

$$\frac{\partial v_\theta}{\partial t} = -\left\{ v_r \frac{\partial v_\theta}{\partial r} + \frac{v_\theta}{r} \frac{\partial v_\theta}{\partial \theta} - \frac{v_r v_\theta}{r} + v_z \frac{\partial v_\theta}{\partial z} \right\} - \frac{1}{\rho} \left\{ \frac{1}{r^2} \frac{\partial}{\partial r}((r^2 \tau_{r\theta})) + \frac{1}{r} \frac{\partial \tau_{\theta\theta}}{\partial \theta} + \frac{\partial \tau_{\theta z}}{\partial z} \right\}$$
$$- \frac{1}{\rho r} \frac{\partial P}{\partial \theta} + g_\theta,$$

z-component

$$\frac{\partial v_z}{\partial t} = -\left\{ v_r \frac{\partial v_z}{\partial r} + \frac{v_\theta}{r} \frac{\partial v_z}{\partial \theta} + v_z \frac{\partial v_z}{\partial z} \right\} - \frac{1}{\rho} \left\{ \frac{1}{r} \frac{\partial}{\partial r}((r\tau_{rz})) + \frac{1}{r} \frac{\partial \tau_{\theta z}}{\partial \theta} + \frac{\partial \tau_{zz}}{\partial z} \right\} - \frac{1}{\rho} \frac{\partial P}{\partial z} + g_z.$$

(7.115)

The general momentum equations in spherical coordinates are:

r-component

$$\frac{\partial v_r}{\partial t} = -\left\{ v_r \frac{\partial v_r}{\partial r} + \frac{v_\theta}{r} \left(v_\theta + \frac{\partial v_r}{\partial \theta} \right) + \frac{v_\phi}{r \sin \theta} \left(\frac{\partial v_r}{\partial \phi} - v_\phi \sin \theta \right) \right\} - \frac{1}{\rho} \frac{\partial P}{\partial r} + g_r$$
$$- \frac{1}{\rho} \left\{ \frac{1}{r^2} \frac{\partial}{\partial r}(r^2 \tau_{rr}) + \frac{1}{r \sin \theta} \frac{\partial}{\partial \theta}(\tau_{r\theta} \sin \theta) + \frac{1}{r \sin \theta} \frac{\partial \tau_{r\phi}}{\partial \phi} - \left(\frac{\tau_{\theta\theta} + \tau_{\phi\phi}}{r} \right) \right\},$$

7.15 Constitutive Relationships for Non-Newtonian Fluids

θ-component

$$\frac{\partial v_\theta}{\partial t} = -\left\{ v_r \frac{\partial v_\theta}{\partial r} + \frac{v_\theta}{r}\left(v_r + \frac{\partial v_\theta}{\partial \theta}\right) + \frac{v_\phi}{r\sin\theta}\left(\frac{\partial v_\theta}{\partial \phi} - v_\phi \cos\theta\right)\right\} - \frac{1}{\rho r}\frac{\partial P}{\partial \theta} + g_\theta$$
$$-\frac{1}{\rho}\left\{\frac{1}{r^2}\frac{\partial}{\partial r}\left(r^2 \tau_{r\theta}\right) + \frac{1}{r\sin\theta}\frac{\partial}{\partial \theta}(\tau_{\theta\theta}\sin\theta) + \frac{1}{r\sin\theta}\frac{\partial \tau_{\theta\phi}}{\partial \phi} + \frac{\tau_{r\theta}}{r} - \frac{\tau_{\phi\phi}\cot\theta}{r}\right\},$$

ϕ-component

$$\frac{\partial v_\phi}{\partial t} = -\left\{ v_r \frac{\partial v_\phi}{\partial r} + \frac{v_\theta}{r}\frac{\partial v_\phi}{\partial \theta} + \frac{v_\phi}{r\sin\theta}\left(\frac{\partial v_\phi}{\partial \phi} + v_r \sin\theta + v_\theta \cos\theta\right)\right\} - \frac{1}{\rho r\sin\theta}\frac{\partial P}{\partial \phi}$$
$$+ g_\phi - \frac{1}{\rho}\left\{\frac{1}{r^2}\frac{\partial}{\partial r}\left(r^2 \tau_{r\phi}\right) + \frac{1}{r}\frac{\partial \tau_{\theta\phi}}{\partial \theta} + \frac{1}{r\sin\theta}\frac{\partial \tau_{\phi\phi}}{\partial \phi} + \frac{\tau_{r\phi}}{r} + \frac{2\tau_{\theta\phi}\cot\theta}{r}\right\}.$$

(7.116)

Before these equations can be solved to determine the velocity components, it is necessary to understand how the stress components are related to velocity gradients. Common constitutive relationships applicable to biomedical engineering are discussed in the next section.

7.15 Constitutive Relationships for Non-Newtonian Fluids

We introduced the general expression for non-Newtonian flow of an incompressible, isotropic fluid in Sect. 7.5. The apparent viscosity of a fluid is defined by (7.30)

$$\tau_{ij} = f(D_{ij}) = -2\eta D_{ij} = -\eta\left(\frac{\partial v_j}{\partial x_i} + \frac{\partial v_i}{\partial x_j}\right). \tag{7.117}$$

In Chap. 4, we introduced models for four non-Newtonian fluids: the Power Law fluid, the Bingham fluid, the Casson fluid, and the Herschel–Bulkley fluid. The models were developed for parallel flow situations. In this section, we present the more general three-dimensional expressions for the apparent viscosity of each of these fluids. Let us define two factors $\dot{\gamma}^2$ and T^2 which are related to the sum of the squares of all the rate of deformation components and viscous stress components, respectively:

$$\dot{\gamma}^2 = 2\sum_i \sum_j D_{ij}^2 = 2\left[D_{11}^2 + D_{22}^2 + D_{33}^2\right] + 4\left[D_{12}^2 + D_{13}^2 + D_{23}^2\right], \tag{7.118}$$

$$T^2 = \frac{1}{2}\sum_i \sum_j \tau_{ij}^2 = \frac{1}{2}\left[\tau_{11}^2 + \tau_{22}^2 + \tau_{33}^2\right] + \tau_{12}^2 + \tau_{13}^2 + \tau_{23}^2. \tag{7.119}$$

The indices (1, 2, 3) take on the values (x, y, z) in rectangular coordinates, (r, θ, z) in cylindrical coordinates, and (r, θ, ϕ) in spherical coordinates. T has

Table 7.4 Rate of deformation components in terms of velocity gradients

Rectangular	D_{xx}	$\dfrac{\partial v_x}{\partial x}$
	D_{yy}	$\dfrac{\partial v_y}{\partial y}$
	D_{zz}	$\dfrac{\partial v_z}{\partial z}$
	$D_{xy} = D_{yx}$	$\dfrac{1}{2}\left(\dfrac{\partial v_y}{\partial x} + \dfrac{\partial v_x}{\partial y}\right)$
	$D_{xz} = D_{zx}$	$\dfrac{1}{2}\left(\dfrac{\partial v_z}{\partial x} + \dfrac{\partial v_x}{\partial z}\right)$
	$D_{yz} = D_{zy}$	$\dfrac{1}{2}\left(\dfrac{\partial v_y}{\partial z} + \dfrac{\partial v_z}{\partial y}\right)$
Cylindrical	D_{rr}	$\dfrac{\partial v_r}{\partial r}$
	$D_{\theta\theta}$	$\dfrac{1}{r}\dfrac{\partial v_\theta}{\partial \theta} + \dfrac{v_r}{r}$
	D_{zz}	$\dfrac{\partial v_z}{\partial z}$
	$D_{r\theta} = D_{\theta r}$	$\dfrac{1}{2}\left(r\dfrac{\partial}{\partial r}\left(\dfrac{v_\theta}{r}\right) + \dfrac{1}{r}\dfrac{\partial v_r}{\partial \theta}\right)$
	$D_{rz} = D_{zr}$	$\dfrac{1}{2}\left(\dfrac{\partial v_r}{\partial z} + \dfrac{\partial v_z}{\partial r}\right)$
	$D_{\theta z} = D_{z\theta}$	$\dfrac{1}{2}\left(\dfrac{1}{r}\dfrac{\partial v_z}{\partial \theta} + \dfrac{\partial v_\theta}{\partial z}\right)$
Spherical	D_{rr}	$\dfrac{\partial v_r}{\partial r}$
	$D_{\theta\theta}$	$\dfrac{1}{r}\dfrac{\partial v_\theta}{\partial \theta} + \dfrac{v_r}{r}$
	$D_{\phi\phi}$	$\dfrac{1}{r\sin\theta}\dfrac{\partial v_\phi}{\partial \phi} + \dfrac{v_r}{r} + \dfrac{v_\theta \cot\theta}{r}$
	$D_{r\theta} = D_{\theta r}$	$\dfrac{1}{2}\left(r\dfrac{\partial}{\partial r}\left(\dfrac{v_\theta}{r}\right) + \dfrac{1}{r}\dfrac{\partial v_r}{\partial \theta}\right)$
	$D_{r\phi} = D_{\phi r}$	$\dfrac{1}{2}\left(\dfrac{1}{r\sin\theta}\dfrac{\partial v_r}{\partial \phi} + r\dfrac{\partial}{\partial r}\left(\dfrac{v_\phi}{r}\right)\right)$
	$D_{\theta\phi} = D_{\phi\theta}$	$\dfrac{1}{2}\left(\dfrac{\sin\theta}{r}\dfrac{\partial}{\partial \theta}\left(\dfrac{v_\phi}{\sin\theta}\right) + \dfrac{1}{r\sin\theta}\dfrac{\partial v_\theta}{\partial \phi}\right)$

dimensions of stress and $\dot{\gamma}$ has dimensions of strain rate. Expressions for D_{ij} and $\dot{\gamma}^2$ are given in Tables 7.4 and 7.5 in terms of velocity gradients for rectangular, cylindrical, and spherical coordinate systems. The relationship between stress components and velocity gradients is shown in Table 7.6. The apparent viscosity of the non-Newtonian fluids of importance in biomedical engineering can be written in terms of $\dot{\gamma}$, and limits for the validity of non-Newtonian constitutive relationships can be expressed in terms of T. Examples are provided in the following sections.

7.15 Constitutive Relationships for Non-Newtonian Fluids

Table 7.5 Total strain rate squared, $\dot{\gamma}^2$ (7.118)

Rectangular	$\dot{\gamma}^2 = 2\left[\left(\dfrac{\partial v_x}{\partial x}\right)^2 + \left(\dfrac{\partial v_y}{\partial y}\right)^2 + \left(\dfrac{\partial v_z}{\partial z}\right)^2\right]$ $+ \left[\left(\dfrac{\partial v_y}{\partial x} + \dfrac{\partial v_x}{\partial y}\right)^2 + \left(\dfrac{\partial v_z}{\partial x} + \dfrac{\partial v_x}{\partial z}\right)^2 + \left(\dfrac{\partial v_y}{\partial z} + \dfrac{\partial v_z}{\partial y}\right)^2\right]$
Cylindrical	$\dot{\gamma}^2 = 2\left[\left(\dfrac{\partial v_r}{\partial r}\right)^2 + \left(\dfrac{1}{r}\dfrac{\partial v_\theta}{\partial \theta} + \dfrac{v_r}{r}\right)^2 + \left(\dfrac{\partial v_z}{\partial z}\right)^2\right]$ $+ \left[\left(r\dfrac{\partial}{\partial r}\left(\dfrac{v_\theta}{r}\right) + \dfrac{1}{r}\dfrac{\partial v_r}{\partial \theta}\right)^2 + \left(\dfrac{1}{r}\dfrac{\partial v_z}{\partial \theta} + \dfrac{\partial v_\theta}{\partial z}\right)^2 + \left(\dfrac{\partial v_r}{\partial z} + \dfrac{\partial v_z}{\partial r}\right)^2\right]$
Spherical	$\dot{\gamma}^2 = 2\left[\left(\dfrac{\partial v_r}{\partial r}\right)^2 + \left(\dfrac{1}{r}\dfrac{\partial v_\theta}{\partial \theta} + \dfrac{v_r}{r}\right)^2 + \left(\dfrac{1}{r\sin\theta}\dfrac{\partial v_\phi}{\partial \phi} + \dfrac{v_r}{r} + \dfrac{v_\theta \cot\theta}{r}\right)^2\right]$ $+ \left[\left(r\dfrac{\partial}{\partial r}\left(\dfrac{v_\theta}{r}\right) + \dfrac{1}{r}\dfrac{\partial v_r}{\partial \theta}\right)^2 + \left(\dfrac{\sin\theta}{r}\dfrac{\partial}{\partial \theta}\left(\dfrac{v_\phi}{\sin\theta}\right) + \dfrac{1}{r\sin\theta}\dfrac{\partial v_\theta}{\partial \phi}\right)^2\right.$ $\left. + \left(\dfrac{1}{r\sin\theta}\dfrac{\partial v_r}{\partial \phi} + r\dfrac{\partial}{\partial r}\left(\dfrac{v_\phi}{r}\right)\right)^2\right]$

7.15.1 Power Law Fluid

The apparent viscosity of a Power Law fluid (Ostwald–de Waele model) is:

$$\eta = K\left(\sqrt{\dot{\gamma}^2}\right)^{n-1} = K|\dot{\gamma}|^{n-1}. \tag{7.120}$$

The parameters K and n are constants. The exponent n is known as the *behavior index* or *power law index*. The proportionality factor K is known as *the flow consistency index*. The absolute value is used in the second expression in (7.120) to ensure that the positive root of $\dot{\gamma}^2$ is used. The constitutive relationships for a power law fluid can be written:

$$\tau_{ij} = -K|\dot{\gamma}|^n \left(\frac{2D_{ij}}{|\dot{\gamma}|}\right), \tag{7.121}$$

where the rate of deformation components D_{ij} are found in Table 7.4.

> **Example 7.15.1.1 Power Law Fluid in Parallel Flow.**
> Show that (7.120) reduces to (4.27) in the case of steady flow of a Power Law fluid between two plates.

Table 7.6 Non-Newtonian stress–strain rate relationships

Rectangular	$\tau_{xx} = -2\eta(\dot{\gamma})\dfrac{\partial v_x}{\partial x}$
	$\tau_{yy} = -2\eta(\dot{\gamma})\dfrac{\partial v_y}{\partial y}$
	$\tau_{zz} = -2\eta(\dot{\gamma})\dfrac{\partial v_z}{\partial z}$
	$\tau_{xy} = \tau_{yx} = -\eta(\dot{\gamma})\left(\dfrac{\partial v_y}{\partial x} + \dfrac{\partial v_x}{\partial y}\right)$
	$\tau_{xz} = \tau_{zx} = -\eta(\dot{\gamma})\left(\dfrac{\partial v_z}{\partial x} + \dfrac{\partial v_x}{\partial z}\right)$
	$\tau_{yz} = \tau_{zy} = -\eta(\dot{\gamma})\left(\dfrac{\partial v_y}{\partial z} + \dfrac{\partial v_z}{\partial y}\right)$
Cylindrical	$\tau_{rr} = -2\eta(\dot{\gamma})\dfrac{\partial v_r}{\partial r}$
	$\tau_{\theta\theta} = -2\eta(\dot{\gamma})\left[\dfrac{1}{r}\dfrac{\partial v_\theta}{\partial \theta} + \dfrac{v_r}{r}\right]$
	$\tau_{zz} = -2\eta(\dot{\gamma})\dfrac{\partial v_z}{\partial z}$
	$\tau_{r\theta} = \tau_{\theta r} = -\eta(\dot{\gamma})\left(r\dfrac{\partial}{\partial r}\left(\dfrac{v_\theta}{r}\right) + \dfrac{1}{r}\dfrac{\partial v_r}{\partial \theta}\right)$
	$\tau_{rz} = \tau_{zr} = -\eta(\dot{\gamma})\left(\dfrac{\partial v_r}{\partial z} + \dfrac{\partial v_z}{\partial r}\right)$
	$\tau_{z\theta} = \tau_{\theta z} = -\eta(\dot{\gamma})\left(\dfrac{1}{r}\dfrac{\partial v_z}{\partial \theta} + \dfrac{\partial v_\theta}{\partial z}\right)$
Spherical	$\tau_{rr} = -2\eta(\dot{\gamma})\dfrac{\partial v_r}{\partial r}$
	$\tau_{\theta\theta} = -2\eta(\dot{\gamma})\left[\dfrac{1}{r}\dfrac{\partial v_\theta}{\partial \theta} + \dfrac{v_r}{r}\right]$
	$\tau_{\phi\phi} = -2\eta(\dot{\gamma})\left[\dfrac{1}{r\sin\theta}\dfrac{\partial v_\phi}{\partial \phi} + \dfrac{v_r}{r} + \dfrac{v_\theta \cot\theta}{r}\right]$
	$\tau_{r\theta} = \tau_{\theta r} = -\eta(\dot{\gamma})\left(r\dfrac{\partial}{\partial r}\left(\dfrac{v_\theta}{r}\right) + \dfrac{1}{r}\dfrac{\partial v_r}{\partial \theta}\right)$
	$\tau_{r\phi} = \tau_{\phi r} = -\eta(\dot{\gamma})\left(\dfrac{1}{r\sin\theta}\dfrac{\partial v_r}{\partial \phi} + r\dfrac{\partial}{\partial r}\left(\dfrac{v_\phi}{r}\right)\right)$
	$\tau_{\phi\theta} = \tau_{\theta\phi} = -\eta(\dot{\gamma})\left(\dfrac{\sin\theta}{r}\dfrac{\partial}{\partial \theta}\left(\dfrac{v_\phi}{\sin\theta}\right) + \dfrac{1}{r\sin\theta}\dfrac{\partial v_\theta}{\partial \phi}\right)$

Solution. Taking the direction of flow to be in the x-direction and the plates to be located at $y = h$ and $y = -h$, the only nonzero component of velocity will be v_x and its magnitude will vary with y. In this case, $\dot{\gamma}^2$ from (7.118) will be:

$$\dot{\gamma}^2 = 4\left[D_{xy}^2\right] = 4\left[\frac{1}{2}\left(\frac{\partial v_x}{\partial y}\right)\right]^2 = \left(\frac{\partial v_x}{\partial y}\right)^2.$$

Taking the positive root:

$$|\dot{\gamma}| = \left|\frac{\partial v_x}{\partial y}\right|.$$

7.15 Constitutive Relationships for Non-Newtonian Fluids

Substituting this into (7.120):

$$\eta = K \left| \frac{\partial v_x}{\partial y} \right|^{n-1},$$

which is the same as (4.27).

7.15.2 Bingham Fluid

A Bingham fluid behaves as a solid as long as the yield stress τ_y is not exceeded:

$$D_{ij} = 0, \quad T^2 < \tau_y^2. \tag{7.122}$$

Otherwise, it behaves as a fluid with apparent viscosity, η:

$$\eta = \mu_0 + \frac{\tau_y}{\sqrt{\dot{\gamma}^2}} = \mu_0 + \frac{\tau_y}{|\dot{\gamma}|}, \quad T^2 \geq \tau_y^2. \tag{7.123}$$

Equation (7.122) represents the constitutive relationships for a Bingham fluid when the yield stress is not exceeded. If the yield stress is exceeded, the constitutive relationships for a Bingham fluid can be written:

$$\tau_{ij} = -\left(\mu_0 |\dot{\gamma}| + \tau_y\right)\left(\frac{2D_{ij}}{|\dot{\gamma}|}\right), \quad T^2 \geq \tau_y^2. \tag{7.124}$$

Example 7.15.2.1 Bingham Fluid in Parallel Flow.
Show that the relationship between shear stress and velocity gradient for parallel flow of a Bingham fluid is that given by (4.28)–(4.30).

Solution. First, we compute the quantities T^2 and $\dot{\gamma}^2$ from (7.118) and (7.119):

$$T^2 = \tau_{yx}^2,$$

$$\dot{\gamma}^2 = \left(\frac{\partial v_x}{\partial y}\right)^2.$$

From Table 7.4:

$$D_{yx} = \frac{1}{2}\left(\frac{\partial v_x}{\partial y}\right).$$

Substituting these into (7.124), we find:

$$\tau_{yx} = -\left(\mu_0 \left|\frac{\partial v_x}{\partial y}\right| + \tau_y\right) \frac{\left(\frac{\partial v_x}{\partial y}\right)}{\left|\frac{\partial v_x}{\partial y}\right|}, \quad \tau_{yx}^2 \geq \tau_y^2.$$

This can be written:

$$\tau_{yx} = \tau_y + \mu_0\left(-\frac{\partial v_x}{\partial y}\right), \quad \tau_{yx}^2 \geq \tau_y^2, \quad \frac{\partial v_x}{\partial y} < 0,$$

$$\tau_{yx} = -\tau_y - \mu_0\left(\frac{\partial v_x}{\partial y}\right), \quad \tau_{yx}^2 \geq \tau_y^2, \quad \frac{\partial v_x}{\partial y} > 0,$$

$$\frac{\partial v_x}{\partial y} = 0, \quad \tau_{yx}^2 < \tau_y^2.$$

These are identical to (4.28)–(4.30). Therefore, the general constitutive relations for a Bingham fluid are consistent with the expressions used in Sect. 4.8.2.2 for parallel flow of a Bingham fluid. The apparent viscosity, from (7.123) is:

$$\eta = \mu_0 + \frac{\tau_y}{\sqrt{\left(\frac{\partial v_x}{\partial y}\right)^2}}, \quad \tau_{yx}^2 > \tau_y^2.$$

Even though the velocity gradient can be either positive or negative, η will always be positive.

7.15.3 Casson Fluid

Like a Bingham fluid, a Casson fluid behaves as a solid until the yield stress τ_y is exceeded:

$$D_{ij} = 0, \quad T^2 < \tau_y^2. \tag{7.125}$$

When the yield stress is exceeded, the apparent viscosity of a Casson fluid is:

$$\eta = \left(S + \sqrt{\left(\frac{\tau_y}{|\dot\gamma|}\right)}\right)^2, \quad T^2 \geq \tau_y^2. \tag{7.126}$$

7.15 Constitutive Relationships for Non-Newtonian Fluids

Here, S is a material property with dimensions the same as the square root of viscosity. We leave it as an exercise (Problem 7.19.13) to show that the relationship between shear stress and velocity gradient reduces to (4.31)–(4.33) for parallel flow between flat plates. Equation (7.122) represents the constitutive relationships when the yield stress is not exceeded. When the yield stress is exceeded the constitutive equations for a Casson fluid can be written:

$$\tau_{ij} = -\left(S\sqrt{|\dot{\gamma}|} + \sqrt{\tau_y}\right)^2 \left(\frac{2D_{ij}}{|\dot{\gamma}|}\right), \quad T^2 \geq \tau_y^2. \tag{7.127}$$

It is often convenient to take the square root of each side for cases when D_{ij} is negative and when D_{ij} is positive:

$$\sqrt{\tau_{ij}} = \left(S\sqrt{|\dot{\gamma}|} + \sqrt{\tau_y}\right)\sqrt{\frac{-2D_{ij}}{|\dot{\gamma}|}}, \quad T \geq \tau_y, \quad D_{ij} < 0, \tag{7.128}$$

$$\sqrt{-\tau_{ij}} = \left(S\sqrt{|\dot{\gamma}|} + \sqrt{\tau_y}\right)\sqrt{\frac{2D_{ij}}{|\dot{\gamma}|}}, \quad -\tau_y \geq -T, \quad D_{ij} > 0. \tag{7.129}$$

For a simple parallel flow situation where the velocity in the x-direction is a function of y, then $|2D_{ij}| = |\dot{\gamma}|$, and (7.122), (7.128), and (7.129) reduce to (4.31)–(4.33).

7.15.4 Herschel–Bulkley Fluid

A Herschel–Bulkley fluid is similar to a power law fluid, but also exhibits a yield stress. Consequently, it will not move if the applied shear stress is less than the yield stress:

$$D_{ij} = 0, \quad T^2 < \tau_y^2. \tag{7.130}$$

When the yield stress is exceeded, the apparent viscosity obeys the following relation:

$$\eta = \frac{\tau_y}{|\dot{\gamma}|} + K|\dot{\gamma}|^{n-1}, \quad T^2 > \tau_y^2. \tag{7.131}$$

Dimensions of the material coefficient K depend on the value of the exponent n. The Herschel–Bulkley fluid is a three-parameter fluid model (K, n, τ_y). When the yield stress is exceeded, the constitutive relationships can be expressed as follows:

$$\tau_{ij} = -2\left(\tau_y + K|\dot{\gamma}|^n\right)\frac{D_{ij}}{|\dot{\gamma}|}, \quad T^2 > \tau_y^2. \tag{7.132}$$

7.16 Setting Up and Solving Non-Newtonian Problems

The following procedure can be used to set up non-Newtonian problems:

1. Determine $\dot{\gamma}^2$ from Table 7.5 and T^2 from (7.119)
2. Find the apparent viscosity for the appropriate non-Newtonian fluid by using the relationships given in Sects. 7.15.1–7.15.4.
3. Substitute the values for apparent viscosity into the appropriate stress–strain rate relationships listed in Table 7.6.
4. Substitute the stress components into the momentum equations, (7.114) (rectangular), (7.115) (cylindrical), or (7.116) (spherical).
5. Apply boundary conditions and solve.

Other variations on this procedure can also be fruitful. For instance, it is sometimes possible to solve for one or more components of stress using the momentum equation, then apply the constitutive relationships to determine the velocity. The remainder of this section is devoted to examples of non-Newtonian problems in rectangular, cylindrical, and spherical coordinate systems.

> **Example 7.16.1 Flow of a Bingham Fluid Between Horizontal, Parallel Plates.**
> An incompressible biological fluid with Bingham fluid properties is forced at a constant rate through a wide, horizontal slit of height h under the influence of a constant pressure difference (Fig. 7.26). We wish to solve for the velocity profile and flow rate.

Solution. *Initial considerations:* First, we must realize that we cannot apply the Navier–Stokes equations because the fluid under investigation is not a Newtonian fluid.

System definition and environmental interactions: The system of interest is the Bingham fluid between the two parallel plates. A pressure gradient is applied in the axial direction and the walls exert a shear stress on the fluid, which opposes motion of the fluid.

Apprising the problem to identify governing relationships: Our starting equations are the momentum equations in terms of viscous stress, the continuity equation and the constitutive relations for a Bingham fluid.

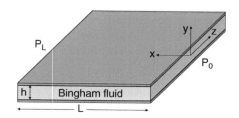

Fig. 7.26 Parallel plates

7.16 Setting Up and Solving Non-Newtonian Problems

Analysis: The continuity equation for an incompressible fluid is:

$$\frac{\partial v_x}{\partial x} + \frac{\partial v_y}{\partial y} + \frac{\partial v_z}{\partial z} = 0.$$

Since the slit is wide in one direction perpendicular to the flow (z-direction) and narrow in the other direction perpendicular to flow (y-direction), we postulate the following:

1. Parallel flow: $v_y = 0$.
2. Parallel flow: $v_z = 0$.
3. No variations in z-direction: $\partial(\ )/\partial z = 0$.
4. Steady-state: $\partial(\ )/\partial t = 0$.

Applying the first three conditions to the continuity equation, leads to:

5. $\partial v_x/\partial x = 0$, or, from (3), (4) and (5): $v_x = v_x(y)$.

The only nonzero velocity gradient is dv_x/dy. Therefore, from Table 7.4, the rate of deformation components are:

6. $D_{xz} = D_{yz} = D_{xx} = D_{yy} = D_{zz} = 0$; $D_{yx} = (1/2)(dv_x/dy)$.

Consequently, the following viscous stresses must be zero from Table 7.6:

7. $\tau_{xz} = \tau_{yz} = \tau_{xx} = \tau_{yy} = \tau_{zz} = 0$.

The only nonzero viscous stress is τ_{yx}, and this can only depend on y:

8. $\tau_{yx} = \tau_{yx}(y)$.

We can calculate T^2 from (7.119):

$$T^2 = \tau_{yx}^2,$$

and the shear rate squared, from Table 7.5 is:

$$\dot{\gamma}^2 = \left(\frac{\partial v_x}{\partial y}\right)^2,$$

so the absolute value of the shear rate is:

$$|\dot{\gamma}| = \left|\frac{\partial v_x}{\partial y}\right|.$$

The Bingham fluid constitutive relationships, (7.122) and (7.124) are:

$$D_{yx} = \frac{1}{2}\left(\frac{dv_x}{dy}\right) = 0, \quad \tau_{yx}^2 < \tau_y^2,$$

$$\tau_{yx} = -(\mu_0|\dot{\gamma}| + \tau_y)\left(\frac{2D_{ij}}{|\dot{\gamma}|}\right) = -\left(\mu_0\left|\frac{dv_x}{dy}\right| + \tau_y\right)\left(\frac{2\left(\frac{1}{2}\left(\frac{dv_x}{dy}\right)\right)}{\left|\frac{dv_x}{dy}\right|}\right), \quad \tau_{yx}^2 \geq \tau_y^2.$$

The absolute value signs can be eliminated by considering separately the cases, where the velocity gradient is negative and when it is positive:

$$\tau_{yx} = -\mu_0 \frac{dv_x}{dy} + \tau_y, \quad \tau_{yx}^2 \geq \tau_y^2, \quad \frac{dv_x}{dy} < 0,$$

$$\tau_{yx} = -\mu_0 \frac{dv_x}{dy} - \tau_y, \quad \tau_{yx}^2 \geq \tau_y^2, \quad \frac{dv_x}{dy} > 0.$$

We can use conditions (1)–(8) to eliminate terms in the momentum equation, (7.13), as shown below:

X – momentum:

$$\rho \frac{\partial v_x}{\partial t}^{\,4} = -\frac{\partial \wp}{\partial x} - \rho v_x \frac{\partial v_x}{\partial x}^{\,5} - \rho v_y \frac{\partial v_x}{\partial y}^{\,1} - \rho v_z \frac{\partial v_x}{\partial z}^{\,2} - \frac{\partial \tau_{xx}}{\partial x}^{\,3} - \frac{\partial \tau_{yx}}{\partial y}^{\,6} - \frac{\partial \tau_{zx}}{\partial z}^{\,6,3}$$

Y – momentum:

$$\rho \frac{\partial v_y}{\partial t}^{\,4} = -\frac{\partial \wp}{\partial y} - \rho v_x \frac{\partial v_y}{\partial x}^{\,1} - \rho v_y \frac{\partial v_y}{\partial y}^{\,1} - \rho v_z \frac{\partial v_y}{\partial z}^{\,2} - \frac{\partial \tau_{xy}}{\partial x}^{\,2,3} - \frac{\partial \tau_{yy}}{\partial y}^{\,7} - \frac{\partial \tau_{zy}}{\partial z}^{\,6,3}$$

Z – momentum:

$$\rho \frac{\partial v_z}{\partial t}^{\,4} = -\frac{\partial \wp}{\partial z} - \rho v_x \frac{\partial v_z}{\partial x}^{\,2} - \rho v_y \frac{\partial v_z}{\partial y}^{\,1} - \rho v_z \frac{\partial v_z}{\partial z}^{\,2} - \frac{\partial \tau_{xz}}{\partial x}^{\,2,3} - \frac{\partial \tau_{yz}}{\partial y}^{\,6} - \frac{\partial \tau_{zz}}{\partial z}^{\,6,3}$$

The y and z-momentum equations reduce to $\partial \wp/\partial y = \partial \wp/\partial z = 0$, so $\wp = \wp(x)$ alone. Therefore, the x-momentum equation reduces to:

$$\frac{d\wp}{dx} = -\frac{d\tau_{yx}}{dy}.$$

This is valid for any fluid, Newtonian or non-Newtonian. Since the left-hand side is a function of x and the right-hand side is a function of y, both terms must be constant. Applying the boundary conditions for pressure: $\wp(x=0) = P(x=0) = P_0$, $\wp(x=L) = P(x=L) = P_L$:

$$P(x) = P_0 + \left(\frac{P_L - P_0}{L}\right)x = P_0 - \Delta P\left(\frac{x}{L}\right).$$

The pressure varies linearly with x. Taking $y = 0$ at the midpoint between the two walls of the slit, the boundary condition at the center, by symmetry, is that the momentum flux is zero at the center ($\tau_{yx}(0) = 0$). Therefore, the solution for $\tau_{yx}(y)$ is also linear, but linear with y:

7.16 Setting Up and Solving Non-Newtonian Problems

$$\tau_{yx}(y) = \left(\frac{\Delta P}{L}\right) y. \qquad (7.133)$$

Note that this equation is valid for any fluid, Newtonian or non-Newtonian. At this point, we must introduce the constitutive relationships derived previously. To apply the constitutive relationships, we must first find where the viscous stress is equal to the yield stress. Setting $y = y_y$ when $\tau_{yx} = \tau_y$:

$$y_y = \frac{\tau_y L}{\Delta P}.$$

Thus for $-y_y \leq y \leq y_y$, dv_x/dy will be zero and the velocity will be constant. Assuming $P_0 > P_L$, so flow is in the positive x-direction, dv_x/dy will be positive for $-h/2 \leq y \leq -y_y$ and dv_x/dy will be negative for $y_y \leq y \leq h/2$. Substituting $\tau_{yx} = \Delta P(y/L)$ into the constitutive relationships and integrating, we obtain:

$$v_x = \frac{\tau_y}{\mu_o} y - \frac{\Delta P}{2\mu_o L} y^2 + C_1, \quad \frac{\tau_y L}{\Delta P} \leq y \leq \frac{h}{2},$$

$$v_x = -\frac{\tau_y}{\mu_o} y - \frac{\Delta P}{2\mu_o L} y^2 + C_2, \quad -\frac{h}{2} \leq y \leq -\frac{\tau_y L}{\Delta P}.$$

where C_1 and C_2 are constants of integration. C_1 can be found by applying the no-slip boundary condition at $y = h/2$ and C_2 can be found by applying the same condition at $y = -h/2$. The velocity of the plug flow region can be found by finding the velocity at either $y = -y_y$ or $y = +y_y$. The complete solution for the velocity profile is:

$$v_x = \frac{\Delta P h^2}{2\mu_o L}\left[\frac{1}{4} - \frac{y^2}{h^2}\right] - \frac{\tau_y h}{\mu_o}\left[\frac{1}{2} - \frac{y}{h}\right], \quad \frac{\tau_y L}{\Delta P} \leq y \leq \frac{h}{2},$$

$$v_x = \frac{\Delta P h^2}{2\mu_o L}\left[\frac{1}{4} - \left(\frac{\tau_y L}{\Delta P h}\right)^2\right] - \frac{\tau_y h}{\mu_o}\left[\frac{1}{2} - \left(\frac{\tau_y L}{\Delta P h}\right)\right], \quad -\frac{\tau_y L}{\Delta P} \leq y \leq \frac{\tau_y L}{\Delta P},$$

$$v_x = \frac{\Delta P h^2}{2\mu_o L}\left[\frac{1}{4} - \frac{y^2}{h^2}\right] - \frac{\tau_y h}{\mu_o}\left[\frac{1}{2} + \frac{y}{h}\right], \quad -\frac{h}{2} \leq y \leq -\frac{\tau_y L}{\Delta P}.$$

Examining and interpreting the results: The maximum stress will occur at $y = h/2$ and the minimum stress at $y = -h/2$. If the magnitude of the shear stress at either wall is less than the yield stress, the velocity gradient will be zero everywhere in the fluid, and since the fluid velocity at the walls are always zero, the fluid will remain stationary everywhere. When the shear stress at the wall exceeds the yield stress, the fluid will begin to move. According to (7.133), this will occur when

$$\tau_{yx}\left(\frac{h}{2}\right) = \tau_y = \left(\frac{\Delta P}{L}\right)\frac{h}{2}.$$

Therefore, the minimum pressure difference required to move the Bingham fluid through the tube is:

$$\Delta P_{\min} = 2\tau_y \frac{L}{h}.$$

If the pressure difference is just slightly greater than the minimum, the bulk of the fluid will move as a plug between the parallel plates. The stress at $y = 0$ is always zero, which is less than the yield stress, so once the minimum pressure difference is exceeded, there will always be a plug region in the range $-\frac{\tau_y L}{\Delta P} \leq y \leq \frac{\tau_y L}{\Delta P}$. The velocity profiles above and below these boundaries will be symmetric in position y. We leave the flow calculation as an exercise in Problem 7.19.15.

> **Example 7.16.2 Couette Flow of Cytoplasm from Neutrophils.**
> Tsai et al. (1993) have shown that the cytoplasm of human neutrophils behaves as a Power law fluid with $n = 0.48$ and $K = 130$ Pa s$^{0.52}$. This fluid is placed in a Couette viscometer with inner radius $R_i = 5$ cm, outer radius $R_o = 5.1$ cm, and height $h = 10$ cm. The inner cylinder is held stationary, while the outer cylinder is rotated at angular velocity Ω. Find the force required to hold the inner cylinder stationary, when $\Omega = 1$ radian/s and $\Omega = 10$ radians/s.

Solution. *Initial considerations:* For steady-state rotation of the outer cylinder, the only nonzero component of fluid velocity in cylindrical coordinates is the component in the tangential direction. Because of the no slip condition, we can postulate:

$$v_\theta = v_\theta(r), \quad v_r = v_z = 0.$$

System definition and environmental interactions: The system of interest is the cytoplasm fluid contained in the Couette viscometer.

Apprising the problem to identify governing relationships: Since the fluid is a power law fluid, the governing equations will be the continuity equation, the momentum equation in cylindrical coordinates, and the constitutive relationship for a power law fluid.

Analysis: Examining the relationships between shear stress and velocity gradients in Table 7.6, we find that only one component of shear stress is nonzero:

$$\tau_{rr} = \tau_{\theta\theta} = \tau_{zz} = \tau_{rz} = \tau_{z\theta} = 0,$$

$$\tau_{r\theta} = -\eta(\dot{\gamma})\left[r\frac{\partial}{\partial r}\left(\frac{v_\theta}{r}\right)\right].$$

Reducing the expression for $\dot{\gamma}^2$ in cylindrical coordinates (Table 7.5):

7.16 Setting Up and Solving Non-Newtonian Problems

$$\dot{\gamma}^2 = \left[r\frac{\partial}{\partial r}\left(\frac{v_\theta}{r}\right)\right]^2,$$

$$|\dot{\gamma}| = \left|r\frac{\partial}{\partial r}\left(\frac{v_\theta}{r}\right)\right|.$$

Substituting this into the relationship for the apparent viscosity of a Power Law fluid (7.120):

$$\eta = K|\dot{\gamma}| = K\left|r\frac{\partial}{\partial r}\left(\frac{v_\theta}{r}\right)\right|^{n-1}.$$

The velocity gradient will be positive in this case, so the absolute value is the same as the value. Substituting the expression for η back into the expression for the shear stress and recognizing that the partial derivate is a total derivative in this case:

$$\tau_{r\theta} = -K\left[r\frac{d}{dr}\left(\frac{v_\theta}{r}\right)\right]^n.$$

A second expression for $\tau_{r\theta}$ can be found by examining the θ-component of the momentum equation in cylindrical coordinates (7.115). Assuming the pressure to be independent of angular position at any fixed value or r and z, this reduces to:

$$\frac{d}{dr}\left(r^2 \tau_{r\theta}\right) = 0.$$

Integrating this expression and setting it equal to the power law expression above:

$$\tau_{r\theta} = \frac{C_1}{r^2} = -K\left[r\frac{d}{dr}\left(\frac{v_\theta}{r}\right)\right]^n,$$

where C_1 is a constant of integration. This can be rearranged and integrated to solve for velocity in terms of a second constant of integration, C_2:

$$v_\theta = C_2 r - \frac{n}{2}\left(-\frac{C_1}{K}\right)^{\frac{1}{n}} r^{(n-2)/n}.$$

The constants C_1 and C_2 can be found by applying the boundary conditions at $r = R_i$ and $R = R_o$:

$$v_\theta(r = R_i) = 0,$$

$$v_\theta(r = R_o) = \Omega R_o.$$

From which, we find:

$$C_1 = -KR_i^2\left(\frac{2\Omega}{n\left(1-\left(\frac{R_o}{R_i}\right)^{-\frac{2}{n}}\right)}\right)^n, \quad C_2 = \left(\frac{\Omega}{1-\left(\frac{R_o}{R_i}\right)^{-\frac{2}{n}}}\right).$$

Substituting these constants into the expressions for velocity and shear stress, we obtain:

$$v_\theta = \Omega r \left[\frac{1-\left(\frac{r}{R_i}\right)^{-\frac{2}{n}}}{1-\left(\frac{R_o}{R_i}\right)^{-\frac{2}{n}}}\right],$$

$$\tau_{r\theta} = -K\left(\frac{R_i}{r}\right)^2 \left(\frac{2\Omega}{n\left(1-\left(\frac{R_o}{R_i}\right)^{-\frac{2}{n}}\right)}\right)^n.$$

Examining and interpreting the results: The velocity is directly proportional to the angular velocity, as expected. The velocity profile depends on the power law index n, but is independent of K.

The force F which must be applied to the inner cylinder to keep it stationary is equal in magnitude and opposite in sign to the stress applied to the wall by the fluid multiplied by the contact area between fluid and the wall:

$$F = 2\pi R_i h(-\tau_{r\theta}) = 2\pi h R_i K \left(\frac{2\Omega}{n\left(1-\left(\frac{R_o}{R_i}\right)^{-\frac{2}{n}}\right)}\right)^n.$$

For an angular velocity of 1 radian/s, the force is:

$$F = 2\pi(0.05\,\mathrm{m})(0.1\,\mathrm{m})\left(130\frac{\mathrm{N\,s^{0.48}}}{\mathrm{m^2}}\right)\left(\frac{2(1\,\mathrm{s^{-1}})}{(0.48)\left(1-\left(\frac{0.051\,\mathrm{m}}{0.05\,\mathrm{m}}\right)^{-\frac{2}{0.48}}\right)}\right)^{0.48} = 27.4\,\mathrm{N}.$$

7.16 Setting Up and Solving Non-Newtonian Problems

The force applied to the viscometer wall by the fluid at 10 radians/s will be $(27.4 \text{ N})[(10 \text{ radian/s})/(1 \text{ radian/s})]^{0.48} = 82.6 \text{ N}$. Although the angular velocity is ten times higher, the wall force is only three times higher. Thus, the apparent viscosity is reduced at higher shear rates. It is instructive to examine the factors that contribute to the apparent viscosity in this problem

$$\eta = K|\dot{\gamma}|^{n-1} = K \left| r \frac{d}{dr}\left(\frac{v_\theta}{r}\right) \right|^{n-1} = K \left| \frac{\left(\frac{2\Omega}{n}\right)\left(\frac{r}{R_i}\right)^{-\frac{2}{n}}}{1 - \left(\frac{R_o}{R_i}\right)^{-\frac{2}{n}}} \right|^{n-1}.$$

The apparent viscosity is directly proportional to the flow consistency index K, as expected, but depends on the power index n in a complex manner. The apparent viscosity also depends on the radial position r, the ratio of the outside to the inside radii of the viscometer, and the angular velocity of the outer wall. For $0 < n < 1$, the apparent viscosity decreases with increasing angular velocity as Ω^{n-1}.

Additional comments: The pressure distribution in the fluid can be found by applying the z- and r-components of the momentum equation, along with the solution for $v_\theta(r)$ given above. The pressure gradient is found to be directly proportional to the angular velocity squared, and is independent of K. We leave this derivation as an exercise for the student (Problem 7.19.16).

Example 7.16.3 Cone and Plate Viscometer.
A disadvantage of capillary tube viscometers and Couette viscometers is that the shear rate varies with radial position within these devices. When trying to characterize a non-Newtonian fluid, it would be highly desirable to expose all fluid elements within the device to the same shear rate. The cone and plate viscometer shown in Fig. 7.27 provides a nearly constant shear rate on all portions of the fluid, as long as the cone angle α is very small (1° or less). A small amount of the fluid to be tested is placed on the plate and the cone is lowered into the fluid until the tip of the cone just touches the plate. The torque required to rotate the cone at constant angular velocity is measured, and the effective viscosity of the fluid can be computed from the device geometry, the angular velocity, and the torque. Show that the shear rate applied to a fluid within a cone and plate viscometer is nearly independent of position.

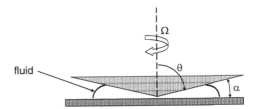

Fig. 7.27 Cone and plate viscometer

Solution. *Initial considerations:* We begin by compiling an appropriate list of assumptions:

1. The fluid is incompressible.
2. Flow is steady ($\partial(\ )/\partial t = 0$).
3. Velocity in the radial direction is zero.
4. Velocity in the θ-direction is zero.
5. There are no changes in any variable in the ϕ-direction ($\partial(\ )/\partial \phi = 0$).
6. Neglect body forces.
7. Flow is laminar and very slow, so $Re \ll 1$, and inertial terms are negligible.

System definition and environmental interactions: The system of interest in a cone and plate viscometer is the fluid contained between the cone and the plate, as shown in Fig. 7.27. The rotating cone forces the fluid to rotate, and motion is opposed by the fluid in contact with the stationary plate.

Apprising the problem to identify governing relationships: Since the fluid is not necessarily a Newtonian fluid, it is inappropriate to use the Navier–Stokes equation. Instead, we must use the momentum equation in terms of shear stress, since this is independent of the constitutive relationship for the fluid being tested. In addition, we will need to use the continuity equation and the definition of apparent viscosity. The geometry of the device also suggests the use of a spherical coordinate system.

Analysis: Based on the assumptions, every term in the continuity equation is zero. The only nonzero component of velocity is v_ϕ, and according to assumptions (2) and (5) this component can depend only on r and θ. Fluid at $\theta = \pi/2 - \alpha$ will rotate at the angular velocity of the cone, while fluid at $\theta = \pi/2$ will be stationary. At any other angle θ between the two walls, we postulate that the fluid rotates at an angular velocity ω that lies between 0 and Ω, and which depends only on θ. Consequently, the velocity in the fluid can be written:

8. $v_\phi(r, \theta) = \omega(\theta) r$

Based on the assumptions listed above, the shear rate squared can be determined from Table 7.5:

$$\dot{\gamma}^2 = \left[\left(\frac{\sin\theta}{r} \frac{\partial}{\partial \theta} \left(\frac{v_\phi}{\sin\theta} \right) \right)^2 + \left(r \frac{\partial}{\partial r} \left(\frac{v_\phi}{r} \right) \right)^2 \right].$$

However, from assumption (8), $v_\phi/r = \omega$, which is just a function of theta. Consequently, this reduces to:

$$|\dot{\gamma}| = \left| \frac{\sin\theta}{r} \frac{\partial}{\partial \theta} \left(\frac{v_\phi}{\sin\theta} \right) \right| = \left| \sin\theta \frac{\partial}{\partial \theta} \left(\frac{v_\phi/r}{\sin\theta} \right) \right| = \left| \sin\theta \frac{d}{d\theta} \left(\frac{\omega(\theta)}{\sin\theta} \right) \right|.$$

Since the angle α is small, we can approximate the derivative in this expression for shear rate as follows:

7.16 Setting Up and Solving Non-Newtonian Problems

$$\frac{d}{d\theta}\left(\frac{\omega(\theta)}{\sin\theta}\right) \approx \frac{\left(\dfrac{\omega\left(\dfrac{\pi}{2}\right)}{\sin\left(\dfrac{\pi}{2}\right)} - \dfrac{\omega\left(\dfrac{\pi}{2} - \alpha\right)}{\sin\left(\dfrac{\pi}{2} - \alpha\right)}\right)}{\dfrac{\pi}{2} - \left(\dfrac{\pi}{2} - \alpha\right)} = -\frac{\Omega}{\alpha \sin\left(\dfrac{\pi}{2} - \alpha\right)}.$$

Therefore, $|\dot{\gamma}|$ can be written:

$$|\dot{\gamma}| = \left(\frac{\Omega}{\alpha}\right)\left(\frac{\sin\theta}{\sin\left(\dfrac{\pi}{2} - \alpha\right)}\right).$$

But if α is very small, then the angle theta is nearly $\pi/2$ and $\sin(\theta) \approx \sin(\pi/2 - \alpha) \approx \sin(\pi/2) = 1$. Therefore, the shear rate is very nearly constant throughout the entire fluid:

$$|\dot{\gamma}| \approx \left(\frac{\Omega}{\alpha}\right).$$

The rate of deformation components from Table 7.4 in spherical coordinates are found to be:

$$D_{rr} = D_{\theta\theta} = D_{\phi\phi} = D_{r\theta} = 0$$

$$D_{r\phi} = \frac{r}{2}\frac{\partial}{\partial r}\left(\frac{v_\phi}{r}\right) = \frac{r}{2}\frac{\partial}{\partial r}(\omega(\theta)) = 0$$

$$D_{\theta\phi} = \frac{1}{2}\left(\frac{\sin\theta}{r}\right)\frac{\partial}{\partial\theta}\left(\frac{v_\phi}{\sin\theta}\right) = \left(\frac{\sin\theta}{2}\right)\frac{d}{d\theta}\left(\frac{\omega(\theta)}{\sin\theta}\right).$$

Applying the same approximation for the derivative applied above, $D_{\theta\phi}$ can be written:

$$D_{\theta\phi} = -\left(\frac{\Omega}{2\alpha}\right)\left(\frac{\sin\theta}{\sin\left(\dfrac{\pi}{2} - \alpha\right)}\right) \approx -\left(\frac{\Omega}{2\alpha}\right).$$

Since the rate of deformation components are all zero except for $D_{\theta\phi}$, the only nonzero viscous stress component will be $\tau_{\theta\phi}$. Since $D_{\theta\phi}$ depends only on θ, then $\tau_{\theta\phi}$ will also be a function of θ alone. The ϕ-component of the momentum equation reduces to:

$$0 = \frac{d\tau_{\theta\phi}}{d\theta} + 2\tau_{\theta\phi}\cot\theta.$$

Separating variables and integrating, we find the solution for $\tau_{\theta\phi}$ to be:

$$\tau_{\theta\phi} = \frac{C}{\sin^2\theta},$$

where C is a constant of integration. If the cone angle α is small, then $\sin(\theta)$ will be nearly unity throughout the fluid and $\tau_{\theta\phi}$ will equal a constant C. This is consistent with $D_{\theta\phi}$ being nearly constant throughout the fluid:

$$\tau_{\theta\phi} = -2\eta D_{\theta\phi} = \eta\left(\frac{\Omega}{\alpha}\right) = C.$$

Thus, if $\tau_{\theta\phi}$ can be measured at a given value of angular velocity Ω, then both the shear rate and the apparent viscosity can be computed. In practice, the shear stress is computed from the torque required to rotate the cone at constant angular velocity. The measured torque $\mathfrak{J}$ is related to the shear stress as follows:

$$\mathfrak{J} = \int_A r\tau_{\theta\phi}dA = \tau_{\theta\phi}\int_0^{2\pi}\int_0^R r^2 dr d\phi = \frac{2\pi R^3 \tau_{\theta\phi}}{3},$$

where R is the distance from the cone apex to the outside surface of the fluid. Equating shear stresses in the last two equations:

$$\tau_{\theta\phi} = \frac{3\mathfrak{J}}{2\pi R^3} = \frac{\eta\Omega}{\alpha}. \tag{7.134}$$

Apparent viscosity of the fluid at a shear rate Ω/α can be computed from the geometry of the viscometer, the measured torque and the angular velocity as follows:

$$\eta = \frac{3\alpha\mathfrak{J}}{2\pi\Omega R^3}.$$

Examining and interpreting the results: The cone and plate viscometer provides a relatively uniform shear rate in the fluid, making apparent viscosity measurements possible at different shear rates. The apparent viscosity is directly proportional to the applied torque, and is highly sensitive to the contact radius, R.

7.17 Summary of Key Concepts

Equations of motion. In this chapter, we use the shell balance approach to derive the general governing equations for multidimensional, unsteady-state fluid motion. These are known as the equations of motion, consisting of (1) conservation of mass,

7.17 Summary of Key Concepts

Table 7.7 Equations of motion for Newtonian fluids

	Rectangular	Cylindrical	Spherical
Continuity equation	(7.4)	(7.69)	(7.70)
Momentum equations	(7.13) or (7.114)	(7.115)	(7.116)
Newtonian fluid constitutive relationships	(7.32)	(7.73)	(7.74)
Navier–Stokes equations	(7.33)	(7.75)	(7.76)

(2) conservation of momentum, and (3) the constitutive relationships for the fluid in the system. For Newtonian fluids, the six constitutive relationships can be substituted into the momentum equation to produce a set of three scalar equations, known as the Navier–Stokes equations. Table 7.7 summarizes the appropriate relationships to be used for Newtonian fluid systems that are analyzed in rectangular, cylindrical, and spherical coordinate systems.

Solving Newtonian flow problems. The Continuity and Navier–Stokes equations form the basis for analysis of any problem involving a Newtonian fluid. The first step in solving this set of equations is to reduce the complexity of the equations by eliminating terms that are identically zero or are very small. Begin by making a numbered list of all of the assumptions that apply to the situation under analysis. For instance, if flow is in the x-direction only, then (1) $v_y = 0$ and (2) $v_z = 0$, and consequently, all derivatives of v_y and v_z will also be zero. Based on your list, eliminate terms in the continuity and N–S equations by drawing a line through unimportant terms. List the assumption number(s) justifying elimination above each term that is dropped. Start with the continuity equation, and use results from the continuity equation to further simplify the N–S equations. For the example above, continuity provides $\partial v_x/\partial x = 0$, allowing us to eliminate $\partial v_x/\partial x$ and $\partial^2 v_x/\partial x^2$ from the N–S equations.

Once the equations of motion are reduced to the minimum number of terms, the appropriate auxiliary conditions must be specified. Since the N–S equations are first order in time, a single initial condition for each nonzero velocity component must be specified in unsteady-state problems. Velocity components in the NS equations are second order with respect to each coordinate direction. Consequently, for each direction in which a velocity gradient exists, we must specify two boundary conditions for each nonzero velocity component. Before attempting to solve a problem, be sure you have specified all of the necessary initial conditions and boundary conditions that ensure a well-posed problem. Even then, many multidimensional Newtonian fluid problems cannot be solved analytically, so numerical methods are often applied. We have selected a few example problems, which analyze different types of solution methods.

Solving non-Newtonian flow problems. The approach to solving problems involving the flow of non-Newtonian fluids is similar to that outlined above. However, the Navier–Stokes equations *cannot* be used to analyze the flow of non-Newtonian fluids. Such problems must use the continuity equation, momentum equations (7.114)–(7.116), and the constitutive relationship for the non-Newtonian fluid. The apparent viscosity of non-Newtonian fluids (Table 7.6) is a function of the shear rate (Table 7.5), which depends on the rate of deformation components

(Table 7.4). Constitutive relationships are provided for Power Law fluids (7.120), Bingham fluids (7.122)–(7.124), Casson fluids (7.125)–(7.127), and Herschel–Bulkley fluids (7.130)–(7.132). The procedure for solving problems involving non-Newtonian fluids is discussed in detail in Sect. 7.16.

7.18 Questions

7.18.1 The continuity equation is another name for which conservation law?
7.18.2 The substantial derivative is a special case of which time derivative?
7.18.3 Explain the difference between dv/dt, $\partial v/\partial t$, and Dv/Dt.
7.18.4 What is the advantage of using the modified pressure?
7.18.5 The stream function automatically satisfies which equation?
7.18.6 The Navier–Stokes equation is based on which assumptions?
7.18.7 Explain the meaning of "fully developed flow" and "entry length."
7.18.8 Explain why the pressure gradient is higher in the entry region of a tube than it is far downstream.
7.18.9 What boundary conditions apply at the interface between two immiscible fluids?
7.18.10 What boundary conditions apply for fluid in a Couette viscometer, with inside radius stationary and outside radius rotating at constant angular velocity?
7.18.11 What boundary conditions apply for flow in a rectangular duct full of fluid?
7.18.12 What boundary conditions apply for flow in a half–full rectangular duct, tilted at an angle β with the horizontal? What is the axial pressure gradient in this case?
7.18.13 What terms in the Navier–Stokes equations disappear when the "creeping flow" assumption is made?
7.18.14 Stokes law relates the drag force on a sphere to what other quantities? What restrictions apply to its derivation?
7.18.15 Can the Navier–Stokes equation be used to solve for the flow of a Bingham fluid in a tube? Explain.
7.18.16 Explain why blood velocity in arteries is in-phase with the pressure gradient oscillations for low values of the Womersley Number, but is 90° out of phase when Womersley Number is high.
7.18.17 Why does the velocity profile near the center of the tube tend to become flatter as the Womersley number increases in periodic flow?
7.18.18 How does the addition of a steady-state pressure gradient influence the flow and velocity profile in pulsatile flow?
7.18.19 What is the purpose of scaling the equations of motion?
7.18.20 What constitutes the equations of motion for a non-Newtonian fluid?
7.18.21 The stress component τ_{ij} for any fluid is a function of which rate of deformation component?
7.18.22 The apparent viscosity of a non-Newtonian fluid is a function of what key factor?
7.18.23 What procedure should be followed in analyzing the flow of a non-Newtonian fluid?

7.19 Problems

7.19.1 Falling Film

Find the velocity distribution and shear stress distribution by first making a list of appropriate assumptions, then by reducing the continuity and Navier–Stokes equations for the falling film problem discussed in Sects. 6.2–6.2.5 (Fig. 6.1).

7.19.2 Immiscible Fluids

Find the velocity distribution and shear stress distribution by first making a list of appropriate assumptions, then by reducing the continuity and Navier–Stokes equations for the problem in Example 6.2.6.1 (flow of two immiscible fluids down an inclined plane).

7.19.3 Cylindrical Tube

Find the velocity distribution and shear stress distribution by first making a list of appropriate assumptions, then by reducing the continuity and Navier–Stokes equations for flow of a Newtonian fluid in a cylindrical tube, as discussed in Sect. 6.3.1.

7.19.4 Annulus Flow

Find the velocity distribution and shear stress distribution by first making a list of appropriate assumptions, then by reducing the continuity and Navier–Stokes equations for flow of a Newtonian fluid in an annulus with inner wall moving, as discussed in Sect. 6.3.2.

7.19.5 Start-Up Flow in a Circular Tube

Simplify the Navier–Stokes equations for flow of a Newtonian fluid through a tube in which the fluid is initially at rest and a constant pressure gradient is suddenly imposed at time $t = 0$. Provide initial and boundary conditions needed to solve the problem. (Do not solve).

7.19.6 Pressure Drop Across a Red Blood Cell

Consider the pressure drop across a single red cell. Adopt a coordinate system that moves at the same constant velocity as a red cell. Assume the cell has radius $R_c < R_t$ and length L. Assume that plasma is a Newtonian fluid and the fluid velocity between the red cell wall and the capillary wall is fully developed.

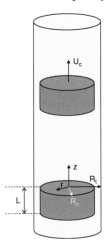

(a) What system is to be analyzed?
(b) What equations are necessary to find the pressure drop across a red cell?
(c) What constitutive relationship is appropriate?
(d) What constraints are imposed on the system by the environment?
(e) Derive an expression that can be used to predict the pressure difference that must be applied across the red cell to move it at constant velocity U_c.
(f) Would you expect the actual measured pressure difference to be greater or less than your computed value? Explain.

7.19.7 Addition of Pulsations to a Bioreactor

The walls of a parallel plate bioreactor are produced by growing a monolayer of cells on thin, porous membranes. The cells produce a substance used as an ingredient in a new drug. The cells have been shown to secrete the substance at much higher rates when they are exposed to oscillating shear stresses with magnitudes between 25 and 50 dynes/cm^2. The cells suffer irreparable damage when subjected to shear stresses above 100 dynes/cm^2.

Each plate is 10 cm long by 10 cm wide. Once the cells are in place, the plates are arranged in parallel and perfused with an oxygen-rich and nutrient-rich fluid. The fluid is Newtonian with viscosity of 2 cp and the fluid density is 1.1 g/ml.

The spacing between plates is 0.04 cm and the flow rate needed to keep the cells viable is 160 ml/min per channel.

A roller pump is used to perfuse the bioreactor. It can produce a pressure gradient across the device of:

$$-\frac{\partial P}{\partial z} = A_0 + A_1 \cos(\omega t).$$

The period for each pump cycle is 1 s. Our goal is to propose appropriate values for A_0 and A_1 when downstream pressure is held at zero (relative to atmospheric pressure). (a) Find the steady component A_0 required to provide the minimum required flow per channel. (b) Provide a detailed step by step procedure of how you would go about finding A_1. Begin by reducing the Navier–Stokes equations and specify all boundary conditions necessary to solve the resulting partial differential equation, but do not attempt to solve it.

7.19.8 Blood Flow in a Slightly Convergent Channel

Blood is forced through a channel with width W, length L, and height $2h$. The half height h decreases linearly with axial position as shown in the figure. Pressures P_0 and P_L are imposed at $z = 0$ and $z = L$, respectively.

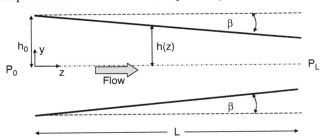

You can make the following assumptions:

1. Blood is incompressible.
2. Blood can be considered a Newtonian fluid.
3. Flow is steady.
4. $v_x = 0$.
5. $\partial(\ )/\partial x = 0$.
6. $W \gg h_0$.
7. $L \gg h_0$.
8. The Reynolds number $\langle v \rangle h_0 / \nu$ is small.
9. Flow is fully developed at $z = 0$.
10. The angle β is small, so $\beta \approx \tan(\beta)$.

(a) Use scaling to find appropriate scale factors for v_y and P.
(b) Neglect small terms and find an expression for v_z in terms of y, $h(z)$ and pressure gradient.
(c) Derive an expression for v_y in terms of y, $h(z)$ and derivates of P.
(d) Find an expression for pressure as a function of $h(z)$ or z.
(e) For the same pressure difference, compare flow through the converging channel to blood flow through a channel with the same length and width, but with horizontal walls a distance $2h_0$ apart.

7.19.9 Blood Flow in Alveolar Wall of the Lung

Pulmonary capillaries are modeled as sheet-like structures with width W. Blood is confined between two thin alveolar membranes as shown. The pressure on the gas side of each membrane is constant and equal to alveolar pressure, P_{alv}. As blood flows through the capillary the transmural pressure decreases and the distance between membranes $2h(z)$ decreases in a nonlinear fashion.

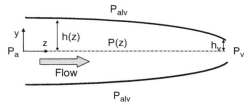

The relationship for the half height h in terms of transmural pressure is:

$$h(z) = h_0 + \frac{\alpha}{2}(P(z) - P_{\text{alv}}), \quad P(z) - P_{\text{alv}} \geq 0,$$
$$h(z) \approx 0, \quad P(z) - P_{\text{alv}} < 0,$$

where α is a constant compliance coefficient and h_0 is the half height at zero transmural pressure. Assumptions (1)–(9) from problem 7.19.8 apply to this case.

(a) Find an expression for blood flow through the capillary sheet in terms of arterial inlet pressure P_a, surrounding alveolar pressure P_{alv} and downstream venous pressure P_v. Assume $P_{\text{alv}} < P_v$.
(b) What is the driving force and blood flow rate through a partially collapsed capillary sheet where $P_{\text{alv}} > P_v$ and $h_v \approx 0$?

7.19.10 Scaling

Use scaling to show that the inertial terms can be neglected for the flow of a Newtonian fluid past a sphere when the Reynolds number based on the sphere diameter is small.

7.19.11 Scaling and Blood Flow in the Lung Microcirculation

Let blood flow in the lung microvessels be modeled as flow between two parallel plates separated by a constant distance h. The width of the channel is approximately equal to the channel length L, and L is much larger than h. Flow can be assumed to be steady and inertial effects are negligible. A top view (xy-plane) and a side view (x–z plane) are shown below. The pressures at the inlet and outlet of the microvessel are known.

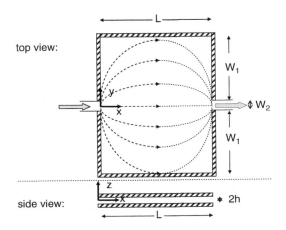

(a) Use scaling arguments to show that the pressure gradient in the x or y directions are greater than the pressure gradient across the height of the channel (z-direction).
(b) Provide a set of equations that can be used to solve this problem for v_x, v_y and P.
(c) Provide all the boundary conditions that are necessary to solve this problem.

7.19.12 Flow of a Viscous Fluid Through a Compliant Tube

The following linear relationship between vessel cross-sectional area A and intravascular pressure P is measured in an artery:

$$A - A_o = \gamma(P - P_o),$$

where A_o and P_o are constants and the cross-section is circular. Derive a relationship between tube flow and the upstream and downstream pressures. Show that this reduces to Poiseuille flow as γ becomes very small.

7.19.13 Casson Fluid

Show that (7.125) and (7.127) for a Casson fluid are consistent with (4.31)–(4.33) for steady, fully developed flow between two parallel plates.

7.19.14 Herschel–Bulkley Fluid

Show that (7.130) and (7.132) for a Herschel–Bulkley fluid are consistent with (4.34)–(4.36) for steady, fully developed flow between two parallel plates.

7.19.15 Flow of Bingham Fluid Between Parallel Plates

Derive an expression for the flow rate of a Bingham fluid between parallel plates from velocity expressions given in Example 7.16.1.

7.19.16 Power Law Fluid in Couette Viscometer

Find an expression for the distribution of pressure $P(r, z)$ in a Couette viscometer filled with a power law fluid, as presented in Example 7.16.2. Take $z = 0$ at the bottom of the viscometer, and the pressure above the fluid at $z = h$ and $r = R_i$ is atmospheric pressure, P_{atm}.

7.19.17 Blood as a Casson Fluid

Anticoagulant is added to blood that obeys the Casson constitutive relationship. The blood is poured onto a flat surface, where it comes to rest as a pool with height h, width W and length L. Pressure above the pool is constant. The surface is slowly tilted until the blood just begins to flow due to gravity down the surface.

(a) Relate the angle where flow is initiated to properties of the blood and height of the pool.
(b) Derive an expression for the velocity of the blood $v_z(y)$ if the surface is tilted at an angle that is greater than the angle found in part a. Assume flow is steady state and the thickness of the blood layer remains constant (i.e., blood is supplied at a constant rate at $z = 0$). Sketch $v_z(y)$.

7.19.18 Bingham Fluid in a Couette Viscometer

A Bingham fluid is placed between the inner and outer walls of a Couette viscometer to a height h (perpendicular to plane of paper). A torque $\mathfrak{J}$ is applied to the outer wall. The outer wall can rotate but the inner wall is stationary.

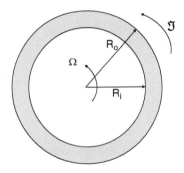

(a) Find an expression for shear stress in the fluid as a function of the applied torque and radial position.
(b) What minimum torque must be applied to the outer wall for the wall to just start rotating?
(c) What minimum torque must be applied to the outer wall for the shear stress to be greater than or equal to the yield stress everywhere in the fluid?
(d) When the applied torque lies between the values found in parts (b) and (c), will the "fluid" or the "solid" region of the Bingham fluid be closer to the inner wall? Explain.
(e) Find a relationship between angular velocity of the wall Ω as a function of applied torque when the torque is greater than the minimum torque found in part (c).
(f) Find the velocity profile in the fluid for conditions (d) and (e) above.

7.19.19 Flow Past a Cylinder

A swimmer holds his/her arm out in a stream perpendicular to the oncoming velocity. We would like to predict the velocity of the fluid as it flows around his/her arm.

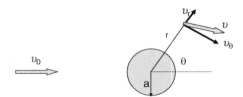

(a) Simplify the continuity and Navier–Stokes equations by applying the following assumptions:
  1. The fluid is an incompressible, Newtonian fluid with kinematic viscosity, ν.
  2. The approach velocity of the stream v_0 is steady and uniform.
  3. There are no changes in velocity in the z-direction.
  4. Her arm is modeled as a circular cylinder with radius a.
  5. Neglect gravitational effects.
  6. Pressure in the fluid is constant.
  7. Velocity in the z-direction is zero.
(b) Use scaling to further simplify the equations obtained above for the case where $v_0 a/\nu \ll 1$.
(c) What boundary conditions would you use to solve for v_r and v_θ? (*Do not solve*).

7.20 Challenges

7.20.1 Laminar Flow in Noncircular Tubes

Figure 5.21 provides the dimensionless parameter M_0 used to determine the pressure drop for laminar flow in noncircular tubes. *Challenge:* How were these relationships for M_0 vs. the ratio of the minor axis to major axis determined for laminar flow of a Newtonian fluid in an elliptical tube? *Generate ideas:* What is different about velocity profiles in noncircular vessels relative to the velocity profile in a tube or between wide parallel plates? What equations govern flow through the elliptical vessel? What assumptions would you use? What boundary conditions are appropriate? What procedures would you follow to find M_0?

References

Blasius H (1908) Grenzschichten in Flussigkeiten mit kleiner Reibung. Z Math Phys 56:1–37

Darbandi M, Hosseinizadeh SF (2004) Remarks on numerical prediction of wall shear stress in entry flow problems. Commun Numer Meth Eng 20:619–625

Mokheimer EMA (2002) Simplified solution of developing laminar flow between parallel plates. J Energy Res 26:399–411

Richardson SM (1989) Fluid Mechanics, Hemisphere Publishing

Tsai MA, Frank RS, Waugh RE (1993) Passive mechanical behavior of human neutrophils: power-law fluid. Biophys J 65:2078–2088

Womersley JR (1955) Method for the calculation of velocity, rate flow, and viscous drag in arteries when the pressure gradient is known. J Physiol 127(3):553–563

Part IV
Bioheat Transport

Chapter 8
Heat Transfer Fundamentals

8.1 Introduction

Heat transfer describes the exchange of energy between materials as a consequence of a difference in temperature. Heat is transmitted by three distinct physical mechanisms: conduction, convection, and radiation. Although heat transfer analysis makes a clear distinction among these three mechanisms, in nature it is common that more than one mechanism occurs simultaneously. Thus, the solution of real problems often involves dealing with more than just a single mechanism. It is convenient for pedagogical purposes to consider the mechanisms of conduction, convection, and radiation separately, and it is important to learn how to identify and solve for the combined effects of these mechanisms.

The analysis of heat transfer seeks to determine *how much* thermal energy will be exchanged under specified conditions and *how fast* the exchange will occur. All living species are directly dependent on heat transfer for the maintenance of a healthy life state. An uncountable number of devices that depend on heat transfer for their function are used for diagnosis and treatment of diseases. Thus, the ability to understand and apply the methods of heat transfer analysis is an important component of the working arsenal of a biomedical engineer. In this chapter, we introduce the fundamental principles that govern these mechanisms of heat transfer and provide examples of how they may be applied in practical problems in biological and medical systems.

8.2 Conduction

Energy can be transmitted through materials via diffusion under the action of an internal temperature gradient. This process is called *heat conduction*, and it results from a transfer of energy from greater to lesser energetic molecules via their random interactions within the structural constraints of the material. There is no net transfer of material during thermal conduction, only energy of molecular motion.

Conduction occurs in all phases of material: solid, liquid, and gas, although the dimensions of molecular motion vary tremendously across the phases. Likewise, the effectiveness of the different phases (solid, liquid, and vapor) in transmitting thermal energy can vary dramatically as a function of the freedom of their molecules to interact with nearest neighbors. The measure of this effectiveness is a property called the *thermal conductivity*. Likewise, the property *temperature* is a measure of the level of energy that determines the potential for transmitting thermal energy. The conductivity and temperature of a material are key parameters used to describe the process by which a material may be engaged in heat conduction. In this chapter, we will develop and apply analysis tools that allow us to study processes that involve heat conduction in living tissues.

The fundamental constitutive expression that describes the one-dimensional conduction of heat is called Fourier's law, as was presented in Chap. 2.

$$q_x = \frac{\dot{Q}_x}{A} = -k\frac{dT}{dx}. \tag{8.1}$$

The heat flow, $\dot{Q}_x$ (W), is a measure of the magnitude of rate at which energy is moving along the direction of an applied gradient in temperature, dT/dx (K/m), which, in this case, is along the x coordinate. The thermal conductivity, k (W/(m K)), is an indicator of the propensity of a material to support the flow of heat when a temperature gradient is imposed across it. The greater the conductivity, the larger will be the heat flow for a given magnitude of temperature gradient. Also, the cross-sectional area, A (m^2), normal to the temperature gradient provides the pathway for the energy flow. The heat flow per unit area is the heat flux, q_x (W/m^2).

The property thermal conductivity, k, is fundamental to understanding the conduction of heat through materials. Values for the property k have been measured for a large spectrum of materials and states, as may be affected by temperature and pressure. The numerical value of k can vary by five or six orders of magnitude from highly conductive metals to insulating materials to liquids and to gases. An abbreviated collection of the thermal conductivity for frequently encountered materials is presented in Appendix C. Values for biological materials are also presented. In general, it is very difficult to measure the thermal conductivity of living tissue as compared with inanimate materials, making this data relatively sparse. The authors have attempted to assemble one of the most complete compilations available of this information.

Consider the system in Fig. 8.1, composed of a solid slab in which the height and width are much greater than the length L. The temperature of the side at $x = 0$ is maintained at T_0 and the temperature of the side at $x = L$ is maintained at T_L. Temperature gradients in the y and z directions are negligible, so the heat flux can be assumed to be restricted to the x-direction. After a long period of time, a steady-state temperature gradient will be established in the material, and heat will flow from the region of greater temperature to the region with lower temperature.

8.2 Conduction

Fig. 8.1 Constant temperature gradient for steady-state heat flow in a finite-sized system

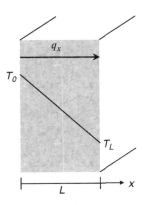

Integration of Fourier's law from $x = 0$ to $x = L$ yields the following relationship for heat flow through the slab:

$$q_x = \frac{\dot{Q}_x}{A} = \frac{k}{L}(T_0 - T_L). \tag{8.2}$$

Thus, the conduction heat flux is directly proportional to the temperature difference across the slab, directly proportional to the thermal conductivity of the slab, and inversely proportional to the thickness of the slab. The right side of (8.2) has the same form as does (2.34) in which a flux is defined in terms of a constitutive property and a difference in driving potential that is applied across a distance. If we integrate Fourier's law from $x = 0$ to an arbitrary position x where the temperature is $T(x)$, we find:

$$q_x x = k(T_0 - T(x)). \tag{8.3}$$

Substituting (8.2) for q_x into (8.3) provides the following linear relationship for the temperature as a function of position in the slab:

$$T(x) = (T_L - T_0)\frac{x}{L} + T_0. \tag{8.4}$$

In biological applications, cylindrical geometries are frequently encountered. Figure 8.2 illustrates heat flow through a hollow cylinder such as may be encountered in the tissue surrounding a length of vessel with blood flowing through it or in a bronchiole with air flowing through it. For steady-state conditions, the radial heat flow will be the same at every position r since the temperature remains constant over time because there is no change in energy storage. The heat flow in the radial direction occurs through a cross-sectional area of $A_r = 2\pi r L$. Note that unlike the slab geometry, the cross-sectional area changes with radial position. Thus, Fourier's Law, (8.1), becomes.

$$\dot{Q}_r = -kA_r\frac{dT}{dr} = -2\pi k r L\frac{dT}{dr}. \tag{8.5}$$

Fig. 8.2 One-dimensional steady-state heat flow through a hollow cylinder

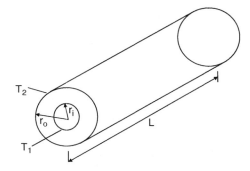

Temperature boundary conditions are specified at the inner and outer radii of the cylinder.

$$T = T_1 \quad \text{at} \quad r = r_i; \quad T = T_2 \quad \text{at} \quad r = r_o.$$

The solution of (8.5) for these boundary conditions is,

$$\dot{Q}_r = \frac{2\pi k L(T_1 - T_2)}{\ln\left(\frac{r_o}{r_i}\right)}. \tag{8.6}$$

Heat flow is directly proportional to the temperature difference and thermal conductivity, as was found in the case of the slab. In addition, heat flow is proportional to the length of the cylinder and inversely proportional to the logarithm of the ratio of the outside radius to inside radius. The temperature distribution in the cylinder can be found by integrating (8.5) from $r = r_i$ to an arbitrary radius, r, and substituting (8.6) for heat flow:

$$\frac{T(r) - T_1}{T_2 - T_1} = \frac{\ln\left(\frac{r}{r_i}\right)}{\ln\left(\frac{r_o}{r_i}\right)}. \tag{8.7}$$

The temperature varies in a nonlinear fashion with radial position.

8.2.1 Thermal Resistance in Conduction

Using the analogy with Ohm's law for the flow of electricity through a conductor, the thermal resistance to conduction of a slab, $\Re_{T,\text{cond}}$, having a specific cross-sectional area A available for flow is given by (2.33) and (8.2)

$$\Re_{T,\text{cond}} = \frac{L}{kA} = \frac{T_0 - T_L}{\dot{Q}_x}. \tag{8.8}$$

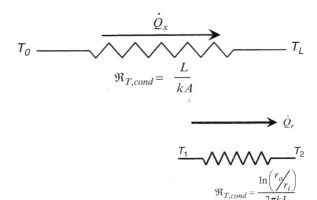

Fig. 8.3 Equivalent electrical network for heat conduction through a single material

Fig. 8.4 An equivalent electrical network for heat flow through a hollow cylinder

The thermal conduction process can be understood in analogous terms of a flow occurring in response to an applied potential difference against a material resistance, as illustrated in Fig. 8.3. In this case, the potential is temperature, and the flow is heat. Similarly, the thermal resistance to conduction for a hollow cylinder can be found by substituting (8.6) into the definition of thermal resistance, (2.33):

$$\Re_{T,\text{cond}} = \frac{\ln\left(\frac{r_o}{r_i}\right)}{2\pi k L}. \tag{8.9}$$

This is illustrated in Fig. 8.4. Applications of the concept of thermal resistance are discussed in Chap. 9.

8.3 Convection

Heat convection describes the exchange of heat associated with relative motion between a fluid and a solid substrate. The fluid may be encountered in either the gaseous or the liquid phase, although for some special cases both a liquid and a gas may coexist. When studying conduction processes, convection is viewed as a boundary condition that contributes to the internal diffusion of heat in the solid. We will focus on this heat transfer at the fluid/solid boundary as it is manifested in the material of the fluid.

The primary objective of convection analysis is to determine an appropriate value of the convective heat transfer coefficient, h. Newton's law of cooling describes the convective flow at the surface, $\dot{Q}_s$ (W), in terms of h, the interface or surface area, S, between the fluid and solid, and the surface and bulk fluid temperatures, (T_s) and (T_∞).

$$\dot{Q}_s = hS(T_s - T_\infty). \tag{8.10}$$

Since convection always involves a relative motion between the fluid and solid, it will be necessary to apply the principles of fluid flow analysis developed earlier in Chaps. 4–7 to describe the characteristics of fluid movement that govern the temperature gradient and heat transfer in the fluid. Thus, analysis of fluid flow and convective heat transfer are very closely related.

8.3.1 Four Principle Characteristics of Convective Processes

There are four distinguishing characteristics of convective flow systems that determine the nature and intensity of the resulting heat transfer process. It is necessary to evaluate each of these characteristics to calculate the value for the convective heat transfer coefficient, h. These characteristics and the various options they may take are:

1. The source of relative motion between the fluid and solid, resulting in *forced* or *free* convection.
2. The geometry and shape of the boundary layer region of the fluid in which convection occurs, producing *internal* or *external* flow. In addition, for free convection the orientation of the fluid/solid interface in the gravitational field is important.
3. The *boundary layer* flow domain, being *laminar* or *turbulent*.
4. The *chemical composition* and *thermodynamic state* of the fluid in the boundary layer, which dictate numerical values for the constitutive properties relevant to the convective process.

When a new convective transport problem is encountered, it will be necessary to address the influence of each of these four characteristics of the process. Together the understanding obtained by evaluating these characteristics provides a specific guide for determining and applying the most appropriate analysis tool for a particular convective problem. Although it is possible to perform a comprehensive study of some convection processes based on a full fluid flow analysis, the mathematical process is quite tedious and lengthy, and many types of convection are too complex to allow this approach. Alternatively, a very broad range of experimental studies have been conducted to measure convective heat transfer as a function of the four primary characteristics. The results of these experiments are presented in terms of *empirical correlation equations* written with *dimensionless parameters* that enable application for convective processes that differ from the exact physical conditions of the experiment. We have adopted this approach for convective analysis, which generally provides a quicker and easier route to determining a value for h while incorporating many elements of a fundamental understanding of the governing physical processes. Much of this section is organized around exploring how the four principle characteristics individually affect the convection process and how they combine to provide a full description of a specific convection condition.

8.3 Convection

The relative motion between a fluid and solid may be caused by differing kinds of energy sources. Perhaps most obviously, an external force can be applied to the fluid or solid to produce the motion (which is termed forced convection). This force is most frequently a mechanical force to move the solid or a pressure gradient on the fluid. However, in the absence of an external motivational force, the heat transfer process itself will cause relative motion. Owing to the constitutive properties of fluids, the existence of a temperature gradient produces a concomitant density gradient. When the fluid is in a force field such as gravity or centrifugation, the density gradient causes internal motion within the fluid by buoyancy effects as the less dense fluid rises and the more dense fluid falls under the action of the force field. This phenomenon is called free convection since no external energy source is applied to cause the motion. Any time there is a temperature gradient in a fluid, the potential exists for having free convection heat transfer.

As can be anticipated, the fluid flow patterns for forced and free convection are very different, and therefore forced and free convection produce quite disparate heat transfer effects. Also, analyses of the fluid flow characteristics in forced and free convection are unique because of the differing patterns of motion. Usually, the magnitude of forced convection effects is much larger than for free convection. Thus, although the potential for free convection will be present whenever a temperature field exists in a fluid, if there is also an imposed forced source of fluid motion, the free convection effects will be masked since they are much smaller, and they can be neglected. We will discuss free and forced convection phenomena separately since they are so distinctive.

8.3.2 Fundamentals of Convective Processes

Convection involves a combination of concurrent mass, energy, and momentum flows. Therefore, convective analysis requires that we adapt simultaneous considerations of the influence of each of these domains. The conservation laws in each of these domains contribute to understanding convective phenomena, plus there are many special constitutive equations that apply. In the following sections, we will discuss the physical nature of convective flows and apply analysis tools to develop methods to determine values of the convective heat transfer coefficient, h, for specific types of flow geometries and conditions.

8.3.2.1 The Constitutive Equation for Convection: Newton's Law of Cooling

Newton's law of cooling relates the magnitude of heat flux to the difference in the temperatures of the solid substrate and at a representative state in the fluid. This equation is stated as:

$$q_S = \frac{\dot{Q}_S}{S} = h(T_S - T_\infty). \tag{8.11}$$

The magnitude of the convection heat transfer is scaled in proportion to the system property h. The quantitative influence of all four principle characteristics of the convection process are embodied in the h value. The convective heat transfer is also directly proportional to the area, S, of the solid–fluid interface.

Although (8.11) is written in a very simple format, determining an appropriate value for h can be a complex undertaking. The influence of each of the four principle characteristics must be evaluated individually and collectively, and the value determined for h may vary over many orders of magnitude, depending on the combined effects of the characteristics. Table 8.1 presents the range of typical values for h for various combinations of convective conditions that may be encountered.

The convection process consists of the sum of two separate effects. First, when there is a temperature gradient in a fluid, heat conduction will occur consistent with the thermal conductivity of the chemical species and its state. The conduction effect can be very large in a liquid metal or very small in a low density vapor. The conduction occurs via microscopic scale interactions among atoms and molecules, with no net translation of mass. Second, there will be transport of energy associated with the bulk movement of a flowing fluid. The component due to only bulk motion is referred to as *advection*. Convection involves a net aggregate motion of the fluid thereby carrying the energy of the molecules from one location to another. These conduction and advection effects are additive and superimposed.

Under certain circumstances, a third effect can contribute to convection, that being when a material undergoes a phase change at the interface surface. In biological applications, this phenomenon is most often associated with the evaporation of water, such as occurs during sweating. If the chemical potential (expressed as a partial pressure) of liquid water on the skin is greater than in the vapor phase in surrounding air, liquid water will evaporate. The consequences are that there will be a mass flow away from the surface, and there will be an energy loss from the solid substrate to provide the latent heat necessary to drive the phase change from the liquid to the vapor state. This process provides a cooling effect to the underlying tissue that can be a direct benefit of sweating to the thermoregulatory process. Thus, the energy balance at the interface surface will involve not only the sensible components associated with the temperature gradient, but also a latent component due to the phase change. The result is an enhancement of convection, which in some instances can be very important in the boundary conditions. Another important

Table 8.1 Range of values for h as encountered for various combinations of convective transport process characteristics

Process characteristics	Range of h (W/(m² K))
Free convection	
Vapors	3–25
Liquids	20–1000
Forced convection – internal/external	
Vapors	10–500
Liquids	100–15,000
Phase change	
Between liquid and vapor	5,000–100,000

8.3 Convection

phase change enhancement of convection occurs in the respiratory pathways as water alternately evaporates and condenses during the breathing cycle.

Next, we will consider various specific methods to evaluate the four principle characteristics of convection to determine numerical values for h. An important point to remember is that the study of convection deals with methods to determine values for h that can be used in specifying the boundary conditions for thermal conduction in solids.

8.3.2.2 Temperature and Velocity Boundary Layers

A fundamental aspect of convection heat transfer is that the processes produce both velocity and temperature boundary layers in the fluid adjacent to a solid interface. Illustrations of these boundary layers are shown in Fig. 8.5.

The velocity and temperature boundary layers have similar features, which have previously been described when we discussed fluid flow. Both define a layer in the fluid adjacent to a solid in which a property gradient exists. The temperature boundary layer develops because there is a temperature difference between the fluid in the free stream, T_∞, and the solid surface, T_s. A thermal gradient exists between the free stream and the surface, which has the largest slope at the surface and diminishes to zero at the outer limit of the boundary layer at the free stream.

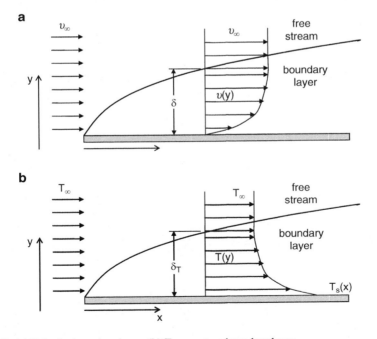

Fig. 8.5 (a) Velocity boundary layer. (b) Temperature boundary layer

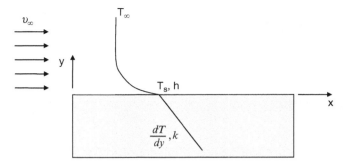

Fig. 8.6 Convective boundary condition at the surface of a conducting solid

The temperature gradient at the surface defines the thermal boundary condition for conduction in the solid substrate. The boundary condition can be written by applying conservation of energy at the surface, which, since it has no thickness, has no mass and is therefore incapable of energy storage. Thus, the conductive energy outflow from the solid is equal to the convective inflow to the fluid as illustrated in Fig. 8.6.

An important feature of the convection interface is that there is continuity of both temperature and heat flux at the surface, the latter of which is expressed in (8.12).

$$-k_f \frac{dT}{dy}\bigg|_{y=0} = h(T_s - T_\infty). \tag{8.12}$$

These two boundary conditions (continuity of flow and potential) are encountered at the interface between differing materials for all modes of transport (fluid, heat, and mass). The equality of temperatures in the fluid and solid at the surface is equivalent to the no-slip velocity condition for fluid flow. You should become familiar with this set of boundary conditions since they will be encountered repeatedly as you analyze transport problems.

Despite so many similarities, we should point out one way in which the velocity and temperature boundary layers differ, that being in their thicknesses, δ and δ_T. The relative rates of momentum and energy diffusion in a fluid may be very different, depending on the properties of viscosity, thermal conductivity and heat capacity. These properties can be combined into a single dimensionless parameter called the *Prandtl number*, Pr, which provides a means of comparing the effectiveness of momentum and energy diffusion in a fluid. Pr is defined as:

$$Pr = \frac{c_p \mu}{k} = \frac{c_p \mu \rho}{k \rho} = \frac{\nu}{\alpha}. \tag{8.13}$$

The last term is the ratio of the momentum diffusivity (kinematic viscosity) to the thermal diffusivity. The value of Pr can be very sensitive to the temperature, especially because of the property viscosity, which can change substantially with temperature. Also, there can be quite large differences in the value of Pr as a function of the chemical composition of a fluid and whether it is a liquid or vapor. For example, at physiological conditions, water has a Pr of about 4.6 as a liquid and 0.87 as a vapor. As the temperature drops to near freezing, the liquid value increases to near 13, and the vapor drops slightly to about 0.82. Viscous liquids such as oil and glycerol can have Pr values in the range $10^4 - 10^5$. Liquids with a high thermal conductivity, including liquid metals such as mercury, have Pr values as low as 10^{-2}. Many vapors have Pr values close to 1 or slightly lower.

The magnitude of Pr is an indicator of the relative growth rates of the velocity and thermal boundary layers, the direct implication being that the thicknesses of the layers can be very different in a given convective flow system. For small Pr, $\delta \ll \delta_T$, and for large Pr, $\delta \gg \delta_T$. For a liquid metal, the thermal effects at the surface can penetrate far into the fluid flow stream; for an oil, the viscous effects penetrate far from the wall in the fluid. For vapors, $\delta \approx \delta_T$ so that the extent of the thermal and viscous boundary effects is similar.

Another important feature of convective heat transfer can be identified by inspection of Fig. 8.5b. For the flow system illustrated, the temperature boundary layer is growing in thickness as the fluid flows further along the surface from the leading edge. Therefore, the convective transport distance between the surface and the freestream changes with position along the length of the surface. In general, the magnitude of convective transport varies inversely with the thickness of the temperature boundary layer. The thicker the boundary layer, the greater is the distance that energy must be transported across a fixed temperature differential, $T_s - T_\infty$. Therefore, the convective resistance is larger, and the convection coefficient, h, is proportionately smaller. The consequence is that the value of h will vary with position under conditions for which the temperature boundary layer is growing, giving rise to a local convection heat transfer coefficient h_x. For some types of problems, local variations in convection may be quite important and must be accounted for mathematically in the analysis procedure. Processes for which a local peak in convective transport can lead to a limiting response require the localized approach. Alternatively, for many problems it is sufficient to use a single convection coefficient that is averaged over the entire participating surface area, $\bar{h}$. The relationship between the local and average coefficients is:

$$\bar{h} = \frac{1}{S} \int_S h_x \, dS, \tag{8.14}$$

where S is the total surface area for a convection interface. The total convective heat flow over an entire surface area S is then given by:

$$\dot{Q}_s = \bar{h} S (T_s - T_\infty). \tag{8.15}$$

Values for $\bar{h}$ and h_x are calculated by application of correlation equations that are expressed in terms of dimensionless parameters. These parameters can and have been developed by application of the Pi theorem as was presented in Chap. 3. A number of dimensionless parameters are fundamental to the expression of these correlation equations, and they are relevant to an impressively large diversity of physical convection processes. The most important and general of these parameters are discussed in the next section.

8.3.2.3 Dimensionless Parameters of Convection

We will consider a dimensional analysis of convective heat transfer phenomena. The convective heat transfer coefficient between a solid body with two characteristic lengths, d and L, surrounded by a flowing fluid might be expected to depend on the following variables:

Geometry of the solid: d, L
Thermal conductivity of the solid: k_s
Properties of the fluid: $\rho_f, c_{pf}, k_f, \mu, \beta g\,(T_s - T_\infty)$ (free convection)
Average fluid velocity: $\langle v \rangle$ (forced convection)

Therefore, in general, one can write:

$$h = h(d, L, k_s, \rho_f, c_{pf}, k_f, \mu, \beta g(T_s - T_\infty), \langle v \rangle). \tag{8.16}$$

Four fundamental dimensions (M, L, T, Θ) are included in these 10 variables. According to the Buckingham Pi theorem, we would expect six dimensionless groups. By inspection, two of these are k_s/k_f and d/L. Choosing k_f, L, μ and ρ_f as the core variables, we find the dimensionless heat transfer coefficient to be a function of the following five dimensionless groups:

$$\frac{hL}{k_f} = f\left(\frac{c_p\mu}{k_f}, \frac{\rho_f \langle v \rangle L}{\mu}, \frac{\rho_f^2 L^3 g \beta (T_s - T_\infty)}{\mu^2}, \frac{d}{L}, \frac{k_s}{k_f}\right). \tag{8.17}$$

The first four dimensionless groups are given special names and symbols, as indicated in Table 8.2. We will describe the importance of each of these dimensionless groups shortly. Thus, we can rewrite (8.17) as:

$$Nu_L = f\left(Pr, Re_L, Gr_L, \frac{d}{L}, \frac{k_s}{k_f}\right). \tag{8.18}$$

The dimensionless heat transfer coefficient hL/k_f is known as the *Nusselt number*, Nu_L, based on the dimension L. Any two dimensionless groups can be multiplied together to form a third dimensionless group. However, the new group will not be independent of the other two. For instance, we could multiply Nu_L by d/L to obtain Nu_d, the Nusselt number based on the characteristic length d. We could then replace Nu_L with Nu_d in (8.18). Alternatively, we could multiply Nu_L by k_f/k_s to

8.3 Convection

Table 8.2 Dimensionless parameters encountered most commonly in biotransport applications

Dimensionless parameter	Definition	Explanation
Bi – Biot number	$Bi = \dfrac{hL}{k_s}$	Ratio of internal thermal conduction resistance to external convective flow resistance
Bi_A – Biot number for mass transfer	$Bi_A = \dfrac{k_{Ae}L}{D_{Ai}\Phi_{Aie}}$	Ratio of internal to external resistance to mass transfer
Fo – Fourier number	$Fo = \dfrac{\alpha t}{L^2}$	Ratio of rates of heat conduction to thermal energy storage in a solid
Gr – Grashof number	$Gr_L = \dfrac{g\beta(T_s - T_\infty)L^3}{\nu^2}$	Ratio of buoyant to viscous forces
Gz – Graetz number	$Gz_d = \dfrac{L}{d} Re_d Pr$	Ratio of advection to pure conduction for the entrance length
Nu – Nusselt number	$Nu_L = \dfrac{hL}{k_f}$	Ratio of net convection to pure conduction in a fluid
Pe – Peclet number	$Pe = \dfrac{vL}{\alpha} = Re_L Pr$	Ratio of advection to pure conduction in a fluid
Pr – Prandtl number	$Pr = \dfrac{c_p \mu}{k_f} = \dfrac{\nu}{\alpha}$	Ratio of momentum and thermal diffusivities in a fluid
Ra – Rayleigh number	$Ra_L = \dfrac{g\beta(T_s - T_\infty)L^3}{\alpha \nu} = Gr_L Pr$	Ratio of free convection and pure conduction effects
Re – Reynolds number	$Re_L = \dfrac{\rho v L}{\mu}$	Ratio of inertial flow effect to viscous retardation forces
Sc – Schmidt number	$Sc = \dfrac{\nu}{D_{Am}}$	Ratio of the momentum and mass diffusivities
Sh – Sherwood number	$Sh_L = \dfrac{k_{Am}L}{D_{Am}}$	Ratio of convective to diffusive mass transport
We – Weber number	$We = \dfrac{\rho v^2 L}{\sigma}$	Ratio of inertia effects to surface tension forces

obtain hL/k_s, the Biot number, also found in Table 8.2. Although dimensional analysis indicates that the Nusselt number can be a function of as many as five independent dimensionless groups, in most cases we find the dependence to be limited to two or three dimensionless groups, depending on the situation.

The fundamental form of the convection correlation equations is that they are written entirely in terms of dimensionless parameters. The primary advantage of this format is that the equations can be scaled for applicability across a wide range of physical dimensions and property values. In this context, the convection coefficient, h, is expressed in terms of the Nusselt number, Nu.

$$Nu = \frac{hL}{k_f}. \qquad (8.19)$$

L is a characteristic physical dimension of the interface surface, and k_f is the thermal conductivity of the fluid evaluated at an average temperature within the boundary layer. You should be careful to note that although the Nu has the same combination of properties as the Biot number Bi, these are two completely different parameters. The thermal conductivity in Bi, refers to the solid substrate, but k_f in

Nu refers to the thermal conductivity of the fluid. Thus, Nu is the ratio of convection to pure conduction within the fluid and is a measure of the contribution of fluid movement to the overall convective process. In the limit, the smallest value Nu can have is 1.0 for the case in which there is no fluid motion at all, and heat transfer in the fluid occurs entirely via conduction.

Analysis and experiments have demonstrated that the Nusselt number is a function of the four primary characteristics of the convection process, all of which can be expressed in terms of appropriate dimensionless parameters. Thus, for forced convection,

$$Nu = f(Re, Pr) \tag{8.20}$$

and for free convection,

$$Nu = f(Gr, Pr). \tag{8.21}$$

The Prandtl number, Pr, was defined previously in (8.13). It is a function of the thermodynamic state and transport properties of the fluid chemical species. The *Reynolds number* was defined earlier relating to fluid flow (3.7) and applies for forced convection. It describes the ratio of inertial force driving a flow and of the viscous retardation against the flow. Recall that there are different definitions of the geometric dimension in Re for internal and external flows. The magnitude of Re identifies whether a flow is in the laminar or turbulent domain. The transition conditions are 2,200 for internal flow and 5×10^5 for external flow.

The equivalent parameter for free convection is the *Grashof number*, which is the ratio of buoyant effects that drive the fluid flow and of viscous retardation of the flow.

$$Gr = \frac{g\beta L^3 (T_s - T_\infty)}{v^2}, \tag{8.22}$$

where g is the acceleration of gravity and β is the coefficient of thermal expansion of the fluid. For an ideal gas, the value of β equals the reciprocal of the temperature expressed in absolute units (K).

For both forced and free convection, the function f in (8.20) and (8.21) embodies a dependency on each of the four primary characteristics of convection. Therefore, identification of the most appropriate correlation to use in a particular application depends on correctly specifying the combination of characteristics unique to the process of interest.

Table 8.2 presents a listing of the most common dimensionless parameters that are used in biotransport analysis. Note that there are parameters for mass transport that are analogous to those for heat transport. These will be discussed in later chapters, but are included here to provide a concise point of reference for these parameters.

Upon encountering a convection heat transfer process, the first step in analysis is to assess it in light of the four principle characteristics of convection. The initial considerations are whether the source of flow is forced or free and whether the geometry results in external or internal flow. Thereafter, it will be necessary to determine values of thermodynamic and transport properties to calculate the flow domain, whether it be laminar or turbulent. This information should be sufficient to direct you to an appropriate correlation equation by which Nu and h can be calculated. The following sections present correlation equations, written in terms of the above dimensionless parameters, that provide a basis for evaluating a wide range of convective heat transfer processes.

8.3.3 Forced Convection Analysis

Forced convection is the most frequently encountered convective mechanism in biological environments. External flows occur between the body surface and surrounding air, or sometimes water. Internal flows occur within the body as both liquids and vapors move through various lumens. All of these processes can be modeled with existing correlations developed many years previously, primarily for industrial and military applications. Correlation equations are identified according to the four primary characteristics of convection heat transfer.

8.3.3.1 Internal Flow Geometries

Introductory Concepts and Background

Internal flows occur in two modes. When a flow is introduced into the entrance of a lumen, the boundary layer will develop progressively along the length of the flow passage. In this region, the boundary layer does not yet fill the entire cross section of the lumen so that there is not a velocity gradient in the central portion of the diameter. Viscous drag does not play a role in this central portion of the flow, and it moves as though the fluid were inviscid. The initial portion of the lumen over which the boundary layer develops is called the *entrance length*. Eventually, if the lumen is long enough and does not have geometric perturbations, a *fully developed* region will be reached in which there are no further changes in the boundary layer with position deeper into the passage. In the fully developed region, the boundary layers on opposing surfaces have grown to the extent that they come together along the center line so that a velocity gradient exists throughout the entire flow cross section. These two regions are illustrated in Fig. 8.7.

Note that if a fully developed flow never occurs no matter how long the lumen is, then the geometry falls into the external flow classification. Fully developed internal flows will always have a velocity gradient extending over the entire lumen cross section.

When a lumen goes around a bend, splits into two smaller lumens, or combines with another lumen to form a single larger passage, the boundary layer will be disturbed and will have to be reestablished, leading to another entrance length. Usually, these flow patterns are complex and are challenging to analyze according

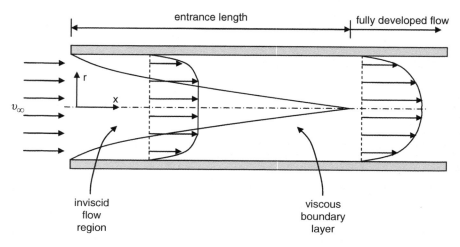

Fig. 8.7 Entrance and fully developed velocity boundary layer regions for flow in a lumen

to the specific flow geometry. If the upstream lumen contains a fully developed flow, then following the geometric change there will likely still be a velocity gradient throughout the lumen cross section, but it may be highly asymmetric, which presents a major challenge to analysis. For our purposes, in the absence of subject-specific information, internal flows will be treated as occurring in simple, straight lumens to which existing correlation equations can be applied directly. The complex geometries can be analyzed using more sophisticated methods such as the finite element method or computational fluid dynamics, but such methods are beyond the introductory scope of this text.

When the source of motion is identified as forced convection, and it is anticipated that the geometry is internal, the Reynolds number should be calculated to determine whether the flow domain is laminar or turbulent. The Reynolds number must be determined using a mean value of the flow velocity within the boundary layer.

$$Re_d = \frac{\rho \langle v \rangle d}{\mu}, \qquad (8.23)$$

where $\langle v \rangle$ is the velocity averaged over the lumen diameter d. The mean velocity is calculated from the total mass flow rate w through the lumen with cross-sectional area A_c

$$w = \rho \langle v \rangle A_c, \qquad (8.24)$$

which for a round lumen results in:

$$Re_d = \frac{4w}{\pi d \mu}. \qquad (8.25)$$

8.3 Convection

For flow in a straight, round lumen turbulent flow will occur for values of Re greater than about 2,200 in fully developed flow. In some cases, the flow conditions may lead to a transition to turbulent flow within the entrance length before the fully developed condition is reached. The length of the entrance region, L_e, for laminar flow from (5.70) with $Re_d > 50$ is given by:

$$\left(\frac{L_e}{d}\right)_{lam} \approx 0.0567\, Re_d. \tag{8.26}$$

For turbulent flow, the entrance length is:

$$\left(\frac{L_e}{d}\right)_{turb} \approx 10. \tag{8.27}$$

For convection heat transfer, a temperature boundary layer will develop in a lumen in addition to the velocity boundary layer. Since the motion of the fluid has a major effect on the convection process, the fluid flow domain and the shape of the velocity boundary layer will have a major influence on the shape of the temperature boundary layer. The concept of a temperature boundary layer is illustrated in Fig. 8.8.

The thermal boundary condition for the fluid at the lumen wall may be defined in terms of a specified temperature or heat flux, and for the simplest cases is constant along the length. One important difference in comparison with the velocity boundary layer is that whereas the velocity remains constant after the fully developed region is established (consider the implications of conservation of mass for the flow process), since heat is being added continuously along the length of the lumen, the average fluid temperature will also rise proportionately.

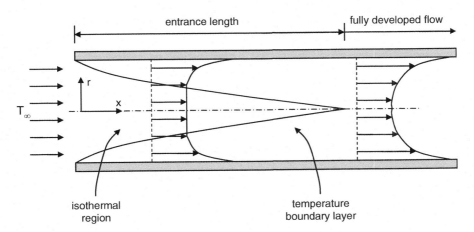

Fig. 8.8 Entrance and fully developed temperature boundary layer regions for flow in a lumen with a heated wall

The thermal entrance length, L_{et}, for laminar flow is given by:

$$\left(\frac{L_{et}}{d}\right)_{lam} \approx 0.0567\, Re_d\, Pr. \tag{8.28}$$

Note that for $Pr = 1$ the temperature and velocity entrance lengths will be identical. Otherwise, they are not. For turbulent flow, the entrance length is nearly independent of Pr so that:

$$\left(\frac{L_{et}}{d}\right)_{turb} \approx 10 \tag{8.29}$$

and the velocity and temperature boundary layers grow together in the entrance length.

When Newton's law of cooling (8.10) is applied to calculate a convective heat flux it is necessary to specify a value for the temperature of the fluid, T_∞. For an internal flow geometry there is not a value for the free stream temperature since the entire flow is occupied by a boundary layer. Therefore, a mean value of fluid temperature must be used to represent the overall state within the lumen volume. The mean temperature, T_m, is defined as the flow-averaged enthalpy per unit volume relative to the flow averaged heat capacity per unit volume averaged over the cross section of the lumen, A_c

$$T_m = \frac{\int_{A_c} (\rho c_p T)\, v\, dA_c}{\int_{A_c} (\rho c_p)\, v\, dA_c}. \tag{8.30}$$

For an incompressible fluid with constant c_p flowing in a circular lumen of radius r_0 this expression becomes:

$$T_m = \frac{2}{\langle v \rangle r_0^2} \int_0^{r_0} v(r) T(r)\, r\, dr, \tag{8.31}$$

where $\langle v \rangle$ is the mean velocity in the lumen. If the flow is incompressible (nearly all biological flows meet the incompressibility requirement of a Mach number less than 0.3, i.e., the velocity is less than 0.3 times the speed of sound in the flow medium), then the mean velocity (5.7) is calculated by integrating the velocity profile over the cross sectional area in a manner similar to what was done above for temperature.

$$\langle v \rangle = \frac{w}{\rho A_c} = \frac{\int_{A_c} \rho v(r)\, dA_c}{\rho A_c} = \frac{2\pi \rho}{\rho \pi r_0^2} \int_0^{r_0} v(r)\, r\, dr = \frac{2}{r_0^2} \int_0^{r_0} v(r)\, r\, dr. \tag{8.32}$$

T_m represents the mean temperature that would result from collecting the fluid flowing through at a particular cross section and mixing it together to achieve an average, uniform state. Newton's law of cooling for internal flow is then written as:

$$q_s = h(x)[T_s - T_m(x)]. \tag{8.33}$$

Note that both the mean fluid temperature, T_m, and the convective coefficient, h, may vary with position along the length of the lumen.

Correlation Equations for Internal Convection Flow

The practical solution of many types of convection problems is dependent on determining numerical values for the convective coefficient, h. Correlation equations by which h may be calculated are written in terms of the relevant dimensionless parameters defined earlier. These equations are derived from many different analytical and empirical sources, and a comprehensive review of them is beyond the scope of this text. There are advanced treatises that present this material in well-organized formats and serve as authoritative references (Bejan 2004; Burmeister 1993; Kays et al. 2005). The correlation equations can be logically organized according to how they match the primary characteristics of convection processes, and they are presented here in that format.

After a given convection condition is established as having a forced source of motion and an internal geometry, the next characteristics to determine are whether the flow domain is laminar or turbulent and what are the further details of the geometry, such as the shape of the cross-sectional flow area and if entrance length effects are important to consider. Laminar flow patterns are much more orderly than are turbulent and have therefore been subjected to theoretical modeling to develop correlation equations. The more complex turbulent flow patterns have issued an adoption of empirical correlations fit to experimental data. The mathematical structure of the equations reflects these two types of development. We will present only a basic set of correlation coefficients for laminar and turbulent flows.

Convection in Fully Developed Laminar Flow

For steady state, fully developed laminar flow through a lumen of circular cross section having a diameter d and a constant heat flux q_s at the inner wall surface, the Nusselt number is given by:

$$Nu_d = \frac{hd}{k_{T_m}} = \frac{48}{11} = 4.36 \quad q_s \text{ is constant}, \tag{8.34}$$

where k_{T_m} is the thermal conductivity of the flowing fluid evaluated at the mean temperature (8.30).

Under these conditions, h does not vary along the length of the lumen, and there is no need to integrate along the flow axis to obtain an averaged value. When the fluid interface with the lumen wall is defined by a constant temperature T_s, the Nusselt number is given by:

$$Nu_d = \frac{hd}{k_{T_m}} = 3.66 \quad T_s \text{ is constant.} \tag{8.35}$$

The basis for the simplicity of these equations is due to the flow being fully developed wherein there is no radial component of velocity. Therefore, the only radial component of energy flow is due to conduction. We will see later that the analysis becomes more intricate in the entrance length and for turbulent flow.

Convection in the Laminar Flow Entrance Length

Convection in the entrance length of a lumen where either or both the velocity and temperature profiles are developing involves a more complex flow pattern than in the fully developed region. Two different types of entrance flow may be encountered: the velocity profile is already fully developed, and only the temperature profile is changing; and, both the velocity and the temperature profiles are developing simultaneously. The first condition occurs when the Prandtl number of the fluid is large or if flow is already established in a lumen before wall convection is initiated downstream of the entrance. The latter condition is called the *combined entry length problem*. Correlations have been available for many years for the average convection transfer that occurs for both of these entry flow problems with a constant wall surface temperature (Hausen 1943).

$$\overline{Nu_d} = 3.66 + \frac{0.0668\left(\frac{d}{L}\right)Re_d Pr}{1 + 0.04\left[\left(\frac{d}{L}\right)Re_d Pr\right]^{\frac{2}{3}}} \quad T_s \text{ is constant.} \tag{8.36}$$

The combination of dimensionless parameters $(d/L)\, Re_d\, Pr$ is known as the Graetz number, Gz_d, which represents the ratio of advection to conduction transport of heat normalized to the relative entrance length. Thus:

$$\overline{Nu_d} = 3.66 + \frac{0.0668 Gz_d}{1 + 0.04 Gz_d^{\frac{2}{3}}} \quad T_s \text{ is constant.} \tag{8.37}$$

For combined entry flow problems with the Prandtl number on the order of one, an alternate correlation is given below (Sieder and Tate 1936), where the fluid state must satisfy the indicated conditions:

$$\overline{Nu_d} = 1.86 Gz_d^{\frac{1}{3}}\left(\frac{\mu_{T_m}}{\mu_s}\right)^{0.14} \quad T_s \text{ is constant,} \tag{8.38}$$

$$0.60 \leq Pr \leq 5,$$

$$0.0044 \leq \frac{\mu_{T_m}}{\mu_s} \leq 9.75,$$

8.3 Convection

where μ_{T_m} and μ_s are fluid viscosities evaluated at the mean temperature and surface temperature, respectively.

More detailed analyses have shown that the Nusselt number decreases progressively along the entrance length from a maximum value at a lumen entrance to a minimum at the start of fully developed flow conditions. This result is anticipated since the resistance to convective heat transport will increase as the thickness of the thermal boundary layer grows.

Convection in Turbulent Internal Flow

Empirically derived correlation equations can be applied for fully developed turbulent flow in circular tubes. A number of alternative equations are available which have defined domains of applicability. The most general equation is:

$$Nu_d = 0.023\, Re_d^{4/5}\, Pr^{1/3}. \tag{8.39}$$

A refined form of this equation is (Dittus and Boelter 1930), with the indicated range of validity.

$$Nu_d = 0.0265\, Re_d^{4/5}\, Pr^{0.3} \quad \text{for cooling,} \tag{8.40}$$

$$Nu_d = 0.0243\, Re_d^{4/5}\, Pr^{0.4} \quad \text{for heating,} \tag{8.41}$$

$$0.7 \leq Pr \leq 160,$$

$$Re_d \geq 10^4,$$

$$\frac{L}{d} \geq 10.$$

For processes in which there may be large internal variations in key properties such as density, a further refinement of the equation is available (Seider and Tate 1936).

$$Nu_d = 0.027\, Re_d^{4/5}\, Pr^{1/3} \left(\frac{\mu_{T_m}}{\mu_s}\right)^{0.14}, \tag{8.42}$$

$$0.7 \leq Pr \leq 1.67 \times 10^4,$$

$$Re_d \geq 10^4,$$

$$\frac{L}{d} \geq 10.$$

For all of these equations, the fluid properties are evaluated at temperature T_m.

The entrance length in turbulent flow tends to be shorter (a rough estimate is less than sixty tube diameters) than for laminar flow due to aggressive mixing along the radial axis. Therefore, it is more likely that entrance effects will not need to be accounted for in turbulent flow. It is usually acceptable to assume that the average Nusselt number is equal to that measured in the fully developed zone.

8.3.3.2 External Flow Geometries

External flow convection occurs when the velocity and temperature boundary layers that develop as a fluid moves past a solid substrate at a dissimilar temperature can grow without any imposed physical constraints. Typically, a boundary layer will start out at the leading edge where the fluid first encounters a solid substrate in the laminar domain, and, if the interface flow length is sufficient, will reach conditions for which a transition to the turbulent domain will occur. As in internal flow, velocity and temperature boundary layers may grow separately or simultaneously, depending on the magnitudes of the diffusion properties as reflected in the Prandtl number. For the laminar domain, the flow pattern is well defined, and theoretical methods of analysis can be applied. For turbulent flow, empirical correlations prevail.

In external flow geometries, there will be a specific free stream temperature, T_∞, in the region outside the boundary layer and a temperature T_S at the solid surface. The properties of the fluid in the boundary layer are evaluated at an average fluid temperature, T_f, also sometimes called the film temperature, (Fig. 8.9)

$$T_f = \frac{T_s + T_\infty}{2}. \tag{8.42a}$$

There are two primary classes of geometric shapes that are encountered in external convection: flat surfaces that are parallel or oblique to the free stream flow, and curved surfaces. External convection analysis will be organized according

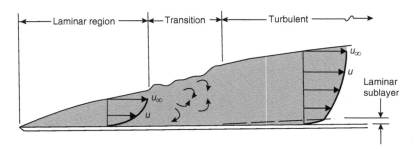

Fig. 8.9 Drawing of external flow boundary layer showing transition to turbulent flow (Holman 1972, with permission)

8.3 Convection

to whether the flow is laminar, turbulent, or mixed (laminar and turbulent regions on the same surface) and the interface geometry (flat or curved).

Laminar External Flow over a Flat Plate

Boundary layer analysis has been applied to develop a description of convective heat transfer in the laminar region of a boundary layer for flow on a flat plate. The convective heat transfer coefficient is a function of the distance x from the leading edge of the plate.

$$Nu_x = \frac{h_x x}{k} = 0.332\, Re_x^{1/2}\, Pr^{1/3}, \qquad (8.43)$$

$$Pr \geq 0.6.$$

The convection process can be integrated over the entire length L of the laminar region on the plate to obtain an averaged value for the Nusselt number.

$$\bar{Nu}_L = 0.664\, Re_L^{0.5}\, Pr^{0.33}, \qquad (8.44)$$

$$Pr \geq 0.6.$$

Turbulent External Flow over a Flat Plate

For flow conditions in which there is a transition from laminar to turbulent flow, the local value for the Nusselt number in the turbulent region is given by

$$Nu_x = 0.0296\, Re_x^{4/5}\, Pr^{1/3}, \qquad (8.45)$$

$$Re_x \leq 10^8,\ 0.6 \leq Pr \leq 60.$$

The convection process can be integrated over the combined laminar and turbulent regions of flow for the complete length L of the flat plate to obtain an averaged value for the Nusselt number. When the convection conditions support a transition from laminar to turbulent flow, the net combined flow is called *mixed*.

$$\bar{Nu}_L = \left(0.037\, Re_L^{0.8} - 871\right) Pr^{0.33}, \qquad (8.46)$$

$$Re_x \leq 10^8,\ 0.6 \leq Pr \leq 60.$$

External Flow over a Perpendicular Cylinder

Convection heat transfer during external flow over the outside of a heated cylinder is illustrated in Fig. 8.10. The interferometer fringe lines represent isotherms in the

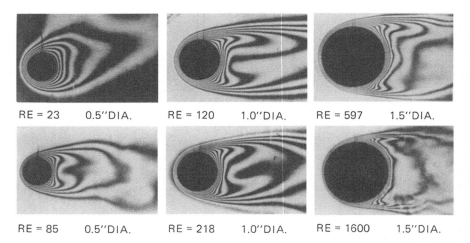

Fig. 8.10 Interferometer photos showing isotherms in the thermal boundary layer for external flow over a heated cylinder (Holman 1972, with permission)

thermal boundary layer. Note the large change in the geometry of the boundary layer as the Reynolds number increases.

The convection process is averaged across the entire circumferential surface around a cylinder of diameter d in perpendicular flow,

$$\bar{Nu}_d = CRe_d^n Pr^{0.33}, \qquad (8.47)$$

$$0.4 \leq Re_d \leq 4 \times 10^5, 0.7 \leq Pr,$$

where the values of C and n are functions of Re_d as given below (Hilpert 1933).

Re_d	C	n
0.4–4	0.989	0.330
4–40	0.911	0.385
40–4,000	0.683	0.466
4,000–40,000	0.193	0.618
40,000–400,000	0.027	0.805

For flow that is normal to the front, the leading position on a cylinder starts as laminar and progresses to turbulent transition at a further angular position around the circumference of the surface. The result is a mixed flow.

External Flow over a Sphere

The convection is averaged across the entire surface around a sphere of diameter d with properties based on T_∞ (Whitaker 1972).

$$\bar{N}u_\mathrm{d} = 2 + \left(0.4\,Re_\mathrm{d}^{0.5} + 0.06\,Re_\mathrm{d}^{0.67}\right)Pr^{0.4}\left(\frac{\mu}{\mu_\mathrm{s}}\right)^{0.25}, \tag{8.48}$$

$$3.5 \leq Re_\mathrm{d} \leq 7.6 \times 10^4,\ 0.71 \leq Pr \leq 380.$$

The flow domain around the surface of a sphere is mixed.

Impingent Flow from a Round Jet onto a Planar Perpendicular Surface

Many types of diagnostic and therapeutic devices make use of a fluid jet at a hot or cold temperature directed onto the skin or other tissue surface. This arrangement is applied to ensure a highly effective convective heat transfer process. The convection due to an impingent fluid jet from a round nozzle normal to a solid planar surface with the geometry as defined in Fig. 8.11 is calculated by (8.49) for limiting conditions of: $2{,}000 \leq Re \leq 400{,}000;\ 2 \leq L/d \leq 12;\ 0.004 \leq A_r \leq 0.04$ (Martin 1977):

$$\bar{N}u_\mathrm{d} = 2 + \left(0.4\,Re_\mathrm{d}^{0.5} + 0.06\,Re_\mathrm{d}^{0.67}\right)Pr^{0.4}\left(\frac{\mu}{\mu_\mathrm{s}}\right)^{0.25}, \tag{8.49}$$

where the relative nozzle area, A_r, is the ratio of the nozzle exit cross-section to the surface area of the target surface.

$$A_r = \frac{A_\mathrm{nozzle}}{A_\mathrm{imping}} = \frac{d^2}{4r^2}. \tag{8.50}$$

Forced convection correlation relations for numerous other geometries and conditions are available in various primary sources in the heat transfer literature.

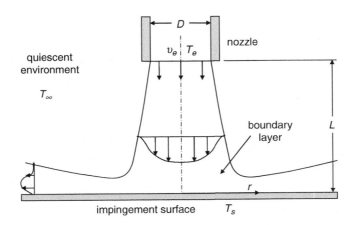

Fig. 8.11 Impingement of a circular fluid jet onto a planar surface

8.3.4 Free Convection Processes

Since free convection processes are driven by buoyant effects, determination of the relevant correlation relations to determine the convection coefficient must start with analysis of the shape and orientation of the fluid/solid interface. This effect is illustrated in the free convection boundary layer adjacent to a vertical cooled flat plate shown in Fig. 8.12. Note that the flow velocity is zero at both the inner and the outer extremes of the boundary layer, although the gradient is finite at the solid interface owing to viscous drag of the fluid. The environment is assumed to be quiescent so that there is no viscous shearing action at the outer region of the boundary layer. The following are some of the most commonly applied free convection correlation relations. Many make use of a dimensionless constant, the *Rayleigh number*, $Ra = Gr \cdot Pr$.

8.3.4.1 Free Convection over a Vertical Plate

The free convection process is averaged over a vertical plate of length L, including both the laminar and the turbulent flow regions over the entire range of Ra (Churchill and Chu 1975b),

$$\bar{Nu}_L = \left\{ 0.825 + \frac{0.387 \, Ra_L^{1/6}}{\left[1 + \left(\frac{0.492}{Pr}\right)^{9/16}\right]^{8/27}} \right\}^2. \tag{8.51}$$

If the entire flow is in the laminar regime and convection averaged over the length L for $10^4 \leq Ra_L \leq 10^9$,

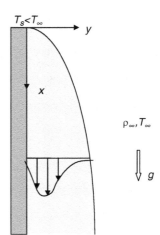

Fig. 8.12 Growth of a free convection boundary layer on a vertical cooled plate

$$\bar{Nu}_L = 0.59\, Ra_L^{1/4}. \tag{8.52}$$

For a turbulent flow with convection averaged over a vertical plate of length L for $10^9 \leq Ra_L \leq 10^{13}$,

$$\bar{Nu}_L = 0.1\, Ra_L^{1/3}. \tag{8.53}$$

8.3.4.2 Free Convection over a Horizontal Plate

The characteristic length for free convection with a horizontal plate is defined by the ratio of the area to the perimeter.

$$L = \frac{A}{P}.$$

For free convection averaged over the upper surface of a hot plate or lower surface of a cold plate having a dimension L; $10^4 \leq Ra_L \leq 10^7$ the Nusselt number is given by (Goldstein et al. 1973),

$$\bar{Nu}_L = 0.54\, Ra_L^{1/4}. \tag{8.54}$$

For free convection averaged over the upper surface of a hot plate or lower surface of a cold plate having a dimension L, $10^7 \leq Ra_L \leq 10^{11}$,

$$\bar{Nu}_L = 0.15\, Ra_L^{1/3}. \tag{8.55}$$

For free convection averaged over the lower surface of a hot plate or upper surface of a cold plate having a dimension L; $10^5 \leq Ra_L \leq 10^{10}$,

$$\bar{Nu}_L = 0.27\, Ra_L^{1/4}. \tag{8.56}$$

8.3.4.3 Free Convection over a Horizontal Cylinder

For free convection averaged over the entire circumferential surface of a horizontal cylinder having an isothermal surface and a diameter d; $Ra_d \leq 10^{12}$ (Churchill and Chu 1975a),

$$\bar{Nu}_d = \left\{ 0.6 + \frac{0.387\, Ra_d^{1/6}}{\left[1 + \left(\dfrac{0.559}{Pr}\right)^{9/16}\right]^{8/27}} \right\}^2. \tag{8.57}$$

Alternatively, free convection averaged over the entire circumferential surface of a horizontal cylinder having an isothermal surface and a diameter d (Morgan 1975),

$$\bar{Nu}_d = CRa_d^n, \qquad (8.58)$$

where the values of C and n are functions of Ra_d as given in the table below.

Ra_d	C	n
10^{-10}–10^{-2}	0.675	0.058
10^{-2}–10^2	1.02	0.148
10^2–10^4	0.85	1.88
10^4–10^7	0.48	0.25
10^7–10^{12}	0.125	0.333

8.3.4.4 Free Convection over a Sphere

For free convection averaged over the entire surface of a sphere having an isothermal surface and a diameter d; $Ra_d \leq 10^{11}$; $Pr \geq 0.7$ (Churchill 2002).

$$\bar{Nu}_d = 2 + \frac{0.589\,Ra_d^{1/4}}{\left[1 + \left(\dfrac{0.469}{Pr}\right)^{9/16}\right]^{4/9}}. \qquad (8.59)$$

8.3.4.5 Free Convection Inside Closed Cavities

There are many situations in which free convective flows are set up in the interior of enclosed cavities. In these cases, the flow pattern is contained within an enclosure and limited by the presence of physical barriers, analogously to internal forced convection. These convective flows embody added complexity, often having recirculating patterns of movement. The flows are a function of the orientation of the surface being heated or cooled and of the overall geometry of the surfaces that constrict the fluid movement. Frequently, there is active heating or cooling on one or more of the enclosure surfaces, and the resulting convective flow causes heat transfer with other surfaces of the cavity. The constitutive correlation equations are organized according to the cavity geometry and heating pattern.

Horizontal Concentric Cylinders

When two long horizontal cylinders are positioned with one inside the other and held at different temperatures, a free convection flow pattern may be established in the fluid between the cylinders. For a heated outer cylinder as shown in Fig. 8.13,

8.3 Convection

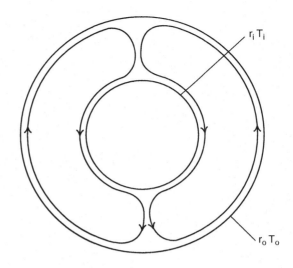

Fig. 8.13 Free convection in the fluid between two concentric long horizontal cylinders or two concentric spheres held at different temperatures. The flow pattern shown will occur when the outer cylinder temperature is higher than for the inner cylinder

fluid adjacent to it will become warmer and tend to rise in the gravity field. Fluid adjacent to the cooler inner cylinder tends to fall, thereby setting up symmetric convection loops in the annular volume. The circulation of fluid from the outer cylinder to the inner cylinder will pump heat inwardly, thereby enhancing the overall transport process in comparison with what would occur by pure conduction.

The influence of convective fluid movement can be described in terms of an *effective thermal conductivity* of the fluid, k_{eff}. The equation for steady-state heat conduction within a cylindrical annulus is modified to become (Raithby and Hollands 1975),

$$\dot{Q} = \frac{2\pi L k_{\text{eff}}(T_i - T_o)}{\ln\left(\dfrac{r_o}{r_i}\right)}, \tag{8.60}$$

where the correlation equation for the effective conductivity is

$$k_{\text{eff}} = 0.386 k \left(\frac{Pr}{0.861 + Pr}\right)^{\frac{1}{4}} Ra_r^{\frac{1}{4}} \tag{8.61}$$

with the Rayleigh number based on a length defined as

$$L_r = \frac{2\left[\ln\left(\dfrac{r_o}{r_i}\right)\right]^{\frac{4}{3}}}{\left[\left(\dfrac{1}{r_i}\right)^{\frac{3}{5}} + \left(\dfrac{1}{r_o}\right)^{\frac{3}{5}}\right]^{\frac{5}{3}}}. \tag{8.62}$$

The fluid properties are evaluated at the mean temperature, T_m

$$T_m = \frac{T_i + T_o}{2}.$$

Equation (8.61) is valid in the range of $0.7 \leq Pr \leq 6{,}000$ and $Ra_r \leq 10^7$.

Concentric Spheres

Free convection between two concentric spheres is similar to that for horizontal cylinders except that the specific geometric influences are altered somewhat. The set of equations is (Raithby and Hollands 1975)

$$\dot{Q} = \frac{4\pi L k_{\text{eff}}(T_i - T_o)}{\dfrac{1}{r_i} - \dfrac{1}{r_o}}, \tag{8.63}$$

where the correlation equation for the effective conductivity is

$$k_{\text{eff}} = 0.74\, k \left(\frac{Pr}{0.861 + Pr}\right)^{\frac{1}{4}} Ra_{L_s}^{\frac{1}{4}} \tag{8.64}$$

with the Rayleigh number based on a length defined as

$$L_s = \frac{\left(\dfrac{1}{r_i} - \dfrac{1}{r_o}\right)}{2^{\frac{1}{3}}\left[\left(\dfrac{1}{r_i}\right)^{\frac{7}{5}} + \left(\dfrac{1}{r_o}\right)^{\frac{7}{5}}\right]^{\frac{5}{3}}}. \tag{8.65}$$

Equation (8.64) is valid in the range of $0.7 \leq Pr \leq 4{,}000$ and $Ra_s \leq 10^4$.

Enclosed Straight Sided Spaces

Free convection can occur when a fluid is contained within a volume having straight sides, such as a rectangular cross section. The potential for convection is highly dependent on geometric properties such as the aspect ratio of the volume (height vs width) and the orientation within the gravity field, plus the pattern of heating or cooling of various walls of the volume. Correlation equations for many different combinations of these parameters may be found in the heat transfer literature.

Example Problem 8.3.1 Cold Exposure Weather Advisory.
One of the local television stations has issued an advisory to alert the population about the potential for hypothermia by cold water exposure. The

8.3 Convection

advisory states that at a temperature of 0°C the heat loss from the body is 50 times faster in stagnant water than in stagnant air. Is this statement accurate, and if not, what is an estimate of the water to air risk ratio?

Solution. *Initial considerations*: The risk of exposure to a cold environment is directly proportional to the rate at which heat may be lost from the body to the surroundings. This heat loss will depend on the temperature of the immediate environment and the effectiveness of heat transfer with that environment. In this case, the heat transfer is via free convection to a stagnant fluid. We can calculate the relative values for the convection coefficient to water and air, which will give the ratio of heat transfers. The temperature differences are identical for both cases. During the time of exposure, the temperature of the human will drop as heat is lost to the environment. We can consider the initial rate of heat loss, which will occur for the maximum temperature difference.

System definition and environmental interactions: The system of interest is the human body, and the process is losing heat to a cold environment. Since this is a free convection problem, it is necessary to specify the geometry and orientation in the gravity field. The most general assumption is that the human is vertical, i.e., having longitudinal axis aligned with gravity. As a first-order approximation, the geometry is a circular cylinder.

Apprising the problem to identify governing equations: The heat transfer will be calculated by Newton's law of cooling, (8.10) and free convection from a hot vertical surface by the correlation equation, (8.51).

$$\dot{Q}_s = hS(T_s - T_\infty), \quad (8.66)$$

$$\bar{Nu}_L = \left\{ 0.825 + \frac{0.387 Ra_L^{1/6}}{\left[1 + \left(\frac{0.492}{Pr}\right)^{9/16}\right]^{8/27}} \right\}^2. \quad (8.67)$$

Analysis: Evaluation of the convective heat transfer coefficient from the correlation equation requires values for the properties of air and water. The average skin temperature for humans is about 34°C. Therefore, the fluid temperature for which properties are evaluated is the film temperature (8.42a)

$$T_f = \frac{0 + 34}{2} = 17°C; \quad 290 K.$$

From Appendix C, for air: $k = 0.0293$ W/mK; $\nu = 19.91 \times 10^{-6}$ m²/s; $\alpha = 28.4 \times 10^{-6}$ m²/s; $Pr = 0.709$. For water: $k = 0.598$ W/mK; $\nu = \mu\rho_f = 1.08 \times 10^{-6}$ m²/s; $\alpha = k/\rho_f c_p = 1.43 \times 10^{-6}$ m²/s; $Pr = 7.56$; $\beta_f = 174 \times 10^{-6}$ 1/K.

The dimensions of the person are taken as $d = 0.25$ m and $L = 1.7$ m. The coefficient of thermal expansion for air is $\beta_f = 1/T_f = 1/290 \text{ K} = 3.45 \times 10^{-3}$ 1/K. The Rayleigh numbers for air and water environments are computed as follows:

$$Ra_{L,a} = \frac{g\beta\Delta TL^3}{\alpha\nu} = \frac{9.8 \text{ m/s}^2 \times 3.45 \times 10^{-3} \text{K}^{-1} \times (34-0)\text{K} \times (1.7\text{m})^3}{28.4 \times 10^{-6}\text{m}^2/\text{s} \times 19.91 \times 10^{-6}\text{m}^2/\text{s}} = 9.98 \times 10^9,$$

$$Ra_{L,w} = \frac{g\beta\Delta TL^3}{\alpha\nu} = \frac{9.8 \text{ m/s}^2 \times 1.74 \times 10^{-4} \text{K}^{-1} \times (34-0)\text{K} \times (1.7\text{m})^3}{1.43 \times 10^{-6}\text{m}^2/\text{s} \times 1.08 \times 10^{-6}\text{m}^2/\text{s}} = 1.84 \times 10^{11}.$$

The Nusselt numbers are computed as

$$\bar{Nu}_{L,a} = \left\{ 0.825 + \frac{0.387(9.98 \times 10^9)^{1/6}}{\left[1 + \left(\frac{0.492}{0.709}\right)^{9/16}\right]^{8/27}} \right\}^2 = 252,$$

$$\bar{Nu}_{L,w} = \left\{ 0.825 + \frac{0.387(1.84 \times 10^{11})^{1/6}}{\left[1 + \left(\frac{0.492}{7.56}\right)^{9/16}\right]^{8/27}} \right\}^2 = 806.$$

The convective heat transfer coefficients are computed from the Nusselt numbers.

$$h_a = \frac{\bar{Nu}_{L,a} \times k_a}{L} = \frac{252 \times 0.0293 \text{ W/m K}}{1.7 \text{ m}} = 4.34 \text{ W/m}^2 \text{ K},$$

$$h_w = \frac{\bar{Nu}_{L,w} \times k_w}{L} = \frac{806 \times 0.598 \text{ W/m K}}{1.7 \text{ m}} = 283 \text{ W/m}^2 \text{ K}.$$

The ratio of heat transfer to water and air equals the ratio of the convective heat transfer coefficients.

$$\frac{q_w}{q_a} = \frac{h_w(T_\infty - T_s)}{h_a(T_\infty - T_s)} = \frac{h_w}{h_a} = \frac{283}{4.34} = 65.2.$$

Examining and interpreting the results: Our calculated ratio of heat loss to water and air is of the same order of magnitude, but somewhat larger, than the advisory to the public. There are a number of possible sources for this difference. One source is that we have assumed the only mechanism of heat loss is through convection. This assumption is quite reasonable for water, but in air radiation heat transfer to the environment will be important. (Later in this chapter, we will learn how to make

8.3 Convection

radiation heat transfer calculations. We will revisit further refinement of our calculations there.) On a relative basis, we expect radiation to increase heat loss more in air than in water, which will reduce the ratio and perhaps bring it closer to the advisory value. An additional issue is that we do not know how the heat loss ratio in the advisory was determined. It could well have been by a method less rigorous than the calculations we have followed. Certainly, the analysis process was different.

> **Example Problem 8.3.2 External Flow Heat Transfer.**
> A very commonly encountered condition is air flowing over a flat surface such as the body, a fixed structure, or an assembly moving through the atmosphere. Since the boundary layer will be growing in the directions of flow, it is expected that the local convective heat transfer coefficient will be a function of position along the plate. Sometimes, it is desirable to determine an averaged value of convection over the entire length of a surface, and other times it is necessary to calculate local variations in convection. We will consider in this example a flat plate at 37°C with 20°C air flowing over it at 1.5 m/s. Determine the convective heat transfer coefficient at positions 150 mm and 300 mm from the leading edge and the average value over the entire surface.

Solution. *Initial considerations*: In this problem we are concerned only with determining the convective heat transfer coefficient for the specific conditions indicated. Thus, it will be necessary only to identify and evaluate the appropriate correlation equations.

System definition and environmental interactions: The system is the interface between the solid and the air where the convection occurs. There is no energy storage since there is no mass at the interface. The conduction heat flow from the solid equals the convection heat flow into the air.

Apprising the problem to identify governing equations: We must calculate the Reynolds number based on flow distance along the surface so that an appropriate correlation equation can be identified. The fluid temperature is $T_f = 28.5°C$, which is about 302 K, which we will round off to 300 K for simplicity. The required property values are: $\rho = 1.16$ kg/m^3; $\mu = 1.85 \times 10^{-5}$ Ns/m^2; $\nu = 1.59 \times 10^{-5}$ m^2/s; $k = 0.0263$ W/(mK); $Pr = 0.707$. The Reynolds numbers for the two lengths are

$$Re_{150} = \frac{\rho v x}{\mu} = \frac{v x}{\nu} = \frac{1.5 \, \text{m/s} \times 0.15 \, \text{m}}{1.59 \times 10^{-5} \text{N s/m}^2} = 1.42 \times 10^4,$$

$$Re_{300} = \frac{1.5 \, \text{m/s} \times 0.3 \, \text{m}}{1.59 \times 10^{-5} \text{N s/m}^2} = 2.83 \times 10^4.$$

Both these Reynolds numbers are less than the transition value of 5×10^5 so that the flow is totally in the laminar regime. The correlation equations for laminar external flow are (8.43) and (8.44) for the local and average values of h.

Analysis: Since $Pr > 0.6$, the heat transfer coefficients are calculated at 150 and 300 mm from the leading edge as

$$Nu_{150} = 0.332\, Re_{150}^{1/2} Pr^{1/3} = 0.332 \cdot (1.42 \times 10^4)^{\frac{1}{2}} \times (0.707)^{\frac{1}{3}} = 35.2,$$

$$Nu_{300} = 0.332\, Re_{300}^{1/2} Pr^{1/3} = 0.332 \times (2.83 \times 10^4)^{\frac{1}{2}} \times (0.707)^{\frac{1}{3}} = 49.8,$$

$$h_{150} = \frac{Nu_{150} k}{x_{150}} = \frac{35.2 \times 0.0263\, \text{W/m K}}{0.15\, \text{m}} = 6.18\, \text{W/m}^2\, \text{K},$$

$$h_{300} = \frac{49.8 \times 0.0263\, \text{W/m K}}{0.3\, \text{m}} = 4.37\, \text{W/m}^2\, \text{K}.$$

The average convection coefficient over the entire length is

$$\bar{Nu}_{300} = 0.664\, Re_{300}^{1/2} Pr^{1/3} = 0.664 \times (28.3 \times 10^4)^{\frac{1}{2}} \times (0.707)^{\frac{1}{3}} = 99.6,$$

$$\bar{h}_{300} = \frac{99.6 \times 0.0263\, \text{W/m K}}{0.3\, \text{m}} = 8.73\, \text{W/m}^2\, \text{K} = 2 \times h_{300}.$$

Examining and interpreting the results: The heat transfer coefficient decreases along the length of the plate because the boundary layer continually grows thicker resulting in a progressively greater resistance to heat flow between the plate and the free stream fluid. The average value of the heat transfer coefficient is larger than at the midpoint position along the plate. This result is expected because the growth of the boundary layer is not a linear function along the length of the plate, i.e., the outer limit of the boundary layer is not a straight line. Rather, it increases in thickness most rapidly near the leading edge. Thus, the region of thinnest boundary layer and therefore with the highest convective heat transfer coefficient is biased toward the front edge of the plate. This is the region where the greatest heat transfer will occur.

8.3.5 Thermal Resistance in Convection

Newton's law of cooling can be written in terms of a thermal resistance $\Re_{T,\text{conv}}$:

$$\dot{Q} = \frac{T_S - T_\infty}{\Re_{T,\text{conv}}}, \qquad (8.68)$$

where:

$$\Re_{T,\text{conv}} = \frac{1}{hS}. \qquad (8.69)$$

The analogy to Ohm's law is shown in Fig. 8.14.

Fig. 8.14 Convective thermal resistance

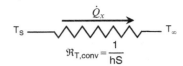

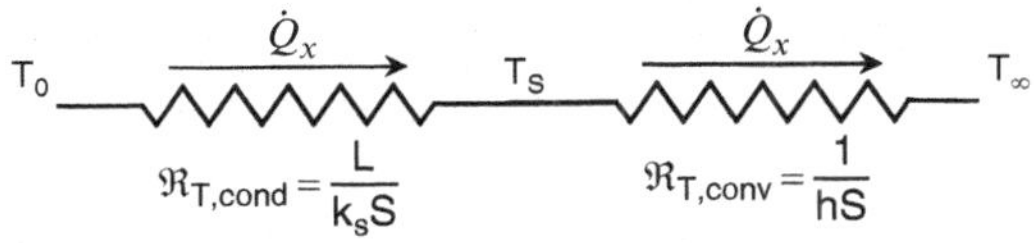

Fig. 8.15 Heat flow from a solid slab into a flowing fluid

8.3.6 Biot Number

Consider heat transfer from a slab that is heated at a constant rate on one side and is in contact with a moving fluid on the other side. Heat flows by conduction from the slab interior to the surface and then by convection away from the surface into the fluid. If the heat transfer rate is constant, we may model the combined resistance as shown in (2.33). Fig. 8.15.

Setting conductive heat flow to the surface equal to convective heat flow from the surface, we have:

$$\dot{Q} = \frac{T_0 - T_S}{\mathfrak{R}_{T,cond}} = \frac{T_S - T_\infty}{\mathfrak{R}_{T,conv}}. \tag{8.70}$$

Consequently, the ratio of thermal resistance in the solid relative to thermal resistance in the fluid is:

$$\frac{\mathfrak{R}_{T,cond}}{\mathfrak{R}_{T,conv}} = \frac{hL}{k_s} = \frac{T_0 - T_S}{T_S - T_\infty} = \frac{\Delta T_{solid}}{\Delta T_{fluid}}. \tag{8.71}$$

The dimensionless number hL/k_s was defined as the Biot number in Sect. 8.3.2.3 and Table 8.1. Equation (8.71) shows that this ratio represents the resistance to conduction in the slab relative to resistance to convection heat transfer in the fluid. If this number is small, (8.71) indicates that the temperature difference in the solid will be small relative to the temperature difference in the fluid. In that case, nearly all of the resistance to heat transfer resides in the fluid phase, and we can treat the solid as though it has a uniform temperature. The resulting analysis is known as a 'lumped' analysis, and is used for $Bi < 0.1$, even for transient applications where the heat flux is not constant. We will return to the importance of the Biot number in Chap. 9. For steady heat flow through a slab, the surface temperature can be found from (8.71):

$$T_S = \frac{T_0 + Bi(T_\infty)}{(1 + Bi)}. \tag{8.72}$$

Thus, the surface temperature tends to be close to T_0 at low values of Bi and close to T_∞ for high values of Bi.

8.4 Thermal Radiation

Thermal radiation is primarily a surface phenomenon as it interacts with a solid, or in some cases with a fluid through which it passes. Thermal radiation is important in many types of heating, cooling, and drying processes. In the out-of-doors, environment solar thermal radiation can have a significant influence on the overall heat load on the skin.

Thermal radiation occurs via the propagation of electromagnetic waves. It does not require the presence of a transmitting material as do conduction and convection. Therefore, thermal radiation can proceed in the absence of matter, such as in the radiation of heat from the sun to earth. All materials are continuously emitting thermal radiation from their surfaces as a function of their temperature and radiative constitutive properties. All surfaces also are continuously receiving thermal energy from their environment. The balance between radiation lost and gained defines the net radiation heat transfer for a body. The wavelengths of thermal radiation extend across a spectrum from about 0.1 μm to 100 μm, embracing the entire visible spectrum. It is for this reason that some thermal radiation can be observed by the human eye, depending on the temperature and properties of the emitting surface.

8.4.1 Three Governing Characteristics of Thermal Radiation Processes

The foregoing observations indicate that there are three characteristics of a body (i.e., a system) and its environment that govern the rate of radiation heat transfer:

- Surface temperature
- Surface radiation properties
- Geometric sizes, shapes, separation, and orientation of the body surface in relation to the aggregate surfaces in the environment.

Each of these three effects can be quantified and expressed in equations used to calculate the magnitude of radiation heat transfer. The objective of this presentation is to introduce and discuss how each of these three factors influences radiation processes and to show how they can be grouped into a single approach to analysis.

8.4.2 The Role of Surface Temperature in Thermal Radiation

The first property to consider is temperature. The relationship between the temperature of a perfect radiating (black) surface and the rate at which thermal radiation is emitted is known as the Stefan–Boltzmann law,

8.4 Thermal Radiation

$$E_b = \sigma T^4, \tag{8.73}$$

where E_b is the blackbody emissive power [W/m^2], and σ is the Stefan–Boltzmann constant, which has the numerical value

$$\sigma = 5.67 \times 10^{-8} \text{ W/m}^2 \cdot \text{K}^4.$$

Note that the temperature must be expressed in absolute units [K].

E_b is the rate at which energy is emitted diffusely (without directional bias) from a surface at temperature T [K] having perfect radiation properties. It is the summation of radiation emitted at all wavelengths from a surface. A perfect radiating surface is termed *black* and is characterized by emitting the maximum possible radiation at any given temperature. The blackbody monochromatic (at a single wavelength, λ) emissive power is calculated from the Planck distribution as

$$E_{\lambda,b}(\lambda, T) = \frac{2\pi h c_o^2}{\lambda^5 \left[\exp\left(\frac{hc_o}{\lambda k_B T}\right) - 1\right]} \text{ [W/m}^2 \cdot \mu\text{m]}, \tag{8.74}$$

where $h = 6.636 \times 10^{-34}$ [J·s] is the Planck constant, $k_B = 1.381 \times 10^{-23}$ [J/K] is the Boltzmann constant, and $c_o = 2.998 \times 10^8$ [m/s] is the speed of light in vacuum. The Planck distribution can be plotted showing $E_{\lambda,b}$ as a function of λ for specific constant values of absolute temperature, T. The result is the nest of spectral emissive power curves in Fig. 8.16. For each temperature, there is an intermediate wavelength for which $E_{\lambda,T}$ has a maximum value, and this maximum increases monotonically in magnitude and occurs at shorter wavelengths with increasing temperature. Wien's displacement law, (8.75), describes the relationship between the absolute temperature and the wavelength at which maximum emission occurs.

$$\lambda_{\max} T = 2,898 \, [\mu\text{m} \cdot \text{K}]. \tag{8.75}$$

Equation (8.74) is integrated over the entire emission spectrum to obtain the expression for the total emitted radiation, the black body emissive power, E_b, (8.73).

$$E_b(T) = \int_0^\infty \frac{2\pi h c_o^2}{\lambda^5 \left[\exp\left(\frac{hc_o}{\lambda k_B T}\right) - 1\right]} d\lambda = \sigma T^4. \tag{8.76}$$

Equation (8.76) represents the area under an isothermal curve in Fig. 8.16 depicting the maximum amount of energy that can be emitted from a surface at a specified temperature. This set of equations provides the basis for quantifying the temperature effect on thermal radiation. It applies to idealized, black surfaces.

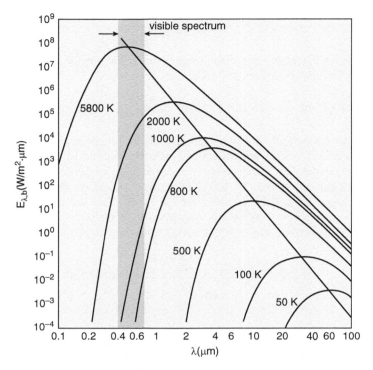

Fig. 8.16 Spectral blackbody emissive power as a function of surface temperature and wavelength. The spectrum visible to humans between about 0.3 and 0.7 μm is indicated

It is helpful to keep in mind the following characteristics of the thermal radiation emitted from a perfect (black) surface:

- As the temperature of the surface changes so does the total amount (rate) of radiation emitted.
- Higher temperature surfaces emit more radiation according to the T^4 law. Thus, radiation is a highly nonlinear function of temperature.
- A surface emits thermal radiation at many wavelengths across a spectrum defined by the temperature.

> **Example 8.4.1 Emissive Power of the Sun.**
> What must the emissive power of the sun be in order to produce the measured solar radiation flux incident at the earth's surface of 1353 W/m². If the sun's surface can be considered as black body, what surface temperature of the sun is necessary to produce the calculated flux?

8.4 Thermal Radiation

Solution. *Initial considerations*: The goal is to determine what the strength of the thermal emission flux (W/m²) is from the sun's surface to produce the measured flux on the surface of the earth. For this purpose, we will need some geometric information such as the distance between the sun and the earth and their diameters. These data are: $d_{sun} = 1.4 \times 10^9$ m; $d_{earth} = 1.3 \times 10^7$ m; $R_{s-e} = 1.5 \times 10^{11}$ m. Also, since we have no information to the contrary, we will assume that the surface of the sun behaves as a black body and that there is no attenuation of the emitted energy between the sun and earth, and in particular in the earth's atmosphere. The black body emissive power of the sun is related directly to its absolute temperature. Thus, if we can determine E_b, the solution is straightforward.

System definition and environmental interactions: The process of interest is the radiation transport of thermal energy from the sun to the earth. The radiation flux that defines this process can be identified at the surfaces of the sun and the earth. Therefore, it is effective to define a system with boundaries at these two surfaces. Such a system consists of the volume between two concentric spheres, the inner having the radius of the sun, R_s, and the outer having the radius of the sun plus the distance between the sun and earth, $R_s + R_{s-e}$. See Fig. 8.17 for layout features.

Apprising the problem to identify governing equations: The conservation of energy is applied to the system, which is quite reasonably assumed to be at steady state. Therefore, the radiant energy entering the system at the inner surface, corresponding to the radius of the sun, equals the energy leaving at the outer surface, which cuts through the center of the earth. Note that only a very minor fraction of the energy flux at the outer surface is incident onto the earth. Conservation of energy in this case can be written:

$$0 = \dot{Q}_{in} - \dot{Q}_{out} = E_s A_{in} - q_s A_{out},$$

where E_s is the outward radiation flux at the sun's surface and q_s is the incident flux from the sun at the outer surface.

Analysis: The black body emissive power, E_b, for the surface of the sun defines the radiation flux outward at the surface, E_s. For the measured solar flux at the earth and the geometry as defined, the radiation flux at the sun is calculated from the conservation of energy equation.

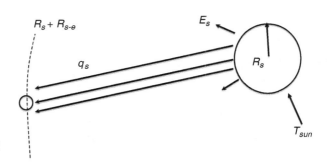

Fig. 8.17 Flux of radiant energy from sun to earth. Note that the dimensions are not to scale

$$E_s = q_s \frac{A_{\text{out}}}{A_{\text{in}}} = q_s \frac{4\pi(R_s + R_{s-e})^2}{4\pi R_s^2}$$

$$= 1353 \,\text{W/m}^2 \frac{(1.5 \times 10^{11} + 7.0 \times 10^8)^2}{(7.0 \times 10^8)^2} = 6.3 \times 10^7 \,\text{W/m}^2.$$

The temperature of the sun can now be calculated via (8.73) for black body emissive power.

$$T_s = \left(\frac{E_s}{\sigma}\right)^{\frac{1}{4}} = \left(\frac{6.3 \times 10^7 \,\text{W/m}^2}{5.67 \times 10^{-8}\,\text{W/m}^2\,\text{K}^4}\right)^{\frac{1}{4}} = 5,774\,\text{K}.$$

Examining and interpreting the results: This value for the temperature of the sun is in very close agreement with other independent measures. The solar flux is incident on the earth over the hemisphere facing the sun. This analysis assumed implicitly that the solar flux was measured at an equator most distant from the sun. However, the radius of the earth is so small in comparison with the other dimensions involved that the effect is less than the significant figures of the calculation.

Additional comments: With some further assumptions, it should be possible to estimate the temperature of the surface of the earth. As with the sun, if we can determine the radiation flux emitted from the earth's surface, then its temperature is determined from (8.76). For this purpose, we will define a system consisting of the earth, which is at thermal equilibrium. The solar flux is absorbed over a projected area equal to the cross section of the earth at the equator. Thermal radiation is assumed to be emitted uniformly and continuously over the entire surface of the earth. Conservation of energy is written for the system as shown in Fig. 8.18.

$$0 = q_s A_{\text{in}} - E_e A_{\text{out}}, \quad E_e 4\pi R_e^2 = q_s \pi R_e^2, \quad E_e = \frac{q_s}{4},$$

$$T_e = \left(\frac{q_s}{4\sigma}\right)^{\frac{1}{4}} = \left(\frac{1,353\,\text{W/m}^2}{4 \times 5,67 \times 10^{-8}\,\text{W/m}^2\,\text{K}^4}\right)^{\frac{1}{4}} = 278\,\text{K}.$$

We know in actuality that the surface of the earth is warmer on average than 278K (5°C). Clearly, one of our assumptions must be flawed. The primary culprit is

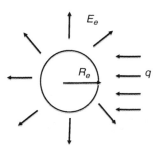

Fig. 8.18 Incident and emitted radiation fluxes for the earth

that the atmosphere of the earth does not participate in the radiation exchange. It is relatively transparent to the incident solar radiation that is predominately at short wave lengths associated with the high temperature of the sun's surface. However, the much cooler surface of the earth results in radiation at longer (infrared) wave lengths as seen in Fig. 8.16. The atmosphere is much more interactive with infrared radiation, absorbing and reflecting much of it. As a consequence, a considerable portion of the thermal radiation emitter from the earth's surface is trapped in the atmosphere, altering the balance on the conservation of energy equation for the earth and causing the average surface temperature to be higher. This phenomenon is the well-publicized greenhouse effect that has been increasing in intensity in recent years as more "greenhouse gases" are added to the atmosphere.

8.4.3 *The Role of Surface Properties in Thermal Radiation*

Next, we will consider the effect of real, rather than idealized, surface properties on thermal radiation exchange. Real surfaces emit less than black body radiation at a given temperature. The ratio of real to black radiation levels defines a property called the *emissivity*, ε. In general, radiation properties are functions of the radiation wavelength and for many practical systems can change significantly over the thermal spectrum. Thus,

$$\varepsilon_\lambda(\lambda, T) = \frac{E_\lambda(\lambda, T)}{E_{\lambda,b}(\lambda, T)}. \tag{8.77}$$

E_λ is the monochromatic blackbody emissive power at a specific wavelength λ.

An idealized real surface has radiant properties that are wavelength independent and is termed a gray surface. For these conditions,

$$\varepsilon(T) = \frac{E(T)}{E_b(T)} = \frac{E(T)}{\sigma T^4}. \tag{8.78}$$

A gray surface has the effect of decreasing the magnitude of the curves in Fig. 8.16 by a constant factor over all wavelengths.

In addition to emission, surfaces are continually bombarded by thermal radiation from their environments. The net radiant flux at a surface is the difference between the energies received and lost. As with emission, the surface radiation properties play an important role in determining the amount of energy absorbed by a surface. A black surface absorbs all incident radiation, whereas real surfaces absorb only a fraction of the incident radiation. The total radiant flux onto a surface from all sources is called the *irradiation* and is denoted by the symbol G [W/m^2]. In general, the incident radiation will be composed of many wavelengths, denoted by G_λ.

A surface can have three modes of response to incident radiation: the radiation may be absorbed, reflected, and/or transmitted. The fractions of incident radiation that undergo each of these responses are determined by three dimensionless properties: the coefficients of *absorption*, α, *reflection*, ρ, and *transmission*, τ. Conservation of energy applied at a surface dictates that the relationship among these three properties must be

$$\alpha + \rho + \tau = 1. \tag{8.79}$$

Figure 8.19 illustrates these phenomena for radiation incident onto a surface that is translucent, allowing some of the radiation to pass through. All three of the properties are wavelength dependent.

The three properties are defined according to the fraction of irradiation that is absorbed, reflected, and transmitted.

$$\alpha_\lambda(\lambda) = \frac{G_{\lambda,\text{abs}}(\lambda)}{G_\lambda(\lambda)}, \tag{8.80}$$

$$\rho_\lambda(\lambda) = \frac{G_{\lambda,\text{ref}}(\lambda)}{G_\lambda(\lambda)}, \tag{8.81}$$

$$\tau_\lambda(\lambda) = \frac{G_{\lambda,\text{tr}}(\lambda)}{G_\lambda(\lambda)}. \tag{8.82}$$

It is well known that there is a very strong spectral (wavelength) dependence of these properties. For example, the greenhouse effect occurs because glass has a high transmissivity (τ) at relatively short wavelengths in the visible spectrum that are characteristic of the solar flux. However, the transmissivity of glass is very small in the infrared spectrum in which terrestrial emission occurs. Therefore, heat from the sun readily passes through glass and is absorbed by interior objects. In contrast, radiant energy emitted by these interior objects is reflected back to the source. The net result is a warming of the interior of a system that has a glass surface exposed to

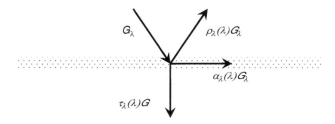

Fig. 8.19 Absorption, reflection, and transmission phenomena for a surface irradiated with a multi-wavelength incident radiation, G_λ

8.4 Thermal Radiation

the sun. The lens of a camera designed to image terrestrial sources of thermal radiation, predominantly in the infrared spectrum, must be fabricated from a material that is transparent at those wavelengths. It is important to verify whether the spectral dependence of material surface properties is important for specific applications involving thermal radiation.

An additional important surface property relationship is defined by Kirchhoff's law, which applies to a surface which is in thermal equilibrium with its environment. Most thermal radiation analyses are performed for processes that are in a steady state. For the surface of a body n having a surface area A_n, at steady state the radiation gained and lost is balanced so that the net exchange is zero. To illustrate, we may consider a large isothermal enclosure at a temperature, T_s, containing numerous small bodies, each having unique properties and temperature. See Fig. 8.20.

Since the surface areas of the interior bodies are very small in comparison with the enclosure area, their individual effects on the radiation field are negligible. The radiation flux to the interior bodies is a combination of emission and reflection from the outer enclosure surface. This total radiant energy leaving the enclosure surface per unit area is known as the *radiosity*, J (W/m²). In this case, the radiosity from the enclosure surface equals the irradiance G on all of the enclosed bodies.

For conditions of thermal equilibrium within the cavity, the temperatures of all surfaces must be equal. $T_{n-1} = T_n = T = T_s$. A steady-state energy balance between absorbed and emitted radiation on one of the interior bodies, designated n, yields

$$\alpha_n G A_n - E_n(T_s) A_n = 0. \tag{8.83}$$

If we were to replace the same interior body with a black body having the same temperature, shape, and orientation, then $\alpha_n = 1$ and $E_n(T_s) = E_b$, and (8.83) becomes:

$$G = E_b(T_s). \tag{8.84}$$

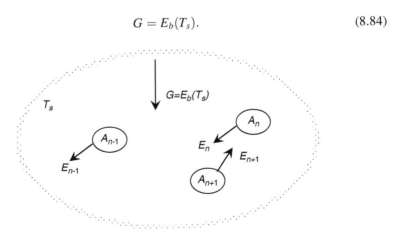

Fig. 8.20 Steady-state thermal radiation within a large isothermal enclosure containing multiple small bodies

Since the surface area of the enclosure is much larger than all other surfaces combined, all incident radiation is eventually absorbed, even if there are multiple reflections in the process. As a consequence, the very large enclosure behaves as if it were a black surface, and the irradiance from the enclosure surface equals the emissive power of a black body at T_s. The net effect is that the enclosure acts as a blackbody cavity regardless of its surface properties.

The term G may be eliminated between the two foregoing equations.

$$E_b(T_s) = \frac{E_n(T_s)}{\alpha_n}. \tag{8.85}$$

This relationship holds for all of the interior bodies. Comparison of (8.78) and (8.85) shows that the emissivity and absorptivity are equal.

$$\alpha = \varepsilon, \quad \text{or}, \quad \alpha_\lambda = \varepsilon_\lambda. \tag{8.86}$$

Equation (8.86) is known as *Kirchhoff's identity*. The general statement of this relationship is that for a gray surface; the emissivity and absorptivity are equal and independent of spectral conditions.

8.4.4 The Role of Geometric Sizes, Shapes, Separation, and Orientation in Thermal Radiation

The third factor influencing thermal radiation transfer is the geometric sizes, shapes, separation, and orientation of body surfaces in relation to the aggregate surfaces in the environment. This effect is quantified in terms of a property called the *shape factor*, which is solely a function of the geometry of a system and its environment. By definition, the shape factor is determined for multiple bodies, and it is related to the size, shape, separation, and orientation of the bodies. The shape factor $F_{m \to n}$ is defined between two surfaces, m and n, as the fraction of energy that leaves surface m that is incident onto the surface n. It is very important to note that the shape factor is directional. The shape factor from body m to n is usually not equal to that from body n to m except for special arrangements.

Values for shape factors have been compiled for a broad range of combinations of size, shape, separation, and orientation and are available as figures, tables, and equations (Howell 1982; Siegel and Howell 2002). There are a number of simple relations that govern shape factors and that are highly useful in working many types of problems. One is called the *reciprocity relation*.

$$A_n F_{n \to m} = A_m F_{m \to n}. \tag{8.87}$$

The recipocity relation is used for calculating the value of a second shape factor between two bodies if the first is already known.

8.4 Thermal Radiation

A second relation is called the *summation rule*, which says the sum of shape factors for the complete environment of a body equals 1.0.

$$\sum_{k=1}^{n} F_{m \to k} = 1.0. \tag{8.88}$$

The summation rule accounts for the entire environment for an object.

A limiting case is shown in Fig. 8.20 in which $A_s \gg A_n$. For this geometry, the reciprocity relation dictates that $F_{s \to n}$ be vanishingly small since only a very small fraction of the radiation leaving the surface s will be incident onto n. The summation rule then shows that effectively $F_{n \to s} = 1$.

The third geometric relationship states that the shape factors for each component of a surface are additive. If a surface n is divided into l components, then the *additive rule* states

$$F_{m \to (n)} = \sum_{j=1}^{l} F_{m \to j}, \tag{8.89}$$

where

$$A_n = \sum_{j=1}^{l} A_j.$$

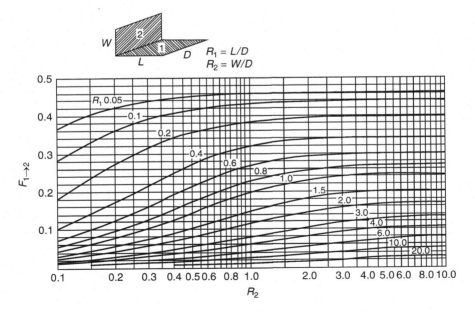

Fig. 8.21 Shape factors for radiation between aligned, adjacent, perpendicular rectangles as a function of their relative sizes and separations

Equation (8.89) is useful for calculating the shape factor for complex geometries that can be subdivided into an assembly of more simple shapes. For example, if two adjacent surfaces 1 and 2 exchange radiation with a separate surface 3, then the additive rule for view factors is

$$A_{1+2}F_{1+2\to3} = A_1 F_{1\to3} + A_2 F_{2\to3}. \tag{8.90}$$

The reader is directed to comprehensive compendia of data for determination of a wide array of shape factors, such as the foregoing Howell references, for detailed information. Some of the most general examples are included in this chapter in graphical format for simple parallel and perpendicular geometries, as shown in the accompanying figures.

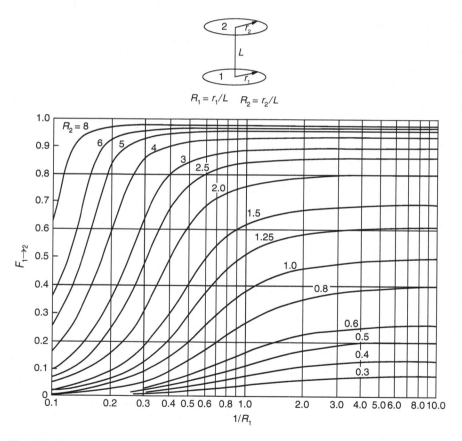

Fig. 8.22 Shape factors for radiation between aligned, parallel, circular, and equal-sized discs as a function of their relative separations

8.4 Thermal Radiation

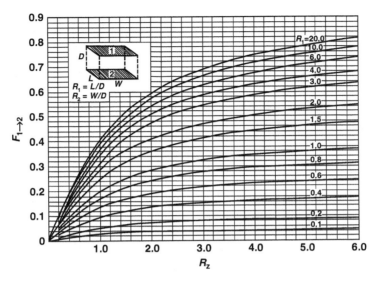

Fig. 8.23 Shape factors for radiation between aligned, parallel rectangles as a function of their relative sizes and separations

Example Problem 8.4.2 Shape Factor Calculation.
Determine the shape factors from surface 2 to 1 for the geometric configurations shown below.

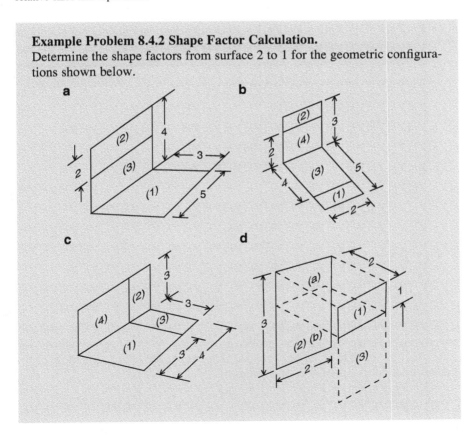

Solution. *Initial considerations*: This type of calculation is based solely on geometric properties. All of the configurations embody perpendicular adjoining planar surfaces. Although the exact geometry of the configuration may not be apparent on the shape factor graphs, it is often possible to apply the shape geometry equations to calculate the target geometric relationship.

System definition and environmental interactions: For these problems, the systems that interact consist of the defined surfaces that have specific size, shape, separation, and configuration properties. It is helpful to set up a coordinate system that matches the dimensions of the shape factor graph to make the analysis more organized.

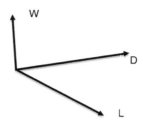

Apprising the problem to identify governing equations: The governing equations for shape factors are embedded in the graphs of shape factor relationships. Although these graphs are based on equations that enable calculation of the specific geometric functions, it is easier and faster to read the results off the graphs.

Analysis:

(a) The shape factors are identified in terms of ratios of the graphic dimensions.

$$R_1 = \frac{L}{D} = \frac{3}{5} = 0.6 \quad R_2 = \frac{W_{1+3}}{D} = \frac{4}{5} = 0.8 \quad R_2 = \frac{W_3}{D} = \frac{2}{5} = 0.4.$$

From Fig. 8.21, $F_{1 \to 2+3} = 0.25$ and $F_{1 \to 3} = 0.19$. From (8.89),

$$F_{1 \to 2+3} = F_{1 \to 3} + F_{1 \to 2} \quad \text{so that}$$
$$F_{1 \to 2} = F_{1 \to 2+3} - F_{1 \to 3} = 0.25 - 0.19 = 0.06.$$

From (8.87),

$$A_1 F_{1 \to 2} = A_2 F_{2 \to 1} \quad F_{2 \to 1} = \frac{A_1}{A_2} F_{1 \to 2} = \frac{15}{10} 0.06 = 0.09.$$

Examining and interpreting the results: $F_{2 \to 1}$ Is larger than $F_{1 \to 2}$ due primarily because A_1 is larger than A_2. The smaller A_2 is able to intercept only a rather small fraction of the diffuse radiation from A_1. Conversely, the larger A_1 will receive a greater fraction of radiation emitted from A_2.

8.4 Thermal Radiation

(b) We can begin by applying (8.89) and (8.90) repeatedly for this geometry.

$$(A_1 + A_3)F_{1+3 \to 2} = A_1 F_{1 \to 2} + A_3 F_{3 \to 2} \quad \text{so that} \quad F_{1 \to 2} = \frac{A_1 + A_3}{A_1} F_{1+3 \to 2} - \frac{A_3}{A_1} F_{3 \to 2},$$

$$F_{1+3 \to 2+4} = F_{1+3 \to 2} + F_{1+3 \to 4} \quad \text{so that} \quad F_{1+3 \to 2} = F_{1+3 \to 2+4} - F_{1+3 \to 4},$$

$$F_{3 \to 2+4} = F_{3 \to 2} + F_{3 \to 4} \quad \text{so that} \quad F_{3 \to 2} = F_{3 \to 2+4} - F_{3 \to 4}.$$

These relations may be combined to yield an expression for $F_{2 \to 1}$.

$$F_{1 \to 2} = \frac{A_1 + A_3}{A_1} (F_{1+3 \to 2+4} - F_{1+3 \to 4}) + \frac{A_3}{A_1} (F_{3 \to 4} - F_{3 \to 2+4}).$$

All terms on the right side of this equation may be determined from the data in Fig. 8.21 since all surfaces share a common border.

$$R_1 = \frac{L_{1+3}}{D} = \frac{5}{2} = 2.5 \quad R_1 = \frac{L_3}{D} = \frac{4}{2} = 2.0 \quad R_2 = \frac{W_{2+4}}{D} = \frac{3}{2} = 1.5 \quad R_2 = \frac{W_4}{D} = \frac{2}{2} = 1.0.$$

From Fig. 8.21, $F_{1+3 \to 2+4} = 0.118$; $F_{1+3 \to 4} = 0.098$; $F_{3 \to 4} = 0.118$; $F_{3 \to 2+4} = 0.137$. Also, $A_1 = 2$, and $A_3 = 8$. The units are arbitrary. Substituting these values into the preceding equation yields

$$F_{1 \to 2} = \frac{2+8}{2}(0.118 - 0.098) + \frac{8}{2}(0.118 - 0.137) = 0.024.$$

Given that $A_1 = A_2$ and that $A_1 F_{1 \to 2} = A_2 F_{2 \to 1}$, $F_{2 \to 1} = F_{1 \to 2} = 0.024$.

Examining and interpreting the results: Even though surfaces 1 and 2 appear to be reasonably close, only about 2% of the radiation is exchanged. In this problem, we have made calculations based on small differences in numbers that are difficult to read accurately off the shape factor graph. It may well not be justified to read the F values to an accuracy of 0.001. If the values are reduced by one significant digit, the shape factors become $F_{1+3 \to 2+4} = 0.12$; $F_{1+3 \to 4} = 0.10$; $F_{3 \to 4} = 0.12$; $F_{3 \to 2+4} = 0.14$ and the result for radiation exchange between surfaces 1 and 2 is $F_{2 \to 1} = F_{1 \to 2} = 0.02$.

(c) Surfaces 1 and 2 are perpendicular and touch only at a common corner. The basic geometry matches that of the shape factor graph in Fig. 8.21. Adding the two hypothetical surfaces 3 and 4 results in two perpendicular flat surfaces that share a common edge. Equation (8.89) may be applied as

$$A_{1+3} F_{1+3 \to 2+4} = A_1 (F_{1 \to 2} + F_{1 \to 4}) + A_3 (F_{3 \to 2} + F_{3 \to 4}).$$

The shape factors in this expression that cannot be evaluated directly from Fig. 8.21 are $F_{1\to2}$ and $F_{3\to4}$. The conditions of reciprocity and symmetry show

$$A_1 F_{1\to2} = A_2 F_{2\to1} \quad \text{and} \quad A_4 F_{4\to3} = A_3 F_{3\to4} \quad \text{and} \quad A_4 F_{4\to3} = A_2 F_{2\to1},$$

so that

$$A_1 F_{1\to2} = A_4 F_{4\to3} \quad \text{and} \quad A_1 F_{1\to2} = A_3 F_{3\to4}.$$

Substituting into the initial equation yields an expression for which all shape factors can be determined except for $F_{1\to2}$.

$$F_{1\to2} = \frac{1}{2A_1}[A_{1+3}F_{1+3\to2+4} - A_1 F_{1\to4} - A_3 F_{3\to2}].$$

The geometry ratios for Fig. 8.21 are calculated for this problem as

$$R_1 = \frac{L_1}{D_1} = \frac{3}{3} = 1.0 \quad R_1 = \frac{L_{1+3}}{D_{1+3}} = \frac{3}{4} = 0.75 \quad R_1 = \frac{L_3}{D_3} = \frac{3}{1} = 3.0,$$

$$R_2 = \frac{W_2}{D_2} = \frac{3}{1} = 3.0 \quad R_2 = \frac{W_{2+4}}{D_{4+4}} = \frac{3}{4} = 0.75 \quad R_2 = \frac{W_4}{D_4} = \frac{3}{3} = 1.0.$$

The resulting shape factors are: $F_{1+3\to2+4} = 0.215$; $F_{1\to4} = 0.20$; $F_{3\to2} = 0.12$. Substituting these values yields

$$F_{1\to2} = \frac{1}{2 \times 3 \times 3}[3 \times 4 \times 0.215 - 3 \times 3 \times 0.2 - 1 \times 3 \times 0.12] = 0.05.$$

(d) This geometry matches the basic configuration shown in Fig. 8.22 for parallel facing rectangles. Since the areas of surfaces 1 and 2 are unequal, a virtual surface 3 can be added to 1 to equal the total area of 2. Area 2 may be divided into two components a and b, of which a equals the geometry of surface 1. Our objective is to obtain an expression for $F_{1\to2}$ in terms of other shape factors that can be determined directly from Fig. 8.22. We may apply shape factor geometry rules to write

$$A_{1+3}F_{1+3\to2} = A_1 F_{1\to2} + A_3(F_{3\to2a} + F_{3\to2b}) \quad \text{and} \quad F_{2a\to1+3} = F_{2a\to1} + F_{2a\to3}.$$

Applying symmetry and reciprocity yields

$$A_{2a}F_{2a\to1+3} = A_1 F_{1\to2} \quad \text{and} \quad A_{2a}F_{2a\to3} = A_3 F_{3\to2a}.$$

We can eliminate the terms $A_3 F_{3\to2a}$ and then $A_{2a}F_{2a\to3}$ as follows:

$$A_{1+3}F_{1+3\to2} = A_1 F_{1\to2} + A_{2a}F_{2a\to3} + A_3 F_{3\to2b},$$

8.4 Thermal Radiation

$$A_{1+3}F_{1+3\to 2} = A_1 F_{1\to 2} - A_{2a}F_{2a\to 1} + A_1 F_{1\to 2} + A_{3+}F_{3\to 2b}.$$

All shape factors except for $F_{1\to 2}$ are now known, so that

$$F_{1\to 2} = \frac{1}{2A_1}(A_2 F_{1+3\to 2} - A_3 F_{3\to 2b} + A_{2a}F_{2a\to 1}).$$

We next compute the areas and geometric ratios as defined in Fig. 8.22.

$$R_1 = \frac{L}{D} = \frac{2}{2} = 1.0 \quad R_2 = \frac{W_1}{D} = \frac{1}{2} = 0.5 \quad R_2 = \frac{W_{1+3}}{D} = \frac{3}{2} = 1.5 \quad R_2 = \frac{W_2}{D} = \frac{3}{2} = 1.5,$$

$$R_2 = \frac{W_{2a}}{D} = \frac{1}{2} = 0.5 \quad R_2 = \frac{W_{2b}}{D} = \frac{2}{2} = 1.0.$$

The shape factors are determined to be: $F_{1+3\to 2} = 0.25$; $F_{3\to 2b} = 0.20$; $F_{2a\to 1} = 0.11$. These values are applied to calculate $F_{1\to 2}$.

$$F_{1\to 2} = \frac{1}{2\times 1\times 2}(2\times 3\times 0.25 - 2\times 2\times 0.20 + 1\times 2\times 0.11) = 0.23.$$

Additional comments: The shape factor rules are few and simple, but they must be applied thoughtfully in order to make accurate calculations.

8.4.5 *Electrical Resistance Model for Radiation*

Evaluation of the temperature, surface property, and geometry effects can be combined to calculate the magnitude of radiation exchange among a system of surfaces. The most simple approach is to represent the radiation process in terms of an equivalent electrical network. For this purpose, two special properties are used: the irradiation, G, which is the total radiation incident onto a surface per unit time and area, and the radiosity, J, which is the total radiation that leaves a surface per unit time and area. Also, for present purposes it is assumed that all surfaces are opaque (no radiation is transmitted, $\tau = 0$), and the radiation process is steady state. Thus, there is no energy storage within any components of the radiating portion of the system.

The radiosity can be written as the sum of radiation emitted and reflected from a surface, which is expressed as

$$J = \varepsilon E_b + \rho G. \tag{8.91}$$

The net energy exchanged by a surface is the difference between the radiosity and the irradiation. For a gray surface with $\alpha = \varepsilon$, so that $\rho = 1 - \varepsilon$,

Fig. 8.24 Electrical resistance model for the drop in radiation potential due to a gray surface defined by the property ε

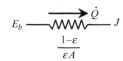

$$\frac{\dot{Q}}{A} = q = J - G. \qquad (8.92)$$

Eliminating G from (8.91) and (8.92), we obtain:

$$\dot{Q} = \frac{E_b - J}{(1 - \varepsilon)/\varepsilon A}. \qquad (8.93)$$

The format of (8.93) is in terms of a flow that equals a difference in potential divided by a resistance. In this case, the equation represents the drop in potential from a black to a gray surface associated with a finite surface radiation resistance. The equation can be represented graphically in terms of a steady state resistance, as shown in Fig. 8.24.

This resistance applies at every surface within a radiating system, which has nonblack radiation properties. Note that for a black surface for which $\varepsilon = 1$, the resistance goes to zero.

A second type of radiation resistance is due to the geometric shape factors among multiple radiating bodies. The apparent radiation potential of a surface is the radiosity. For the exchange of radiation between two surfaces A_1 and A_2, the net energy flow equals the sum of the flows in both directions. The radiation leaving surface 1 that is incident on surface 2 is

$$J_1 A_1 F_{1 \to 2}$$

and in like manner the radiation from 2 to 1 is

$$J_2 A_2 F_{2 \to 1}.$$

The net interchange between surfaces 1 and 2 is then the difference between these two flows, with reciprocity applied.

$$\dot{Q}_{1 \to 2} = J_1 A_1 F_{1 \to 2} - J_2 A_2 F_{2 \to 1} = \frac{J_1 - J_2}{1/A_1 F_{1 \to 2}}. \qquad (8.94)$$

This process can also be modeled through an electrical network as shown in Fig. 8.25.

These two types of resistance elements can be coupled to model the steady-state interactions among systems of radiating bodies. When equivalent resistors are joined together, Kirchhoff's current law applies at each junction, and the sum of the current flows at the junction is zero. Translated to an equivalent radiation circuit, this means that the net radiation flows are zero, or, that energy is conserved. This relationship is very useful for computing heat flows in radiation processes.

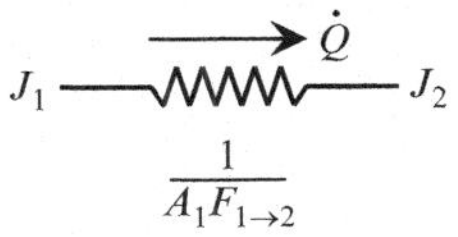

8.4 Thermal Radiation

Fig. 8.25 Electrical resistance model for the radiation exchange between two surfaces with a shape factor $F_{1 \to 2}$

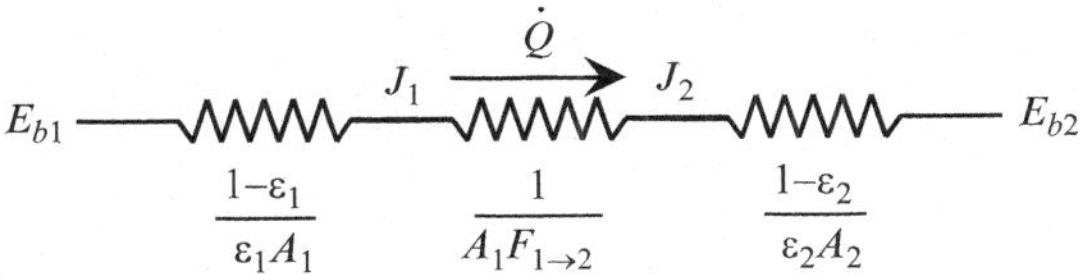

Fig. 8.26 Electrical resistance model for the radiation exchange between two surfaces with a shape factor $F_{1 \to 2}$

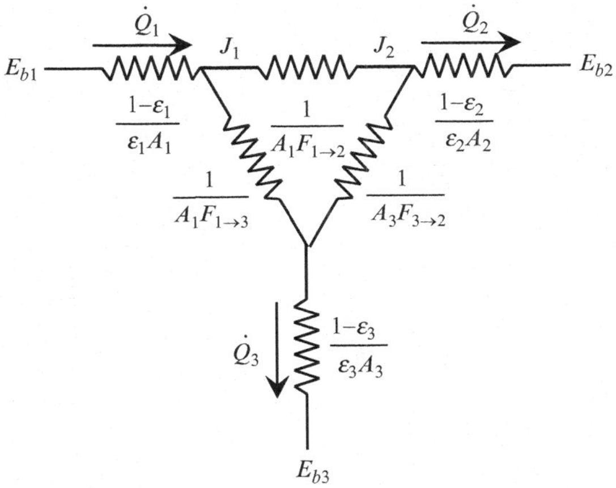

Fig. 8.27 Electrical resistance model for the radiation exchange among the surfaces of a three-body system

Consider two opaque bodies that exchange radiation only with each other. This problem is characterized by the network shown in Fig. 8.26. This network can be solved to determine the radiation heat flow in terms of the temperatures (T_1, T_2), surface properties $(\varepsilon_1, \varepsilon_2)$, and system geometry $(A_1, A_2, F_{1 \to 2})$.

$$\dot{Q}_{1 \to 2} = \frac{E_{b_1} - E_{b_2}}{\frac{1-\varepsilon_1}{\varepsilon_1 A_1} + \frac{1}{A_1 F_{1 \to 2}} + \frac{1-\varepsilon_2}{\varepsilon_2 A_2}} = \frac{\sigma\left(T_1^4 - T_2^4\right)}{\frac{1-\varepsilon_1}{\varepsilon_1 A_1} + \frac{1}{A_1 F_{1 \to 2}} + \frac{1-\varepsilon_2}{\varepsilon_2 A_2}}. \quad (8.95)$$

Note that the second expression in this equation has a linear differential in the driving potential, whereas the third expression has a fourth power differential. A major advantage of the electrical circuit analogy is that a radiation problem can be expressed as a simple linear network as compared to the thermal formulation in which temperature must be raised to the fourth power.

Given common network modeling tools, it becomes straightforward to describe radiation exchange among the components of an n bodied system. A three-bodied system can be used to illustrate this approach, as shown in Fig. 8.27. In this system, each of the three bodies experiences a unique radiant heat flow. For the special case

Fig. 8.28 Electrical resistance model for the radiation exchange among the surfaces of a three-body system in which surface 3 is perfectly insulated

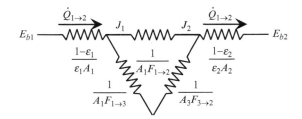

in which one surface, such as 3, is perfectly insulated, meaning that all incident radiation is reradiated, then the diagram is simplified to a combined series/parallel exchange between surfaces 1 and 2 as shown in Fig. 8.28. The same diagram will apply if surface 3 is black, but there will be an energy flow at node 3.

Although surface 3 has no net heat transfer, its radiation properties affect the heat flow for this system between surfaces 1 and 2. The equation for heat flow is written for the equivalent electrical network as a function of the system properties, with the shape factor reciprocity relation applied for $A_n F_{n \to m} = A_m F_{m \to n}$, as

$$\dot{Q}_{1 \to 2} = \frac{\sigma A_1 \left(T_1^4 - T_2^4 \right)}{\frac{1}{A_1}\left(\frac{1}{\varepsilon_1} - 1\right) + \frac{1}{A_1 F_{1 \to 2} + \left(\frac{1}{A_1 F_{1 \to 3}} + \frac{1}{A_2 F_{2 \to 3}}\right)^{-1}} + \frac{1}{A_2}\left(\frac{1}{\varepsilon_2} - 1\right)}. \quad (8.96)$$

> **Example Problem 8.4.3 Radiation to an Absorbing Environment.**
> Consider the performance of a special scientific apparatus consisting of a metal plate 2.5 cm thick that has a 2.5 cm hole drilled through it. By means of a heater, the plate is held at a constant uniform temperature of 260°C. The emissivity of the plate material is 0.07. One end of the hole is covered by a heater with an emissivity of 0.5 that maintains a constant surface temperature of 425°C. The other end of the hole is open to the room in which the plate is placed, having a temperature of 25°C. How much heat will be lost through the hole?

Solution. *Initial considerations*: This process is at steady state and can be represented by an analogous circuit diagram. The radiation properties of the room are not specified. However, all radiation that leaves the open hole will eventually be absorbed by the surfaces of the room after multiple partial absorptions and reflections. As a result, the open end of the hole will behave as a black surface since no incident radiation will be reflected back.

System definition and environmental interactions: The system consists of three surfaces that exchange radiation. The configuration is shown in the accompanying figure, and the equivalent circuit diagram is shown in Fig. 8.28, since the radiation resistance of the hole is zero.

8.4 Thermal Radiation

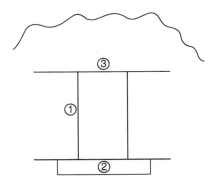

Apprising the problem to identify governing equations: The governing equations are embedded in the circuit diagram. Kirchhoff's current law will be applied at nodes to perform conservation of energy calculations.

Analysis: The conservation of energy applied at the nodes for J_1 and J_2 is

$$0 = \frac{E_{b_1} - J_1}{1 - \varepsilon_1/\varepsilon_1 A_1} + \frac{J_2 - J_1}{1/A_1 F_{1 \to 2}} + \frac{E_{b_3} - J_1}{1/A_1 F_{1 \to 3}}$$

$$0 = \frac{E_{b_2} - J_2}{1 - \varepsilon_2/\varepsilon_2 A_2} + \frac{J_1 - J_2}{1/A_2 F_{2 \to 1}} + \frac{E_{b_3} - J_2}{1/A_2 F_{2 \to 3}}.$$

The blackbody emissive powers and areas can be calculated directly from the given data.

$$E_{b_1} = \sigma T_1^4 = 5.67 \times 10^{-8}\, \text{W/m}^2\, \text{K}^4 \times 533^4 \text{K}^4 = 4.58 \times 10^3\, \text{W/m}^2,$$

$$E_{b_2} = \sigma T_2^4 = 5.67 \times 10^{-8}\, \text{W/m}^2\, \text{K}^4 \times 698^4 \text{K}^4 = 1.35 \times 10^4\, \text{W/m}^2,$$

$$E_{b_3} = \sigma T_3^4 = 5.67 \times 10^{-8}\, \text{W/m}^2\, \text{K}^4 \times 298^4 \text{K}^4 = 4.47 \times 10^2\, \text{W/m}^2,$$

$$A_1 = \pi d L = 3.14 \times 0.025\, \text{m} \times 0.025\, \text{m} = 1.96 \times 10^{-3}\, \text{m}^2,$$

$$A_2 = A_3 = \frac{\pi d^2}{4} = \frac{3.14 \times 0.025^2\, \text{m}^2}{4} = 4.90 \times 10^{-4}\, \text{m}^2.$$

The shape factor is determined from Fig. 8.22.

$$R_1 = \frac{r_1}{L} = \frac{0.0125}{0.025} = 0.5 \quad R_2 = \frac{r_2}{L} = \frac{0.0125}{0.025} = 0.5 \quad F_{2 \to 3} = 0.06.$$

The shape factor summative law yields, $F_{2 \to 1} = 1 - F_{2 \to 3} = 0.94$. Symmetry and the reciprocal law yield

$$F_{1\to 2} = F_{1\to 3} = \frac{A_2}{A_1} F_{2\to 1} = \frac{4.90 \times 10^{-4}}{1.96 \times 10^{-3}} \cdot 0.94 = 0.235.$$

These property values may be substituted into the two node equations to solve simultaneously for J_1 and J_2, which are applied to determine the heat flow at node E_{b3}.

$$0 = \frac{4.58 \times 10^3 - J_1}{1 - 0.07/0.07 \times 1.96 \times 10^{-3}} + \frac{J_2 - J_1}{1/1.96 \times 10^{-3} \times 0.235} + \frac{4.47 \times 10^2 - J_1}{1/1.96 \times 10^{-3} \times 0.235},$$

$$0 = \frac{1.35 \times 10^4 - J_2}{1 - 0.5/0.5 \times 4.90 \times 10^{-4}} + \frac{J_1 - J_2}{1/4.90 \times 10^{-4} \times 0.94} + \frac{4.47 \times 10^2 - J_2}{1/4.90 \times 10^{-4} \times 0.06}.$$

The values for J_1 and J_2 are

$$J_1 = 4.68 \times 10^3 \text{ W/m}^2 \quad J_2 = 8.94 \times 10^3 \text{ W/m}^2.$$

The heat loss through the hole is the sum of the flows from nodes 1 and 2.

$$q_3 = \frac{J_1 - E_{b3}}{1/A_1 F_{1\to 3}} + \frac{J_2 - E_{b3}}{1/A_2 F_{2\to 3}} = 2.20 \text{ W}.$$

Examining and interpreting the results: An alternative approach to calculating the heat flow through the open end of the hole is to calculate the radiation flows for surfaces 1 and 2 and take their sum. Because the system is at steady state, conservation of energy dictates that the net heat flow be zero. We can write the radiation flow equation for each surface in the format that a flow to the surface is positive. This is the format in which q$_3$ was calculated above. It is important that our notation be consistent for all of the surfaces when we apply the calculated flows to the conservation of energy equation.

$$q_1 + q_2 + q_3 = 0,$$

$$q_1 = \frac{J_1 - E_{b_1}}{1 - \varepsilon_1/\varepsilon_1 A_1} = \frac{(4.68 \times 10^3 - 4.58 \times 10^3) \text{ W/m}^2}{1 - 0.07/0.07 \times 1.96 \times 10^{-3} \text{ m}^2} = 1.51 \times 10^{-2} \text{ W},$$

$$q_2 = \frac{J_2 - E_{b2}}{1 - \varepsilon_2/\varepsilon_2 A_2} = \frac{(8.94 \times 10^3 - 1.35 \times 10^4) \text{ W/m}^2}{1 - 0.5/0.5 \times 4.90 \times 10^{-4} \text{m}^2} = -2.22 \text{ W}.$$

The radiation heat flow away from the bottom heated hole supplies a net energy to both the side walls and the open hole. Comparison of the radiosities for surfaces 1 and 2 shows the potential drop owing to a nonideal surface in the direction in which the flow occurs. Thus, $J_1 > E_{b1}$ in conjunction with a net flow to surface 1.

8.4 Thermal Radiation

A final radiation process to be discussed addresses the effect of an absorbing and transmitting medium included in a radiating system. Glasses and some gases are examples of this type of media. Nonpolar gaseous molecules such as O_2 and N_2, the primary constituents of air, are largely transparent and do not interact with thermal radiation. The same is not true for polar gaseous molecules commonly encountered environmentally such as CO_2 and H_2O. These media may have a significant radiation interaction, depending on the radiation properties of the molecule at the specific wavelength, the gas pressure, and the path length of the radiation through the medium. When these media are between two radiating surfaces they may have a significant effect on the energy exchange. When a surface is in a very large gaseous environment, such as your body positioned out of doors, then the effective emissivity of the gas approaches 1.0 because the area of your body is small compared with the environment.

If the medium is nonreflective, then

$$\alpha_m + \tau_m = \varepsilon_m + \tau_m = 1. \tag{8.97}$$

Radiation that is absorbed by the medium is transmitted and then emitted to its environment. Unlike a vacuum which does not interact with radiation and therefore transmits 100%, an absorbing medium will not transmit all incident radiation. It thereby presents a resistance to the flow of radiation and can be modeled as such. For a system consisting of two surfaces, 1 and 2, that see only each other, plus an intervening medium m, the net energy leaving surface 1 that is transmitted through the medium and arrives at surface 2 is

$$J_1 A_1 F_{1 \to 2} \tau_m.$$

Likewise, the energy flow in the opposite direction is

$$J_2 A_2 F_{2 \to 1} \tau_m.$$

The net interchange between surfaces 1 and 2 via transmission through the medium is then the sum of these two flows

$$\dot{Q}_{1 \to 2} = J_1 \tau_m A_1 F_{1 \to 2} - J_2 \tau_m A_2 F_{2 \to 1} = \frac{J_1 - J_2}{1/A_1 F_{1 \to 2}(1 - \varepsilon_m)}. \tag{8.98}$$

The effect of the medium can be represented by a radiation network element as shown in Fig. 8.29.

This circuit element can be included in a radiation network model as appropriate to represent the effect of an interstitial medium between radiating surfaces.

For the radiation analysis that we have considered, all of the equivalent electrical networks contain only resistors, and specifically there are no capacitors. The explicit interpretation of this arrangement is that all of the radiation processes are

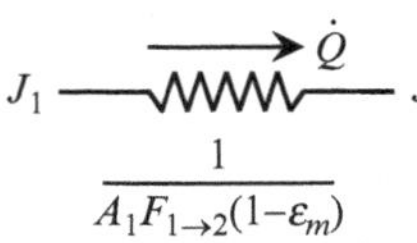

$$J_1 \xrightarrow{} \text{WWW} \xrightarrow{\dot{Q}} J_2$$
$$\frac{1}{A_1 F_{1\to 2}(1-\varepsilon_m)}$$

Fig. 8.29 Electrical resistance model for the effect of an absorbing and transmitting medium on the thermal radiation between two surfaces

at steady state such that no energy storage occurs. For our present analysis, the mass of all radiating bodies has been neglected. Since many thermal radiation processes are approximated as surface phenomena, this is a reasonable assumption. Under conditions that demand more comprehensive and sophisticated analysis, this assumption may have to be relaxed, which leads to a significant increase in the complexity of the thermal radiation analysis.

Thermal radiation effects can be distinguished from optical irradiation during laser/tissue interactions. The laser energy is absorbed by the tissue as a spatially and temporally distributed heating source, $\dot{Q}_{gen}(r,t)$. Thermal radiation is emitted and absorbed from the surface of the tissue as a function of the temperature and radiative properties of the tissue and the environment (Welch and van Gemert 2011). These are different physical phenomena, and they likewise are distinguished mathematically when modeling the system behavior.

Example Problem 8.4.4 Cold Exposure Weather Advisory: Revisited.
This is a continuation of example problem 8.3.1 in which we calculated a ratio for heat loss via free convection to stagnant water and air at 0°C of 65.2, whereas the National Weather Service advertised it to be 50. The initial calculations did not account for the effects of radiation loss in air. The problem to solve in this example is calculation of the added effect of radiation.

Solution. *Initial considerations*: Since free convection in air is smaller than in water, augmentation of the heat loss by radiation will provide a much greater relative increment for air than water. Because the radiation exchange is with the environment, the effective emissivity can be taken as 1.0. Radiation and free convection can be assumed to occur in parallel and independently. This situation is termed multimode heat transfer.

System definition and environmental interactions: The system of interest is defined the same as in Example 8.3.1, that being the human body. Radiation is added to free convection as a mechanism of interaction with the surroundings.

Apprising the problem to identify governing equations: Equation (8.95) can be applied for the radiation heat exchange with $A_1 = 1.0$ and $\varepsilon_2 = 1.0$.

Analysis: The emissivity of the skin is $\varepsilon_1 = 0.95$.

$$\dot{Q}_{1\to 2} = \frac{E_{b_1} - E_{b_2}}{\frac{1-\varepsilon_1}{\varepsilon_1 A_1} + \frac{1}{A_1 F_{1\to 2}} + \frac{1-\varepsilon_2}{\varepsilon_2 A_2}} = \frac{\sigma(T_1^4 - T_2^4)}{\frac{1-\varepsilon_1}{\varepsilon_1 A_1} + \frac{1}{A_1 \cdot 1} + \frac{1-1}{\varepsilon_2 A_2}} = \varepsilon_1 A_1 \sigma(T_1^4 - T_2^4),$$

$$q_{1\to 2} = \varepsilon_1 \sigma(T_1^4 - T_2^4).$$

This equation can be written in a format to compare directly with the convection heat transfer as

$$q_{1\to 2} = h_{\text{rad}}(T_s - T_\infty) = \varepsilon_s \sigma(T_s^4 - T_\infty^4).$$

An equivalent radiation heat transfer coefficient is then calculated as

$$h_{\text{rad}} = \frac{\varepsilon_s \sigma(T_s^4 - T_\infty^4)}{T_s - T_\infty} = \frac{0.95 \times 5.67 \times 10^{-8}\,\text{W/m}^2\,\text{K}^4 (307^4 - 273^4)\,\text{K}^4}{(307 - 273)\,\text{K}}$$
$$= 5.27\,\text{W/m}^2\,\text{K}.$$

Recall from Example 8.3.1 that we calculated for free convection: $h_a = 4.34$ W/(m² K) and $h_w = 283$ W/(m² K). If it is valid to assume that the radiation effects are identical for air and water, then the ratio of combined heat losses is

$$\frac{h_w + h_{\text{rad}}}{h_a + h_{\text{rad}}} = \frac{283 + 5.27}{4.34 + 5.27} = 30.0.$$

Examining and interpreting the results: Inclusion of radiation effects reduces the ratio of heat loss for water and air by a factor of more than two. This result points out that we need to give careful consideration for some problems as to whether multiple heat transfer mechanisms need to be evaluated. When we were working out a solution to Example 8.3.1, it was not obvious that for free convection in air we were omitting an effect that was more important than that which we were calculating.

When the effects of radiation are included, the ratio of heat loss to water and air becomes less than the value of 50 that was advertised in the popular media. We can have a greater level of confidence in the calculations that include the effects of both convection and radiation heat transfer. If you are interested in tracking down the reason for the difference in our calculations compared to the advisory circulated by the television station, you may want to invest the time and effort required. However, the source of data presented in the media is often difficult to identify.

8.5 Common Heat Transfer Boundary Conditions

There are several kinds of boundary conditions that are encountered most frequently in bioheat transfer phenomena. These come under the categories of specified temperature, convection heat exchange conditions, and heat flow, which most often is associated with a radiation (perhaps from the sun), phase change (such as evaporation of sweat), or insulation (heat flow is zero) processes. The effects of these types of boundary condition on the transient development of an internal temperature distribution are illustrated in Fig. 8.30.

The simplest specification is that the temperature or flux remains constant at the surface over time, which leads to the simplest mathematical expression. However, if the actual process of interest precludes using a constant boundary condition, as may be encountered in biological applications, then a more complex specification is required that may lead to a numerical solution.

An important class of boundary conditions occurs when a material interface exists interior to a system, as illustrated in Fig. 8.31. The basis for deriving the

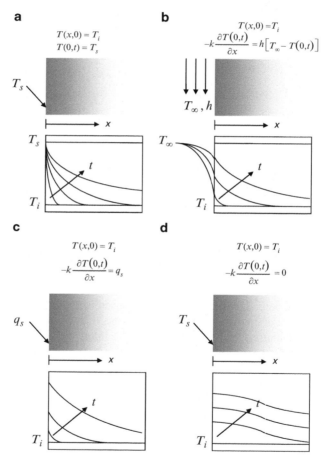

Fig. 8.30 Types of thermal boundary conditions for heat conduction in a material: (**a**) constant temperature, (**b**) convection with a constant temperature fluid, (**c**) constant applied surface energy flux, (**d**) insulated surface (zero heat flux). The latter case requires an internal heat source to produce a transient temperature field

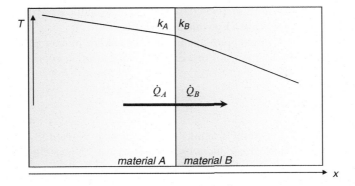

Fig. 8.31 Specification of boundary conditions at a material interface. Note that there is continuity of both potential and flow

boundary conditions was presented in Sect. 8.3.3.2. Physical constraints demand that two independent conditions occur at the interface, which can be interpreted directly as two boundary conditions. One constraint is that the flux be continuous at the interface of the materials. Since there is no mass associated with an interface, there is no capacity for storing energy. Conservation of energy then states that the flow into the interface must equal the flow out of the interface.

$$\dot{Q}_A = \dot{Q}_B. \tag{8.99}$$

Assuming there is no convective heat flux at the interface, Fourier's law applies for the conduction in both materials, which are anticipated to have different values for their thermal conductivities. Equation (8.99) becomes at the interface

$$k_A \frac{\partial T_A}{\partial x} = k_B \frac{\partial T_B}{\partial x}. \tag{8.100}$$

The second constraint is that the temperatures be continuous at the interface.

$$T_A = T_B. \tag{8.101}$$

Note, however, that differing thermal conductivities of the interface materials means that the temperature slopes will not be continuous at the interface. It will be very helpful in solving transport problems to realize that this set of boundary conditions applies at all internal material interfaces.

8.6 Summary of Key Concepts

Heat transfer is one of the three primary transport phenomena covered in this text. All three are related mechanistically, but each also has unique characteristics. There are three distinctly different modes by which heat flows: conduction, convection, and radiation. There is a distinctive constitutive equation that describes each mode that represents the mechanism by which heat moves.

Conduction Heat Transfer. For conduction, the governing constitutive equation is the Fourier law. It is written for one-dimensional Cartesian coordinates as

$$q_x = \frac{\dot{Q}_x}{A} = -k\frac{dT}{dx}. \qquad (8.1)$$

This expression says that if a temperature gradient is applied in space across a medium that has a thermal conductivity, k, there will be a heat flow by diffusion along the direction of the gradient. The magnitude of the flow is proportional to the area available for the process, normal to the temperature gradient. The negative sign means that heat must flow from a higher temperature region to one of lower temperature, in accordance with the second law of thermodynamics. Thus, the temperature must be decreasing along the x coordinate to generate a diffusion of heat toward the positive coordinate direction. Heat conduction is most commonly encountered in solid phase systems, and very often the boundary conditions are determined by convection and/or radiation with the surroundings. In this way, conduction usually occurs in a series configuration with convection and radiation.

Convection Heat Transfer. For convection, the governing constitutive equation is Newton's law of cooling. It is written as

$$\dot{Q}_s = hS(T_s - T_\infty). \qquad (8.10)$$

Convection occurs when there is a relative motion between a fluid, which may be a gas or liquid, and an adjacent solid substrate. The magnitude of the convection is proportional to the surface area at the solid–fluid interface, the convective coefficient, h, and the difference between the temperatures of the substrate and the fluid. The latter property is assessed at different locations in the fluid, depending on the geometry and conditions of fluid flow. Most convective problems focus on determining a numerical value for the convective heat transfer coefficient. There are four key factors that must be taken into consideration to make this calculation. If you follow a logical procedure to evaluate these factors, the analysis will lead you directly to an appropriate correlation equation to apply to calculate h. The key factors are:

1. The source of relative motion between the fluid and solid, resulting in *forced* or *free* convection.
2. The geometry and shape of the boundary layer region of the fluid in which convection occurs, producing *internal* or *external* flow. In addition, for free convection the orientation of the fluid/solid interface in the gravitational field is important.
3. The *boundary layer* flow domain, being *laminar* or *turbulent*.
4. The *chemical composition* and *thermodynamic state* of the fluid in the boundary layer, which dictate numerical values for the constitutive properties relevant to the convective process.

Convection analysis is based heavily on equations written in terms of dimensionless parameters. These parameters have been derived so that they provide physical insight into how the convection process occurs for a problem being analyzed. The compilation of these parameters in Table 8.2 is important information to understand and have available while you are working convection problems.

Radiation Heat Transfer. For radiation, the governing constitutive equation is the Stefan–Boltzmann law. It is written

$$E_b = \sigma T^4. \tag{8.73}$$

There are three key categories of governing properties to be evaluated in performing analysis of radiation heat transfer. These are:

1. Temperature of the surface expressed in absolute units (K).
2. Radiation properties of the surface as defined by absorption, reflection, and transmission.
3. The geometric configuration of multiple surfaces that exchange radiation as defined by the sizes, shapes, separation, and orientation.

Because the Stefan–Boltzmann law is written in terms of the fourth power of temperature which results in equations that are nonlinear in the thermal dimension, it is most practical to perform radiation analysis based on the electrical circuit analogy. This method is well defined and provides a simple calculation technique.

Formulating Boundary and Initial Conditions. One of the skills that is most important to develop for solving heat transfer problems is to be able to recognize and apply the boundary and initial conditions associated with a system and environmental interaction process. These conditions are applied in conjunction with the solution of governing equations that you develop for the problem solution. In general, you need one independent condition for each order of the differential equation applied to each dimension of your system. In addition, each unique material domain in your system will require two additional spatial boundary conditions applied at a material interface. For example, a transient conduction process that is described by a partial differential equation that is first order in time and second order in space, and occurs in one dimension through a two-layer skin system will require one condition in time (generally evaluated at $t = 0$, the initial condition), plus four spatial conditions. These conditions are evaluated one each at the inner and outer surfaces and two at the interface between the two skin layers.

8.7 Questions

8.7.1. How large is the range of thermal conductivities that you may encounter as a biomedical engineer?

8.7.2. How do you know how many spatial boundary conditions must be identified to solve a particular heat transfer problem?

8.7.3. Do the number of boundary conditions depend on whether convection or radiation is involved in the process?
8.7.4. How do you describe an insulated surface mathematically?
8.7.5. Why is the Reynolds number important in heat transfer?
8.7.6. What is the meaning of a thermal boundary layer?
8.7.7. What are the fundamental differences between forced and free convection? What happens when they occur simultaneously?
8.7.8. Define the Prandtl number and tell why it is important to heat transfer.
8.7.9. What is the bulk temperature, and how is it used? What is the film temperature?
8.7.10. In what ways are the Grashof and Reynolds numbers similar, and in what ways are they different?
8.7.11. What is the Stefan–Boltzmann law?
8.7.12. What is a gray body, and what are its unique features? How is the radiation for a gray body different than that for a black body?
8.7.13. Define irradiation and radiosity. How to these properties relate to practical thermal radiation processes of the human body and for the design of biomedical devices?
8.7.14. What is Kirchhoff's current law as used in radiation analysis? Under what conditions does it apply?
8.7.15. Explain the greenhouse effect and why it occurs. What are the major physical factors that affect this phenomenon? Name as many as possible common occurrences that you have observed.
8.7.16. What are some of the unique difficulties in designing a therapeutic device based on cooling tissue as compared with a heating device?
8.7.17. Explain the meaning of a black body surface for both radiation and absorption processes.
8.7.18. What is the meaning of a radiation shape factor?
8.7.19. Why do surfaces absorb solar and earthbound radiation differently?
8.7.20. What is meant by a thermal resistor?
8.7.21. In what ways are internal and external flows the same, and in what ways are they different? Name biological examples of both.

8.8 Problems

8.8.1 Radiation During Barbequing

A barbeque grill is built from bricks which act as insulating surfaces, and it is large enough (about 2 feet in diameter) to be able to cook 25 or more hamburgers simultaneously. The rack on which the charcoal is placed is fixed on the grill, and the rack on which the burgers are placed is adjustable in height increments of 7 cm, starting at a minimum of 10 cm. When the grill is initially set up for cooking, the charcoal rack is completely covered with briquettes 2 cm in diameter. The briquettes are very black, and they reach a temperature of 700°C after being ignited.

8.8 Problems

After the charcoal reaches its highest temperature, the hamburgers are positioned on the grill as they are removed from the refrigerator at 4°C, and their absorption coefficient can be assumed to be 0.8. The objective of this problem is to compare the performance of the grill with the meat rack placed at the closest and the third to closest settings. Assume that the rack itself does not participate in or affect the heat transfer process. Analyze the process for one 10 cm diameter hamburger placed in the middle of the grill. What is the initial rate of heat transfer to the surface of the hamburger for the two grill settings?

8.8.2 Heat Transfer Through an Insulated Window

Compare the heat loss from two windows of a room, one having a single sheet of glass 5 mm thick while the other is composed of two sheets of glass, each 2.5 mm thick, separated by 2.5 mm of air. The second window is called a thermopane. The window sizes are identical at 1 m × 1 m. The air temperature in the room is 20°C and the outside air temperature is −5°C. The average value of the thermal conductivity of glass is 0.85 W/(m K) and that of air is 0.025 W/(m K) over the temperature range of interest. The convective heat transfer coefficients are 20 W/(m² K) and 200 W/(m² K) inside the room and outside, respectively. Consider the thin layer of air between the glass sheets to be stagnant.

8.8.3 Steady-State Temperature Distribution in the Arm

Consider the arm to be a cylinder composed of bone ($r < R_i$) and soft tissue ($R_i < r < R_o$). Heat is produced in the soft tissue at a constant rate per unit volume Q. Assume no heat is produced in the bone, and Newton's Law of Cooling applies at the outside surface. Find: (a) equations for the steady-state radial temperature profile and (b) the temperature of the bone. Assume constant tissue properties. What important mechanism is ignored in this problem?

8.8.4 Blood Bank Refrigerator

Blood is sometimes stored for refrigeration in cylindrical containers of dimensions 150 mm long and 60 mm in diameter. After processing subsequent to donation the blood and container are at a uniform temperature of 27°C. The container is then placed into the refrigerator in the blood bank at a temperature of 4°C. A question has arisen among the blood bank staff as to whether the containers should be placed standing up vertically or laying down horizontally in order to achieve the most rapid cooling after being placed into the refrigerator. It is assumed that no significant cooling occurs from the ends of the container. Only the sides are cooled. Direct contact with the shelving has no effect on the cooling process. What is your recommendation, and how much of an advantage does the best method have over the alternative?

8.8.5 Cooling Coffee When It Is Poured into a Cup

Many burn injuries due to hot beverage (such as coffee) spills occur at the drive-thru windows of fast food restaurants. Coffee is poured from a holding container into a cup and is served to the customer. One issue in this scenario is how much the coffee is cooled during the process of being poured from the holding container into the serving cup as the coffee and cup come to thermal equilibrium. The objective of this problem is to calculate the amount of cooling that occurs for two different styles of cups.

A typical holding temperature for coffee at a fast food restaurant is 185°F (85°C). Coffee is poured into a cup that is initially at room temperature, which can be assumed as 25°C. Since the coffee is hotter than the cup, some of its energy is transferred to heat up the cup, thereby cooling the coffee. The coffee will also be cooled by convection and radiation to the environment during the pouring process, but you can neglect these factors for the purposes of this calculation. You are to determine the cooling for a Styrofoam cup and for a ceramic cup. The amount of liquid poured into the cup is 250 ml. The cup dimensions are 70 mm diameter and 90 mm height. The wall and bottom thickness of the Styrofoam cup is 2mm and of the ceramic cup is 7.1 mm. The thermal properties of these materials are given below.

Material	Conductivity (W/mK)	Density (kg/m^3)	Specific heat (J/kgK)
Ceramic	1.3	2083	835
Styrofoam	0.026	70	1045
Coffee	0.674	980	4200

How much will the temperature of the coffee be reduced in coming to thermal equilibrium with the cup? For this calculation, you can assume the final temperature distribution in the material of the cup is the same for both the wall and the bottom and that the outer and inner surface temperatures are identical to the surrounding fluids (air on the outside and coffee on the inside). What is the ratio of the change in coffee temperature when poured into Styrofoam and ceramic cups?

8.9 Challenges

8.9.1 Debunking Brain Freeze Myth from Reality

A common cause of headache pain is taking a big bite of ice cream, occurring in one third of a randomly selected population (1). It occurs regardless of whether someone suffers from other types of headache. Children know all about ice cream headache, generally describing it by the term "brain freeze." The pain begins a few seconds after the rapid ingestion of very cold foods or beverages. The pain is usually located in the midfrontal area, but can be unilateral in the temporal, frontal, or retro-orbital region. It is a stabbing or aching type of pain that recedes 10–20 seconds after its onset (2).

There has been an ongoing discussion in the medical literature about how and why brain freeze occurs when eating ice cream rapidly (3). The phenomenon is obviously initiated by heat transfer associated with holding a cold mass in the mouth. Three primary explanations have been set forth, none of which has been proved rigorously: the ice cream cools the brain (resulting in pain) by direct conduction; the ice cream cools blood circulating through the palate, which then is circulated to the brain where it induces cooling; and the ice cream cools local nerves in the palate, which cause sudden vasodilation in the brain and a painful hyperperfusion. There is no thermal analysis of this phenomenon reported in the biomedical literature that provides a quantitative basis for resolving the debate. A well-conceived and executed bioheat transfer analysis of this phenomenon would likely be worthy of a new archival journal publication in the medical literature as well as providing the motivation for a novel discussion during your next visit to the local ice cream shoppe.

Your job is to develop a strategy for formulating a model of brain freeze that could be used to understand and describe quantitatively how heat transfer could induce the painful sensation sometimes associated with eating ice cream. Present your answer in a full Generate Ideas Model format.

1. Raskin NH. *Headache*. 2nd ed. London: Churchill Livingstone, 1988.
2. Bird N, MacGregor A, Wilkinson MIP. Ice cream headache–site, duration, and relationship to migraine. *Headache* 1992;32:35–38.
3. Hulihan, J. Ice cream headache. *BMJ* 1997;314:1364.

8.9.2 Jet Impingement Cooling of Skin During Laser Surgery

Laser irradiation is an effective tool for the destruction or denaturation of subsurface structure below the surface of the skin. Typical processes include ablation of the large and profuse blood vessels that cause port wine stain birth marks and thermal reshaping of cartilage to obviate congenital, traumatic, and oncologic defects in the head and neck. The laser light is irradiated onto the surface of the skin, and it is propagated progressively into the interior of the skin. At each depth of the skin, some of the energy is absorbed, and the remainder is transmitted to lower depths. Eventually, all the incident laser energy is absorbed. The result is a heat source that is distributed over a considerable depth of the skin, with the distribution pattern being a function of the absorptivity of different structures in the skin to the specific wave length of laser light applied. For example, to destroy blood vessels a laser is chosen with a wavelength that is most strongly absorbed by the hemoglobin in red blood cells so that the energy source is selectively concentrated on the deformed blood vessels. A limiting factor of this technique is that the laser energy absorbed at the most superficial levels (near the surface) may be enough to cause second- or third-degree burns, producing scarring. Since most port wine stains that are surgically treated are on the face, this is an unacceptable result.

A biomedical engineer has invented an ingenuous method to eliminate the injury to superficial tissue during laser irradiation. It is based on spraying a cryogenic refrigerant fluid onto the surface of the skin coincidently with the laser irradiation. Thus, there are two simultaneous heat transfer processes: convective cooling of the skin surface and deep heating of the skin as a function of the progressive absorption of the light with depth into the tissue and according to the absorption coefficients of different skin structures. This technique has been tremendously successful and has been rapidly adopted for many types of plastic surgery procedures.

The challenge problem you are to address is to develop a strategy for building a model of laser irradiation of the skin with simultaneous convective cooling of the skin surface. Use the standard Generate Ideas Model. Include as much information as you deem necessary to describe your approach. The constitutive expression for absorption of laser energy in tissue is quite complex, and for our purposes in this analysis, it can be represented as a simple function of time and depth, with no further details necessary.

References

Bejan A (2004) Convection heat transfer, 3rd edn. Wiley, Hoboken, NJ
Burmeister LC (1993) Convective heat transfer, 2nd edn. Wiley, Hoboken, NJ
Churchill SW (2002) Free convection around immersed bodies. In: Hewitt GF (ed) Heat exchanger design handbook. Begel House, New York
Churchill SW, Chu HHS (1975a) Correlating equations for laminar and turbulent free convection from a horizontal cylinder. Int J Heat Mass Transf 18:1049–1053
Churchill SW, Chu HHS (1975b) Correlating equations for laminar and turbulent free convection from a vertical plate. Int J Heat Mass Transf 18:1323–1329
Dittus FW, Boelter LMK (1930) Univ. Calif. Berkeley Pub. Eng. 2, 443
Goldstein RJ, Sparrow EM, Jones DC (1973) Natural convection mass transfer adjacent to horizontal plates. Int J Heat Mass Transf 16:1025–1035
Hausen H (1943) Darstellung des Warmeuberganges in Rohren durch veralgemeinerte Potenzbeziehungen. Zeitschrift des Vereines Deutscher Ingenieure 4:91
Hilpert R (1933) Warmeabgabe von geheizten Drahten und Rohren im Luftstrom. Forsch Gebiete Ingenieurw 4:215–220
Holman JP (1972) Heat transfer, McGraw-Hill, New York
Howell JR (1982) A catalog of radiation configuration factors. McGraw-Hill, New York
Kays W, Crawford M, Weigand B (2005) Convective heat and mass transfer, 4th edn. McGraw-Hill, New York
Martin H (1977) Heat and mass transfer between impinging gas jets and solid surfaces. Adv Heat Transfer 13:1–60
Morgan VT (1975) The overall convective heat transfer from smooth circular cylinders. Adv Heat Transfer 11:199–264

Raithby GD, Hollands KGT (1975) A general method for obtaining approximate solutions in laminar and turbulent free convection problems. Adv Heat Transfer 11:265–315

Sieder EN, Tate CE (1936) Heat transfer and pressure drop of liquids in tubes. Ind Eng Chem 28:1429

Siegel R, Howell JR (2002) Thermal radiation heat transfer, 4th edn. Taylor and Francis, New York

Welch AJ, van Gemert MJC (2011) Optical-thermal response of laser-irradiated tissue. Springer Press, New York

Whitaker S (1972) Forced convection heat transfer correlation for flow in pipes, past flat plates, single cylinders, single spheres, and for flow in packed beds and tube bundles. AIChE J 18:361–371

Chapter 9
Macroscopic Approach to Bioheat Transport

9.1 Introduction

Bioheat transport principles can be used to find heat flux and temperature in biological systems and devices as a function of both position and time. If we are interested in spatial variations of heat flux or temperature in a system, then it is necessary to use a microscopic approach, as presented in Chaps. 10 and 11. However, in many situations spatial variations are not of interest. Instead, we might be interested in predicting the rate or the amount of energy that flows in or out of the system as a whole, or we might like to know how the total energy varies with time inside the system. In such cases, a macroscopic approach can be used to describe the accumulation of energy within the system or the flow of energy through a system. Take, for example, the flow of heat from a human arm through a layer of clothing. We found in Chap. 8 that the temperature in the layer varies with radial position in a nonlinear fashion. However, we are usually interested in the heat flux from the arm, not the temperature distribution in the clothing. To compute the heat flux we only need to know the thermal resistance of the clothing layer and the temperatures at the inside and outside surface. If an additional layer, with known thermal resistance, is added to the clothing ensemble, its effect on heat loss can be computed using a macroscopic approach.

9.2 General Macroscopic Energy Relation

The general macroscopic conservation of energy, also known as the first law of thermodynamics, was introduced in Sect. 5.4. In most bioheat applications, changes in potential energy and kinetic energy are small in comparison with changes in internal energy, and frictional work is negligible. Under these circumstances, (5.42), which is applicable to a system consisting of multiple inlets and outlets, reduces to:

$$\frac{dE}{dt} = \sum_{i=1}^{N_{\text{inlets}}} w_i \hat{H}_i - \sum_{j=1}^{N_{\text{outlets}}} w_j \hat{H}_j + \dot{Q}_s + \dot{Q}_{\text{gen}} - \dot{W}_s. \tag{9.1}$$

The left side of (9.1) represents the accumulation of energy within the system. Energy can accumulate if heat enters through the system boundaries or if heat is generated within the system, and energy will be lost as the system performs work on the surroundings. The net rate at which energy enters the system through the system boundaries is given by the first three terms on the right side of (9.1). The first two terms account for the rate at which energy is transported by convection into and out of the system through multiple conduits. The term $\dot{Q}_S$ refers to the rate heat enters through the surface of the system by all other mechanisms. This can include conduction and radiation at the conduit surfaces, and heat transfer by all mechanisms through the nonconduit surfaces of the system. The last two terms represent the rate at which energy is gained by heat generation within the system, and the rate energy is lost when the system performs work on the surroundings. The last term is often referred to as "shaft work" which results when fluid inside a system forces an object, such as an impeller, to rotate and perform work on the surroundings.

Changes in system energy ordinarily reflect changes in internal energy (i.e., $E = U$). In addition, most of the systems that we will deal with in bioheat transport perform little or no work on the surroundings. Finally, as we showed in Example 5.4.1, the rate of change of enthalpy at system inlets and outlets is dictated nearly completely by the temperature of the mass crossing the boundary. With these assumptions, (9.1) simplifies to:

$$\frac{d\bar{U}}{dt} = \frac{d}{dt}\left(mc_p\bar{T}\right) = \sum_{i=1}^{N_{\text{inlets}}} w_i c_{pi} T_{m,i} - \sum_{j=1}^{N_{\text{outlets}}} w_j c_{pj} T_{m,j} + \dot{Q}_S + \dot{Q}_{\text{gen}}. \quad (9.2)$$

The overbar indicates an average internal energy in the system. $T_{m,k}$ is the flow-averaged or mixing cup temperature of fluid stream k which enters or leaves the system. If the system is a solid with constant mass (a closed system) and the temperature is assumed to be uniform throughout the solid, then (9.2) becomes:

$$mc_p \frac{dT}{dt} = \dot{Q}_S + \dot{Q}_{\text{gen}}. \quad (9.3)$$

This can be used to estimate heat transfer from solids with low internal thermal resistance relative to the resistance to heat transfer with the surroundings at the surface of the solid. This is a simplified equation, and the test for when it can be applied is whether the Biot number $Bi < 0.1$, as is discussed in Sect. 8.3.6.

If the system consists of a well-mixed fluid with constant mass and multiple inlets and outlets, then all of the outlet temperatures will equal the system temperature T, and (9.2) becomes:

$$mc_p \frac{dT}{dt} + c_p \left(\sum_{j=1}^{N_{\text{outlets}}} w_j\right) T = \sum_{i=1}^{N_{\text{inlets}}} w_i c_{pi} T_{m,i} + \dot{Q}_S + \dot{Q}_{\text{gen}}. \quad (9.4)$$

9.3 Steady-State Applications of the Macroscopic Energy Balance

Equations (9.3) and (9.4) are the starting points for the macroscopic treatment of many solid and fluid systems, respectively. The remainder of this chapter will examine relevant problems in bioheat transfer that can be solved by applying a macroscopic energy balance. We begin with steady-state applications, move to situations where system temperature changes with time, and finally consider cases where the system is divided into interacting subsystems.

9.3 Steady-State Applications of the Macroscopic Energy Balance

In steady-state applications, the total energy in the system does not change and (9.2) reduces to:

$$0 = \sum_{i=1}^{N_{\text{inlets}}} w_i c_{\text{p}i} T_i - \sum_{j=1}^{N_{\text{outlets}}} w_j c_{\text{p}j} T_j + \dot{Q}_S + \dot{Q}_{\text{gen}}. \tag{9.5}$$

This expression is appropriate for fluid systems with multiple inlets and outlets. It states that the net flow of energy into the system via conduits, through the system boundaries or via heat generation, must be balanced by the net flow of energy out of the system. If the system is a solid, the mass flow terms are all zero and (9.5) reduces to:

$$0 = \dot{Q}_S + \dot{Q}_{\text{gen}}. \tag{9.6}$$

Recall that $\dot{Q}_S$ is the rate at which heat enters the system through the boundaries, and $\dot{Q}_{\text{gen}}$ is the rate that heat is generated in the system. If $\dot{Q}_{\text{gen}}$ is positive, then $\dot{Q}_S$ must be negative. If no heat is generated in the system, then there will be no net heat exchanged with the surroundings for steady-state heat transfer. Therefore, if heat enters the solid through one portion of the surface, it must leave at the same rate through another portion of the surface.

9.3.1 Thermal Resistances

We introduced the concept of thermal resistance in Chap. 8, which emphasizes the analogy between Fourier's law and Ohm's law in steady-state heat transport. In many biological systems, the tissue is not composed of a homogeneous material, but of a layered structure which may have series and/or parallel configurations of materials, each having unique thermal transport properties. These are called composite systems, and they must be accommodated frequently when modeling biological tissues. Figure 9.1a, b shows examples of a simple series system, such as the epidermis with an underlying dermis and subcutaneous fat layer, and a more complex arrangement in which both series and parallel flow paths exist.

If heat generation and blood flow to the layers are negligible, the macroscopic energy equation reduces to $\dot{Q}_S = 0$. If $\dot{Q}_x$ is the rate at which heat flows into the composite system at $x = 0$, then in the steady state $\dot{Q}_x$ must also equal the rate that heat flows out at the other end. If we were to consider each layer as a separate

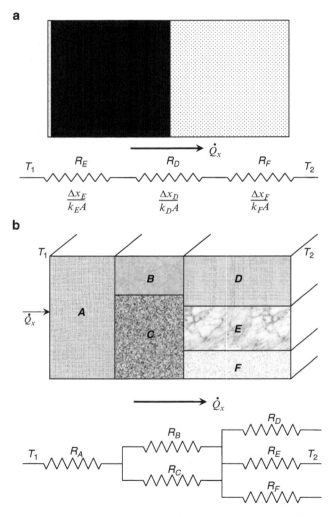

Fig. 9.1 (a) Heat conduction through the skin, consisting of a thin epidermis (E) and much thicker dermis (D), and subcutaneous fat (F), with an equivalent electrical network. (b) Heat conduction in a more complex structure with both series and parallel flow paths and the equivalent electrical network

system, the heat flow into that layer must equal the heat flow out of the same layer. Therefore, an analogous resistance network can be constructed to represent the heat conduction patterns, and the network can be solved to calculate an overall heat flow by direct analogy to the Kirchhoff laws that are used for electrical circuits. Thus, the heat flow through the skin and fat composite system must be identical for every layer since there is no storage of energy for a steady-state process.

$$\dot{Q}_x = -k_E A \frac{T_a - T_1}{\Delta x_E} = -k_D A \frac{T_b - T_a}{\Delta x_D} = -k_F A \frac{T_2 - T_b}{\Delta x_F}$$

$$= \frac{T_1 - T_2}{\Delta x_E/k_E A + \Delta x_D/k_D A + \Delta x_F/k_F A}. \tag{9.7}$$

The flow of heat can be written in terms of the overall applied temperature difference and the net thermal resistance of the composite structure.

$$\dot{Q}_x = \frac{\Delta T_{\text{overall}}}{\sum \Re_T}. \tag{9.8}$$

Example 9.3.1.1 Heat Flux Through Skin.
It is desired to know the heat flux that must be applied to the skin and fat structure shown in Fig. 8.7a to maintain a steady temperature difference of 20 K across the entire structure. The thicknesses of the layers are: epidermis -80 μm, dermis -2 mm, and fat -3 mm. What are the temperature drops across each layer under these conditions?

Solution. *Initial considerations:* The goal is to perform an analysis to determine the magnitude of steady-state heat flux that must be supplied through the physiological structure to support the specified temperature drop between opposing boundaries.

System definition and environmental interactions: The system is defined as a serial layering of materials as illustrated in Fig. 9.1a, including its interaction with the environment in terms of the heat flow at the skin surface.

Apprising the problem to identify governing equations: The following assumptions are made to define conditions for application of the conservation laws, constitutive equations, and boundary conditions.

1. *The heat flow through the skin is one-dimensional along a Cartesian coordinate, x.*
2. *The process is at steady state for the period of analysis.*
3. *The base temperature deep within the tissue below the fat and into which the heat flows is 37°C.*
4. *The material properties are not dependent on temperature, and although the temperatures are high enough to cause tissue damage, those effects can be neglected.*

5. *The applied heat flux comes from a source that has adequate power to maintain the temperature differential across the tissues.*
6. *There is perfect thermal contact between all of the tissue layers with no interfacial temperature drop.*

Analysis: The steady-state heat flux is based on (9.7) with the heat flow written per unit cross-sectional area.

$$\frac{\dot{Q}_x}{A} = q_x = \frac{T_1 - T_2}{\Delta x_E/k_E + \Delta x_D/k_D + \Delta x_F/k_F}.$$

The values for thermal properties are given in Appendix C:
$k_E = 0.209$ (W/(m K)); $k_D = 0.30$ (W/(m K)); $k_F = 0.21$ (W/(m K))
Substituting the given constitutive and property values into the heat flux equation, it follows that:

$$q_x = \frac{20(K)}{(0.080/0.209) + (2.0/0.30) + 3.0/0.21 \left(\frac{m \cdot 10^{-3}}{W/(mK)}\right)} = 937 \frac{W}{m^2}.$$

The percentage of the total thermal resistance per unit cross-sectional flow area that occurs in each conducting layer are calculated from the denominator of the above equation.

$$\Re_{TE} = 0.383 \times 10^{-3} \text{ (K/W)}; \; \Re_{TD} = 6.67 \times 10^{-3} \text{ (K/W)};$$
$$\Re_{TF} = 14.3 \times 10^{-3} \text{ (K/W)},$$

$$\Re_T = \Re_{TE} + \Re_{TD} + \Re_{TF} = 21.3 \times 10^{-3} (\text{K/W}),$$

$$\Re_{TE}/\Re_T = 1.8\%; \; \Re_{TD}/\Re_T = 31.2\%; \; \Re_{TF}/\Re_T = 67.0\%.$$

As shown in (9.7), the temperature drop across each material is proportional to the percent of the total thermal resistance for the series composite. The temperature drops across each material are calculated as:
$\Delta T_E = 0.36$ K; $\Delta T_D = 6.25$ K; $\Delta T_F = 13.39$ K.

Examining and interpreting the results: Nearly a kilowatt per square meter must be applied to hold a 20 K temperature difference across the skin and fat combined. Compare this with the normal heat flux through the skin of 30–60 W/m² resulting from basal metabolic heat loss, which is one and one-half orders of magnitude less. As would be anticipated intuitively, the largest component of the temperature drop occurs across the fat layer because of its higher thermal resistance. Although the fat and epidermis have equivalent thermal conductivities, only a minor portion of the total temperature drop occurs across the epidermis because of its very small thickness.

9.3 Steady-State Applications of the Macroscopic Energy Balance

Additional comments: The boundary conditions assumed for this problem are expressed as defined temperatures. Other types of boundary conditions are frequently encountered with biological systems, such as convection with a flowing fluid or a heat flux associated with exposure to a radiation source. Developing a model that embodies these types of boundary conditions is discussed and illustrated in other text locations and problems. It is important to learn to work with all three primary types of thermal boundary conditions: temperature, convection, and heat flux.

Example 9.3.1.2 Heat Loss in a Heart Lung Machine.

The heat exchanger in a heart lung machine is constructed so that blood flows through an array of tubes with controlled temperature water circulated through an outer exterior shell. Although the temperature of the blood varies as it passes through the heat exchanger, during the cool down cycle when a patient is being brought to a state of hypothermia in preparation for heart surgery, the average temperature difference between the blood and water within the heat exchanger is targeted to be 17°C. The effectiveness of the heat exchanger is dependent on a number of factors including the convective flow coefficients between the blood and water and the adjacent walls of the tube, plus the thermal conductivity and geometry of the tube. Determine the average heat flow per unit surface area of the heat exchanger for the following specific operating conditions.

The inner diameter of the heat exchanger tubing is 10 mm, and the wall thickness is 1 mm. The material of the tubing is 304 stainless steel, which has a thermal conductivity of 14.9 W/m K. The convective heat transfer coefficient on the interior of the tube is 750 W/m^2 K and on the exterior is 2,250 W/m^2 K. As a further design consideration, it is proposed to alter the design of the heat exchanger by substituting tubing fabricated from treated aluminum alloy with a thermal conductivity of 175 W/m K. How do you predict this change will affect the performance of the heat exchanger?

Solution. *Initial considerations:* The goal is to explore different options for achieving a target level of performance of a heat exchanger designed to be an integral component of a heart lung machine. In this phase of the analysis, it is only necessary to consider the local thermal performance of a single tube within the heat exchanger. A more comprehensive analysis would involve integrating over the entire heat exchange surface of the device.

System definition and environmental interactions: The system that embodies the process of interest is the tubing, and the interactions with the environment that drive the process are the interior and exterior surface convective heat transfers with blood and water. For purposes of this analysis, the system is considered as closed, with the only energy flows of interest being the convection with blood and water as coupled via diffusion through the tube wall. Note that the tube could also be defined as an open system in which there is a change of enthalpy of the blood as it enters and leaves a defined length of the tube as a result of the convective exchange with the tube wall. If we define the system this way, we will lose the ability to analyze the convection between the blood and tube wall as it flows through the exchanger. This type of analysis will be addressed in a later section. The system diagram is shown in Fig. 9.2. This problem involves a system with two convection boundary conditions in contrast to the previous example for which the boundary conditions were specified by defined temperatures.

Apprising the problem to identify governing equations: The following assumptions are made to define conditions for application of the conservation laws, constitutive equations, and boundary conditions.

1. *The heat flow through the tube is one-dimensional along a radial coordinate, r.*
2. *The internal and external convective heat exchanges are also one-dimensional in the radial direction.*
3. *The heat exchanger is operating at a steady-state condition.*
4. *The temperature of the blood will change continuously through the heat exchanger as it flows from the inlet to the exit. Nonetheless, a single representative temperature drop between the blood and water can be analyzed.*
5. *If the water is well circulated and mixed in the shell that encloses the tubing, which is a common design objective for a heat exchanger, then its temperature can be assumed to be uniform over the entire exterior tubing surface.*
6. *A unit length of the tubing along the blood flow direction is considered.*

Analysis: Since the system is modeled as being at steady state, there is no change in the energy stored during the process. Therefore the conservation of energy dictates that the sum of the heat flows into the tube is equal to the heat flow out. For this type of process, an equivalent electrical resistance network can be used to represent the serial flow of heat through the system (Fig. 9.3).

The radial equivalent for the flux equation (9.8) can be applied to describe the temperature drop and overall thermal resistance.

$$\dot{Q}_r = \frac{\Delta T_{\text{overall}}}{\sum \Re_T}.$$

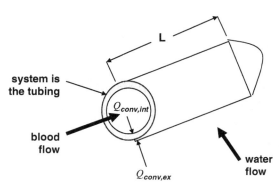

Fig. 9.2 A representative length of tubing from a heart lung machine heat exchanger with blood flowing internally and water at a controlled temperature circulated externally

9.3 Steady-State Applications of the Macroscopic Energy Balance

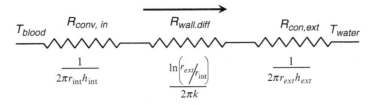

Fig. 9.3 Equivalent electrical resistance network for steady-state radial heat flow through a tube wall with internal and external convection boundary conditions

The total thermal network resistance has three terms arranged in series representing the internal convection, wall diffusion, and external convection.

$$\sum_i \Re_{T,i} = \Re_{T,\text{conv,int}} + \Re_{T,\text{wall,diff}} + \Re_{T,\text{conv,ext}}$$

The constitutive relation can be substituted directly for each of the equivalent resistance terms per unit tube length in the network. These relations from (8.9) and (8.69) are as follows:

$$\Re_{T,\text{conv,int}} = \frac{1}{2\pi r_{\text{int}} h_{\text{int}}},$$

$$\Re_{T,\text{wall,diff}} = \frac{\ln(r_{\text{ext}}/r_{\text{int}})}{2\pi k},$$

$$\Re_{T,\text{conv,ext}} = \frac{1}{2\pi r_{\text{ext}} h_{\text{ext}}}.$$

The overall expression for the heat flow per unit length is:

$$\frac{\dot{Q}_r}{L} = \frac{T_{\text{int}} - T_{\text{ext}}}{(1/2\pi r_{\text{int}} h_{\text{int}}) + (\ln(r_{\text{ext}}/r_{\text{int}})/2\pi k) + 1/2\pi r_{\text{ext}} h_{\text{ext}}}.$$

Hence,

$$\frac{\dot{Q}_r}{L} = \frac{17 K}{\frac{1}{2\pi (0.01\,\text{m}) \cdot (175\,\text{W/m}^2 \cdot K)} + \frac{\ln(0.011/0.01)}{2\pi (14.9\,\text{W/m} \cdot K)} + \frac{1}{2\pi (0.011\,\text{m}) \cdot (2,250\,\text{W/m}^2 \cdot K)}} = 593 \frac{\text{W}}{\text{m}}.$$

Examining and interpreting the results: When aluminum is substituted for stainless steel as the tube material, the only alteration in the thermal performance is the magnitude of the conduction resistance of the wall. Therefore, the thermal effect of the new material will be dependent on the relative magnitudes of the resistance terms. These are calculated from the given data as follows:

$$\Re_{T,\text{conv,int}} = 0.0212 \, \frac{\text{mK}}{\text{W}}; \Re_{T,\text{wall,ss}} = 0.00102 \, \frac{\text{mK}}{\text{W}};$$

$$\Re_{T,\text{wall,al}} = 0.000067 \, \frac{\text{mK}}{\text{W}}; \; \Re_{T,\text{conv,ext}} = 0.00643 \, \frac{\text{mK}}{\text{W}}.$$

The percentage of overall thermal resistance attributable to conduction through the stainless steel wall is about 3.6%. When the material is changed to aluminum, this drops to 0.3%. In either case the conduction resistance of the wall is a small fraction of the total thermal resistance to the flow of heat between the blood and water. Therefore, there is only a marginal benefit in the thermal performance for aluminum tubing, with a new value of $\dot{Q}_r/L = 613 \, \text{W/m}$.

Additional comments: This problem illustrates the fact that for many practical cases in which there is a convection boundary condition acting in concert with conduction through an adjacent solid, the convective resistance can dominate the heat flow process. We will evaluate this effect more fully when we study the behavior of transient systems in which the temperature of the conducting material changes with time.

9.3.2 Heat Transfer Coefficients

In Chap. 8, we provided relationships for the dimensionless heat transfer coefficient, Nu, obtained for objects having various geometries studied under many different experimental conditions. Taking a solid object as our system, heat loss from the various surfaces of the object is governed by Newton's law of cooling. For steady-state external flow problems where there are no internal conduits entering or leaving the system and N different surfaces available for energy exchange with the surroundings, the macroscopic energy equation becomes:

$$0 = \sum_{n=1}^{N} h_n S_n (T_\infty - T_{Sn}) + \dot{Q}_{\text{gen}}. \tag{9.9}$$

For surface n, S_n is the surface area, h_n is the convective heat transfer coefficient, and T_{Sn} is the temperature, which is assumed to be uniform. If fluid flows through a solid system via a single conduit in which heat transfer at the outside surface obeys Newton's law of cooling, the surface temperature is constant, and no heat generation occurs in the system, the macroscopic energy equation becomes:

$$0 = wc_p (T_{m,\text{in}} - T_{m,\text{out}}) + hS(T_\infty - T_S). \tag{9.10}$$

In many situations, the surface temperature varies with position along the surface, so these equations must be modified to account for this variation. Under

9.3 Steady-State Applications of the Macroscopic Energy Balance

these circumstances (9.10) would be modified to account for changes in heat flux with position on the surface of the system:

$$0 = wc_p(T_{m,in} - T_{m,out}) + \int_S h(T_\infty - T_S)dS. \tag{9.11}$$

In the sections which follow, we provide examples of how these expressions are applied in practical problems.

Example 9.3.2.1 Hot Wire Anemometer.

A hot wire anemometer is used to measure air velocity in respiratory studies. The device consists of a small diameter wire that is heated electrically. The electrical power supplied to the wire is measured along with the temperature of the wire. The greater the air velocity, the lower will be the wire temperature for a given power. If the power supplied to the wire and the wire temperature are measured, how might we compute the fluid velocity?

Solution. The power supplied to the wire is converted to heat, so $\dot{Q}_{gen} = I^2 \Re_e$, and in the steady-state this heat is lost by convection at the surface according to Newton's law of cooling. Conservation of energy in this case is:

$$h\pi dL[T - T_\infty] = I^2 \Re_e,$$

where πdL is the surface area of the wire. The heat transfer coefficient for flow past a cylinder is given by (8.47):

$$Nu_d = \frac{hd}{k_f} = CRe_d^n Pr^{0.33}.$$

Substituting this expression for h in Newton's law of cooling, applying the definition of the Reynolds number, and solving for the velocity, we have:

$$v = \frac{\nu}{d} \left[\frac{I^2 \Re_e}{C\pi L k_f (T - T_\infty) Pr^{.33}} \right]^{1/n}.$$

All of the parameters on the right side of the equation are either measured or known except n and C, which must be determined from the value of the Reynolds number. Therefore, since we cannot compute v directly, an iterative scheme must be used in which an initial guess is made for the values of C and n, and these are used to calculate v. This value of v is then applied to calculate Re_d, and the values assumed for C and n are checked to see whether they are appropriate for the value of Re_d. If these are appropriate, we have found the correct velocity, if not we must try a new set of C and n from the table accompanying (8.47) that corresponds to the Reynolds number just computed and repeat the iterative analysis.

Example 9.3.2.2 Wind Chill Factor.
The wind chill factor (WCF) is used to describe quantitatively the feeling of how cold an environment is. It is based on the combined effects of air temperature and air velocity on cooling the surface of a human. Although the commonly used WCF is based on a complex empirical formula, it should be possible to calculate the difference in WCF for varying environmental conditions based on a simple and fundamental convection analysis. As the air velocity increases, the convective heat transfer also increases, making the environmental temperature feel colder (assuming that air temperature is lower than skin temperature.) The WCF compensates for the larger air velocity by determining a lower temperature that would have an equivalent convective heat transfer for an air velocity of zero.

Your job on this problem is to calculate how the WCF changes under different wind velocities. A thermometer has measured air temperature at 0°C, which has remained constant for a period of 10 h. Then the wind velocity increases from 0 to 10 miles/h. How much will the WCF change? Some relevant properties may include:

Property at 0°C	Value	Units
Air thermal diffusivity, $\alpha = k/\rho c$	18.6×10^{-6}	m^2/s
Air kinematic viscosity, $\nu = \mu/\rho$	13.3×10^{-6}	m^2/s
Prandtl No., Pr	0.709	
Air thermal conductivity, k	0.024	W/m K

Solution. *Initial considerations:* The wind chill calculation is based on the assumption that the subjective sensation of the relative coldness of an environment is based on the rate at which heat is lost from the body surface. The rate of loss of heat by convection to the surroundings depends primarily on two factors: the temperature of the air and the convective heat transfer coefficient between the air and the clothing or skin surface. The heat loss can be increased by either lowering the temperature or increasing the convection. The wind chill temperature will be lower than the actual temperature if the air velocity is greater than what would occur by natural convection when the wind is not blowing. For this problem, we will need to calculate the heat loss for forced convection at 0°C and then determine how low the air temperature would need to be for natural convection conditions to produce the same rate of heat transfer. This latter temperature should define the wind chill value for an air velocity of 10 miles/h at 0°C. The natural convection calculation will be somewhat challenging since it involves what is called an *inverse solution*. We will start with a known value for the rate of heat transfer as determined by the forced convection analysis. Then we will need to work backwards to determine what

9.3 Steady-State Applications of the Macroscopic Energy Balance

temperature and natural convection coefficient combined to produce this heat transfer. The complicating factor is that the natural convection coefficient is a function of the air temperature; so our solution procedure will need to be iterative.

System definition and environmental interactions: The process on which this problem is focused is the convective heat transfer between the human and the surrounding air. Therefore, the system should consist of the human substrate, including its surface. The system geometries can be assumed to be vertical to accommodate the greatest natural convection flow and to be parallel to the direction the wind is blowing; so the forced convection can be represented as external flow over a flat plate. The only environmental interaction of consequence to this problem is the convective heat loss from the human to the environment.

Apprising the problem to identify governing equations: We need to identify the appropriate natural and forced convection correlation equations that will be applied for heat transfer at the human/air interface. It will be necessary to analyze the flow domain to determine whether it is laminar or turbulent before we can select correlation equations appropriate for this problem.

Analysis: Both the forced and natural convection flows will be external. We can choose characteristic dimensions of the body trunk for these processes: 0.3 m laterally for forced convection and 0.5 m vertically for natural convection. The Reynolds number for forced convection is:

$$Re_L = \frac{\rho v L}{\mu} = \frac{vL}{\nu} = \frac{10\,\text{mi/h} \cdot 1.61 \times 10^3\,\text{m/mi} \cdot 1/3{,}600\,\text{s/h} \cdot 0.3\,\text{m}}{13.3 \times 10^{-6}\,\text{m}^2/\text{s}}$$
$$= 9.32 \times 10^4.$$

This Reynolds number is substantially less than the transition value of 5×10^5; so a laminar flow correlation is appropriate with (9.12).

$$\bar{Nu}_L = 0.664 Re_L^{0.5} Pr^{0.33} = 0.664 \cdot \sqrt{9.32 \times 10^4} \cdot 0.709^{\frac{1}{3}} = 181, \qquad (9.12)$$

$$Pr \geq 0.6.$$

The convective heat transfer coefficient is calculated from the Nusselt number with (8.19).

$$\bar{h}_f = \frac{Nu_L k_f}{L} = \frac{181 \times 0.024\,\text{W/mK}}{0.3\,\text{m}} = 15.2\,\text{W/m}^2\,\text{K}.$$

In order to calculate what the forced convection heat flux will be, it is necessary to choose a value for the temperature of the clothing or skin surface. We will select 4°C and revisit this choice later if necessary. The heat flux associated with forced convection then is 60.9 W/m².

$$q_f = \bar{h}_f(T_s - T_a) = 15.2\,\text{W/m}^2\,\text{K} \cdot (4-0)\,\text{K} = 60.9\,\text{W/m}^2.$$

The next step is to determine what natural convection conditions will produce the same heat flux. There are two correlations for natural convection on a flat plate: Equations (9.13) and (9.14). We can compare the calculation of the natural convection coefficient by the following two equations:

$$\bar{Nu}_L = \left\{ 0.825 + \frac{0.387 Ra_L^{1/6}}{\left[1 + (0.492/Pr)^{9/16}\right]^{8/27}} \right\}^2, \qquad (9.13)$$

$$\bar{Nu}_L = 0.59 Ra_L^{1/4}. \qquad (9.14)$$

First the Rayleigh number must be computed, and it will be at a lower temperature because of the effect of the WCF. The properties k, α, and ν for air are all temperature dependent. We can select an initial estimate of the wind chill temperature as $-10°C$ and interpolate the values of the air properties from Appendix C for the film temperature

$$T_f = \frac{T_s + T_a}{2} = \frac{4 - 10}{2} = -3°C = 270.15\,K,$$

$$Ra_L = Gr_L Pr = \frac{g\beta(T_s - T_a)L^3}{\alpha\nu} = \frac{9.8\,m/s^2 \cdot 1/270.15\,K \cdot (4 - 0)\,K \cdot 0.5^3\,m^3}{1.85 \times 10^{-5}\,m^2/s \cdot 1.32 \times 10^{-5}\,m^2/s}$$
$$= 2.59 \times 10^8.$$

This value is in the laminar range, so both correlation equations are applicable.

$$\bar{Nu}_L = \left\{ 0.825 + \frac{0.387 \cdot \left[2.59 \times 10^8\right]^{1/6}}{\left[1 + (0.492/0.709)^{9/16}\right]^{8/27}} \right\} = 74.9,$$

$$\bar{Nu}_L = 0.59 \cdot \left(2.59 \times 10^8\right)^{1/4} = 73.0.$$

There is no significant difference in these two Nusselt numbers, from which the convective heat transfer coefficient can be calculated directly.

$$\bar{h}_n = \frac{Nu_L k_f}{L} = \frac{74.9 \cdot 0.024\,W/mK}{0.5\,m} = 3.58\,W/m^2\,K.$$

We can now check our assumption of the wind chill temperature of $-10°C$ by applying the same heat flux as determined for forced convection at $0°C$ in combination with the natural convection coefficient to see whether it predicts the same wind chill temperature.

9.3 Steady-State Applications of the Macroscopic Energy Balance

$$T_\text{a} = T_\text{s} - \frac{q_\text{f}}{h_n} = 4°\text{C} - \frac{60.9\,\text{W/m}^2}{3.58\,\text{W/m}^2} = -13.03°\text{C}.$$

Our assumed wind chill temperature was higher than we calculated. An intermediate value is likely to be correct, and after iterative calculations we find that an air temperature of $-12.38°$C yields forced and natural convection heat fluxes that are equal.

Examining and interpreting the results: We should check with standard wind chill tables to test the accuracy of our calculation. Reference to these tables (which in the USA are presented in units of temperature in °F and wind velocity in miles per hour) shows that for 32°F and 10 mph the wind chill temperature is 18°F $= -7.8°$C. This temperature is off by quite a bit from our calculated value; so it behooves us to explore a little more to determine the reason for the difference. One useful piece of information is the algorithm by which wind chill is calculated, which is:

$$\text{wind chill}(°\text{F}) = 35.74 + 0.6215 \cdot T(°\text{F}) - 35.75 \cdot v^{0.16}(\text{mph}) + 0.4275 \cdot T(°\text{F})$$
$$\times\, v^{0.16}(\text{mph}).$$

This expression is obviously derived by being fit to empirical data in comparison to our analysis that is based on convective heat transfer principles (note, however, that our convective coefficients are also derived by a fit to empirical data). We have only calculated a single point of comparison. It would be instructive in evaluating our method to investigate some other wind velocities to see how they compare to the empirical data. Once we have a computational program set up to make the initial calculations, it is relatively simple to insert alternate wind velocities. At 20 and 30 mph, the wind chill temperatures are $7°$F $= -13.9°$C and $1°$F $= -17.2°$C. The corresponding calculated numbers are $-17.5°$C and $-21.3°$C, respectively. All of these values are lower than the empirical data, which probably indicates that there is a systematic problem with our calculations. The most obvious point in our calculations to consider changing is the value chosen for the surface temperature from which heat is lost to the surrounding air. We started by assuming it is 4°C, and it is possible to iterate our calculations to look at other values. Table 9.1 below presents the results of these further calculations.

Table 9.1 Comparison of computed wind chill temperatures (°C) based on convective heat transfer analysis and empirical data for air temperature of 0°C

Wind velocity (mph)	10	20	30
Empirical wind chill	−7.8	−13.9	−17.2
T_s (2°C)	−7.4	−8.4	−12.2
T_s (3°C)	−10.1	−13.1	−16.9
T_s (4°C)	−13.0	−17.5	−21.3
T_s (5°C)	−14.7	−21.7	−25.4

As we examine and evaluate these results, a number of trends are apparent. Perhaps most obviously, our computational method does not exactly match the empirical algorithm for wind chill. Nonetheless, the changes with wind velocity are in the correct direction. The calculated wind chill values are very sensitive to the temperature chosen for the heat loss surface. Among the values we have explored, 3°C gives the best match, although the sensitivity to wind velocity is less than the empirical data. In addition, we have not performed calculations for air temperatures other than 0°C, and the 3°C may not be uniformly best for other air temperatures. We could readily perform these calculations, but in the context of presenting an example problem, the present state of the analysis is likely complete enough to achieve the intended learning benefit.

Additional comments: One value of going through this rather lengthy example problem is to realize that it is necessary to be continually thinking about the meaning and interpretation of the results we are obtaining while in the process of performing calculations. In this case, some of our initial assumptions were off the mark and needed to be adjusted. In addition, we needed to consult outside reference materials, readily available on the web, to evaluate our results.

9.3.3 Convective Heat Transport

There are many problems in bioheat transfer where convective heat transport is the dominant mode of energy exchange between a system and its surroundings. If energy exchange is steady without heat generation in the system, (9.5) reduces to:

$$0 = \sum_{i=1}^{N_{\text{inlets}}} w_i c_{pi} T_{m,i} - \sum_{j=1}^{N_{\text{outlets}}} w_j c_{pj} T_{m,j} + \dot{Q}_S. \tag{9.15}$$

For steady flow through a system with a single inlet and a single outlet, the mass flow rate will be constant, and if temperature variations are not large, the heat capacity per unit mass will also be constant. Under these circumstances, the energy equation becomes:

$$0 = w c_p (T_{m,\text{in}} - T_{m,\text{out}}) + \dot{Q}_S. \tag{9.16}$$

This expression is used extensively in heat exchanger design.

> **Example 9.3.3.1 Energy Exchange in a Blood Vessel.**
> Blood with a mean temperature of 37°C is cooled to 35°C as it passes through a blood vessel in skin. Flow rate is 10 ml/min. What is the rate of heat loss to the skin? Blood density is 1,050 kg/m^3 and specific heat is 3,740 J kg^{-1} K^{-1}.

9.3 Steady-State Applications of the Macroscopic Energy Balance

Solution. This problem involves a direct application of the foregoing equation in which all of the information requisite to making a calculation is given. The system for analysis is the blood vessel through which the blood flows and with which there is convective heat transfer with the blood. In the steady state, the rate of heat added through the walls of the vessel plus the rate enthalpy enters at the inlet minus the rate enthalpy leaves at the outlet of the vessel must sum to zero:

$$0 = wc_p(T_{m,in} - T_{m,out}) + \dot{Q}_{blood}.$$

In this case, the rate at which heat is added to the blood is negative and is equal to heat lost to the skin:

$$-\dot{Q}_{blood} = \dot{Q}_{skin} = wc_p(T_{m,in} - T_{m,out})$$

$$\dot{Q}_{skin} = \left(1{,}050\frac{kg}{m^3}\right)\left(10\frac{ml}{min}\right)\left(3{,}740\frac{J}{kg°K}\right)(2°K)\left(10^{-6}\frac{m^3}{ml}\right)\left(\frac{1min}{60s}\right) = 1.31 W.$$

Example 9.3.3.2 Mixing of Streams with Different Temperatures.
Derive an expression for the temperature of fluid flowing in a vessel formed by the merging of two inlet vessels with streams at different temperatures, flow rates, and specific heats. Assume no energy is exchanged through the walls of the vessels.

Solution. The relevant system is the vessel with two inlets and one outlet. The vessel is insulated, so no heat transfer occurs. The only interactions with the environment are the enthalpy flows across the boundary. We begin with (9.15) for two inlet streams 1 and 2 and a single outlet stream and with $\dot{Q}_s = 0$:

$$0 = w_1 c_{p1} T_{m1} + w_2 c_{p2} T_{m2} - w_{out} c_{p,out} T_{m,out},$$

which can be rearranged to solve for the outlet mean temperature, recognizing that the mass flow rate out of the system must equal the sum of the mass flow rates into the system:

$$T_{m,out} = \frac{w_1 c_{p1} T_{m1} + w_2 c_{p2} T_{m2}}{(w_1 + w_2) c_{p,out}}.$$

If the specific heat of each of the three fluids is the same, then the mean temperature of the merged streams is the mass flow-weighted average of the temperatures of the two merging streams.

9.3.4 Biomedical Applications of Thermal Radiation

Thermal radiation is an important factor in many applications involving living systems. Although it is not possible to present a comprehensive review of these applications, many will be discussed to illustrate the utility of radiation heat transfer analysis.

9.3.4.1 Steady-State Radiation Exchange

> **Example Problem 9.3.4.1.1 Radiation in Thermal Equilibrium.**
> A person dressed in black clothing and a white cat have been together in a large interior room for a long period of time. The net radiation flux from the isothermal walls, floor, and ceiling of the room onto both the person and cat is 500 W/m^2. The person and the cat absorb incident radiation at rates of 400 and 200 W/m^2, respectively. Given this information, find the coefficients of absorption, emissivities, net heat fluxes, emissive powers, and temperatures of the person and cat.

Solution. *Initial considerations*: A key issue relating to this problem is that the person and cat have been in the room for a long enough time to come to thermal equilibrium with their environment. All initial transients are completed so that the net heat fluxes from both the person and cat will be zero. Further, since the surface area of the room is much larger than that of either the person or cat, it will behave as if it was black. All incident radiation onto room surfaces will eventually be absorbed, even if multiple partial reflection iterations are required. The shape factors between the person and cat are very small so that they experience no significant radiation exchange. When the person and cat first enter the room, their temperatures will adjust under the action of the radiation flux from the room until a condition of equilibrium is achieved. For our present analysis considerations, modes of heat transfer other than radiation will be neglected. In most real situations, convection with air in the room will also exert an influence on the final equilibrium state. Another simplifying assumption is that the internal metabolism of the person and cat is neglected, so there is no net radiation loss in the living animals. If only radiation interactions are allowed, then the equilibrium state must be characterized by equal temperatures for the room, person, and cat so that there remains no potential difference for further heat transfer. This is the equilibrium condition that would exist if the person and cat were no longer alive.

9.3 Steady-State Applications of the Macroscopic Energy Balance

System definition and environmental interactions: The overall system consists of three surfaces: the room, person, and cat. Since we are required to perform calculations on the properties and heat exchanges for the person and cat, each of these must also be identified as an individual system for analysis. See the figure for layout features.

Apprising the problem to identify governing equations: The conservation of energy is applied for both the person and the cat.

$$0 = \dot{Q}_s = \dot{Q}_{in} - \dot{Q}_{out} = G - E.$$

The surface temperatures for the room, person, and cat are all identical.

$$T_r = T_{pers} = T_{cat}.$$

The surface area for the room is much larger than for the person and cat so that all incident radiation is eventually absorbed making it effectively black.

$$A_r \gg A_{pers}, \ A_r \gg A_{cat}, \ \alpha_r = \varepsilon_r \approx 1.$$

Analysis: The radiation flux incident on the person and cat is the same as the emissive power from the room, which can be applied to calculate the temperature of the room and therefore also the person and cat.

$$G = E_r = \sigma T_r^4,$$

$$T_r = \left(\frac{G}{\sigma}\right)^{1/4} = \left(\frac{500 \text{ W/m}^2}{5.67 \times 10^{-8} \text{ W/m}^2 \text{ K}^4}\right)^{1/4} = 306 \text{ K} = T_{pers} = T_{cat}.$$

The fraction of incident radiation absorbed is defined as the coefficient of absorption, which is assumed to equal the emissivity. Therefore, for the person and cat, the values of this property are calculated as:

$$\alpha_{\text{pers}} = \frac{G_{\text{abs,pers}}}{G} = \frac{400 \text{ W/m}^2}{500 \text{ W/m}^2} = 0.8, \ \alpha_{\text{cat}} = \frac{G_{\text{abs,cat}}}{G} = \frac{200 \text{ W/m}^2}{500 \text{ W/m}^2} = 0.4.$$

The conservation of energy requires that at equilibrium the emissivity equals the incident rate of radiation. Since the values of irradiance, G, are given:

$$E_{\text{pers}} = G_{\text{abs,pers}} = 400 \text{ W/m}^2, \ E_{\text{cat}} = G_{\text{abs,cat}} = 200 \text{ W/m}^2.$$

Examining and interpreting the results: Even though the person and cat have different surface radiation properties, at equilibrium they come to identical temperatures. The difference in their properties make a more significant difference during the transient period of the heat transfer process during which equilibrium is achieved. The lower absorptivity (and therefore higher reflectivity) for the cat means that it will have a larger resistance to radiation and will therefore take longer to equilibrate with the room than the person.

Additional comments: Calculating the transient portion of the heat transfer process would involve considerably greater complexity since the energy storage term in the conservation of energy equation will not be zero, by definition. For these conditions, the radiation gain and loss on the boundary will provide flux style boundary conditions for a transient internal conduction problem.

Example Problem 9.3.4.1.2.

Three long flat surfaces are arranged as an equilateral triangular. Surface 1 is at a temperature of 500 K; surface 2 is at 350 K; surface 3 is insulated and therefore at a floating temperature. The emissivities of the three surfaces are 0.25, 0.75, and 0.5, respectively. What is the heat flux from surface 1 and the temperature of surface 3?

Solution. *Initial considerations*: Since the plates are long, end effects will be neglected. The process is assumed to have reached a steady state. Surfaces 1 and 2 will have communicating heat fluxes with the underlying substrate. These can be neglected for the present analysis with the exception of assuming that the heat flow is adequate to maintain the surfaces at the indicated constant temperatures.

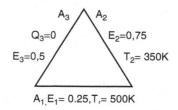

9.3 Steady-State Applications of the Macroscopic Energy Balance

System definition and environmental interactions: The system consists of the three radiation communicating surfaces 1 and 2 which have net heat fluxes due to conduction in the underlying material, and 3 which is insulated and therefore has no net flux.

Apprising the problem to identify governing equations: The conservation of energy can be represented by an equivalent electrical diagram as shown in Fig. 8.24. The steady-state heat flow between surfaces 1 and 2 is given by (8.95), which for $A_1 = A_2$ and, because of geometric symmetry, $F_{13} = F_{23} = F_{12} = 0.5$, becomes:

$$\dot{Q}_{1 \to 2} = \frac{\sigma A_1 \left(T_1^4 - T_2^4\right)}{\dfrac{1}{\varepsilon_1} + \dfrac{1}{\varepsilon_2} - 2 + \dfrac{1}{(1/2) + (1/4)}} = \frac{\sigma A_1 \left(T_1^4 - T_2^4\right)}{\dfrac{1}{\varepsilon_1} + \dfrac{1}{\varepsilon_2} - \dfrac{2}{3}}.$$

The area of each side is given by the common length L multiplied by the very large depth, D.

Analysis: The heat flow can be calculated as normalized to the area of one of the equal sides, since we are not given the specific dimensions. The property values are substituted into this equation as follows:

$$q_{1 \to 2} = \frac{5.67 \times 10^{-8} \text{ W/m}^2 \text{ K}^4 \left(500^4 - 350^4\right) \text{ K}^4}{\dfrac{1}{0.25} + \dfrac{1}{0.75} - \dfrac{2}{3}} = 577 \text{ W/m}^2.$$

The temperature of the insulating surface is found by determining the value of the potential at the network node corresponding to surface 3. The potential drops across circuit resistances are determined as the product of the flow and resistance. For example, the effective radiation potentials for surfaces 1 and 2 involve a drop from the black body emissive powers owing to the imperfect surface radiation properties. These potentials are the radiosities, J_1 and J_2. The conservation of energy can be written for surface 3, which is at steady state and has no net heat flow. Thus, the radiation inward to surface 3 must equal the radiation outward.

$$q_3 = 0 = q_{13} + q_{12} = \frac{E_{b3} - J_1}{\dfrac{1}{F_{13}}} + \frac{E_{b3} - J_2}{\dfrac{1}{F_{23}}}.$$

For equal shape factors:

$$E_{b,3} = \frac{J_1 + J_2}{2}.$$

The radiosities J_1 and J_2 are determined by writing the conservation of energy for surfaces 1 and 2, respectively.

$$q_1 = \frac{E_{b,1} - J_1}{\dfrac{1-\varepsilon_1}{\varepsilon_1}}, \quad q_2 = \frac{E_{b,2} - J_2}{\dfrac{1-\varepsilon_2}{\varepsilon_2}},$$

which can be solved for the radiosities as:

$$J_1 = E_{b,1} - q_1 \frac{1-\varepsilon_1}{\varepsilon_1} = 5.67 \times 10^{-8} \cdot 500^4 - 577 \frac{1-0.25}{0.25} = 1,813 \text{ W/m}^2,$$

$$J_2 = E_{b,2} + q_2 \frac{1-\varepsilon_2}{\varepsilon_2} = 5.67 \times 10^{-8} \cdot 350^4 + 577 \frac{1-0.75}{0.75} = 1,043 \text{ W/m}^2.$$

The black body emissive power of the insulated surface may now be computed, from which the temperature is determined.

$$E_{b,3} = \frac{1,813 + 1,043}{2} = 1,428 \text{ W/m}^2, \quad T_3 = \left(\frac{E_{b,3}}{\sigma}\right)^{1/4} = 398 \text{ K}.$$

Examining and interpreting the results: Note that the problem statement asked for the heat flux from surface 1. Since there is no net heat flux to surface 3 because it is insulated, the total flux from 1 must be exchanged with 2. Even though the system of surfaces is geometrically symmetric, the temperature of the insulated surface is not the average of those of the two heat exchanging surfaces because the emissivities are not also symmetric. The intermediate temperature is closest to that of the surface with the highest emissivity and absorptivity.

9.3.4.2 Environmental Radiation Load on the Human Body

Radiation heat transfer can comprise an important component of the total heat load on a human. For example, on a hot summer day under clear skies, walking from the shade into the sun causes an immediate subjective sensation of being under a much larger thermal burden, and your thermoregulatory system may start to compensate within a few minutes by increasing the rates of blood perfusion to the skin and of sweating. Conversely, during very cold winter months, you will likely experience a subjective difference in body heat loss when sitting close to a closed window (which may have a relatively low interior surface temperature if it is not well insulated from the exterior weather) as compared with sitting in the middle of the room. In both cases, the air temperature for convective effects will be approximately equivalent, but the radiation exchange with a cold wall will vary with separation, which will alter the shape factor between your body and the wall. You will probably feel a perceptible heat loss when seated close to a large and cold surface.

9.3 Steady-State Applications of the Macroscopic Energy Balance 581

The primary component of environmental radiation loading on the body is due to the solar flux from the sun. The solar radiation entering the earth's atmosphere is about 340 W/m². As radiation passes through the atmosphere to the earth's surface, there is some attenuation owing to interaction with gases and particulate matter. The degree of attenuation may vary considerably as a function of global location and transient atmospheric conditions. The converse situation occurs at nighttime when the radiation temperature of the sky can be as low as $-100°C$ on a clear night resulting in a terrestrial heat loss. As during the daytime, atmospheric conditions can have a major effect in reducing atmospheric radiation exchange. Thus, the temperature of objects at ground level may become considerably cooler than the surrounding air temperature if radiation loss to the sky is significant. This phenomenon is illustrated in the following example.

Example Problem 9.3.4.2.1 Simultaneous Convection and Radiation.
On a very warm summer evening, you have decided to set up your $8''$ telescope at a country location to engage in some star gazing. The air temperature is 100°F (37.8°C), and there is no detectable wind velocity. The effective radiation temperature of the sky on a very clear summer night is $-100°C$. Your telescope is a passive device that will eventually come to an equilibrium temperature based on convection with the air balanced by radiation with the sky. What will be the equilibrium temperature? You can assume that it is positioned vertically during observation. In contrast, your skin temperature may not change appreciably because of the capability of your thermoregulatory system to actively modulate the body's interaction with the environment. Under these conditions, the primary thermoregulatory mechanism would be via control of blood perfusion through the skin. Thus, there may be a net heat loss or gain while the skin temperature is maintained at 34°C. The environmental radiation temperature is colder, and the convection temperature is warmer. Because of the very warm temperature, you decided to wear your swim suit with the consequence that there is very little intermediate clothing to modify the thermal interaction between the skin and the environment. What will be the net heat gain or loss from the body, and what will be the convective and radiative contributions? For simplicity of calculation in solving this problem and for your physical comfort, assume that you have set your telescope so that you can be reclined while making your observations.

Solution. *Initial considerations*: This problem deals with convective and radiation inputs from the surroundings to a passive mass (the telescope) and to a living body (the human) having an internal thermoregulatory system. When the telescope reaches its equilibrium temperature, the rate of heat gain by convection will be exactly offset by the rate of heat loss by radiation. For the human, there are many complex features of the thermoregulatory system such as water evaporation from the skin surface (which is affected by the relative humidity of the air), respiratory energy exchange, and others, and we will not attempt to include any of these effects in this analysis. The most simple approach that will still provide useful information is to assume that the convection and radiation processes occur simultaneously and independently.

System definition and environmental interactions: The human system consists of the skin surface that exchanges heat with the environment via natural convection and radiation. Because these are both boundary effects and thermoregulation is assumed to hold the skin temperature constant, we can view the system as being mass-less, with no capacity for energy storage. The lens will be the system for the telescope. Its temperature will be transient, and the convection and radiation boundary conditions will combine to drive that process.

Apprising the problem to identify governing equations: We apply directly the appropriate constitutive equations to calculate the convection and radiation heat transfers. Radiation will be calculated by (8.94) and convection by correlation equations that are dependent on geometry and the range of operating conditions.

The constitutive equation for radiation exchange between the skin or the lens and the sky is given by (8.94). In this application, the emissivity of the sky is 1.0. None of the radiation to it is reflected to the lens; the sky is truly black. Also, the shape factor for both the upward facing skin surface and the lens toward the sky can be taken as 1.0. Thus, the constitutive equation is based on two radiation resistances in series.

$$\dot{Q}_r = \frac{\sigma(T_t^4 - T_s^4)}{\frac{1-\varepsilon_1}{\varepsilon_1 A_1} + \frac{1}{A_1}} = \varepsilon_1 A_1 \sigma (T_t^4 - T_s^4).$$

The emissivity of skin is about 0.95 and of the glass lens about 0.9. The telescope temperature remains to be determined.

Analysis: We will start first with the human since that problem is easier to solve. The skin temperature is specified and does not change; so we know what the thermal driving potentials are for all the heat flows. The air temperature is higher than that of the skin. Thus, the geometry dictates that this is a process of heating the upper horizontal surface of a body. The air will be cooled as it contacts the skin, thereby increasing in density and spilling downward over the sides of the body under action of the gravity field. Equation (8.56) may be applied to this process. First, it is necessary to calculate the Rayleigh number. We can take the characteristic length for the upper body surface as 0.5 m.

$$Ra = Gr_L Pr = \frac{g\beta(T_t - T_a)L^3}{\alpha \nu}$$

$$= \frac{9.8 \, \text{m/s}^2 \cdot 1/310.95 \, \text{K} \cdot (310.95 - 300) \, \text{K} \cdot (0.5)^3 \, \text{m}^3}{2.41 \times 10^{-7} \, \text{m}^2/\text{s} \cdot 1.69 \times 10^{-7} \, \text{m}^2/\text{s}} = 3.68 \times 10^7.$$

The Nusselt number is:

$$\bar{Nu}_L = 0.27 Ra_L^{1/4} = 21.0,$$

9.3 Steady-State Applications of the Macroscopic Energy Balance

and the convective heat transfer coefficient is:

$$h_L = \frac{\bar{Nu}_L k_{\text{air}}}{L} = \frac{21.0 \cdot 2.71 \times 10^{-2}\,\text{W/mK}}{0.5\,\text{m}} = 1.14\,\text{W/m}^2\,\text{K}.$$

The convective heat flux is:

$$q_{\text{conv}} = \frac{\dot{Q}_{\text{conv}}}{A} = h_L(T_a - T_{\text{skin}}) = 1.14\,\text{W/m}^2\,\text{K} \cdot (310.95 - 307.15)\,\text{K}$$
$$= 4.33\,\text{W/m}^2.$$

The radiation flux may also be calculated directly from the constitutive equation:

$$q_{\text{rad}} = \frac{\dot{Q}_r}{A} = \varepsilon_{\text{skin}} \sigma \left(T_{\text{skin}}^4 - T_{\text{sky}}^4\right) = 431\,\text{W/m}^2.$$

As anticipated, the heat loss by radiation to a dark and cold sky far exceeds the slight warming by natural convection. Note that both the radiation and convection occur via the same surface area. The thermoregulation system accommodates for the deficit in boundary heat loss via the supply of warm blood perfusing the skin.

Next we will consider the telescope and its geometry relative to the sky and the gravity field. The telescope is our system, and it interacts with the surroundings via convective heat gain and radiative heat loss. The primary shape of the telescope is a short cylinder (0.25 m diameter and 0.45 m length); so it may be necessary to consider heat transfer for all surfaces. Its orientation when viewing the heavens will likely be toward vertical. In the absence of other information, we will assume that the sides are vertical, and the lens is horizontal.

Further, we assume that the radius is large in comparison to the thickness of the convection boundary layer that is set up. The lens will be cooler than the surrounding air so that convection will increase the density of the air, causing it to spill downward over the sides under the action of the gravity field. The air will also be cooled by the sides of the telescope and flow downward. We will assume that the air flow from the lens does not affect the boundary layer that forms along the sides. The first step in applying the constitutive equations is to calculate the Rayleigh number for both the horizontal and the vertical services. Since the Rayleigh number is directly proportional to the difference in temperatures of the telescope and surrounding air, it will be necessary to assume the final solution temperature and use that value to perform the calculations. We will start with a value of 300K. A further important assumption is that the telescope is isothermal, with implications that all external heat transfer surfaces are at the same temperature even though they may experience different magnitudes of heat flux. For the top and bottom surfaces:

$$Ra_{\text{horiz}} = Gr_D Pr = \frac{g\beta(T_t - T_a)D^3}{\alpha \nu}$$
$$= \frac{9.8\,\text{m/s}^2 \cdot 1/310.95\,\text{K} \cdot (310.95 - 300)\,\text{K} \cdot (0.25)^3\,\text{m}^3}{2.41 \times 10^{-7}\,\text{m}^2/\text{s} \cdot 1.69 \times 10^{-7}\,\text{m}^2/\text{s}} = 1.32 \times 10^7.$$

For the vertical surface:

$$Ra_{\text{vert}} = Gr_L Pr = \frac{g\beta(T_t - T_a)L^3}{\alpha \nu}$$

$$= \frac{9.8 \text{ m/s}^2 \cdot 1/310.95 \text{ K} \cdot (310.95 - 300) \text{ K} \cdot (0.75)^3 \text{ m}^3}{2.41 \times 10^{-7} \text{ m}^2/\text{s} \cdot 1.69 \times 10^{-7} \text{ m}^2/\text{s}} = 7.72 \times 10^7.$$

Given the values of Ra for each of these cases, the correlation equations for calculation of the Nusselt numbers are: top (8.56); bottom (8.55); vertical sides (8.52). Thus:

$$\bar{N}u_{\text{top}} = 0.27 Ra_{\text{horiz}}^{1/4} = 16.3,$$

$$\bar{N}u_{\text{bot}} = 0.15 Ra_{\text{horiz}}^{1/3} = 33.6,$$

$$\bar{N}u_{\text{vert}} = 0.59 Ra_{\text{vert}}^{1/4} = 55.3.$$

The convective heat transfer coefficients and corresponding surface heat flows are calculated for these Nusselt numbers.

$$h_{\text{top}} = \frac{\bar{N}u_{\text{top}} k_{\text{air}}}{D} = \frac{16.3 \cdot 2.71 \times 10^{-2} \text{ W/mK}}{0.25 \text{ m}} = 1.77 \text{ W/m}^2 \text{ K},$$

$$h_{\text{bot}} = \frac{\bar{N}u_{\text{bot}} k_{\text{air}}}{D} = \frac{33.6 \cdot 2.71 \times 10^{-2} \text{ W/mK}}{0.25 \text{ m}} = 3.64 \text{ W/m}^2 \text{ K},$$

$$h_{\text{vert}} = \frac{\bar{N}u_{\text{vert}} k_{\text{air}}}{L} = \frac{55.3 \cdot 2.71 \times 10^{-2} \text{ W/mK}}{0.75 \text{ m}} = 3.33 \text{ W/m}^2 \text{ K},$$

$$\dot{Q}_{\text{top}} = h_{\text{top}} A_{\text{top}} (T_a - T_t) = 1.77 \text{ W/m}^2 \text{ K} \cdot \frac{\pi}{4}(0.25)^2 \text{ m}^2 \cdot (310.95 - 300) \text{ K}$$
$$= 0.95 \text{ W},$$

$$\dot{Q}_{\text{bot}} = h_{\text{bot}} A_{\text{bot}} (T_a - T_t) = 3.64 \text{ W/m}^2 \text{ K} \cdot \frac{\pi}{4}(0.25)^2 \text{ m}^2 \cdot (310.95 - 300) \text{ K}$$
$$= 1.96 \text{ W},$$

$$\dot{Q}_{\text{vert}} = h_{\text{vert}} A_{\text{vert}} (T_a - T_t) = 3.33 \text{ W/m}^2 \text{ K} \cdot \pi \cdot 0.25 \cdot 0.75 \text{ m}^2 \cdot (310.95 - 300) \text{ K}$$
$$= 12.9 \text{ W}.$$

9.3 Steady-State Applications of the Macroscopic Energy Balance

The total convective heat transfer is the sum of all three surfaces. $\dot{Q}_{\text{conv}} = \dot{Q}_{\text{top}} + \dot{Q}_{\text{bot}} + \dot{Q}_{\text{vert}} = 15.8$ W. This value will be compared with the radiation heat flow calculated for the same telescope temperature.

$$\dot{Q}_r = \varepsilon_1 A_1 \sigma \left(T_t^4 - T_s^4 \right)$$
$$= 0.9 \cdot \pi \cdot 0.25 \text{ m} \cdot 0.75 \text{ m} \cdot 5.67 \times 10^{-8} \text{ W/m}^2 \text{ K}^4 \left(310.95^4 - 300^4 \right) \text{ K}^4$$
$$= 18.1 \text{ W}.$$

These calculations show that our initial estimate of 300 K for the telescope temperature is very close to the final answer, being just slightly high. As this value is diminished, the convection will increase and radiation will decrease. By iterative calculations, we find that the two heat flows are equal at 299 K.

Examining and interpreting the results: The telescope equilibrates at a temperature somewhat lower than does the skin. And, while the convection and radiation fluxes for the telescope are equal, for the skin radiation exceeds convection by two orders of magnitude. At first this difference may seem somewhat incongruous, but we should expect that there is a logical explanation. Actually, there are a number of factors that contribute to this difference. For the skin the radiation and convection areas are identical, and for the telescope the convection area is more than seven times larger than for radiation. The temperature differential driving convection is more than three times larger for the telescope than for the skin. The convection coefficients for most of the telescope surfaces are about three times larger than for the skin. In combination, these effects account for approximately a two order magnitude differential in the relative convective and radiative heat flows for the skin and the telescope, which is in line with the numerical results we computed.

9.3.4.3 Radiation Insulation

There are many situations in which it is necessary to provide insulation against radiation heat transfer, as for example in protecting humans from an excessively hot or cold environment. Because radiation occurs by a mechanism entirely different than conduction and convection, insulation requires a totally different strategy. Thus, devices that provide excellent radiation insulation are often poor insulators for thermal conduction.

The radiation flux between two surfaces can be reduced by introducing an intermediate material having highly reflective surface properties. A common example is the "space blanket" which consists of a thin sheet of plastic that has a highly reflective metallic coating, often aluminum, on both sides. This device was developed in the early 1960s by NASA to limit heat loss by radiation under the conditions in space for which there can be very large temperature differentials with the environment with ambient pressure near that of a vacuum. Space blankets can be used in multiple layers to provide greater insulation. When this arrangement is used the materials may be purposefully wrinkled, so that the contact area between adjacent sheets is minimized, limiting the opportunity for conduction exchange and helping maintain the largest temperature difference between the sheets.

The effectiveness of an insulating radiation barrier can be evaluated by building a model for the process. The configuration is shown in Fig. 9.4.

The network can be translated into a heat flow equation. For conditions of all three areas equal and all shape factors unity, the equation is:

$$q_{12} = \frac{\sigma(T_1^4 - T_2^4)}{\frac{1-\varepsilon_1}{\varepsilon_1} + 1 + \frac{1-\varepsilon_{3,1}}{\varepsilon_{3,1}} + \frac{1-\varepsilon_{3,2}}{\varepsilon_{3,2}} + 1 + \frac{1-\varepsilon_2}{\varepsilon_2}}$$
$$= \frac{\sigma(T_1^4 - T_2^4)}{\frac{1}{\varepsilon_1} + \frac{1-\varepsilon_{3,1}}{\varepsilon_{3,1}} + \frac{1-\varepsilon_{3,2}}{\varepsilon_{3,2}} + \frac{1}{\varepsilon_2}}. \quad (9.17)$$

A radiation barrier is most effective when the surfaces are shinny, i.e., when they have a high reflectivity (which requires low emissivity). The following example illustrates this effect.

Example Problem 9.3.4.3.1.
By what factor is the radiant heat transfer between two large parallel flat surfaces with temperatures and emissivities of $T_1 = 500\,\text{K}$, $T_2 = 310\,\text{K}$, $\varepsilon_1 = 0.85$, and $\varepsilon_2 = 0.65$ decreased by introducing an intermediate radiation barrier having an emissivity of 0.03 on both sides?

Solution. *Initial considerations*: The factor by which radiation is reduced by introduction of a radiation barrier is determined by calculating the heat flows with and without the barrier for the same surface temperatures. The potential difference for

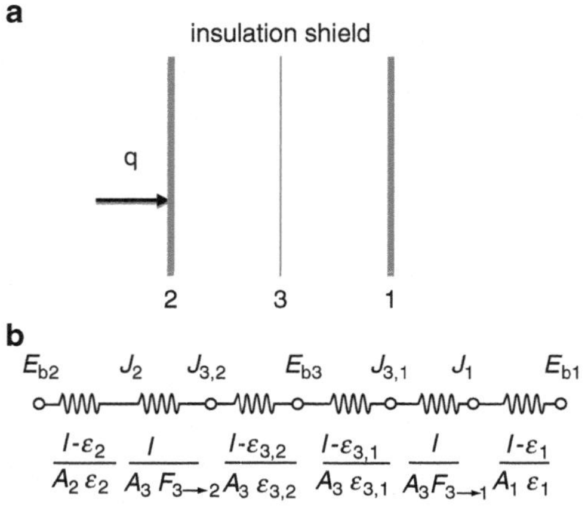

Fig. 9.4 Radiation heat transfer between two large, parallel flat surfaces (1 and 2) separated by a radiation shield (3). (**a**) Schematic of physical configuration. (**b**) Equivalent network representation

radiation transport ($E_{b1} - E_{b2}$) is the same for both cases, so heat flow will be altered only by changes in the resistance caused by adding the radiation barrier.

System definition and environmental interactions: The system consists of the two primary surfaces what exchange energy via thermal radiation, plus the barrier when it is added. The system matches that shown in Fig. 9.4.

Apprising the problem to identify governing equations: The heat fluxes with and without the barrier in place are compared as a ratio of the resistances in the denominator of (9.17).

Analysis: Introduction of a simple radiation barrier reduces the radiation flux by 97.5%.

$$\frac{q_{12,\text{with}}}{q_{12,\text{without}}} = \frac{\frac{1}{\varepsilon_1}+\frac{1}{\varepsilon_2}-1}{\frac{1}{\varepsilon_1}+\frac{1}{\varepsilon_2}+\frac{1-\varepsilon_{31}}{\varepsilon_{31}}+\frac{1-\varepsilon_{32}}{\varepsilon_{32}}} = \frac{\frac{1}{0.85}+\frac{1}{0.65}-1}{\frac{1}{0.85}+\frac{1}{0.65}+\frac{1-0.03}{0.03}+\frac{1-0.03}{0.03}}$$

$$= 0.025.$$

Examining and interpreting the results: For many biomedical applications, a reduction in radiation by a factor of nearly two orders of magnitude provides adequate insulation. In cases where higher performance is required, additional barriers can be used to accomplish similar gains in insulating capacity.

Additional comments: Radiation barriers have long been used for purposes of enhancing thermal comfort. For example, in cold climates where some people either lived in or gathered in large structures constructed of stone or brick (which can have emissivities on the order of 0.9 and thereby function as effective radiation heat sinks for exchange with humans), it was not practical to attempt to centrally heat the entire building volume as is commonly done today. Rather, one widely practiced strategy was to hang large tapestries on the walls that served as a radiation barrier. The tapestries did not have a high insulating efficiency as does a space blanket, but even with a relatively poor coefficient of reflectivity of about 0.5 they would reduce the radiation flux (as calculated in the foregoing example) to about one-third of that without the tapestry. Undoubtedly this gain in radiation insulation was enough to affect benefits in human thermal comfort, and it gave rise to the fabrication of the great tapestries of past centuries that are primarily viewed in present times only as works of art.

9.3.5 Heat Transfer with Phase Change

Phase change is a process in which the molecular structure of a material is altered. The most commonly encountered phases of materials are solid, liquid, and vapor, but other phenomena such as a protein denaturation may also be grouped with this category of processes. There are two important characteristics of a phase change that relate directly to heat transfer. First, when the phase of a material is altered, the stored internal energy is changed. Second, a phase change does not require an associated change in temperature. As a consequence, there needs to be an additional term added to the left side of the conservation of energy equation that is a function of the phase state of the system but is not temperature dependent. This term is called the *latent heat*, and it is a measure of the amount of energy that must be added or removed from a material to alter its state. In contrast, changes in the internal energy of a material associated with changes in temperature denote the *sensible heat*. We have seen that the constitutive equation for sensible heat involves a specific heat unique to the chemical composition and phase of a material multiplied by a temperature differential. The constitutive equation for latent heat is simply the amount of energy per unit mass necessary to cause a transition from one phase to another. For example, for a system pressure of one atmosphere water can be altered between the liquid and vapor phases when the temperature is $100°C$ and between the solid and liquid phases when the temperature is $0°C$. Under these conditions, when energy is added or removed to a material, the phase rather than the temperature is altered. At thermodynamic states away from the phase temperature when energy is added or removed, the result is a change in material temperature. Phase changes are very important for some types of bioheat transfer processes. Examples are the alternating evaporation and condensation of water into air during the respiratory cycle, the denaturation of protein molecules under a variety of types of stress, the explosive boiling of tissue during laser surgery, and the use of controlled freezing for cryopreservation of tissues or for targeted tissue destruction during cryosurgery. As is often the case for biological materials, phase changes in these systems are accompanied by increased complexity in comparison with inanimate materials. Delayed phase nucleation may result in undercooling or supercooling from the phase change temperature, causing a state of metastable equilibrium to develop from there may a sudden and large release of latent energy when nucleation in finally triggered. The fact that biological materials are mixtures rather than a pure chemical substance means that many phase changes occur over a range of temperatures rather than a single value. As a consequence, the process will combine simultaneous latent and sensible energy storage.

The required mathematical analysis of phase change heat transfer is among the most complex in the field of biotransport; so treatment of these phenomena is generally omitted from introductory texts. Thus, we will not give this topic further consideration.

9.4 Unsteady-State Macroscopic Heat Transfer Applications

In steady-state applications, the accumulation of energy in the system is zero. However, there are many situations where the system temperature and energy transfer to the surroundings are not constant, but change with time. In the following sections, we will apply the unsteady-state macroscopic energy balance (9.2) to systems that are not in the steady state.

9.4.1 Lumped Parameter Analysis of Transient Diffusion with Convection

The simplest form of analysis for a transient diffusion problem occurs when there is a sudden change in convective transport with the surroundings, and the internal resistance to heat conduction is small compared to the convective resistance at the boundary. This situation may be encountered when a solid object is quenched into a large body of a fluid. If there is an imbalance of heat flow resistance toward the convective boundary, then the conduction temperature gradient in the interior of the solid will be negligible, and at any instant of time the solid will appear as isothermal. Under these conditions, the thermal state of the solid can be described by a single temperature value which will be a function only of time.

We can consider the system shown in Fig. 9.5 to illustrate this type of process. A solid object with a characteristic dimension of D is the system of interest and is initially at a temperature T_i. At time $t = 0$, it is rapidly immersed into a new environment consisting of a large fluid bath at a temperature T_∞. For this case, we assume that the solid will cool with a spatially uniform temperature until it comes to equilibrium with the fluid at T_∞ via convection at its surface. The magnitude of convection heat transfer is described by the convective coefficient, h, and at any given time is in proportion to the temperature difference between the solid and the fluid, $T(t) - T_\infty$.

To analyze this process, we apply the conservation of energy to the solid with volume V and surface area S. There are two nonzero terms in the conservation equation: the rate of change of energy stored in the solid on the left side, and the rate of convective heat loss to the fluid on the right side.

$$\frac{dE}{dt} = -\dot{Q}_{conv}. \qquad (9.18)$$

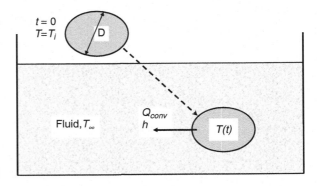

Fig. 9.5 Cooling of a hot solid by quenching into a fluid

Upon substitution of the appropriate constitutive equations, (9.18) becomes:

$$\rho c_P V \frac{dT}{dt} = -hS(T - T_\infty). \tag{9.19}$$

The negative sign exists on the right side of the equation since a positive change of energy stored in the system requires that the environmental temperature exceed that of the system. Equation (9.19) may be solved to obtain an expression for temperature of the solid as a function of time by separation of variables and then integrating the resulting expression over the duration of the process. The representation of temperature is simplified if it is rescaled to that of the fluid: $\theta(t) = T(t) - T_\infty$

$$\int_{\theta_i}^{\theta} \frac{d\theta}{\theta} = -\frac{hS}{\rho c_P V} \int_0^t dt, \tag{9.20}$$

where it is assumed that no constitutive properties change with time and temperature.

$$\ln \frac{\theta}{\theta_i} = -\frac{hS}{\rho c_P V} t, \tag{9.21}$$

or

$$\frac{\theta}{\theta_i} = \frac{T - T_\infty}{T_i - T_\infty} = e^{-\left(\frac{hS}{\rho c_P V}\right)t}. \tag{9.22}$$

Equation (9.22) describes a classic exponential decay process in which a system responds to a step change in the environmental state by an asymptotic approach to a new equilibrium condition. An analogous process can be observed in many physical domains, with perhaps the most familiar being the discharge of an electrical RC circuit. The thermal analogy is based on a thermal resistance associated with the convective heat flow,

$$\mathfrak{R}_T = \mathfrak{R}_{T,\text{conv}} = \frac{1}{hS}, \tag{9.23}$$

and a thermal capacitance associated with the change in energy storage,

$$C_T = \rho c_P V. \tag{9.24}$$

These two terms are combined with (9.22) to describe an equivalent thermal RC series circuit behavior.

$$\frac{\theta}{\theta_i} = \frac{T - T_\infty}{T_i - T_\infty} = e^{-\frac{t}{\mathfrak{R}_T C_T}}. \tag{9.25}$$

9.4 Unsteady-State Macroscopic Heat Transfer Applications

The resistance × capacitance product defines a time constant for the thermal decay process,

$$\Re_T C_T = \tau_T, \quad (9.26)$$

so that the transient temperature history of the solid is most generally described by:

$$\frac{\theta}{\theta_i} = e^{-\frac{t}{\tau_T}}. \quad (9.27)$$

The behavior of this system is illustrated in Fig. 9.6 depicting the rate of discharge of a thermal RC circuit in terms of the constitutive and geometric properties of the system and environment.

This analysis of the transient temperature in a convectively cooled solid is very simple and clean. But it is applicable for only a very limited range of processes as defined by a small ratio of the resistances for internal conduction to surface convection. This ratio can be quantified in terms of a dimensionless parameter called the Biot number (Bi). It is used as a test to determine when a lumped analysis is appropriate. The convective resistance is already defined in (9.23). The conductive resistance is defined based on Fourier's law, (2.9). The characteristic distance L_c across which the temperature gradient occurs and which represents the conduction path length within the solid is often expressed as the ratio of the volume to the surface area.

$$L_c = \frac{V}{S}. \quad (9.28)$$

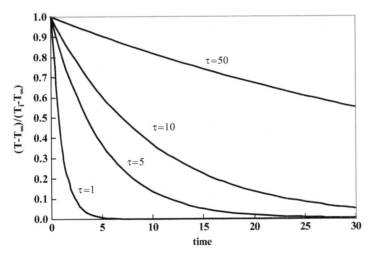

Fig. 9.6 Cooling of a solid immersed into a fluid when the internal conduction resistance is negligible in comparison to convective resistance at the surface

For example, the characteristic length of a sphere of radius r_c is $r_c/3$ and for a cylinder is $r_c/2$. The conduction resistance is then written as:

$$\mathfrak{R}_{T,cond} = \frac{L_c}{kS}. \quad (9.29)$$

The Biot number is obtained by combining (9.23) and (9.29).

$$Bi = \frac{\mathfrak{R}_{T,cond}}{\mathfrak{R}_{T,conv}} = \frac{L_c/kS}{1/hS} = \frac{hL_c}{k}. \quad (9.30)$$

Note that the thermal conductivity is that of the solid, which is different than the conductivity of the surrounding fluid. In the analysis of convection processes, a similar parameter called the Nusselt number, Nu, is developed based on an identical set of properties, but with the conductivity equal to that of the fluid. These two parameters are defined to be applied for different purposes, and they should not be confused.

The Biot number provides an effective measure to determine whether the lumped parameter method may be applied for the analysis of any given system and process. Small values for Bi indicate that the internal conduction resistance of a solid is relatively small during a convective interaction with its environment. Therefore, the lumped parameter analysis can be considered valid to use for circumstances for which Bi is less than 0.1.

$$Bi = \frac{hL_c}{k} < 0.1 \quad \text{to use lumped analysis.} \quad (9.31)$$

The lumped analysis method is much simpler to apply and to interpret than the more comprehensive approaches for transient heat transfer. Thus, when starting the analysis for a new transient conduction problem with convection boundary conditions, after the system and environmental interactions are defined, the Bi value should be calculated to see whether the lumped method is applicable.

Example 9.4.1.
A component of a prosthetic implant is fabricated of stainless steel and has a spherical geometry of 3 cm diameter. During the manufacturing process, it is necessary to subject the sphere to a controlled annealing process from a high temperature in a gaseous nitrogen environment. For the purposes of process design, it is necessary to predict the time that will be required to cool the spheres from 1,050 to 125°C in a circulating gaseous environment at 25°C. The circulating system has an operating state for which the convective heat transfer coefficient over the surface of the sphere is 150 W/m² K.

Solution. *Initial considerations:* This is a quenching process of introducing a heated solid into a cold gaseous environment. Since the temperature of the sphere will be changing during the process, a transient analysis is required.

9.4 Unsteady-State Macroscopic Heat Transfer Applications

System definition and environmental interactions: The system consists of a single solid mass of stainless steel with a perfectly spherical geometry. The interaction with a gaseous nitrogen environment causes cooling of the sphere by convection. No other thermally significant interactions occur.

Apprising the problem to identify the governing equations: The governing equation for this process is the conservation of energy. There will be only the storage term and a single convective heat transfer with the environment as per (9.18).

$$\frac{dE}{dt} = -\dot{Q}_{conv}.$$

Several aspects of the interaction between the system and environment warrant detailed consideration for simplifying assumptions.

1. *The gaseous nitrogen environment has very large dimensions in comparison to the diameter of the sphere so that it remains constant during the process.*
2. *The convective coefficient is uniform over the entire surface of the sphere.*
3. *The sphere is isolated in the nitrogen environment so that the convection is not influenced by the proximity of other objects.*
4. *The sphere enters the nitrogen very quickly, approximating a step-wise manner.*

Analysis: Since this is a transient conduction process with a convective environment, the first step is to check whether a lumped analysis model is valid. For this purpose, we must calculate the value of the Biot number. The convective heat transfer coefficient is given, the characteristic length is one-sixth of the diameter, and the thermal conductivity of stainless steel is found from the property tables to be 14.2 W/m K.

$$Bi = \frac{hL_c}{k_{solid}} = \frac{150 \, \text{W/m}^2 \, \text{K} \cdot (3 \times 10^{-2} \, \text{m}/6)}{14.2 \, \text{W/m K}} = 5.28 \times 10^{-2} < 0.1.$$

The Biot number shows that the sphere will remain nearly isothermal as its temperature drops during quenching. Therefore, it is appropriate to use the simplified lumped parameter analysis of (9.21) to determine the process time. For a sphere of radius R, this expression is written as:

$$\ln\frac{\theta}{\theta_i} = -\frac{hS}{\rho c_P V}t = -\frac{3h}{\rho c_P R}t.$$

Given the properties for density and specific heat of stainless steel plus the process data, this equation can be solved for the time to reduce the temperature to 125°C.

$$t = -\frac{\rho c_P R}{3h}\ln\frac{\theta}{\theta_i} = -\frac{8.24 \times 10^3 \, \text{kg/m}^3 \cdot 468 \, \text{J/kgK} \cdot 1.5 \times 10^{-2} \, \text{m}}{3 \cdot 150 \, \text{W/m}^2 \, \text{K}} \cdot \ln\frac{125 - 25}{1025 - 25}$$

$$= 296 \, \text{s}.$$

The time to cool the sphere to the specified temperature is approximately 5 min.

Examining and interpreting the results: Even though a substantial time is required to cool the sphere to from 1,025 to 125°C, the internal temperature gradient remains minimal because the relative convective resistance on the surface remains much larger than the internal conductive resistance.

Additional comments: We can examine the cooling process as it continues further toward a final equilibrium state with the nitrogen environment. For example, the time to reduce the temperature to 35°C doubles to 592 s. Our initial calculation for cooling to 125°C covered 90% of the complete temperature range, leaving 10^{-1} of the process remaining. Cooling to 35°C covers 99% of the range, leaving 10^{-2} of the process remaining. Given the exponential relationship between time and temperature in (9.21) and (9.22), the ratio of the fractions of the temperature ranges covered matches the difference in the required times of a fraction of 2.

One potential concern with our analysis is that we have assumed that the convective heat transfer coefficient is constant over the entire process. Reference to the property table for the nitrogen gas shows that over the temperature range of the cooling process, the thermal conductivity of N_2 is reduced by a factor of 3. Accordingly, the film temperature in the boundary layer will change by about 1.5. Since h is directly proportional in the Nusselt number to the conductivity of the gas, the assumption of constant h during a process over such a large span of temperatures can be called into question. If we are concerned with achieving a highly accurate analysis, this issue should be investigated further.

Example 9.4.2.
Show an example in which the Bi calculation indicates lumped analysis is not appropriate. The example could subsequently be solved using a distributed technique as discussed in Sect. 10.6.

Solution. *Initial considerations:* This problem involves what is called an *inverse solution*. We start with the end result defined and then work backward to determine what kind of a problem statement will produce those results. The criterion that must be satisfied is that $Bi > 0.1$. There are nearly an infinite number of problems that can meet this criterion. From the definition of the Biot number:

$$Bi = \frac{hL_c}{k_{\text{solid}}},$$

we can see that common characteristics include a large convective heat transfer coefficient at the boundary and/or large dimensions of the solid and/or poor thermal conductivity of the solid.

System definition and environmental interactions: We can consider a system similar to that of Example 9.4.1, but with different materials and operating characteristics. We will assume that the 3 cm sphere is made of Teflon and the quenching fluid is liquid water.

Apprising the problem to identify the governing equations: The value of Bi is calculated based on its definition as shown above.

Analysis: The convective heat transfer coefficient is determined to be 1,500 W/m² K, and the thermal conductivity of Teflon is 0.23 W/m K. The value of Bi is calculated to be 36.

Comments: Although this example is simple, inverse problems are often an important part of engineering analysis. For example, the design of a particular device may require that specified performance criteria are met. This condition will drive the design of the features of the device, and an inverse solution may be applied to calculate particular aspects of the device structure and operational modes.

9.4.2 Thermal Compartmental Analysis

In this section, we will consider the unsteady-state analysis of fluid systems governed by the macroscopic energy equation (9.2). If the system is well mixed, then the temperature in the system is uniform throughout. This means that all fluid streams emerging from the system will have the same temperature and enthalpy as the system; so the macroscopic energy equation becomes:

$$\frac{d}{dt}(mc_pT) = \sum_{i=1}^{N_{inlets}} w_i c_{pi} T_{m,i} - \left(\sum_{j=1}^{N_{outlets}} w_j\right) c_p T + \dot{Q}_s + \dot{Q}_{gen}. \quad (9.32)$$

Example 9.4.2.1 Well-Mixed Compartment.
An insulated reservoir with volume V is filled with dialysis fluid at temperature T_0. At time $t = 0$, dialysis fluid at temperature T_1 is introduced via a single inlet with a constant mass flow rate w_1. Fluid leaves the reservoir at the same rate through outlet channels. The reservoir is well mixed. Derive an expression for the rate of change of temperature with time in the reservoir.

Solution. *Initial considerations:* Because the fluid flow rates in and out are identical, the reservoir mass remains constant. Since the temperature of the entering fluid is different than the initial reservoir temperature, the process is transient with respect to energy, but not with respect to mass.

System definition and environmental interactions: The system will consist of the reservoir, and it is open because mass crosses the boundary. For a rigid, insulated container, the only method of environment interaction is via mass flow through the inlet and outlet ports.

Apprising the problem to identify the governing equations: Equation (9.32) is directly applicable.

Analysis: For an insulated system, $\dot{Q}_S = 0$. The system mass is constant and equal to ρV. The sum of all outlet mass flows must equal the inlet mass flow w_1. It is quite reasonable to assume that no heat is generated in the fluid and that the specific heat is constant. With these restrictions, (9.32) reduces to:

$$\rho V \frac{dT}{dt} = w_1(T_1 - T).$$

The solution with initial condition $T(0) = T_0$ is:

$$\frac{T - T_1}{T_0 - T_1} = \exp\left(-\frac{w_1}{\rho V}t\right).$$

Comments: As a check we see that at $t = 0$ the temperature is T_0, and as t approaches infinity the temperature approaches the inlet temperature. The time constant of the system is the mass of the system divided by the mass flow rate through the system.

Note that an important unstated assumption of this analysis is that the incoming fluid is thoroughly mixed with the reservoir fluid before it exits. When you encounter a seemingly simple and tightly prescribed problem such as this one, it is often the case that implicit assumptions that may have significant physical consequences are not stated explicitly. Thereby it is easy to miss critical aspects of solving a real world problem that would need to be considered if you had the responsibility of formulating the problem on your own.

Example 9.4.2.2 Thermal Dilution Cardiac Output Measurement.
The thermal dilution technique is used clinically to measure the flow output of the heart. Show how cardiac output can be estimated from a procedure in which cold saline solution is injected as a bolus into the pulmonary artery and the temperature of blood measured in the ascending aorta as a function of time.

Solution. *Initial considerations:* The cardiac output is the volume of blood pumped by the heart in 1 min. The thermal dilution process consists of inserting a catheter into the pulmonary artery and injecting a small volume of cold saline at a known temperature. The temperature is measured by a sensor on the catheter at a known distance downstream of the injection site. The distance is typically 6–10 cm, depending on the configuration of the catheter. The volume flow rate of blood will affect how rapidly the temperature at the measurement site changes. High flow will cause a more rapid change in temperature, whereas low flow will result in a slower change. A major drawback of this technique is its invasiveness, as the catheter must be threaded through the right ventricle to reach the pulmonary artery.

9.4 Unsteady-State Macroscopic Heat Transfer Applications

System definition and environmental interactions: The system of interest consists of the pulmonary vasculature from the injection site to the measurement site. The fluids flowing in this problem are the blood entering and leaving plus the cold injected solution.

Apprising the problem to identify the governing equations: We can apply the general macroscopic energy balance, (9.2), to this system for a single inlet (pulmonary artery) and a single outlet (ascending aorta).

Analysis: We would the mass contained within the boundary of the system and specific heat of the system to remain constant during the course of a thermal dilution measurement, since this section of vasculature will maintain a constant volume. The rate of internal heat generation is negligible. Also, we can assume that there is no net exchange of heat with the tissues in which the vasculature is embedded during the procedure. Consequently, (9.2) becomes:

$$\frac{dU}{dt} = w\left[c_{p,in}T_{in} - c_{p,out}T_{out}\right].$$

The mass flow rate w is equal to the product of blood flow Q_b and blood density. The inlet temperature is body temperature, T_b. Since the saline solution is diluted significantly by the time it reaches the detection site, the specific heat at the outlet is approximately equal to the specific heat at the inlet, which is that of blood, c_{pb}. Multiplying both sides of the equation by dt, integrating from $t = 0$ to $t = \infty$, and assuming flow is constant:

$$U(\infty) - U(0) = \rho_b Q_b c_{pb} \int_0^\infty [T_b - T_{out}]dt.$$

The difference in internal energy between when the system is filled with blood at T_b and just after a mass of cold saline, m_s, at temperature T_s is introduced is:

$$U(\infty) - U(0) = m_s c_{ps}[T_b - T_s] = \rho_s V_s c_{ps}[T_b - T_s].$$

Note that the mass of saline injected replaces an exactly equivalent mass of blood within the system boundary. Equating the two foregoing equations yields an expression for blood flow rate.

$$Q_b = \frac{\rho_s c_{ps}}{\rho_b c_{pb}} \frac{V_s[T_b - T_s]}{\int_0^\infty [T_b - T_{out}]dt}.$$

Comments: To estimate cardiac output, approximately 5 ml of ice cold saline is injected into the pulmonary artery, and temperature is measured in the ascending aorta for about 20 s. The area under the $T_b - T_{out}(t)$ curve over this time is measured, representing the integral in the denominator. Given the physical properties of blood and saline, plus the volume and temperature of saline injected, the volume flow rate of blood can be calculated. An advantage of the thermal dilution method over the dye dilution method is that the cooled blood is reheated as it passes through the systemic circulation, reducing the effects of recirculation observed with the dye method. Blood flow is not actually constant during a thermal dilution measurement since the heart is beating, but it has been shown that little error is introduced by making this assumption.

9.5 Multiple System Interactions

Many systems can be broken down into interacting subsystems that can be analyzed with the macroscopic approach. In the following sections, we provide examples of interacting subsystems involving conduction, convection, and radiation. These examples are not meant to provide an exhaustive set of applications, but instead are presented as being representative of the method of analysis.

9.5.1 Convection: Multiple Well-Mixed Compartments

Compartmental analysis involves the exchange of mass and energy between well-mixed compartments. This type of analysis will be applied for many mass transfer applications in Chap. 13 for which tracers or drugs are exchanged between interacting compartments linked in series or in parallel. Here we will consider a heat transfer problem that illustrates how thermal dilution methods can be used to estimate cardiac chamber volumes. The overall system consists of an isolated heart preparation that is perfused with blood at a constant volume flow rate Q_b. A bolus of cold blood is injected into the left atrium at time $t = 0$, and the temperature of the blood is monitored in the left ventricle as a function of time. We would like to use the temperature/time data for the left ventricle to estimate the volumes of the left atrium and left ventricle.

The blood in each chamber will be assumed to be well-mixed system, so that each is isothermal. The only energy exchange with the environment is via the flow of blood from the adjacent compartment. Applying a macroscopic energy balance to the left atrium (LA):

$$\frac{d}{dt}(\rho_b V_{LA} T_{LA}) = \rho_b Q_b [T_b - T_{LA}].$$

The entering blood will always be at body temperature, including after the bolus is introduced. Assuming that the blood volume in the left atrium remains constant, the equation can be solved to find an expression for left atrial temperature if the initial temperature T_{LA0} is known:

9.5 Multiple System Interactions

$$T_{LA} - T_b = (T_{LA0} - T_b)\exp\left\{-\frac{Q_b}{V_{LA}}t\right\}.$$

If we monitor the temperature in the left atrium, we can estimate the volume of the left atrium by plotting the logarithm of $T_{LA} - T_b$ vs. time. After a short period of time, the slope will be constant, and V_{LA} can be estimated from the slope and the known blood flow Q_b. Extrapolation of $T_{LA} - T_b$ back to zero time defines an intercept on the temperature axis, from which T_{LA0} can be computed. The initial temperature in the left atrium can be computed from the volume of the injected blood V_0 and its temperature T_0:

$$\frac{T_{LA0} - T_b}{T_0 - T_b} = \frac{V_0}{(V_{LA} + V_0)}.$$

Therefore, V_{LA} can be determined from the slope and from the intercept of the ln $(T_{LA} - T_b)$ vs. time data, and the two estimates compared. Unfortunately, temperature is not measured in the left atrium, but rather in the left ventricle. Still, this provides some insight on how data collected from the left ventricle might be analyzed using graphical methods.

The conservation of energy applied to the left ventricle yields:

$$\frac{d(\rho_b V_{LV} T_{LV})}{dt} = \rho_b Q_b [T_{LA} - T_{LV}].$$

Assuming left ventricular volume to be constant in this system, the energy equation can be rewritten in terms of the expression found for the left atrium temperature.

$$\frac{dT_{LV}}{dt} + \left(\frac{Q_b}{V_{LV}}\right) T_{LV} = \frac{Q_b}{V_{LV}}\left[T_b - (T_b - T_{LA0})\exp\left\{-\frac{Q_b}{V_{LA}}t\right\}\right].$$

The solution for left ventricle temperature, using the initial condition $T_{LV}(0) = T_b$, is:

$$\frac{T_{LV} - T_b}{T_{LA0} - T_b} = \left(\frac{V_{LA}}{V_{LV} - V_{LA}}\right)\left(e^{-\left(\frac{Q_b}{V_{LV}}\right)t} - e^{-\left(\frac{Q_b}{V_{LA}}\right)t}\right).$$

Substituting for T_{LA0} and defining a dimensionless temperature, θ:

$$\frac{T_{LV} - T_b}{T_0 - T_b} \equiv \theta(t) = \left(\frac{V_{LA} + V_0}{V_{LV} - V_{LA}}\right)\left(\frac{V_{LA}}{V_0}\right)\left(e^{-\left(\frac{Q_b}{V_{LV}}\right)t} - e^{-\left(\frac{Q_b}{V_{LA}}\right)t}\right).$$

This equation can be rewritten as:

$$\theta(t) = ae^{-\frac{t}{\tau_{LV}}} - ae^{-\frac{t}{\tau_{LA}}} = \theta_1(t) - \theta_2(t),$$

where $\theta_1 = a\exp(-t/\tau_{LV})$, $\theta_2 = a\exp(-t/\tau_{LA})$, $\tau_{LV} = V_{LV}/Q_V$, $\tau_{LA} = V_{LA}/Q_V$, and

$$a = \left(\frac{V_{LA} + V_0}{V_{LV} - V_{LA}}\right)\left(\frac{V_{LA}}{V_0}\right).$$

Simulated $\theta(t)$ data are plotted on a log temperature scale vs. time in Fig. 9.7. Cardiac volumes can be found by analyzing the plotted data. In this case, 5 ml of blood at 0°C was injected into the left atrium at $t = 0$. Entering blood temperature was 37°C, and blood flow rate was 60 ml/s. After about 3 s the slope is constant, reflecting the longer time constant of the larger volume (i.e., the left ventricle). Left ventricular volume computed from the measured slope of -0.6 s^{-1} is 100 ml. Extrapolation of the curve with that slope back to zero time gives an intercept of $a = 4.94$. One could use the measured intercept to compute the volume of the left atrium. However, another method, known as *peeling off exponentials*, would be to compute θ_2 by subtracting θ from θ_1 and plotting θ_2 vs. time. This is also shown in Fig. 9.7. Left atrial volume can be computed by dividing the flow rate by the negative slope of the $\ln(\theta_2)$ vs. time graph, 1.622 s^{-1}, to obtain a volume of 37 ml.

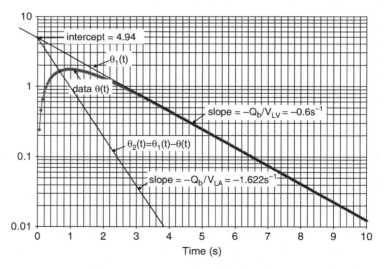

Fig. 9.7 Graphical determination of cardiac volumes from temperature measured in the left ventricle

9.5 Multiple System Interactions

Substituting the values of 37 and 100 ml for the cardiac volumes into the expression for the intercept a gives 4.94, which is consistent with the measured intercept on the graph.

9.5.2 Combined Conduction and Convection

In Sect. 9.4.1, we considered the cooling of a solid when it was placed in a fluid at a different temperature. We assumed that the fluid volume was very large, so the heat transferred from the solid had a negligible effect on the temperature of the fluid. Let us now consider a case where the fluid volume V_f is finite, well mixed (and therefore isothermal), and insulated from the surroundings. The solid, initially at temperature T_{s0}, is introduced into the fluid at $t = 0$, and we wish to find both the temperatures of the solid and of the fluid as functions of time. The Biot number is very small, so a lumped parameter representation of the solid is valid.

The solid and the fluid are each a separate system, and they interact only with each other during the quenching process. We start by applying the conservation of energy as macroscopic energy balances for both the solid and the fluid:

$$\text{Solid}: m_s c_{ps} \frac{dT_s}{dt} = hS(T_f - T_s), \tag{9.33}$$

$$\text{Fluid}: m_f c_{pf} \frac{dT_f}{dt} = hS(T_s - T_f). \tag{9.34}$$

Solving the (9.33) for T_f:

$$T_f = T_s + \frac{m_s c_{ps}}{hS} \frac{dT_s}{dt}. \tag{9.35}$$

Substituting this expression into (9.34) provides a single second-order equation for the solid temperature.

$$\frac{d^2 T_s}{dt^2} + \frac{hS}{m_s c_{ps}} \left(1 + \frac{m_s c_{ps}}{m_f c_{pf}}\right) \frac{dT_s}{dt} = 0. \tag{9.36}$$

The general solution to this equation is:

$$T_s = C_1 \exp\left[-\frac{hS}{m_s c_{ps}} \left(1 + \frac{m_s c_{ps}}{m_f c_{pf}}\right) t\right] + C_2. \tag{9.37}$$

The initial conditions are $T_s(0) = T_{s0}$ and $T_f(0) = T_{f0}$. Therefore, $C_1 + C_2 = T_{s0}$. Applying the initial conditions in (9.35), we find:

$$\left.\frac{dT_s}{dt}\right|_{t=0} = \frac{hS}{m_s c_{ps}}[T_{f0} - T_{s0}] = -\frac{hS}{m_s c_{ps}}\left(1 + \frac{m_s c_{ps}}{m_f c_{pf}}\right)C_1. \quad (9.38)$$

The coefficients C_1 and C_2 are:

$$C_1 = \left(\frac{m_f c_{pf}}{m_f c_{pf} + m_s c_{ps}}\right)[T_{s0} - T_{f0}], \quad C_2 = T_{s0} - C_1. \quad (9.39)$$

The solution for the solid temperature is:

$$\frac{T_s - T_{s0}}{T_{f0} - T_{s0}} = \left(\frac{m_f c_{pf}}{m_f c_{pf} + m_s c_{ps}}\right)\left\{1 - \exp\left[-\frac{hS}{m_s c_{ps}}\left(1 + \frac{m_s c_{ps}}{m_f c_{pf}}\right)t\right]\right\}. \quad (9.40)$$

Note that if m_f is very large and $T_{f0} = T_\infty$, then the solution is:

$$\frac{T_s - T_{s0}}{T_\infty - T_{s0}} = \left\{1 - \exp\left[-\left(\frac{hS}{m_s c_{ps}}\right)t\right]\right\}. \quad (9.41)$$

This can be shown to be identical to the solution found in Sect. 9.4.1 for the solid cooling in an infinitely large fluid volume (9.22). The actual time constant is smaller, so it will take less time for the solid to equilibrate with a finite fluid volume than with an infinitely large fluid volume. Because the fluid temperature will rise as the solid cools, the result is that the final temperature of the solid will be higher. The temperature in the fluid can be found by substituting the solution for the solid into (9.35):

$$\frac{T_f - T_{s0}}{T_{f0} - T_{s0}} = \frac{m_f c_{pf} + m_s c_{ps} \exp\left\{-\frac{hS}{m_s c_{ps}}\left(1 + \frac{m_s c_{ps}}{m_f c_{pf}}\right)t\right\}}{m_f c_{pf} + m_s c_{ps}}. \quad (9.42)$$

Note that if $m_f \gg m_s$, then the fluid temperature remains constant at T_{f0}. Note also that as time approaches infinity, both the solid and the fluid approach the same equilibrium temperature:

$$T_{f\infty} = T_{s\infty} = T_{s0} + \frac{m_f c_{pf}[T_{f0} - T_{s0}]}{m_f c_{pf} + m_s c_{ps}}. \quad (9.43)$$

9.5.3 Radiation: Flame Burn Injury

An important class of radiation sources results from combustion processes and other large high temperature systems, such as a blast furnace, that can cause heat transfer to the skin at a rate sufficient to cause burns. The level of danger is a function primarily of the source temperature, the distance of separation from the source, and the presence of protective garments. Under moderate conditions, a fire can be used to create a situation of comfort in a cold environment. Contrastingly, under extreme conditions a fire can cause a severe burn injury, and this process has been examined extensively with the objective of providing safety for personnel who must work proximal to combustion or high temperature processes and to persons who function in environments that may expose them to present or potential fire hazards, such as fire fighters, military pilots, and race car drivers.[1] A distinguishing factor of this type of radiation is that it is of a much larger magnitude and typically of a shorter duration than occurs by passive environmental sources. Because the magnitude of the boundary heat flux is much greater, the intensity of the response in the tissue is proportionately greater, resulting in injury to the tissue in a very short time.

The intensity of the source can vary widely depending on the type of combustion process and whether it is in a closed or open environment. A sustained burning of cellulosic and wood products requires a minimum local flux in the range of 10–25 kW/m^2. At the base of the fire the temperature is about 900°C, and it decreases to about 350°C outwardly in the plume to the visible tips of the flame. The temperature for combustion of methane vapor and liquid fuels can be much higher, in a range of 1,100–1,200°C. In any actual fire, the temperature will be strongly dependent on the supply rates of fuel and oxygen. In some types of industrial combustion such as a gas fired burner or a blast furnace, temperatures can be as high as 1,500–2,000°C. Needless to say, the radiation emitted by these sources is extremely large and highly dangerous to proximal humans. For example, the black body emissive power at 2,000°C (2,273 K) is $E_{b,2273} = 1.5 \times 10^6$ W/m^2.

The intensity of radiation from a fire that is incident on the skin also can be controlled by adjusting the shape factor and the level of thermal insulation covering the skin. The easiest way to control the shape factor is to change the distance from the radiation source and/or to alter the geometric relationship with the source. The shape factor decreases rapidly with increasing distance. Also, the geometric configuration can be changed to either heat or protect specific areas of the body. For example, during and shortly following the Second World War, much research was conducted to understand the action of bombs creating large fire storms and how to provide protection from the huge fires. Buettner (1952) performed an analysis of thermal radiation energy from a bomb and recommended that a person exposed to the ignition of a large bomb would be well served to spin one's body about an axis perpendicular to the normal from the source to provide a continuously changing shape factor with the source thereby reducing the heat load on any one area of the body. The primary source of thermal insulation is provided by various articles of

---

[1] Many of the pioneering studies in this area were conducted by Alice Stoll (Stoll and Greene 1959; Stoll 1967) during the 1950s and 1960s with the objective of developing protective gear for pilots landing planes on aircraft carriers during which a wreck resulted in an intense fuel fire. The outcome of her studies was the development of the thermally protective fabric Nomex® (Stoll et al. 1971).

clothing. Some clothing materials can be designed specifically to provide effective thermal insulation for industrial and military applications.

Alice Stoll (Stoll and Greene 1959) demonstrated that the pain associated with a burn is proportional to the rate of injury. Therefore, exposure to a strong source of radiation such as a fire can cause a highly painful experience, and a conscious person will rapidly take action to reduce the injury process by moving further away to decrease the shape factor and thereby the radiation flux.

Example Problem 9.5.3.1 Design a Fire Fighter Burnover Shelter.

Fighting large forest fires and wildfires is a physically taxing and extremely dangerous task. One of the primary dangers that fighters operating on the ground face is being overcome by a rapidly moving fire front that advances faster than they can evacuate, trapping them with sometimes fatal consequences. In recent times, a personal safety device has become available for fighters to carry to provide a limited period of protection from getting burned. Only a couple of minutes may be required to allow for a fire front to burn past. The device consists of a personal shelter in which a fighter can be enclosed during a burnover event. The picture below shows a number of fire fighters enclosed in their personal shelters while a brush fire burns past their location (source: Storm King Mountain Technologies©).

Fig. 9.8 Personal shelters

If you were given the assignment of designing a personal protective shelter that could be carried by a fire fighter, what are the most important features that you would incorporate into it? Demonstrate some of the most critical first iteration performance calculations that you would make.

9.5 Multiple System Interactions

Solution. *Initial considerations*: This is a challenge that embodies many complex and detailed aspects. As we will see, there are many decision points and assumptions to be made in arriving at an analysis that can be applied to the design of a shelter. In writing the solution for this problem, for the sake of modeling the process by which an engineer can work through the complexities of analyzing a very practical challenge, the authors have elected to provide extensive commentary; in essence we have attempted to put into writing much of the thinking that we have followed in developing a solution. We will use the GIM to provide direct guidance to our thinking process.

For the *first iteration* solution, we should be focused only on identifying and characterizing the governing system, subsystems, and processes that have the greatest influence on the phenomenon. Secondary effects should be put off for consideration in subsequent more refined analyses. We are most concerned with what happens inside the shelter since that will determine the outcome for the human being. How the fire interacts with the shelter will have a major impact on what happens to the human inside. The heat transfer properties (and probably also the physical structure) of the shelter will determine its effectiveness in protecting the fire fighter. We are dealing with a transient process, and the property of greatest interest is the transient thermal state of the human inside the shelter.

Before attempting to define the system, we should consider what are the most important energy transport and storage processes and where they are occurring so that we can be sure the system and boundary definitions enable us to write equations to describe these events. Outside the shelter, the fire will be a source of high temperature and heat transfer by both radiation and convection. For the initial iteration, we can assume that radiation predominates. The validity of this assumption may be checked in subsequent more detailed solution iterations that include the contributions of convection. The temperature of the radiation source will vary as a function of the intensity of the fire and the material being combusted. An investigation of the combustion literature reveals that the ignition heat flux for cellulosic materials is between 10 and 25 kW/m^2, so the radiative heat flux should be in that range. In compartment fires, an upper layer temperature of about 900 K corresponds to the lower temperature range that will support a flashover scenario. Thus, we can take 900 K as our source temperature and then compare the radiation flux with the known range of radiation required to sustain combustion so as to verify whether our model is consistent with independent physical data.

The primary method of isolation from the fire is via a radiation shield, usually a very thin sheet of plastic with a shiny metallic coating on both sides. The shield can be light weight and folded into a small package for carrying without losing its effectiveness, an attribute that matches well with the requirement that the shelter not be large and heavy for the fighter to carry when not in use. The radiation barrier will come to a temperature between that of the fire and the shelter interior. Accordingly, since it will be relatively hot, contact between the human and the barrier should be minimized to reduce the risk of heating via conduction or of a direct burn. A light weight, rapidly deployed internal frame similar to that used in many small tents could serve to hold the radiation barrier away from the human on the interior. The material must also be able to withstand the maximum temperatures to which it will be exposed during a fire.

The three potential major thermal insults to the human that need to be avoided are surface burns, inhalation of heated air, and heat stress associated with the body core becoming hyperthermic. Many unprotected burn fatalities occur by the first mechanism, which the shelter should protect against. We must design the shelter to protect against thermal radiation burns. The time constant for the body to become hyperthermic via conduction of heat from the skin to the core is anticipated to be large in comparison to the several minutes of protection for which the shelter is designed. Thus, it will be most relevant to focus our analysis on whether the shelter will provide protection from inhalation burn injury.

A detailed analysis of convective heat transfer with heated air flowing through the respiratory tract leading to an inhalation burn is highly complex and well beyond the scope of this text. However, if we neglect the internal details of the heat transfer process and just consider the overall energy balance between the air and lung tissues, it should be possible to determine the rate at which the lungs are heated during respiration to identify the survival time for a human inside the shelter.

The net temperature difference between the fire and the human may be large enough that there would be benefit to having a multilayered space blanket structure for the shelter. The more the layers of insulation, the lower should be the temperature on the interior of the shelter. Thus, there should be a tradeoff between the complexity of the shelter structure and the effectiveness by which it affords protection to a firefighter.

One could ask whether it is likely that the fire would consume oxygen at the rate that would compromise the firefighter's ability to breathe. The type of fire we are considering is usually driven by high wind currents. Thus, we will make the assumption that there is plenty of oxygen available for both the fire to burn and the firefighter to breathe.

With these thoughts in mind, we can proceed with identifying systems for analysis that will allow us to write equations to describe the foregoing processes.

System definition and environmental interactions: The primary element of interest in this problem is the changing thermal state of the human while there is a heat interaction with the environment. Therefore, it will be important to define the human as a system so that heat flow across the boundary can be identified. The most significant component of the environment that can affect the human is the fire, which will communicate with the human by means of the intermediate shelter consisting of an as-yet undetermined number of layers. Therefore, it is also useful to define each of the shelter layers as a subsystem. The shelter will have a negligible thermal mass (defined as a mass having a capacity for storing thermal energy, i.e., via a change in temperature) because of the light weight and very thin dimension of the radiation layers, and it will communicate externally via radiation with the fire and internally via convection with the human. These systems and interactions are shown in Fig. 9.9. As an initial approximation, the systems are taken as

9.5 Multiple System Interactions

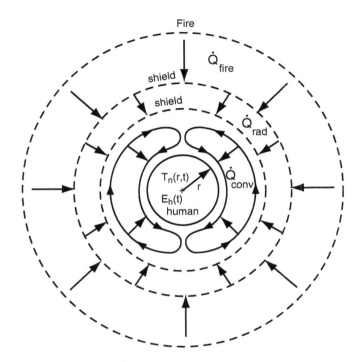

Fig. 9.9 System interactions

geometrically symmetric and one-dimensional, with property variations only in the r coordinate, and with time as appropriate. Although somewhat inaccurate, this assumption enables the problem to be modeled using closed form mathematical representations. If there is concern that asymmetry effects need to be included, a subsequent more complex analysis can be performed.

One possible mechanism of interaction of the human with the environment within the shelter is via convection heat transfer to the skin surface. The highest heat flow to the skin will occur in areas not covered by clothing. The result will be a surface burn.

A second mechanism of interaction of the human with the heated air inside the shelter is by breathing air into and out of the lungs resulting in an inhalation burn. For this system energy exchange with the environment is dominated by mass flow into and out of the lungs, with other heat transfer mechanisms playing a secondary role.

Apprising the problem to identify governing equations: The conservation of energy equation can be used to describe the heat flow for all systems. For the human, this is a transient process resulting in a temperature change. For all other components, the temperature change during the period while the firefighter is within the shelter will be small and/or the thermal masses of the components are small, with the results that steady state can be assumed (no significant change in the stored energy). Thus, initial transients in the shelter and interior air are neglected. Although the temperature at the surface of the human will increase with time, and that change may be significant with respect to the physiological state and to thermal injury processes, the change will be small in comparison to the overall temperature differential driving the heat transfer process between the fire and the human. Therefore, the surface temperature for the human is assumed to be constant with respect to effects on the heat transfer in the human's environment, so that the environmental constitutive properties can be taken as constant during the time of occupancy in the shelter. Note that the foregoing simplifying assumptions for the initial solution iteration are chosen with a realization that they are all related to second-order effects on the overall heat transfer process.

The net heat flow through the series of subsystems from the fire to the surface of the human is illustrated in the equivalent network diagram (Fig. 9.10). Note that we are assuming that radiation between the inner shelter surface and the human in parallel with the convection is negligible. The higher the shelter inner surface temperature, the more compromising is this assumption. If there is a second iteration on the solution, this would likely be an important effect to evaluate.

The conservation of energy is written for the human, the air inside the shelter, and each of the shelter layers. The number of layers assumed initially is arbitrary, although you want to make the number reasonable: not too large to be unnecessarily burdensome to calculations. Based on our intuition and experience, we will select two layers for initial evaluation. If that number does not work, it can be adjusted for subsequent iterations in the analysis. We will assume that the most probable mechanism of injury is an inhalation burn associated with breathing hot air.

$$\text{for the human,} \quad \frac{dE_{\text{hum}}(t)}{dt} \ (\text{W}) = w_{\text{resp}} \left(\frac{\text{kg}}{\text{s}}\right) c_{\text{air}} \left(\frac{\text{kJ}}{\text{kg K}}\right) (T_{\text{in}} - T_{\text{out}}) \ (\text{K}),$$

$$\text{for the air inside the shelter,} \quad \frac{dE_{\text{air}}(t)}{dt} = 0 = \dot{Q}_{\text{conv}} - w_{\text{resp}} c_{\text{air}} (T_{\text{in}} - T_{\text{out}}),$$

$$\text{for the inner shelter layer,} \quad \frac{dE_{\text{shltr}}(t)}{dt} = 0 = \dot{Q}_{\text{rad}} - \dot{Q}_{\text{conv}}.$$

$$\text{for the outer shelter layer,} \quad \frac{dE_{\text{shltr}}(t)}{dt} = 0 = \dot{Q}_{\text{fire}} - \dot{Q}_{\text{rad}}.$$

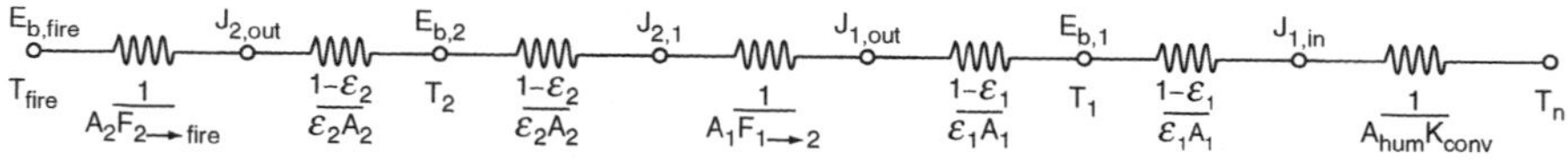

Fig. 9.10 Equivalent network for heat exchange

9.5 Multiple System Interactions

The conservation of energy equation may be replicated for each of the shelter layers for as many layers as may be used.

The subsystems exchange heat in series so that the intermediate energy exchanges are defined by continuity of flow. The equivalent electrical network for a two layer shelter shows the flow of heat continuously through the various systems of the overall process.

Analysis: Although the heat flow from the fire to the surface of the fire fighter is continuous as shown in the equivalent electrical circuit, the potentials that cause the flow are different for the radiation component (black body emissivity – proportional to T^4) and the convection component (proportional to T^1). Therefore, it is not possible to solve for the heat flow across the entire circuit with a single calculation since the end potentials do not match. In a situation like this, it is necessary to estimate the temperature value at the interface between radiation and convection and solve for the radiation and convection flows using the assumed value. If the two flows match, the assumed value was valid. If they do not match, then it is necessary to iterate with a new assumed intermediate temperature value until the calculated radiation and convection flows match to within an acceptable tolerance. With thoughtful interpretation of the calculation results, it should be possible to converge to an acceptable value for the intermediate temperature with no more than a few iterations.

The fire can be treated as a black body at a temperature of 900 K. The emissive power is:

$$E_{b,\text{fire}} = \sigma T_{\text{fire}}^4 = 5.67 \times 10^{-8}\,\text{W/m}^2\,\text{K}^4 \times 900^4\,\text{K}^4 = 3.72 \times 10^4\,\text{W/m}^2,$$

which is a very large heat flux.

The radiation resistances are determined by the emissivities of the shelter layers. For such a high heat load, an efficient radiation reflector will be a great benefit. The best value which can be achieved without incurring excessive materials costs is about $\varepsilon = 0.025$. The reflective insulation will act on each side of every layer of insulating material. The effectiveness of the insulation can be increased most easily by adding more layers of foil. The layers are so thin that increasing the number does not incur a large cost in weight. The shape factors from the shelter to the fire can be taken as about 0.5 and between the inner shelter layers as 1.0. Thus, as can be extrapolated from the equivalent resistance diagram for radiation insulation, for n layers of insulation in the present configuration, the thermal resistance per unit cross-sectional area will be:

$$R_{\text{rad},n} = 2 + (n-1) + 2n\frac{1-\varepsilon}{\varepsilon}.$$

The total radiation heat flow will be proportional to the cross-sectional area presented by the shelter. The area of the human can be approximated as a cylinder of length $L = 2$ m with a radius of $R_{\text{hum}} = 0.25$ m, with end effects neglected.

The radius of the shelter can be taken as twice that of the human, with all reflective layers approximately equal in size. $R_{\text{shltr}} = 0.5$ m. Thus, the area of the shelter is:

$$A_{\text{shltr}} = 2\pi R_{\text{shltr}} L_{\text{shltr}} = 2\pi \cdot 0.5(\text{m}) \cdot 2(\text{m}) = 6.28\,(\text{m}^2).$$

The intended purpose of the radiation shield is to impose most of the temperature drop between the fire and the inner surface of the shelter that convects heat to the air breathed by the firefighter. Thus, we can start with an initial guess for the effective temperature of the inner surface of the shelter as 360 K in comparison with the approximate skin temperature of 310 K.

The radiation portion of the equivalent network is calculated as a function of the number of shields in the shelter. Consider the cases of 2 and 4 shields. Thus:

$$\dot{Q}_{\text{rad},2} = \frac{E_{b,\text{fire}} - J_{1,\text{in}}}{\sum R_{\text{rad}}} = \frac{\sigma(900^4 - 360^4)\,\text{W/m}^2}{\dfrac{1}{2\pi \cdot 2\,\text{m} \cdot 0.5\,\text{m}}\left[3 + 4\dfrac{1 - 0.025}{0.025}\right]} = 1{,}432\,\text{W},$$

$$\dot{Q}_{\text{rad},4} = \frac{E_{b,\text{fire}} - J_{1,\text{in}}}{\sum R_{\text{rad}}} = \frac{\sigma(900^4 - 360^4)\,\text{W/m}^2}{\dfrac{1}{2\pi \cdot 2\,\text{m} \cdot 0.5\,\text{m}}\left[5 + 8\dfrac{1 - 0.025}{0.025}\right]} = 718\,\text{W}.$$

Note that the radiation heat transfer scales very closely with the number of radiation shields. Also, the radiation emissivity of the fire is far greater than that for the inner surface of the shelter, with a ratio greater than 200. The net radiation flux is not affected very strongly by the value of the temperature on the inside of the shelter. Therefore, as we alter the assumed temperature on the inside of the shelter, it will not have much influence on the net radiation heat transfer. This information should be quite useful to our design process.

Next, we will calculate the convection heat transfer between the inner surface of the shelter and the firefighter for our assumed temperature. The convection heat transfer process is characterized in terms of free convection on the interior of the annular space between two horizontal concentric cylinders. Equations (8.60)–(8.62) provide a basis for determining the overall convective transport between the inside of the shelter and the outer surface of the firefighter. The convective heat flow is calculated by:

$$\dot{Q}_{\text{conv}} = \frac{2\pi L k_{\text{eff}}(T_i - T_o)}{\ln(r_o/r_i)}.$$

The only unidentified term in this equation is k_{eff}, which is given by:

$$k_{\text{eff}} = 0.386 k \left(\frac{Pr}{0.861 + Pr}\right)^{1/4} Ra_r^{1/4} \quad \text{with } L_r = \frac{2[\ln(r_o/r_i)]^{4/3}}{\left[(1/r_i)^{3/5} + (1/r_o)^{3/5}\right]^{5/3}}.$$

9.5 Multiple System Interactions

The properties of air are evaluated at the film temperature $T_f = (360 + 310)/2 = 335$ K. By interpolating for temperature in the table of air properties, we find $Pr = 0.701$. $k = 2.94 \times 10^{-2}$ W/m K. $\alpha = 28.7 \times 10^{-6}$ m²/s. $\nu = 20.1 \times 10^{-6}$ m²/s. Also, $g = 9.8$ m/s². $\beta = 1/T_f = 2.92 \times 10^{-3}$ K^{-1}. The characteristic radial length is:

$$L_r = \frac{2(\ln 2)^{4/3}}{\left[\left(\frac{1}{0.25}\right)^{3/5} + \left(\frac{1}{0.5}\right)^{3/5}\right]^{5/3}} = 0.132 \text{ m}.$$

The Rayleigh number is calculated as:

$$Ra_r = \frac{g\beta(T_s - T_\infty)L_r^3}{\nu\alpha}$$

$$= \frac{9.8 \text{ m/s}^2 \cdot 2.99 \times 10^{-3} \text{ K}^{-1} \cdot (360 - 310) \text{ K} \cdot (0.132)^3 \text{ m}^3}{27.7 \times 10^{-6} \text{ m}^2/\text{s} \cdot 19.4 \times 10^{-6} \text{ m}^2/\text{s}} = 6.27 \times 10^6.$$

The effective thermal conductivity is:

$$k_{\text{eff}} = 0.386 \cdot 2.89 \times 10^{-2} \text{W/m K} \left(\frac{0.701}{0.861 + 0.701} 6.27 \times 10^6\right)^{1/4}$$

$$= 0.457 \text{ W/m K}.$$

Internal convective circulation enhances the heat flow between the firefighter and the inside of the shelter by a factor on the order of 15 above that for pure conduction. The convective heat transfer is:

$$\dot{Q}_{\text{conv}} = \frac{2\pi L k_{\text{eff}}(T_o - T_i)}{\ln(r_o/r_i)} = \frac{2 \cdot 3.14 \cdot 2 \text{ m} \cdot 0.457 \text{ W/mK}(360 - 310) \text{ K}}{\ln 2} = 414 \text{ W}.$$

We know that the radiation and convection flows need to be equal at the inner surface of the shelter to satisfy the requirements of conservation of energy. The magnitude of convection is approximately one-half that of the radiation with four shield layers, and one-fourth that with two layers. Thus, the calculated radiation could be reduced to be approximately equal to that of convection for an inner shield temperature of 360 K by increasing the number of shields to eight. That is probably not a practical solution. Thus, we should adjust our estimated temperature upward to balance the convective and radiative flows. Note that the convection will scale approximately with the difference between the skin and the inner shield temperatures.

This path of calculation will show that for an inner shield temperature of 387 K the convection heat flow is 706 W, and the radiation flow with four shield layers is 712 W and with two layers is 1,419 W. Alternatively, for an inner shield temperature of 442 K, the convective flow is 1,380 W, and the radiation flow for two layers is 1,384 W. A single shield produces a radiation flow of 2,574 W and a convection flow of 2,577 W with an inner shield temperature of 528 K. These calculations show the performance of three different shelter designs, the primary practical difference being the temperature of the inner surface of the shelter.

The next step in the analysis is to see whether any of these air temperatures inside the shelter will be hot enough to produce an inhalation burn. The flow of hot air into the respiratory tract has a thermal behavior similar to that of the perfusion of blood through the cardiovascular system. As the flow network branches into progressively smaller diameter flow lumens, the ratio of net surface area to volume increases dramatically with the result that there is total equilibration between the flowing fluid and the lumen tissue. The heated air will equilibrate to the local temperature of the lung tissue, and all of the excess energy of the inhaled air represented by its elevated temperature will be added to the energy stored in the lungs. This calculation will be based on the conservation of energy for the human.

$$\frac{dE_{\text{hum}}(t)}{dt} = w_{\text{air}} c_{\text{air}} (T_{\text{in}} - T_{\text{out}}).$$

The mass flow rate w_{air} is the product of the respiratory rate (12/min), the tidal volume (500 ml), and the density of air as a function of temperature. The fluid temperature within the shelter (the average of the inner shelter surface and the human outer surface temperature) varies with the number of radiation shields. The densities, specific heats, and resulting air flow rates corresponding to the three shelter designs are shown below. For example, for the four shield design:

$$w_{\text{air}} = 12(1/\text{min}) \frac{1}{60(\text{s/min})} 5 \times 10^{-4} \, (\text{m}^3) 0.995 \, (\text{kg/m}^3) = 9.95 \times 10^{-5} \, (\text{kg/s}).$$

Number of shields	1	2	4
T_f in shelter (K)	419	376	349
Density of air (kg/m³)	0.83	0.933	0.995
c (kJ/kg K)	1.017	1.012	1.009
w_{air} (kg/s)	8.3×10^{-5}	9.3×10^{-5}	9.95×10^{-5}

The specific heat of the lung tissue is 4.178 (kJ/kg K), and physiological data show that the lungs can be taken to be 1% of the total body mass. For a 70 kg firefighter, the lung mass is $m_{\text{lung}} = 0.7$ kg. We should consider how much of the total body mass is warmed by breathing heated air into the lungs (i.e., what storage mass do we use when writing the conservation of energy for the body?). This is a very complicated issue to address, well beyond the scope of our current problem. Thus, we should identify our most reasonable approximation to apply in this case. In one extreme, we could assume that the entire body is warmed uniformly as a result of breathing hot air. Based on our common human experiences of being in hot and cold environments, this assumption is illogical. The other extreme assumption is that all the heat added to the body is confined to the lungs, which is probably much closer to what really happens. Some heat will be lost to surrounding visceral

9.5 Multiple System Interactions

tissues and organs, primarily via convection by flowing blood. However, when appraising the safety of a shelter design, the most conservative assumption is that all the heat added to the body remain in the lungs, resulting in the most rapid and greatest rise in lung temperature and the highest risk of injury. Therefore, we will assume that the effective mass of the body corresponds to that of the lungs with the realization that this is a worst case scenario.

The inlet temperature of the air is the average value from within the shelter, which is the fluid temperature, T_f. These constitutive values are applied in the conservation of energy equation to determine the time rate of change of lung tissue temperature, which can then be used to determine the time to produce an 8°C change in lung temperature to increase from normothermia at 37°C to the threshold for thermal injury at 43°C.

Number of shields	1	2	4
$dT_{\text{lung}}/dt \left(\dfrac{K}{s}\right)$	3.15×10^{-2}	2.13×10^{-2}	1.34×10^{-2}
Time to increase 8°C (min)	42.4	62.6	99.6

Examining and interpreting the results: These results clearly show that even with a single radiation shield, the shelter should be more than adequate to protect a fire fighter from an inhalation burn for a time adequate to allow a fire to burn over. As long as the inside of the shelter is not touched causing a possible contact burn, the thermal design of the shelter should be thermally adequate.

Additional comments: This is a complex problem in terms of both the heat transfer processes and the physiological response to the thermal stress. The initial iteration we have performed on designing a protective shelter system indicates that the concept has promise and is worthy of additional study and development unless nonthermal factors dictate otherwise. The number of additional analysis iterations worth pursuing is dependent on the level of thoroughness necessary to develop an acceptable design for the shelter.

It would be interesting to simulate what would happen thermally to a firefighter if a protective shelter was not available. Under these conditions, the greatest risk would be associated with a direct radiation exposure to the fire. The skin would act like a semi-infinite medium with a flux governed boundary condition. The equation that describes the transient temperature in tissue for these conditions is (10.146).

$$T(x,t) - T_i = \frac{2\dot{Q}_s \sqrt{\alpha t/\pi}}{k} e^{-x^2/4\alpha t} - \frac{\dot{Q}_s x}{k}\left(1 - \text{erf}\frac{x}{2\sqrt{\alpha t}}\right).$$

The heat flux from the fire is calculated from (8.95) for a temperature of 900°C and for a shape factor of 0.5. Thus:

$$\dot{Q}_s = \sigma AF\left(T_{\text{fire}}^4 - T_{\text{skin}}^4\right) = 5.67 \times 10^{-8} \; (\text{W/m}^2\,\text{K}^4) \cdot 0.5\left(900^4 - 310^4\right) (\text{K}^4)$$
$$= 1.83 \times 10^4 \,(\text{W/m}^2).$$

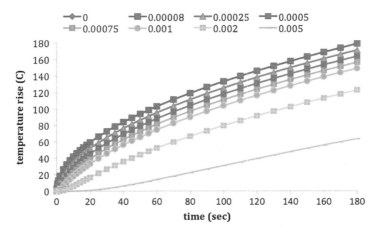

Fig. 9.11 Tissue temperature vs. time for various tissue depths (m)

The temperature as a function of time and position can be calculated for specific values of x and t. We choose to perform these calculations at the base of the dermis (0.00008 m), at incremental positions through the dermis (0.25, 0.5, 0.75, 1, and 2 mm), and at a subcutaneous point (5 mm). The process was simulated for a period of 3 min. Temperature rises greater than 11°C above the initial dermal value of 34°C are considered injurious. Figure 9.11 shows that within about 15 s, a full third-degree burn will have occurred (injury at the base of the dermis, 2 mm), and after only a minute the injury will have penetrated deep into the muscle, undoubtedly indicative of a fatal event. The availability of a protective shelter can have a life saving role for a fire fighter caught in a burnover situation.

9.5.4 Human Thermoregulation

Humans possess an elaborate thermal control (thermoregulation) system that ensures maintenance of internal body temperatures near a physiological set point of about 37°C under a large spectrum of environmental conditions and metabolic rate activities. Slight variations of the core temperature from the normal set point value can result in serious health consequences including lowered resistance to infection, difficulties in coagulation, cardiac anomalies, and even death. It is not surprising then that the regulation of core body temperature is among the most complex and precise of all physiological functions, being held to within a few tenths of a degree Celsius even while the set point can vary as much as 1°C during the circadian cycle. To achieve this objective, it is necessary to be able to: protect the body core mass (primarily the visceral organs and the brain, amounting to about 10% of the total body mass) from external thermal fluctuations, and dissipate excess heat that is generated internally, such as during exercise. The human body is very effectively equipped to meet this requirement.

9.5 Multiple System Interactions

The body incorporates a number of energy production and dissipation mechanisms to preserve thermal homeostasis. Examples of important thermoregulatory processes that are governed by feedback loops are sweating, shivering, and variable localized blood flow, especially to the skin. To a large extent, these mechanisms are controlled by feedback signals based on specific physiological states. The quantitative functional characteristics of the controller are perhaps the most challenging aspects of thermoregulation to understand. The control processes tend to be affected by multiple inputs that are nonlinear over typical signal ranges and that may be coupled with other inputs.

Transient blood perfusion provides an effective means for adjusting the thermal conductance between the body core and surface. Direct thermal conduction provides an ineffective pathway for moving heat between the core and periphery because of the relatively low thermal conductivity of tissues and the long conduction distances. Alternatively, blood is circulated rapidly between the core and the skin, providing a convective transport medium for quickly moving heat large distances. Blood flow distribution within the body is controlled according to metabolic needs of the body, local tissue temperature, and the need to maintain the appropriate core temperature. When the core becomes too hot, the blood vessels in the skin dilate to allow increased perfusion to the body surface. The blood is cooled by the environment, and the cooler blood is returned to the core. Increased blood flow to the skin surface also enables greater sweat production, adding to the cooling process. In contrast, when the core becomes too cold, blood flow to the skin is constricted to conserve the body's internal energy.

As the operational mechanisms of thermoregulation have been better understood, it has been possible to model these functions in an attempt to simulate and predict thermoregulatory behavior, and to be able to design systems to interact thermally with the human body (such as an environmental protective garment or a space suit) without compromising the health and safety of the subject. There are two different levels of modeling resolution that are applied to human thermoregulation. At the macroscopic scale the body is viewed as a single uniform mass that interacts with the environment via a menu of specific energy exchange mechanisms. This approach will be discussed in the following section. At the microscopic level, the internal thermoregulatory processes are identified and characterized as they occur locally within the body, including individual afferent inputs to and efferent outputs from the central thermal controller as well as local physiological signals. This multitude of operations is integrated to produce a mechanistic simulation of whole body thermoregulation. This approach can be exceptionally complicated to represent mathematically, and after decades of development some of the most comprehensive and effective models are still a work in progress (e.g., see Wissler 1961, 1985, 2008). These models are complex beyond what can be covered in an introductory text (e.g., the current Wissler thermoregulation model computes simultaneous energy and mass balances for about 5,000 individual volume elements that represent a fully clothed human, at 1.5 s simulation intervals), but their development and application remain highly fertile areas of research in biotransport.

9.5.4.1 Conservation of Energy

Conservation of energy for the human body Applied to the Human Body must account for internal energy production as a byproduct of cellular metabolism plus multiple mechanisms of mass, heat, and work exchange across the body surface. The most important contributions to the energy balance for the body are represented in the following conservation statement:

$$\frac{dE_{body}(t)}{dt} = \dot{M} - \left(\dot{W} + \dot{Q}_{conv} + \dot{Q}_{cond} + \dot{Q}_{rad} + \dot{Q}_{evap} + \dot{Q}_{resp}\right) + \dot{H}_{mass}, \quad (9.44)$$

where E_{body} is the energy stored in the body; $\dot{M}$ = the rate of metabolic energy production; $\dot{W}$ = the rate at which work is performed on the environment; $\dot{Q}_{conv}$ = rate of heat loss by convection; $\dot{Q}_{cond}$ = rate of heat loss by conduction; $\dot{Q}_{rad}$ = rate of heat loss by radiation; $\dot{Q}_{evap}$ = rate of heat loss by evaporation; $\dot{Q}_{resp}$ = rate of respiratory heat loss; and $\dot{H}_{mass}$ = rate of enthalpy exchange with added or lost mass.

Energy is produced in the body by basal, or resting, metabolism, defined as the minimal metabolism measured at a temperature of thermal neutrality in a resting homeotherm, with normal body temperature several hours after a meal and not immediately after hypothermia. As with chemical reactions in general, body metabolism is a function of temperature, increasing at a rate of about 6–9% per degree Celsius. Increased energy generation is due to muscle activity, including physical exercise and shivering, and by food intake and digestion. Therefore, the total energy production in the body is determined by the energy needed for basic body processes plus any external work. Since the body operates with less than 100% efficiency, only a fraction of the metabolic rate is applied to work, with the remainder dissipated as heat. The mechanical efficiency, η, associated with metabolic energy utilization, is zero for most activities except when the person is performing external mechanical work such as walking upstairs, lifting a mass to a higher level, or pedaling a bicycle. Work transferred from the environment is dissipated as heat in the human body, resulting in a negative η (work output is negative, but waste heat is still generated). An example of this case is walking downstairs. Thus, for many work-producing activities there are joint contributions to the energy balance from both the $\dot{M}$ and $\dot{W}$ terms.

Convection, radiation, conduction, and evaporation of sweat at the skin surface all facilitate heat transfer from the body. The magnitudes of these phenomena are dependent on autonomic physiological controls plus environmental conditions including temperature, humidity, air velocity, cloud cover, and time of day. Humans may also make adjustments to regulate environmental heat transfer by diverse methods including clothing ensembles with appropriate insulating properties, postural adjustments, movement to a position of different radiation flux, and altering the temperature, humidity, and air velocity of their immediate environment. Heat transfer occurs within the respiratory tract and lungs as the temperature of air is altered during inspiration and expiration, although this effect is minor in human beings as compared with many other mammalian and reptile species. The energy storage of the body is altered as mass is added and removed during normal life processes including eating and drinking, elimination, surface evaporation including sweating, and net evaporation within the respiratory tract as air is humidified as it flows into the lungs.

We have described the mechanistic basis for calculating each of the foregoing phenomena in other sections of this book. Appropriate constitutive equations can be identified and substituted for each term in the energy conservation equation (9.44) to develop an expression for how the energy of the body changes with time based on the level of physical activity, environmental conditions, and other personal factors. As an alternative, the heating, ventilating and air-conditioning (HVAC) industry has long had a mission to design and implement environmental systems that provide for human thermal comfort for a very wide range of building and activity combinations. This work has led to an array of empirical correlation functions for the various terms in (9.44) that provide quantitative guidelines for the individual environmental components that contribute to thermoregulatory interactions. This large body of knowledge is compiled in design handbooks such as the ASRHAE Handbook (2009) (from the American Society of Refrigeration, Heating and Air-Conditioning Engineers, Inc.). We do not attempt to cover this literature in an introductory textbook other than referring to it for specific constitutive expressions in example problems.

9.5.4.2 Interactive Garments: Space Suits and Protective Clothing

One very important area of application of thermoregulation models is to predict the interaction between a human and an engineered environment such as a space suit or a hazmat protective garment. These devices constitute a miniature and portable environment for a human that must be capable of interacting compatibly with all living functions to ensure that a life-threatening condition is not created. From a thermoregulatory perspective, the device must be able to remove excess heat when the work and metabolic rates are high and to provide a warm environment when the metabolic rate is insufficient to maintain the core temperature. Thus, the suits must have their own control system that is able to interact with the human so that the thermoregulatory control is able to function normally. Often the control of these devices is left to the wearer to adjust and maintain manually with the result that on occasions the thermal state of the human is allowed to drift to a condition that is dangerous and incapacitating. The alternative is to provide an automated control for the device (Nyberg et al. 2000). In principle this approach should be superior. In fact, it is enormously challenging to design a microenvironment for a human that performs in a manner that is in coordination with the frequently changing operations of the thermoregulatory system. The key to success is to have a good simulation model to represent the thermoregulatory system function to use for the design of the thermal suit. The level of sophistication of a model for thermoregulation must be at a level for which the internal energy flows and gradients within the body are represented since these play key roles in human thermostasis.

9.6 Summary of Key Concepts

Many processes are compatible with viewing the system from a macroscopic perspective for which it is not necessary to be concerned with details of what happens inside the system boundary to understand the net effect produced by the process. The key equation for the entire chapter is the conservation of energy, (9.1).

$$\frac{dE}{dt} = \sum_{i=1}^{N_{\text{inlets}}} w_i \hat{H}_i - \sum_{j=1}^{N_{\text{outlets}}} w_j \hat{H}_j + \dot{Q}_S + \dot{Q}_{\text{gen}} - \dot{W}_s. \tag{9.1}$$

The energy storage term on the left side is balanced by the sum of all energy flows across the boundary. These include the enthalpies of mass flow in and out, all types of boundary heat flows, which are usually either convection or radiation, but can also include phenomena such as evaporation during sweating, work exchange across the boundary, and internal energy generation. The internal energy generation is special compared to the other boundary interactions since it represents a source distributed internally within the system. This term includes heat released as a consequence of metabolism and muscle action. Other examples are energy dissipated in tissues during diagnostic and therapeutic procedures that use sources such as lasers, microwaves, ultrasound, and radiofrequency. Energy storage within the system determines whether a process is steady state (constant temperature over time) or transient (time varying temperature).

Steady-State Processes. The primary feature of steady-state processes are that the energy flows into the system are exactly balanced by the outward flows so that the net change in stored energy is zero. When analyzing a heat transfer problem, one of the first issues to be evaluated during the *Initial Considerations* step is whether the process is steady state or transient. A steady-state process can occur for both open and closed systems. For a closed system, the rate of work flow is balanced by the net heat flow. In living systems, an important component of heat flow can be internal metabolic energy generation, which in turn gives rise to heat rejection from the body surface to maintain a state of homeostasis.

Analogous electrical circuit diagrams are an effective mechanism for representing and solving steady-state heat transfer problems, and especially so for radiation heat transfer phenomena. Since steady-state processes do not involve changes in energy storage, the equivalent electrical circuits have no capacitance elements. The circuits are simple arrangements of resistors in series and parallel configurations. Learning this method of modeling is well worth the effort to have a very helpful tool for visualizing heat transfer processes in an easily interpreted graphical format.

Transient Processes. In order to analyze a transient heat transfer process from a macroscopic perspective, the system is assumed to be isothermal. Therefore, as the stored energy of the system is changing, its state can be followed since the temperature is the same everywhere. A microscopic perspective is applied to be able to analyze processes in which the internal storage of energy may be distributed nonuniformly throughout a system, and will be discussed in the following chapters.

The criteria for being able to apply macroscopic analysis to a transient heat transfer process are that either a fluid must be well mixed or a solid must be able to distribute heat internally by conduction much more rapidly than it is being received or lost at the boundary. The latter criterion is evaluated by calculating the Biot number (9.30), but only if the heat exchange at the boundary occurs via convection.

$$Bi = \frac{\Re_{T,\text{cond}}}{\Re_{T,\text{conv}}} = \frac{hL_c}{k_{\text{solid}}}. \tag{9.30}$$

In evaluating the Biot number, it is important to remember that the convection coefficient, h, applies to heat transfer in the fluid, and the thermal conductivity, k, applied to heat transfer in the solid. When the value of the Biot number is small, i.e., $Bi < 0.1$, the relative resistance to internal conduction within the solid is small in comparison to the convective resistance at the boundary. Under these conditions, there is adequate time for heat to be distributed uniformly within the solid relative to the rate at which it is exchanged at the exterior boundary. If the mechanism of heat exchange at the boundary is via radiation, the same principle applies for being able to assume that the solid remains essentially isothermal during a process: the rate at which heat can be dispersed by conduction internally must be rapid in comparison to the rate of radiation exchange at the boundary. However, there does not exist a convenient quantitative measure of the ratio of conductive and radiative resistances analogous to the Biot number for a convective boundary condition.

9.7 Questions

9.7.1 What is the physical interpretation of the Biot number?

9.7.2 Explain why a lumped analysis of a conduction problem cannot be used when $Bi > 0.1$.

9.7.3 How can an electrical resistance be applied to represent conduction, convection, and radiation processes?

9.7.4 Explain how you would deal with a problem for which a solid experienced simultaneous convection and radiation at its boundary.

9.7.5 What characteristics of a radiation insulator make it effective and why?

9.7.6 Discuss situations in which additional terms may be included in the conservation of energy (9.1) beyond what appear as it is currently written.

9.7.7 What are the boundary conditions for heat flow at the interface of differing materials?

9.7.8 Explain how to define the boundary conditions at the surface of a solid that is undergoing convection and radiation with the environment.

9.7.9 What is the meaning of the time constant for a lumped analysis conduction/convection process?

9.7.10 What is the physical interpretation of the radiosity property of a surface?

9.7.11 Equation (9.16) relates heat transfer between blood and vessel wall to the change in enthalpy over the inlet and outlet of a vessel segment. What type of assumptions would you need to make to perform an analysis of convective process at the wall?

9.8 Problems

9.8.1 Conservation of Energy

Here we will consider an alternate method of solving Problem 3.6.3. Blood at a hematocrit of 0.45 in the left reservoir is separated from plasma in the right reservoir by a short tube that is clamped to prohibit any flow. Initially the blood level is higher than the plasma. When the clamp is released, there will be a flow between the two reservoirs until equilibrium is reached. Show that application of the conservation of energy relationship to the blood in the right reservoir after the clamp is released leads to an analogous equation to that derived using the conservation of red blood cells in the reservoir. Is there any essential information missing from this problem statement?

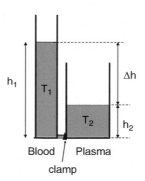

9.8.2 Scalding Water in a Bath Tub

Water at 100°F flows into a bathtub at 1 gallon/min and flows out through the drain at the same rate. Twenty gallons of water at 100°F are initially in the tub. At time $t = 0$, the inlet water temperature is changed to 140°F, but the flow rate is unchanged. If the water in the tub is well mixed, how long will it take for the water temperature to increase to 120°F? Are there any hidden assumptions built into this problem statement, and if so, what are they?

9.8.3 Insulating Properties of Clothing

The heat flux through a 1-mm thick layer of skin is 1.05×10^4 W/m^2. The temperature at the inside surface is 37°C and the temperature at the outside surface is 30°C.

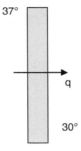

(a) What is the thermal conductivity of skin?
(b) A layer of clothing material with half the thermal conductivity of skin and twice the thickness of skin is placed on the outside surface of the skin. If the outside surface of the clothing is maintained at 30°C, what is the new heat flux from the skin and what is the temperature at the skin–insulation interface?

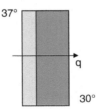

9.8.4 Heat Exchange Between Man and Environment

Heat generated by a naked man standing in a room is 100 kcal/h. Eighty percent of the generated heat is lost by a combination of forced convection (1 m/s wind) and radiation to the walls of the room. The man can be modeled as a circular cylinder with height 1.7 m and diameter of 0.34 m. Only 77% of the man's surface area is involved in radiation exchange with the surroundings because of shape factor effects. If the temperature of the room and surrounding air is 25°C, what is the skin temperature of the man? Be sure that you state all assumptions that you find necessary to solve this problem.

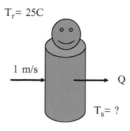

9.8.5 Heat Transfer in the Left Ventricle of a Mouse

A small catheter is inserted into the left ventricle of a mouse (ventricular volume = V). At time zero, blood at temperature T_1 is infused at a constant rate Q_1 through the catheter into the ventricle. This mixes with blood entering from the left atrium at a flow of Q_0 and body temperature T_0. Assume the blood in the left ventricle is well-mixed, ventricular volume remains constant, and there is no heat transfer to the wall of the ventricle.

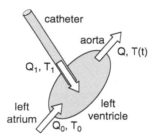

(a) Use conservation principles to find total blood flow emerging from the ventricle, and the transient temperature of blood leaving the ventricle from the time that the flow is started from the catheter.
(b) Find the final steady-state temperature of blood emerging from the ventricle after a long period of time.

9.8.6 Blood Warmer

Blood for transfusion is warmed with a heater in a well-mixed reservoir to a steady-state value of 37°C. Air temperature surrounding the reservoir is 20°C. The overall heat transfer coefficient from reservoir walls to air is 100 W m^{-2} K^{-1}, the reservoir volume is 1×10^{-3} m^3, and the surface area of the reservoir is 0.01 m^2.

(a) What is the heat transfer rate from the heater in watts?
(b) At time $t = 0$, the heater is turned off. How long will it take for the temperature to reach 30°C?

9.8.7 Steady Conduction Through Multilayered Skin

Consider steady-state heat transfer through skin, which is composed of two layers, the dermis and the epidermis. The thicknesses of the two layers are δ_d and δ_e, respectively, and the thermal conductivities are k_d and k_e. The epidermis surface temperature is maintained at T_2 and the dermis surface temperature is maintained at T_1.

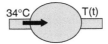

(a) Find algebraic expressions for the heat flux q through the skin and the temperature at the interface between dermis and epidermis.
(b) An overall heat transfer coefficient U is defined from the following equation: $q = U(T_1 - T_2)$. Find U in terms of k_e, k_d, δ_d, and δ_e.

9.8.8 Thermal Mixing of Blood

Blood at 37°C enters the left ventricle (LV) of a skier at a rate of 100 ml/s. At time $t = 0$, the skier steps outside and begins to breathe in very cold air. Consequently, the temperature of blood entering the ventricle is suddenly reduced to 34°C. Assume the ventricle has a constant volume of 100 ml, the ventricle is well-mixed, and the specific heats and densities of blood are independent of temperature.

(a) Derive an algebraic expression for the temperature of blood exiting from the ventricle as a function of time.
(b) What will be the temperature of blood emerging from the ventricle after a long period of time?

9.8.9 Postmortem Interval

Body cooling is used to estimate the time of death. Compute the Biot Number associated with heat transfer from the surface of the body modeled as a cylinder with radius $R = 20$ cm and $k_{body} = 0.5$ W/(m°C) for the following two cases:

(a) Air: $h = 2.46$ W/(m²°C)
(b) Water: $h = 100$ W/(m²°C)
(c) Based on these computations and assuming the body cools like a cylinder, do we need to account for internal thermal resistance in estimating postmortem interval in either case? Why or why not?
(d) What analysis would you use to get a better estimate of the time of death?

9.9 Challenges

9.9.1 Postmortem Interval Analysis

You have been called to testify as a biomedical engineering expert witness on behalf of the defendant who is accused of murder. The body of the defendant's boyfriend was found at 5:30 a.m. in a creek behind her house, and there was evidence of foul play. The forensic scientist working with the prosecutor places the time of death at about 2:45 a.m. The defendant has an unshakeable alibi after 4 a.m. when she checked into work as a nurse at the local hospital, but testifies that she was home alone after last being seen with the victim the previous night at 10 p.m.

The forensic scientist based the estimate of postmortem interval (PMI) on the change in body temperature as it came toward equilibrium with the environment, using the standard forensic empirical equation: $dT/dt = -\kappa(T - T_a)$, where T is body temperature, κ is a constant, and T_a = ambient temperature = 60°F. The 2:45 a.m. estimate is based on two measurements of rectal body temperature [T(5:30 a.m.) = 90°F, T(7:30 a.m.) = 85°F] after it was removed from the creek but retained at the scene. The temperature at the time of death was assumed to be 98.6°F.

Your job is to evaluate whether the analysis of PMI by the forensic scientist is credible, and, if not, what corrections need to be made and/or what additional information needs to be considered in determining the culpability of the defendant.

Your initial investigation uncovers additional information that may or may not be relevant. The body was nearly submerged in the creek when found and pulled from the water at 5:30 a.m. Water temperature was the same as ambient air, 60°F, and the water velocity was essentially zero. The mass of the victim was 80 kg, and surface area was 1.7 m². The specific heat of the body can be taken as $c_p = 4.2$ kJ/kg K and the thermal conductivity of skin is 0.4 W/m K. You also find that typical heat transfer coefficients in stagnant fluids are in the range of 5 W/(m²°C) for air and 50 W/(m²°C) for water.

9.9 Challenges

Based on this new information and your understanding of how the forensic scientist estimated the PMI, perform the following analyses:

1. Estimate what the value of κ in the forensic scientist's equation would be for heat transfer in stagnant water.
2. Can you do an inverse calculation to determine the value of κ for air used by the forensic scientist? How does this value compare with a calculation of κ from constitutive properties?
3. Use the value for κ in water to estimate the time of death.
4. Evaluate this method of analysis of PMI. Do you think your estimate in the foregoing calculations is an accurate representation of the actual time of death, and why? Will your testimony as it stands right now be sufficient to provide reasonable doubt of the defendant's guilt, or do you think there is further work to do?

9.9.2 Heart–Lung Machine Perfusion of Patient During Heart Surgery

A 37-year-old female was seen by her physician for signs of hypertension. An extra heart sound and subsequent echocardiogram revealed a large tumor mass on the right atrium through the tricuspid valve into right ventricle. A further CT scan showed a possible abdominal tumor extending up the inferior vena cava and through the heart. A decision was made to operate immediately using cardiopulmonary bypass with deep hypothermic circulatory arrest. The heart lung machine was operated and monitored continuously by a perfusion technologist. The patient core temperature was monitored via a probe inserted into the bladder.

The patient's core temperature was lowered by controlling the temperature of the solution perfused through her cardiovascular system while the head was packed in ice. Primary data recorded were the time, the perfusate flow rate and temperature (measured in the HLM), and the bladder probe temperature. The target core temperature was 15°C, at which state circulatory arrest was initiated and surgery performed on the heart for 22 min to identify and remove a tumor originating in the uterus and growing up through the vena cava into the heart. Next, HLM perfusion was reinstated with warm solution to actively rewarm the patient to 25°C. Over the next 10 days her neurological condition deteriorated, and she remains in a state of permanent severe brain damage. A malpractice lawsuit has ensued alleging that the surgical team stopped the perfusion to lower body core temperature before it reached 15°C. You have been assigned the task of reviewing the available data to determine whether it supports or negates the claim of the surgical team that they followed protocol and ensured that the patient's core temperature was lowered to 15°C before the perfusion was stopped.

The following data taken directly from the perfusionist's log sheet are provided to you for analysis. All of the data were entered by hand during the procedure. When the cardiopulmonary bypass procedure began, both the bladder probe and the perfusate return temperatures from the HLM read 35.5°C. The patient was then perfused for 15 min at a rate of 3.1 L/min, after which the bladder and perfusate temperatures read 24.6 and 19.8°C, respectively. The perfusion rate was then slowed to 2.5 L/min for an additional 7 min (22 min total of perfusion), after which circulatory arrest was started. The data sheet shows the bladder temperature as 15°C. The surgical procedure lasted 22 min, at the end of which the bladder and perfusate return temperature values were 21.0 and 14.7°C. The next data recording was after 16 min of perfusion at 1.5 L/min, with bladder and perfusate temperatures of 24.0 and 27.8°C.

It is alleged that the cause of the brain damage was that the perfusion was terminated at a temperature above 15°C, and even above 20°C, and therefore that the perfusionist's data log was falsified to cover the error. This allegation presumes that it would not have been possible to cool the patient's core temperature to 15°C for the recorded time of perfusion and the perfusion rates used. Your job is to analyze the available data and to determine if it is possible to prove whether the data were falsified. What type of information will you need to perform this analysis, and what assumptions will you be justified in making to perform the analysis? Be ready to defend your analysis in court before a hostile lawyer.

9.9.3 Design of a Combined Heat and Mass Exchanger

Your boss asks you to design a hollow fiber heat exchanger with air as the coolant. The idea is to use the same fibers for both heat exchange and gas exchange in a heart–lung machine. The heat exchanger is to remove heat at a rate of 2.27 kW, with a blood temperature difference of 10°C and an air temperature difference of 1°C. Blood flow rate through the device is 3.5 L/min. Your job is to analyze this potential design for feasibility and report results to your boss. What recommendation(s) will you make to your boss in relation to the initial thought of combining heat and mass exchange?

References

ASRHAE Handbook – Fundamentals (2009) ASRHAE, Inc., Atlanta
Buettner K (1952) Numerical analysis and pilot experiments of penetrating flash radiation effects. J Appl Physiol 5:207–222
Nyberg KL, Diller KR, Wissler EH (2000) Automatic control of thermal neutrality for space suit applications using a liquid cooling garment. Aviat Space Environ Med 71:904–915
Stoll AM (1967) Heat transfer in biotechnology. Adv Heat Transf 4:61–141

Stoll AM, Greene LC (1959) Relationship between pain and tissue damage due to thermal radiation. J Appl Physiol 14:373–382

Stoll AM, Chianta MA, Judge LB (1971) Development of practical high-intensity thermal protection systems. Aviat Med 42:54–58

Wissler EH (1961) Steady-state temperature distribution in man. J Appl Physiol 16:734–740

Wissler EH (1985) Mathematical simulation of human thermal behavior using whole body models. In: Shitzer A, Eberhart RC (eds) Heat transfer in medicine and biology: analysis and applications, vol 1. Plenum, New York, pp 325–373, Chap. 13

Wissler EH (2008) A quantitative assessment of skin blood flow in humans. Eur J Appl Physiol 104:145–157

Chapter 10
Shell Balance Approach for One-Dimensional Bioheat Transport

10.1 Introduction

Analysis methods in Chap. 9 were based on a macroscopic approach to bioheat transport, in which the system of interest was assumed to have a relatively uniform temperature. The temperature might change with time, but spatial variations within the system were assumed to be negligible or unimportant. In this chapter we will turn our interest to problems where there is an important spatial gradient in temperature. The spatial temperature gradient will cause a conduction heat flux in accordance with Fourier's law (2.9). Here we will restrict the scope of problems to those that are dimensional or nearly one-dimensional. Our analysis in each of these problems will begin by applying conservation of energy on an infinitesimally small portion of the system, known as a *shell*. A feature which distinguishes this approach from the more general approach to be applied in Chap. 11 is that the shells used in this chapter will shrink only in one dimension, so they might include a portion of the system boundary, while the shells used in the general approach shrink around an interior point in the system and do not include any portion of the system boundary. Consequently, energy that enters the system through the portion of the shell that includes the system boundary will be treated in this chapter as a term in the conservation equation rather than as a boundary condition. In reality, heat that enters through the system boundary often enters in a direction that is perpendicular to the assumed direction of energy flow. Although energy flow is not truly one-dimensional in such cases, the shell balance approach allows us to obtain realistic approximate solutions in which a more rigorous multidimensional approach would greatly increase the complexity but add little to the understanding or accuracy of the solution.

10.2 General Approach

Our approach to one-dimensional heat transfer problems will be similar to the approach outlined in Sect. 6.2 for momentum transport. Once a system has been identified, the first step is to construct a shell within the system. Next, if energy is assumed to flow in the x-direction, select an arbitrary point inside the system at a position x and draw a plane through the point perpendicular to the x-direction. Second, draw another plane parallel to the first through a point at $x + \Delta x$. The volume between the two planes constitutes the shell. Third, apply conservation of energy to the shell. Fourth, divide the resulting energy equation by the shell volume, and take the limit as Δx approaches zero. This step will result in an ordinary differential equation in the case of steady-state transport, or a partial differential equation in the case of unsteady–state transport. Fifth, solve the differential equation, subject to the relevant boundary conditions and, if an unsteady–state problem, the initial condition. Note that this approach closely follows the steps of the GIM.

10.3 Steady-State Conduction with Heat Generation

There are various types of bioheat transfer processes in which the distributed internal generation of energy plays an important role. These include metabolism, energy-based surgical procedures, and a wide variety of diagnostic and therapeutic procedures. The common characteristic of each of these processes is that energy is produced in the interior of the medium, thereby contributing to the heat that is conducted internally. The spatial pattern of energy generation influences the temperature profile and therefore the pattern of heat flow. In accordance with the principle of conservation of energy, heat that is generated within a material must either contribute to the local storage, which would cause the process to be transient since the thermal state of the system would be changing, or it must be carried away via some transport mechanism such as conduction or internal convection to blood. Many biologically significant processes involving internal energy generation occur under transient conditions and will be discussed later in this chapter.

At steady state, by definition, the net energy storage effects are zero. The mathematically simplest case is for a generation process that is uniform and invariant throughout the material, as may be encountered in conjunction with metabolism.

10.3.1 Steady-State Conduction with Heat Generation in a Slab

Let us begin by considering the slab of tissue in Fig. 10.1 with uniform heat generation per unit volume, $\dot{q}_{met}$. The top and bottom surfaces are insulated so heat is transferred only in the x-direction, and the temperature is expected to be symmetrical about the center of the slab ($x = 0$).

A shell is constructed by extending planes through the tissue at x and $x + \Delta x$, as shown in Fig. 10.1. If the cross-sectional area of the tissue is A_c, the volume of the shell is $A_c \Delta x$. Conservation of energy applied on the shell is:

10.3 Steady-State Conduction with Heat Generation

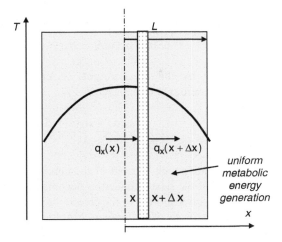

Fig. 10.1 Heat conduction in a one-dimensional symmetric Cartesian system with a uniform internal heat generation

$$\left\{\begin{array}{c}\text{rate at which}\\\text{energy accumulates}\\\text{in shell}\end{array}\right\} = \left\{\begin{array}{c}\text{rate at which}\\\text{energy enters}\\\text{shell}\end{array}\right\} - \left\{\begin{array}{c}\text{rate at which}\\\text{energy leaves}\\\text{shell}\end{array}\right\}$$

$$+ \left\{\begin{array}{c}\text{rate at which}\\\text{energy is}\\\text{produced in shell}\end{array}\right\}. \quad (10.1)$$

Since we are dealing with steady-state heat transfer, there is no accumulation of energy in the shell, so the left-hand side of this expression is zero. Energy enters and leaves the shell by conduction, so these terms are written as:

$$\left\{\begin{array}{c}\text{rate at which}\\\text{energy enters}\\\text{shell}\end{array}\right\} - \left\{\begin{array}{c}\text{rate at which}\\\text{energy leaves}\\\text{shell}\end{array}\right\} = q_x(x)A_c - q_x(x+\Delta x)A_c. \quad (10.2)$$

Note that if the top and bottom of the shell were not insulated, we would need to account for heat entering and leaving through these surfaces, and the problem would be multidimensional. Examples where heat is transferred through the shell surface will be given in the later sections. The rate of production of energy is equal to the rate of production per unit volume multiplied by the shell volume:

$$\left\{\begin{array}{c}\text{rate at which}\\\text{energy is}\\\text{produced in shell}\end{array}\right\} = \dot{q}_{\text{met}} A_c \Delta x. \quad (10.3)$$

Combining these terms in the conservation of energy expression results in:

$$0 = q_x(x)A_c - q_x(x + \Delta x)A_c + \dot{q}_{met}A_c\Delta x. \tag{10.4}$$

Dividing by the shell volume $A_c\Delta x$ and taking the limit as the volume approaches zero, we find:

$$0 = \lim_{\Delta x \to 0}\left(\frac{q_x(x) - q_x(x + \Delta x)}{\Delta x}\right) + \dot{q}_{met} = -\frac{dq_x}{dx} + \dot{q}_{met}. \tag{10.5}$$

The negative sign in front of the conduction term results from the fact that for increasing x, the heat conduction becomes larger as more accumulated metabolic heat must be transported to the free surface of the material. As the heat conduction increases, the slope of the temperature gradient becomes increasingly more negative, reaching a maximum at the surface. At the center plane the temperature gradient and heat conduction are both zero since the boundary conditions are identical at each end, resulting in symmetry about the center. This situation is depicted in Fig. 10.1.

Since the heat flux is by conduction, we can introduce Fourier's law for q_x:

$$0 = \frac{d}{dx}\left(-k\frac{dT}{dx}\right) + \dot{q}_{met}. \tag{10.6}$$

Finally, if the thermal conductivity is independent of the position,

$$0 = \frac{d^2T}{dx^2} + \frac{\dot{q}_{met}}{k}. \tag{10.7}$$

This second-order differential equation requires two boundary conditions, which are identified at the centerline, $x = 0$, where the slope of the temperature gradient is zero due to combined geometric and thermal symmetry, and the outer surfaces of the material, $x = \pm L$, where standard conditions of temperature, heat flux or convection may be specified. If the environment results in the temperature being fixed, then the boundary conditions are described as

$$T = T_L \text{ at } x = \pm L \text{ and } \frac{dT}{dx} = 0 \text{ at } x = 0.$$

Integrating twice yields the general solution of (10.7),

$$T(x) = -\frac{\dot{q}_{met}}{2k}x^2 + C_1 x + C_2. \tag{10.8}$$

The boundary condition at $x = 0$ dictates that $C_1 = 0$. C_2 is evaluated at $x = L$,

$$T_L = -\frac{\dot{q}_{met}L^2}{2k} + C_2 \text{ or } C_2 = T_L + \frac{\dot{q}_{met}L^2}{2k}.$$

10.3 Steady-State Conduction with Heat Generation

Thus, the full solution for the temperature distribution is

$$T(x) - T_L = \frac{\dot{q}_{\text{met}}}{2k}(L^2 - x^2). \tag{10.9}$$

The maximum temperature difference occurs between the center and the surface, for which

$$T_0 - T_L = \frac{\dot{q}_{\text{met}}}{2k}L^2. \tag{10.10}$$

By subtracting (10.10) from (10.9), the temperature can be scaled to the maximum value.

$$T(x) - T_0 = -\frac{\dot{q}_{\text{met}}}{2k}x^2. \tag{10.11}$$

The temperature can be normalized to the surface value to show the parabolic distribution.

$$\frac{T(x) - T_0}{T_L - T_0} = \left(\frac{x}{L}\right)^2. \tag{10.12}$$

At steady state the total heat generated within the material per unit of cross-sectional area, A_c, must equal the rate of conduction at the surfaces

$$2LA_c\dot{q}_{\text{met}} = -2kA_c\frac{dT}{dx}\bigg)_{x=L} = -2kA_c(T_L - T_0)\frac{2}{L}$$

$$\text{or } \dot{q}_{\text{met}} = \frac{2k}{L^2}(T_0 - T_L), \tag{10.13}$$

which could alternatively be written directly from (10.11) as evaluated over the entire system volume.

10.3.2 Steady-State Conduction with Heat Generation in a Cylinder

An analogous analysis can be performed for a system characterized in cylindrical coordinates as shown in Fig. 10.2. In this case the shell is the volume between two cylinders of constant r and $r + \Delta r$. Applying conservation of energy to the cylindrical shell with surface area $2\pi rL$ at position r, $2\pi(r + \Delta r)L$ at position $r + \Delta r$ and shell volume of $2\pi rL\Delta r$, where L is the length of the shell:

$$0 = q_r(r)(2\pi rL) - q_r(r + \Delta r)(2\pi(r + \Delta r)L) + \dot{q}_{\text{met}}(2\pi rL\Delta r). \tag{10.14}$$

Fig. 10.2 Heat conduction in a one-dimensional cylindrical system with a uniform internal heat generation

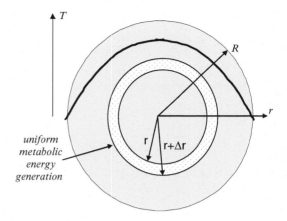

Noting that the product $(r + \Delta r)q_r(r + \Delta r)$ is simply rq_r evaluated at $r + \Delta r$, we can divide (10.14) by the shell volume and let the Δr approach zero:

$$0 = \lim_{\Delta r \to 0} \left(\frac{(rq_r)|_r - (rq_r)|_{r+\Delta r}}{r\Delta r} \right) + \dot{q}_{\text{met}} = -\frac{1}{r}\frac{d}{dr}(rq_r) + \dot{q}_{\text{met}}. \quad (10.15)$$

Introducing Fourier's law for q_r and assuming k is constant, (10.15) can be rearranged as:

$$0 = \frac{1}{r}\frac{d}{dr}\left(r\left(\frac{dT}{dr}\right)\right) + \frac{\dot{q}_{\text{met}}}{k}. \quad (10.16)$$

The two required boundary conditions may be identified at the centerline, where the slope of the temperature gradient is zero due to combined geometric and thermal symmetry, and known temperature the outer surface of the material, analogous to the conditions for Cartesian coordinates. For steady state the conservation of energy requires that the cumulative rate of internal energy generation must equal the rate of conduction at the surface to the environment. These boundary conditions are described as

$$T = T_R \text{ at } r = R \text{ and } \frac{dT}{dr} = 0 \text{ at } r = 0 \text{ or } \dot{q}_{\text{met}}\pi R^2 L = -2\pi RLk\left.\frac{dT}{dr}\right)_{r=R}.$$

The latter condition can be simplified as

$$\left.\frac{dT}{dr}\right)_{r=R} = -\frac{\dot{q}_{\text{met}}R}{2k}.$$

The general solution of (10.16) is obtained by integrating twice with respect to r.

$$T(r) = -\frac{\dot{q}_{\text{met}}}{4k}r^2 + C_1 \ln r + C_2. \quad (10.17)$$

10.3 Steady-State Conduction with Heat Generation

The second boundary condition dictates that $C_1 = 0$. C_2 is evaluated at $r = R$.

$$T_R = -\frac{\dot{q}_{met} R^2}{4k} + C_2 \quad \text{or} \quad C_2 = T_R + \frac{\dot{q}_{met} R^2}{4k}.$$

Thus, the full solution for the temperature distribution is

$$T(r) - T_R = \frac{\dot{q}_{met}}{4k}\left(R^2 - r^2\right). \tag{10.18}$$

The maximum temperature difference occurs between the center and the surface, for which

$$T_0 - T_R = \frac{\dot{q}_{met}}{4k} R^2. \tag{10.19}$$

By subtracting (10.19) from (10.18), the temperature can be scaled to the maximum value.

$$T(r) - T_0 = -\frac{\dot{q}_{met}}{4k} r^2. \tag{10.20}$$

The temperature can be normalized to the surface value to show the parabolic distribution.

$$\frac{T(r) - T_0}{T_R - T_0} = \left(\frac{r}{R}\right)^2. \tag{10.21}$$

Example 10.3.1 Electrical Current Treatment for Pain.
A novel method of treating pain is to direct an electrical current to the tissue that is the source of the signal. This therapy is accomplished by inducing a small electrical current through tissue for times that are long enough for the system to reach steady state. Although the therapeutic effects have been demonstrated, there is a lingering concern that the procedure could cause a burn to the patient. The protocol specifies the use of currents as large as 15 mA for 45 min. The criterion for avoiding a burn under these conditions is that the temperature should not exceed 43°C. This procedure is normally performed with no active control of the surface boundary condition, which by default is natural convection with air in the clinic. Figure 10.3 shows application of this technique to a finger. The operating conditions under this state are to be determined. It has been suggested that proactive intervention for surface cooling should be implemented to ensure that burns cannot occur. A possible control scheme would be to apply a cooling cuff to a finger being treated with a circulating refrigerant capable of dictating the surface temperature on the skin. If this latter action were taken, what skin temperature would you recommend to meet the safety criterion?

Fig. 10.3 Heat generation in a finger during electrical current therapy

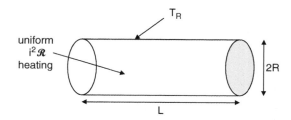

Solution. *Initial considerations:* The objective is to determine the conditions under which the electrical current therapy technique can be applied to the finger without the danger of causing thermal injury. The mechanism of concern is electrical ($I^2 \Re_e$) heating of the tissues. It will be necessary to identify the quantitative operational parameter values for the process so that they can be implemented for specific patients.

System definition and environmental interactions: The relevant system is the finger as shown in Fig. 10.3. At steady state, the finger loses heat to the environment from the skin surface at the same rate that it is generated internally by the therapeutic current.

Apprising the problem to identify governing equations: For purposes of this problem several assumptions can be made:

- The rate of metabolic heat generation in the finger is negligible in comparison to that for the therapeutic electrical current.
- The finger is treated as a homogeneous medium from the aspects of both electrical energy generation via $I^2\Re_e$ dissipation and heat conduction. The electrical and thermal properties, and the heat generation are uniform throughout the entire finger and assumed to be that of muscle.
- The primary temperature gradient occurs only in the radial direction, and heat loss from the tip is negligible.
- The entire process occurs with a geometric symmetry, including convection from the skin surface to surrounding air.
- Passive heat transfer from skin surface to surrounding air occurs by natural convection, with a coefficient of 2 W/m² K.
- The ambient air temperature is 25°C.
- Blood perfusion through the finger does not have a significant effect on the temperature distribution during the therapeutic protocol.

Application of the conservation of energy and the relevant constitutive equations has already resulted in the definition of the general solution for the temperature distribution given in (10.17).

$$T(r) = -\frac{\dot{q}_e}{4k}r^2 + C_1 \ln r + C_2. \qquad (10.22)$$

10.3 Steady-State Conduction with Heat Generation

Analysis: For this problem the energy generation per unit volume $\dot{q}_e$ is caused by electrical energy dissipation rather than metabolism. However, a different set of boundary conditions is applicable to this problem than was used previously for deriving a solution. Since the maximum temperature will occur at the center line of the finger, this is the location where the burn prevention criterion should be applied. Therefore, the centerline temperature should not exceed 43°C, which means that in the extreme case it would equal this value. Also, due to conditions of symmetry, we know that at the center the temperature gradient will be zero irrespective of the value of the temperature.

When natural convection occurs from the skin surface, the boundary condition states that the conductive and convective fluxes at the surface are continuous (i.e., equal), and they must also equal the net rate of internal heat generation.

$$-k2\pi RL\frac{dT}{dr}\bigg)_{r=R} = h2\pi RL(T_R - T_\infty) = \pi R^2 L \dot{q}_e.$$

Thus, the temperature at the surface of the skin is related to the combined convection and heat-generation processes.

$$T_R = T_\infty + \frac{R}{2h}\dot{q}_e.$$

The heat generation in tissue with volume V occurs as a result of $I^2\mathfrak{R}_e$ dissipation in the tissue.

$$V\dot{q}_e = I^2\mathfrak{R}_e.$$

The heating resistance $\mathfrak{R}_e$ is dependent on the intrinsic resistivity $\mathfrak{R}'_e$ and on the geometry of the finger,

$$\mathfrak{R}_e = \frac{\mathfrak{R}'_e L}{A_c} = \frac{\mathfrak{R}'_e L}{\pi R^2},$$

where A_c is the cross-sectional area through which the electrical current flows. The electrical resistivity is taken as that of muscle at 2.5 Ω m which yields an overall resistance of 636 Ω distributed uniformly throughout the finger. Thus, for a current of 15 mA, the energy generation per unit volume by electrical dissipation is

$$\dot{q}_e = \frac{I^2\mathfrak{R}_e}{V} = \frac{I^2\mathfrak{R}'_e L}{\pi R^2 \cdot \pi R^2 L} = \frac{I^2\mathfrak{R}'_e}{(\pi R^2)^2} = \frac{(1.5 \times 10^{-2}\,\text{A})^2 \times 2.5\,\Omega\,\text{m}}{(3.14 \times 10^{-4}\,\text{m}^2)^2}$$

$$= 5.70 \times 10^3 \left(\frac{\text{W}}{\text{m}^3}\right).$$

This energy source is balanced by the convective heat loss from the skin at steady state, which thereby determines the surface temperature.

$$T_R = 25°C + \frac{10^{-2} \text{ m}}{2 \times 2 \text{ W/m}^2 \text{ K}} \times 5.70 \times 10^3 \text{ W/m}^3 = 39.25°C.$$

The temperature difference between the surface and the center line is determined by (10.19), which will identify the locus of highest temperature at the geometric center. The thermal conductivity of muscle is 0.52 W/m K.

$$T_0 = T_R + \frac{\dot{q}_e}{4k} R^2 = 39.25°C + \frac{5.70 \times 10^3 \text{ W/m}^3}{4 \times 0.52 \text{ W/m K}} \times (10^{-2} \text{ m})^2 = 39.52°C.$$

The temperature difference between the center and surface of the finger is only 0.27°C.

Examining and interpreting the results: The analysis is directed toward evaluating the potential for causing a burn and identifying the conditions that will prevent a burn. The greatest risk for a burn occurs at the highest temperature. Although the protocol must have a transient phase initially during which the tissue temperature range rises from physiological to therapeutic, the highest temperatures will occur only after a steady state is attained. Therefore, a steady-state analysis will identify the conditions that define the maximum risk of a burn, and it is not necessary to conduct the more complex analysis of the transient phase of the protocol. At steady state the finger loses heat to the environment at the same rate that it is generated internally by the therapeutic current.

It is a valid concern that the electrical current therapy could cause a tissue burn. However, under the assumptions exercised for this problem, the analysis shows that temperature rise is well below the conservative threshold that is needed to cause a thermal injury at 43°C. The implication is that this therapy should be considered safe for the conditions considered.

The temperature calculations were performed on the assumption that the thermal field was dictated only by considerations of the internal energy generation and the convection from the skin. In actuality, under normothermia there are numerous factors that contribute to determining the finger temperature, including conduction from the hand and convection with perfused blood. Both of these phenomena will have the tendency to cause the tissue temperature to be lower than was calculated, thereby increasing the margin of safety. Although a more complicated analysis would be required to investigate these added effects, it is reasonable to neglect them when assessing the safety of the proposed therapeutic protocol.

Additional comments: There still may be an additional effect that could negatively influence the safety of the protocol owing to the influence of heating in the bone not being included. The intrinsic electrical resistance of bone is about 40 times that of muscle (Lee et al. 2000). Thus, it could be possible that the same current density in the bone would result in an injurious elevation of temperature. The analysis of this

10.3.3 Steady-State Conduction with Heat Generation in a Sphere

Metabolic heat generation and conduction in a spherically shaped object, such as a spherical cell, can be approached in the same manner as for the slab or cylinder. Here we will assume that the heat sources are evenly distributed throughout the interior of the cell and that heat flows in the radial direction, r. In the steady state, energy does not accumulate in the cell, so all of the heat generated ($\dot{q}_{met} 4\pi R^3/3$) is lost through the surface of the cell ($4\pi R^2 q_r|_R$). Therefore, an appropriate boundary condition at the cell surface is that the rate at which heat is generated within the cell equals the rate at which it is lost at the surface.

$$q_r|_{r=R} = \frac{\dot{q}_{met} R}{3}. \qquad (10.23)$$

To find the temperature distribution in the cell we first select shells at r and $r + \Delta r$, similar to the approach taken with the cylinder. The difference here is that there is no exposed edge in a nonradial dimension of the shell as was the case of the cylinder and plane. The surface area of the inner shell is $4\pi r^2$, the surface area of the outer surface of the shell is $4\pi(r + \Delta r)^2$, and the shell volume is $4\pi r^2 \Delta r$. The steady-state conservation of energy statement for the shell is:

$$0 = q_r(r)(4\pi r^2) - q_r(r + \Delta r)\left(4\pi(r + \Delta r)^2\right) + \dot{q}_{met}(4\pi r^2 \Delta r). \qquad (10.24)$$

Noting that the product $(r + \Delta r)^2 q_r(r + \Delta r)$ is simply $r^2 q_r$ evaluated at $r + \Delta r$, we can divide (10.24) by the shell volume and let the Δr approach zero:

$$0 = \lim_{\Delta r \to 0} \left(\frac{(r^2 q_r)|_r - (r^2 q_r)|_{r+\Delta r}}{r^2 \Delta r} \right) + \dot{q}_{met} = -\frac{1}{r^2} \frac{d}{dr}(r^2 q_r) + \dot{q}_{met}. \qquad (10.25)$$

Integrating (10.25) from r to R, and using (10.23) as a boundary condition at the cell surface, we find that the heat flux increases linearly with the radial position:

$$q_r = \frac{\dot{q}_{met} r}{3}. \qquad (10.26)$$

Applying Fourier's law:

$$-k\frac{dT}{dr} = \frac{\dot{q}_{met} r}{3}.\tag{10.27}$$

Integrating and applying the boundary condition of a known temperature T_R at the outside surface, we find the following parabolic temperature distribution in the cell:

$$T = T_R + \frac{\dot{q}_{met} R^2}{6k}\left(1 - \left(\frac{r}{R}\right)^2\right).\tag{10.28}$$

The temperature is maximum at the center of the cell, and the difference between the temperature at the center of the cell and the surface temperature is $\dot{q}_{met} R^2/6k$.

10.4 Steady-State One-Dimensional Problems Involving Convection

There are many instances in which a fluid is heated or cooled as it flows through a conduit. Energy exchange between the surroundings and the fluid flowing through the conduit is classified as an internal convection problem. Examples in biomedical engineering include devices, such as blood heat exchangers, and in vivo heat transfer in airways and blood vessels.

For an internal flow of a fluid through a lumen of finite length, conservation of energy applied for steady state relates the convective heat flow at the lumen inner surface to the difference in the mean inlet and outlet temperatures. This type of process occurs very frequently in internal convective flows, and is illustrated in Fig. 10.4.

The system in this case consists of the fluid identified when it is between a plane at x and a second plane at $x + \Delta x$, as shown in Fig. 10.4. The temperature of the fluid, T_m, is defined as a mixed value representative of the state of the entire flow cross-section. Note that the fluid in contact with the wall forms part of the boundary of the system. Therefore, a statement of conservation of energy will include not only the rate at which energy enters the shell at x and leaves at $x + \Delta x$, but must also include the amount of heat $\Delta \dot{Q}_{conv}$ added at the boundary. Under these conditions, the rate at which energy is added by convection at the periphery of the lumen will equal the difference in the rates at which enthalpy enters and leaves the shell at x and $x + \Delta x$. For steady-state heat transfer, the accumulation term is zero and an energy balance on the shell is:

$$0 = w\hat{H}_m\big|_x + \Delta\dot{Q}_{conv} - w\hat{H}_m\big|_{x+\Delta x}.\tag{10.29}$$

10.4 Steady-State One-Dimensional Problems Involving Convection

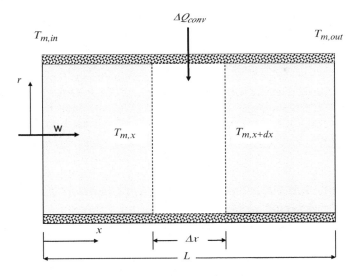

Fig. 10.4 System for analysis of a small length, Δx, during convective flow through a lumen

For physiological levels of pressure, the mean enthalpy per unit mass $\hat{H}_m$ is nearly equal to the mean internal energy per unit mass, $c_p \Delta T_m$.

$$\hat{H}_m\big|_x - \hat{H}_m\big|_{x+\Delta x} = c_p\left(T_m\big|_x - T_m\big|_{x+\Delta x}\right). \tag{10.30}$$

In addition, the heat flow by convection at the periphery, P, of the lumen is given by:

$$\Delta \dot{Q}_{conv} = q_s P \Delta x, \tag{10.31}$$

where q_s is the surface flux at the boundary, P is the peripheral dimension of the lumen and $P\Delta x$ is the peripheral surface area of the fluid shell. Substituting (10.30) and (10.31) into (10.29), dividing by Δx, and taking the limit as Δx approaches zero yields:

$$0 = P q_s - w c_p \frac{dT_m}{dx}. \tag{10.32}$$

Newton's law of cooling (8.103) can be applied to represent the local convective heat flux, q_s.

$$\frac{dT_m}{dx} = \frac{P}{w c_p} h(x)(T_s - T_m). \tag{10.33}$$

The solution to (10.33) will describe how the mean temperature of a fluid will change as it flows through a lumen with convective heat transfer at the wall. The shape of this temperature profile will depend on the specific boundary conditions at the wall of the lumen. The two most simple boundary conditions are for the surface temperature, T_s, or the heat flux, q_s, to be constant along the length of the lumen. Further, the convective heat transfer coefficient may be constant with position if the flow pattern in the lumen is fully developed. We will analyze the convection process for conditions of constant temperature and constant heat flux imposed at the wall surface.

10.4.1 Internal Flow Convection with a Constant Temperature Boundary Condition

For the case of a *constant surface temperature*, it is convenient to redefine the temperature scale as $\theta = T_s - T_m$ so that $d\theta/dx = -dT_m/dx$, since T_s is invariant and its derivative must be zero. Then (10.33) becomes

$$-\frac{d\theta}{dx} = \frac{P}{wc_p} h(x)\, \theta. \tag{10.34}$$

The variables can be separated, and, for a steady-state flow through a lumen of constant cross-section, the resulting equation integrated from the lumen inlet to outlet is:

$$\int_{\theta_{in}}^{\theta_{out}} \frac{d\theta}{\theta} = \ln \frac{\theta_{out}}{\theta_{in}} = -\frac{P}{wc_p} \int_0^L h(x)dx = -\frac{PL}{wc_p} \bar{h}_L, \tag{10.35}$$

where $\bar{h}_L$ is the average convective heat transfer coefficient over the entire surface area of the lumen. Equation (10.35) can be rewritten as an exponential function,

$$\frac{\theta_{out}}{\theta_{in}} = \frac{T_s - T_{m,out}}{T_s - T_{m,in}} = \exp\left(-\frac{PL}{wc_p}\bar{h}_L\right)_{T_s} = \exp\left(-\frac{S}{wc_p}\bar{h}_L\right)_{T_s}, \tag{10.36}$$

where $S = PL$ is the surface area of the lumen and the subscript T_s indicates constant surface temperature. The temperature at any position x along the length of the lumen is given by:

$$\frac{\theta_x}{\theta_{in}} = \frac{T_s - T_{m,x}}{T_s - T_{m,in}} = \exp\left(-\frac{Px}{wc_p}\bar{h}_x\right)_{T_s}. \tag{10.37}$$

This equation shows that the difference between the wall surface and the mean fluid temperatures decreases exponentially from the inlet of the lumen.

10.4 Steady-State One-Dimensional Problems Involving Convection

Conservation of energy shows that the net total convection between the wall surface and the fluid is equal to the overall mean enthalpy change of the fluid as it flows from the lumen inlet to outlet:

$$\dot{Q}_{conv} = w\left(\hat{H}_{m,in} - \hat{H}_{m,out}\right) = wc_p\left(T_{m,in} - T_{m,out}\right), \qquad (10.38)$$

which can be written in terms of θ as:

$$\dot{Q}_{conv} = wc_p\left(\theta_{m,in} - \theta_{m,out}\right). \qquad (10.39)$$

An expression for wc_p can be introduced by taking the logarithm of (10.36). Inserting this into (10.39), for a constant surface temperature, T_s:

$$\dot{Q}_{conv} = \bar{h}_L S \frac{\theta_{out} - \theta_{in}}{\ln(\theta_{out}/\theta_{in})} = \bar{h}_L S \theta_{lm}, \qquad (10.40)$$

where θ_{lm} is defined as the *log mean temperature difference*.

$$\theta_{lm} = \frac{\theta_{out} - \theta_{in}}{\ln(\theta_{out}/\theta_{in})} = \frac{(T_s - T_{m,out}) - (T_s - T_{m,in})}{\ln((T_s - T_{m,out})/(T_s - T_{m,in}))}. \qquad (10.41)$$

In some types of application, a fluid flows through a tube that is surrounded by a second external fluid at a temperature T_∞ that exchanges heat with the exterior surface via convection. An example of this arrangement is a shell and tube heat exchanger such as might be applied to regulate the temperature of blood during a cardiovascular bypass procedure. Blood would flow through the inner tubes to have its temperature controlled to a selected value, and water would flow in relatively large amounts through the external shell outside the tube at the control temperature T_∞. If the relative flow rate of the external fluid is large so that it is continually being renewed at the temperature T_∞, then it can act to produce a boundary condition for the internal fluid that approximates a constant wall temperature. In heat exchanger analysis the net heat exchange coefficient between the external fluid at T_∞ and the internal fluid at T_m can be represented by an *average overall heat transfer coefficient*, $\bar{U}$. $\bar{U}$ embodies the effects of the average internal and external convection coefficients plus the conduction through the lumen wall thickness. Then, the equation for net convective transfer for the fluid flowing through the lumen is written with the reference temperature of the external fluid rather than that of the lumen wall.

$$\frac{\theta_{out}}{\theta_{in}} = \frac{T_\infty - T_{m,out}}{T_\infty - T_{m,in}} = \exp\left(-\frac{\bar{U}S}{wc_p}\right)_{T_s} \qquad (10.42)$$

and

$$\dot{Q}_{\text{conv}} = \bar{U}S\theta_{\text{lm}}. \tag{10.43}$$

We will take a more comprehensive look at this type of problem later in this chapter when heat exchangers are discussed.

> **Example 10.4.1 Cooling a Solution in a Research Lab.**
> A laboratory procedure requires that a large volume of an aqueous solution be cooled to a temperature of 4°C or lower after it is prepared and before it is used in an experiment. This process is to be accomplished via a single pass through a copper coil immersed in an ice water bath. The solution is prepared at room temperature, 22°C. The copper tube available to fabricate a cooling coil has a nominal diameter of 1 in., which has an inner diameter of 0.995 in. (25.3 mm). Separate conditions dictate that the flow rate be 0.05 kg/s. It has been independently determined that the convective heat transfer coefficient for the solution flowing through the tube is 4,000 W/m² K. How long should the coil be to achieve the required temperature drop of the flowing solution?

Solution. *Initial considerations:* As the aqueous solution flows through the copper tube heat will be lost to the ice water bath. We may assume that the procedure is run in a manner such that not all of the ice is melted. Therefore, the outside of the copper tube will be maintained along its entire length at 0°C. The thermal conductivity of copper is quite large, meaning that the thermal resistance is small. For our initial calculations we can assume that the internal convective resistance between the tube and the flowing water completely dominates heat transfer from the solution to the ice water bath. The copper tube will have a temperature of 0°C along its entire length.

System definition and environmental interactions: The process of interest is the loss of heat from the aqueous solution as it flows through the copper tube, causing the outlet temperature to be lower than the inlet temperature. To describe and analyze this process, the most appropriate system is the copper tube with open boundaries at the inlet and outlet. Interactions with the environment occur via convection with the solution at the inner wall and via enthalpy flows in and out.

Apprising the problem to identify governing equations: We can apply (10.39) and (10.40) to calculate the required length of tubing.

$$\dot{Q}_{\text{conv}} = wc_{\text{p}}\left(\theta_{\text{m,in}} - \theta_{\text{m,out}}\right), \tag{10.44}$$

$$\dot{Q}_{\text{conv}} = \bar{h}_{\text{L}}S\frac{\theta_{\text{out}} - \theta_{\text{in}}}{\ln(\theta_{\text{out}}/\theta_{\text{in}})} = \bar{h}_{\text{L}}S\theta_{\text{lm}}. \tag{10.45}$$

10.4 Steady-State One-Dimensional Problems Involving Convection

The convective heat flow can be eliminated between these two equations, leaving the surface area of the tube interior as the only unknown parameter. Given the tube diameter, the length can be calculated from the surface area.

Analysis: Equations (10.39) and (10.40) are combined to obtain an expression for the surface area of the tube

$$\bar{h}_L S \frac{\theta_{out} - \theta_{in}}{\ln(\theta_{out}/\theta_{in})} = -wc_p(\theta_{m,in} - \theta_{m,out})$$

or

$$S = -\frac{wc_p}{\bar{h}_L} \ln(\theta_{out}/\theta_{in}).$$

From Appendix C, the specific heat of water is $c_p = 4.18 \times 10^3$ J/kg K, which can be taken to approximate the value for the aqueous solution. The surface area is computed from the known property values,

$$S = -\frac{0.05 \text{ kg/s} \cdot 4.18 \times 10^3 \text{ J/kgK}}{4,000 \text{ W/m}^2 \text{ K}} \ln((4-0)/(22-0)) = 0.0891 \text{ m}^2.$$

The tube length is determined from the surface area,

$$L = \frac{S}{d} = \frac{0.0891 \text{ m}^2}{0.0253 \text{ m}} = 3.52 \text{ m}.$$

Examining and interpreting the results: The length of the tube is based on the assumption that there is no significant thermal resistance associated with heat conduction through the copper tube or heat transfer to the ice water bath. In actuality, these resistances must be nonzero since a practical perfect conductor does not exist. Therefore the rate of heat loss will be less than we calculated, with the result that the temperature drop of the aqueous solution will be less than the desired amount. To compensate, we can lengthen the tube. We could do a more comprehensive calculation, taking into account conduction through the copper tube and convection to the ice water bath. Alternatively, if we are confident that these effects are minimal, we could simply lengthen the tube to 4 m (an additional 14%), which should be adequate.

Additional Comments: It may be helpful to know the rate at which ice will be melted during the process. Either (10.39) or (10.40) may be applied to calculate the value of $\dot{Q}_{conv}$. Given that the latent heat of fusion for water is $\Lambda = 333.5$ J/g, the melt rate can be determined.

$$\dot{Q}_{conv} = 0.05 \text{ kg/s} \cdot 4.18 \times 10^3 \text{ J/kg K}(22-4) \text{ K} = 3.76 \times 10^3 \text{ W}.$$

Then, for a melt rate of w_m kg/s,

$$\dot{Q}_{conv} = 3.76 \times 10^3 \text{ W} = (w_m \text{ kg/s})(\Lambda \text{ J/kg})$$

or

$$w_m = \frac{\dot{Q}_{conv}}{\Lambda} = \frac{3.76 \times 10^3 \text{ W}}{333.5 \text{ J/g}} = 11.3 \text{ g/s}.$$

The ratio of the rate of melting of water to the rate of flow of an aqueous solution changing temperature from 22°C to 4°C is dimensionless,

$$\frac{w_m}{w_{flow}} = \frac{11.3 \text{ g/s}}{50 \text{ g/s}} = 0.226.$$

We calculate that 0.226 g of ice melts for each gram of aqueous solutions that is cooled from 22°C to 4°C. This ratio is reflective of the relative magnitudes of latent and sensible heats. The ratio of sensible to latent heat for a process is defined as the *Stefan number*. It is quite useful for the analysis of freezing and thawing problems,

$$Ste = \frac{c_p \Delta T}{\Lambda}.$$

10.4.2 Internal Flow Convection with a Constant Heat Flux Boundary Condition

The alternate boundary condition inside the lumen wall is for a *constant surface heat flux*. Starting with (10.32), the heat flux term, q_s, will be the same at all the positions along the lumen, dictating that the temperature gradient also will not change with the position along the length of the lumen,

$$\frac{dT_m}{dx} = \left(\frac{q_s P}{W c_p}\right)_{q_s}. \tag{10.46}$$

Therefore, the temperature will increase linearly as the fluid flows through the lumen. Likewise, the total convective heat transfer along the length of the lumen is simply the product of the wall area and the uniform heat flux,

$$\dot{Q}_{cond} = q_s S. \tag{10.47}$$

10.4 Steady-State One-Dimensional Problems Involving Convection

Example 10.4.2 Uniform Convective Heating of a Fluid Flowing Inside a Pipe.
A bioprocessing plant uses a heater to warm an aqueous protein solution during manufacturing from 10°C to 42°C. The heater consists of a straight, thin-walled tube with an inner diameter of 15 mm. The heating process is accomplished by an electric resistance element that is affixed to the outside surface of the tube. The total length of the tube is 5 m, and the average flow velocity of the solution is 0.25 m/s. As the solution enters the tube, the temperature is uniform, and the velocity profile can be considered to be fully developed. What must the rate of heating be to produce the targeted temperature increase?

Solution. *Initial considerations:* The goal is to calculate the rate of heat flux that must be applied for the flow conditions and geometry described to increase the temperature of a flowing solution from 10°C to 42°C. The resistance heater applied at the wall enforces that the convection boundary condition is a constant heat flux. The highest temperature reached by the solution must be determined so that the conditions can be avoided that cause denaturation of the protein product. For this particular product, denaturation can begin at temperatures above 43°C. Designing for a maximum temperature of 42°C will provide a buffer against damaging the product, while still allowing for a good level of performance at the maximum safe operating temperature.

System definition and environmental interactions: An open system is defined with the protein solution flowing through it. The boundary of the system is the inner walls of the heat exchanger tube, plus the flow cross-sectional areas at the inlet and outlet. The interactions are convective heat flux at the walls and the mass flow across the inlet and outlet cross-sections.

Apprizing the problem to identify governing equations: The conservation of energy is embedded in (10.46), which is applicable to this problem,

$$\frac{dT_m}{dx} = \left(\frac{q_s P}{W c_p}\right)_{q_s}.$$

Analysis: Since the surface heat flux is uniform over the entire length of the tube, all the terms in this equation are constant. The temperature gradient is constant, and the temperature increases linearly with the position along the length of the tube,

$$\frac{dT_m}{dx} = \frac{T_{out} - T_{in}}{L} = \frac{42 - 10}{5}\frac{K}{m} = 6.4\,K/m.$$

The flow is given in terms of the solution velocity. It is necessary to identify a density of the solution to determine the mass flow rate. The density should be close to that of water. Thus,

$$w = \rho \cdot A \cdot <v> = 1 \times 10^3 \text{kg/m}^3 \cdot \frac{\pi}{4}(0.015)^2 \text{ m}^2 \cdot 0.25 \text{ m/s} = 4.42 \times 10^{-2} \text{ kg/s}.$$

The requisite heat flux is then calculated as

$$q_s = \frac{T_{\text{out}} - T_{\text{in}}}{L} \frac{wc_p}{P} = 6.4 \text{ K/m} \frac{4.42 \times 10^{-2} \text{ kg/s} \cdot 4.36 \times 10^3 \text{ J/kgK}}{\pi \cdot 0.015 \text{ m}}$$
$$= 26.2 \text{ kW/m}^2.$$

Examining and interpreting the results: The total heat transfer can be determined by multiplying the heat flux and the convective transfer area of the tube,

$$\dot{Q}_x = q_s \cdot S = q_s \cdot P \cdot L = 6.16 \times 10^3 \text{ W}$$

The actual capacity of the electric heater would need to be greater than this value since not all of the energy dissipated by it will be transferred to the protein solution flowing through the tube. Some energy will be lost to the other components of the heater environment. Therefore, if the entire assembly has a reasonably good outer insulation, a 10 kW specification for the electric heater should provide acceptable performance.

10.4.3 Heat Exchangers

One of the most common applications of the principles of heat transfer is in the design of devices to produce a transfer of heat between fluids at different temperatures. These devices are typically called *heat exchangers*, and there is a long history of detailed research and development that has been directed toward this area of heat transfer. In most instances there is a network of rigid flow passages that serve to isolate the two exchanging fluids. The fluid that is the target for having its temperature changed via the heat transfer is often contained within an inner-most flow passage, and the second fluid may be in an outer shell or may be unconstrained to include the entire environment. An example of the latter arrangement is the radiator of an automobile. The function of the radiator is to reduce the temperature of the coolant solution that is circulated to and from the engine. As the coolant is circulated through the inner core tubing of the radiator, a second cooling fluid, environmental air, flows over the external surfaces of the tubes to receive the heat lost by the coolant. The coolant fluid is circulated repeatedly between the engine

10.4 Steady-State One-Dimensional Problems Involving Convection

and the radiator, whereas the air passes only one time through the radiator. New makeup air is supplied continuously to the radiator. A radiator is designed with an external mesh of fins connecting the coolant tubes to provide a larger contact surface area for the air. This design feature makes the overall thermal efficiency of the radiator greater by matching the convection transport of the coolant and air. Since the air is in vapor phase, its convective heat transfer coefficient will be less than that of the coolant, which is in the liquid phase. By providing a larger transfer surface for the air, the products of $h \cdot A$ for the air and coolant are brought into a closer match, and there can be a larger net heat exchange between the fluids consistent with the confines of the radiator's physical size.

In the example of an automotive radiator, one of the fluids, the coolant, is more valuable than the other fluid, the air. Therefore, the coolant is conserved within the engine and the radiator plumbing system, whereas the air is used once and then discarded. There are numerous other designs for heat exchangers, many more than can be covered in this text. We will focus on those most commonly encountered by a biomedical engineer. These include the shell and tube geometry, in which one fluid flows through a tube and the second through an outer shell, and naturally occurring heat exchangers in the body.

The design of heat exchangers has been studied extensively, and there is a rich literature on this subject [for example, see Kays and London (1984), Hewitt (2002), and Shah and Sekulic (2003)]. Most heat exchanger designs are optimized for one or more performance criteria. Examples are the magnitude of heat exchanged between the two fluids as normalized to size of the device, weight of the device, and pressure required to pump one or both of the fluids through the device, cost, corrosion resistance, avoidance of solute deposits from one of the fluids, ease of manufacture, and others. In our treatment we will focus on thermal performance, but if you become involved in the design of a practical heat exchanger, you should be aware that there may be many factors beyond the basic issues of heat transfer to be considered.

The starting point for analysis of a heat exchanger is to consider the heat transfer pathway between two fluids across the wall of a tube. There will be convection on each side of the wall, and conduction through the wall, as illustrated in Fig. 10.5.

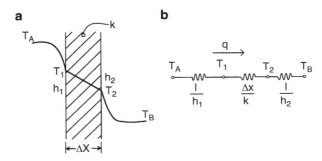

Fig. 10.5 Overall heat transfer through the wall of a heat exchanger tube: (a) convection, conduction, and convection fluxes in series and (b) equivalent electrical network diagram

We will consider only steady-state operation of a heat exchanger. The start up and shut down processes are usually not important phases of heat exchanger function. For these conditions, there is no change in the energy stored in the exchanger, and heat only flows between the fluids as they pass through the exchanger. The series heat transfer from the hotter fluid A to the colder fluid B through a plane wall is expressed by

$$q = h_1(T_A - T_1) = \frac{k}{\Delta x}(T_1 - T_2) = h_2(T_2 - T_B). \tag{10.48}$$

The equation for the equivalent electrical resistance network is

$$q = \frac{T_A - T_B}{1/h_1 + \Delta x/k + 1/h_2} = U \Delta T_{\text{overall}}, \tag{10.49}$$

where U is the *overall heat transfer coefficient*. A similar expression can be written for the heat conduction through a cylinder rather than a planar wall,

$$\dot{Q} = \frac{T_A - T_B}{1/h_i A_i + (\ln(r_o/r_i)/2\pi kL) + 1/h_o A_o} = U_{i/o} A_{i/o} \Delta T_{\text{overall}}. \tag{10.50}$$

In this case, the overall heat transfer coefficient is expressed in terms of cylindrical heat transfer and is normalized to either the inside or outside surface area of the tube, where

$$U_i = \frac{1}{1/h_i + (A_i \ln(r_o/r_i)/2\pi kL) + (A_i/A_o)(1/h_o A_o)}, \tag{10.51}$$

$$U_o = \frac{1}{(A_o/A_i)(1/h_i) + (A_o \ln(r_o/r_i)/2\pi kL) + 1/h_o}. \tag{10.52}$$

The overall heat transfer coefficient is a very useful tool for analyzing the transport of heat between two fluids in an exchanger.

10.4.3.1 Cocurrent and Counter-Current Heat Exchangers

The analyses presented in this section are analogous to those for cocurrent and counter-current mass exchangers presented in Sects. 14.5.2.2 and 14.5.2.3. The fundamental flow geometries of the hot and cold fluids are illustrated in Fig. 10.6 for these two types of heat exchangers.

In a cocurrent arrangement both the hot and cold fluids enter the exchanger at the same end and exit from the opposite end. The temperature difference between the two fluids is maximum at the inlet and decreases continuously to a minimum value

10.4 Steady-State One-Dimensional Problems Involving Convection

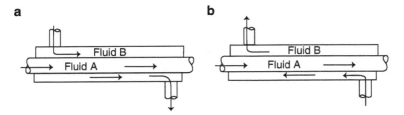

Fig. 10.6 Shell and tube heat exchanger flow geometries: (**a**) cocurrent and (**b**) counter-current

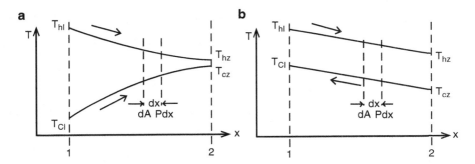

Fig. 10.7 Relative temperature profiles as the hot and cold fluids flow through a heat exchanger: (**a**) cocurrent and (**b**) counter-current

at the outlet. Therefore, the heat transfer between the fluids will be highly nonuniform along the length of the exchanger.

In contrast, for a counter-current arrangement the fluids enter at the opposite ends of the exchanger and flow through it in opposing directions. As a consequence, the temperature difference between the two fluids remains much more uniform throughout the exchanger, and in many cases the overall thermal performance is higher than for a cocurrent system. The relative temperature profiles for cocurrent and counter-current heat exchangers are illustrated in Fig. 10.7. The net temperature changes of the hot and cold fluids between their inlet and outlet states are not necessarily equal, depending on their relative mass flow rates and specific heats.

Given these initial considerations, we will now develop a method of analyzing the thermal performance of a heat exchanger based on the physical parameters of its design and the thermodynamic and flow properties of the two fluid streams moving through it. The end point of this analysis will be the definition of the log mean temperature difference as a representative of the overall thermal potential differential between the two fluids. A fundamental assumption that we will make for this analysis is that we do not need to account for the gradients in temperature that will exist in the hot and cold fluids perpendicular to their primary flow direction. At any flow cross-section of the exchanger, the hot and cold fluids are thermally well mixed. Embedded within this assumption is the implicit realization that there will be temperature gradients at least in the boundary layers near the wall where convection occurs.

Nonetheless, by lumping these effects into the overall heat transfer coefficient it is possible to derive an analysis model that is highly useful for the design of heat exchangers and understanding their performance behavior. An additional assumption is that the convective heat transfer coefficients do not vary along the length of the exchanger. This assumption greatly simplifies the analysis, but we have seen earlier in Sect. 10.4 that the convective heat transfer coefficient can change by a significant factor along the length of a flow surface. In cases where a more detailed analysis is required, numerical analysis methods may be applied. Nonetheless, the basic principles we will use for this simplified analysis can be adopted to develop a more comprehensive model.

First a small length, dx, along the exchanger will be defined, and associated with it there will be a small area, dA, for heat transfer between the two fluids. At steady state there will be no change in the energy stored in the body of the exchanger. Therefore, for a system defined by the area dA, the conservation of energy requires that the energy lost by the hot fluid equals the energy gained by the cold fluid,

$$d\dot{Q} = -w_h c_h dT_h = w_c c_c dT_c. \tag{10.53}$$

These two equations may be rewritten and solved for the change in the difference between the hot and cold fluids,

$$dT_h = -\frac{d\dot{Q}}{w_h c_h} \quad \text{and} \quad dT_c = \frac{d\dot{Q}}{w_c c_c},$$

so that

$$dT_h - dT_c = d(T_h - T_c) = -d\dot{Q}\left(\frac{1}{w_h c_h} + \frac{1}{w_c c_c}\right). \tag{10.54}$$

An alternate expression for the heat flow is based on the overall heat transfer coefficient,

$$d\dot{Q} = U dA (T_h - T_c). \tag{10.55}$$

Next, we eliminate the heat flow term between the last two equations

$$\frac{d(T_h - T_c)}{T_h - T_c} = -U dA \left(\frac{1}{w_h c_h} + \frac{1}{w_c c_c}\right). \tag{10.56}$$

This equation can be integrated over the length of the exchanger from section 1 to 2.

For *cocurrent* flow geometry,

$$\ln \frac{T_{h,2} - T_{c,2}}{T_{h,1} - T_{c,1}} = \ln \frac{T_{h,\text{out}} - T_{c,\text{out}}}{T_{h,\text{in}} - T_{c,\text{in}}} = -UA\left(\frac{1}{w_h c_h} + \frac{1}{w_c c_c}\right). \tag{10.57}$$

10.4 Steady-State One-Dimensional Problems Involving Convection

Equation (10.53) may also be integrated from section 1 to 2 to obtain

$$w_h c_h = \frac{\dot{Q}}{T_{h,in} - T_{h,out}} \quad \text{and} \quad w_c c_c = \frac{\dot{Q}}{T_{c,out} - T_{c,in}}. \tag{10.58}$$

The mass flow time-specific heat terms can be eliminated from the last two equations to obtain

$$\dot{Q} = UA \frac{(T_{h,out} - T_{c,out}) - (T_{h,in} - T_{c,in})}{\ln\left[(T_{h,out} - T_{c,out})/(T_{h,in} - T_{c,in})\right]} = UA\Delta T_{lm}. \tag{10.59}$$

For *counter-current* flow geometry,

$$\ln \frac{T_{h,2} - T_{c,2}}{T_{h,1} - T_{c,1}} = \ln \frac{T_{h,out} - T_{c,in}}{T_{h,in} - T_{c,out}} = -UA\left(\frac{1}{w_h c_h} + \frac{1}{w_c c_c}\right). \tag{10.60}$$

Equation (10.53) may also be integrated from section 1 to 2 to obtain

$$w_h c_h = \frac{\dot{Q}}{T_{h,out} - T_{h,in}} \quad \text{and} \quad w_c c_c = \frac{\dot{Q}}{T_{c,out} - T_{c,in}}. \tag{10.61}$$

The mass flow time-specific heat terms can be eliminated from the last two equations to obtain

$$\dot{Q} = UA \frac{(T_{h,out} - T_{c,in}) - (T_{h,in} - T_{c,out})}{\ln\left[(T_{h,out} - T_{c,in})/(T_{h,in} - T_{c,out})\right]} = UA\Delta T_{lm}. \tag{10.62}$$

Note that for both the cocurrent and counter-current geometries the *log mean temperature difference* can be written as an identical term.

$$\Delta T_{lm} = \frac{(T_{h,2} - T_{c,2}) - (T_{h,1} - T_{c,1})}{\ln\left[(T_{h,2} - T_{c,2})/(T_{h,1} - T_{c,1})\right]}. \tag{10.63}$$

This matches (10.41) which was derived for a more restrictive system.

The heat exchangers we have considered to this point are manufactured. They often have a uniform geometry, including many symmetries, and controlled flow conditions. In contrast, there are many examples of naturally occurring heat exchangers in animals, including humans. In most cases the functions of these structures are to conserve energy, either to reduce the metabolic load on the animal to maintain homeostasis or to protect vital organs in the core and brain from hypothermic or hyperthermic injury. Schmidt-Nielsen (1997) has made an extensive study of physiological heat exchanger systems and of their operational characteristics. Most frequently the geometry of these heat exchangers is counter-current. The fluids are usually arterial and venous blood, although in some fish species there are water and blood heat exchangers. The versatility and breadth of physiological heat exchangers are illustrated in Figs. 10.8–10.13.

Although many more examples of physiological heat exchangers are available, those illustrated above should suffice to show that: (1) they are prevalent throughout the animal kingdom, (2) they occur in many various configurations, and (3) the geometries are exceedingly complex in comparison with heat exchangers designed and manufactured by humans. Another important factor not seen directly from the figures is that the flow conditions often change much more frequently than is typical for fabricated heat exchanger operation. Therefore, transient behavior can be significant in some cases.

Fig. 10.8 Counter-current heat flow in the flipper of a porpoise. A central artery with distal blood flow is completely surrounded by veins with proximal flow. Blood circulating through the flipper is cooled as it loses heat to the cold water environment. The warm arterial blood loses heat to the cooler returning venular blood, pre-heating it before it reenters the core region and conserving heat stored in the core tissues. Also, heat loss by arterial blood to the surrounding cold water is reduced if its temperature in the peripheral circulation is lower (Schmidt-Nielsen 1972), with permission

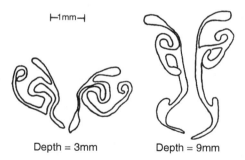

Fig. 10.9 Cross-sectional flow geometry of the respiratory tract in the kangaroo rat at the indicated distances from the inlet. Of note is the extremely tortuous shape of the flow channel that results in a huge ratio of surface area-to-volume, issuing in very effective convection of inhaled and exhaled air with the nasal passage tissue. As a consequence, the temperature of the air may change by as much as 20°C between the tip of the nose and the lungs. This arrangement is highly conservative of both energy and water for an animal that lives in an arid environment (Jackson and Schmidt-Nielsen 1964), with permission

10.4 Steady-State One-Dimensional Problems Involving Convection

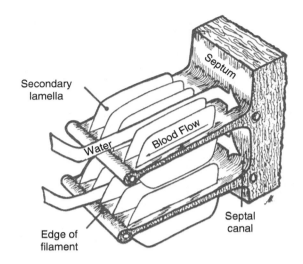

Fig. 10.10 Counter-current flows of blood and water in the gill of a shark. This structure exchanges both heat and oxygen between the blood and water (Grigg 1970), with permission

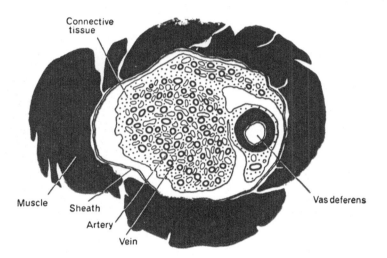

Fig. 10.11 The vascular structure to the testes of a wallaby. The arteries (*thick walls*) and veins (*thin walls*) are interspersed throughout the supporting tissue. The counter-current flow heat exchange serves to cool arterial blood before it reaches the testes. As a consequence, the testes can remain at a temperature below that of the core, which is necessary for the formation of fertile sperm (Barnett et al. 1958), with permission

As may be anticipated, the analysis of physiological heat exchangers is much more complicated than for manufactured exchangers. Nonetheless, because physiological heat exchanger function is so important to homeostasis, including human thermoregulation, many engineers have developed analytical models for the effect of blood flow heat transfer over the past half century. The complexity of nearly all of these models is of necessity beyond the scope of an introductory text such as this

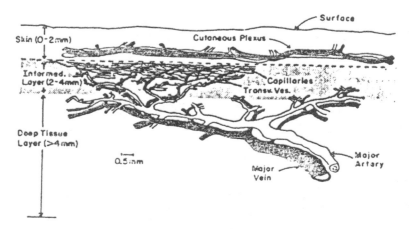

Fig. 10.12 Schematic of the vascular network in the skin and subcutaneous tissue of a rabbit. Important features are that the feeder arteries and veins are paired in counter-current heat transfer geometries that are separate from the vascular supply to muscle. The counter-current arrangement prevails over 5 or 6 branching generations until the diameters become thermally insignificant in the intermediate layer. As the vessels branch their number increases and size decreases. Bleed off of flow to the local capillary networks can affect the overall heat balance (Weinbaum et al. 1984), with permission

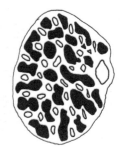

Fig. 10.13 Drawing of the cross-section of a femoral vascular bundle from a three-toed sloth. Veins are *black* and arteries are *white*. The arteries and veins are thoroughly mingled in a counter-current structure of about 1 cm diameter (Scholander and Krog 1957), with permission

one. An excellent summary, comparison, and commentary on many of these models were prepared by Charney (1992). More recent, but briefer, summaries are also available (Diller et al. 2005). This topic remains an area where biomedical engineers can make a significant contribution. There is a short presentation in Sect. 10.5.2 concerning the Pennes model of bioheat transfer, which is the most simple, and also most widely adopted, model for the thermal interaction between flowing and tissue (Pennes 1948).

10.4.3.2 Effectiveness–NTU Analysis Method

If all of the inlet and outlet temperatures are specified or readily determined, the log mean temperature difference analysis is directly applicable to calculate heat

10.4 Steady-State One-Dimensional Problems Involving Convection

exchanger parameters such as the heat transfer performance and required surface area. However, in many cases it is more advantageous to apply alternative analysis procedure known as the effectiveness–NTU method. This approach has been developed by Kays and London (1984 and earlier editions). Many details of the analysis, further examples and data for specific design applications may be found there.

The effectiveness of a heat exchanger is defined as

$$\text{effectiveness} = \varepsilon = \frac{\text{actual heat transfer}}{\text{maximum possible heat transfer}}.$$

The actual heat transfer may be calculated as either the heat lost by the hot fluid or the heat gained by the cold fluid. Retaining the notation of Fig. 10.7 for the inlets and exits of an exchanger, the actual heat transfer for a cocurrent exchanger is

$$\dot{Q} = w_h c_h (T_{h_1} - T_{h_2}) = w_c c_c (T_{c_2} - T_{c_1}), \tag{10.64}$$

and for a counter-current exchanger is

$$\dot{Q} = w_h c_h (T_{h_1} - T_{h_2}) = w_c c_c (T_{c_1} - T_{c_2}). \tag{10.65}$$

The maximum possible heat transfer is defined by the maximum temperature difference that exists within the exchanger, i.e., the difference between the inlet temperatures of the hot and cold fluids. Since conservation of energy for a heat exchanger operating at steady state requires that the heat gained by the cold fluid equals the heat lost by the hot fluid, the maximum possible temperature difference would need to be associated with the fluid that has the minimum value of the product wc. If we were to allow the fluid with the larger value of wc to experience the maximum possible temperature difference, conservation of energy would require that the other fluid undergo a temperature difference greater than the maximum, which is impossible. Thus, the maximum possible heat transfer is defined as

$$\dot{Q}_{\max} = (wc)_{\min}(T_{h_{in}} - T_{c_{in}}). \tag{10.66}$$

Based on this definition, the effectiveness is written in terms of the fluid that has the minimum product of wc. Thus, for a cocurrent exchanger,

$$\varepsilon_{h_{\min}} = \frac{w_h c_h (T_{h_1} - T_{h_2})}{w_h c_h (T_{h_1} - T_{c_1})} = \frac{(T_{h_1} - T_{h_2})}{(T_{h_1} - T_{c_1})} \tag{10.67}$$

or

$$\varepsilon_{c_{\min}} = \frac{w_c c_c (T_{c_2} - T_{c_1})}{w_c c_c (T_{h_1} - T_{c_1})} = \frac{(T_{c_2} - T_{c_1})}{(T_{h_1} - T_{c_1})}. \tag{10.68}$$

For a counter-current exchanger the effectiveness equation is

$$\varepsilon_{h_{min}} = \frac{w_h c_h(T_{h_1} - T_{h_2})}{w_h c_h(T_{h_1} - T_{c_1})} = \frac{(T_{h_1} - T_{h_2})}{(T_{h_1} - T_{c_1})} \quad (10.69)$$

or

$$\varepsilon_{c_{min}} = \frac{w_c c_c(T_{c_1} - T_{c_2})}{w_c c_c(T_{h_1} - T_{c_1})} = \frac{(T_{c_1} - T_{c_2})}{(T_{h_1} - T_{c_1})}. \quad (10.70)$$

The effectiveness function is most useful for analysis purposes when it is related to the flow and thermal properties of the fluids (wc) and the design properties of the heat exchanger (U and A). We can illustrate how to derive this relationship, starting with a cocurrent exchanger with the cold fluid defining the minimum value of wc. (10.57) may be rewritten as

$$\ln \frac{T_{h_2} - T_{c_2}}{T_{h_1} - T_{c_1}} = -UA\left(\frac{1}{w_h c_h} + \frac{1}{w_c c_c}\right) = -\frac{UA}{w_c c_c}\left(1 + \frac{w_c c_c}{w_h c_h}\right) \quad (10.71)$$

or

$$\frac{T_{h_2} - T_{c_2}}{T_{h_1} - T_{c_1}} = \exp\left[-\frac{UA}{w_c c_c}\left(1 + \frac{w_c c_c}{w_h c_h}\right)\right]. \quad (10.72)$$

Equation (10.58) can be written as

$$T_{h_2} = T_{h_1} + \frac{w_c c_c}{w_h c_h}(T_{c_1} - T_{c_2}). \quad (10.73)$$

The temperature ratio in (10.72) may be written in terms of (10.73).

$$\frac{T_{h_2} - T_{c_2}}{T_{h_1} - T_{c_1}} = \frac{T_{h_1} + (w_c c_c/w_h c_h)(T_{c_1} - T_{c_2}) - T_{c_2}}{T_{h_1} - T_{c_1}}. \quad (10.74)$$

The numerator of the right-hand term can be restated by adding and subtracting T_{c_1} so that it can be expressed in terms of the effectiveness,

$$\frac{(T_{h_1} - T_{c_1}) + (w_c c_c/w_h c_h)(T_{c_1} - T_{c_2}) - (T_{c_1} - T_{c_2})}{T_{h_1} - T_{c_1}} = 1 - \varepsilon\left(1 + \frac{w_c c_c}{w_h c_h}\right). \quad (10.75)$$

Combining (10.72), (10.74), and (10.75) yields the desired expression for the effectiveness,

$$\varepsilon = \frac{1 - \exp\left[-\frac{UA}{w_c c_c}\left(1 + \frac{w_c c_c}{w_h c_h}\right)\right]}{1 + \frac{w_c c_c}{w_h c_h}} = \frac{1 - \exp\left[-\frac{UA}{C_{min}}\left(1 + \frac{C_{min}}{C_{max}}\right)\right]}{1 + \frac{C_{min}}{C_{max}}}, \quad (10.76)$$

10.4 Steady-State One-Dimensional Problems Involving Convection

where the heat capacity rate, C, is defined as $C = wc$. It can be shown that for a cocurrent heat exchanger, if the hot fluid has the minimum value of C, the same equation as (10.76) results.

For a counter-current heat exchanger the expression for effectiveness is

$$\varepsilon = \frac{1 - \exp\left[-\dfrac{UA}{C_{min}}\left(1 - \dfrac{C_{min}}{C_{max}}\right)\right]}{1 - \left(\dfrac{C_{min}}{C_{max}}\right)\exp\left[-\dfrac{UA}{C_{min}}\left(1 - \dfrac{C_{min}}{C_{max}}\right)\right]}. \tag{10.77}$$

Finally, the *number of transfer units*, NTU, is defined as

$$\text{NTU} = \frac{UA}{C_{min}} \tag{10.78}$$

and the *heat capacity rate ratio*, C_r, as

$$C_r = \frac{C_{min}}{C_{max}}. \tag{10.79}$$

The effectiveness equation is written in the format

$$\varepsilon = f(\text{NTU}, C_r).$$

For a cocurrent concentric tube heat exchanger this relationship is

$$\varepsilon = \frac{1 - \exp[-\text{NTU}(1 + C_r)]}{1 + C_f} \tag{10.80}$$

and for a counter-current heat exchanger it is

$$\varepsilon = \frac{1 - \exp[-\text{NTU}(1 - C_r)]}{1 - C_r \exp[-\text{NTU}(1 - C_r)]}, C_r < 1. \tag{10.81}$$

Special cases are for a counter-current exchanger for which $C_r = 1$,

$$\varepsilon = \frac{\text{NTU}}{1 + \text{NTU}}, \ C_r = 1, \text{ counter-current flows} \tag{10.82}$$

and for $C_r = 0$ for any flow geometry,

$$\varepsilon = 1 - \exp(-\text{NTU}), \ C_r = 0, \text{ all flow geometries.} \tag{10.83}$$

This case typically occurs when (1) there is a phase change process on one side of the exchanger, as in a boiler or condenser or (2) the volume of the external fluid is very large and well stirred so that heat transfer with the fluid inside a flow tube has no significant effect on its temperature. Under these conditions the performances of cocurrent and counter-current heat exchangers merge to the same point.

Alternatively, the heat exchanger performance can be expressed in a format in which NTU rather than ε is calculated:

$$\text{NTU} = f(\varepsilon, C_r).$$

After algebraic manipulation, the foregoing two equations can be rewritten to solve for NTU. For a cocurrent exchanger,

$$\text{NTU} = -\frac{\ln[1 - \varepsilon(1 + C_r)]}{1 + C_f}. \tag{10.84}$$

For a counter-current exchanger,

$$\text{NTU} = \frac{1}{C_r - 1} \ln\left(\frac{\varepsilon - 1}{\varepsilon C_r - 1}\right), \; C_r < 1, \tag{10.85}$$

$$\text{NTU} = \frac{\varepsilon}{1 - \varepsilon}, \; C_r = 1. \tag{10.86}$$

For all flow conditions for which $C_r = 0$,

$$NTU = -\ln(1 - \varepsilon), C_r = 0. \tag{10.87}$$

Some of the Kays and London (1984) analyses are available in graphical format. Figures 10.14 and 10.15 present graphs of effectiveness–NTU relationships for cocurrent and counter-current concentric tube heat exchangers.

Kays and London (1984) present analytical and graphical data for other heat exchanger arrangements including multiple pass geometries for both the shell and tube flows, plus cross-flow in which the hot and cold fluids move along perpendicular vectors. They also present analyses of the pressure drop associated with fluid flow through specific heat exchanger configurations, which is directly related to the work expenditure required to operate the device. This is an important factor in practical heat exchanger design, but it is not covered in this presentation.

Example Problem 10.4.3 Extracorporeal Blood Cooler and Warmer.
Many surgical procedures require the ability to cool or heat blood as it is circulated through an external heat exchanger. A fraction of the cardiac output is diverted to the heat exchanger where its temperature can be altered as it is circulated through coiled tubing exposed to external water at a controlled temperature. In this problem we will consider only the cooling process. An important limitation of the process is that the temperature of the blood as it

10.4 Steady-State One-Dimensional Problems Involving Convection

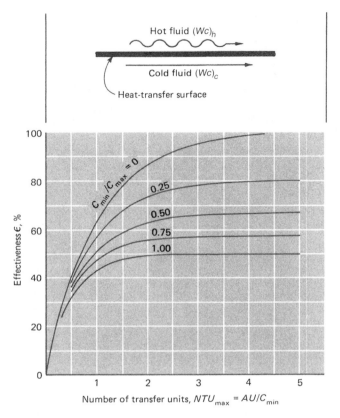

Fig. 10.14 Effectiveness–NTU relationship for co-current concentric tube heat exchanger (Kays and London, 1984, with permission)

reenters the body be no more than 10°C lower than when it left so as to avoid a massive vasoconstrictive reaction. The minimum temperature to which the blood would be cooled is about 5°C to achieve a 15°C core body temperature. You can assume that the flow rate of blood is 3 L/min, and the thermal resistance associated with convection during the flow of blood through the coiled tube is much larger than that of conduction through the tube wall and of convection of water circulated with a large agitation on the outside of the tube. The inner diameter of the tube is 10 mm. What tube length is needed to achieve a 10°C drop in temperature?

Solution. *Initial considerations:* This is a shell and tube heat exchanger arrangement. The effectiveness–NTU analysis method should be applicable to calculate the size of the exchanger needed to produce the required temperature drop in the blood

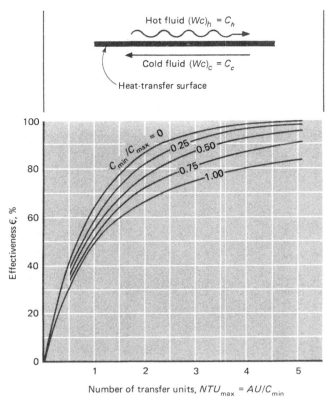

Fig. 10.15 Effectiveness–NTU relationship for counter-current concentric tube heat exchanger (Kays and London, 1984, with permission)

during a single pass through. The water is circulated through the external shell and will likely have a volume that is much larger than that of the coil through which the blood flows. Also, baffles and other flow enhancers can be introduced into the shell flow pathway to agitate the water and make convection with the outside of the blood coil more effective. In general, such convection enhancement methods are not used on the blood side of the exchanger so as to avoid inducing coagulation and cell damage. We will assume that the radius into which the tube is coiled is large in comparison with the radius of the tube so that the tube is essentially straight from a heat transfer perspective.

At the beginning of the cooling process the blood will enter the heat exchanger at 37°C, and the cooling will persist until the core temperature is reduced to 15°C. We can choose any point in the intermediate temperature range for our analysis. As the cooled blood returns to the body it will progressively cool the core. Thus, the temperature entering the heat exchanger will drop over time, making the process transient. We should consider whether it makes any difference if we analyze the heat exchanger performance near the beginning or the end of the cooling process.

10.4 Steady-State One-Dimensional Problems Involving Convection

As a general principle we should design the device for the most limiting case of performance. As the blood temperature drops its viscosity will increase. We know from convection analysis that the heat transfer coefficient and Nusselt number will be proportional to the Reynolds number, which contains the viscosity in its denominator. Therefore, higher viscosity translates to lower convective transport. Conversely, the Prandtl number increases with lower temperatures, offsetting the decreasing Reynolds number. We may be well served by calculating the heat transfer coefficient at both the extreme high- and low-operating temperature ranges.

As an initial approximation, we will neglect the rate of change of energy of the mass stored in the exchanger as the blood cools down as being small in comparison with the rate at which heat crosses the wall of the exchanger. This assumption is a consequence of the mass of blood contained within the heat exchanger being considerably smaller than the mass of the body that is being cooled by the returning blood. It is necessary to cool a small amount of blood over and over in order to cool the large body once. We can revisit this assumption after the calculations are completed to judge whether it is valid. We also assume constant properties for the blood, and that it is incompressible.

Since the water acts as a very large reservoir to receive heat from the blood, its heat capacity rate is much larger than for that of blood. Therefore, the heat capacity rate ratio will be zero, $C_r = 0$. Because this problem asks us to determine the size of the exchanger needed to achieve the target blood-cooling performance, our objective will be to calculate the number of transfer units, NTU, from which the surface area can be computed.

Many aspects of this problem have not been defined so that of necessity we will need to make some design decisions as we proceed with the solution. For example, the materials and wall thickness of the cooling coil are not specified, and the intended effectiveness for the exchanger is not given.

System definition and environmental interactions: It will be helpful to define two different systems for analysis as shown in the figure: (a) one is the overall heat exchanger and (b) one is the blood-cooling coil. Note that we anticipate that the difference between the inlet and outlet temperatures for the water will be much less than for blood because we will have a significantly larger flow rate of water through the exchanger. The overall exchanger thermal performance will be evaluated using the effectiveness–NTU method. The cooling coil will serve for analysis of the convective heat transfer from the blood to the inside wall of the tube, which we anticipate will be the limiting aspect of the exchanger performance. The heat exchanger communicates with the environment only via the flow of blood and water enthalpies across the inlets and outlets. The cooling coil has a wall convection and blood flow enthalpy at the inlet and outlet. A sketch of the anticipated temperature profile of the blood along the length of the coil is shown (c). Note that unless the tube is exceptionally long, the temperature of the blood at exit will not equal that of the water. This relationship illustrates that to achieve a higher effectiveness, a larger NTU is required. For increasingly high values of the efficiency, eventually such a configuration becomes highly uneconomical of materials and space resources, plus the work required to pump the blood through a longer coil is proportionately greater.

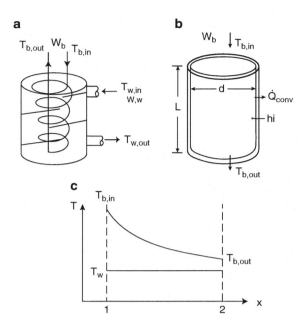

Apprising the problem to identify governing equations: The conservation of energy is applied to the cooling coil to relate the change in enthalpy of blood flowing in and out of the convective heat transfer at the wall (10.38),

$$\dot{Q}_{\text{conv}} = wc_{\text{p}}\left[T_{\text{m,out}} - T_{\text{m,in}}\right].$$

The effectiveness–NTU analysis can be performed graphically using either Fig. 10.14 or 10.15 since the curve for $C_{\text{r}} = 0$ is identical on both.

Analysis: The first step in the effectiveness–NTU analysis is to choose a value for ε. This choice is somewhat arbitrary, but we will pick a value for which the slope of the $C_{\min}/C_{\max} = 0$ curve on Figs. 10.14 and 10.15 is rather large. Note that for values of ε greater than about 0.80, further increments in the efficiency require substantially larger values of NTU. Thus, there is a state of diminishing returns beyond which small increases in performance require large expenditures in size and cost of the heat exchanger. For $\varepsilon = 0.80$, the effectiveness–NTU charts give a value of NTU of approximately 1.6.

Since we are neglecting the thermal resistances of convection on the outside of the tube and of thermal conductivity through the tube wall, the overall heat transfer coefficient is simply equal to the convective heat transfer coefficient on the inside of the tube,

$$U = \bar{h}_{\text{L}}.$$

10.4 Steady-State One-Dimensional Problems Involving Convection

We must calculate the Reynolds number to determine the correlation equation for h, which requires property values for blood to be identified. The values needed are outside the range of temperatures presented in Appendix C. However, it is not difficult to track down the necessary values with a little exploring on the internet. This is a situation you are likely to encounter often in solving practical problems, and it is useful to develop the skill of finding appropriate constitutive property values on your own. The properties for this problem are:

@$T = 10°C$, $\mu = 4.97 \times 10^3$ Pa s; @$T = 32°C$, $\mu = 3.68 \times 10^3$ Pa s;
$\rho = 1.05 \times 10^3$ kg/m^3; $k = 5.07 \times 10^{-1}$ W/m × K; $c_p = 3.74 \times 10^3$ J/kg K.

The velocity is calculated from the volume flow rate as:

$$\langle v \rangle = \frac{Q}{A} = \frac{4Q}{\pi d^2} = \frac{4 \times 3\,\text{L/min} \cdot 10^3\,\text{cm}^3/\text{L} \cdot 10^{-6}\,\text{m}^3/\text{cm}^3}{\pi \cdot (1 \times 10^{-2})^2\,\text{m}^2 \cdot 60\,\text{s/min}} = 6.37\,10^{-1}\,\text{m/s}.$$

The mass flow rate is:

$$w = \rho Q = 1.05 \times 10^3\,\text{kg/m}^3 \cdot 3\,\text{L/min} \cdot \frac{10^3\,\text{cm}^3/\text{L} \cdot 10^{-6}\,\text{m}^3/\text{cm}^3}{60\,\text{s/min}}$$

$$= 5.25 \times 10^{-2}\,\text{kg/s}.$$

The Reynolds numbers at 10°C and 32°C are:

$$Re_{10°C} = \frac{\rho \langle v \rangle d}{\mu} = \frac{1.05 \times 10^3\,\text{kg/m}^3 \cdot 6.37 \times 10^{-1}\,\text{m/s} \cdot 10^{-2}\,\text{m}}{4.97 \times 10^3\,\text{kg/m s}} = 1.34 \times 10^{-3},$$

$$Re_{32°C} = \frac{\rho \langle v \rangle d}{\mu} = \frac{1.05 \times 10^3\,\text{kg/m}^3 \cdot 6.37 \times 10^{-1}\,\text{m/s} \cdot 10^{-2}\,\text{m}}{3.68 \times 10^3\,\text{kg/m s}} = 1.82 \times 10^{-3}.$$

In both the cases, the Reynolds number is very low, deep in the laminar flow region. For a constant temperature wall condition and fully developed laminar flow, the convection correlation equation is (8.35),

$$Nu_d = 3.66 = \frac{hd}{k},$$

from which the convective heat transfer coefficient can be calculated. Note that for this correlation equation the convection coefficient will be the same for both Reynolds numbers and is independent of the Prandtl number,

$$h = 3.66 \frac{k}{d} = 3.66 \frac{0.507\,\text{W/mK}}{10^{-2}\,\text{m}} = 186\,\text{W/m}^2\,\text{K} = U.$$

The heat capacity rate for the blood side of the heat exchanger is calculated as:

$$C_{bld} = C_{min} = wc_p = 5.25 \times 10^{-2}\,\text{kg/s} \cdot 3.74 \times 10^3\,\text{J/kg K} = 1.96 \times 10^2\,\text{W/K}.$$

For an NTU of 1.6, the surface convection area is

$$A = \frac{NTU \cdot C_{min}}{U} = \frac{1.6 \cdot 1.96 \times 10^2\,\text{W/K}}{1.86 \times 10^2\,\text{W/m}^2\,\text{K}} = 1.69\,\text{m}^2,$$

from which the length of the exchanger flow tube is calculated.

$$L = \frac{A}{\pi d} = \frac{1.69\,\text{m}^2}{3.14 \times 10^{-2}\,\text{m}} = 53.9\,\text{m}.$$

A heat exchanger in which blood flows through a single tube 10 mm in diameter with a total length in excess of 50 m is totally impractical. Therefore, we should reevaluate the initial assumptions that were used to design the heat exchanger. We can consider some options for reevaluation.

One parameter that could be changed is the diameter of the blood flow tube. However, inspection of the equations for calculating h and L shows that the diameter will balance out changes in these equations. Repeating the above calculations for larger and smaller tube diameters shows that there is no effect on the tube length.

Another parameter to evaluate is the efficiency. A lower value of ε will correlate with a lower NTU, which in turn leads to a smaller area and shorter length. The tradeoff, which we will see below, is that with a smaller ε, the temperature of the cooling water must be decreased in order to obtain the requisite 10°C temperature drop for the blood as it flows through the exchanger. Decreasing ε to 0.66 or 0.43 reduces the tube length to 34 and 17 m, respectively. Neither length is satisfactorily small, so a different solution must be found.

An alternative design that will produce an acceptable solution is to divide the blood flow into a number of parallel tubes, thereby reducing the mass flux and heat capacity rate, C_{min}, through any single tube. As C_{min} is decreased, the required convection area of the tube is reduced in direct proportion. This design approach is used in a majority of practical heat exchangers. The blood inflow can be collected into a header from which it is divided into a large number of parallel individual tubes, from which it is recollected into an outlet header. For example, if there are 10 flow tubes, the length of a tube is reduced by a factor of 10 to 5.39 m. Twenty-five tubes reduce the length to 2.16 m, and 100 tubes to 0.539 m. One obvious tradeoff in this design is that as the number of parallel tubes increases, the manufacturing complexity and expense increase. Also, it may be more difficult to produce a large convection coefficient with water on the outside surface of all of the tubes.

10.4 Steady-State One-Dimensional Problems Involving Convection

Examining and interpreting the results: The initial statement of this problem led us to a design for a heat exchanger that is totally impractical to build and to use. It was necessary for us to modify the original specifications to obtain a design that is practical. Even though our calculation for the single tube heat exchanger with a length of 53 m is mathematically correct, it is important to examine the results beyond merely checking that the computations are correct, to see if our solution to the problem is acceptable in the total context of its application. In this case, substituting an array of shorter parallel flow tubes for a single long tube provides a much more workable design, and, indeed, this approach is standard in the heat exchanger industry.

Further considerations: The practical implementation of this heat exchanger design will require some further calculations to define its operating characteristics. For example, we will need to know what temperature of water circulating on the outside of the tubes is necessary to ensure that the blood will be cooled adequately. We can make this calculation directly from the definition of the heat exchanger efficiency, ε.

$$\varepsilon = \frac{T_{bld,in} - T_{bld,out}}{T_{bld,in} - T_{water}} = \frac{15 - 5}{15 - T_{water}}.$$

In the foregoing analysis, we considered the thermal performance of the heat exchanger for three different efficiencies, 0.80, 0.66, and 0.43, which from Figs. 10.14 and 10.15 correspond to NTUs of 1.6, 1.0, and 0.5. The required cooling water temperature in each of these cases is

$$T_{water} = T_{bld,in} - \frac{T_{bld,in} - T_{bld,out}}{\varepsilon} = 15°C - \frac{15 - 5°C}{\varepsilon},$$

which yields

ε	T_{water} (°C)
0.80	2.5
0.66	−0.2
0.43	−8.3

If the operating conditions require that the blood should exit the heat exchanger at a temperature as low as 5°C, then the design should embody an efficiency that is high enough so that the water temperature is above its freezing point. Given that the water will need to be chilled in its own heat exchanger, it is probably not reasonable to specify its temperature to be below 2.5°C. Thus, our best choice for ε among the three options considered is 0.80.

It would also be important to know the magnitude of heat transfer that occurs in the exchanger. This quantity can be calculated by applying the conservation of energy to the blood flowing through the exchanger tubes (10.38).

$$\dot{Q}_{\text{conv}} = wc_{\text{p}}(T_{\text{out}} - T_{\text{in}})$$
$$= \frac{3\,\text{l/min} \cdot 1.05 \times 10^3\,\text{kg/m}^3}{60\,\text{s/min} \cdot 10^6\,\text{ml/m}^3} 3.74 \times 10^3\,\text{J/kg K}\,(15-5)^\circ\text{C}$$
$$= 1.96 \times 10^3\,\text{W}.$$

We can compare the amount of heat removed from the blood during one pass through the exchanger with the total amount of energy to be removed from the patient's body to achieve the desired state of hypothermia for the surgical procedure. The energy will be dependent on the mass of the patient and the fraction of the body to be cooled. The blood circulation is focused in the visceral organs, and it is most important that these tissues receive the protection of hypothermia during surgery. The viscera compose about 10% or the total body mass. It can be assumed that there will be a parasitic heat flow from the surrounding tissues amounting to another 5% of the body mass. For a 75 kg patient, the total energy loss during cooling is estimated by writing the conservation of energy equation for 15% of the total mass,

$$\Delta E_{\text{body}} = m_{\text{body}} f_{\text{viscera}} c_{\text{p}} (T_{\text{initial}} - T_{\text{final}})$$
$$= 75\,\text{kg} \cdot 0.15 \cdot 3.74 \times 10^3\,\text{J/kg K}(37-15)^\circ\text{C} = 9.26 \times 10^5\,\text{J}.$$

The time to cool the body at the start of surgery is calculated as the total energy that must be removed divided by the rate at which heat is extracted from blood in the exchanger.

$$t_{\text{cool}} = \frac{\Delta E_{\text{body}}}{\dot{Q}_{\text{conv}}} = \frac{9.26 \times 10^5\,\text{J}}{1.96 \times 10^3\,\text{W}} = 471\,\text{s} = 7.86\,\text{min}.$$

Our analysis has not accounted for start up conditions for the cooling process, which will extend the time of the procedure. However, even a doubling of the cooling period to 15 min would likely be within the expectations of the surgical team. Thus, the heat exchanger thermal design should be adequate for its intended purpose.

We should determine the length of time that the blood will be resident within the heat exchanger for a single pass. That time is simply the length of the tube divided by the average flow velocity,

$$t_{\text{res}} = \frac{L}{\langle v \rangle} = \frac{53.9\,\text{m}}{0.637\,\text{m/s}} = 84.7\,\text{s}.$$

Note that the residence time is the same, irrespective of the number of parallel tubes into which the flow is divided. As the number of tubes increases, both the length and the velocity decrease in direct proportion.

The number of passes of blood through the exchanger is the ratio of the total cooling period to the residence time of blood during a single pass,

$$\text{no. passes} = \frac{t_{\text{cool}}}{t_{\text{res}}} = \frac{471 \text{ s}}{84.7 \text{ s}} = 5.57.$$

Our initial assumption that the rate at which the temperature of the cooling coil is reduced (indicative of the energy stored within the heat exchanger) is small compared to the rate at which blood passes through the heat exchanger (related to the convective heat transfer) is reasonable, but the ratio is less than an order of magnitude. Therefore, if we were to pursue a further refinement of the design, we might consider the heat exchanger as a transient rather than a steady-state device.

10.5 One-Dimensional Steady-State Heat Conduction

There are many conditions in biological systems that are at or close to a thermal steady state. For instances in which the internal thermal resistance is similar to that of the boundary thermal resistance associated with either a convection or radiation process there will be a significant internal temperature gradient within the tissue (this is in contrast with there being a very small internal thermal resistance for which $Bi < 0.1$; see Sect. 9.4.1). One such instance that is special to living tissue occurs in conjunction with the perfusion of blood through a highly organized network of blood vessels. The result is a boundary heat transfer that is distributed through the interior of a thermally conducting medium (the interstitial tissue). In this section, we address some geometrically explicit examples of steady-state heat conduction.

10.5.1 Heat Conduction with Convection or Radiation at Extended Surfaces

Extended surfaces refer to systems in which the geometry of the boundary is expanded to create a larger surface-to-volume ratio resulting in an enhanced ability to exchange heat with the environment via convection and/or radiation. In practice, the extended surfaces are frequently referred to as fins and are seen to protrude from a planar base surface of a solid substrate material. Countless industrial and commercial applications have been developed for fins, and much excellent research has been conducted on this subject over many decades. Fins provide a challenging problem area for design based on optimization of multiple factors such as heat flow achieved per unit area in the basal substrate plane, enhancement of forced or free convection flow, radiation surface shape configuration, weight, manufacturing complexity and cost, compatibility with maintenance and cleaning procedures, etc. Fin design has become a sophisticated and advanced area of heat transfer engineering. In many instances, fins are evaluated in the context of steady-state heat transfer since they exist in a steady state during their primary mode of operation. It is relevant to consider fin thermal performance in the context of biotransport because some biological structures, such as fingers, can behave somewhat like fins in heat exchange with the environment. (It may seem that fingers are totally distinctive from manufactured fins in that although they have the geometry of a fin, they are perfused with blood, which gives them a unique mode of internal heat transfer that is different from an inanimate object. However, there are parallels in manufactured fins, such as the blades on a gas turbine engine, that are kept from becoming too hot by the internal circulation of a cooling fluid. The combination of extended surface heat transfer and internal convective cooling becomes complex very easily, and we will only treat these two phenomena separately.) Some devices used in a biological context also depend on extended surfaces for heat transfer enhancement.

Fins may have a constant or variable cross-sectional thickness as distance from the substrate increases. They may have planar or cylindrical shapes, or they may occupy discrete positions on the substrate with a pin-like configuration. Figure 10.16 presents some standard geometries of finned surfaces.

From among these commonly encountered geometries, the pin fin is selected for analysis because of its geometric similarity to a finger. The primary mechanism of heat exchange with the environment from the surface of the pin is convection, as characterized by the coefficient, h. The starting point for the analysis is to consider an energy balance for a short length of the fin as shown in Fig. 10.17.

An important inherent assumption for this analysis is that the temperature varies only along the length of the fin, so that it is not a function of radius or rotational angle at any longitudinal position. For a steady-state process, the heat flow into the short length of the fin must equal the flow out. The conservation of energy relationship is expressed as:

$$0 = \dot{Q}_x - \dot{Q}_{x+dx} - \dot{Q}_{conv}. \tag{10.88}$$

Writing the heat loss $\Delta \dot{Q}_{conv}$ from the fin in terms of Newton's law of cooling:

$$\dot{Q}_{conv} = hP\Delta x(T - T_\infty). \tag{10.89}$$

Substituting (10.89) into (10.88), dividing by Δx and taking the limit as Δx approaches zero:

$$\frac{d\dot{Q}}{dx} = -hP(T - T_\infty). \tag{10.90}$$

10.5 One-Dimensional Steady-State Heat Conduction

Fig. 10.16 Examples of geometric configurations of finned surfaces: (**a**) Planar, (**b**) cylindrical, (**c**) pin, and (**d**) variable cross-section along the length. In each case the most important dimensions are indicated

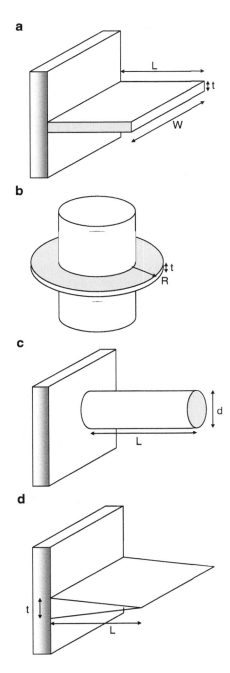

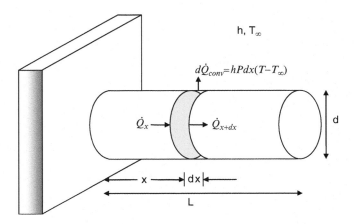

Fig. 10.17 Conduction and convection for a short length, Δx, of a pin fin

Equation (10.90) contains a negative sign because the heat conduction along the fin decreases with the distance from the base. Substituting Fourier's law for conduction in the fin:

$$\frac{d}{dx}\left(-kA_c \frac{dT}{dx}\right) = -hP(T - T_\infty). \tag{10.91}$$

Note that the cross-sectional area, A_c, and peripheral surface area are different. For a constant cross-sectional area and constant thermal conductivity,

$$kA_c \frac{d^2T}{dx^2} - hP(T - T_\infty) = 0. \tag{10.92}$$

For a circular cross-section, $P/A_c = 4/d$. The temperature can be scaled to the fluid state to provide a simpler notation, $\theta = T - T_\infty$. Then,

$$\frac{d^2\theta}{dx^2} - \frac{4h}{kd}\theta = \frac{d^2\theta}{dx^2} - m^2\theta = 0, \tag{10.93}$$

where $4h/kd$ is written as m^2 to be more convenient for defining the solution. Since the equation is second order along the x coordinate, the solution requires two boundary conditions. One is identified at $x = 0$ where the temperature is equal to that of the base substrate material. Several cases may be considered for the second condition.

(a) If the fin is very long, thereby providing a large surface for convection, then the temperature at the tip will approach that of the surrounding fluid. Thus, $T(L) = T_\infty$ or $\theta(L) = 0$.

(b) The fin is of a finite length so that there will be appreciable convective heat exchange on the surface of the tip. In this case the conduction into the tip equals the convection exchange from it.

10.5 One-Dimensional Steady-State Heat Conduction

$$hA_c[T(L) - T_\infty] = -kA_c \frac{dT(L)}{dx} \quad \text{or} \quad hA_c \theta(L) = -kA_c \frac{d\theta(L)}{dx}.$$

(c) The tip of the fin is insulated so that $dT(L)/dx = 0$ or $d\theta(L)/dx = 0$. This condition may arise at the mid plane of a fin when there is a symmetric geometry in which the fin joins to identical substrate surfaces. Because of geometric and thermal symmetry, there will be no heat flow across the midplane, corresponding to the conditions for perfect insulation.

The general solution of (10.93) is:

$$\theta(x) = C_1 e^{-mx} + C_2 e^{mx}. \tag{10.94}$$

Each of the three different boundary conditions issues in a unique set of values for the coefficients C_1 and C_2.

In the first case for thermal equilibration at the tip,

$$\frac{\theta(x)}{\theta_0} = \frac{T(x) - T_\infty}{T_0 - T_\infty} = e^{-mx}. \tag{10.95}$$

In the second case for convection from the tip,

$$\frac{\theta(x)}{\theta_0} = \frac{\cosh m(L-x) + (h/mk)\sinh m(L-x)}{\cosh mL + (h/mk)\sinh mL}. \tag{10.96}$$

In the third case for an insulated tip,

$$\frac{\theta(x)}{\theta_0} = \frac{\cosh[m(L-x)]}{\cosh mL}. \tag{10.97}$$

It is often of interest to determine the amount of heat exchanged via convection from the total surface area of a fin. There are two approaches that can be used for this calculation. One is to apply Fourier's law to the appropriate temperature distribution expression to calculate the total heat flow through the basal plane of the fin ($x = 0$). By applying conservation of energy to the entire fin, it is seen that all of the heat that is convected from the surface of the fin must also be conducted into the fin through its base.

$$\dot{Q}_{total} = -kA \frac{dT}{dx}\bigg|_{x=0}. \tag{10.98}$$

Alternatively, the total heat transfer from the fin can be calculated by integrating the convection over the entire surface exposed to a surrounding fluid.

$$\dot{Q}_{total} = \int_0^L hP(T - T_\infty)dx = \int_0^L hP\theta dx. \tag{10.99}$$

Although these two expressions are equivalent, usually (10.98) is the easier to apply.

After the heat transferred from the fin surface is known, it is possible to compare it with the amount of heat that would have been transferred from the basal area of the substrate had the fin not been present. This comparison is a measure of how much the fin enhances the heat transfer between the material and the surrounding fluid. The ratio of these two quantities is known as the *effectiveness* of the fin, ε_{fin},

$$\varepsilon_{\text{fin}} = \frac{\dot{Q}_{\text{total}}}{hA_c(T_0 - T_\infty)} = \frac{\dot{Q}_{\text{total}}}{hA_c\theta_0}. \tag{10.100}$$

Although it is not intuitively obvious, there is no guarantee that the presence of a fin will result in a value of ε_{fin} greater than 1.0. It is possible that the material and geometry of the fin could cause a lower net heat transfer than would have occurred from the plane substrate. The application of fin analysis is demonstrated in the following example.

Example 10.5.1.
One of the primary situations in which the thermal behavior of a finger as a pin fin may be important occurs under conditions of prolonged exposure in a hostile cold environment. In this context it is useful to be able to predict how cold a finger may become under specific operational conditions. Consider a finger with nominal dimensions of 20 mm diameter and 80 mm length. Determine the environmental temperature that will cause the temperature at the tip of the finger (where it will be coldest) to be 0°C, for which frost bite might be an imminent danger. What will be the temperature at the midpoint of the finger under these conditions? Consider two levels of convection with air for which the environment is quiescent ($h = 2$ W/m^2 K) and is blustery ($h = 50$ W/m^2 K).

Solution. *Initial considerations:* The goal is to quantify various combinations of environmental conditions that can be tolerated on an exposed finger without introducing a risk of frostbite. The approach is to model the finger as a pin fin through which heat is conducted from the palm of the hand and lost via convection from the skin surface to the surrounding air. The temperature distribution along the length of the finger must be determined in order to complete this analysis.

System definition and environmental interactions: The system is closed, consisting of the finger having right circular cylindrical geometry as shown in Fig. 10.16c. Conduction occurs internal to the system longitudinally along its length. Thermal interactions with the environment are by conduction with the palm of the hand at the base of the finger and via convection between the skin surface and surrounding air.

10.5 One-Dimensional Steady-State Heat Conduction

Apprising the problem to identify governing equations: The following assumptions are made to define conditions for application of the conservation laws, constitutive equations, and boundary conditions appropriate to a pin fin analysis:

1. The geometry of the finger is simplified to be a right circular cylinder, with a flat tip. The proximity of adjacent fingers has no effect on the heat transfer for the finger of interest.
2. The heat flow is one-dimensional along the length of the finger, designated by a coordinate, x.
3. The convective heat transfer coefficient is uniform around the entire surface of the finger, as designated by a uniform value of h.
4. Convection occurs on the tip of the finger.
5. The temperature distribution is evaluated after the system has reached steady state.
6. The temperature at the tip of the finger is 0°C.
7. The temperature at the base of the finger is 34°C.
8. The average thermal conductivity of the cross-section of the finger is taken to be 0.5 W/m K, based on the values for skin, muscle, and bone.
9. Blood perfusion effects are not thermally significant in the finger. Since the finger will be very cold, the vascular system will be vasoconstricted, substantiating this assumption.

Analysis: For this system and boundary conditions we have shown that the steady-state temperature distribution along the length of the finger is given by (10.96)

$$\frac{T(x) - T_\infty}{T_0 - T_\infty} = \frac{\theta(x)}{\theta_0} = \frac{\cosh m(L-x) + (h/mk)\sinh m(L-x)}{\cosh mL + (h/mk)\sinh mL}.$$

The temperature is a function of the location along the length of the finger and of the heat transfer properties for the system and environment as used to calculate the values of the hyperbolic functions. For the temperature specified at the finger tip, the only unknown in this equation is the value of the environmental air temperature, T_∞. The value of the dimensionless temperature ratio at the finger tip can be designated by u_L for simplicity of notation. Then,

$$u_L = \frac{T_L - T_\infty}{T_0 - T_\infty} \quad \text{or} \quad T_\infty = \frac{u_L T_0 - T_L}{u_L - 1}.$$

The value of m must be determined to complete the calculations, and it is a function of the magnitude of the convective heat transfer coefficient. Therefore there is a unique value for each of the environmental conditions specified.

$$m^2 = \frac{hP}{kA_c} \quad \text{or, for the pin fin } m = 2\sqrt{\frac{h}{kd}}.$$

Thus,

$$m_2 = 2\sqrt{\frac{2\,\text{W/m}^2\,\text{K}}{0.5\,\text{W/m K} \times 0.02\,\text{m}}} = 28.3\,\text{m}^{-1}$$

and

$$m_{50} = 2\sqrt{\frac{50\,\text{W/m}^2\,\text{K}}{0.5\,\text{W/m K} \times 0.02\,\text{m}}} = 141.4\,\text{m}^{-1}$$

When the geometric and constitutive values are substituted into (10.96), the values for u_L and T_∞ are computed to be

$$u_{L-2} = 0.181 \quad u_{L-50} = 0.0000143,$$

$$T_{\infty-2} = -7.51°\text{C} \quad T_{\infty-50} = -0.000486°\text{C}.$$

Given these values for the environmental temperature, (10.96) can be applied to calculate the temperature at any distance along the finger. For the half distance, $x = L/2$.

$$T(L/2) = T_\infty + (T_0 - T_\infty)\frac{\cosh m(L/2) + (h/mk)\sinh m(L/2)}{\cosh mL + (h/mk)\sinh mL}.$$

The mid-finger temperature values for the two convective environments are

$$T_{L/2-2} = 6.81°\text{C} \quad T_{L/2-50} = 0.118°\text{C}.$$

Examining and interpreting the results: The strength of the environmental heat loss by convection has a large effect on the thermal state of the finger. With a diminished convective heat transfer coefficient, a much lower environmental temperature is required to cool the tip of the finger to freezing conditions. Also, the overall temperature of the finger is higher for smaller convective cooling. The temperature at the midpoint of the finger is not midway between the base and tip temperatures. The data show that the greatest cooling occurs near the base where the temperature difference between the skin and air is largest.

Additional Comments: It should be easy to learn more about the behavior of this system by performing a few additional calculations using the analysis that is already set up. For example, we can investigate an intermediate value of the convective heat transfer coefficient, 10 W/m^2 K, and we can evaluate the temperatures at locations

10.5 One-Dimensional Steady-State Heat Conduction

closer to the base, $x = L/5$ and $x = L/10$, where the temperature gradient should be larger. The ambient air and finger temperatures are calculated for each of these conditions and are plotted below.

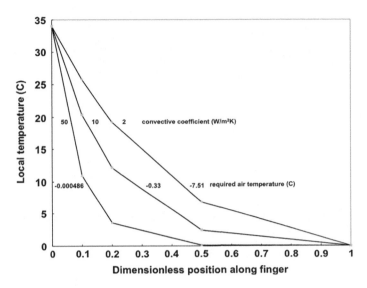

The added calculations provide further refinement to our understanding of how the finger cools under the differing convective conditions. Although the tip temperatures are identical for each scenario, for larger convection effects, the finger is much colder. The net heat stored in the finger is equal to the specific heat times the temperature at each location. Thus, the area under each of the three curves is proportional to the total energy in the finger. Complementarily, the area between the 34°C isotherm and each curve represents the heat lost by the finger to the environment, assuming that it was originally at a uniform temperature.

The temperature plots along the length of the finger show that the slope decreases as the tip is approached, and the larger the convection coefficient, the smaller is the tip slope. A slope that becomes vanishingly small means that a state of thermal equilibrium is approached at the tip, satisfying the boundary criteria for (10.95),

$$\frac{\theta(x)}{\theta_0} = \frac{T(x) - T_\infty}{T_0 - T_\infty} = e^{-mx}. \tag{10.101}$$

The temperatures at the identical locations can be computed with this equation and compared with the values for the convective tip assumption as shown in the plot below.

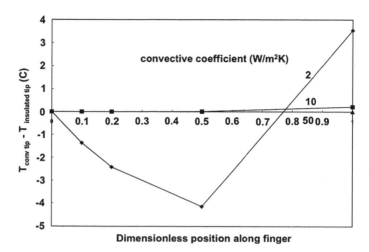

The difference between the temperatures calculated with the two assumptions is negligible for the largest convection coefficient and is significant for the smallest convection coefficient. This result should be anticipated since a more effective heat transfer between the finger and the environment will provide a more complete equilibration toward the tip.

It should be noted for the second set of calculations, that the assumption of equilibrium with the environment at the tip in conjunction with the stated constraint that the tip temperature be $0°C$, required that the air temperature also be set to $0°C$. For the first set of calculations there was a finite temperature difference between the skin at the finger tip and the surrounding air. In a quiescent environment the comparative analysis shows that this effect is important, and it is not wise to assume that thermal equilibrium is achieved at the finger tip.

Some of the assumptions invoked for this analysis may be questioned. In general, after completing a problem analysis, it is always a good practice to review all of the assumptions for appropriateness and for their effect on the results. For example, the convection between the skin and air was assumed not to be influenced by the proximity of adjacent fingers. Actually, the individual fingers are not free standing with respect to environmental air. This geometric effect is largest near the hand where the analysis points to the convective transport being greatest. The anticipated result should be that the convective loss is reduced, and a lower environmental temperature is required to cool the tip to $0°C$.

10.5.2 Heat Exchange in Tissue: Transient and Steady-State Pennes Equation

Bioheat transfer processes in living tissues are often influenced by blood perfusion through the vascular network. When there is a significant difference between the temperature of blood and the tissue through which it flows, convective heat transport will occur, altering the temperatures of both the blood and the tissue. Perfusion-based heat transfer interaction is critical to a number of physiological processes such as thermoregulation and inflammation. The blood/tissue thermal interaction is a function of several parameters including the rate of perfusion and the vascular anatomy, which vary widely among the different tissues, organs of the body, and pathology. Appendix C contains an extensive compilation of perfusion rate data for many tissues and organs and for many species.

The rate of perfusion of blood through different tissues and organs varies over the time course of a normal day's activities, depending on the factors such as physical activity, physiological stimulus, and environmental conditions. Further, many disease processes are characterized by alterations in blood perfusion, and some therapeutic interventions result in either an increase or decrease in blood flow in a target tissue. For these reasons, it is very useful in a clinical context to know what the absolute level of blood perfusion is within a given tissue. There are numerous techniques that have been developed for this purpose over the past several decades. In some of these techniques, the coupling between vascular perfusion and local tissue temperature is applied to advantage to assess the flow through local vessels via inverse solution of equations which model the thermal interaction between perfused blood and the surrounding tissue.

Pennes (1948) published the seminal work on developing a quantitative basis for describing the thermal interaction between tissue and perfused blood. His work consisted of a series of experiments to measure temperature distribution as a function of radial position in the forearms of nine human subjects. A butt-junction thermocouple was passed completely through the arm via a needle inserted as a temporary guideway, with the two leads exiting on opposite sides of the arm. The subjects were unanesthetized so as to avoid the effects of anesthesia on blood perfusion. Following a period of normalization, the thermocouple was scanned transversely across the mediolateral axis to measure the temperature as a function of radial position within the interior of the arm. The environment in the experimental suite was kept thermally neutral during the experiments. Pennes' data showed a temperature differential of 3–4°C between the skin and the interior of the arm, which he attributed to the effects of metabolic heat generation and heat transfer with arterial blood perfused through the microvasculature.

Pennes proposed a model to describe the effects of metabolism and blood perfusion on the energy balance within tissue. The shell balance on uniformly perfused tissue can be used to incorporate these two effects, along with simple diffusion, to produce the following relation:

$$\rho c \frac{\partial T}{\partial t} = \nabla \cdot k \nabla T + (\rho c)_b \omega_b (T_a - T) + \dot{q}_{met}. \qquad (10.102)$$

This equation is written to describe the thermal effects of blood flow through a local region of tissue having a temperature T. It contains the familiar energy storage and conduction terms, plus terms to account for convection with perfused blood and metabolic heat generation. The blood is perfused at a rate ω_b (ml blood flow/ml tissue; units 1/s). The temperature of perfused blood entering into a tissue region is that of the arterial supply, T_a, and the leaving temperature is T because the relatively small volume of flowing blood completely equilibrates with the surrounding tissue via the very large surface area to volume ratio of the microvasculature through which it flows. If the local tissue is at steady state, which is often the case, then the storage term on the left side of the equation is zero.

$$0 = \nabla \cdot k\nabla T + (\rho c)_b \omega_b (T_a - T) + \dot{q}_{met}. \quad (10.103)$$

The Pennes model contains no specific information about the morphology of the vasculature through which the blood flows. The somewhat simple assumption is that the fraction of blood flowing through a tissue that is diverted through the microvasculature comes to thermal equilibration with the local tissue as it passes to the venous return vessels. A major advantage of the Pennes model is that the term that accounts for perfusion heat transfer is linear in temperature, which facilitates the solution of (10.102) and (10.103). Since the publication of this work, the Pennes model has been adapted by many researchers for the analysis of a variety of bioheat transfer phenomena. These applications vary in physiological complexity from a simple homogeneous volume of tissue to thermal regulation of the entire human body. As more scientists have evaluated the Pennes model for application in specific physiological systems, it has become increasingly clear that some of the assumptions foundational to the model are not valid for some vascular geometries that vary greatly among the various tissues and organs of the body.

Given that the validity of the Pennes model has been questioned for many applications, Wissler (1998) has revisited and reanalyzed Pennes' original data. Given the hindsight of five decades of advances in bioheat transfer plus greatly improved computational tools and better constitutive property data, Wissler's analysis pointed out flaws in Pennes' work which had not been appreciated previously. However, he also showed that much of the criticism that has been directed toward the Pennes model is not justified, in that his improved computations with the model demonstrated a good standard of agreement with the experimental data. Thus, Wissler's conclusion is that "those who base their theoretical calculations on the Pennes model can be somewhat more confident that their starting equations are valid." The quantitative analysis of the effects of blood perfusion on the internal temperature distribution in living tissue remains a topic of active research after a half century of study.

10.6 Transient Diffusion Processes with Internal Thermal Gradients

When the diffusion of heat inside a material occurs relatively slowly, significant spatial temperature gradients will develop between the surface and the interior

10.6 Transient Diffusion Processes with Internal Thermal Gradients

areas. Under these conditions the effects of the limited rate of internal conduction of heat must be accounted for when predicting how the temperature will change with time. In such cases energy can accumulate in a small element of material as a function of time. The rate of accumulation of energy will equal the rate of change of internal energy. For a shell with volume $A_c \Delta x$, such as shown in Fig. 10.1, the rate of accumulation is:

$$\left\{ \begin{array}{c} \text{rate at which} \\ \text{energy accumulates} \\ \text{in shell} \end{array} \right\} = \frac{\partial}{\partial t}(m\hat{U}) = \frac{\partial}{\partial t}\left(\rho A_c \Delta x c_p [T - T_{\text{Ref}}]\right)$$

$$= \rho A_c \Delta x c_p \frac{\partial T}{\partial t}. \tag{10.104}$$

This term was zero for steady-state transport, but for unsteady-state transport in a slab, (10.4) would be modified to include energy accumulation:

$$\rho A_c \Delta x c_p \frac{\partial T}{\partial t} = q_x(x) A_c - q_x(x + \Delta x) A_c + \dot{q}_{\text{met}} A_c \Delta x. \tag{10.105}$$

Dividing by the volume of the shell and letting Δx approach zero results in the following partial differential equation:

$$\rho c_p \frac{\partial T}{\partial t} = -\frac{\partial q_x}{\partial x} + \dot{q}_{\text{met}}. \tag{10.106}$$

Finally, introducing Fourier's law with constant thermal conductivity:

$$\frac{\partial T}{\partial t} = \alpha \frac{\partial^2 T}{\partial x^2} + \frac{\dot{q}_{\text{met}}}{\rho c_p}, \tag{10.107}$$

where $\alpha = k/\rho c_p$ is the thermal diffusivity. This corresponds to the Pennes equation (10.102) with the term for internal convection with perfused blood dropped. The solution in the spatial domain involves specifying two boundary conditions for each distinct material through which heat diffuses. In addition, an initial temperature distribution is also needed in each material, $T(0, x)$.

The geometry of the system will influence the form of the one-dimensional conduction relationship. For the radial flow of heat through a material with cylindrical geometry, such as Fig. 10.2, the rate of accumulation of energy is:

$$\left\{ \begin{array}{c} \text{rate at which} \\ \text{energy accumulates} \\ \text{in shell} \end{array} \right\} = \frac{\partial}{\partial t}\left(\rho 2\pi r L \Delta r c_p [T - T_{\text{Ref}}]\right)$$

$$= 2\pi \rho c_p L r \Delta r \frac{\partial T}{\partial t}. \tag{10.108}$$

Inclusion of energy accumulation in (10.14) for a cylindrical shell leads to:

$$2\pi \rho c_\text{p} L r \Delta r \frac{\partial T}{\partial t} = q_r(r)\,(2\pi r L) - q_r(r+\Delta r)\,(2\pi(r+\Delta r)L)$$
$$+ \dot{q}_\text{met}(2\pi r L \Delta r). \tag{10.109}$$

Dividing by the volume of the shell and letting Δr approach zero results in the following partial differential equation:

$$\rho c_\text{p} \frac{\partial T}{\partial t} = -\frac{1}{r}\frac{\partial (r q_r)}{\partial r} + \dot{q}_\text{met}. \tag{10.110}$$

Finally, introducing Fourier's law with constant thermal conductivity:

$$\frac{\partial T}{\partial t} = \frac{\alpha}{r}\frac{\partial}{\partial r}\left(r\frac{\partial T}{\partial r}\right) + \frac{\dot{q}_\text{met}}{\rho c_\text{p}}. \tag{10.111}$$

We leave it as an exercise to show that for objects with a spherical geometry, such as a spherically shaped white blood cell, a shell balance can be used to derive the following one-dimensional conduction equation:

$$\frac{\partial T}{\partial t} = \frac{\alpha}{r^2}\frac{\partial}{\partial r}\left(r^2\frac{\partial T}{\partial r}\right) + \frac{\dot{q}_\text{met}}{\rho c_\text{p}}. \tag{10.112}$$

Relevant boundary conditions were discussed in Chap. 8. Specific examples of solutions of (10.107), (10.111), and (10.112) are treated in the remainder of this chapter.

10.6.1 Symmetric Geometries: Exact and Approximate Solutions for Negligible Heat Generation

Systems and processes that exhibit both geometric and thermal symmetry fall into a special class of analysis for which well-known exact and approximate mathematical solutions exist. By definition, these systems must have finite dimensions. Most frequently, they are encountered in one-dimensional format as a plane wall (with temperature varying along the x coordinate), a long cylinder, or a sphere (in both the cases the temperature varies only with radius). These geometries are illustrated in Fig. 10.18 for thermal interactions with the environment via convection.

For the simplified case in which the initial temperature in the system is uniform, there is no significant internal energy generation, and blood perfusion effects can be neglected; at a defined starting time, the environment undergoes a step change to

Fig. 10.18 A plane wall (a), infinite cylinder (b) and sphere (c) showing symmetry of geometry in one dimension and convection thermal boundary conditions

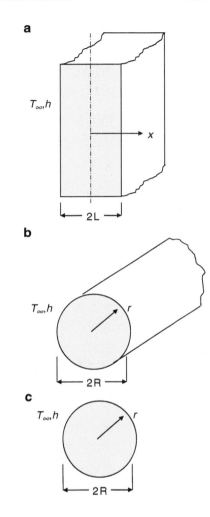

a new value. The analytical solution for this problem is in the form of an infinite series. As will become apparent, it is advantageous to write the problem statement and solution in terms of dimensionless variables. We will first consider the solution for a system in rectangular coordinates, after which the cylindrical and spherical solutions are obtained by straightforward mathematical operations.

The temperature is scaled to the environmental value as $\theta = T - T_\infty$, and is normalized to the initial value.

$$\theta^* = \frac{\theta}{\theta_i} = \frac{T - T_\infty}{T_i - T_\infty}. \tag{10.113}$$

Likewise, the independent variables for position and time are normalized to the size and thermal time constant of the system

$$x^* = \frac{x}{L}, \tag{10.114}$$

where L is the half width of the system along the primary thermal diffusion vector

$$t^* = Fo = \frac{\alpha t}{L^2}, \tag{10.115}$$

where Fo is called the *Fourier number*, representing a dimensionless time. It is the ratio of the actual process time compared to the thermal diffusion time constant for the system.

The heat diffusion equation in one dimension (10.107) can be written in terms of these dimensionless variables

$$\frac{\partial \theta^*}{\partial Fo} = \frac{\partial^2 \theta^*}{\partial x^{*2}}, \tag{10.116}$$

for which the initial and boundary conditions are written as

$$\theta^*(x^*, 0) = 1, \tag{10.117}$$

$$\left. \frac{\partial \theta^*}{\partial x^*} \right|_{x^*=0} = 0, \tag{10.118}$$

which is a result of thermal and geometric symmetry

$$-\left. \frac{\partial \theta^*}{\partial x^*} \right|_{x^*=1} = Bi\, \theta^*(1, Fo), \tag{10.119}$$

where Bi is the Biot number defined in Sect. 8.3.2.3:

$$Bi = \frac{hL}{k_s}. \tag{10.120}$$

(You should be aware that a more general presentation of this analysis is given in Appendix D that covers applications to both conduction of heat and diffusion of mass. For sake of generality, an alternative nomenclature is used in which $\theta^* = Y$, $x^* = n$, $Fo = X$, and $Bi = 1/m$. The material presented in Appendix D should be an enlightening complement to your understanding of the present analysis.)

10.6 Transient Diffusion Processes with Internal Thermal Gradients

The solution for this problem is in the form of an infinite series

$$\theta^* = \sum_{n=1}^{\infty} C_n e^{-\lambda_n^2 Fo} \cos(\lambda_n x^*), \qquad (10.121)$$

where C_n satisfies the following relationship for each value of n

$$C_n = \frac{4 \sin \lambda_n}{2\lambda_n + \sin(2\lambda_n)}, \qquad (10.122)$$

and the eigenvalues λ_n are defined as the positive roots of the transcendental equation

$$\lambda_n \tan \lambda_n = Bi. \qquad (10.123)$$

There are unique values of C_n and λ_n for each value of Bi.

Although the exact solution takes the form of an infinite series, for many problems it is adequate to use only a limited number of terms and still maintain an acceptable level of accuracy. If the analysis can be restricted to portions of the process following the initial transient for which $Fo > 0.2$, then only the first term is required. This effect will be demonstrated in Sect. 10.6.3, where graphical results are discussed. The closer the analysis must approach the beginning of the process, the more terms must be included in the calculation. In these cases, the exact solution still can be computed in a relatively straightforward manner (Diller 1990a, b, c), although the detail that must be included increases with each additional term. Unfortunately, in many classes of biomedical processes, information concerning the initial transient behavior is of greatest interest, and it is not possible to use the single-term approximation. Solutions in Appendix D use the first 30 terms of the series.

A fully analogous analysis can be applied for systems modeled in cylindrical and spherical coordinates. For the cylindrical geometry defined in Fig. 10.18, the dimensionless temperature is given by

$$\theta^* = \sum_{n=1}^{\infty} C_n e^{-\lambda_n^2 Fo} J_0(\lambda_n r^*), \qquad (10.124)$$

where $Fo = \alpha t/R^2$. C_n satisfies for each value of n

$$C_n = \frac{2 J_1(\lambda_n)}{\lambda_n J_0^2(\lambda_n) + J_1^2(\lambda_n)}, \qquad (10.125)$$

and the eigenvalues λ_n are defined as the positive roots of the transcendental equation

$$\lambda_n \frac{J_1(\lambda_n)}{J_0(\lambda_n)} = Bi, \qquad (10.126)$$

where $Bi = hR/k$.

Likewise, for the spherical geometry defined in Fig. 10.18, the dimensionless temperature is given by

$$\theta^* = \sum_{n=1}^{\infty} C_n e^{-\lambda_n^2 Fo} \frac{1}{\lambda_n r^*} \sin(\lambda_n r^*), \tag{10.127}$$

where $Fo = \alpha t / R^2$. C_n satisfies for each value of n

$$C_n = \frac{4[\sin(\lambda_n) - \lambda_n \cos(\lambda_n)]}{2\lambda_n + \sin(2\lambda_n)}, \tag{10.128}$$

and the eigenvalues λ_n are defined as the positive roots of the transcendental equation

$$1 - \lambda_n \cot \lambda_n = Bi, \tag{10.129}$$

where $Bi = hR/k$.

The one-term approximate solutions for these three coordinate systems are all written in a format consisting of the product of two terms: one is the temperature at the geometric center, which is a function of time, and the other is the spatial distribution of the temperature between the center and the surface, which does not change with time. Thus, the total temperature solution is given in the form of

$$\theta^*(Fo, x^*) = \theta_0^*(Fo, 0)\theta_x^*(x^*) \text{ and } \theta^*(Fo, r^*) = \theta_0^*(Fo, 0)\theta_r^*(r^*), \tag{10.130}$$

which is a short-hand notation for

$$\theta^* = \frac{T - T_\infty}{T_i - T_\infty} = \theta_0^* \cdot \theta_x^* = \frac{T_0 - T_\infty}{T_i - T_\infty} \cdot \frac{T - T_\infty}{T_0 - T_\infty} \text{ and}$$
$$\theta^* = \frac{T - T_\infty}{T_i - T_\infty} = \theta_0^* \cdot \theta_r^* = \frac{T_0 - T_\infty}{T_i - T_\infty} \cdot \frac{T - T_\infty}{T_0 - T_\infty}. \tag{10.131}$$

The transient temperature at the center and the constant spatial temperature distribution are calculated independently and then multiplied together to determine the full solution as a function of time and space.

Since λ_1 is a function of Bi, C_1 must also depend on Bi. Therefore, to determine the single-term approximate solution, it is necessary to first calculate the value of Bi, from which the pre-exponential and transcendental constants are determined. The values of these two constants for the three coordinate systems are compiled in Table 10.1 for discrete increments of Bi values. It is logical that the approximate solution is not valid for short times after a change in the boundary condition since it will require a finite period for a change effected at the surface to penetrate to the center. The added terms in the series solution provide further information about the diffusion of the heat wave progressively closer to the geometric center.

10.6 Transient Diffusion Processes with Internal Thermal Gradients

Table 10.1 Coefficients for the single-term approximation for the series solutions to one-dimensional transient conduction problems

Bi^a	Plane wall		Infinite cylinder		Sphere	
	λ_1 (rad)	C_1	λ_1 (rad)	C_1	λ_1 (rad)	C_1
0.01	0.0998	1.0017	0.1412	1.0025	0.1730	1.0030
0.02	0.1410	1.0033	0.1995	1.0050	0.2445	1.0060
0.03	0.1723	1.0049	0.2440	1.0075	0.2991	1.0090
0.04	0.1987	1.0066	0.2814	1.0099	0.3450	1.0120
0.05	0.2218	1.0082	0.3143	1.0124	0.3854	1.0149
0.06	0.2425	1.0098	0.3438	1.0148	0.4217	1.0179
0.07	0.2615	1.0114	0.3709	1.0173	0.4551	1.0209
0.08	0.2791	1.0130	0.3960	1.0197	0.4860	1.0239
0.09	0.2956	1.0145	0.4195	1.0222	0.5150	1.0268
0.10	0.3111	1.0161	0.4417	1.0246	0.5423	1.0298
0.15	0.3779	1.0237	0.5376	1.0365	0.6609	1.0445
0.20	0.4328	1.0311	0.6170	1.0483	0.7593	1.0592
0.25	0.4801	1.0382	0.6856	1.0598	0.8447	1.0737
0.30	0.5218	1.0450	0.7465	1.0712	0.9208	1.0880
0.4	0.5932	1.0580	0.8516	1.0932	1.0528	1.1164
0.5	0.6533	1.0701	0.9408	1.1143	1.1656	1.1441
0.6	0.7051	1.0814	1.0184	1.1345	1.2644	1.1713
0.7	0.7506	1.0919	1.0873	1.1539	1.3525	1.1978
0.8	0.7910	1.1016	1.1490	1.1724	1.4320	1.2236
0.9	0.8274	1.1107	1.2048	1.1902	1.5044	1.2488
1.0	0.8603	1.1191	1.2558	1.2071	1.5708	1.2732
2.0	1.0769	1.1785	1.5994	1.3384	2.0288	1.4793
3.0	1.1925	1.2102	1.7887	1.4191	2.2889	1.6227
4.0	1.2646	1.2287	1.9081	1.4698	2.4556	1.7202
5.0	1.3138	1.2402	1.9898	1.5029	2.5704	1.7870
6.0	1.3496	1.2479	2.0490	1.5253	2.6537	1.8338
7.0	1.3766	1.2532	2.0937	1.5411	2.7165	1.8673
8.0	1.3978	1.2570	2.1286	1.5526	2.7654	1.8920
9.0	1.4149	1.2598	2.1566	1.5611	2.8044	1.9106
10.0	1.4289	1.2620	2.1795	1.5677	2.8363	1.9249
20.0	1.4961	1.2699	2.2881	1.5919	2.9857	1.9781
30.0	1.5202	1.2717	2.3261	1.5973	3.0372	1.9898
40.0	1.5325	1.2723	2.3455	1.5993	3.0632	1.9942
50.0	1.5400	1.2727	2.3572	1.6002	3.0788	1.9962
100.0	1.5552	1.2731	2.3809	1.6015	3.1102	1.9990
∞	1.5708	1.2733	2.4050	1.6018	3.1415	2.0000

The two-component, single-term approximate solutions for the three coordinate systems are given in Table 10.1.

For a plane wall the transient temperature at the center plane is

$$\theta_0^*(Fo, 0) = C_1 e^{-\lambda_1^2 Fo} \tag{10.132}$$

and the spatial temperature distribution is

$$\theta^*(x^*) = \cos(\lambda_1 x^*). \tag{10.133}$$

The total solution based on the first term of the infinite series for a slab is

$$\theta_0^*(Fo, x^*) = C_1 e^{-\lambda_1^2 Fo} \cos(\lambda_1 x^*). \tag{10.134}$$

For a cylinder the transient temperature at the center line is

$$\theta_0^*(Fo, 0) = C_1 e^{-\lambda_1^2 Fo} \tag{10.135}$$

and the spatial temperature distribution is

$$\theta^*(r^*) = J_0(\lambda_1 r^*), \tag{10.136}$$

The total solution based on the first term of the infinite series for a cylinder is

$$\theta_0^*(Fo, r^*) = C_1 e^{-\lambda_1^2 Fo} J_0(\lambda_1 r^*). \tag{10.137}$$

For a sphere the transient temperature at the center point is

$$\theta_0^*(Fo, 0) = C_1 e^{-\lambda_1^2 Fo} \tag{10.138}$$

and the spatial temperature distribution is

$$\theta^*(r^*) = \frac{1}{\lambda_1 r^*} \sin(\lambda_1 r^*). \tag{10.139}$$

The total solution based on the first term of the infinite series for a sphere is

$$\theta_0^*(Fo, r^*) = C_1 e^{-\lambda_1^2 Fo} \frac{1}{\lambda_1 x^*} \sin(\lambda_1 x^*). \tag{10.140}$$

Note that the expression for calculating the transient center temperature is identical for all geometries, but the constants are different so that the numerical solutions also differ. Graphical representations of the foregoing solutions, called Heisler charts, are presented in Sect. 10.6.3.2.

Example 10.6.1 Tissue Storage in the Vitrified State.
The storage of human tissues and organs at ultra low temperatures offers a method of accumulating a large bank of materials available for transplantation, thereby alleviating one of the primary difficulties in matching recipients with an adequate number of donors. Although researchers have been working for decades to perfect effective cryopreservation methods, there remain significant problems that have blocked the broad adoption of low-temperature banking methods. A major problem is irreversible injury caused by ice crystal formation

when the temperature is lowered to a value of $-100°C$ or below that is necessary for long-term stability. One approach to avoid complications caused by ice crystals is to chemically modify the biological tissue so that the temperature can be lowered very rapidly to $-120°C$ (the glass transition temperature of water) or lower to avoid ice nucleation, thereby achieving a state for which the water will vitrify. This process is equivalent to turning the tissue into a glass, but it is a glass based on H_2O rather than SiO_2 that we are familiar with. However, this solution to the ice crystal injury problem creates a different problem due to internal thermal stresses that are locked into the glassy tissue when it vitrifies. As a consequence, a vitrified organ such as a kidney could crack into two or more pieces if the internal stresses are greater than the strength of the vitrified material, rendering it useless for transplantation when rewarmed to physiological temperature. One possible solution to this problem is to apply the standard method used in the metal working and ceramics industries to relieve internal stresses, which is the process of annealing. The difference for the present case is that the temperature range and the thermal and mechanical properties are quite different for a vitrified aqueous system than for a ceramic or metal. Your job on this problem is to calculate a thermal protocol for annealing a vitrified human kidney. The following background information is available for you to consider using for your analysis.

- The kidney can be treated geometrically as a sphere 7 cm in diameter.
- The vitrification process consists of cooling the organ in a quiescent vapor phase above a pool of liquid nitrogen to equilibrate at a long-term holding temperature of $-150°C$, which is well below the temperature of $-120°C$ at which ice crystals can form spontaneously.
- Annealing consists of warming the kidney from $-150°C$ to $-125°C$ to allow the internal stresses to relax. The annealing process is conducted by blowing a stream of nitrogen gas over the surface of the kidney at a temperature of $-123°C$.
- To assure an adequate state of annealing, the process should last long enough so that the center of the kidney is warmed to at least $-125°C$ for a minimum time of 5 h.
- The manufacturer of the refrigeration hardware is able to supply you with a unit that will blow the $-123°C$ nitrogen gas over the surface of the kidney at a minimum velocity of 1 m/s for the duration of the annealing process.
- For purposes of your calculations it is acceptable to approximate the properties of the vitrified kidney by those of ice.

How long should the annealing process be run from start to finish (warming the center of the kidney to $-125°C$, then holding for another 5 h) to achieve the target state necessary to reduce the probability of fracturing the kidney according to the criteria given above?

Solution. *Initial considerations:* This problem requires calculation of the time that a vitrified kidney with a uniform initial temperature of $-150°C$ would have to be warmed via a nitrogen stream at $-123°C$ to bring the center temperature to a value of $-125°C$. There will be a convective warming on the surface of the kidney in series with heat conduction through the interior. Since this is a transient problem, the starting point of the analysis will be to calculate Bi to determine if a simple lumped model can be applied.

System definition and environmental interactions: The system consists of a spherical frozen kidney that interacts with the environment via convection, as shown in Fig. 10.19. The governing environmental interaction is convective warming by external flowing nitrogen gas. The process is transient, and the duration is determined by 5 h after the center of the kidney is warmed from $-150°C$ to $-125°C$.

This problem statement is rather highly prescribed in that many of the important assumptions necessary to proceed with the analysis are described explicitly under the assumption that they may be too subtle and field-specific for the reader to anticipate. Some of the analysis involves application of convective heat transfer principles covered later in the text. These can be applied simply in the present context without further comment.

Apprising the problem to identify the governing equations: The conservation of energy states that the convection of heat to the kidney equals the change in the internal energy.

$$\frac{dE}{dt} = \dot{Q}_{conv}.$$

The first question to be addressed is whether the Biot number is small enough so that a lumped parameter analysis can be applied. This calculation will require that the convective heat transfer coefficient be determined.

Analysis: The flow pattern is externally around the kidney. Relevant properties are outside the temperature range found in Appendix C, but can be found in other handbooks. These are tabulated below.

Material	Property	Value	Units
N_2 gas (150 K)	Density	2.26	kg/m^3
N_2 gas (150 K)	Viscosity	1.006×10^{-5}	N s/m^2
N_2 gas (125 K)	Viscosity	0.85×10^{-5}	N s/m^2
N_2 gas (150 K)	Prandtl no.	0.759	
N_2 gas (150 K)	Conductivity	1.39×10^{-2}	W/m K

(*continued*)

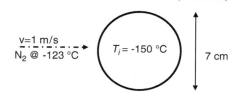

Fig. 10.19 A closed system consisting of a vitrified kidney that is exchanging heat with the environment via forced convection

10.6 Transient Diffusion Processes with Internal Thermal Gradients

Material	Property	Value	Units
N_2 gas (150 K)	Thermal diffusivity	5.86×10^{-6}	m^2/s
Ice (253 K)	Conductivity	2.03	W/m K
Ice (253 K)	Density	920	Kg/m^3
Ice (253 K)	Specific heat	1,945	J/kg K

For external flow over a sphere the Nusselt number can be calculated from (8.48), for which the Reynolds number must first be computed,

$$Re_D = \frac{\rho V D}{\mu} = \frac{2.26 \,\text{kg/m}^3 \cdot 1 \,\text{m/s} \cdot 7 \cdot 10^{-2}}{1.006 \,\text{N s/m}^2} = 1.57 \cdot 10^4$$

and

$$\overline{Nu}_D = 2 + \left(0.4 Re_D^{1/2} + 0.06 Re_D^{2/3}\right) Pr^{0.4} \left(\frac{\mu}{\mu_s}\right)^{1/4} = 84.1.$$

The convection coefficient is calculated from the Nusselt number,

$$\bar{h} = \frac{\overline{Nu} \cdot k_n}{D} = 16.7 \left(\frac{W}{m^2 K}\right).$$

Next, the Biot number is computed with the characteristic conduction length for a sphere equal to the volume-to-area ratio $L_c = \frac{V}{S} = \frac{4/3 \pi R^3}{4 \pi R^2} = \frac{R}{3}$,

$$Bi = \frac{h L_c}{k_i} = \frac{16.7 \,(W/m^2 \,K) \cdot 7/6 \cdot 10^{-2} \,(m)}{2.03 \,(W/m \,K)} = 0.096.$$

The lumped analysis method is just at the boundary of being usable for this problem. A distributed analysis can be used and the single-term simplification is justified since only longer times in the process are of interest. For this purpose the coefficients for the solution need to be identified based on the Biot number defined by $hR/k = 0.288$ (*note*: for purposes of analysis of the radial temperature diffusion process, the Biot number is defined in terms of the actual radius of the sphere, not the characteristic length).

The dimensionless temperature at the center of the sphere for the defined conditions is given by

$$\theta_0^* = \frac{T_0 - T_\infty}{T_i - T_\infty} = \frac{125 - 123}{150 - 123} = 7.41 \cdot 10^{-2}.$$

The center temperature is related to the time, properties and environmental state by

$$\theta_0^* = C_1 \exp(-\lambda_1^2 Fo),$$

where for $Bi = 0.288$ the constants are $C_1 = 1.080$ and $\lambda_1 = 0.877$. This equation can be solved for the Fourier number, Fo, which will identify the time to achieve a center temperature of $-125°C$.

$$-\lambda_1^2 Fo = \ln \frac{\theta_0^*}{C_1} \quad Fo = \frac{1}{\lambda_1^2} \ln \frac{C_1}{\theta_0^*} = \frac{1}{0.877^2} \ln \frac{1.080}{0.0741} = 3.49.$$

The time for the center to reach $-125°C$ is calculated from the Fo.

$$t_o = Fo \frac{R^2}{\alpha} = 3.49 \frac{(7/2 \cdot 10^{-2})^2 \text{m}^2}{5.86 \cdot 10^{-6} \text{ (m}^2/\text{s)}} = 729 \text{ (s)} = 12.1 \text{ (min)}.$$

Comments: Since $3.49 = Fo \gg 0.2$, the assumption that using the single-term approximate solution is validated. The time required to warm the kidney to the annealing temperature (about 12 min) is essentially insignificant in comparison to the total annealing time (5 h). The time constant for thermal diffusion is much smaller than the time constant for the relaxation of internal elastic stresses. This situation may not be true for all processes of this nature since the time constants are dependent on the constitutive thermal and elastic properties and the geometry, which will usually change among various systems.

10.6.2 Semi-Infinite Geometry

A semi-infinite geometry is defined for a system when the dimension in the primary direction of heat flow is large enough that the thermal effects on one surface are not propagated to the opposite side. Thus, the system behaves as if it is infinitely thick as one moves from the surface to the interior; hence the description as being semi-infinite. A common example of a semi-infinite system is the earth as it interacts with the atmosphere via changing boundary conditions. Changes in the environment will continue to be propagated further into the system with time, but with a diminished magnitude as the depth increases. Since a defined stable state is not achieved, processes in semi-infinite coordinates are most frequently modeled as transient. Figure 10.20 shows a typical representation for a semi-infinite system in Cartesian coordinates. Although many surface geometries are not perfectly flat in a Cartesian sense, if the radius of curvature is large in comparison with the effective penetration

10.6 Transient Diffusion Processes with Internal Thermal Gradients

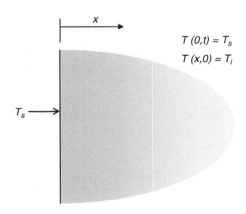

Fig. 10.20 A one-dimensional coordinate system for a semi-infinite geometry

depth of the boundary perturbation, it is acceptable to assume simple Cartesian semi-infinite coordinates.

For the sake of simplicity the effects of blood perfusion and internal energy generation will be omitted for the present development, although they can easily be added to the statement of conservation of energy if necessary. The governing equation for the one-dimensional transient temperature distribution in a homogeneous medium (10.107) reduces to:

$$\frac{\partial T}{\partial t} = \alpha \frac{\partial^2 T}{\partial x^2}. \tag{10.141}$$

The initial and boundary conditions are defined with the following considerations. The simplest initial condition is that the temperature be uniform everywhere.

$$T(x.0) = T_i.$$

In many cases, if this criterion is not met, a closed form analytical solution is precluded. The first spatial boundary condition is defined at a large depth into the material beyond which any changes implemented on the surface are not propagated over the entire time course of the process of interest. Thus, at a large depth the temperature never changes from the initial value,

$$T(\infty, t) = T_i.$$

The second spatial boundary condition is identified at the surface, and any of the three examples defined in Sect. 8.5 may be encountered:

$$T(0,t) = T_s \text{ for a specified temperature,}$$

$$-k\frac{\partial T(0,t)}{\partial x} = h[T(0,t) - T_\infty] \text{ for convection,}$$

$$-k\frac{\partial T(0,t)}{\partial x} = q_s \text{ for a heat flux, such as radiation.}$$

For the case of a specified step change in the surface temperature from the initial value to a new and constant value, the solution is the well-known Gaussian error function

$$\frac{T(x,t) - T_s}{T_i - T_s} = erf\frac{x}{2\sqrt{\alpha t}}, \tag{10.142}$$

where the error function is defined as

$$erf\frac{x}{2\sqrt{\alpha t}} = \frac{2}{\sqrt{\pi}}\int_0^{x/2\sqrt{\alpha t}} e^{-\eta^2}d\eta. \tag{10.143}$$

Note that the symbol η is a dummy variable used in the integration, and the value of the integral is a function of its upper limit. The error function is available as an embedded function in most common computational software programs that run on a personal computer.

The instantaneous heat flux can be computed at any depth by applying Fourier's law (2.9) in conjunction with (10.142).

$$\dot{Q}_x = -kA\frac{\partial T}{\partial x} = kA(T_s - T_i)\frac{2}{\sqrt{\pi}}e^{-x^2/4\alpha t}\frac{\partial}{\partial x}\left(\frac{x}{2\sqrt{\alpha t}}\right)$$

$$= kA\frac{T_s - T_i}{\sqrt{\pi\alpha t}}e^{-x^2/4\alpha t}. \tag{10.144}$$

Likewise, the solutions for temperature distribution for the convection and heat flow boundary conditions are, respectively:

for surface convection: $-k\left.\frac{\partial T}{\partial x}\right|_{x=0} = h[T_\infty - T(0,t)],$

$$\frac{T(x,t) - T_i}{T_\infty - T_i} = 1 - erf\frac{x}{2\sqrt{\alpha t}}$$
$$- e^{\left(\frac{hx}{k} + \frac{h^2\alpha t}{k^2}\right)}\left[1 - erf\left(\frac{x}{2\sqrt{\alpha t}} + \frac{h\sqrt{\alpha t}}{k}\right)\right], \tag{10.145}$$

for a constant surface heat flux: $-k\left.\frac{\partial T}{\partial x}\right|_{x=0} = q_s,$

10.6 Transient Diffusion Processes with Internal Thermal Gradients

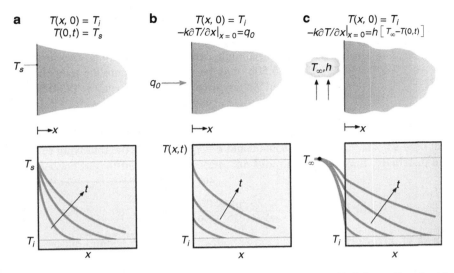

Fig. 10.21 Examples of the transient temperature distributions in a semi-infinite medium for (**a**) constant temperature, (**b**) constant heat flux, and (**c**) convection boundary conditions

$$T(x,t) - T_i = \frac{2q_s\sqrt{\alpha t/\pi}}{k} e^{-x^2/4\alpha t} - \frac{q_s x}{k}\left(1 - \operatorname{erf} \frac{x}{2\sqrt{\alpha t}}\right). \tag{10.146}$$

Figure 10.21 illustrates the differences in the surface and internal temperature distributions that occur with these three types of boundary conditions. An interesting limiting case occurs when the convective heat transfer coefficient becomes very large ($h \to \infty$), the convective condition becomes identical to an imposed constant surface temperature.

An interesting and very relevant special case is defined by bringing two semi-infinite media into direct contact, as shown in Fig. 10.22. This case approximates the event in which human flesh contacts a surface that is either hotter or colder than the skin temperature. The problem has been studied to determine the "*safe touch temperature*" of specific types of materials as a way to define safety standards for how hot exposed surfaces can be without the risk of causing contact burns. The safe touch temperature is a function of both the actual material temperature and the material thermal properties. For example, a sheet of aluminum and a sheet of low-density polymer at the same elevated temperature will "feel" quite different to the touch and accordingly have different levels of capability for causing a contact burn.

Since this problem involves conduction in a composite system, there must be two boundary conditions identified at the interface: one related to continuity of temperature and one to continuity of heat flow, in accordance with (8.100) and (8.101). If there is no interfacial thermal resistance at the contacting surfaces, then the temperatures of the two materials will be equal on a continuing basis,

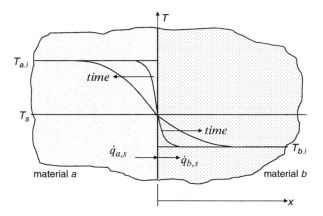

Fig. 10.22 Contact between two semi-infinite media, denoted by *a* and *b*, at different initial temperatures and with different thermal properties

$$T_a(0,t) = T_b(0,t) = T_s. \qquad (10.147)$$

The surface temperature remains constant over time following an initial step change from the initial value to an intermediate interface value, T_s. Likewise, the heat flows must be equal since the interface has no thermal mass in which energy could be stored. Thus, the heat flow into the interface must be equal in magnitude and opposite in sign to that flowing out, as shown by the assumed directional arrows in Fig. 10.22,

$$\dot{q}_{a,s} = \dot{q}_{b,s}. \qquad (10.148)$$

This process corresponds to a step change in the surface temperature. Equation (10.144) may be substituted on both the sides of this relationship resulting in

$$\frac{-k_a(T_s - T_{a,i})}{\sqrt{\pi \alpha_a t}} = \frac{-k_b(T_s - T_{b,i})}{\sqrt{\pi \alpha_b t}}, \qquad (10.149)$$

from which the surface temperature may be solved,

$$T_s = \frac{\sqrt{(k\rho c_P)_a} T_{a,i} + \sqrt{(k\rho c_P)_b} T_{b,i}}{\sqrt{(k\rho c_P)_a} + \sqrt{(k\rho c_P)_b}}. \qquad (10.150)$$

The property product $k\rho c_P$ is known as the *thermal inertia*. It is a measure of how effectively a material can exchange thermal energy with its environment. The higher the thermal inertia of a material, the more the interface temperature will be biased toward its initial value when touched by human skin and therefore the greater the potential for causing a burn.

10.6 Transient Diffusion Processes with Internal Thermal Gradients

Example 10.6.2 Safe Touch Temperature Analysis for Different Materials.
Consider the relative susceptibility of receiving a burn when touching different solid surfaces that are initially at 90°C. The materials are aluminum, maple wood and rigid polystyrene insulation. How long is required for the temperature in the skin at a depth of 500 μm to reach 45°C in each case?

Solution. *Initial considerations:* When human skin contacts the heated solid material, the temperature at the interface will come to an intermediate value. If the solid substrate is sufficiently thick, both the skin and the solid will experience a transient heat transfer in a semi-infinite geometry.

System definition and environmental interactions: The system is defined as skin in contact with the hot substrate material, as shown in Fig. 10.22.

Apprising the system to identify governing equations: The interface temperature between two semi-infinite solids at different temperatures brought into contact is given by (10.150)

$$T_s = \frac{\sqrt{(k\rho c_P)_a} T_{a,i} + \sqrt{(k\rho c_P)_b} T_{b,i}}{\sqrt{(k\rho c_P)_a} + \sqrt{(k\rho c_P)_b}}.$$

Analysis: The surface temperature is calculated with the appropriate property values from the table below.

Material	Conductivity, k (W/m K)	Density, ρ (kg/m³)	Specific heat, c_P (J/kg K)	Thermal inertia, $k\rho c_P$ (J²/s m⁴ K²)
Skin	0.37	1,000	4,217	1.56×10^6
Aluminum	237	2,702	903	5.78×10^8
Maple wood	0.16	720	1,255	1.44×10^5
Polystyrene	0.027	56	1,210	1.83×10^3

The initial temperatures for the skin and the hot material are 34°C and 90°C. Substitution of the above property values gives the following set of interface temperatures. Note that since the values of temperatures are being multiplied, it is necessary to express the temperatures using absolute units. As an example, for aluminum,

$$T_s = \frac{\sqrt{(k\rho c)_{al}} T_{al,i} + \sqrt{(k\rho c)_{sk}} T_{sk,i}}{\sqrt{(k\rho c)_{al}} + \sqrt{(k\rho c)_{sk}}}$$

$$= \frac{\sqrt{5.78 \times 10^8} \cdot 363.16\,\text{K} + \sqrt{1.56 \times 10^6} \cdot 307.16\,\text{K}}{\sqrt{5.78 \times 10^8} + \sqrt{1.56 \times 10^6}} = 360.4\,\text{K} = 87.2°\text{C}.$$

Material	Interface temperature, T_s (°C)
Aluminum	87.2
Maple wood	47.1
Polystyrene	35.9

The times required to reach 45°C at a depth of 500 μm are calculated from (10.142) by solving for the time that yields these specified results. Thus, this calculation involves an inverse solution of this equation to identify the time,

$$\frac{T(x,t) - T_s}{T_i - T_s} = erf \frac{x}{2\sqrt{\alpha t}}.$$

The calculations show that for aluminum the touch time is 0.892 s, and for maple wood it is 35.7 s. The time for aluminum is on the same order as the response time constant for a human touching a hot surface, whereas the time for maple wood is so long that it would take a concerted effort to maintain a continuous contact with the hot surface for the entire period. Since the interface temperature for polystyrene is below 45°C, that temperature will never be reached in the skin.

Thus, under these conditions, at a temperature of 90°C aluminum should be considered dangerous, maple wood as relatively safe and polystyrene as totally harmless.

Examining and interpreting the results: As the skin contacts the hot surface, the temperature of the skin surface will rise, and the temperature of the solid surface will decrease. The amount of decrease will depend on how readily the solid is able to conduct heat to the skin touching it. When the solid is more capable of conducting heat, the temperature at the interface will be greater. Our calculations show that the surface of the skin does not actually reach the value of the solid substrate. This explains why a hot solid with a low thermal inertia feels cooler than does a solid at exactly the same temperature, but with a high thermal inertia, because the resulting interface temperature is lower. Since the high thermal inertia solid raises the skin surface temperature to a higher value, the risk for causing a burn is greater. We are subjectively aware of this phenomenon based on our human experiences.

10.6.3 Graphical Methods

In the past years, graphical methods were a very convenient means to obtain approximate solutions to some classes of transient conduction problems. Modern computation tools have largely replaced graphical methods in most practical applications. However, it is useful to briefly consider these methods since they provide considerable intuitive insight into how the temperature will change within a system with internal diffusion and a convection boundary condition.

10.6 Transient Diffusion Processes with Internal Thermal Gradients

As would be anticipated, the simplest graphical solution methods have been developed for problems corresponding to the simplest mathematical solutions. These systems generally allow temperature to be a function of time and only a single spatial coordinate, have uniform initial temperature, have symmetric or planar geometry and experience a step change in boundary conditions. We will consider two types of graphical solution methods: one for semi-infinite geometry (Schmidt plot) and one for symmetric finite dimensioned systems (Heisler charts).

10.6.3.1 Schmidt Plot

The *Schmidt plot* is a discrete approximation of a continuous function to obtain a graphical solution for a transient heat transfer problem (Kreith 1958). The most common system for which it is applied is a semi-infinite solid with a planar surface, a homogeneous interior composition, and a uniform initial temperature which is subjected to a step change in the surface temperature that is subsequently maintained constant. This problem is rather easily solved mathematically with the error function as in (10.142), but the graphical solution provides interpretive insights that are not readily apparent for a purely mathematical solution. Other variations of this problem such as a convective and time-dependent boundary conditions, nonuniform initial temperature, and composite materials interfaces can be addressed by the Schmidt plot, but require added complexity without providing much added insight into the solution method, and are not included in this discussion. A more detailed discussion of this method is available in Kreith (1958).

The starting point is to divide the physical system into a set of equally spaced spatial increments or nodes, Δx. The time scale is also divided into a series of uniform increments, Δt. Integer counters m and p are used to keep track of the position and time, respectively,

$$x = m\Delta x \quad \text{and} \quad t = p\Delta t. \tag{10.151}$$

Changes in temperature are identified only at node points. Therefore, the magnitude of Δx defines the limit of spatial resolution possible with this method. Likewise, Δt defines the temporal resolution.

The mid-plane is identified between any pair of adjacent nodes, extending a distance $\Delta x/2$ in both directions as shown in Fig. 10.23. A subsystem is defined associated with the mass extending $\Delta x/2$ on both sides of a node and interacting via heat conduction with each of its nearest neighbor nodes. Although in actuality there will be a continuous temperature gradient across the mass associated with node m, for this analysis there will be only a single value recognized at the center position. At a particular time $t = p\Delta t$, this temperature is denoted by T_m^p. The conservation of energy can be written for this node as it exchanges heat via conduction with the neighboring nodes and alters the amount of energy stored internally. At the start of a time step p, the temperature gradients into and out of the node (in the positive x direction) are $(T_{m-1}^p - T_m^p)/\Delta x$ and $(T_m^p - T_{m+1}^p)/\Delta x$. The conservation of energy

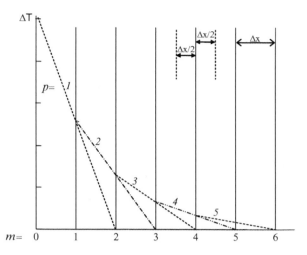

Fig. 10.23 Schmidt plot graphical solution for a step change of surface temperature in a homogeneous semi-infinite material

relationship applied for the node states that the difference between the heat flows in and out equals the rate of internal storage over time Δt. The relationship is expressed for the subsystem having a unit cross-section area normal to the x-axis as

$$\frac{\rho c A \Delta x \left(T_m^{p+1} - T_m^p\right)}{\Delta t} = kA \frac{\left(T_{m-1}^p - T_m^p\right)}{\Delta x} - kA \frac{\left(T_m^p - T_{m+1}^p\right)}{\Delta x}, \qquad (10.152)$$

which may be simplified to

$$\frac{\rho c_P \Delta x^2 \left(T_m^{p+1} - T_m^p\right)}{2k\Delta t} = \frac{\left(T_{m-1}^p + T_{m+1}^p\right)}{2} - T_m^p. \qquad (10.153)$$

Next, we specify that the values of Δt and Δx be scaled so that

$$\frac{\rho c_P \Delta x^2}{2k\Delta t} = \frac{\Delta x^2}{2\alpha \Delta t} = 1, \qquad (10.154)$$

in which case (10.153) becomes

$$T_m^{p+1} = \frac{\left(T_{m-1}^p + T_{m+1}^p\right)}{2}. \qquad (10.155)$$

This assumption corresponds to setting the Fourier number equal to ½.

With the T_m^p term now eliminated from the equation by a judicious choice of the relative values of Δt and Δx, the value for the temperature of node m one time step

10.6 Transient Diffusion Processes with Internal Thermal Gradients

$p + 1$ into the future is simply the average of the nearest neighbor values at the current time step p. Equation (10.155) can be constructed graphically to solve for the temperature T_m^{p+1} by drawing a straight line on a plot of temperature versus position from the temperature values at nodes $m - 1$ and $m + 1$. This graphical solution process is illustrated in Fig. 10.23 for a series of five sequential time steps affecting the first five nodes inward from the surface. The current temperature for any time step consists of the uppermost set of connected straight lines. This stepwise approximation to the actual continuous temperature function is considered to be most accurate at the central position of any node. That is, the slope of the actual temperature curve can be expected to be closest to the straight line approximation near the center of a node subsystem.

The fact that the Schmidt plot is based on a discrete analysis for which the solution is identified and evaluated only at finite increments in time and space, rather than as a continuous function as would be the case for a mathematical solution, gives the solution some distinctive features. In Sect. 11.2 we will present an introduction to numerical solutions via the finite difference method. This solution method also is discrete and therefore shares some of the features of the Schmidt plot. It is important to be aware of these features in order to understand the limitations of discrete solutions.

Perhaps the most striking and limiting characteristic of discrete solution methods is that the temperature resolution is limited in both time and space to Δx and Δt, and it is not possible to represent the temperature on a smaller scale than was used to establish the plot. In many cases it is desirable to make Δx as small as possible in the Schmidt plot to have good spatial resolution and to make Δt as large as possible in order to limit the number of solution iterations required to follow a transient process over a specified time period. As was shown in (10.154) the magnitudes of Δx and Δt are not independent, so that decreasing one demands that the other also be diminished, and vice versa. Good spatial representation comes only at a price of greater calculational burden. When it is necessary to analyze a process for an extended period of time, the number of time steps is increased in inverse proportion to how much the time step increment must be reduced. Obviously, in planning a solution strategy using a discrete approximation method, it is advisable to calculate and plan beforehand what the calculation burden will be as a function of how the time and spatial steps are set up.

There are other features that can be observed from the Schmidt plot. The effect of changing the boundary condition at $t = 0$ is propagated into the material one additional node for each further time step. Thus, there will be a finite passage of time required before a change made at the surface can be sensed at the interior. The greater the depth, the larger will be the time lapse until the boundary change can be propagated to that position.

Another characteristic is that the temperature at each node is updated only for alternate time steps. In Fig. 10.23 alternating time steps are denoted with distinctive line patterns to highlight this feature. An additional characteristic is that the magnitude of the temperature increments diminishes progressively with distance away from the surface. A limiting aspect of the graphical solution method is that with increasing large values of position and time, the temperature changes rapidly become smaller and more difficult to resolve spatially on the graph, and accuracy is lost.

The exercise of building a Schmidt plot provides a useful background for beginning the study of finite difference analysis. Many of the lessons learned therein are directly applicable to designing a finite difference grid.

10.6.3.2 Heisler Charts

The Heisler charts, first published in 1947 (Heisler 1947), are based on the single-term approximation solution to the transient heat conduction problem presented in Sect. 10.6.1. (You should also be aware of Appendix D in which is presented a more complete graphical solution of the transient conduction based on 30 terms, not just one. The Heisler charts have limitations that Appendix D charts do not.) Equations (10.134), (10.137) and (10.140) show that the solution for temperature, based on the first term of the infinite series, is the product of two terms: one for the transient behavior and one for the internal spatial distribution. Correspondingly, a set of two Heisler charts may be applied to develop a complete solution for a transient conduction problem. The center temperatures are calculated from (10.132), (10.135), and (10.138). The spatial distributions are calculated from (10.133), (10.136), and (10.139). Since the values of C_1 and λ_1 depend on the magnitude of Bi for the problem to be solved, each Bi results in a unique curve for temperature as a function of time or position. Therefore, one of the Heisler charts contains a nest of curves for how the temperature at the geometric center of a system changes with elapsed time from when the boundary condition was changed. Each curve represents a unique value for Bi. Since only a limited number of Bi curves can be plotted, it may be necessary to interpolate between curves if the Bi for the system does not appear on the chart. The second Heisler chart contains a nest of curves for the variation in temperature internal to the system between the center and surface as a function of the Bi. Each curve denotes the temperature at a specific interior location. Note that for very small values of Bi the temperatures at all interior locations merge into a single value consistent with internal conductance resistance becoming negligible in comparison to convective resistance at the surface. We show the single-term Heisler charts for a slab in Fig. 10.24, for a cylinder in Fig. 10.25, and for a sphere in Fig. 10.26. A more complex set of charts based on the first 30 terms of the infinite series governing a slab (10.121), a cylinder (10.124) and a sphere (10.127) are provided in Diller (1990b). As will be illustrated in the following example, these charts are used in pairs to solve for the temperature as a function of both position and time. The values identified from the (a) and (b) charts are multiplied together to calculate temperature as a function of position and time according to (10.134), (10.137), and (10.140).

10.6 Transient Diffusion Processes with Internal Thermal Gradients

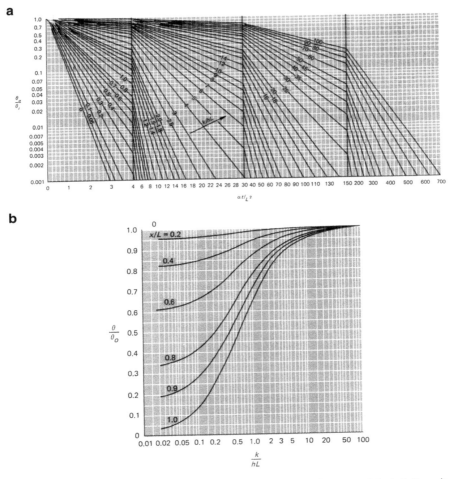

Fig. 10.24 Heisler charts for a slab of thickness $2L$ based on the first term of the infinite series ((10.132) and (10.133)): (**a**) temperature as a function of time at the midplane and (**b**) temperature vs. position for a step change in convective boundary condition (redrawn from Holman, 1972, with permission)

Example 10.6.3 Cooling a Long Aluminum Rod Exposed to a Cold Fluid.
Consider the process of cooling a long heated aluminum rod initially at a uniform temperature of 220°C when it is exposed to a fluid at 65°C. The convection heat transfer coefficient is determined independently to be 2,500 W/m² K. The diameter of the cylinder is 5 cm. Use the Heisler charts to calculate the temperature after 30 s at a radius of 2 cm.

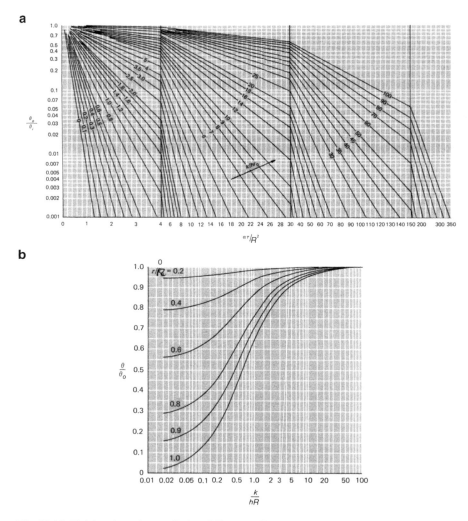

Fig. 10.25 Heisler charts for a cylinder of diameter d based on the first term of the infinite series ((10.132) and (10.133)): (**a**) temperature as a function of time at the centerline and (**b**) temperature vs. position for a step change in convective boundary condition (redrawn from Holman, 1972, with permission)

Solution. *Initial considerations:* The rod is initially at a uniform temperature, it is exposed to a new convective environment in a stepwise manner and, owing to a symmetric boundary and long length, the temperature will vary only along the radial dimension. Therefore, the conditions requisite for applying the Heisler charts are satisfied.

System definition and environmental interactions: The system is defined as the long rod that is losing heat via convection to its environment.

Apprising the system to identify governing equations: For a cylindrical geometry Fig. 10.25 is the appropriate set of Heisler charts.

10.6 Transient Diffusion Processes with Internal Thermal Gradients

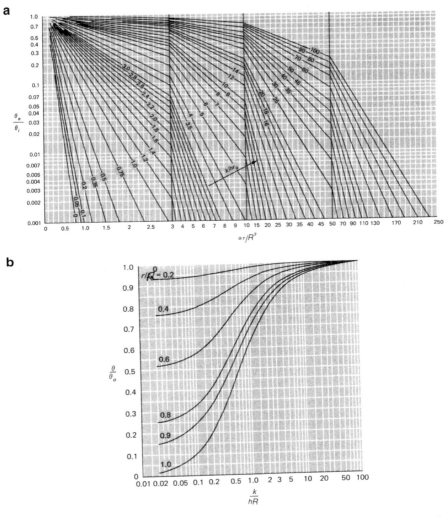

Fig. 10.26 Heisler charts for a sphere of diameter d based on the first term of the infinite series ((10.132) and (10.133)): (**a**) temperature as a function of time at the midpoint and (**b**) temperature vs. position for a step change in convective boundary condition (redrawn from Holman, 1972, with permission)

Analysis: Application of the Heisler charts requires that particular dimensionless property ratios be calculated as below. For this purpose it is necessary to obtain values for material properties. From Appendix C, the thermal conductivity of aluminum is 237 W/m K, density is 2,700 kg/m^2, specific heat is 900 J/kg K, and thermal diffusivity is 9.8×10^{-5} m^2/s:

$$\frac{r}{R} = \frac{2}{2.5} = 0.8,$$

$$Bi = \frac{hR}{k} = \frac{2{,}500 \text{ W/m}^2 \text{ K} \cdot 0.025 \text{ m}}{237 \text{ W/m K}} = 0.264 \quad \frac{1}{Bi} = \frac{1}{0.264} = 3.79.$$

Note that $Bi > 0.1$ so that the simplified lumped analysis method is not applicable for this problem.

$$Fo = \frac{\alpha t}{R^2} = \frac{9.8 \times 10^{-5} \text{ W/m K} \cdot 30 \text{ s}}{(0.025 \text{ m})^2} = 4.68.$$

From Fig. 10.25a we determine the temperature as a function of time along the centerline of the rod.

$$\frac{\theta_0}{\theta_i} = 0.13 \quad \theta_0 = T_0 - T_\infty = 0.13(220 - 65)°\text{C} = 20.2°\text{C}.$$

From Fig. 10.25b we determine the temperature distribution at any given time within the interior of the rod.

$$\frac{\theta}{\theta_0} = 0.93 \quad \theta = T - T_\infty = 0.93 \cdot 20.2°\text{C} = 18.7°\text{C}.$$

Alternatively written,

$$\theta = \frac{\theta_0}{\theta_i} \cdot \frac{\theta}{\theta_0} \cdot \theta_i = 0.13 \cdot 0.93 \cdot (220 - 65)°\text{C} = 18.7°\text{C}.$$

The temperature at a radius of 2 cm after 30 s of cooling is:

$$T(2 \text{ cm}, 30 \text{ s}) = \theta + T_\infty = 18.7 + 65 = 83.7°\text{C}.$$

Examining and interpreting the results: Further inspection of the Heisler charts can tell us more information about this process. From Fig. 10.25b for the spatial temperature distribution, we can observe that for the $Bi^{-1} = 3.79$ that applies to this problem, for $r/R = 1.0$, corresponding to the surface of the cylinder, $\theta/\theta_0 = 0.89$. This means that only 11% of the total temperature drop at any time exists between the surface and the centerline, and 89% is due to convective resistance on the surface with the environment. Given that the value of $Bi = 0.264$ is only marginally greater than the threshold of $Bi < 0.1$ for using a lumped analysis for which the solid is assumed to be isothermal, it is not surprising that only a small fraction of the temperature drop is due to conduction. From Fig. 10.25a we can see that for $Bi^{-1} = 3.79$ the smallest value of Fo for which a temperature can be resolved is about $Fo = 0.5$. This is approximately 1/10

10.6 Transient Diffusion Processes with Internal Thermal Gradients

of the value we calculated for a time of 30 s. The implication is that for this problem, the smallest process time for which temperature can be determined from the Heisler charts is around 3 s. To analyze the temperature for shorter times it would be necessary to use the infinite series solution with multiple terms as given in (10.124).

In the foregoing example, we conveniently embedded into the problem statement that the cylinder was very long, which led to the assumption that the temperature distribution was dependent on only a single spatial dimension, the radius. In many practical cases this simplifying assumption is not valid. For example, if the cylinder was quite short, the temperature would vary with both the radial and axial locations, requiring a two-dimensional solution. We treat multidimensional heat transfer problems in Chap. 11. However, for a linear problem like this one, the Heisler charts can be used, based on the principle of superposition. This analysis is based on the fact that the solution of the partial differential equation for the temperature can be separated into independent solutions, each of which is dependent on only one of the spatial variables. The complete solution is then the product of the single-dimension solutions. Thus, to illustrate for a two-dimensional Cartesian system, the governing partial differential equation is

$$\frac{1}{\alpha}\frac{\partial T}{\partial t} = \frac{\partial^2 T}{\partial x^2} + \frac{\partial^2 T}{\partial y^2}. \tag{10.156}$$

The solution of this equation can be written as

$$T(x,y,t) = X(x) \cdot Y(y) \cdot \Theta(t). \tag{10.157}$$

An equivalent solution is that for two orthogonal infinite plates of thicknesses $2L_1$ and $2L_2$, and denoted as T_1 and T_2. For the two plates, the corresponding differential equations are:

$$\frac{1}{\alpha}\frac{\partial T_1(x,t)}{\partial t} = \frac{\partial^2 T_1(x,t)}{\partial x^2}, \tag{10.158}$$

$$\frac{1}{\alpha}\frac{\partial T_2(y,t)}{\partial t} = \frac{\partial^2 T_2(y,t)}{\partial y^2}, \tag{10.159}$$

and the corresponding product solution is

$$T(x,y,t) = T_1(x,t) \cdot T_2(y,t). \tag{10.160}$$

The solution can be written in terms of the Heisler chart variables,

$$\left(\frac{T(x,y,t) - T_\infty}{T_i - T_\infty}\right) = \left(\frac{T_1(x,t) - T_\infty}{T_i - T_\infty}\right)_{2L_1} \cdot \left(\frac{T_2(y,t) - T_\infty}{T_i - T_\infty}\right)_{2L_2}. \tag{10.161}$$

Thus, a multidimensional problem is solved using the Heisler charts by solving for the temperature in each coordinate independently and then multiplying the individual solutions together. See Sect. 15.3.2 for a detailed treatment of superposition.

Example 10.6.4 Cooling a Short Aluminum Rod Exposed to a Cold Fluid. Consider the same process as for Example 10.6.3, except with the rod only 5 cm long. Determine the temperature at a position in the rod 0.5 cm from the outer radial surface and 0.5 cm from a flat end after 30 s of cooling.

Solution. *Initial considerations:* The radial analysis will be identical to that of the prior solution. The axial analysis will be performed independently in Cartesian coordinates and multiplied with the radial solution. Since heat will flow from the point of analysis to the surface along both radial and axial coordinates, we anticipate that cooling process will be further toward completion at a given time than for a very long rod in which heat flow is only radial.

System definition and environmental interactions: The system is defined as the short rod that is losing heat via convection to its environment from both the radial and axial surfaces.

Apprising the system to identify governing equations: For a flat plate geometry Fig. 10.24 is the appropriate set of Heisler charts.

Analysis: We will start by calculating the values of the dimensionless properties required to use the Cartesian Heisler charts:

$$\frac{x}{L} = \frac{2}{2.5} = 0.8,$$

$$Bi = \frac{hL}{k} = \frac{2,500 \text{ W/m}^2 \text{ K} \cdot 0.025 \text{ m}}{237 \text{ W/m K}} = 0.264 \quad \frac{1}{Bi} = \frac{1}{0.264} = 3.79,$$

$$Fo = \frac{\alpha t}{L^2} = \frac{9.8 \times 10^{-5} \text{ W/mK} \cdot 30 \text{ s}}{(0.025 \text{ m})^2} = 4.68.$$

From Fig. 10.24a and previously from Fig. 10.25a,

$$\left(\frac{\theta_0}{\theta_i}\right)_{\text{axial}} = 0.28 \quad \left(\frac{\theta_0}{\theta_i}\right)_{\text{radial}} = 0.13.$$

From Fig. 10.25b,

$$\left(\frac{\theta}{\theta_0}\right)_{\text{axial}} = 0.93 \quad \left(\frac{\theta}{\theta_0}\right)_{\text{radial}} = 0.93.$$

The net effect of axial and radial heat flows is calculated by multiplying the combined spatial and temporal factors times the overall temperature drop, θ_i.

$$\theta = \theta_i \cdot 0.28 \cdot 0.13 \cdot 0.93 \cdot 0.93 = 3.15 \times 10^{-2}\theta_i = 4.88°C,$$

$$T(2\,\text{cm}, 2\,\text{cm}, 30\,\text{s}) = 65 + 4.88 = 69.9°C.$$

Examining and interpreting the results: Adding an axial dimension of heat flow results in a further reduction in the temperature on the interior of the rod. We could use the Heisler charts to determine the cooling time for a short rod to reach the same temperature as required for a long rod. Since the two-dimensional temperature is determined as the product of the axial and radial solutions, it is necessary to iterate to determine desired time. We are looking for a time that yields the same centerline dimensionless temperature that occurs in the long cylinder, that is at $t = 30$ s,

$$\left(\frac{\theta_0}{\theta_i}\right)_{\text{radial,1D}} = 0.13.$$

After calculating the two-dimensional product for the center temperature for differing times, we can find that at $t = 20$ s, for which $Fo = 3.12$,

$$\left(\frac{\theta_0}{\theta_i}\right)_{\text{axial}} \cdot \left(\frac{\theta_0}{\theta_i}\right)_{\text{radial,2D}} = 0.55 \cdot 0.24 = 0.13.$$

Thus, providing for axial heat flow reduces the cooling time by about 1/3 for the conditions of this problem.

10.7 Summary of Key Concepts

We have seen repeatedly in this chapter that our basic approach to problem solving according to the Generate Ideas Method (defining a system consistent with our geometry and process of interest and how the system interacts with the environment; writing the governing conservation equations for the process; developing the differential equation(s) to model the process using appropriate constitutive relations; and applying initial and boundary conditions along with material properties to obtain a numerical result) can be applied in a very wide range of problem analysis situations. Indeed, this approach will be valid for nearly any type of problem you may encounter. Thus, we have been illustrating and practicing it over and over. It is our intention that the readers of this text will be able to commit the GIM to a level of enduring understanding in their arsenal of engineering tools. Then it will be possible to apply this analysis creatively in many different problem solving contexts.

Conduction with Internal Energy Generation. Heat transfer with internal energy generation is particularly important for biological systems because of the effects of metabolism, surgical procedures that use dissipative energy sources such as lasers, radio frequency, microwave and ultrasound, and diagnostic techniques. These phenomena produce internal temperature gradients within tissue that can be analyzed by defining a shell system and applying the principles of conservation. Each term in a conservation equation is represented by a constitutive relation, leading to a differential equation that is suitable to describe the temperature distribution. In this chapter we have only considered simple heat-generation equations. In the following chapter we will illustrate the more complex analysis that results from heating tissue with a laser having a nonlinear energy absorption pattern.

Heat Exchangers. Heat exchangers occur both within the body and in medical devices in which heat is transferred between fluids flowing proximally, but without mixing, at different temperatures. The geometry can be arranged so that at the point of transfer the fluids flow in the same direction (cocurrent), in opposite directions (counter-current) or in perpendicular directions (cross-current). Only the first two cases are examined in this chapter. Two alternative analysis procedures can be applied: the log mean temperature difference (LMTD) in combination with an overall heat transfer coefficient and the effectiveness–number of transfer units (NTU) method. Extensive experimental data on heat exchanger performance provides guidance for practical system designs.

Fins. The efficacy of heat exchange between a base material and the surroundings can be enhanced by extending the exchange surface in the geometry of a fin. The increase in heat transfer resulting from a greater surface area is described by the fin effectiveness,

$$\varepsilon_{\text{fin}} = \frac{\dot{Q}_{\text{total}}}{hA_c(T_0 - T_\infty)} = \frac{\dot{Q}_{\text{total}}}{hA_c\theta_0}. \tag{10.100}$$

The Effect of Blood Perfusion on Heat Transfer in Tissues. The Pennes equation provides a simple but widely applicable tool for describing the convective heat transfer that occurs when blood is perfused through tissue at a different temperature,

$$\rho c \frac{\partial T}{\partial t} = \nabla \cdot k\nabla T + (\rho c)_b \omega_b (T_a - T) + \dot{q}_{\text{met}}. \tag{10.102}$$

Convective heat exchange between blood and tissue is highly dependent on the geometry of the vascular network through which blood flows. It constitutes one of the most important components of the human thermoregulatory system.

Transient Diffusion Analysis by Analytical Methods. Mathematical descriptions of the transient diffusion of heat in a material are dependent on the geometry and boundary conditions, even for the simplest processes that start with a uniform initial temperature and homogeneous properties. When the geometry is finite in size and symmetric with respect to shape and boundary conditions, the temperature field is described by an infinite series that is unique for Cartesian, cylindrical, and spherical coordinates, as given below:

$$\text{Cartesian } \theta^* = \sum_{n=1}^{\infty} C_n e^{-\lambda_n^2 Fo} \cos(\lambda_n x^*), \qquad (10.121)$$

$$\text{cylindrical } \theta^* = \sum_{n=1}^{\infty} C_n e^{-\lambda_n^2 Fo} J_0(\lambda_n r^*), \qquad (10.124)$$

$$\text{spherical } \theta^* = \sum_{n=1}^{\infty} C_n e^{-\lambda_n^2 Fo} \frac{1}{\lambda_n r^*} \sin(\lambda_n r^*). \qquad (10.127)$$

When the process is evaluated only for longer times corresponding to $Fo > 0.2$, the temperature solution may be represented by only the first term of the series. When the geometry is semi-infinite, the temperature solution is given by an error function. For a constant temperature boundary condition at the surface,

$$\frac{T(x,t) - T_s}{T_i - T_s} = erf \frac{x}{2\sqrt{\alpha t}}. \qquad (10.142)$$

Transient Diffusion Analysis by Graphical Methods. Although most transient diffusion problems are now solved either analytically or by numerical approximation, graphical methods that were developed long before the widespread use of computers provide considerable insight into the governing behavior of transport processes. The Schmidt plot is a discrete approximation of the continuous temperature function that shares fundamental features with finite difference analysis. Heisler charts provide a graphical representation to the single-term approximation of the infinite series solution for finite, symmetric systems. Both are simple and fast to implement and offer explicit visual understanding of the heat transfer process.

10.8 Questions

10.8.1. Explain the significance of differences between transient and steady-state conduction processes.

10.8.2. How do conduction processes typically relate to convection and radiation processes?

10.8.3. For a transient conduction process how does the approach to analysis change as the internal temperature gradient becomes larger?

10.8.4. Describe how one-dimensional transient solutions may be used to solve two- and three-dimensional problems.

10.8.5. How do the physical characteristics of cocurrent and counter-current heat exchangers differ, and how do they affect the thermal performance of a heat exchanger?

10.8.6. What are the advantages and limitations of the Schmidt plot and Heisler charts for solving transient heat transfer problems?

10.8.7. Define the effectiveness of a heat exchanger. How can it be used for heat exchanger design?

10.8.8. Explain how internal heat generation affects the temperature distribution within a conducting material.

10.9 Problems

10.9.1 Cryotherapy Safety

Oftentimes, in treating injuries a cold pack is immediately applied to the skin of the affected area. Penetration of the cooling effect to the underlying target tissue is essential to derive the desired therapeutic benefit. A major concern relating to using the pack is that it should be cold enough to produce a therapeutic effect and not be so cold as to cause thermal injury of the skin at the site of application. If the surface temperature of the skin should not be lowered to less than 5°C, how cold can the pack be pre-cooled safely? Also, what will be the rate of heat transfer from the skin to the cold pack during the first 10 min of treatment? The gel within a typical cold pack consists of a solution of water and salt (to reduce the freezing temperature) with a small amount of cellulose added to increase the viscosity. The eutectic temperature for a sodium chloride (23.3% by weight) and water mixture is $-21.2°C$.

10.9.2 Warming a Three-Dimensional Stick

In USA, butter is generally sold by pounds, with four sticks per pound and each stick being 11.45 cm in length.

(a) If each stick is in the shape of a rectangle with square cross-section, find the dimensions of each stick (cm). For butter: $k = 0.2$ W/(m K); $\rho = 998$ kg/m^3; $c_p = 2{,}300$ J/(kg K).

(b) Suppose that a stick of butter is maintained for a week at 280 K. It is then placed in the surroundings at room temperature (293 K) so that all the six sides are exposed to air with $h = 8$ W/(m^2 K). Find the temperature (K) at the center of a single stick after 1 h.

10.9.3 Heating in a Muscle Cell

A long, cylindrical skeletal muscle cell with radius R produces heat at a constant rate per unit volume, q_{met}.

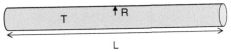

(a) The cell surface is insulated so no heat escapes from the cell. Assuming radial variations in the cell to be negligible, how long will it take for the temperature inside the cell to increase by 1°C?
(b) When the cell dies ($q_{met} = 0$), how long will it take to cool from 49°C to 48°C? Assume that Newton's law of cooling governs heat transfer at the outer surface of the cell, with surrounding temperature $T_\infty = 37$°C and $h = 200$ W/m^2°C. Is a lumped analysis justified in this case? Explain.

10.9.4 Heat Conduction from the Body

Heat is lost at a constant rate from the body by conduction through a layer of skin of thickness L_s, and then through a layer of clothing of thickness L_c. The skin has a thermal conductivity k that is twice the thermal conductivity of the clothing.

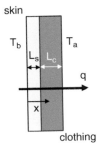

(a) If the temperature at the inner surface of the skin is T_b and the temperature at the outer surface of the clothing is T_a, derive expressions showing how the temperature varies with x in the skin and in the clothing.
(b) What is the temperature at the skin–clothing interface?

10.9.5 Thanksgiving Festivities

A turkey at room temperature is placed in a convection oven at 350°F. The turkey can be assumed to be a sphere with radius 10 cm and thermal conductivity

0.5 W/(m K). The heat transfer coefficient between turkey and air in the oven is 5 W/(m² K). Will a temperature reading taken at a radial position of 5 cm be a good approximation to the temperature in the center of the turkey? Provide a quantitative explanation.

10.9.6 Hot Surface Burn

A human forearm accidentally comes into contact with a hot oven rack. The temperature throughout the skin and tissue is initially uniform at 33°C before contact with the oven. Skin surface temperature instantaneously reaches 200°C upon oven contact and remains at that temperature during exposure to the hot surface. $\alpha_{skin} = 2.5 \times 10^{-7}$ m²/s. Estimate the time required to produce tissue damage to a depth of 1 mm if 44°C is the threshold for tissue damage.

10.9.7 Heat Loss by Walking on the Floor

We wish to reduce the heat lost by conduction from bare feet to a tiled bathroom floor by installing a carpet. Heat loss without the carpet is 10 kcal/h.

(a) How thick should the carpet be if this heat loss is to be cut in half? The floor temperature is 25°C and skin temperature is 32°C. The sole of each foot is approximately 30 cm long and 8 cm wide. The thermal conductivity of the carpet is reported to be 0.15 kcal/(m h C).
(b) A subcarpet is placed between the floor and the carpet. It has a thickness of 0.5 cm and a thermal conductivity of 0.02 kcal/(m h C). What is the conduction loss from the feet? What is the temperature at the carpet–subcarpet interface?

10.9.8 Transient Temperature Caused by Quenching

A thin slice of tissue initially at 40°C is immersed in a saline solution held at 10°C. After 100 s the temperature at the center is 25°C. We wish to use this information along with the *known* tissue geometry x_1, x_2, and x_3, and properties k, ρ, and C_p to estimate the heat transfer coefficient h. Neglect tissue heat generation.

(a) Show how you would use a lumped analysis to estimate h.
(b) Outline how you would use the Heisler chart provided to estimate h. Which value of k is used in computing Bi on the chart – tissue or saline?
(c) Which method would you expect to give the best estimate of h? Explain.

10.9 Problems

10.9.9 Cooling the Skin

An ice cube $T_s = 0°C$ is brought in contact with human skin having $\rho C_p = 3.7 \times 10^6$ J/m³·C, $k = 0.293$ W/m·C, and $T(t = 0) = 37°C$ for a total elapsed time of $t = 300$ s. The surface area of heat transfer is 100 cm². Use the solution for transient conduction through a semi-infinite slab with prescribed surface temperature to determine the amount (J) of heat transfer from the skin to the ice during this elapsed time.

10.9.10 Contact Burn

When human skin is brought in contact with a hot surface, it burns. The degree of burn is characterized by the temperature of the contact material T_s and the contact time t. A first-degree burn displays no blisters and produces reversible damage. A second-degree burn is moist, red, and blistered and produces partial skin loss. A third-degree burn is dry, white, leathery, and blisterless and produces whole skin loss. A pure copper pipe with constant temperature $T_s = 80°C$ is brought in contact with human skin having $\rho c_p = 3.7 \times 10^6$ J/m³·C, $k = 0.293$ W/m·C, and $T(t = 0) = 37°C$ for a total elapsed time of $t = 300$ s.

Use the solution for transient conduction through a semi-infinite slab with prescribed surface temperature to plot the temperature distribution $T(x, t)(°C)$ as a function of position x(mm) at elapsed times $t = 1, 10, 20, 40, 50, 100, 150, 200, 220, 240, 280,$ and 300 s.

10.9.11 Cooling an Isolated Muscle

An isolated muscle with cylindrical shape is stimulated repeatedly in a laboratory experiment until its temperature reaches 30°C. At that point, stimulation ceases and the muscle is allowed to cool in air at 20°C.

Muscle radius = 0.5 cm
Muscle length = 30 cm
Heat transfer coefficient between muscle surface and air: $h = 100$ W/(m²·K)

Thermal properties	Muscle	Air
Thermal conductivity, k (W/(m°K))	0.5	2.5×10^{-2}
Thermal diffusivity, α (m²/s)	2.5×10^{-7}	2.0×10^{-5}
Density, ρ (kg/m³)	1,000	1.25
Heat capacity, C_P (J/kg°K)	5,000	1,000

(a) How many seconds will it take for the temperature at the center of the muscle to reach 21°C?
(b) Find the initial rate of heat loss from the muscle (in watts).

10.9.12 Brewing a Coffee Bean

A spherical coffee bean with a radius of 7×10^{-3} m and temperature of 20°C is dropped into hot water at 90°C. How long will it take for the center of the bean to reach 80°C? The heat transfer coefficient between bean and water is 90 W/m²°C, the thermal conductivity of the bean is 0.15 W/m°C, the thermal diffusivity of the bean is 2×10^{-7} m²/s, specific heat of the bean is 0.577 kJ/kg°C, and the density of the bean is 1,300 kg/m³.

10.9.13 Protecting an Orange from a Freeze

An orange grove in Florida is exposed to a cold front causing a sudden change in air temperature from 18°C to 0°C. How long will it take for the oranges to reach 3°C in the center? Assume that the oranges are spherical with a radius of 4 cm, the thermal conductivity of the orange is 0.6 W/m°C, specific heat of the orange is 4.18 kJ/kg°C, the density of the orange is 1,000 kg/m³, the thermal conductivity of the air is 0.027 W/m°C, the density of air is 1.17 kg/m³, the specific heat of air is 1 kJ/kg°C, and the heat transfer coefficient between orange and still air is 0.675 W/m²°C.

10.9.14 Postmortem Interval Case Revisited

In a prior challenge (9.8.1) we considered the standard approach in determining the time of death according to a lumped mass model, as published prominently in the forensic science literature. After performing a heat transfer analysis, we realize that this model is inappropriate for the case in question. The prosecutor is astute enough during cross-examination to realize that this is one of the prime conclusions of our analysis. Consequently, the prosecutor gets wise and hires a bioengineer who testifies as follows:

- Your model prediction is criticized because a lumped analysis (macroscopic) was used.
- The witness states that internal thermal resistance in the body cannot be neglected.
- They claim the body takes longer to cool than you predicted.
- They present experimental evidence that body temperature varies with position and time.

10.9 Problems

1) Would you agree with the prosecutor's witness? Explain why or why not.
2) Model the body as a cylinder of tissue with length 1.8 m and surface area (excluding the ends) of 1.7 m². Neglect heat transfer from the ends of the cylinder. Use the Heisler chart for a cylinder to estimate the time it takes the centerline temperature T_c (core temperature) to cool from initial body temperature T_0 to the estimated temperature T at 5:30 AM. Does your estimate suggest that the defendant is guilty or innocent? Are there other factors that may change your estimate of the time of death? In particular, discuss the effect of creek velocity on heat transfer coefficient. Write this discussion as a simple explanation to the jury.

10.9.15 Hibernation Temperature

During hibernation of warm-blooded animals (homeotherms), the heart beat and the body temperature are lowered and in some animals the body waste is recycled to reduce energy consumption. Up to 40% of the total weight may be lost during the hibernation period. The nesting chamber of the hibernating animals is at some distance from the ground surface, as shown in figure (i). The heat transfer from the body is reduced by the reduction in the body temperature T_1 and by the insulating effects of the body fur and the surrounding air (assumed stagnant). A simple thermal model for the steady-state, spherical, one-dimensional heat transfer is given in figure (ii). The thermal resistances of air and soil can be determined from

$$R_{k,\text{air}} = \frac{1 - \dfrac{R_a}{2L}}{4\pi k_s R_a}, \quad R_{k,\text{soil}} = \frac{L - R_a}{4\pi k_s R_a L}.$$

An average temperature T_2 is used for the ground surrounding the nest. The air gap size $R_a - R_f$ is an average taken around the animal body.

Determine the heat loss from the body for (1) $L = 2.5 R_a$ and (2) $L = 10 R_a$. $R_1 = 10$ cm, $R_f = 11$ cm, $R_a = 11.5$ cm, $T_1 = 20°C$, and $T_2 = 0°C$. For air $k_a = 0.0267$ W/m K, for fur $k_f = 0.036$ W/m K, and for soil $k_s = 0.52$ W/m K.

(i) Diagram of Woodchuck Home

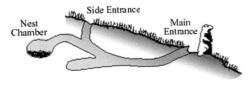

(ii) Simple Thermal Model

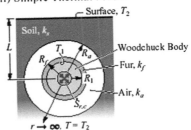

10.9.16 Frostbite to a Mountain Climber

On a hiking expedition in the Himalayas, a climber has an accident, losing his boot and sock, and his bare foot becomes wedged into a crevasse that is full of snow. The climber and his comrades have a radio and are able to call for help to come with equipment to release his foot. Help is expected to arrive in 30 min. Although his ultimate survival is not in doubt, there remains a concern as to whether he will suffer frost bite in his exposed skin. The conditions for frostbite are for the temperature be reduced to $-3°C$ at the base of the dermis, which is 2 mm thick. The temperature of the snow is $-15°C$, and the initial temperature of his skin at the time of the accident was $34°C$. What can you tell the trapped climber about his prospects for avoiding frost bite?

The following information is available for your use.

Density of skin	1,040	kg/m^3
Density of snow	500	kg/m^3
Specific heat of skin	4.0	kJ/kg K
Thermal conductivity of skin	0.21	W/m K
Thermal conductivity of snow	0.19	W/m K

The skin of the climber is in direct contact with the snow. Thus, there is an imposed fixed temperature boundary condition on the tissue surface. It is reasonable to assume that the mass and thermal properties of the snow are such that as they receive heat from the warm foot of the climber their temperature does not change significantly. The geometry of this system is best approximated as a semi-infinite solid, with the skin treated as a homogeneous material. At this low temperature blood perfusion will not have any significant effect, plus there are no vessels in the superficial layer of skin of greatest interest. The thermal consequences of the tissue freezing should not be included in your analysis, as they lead to an extremely difficult mathematical problem.

10.9.17 Heat Exchanger

A swimmer with body mass 50 kg is pulled from the ocean with hypothermia. Her body temperature is $27°C$. She is connected to a shell and tube heat exchanger. Blood flow through the device is kept constant at 1,000 ml/min. The perfusionist adjusts the flow rate in the perfusion fluid $Q_s(t)$; so blood leaving the heat exchanger always has a constant temperature of $37°C$. The inlet fluid temperature to the shell is kept constant at $45°C$. At time $t = 0$, the shell flow rate is 5,000 ml/min. Density and specific heats of blood, tissue, and perfusion fluid are the same: $\rho = 1$ g/ml; $c_p = 4$ J g^{-1} K^{-1}.

10.10 Challenges

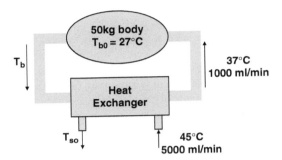

(a) What is the outlet shell temperature T_{so} at $t = 0$?
(b) What minimum product of overall heat transfer coefficient and exchange area must the heat exchanger have to allow the initial amount of heat exchange?
(c) Perform an energy balance on the body to determine how long it will take to heat the body to a temperature 35°C.

10.9.18 Steady-State Temperature Distribution in the Boundary Layer

Consider the head to be a sphere with radius 10 cm. Surface temperature of the skin is maintained at 35°C. The air surrounding the head is still (zero velocity) and the temperature of air far away ($r \to \infty$) from the head is 15°C.

Thermal properties of air	
Thermal conductivity, k (W m^{-1} K^{-1})	2.5×10^{-2}
Thermal diffusivity, α (m^2 s^{-1}) = $k/\rho C_p$	2.0×10^{-5}
Density, ρ (kg m^{-3})	1.25
Heat capacity, C_P (J kg^{-1} K)	1,000

(a) Derive an expression for the distribution of temperature $T(r)$ *in the air* for $r \geq 10$ cm.
(b) Find the heat loss from the head (in watts).
(c) Use your solution and Newton's law of cooling to compute a heat transfer coefficient for heat loss from the head to the air (W m^{-2} K^{-1}).

10.10 Challenges

10.10.1 Heat Transfer with Blood as it Flows Through Vessels

One of the most important components of the body's thermoregulatory function is based on convective heat transfer that occurs when the blood flows through vessels that are of different temperatures. In most situations the distribution of heat around the body via convection is much more effective than by conduction pathways. Quantitative knowledge of the convection effect is critical to the design of many types of diagnostic and therapeutic devices. A significant component of this knowledge base is an understanding of which elements of the vascular network function best for convective heat transfer. We may anticipate that the answer is dependent on the flow rate, length and diameter of the various vascular elements. In 1980 both Chato (1980) and Chen and Holmes (1980) performed a classic analysis of where in the vascular tree thermal equilibration occurs between blood and tissue. The topic has been revisited much more recently by Shrivastava and Roemer (2006). Their analysis showed that as the blood flows through progressively smaller and shorter branches of the arterial network, at a level of the approximately 60 μm arterioles, the blood will reach the same temperature as the tissue through which it is flowing. This result is quite different than that for chemical equilibration between blood and tissue, which occurs primarily in the much smaller capillaries where the lumen permeability allows for a transport of many molecular species. Heat flow is not limited by the vascular permeability, so that thermal equilibration can occur much earlier in the circulatory network than can chemical equilibration.

In this challenge, we investigate the convective heat transfer for blood flow through an artery, an arteriole, and a capillary having the geometries and flow states shown below (Whitmore 1968). Note that Whitmore calculated the Reynolds number for each of the flow conditions based on a blood viscosity of 0.03 poise. How would you develop a model for the thermal equilibration process that is at the level of mathematical sophistication that matches the material in this book?

Vessel	Diameter (μm)	Length (mm)	Velocity (mm/s)	Re_d
Large artery	3,000	200	130	130
Small artery	600	10	60	12
Arteriole	20	2	3	0.02
Capillary	8	1	0.7	0.002

Although these data are for dogs, it is in agreement with fewer but similar measurements for humans (Whitmore 1968).

References

Barnett CH, Harrison RJ, Tomlinson JDW (1958) Variations in the venous system of mammals. Biol Rev 22:442–487
Charney CK (1992) Mathematical models of bioheat transfer. Advances in heat transfer, vol 22. Academic, New York, pp 19–156
Chato JC (1980) Heat transfer in blood vessels. J Biomech Eng 102:110–118
Chen MM, Holmes KR (1980) Microvascular contributions in tissue heat transfer. Ann N Y Acad Sci 335:137–143
Diller KR (1990a) A simple procedure for determining the spatial distribution of cooling rates within a specimen during cryopreservation I. Analysis. Proc Inst Mech Eng J Eng Med 204:179–187

References

Diller KR (1990b) A simple procedure for determining the spatial distribution of cooling rates within a specimen during cryopreservation II. Graphical solutions. Proc Inst Mech Eng J Eng Med 204:188–197

Diller KR (1990c) Coefficients for solution of the analytical freezing equation in the range of states for rapid solidification of biological systems. Proc Inst Mech Eng J Eng Med 204:199–202

Diller KR, Valvano JW, Pearce JA (2005) Bioheat transfer. In: Kreith F, Goswami Y (eds) The CRC handbook of mechanical engineering, 2nd edn. CRC, Boca Raton, pp 4-282–4-361

Grigg GC (1970) Water flow through the gills of Port Jackson sharks. J Exp Biol 52:565–568

Heisler MP (1947) Temperature charts for induction and constant temperature heating. Trans ASME 69:227–236

Hewitt GF (2002) Heat exchanger design handbook 2002 (5 volumes). Begell House, New York

Jackson DC, Schmidt-Nielsen K (1964) Countercurrent heat exchange in the respiratory passages. Proc Natl Acad Sci USA 51:1192–1197

Kays WM, London AL (1984) Compact heat exchangers, 3rd edn. McGraw-Hill, New York

Kreith FM (1958) Principles of heat transfer. International Textbook Co., Scranton

Lee RC, Zhang D, Hannig J (2000) Biophysical injury mechanisms in electrical shock trauma. Annu Rev Biomed Eng 2:477–509

Pennes HH (1948) Analysis of tissue and arterial blood temperatures in the resting forearm. J Appl Physiol 1:92–122 (republished on 50th anniversary in (1998) J Appl Physiol 85:5–34)

Schmidt-Nielsen K (1972) How animals work. Cambridge University Press, Cambridge

Schmidt-Nielsen K (1997) Animal physiology: adaptation and environment, 5th edn. Cambridge University Press, Cambridge

Scholander PF, Krog J (1957) Countercurrent heat exchange and vascular bundles in sloths. J Appl Physiol 10:405–411

Shah RK, Sekulic DP (2003) Fundamentals of heat exchanger design. Wiley, Hoboken

Shrivastava D, Roemer RB (2006) Readdressing the issue of thermally significant blood vessels using a countercurrent vessel network. J Biomech Eng 128:210–216

Weinbaum S, Jiji LM, Lemons DE (1984) Theory and experiment for the effect of vascular microstructure on surface tissue heat transfer. Part I. Anatomical foundation and model conceptualization. J Biomech Eng 106:331–341

Whitmore RL (1968) Rheology of the circulation. Pergamon, Oxford

Wissler EH (1998) Pennes' 1948 paper revisited. J Appl Physiol 85:35–41

Chapter 11
General Microscopic Approach for Bioheat Transport

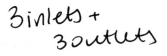
3 inlets + 3 outlets (handwritten annotation)

11.1 General Microscopic Formulation of Conservation of Energy

The flow of energy through biological materials such as blood vessels, airways, or tissue is integral to a very large array of processes in living systems. Thus, it is interesting and important to develop a facility for describing the transport of heat as a function of a standard set of critical parameters. These include:

- The properties of the material, such as ρ, k, c_p
- The temperature gradient
- The velocity field
- The geometry of the system
- Distributed internal energy sources such as metabolism, therapeutic heating, and blood perfusion
- The initial thermal state for a system that experiences a transient process
- The environmental thermal stressors that cause heat flow to occur (which are usually referred to as the boundary conditions)

As we consider the analysis of heat transfer in many types of systems, in general we will need to identify information about each of these parameters as we execute the Generate Ideas analysis method. Indeed, the GIM can be applied to develop the partial differential equation that describes the conduction of heat through a material.

11.1.1 Derivation of Conservation of Energy for Combined Conduction and Convection

For this procedure, the starting point is to define the system of analysis. A homogeneous isotropic material will serve to illustrate the process, although a more complex material could be chosen. The system is defined in Cartesian coordinates as a small cube on the interior of the material as illustrated in Fig. 7.1, consisting of an infinitesimally small volume with sides of length Δx, Δy, and Δz that surround an internal point within the system (refer to Fig. 7.1). We apply conservation of energy to the material contained in this subsystem or shell, which is open so that mass may cross the boundary:

$$\left\{\begin{array}{c}\text{Rate of}\\ \text{accumulation of}\\ \text{energy}\end{array}\right\} = \left\{\begin{array}{c}\text{Rate at which}\\ \text{energy}\\ \text{enters shell}\end{array}\right\} - \left\{\begin{array}{c}\text{Rate at which}\\ \text{energy}\\ \text{leaves shell}\end{array}\right\}$$

$$+\left\{\begin{array}{c}\text{energy trans.}\\ \text{across the}\\ \text{wall}\end{array}\right\} + \left\{\begin{array}{c}\text{Rate of}\\ \text{energy production}\\ \text{within shell}\end{array}\right\} \;\;\text{internal perfusion} \atop \text{metabolism} \quad (11.1)$$

Neglecting contributions from kinetic and potential energy and phase change, the rate of accumulation of energy is equal to the rate of change of internal energy:

$$\left\{\begin{array}{c}\text{Rate of}\\ \text{accumulation of}\\ \text{energy}\end{array}\right\} = \frac{\partial}{\partial t}\left(mc_\text{p}(T - T_\text{R})\right) = \frac{\partial}{\partial t}\left(\rho \Delta x \Delta y \Delta z c_\text{p} T\right). \quad (11.2)$$

Energy enters and leaves the shell through the six surfaces by conduction and convection (via mass flow with the property enthalpy). Assuming the energy flux in each coordinate direction to be positive, energy will enter the shell through surfaces at x, y, and z, and will leave at $x+\Delta x$, $y+\Delta y$, and $z+\Delta z$. Thus, the rate at which energy enters by conduction and convection (2.27) is:

$$\left\{\begin{array}{c}\text{Rate at which}\\ \text{energy}\\ \text{enters shell}\end{array}\right\} = (q_x + \rho v_x c_\text{p}(T - T_\text{R}))\big|_x \Delta y \Delta z + (q_y + \rho v_y c_\text{p}(T - T_\text{R}))\big|_y \Delta x \Delta z$$
$$+ (q_z + \rho v_z c_\text{p}(T - T_\text{R}))\big|_z \Delta x \Delta y. \quad (11.3)$$

Note that the convective flow across the boundary is written in the format: $w = \rho \Delta A v$. Similarly, the rate energy leaves through the other three surfaces is:

$$\left\{\begin{array}{c}\text{Rate at which}\\ \text{energy}\\ \text{leaves shell}\end{array}\right\} = (q_x + \rho v_x c_\text{p}(T - T_\text{R}))\big|_{x+\Delta x} \Delta y \Delta z$$
$$+ (q_y + \rho v_y c_\text{p}(T - T_\text{R}))\big|_{y+\Delta y} \Delta x \Delta z$$
$$+ (q_z + \rho v_z c_\text{p}(T - T_\text{R}))\big|_{z+\Delta z} \Delta x \Delta y. \quad (11.4)$$

11.1 General Microscopic Formulation of Conservation of Energy

Finally, energy can be produced inside the shell by metabolic processes, laser irradiation, chemical reaction, or other spatially distributed sources. Grouping these as a single term, and letting the rate of heat production per unit volume be $\dot{q}_{met}$:

$$\left\{\begin{array}{c} \text{Rate of} \\ \text{energy production} \\ \text{within shell} \end{array}\right\} = \dot{q}_{met} \Delta x \Delta y \Delta z. \tag{11.5}$$

Inserting all terms into the conservation of energy expression, we obtain:

$$\begin{aligned}\frac{\partial(\rho \Delta x \Delta y \Delta z c_p T)}{\partial t} =& \left((\rho c_p v_x T)\big|_x - (\rho c_p v_x T)\big|_{x+\Delta x}\right) \Delta y \Delta z \\ &+ \left((\rho c_p v_y T)\big|_y - (\rho c_p v_y T)\big|_{y+\Delta y}\right) \Delta x \Delta z \\ &+ \left((\rho c_p v_z T)\big|_z - (\rho c_p v_z T)\big|_{z+\Delta z}\right) \Delta x \Delta y \\ &+ (q_x|_x - q_x|_{x+\Delta x}) \Delta y \Delta z + (q_y|_y - q_y|_{y+\Delta y}) \Delta x \Delta z \\ &+ (q_z|_z - q_z|_{z+\Delta z}) \Delta x \Delta y + \dot{q}_{met} \Delta x \Delta y \Delta z. \end{aligned} \tag{11.6}$$

Dividing by the shell volume and letting the volume approach zero, we find:

$$\frac{\partial(\rho c_p T)}{\partial t} = -\underbrace{\frac{\partial(\rho c_p v_x T)}{\partial x} - \frac{\partial(\rho c_p v_y T)}{\partial y} - \frac{\partial(\rho c_p v_z T)}{\partial z}}_{\text{conv. terms}} \underbrace{-\frac{\partial q_x}{\partial x} - \frac{\partial q_y}{\partial y} - \frac{\partial q_z}{\partial z}}_{\text{cond. terms}} + \dot{q}_{met}. \tag{11.7}$$

Introducing Fourier's Law and assuming c_p and ρ are constant:

$$\rho c_p \frac{\partial T}{\partial t} = -\rho c_p \left\{v_x \frac{\partial T}{\partial x} + v_y \frac{\partial T}{\partial y} + v_z \frac{\partial T}{\partial z}\right\} + \frac{\partial}{\partial x}\left(k \frac{\partial T}{\partial x}\right) + \frac{\partial}{\partial y}\left(k \frac{\partial T}{\partial y}\right)$$
$$+ \frac{\partial}{\partial z}\left(k \frac{\partial T}{\partial z}\right) + \dot{q}_{met}. \tag{11.8}$$

Writing this equation in vector notation, we obtain an expression that is applicable for alternate coordinate systems:

$$\rho c_p \frac{\partial T}{\partial t} = -\rho c_p \vec{v} \cdot \vec{\nabla} T + \vec{\nabla} \cdot \left(k \vec{\nabla} T\right) + \dot{q}_{met}. \tag{11.9}$$

Dividing by ρc_p and using the definition of the thermal diffusivity:

$$\frac{\partial T}{\partial t} = -\vec{v} \cdot \vec{\nabla} T + \vec{\nabla} \cdot \left(\alpha \vec{\nabla} T\right) + \frac{\dot{q}_{met}}{\rho c_p}. \tag{11.10}$$

If, in addition, the thermal conductivity and thermal diffusivity are constant, this can be written in terms of the substantial derivative (Sect. 7.6):

$$\frac{DT}{Dt} = \left[\frac{\partial T}{\partial t} + \vec{v}\cdot\vec{\nabla}T\right] = \alpha\nabla^2 T + \frac{\dot{q}_{met}}{\rho c_p}. \tag{11.11}$$

In the rectangular Cartesian coordinate system, the general microscopic thermal energy equation for constant ρ, c_p and k is:

$$\rho c_p \frac{\partial T}{\partial t} = -\rho c_p \left\{ v_x \frac{\partial T}{\partial x} + v_y \frac{\partial T}{\partial y} + v_z \frac{\partial T}{\partial z}\right\} + k\left(\frac{\partial^2 T}{\partial x^2} + \frac{\partial^2 T}{\partial y^2} + \frac{\partial^2 T}{\partial z^2}\right) + \dot{q}_{met}. \tag{11.12}$$

In the cylindrical coordinate system, the general expression for constant ρ, c_p, and k is:

$$\rho c_p \frac{\partial T}{\partial t} = -\rho c_p \left\{ v_r \frac{\partial T}{\partial r} + \frac{v_\theta}{r}\frac{\partial T}{\partial \theta} + v_z \frac{\partial T}{\partial z}\right\}$$
$$+ k\left\{\frac{1}{r}\frac{\partial}{\partial r}\left(r\frac{\partial T}{\partial r}\right) + \frac{1}{r^2}\frac{\partial^2 T}{\partial \theta^2} + \frac{\partial^2 T}{\partial z^2}\right\} + \dot{q}_{met}. \tag{11.13}$$

In the spherical coordinate system, the general expression for constant ρ, c_p, and k is:

$$\rho c_p \frac{\partial T}{\partial t} = -\rho c_p \left\{ v_r \frac{\partial T}{\partial r} + \frac{v_\theta}{r}\frac{\partial T}{\partial \theta} + \frac{v_\phi}{r\sin\theta}\frac{\partial T}{\partial \phi}\right\}$$
$$+ k\left\{\frac{1}{r^2}\frac{\partial}{\partial r}\left(r^2 \frac{\partial T}{\partial r}\right) + \frac{1}{r^2 \sin\theta}\frac{\partial}{\partial \theta}\left(\sin\theta \frac{\partial T}{\partial \theta}\right) + \frac{1}{r^2 \sin^2\theta}\frac{\partial^2 T}{\partial \phi^2}\right\} + \dot{q}_{met}. \tag{11.14}$$

The full solution of the general microscopic thermal energy equation is accomplished by integrating once in the time domain and twice for each independent spatial domain. Each spatial integration generates functions of the other spatial coordinates which must be evaluated from known values of the solution at defined positions. The conditions by which the functions are evaluated are called the initial and boundary conditions, respectively, as discussed in Chap. 8. They are determined according to: (a) the shape of the system; (b) what the temperature field interior to the system is like at the beginning of the process; (c) the geometry of imposed heat transfer interactions with the environment; and (d) how these environmental interactions may change over time. As an aggregate, these four types of conditions dictate the form and complexity of the mathematical solution to (11.10), and there are many different outcomes that may be encountered. The remainder of this chapter is organized according to specific combinations of these conditions that are often found in biological applications of heat conduction.

11.1.2 Simplifying the General Microscopic Energy Equation

In this section, we will reduce the general energy equation for some of the cases that we have analyzed previously and compare the resulting expressions with equations

11.1 General Microscopic Formulation of Conservation of Energy

and solutions found in Chaps. 9 and 10. This will provide some practice in eliminating terms that are identically zero or negligible, and will also illustrate some of the assumptions made in previous applications.

11.1.2.1 One-Dimensional Steady-State Conduction in Tissue with Heat Generation

For steady-state conduction with metabolic heat generation in tissue in the x-direction and no mass flow, the convection terms are zero (i.e., $v_x = v_y = v_z = 0$), and heat does not flow in the y or z directions. The general microscopic energy equation reduces to:

$$0 = k\frac{\partial^2 T}{\partial x^2} + \dot{q}_{\text{met}}. \tag{11.15}$$

This is identical to the (10.7) derived using the 1D shell balance approach.

11.1.2.2 Cooling of a Cylinder

Let us consider a long solid cylinder at temperature T_0 that is cooled by placing it into a fluid bath with temperature T_∞. The system of interest is the solid cylinder. Velocities inside the cylinder are all zero, and since the cylinder is long we can assume that all the heat transfer occurs in the radial direction. No heat is generated inside the cylinder. In this case, the general energy expression for cylindrical coordinates, (11.13), reduces to:

$$\rho c_p \frac{\partial T}{\partial t} = \frac{k}{r}\frac{\partial}{\partial r}\left(r\frac{\partial T}{\partial r}\right). \tag{11.16}$$

This corresponds to the expression solved in Sect. 10.6. We will evaluate this equation for the initial and boundary conditions $T(0, r) = T_0$, $q_r(0) = 0$, and $q_r(R) = h(T_s - T_\infty)$. Applying Fourier's law, this can be written as:

$$\rho c_p \frac{\partial T}{\partial t} = -\frac{1}{r}\frac{\partial}{\partial r}(rq_r). \tag{11.17}$$

Integrating both sides over the volume of the cylinder:

$$\rho c_p \frac{d}{dt}\int_V T dV = -\int_0^R \frac{1}{r}\frac{\partial}{\partial r}(rq_r)(2\pi rL dr). \tag{11.18}$$

Defining the average temperature in the sphere as follows:

$$\bar{T} = \frac{1}{V}\int_V T dV. \tag{11.19}$$

The integral on the right side of (11.18) can be written in terms of the heat flux at the surface of the cylinder, since conduction must equal convection at the surface boundary:

$$\int_0^R \frac{1}{r}\frac{\partial}{\partial r}(rq_r)(2\pi rL\mathrm{d}r) = 2\pi L \int_0^{Rq_s} \mathrm{d}(rq_r) = 2\pi L R q_S = hS(T_s - T_\infty), \quad (11.20)$$

where $S = 2\pi R L$ and h is the heat transfer coefficient. The heat flux q_s is assumed to be in the positive r-direction at $r = R$. Substituting (11.19) and (11.20) into (11.18), we find:

$$\rho c_p V \frac{\mathrm{d}\bar{T}}{\mathrm{d}t} = hS(T_s - T_\infty). \quad (11.21)$$

If the internal resistance is low ($Bi < 0.1$), then the average temperature and the surface temperature will be nearly equal and (11.21) reduces to the (9.19) which was derived using the macroscopic approach. Note that the heat flux at the surface was a term in the macroscopic conservation equation, but here was determined by applying a boundary condition to the general microscopic equation.

11.1.2.3 Steady Flow Through a Tube with Constant Heat Flux at the Boundary

Consider the steady-state flow of fluid through a tube with radius R and length L in which a constant heat flux q_s is applied at the tube wall. We would like to compare the shell balance approach to the approach where the general equation is simplified. In Sect. 10.4.2, we used a shell balance approach to obtain an expression for the mean temperature in a vessel as a function of axial position. Assuming the vessel is a cylinder with its axis directed in the z-direction, vessel perimeter $P = 2\pi R$ and $w = \rho Q_V$, (10.46) becomes:

$$\frac{\mathrm{d}T_m}{\mathrm{d}z} = \frac{2\pi R q_s}{\rho c_p Q_V}. \quad (11.22)$$

Therefore, the shell balance approach predicts that the mixing cup temperature increases linearly with axial position. Let us now reduce the general thermal energy equation, (11.13), for one-dimensional steady-state transport with $\dot{q}_{met} = 0$ and negligible axial conduction. The only remaining term is for convection:

$$0 = -\rho c_p v_z \frac{\partial T}{\partial z}. \quad (11.23)$$

This result indicates that the temperature remains constant with increasing axial position, which is obviously incorrect. So what is the problem with this approach? The problem is that this is not strictly a one-dimensional problem. Although conduction in the axial direction is indeed small with respect to convection, conduction in the radial direction is not zero. In fact, radial conduction along the temperature gradient in the boundary layer is the mechanism by which heat is transferred from the boundary at R toward the center of the tube. Including radial conduction along with axial convection in the energy balance for this system, we have:

$$0 = -\rho c_p v_z \frac{\partial T}{\partial z} + \frac{k}{r}\frac{\partial}{\partial r}\left(r\frac{\partial T}{\partial r}\right). \tag{11.24}$$

Applying Fourier's law and recognizing that for steady, incompressible flow, v_z is independent of axial position z, this can be rewritten as:

$$0 = -\rho c_p \frac{\partial}{\partial z}(v_z T) - \frac{1}{r}\frac{\partial}{\partial r}(rq_r). \tag{11.25}$$

Integrating both terms over the cross-sectional area of the tube at any position z:

$$0 = -\rho c_p \frac{d}{dz}\int_{A_c}(v_z T)dA_c - \int_0^r \frac{1}{r}\frac{\partial}{\partial r}(rq_r)(2\pi rdr). \tag{11.26}$$

The left integral, by definition of the mixing cup temperature, is $Q_V T_m$. The right integral reduces to:

$$\int_0^r \frac{1}{r}\frac{\partial}{\partial r}(rq_r)(2\pi rdr) = 2\pi \int_{rq_r=0}^{rq_r=-Rq_s} d(rq_r) = -2\pi Rq_s. \tag{11.27}$$

The heat flux at the wall is directed in the negative r-direction, which accounts for the negative sign in (11.27). The final expression for (11.26) becomes:

$$\frac{dT_m}{dz} = \frac{2\pi Rq_s}{\rho c_p Q_V}. \tag{11.28}$$

This is identical to the expression derived with the shell balance approach. The primary difference in the derivations is that the wall flux term was included in the conservation expression for the shell balance but entered into the problem as a boundary condition when integrating the general microscopic energy equation.

11.2 Numerical Methods for Transient Conduction: Finite Difference Analysis

In many cases involving biotransport, it is not possible to obtain a closed form analytical solution for the governing differential equation that describes a process. Effects such as a complicated geometry, layered materials having significantly different transport properties, or a complex spatial and/or temporal pattern of energy deposition can cause this situation. Nonetheless, it is still possible to achieve a solution to the governing equation by numerical approximation methods. All of the limiting factors listed above can be accommodated numerically, but with a tradeoff of sacrificing the accuracy and spatial and temporal resolution inherent to analytical solutions. The most common numerical analysis methods are finite difference and finite element. In both cases, the continuous function of temperature in both time and space is replaced by a mathematical formulation that varies in value only at discrete increments such as Δt and Δx. In general, as Δt and Δx are made smaller, the numerical analysis more closely approximates the complete and continuous solution, but a larger computational burden is also incurred since the system is divided into a larger number of finite chunks. Thus, setting up a numerical analysis of a problem involves weighing the balance between the need for resolution in time and/or space against the capacity and capability of the computational resources available, and how long it will take to obtain a solution. As advances in computational hardware and software have been achieved, the utility and application of numerical analyses have also increased correspondingly.

Although both the finite difference and finite element methods are based on discrete approximations, there are significant practical differences. The finite element method is more versatile and powerful, and many commercial codes have been developed that can be applied to the solution of a very broad range of problems. However, the initial development of a finite element code requires relatively sophisticated techniques and can be quite time consuming. Conversely, finite difference methods lack versatility, generally being applicable to only the single specific problem for which they are written. The major advantage is that code development is relatively straightforward and can be accomplished with a small investment of time.

In this text only finite difference methods will be discussed. Finite difference solutions are more frequently developed by individual analysts, whereas for finite element analyses commercial codes are usually applied. Also, in a general course that covers a broad range of transport topics, it is not possible to devote the coverage that would be required to introduce finite element analysis to the extent that a student could develop an independent proficiency. That is not the case for finite differences. With little effort, it is possible to develop the ability to write an appropriate set of finite difference equations for a specific problem and to implement them for solution on a personal computer.

The starting point for developing a finite difference analysis is to convert a system representation in which the properties vary continuously to one in which only incremental changes are allowed. The objective is to be able to write an equivalent set of algebraic difference equations as an alternative to a single partial differential equation, with the anticipation that the set of algebraic equations will be easier to solve than will the PDE for the accompanying boundary conditions. To illustrate the process, a two-dimensional system will be used, although it is easily extended to three dimensions. Figure 11.1 shows how a two-dimensional physical system can be divided into finite increments. Although it is not necessary,

11.2 Numerical Methods for Transient Conduction: Finite Difference Analysis

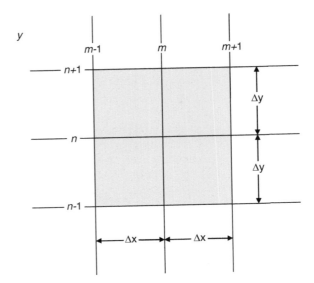

Fig. 11.1 Layout of a two-dimensional finite difference grid in a Cartesian coordinate system

for simplicity the size of the increments is the same for both the x and the y dimensions.

The grid lines are spaced at equal intervals Δx and Δy. The intersections of grid lines are called nodes. Values of the temperature can be identified only at grid line intervals, meaning that rather than being a point function, the temperature is taken to represent the average for a region associated with the mass surrounding an x and y grid line location. We will discuss how to account for this averaging effect later. The magnitudes of Δx and Δy define the limiting spatial resolution at which the temperature can be characterized, and likewise Δt defines the temporal resolution of the analysis.

Index counters are used to keep track of the coordinate node location and the elapsed time. As shown in Fig. 11.1, the indices m and n define the location in the Δx and Δy grid. Therefore, the absolute coordinate positions in x and y are given as:

$$x = m\Delta x \quad \text{and} \quad y = n\Delta y. \tag{11.29}$$

The index p is used to count the number of time increments so that:

$$t = p\Delta t. \tag{11.30}$$

The position indices are written as subscripts to temperature, and the time index as a superscript. Thus, the temperature at location m, n and time p is denoted as $T_{m,n}^p$.

The next step is to develop difference expressions to represent the partial differentials encountered in differential equations. These individual expressions are shown below. The differences are written for a location centered on node m, n. The temperature gradient is defined between adjacent nodes and therefore represents the value intermediate between the nodes.

$$\left.\frac{\partial T}{\partial x}\right|_{m-\frac{1}{2},n} \approx \frac{T_{m,n} - T_{m-1,n}}{\Delta x}, \tag{11.31}$$

$$\left.\frac{\partial T}{\partial x}\right|_{m+\frac{1}{2},n} \approx \frac{T_{m+1,n} - T_{m,n}}{\Delta x}, \tag{11.32}$$

$$\left.\frac{\partial T}{\partial y}\right|_{m,n-\frac{1}{2}} \approx \frac{T_{m,n} - T_{m,n-1}}{\Delta y}, \tag{11.33}$$

$$\left.\frac{\partial T}{\partial y}\right|_{m,n+\frac{1}{2}} \approx \frac{T_{m,n+1} - T_{m,n}}{\Delta y}. \tag{11.34}$$

The second-order difference is defined as the gradient between adjacent gradients, which means that it is associated directly with the node.

$$\left.\frac{\partial^2 T}{\partial x^2}\right|_{m,n} \approx \frac{\left.\frac{\partial T}{\partial x}\right|_{m+\frac{1}{2},n} - \left.\frac{\partial T}{\partial x}\right|_{m-\frac{1}{2},n}}{\Delta x} = \frac{T_{m+1,n} + T_{m-1,n} - 2T_{m,n}}{(\Delta x)^2}, \tag{11.35}$$

$$\left.\frac{\partial^2 T}{\partial y^2}\right|_{m,n} \approx \frac{\left.\frac{\partial T}{\partial y}\right|_{m,n+\frac{1}{2}} - \left.\frac{\partial T}{\partial y}\right|_{m,n-\frac{1}{2}}}{\Delta y} = \frac{T_{m,n+1} + T_{m,n-1} - 2T_{m,n}}{(\Delta y)^2}. \tag{11.36}$$

The temperature difference with respect to time can be written in either of two formats that are termed *forward* and *backward difference*, based on whether one looks into the future or at the past with respect to the present value of temperature. These two approaches require very different solution methods for the resulting set of algebraic equations. The forward difference format is:

$$\frac{\partial T}{\partial t} \approx \frac{T_{m,n}^{p+1} - T_{m,n}^p}{\Delta t} \tag{11.37}$$

and the backward difference format is:

$$\frac{\partial T}{\partial t} \approx \frac{T_{m,n}^p - T_{m,n}^{p-1}}{\Delta t}. \tag{11.38}$$

We will present the solutions for both formats, starting with the forward (explicit) difference.

11.2.1 Forward Finite Difference Method

The relevant partial differential equation to be approximated is:

$$\frac{\partial T}{\partial t} = \alpha \left(\frac{\partial^2 T}{\partial x^2} + \frac{\partial^2 T}{\partial y^2} \right). \tag{11.39}$$

Substituting (11.36), (11.37), and (11.38) for an interior node that is surrounded by like nodes yields:

$$\frac{T_{m,n}^{p+1} - T_{m,n}^p}{\Delta t} = \alpha \left[\frac{T_{m+1,n}^p + T_{m-1,n}^p - 2T_{m,n}^p}{(\Delta x)^2} + \frac{T_{m,n+1}^p + T_{m,n-1}^p - 2T_{m,n}^p}{(\Delta y)^2} \right]. \tag{11.40}$$

This equation could also be derived by applying the conservation of energy to a system consisting of the mass that is averaged over the region surrounding node m, n.

Since Δx and Δy are equal for this derivation, they may be combined. With minor rearrangement of terms, the foregoing equation becomes:

$$T_{m,n}^{p+1} = Fo\left(T_{m+1,n}^p + T_{m-1,n}^p + T_{m,n+1}^p + T_{m,n-1}^p\right) + (1 - 4Fo)T_{m,n}^p, \tag{11.41}$$

where the dimensionless property grouping is defined as the *Fourier number* in difference format. It is a measure of the thermal diffusion time constant for the system, based on its thermal properties and size, in comparison to the time stepping increment defined for the finite difference analysis.

$$Fo = \frac{\alpha \Delta t}{(\Delta x)^2}. \tag{11.42}$$

Before proceeding, several important features of (11.41) should be pointed out. First, if the full initial temperature distribution is known (at $p = 0$), then henceforth the temperature at the next time increment can be calculated independently for each node using this equation. When the value of p is incremented by 1 after the full temperature distribution is calculated, based on its prior known value, another complete set of calculations is performed to advance the solution one additional increment in time. This process is repeated as often as necessary to reach the targeted elapsed time.

Second, the future value of temperature at a node is determined as a function of the thermal and finite difference incremental space and time sizes in conjunction with the current temperatures at the node and its nearest neighbors. For a one-dimensional system there are two neighbors, for two dimensions there are four, and for three dimensions there are six, with (11.41) adjusted accordingly.

Third, the selection of the magnitudes of Δx and Δt is not arbitrary, but the values are strongly interdependent. The criterion that links these two values is that the coefficient for the $T_{m,n}^p$ term has to be nonnegative to avoid having the computation process become numerically unstable. If the size of the forward time step is too large, the solution will diverge from the correct value, resulting in the calculations diverging from the final steady-state value. For (11.41) the stability criterion is:

$$1 - 4Fo \geq 0 \quad \text{or} \quad Fo = \frac{\alpha \Delta t}{(\Delta x)^2} \leq \frac{1}{4}. \tag{11.43}$$

Equation (11.43) fixes the limiting relationship between Δt and Δx. In general, it is desirable to set Δt as large as possible to minimize the number of computational iterations to complete a simulation of a given duration. However, as Δt is enlarged, Δx must also grow, thereby diminishing the spatial resolution with which the temperature can be calculated. In most cases spatial resolution is more useful than temporal resolution, meaning it is desirable to have Δx as small as possible. This competition between setting the magnitudes of Δt and Δx is an issue that must be resolved in building most forward difference models. A further consideration is that if not all the node equations are the same for a system, the stability criterion (the coefficient of every $T_{m,n}^p$ term must be nonnegative) is set by the node for which the most stringent limitation applies to the relationship between Δt and Δx. This condition often exists at a boundary node rather than an interior node and is discussed below.

Fourth, (11.41) holds only for an internal node. If a node has a neighbor that is at a boundary, either on the surface or at a materials interface, then some of the terms have to be adjusted in accordance with the local grid geometry and how each adjacent node contributes to the conservation of energy for the subject node. Examples for some of these situations are derived below. As long as a logical process is followed, based on proper definition of the nodal system and its environmental interactions and application of conservation of energy, derivation of each individual nodal equation is quite straightforward. Figure 11.2 shows the grids for a number of commonly encountered boundary conditions. A planar surface node with convection (a) will be selected for detailed analysis. Conservation of energy applied to the surface node is different from the interior node in that there is only half of the mass surrounding the node to participate in energy storage, and the conduction along the y axis occurs via a cross-sectional area only half that of an interior node. This characteristic is important when writing the conservation of energy in terms of the difference equation.

Consider the energy balance for the mass associated with node $m, 0$ in Fig. 11.2a as shown in Fig. 11.3. Since the surface is defined normal to the y axis, a logical index numbering scheme is to start with the surface node indicated by 0, with each successive node toward the interior increasing by 1.

$$\dot{E}_{st} = \sum_{\text{boundary}} \left(\dot{E}_{in} - \dot{E}_{out} \right). \tag{11.44}$$

All energy exchanges with the environment are arbitrarily shown as positive into the system. Application of the conservation of energy yields:

11.2 Numerical Methods for Transient Conduction: Finite Difference Analysis

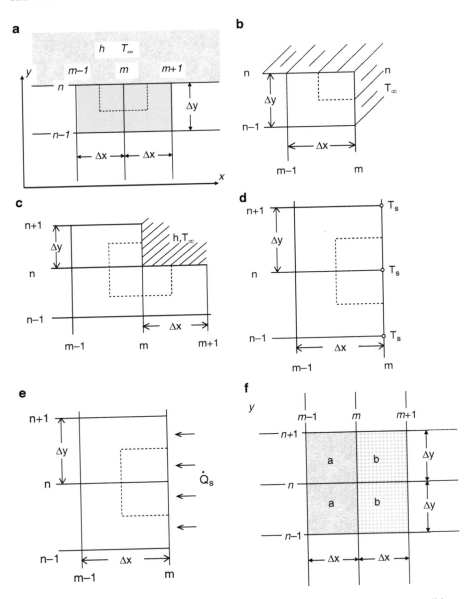

Fig. 11.2 Finite difference grid geometries for nodes with various typical boundary conditions: (**a**) planar surface with convection, (**b**) outside corner with convection, (**c**) inside corner with convection, (**d**) fixed surface temperature, (**e**) defined surface heat flux, (**f**) interior material interface

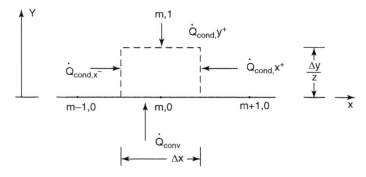

Fig. 11.3 Conservation of energy for node $m, 0$ with surface convection with the environment

$$\dot{E}_{st} = \dot{Q}_{cond,x^+} + \dot{Q}_{cond,x^-} + \dot{Q}_{cond,y^+} + \dot{Q}_{conv}. \tag{11.45}$$

The constitutive expressions for each term in this equation can be substituted in the finite difference format, taking note of the altered geometry of the surface node. In particular, the mass associated with surface nodes is only half that of an interior node, which affects the capacity for storing energy and the heat conduction cross-section along the x axis. The term Z is used to denote the unit length in the dimension for which temperature does not vary in a two-dimensional system.

$$\rho c Z \Delta x \frac{\Delta y}{2} \frac{T_{m,0}^{p+1} - T_{m,0}^p}{\Delta t} = \frac{kZ\Delta y}{2\Delta x}\left(T_{m-1,0}^p - T_{m,0}^p\right) + \frac{kZ\Delta y}{2\Delta x}\left(T_{m+1,0}^p - T_{m,0}^p\right) \\ + \frac{kZ\Delta x}{\Delta y}\left(T_{m,1}^p - T_{m,0}^p\right) + hZ\Delta x\left(T_\infty - T_{m,0}^p\right). \tag{11.46}$$

For $\Delta x = \Delta y$, the temperature for the central surface node is written as:

$$T_{m,0}^{p+1} = \frac{\alpha \Delta t}{\Delta x^2}\left(T_{m-1,0}^p + T_{m+1,0}^p + 2T_{m,1}^p\right) + \frac{2h\Delta t}{\rho c \Delta x}T_\infty + \left(1 - 4\frac{\alpha \Delta t}{\Delta x^2} - 2\frac{h\Delta t}{\rho c \Delta x}\right)T_{m,0}^p. \tag{11.47}$$

As with the Fourier number, the *Biot number* may also be defined for a discrete system as:

$$Bi = \frac{h\Delta x}{k}. \tag{11.48}$$

Substitution of Bi and Fo into (11.47) yields:

$$T_{m,0}^{p+1} = Fo\left(T_{m-1,0}^p + T_{m+1,0}^p + 2T_{m,1}^p\right) + 2BiFoT_\infty + (1 - 4Fo - 2BiFo)T_{m,0}^p. \tag{11.49}$$

11.2 Numerical Methods for Transient Conduction: Finite Difference Analysis

The stability criterion as defined for node m, 0 demands that the coefficient for $T_{m,0}^p$ be nonnegative.

$$1 - 4Fo - 2BiFo \geq 0. \qquad (11.50)$$

Thus,

$$\frac{1}{4} \geq Fo\left(1 + \frac{Bi}{2}\right) \quad \text{or} \quad Fo \leq \frac{1}{4}\left(\frac{2}{2+Bi}\right). \qquad (11.51)$$

Note that the stability criterion for the surface node is the same (if $Bi = 0$) or more stringent (if $Bi > 0$) than for an interior node as defined in (11.43) since Bi must always have a nonnegative value. As a consequence, when developing a forward (or explicit) finite difference analysis for a system, it is necessary to check the stability criterion for all types of nodes present to determine which one places the most limiting restrictions on the relative values chosen for Δx and Δt. That criterion must then be applied for establishing the magnitudes of all finite increments for the analysis.

The principle of conservation of energy can be applied as illustrated in the foregoing explanation to derive the governing node equations for all of the conditions shown in Fig. 11.2.

Example 11.2.1 Heating of a Cool Plate.
A plate-shaped material initially at a uniform temperature of $10°C$ is rapidly immersed in a stirred fluid with temperature $90°C$. The thickness of the plate is 1 cm and is much smaller than its other dimensions. Key properties are a material thermal diffusivity of $5.1 \times 10^8 \, m^2/s$ and a convection heat transfer coefficient between the material and the fluid that is very large. Perform an explicit finite difference analysis to predict what the temperature distribution will be within the material at 3 min after the immersion.

Solution. *Initial considerations*: The specified dimensions of the material dictate that this process can be assumed to be one-dimensional. Because the convective heat transfer coefficient is very large, $Bi \gg 0.1$, the simplified lumped analysis solution method cannot be applied. There are actually a number of alternative methods by which this problem could be solved. One approach is to apply an analytical solution, as in the infinite series of (10.121), or the one term approximation of (10.134). Another approach would be to apply the Heisler charts in Fig. 10.24. Here we will follow a finite difference analysis to illustrate the method, even though for this problem it will probably involve more work than the foregoing alternatives.

Given that the convection coefficient is so large, it is reasonable to assume that the surface of the material adopts the temperature of the surrounding fluid as soon as it is immersed. This then becomes a constant boundary temperature problem.

System definition and environmental interactions: The system is a homogeneous material with a one-dimensional internal temperature gradient caused by exposure to a surrounding fluid at a different temperature. For the finite difference analysis, we will define a system with a uniform overlaid grid as shown below. We will define nine equally spaced nodes with index m, starting and ending at the boundaries. The convection interaction at the boundaries is sufficient to maintain a constant value for the surface temperature.

Each node is separated by an equal distance, Δx, and represents the property values of the mass for a distance $\Delta x/2$ on each side.

Apprising the problem to determine the governing equations: The governing equation for the temperature distribution is the one-dimensional equivalent of (11.41).

$$T_m^{p+1} = Fo\left(T_{m+1}^p + T_{m-1}^p\right) + (1 - 2Fo)T_m^p.$$

Analysis: The stability criterion that is applied to determine the maximum ratio of $\Delta t/\Delta x$ is that the coefficient of T_m^p be nonnegative.

$$0 \leq 1 - 2Fo = 1 - 2\frac{\alpha \Delta t}{\Delta x^2}$$

$$\frac{\alpha \Delta t}{\Delta x^2} \leq \frac{1}{2}$$

$$\Delta t \leq \frac{\Delta x^2}{2\alpha} = \frac{(1.25 \times 10^{-3}\text{m})^2}{2 \cdot 5.1 \times 10^{-8}\frac{\text{m}^2}{\text{s}}} = 15.3 \text{ s}.$$

$\Delta t = 15$ s satisfies the stability criterion and provides an integer number of iterations to 3 min. The value for Fo is:

$$Fo = \frac{\alpha \Delta t}{\Delta x^2} = \frac{5.1 \times 10^{-8}\frac{\text{m}^2}{\text{s}} \cdot 15 \text{ s}}{(1.25 \times 10^{-3}\text{m})^2} = 0.49.$$

Owing to the boundary conditions, the surface node temperatures are fixed at:

$$T_0^p = T_8^p = 90°C,$$

starting with the immersion event at $p = 0$. All of the interior nodes obey the first difference equation above. Because the system is geometrically and thermally symmetric, it is necessary to calculate temperatures for only one-half space consisting of nodes 0–4.

11.2 Numerical Methods for Transient Conduction: Finite Difference Analysis

$$T_m^{p+1} = 0.49\left(T_{m+1}^p + T_{m-1}^p\right) + 0.0208 T_m^p.$$

The table below presents the temperatures for each node in the material half space for time steps to 5 min.

Time step		Node/temperature (°C)				
p	t (s)	0	1	2	3	4
0	0	90	10	10	10	10
1	15	90	49.2	10	10	10
2	30	90	50.0	29.2	10	10
3	45	90	59.5	30.0	19.4	10
4	60	90	60.1	39.3	20.0	10
5	75	90	64.6	40.1	29.1	19.3
6	90	90	65.1	46.7	30.0	20.0
7	105	90	68.4	47.6	37.7	28.9
8	120	90	68.8	53.0	38.8	30.0
9	135	90	71.5	53.9	45.2	37.6
10	150	90	72.0	58.3	46.4	38.8
11	165	90	74.2	59.2	51.6	45.1
12	180	90	74.6	62.9	52.8	46.4
13	195	90	76.5	63.8	57.2	51.5
14	210	90	76.9	66.8	58.3	52.8
15	225	90	78.4	67.7	61.9	57.1
16	240	90	78.9	70.2	63.0	58.3
17	255	90	80.1	71.0	66.0	61.9
18	270	90	80.6	73.1	67.1	63.0
19	285	90	81.6	73.8	69.6	66.0
20	300	90	82.0	75.6	70.5	67.1
						69.6

The same data can be plotted for a more comprehensive view of the evolution of temperature in time and space.

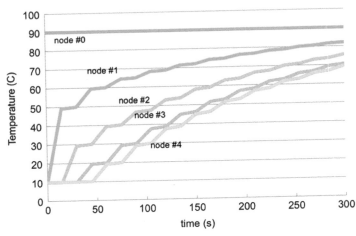

Examining and interpreting the results: The results for the finite difference analysis appear quite similar to that of the Schmidt plot method in Sect. 10.6.3.1.

The change in boundary temperature is propagated into the material interior one additional node for each time step, and interior node temperatures change only on alternate steps. However, in this present analysis we can see a very small change on the alternate steps rather than no change at all. This behavior is a consequence of our choosing a time step so that the coefficient of the T_m^p term is nonzero, providing it at least a small input value of each time step. Comparison of the analysis equation immediately preceding the table of computed temperatures to (10.155) shows a similarity in structure, but with a small added term for the node temperature for which the equation is written. Had we chosen the time step to be $\Delta t = 15.3$ s, this added term would become zero, and the temperature change on alternate time steps would have been zero instead of near to zero. This choice has no effect on the accuracy of our solution. However, both the spatial and temporal steps are rather large for this analysis, so that the resolution of temperature is quite coarse. We could improve this resolution using smaller values for Δx and Δt to obtain a smoother temperature curve, but at the expense of a larger number of computational steps to perform a simulation to a given elapsed process time.

Example 11.2.2 Temperature in Skin During a Burn.
Develop a finite difference model for the transient temperature field in the skin during a burn injury. Two types of thermal insult are to be evaluated: a prolonged contact with a hot surface that maintains a constant boundary temperature, and convection with a hot fluid via a convective heat transfer coefficient of 8,400 W/m² K. The initial temperature of the skin is 34°C, and the environmental temperature is 60°C.

Solution. *Initial considerations:* The burn will occur by contact with a heated substrate or convective flow over a finite sized area of the skin. The most severe burn will arise in the center of the insult area, with the severity diminishing as the edge is approached where the temperature will not be so high. In most cases, the interest is focused on the region of greatest injury in the central area. In the center, the lateral temperature gradient can be assumed to be much smaller than the gradient directed into the skin. Therefore, a one-dimensional analysis can be applied for evaluating the worst case of a burn in the center of the affected tissue. For simplicity, the initial temperature in the skin will be assumed to be a uniform value, T_0, and the temperature on the surface or the temperature of the surrounding air to be raised instantaneously to T_∞. The skin and subcutaneous tissue are a multilayered structure, with each layer having unique dimensions and thermal properties.

System definition and environmental interactions: The system is defined along a one-dimensional Cartesian coordinate directed into the skin, with a unit cross-sectional area sufficient in size to ensure that all property values across it are uniform. In this case, the epidermis has a thickness of 100 μm, the dermis 2 mm, and the subcutaneous fat 1 mm. A representation of this composite system is shown below.

11.2 Numerical Methods for Transient Conduction: Finite Difference Analysis

Note that the epidermis is very thin in comparison to the dermis and fat. However, it may be thermally significant since it controls the thermal interaction with the environment at the surface of the skin. In order to apply the finite difference method, it is necessary to divide the system into a series of increments. Because the epidermis is the thinnest tissue in the system, its dimensions will dictate the size of the smallest spatial increment. The grid is established near the surface of the skin with first interior node located at the epidermal/dermal interface so that $\Delta x_e = 100$ µm ($x_2 - x_1 = 100$ µm). Equally spaced grid nodes can be propagated into the dermis and fat so that $\Delta x_d = \Delta x_f = 100$ µm. The uniform mass associated with each increment extends a distance $\Delta x/2$ to each side of the node. Note that for node 2, one-half of the mass is epidermis and one-half is dermis.

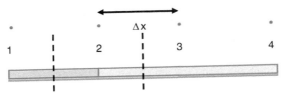

We assume that there is not an active blood flow through the system during the actual burning process, and that metabolic energy generation is negligible in magnitude in comparison to heat diffusion caused by the thermal insult on the surface of the epidermis. Thermal property values will be taken as constant.

Apprising the problem to determine the governing equations: The governing equation for the temperature distribution for an interior is the one-dimensional equivalent of (11.41).

$$T_m^{p+1} = Fo\left(T_{m+1}^p + T_{m-1}^p\right) + (1 - 2Fo)T_m^p.$$

For convenience, we can write the finite difference equations in terms of a dimensionless temperature defined as a function of the initial and environmental values.

$$\theta_m^{p+1} = Fo\left(\theta_{m+1}^p + \theta_{m-1}^p\right) + (1 - 2Fo)\theta_m^p \quad \text{where} \quad \theta_m^p = \frac{T_m^p - T_\infty}{T_i - T_\infty}.$$

Also, since there are unique thermal properties for epidermis, dermis, and fat, the Fourier number will be defined specially to each material.

$$Fo_e = \frac{k_e \Delta t}{\rho_e c_e \Delta x^2} \quad Fo_d = \frac{k_d \Delta t}{\rho_d c_d \Delta x^2} \quad Fo_f = \frac{k_f \Delta t}{\rho_f c_f \Delta x^2}.$$

The initial condition for all nodes except $m = 1$ at the surface is $\theta_m^0 = 1$. The boundary nodes at the surface and the interface between different materials require the formulation of special equations. For a constant surface temperature, node $m = 1$ is written $\theta_1^p = 0$. For a convective heat flow on the surface, (11.49) is modified for a one-dimensional grid as:

$$T_1^{p+1} = 2Fo_e(T_2^p + Bi_e T_\infty) + (1 - 2Fo_e - 2Bi_e Fo_e)T_1^p$$

or

$$\theta_1^{p+1} = 2Fo_e \theta_2^p + (1 - 2Fo_e - 2Bi_e Fo_e)\theta_1^p,$$

where the Biot number is written in terms of the grid spacing and properties of the epidermis.

$$Bi_e = \frac{h\Delta x}{k_e}.$$

Nodes at the interface between two different materials require a special examination. For example, the diagram above shows that the mass associated with node 2 consists of half an epidermal element and half a dermal element. We know that the two boundary conditions that must be satisfied at a material interface are that both the temperature and heat flow are continuous. Writing conservation of energy for a node that embodies mass from two different materials requires finite difference expressions for energy storage and for the second-order spatial derivative of temperature for a composite system. The composite property values are based on weighting according to the contributions of each material, which may vary if the grid spacing is nonuniform (Minkowycz et al. 1988). To illustrate, for a uniform grid spacing, at node 2 these expressions are:

$$\rho c \frac{\partial \theta}{\partial t} \approx \frac{\rho_e c_e(x_2 - x_1) + \rho_d c_d(x_3 - x_2)}{(x_2 - x_1) + (x_3 - x_2)} \cdot \frac{\theta_2^{p+1} - \theta_2^p}{\Delta t} = \frac{\rho_e c_e + \rho_d c_d}{2} \cdot \frac{\theta_2^{p+1} - \theta_2^p}{\Delta t},$$

$$k\frac{\partial^2 \theta}{\partial x^2} \approx \frac{2}{(x_2 - x_1) + (x_3 - x_2)}\left[k_d \frac{\theta_3^p - \theta_2^p}{x_3 - x_2} - k_e \frac{\theta_2^p - \theta_1^p}{x_2 - x_1}\right]$$

$$= k_d \frac{\theta_3^p - \theta_2^p}{\Delta x^2} - k_e \frac{\theta_2^p - \theta_1^p}{\Delta x^2}.$$

These two equations are equated in writing conservation of energy.

$$\frac{\theta_2^{p+1} - \theta_2^p}{\Delta t} = \frac{2}{\rho_e c_e + \rho_d c_d}\left[k_d \frac{\theta_3^p - \theta_2^p}{\Delta x^2} - k_e \frac{\theta_2^p - \theta_1^p}{\Delta x^2}\right].$$

Further dimensionless property ratios may be defined as:

$$a = \frac{k_d}{k_e} \quad b = \frac{\rho_d c_d}{\rho_e c_e} \quad g = \frac{k_f}{k_d} \quad j = \frac{\rho_f c_f}{\rho_d c_d},$$

11.2 Numerical Methods for Transient Conduction: Finite Difference Analysis

which are applied to write the conservation of energy for interface node 2.

$$\theta_2^{p+1} = \frac{2Fo_e}{(1+b)}[a\theta_3^p + \theta_1^p] + \left[1 - 2Fo_e\frac{1+a}{1+b}\right]\theta_2^p.$$

Node 22 is at the interface between dermis and subcutaneous fat, and its mass is split between these two materials. The conservation of energy equation is:

$$\theta_{22}^{p+1} = \frac{2Fo_d}{1+j}[g\theta_{23}^p + \theta_{21}^p] + \left[1 - Fo_d\frac{1+g}{1+j}\right]\theta_{22}^p.$$

The final boundary for the system is at the innermost node, 32, of the subcutaneous fat. If it is assumed to be insulated, geometric and thermal symmetry will hold.

$$\theta_{32}^{p+1} = 2Fo_f\theta_{31}^p + (1 - 2Fo_f)\theta_{32}^p.$$

Analysis: The size of the time step will be determined by the most stringent of all stability criteria for the system. The criterion for each node m is that the coefficient of T_m^p be nonnegative. Owing to the various combinations of material properties that appear among the node equations, there are numerous criteria. Δt must be chosen to not be greater than the criterion that specifies the smallest time step. In order to compute the stability criteria, it is necessary to identify constitutive property values for epidermis, dermis, and fat. The following property values are found in Appendix C.

Material	$T(°C)$	ρ (kg/m³)	$k \times 10^2$ (W m⁻¹ °K⁻¹)	$c_p \times 10^{-3}$ (J kg⁻¹ °K⁻¹)	$\alpha \times 10^6$ (m²/s)
Epidermis	37	1,200	20.9	3.60	0.048
Dermis	37	1,200	29.3	3.22	0.076
Fat	37	916	23	2.30	0.109

The table below presents a summary of the numerical solution stability criteria for each type of node subsystem.

Node	Stability equation	Stability criterion	Time step (s)
1	$Fo_e(1+Bi_e) \leq \frac{1}{2}$	$\Delta t \leq \frac{\Delta x^2}{2\alpha_e}\left(\frac{1}{1+Bi_e}\right)$	$\Delta t \leq 2.06 \times 10^{-2}$ s
2	$Fo_e \leq \frac{1}{2}\frac{1+b}{1+a} = 0.394$	$\Delta t \leq 0.394\frac{\Delta x^2}{\alpha_e}$	$\Delta t \leq 8.15 \times 10^{-2}$ s
3–21	$Fo_d \leq \frac{1}{2}$	$\Delta t \leq \frac{\Delta x^2}{2\alpha_d}$	$\Delta t \leq 6.59 \times 10^{-2}$ s
22	$Fo_d \leq \frac{1}{2}\frac{1+j}{1+g} = 0.440$	$\Delta t \leq 0.440\frac{\Delta x^2}{\alpha_d}$	$\Delta t \leq 5.79 \times 10^{-2}$ s
23–32	$Fo_f \leq \frac{1}{2}$	$\Delta t \leq \frac{\Delta x^2}{2\alpha_f}$	$\Delta t \leq 4.59 \times 10^{-2}$ s

Note that the stability criterion given for node 1 is for a convective boundary condition. For a constant temperature boundary condition, the equation is inherently stable for any time step chosen. From this analysis, we see that the step size is determined by the conditions for the nodes in the fat because of its higher thermal diffusivity. To this end, we will set the time step to be 0.04 s. This means that it will require 25 rounds of calculations for all 32 nodes to simulate 1 s of the burn process. Since a change in boundary conditions is propagated into the tissue at a rate of only one node per time step, after 1 s there will still not be an effect of what has happened on the surface realized at the innermost boundary.

A process was simulated consisting of 5 s of exposure at an imposed surface temperature of 60°C, followed by 3 s of surface cooling at 34°C.

Examining and interpreting the results: Plots of the temperature as a function of time and of position are shown below.

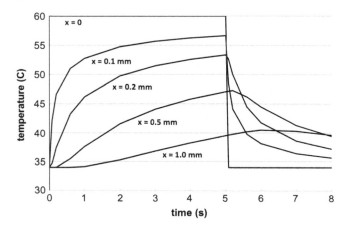

The boundary node has a step increase and decrease in temperature, and at progressively deeper positions in the skin the changes are reduced in magnitude and delayed as a consequence of the heat diffusion and energy storage processes. Note that at 0.5 mm and greater locations, the temperature continues to rise for a period of time after the surface temperature is reduced.

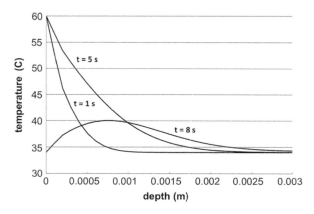

11.2 Numerical Methods for Transient Conduction: Finite Difference Analysis

The temperature gradient shows the gradual diffusion of heat deeper into the skin with time. Note that at any given time the gradient slope is maximum at the surface and decreases continuously with depth. Application of Fourier's law, (2.9) to the $t = 1$ s and $t = 5$ s curves shows that the diminishing slope indicates a proportionately smaller conduction heat flow. After the surface temperature is reduced, a point of inflection develops in the gradient as heat is conducted both toward the surface and further into the skin. Even though deeper locations in the skin continue to be heated for some time after the hot surface is removed from the surface, temperatures are very quickly brought below the threshold value (43°C) at which burn injury occurs.

Further considerations: The problem was also solved for the convection boundary condition being programmed for the temperature history of node 1. We anticipate that the heating process will be slower for a convection boundary than for a constant temperature boundary since the thermal resistance with the surroundings will be larger. The plot below shows the calculated temperature difference over time at progressive locations into the skin for constant temperature and convection boundary conditions. The temperature difference is largest closest to the skin and is diminished at greater depths that are less influenced by what happens on the surface. However, because of the larger thermal resistance at the surface, there is a persistent lower temperature throughout the convectively heated skin.

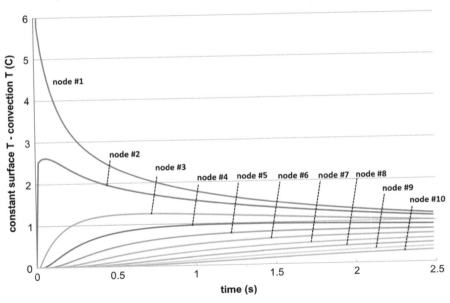

Another analysis of interest is to compare the finite difference simulation of the temperature history with the exact analytical solution for the temperature field in a semi-infinite solid for a step change in surface temperature. We know from (10.167) that the analytical solution takes the form of the error function.

$$\frac{T(x,t) - T_s}{T_i - T_s} = \text{erf} \frac{x}{2\sqrt{\alpha t}}.$$

Since the analytical solution assumes that the material has homogeneous properties, it is necessary to also apply this assumption to the finite difference model to be able to compare the two methods (here we assume the properties of dermis since it constitutes a majority of the skin). Plus, the two methods can only be compared for the heating portion of the process since at the start of cooling the internal temperature is nonuniform, and the analytical solution cannot match that condition. The calculated temperature difference between the finite difference and error function solutions is shown below.

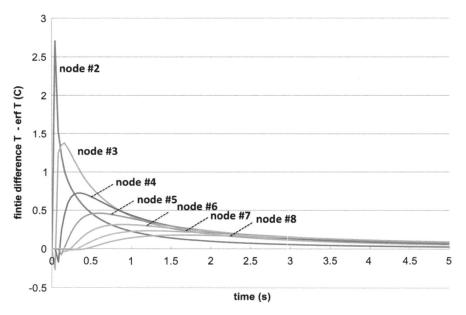

Since the finite difference model is a discrete approximation of a continuous function, it is predictable that numerical results will not be absolutely accurate. This difference is often referred to as truncation error. The difference is greatest near the surface and closest to the starting conditions. In general, the difference can be reduced by decreasing the size of the grid spacing and time steps, but the number of computation iterations to simulate a given duration of process will increase. When you are designing a finite difference model for a process, it is often necessary to consider the balance between achieving high resolution and accuracy in your calculations and avoiding an unnecessarily large number of computational iterations. A comprehensive discussion of the important issues to be considered in the design of a finite difference analysis scheme is presented in Minkowycz et al. (1988).

11.2.2 Backward Finite Difference Method

The backward (implicit) finite difference method can be used to formulate an analysis, as a complement to the explicit methods shown in Sect. 11.2.1. An interior node is described as:

$$\frac{T_{m,n}^{p+1} - T_{m,n}^p}{\Delta t} = \alpha \left[\frac{T_{m+1,n}^{p+1} + T_{m-1,n}^{p+1} - 2T_{m,n}^{p+1}}{(\Delta x)^2} + \frac{T_{m,n+1}^{p+1} + T_{m,n-1}^{p+1} - 2T_{m,n}^{p+1}}{(\Delta y)^2} \right]. \quad (11.52)$$

Note that the values for temperatures at all surrounding nodes are written in terms of the future state as opposed to the current (known) state. Thus, the algebraic equation for temperature solution has unknown values for not only the central node of interest, but also all of the surrounding nodes. This equation cannot be solved explicitly since there is more than one unknown. Multiple equations must be solved simultaneously for the complete set of node temperatures, using methods such as matrix algebra. To follow our standard convention of writing equations with unknown parameters on the left side, (11.52) becomes:

$$(1 + 4Fo)T_{m,n}^{p+1} - FoT_{m+1,n}^{p+1} - FoT_{m-1,n}^{p+1} - FoT_{m,n+1}^{p+1} - FoT_{m+1,n-1}^{p+1} = T_{m,n}^p. \quad (11.53)$$

A similar equation is written for each node resulting in a matrix of $m \times n$ equations which must be solved simultaneously to determine the future temperature distribution at one time step ahead. Although this solution method is more complex than for the explicit formulation, it does have a major advantage. Since Fo is always non-negative, the coefficient for the term $T_{m,n}^{p+1}$ must be positive, meaning that the solution is stable regardless of the choice of values for Δx and Δt. Therefore, arbitrarily large values may be chosen for Δt, thereby greatly reducing the number of computational iterations in comparison with the explicit method to complete an analysis to a predetermined elapsed time. This becomes an important consideration when a process of long duration must be simulated.

Example 11.2.3 Comparison of Forward and Backward Difference Methods.
Consider a section through a muscle that has a uniform thickness of 2 mm. Owing to its placement in the body and regional blood perfusion, one side of the muscle is at 36.9°C and the other side at 37°C. The longitudinal and lateral dimensions of the muscle are quite large in comparison to its thickness so that the temperature field can be presumed to be one dimensional. From this initial condition, the rate of internal energy generation within the muscle is suddenly increased to $\dot{q}_m = 5 \times 10^4 \text{W/m}^3$ by means of electrical stimulation. Develop and compare forward and backward finite difference models for the transient thermal response of the muscle to the stimulation.

Solution. *Initial considerations*: There is an initial linear temperature gradient within the muscle because of the imposed boundary conditions. In the absence of other information, we can assume that imposed fixed temperature boundary conditions are maintained throughout the process. With an added internal generation of energy, we anticipate that the temperature in the muscle will increase. The muscle will transition from an initial steady-state linear temperature gradient to a final steady state in which the gradient is nonlinear. The internal heat generation during stimulation is assumed to be uniform throughout the muscle. Further, we will assume the muscle to be homogeneous and to have constant properties and dimensions. We anticipate that the initial and final temperature distributions will have the geometries shown below.

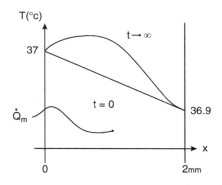

System definition and environmental interactions: The system is defined with a one-dimensional Cartesian coordinate through the muscle in the direction of the temperature gradient as indicated. A finite difference grid is overlaid onto the muscle. The most simple grid, from a computational perspective, is to establish a node on each surface and a node in the geometric center of the muscle. The node spacing, Δx, is one-half of the thickness, 1 mm. The mass associated with each node is indicated by the dashed lines, with heat flow interactions between adjacent nodes shown at the boundaries.

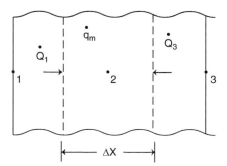

Apprising the problem to determine the governing equations: Since the temperatures of the surface nodes, 1 and 3, are fixed over time, the only node which

11.2 Numerical Methods for Transient Conduction: Finite Difference Analysis

must be analyzed is node 2. Applying conservation of energy for a unit cross-sectional area normal to the temperature gradient:

$$\frac{dE_2}{dt} = \dot{Q}_1 + \dot{Q}_3 + \dot{q}_m \Delta x.$$

The governing forward and backward difference equations are written directly from the conservation of energy equation.

$$\text{forward}: \quad \rho c \Delta x \frac{T_2^{p+1} - T_2^p}{\Delta t} = k \frac{T_1^p - T_2^p}{\Delta x} + k \frac{T_3^p - T_2^p}{\Delta x} + \dot{q}_m \Delta x.$$

$$\text{backward}: \quad \rho c \Delta x \frac{T_2^{p+1} - T_2^p}{\Delta t} = k \frac{T_1^{p+1} - T_2^{p+1}}{\Delta x} + k \frac{T_3^{p+1} - T_2^{p+1}}{\Delta x} + \dot{q}_m \Delta x.$$

These equations may be written in terms of the Fourier number: $Fo = \dfrac{k \Delta t}{\rho c \Delta x^2}$

$$\text{forward}: \quad T_2^{p+1} = Fo(T_1^p + T_3^p) + (1 - 2Fo) T_2^p + \frac{\dot{q}_m \Delta t}{\rho c}.$$

$$\text{backward}: \quad T_2^{p+1} = \frac{Fo(T_1^{p+1} + T_3^{p+1}) + T_2^p + \dfrac{\dot{q}_m \Delta t}{\rho c}}{1 + 2Fo}.$$

Analysis: The magnitude of Δx has already been set. The size of Δt is arbitrary for the backward difference equation, but must satisfy the stability criterion for the forward difference equation, $0 \leq 1 - 2Fo$. From Appendix C, for skeletal muscle $\alpha = 1.16 \times 10^{-7} \text{ m}^2/\text{s}$.

$$\frac{1}{2} \geq Fo = \frac{\alpha \Delta t}{\Delta x^2} \quad \Delta t \leq \frac{1}{2} \frac{\Delta x^2}{\alpha} = \frac{1}{2} \frac{(0.001\text{m})^2}{1.16 \times 10^{-7} \dfrac{\text{m}^2}{\text{s}}} = 4.31 \text{ s}.$$

The Fourier number must be calculated to apply in both difference equations. A time step that satisfies the stability criterion was chosen, $\Delta t = 4$ s.

$$Fo = \frac{\alpha \Delta T}{\Delta x^2} = \frac{1.16 \times 10^{-7} \dfrac{\text{m}^2}{\text{s}} \cdot 4 \text{ s}}{(10^{-3}\text{m})^2} = 4.64 \times 10^{-1}.$$

The forward and backward difference equations were applied to the process, with the results shown in the table below.

Step (p)	Time (s)	T_1 (°C)	T_2 (°C)		T_3 (°C)
			Backward	Forward	
0	0	37	36.9500	36.9500	36.9
1	4	37	36.9762	37.0004	36.9
2	8	37	36.9897	37.0041	36.9
3	12	37	36.9968	37.0043	36.9
4	16	37	37.0004	37.0043	36.9
5	20	37	37.0023	37.0043	36.9
6	24	37	37.0033	37.0043	36.9
7	28	37	37.0038	37.0043	36.9
8	32	37	37.0041	37.0043	36.9
9	36	37	37.0042	37.0043	36.9
10	40	37	37.0043	37.0043	36.9

Examining and interpreting the results: Both difference methods provide essentially identical final temperatures for the center node after ten time step calculations. However, the forward difference equation predicts a much more rapid transition between the initial and final states than does the backward difference equation, completing the entire process in only three time steps.

Further considerations: The grid spacing and time steps are both very coarse, providing very little resolution for the transient process. We may learn more about the characteristics of the forward and backward difference estimations of the transient process by reducing the size of both Δx and Δt. If the muscle is overlaid with a grid having eight equal increments, then $\Delta x = 0.25$ mm. For a smaller Δx, the stability criterion will require a correspondingly smaller time step, Δt.

$$\Delta t \leq \frac{1}{2}\frac{\Delta x^2}{\alpha} = \frac{1}{2}\frac{(0.00025 \text{ m})^2}{1.16 \times 10^{-7} \frac{\text{m}^2}{\text{s}}} = 0.269 \text{ s}.$$

Accordingly, a new time step of $\Delta t = 0.25$ s was chosen. The value of Fo is unchanged from the initial set of calculations. The results for the forward and backward difference calculations for the coarser and finer resolution analyses are plotted below.

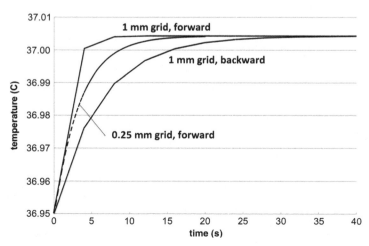

It is immediately apparent that reducing the sizes of the grid spacing and the time step produces a much smoother simulation of the continuous temperature change process through the muscle. Also, the discrepancy between the forward and backward difference calculations is reduced appreciably. The initial and final temperature distributions through the muscle for the coarser and finer grids are plotted below.

It is obvious that the finer grid and smaller time steps yield a more accurate representation of the actual temperature history within the muscle. However, the more pleasing results come at a price of greater computational effort. This example addresses a simple problem, so the added computation is inconsequential. For more complex problems this may not be the case, and the tradeoff between increased spatial and temporal resolution and a larger number of computational steps may need to be weighed.

11.3 Thermal Injury Mechanisms and Analysis

Exposure to temperatures above normal physiologic ranges (>43°C) can result in measurable irreversible changes in tissue structure or function. Cell death or tissue alterations may be detrimental – for example, skin burns – or beneficial, as in vessel sealing or tumor destruction. Tissues of the central nervous system are the most thermally sensitive, exhibiting irreversible changes for long-term exposures above about 42°C. Thermal kinetic models for cell and tissue injury as a function of exposure temperature and time provide helpful insights into the underlying principles of tissue thermal damage.

11.3.1 Burn Injury

The original work on the application of rate process models to describe a burn injury was reported by Moritz and Henriques in a series of seminal papers in 1947 (Henriques 1947; Henriques and Moritz 1947; Moritz and Henriques 1947). They applied flowing water at elevated temperatures to pig and human skin and measured exposure times required to create threshold first and second degree burns. In their work, the damage was quantified using a single parameter, Ω, which is calculated from an Arrhenius integral. Accordingly, the development of a thermal injury is treated as a first-order rate process as a function of the local temperature. For a one-dimensional Cartesian system, this expression is written as:

$$\frac{d\Omega(x,t)}{dt} = Ae^{-\frac{\Delta E}{RT(x,t)}}, \tag{11.54}$$

where Ω is a quantitative measure of the extent of injury, A a frequency factory (s^{-1}), ΔE an activation energy barrier for the injury process (J/mol), R the universal gas constant (8.32 J/mol K), and T the temperature as a function of time and position in the heated tissue, expressed in absolute units (K). The total injury accrued over a total duration of exposure of τ is obtained by integrating the injury rate function over the entire process.

$$\Omega(x,\tau) = \int_0^\tau A e^{-\frac{\Delta E}{RT(x,t)}} dt. \quad (11.55)$$

Subsequently, other investigators added further data and analysis to the understanding of this problem, including defining the conditions for a third degree burn in terms of Ω (Stoll 1960; Takata 1974; Lawrence and Bull 1976). Table 11.1 shows the values of Ω for threshold levels of burn injury.

The most striking feature of the Ω function is that it is highly nonlinear with respect to the degree of injury. This nonlinearity is a result of the greatly increasing extent of damage that is associated progressively with first, second, and third degree burns.

Burns are related directly to the structure of skin and how it is affected by the injury process (Fig. 11.4).

The primary feature of skin that pertains to burn injuries is that it is a multilayered structure. The outermost layer is a thin epidermis that has no blood perfusion. It is underlayed by a much thicker dermis that has a rich microvascular network. Beneath the dermis is subcutaneous fat. The thickness of skin is highly variable among the

Table 11.1 Values of Ω corresponding to threshold levels of thermal injury

Degree of burn injury	Ω
First	0.53
Second	1.0
Third	10^4

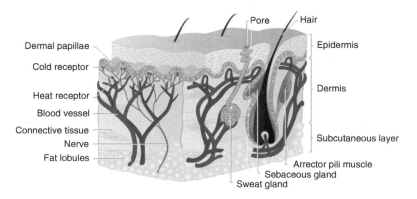

Fig. 11.4 Structure of human skin showing the primary elements and vascular network

11.3 Thermal Injury Mechanisms and Analysis

population and over the body surface of any given individual. Many different methods have been applied to measure skin thickness, including histology and ultrasound, and they give somewhat differing results (Southwood 1955; Seidenari et al. 2000; Moore et al. 2003). The most widely applied value for the thickness of epidermis is 80 μm and dermis is 2 mm. However, thicknesses can vary by more than a factor of two across different areas of the body. The thicknesses are greatest on the soles of the feet and palms of the hands and much thinner on the eyelids and creases of the elbows. Further, skin is much thinner on the very young and the elderly in comparison to mature adults.

A first degree injury is quite superficial, producing no irreversible physiological damage, and is characterized primarily by transient hyperemia (an accumulation of blood in the skin vessels resulting in a reddened appearance in the affected area). Second degree burns cover a very wide range of injuries, from superficial to deep. An important aspect of second degree burns is that some level of perfusion of blood is maintained. Moritz and Henriques characterized a threshold second degree burn, the mildest superficial level, as one for which total epidermal necrosis occurs. Injury to the vascular bed causes an inflammatory response with attendant increased perfusion and vascular permeability. Consequently, a clinical manifestation of a second degree burn is the formation of edema and blistering. Since the nerve endings are still alive, the level of pain is generally quite high. Applied pressure will cause blanching due to a temporary suppression of the enhanced flow through the surface capillaries. Healing time may be on the order of 2 weeks. A deep second degree burn will have a more mottled appearance since the more superficial areas of dermis are destroyed, eliminating capillary perfusion. Healing time may be extended to a month or more, but the retention of an active blood perfusion and live skin cells at the base of the dermis will allow the tissue to self-heal. A third degree burn occurs when the penetration of heat from the surface is deep enough to destroy the entire thickness of the dermis. Since there are no active nerves and circulatory elements, clinically, third degree burns are dry, colorless, and painless. In the absence of live skin cells, a third degree burn is not self healing, and placement of a skin graft is required to enable non-scarred skin to cover the affected area.

From this brief discussion of burn injuries, it should be apparent that the extent and volume of tissue damage are significantly larger for a third degree burn than for a second degree, and likewise for a first degree. Thus, the disproportionate increments in Ω are to be expected.

The values of the empirical constants A and ΔE in (11.54) were determined by Moritz and Henriques by a graphical fit of the Arrhenius function to their experimental data for a threshold second degree burn (Fig. 11.5). Their experimental data cover the temperature range of 44–70°C. It is not valid to extrapolate their analysis model to temperatures beyond this range since the tissue response may vary significantly from the experimental data. The slope (ΔE) and intercept (A) of the plotted Arrhenius function yielded values of $A = 3.1 \times 10^{98}$ s^{-1} and $\Delta E = 6.03 \times 10^5$ J/mole. These values can be applied in (11.54) to predict the rate at which injury will accrue in skin exposed to a given surface temperature.

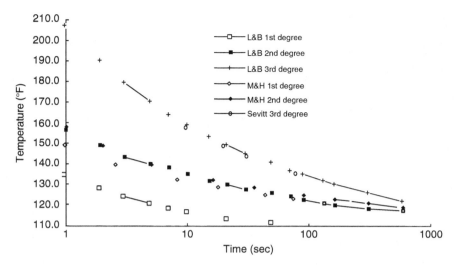

Fig. 11.5 Experimental data for combinations of exposure time and constant surface temperature to cause threshold first, second, and third degree skin burns. Replotted from data in Moritz and Henriques (1947), Sevitt (1957), and Lawrence and Bull (1976)

An alternative interpretation of the injury function Ω is that it is related to the extent of molecular-scale damage in the affected tissue. This relationship is expressed as:

$$\Omega(x,\tau) = \ln\left\{\frac{C(0)}{C(\tau)}\right\}, \qquad (11.56)$$

where $C(\tau)$ is defined as the concentration of undamaged (native state) molecules in the tissue (Pearce 2009; Diller et al. 2005). $C(0)$ is the concentration before heating and $C(\tau)$ after an elapsed time τ. The advantage of this Ω formulation is that it can be related directly to quantitative laboratory measures rather than clinical outcomes. Typical quantitative pathologic end points include birefringence loss in collagen and muscle, collagen Hyaline damage, leakage of fluorescent dyes, and cell survival in culture (Diller et al. 2005). Combining (11.55) and (11.56) enables the determination of values for A and ΔE for molecular, cellular, and tissue-based scales of resolution as a complement to clinically derived assessments. Combined values for A and ΔE have been reported for thermal injury in many different types of biological systems and have been compiled by He and Bischof (2003) and Pearce (2009), as summarized in Table 11.2.

He and Bischof (2003) have plotted A and ΔE for more than 100 molecular, cellular, and tissue systems and have found a very tight correlation between these two properties for a very wide diversity of biological materials. This data plot is shown in Fig. 11.6. In addition to incorporating many different biological materials, this data also represent a broad diversity of injury end point assessment methods. The tight correlation between A and ΔE indicates an intrinsic interdependence

11.3 Thermal Injury Mechanisms and Analysis

Table 11.2 Kinetic coefficients A and ΔE for thermal injury over the indicated temperature ranges (He and Bischof 2003; Pearce 2009). This is a small sampling of the data available in the source references

Material	A (1/s)	ΔE (J/mol)	T range °C	Reference
DNA		1.00×10^5		Eigner et al. (1961)
RNA		7.12×10^4		Eigner et al. (1961)
Egg albumin	2.0×10^{81}	5.50×10^5	65	Eyring and Stearn (1938)
HeLa cells	7.37×10^{277}	1.70×10^6	$41 < T < 44$	Landry and Marceau
	1.19×10^{113}	7.06×10^5	$44 < T < 49$	(1978)
	1.0×10^{16}	1.08×10^5	$49 < T < 55$	
CHO cells	1.7×10^{249}	1.53×10^6	$41.5 < T < 43.5$	Dewey et al. (1977)
	3.1×10^{100}	6.20×10^5	$43.5 < T < 46.5$	
CHO cells	2.84×10^{99}	6.19×10^5	$T \geq 43$	Sapareto (1982)
Erythrocytes	1×10^{31}	2.12×10^5		Flock et al. (1993)
Erythrocytes	7.6×10^{27}	1.94×10^5	$44 < T < 60$	Moussa et al. (1979)
Retina	3.1×10^{98}	6.28×10^5		Welch and Polhamus (1984)
Collagen	1.606×10^{45}	3.06×10^5		Pearce et al. (1993)
Collagen	1.77×10^{56}	3.676×10^5		Maintland and Walsh (1997)
Muscle	2.94×10^{39}	2.596×10^5		Jacques and Gaeeni (1989)
Skin	3.1×10^{98}	6.28×10^5	$44 < T < 70$	Henriques (1947)
Skin	1.3×10^{95}	6.04×10^5	$T \leq 50$	Diller and Klutke (1993)
Skin	2.185×10^{124}	7.82×10^5	$T \leq 50$	Weaver and Stoll (1969)
	1.823×10^{51}	3.27×10^5	$T > 50$	
Skin	8.82×10^{94}	6.028×10^5	$T \leq 53$	Pearce (2009)
	1.297×10^{31}	2.04×10^5	$T > 53$	
Joint capsule	4×10^5	8.1×10^4	$44 < T < 60$	Moran et al. (2000)
	1.85×10^{32}	2.34×10^5	$60 < T < 70$	
Prostate tumor AT-1	1.7×10^{91}	5.68×10^5	$40 < T < 70$	Bhowmick et al. (2004)
Kidney	3.27×10^{38}	2.57×10^5	$37 < T < 57$	He et al. (2004)
Intestine	1.6×10^{93}	5.82×10^5	$42.5 < T < 44.5$	Milligan et al. (1984)

based on common characteristics among thermal injury processes involving all biological materials.

Although the differences in the activation energy appear to be quite small, in actuality, since ΔE appears as an exponential term in the injury rate equation, even minute changes can have a significant effect. To wit, even though the Henriques kinetic coefficient values have been applied widely for modeling burn processes for well over a half century, it was discovered many decades after publication that their own coefficients are not a good fit to their experimental data (Diller and Klutke 1993). The source of this error is undoubtedly the methods that were available at the time for reducing experimental data, which consisted of adjusting a straight edge placed on the plotted data to obtain an estimated best fit. A line was drawn along the edge, and the slope and intercept measured from which A and ΔE were calculated. Also, the kinetic coefficients for skin presented in Table 11.2 do not cover the entire range of temperatures for the Henriques data. The reason is that there is a break point in the data around 50°C or slightly higher, probably because the molecular

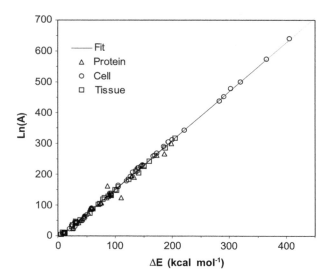

Fig. 11.6 Relationship between activation energy, ΔE (kcal/mol), and the natural log of frequency factor, A (1/s), for thermal injury across a large number of molecular, cellular, and tissue systems (He and Bischof 2003), with permission

mechanism of injury is altered as the insult temperature exceeds a threshold value. A comprehensive discussion of this phenomenon is found in Pearce (2009).

> **Example 11.3.1 Rate of Burn Injury Varies with Temperature.**
> In evaluating the potential for a heated source to cause a burn injury, it is important to be able to determine the effect of changing the temperature, either if it is raised thereby increasing the risk factor, or lowered to reduce the risk. The objective of this problem is to quantify the influence of changing the temperature on altering the rate at which a burn occurs.

Solution. *Initial considerations*: Quantifying the temperature governed rate at which a burn proceeds requires a model for the process, such at the Arrhenius model in (11.54). It will be possible to compare the values of $d\Omega/dt$ at different temperatures to make this calculation. We will need to consider the effect of the reference temperature that we choose because the injury rate is a function of temperature. Thus, the ratio may depend on the reference state that we select.

System definition and environmental interactions: For this analysis our system will be a volume of tissue for which the temperature is known at a specific point in time. It could have been determined independently via (11.12) or (11.13) or (11.14). $T(x,t)$ is supplied as an independently determined value for this calculation.

Apprising the problem to identify governing equations: We may apply (11.54) to calculate the rate at which injured tissue is accruing at a given temperature.

11.3 Thermal Injury Mechanisms and Analysis

$$\frac{d\Omega(x,t)}{dt} = A e^{-\frac{\Delta E}{RT(x,t)}}.$$

Analysis: The ratio of injury rates at two different temperatures is given by:

$$\frac{\left.\frac{d\Omega}{dt}\right|_{T_2}}{\left.\frac{d\Omega}{dt}\right|_{T_1}} = \text{Rate Ratio}\left(\frac{T_2}{T_1}\right) = \frac{e^{-\frac{\Delta E}{RT_2}}}{e^{-\frac{\Delta E}{RT_1}}}.$$

The following table gives the ratios of injury rate for 1°C increments across the range of temperatures for Moritz and Henriques' experimental data. Note that the experimental temperatures in °C have been converted to K by adding 273.15 to perform the injury rate calculations.

T_1 (°C)	T_1 (K)	$T_2 = T_1 + 1$ (K)	Injury rate ratio (T_2/T_1)
44	317.15	318.15	2.11
45	318.15	319.15	2.10
46	319.15	320.15	2.09
47	320.15	321.15	2.08
48	321.15	322.15	2.07
49	322.15	323.15	2.06
50	323.15	324.15	2.05
51	324.15	325.15	2.05
52	325.15	326.15	2.04
53	326.15	327.15	2.03
54	327.15	328.15	2.02
55	328.15	329.15	2.01
56	329.15	330.15	2.00
57	330.15	331.15	1.99
58	331.15	332.15	1.99
59	332.15	333.15	1.98
60	333.15	334.15	1.97
61	334.15	335.15	1.96
62	335.15	336.15	1.95
63	336.15	337.15	1.95
64	337.15	338.15	1.94
65	338.15	339.15	1.93
66	339.15	340.15	1.92
67	340.15	341.15	1.92
68	341.15	342.15	1.91
69	342.15	343.15	1.90
70	343.15	344.15	1.89

Examining and interpreting the results: Over the range of temperatures for which a majority of skin burns occur, an increase of 1°C causes approximately a twofold increase in the rate of injury. Therefore, in considering the risk of a thermal injury hazard for a system, even small alterations in temperature can translate into large effects in burn outcome. For example, a reduction in temperature of 10°C will produce a drop in the injury rate of about 2^{10}, which corresponds to a reduced risk on the order of 1,000. The ratio of injury rates at 54°C and 44°C is 1.43 × 10³, and at 64°C and 44°C it is 1.34 × 10⁶.

The other obvious feature of the injury rate ratio calculations is that the ratio decreases slightly as the temperature increases. Although the absolute value of the injury rate is amplified exponentially as the temperature becomes higher, the relative change is diminished slightly. The most important practical consequence of the temperature dependence of injury rate is that during an actual burning scenario, when the temperature is reduced even a few degree Celsius from the maximum value, the rate of injury becomes inconsequential in contributing to further tissue damage. For example, when the tissue is cooled just 3°C from its peak value, the rate of further damage accumulation is diminished by approximately 2^3, or about one order of magnitude.

Further considerations: The kinetic coefficient data in Table 11.2 show that there are many alternative approaches to calculating the rate at which a burn occurs. It would be interesting to explore how much of a difference the choice of coefficients makes. In part, this information will contribute to the level of confidence we can have in injury calculations. Here we can compare the injury rates predicted by the Henriques and the Diller–Klutke coefficients. Both sets of coefficients were derived from the same experimental data set, with the latter using modern curve-fitting optimization methods to more accurately match the data. The table below presents the ratio of the predicted injury rates as a function of temperature. Over the entire temperature range, the Henriques coefficients give a higher injury rate. Of greatest relevance are the results for temperatures below 50°C where the models are thought to match the data most accurately. The ratio of the predicted rates is slightly below 0.5, which is equivalent to a temperature sensitivity of about 1°C. The conditions to which the model is being applied will determine whether this level of sensitivity is acceptable or not.

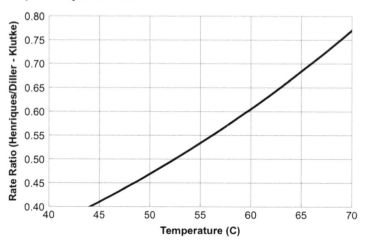

For many heating scenarios that result in a burn injury, the tissue behaves as if it were a semi-infinite solid. The high temperature or heat flux imposed on the surface

11.3 Thermal Injury Mechanisms and Analysis

of the skin penetrates into the underlying tissues without encountering any further physical boundaries. In principle, it should be possible to model the temperature rise process in the skin via the analytical solution of Sect. 10.6.2 in terms of the error function. However, in some cases the assumptions requisite to this analysis, including a homogeneous material composition and properties, uniform initial temperature distribution, and step change to a constant surface boundary condition, do not match the actual conditions that need to be analyzed. The alternative approach to analysis is to apply a numerical method such as finite differences as presented in Sect. 11.2. A grid can be established on which temperatures can be determined as a function of position and time. This same grid can then be applied to use the temperatures to calculate the local rates of thermal injury and to integrate this injury over time to determine the net accrued injury distribution within the tissue caused by a specified period of exposure to a thermal insult. To illustrate this process, we will continue Example 11.3.1 for which a transient temperature field was determined, by adding a further computational step to predict the development of the pattern of injury caused by the elevated temperatures.

> **Example 11.3.2 Computation of the Degree of Burn Injury.**
> In example 11.2.2, we applied the finite difference method to calculate the temperature field that develops in skin during the burn process. The objective of this problem is to apply the temperature time data from that example to compute pattern of thermal injury that is produced.

Solution. *Initial considerations*: The solution of the transient temperature analysis, $T(x,t)$, may be used directly in (11.54) to calculate the rate of injury at any time and location in the skin. An important decision in performing these calculations is choosing which combination of values for A and ΔE to apply for the kinetics of the injury process. Table 11.2 provides many different sets of data. For this example, we will use the Henriques data for skin. It is the most widely adopted burn model. Although the accuracy of the Henriques model is less then some other alternatives, it provides the best basis for comparison with other existing data.

System definition and environmental interactions: The system is identical to that used in Example 11.2.2, including the finite difference grid as defined.

Apprising the problem to identify governing equations: We may apply (11.55) to calculate the degree at which injured tissue is accrued after a time of exposure, τ.

$$\Omega(x,\tau) = \int_0^\tau A e^{-\frac{\Delta E}{RT(x,t)}} dt.$$

Because the finite difference solution for temperature only provides values at discrete time and space intervals, we must adapt the Arrhenius injury model to match the temperature data set, $T(\Delta x, \Delta t)$. Accordingly, at a node position, m, after P time steps have occurred, where $\tau = P\Delta t$:

$$\Omega(m\Delta x, P\Delta t) = \sum_{p=1}^{P} \frac{\partial \Omega(m\Delta x, p\Delta t)}{\partial t} \Delta t = \sum_{p=1}^{P} A e^{-\frac{\Delta E}{RT(m\Delta x, p\Delta t)}} \Delta t.$$

The summation of the injury function will be updated for each node position and time step for the finite difference analysis.

Analysis: Calculation of the injury function is added to the temperature computation algorithm already established for Example 11.2.2. The Henriques injury coefficients applied are: $A = 3.1 \times 10^{98}$ 1/s and $E = 6.28 \times 10^5$ J/mol. A plot of the time-wise development of the injury function, Ω, at selected depths into the skin is shown below.

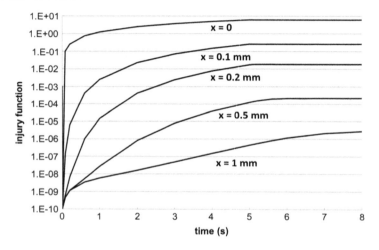

Examining and interpreting the results: When evaluating a burn plot, it is important to keep in mind that the injury axis is plotted on a logarithmic scale. The injury values appearing in the lower portion of the graph have inconsequential clinical effects. The injury data display a number of foreseen characteristics. The manifestation of injury is progressive over time and decreases in severity with distance into the skin. The predicted injury is a good match for the experimental data published by Henriques (1947) for which a 60°C surface temperature produces a first degree injury ($\Omega = 0.53$) in 2.6 s and a second degree injury ($\Omega = 1.0$) in 5 s. Henriques used an epidermal thickness intermediate to the node positions at 0 and 0.1 mm applied in the finite difference model, which provides for an acceptable correlation with his experimental data. It is important to note that when cooling is initiated on the skin surface at 5 s, the injury process ceases immediately in the most superficial region of greatest damage. The practical implication is that removing the source of thermal insult during a burn as soon as possible can have a major benefit in limiting the extent of injury.

11.3.2 Therapeutic Applications of Hyperthermia

In contrast to burn injuries, which normally occur accidentally under uncontrolled conditions, during a surgical procedure tissue temperatures may be raised according to a planned protocol to selectively destroy a target volume of tissue, such as a tumor. Although surface heating may be used for this purpose, most frequently an external energy source, such as a laser, ultrasound, radiofrequency, microwave, and others, is focused to dissipate energy within a target tissue according to a determined pattern in space and time. The design of the procedure may be specific to the needs of a particular patient's diagnosis, with the objective of killing a defined area of tissue with a high level of confidence and limiting the collateral injury effects in the surrounding healthy tissue. It is possible to modify the tissue energy absorption properties by introducing materials such as nanoparticles selectively into the tissue volume targeted for destruction before the procedure. The ability to model and predict the outcome of the heating process enables precise real-time control over the protocol parameters to produce a desired outcome (Fuentes et al. 2009). Thus, the explicit distinction between burns and hyperthermic surgery: in burns the objective is to minimize the extent of injury to the affected tissue; in surgery the objective is to maximize injury to the target tissue. In both instances, the response is governed by the temperature – time history in the tissue of interest, and models for these processes have proven to be highly useful for planning therapeutic procedures and for designing safe environments and operating practices.

Historically, an alternative approach to the Arrhenius model called the *Thermal Dose* has been adopted for simulating induced thermal injury during surgical procedures (Sapareto and Dewey 1984; Dewey 1994, 2009). The thermal dose describes the number of minutes of exposure at 43°C required to produce a thermal effect equivalent to that at a different (usually higher) temperature. The "cumulative equivalent minutes" of exposure at 43°C, CEM 43, can be applied to represent the effects of an entire heating protocol that may include exposures for defined times at many temperatures. Further, since a tumor is not isothermal during a hyperthermia protocol, another useful descriptor is to identify the thermal dose for which a threshold temperature is exceeded in a minimum percentage of the tumor volume. For example, the therapeutic effectiveness of a particular treatment can be described by a thermal dose that is exceeded in 90% of the tumor as (t_{43}T90) (Dewey 2009). Pearce (2009) notes that the CEM 43 can be calculated for a protocol that involves exposures at a series of N temperature steps T_i for times t_i as:

$$CEM43 = \sum_{i=1}^{N} R^{(43-T_i)} t_i, \qquad (11.57)$$

where R is a constant of proportionality determined from experimental injury rate data for a specific time temperature protocol applied to a particular biomaterial. This analysis is illustrated as follows, based on the work of Pearce (2009).

D_0 is the time (in minutes) at which the surviving number of cells has decreased to 1/e of the original population. It corresponds to the conditions for which $\Omega = 1$. If the original population of cells is denoted by N_0, and the surviving number of cells by $N(\tau)$ at time t during exposure to a temperature T that started at time t_0, then the expression for D_0 is:

$$\frac{N(\tau)}{N_0} = e^{\frac{t-t_0}{D_0(T)}} = e^{\frac{\tau}{D_0(T)}}, \tag{11.58}$$

where the lapsed time at T is given by τ. The cell death rate at temperature T then is the reciprocal of D_0.

$$\frac{1}{D_0(T)} = \frac{\partial}{\partial t}\left(1 - e^{\frac{\tau}{D_0(T)}}\right). \tag{11.59}$$

This expression is related to the Arrhenius injury function by:

$$\Omega(\tau) = \ln\left\{\frac{N_0}{N(\tau)}\right\} = \left\{\frac{C(0)}{C(\tau)}\right\} = 1 = \int_0^{D_0} Ae^{-\frac{\Delta E}{RT}}dt. \tag{11.60}$$

The injury data in Fig. 11.7 can be applied to illustrate how to calculate the value of R in (11.57). From Fig. 11.7b, one can read that $D_0(43°C) = 11.1$ min and $D_0(46.35°C) = 1$ min. From (11.57) we can write:

$$\ln(R) = \frac{\ln\left(\frac{t_{43}}{t_i}\right)}{43 - T_i} = \frac{\ln\left(\frac{t_{43}}{t_{46.35}}\right)}{43 - 46.35} = -\frac{\ln(11.1)}{3.35} = -0.719, \tag{11.61}$$

from which $R = 0.487$ for $T \geq 43°C$. In like manner, we can calculate the value for R at temperatures below the breakpoint in injury using the data at 41.5°C. For these conditions, $R = 0.144$ for $T < 43°C$. Finally, the relationship between the Ω burn injury function and the cumulative equivalent minutes at 43°C is:

$$\Omega = \frac{CEM.43}{D_0(43)}. \tag{11.62}$$

As was noted earlier, many different types of energy sources have been adapted to create focused elevated temperatures in tissue for performing hyperthermic surgery. These methods are based on the absorption of energy dissipated in a planned pattern within a target tissue volume. This pattern is a function of the *specific absorption rate* (SAR) for a particular modality used to heat the tissue. Each energy domain source has a unique constitutive equation for determining the magnitude and distribution of energy deposition in tissue. A comprehensive review of this literature is beyond what can be covered in an introductory text. However, to illustrate

11.3 Thermal Injury Mechanisms and Analysis

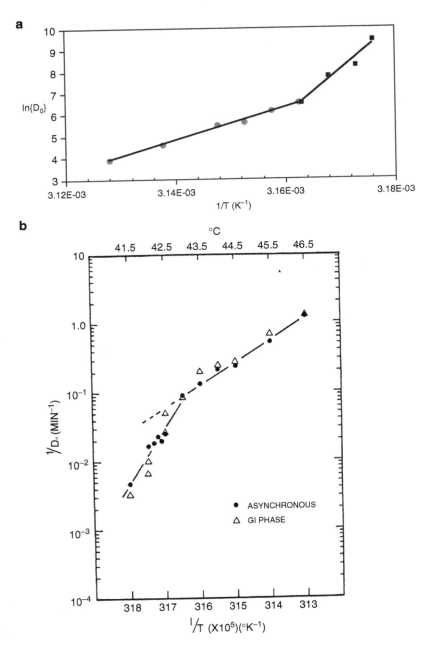

Fig. 11.7 Cell survival data for CHO exposed to hyperthermic stress (Sapareto 1982, with permission): (**a**) Arrhenius plot of cell survival (ln D_0), (**b**) cell survival rate plot ($1/D_0$). Note the breakpoint in the injury curves at 43°C, which typifies a transition in the injury mechanism

the analysis procedure, in the following section we will discuss laser irradiation as an energy source for performing surgery in tissue systems.

11.4 Laser Irradiation of Tissue

Lasers have come to be used in a very broad range of medical procedures. The irradiation of tissue by laser light results in the absorption of energy. Many medical applications involving lasers rely upon the generation of heat within the tissue for the desired therapeutic effect. Since laser energy absorption is a fully dissipative process, the consequence is that the increment in energy is expressed entirely as a heat transfer source distributed within the tissue. In conjunction with the irradiation, there will be an increase in the energy stored locally in the tissue as a function of the geometric pattern of absorption. Two primary mechanisms of energy storage are encountered most frequently during laser irradiation: sensible and latent. Sensible storage results in a change in temperature, and latent storage results in a change in phase. The phase change may be subtle, such as the denaturation of protein molecules, or it may be much more overt, such as when the boiling of water causes an explosive popping as the greatly increased vapor pressure of steam produces a ripping and tearing of tissue structures. The two mechanisms may occur simultaneously or singularly, depending on the initial state of the tissue and the intensity of the irradiation.

Unlike thermal irradiation, which is primarily a surface event, laser light irradiation results in a distributed absorption of energy below the tissue surface. In some surgical procedures, the laser light is introduced to a target volume deep within a tissue by inserting a fiber optic probe into the subject location. Heat generated within tissue is a function of the laser power, the shape, and the size of the incident beam and the optical properties of the tissue at the irradiation wavelength. Key to the calculation of heat source strength is the accurate estimation of the light distribution. The action of irradiated laser light as a distributed energy source in tissue has been reviewed by Welch et al. (1989) and Welch and van Gemert (2011).

11.4.1 Distributed Energy Absorption

Determination of the absorbed light energy in tissue is difficult in many cases. Although UV wavelengths of the excimer laser and 10.6 μm wavelength of the CO_2 laser are absorbed within the first 20 μm of soft tissue, visible and near infrared wavelengths are scattered and absorbed. Typically, the occurrence of multiple scattering events is a significant factor in determining the distribution of light in tissue and the resulting heat source term. Figure 11.8 presents a diagram of the various energy events that may be anticipated in conjunction with laser irradiation

11.4 Laser Irradiation of Tissue

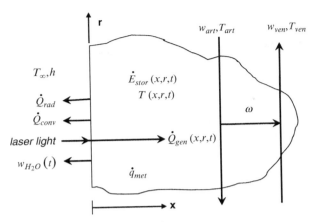

Fig. 11.8 Conservation of energy terms during laser irradiation of a surface tissue that is perfused with blood

of tissue. A cylindrical coordinate system is applied to the tissue system since most laser irradiation is symmetric about a central axis. The resulting energy and temperature fields will be dependent on radial distance from the center of the incident beam and the depth into the tissue. The target tissue is often perfused with blood, so it may be necessary to include the thermal influence of perfusion from the Pennes equation (10.102) in the analysis. Thermal radiation, convection, and evaporation of water at the skin surface can be highly important in some processes. As a consequence of all of these potentially important effects, you should carefully consider each of the initial steps of the GIM, including especially the initial evaluation of what are the important components of a process and how the system and its interactions with the environment should be defined to embody the processes that govern the process to be studied. We will next derive an expression for the heat source term as a function of the laser irradiation and tissue optical properties.

A laser beam of intensity $I_o(r,t)$ that is aligned with the x axis at the surface of the absorbing layer will have an intensity within the tissue:

$$I(r,x,t) = I_o(r,t)e^{-\beta x}, \tag{11.63}$$

at a depth of x. β is the absorption coefficient of the tissue at the wavelength of the laser. The heat source term then is:

$$\dot{Q}_{\text{gen}}(r,x,t) = \beta I_o(r,t)e^{-\beta x}. \tag{11.64}$$

The intensity function $I_o(r,t)$ is not uniform across the laser beam. Rather, the intensity is the strongest along the center line and diminishes with increasing radial distance. A common assumption is that the spatial variation in the intensity along the radius follows a Gaussian distribution pattern. After $\dot{Q}_{\text{gen}}(r,x,t)$ is calculated

Table 11.3 Absorption coefficient (β) values (1/s) for various biological tissues at three common medical laser wavelengths (nm) (Welch and van Gemert 2011, with permission)

Tissue	Wavelength (nm)		
	500 (Argon)	1,100 (Nd-Yag)	10,640 (CO_2)
Water	0.00025	0.36	1,106
Skin-epidermis	55	231	911
Skin-dermis	13	13	
Fat	13	10	
Damaged tissue	19	10	
Eye-pigment epithelium	1545	169	
Eye-choroid	169	107	
Stomach	28	6	
Liver	50	12.5	200
Blood-oxygenated	105	9.9	
Blood-deoxygenated	105	1.8	

for a specific combination of laser irradiation and target tissue, it can then be applied as a term in the conservation of energy equation such as (11.9) to analyze the thermal effects of the radiation process.

The absorption coefficient β is highly dependent on both the wavelength and tissue optical properties. Table 11.3 gives values of β for various tissues and wavelength combinations.

11.4.2 Time Constant Analysis of the Transient Temperature Field

The transient temperature field that develops during laser irradiation of tissue is described by a partial differential equation in both space and time (11.11). For many practical conditions of system geometry, composite tissue structure, boundary interactions, and energy deposition pattern, the solution of this equation is highly challenging and beyond the scope of analytical methods. The most common alternative is numerical methods, which are effective but tend to be time consuming in their implementation. An alternative approach is based on identifying diffusion time constants that can be identified in conjunction with short-term laser irradiation (van Gemert and Welch 1989). The following discussion provides a brief synopsis of this approach.

Heat transfer occurs in tissue by simple diffusion. In the absence of confounding factors such as internal convection by blood perfusion, the diffusion equation (11.11) can be written in terms of dimensionless parameters without the presence of an internal generation source as follows:

$$\frac{\partial \theta^*}{\partial Fo} = \frac{\partial^2 \theta^*}{\partial x^{*2}}, \qquad (11.65)$$

11.4 Laser Irradiation of Tissue

where the dimensionless parameters are defined for a system having a significant dimension L and a reference boundary temperature of T_0.

$$\theta^* = \frac{\theta}{\theta_i} = \frac{T - T_0}{T_i - T_0}, \tag{11.66}$$

$$x^* = \frac{x}{L}, \tag{11.67}$$

$$t^* = Fo = \frac{\alpha t}{L^2}. \tag{11.68}$$

The last parameter, Fo, the Fourier number, defines the relationship among a characteristic dimension of a system, the diffusional properties of the medium, and the elapsed diffusion time. As we have seen in numerous other occasions, these dimensionless combinations of temporal, spatial, and constitutive properties occur frequently in transient transport analysis and provide a basis for identifying a time constant to apply in describing the progression of a diffusion process. If an energy deposition process, having a clearly defined spatial distribution within a system, occurs very rapidly in comparison with the rate at which that energy diffuses into the surrounding medium, a time constant written in the format of the Fourier number can be an effective tool for describing the transient diffusion. Given the conceptual power of a time constant, it can lead to a simplified representation of a geometrically and temporally complex process. Laser irradiation is a particularly appropriate phenomenon to which this method can be applied since the energy deposition process can be extremely rapid thereby creating a well-defined initial temperature pattern which subsequently diffuses into the surrounding tissue.

There have been a number of analyses of laser irradiation in tissue which focused on diffusion in a single spatial coordinate. Many of these analyses have been summarized by McKenzie (1990). Van Gemert and Welch (1989) have extended this concept to combine simultaneous diffusion time constants in two orthogonal coordinates, which provides a basis for a more comprehensive analysis. The result is the ability to predict the spatial propagation of injury from a short duration laser pulse.

The starting point for this analysis is the diffusion equation in radial and axial dimensions, with a heat source owing to optical energy absorption as described by a coefficient μ_a applied to a local two-dimensional fluence rate $\Phi(x,r)$. The temperature scale is shifted with respect to the initial value so that the dependent variable is the temperature rise, ΔT.

$$\frac{\partial \Delta T(x,r,t)}{\partial t} = \frac{\mu_a \Phi(x,r)}{\rho c_p} + \frac{k}{\rho c_p}\left[\frac{\partial^2 \Delta T}{\partial x^2} + \frac{\partial^2 \Delta T}{\partial r^2} + \frac{1}{r}\frac{\partial \Delta T}{\partial r}\right]. \tag{11.69}$$

In order to define axial and radial time constants, τ_x and τ_r, it is necessary to identify relevant characteristic dimensions along the axial and radial coordinates, which are the penetration depths of the temperature and fluence rate, denoted as

x_o and r_o, respectively. The solution is written in terms of differential operators in x and r having eigenfunctions that satisfy the relationships:

$$\frac{k}{\rho c_p} \frac{d^2 X(x)}{dx^2} = -\frac{1}{\tau_x} X(x), \tag{11.70}$$

$$\frac{k}{\rho c_p} \left[\frac{d^2 R(r)}{dr^2} + \frac{1}{r} \frac{dR(r)}{dr} \right] = -\frac{1}{\tau_r} R(r). \tag{11.71}$$

For a cylindrically symmetric irradiation beam in a tissue with an insulated surface and with semi-infinite geometry in both dimensions, the boundary conditions for these equations are:

$$\frac{dX(0)}{dx} = 0, \tag{11.72}$$

$$\frac{dR(0)}{dr} = 0. \tag{11.73}$$

(11.70)–(11.74) will be satisfied if

$$\frac{k}{\rho c_p} \left(\frac{\pi}{2x_o} \right)^2 = \frac{1}{\tau_x}, \tag{11.74}$$

$$\frac{k}{\rho c_p} \left(\frac{2.4}{r_o} \right)^2 = \frac{1}{\tau_r}. \tag{11.75}$$

Note that both of these solutions define time constants in terms of the combination of properties associated with the Fourier number.

An overall time constant for the parallel axial and radial diffusion processes is τ:

$$\frac{1}{\tau} = \frac{1}{\tau_x} + \frac{1}{\tau_r}. \tag{11.76}$$

The diffusion equation with a laser irradiation source can be written as:

$$\frac{d\Delta T(x,r,t)}{dt} = \frac{\mu_a \Phi(x,r)}{\rho c_p} - \frac{\Delta T(x,r,t)}{\tau} \tag{11.77}$$

for which the solution is:

$$\Delta T(x,r,t) = \frac{\tau \mu_a \Phi(x,r)}{\rho c_p} \left(1 - e^{-\frac{t}{\tau}} \right). \tag{11.78}$$

11.4 Laser Irradiation of Tissue

For long times, the steady-state solution is:

$$\Delta T(x, r, t) = \frac{\tau \mu_a \Phi(x, r)}{\rho c_p}. \qquad (11.79)$$

As would be anticipated, a solution of the diffusion problem based on a time-constant analysis takes the form of a decaying exponential function. This solution is an approximation based on the assumption that ΔT is proportional to Φ, which holds greatest validity for the shortest times and is increasingly compromised for longer times. The solution is reflective of the laser beam radius and optical penetration into the tissue, providing an intuitive understanding of the thermal response to the irradiation. This analysis is most accurate for diffusion periods less than three times the time constant.

11.4.3 Surface Cooling During Irradiation

There are instances during the laser irradiation of tissue when other simultaneous interactions with the surroundings can have a significant effect on the temperature distribution. In this section, we will present a practical example of combined laser heating and environmental heat transfer that has a major beneficial impact on the ability to apply laser surgical techniques.

A frequent objective of laser irradiation procedures is to deposit thermal energy (heat) into tissue within a targeted subsurface volume. The absorption characteristics of the tissue for the applied wavelength may be such that to achieve the necessary temperature elevation in the interior volume, an unacceptable temperature rise occurs in the surface tissue resulting in unwanted damage to that tissue. An option for overcoming this limitation is to provide simultaneous cooling of the surface in conjunction with the irradiation to prevent the surface tissue temperature from rising to injurious levels (Anvari et al. 1995; Pikkula et al. 2001). This procedure is based on a process known as *spray cooling* or *impingement jet heat transfer* in which a stream of cool fluid is directed obliquely onto a substrate to produce a local convective cooling effect (Martin 1977). Jet impingement convection is illustrated in Fig. 8.11. The magnitude of the convection cooling is determined by the velocity, direction and pattern of spray, the thermal properties of the spray fluid, and the temperature of the spray. A correlation equation for conditions of a simplified geometry is given in (8.49). This process has a long history of many industrial applications, and its adaptation to control the spatial temperature distribution in tissue during laser irradiation is a novel application that has issued in significantly improved outcomes for numerous laser treatments. Thermal analysis of the process may be additionally complicated if the cooling fluid is a liquid cryogen, in which case the liquid evaporates upon contact with the skin, issuing in a joint sensible and latent heat transfer interaction with the skin and a concomitant two phase external flow phenomenon.

Holman and colleagues (Holman and Kendall 1993) have investigated spray cooling of vertical surfaces by a horizontal stream of subcooled Freon-113. The subcooling has the effect of reducing the complexity of boiling from the impingement heat transfer process, making the analysis and correlation considerably simpler (also the heat transfer is more effectual from a liquid phase coolant). Nonetheless, unless the extent of subcooling is very large, phase change effects will persist at the cooled surface. The primary motivation for this study was the cooling of electronic equipment, which also has a spatially distributed pattern of internal energy generation. Pressurized spray from a nozzle produced a stream of liquid droplets striking the heated surface. From a mechanistic perspective, a large number of physical characteristics of the spray system have been identified that exert a direct influence on the ability to cool a surface. These include the mass flux, spray droplet velocity, droplet diameter, and distance between the spray nozzle and surface. The cooling effect is greater with increasing droplet mass flux, degree of subcooling of the cryogen, and magnitude of the *Weber number*, We, which is a dimensionless parameter defined in terms of the ratio of the inertial and surface tension properties of the spray.

$$We = \frac{\rho v^2 d_d}{\sigma}, \tag{11.80}$$

where d_d is the diameter of the spray droplets and σ is the surface tension between the liquid and vapor phases of the spray fluid. The Weber number characterizes the impact dynamics of the spray droplets with the surface. A large value of We describes physical conditions for which the formation of a vapor layer between the droplets and the surface will be minimized, thereby leading to more efficient cooling of the surface. Experimental observations have shown that the mechanism of heat transfer between the spray droplets and the solid surface is by subcooled boiling which is a combination of convection and liquid–vapor phase change. Extensive experimental data for spray cooling with Freon-113 led to a correlation equation relating the cooling heat flux, q (W m^{-2}) to: We, the flow velocity of the spray, the temperature difference between the spray liquid and the surface, ΔT, and the magnitude of the mass flux of spray, which is linearly related to the separation of the spray nozzle and warm surface, x. For most spray nozzle configurations, the cross-sectional area of the spray pattern will increase in size with distance from the source, thereby reducing the magnitude of the flux (kg s^{-1} m^{-2}). The degree of spray liquid subcooling is embodied in the term ΔT, which is the subcooling of the liquid below the saturation temperature, T_{sat}, plus the differential between the saturation and warm surface temperatures, if indeed a coolant is used which boils at a temperature below that of the surface being cooled. With adequate nucleation conditions present, the spray liquid will begin to boil when it is warmed to the saturation temperature. These relationships are expressed in terms of a correlation equation written with dimensionless parameters (Holman and Kendall 1993).

$$\frac{\dot{q}x}{\mu_f h_{\text{fg}}} = 9.5 We^{0.6} \left(\frac{c_p \Delta T}{h_{\text{fg}}}\right)^{1.5}. \tag{11.81}$$

11.4 Laser Irradiation of Tissue

The thermodynamic properties of the spray are evaluated at the film temperature, T_f, (8.42a).

The temperature difference between the spray liquid and the heated surface is normalized to the latent heat of vaporization between the liquid and vapor states, h_{fg}, with the liquid specific heat, c_p. This ratio compares the capability of the spray liquid to undergo sensible and latent heat transfers as it interacts with the warm surface, which is referred to as the *Stefan number*, *Ste*. The *Reynolds number* may be defined in terms of the droplet diameter, and the droplet diameter is normalized by the nozzle to surface distance. These dimensionless parameters are given by the relations:

$$Ste = \frac{c_p \Delta T}{h_{fg}}, \tag{11.82}$$

$$Re_{d_d} = \frac{\rho v d_d}{\mu_f}, \tag{11.83}$$

$$d_d^* = \frac{d_d}{x}, \tag{11.84}$$

so that the spray cooling heat flux can be written as:

$$\frac{\dot{q}}{\rho v h_{fg}} = 9.5 \frac{We^{0.6}}{Re} d_d^* Ste^{1.5}. \tag{11.85}$$

The heat flux at the surface is normalized to the energy of the approaching spray stream in terms of the momentum per unit volume and the latent heat. As with all empirical heat transfer correlations, the above equation is valid over the range of experimental test conditions for which it was derived.

The spray cooling process in conjunction with laser irradiation of skin has been analyzed from the specific perspective of augmenting laser irradiation of the skin to protect near surface structures from thermal injury (Anvari et al. 1995; Pikkula et al. 2001). Analysis of this process is complicated beyond that for only spray cooling since there are combined simultaneous surface cooling and penetrating heating effects. This process can be simplified for analysis by assuming that the dynamics of the jet impingement and evaporation result in a uniform material boundary on the skin surface consisting of mixed cryogen and ice that grows in thickness $b(t)$ during the spraying process. The geometry for this system is a one-dimensional, multi-layered, semi-infinite medium in Cartesian coordinates as shown in Fig. 11.9.

The transient heat transfer process [which for our case is targeted to destroy port wine stain (PWS) vessels deep within the skin without causing injury to the more superficial tissues that could result in permanent scarring] can be considered in terms of two separate phenomena. One is the spray cooling onto the skin surface which will issue in the penetration of a thermal cooling wave into the underlying tissues. The second is the absorption of laser energy distributed as a heat source on the interior of the tissue as a function of the optical properties of the laser light and

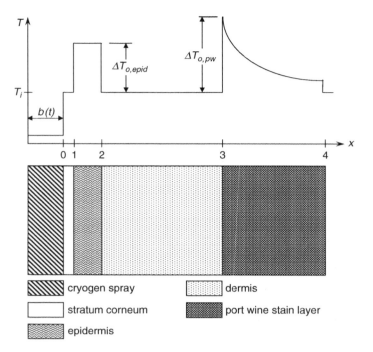

Fig. 11.9 Composite system for analysis of jet impingement cooling of the skin surface with a cryogen spray. A boundary layer consisting of a mixture of cryogen and ice particles grows in thickness $b(t)$ on the skin surface during the spraying process. The indicated temperature distribution depicts the cooling effect at the skin surface and immediately at the completion of the laser-induced heating within the various skin structures, including the PWS vessels [adapted from Anvari et al. (1995 with permission)]

the matching spectral absorption properties of the various tissue components. The heating is assumed to occur very rapidly during a brief laser pulse (e.g., 450 μs) that is short in comparison with the subsequent thermal diffusion processes within the skin. Thus, the spatial temperature profile that is produced at the end of the laser pulse, relative to the starting uniform value, T_i, as shown in Fig. 11.9, can be taken as the initial conditions for the transient heat transfer that determines whether the PWS vessels will be destroyed and the epidermis undamaged. The temperature distribution pattern in the skin is characterized using a simple light absorption model (Anvari et al. 1995) in which heat is generated only in the melanin of the epidermis and in the vasculature of the PWS tissue volume. The fraction of the PWS layer occupied by the vasculature is denoted by f_{area}. The resulting initial temperature rise function within the skin is given by:

$$T(x,0) - T_{\text{i}} = \begin{cases} \Delta T_{0,\text{epid}} \\ \Delta T_{0,\text{PWS}} f_{\text{area}} e^{-[\mu(x-x_{\text{PWS}})]} \\ 0 \end{cases} \text{for} \begin{cases} x_1 \leq x \leq x_2 \\ x_3 \leq x \leq x_4 \\ x<x_1, x_2 \leq x \leq x_3, x>x_4 \end{cases} . \quad (11.86)$$

11.4 Laser Irradiation of Tissue

It is important to note that the cooling boundary condition can be controlled to manipulate the temperature distribution caused by the laser within the skin by applying a differential start time for the spraying and irradiation. In particular, pre-cooling will initiate a cold wave to move into the skin. With proper timing, the penetration depth of the cooling will be limited to an intermediate level so that the epidermis is protected from thermal injury, whereas the PWS remains fully sensitive to the therapeutic effects of the laser irradiation. This type of cooling strategy will alter the initial thermal conditions for the laser heating process to be more complex than defined in (11.86), with the superposition of the preliminary cooling pattern.

Spray cooling of a cryogen at subzero temperatures runs the risk of causing ice formation, either at the exterior boundary to the skin by frost formation from water vapor in the ambient air or additionally within the skin itself. In nearly all aspects, ice formation is an undesirable result and to be avoided. A layer of ice will significantly slow the rate of cooling for interior tissues by sensible heat loss by imposing a layer undergoing phase change at the skin surface. Liberation of the latent heat of fusion from liquid water when it freezes represents a large heat sink to absorb energy flow that would otherwise be cooling the tissue. Because freezing occurs at a constant temperature, the process will enforce an isothermal boundary condition of $0°C$ onto the skin. Freezing of the skin will result in frostbite, with a resultant likelihood of tissue injury and possible scarring. Thus, ice formation during spray cooling is to be avoided. Because of its mathematical complexity, the freezing process is omitted from our analysis. It can be replaced by the more simple condition of avoiding spay conditions that will result in a drop of the surface temperature to $0°C$.

The cooling at the surface is characterized in terms of a simple convective heat transfer coefficient, h, that is uniform across the boundary area of interest. Thus, the boundary condition is expressed as (8.12) during active cooling and is assumed to be insulated before and after.

$$-k_f \frac{dT}{dx}\bigg|_{x=0} = h(T_s - T_\infty). \tag{11.87}$$

Note that the entire effect of the spray cooling is lumped into the convective heat transfer coefficient, h. In conjunction with Newton's law of cooling, (2.40), it should be possible to derive a value for h from a correlation relation such as (8.49).

$$\dot{q} = \frac{\dot{Q}}{S} = h(T_\infty - T_s) = h\Delta T. \tag{11.88}$$

By incorporating the laser heating cycle into the initial conditions and assuming that blood perfusion (i.e., convection) has no significant thermal effect, the transient tissue temperature is described by the basic one-dimensional diffusion equation (11.12) without heat generation, which is written in a single spatial dimension with constant property values as:

$$\frac{\partial T(x,t)}{\partial t} = \alpha \frac{\partial^2 T}{\partial x^2}. \tag{11.89}$$

Several application cases can be considered: thermal response to laser heating with no cooling, with precooling, and with both precooling and postcooling. The general form of the solution to the diffusion problem in a semi-infinite system is the error function. Accordingly, the analytical solutions follow for each of the three cases, based on Carslaw and Jaeger (1959). The time scale is adjusted to set the zero point at the completion of the laser irradiation of duration t_{laser} as defined by $t^* = t - t_{\text{laser}}$. Since there are two spatial temperature sources in the system, the temperature rise at any location in the skin is the superimposed sum of the effects derived from both of the initial areas of temperature rise (in the epidermis and PWS layers). Thus:

$$\Delta T(x,t^*) = \Delta T_{\text{epid}}(x,t^*) + \Delta T_{\text{PWS}}(x,t^*). \tag{11.90}$$

For the first case in which there is no cooling, the boundary condition is insulated. The solutions for the two components of heating in the preceding equation are:

$$\Delta T_{\text{epid}}(x,t^*) = \frac{\Delta T_{0,\text{epid}}}{2} \left\{ \text{erf}\left(\frac{x_i}{2\sqrt{\alpha t^*}} - \frac{x}{2\sqrt{\alpha t^*}}\right) + \text{erf}\left(\frac{x_i}{2\sqrt{\alpha t^*}} + \frac{x}{2\sqrt{\alpha t^*}}\right) \right\} \Big|_{\frac{x_1}{2\sqrt{\alpha t^*}}}^{\frac{x_2}{2\sqrt{\alpha t^*}}}, \tag{11.91}$$

$$\Delta T_{\text{PWS}}(x,t^*) = \frac{\Delta T_{0,\text{PWS}}}{2 f_{\text{area}}} \exp\left[2\mu_a^{\text{blood}}\sqrt{\alpha t^*}\left(\frac{x_3}{2\sqrt{\alpha t^*}} + \mu_a^{\text{blood}}\sqrt{\alpha t^*}\right)\right]$$

$$\cdot \left\{ \exp\left[-2\frac{x}{2\sqrt{\alpha t^*}}\mu_a^{\text{blood}}\sqrt{\alpha t^*}\right] \text{erf}\left(\frac{x_i}{2\sqrt{\alpha t^*}} - \frac{x}{2\sqrt{\alpha t^*}} + \mu_a^{\text{blood}}\sqrt{\alpha t^*}\right) \right.$$

$$\left. + \exp\left[2\frac{x}{2\sqrt{\alpha t^*}}\mu_a^{\text{blood}}\sqrt{\alpha t^*}\right] \text{erf}\left(\frac{x_i}{2\sqrt{\alpha t^*}} + \frac{x}{2\sqrt{\alpha t^*}} + \mu_a^{\text{blood}}\sqrt{\alpha t^*}\right) \right\} \Big|_{\frac{x_3}{2\sqrt{\alpha t^*}}}^{\frac{x_4}{2\sqrt{\alpha t^*}}}, \tag{11.92}$$

where μ_a^{blood} is the optical absorption coefficient of blood.

When the skin is precooled before laser irradiation, the initial conditions are altered as the cooling wave penetrates the skin. For the period before irradiation, time is measured from the beginning of cooling, and the temperature history is described in terms of the classic problem of diffusion in a semi-infinite medium with a convective boundary condition.

$$\Delta T_{\text{cooling}}(x,t) = (T_\infty - T_0)$$

$$\times \left\{ \text{erfc}\left(\frac{x}{2\sqrt{\alpha t}}\right) - \left[\exp\left(\frac{hx}{k} + \frac{h^2 \alpha t}{k^2}\right)\right]\left[\text{erfc}\left(\frac{x}{2\sqrt{\alpha t}} + \frac{h\sqrt{\alpha t}}{k}\right)\right] \right\}. \tag{11.93}$$

11.4 Laser Irradiation of Tissue

The cooling spray is discontinued when the short laser irradiation starts, which imposes the energy distribution pattern shown in Fig. 11.9 onto the preexisting temperature distribution at the end of precooling. An exact analytical function does not exist for this problem, so an approximation has been developed (Anvari et al. 1995), although there may be many options for an approximation beyond that chosen by those authors. The result is that the post-heating temperature is given by a superposition of the combined effects of cooling and the two layers of laser heating where there still exist residual temperature alterations owing to cooling being terminated after that process.

$$\Delta T(x, t^*) = \Delta T_{\text{epid}}(x, t^*) + \Delta T_{\text{PWS}}(x, t^*) + \Delta T_{\text{cooling}}(x, t^*). \tag{11.94}$$

For the conditions of continuous cooling, a superposition solution is given in (11.94) with the individual terms determined for the appropriate initial and boundary conditions. The expression of cooling effect applies throughout the entire process, (11.93). The heating terms are given by:

$$\Delta T_{\text{epid}}(x, t^*) = \frac{\Delta T_{0,\text{epid}}}{2} \left\{ \text{erf}\left(\frac{x_i}{2\sqrt{\alpha t^*}} - \frac{x}{2\sqrt{\alpha t^*}}\right) + \text{erf}\left(\frac{x_i}{2\sqrt{\alpha t^*}} + \frac{x}{2\sqrt{\alpha t^*}}\right) \right.$$
$$\left. - 2\exp\left[\frac{x}{2\sqrt{\alpha t^*}} + \frac{x_i}{2\sqrt{\alpha t^*}}\right]^2 \exp\left[\frac{x}{2\sqrt{\alpha t^*}}\right]^2 \right. \tag{11.95}$$
$$\left. \times \text{erfc}\left(\frac{h}{k}\sqrt{\alpha t^*} + \frac{x_i}{2\sqrt{\alpha t^*}} + \frac{x}{2\sqrt{\alpha t^*}}\right)\right\} \Big|_{\frac{x_1}{2\sqrt{\alpha t^*}}}^{\frac{x_2}{2\sqrt{\alpha t^*}}},$$

$$\Delta T_{\text{PWS}}(x, t^*) = \frac{\Delta T_{0,\text{PWS}}}{f_{\text{area}}} \exp\left[2\mu_a^{\text{blood}}\sqrt{\alpha t^*}\frac{x_3}{2\sqrt{\alpha t^*}}\right] \cdot$$
$$\left\{ \exp\frac{\left[2\mu_a^{\text{blood}}\sqrt{\alpha t^*}\right]^2}{2}\left(\exp\left[-2\mu_a^{\text{blood}}\sqrt{\alpha t^*}\frac{x}{2\sqrt{\alpha t^*}}\right]\text{erf}\left[\frac{x_i}{2\sqrt{\alpha t^*}} - \frac{x}{2\sqrt{\alpha t^*}} + \mu_a^{\text{blood}}\sqrt{\alpha t^*}\right]\right.\right.$$
$$+ \left[\frac{\mu_a^{\text{blood}}\sqrt{\alpha t^*} + \frac{h}{k}\sqrt{\alpha t^*}}{\mu_a^{\text{blood}}\sqrt{\alpha t^*} - \frac{h}{k}\sqrt{\alpha t^*}}\right] \exp\left[2\mu_a^{\text{blood}}\sqrt{\alpha t^*}\frac{x}{2\sqrt{\alpha t^*}}\right]\text{erf}\left[\frac{x_i}{2\sqrt{\alpha t^*}} - \frac{x}{2\sqrt{\alpha t^*}} + \mu_a^{\text{blood}}\sqrt{\alpha t^*}\right]\right)$$
$$+ \left[\frac{\frac{h}{k}\sqrt{\alpha t^*}}{\mu_a^{\text{blood}}\sqrt{\alpha t^*} - \frac{h}{k}\sqrt{\alpha t^*}}\right]\exp\left[-\left(\frac{x}{2\sqrt{\alpha t^*}} + \frac{x_i}{2\sqrt{\alpha t^*}}\right)^2 - 2\frac{x_i}{2\sqrt{\alpha t^*}}\mu_a^{\text{blood}}\sqrt{\alpha t^*}\right] \cdot$$
$$\left.\exp\left[\frac{x_i}{2\sqrt{\alpha t^*}} + \frac{x}{2\sqrt{\alpha t^*}} + \mu_a^{\text{blood}}\sqrt{\alpha t^*}\right]^2 \text{erfc}\left(\frac{x_i}{2\sqrt{\alpha t^*}} + \frac{x}{2\sqrt{\alpha t^*}} + \mu_a^{\text{blood}}\sqrt{\alpha t^*}\right)\right\}\Big|_{\frac{x_3}{2\sqrt{\alpha t^*}}}^{\frac{x_4}{2\sqrt{\alpha t^*}}}.$$
$$\tag{11.96}$$

These transient temperature solutions can be applied with a damage rate model to calculate the level of injury to tissue as a function of position and treatment protocol.

The bioheat transfer analysis in this section is obviously quite complex, beyond what is discussed otherwise in the text. It is not intended as material to be learned or memorized. The material is a good example of "knowledge to be familiar with" (see Sect. 1.8), but that is not a part of your essential working tools for biotransport. Its place in an introductory text is to illustrate a bioheat transfer problem solution that may be typical of what you may encounter in reading the research literature. A good exercise would be for you to go through this analysis and overlay the steps of the GIM to the work presented to gain a clearer understanding of how the analysis of these researchers fits into a logical structure. If you advance to further studies in biotransport, it will be most useful for you to be able to put the research literature that you encounter into a context of understanding according to the GIM approach to problem analysis.

11.5 Summary of Key Concepts

In this chapter, we have presented a generalized approach to the analysis of heat transfer processes in living tissues, and shown how it can be applied to specific problems that may involve varying degrees of complexity. We have also introduced the finite difference numerical analysis method for solving problems that involve complex geometries, composite materials structures, nonlinear properties, or other features that render them unsolvable by applying standard analytical functions. These conditions arise quite often in biological systems.

General microscopic formulation of conservation of energy. The analysis strategy embodied in the Generate Ideas Model can be applied to a microscopic scale system to derive a general expression for application of the conservation of energy. A general system may communicate with its environment across the boundary by flows of both energy and mass. In most biomedical heat transfer applications we can neglect the effects of work transfer (muscle activation being an important exception), so that the primary heat flows occur at the boundaries and via various methods of internal generation. Mass flowing across a system boundary has energy that is expressed as the enthalpy to account for the effect of displacing mass in the environment in conjunction with the flow. The mass flow can be expressed in terms of convection. The resulting conservation of energy equation is:

$$\frac{\partial T}{\partial t} = -\vec{v} \cdot \vec{\nabla} T + \vec{\nabla} \cdot \left(\alpha \vec{\nabla} T \right) + \frac{\dot{q}_{\text{met}}}{\rho c_{\text{p}}}, \qquad (11.10)$$

which can also be written in terms of the substantial derivative that we first encountered in conjunction with the analysis of fluid flow in Chap. 7.

$$\frac{DT}{Dt} = \left[\frac{\partial T}{\partial t} + \vec{v} \cdot \vec{\nabla} T \right] = \alpha \nabla^2 T + \frac{\dot{q}_{\text{met}}}{\rho c_{\text{p}}}. \qquad (11.11)$$

11.5 Summary of Key Concepts

Numerical methods for transient conduction: finite difference analysis. It is possible to develop a set of algebraic equations that approximate the differential equation that describes the flow of heat in a system (as well as for applications in fluid flow, mass flow, and many other types of problems such as the stress distribution in a system). The solution to the differential equation will be an analytical function that has values that vary continuously with both position and time for the process of interest. In order to build a set of algebraic approximation equations, it is necessary to divide the continuous system into discrete increments in both space and time, thereby losing resolution in our description of the process. A separate equation is written for each increment in space and time which, for complex or nonlinear systems, can be more readily solved that can an analytical function. We have illustrated the finite difference method in this text, although alternatives such as the finite element method are available. There are two basic approaches to writing a finite difference approximation for how the temperature changes continuously with time: forward difference and backward difference. For a two-dimensional transient heat transfer process in Cartesian coordinates, the two difference equations for a grid position in the interior of a material are:

forward: $\quad T_{m,n}^{p+1} = Fo\left(T_{m+1,n}^{p} + T_{m-1,n}^{p} + T_{m,n+1}^{p} + T_{m,n-1}^{p}\right) + (1 - 4Fo)T_{m,n}^{p}$

$$(11.41)$$

and

backward: $\quad (1+4Fo)T_{m,n}^{p+1} - FoT_{m+1,n}^{p+1} - FoT_{m-1,n}^{p+1} - FoT_{m,n+1}^{p+1} - FoT_{m+1,n-1}^{p+1} = T_{m,n}^{p},$

$$(11.53)$$

where the Fourier number for a finite difference analysis is written as:

$$Fo = \frac{\alpha \Delta t}{(\Delta x)^2}. \quad (11.42)$$

The size of the time and space increments Δt, Δx, and Δy must be chosen so that the coefficient for the $T_{m,n}^p$ term is positive to ensure that the solution of the equations is stable. The backward difference solution is inherently stable. The forward difference solution requires that the values of Δt, Δx, and Δy be chosen so that:

$$1 - 4Fo \geq 0 \quad \text{or} \quad Fo = \frac{\alpha \Delta t}{(\Delta x)^2} \leq \frac{1}{4}, \quad (11.43)$$

where $\Delta x = \Delta y$. Special difference equations are written for boundary and interface nodes and for particular internal energy generation conditions. Each of these equations will have its own stability criterion, and the most stringent criterion for the entire set of equations must be applied in setting up the grid spacing and time steps.

Thermal injury mechanisms and analysis. When the temperature of living tissue is raised above a threshold value, irreversible changes in structure and function occur. The damage process can be described by a first-order kinetics equation based on a standard Arrhenius type model. The rate at which tissue injury occurs is given by:

$$\frac{d\Omega(x,t)}{dt} = Ae^{-\frac{\Delta E}{RT(x,t)}}, \qquad (11.54)$$

where Ω is a quantitative measure of the degree of injury, and the injury model properties A and ΔE are unique to the tissue type and, in some cases, to the temperature of the injury process. The local temperature history, $T(x,t)$, must be applied in absolute units (K). The injury rate function can be integrated over the entire period for which the temperature is at an elevated temperature to obtain a quantitative representation of the level of damage.

$$\Omega(x,\tau) = \int_0^\tau Ae^{-\frac{\Delta E}{RT(x,t)}} dt. \qquad (11.55)$$

Values of Ω can be correlated with the degree of a clinically evaluated burn injury.

First degree – $\Omega = 0.53$
Second degree – $\Omega = 1.0$
Third degree – $\Omega = 10^4$

Laser irradiation of tissue. Lasers are applied in a wide range of medical procedures. The optical interaction of a laser beam with tissue results in an internal energy generation function. Since the power density of a laser can be very high, it is possible to create extremely large temperature gradients in both time and space. The optical and absorption properties of different tissues are highly dependent on wavelength so that it is possible to use lasers of specific wavelengths to "tune" irradiation protocols to meet special diagnostic and therapeutic needs.

A laser beam of intensity $I_o(r,t)$ that is aligned with the x axis at the surface of the absorbing layer will have an intensity within the tissue

$$I(r,x,t) = I_o(r,t)e^{-\beta x} \qquad (11.63)$$

at a depth of x. β is the absorption coefficient of the tissue at the wavelength of the laser. The heat source term then is:

$$\dot{Q}_{\text{gen}}(r,x,t) = \beta I_o(r,t)e^{-\beta x}. \qquad (11.64)$$

The $\dot{Q}_{\text{gen}}$ term is applied in the conservation of energy equation to solve for the development of a transient temperature field during a defined laser treatment protocol.

11.6 Questions

11.6.1. Explain the derivation of the microscopic scale formulation of the conservation of energy equation in terms of the Generate Ideas Model.

11.6.2. For a steady-state system with internal energy generation, how must the temperature gradients at the system boundaries relate to the rate of energy generation?

11.6.3. What is the physical meaning of the *mixing cup temperature*?

11.6.4. What are the advantages and disadvantages of the forward and backward difference formulations in the unsteady-state numerical method? Under what circumstances might you choose one method over the other?

11.6.5. Explain the tradeoffs involved in choosing the size of grid spacing and magnitude of time step in setting up a finite difference solution to a transient diffusion problem.

11.6.6. Why is a stability criterion imposed on a forward difference solution method but not on a backward difference method? What are the physical consequences and interpretation of violating the stability criterion? Have you ever performed calculations using the forward difference method with the stability criterion violated to see what happens?

11.6.7. What are the fundamental assumptions of the Arrhenius model for thermal injury in living cells and tissues?

11.6.8. What units of temperature are required in the Arrhenius thermal injury model?

11.6.9. Explain the physical and physiological correlation between microscopic and macroscopic scale descriptions of first, second, and third degree burn injuries.

11.6.10. How can the wavelength dependence of the absorption properties of light in different tissues be applied to advantage in the design of laser surgery and therapy procedures?

11.6.11. Explain why simultaneous spray cooling of the skin surface in conjunction with laser irradiation allows a higher rate of energy deposition to be applied during surgical procedures.

11.7 Problems

11.7.1 Vulcanization Process

The final mechanical properties of medical grade rubbery polymers are achieved by a process equivalent to vulcanization for other synthetic rubbers. A short cylinder of polymer 10 cm tall and 20 cm in diameter is originally at a uniform temperature of 290 K and must be heated to a center temperature of 410 K to achieve complete crosslinking and develop the desired physical properties. Heating is accomplished through the uniform application of steam to all surfaces of the polymer cylinder.

(a) Determine the center temperature after 16 h of heating. The following constant physical properties may be used: $k = 0.151$ W/(m-K); $c_P = 200$ J/(kg-K); $\rho = 1201$ kg/m^3; $\alpha = 6.19 \times 10^{-8}$ m^2/s; $h = 16$ W/(m2-K); $T_\infty = 435$ K.
(b) Determine the temperature at position $(r,z) = (5$ cm, 2.5 cm$)$ after 16 h of heating.

11.7.2 Metabolic Heat Generation

A thin layer of tissue generates metabolic heat at a constant rate per unit volume, R. The tissue is insulated on the bottom (i.e., no heat flows out from the bottom) and is kept at a constant temperature, T_1 at the top.

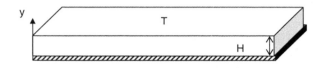

(a) Derive an expression for heat flux as a function of position y.
(b) Derive an expression for temperature as a function of y.

11.7.3 Internal Temperature Gradient

(a) Use a thermal energy microscopic shell balance to derive a *partial differential equation* that describes temperature in the cylinder as a function of both time and radial position. The cylinder is initially at temperature T_0, and heat is transferred to the surroundings at T_∞ via Newton's Law of Cooling. There is no heat produced within the cylinder (i.e., postmortem).
(b) What initial and boundary conditions are needed to actually solve the resulting PDE? *Do not attempt to solve the resulting equation.*

11.7.4 Temperature Gradient in Stagnant Air

(a) Use Fourier's law to find an expression for the temperature distribution $T(r)$ in the air surrounding a solid sphere with diameter d and surface temperature T_s. The air is assumed to be stagnant and the air temperature far from the sphere (i.e., $r \to \infty$) is T_∞. Heat flow $\dot{Q}$ in the air $(r \geq d/2)$ is assumed to be constant (steady-state).

(b) Simplify the general microscopic thermal energy equation in air for this case in spherical coordinates (11.14). Integrate it with the appropriate boundary conditions to show that the result is the same as in part (a).
(c) Apply Newton's law of cooling to your solution in part (a) or (b) to show that the Nusselt number (hd/k_f) for conduction in the air surrounding a sphere is equal to 2.0.
(d) Consider heat loss from a bald head to air as being equivalent to heat loss from a sphere with diameter 18 cm and constant surface temperature of 35°C. The effect of natural convection on heat transfer is given by (8.57). Plot the heat loss from the head relative to heat loss in stagnant air as a function of air temperature as it varies from 35°C to -50°C. Assume constant values for α and ν, but $\beta = 1/T_\infty$, where T_∞ is given in absolute units (K). $\alpha = 1.95 \times 10^{-5}$ m²/s, $\nu = 1.39 \times 10^{-5}$ m²/s.
(e) Now consider heat loss from the head when $T_\infty = 0$°C, and air is blown past the head with various velocities between 0 and 30 m/s. Forced convection from a sphere is given by (8.48) with $\mu/\mu_s = 1$. Plot the heat loss from the head relative to the heat loss for zero velocity as a function of wind velocity. This is known as the wind chill factor.
(f) What assumption is probably invalid in your computations in parts (d) and (e)? How does the body cut down on heat loss from the skin?

11.7.5 Heat Transfer in a Hollow Fiber

Blood flows through a hollow fiber with an inside diameter of 100 μm and outside diameter of 120 μm. The blood is cooled by blowing air at a lower temperature across the outside surface of the fiber. The thermal conductivity of the fiber is 2×10^{-4} kW/m °C. Compute the overall heat transfer coefficient U_o of the fiber based on the outside radius if:

(a) All the thermal resistance is in the fiber wall.
(b) The heat transfer coefficient between blood and the inside fiber wall is given by (8.35) and the heat transfer coefficient between air and the outside surface of the fiber is given by (8.47). Use $k_{blood} = 0.633$ W/m K and $k_{air} = 0.0262$ W/m K, Reynolds number for air $= 200{,}000$ and $Pr_{air} = 0.7$.

How much error is introduced by assuming that all of the thermal resistance is in the fiber wall?

11.7.6 Hyperthermia Therapy for Tumors

There is a growing body of clinical and scientific evidence that mild heating of tumors (a technique called hyperthermia) can provide a significant enhancement of the effectiveness of radiation treatment regimens. An engineering challenge is to design a protocol that will produce temperatures in the therapeutic range within the tumor while minimizing damage to surrounding healthy tissue by burn injury. For this problem consider a tumor growing on the surface of the skin, with a depth of 5 mm and an initial temperature of 34°C. The tumor is heated by placing a 60°C hot pack onto the surface. To simplify the analysis you may assume the tumor to be one dimensional, that all the tissue thermal properties are homogeneous, and that blood perfusion would not influence the heating process. The thermal diffusivity of the tumor and tissue is 8.85×10^{-8} m^2/s. In order for the therapy to be effective, the temperature must reach at least 50°C in the deepest portion of the tumor. In order to minimize damage to surrounding tissues, the temperature should not exceed 45°C at a depth of 8 mm. What are the bounds on the minimum time and the maximum time for which the therapy can satisfy these criteria?

If you want to include the effect of blood perfused through the skin at 37°C on the hyperthermia process, how would you develop the governing equation to model the process? How would you expect blood perfusion to change the two times you have calculated?

11.7.7 Heat Exchanger to Coagulate Blood

Plasma flows between two parallel plates with length L and width w. The distance between plates is 2 h. The inlet temperature is T_0. A constant flux of heat q is applied to the plasma through each wall by solar radiation. Our goal is to determine the length of plates necessary to raise the mean outlet temperature to the coagulation temperature T_c. Simplify (*do not solve*) the continuity, Navier–Stokes, and thermal energy equations for this case, making the following assumptions:

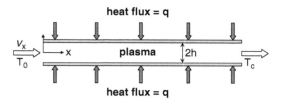

1. $v_y = 0$
2. $v_z = 0$
3. Steady-state, incompressible fluid
4. Neglect gravitational effects
5. No heat production in plasma
6. Nothing varies in the z-direction
7. Neglect axial conduction
8. Temperature is uniform at the inlet ($x = 0$: $T = T_0$)
9. Pressures at $x = 0$ and $x = L$ are known

11.8 Challenges

11.8.1 Kangaroo Care for Enhancing Neonatal Thermoregulatory Function

There is a long established medical literature that advocates skin-to-skin contact between newborn babies and mothers to compensate for frequent deficits in the ability of neonates to thermoregulate, as well as to establish emotional ties and facilitate attachment (see example articles at the end of this statement). Indeed, the American Academy of Pediatrics has recommended that healthy infants should be placed and remain in direct skin-to-skin contact with their mothers immediately after delivery until the first feeding is accomplished (Gartner et al. 2005). The preferred arrangement is skin-to-skin and chest-to-chest placement of the infant between the maternal breasts, sometimes augmented by covering with a preheated blanket, thus the descriptor "kangaroo care." There have been a number of clinical studies that document higher infant average skin and core temperatures during and subsequent to kangaroo care in comparison to babies who have been separated from their mothers. Kangaroo care is advocated as an acceptable and more effective alternative to placement in an incubator to combat hypothermia under normal circumstances.

In view of compelling data for its efficacy, kangaroo care is being adopted ever more widely. Examination of the literature finds that the most rigorous studies of kangaroo care present data on infant skin and core (usually measured rectally) temperatures over post-birth time, and there are some discussions of possible physiological mechanisms. However, a more comprehensive and quantitative understanding (derived via the perspective and methods of an engineer) of the kangaroo care phenomenon would be beneficial to its further development and more optimal and widespread implementation.

Your job in this challenge is to develop a strategy for formulating a model for the thermal effects of kangaroo care. This challenge should provide a rich opportunity for working collaboratively with your classmates and for invoking many of the tools that should now be in your arsenal of bioheat transfer skills. Application of the Generate Ideas Model for a series of iterative analyses should serve you well in addressing this challenge.

Färdig JA (1980) A comparison of skin-to-skin contact and radiant heaters in promoting neonatal thermoregulation. J Nurse Midwifery 25:19–28

Galligan M (2006) Proposed guidelines for skin-to-skin treatment of neonatal hypothermia. Am J Matern Child Nurs 31:298–304

Gartner L, Morton J, Lawrence R, Naylor A, O'Hare D, Schanler R, Eidelman AK (2005) Breastfeeding and the use of human milk. Pediatrics 115:496–506

Klaus J, Jerauld R, Fregers C, McAlpine W, Steffa M, Kennell J (1972) Maternal attachment: importance of first postpartum days. N Engl J Med 28:460–463

Walter MS, Boggs KM, Gudington-Hoe S, Price KM, Morrison B (2007) Kangaroo care at birth for full term infants: a pilot study. Am J Matern Child Nurs 32:375–381

References

Anvari B, Milner TE, Tanenbaum BS, Kimel S, Svaasand LO, Nelson JS (1995) Selective cooling of biological tissues: application for the thermally mediated therapeutic procedures. Phys Med Biol 40:241–252

Bhowmick S, Coad JE, Swanlund DJ, Bischof JC (2004) In vitro thermal therapy of AT-1 dunning prostate tumors. Int J Hyperthermia 20:73–92

Carslaw HS, Jaeger JC (1959) Conduction of heat in solids, 2nd edn. Oxford University Press, London

Dewey WC (1994) Arrhenius relations from the molecule and cell to the clinic. Int J Hyperthermia 10:457–483

Dewey WC (2009) Hyperthermia classic commentary. Int J Hyperthermia 25:21–24

Dewey WC, Hopwood LE, Sapareto SA, Gerweck LE (1977) Cellular response to combinations of hypothermia and radiation. Radiat Biol 123:463–474

Diller KR, Klutke GA (1993) Accuracy analysis of the Henriques model for predicting thermal burn injury. In: Advances in bioheat and mass transfer, ASME, vol HTD 268, pp 117–123

Diller KR, Valvano JW, Pearce JA (2005) Bioheat transfer. In: Kreith F, Goswami Y (eds) The CRC handbook of mechanical engineering, 2nd edn. CRC Press, Boca Raton, pp 4-282–4-361

Eigner J, Boedtker H, Michaels G (1961) The thermal degradation of nucleic acids. Biochim Biophys Acta 51:156–168

Eyring H, Stearn AE (1938) The application of the theory of absolute reaction rates to proteins. In: Proceedings symposium on physical chemistry of proteins, Milwaukee, pp 253–270

Flock S, Smith L, Waner M (1993) Quantifying the effects on blood of irradiation with four different vascular-lesion lasers. In: Proceedings of the SPIE, vol 1882, pp 237–243

Fuentes D, Oden JT, Diller KR, Hazle JD, Elliott A, Shetty A, Stafford RJ (2009) Computational modeling and real-time control of patient-specific laser treatment of cancer. Ann Biomed Eng 37:763–782

He X, Bischof JC (2003) Quantification of temperature and injury response in thermal therapy and cryosurgery. Crit Rev Biomed Eng 31:355–422

He X, Megee S, Coad JE, Schmidlin F, Iaizzo PA, Swanlund DJ, Rudie E, Kluge S, Bischof JC (2004) Investigation of the thermal and tissue injury behavior in microwave thermal therapy using a porcine kidney model. Int J Hyperthermia 20:567–593

Henriques FC (1947) Studies of thermal injury. V. The predictability and signficcance of thermally induced rate processes leading to irreversible epidermal injury. Arch Pathol 23:489–502

Henriques FC, Moritz AR (1947) Studies of thermal injury. I. The conduction of heat to and through skin and the temperature attained therein: a theoretical and experimental investigation. Am J Pathol 23:531–549

Holman JP, Kendall CM (1993) Extended studies of spray cooling with Freon-113. Int J Heat Mass Transf 36:2239–2241

Jacques SL, Gaeeni MO (1989) Thermally induced changes in optical properties of heart. IEEE Eng Med Biol Mag 11:1199–1200

Landry J, Marceau N (1978) Rate-limiting events in hyperthermic cell killing. Radiat Res 75:573–585

Lawrence JC, Bull JP (1976) Thermal conditions which cause skin burns, Proceedings of the Institution of Mechanical Engineers. Eng Med 5:61–63

Maintland DJ, Walsh JT (1997) Quantitative measurements of linear birefringence during heating of native collagen. Lasers Surg Med 20:310–318

Martin H (1977) Heat and mass transfer between impinging gas jets and solid surfaces. Adv Heat Transf 13:1–60

McKenzie AL (1990) Physics of thermal processes in laser-tissue interaction. Phys Med Biol 35:1175–1209

Milligan AJ, Mietz JA, Leeper DB (1984) Effect of interstinal hyperthermia in the Chinese Hamster. Int J Radiat Oncol Biol Phys 10:259–263

References

Minkowycz WJ, Sparrow EM, Pletcher RH, Schneider GE (1988) Overview of basic numerical methods. In: Minkowycz WJ, Sparrow EM, Pletcher RH, Schneider GE (eds) Handbook of numerical heat transfer. Wiley, New York, pp 1–88

Moore TL, Lunt M, McManus B, Anderson ME, Herrick AL (2003) Seventeen-point dermal ultrasound scoring system – a reliable measure of skin thickness in patients with systemic sclerosis. Rheumatology 42:1559–1563

Moran K, Anderson P, Hutcheson J, Flock S (2000) Thermally induced shrinkage of joint capsule. Clin Orthop Relat Res 381:248–255

Moritz AR, Henriques FC (1947) Studies of thermal injury. II. The relative importance of time and surface temperature in the causation of cutaneous burns. Am J Pathol 23:695–720

Moussa NA, Tell EN, Cravalho EG (1979) Time progression of hemolysis of erythrocyte populations exposed to supraphysiological temperatures. J Biomech Eng 101:213–217

Pearce JA (2009) Relationship between Arrhenius models of thermal damage and the CEM 43 thermal dose. In: Energy-based treatment of tissue and assessment, Proceedings of the SPIE, vol 7181, pp 70104:1–15

Pearce JA, Thomsen S, Vijverberg H, McMurray T (1993) Kinetics for birefringence changes in thermally coagulated rat skin collagen. In: Proceedings of the SPIE, vol 1876, pp 180–186

Pikkula BM, Torres JH, Tunnell JW, Anvari B (2001) Cryogen spray cooling: effects of droplet size and spray density on heat removal. Lasers Surg Med 28:103–112

Sapareto SA (1982) The biology of hyperthermia in vitro. In: Nussbaum G (ed) Physical aspects of hyperthermia. American Institute Physics, New York

Sapareto SA, Dewey WC (1984) Thermal dose determination in cancer therapy. Int J Radiat Oncol Biol Phys 10:787–800

Seidenari S, Giusti G, Bertoni L, Magnoni C, Peliacani G (2000) Thickness and echnogenicity of the skin in children as assessed by 20-MHz ultrasound. Dermatology 201:218–222

Sevitt S (1957) Burns: pathology and therapeutic applications. Butterworth, London

Southwood WFW (1955) The thickness of the skin. Plast Reconstr Surg 15:423–429

Stoll AM (1960) A computer solution for determination of thermal tissue damage integrals from experimental data. IRE Trans Med Electron 7:355–358

Takata AN (1974) Development of criterion for skin burns. Aerosp Med 45:634–637

van Gemert MJC, Welch AJ (1989) Time constants in thermal laser medicine. Lasers Surg Med 9:405–421

Weaver JA, Stoll AM (1969) Mathematical model of skin exposed to thermal radiation. Aerosp Med 40:24–30

Welch AJ, Polhamus GD (1984) Measurement and prediction of thermal injury in the retina of Rhesus monkey. IEEE Trans Biomed Eng 31:633–644

Welch AJ, van Gemert MJC (2011) Optical-thermal response of laser-irradiated tissue, 2nd edn. Springer, New York

Welch AJ, Pearce JA, Diller KR, Yoon G, Cheong WF (1989) Heat generation in laser irradiated tissue. J Biomech Eng 111:62–68

Part V
Biological Mass Transport

Chapter 12
Mass Transfer Fundamentals

12.1 Average and Local Mass and Molar Concentrations

When we speak of mass transfer, we are generally referring to the movement of one or more molecular species relative to the others. Before we can describe this relative movement, we need to understand the most common ways of quantifying the presence of each species. Consider the closed system with volume V shown in Fig. 12.1 which contains three different molecular species A, B, and C, represented by three different colors.

The total amount of species A present in the system can be expressed in three ways: (1) the total number of molecules of species A, $\mathbb{N}_A$, (2) the total number of moles of species A, N_A, and (3) the total mass of species A, m_A. However, these are all related, since:

$$N_A = \frac{\mathbb{N}_A}{N_{AV}},$$
$$m_A = M_A N_A = \frac{M_A \mathbb{N}_A}{N_{AV}}, \quad (12.1)$$

where N_{AV} is Avogadro's number (6.02×10^{23} molecules/mole) and M_A is the molecular weight of species A. The average molar concentration of species A in the system is represented by $\bar{c}_A$, where the overbar indicates a volume-averaged value and c_A is the symbol for molar concentration of species A:

$$\bar{c}_A = \frac{N_A}{V}. \quad (12.2)$$

The average mass concentration $\bar{\rho}_A$ of species A can be written in a similar fashion:

$$\bar{\rho}_A = \frac{m_A}{V} = M_A \bar{c}_A. \quad (12.3)$$

Molecular transport of any of the species within the system shown in Fig. 12.1 will occur by diffusion if a concentration gradient exists for that species within the

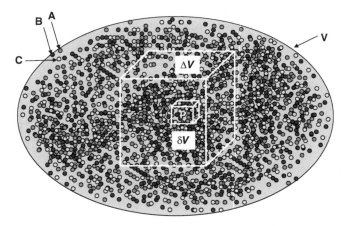

Fig. 12.1 Three molecular species in a closed volume

system. Thus, local concentrations, rather than average concentrations, are of interest in a large fraction of mass transfer applications. Referring to Fig. 12.1, we can define the *local molar concentration of species A* c_A in the region ΔV around a point P as:

$$c_A(x,y,z,t) = \lim_{\Delta V \to \delta V} \left\{ \frac{\Delta N_A}{\Delta V} \right\}, \tag{12.4}$$

where ΔV is the volume surrounding a point $P(x,y,z)$, ΔN_A is the number of moles of species A contained in the volume ΔV, and δV is the smallest volume surrounding P for which the region can be considered a continuum. This local concentration can also be written in terms of the number of molecules $\delta \mathbb{N}_A$ contained in the volume δV:

$$c_A = \frac{\delta \mathbb{N}_A}{N_{AV} \delta V}. \tag{12.5}$$

The local mass density ρ_A can be defined similarly to (12.4) with Δm_A replacing ΔN_A and δm_A replacing δN_A, or can be computed by multiplying (12.5) by the molecular weight of species A:

$$\rho_A(x,y,z,t) = \lim_{\Delta V \to \delta V} \left\{ \frac{\Delta m_A}{\Delta V} \right\} = \frac{\delta m_A}{\delta V} = M_A \frac{\delta c_A}{\delta V} = \frac{M_A \delta \mathbb{N}_A}{N_{AV} \delta V}. \tag{12.6}$$

The total local molar concentration c can be found by adding the molar concentrations of all species contained in the volume δV:

$$c(x,y,z,t) = \sum_{i=1}^{N_{species}} c_i(x,y,z,t). \tag{12.7}$$

12.1 Average and Local Mass and Molar Concentrations

N_{species} is the total number of different species present, which equals three for our particular example in Fig. 12.1. Similarly, the total local mass concentration, or mass density, ρ, can be found by summing the local mass concentrations of the individual species:

$$\rho(x,y,z,t) = \sum_{i=1}^{N_{\text{species}}} \rho_i(x,y,z,t). \tag{12.8}$$

The local average molecular weight for all species M is defined as the ratio of total mass concentration to total molar concentration:

$$M = \frac{\rho}{c}. \tag{12.9}$$

Substituting (12.5) and (12.6) into (12.9), we have:

$$M = \frac{\rho}{c} = \frac{\sum_{i=1}^{N_{\text{species}}} \rho_i}{\sum_{i=1}^{N_{\text{species}}} c_i} = \frac{\sum_{i=1}^{N_{\text{species}}} M_i \delta N_i}{\sum_{i=1}^{N_{\text{species}}} \delta N_i}. \tag{12.10}$$

Thus, one name given to M is the *number-averaged molecular weight*. Another way to write (12.9) is:

$$M = \frac{\rho}{c} = \frac{\sum_{i=1}^{N_{\text{species}}} M_i c_i}{c} = \sum_{i=1}^{N_{\text{species}}} M_i \left(\frac{c_i}{c}\right). \tag{12.11}$$

Consequently, a second name given to M is the *molar-averaged molecular weight*. The local mole fraction of species A is defined as x_A:

$$x_A = \frac{c_A}{c}. \tag{12.12}$$

Similarly, the local mass fraction ω_A of species A is defined as:

$$\omega_A = \frac{\rho_A}{\rho}. \tag{12.13}$$

Writing the mole fraction in terms of the mass fraction:

$$x_A = \frac{c_A}{c} = \frac{\rho_A/M_A}{\rho/M} = \omega_A \frac{M}{M_A}. \tag{12.14}$$

The sum of all mole fractions must equal unity. Adding the mole fractions in (12.14) for all species and solving for M, we find:

$$M = \frac{1}{\sum_{i=1}^{N_{\text{species}}} \frac{\omega_i}{M_i}}. \tag{12.15}$$

Substituting (12.15) into (12.14), we obtain an expression for the mole fraction of species A in terms of the mass fractions and molecular weights of all species:

$$x_A = \frac{\frac{\omega_A}{M_A}}{\sum_{i=1}^{N_{\text{species}}} \frac{\omega_i}{M_i}}. \tag{12.16}$$

This is particularly useful in finding mole fractions of species in liquids, where solution composition is often expressed as weight percentages. We can follow a similar procedure for finding the mass fraction in terms of the mole fractions:

$$\omega_A = \frac{\rho_A}{\rho} = \frac{M_A c_A}{Mc} = \frac{x_A M_A}{M}. \tag{12.17}$$

The sum of all mass fractions must also equal unity. Adding the mass fractions in (12.17) for all species and solving for M, we find:

$$M = \sum_{i=1}^{N_{\text{species}}} x_i M_i. \tag{12.18}$$

This is consistent with (12.11). Substituting (12.18) into (12.17), we obtain an expression for the mass fraction of species A in terms of the mole fractions and molecular weights of all species:

$$\omega_A = \frac{x_A M_A}{\sum_{i=1}^{N_{\text{species}}} x_i M_i}. \tag{12.19}$$

If the volume under investigation contains a gas, the symbol y_A is often used to represent the mole fraction rather than x_A:

$$y_A = (x_A)_{\text{gas}}. \tag{12.20}$$

If the gas is an ideal gas, then the *equation of state* is:

$$P = cRT, \tag{12.21}$$

12.1 Average and Local Mass and Molar Concentrations

where P is the pressure, c is the total molar concentration, R is the universal gas constant and T is the absolute temperature. The SI unit for pressure is the Pascal, which is equal to 1 N/m^2. In physiological applications pressure is often measured in units of Torr, which is equal to 1/760th of an atmosphere, and is very nearly equal to 1 mmHg. According to Dalton's Law, the total pressure is the sum of the partial pressures caused by each species, and thus the partial pressure P_A of species A is related to the molar concentration of species A as follows:

$$P_A = c_A RT. \quad (12.22)$$

Dividing (12.22) by (12.21), we find that the ratio of partial pressure of species A to total pressure is the mole fraction for species A in an ideal gas:

$$\frac{P_A}{P} = \frac{c_A}{c} = y_A. \quad (12.23)$$

Example 12.1.1 Composition of Alveolar Gas.

The partial pressures of the primary components of alveolar gas at 37°C are given below. If alveolar gas is considered to be an ideal gas, find the following quantities: (a) alveolar pressure, (b) mole fraction of each component, (c) the molar-averaged molecular weight, (d) mass fraction of each component, (e) molar concentration of each component, (f) alveolar gas total molar concentration, (g) alveolar gas density, and (h) mass concentration of each component.

Gas	Partial pressure (Torr)	Molecular weight
Nitrogen, N_2	569	28
Oxygen, O_2	104	32
Carbon dioxide, CO_2	40	44
Water vapor, H_2O	47	18

Solution. *Initial considerations:* The objective of this exercise is to apply the definitions introduced in Sect. 12.1 for the specific case of alveolar gas, which is assumed to be well-mixed and to obey the ideal gas law.

System definition and environmental interactions: As the gas is well-mixed and we are seeking only intensive properties, we can take any portion of the gas for our system. Alveolar gas interacts with its environment via gas exchange through the alveolar walls and ventilation at terminal bronchi. However, we are not concerned here with environmental interactions, but rather with the composition of the gas measured at a particular time.

Apprising the problem to identify governing relationships: The governing relationships are the ideal gas law, Dalton's law, and the definitions of species mass concentration, species molar concentration, mass fraction, mole fraction, average molar molecular weight, total molar concentration and mass density.

Analysis:

(a) According to Dalton's law, total pressure is simply the sum of the partial pressures of all of the components, which in this case is 760 Torr.

(b) The mole fraction of a component of an ideal gas, according to (12.23), is simply the ratio of the partial pressure of that component to the total pressure. In the case of N_2, this is $y_{N_2} = (569\,\text{Torr})/(760\,\text{Torr}) = 0.749$. Using the same approach, we find $y_{O_2} = 0.137$, $y_{CO_2} = 0.053$ and $y_{H_2O} = 0.062$.

(c) The molar average molecular weight of alveolar gas M_{alv} can be computed from (12.18):

$$M_{alv} = \sum_{i=1}^{4} y_i M_i = (0.749)(28) + (0.137)(32) + (0.053)(44)$$
$$+ (0.062)(18) = 28.8\,\text{g/mol}.$$

(d) The mass fraction of each species can be computed using (12.17). For N_2:

$$\omega_{N_2} = \frac{x_{N_2} M_{N_2}}{M_{alv}} = \frac{(0.749)(28)}{28.8} = 0.728.$$

This is smaller than the mole fraction of N_2 because the molecular weight of N_2 is lower than the mean molecular weight of the gas mixture. The mass fractions of oxygen and CO_2 will have mass fractions greater than their mole fractions since they have molecular weights greater than the mean: $\omega_{O_2} = 0.152$, $\omega_{CO_2} = 0.081$. The mass fraction of water vapor in alveolar gas is considerably lower than its mole fraction, $\omega_{H_2O} = 0.039$.

(e) The molar concentration of each species in an ideal gas can be computed using (12.22). The universal gas constant is $R = 8.314 \times 10^7\,\text{g cm}^2\,\text{s}^{-2}\,\text{mol}^{-1}\,\text{K}^{-1}$ and the absolute temperature is $T = 273 + 37 = 310\,\text{K}$. Substituting these into (12.22) for nitrogen:

$$c_{N_2} = \frac{P_{N_2}}{RT} = \frac{(569\,\text{Torr})(1333.2\,\text{g cm}^{-1}\,\text{s}^{-2}\,\text{Torr}^{-1})}{(8.314 \times 10^7\,\text{g cm}^2\,\text{s}^{-2}\,\text{mol}^{-1}\,\text{K}^{-1})(310\,\text{K})}$$
$$= 2.94 \times 10^{-5}\,\text{mol/cm}^3.$$

Repeating this calculation for the other species: $c_{O_2} = 5.38 \times 10^{-6}\,\text{mol/cm}^3$, $c_{CO_2} = 2.07 \times 10^{-6}\,\text{mol/cm}^3$ and $c_{H_2O} = 2.43 \times 10^{-6}\,\text{mol/cm}^3$.

(f) The total alveolar molar concentration c_{alv} can be found from (12.7) by adding the molar concentrations of all of the species:

$$c_{alv} = \sum_{i=1}^{4} c_i = (29.4 + 5.38 + 2.07 + 2.43) \times 10^{-6}\,\text{mol/cm}^3$$
$$= 3.93 \times 10^{-5}\,\text{mol/cm}^3.$$

12.2 Phase Equilibrium

(g) Alveolar gas density ρ_{alv} from (12.11) is simply the product of the total molar concentration c_{alv} and the mean molecular weight M_{alv}:

$$\rho_{alv} = c_{alv} M_{alv} = (3.93 \times 10^{-5} \text{ mol/cm}^3)(28.8 \text{ g/mol}) = 1.13 \times 10^{-3} \text{ g/cm}^3.$$

(h) The mass concentration of each species can be computed from the definition of the mass fraction. For nitrogen:

$$\rho_{N_2} = \omega_{N_2} \rho_{alv} = (0.728)(1.13 \times 10^{-3} \text{ g/cm}^3) = 8.23 \times 10^{-4} \text{ g/cm}^3.$$

Using the same procedure, $\rho_{O_2} = 1.72 \times 10^{-4}$ g/cm^3, $\rho_{CO_2} = 9.15 \times 10^{-5}$ g/cm^3 and $\rho_{H_2O} = 4.41 \times 10^{-4}$ g/cm^3.

Examining and interpreting the results: Note that the mass fraction of a species can be greater than, less than, or equal to the mole fraction of that species, depending on the value of the molecular weight of the species relative to the molar average molecular weight. A good check on your computations of mass fractions and mole fractions is to be sure that the sum of all mass fractions and the sum of all mole fractions equal unity. This is true for our computations, within roundoff error. The density of alveolar gas is very similar to the density of air at the same temperature. The greater concentration of CO_2 in alveolar gas tends to raise the density, but the lower concentration of O_2 and the presence of saturated water vapor tend to reduce the density of alveolar gas.

12.2 Phase Equilibrium

The terms "species mass transport" and "mass transfer" imply the movement of one molecular species relative to the others. What are appropriate "driving forces" that cause species to move from one region to another? Before we can understand the factors that cause species to move at different rates, it is useful to first study how species separate between phases under equilibrium conditions. If a system is in equilibrium with its surroundings, then there is no net exchange of mass, momentum or energy with its surroundings. If multiple species are present, then equilibrium can only exist if there is no net mass flow of any of the individual species through the system boundaries. This includes electrically charged species, so no net current flow will exist between two regions that are in equilibrium.

12.2.1 Liquid–Gas Equilibrium

Let us begin by considering the equilibrium between the gas and liquid phases of a single species A. We perform the following experiment. A rigid container is partially filled with pure liquid A, and all gas above the liquid is removed. We measure the pressure in the space above the liquid surface and maintain the container at constant temperature. As time progresses, some of the molecules in the liquid phase will vaporize and the pressure in the gas phase will begin to rise. After a sufficient time has elapsed, the pressure will stabilize and the gas and the liquid will be in equilibrium. The equilibrium pressure established in the gas phase is known as the vapor pressure. If we raise the temperature of the system, the vapor pressure will rise. If we repeat our experiment with a different liquid, we will find a different relationship between the vapor pressure and the equilibrium temperature. Relationships for common liquids are provided in Appendix C.

Can we predict the gas composition that will be in equilibrium with mixtures of different liquid species? If the mixtures are ideal, we can apply *Raoult's Law* that relates the partial pressure of each component to the vapor pressure $P_{vap,A}$ above pure component A at the same temperature and the mole fraction of that component in the liquid:

$$P_A = P_{vap,A} x_A. \tag{12.24}$$

Dividing (12.24) by the total pressure provides an equilibrium relationship between the mole fractions of component A in the gas and liquid phases:

$$y_A = \left(\frac{P_{vap,A}}{P}\right) x_A \tag{12.25}$$

We have used Dalton's law (12.22) in developing (12.25). Therefore, this applies only in the situation where an ideal gas is in equilibrium with an ideal solution. This may be closely approximated in cases where the molecular structures of the components are similar, such as benzene and toluene. However, application of Raoult's Law is inappropriate when the structures are quite different or dissociate into ions in the mixture, such as a mixture of HCl and water. Since biological solutions are aqueous solutions containing many different species, including electrolytes, macromolecules and dissolved gases, it is unlikely that (12.25) is valid in biological applications, except perhaps in very dilute solutions. However, the form of the relationship suggests the use of an empirical relationship between the mole fractions in the gas and liquid phases. This empirical relationship is known as *Henry's Law*:

$$y_A = \phi_A x_A. \tag{12.26}$$

The coefficient ϕ_A is known as the *Henry's Law constant*, which must be determined experimentally at a particular pressure and temperature. Other ways of writing Henry's law are also found in the literature. Two of the most common are:

$$P_{A,gas} = H_A x_A, \tag{12.27}$$

12.2 Phase Equilibrium

$$c_{A,liq} = \alpha_{A,liq} P_{A,gas}. \tag{12.28}$$

The coefficient H_A is also known as a Henry's Law constant and $\alpha_{A,liq}$ is generally referred to as a *solubility coefficient* for gas A in liquid liq. The student needs to pay careful attention to which form of Henry's law is used and to use an appropriate set of units. The coefficients H_A and $\alpha_{A,liq}$ are related to ϕ_A as follows:

$$H_A = \phi_A P, \tag{12.29}$$

$$\alpha_{A,liq} = \frac{c_{liquid}}{\phi_A P}, \tag{12.30}$$

where c_{liquid} is the total molar concentration of the liquid. For biological systems, where pressure is nearly constant, the coefficients in all of the forms of Henry's law remain relatively constant.

In blood gas applications, the amount of gas dissolved in blood plasma or in blood is of interest. This is often expressed as a *volume concentration* or volume fraction C_A^*, defined as the volume of pure dry gas A (ΔV_A) that physically dissolves in a volume of liquid (V_{liquid}) under equilibrium conditions. The number of moles of A that dissolves in the liquid (ΔN_A) is equal to the product of ΔV_A and the total molar concentration in the gas phase, c_{gas}. This can be used to define C_A^* in terms of the molar concentration of A or the mole fraction of A in the liquid:

$$C_A^* \equiv \frac{\Delta V_A}{V_{liquid}} = \frac{\Delta N_A}{c_{gas} V_{liquid}} = \frac{(c_A)_{liquid}}{c_{gas}} = \frac{c_{liquid}}{c_{gas}} x_A. \tag{12.31}$$

If we are dealing with an ideal gas, then $c_{gas} = P/RT$. The total molar concentration at a particular temperature and pressure is independent of the composition of the gas. At standard temperature and pressure conditions (STP), $P = 1$ atm, $T = 273$ K and $c_{gas} = 1$ mol/(22.4 L) $= 0.0464$ mol/L. A form of Henry's Law written in terms of C_A^* at STP is used in blood gas computations:

$$C_A^* = \alpha_{A,liq}^* P_A, \tag{12.32}$$

where $\alpha_{A,liq}^*$ is the *Bunsen solubility coefficient* of gas A in the liquid, with typical units of (ml gas A at STP)(ml liquid)$^{-1}$(atm partial pressure of A)$^{-1}$. This is related to the Henry's Law constant ϕ_A:

$$\alpha_{A,liq}^* = \frac{1}{\phi_A P}\left(\frac{c_{liquid}}{c_{gas}}\right) = \frac{\alpha_{A,liq}}{c_{gas}}. \tag{12.33}$$

The dissolved gas volume computed with (12.32) is the volume of gas dissolved in the liquid corrected to STP. In an ideal liquid, the Henry's Law coefficient ϕ_A is simply the ratio of the vapor pressure of A to total pressure. Since vapor pressure increases with temperature, the solubility of a gas in a liquid will decrease with increasing temperature. We can compare the concentration of A in the liquid phase to the concentration of A in the gas phase by inserting the definition of the mole fraction into Henry's Law:

$$(c_A)_{\text{liquid}} = \left(\frac{1}{\phi_A} \frac{c_{\text{liquid}}}{c_{\text{gas}}}\right)(c_A)_{\text{gas}} = \Phi_{A,\text{liquid,gas}}(c_A)_{\text{gas}}, \quad (12.34)$$

where $\Phi_{A,\text{liqid,gas}}$ is defined as a *partition coefficient*, which represents the equilibrium concentration of species A in the liquid relative to the concentration in the gas. For an ideal gas, the liquid–gas partition coefficient is:

$$\Phi_{A,\text{liquid,gas}} = \left\{\frac{(c_A)_{\text{liquid}}}{(c_A)_{\text{gas}}}\right\}_{\text{equil}} = \frac{\alpha_{A,\text{liq}} P_{A,\text{gas}}}{c_{\text{gas}}(P_{A,\text{gas}}/P)} = \frac{\alpha_{A,\text{liq}}}{c_{\text{gas}}/P} = \frac{\alpha_{A,\text{liq}}}{\left(\frac{P}{RT}\right)\frac{1}{P}} = RT\alpha_{A,\text{liq}}.$$

(12.35)

Therefore, the partition coefficient is directly proportional to the solubility of A in the liquid and the absolute temperature. Alternatively, we can write (12.35) in terms of the Bunsen solubility coefficient, to show $\Phi_{A,\text{liquid,gas}} = \alpha^*_{A,\text{liq}} P$.

Example 12.2.1.1 Plasma-CO_2 Solubility and Partition Coefficient.
Consider plasma to be in equilibrium with alveolar gas at 1 atm and 37°C. The gas has the same composition as given in Example 12.1.1. The Bunsen solubility coefficient for CO_2 in plasma at 37°C is $\alpha^*_{CO_2,\text{plasma}} = 0.57$ (ml CO_2) (ml plasma)$^{-1}$(atmosphere of partial pressure)$^{-1}$. Find (a) the volume fraction of CO_2 in plasma, (b) the equilibrium concentration of CO_2 in plasma, (c) the solubility coefficient for CO_2 in plasma, $\alpha_{CO_2,\text{plasma}}$, (d) the gas–liquid partition coefficient for CO_2, (e) the mole fraction of CO_2 in plasma and (f) the mass fraction of CO_2 in plasma.

Solution. *Initial considerations:* As the molar concentrations of species other than water in plasma are small, we will assume that the total molar concentration of plasma is the same as the total molar concentration of water. We will also assume that alveolar gas and plasma are in equilibrium and that plasma is well mixed, so no spatial variations of solutes exist.

System definition and environmental interactions: Since our goal is to find several intensive properties of plasma, the system of interest is any representative sample of plasma. Since alveolar gas and plasma are in equilibrium, there is no exchange of any species across the gas–liquid interface.

Apprising the problem to identify governing relationships: Henry's law relates the equilibrium concentration of CO_2 in plasma to the partial pressure of the CO_2 in alveolar gas. The other relationships needed to find a solution are the definitions

12.2 Phase Equilibrium

of the solubility coefficient, molar concentration, volume fraction, partition coefficient, mass fraction and mole fraction.

Analysis:

(a) The volume fraction of CO_2 in plasma can be computed using (12.32):

$$C^*_{CO_2} = \alpha^*_{CO_2,plasma} P_{CO_2} = \left(0.57 \text{ ml}_{CO_2} \text{ ml}^{-1}_{plasma} \text{ atm}^{-1}\right) \left(\frac{40 \text{ Torr}}{760 \text{ Torr/atm}}\right)$$

$$= 0.03 \frac{\text{ml}_{CO_2}}{\text{ml}_{plasma}}.$$

(b) The total alveolar gas molar concentration was found in Example 12.1.1 to be:

$$c_{gas} = 3.93 \times 10^{-5} \text{ mol/cm}^3.$$

Substituting the values for c_{gas} and $C^*_{CO_2}$ into (12.31), we can find the molar concentration of CO_2 in the plasma:

$$c_{CO_2,plasma} = c_{gas} C^*_{CO_2} = (3.93 \times 10^{-5} \text{ mol/ml}) (0.03 \text{ ml/ml}_{plasma})$$

$$= 1.18 \times 10^{-6} \text{ mol/cm}^3.$$

(c) The solubility coefficient can be found from (12.28):

$$\alpha_{CO_2,plasma} = \frac{c_{CO_2,plasma}}{P_{CO_2}} = \frac{1.18 \times 10^{-6} \text{ mol/ml}}{40 \text{ Torr}} = 2.95 \times 10^{-8} \text{ mol ml}^{-1} \text{ Torr}^{-1}.$$

(d) In Example 12.1.1 we found the concentration of CO_2 in alveolar gas to be 2.069×10^{-6} mol/cm^3. The gas/liquid partition coefficient for CO_2 is determined from the ratio of equilibrium concentrations in the gas and plasma:

$$\Phi_{CO_2,gas,plasma} = \frac{c_{CO_2,gas}}{c_{CO_2,plasma}} = \frac{2.069 \times 10^{-6} \text{ mol/cm}^3}{1.18 \times 10^{-6} \text{ mol/cm}^3} = 1.75.$$

(e) In order to compute the mole fraction of CO_2 in plasma, we must know the total molar concentration of plasma. We approximate this as the molar concentration of water, which is equal to the mass density of water at 37°C divided by the molecular weight of water:

$$c_{plasma} = c_{H_2O} = \frac{\rho_{H_2O}}{M_{H_2O}} = \frac{0.993 \text{ g/cm}^3}{18 \text{ g/mol}} = 0.05517 \text{ mol/cm}^3.$$

The mole fraction of CO_2 in plasma is:

$$x_{CO_2,plasma} = \frac{c_{CO_2,plasma}}{c_{plasma}} = \frac{1.18 \times 10^{-6} \text{ mol/cm}^3}{0.05517 \text{ mol/cm}^3} = 2.14 \times 10^{-5}.$$

(f) The mass fraction of CO_2 in plasma can be found from the mole fraction using (12.17), assuming that the molecular weight of plasma is approximately the molecular weight of water:

$$\omega_{CO_2,\text{plasma}} = x_{CO_2,\text{plasma}} \frac{M_{CO_2}}{M_{\text{plasma}}} \approx (2.14 \times 10^{-5}) \left(\frac{44}{18}\right) = 5.23 \times 10^{-5}.$$

Examining and interpreting the results: From the partition coefficient, we see that a given volume of alveolar gas will contain 1.75 times more CO_2 than the same volume of plasma. However, since the total number of moles per ml is 1,400 times larger in plasma than in alveolar gas, the mole fraction of CO_2 in plasma is smaller than in alveolar gas by a factor of 4×10^{-4} (i.e., $1/(1.75 \times 1,400)$).

Example 12.2.1.2 Diving and the Bends.
SCUBA divers who surface rapidly after spending significant times at depths greater than 40 feet may experience decompression sickness, or "the bends". This is because the SCUBA apparatus delivers compressed air from a tank to the lungs at a gauge pressure equal to the hydrostatic pressure surrounding the diver. Consider an 80 kg diver who spends a significant time in fresh water at a depth of 100 feet, such that the nitrogen he breathes at that depth equilibrates with body tissues. The Bunsen solubility coefficient for N_2 is 0.012 (ml CO_2)(ml tissue)$^{-1}$(atmosphere of partial pressure)$^{-1}$. How much excess (super saturated) N_2 will be present in the body if the diver rises quickly to the surface? From this analysis, what might cause the pain associated with the bends?

Solution. *Initial considerations:* Since N_2 comprises about 79% of the compressed gas, we will ignore the effects of other gases in our analysis. We will assume that N_2 in the tissue is in equilibrium with N_2 in the airways of the lungs.

System definition and environmental interactions: The system of interest is an arbitrary volume of tissue.

Apprising the problem to identify governing relationships: The concentration of gas that is dissolved in tissue at a gas–liquid interface is given by Henry's law. The total pressure in the lung airway is equal to the hydrostatic pressure at the current depth of the diver.

Analysis: The hydrostatic pressure at 100 feet is:

$$P = \rho g h = \left(1 \frac{\text{g}}{\text{cm}^3}\right) (980 \,\text{cm s}^{-2}) (100 \,\text{ft}) (30.48 \,\text{cm ft}^{-1}) (7.5 \times 10^{-4} \,\text{Torr cm s}^2 \text{g}^{-1}),$$
$$P = 2,240 \,\text{Torr}.$$

12.2 Phase Equilibrium

This is almost three atmospheres. The molar concentration of nitrogen in tissue over and above that at sea level can be computed from (12.31) and (12.32):

$$c_{N_2,\text{tissue}} = c_{\text{gas}} C^*_{N_2} = \left(\frac{P}{RT}\right) \alpha^*_{N_2,\text{tissue}} P_{N_2}.$$

Assuming that the mole fraction of nitrogen is 0.79 in the compressed air tank, the partial pressure of N_2, over and above that at sea level, is 0.79(2,240 Torr) = 1,770 Torr. The additional nitrogen that dissolves in the tissue over and above the amount dissolved at sea level is:

$$c_{N_2,\text{tissue}} = \left(\frac{2,240 \text{ Torr}}{(63,260 \text{ Torr ml mol}^{-1} \text{ K}^{-1})(310 \text{ K})}\right)\left(0.012 \frac{\text{ml}}{\text{ml}_{\text{tissue}} \text{ atm}}\right)$$

$$\times \left(\frac{1 \text{ atm}}{760 \text{ Torr}}\right)(1,770 \text{ Torr}),$$

$$c_{N_2,\text{tissue}} = 3.2 \times 10^{-6} \text{ mol/ml}_{\text{tissue}}.$$

The mass of excess N_2 can be estimated if we assume the body to consist primarily of water, so that the tissue volume of an 80 kg man is about 80,000 ml:

$$\text{mass}_{N_2} = M_{N_2} c_{N_2,\text{tissue}} V_{\text{tissue}} = \left(28 \frac{\text{g}}{\text{mol}}\right)\left(3.2 \times 10^{-6} \frac{\text{mol}}{\text{ml}}\right)(80,000 \text{ ml}) = 7.16 \text{ g}.$$

Examining and interpreting the results: Therefore, over 7 g of excess N_2 will come out of solution in the form of N_2 bubbles if the body is suddenly raised to sea level from a depth of 100 feet. These bubbles will cause tremendous pain, primarily in the joints, headache, dizziness and other symptoms.

Additional comments: Divers avoid these complications by ascending to the surface in stages, waiting long enough at each depth to breathe out excess N_2.

Example 12.2.1.3 Trout Survival in Warm Waters or High Altitude.
The minimum dissolved oxygen concentration (DO) in lakes and streams needed to sustain trout in a healthy state is 5 mg/L. The relationship between solubility of oxygen in water and temperature is given in the graph below. Consider two lakes at different locations that are in equilibrium with the air above them. The mole fraction of oxygen is 0.209 in both the cases. Lake A is at sea level and 20°C, while lake B is at an altitude of 12,000 feet (total pressure = 480 Torr) and 30°C. Will trout have difficulty surviving in either of these lakes?

Solution. *Initial considerations:* We will assume equilibrium between oxygen in the atmosphere and oxygen in the water. The amount of oxygen in the atmosphere is reduced at higher elevations, and the solubility coefficient for oxygen in water declines with temperature.

System definition and environmental interactions: The system of interest is the water in each lake. As we assume that the oxygen is well mixed and in equilibrium, the intrinsic properties of any water sample should be representative of the lake as a whole.

Apprising the problem to identify governing relationships: The relationship between dissolved oxygen in the lake water and partial pressure of oxygen in the atmosphere is given by Henry's law.

Analysis: In both the cases, the saturated oxygen concentration in water can be computed from Henry's law as follows:

$$\rho_{O_2} = M_{O_2} c_{O_2} = M_{O_2} \alpha_{O_2,\text{water}} P_{O_2}.$$

For the sea level case, $P_{O_2} = 0.209(760 \text{ Torr}) = 159$ Torr and the solubility of O_2 in water at 20°C from Fig. 12.2 is 1.8×10^{-6} mmol ml^{-1} Torr^{-1}. The mass concentration of O_2 in water under these conditions is:

$$\rho_{O_2} = \left(32 \frac{\text{mg}}{\text{mmol}}\right) \left(1.8 \times 10^{-6} \frac{\text{mmol}}{\text{ml Torr}}\right) (159 \text{ Torr}) \left(\frac{1,000 \text{ ml}}{\text{L}}\right) = 9.16 \text{ mg/L}.$$

This is above the minimum level of dissolved oxygen for trout survival. For the high altitude case, $P_{O_2} = 0.209(480 \text{ Torr}) = 100$ Torr, and the solubility of O_2 in water at 30°C is 1.5×10^{-6} mmol ml^{-1} Torr^{-1}. The mass concentration of O_2 in water under these conditions is:

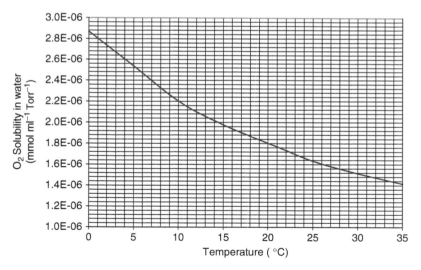

Fig. 12.2 Oxygen solubility in water vs. temperature

12.2 Phase Equilibrium

$$p_{O_2} = \left(32 \frac{mg}{mmol}\right)\left(1.5 \times 10^{-6} \frac{mmol}{ml\, Torr}\right)(100\, Torr)\left(\frac{1{,}000\, ml}{L}\right) = 4.8\, mg/L.$$

Trout would have considerable difficulty surviving under these conditions.

Examining and interpreting the results: The combination of less oxygen in the air at high altitude and the higher temperature both reduce the dissolved oxygen concentration, making the survival of trout at this altitude tenuous.

Additional comments: In reality, the concentration of oxygen in the lakes may be different than the saturation concentration. The metabolic rate of fish increases with temperature, and this tends to reduce the oxygen concentration. Algae and other aquatic plants also consume oxygen, but during daylight hours generally produce more oxygen via photosynthesis than they consume. Another factor tending to favor trout survival is that the temperature decreases and the pressure increases at greater depths. The solubility of oxygen in water increases with increasing pressure, so more oxygen is available in the deeper regions of the lake.

12.2.2 Liquid–Liquid, Gas–Solid, Liquid–Solid, Solid–Solid Equilibrium

Empirical relationships similar to Henry's law for gas–liquid partitioning can be applied to the equilibrium distribution of soluble components between any two phases. Thus, for phase equilibrium of a solute between two immiscible liquids, two solids, a solid and gas, or a solid and liquid, the concentrations in the two phases can usually be related by an expression similar to (12.34):

$$(c_A)_1 = \Phi_{A12}(c_A)_2. \tag{12.36}$$

The partition coefficient or distribution coefficient Φ_{A12} is the equilibrium ratio of the molar concentrations of species A in region 1 relative to the molar concentration of species A in region 2. The subscripts 1 and 2 are usually not included in the definition, but this can lead to difficulty in interpreting solubility data. The concentration of A in region 2 can be expressed in a similar manner by interchanging the subscripts in (12.36):

$$(c_A)_2 = \Phi_{A21}(c_A)_1. \tag{12.37}$$

The partition coefficient Φ_{A21} is the equilibrium ratio of the molar concentrations of species A in region 2 relative to the molar concentration of species A in region 1. Substituting (12.37) into (12.36), we find that Φ_{A21} is simply the inverse of Φ_{A12}:

$$\Phi_{A21} = \frac{1}{\Phi_{A12}}. \tag{12.38}$$

Although (12.36) and (12.37) relate the molar concentrations in the two regions, multiplying both sides by the molecular weight of species A provides similar relationships between the mass concentrations of species A on each side of the interface:

$$(\rho_A)_1 = \Phi_{A12}(\rho_A)_2, \tag{12.39}$$

$$(\rho_A)_2 = \Phi_{A21}(\rho_A)_1. \tag{12.40}$$

There are exceptions to these equilibrium relationships, particularly in solid–fluid interfaces where adsorption of species A may occur at the interface. In such cases, the adsorbed concentration at the interface may be related to the fluid concentration raised to a power n, where n must be determined empirically. In many cases, a linear relationship can be used if concentration changes are small.

Example 12.2.2.1 Equilibrium Between Plasma and an Artificial Membrane.
We wish to find the partition coefficient between oxygen in plasma and oxygen in a polycarbonate membrane under equilibrium conditions with $P = 1$ atm and $T = 37°C$. The partial pressure of oxygen in plasma is 155 Torr. The Bunsen solubility coefficients at 37°C are 0.022 ml (STP) ml^{-1} atm^{-1} for oxygen in plasma and 0.055 ml (STP) ml^{-1} atm^{-1} for oxygen in polycarbonate.

Solution. *Initial considerations:* Since the system is in equilibrium, both the plasma and the polycarbonate membrane behave as though they are in equilibrium with the same gas at 37°C.

System definition and environmental interactions: To find the partition coefficient, we must consider two systems: the plasma and the membrane material. As we are interested in finding the equilibrium concentration in each, an intrinsic property, we do not need to be concerned with the size of either system. We will consider equilibrium between (1) a gas and plasma and (2) the same gas and the membrane material.

Apprising the problem to identify governing relationships: We can use Henry's law to find equilibrium concentrations in each material and take their ratio to find the partition coefficient.

Analysis: Both materials are exposed to the same partial pressure of oxygen. Rewriting (12.28):

$$P_{O_2,\text{gas}} = \frac{c_{O_2,\text{plasma}}}{\alpha_{O_2,\text{plasma}}} = \frac{c_{O_2,\text{membrane}}}{\alpha_{O_2,\text{membrane}}}.$$

Consequently, the partition coefficient is:

$$\Phi_{O_2,\text{plama},\text{membrane}} = \frac{c_{O_2,\text{plasma}}}{c_{O_2,\text{membrane}}} = \frac{\alpha_{O_2,\text{plasma}}}{\alpha_{O_2,\text{membrane}}} = \frac{\alpha^*_{O_2,\text{plasma}}}{\alpha^*_{O_2,\text{membrane}}}.$$

12.2 Phase Equilibrium

Substituting values into this expression, we find:

$$\Phi_{O_2,\text{plama,membrane}} = \frac{0.022 \, \text{ml(STP)ml}^{-1} \, \text{atm}^{-1}}{0.055 \, \text{ml (STP) ml}^{-1} \, \text{atm}^{-1}} = 0.4.$$

Examining and interpreting the results: Consequently, since oxygen is more soluble in polycarbonate, the solid membrane holds 2.5 times more oxygen per unit volume than does the plasma.

Example 12.2.2.2 Decaffeination Process.
The solubility of caffeine in water is 22 mg/ml at 25°C, and the solubility of caffeine in an equal mixture of water and ethanol is 110 mg/ml. We wish to remove caffeine from a 100 ml sample containing 10 mg/ml using a caffeine-permeable membrane that separates this sample from 1,000 ml of a 50% water–ethanol mixture. The compartments on both sides of the membrane are well-mixed. After a long period of time, what will be the concentration of caffeine in each compartment? The membrane is very thin, so the amount of caffeine dissolved in the membrane can be neglected.

Solution. *Initial considerations:* We will assume that the two fluids reach equilibrium after sufficient time has passed.

System definition and environmental interactions: We will analyze the 100 ml system and the 1,000 ml system under equilibrium conditions. No caffeine will be transported across the membrane under equilibrium conditions.

Apprising the problem to identify governing relationships: Since the membrane is permeable to caffeine, we simply need to apply Henry's law to the two fluids when equilibrium is reached, along with conservation of caffeine between the two systems.

Analysis: The partition coefficient between the 50% ethanol–water mixture and water is the ratio of the solubility of caffeine in the two fluids:

$$\Phi_{O_2,\text{mixture,water}} = \frac{110 \, \text{mg/ml}}{22 \, \text{mg/ml}} = 5.0.$$

Therefore, when equilibrium is reached:

$$\rho_{\text{caffeine,mixture}} = 5.0 \left(\rho_{\text{caffeine,water}} \right).$$

If $m_{\text{caffein,water}}$ is the final mass of caffeine in the water compartment and $m_{\text{caffein,mixture}}$ is the final mass of caffeine in the ethanol–water mixture, then at equilibrium:

$$\frac{m_{\text{caffeine,mixture}}}{1,000 \, \text{ml}} = 5.0 \left(\frac{m_{\text{caffeine,water}}}{100 \, \text{ml}} \right) \quad \text{or} \quad m_{\text{caffeine,mixture}} = 50 \left(m_{\text{caffeine,water}} \right).$$

The initial mass of caffeine in the system m_{caffeine} is the amount of caffeine initially confined to the aqueous solution. Conservation of mass states that this must be equal to the sum of the masses of caffeine in each compartment after equilibrium is attained:

$$m_{\text{caffeine}} = (10\,\text{mg/ml})(100\,\text{ml}) = 1\,\text{g} = m_{\text{caffeine,mixture}} + m_{\text{caffeine,water}}$$
$$= 51\left(m_{\text{caffeine,water}}\right).$$

Examining and interpreting the results: Solving the above expression we find, the final distribution of caffeine to be 0.0196 g on the water side and 0.9804 g on the ethanol–water side. When the process is completed, over 98% of the caffeine will be removed from the original solution.

Additional comments: Because of its higher caffeine solubility, the water–ethanol mixture is preferable for removing caffeine. If water had been used to remove the caffeine instead of the water–ethanol mixture, then 9% of the caffeine would remain in the original solution. If we wish to learn how long it takes to reach equilibrium, it will be necessary to know the membrane permeability to caffeine and its surface area.

12.3 Species Transport Between Phases

Let us turn now to the analysis of nonequilibrium situations between two phases in contact. The molar flow of species A across the interface separating the phases W_A (mol/s) will be proportional to the surface area of the interface S and a driving force that represents a departure from equilibrium:

$$W_A \propto (\text{Driving force}) \times S. \tag{12.41}$$

Our task is to identify one or more suitable driving forces that can be used to predict the transfer of a component A from one phase to another. One possible candidate would be the difference in mole fraction between the two regions in contact. However, unless the Henry's Law coefficient ϕ_A is equal to unity, the mole fractions of species A will be different between two phases when they are in equilibrium. Since there is no flow of species A between phases in equilibrium, a difference in mole fraction between phases cannot be considered as a suitable driving force for the movement of species A between phases. Does species A move because there is a difference in concentration between the phases? In general, the partition coefficient Φ_{A12} will not be equal to unity, so there will be a difference in concentrations between phases in equilibrium. Therefore, a molar concentration difference across the interface cannot be an appropriate driving force for species mass transfer. We can use similar arguments to show that mass fractions and mass concentrations are generally different between phases under equilibrium conditions, and they cannot be considered to be appropriate driving forces for mass transfer between phases either.

12.3 Species Transport Between Phases

Let us adopt a coordinate system which is positive in the direction from system 1 to system 2. A suitable driving force must equal zero when the two phases are in equilibrium. A good candidate that arises from the equilibrium analysis in Sect. 12.2 is:

$$\text{Driving force} = (c_A)_1 - \Phi_{A12}(c_A)_2. \tag{12.42}$$

The driving force will be zero if there is no flow of species A, positive if the flow is positive (from region 1 to region 2) and negative if the flow is negative (from region 2 to region 1). Substituting this into (12.41) and introducing a proportionality factor, P_A:

$$W_A = P_A S[(c_A)_1 - \Phi_{A12}(c_A)_2]. \tag{12.43}$$

Multiplying both sides of (12.43) by the molecular weight of species A and defining w_A as the mass flow (e.g., g/s) of species A, we obtain:

$$w_A = P_A S[(\rho_A)_1 - \Phi_{A12}(\rho_A)_2]. \tag{12.44}$$

The coefficient P_A is the same for mass flow and molar flow and is known as the *permeability* of the interface to species A and is also known as an *overall mass transfer coefficient for species A*. This will depend on the relative resistance to mass transfer of species A in each phase, as we will discuss in Sect. 12.4.3. An alternate way to write (12.43) is:

$$W_A = -P_A^* S[(c_A)_2 - \Phi_{A21}(c_A)_1], \tag{12.45}$$

where

$$P_A^* = P_A \Phi_{A12}. \tag{12.46}$$

Φ_{A21} is given by (12.36). If the driving force in (12.45) is positive (i.e., the term in brackets), then the flow will be from region 2 to region 1, which is considered negative.

Example 12.3.1 Driving Force for O_2 Transport from Plasma to Membrane.

The polycarbonate membrane in Example 12.2.2.1 has a surface area of 1 m² and an overall mass transfer coefficient for O_2 of 2×10^{-3} cm/s. The initial concentration of oxygen in the membrane is 3×10^{-4} mol/L. The membrane is brought into contact with plasma having an oxygen concentration of 1.8×10^{-4} mol/L. Will O_2 move out of the membrane or into the membrane? At what rate?

Solution. *Initial considerations:* The direction of flow will depend on the sign of the driving force. The flow rate will depend on the magnitude of the driving force and the permeability-surface area product.

System definition and environmental interactions: We can analyze either the plasma or the membrane. The flow of oxygen out of the plasma will equal flow into the membrane, and vice versa.

Apprising the problem to identify governing relationships: The expression for species molar flow derived in this section can be used to determine the magnitude and direction of oxygen flow.

Analysis: Taking side 1 in (12.43) as plasma and side 2 as the membrane,

$$W_{O_2} = P_{O_2} S \left[(c_{O_2})_{plasma} - \Phi_{O_2,plasma,membrane} (c_{O_2})_{membrane} \right]$$

W_{O2} represents the molar flow of oxygen from plasma to the membrane. From Example 12.2.2.1, we find the partition coefficient $\Phi_{O_2,plasma,membrane} = 0.4$. Substituting the appropriate values:

$$W_{O_2} = \left(2 \times 10^{-3} \frac{cm}{s}\right) (10^4 \, cm^2) \left[\left(1.8 \times 10^{-7} \frac{mol}{cm^3}\right) - 0.4 \left(3 \times 10^{-7} \frac{mol}{cm^3}\right) \right],$$

$$W_{O_2} = 1.2 \times 10^{-6} \, mol/s.$$

Examining and interpreting the results: Therefore, the flow of oxygen across the interface will be from the plasma to the membrane, even though the membrane concentration is higher than the plasma concentration of oxygen.

Additional comments: Because a species can move randomly in a uniform medium, the net diffusive flux of the species always occurs from regions of high concentration to low concentration. Students often tend to apply this same reasoning to diffusion through nonuniform media as well. However, if the solubility of the species in the material is not uniform, then motion in one direction is more restricted than motion in another direction. Consequently, the motion is no longer completely random. Mass flow across an interface, where the solubilities are very different between the two phases, is just one example.

12.4 Species Transport Within a Single Phase

Within a single phase, the concentration of each species will be uniform under equilibrium conditions. The random movement of all the molecular species is responsible for maintaining this uniform distribution, so gradients in species concentration cannot be sustained. In nonequilibrium situations, each species can move at different rates, and concentration gradients can be established within the material. In this section, we will examine the movement of species A and the mechanisms responsible for its movement.

12.4 Species Transport Within a Single Phase

12.4.1 Species Fluxes and Velocities

The flux of any moving quantity is the rate per unit area at which that quantity passes through an area that is perpendicular to the direction of movement. The local molar flux of species A is the number of moles per second of species A that passes through a unit area perpendicular to the movement of species A. Let the local velocity of species A be v_A. Then the distance moved by species A in time Δt will be $\Delta L_A = v_A \Delta t$. The number of moles of species A, ΔN_A, that pass through an area A, perpendicular to the direction of flow of species A, is:

$$\Delta N_A = c_A (\Delta L_A A). \tag{12.47}$$

The *molar flux of species A*, N_A, can be found by dividing the number of moles by $A\Delta t$:

$$N_A = \frac{\Delta N_A}{A \Delta t} = c_A \left(\frac{\Delta L_A}{\Delta t}\right) = c_A v_A. \tag{12.48}$$

Thus, the molar flux of A is simply the product of the molar concentration of A and the local velocity of A. Multiplying each side of (12.48) by a unit vector in the direction of the velocity yields a vector equation for the molar flux:

$$\vec{N}_A = c_A \vec{v}_A. \tag{12.49}$$

We can compute the *mass flux of species A*, $\vec{n}_A$ by multiplying each side of (12.49) by the molecular weight of species A:

$$\vec{n}_A = M_A \vec{N}_A = \rho_A \vec{v}_A. \tag{12.50}$$

Note that molar fluxes will be designated by upper case letters and mass fluxes by lower case letters.

The average velocity of all species can be computed in two ways. The most common average velocity is the *mass average velocity*, $\vec{v}$. The mass average velocity is the sum of the species velocities weighted by their mass concentration and divided by the mass density of all species:

$$\vec{v} \equiv \frac{\sum_{i=1}^{N_{species}} \rho_i \vec{v}_i}{\sum_{i=1}^{N_{species}} \rho_i}. \tag{12.51}$$

Alternate ways of writing (12.51) are:

$$\vec{v} = \frac{\sum_{i=1}^{N_{species}} \vec{n}_i}{\rho} = \sum_{i=1}^{N_{species}} \omega_i \vec{v}_i. \qquad (12.52)$$

The second way we can compute the average velocity is to weight the species velocities by their molar concentrations rather than their mass concentrations. The *molar average velocity* $\vec{v}^*$ is defined as:

$$\vec{v}^* = \frac{\sum_{i=1}^{N_{species}} c_i \vec{v}_i}{\sum_{i=1}^{N_{species}} c_i}. \qquad (12.53)$$

Alternate ways of writing (12.53) are:

$$\vec{v}^* = \frac{\sum_{i=1}^{N_{species}} \vec{N}_i}{c} = \sum_{i=1}^{N_{species}} x_i \vec{v}_i. \qquad (12.54)$$

If the material consists of only a single species, then the mass average velocity will equal the molar average velocity. In previous chapters, we dealt mostly with pure fluids or solids, so it was not necessary to distinguish between these two velocities.

12.4.2 Diffusion Fluxes and Velocities

The velocities defined above are all relative to stationary coordinates. In mass transfer applications it is important to know how a particular species moves relative to either the mass average velocity or the molar average velocity. These are referred to as diffusion velocities. Four different diffusion fluxes can be defined, depending on whether it is a mass flux or molar flux and whether it is measured relative to the mass average velocity or the molar average velocity. The nomenclature used to distinguish the diffusion fluxes from the fluxes relative to a stationary coordinate system and their definitions are defined below.

$\vec{j}_A$: Mass flux of A relative to the mass average velocity

$$\vec{j}_A = \rho_A (\vec{v}_A - \vec{v}) = \vec{n}_A - \omega_A \sum_{j=1}^{N_{species}} \vec{n}_j. \qquad (12.55)$$

12.4 Species Transport Within a Single Phase

$\vec{J}_A^*$: Molar flux of A relative to the molar average velocity

$$\vec{J}_A^* = c_A(\vec{v}_A - \vec{v}^*) = \vec{N}_A - x_A \sum_{j=1}^{N_{species}} \vec{N}_j. \qquad (12.56)$$

$\vec{J}_A$: Molar flux of A relative to the mass average velocity

$$\vec{J}_A = c_A(\vec{v}_A - \vec{v}) = \frac{\vec{j}_A}{M_A} = \frac{1}{M_A}\left[\vec{n}_A - \omega_A \sum_{j=1}^{N_{species}} \vec{n}_j\right]. \qquad (12.57)$$

$\vec{j}_A^*$: Mass flux of A relative to the molar average velocity

$$\vec{j}_A^* = \rho_A(\vec{v}_A - \vec{v}^*) = M_A \vec{J}_A^* = M_A\left[\vec{N}_A - x_A \sum_{j=1}^{N_{species}} \vec{N}_j\right]. \qquad (12.58)$$

The fluxes $\vec{j}_A$ and $\vec{J}_A^*$ are used most commonly in biological mass transfer applications. $\vec{J}_A$ is also used with some frequency. The mass flux relative to the molar average velocity is rarely used and is shown for completeness.

12.4.3 Convective and Diffusive Transport

Equation (12.55) can be rearranged as follows:

$$\vec{n}_A = \rho_A \vec{v} + \vec{j}_A. \qquad (12.59)$$

Thus, the mass flux, $\vec{n}_A$, of species A can be considered to consist of two fundamental components. The first component, $\rho_A \vec{v}$, represents the *convective mass flux* of species A. This flux occurs because species A is carried along with the local mass average velocity of the fluid. This is also known as mass transport of species A resulting from bulk fluid motion. The second component, $\vec{j}_A$, is the diffusion mass flux of species A relative to the mass average velocity of the fluid. We will discuss mechanisms for this flux in more detail in the next section.

The convective and diffusive molar fluxes of species A can be found by dividing (12.59) by the molecular weight of species A:

$$\vec{N}_A = c_A \vec{v} + \vec{J}_A. \qquad (12.60)$$

An alternate way of splitting the molar flux into convective and diffusive components would be relative to the molar average velocity, rather than the mass average velocity. Rearranging (12.56):

$$\vec{N}_A = c_A \vec{v}^* + \vec{J}_A^*. \quad (12.61)$$

However, in most biotransport applications, the *convective molar flux* is based on the mass average velocity, as shown in (12.60).

12.4.4 Total Mass and Molar Fluxes

Total molar and mass fluxes can be found by summing the fluxes of all species. Upon application of the definitions of ρ, c, M, $\vec{v}$ and $\vec{v}^*$, the total fluxes are:

$$\vec{N} = \sum_{i=1}^{N_{species}} \vec{N}_i = c\vec{v}^*, \quad (12.62)$$

$$\vec{n} = \sum_{i=1}^{N_{species}} \vec{n}_i = \rho \vec{v}, \quad (12.63)$$

$$\sum_{i=1}^{N_{species}} \vec{j}_i = 0, \quad (12.64)$$

$$\sum_{i=1}^{N_{species}} \vec{J}_i^* = 0, \quad (12.65)$$

$$\sum_{i=1}^{N_{species}} \vec{J}_i = c(\vec{v}^* - \vec{v}), \quad (12.66)$$

$$\sum_{i=1}^{N_{species}} \vec{j}_i^* = \rho(\vec{v} - \vec{v}^*) = -Mc \sum_{i=1}^{N_{species}} \vec{J}_i. \quad (12.67)$$

Equation (12.62) defines the total molar flux, $\vec{N}$, and (12.63) defines the total mass flux $\vec{n}$. Equation (12.64) indicates that if one were to move with the mass average velocity, then the total mass flux relative to the mass average velocity is zero. Similarly, (12.65) shows that an observer moving with the molar average velocity will measure no net molar flux. The total molar flux measured by an observer moving with the mass average velocity is proportional to the difference between the molar average and mass average velocities. The total mass flux measured by an observer moving with the molar average velocity is proportional to the difference between the mass average and molar average velocities. Mass

12.4 Species Transport Within a Single Phase

fluxes are used in most biotransport applications. Molar fluxes are used primarily when chemical reactions occur in the material.

> **Example 12.4.1 Alveolar Fluxes.**
> Alveolar gas at a given position within an alveolus is composed primarily of O_2, CO_2, H_2O and N_2 at 37°C and 760 mmHg. Partial pressures at that location are: $P_{O2} = 100$ mmHg, $P_{CO_2} = 40$ mmHg and $P_{H_2O} = 47$ mmHg. Total O_2 flow into the lung is 250.3 ml (STP) per minute, total area perpendicular to the direction of gas flow is 36 m^2 and the respiratory quotient $(-N_{CO_2}/N_{O_2})$ is 0.8. Estimate the velocity of O_2 and CO_2 at the position of interest and find the fluxes of all gases relative to stationary coordinates, relative to the mass average velocity and relative to the molar average velocity. Assume alveolar gas is an ideal gas and the velocities of water vapor and nitrogen are zero.

Solution. *Initial considerations:* Our first goal is to estimate the average velocity of oxygen in an alveolus. We are given the total flow of oxygen into the lung and the total area perpendicular to the flow, so we can estimate the average flux of O_2 in lung alveoli. We are also given enough information to compute concentrations of all gases in the lung. Knowing the flux and concentration of oxygen, we can compute the average velocity of oxygen, and from the respiratory quotient we can compute the flux and velocity of CO_2. Next we can compute the mass average and molar average velocities. Finally, the fluxes of each species can be computed relative to each of these velocities.

System definition and environmental interactions: The system is the gas within an alveolus at the plane of interest in Fig. 12.3. We assume that O_2 flux in a single alveolus is the same as the average flux in all alveoli. Consequently, the flux in any alveolus can be estimated by taking the ratio of total oxygen flow to total cross-sectional area of all alveoli.

Apprising the problem to identify governing equations: The governing equations in this case are simply the definitions of the various fluxes, velocities and concentrations presented thus far in this chapter.

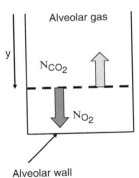

Fig. 12.3 Gas exchange in alveolus

Analysis: At STP, 1 mole occupies 22.4 L. The flow of O_2 (W_{O_2}) in units of mol/s is:

$$W_{O_2} = \left(250.3 \, \frac{\text{ml}}{\text{min}}\right)\left(\frac{1 \, \text{min}}{60 \, \text{s}}\right)\left(\frac{1 \, \text{mol}}{22,400 \, \text{ml}}\right) = 1.862 \times 10^{-4} \, \frac{\text{mol}}{\text{s}}.$$

The flux of oxygen toward the alveolar wall, relative to stationary coordinates, is found by dividing the oxygen flow by the area:

$$N_{O_2} = \frac{W_{O_2}}{A} = \frac{1.86 \times 10^{-4} \, \frac{\text{mol}}{\text{s}}}{36 \, \text{m}^2} = 5.17 \times 10^{-6} \, \frac{\text{mol}}{\text{m}^2 \, \text{s}}.$$

We assume that this flux is uniform throughout the lung. The mass flux of O_2 relative to a stationary coordinate system is:

$$n_{O_2} = M_{O_2} N_{O_2} = \left(32 \times 10^{-3} \, \frac{\text{kg}}{\text{mol}}\right)\left(5.17 \times 10^{-6} \, \frac{\text{mol}}{\text{m}^2 \, \text{s}}\right) = 1.655 \times 10^{-7} \, \frac{\text{kg}}{\text{m}^2 \, \text{s}}.$$

To find the average velocity of oxygen as it moves toward the wall, we use the definition of molar flux (12.48). However, we must first compute the concentration of oxygen from the partial pressure:

$$c_{O_2} = \frac{P_{O_2}}{RT} = \frac{\left(\frac{100}{760} \, \text{atm}\right)}{\left(8.205 \times 10^{-5} \, \frac{\text{m}^3 \, \text{atm}}{\text{mol K}}\right)(310 \, \text{K})} = 5.173 \, \frac{\text{mol}}{\text{m}^3}.$$

The alveolar concentrations of the other gases can be found in the same manner: $c_{CO_2} = 2.069 \, \text{mol/m}^3$, $c_{H_2O} = 2.431 \, \text{mol/m}^3$ and $c_{N_2} = 29.64 \, \text{mol/m}^3$. The total molar concentration can be found by adding the molar concentrations of all species or by applying the ideal gas law: $c = 39.32 \, \text{mol/m}^3$. The velocity of O_2 is computed from the definition of oxygen flux:

$$v_{O_2} = \frac{N_{O_2}}{c_{O_2}} = \frac{5.173 \times 10^{-6} \, \frac{\text{mol}}{\text{m}^2 \, \text{s}}}{5.173 \, \frac{\text{mol}}{\text{m}^3}} = 1 \times 10^{-6} \, \frac{\text{m}}{\text{s}}.$$

Since the respiratory quotient is 0.8:

$$N_{CO_2} = -0.8 N_{O_2} = -4.138 \times 10^{-6} \, \frac{\text{mol}}{\text{m}^2 \, \text{s}}.$$

The negative sign indicates that the flux of CO_2 is in the opposite direction from the O_2 flux, that is, from the alveolar wall toward the alveolar gas. The mass flux of

12.4 Species Transport Within a Single Phase

CO_2 with $M_{CO_2} = 44$ is -1.82×10^{-7} kg m^{-3} s^{-1}. The mass fluxes of water and N_2 are zero. The average velocity of CO_2 is:

$$v_{CO_2} = \frac{N_{CO_2}}{c_{CO_2}} = \frac{-4.138 \times 10^{-6} \frac{\text{mol}}{\text{m}^2 \text{s}}}{2.069 \frac{\text{mol}}{\text{m}^3}} = -2 \times 10^{-6} \frac{\text{m}}{\text{s}}.$$

As the water vapor and nitrogen are stationary, the molar average velocity can be computed from (12.53):

$$v^* = \frac{c_{O_2} v_{O_2} + c_{CO_2} v_{CO_2}}{c} = \frac{\left(5.173 \frac{\text{mol}}{\text{m}^3}\right)\left(10^{-6} \frac{\text{m}}{\text{s}}\right) + \left(2.069 \frac{\text{mol}}{\text{m}^3}\right)\left(-2 \times 10^{-6} \frac{\text{m}}{\text{s}}\right)}{39.32 \frac{\text{mol}}{\text{m}^3}}$$

$$= 2.63 \times 10^{-8} \frac{\text{m}}{\text{s}}.$$

Before we can compute the mass average velocity, we need to compute the mass concentrations for each species. For oxygen:

$$\rho_{O_2} = M_{O_2} c_{O_2} = \left(32 \times 10^{-3} \frac{\text{kg}}{\text{mol}}\right)\left(5.173 \frac{\text{mol}}{\text{m}^3}\right) = 0.1655 \frac{\text{kg}}{\text{m}^3}.$$

Mass concentrations for the other species with $M_{CO_2} = 44$, $M_{H_2O} = 18$ and $M_{N_2} = 28$ are: $\rho_{CO_2} = 0.091$ kg/m^3, $\rho_{H_2O} = 0.0438$ kg/m^3 and $\rho_{N_2} = 0.823$ kg/m^3. The total density is the sum of all mass concentrations: $\rho = 1.13$ kg/m^3. The mass average velocity is:

$$v = \frac{\rho_{O_2} v_{O_2} + \rho_{CO_2} v_{CO_2}}{\rho} = \frac{\left(0.1655 \frac{\text{kg}}{\text{m}^3}\right)\left(10^{-6} \frac{\text{m}}{\text{s}}\right) + \left(0.091 \frac{\text{kg}}{\text{m}^3}\right)\left(-2 \times 10^{-6} \frac{\text{m}}{\text{s}}\right)}{1.13 \frac{\text{kg}}{\text{m}^3}}$$

$$= -1.469 \times 10^{-8} \frac{\text{m}}{\text{s}}.$$

The molar fluxes relative to the molar average velocity are:

$$J^*_{O_2} = c_{O_2}(v_{O_2} - v^*) = \left(5.173 \frac{\text{mol}}{\text{m}^3}\right)\left(1 \times 10^{-6} \frac{\text{m}}{\text{s}} - 2.63 \times 10^{-8} \frac{\text{m}}{\text{s}}\right)$$

$$= 5.037 \times 10^{-6} \frac{\text{mol}}{\text{m}^2 \text{s}},$$

$$J^*_{CO_2} = c_{CO_2}(v_{CO_2} - v^*) = \left(2.069 \frac{\text{mol}}{\text{m}^3}\right)\left(-2 \times 10^{-6} \frac{\text{m}}{\text{s}} - 2.63 \times 10^{-8} \frac{\text{m}}{\text{s}}\right)$$

$$= -4.193 \times 10^{-6} \frac{\text{mol}}{\text{m}^2 \text{s}},$$

$$J^*_{H_2O} = c_{H_2O}(v_{H_2O} - v^*) = \left(2.431 \frac{mol}{m^3}\right)\left(0 - 2.63 \times 10^{-8} \frac{m}{s}\right)$$
$$= -6.4 \times 10^{-8} \frac{mol}{m^2 s},$$

$$J^*_{N_2} = c_{N_2}(v_{N_2} - v^*) = \left(29.64 \frac{mol}{m^3}\right)\left(0 - 2.63 \times 10^{-8} \frac{m}{s}\right) = -7.8 \times 10^{-7} \frac{mol}{m^2 s}.$$

The molar fluxes relative to the mass average velocity are:

$$J_{O_2} = c_{O_2}(v_{O_2} - v) = \left(5.173 \frac{mol}{m^3}\right)\left(1 \times 10^{-6} \frac{m}{s} - \left(-1.469 \times 10^{-8} \frac{m}{s}\right)\right)$$
$$= +5.249 \times 10^{-6} \frac{mol}{m^2 s},$$

$$J_{CO_2} = c_{CO_2}(v_{CO_2} - v) = \left(2.069 \frac{mol}{m^3}\right)\left(-2 \times 10^{-6} \frac{m}{s} - \left(-1.469 \times 10^{-8} \frac{m}{s}\right)\right)$$
$$= -4.168 \times 10^{-6} \frac{mol}{m^2 s},$$

$$J_{H_2O} = c_{H_2O}(v_{H_2O} - v) = \left(2.431 \frac{mol}{m^3}\right)\left(0 - \left(-1.469 \times 10^{-8} \frac{m}{s}\right)\right)$$
$$= +3.572 \times 10^{-8} \frac{mol}{m^2 s},$$

$$J_{N_2} = c_{N_2}(v_{N_2} - v*) = \left(29.64 \frac{mol}{m^3}\right)\left(0 - \left(-1.469 \times 10^{-8} \frac{m}{s}\right)\right)$$
$$= +4.354 \times 10^{-7} \frac{mol}{m^2 s}.$$

Finally, the mass fluxes relative to the mass average velocities are:

$$j_{O_2} = \rho_{O_2}(v_{O_2} - v) = M_{O_2}J_{O_2} = \left(32 \times 10^{-3} \frac{kg}{mol}\right)\left(5.249 \times 10^{-6} \frac{mol}{m^2 s}\right)$$
$$= 1.68 \times 10^{-7} \frac{kg}{m^2 s},$$

$$j_{CO_2} = M_{CO_2}J_{CO_2} = \left(44 \times 10^{-3} \frac{kg}{mol}\right)\left(-4.168 \times 10^{-6} \frac{mol}{m^2 s}\right)$$
$$= -1.808 \times 10^{-7} \frac{kg}{m^2 s},$$

12.4 Species Transport Within a Single Phase

$$j_{H_2O} = M_{H_2O} c_{H_2O} = \left(18 \times 10^{-3} \, \frac{kg}{mol}\right) \left(3.572 \times 10^{-8} \, \frac{mol}{m^2 \, s}\right)$$

$$= 6.43 \times 10^{-10} \, \frac{kg}{m^2 \, s},$$

$$j_{N_2} = M_{N_2} c_{N_2} = \left(28 \times 10^{-3} \, \frac{kg}{mol}\right) \left(4.354 \times 10^{-7} \, \frac{mol}{m^2 \, s}\right) = 1.219 \times 10^{-8} \, \frac{kg}{m^2 \, s}.$$

Examining and interpreting the results: We used a one-dimensional approximation in modeling this problem in Cartesian coordinates. Since alveoli are nearly spherical, a more accurate approach would be to use spherical coordinates. However, this model should give us a good order of magnitude estimate of alveolar gas fluxes, diffusion velocities, etc. The analysis indicates that diffusion velocities of respiratory gases in lung alveoli are expected to be very low, on the order of 1 μm/s. Notice that the molar average velocity is positive (directed from gas toward alveolar wall), while the mass average velocity is negative (directed from wall toward the gas). Although the flux of CO_2 is only 80% of the O_2 flux, its molecular weight is 137.5% of the molecular weight of oxygen. Consequently, the net mass flux is away from the wall, while the net molar flux is toward the wall.

Additional comments: Note that the sum of the molar fluxes relative to the molar average velocities adds to zero, as they should. The same is true for the sum of the mass fluxes relative to the mass average velocity. Even though the net velocity for water and N_2 is zero, the diffusion flux for each of these species is not zero. However, the sum of the diffusive flux, and the convective flux is zero.

12.4.5 Molecular Diffusion and Fick's Law of Diffusion

The mechanism responsible for convective mass transport is relatively easy to understand. We know from our treatment of fluid mechanics that velocity gradients will develop when shear stresses are applied to a fluid. The source of the shear stress could be a moving boundary, a pressure gradient or any of a number of other factors. If different molecular species exist in the fluid, they will be swept along with the solvent, resulting in convective transport of the species.

But what is responsible for a diffusive flux? What would cause one species to move relative to the velocity of all of the species in the mixture? If you carefully open a perfume bottle in the middle of a room full of stagnant air, the scent can be detected in a matter of minutes. Initially, it is detected close to the source, and later can be detected at the edges of the room. At any point in the room, the scent becomes stronger with time, eventually being uniform throughout the room. The flux of perfume is primarily a diffusive flux. To see this, we can compute the molar average velocity from (12.54). If we label perfume as species A, and recognizing that the net velocity of all of the other species is zero, the molar average velocity from (12.53) is:

$$\vec{v}^* = x_A \vec{v}_A. \tag{12.68}$$

The velocity of species A relative to the molar average velocity is:

$$\vec{v}_A - \vec{v}^* = \vec{v}_A[1 - x_A], \tag{12.69}$$

and so the molar flux of A relative to the molar average velocity is:

$$\vec{J}_A^* = c_A \vec{v}_A[1 - x_A] = \vec{N}_A[1 - x_A]. \tag{12.70}$$

However, the mole fraction of perfume x_A is very small, even under equilibrium conditions. Consequently,

$$\vec{v}^* \approx 0 \quad \text{and} \quad \vec{N}_A \approx \vec{J}_A^*. \tag{12.71}$$

Thus, convection is negligible in this case and perfume is transported primarily by diffusion. This flux is caused by the random motion of all of the molecules in the room. A random walk model has been used to model the diffusion process. To understand the mechanism of diffusion, consider an ideal case where molecular motion is restricted to a single dimension. At time t, a molecule of A collides with another molecule of any of the species present. After colliding, the molecule can randomly move forward or backward. If the molecules are uniformly distributed in space, then the times between collisions and the distance traveled between collisions will be relatively constant. Since motion is random, it is highly unlikely that molecules of any one species will tend to congregate spatially as time progresses. In fact, just the opposite will occur. If a region initially contains a high concentration of a species in one region, there will be a larger flux of that species away from that region simply because of random molecular motion.

Therefore, there are two major factors that tend to increase the flux of species A in a mixture. First, the steeper the mole fraction gradient, the greater will be the flux. Second, the larger the total concentration of all species, the greater will be the flux. Adolf Fick (1855) stated this for the one-dimensional diffusion of species A in a binary mixture of species A in species B:

$$J_A^* = -D_{AB} c \frac{dx_A}{dy}, \tag{12.72}$$

where y is the spatial dimension and D_{AB} is the *binary diffusion coefficient* or *diffusivity* for species A in species B. This is known as Fick's Law of diffusion. Note the presence of the minus sign, which indicates that species A moves down the gradient, from regions of high mole fraction to regions of low mole fraction.

12.4 Species Transport Within a Single Phase

Note, also, that Fick's law is written in terms of the molar flux relative to the molar average velocity. The diffusion coefficient increases with temperature and decreases with pressure. The fundamental dimensions of D_{AB} are $L^2 T^{-1}$, which are the same as the fundamental dimensions of kinematic viscosity and thermal diffusivity. Values of the diffusion coefficients for various solutes in biological gases, liquids and solids of interest are provided in Appendix C.

Extending (12.72) to three-dimensional space, we can write the vector form of Fick's Law:

$$\vec{J}_A^* = -D_{AB} c \vec{\nabla} x_A. \tag{12.73}$$

It is easy to show from (12.65) that, for a binary mixture, the diffusion coefficient for species A diffusing through a mixture of A and B (D_{AB}) is the same as the diffusion coefficient for species B diffusing through the mixture (D_{BA}). It is considerably more difficult to show that Fick's Law for a binary mixture can also be written in terms of the mass flux relative to the mass average velocity:

$$\vec{j}_A = -D_{AB} \rho \vec{\nabla} \omega_A. \tag{12.74}$$

The diffusion coefficients in (12.73) and (12.74) are equal. These equations are not strictly valid for mixtures with more than two species. Nevertheless, these expressions for Fick's Law are still used for dilute mixtures, where species A is one of the dilute species.

Substituting Fick's Law into the expression for the total mass flux of species A, (12.59):

$$\vec{n}_A = \omega_A \sum_{i=1}^{N_{species}} \vec{n}_i - D_{AB} \rho \vec{\nabla} \omega_A. \tag{12.75}$$

The total molar flux of species A can be found by substituting Fick's Law into (12.61)

$$\vec{N}_A = x_A \sum_{i=1}^{N_{species}} \vec{N}_i - D_{AB} c \vec{\nabla} x_A. \tag{12.76}$$

Equation (12.76) is a general expression for the flux of species A in any media. It is used primarily for diffusion and convection in gases, where the total molar concentration c cannot be considered constant. For fluids where ρ is constant, (12.75) can be simplified to:

$$\vec{n}_A = \rho_A \vec{v} - D_{AB} \vec{\nabla} \rho_A \text{ (constant } \rho). \tag{12.77}$$

Dividing by the molecular weight of A:

$$\vec{N}_A = c_A \vec{v} - D_{AB} \vec{\nabla} c_A \quad \text{(constant } \rho\text{)}. \tag{12.78}$$

The diffusive fluxes can be written as:

$$\vec{j}_A = -D_{AB} \vec{\nabla} \rho_A \quad \text{(constant } \rho\text{)}, \tag{12.79}$$

$$\vec{J}_A = -D_{AB} \vec{\nabla} c_A \quad \text{(constant } \rho\text{)}. \tag{12.80}$$

For diffusion in solids with constant ρ, the velocity is zero, so

$$\vec{n}_A = \vec{j}_A = -D_{AB} \vec{\nabla} \rho_A \quad \text{(solids)}, \tag{12.81}$$

$$\vec{N}_A = \vec{J}_A = -D_{AB} \vec{\nabla} c_A \quad \text{(solids)}. \tag{12.82}$$

Example 12.4.2 Evaporation of Water Through a Stagnant Gas Film of Air.
Water evaporates at a constant rate from the surface of a lake at atmospheric pressure P and temperature T. The water vapor passes through a film of still, dry air. The diffusivity of water in air is $D_{H_2O,air}$, the partial pressure of water vapor at a height h above the lake is $P_{H_2O}(h)$ and the partial pressure at the lake surface equals the vapor pressure of water at the surface temperature, $P_{H_2O,vap}$. Derive an expression for the flux of water from the surface. How does the partial pressure of water vary with position in the film?

Solution. *Initial considerations:* Since the air is still, the mass average velocity of the film will be lower than the velocity of the water vapor. Consequently, the flux of water vapor through still air will have both convective and diffusive components. Although one may be tempted to apply Fick's law directly to the transport of water vapor through the film, this would ignore the convective contribution. We will retain the convective component and examine its contribution at the end of the analysis.

System definition and environmental interactions: The system of interest in this problem is the layer of air plus water vapor between the lake surface and the top of the film of thickness h. This is a steady-state problem, so the flux of water vapor entering the system at the lake surface is constant at any position in the film.

Apprising the problem to identify governing relationships: As we will include both the convective and diffusive fluxes, we should apply the general expression for the flux of a species in a gas, (12.76).

12.4 Species Transport Within a Single Phase

Analysis: We begin by writing the general expression for the flux of water vapor through the film, (12.76), with $y = 0$ at the surface of the lake:

$$N_{H_2O} = x_{H_2O} \sum_{i=1}^{N_{species}} N_i - D_{H_2O,air} C_{air} \frac{dx_{H_2O}}{dy}.$$

We will assume that the variation in total molar concentration of gas in the film caused by the gradient in water concentration is small, so c_{air} is nearly constant and for an ideal gas equals P/RT. This is not exactly correct, since the water vapor concentration varies with position in the film. However, the change in total molar concentration of air in the film will be small. In addition, since the flux of all species other than water is zero, $\Sigma N_i = N_{H_2O}$, and the expression for water flux becomes:

$$N_{H_2O} = x_{H_2O} N_{H_2O} - D_{H_2O,air} C_{air} \frac{dx_{H_2O}}{dy}.$$

Separating variables:

$$\frac{N_{H_2O}}{D_{H_2O,air} C_{air}} dy = -\frac{dx_{H_2O}}{(1 - x_{H_2O})} = \frac{d(1 - x_{H_2O})}{(1 - x_{H_2O})} = d[\ln(1 - x_{H_2O})]. \qquad (12.83)$$

Integrating from $y = 0$ to $y = h$:

$$\frac{N_{H_2O} h}{D_{H_2O,air} C_{air}} = \ln \frac{(1 - x_{H_2O}(h))}{(1 - x_{H_2O}(0))},$$

where

$$x_{H_2O}(0) = \frac{P_{H_2O,vap}}{P},$$

$$x_{H_2O}(h) = \frac{P_{H_2O}(h)}{P}.$$

The evaporation rate per unit lake surface area is the water flux from the surface:

$$N_{H_2O} = \left(\frac{D_{H_2O,air}}{h}\right) \left(\frac{P}{RT}\right) \ln \frac{\left(1 - \frac{P_{H_2O}(h)}{P}\right)}{\left(1 - \frac{P_{H_2O,vap}}{P}\right)}. \qquad (12.84)$$

Thus, the evaporation rate per unit area could be measured by monitoring the surface temperature and placing a probe that measures P_{H_2O} at a height h in a stagnant layer above the lake surface. The method could only be used when the air was perfectly still. The distribution of water mole fraction in the film can be found by integrating (12.83) from $y = 0$ to arbitrary y:

$$\frac{N_{H_2O}}{D_{H_2O,air}C_{air}} y = \ln\left(\frac{1 - x_{H_2O}(y)}{1 - x_{H_2O}(0)}\right).$$

Substituting the expression in (12.84) for N_{H_2O}:

$$\frac{1 - x_{H_2O}(y)}{1 - x_{H_2O}(0)} = \left(\frac{1 - x_{H_2O}(h)}{1 - x_{H_2O}(0)}\right)^{y/h}.$$

Examining and interpreting the results: Therefore, the steady-state distribution of water vapor is independent of D_{H_2O} and c_{air}, and depends only on y/h and the mole fractions of water at $y = 0$ and $y = h$.

It is not easy to visualize the distribution of water vapor in the film by examining the expression for mole fraction vs. y/h. Let us introduce a concrete example, which will also allow us to examine the convective contribution to the solution. Consider the case where the gas is ideal at atmospheric pressure and a temperature of 37°C. The vapor pressure of water under these circumstances is 47 Torr, so the mole fraction of water at $y = 0$ is 0.0618. If the mole fraction at $y = h$ is kept at zero, the solution for the mole fraction as a function of y/h is:

$$x_{H_2O}(y) = 1 - 0.9382^{(1-y/h)}.$$

A plot of mole fraction vs. y/h is shown in Fig. 12.4. If we ignore convection and simply integrate Fick's law, we find $N_{H_2O} = D_{H_2O,air}c_{air}/x_{H_2O}(0)/h$, and the mole fraction of water is related to position in the film as follows:

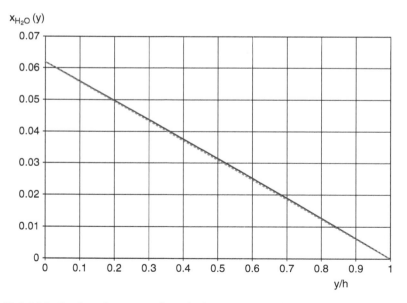

Fig. 12.4 Mole fraction of water vs. dimensionless position in film for two theories: solid line (12.84), dotted line (12.84a)

12.4 Species Transport Within a Single Phase

$$x_{H_2O}(y) = 0.0618\left(1 - \frac{y}{h}\right). \tag{12.84a}$$

This linear distribution is plotted as the dotted line in Fig. 12.4. This is virtually indistinguishable from the full solution.

The water flux from the surface will depend on $D_{H_2O,air}$ and h. The diffusivity if water in air at 37°C is 0.27 cm²/s. Let $h = 10$ cm. The water flux, according to (12.84), will be:

$$N_{H_2O} = \left(\frac{0.27\,\text{cm}^2\,\text{s}^{-1}}{10\,\text{cm}}\right)\left(\frac{1\,\text{atm}}{(82.05\,\text{cm}^3\,\text{atm}\,\text{mol}^{-1}\,\text{K}^{-1})(310\,\text{K})}\right)\ln\left[\frac{(1-0)}{(1-0.0618)}\right],$$

$$N_{H_2O} = 6.77 \times 10^{-8}\,\frac{\text{mol}}{\text{cm}^2\,\text{s}}.$$

We can compare this with the flux based on diffusion alone:

$$N_{H_2O} = \frac{D_{H_2O,air}}{h}\left(\frac{P}{RT}\right)x_{H_2O}(0) = \left(\frac{0.27\,\text{cm}^2\,\text{s}^{-1}}{10\,\text{cm}}\right)$$

$$\times \left(\frac{1\,\text{atm}}{(82.05\,\text{cm}^3\,\text{atm}\,\text{mol}^{-1}\,\text{K}^{-1})(310\,\text{K})}\right)(0.0618),$$

$$N_{H_2O} = 6.56 \times 10^{-8}\,\frac{\text{mol}}{\text{cm}^2\,\text{s}}.$$

Therefore, ignoring convection in this case leads to just a 3% underestimate in flux. Nevertheless, it is advisable to use the flux expression that includes both convection and diffusion components (12.76) as the starting point for species transport in gases.

Example 12.4.3 Diffusion of Gas with Heterogeneous Chemical Reaction. Consider the following heterogeneous chemical reaction that takes place instantaneously on a reactive surface in contact with a gas:

$$2A \rightarrow B$$

Find the steady-state flux of B from the surface if the concentration of A is maintained at $c_A(h)$ at $y = h$, where y is measured from the reactive surface.

Solution. *Initial considerations:* We assume steady-state transport of both species and that the only reaction is the heterogeneous reaction at the surface. We will also assume that all species other than A and B are stationary and that convection cannot be ignored in considering the transport of A and B.

System definition and environmental interactions: The system consists of the gas above the reacting surface. The reacting surface acts as one boundary of the system ($y = 0$) and the surface *ay* $y = h$ with known composition is the other boundary. The flux of species A and the flux of species B are each constant.

Apprising the problem to identify governing relationships: Since we are dealing with transport in a gas, our starting point will be (12.76), which includes the effects of both convection and diffusion. Before this can be integrated, we must specify a relationship between the flux of species A and the flux of species B.

Analysis: For every mole of B produced at the boundary of our system, two moles of A must be transported to the surface. Since A and B move in opposite directions, the relationship between fluxes is:

$$N_A = -2N_B.$$

Since we assume that all of the other species in the gas are stationary, the sum of fluxes is:

$$\sum_{i=1}^{N_{species}} N_i = N_A + N_B = \frac{N_A}{2}.$$

The species continuity equation for species A becomes:

$$N_A = x_A \left(\frac{N_A}{2}\right) - D_{A,gas} c_{gas} \frac{dx_A}{dy}.$$

Separating variables, assuming the total gas concentration is constant, and integrating from $y = 0$ to $y = h$:

$$N_A = \frac{2 D_{A,gas} c_{gas}}{h} \ln\left(\frac{2 - x_A(h)}{2 - x_A(0)}\right).$$

The mole fraction at $y = h$ is given and since the reaction at the surface is instantaneous, we can assume that the mole fraction at the surface is zero. Thus, the final expression for the flux is:

$$N_A = -2N_B = \frac{2 D_{A,gas} c_{gas}}{h} \ln\left(1 - \frac{x_A(h)}{2}\right).$$

Examining and interpreting the results: The logarithm term will be negative, so N_A is negative (toward the surface) and N_B is positive (away from the surface). The higher $x_A(h)$, the greater will be the flux, but the flux will not vary linearly with $x_A(h)$.

12.4 Species Transport Within a Single Phase

Example 12.4.4 Diffusion of Solute Through a Membrane.
The concentrations of drug "A" at each end of a membrane are known to be $C_A(0)$ and $C_A(\delta)$, where δ is the membrane thickness. The diffusion coefficient of A in the membrane is $D_{A,\text{membrane}}$. Find the flux of A through the membrane.

Solution. *Initial considerations:* We will assume that the concentrations at the two ends are maintained constant, so the flux of the drug is constant. We will also assume that the concentration of the drug is small relative to the total molar concentration of the membrane. Therefore, we will neglect convection relative to diffusion.

System definition and environmental interactions: The system to be analyzed is the membrane between the two ends $y = 0$ and $y = \delta$.

Apprising the problem to identify governing relationships: Since convection is assumed to be negligible, we can use Fick's law to determine the flux of drug through the membrane.

Analysis: We start with the diffusion equation for A in a solid (12.82):

$$N_A = J_A = -D_{Am}\frac{dc_{Am}}{dy}.$$

The subscripts "Am" refer to drug A in membrane m. Separating variables and integrating from $y = 0$ to $y = \delta$, we find the following expression for N_A:

$$N_A = \frac{D_{Am}}{\delta}[c_{Am}(0) - c_{Am}(\delta)].$$

Examining and interpreting the results: Consequently, the flux is directly proportional to the concentration difference between the two ends of the membrane, is directly proportional to the diffusion coefficient and inversely proportional to the thickness of the membrane.

Additional comments: The ratio of the diffusion coefficient to membrane thickness is proportional to the membrane permeability. The concentration difference in the expression above is the concentration difference *within* the membrane. If the membrane separates two well-mixed fluids, then we can relate the concentration within the membrane to the concentration in the fluid at each fluid–membrane interface. Assuming local equilibrium at the interface $y = 0$, $c_{Am}(0) = \Phi_{Am0}C_{A0}$, where C_{A0} is the liquid concentration in contact with the membrane at $y = 0$. Likewise, at $y = \delta$, $c_{Am}(\delta) = \Phi_{Am1}C_{A1}$, where C_{A1} is the liquid concentration in contact with the membrane at $y = \delta$. Substituting for membrane concentrations in terms of fluid concentrations at the two ends of the membrane, and recognizing that $\Phi_{Am1}/\Phi_{Am0} = \Phi_{A01}$:

$$N_A = \frac{D_{Am}\Phi_{Am0}}{\delta}[C_{A0} - \Phi_{A01}C_{A1}] = P_A[C_{A0} - \Phi_{A01}C_{A1}].$$

Therefore, for a solid membrane separating two well-mixed fluids, the membrane permeability P_A is equal to the partition coefficient between the fluid at $y = 0$ and the membrane multiplied by the ratio of the diffusion coefficient in the membrane relative to the membrane thickness. If the membrane is porous or if the fluids are not well mixed, then the permeability will depend on additional factors.

Example 12.4.5 Diffusion Through a Cylindrical Vessel Wall.
Consider steady-state diffusion of species A from blood plasma to interstitial fluid through a blood vessel wall having inside radius R_i and outside radius R_o. Solute concentration in blood plasma is C_p and solute concentration in interstitial fluid is C_i. The partition coefficient for species A in the vessel wall relative to plasma is Φ_{Awp} and the partition coefficient between vessel wall and interstitial fluid is Φ_{Awi}. Find the steady-state concentration as a function of radial position in the vessel wall. Find the overall *mass transfer resistance* $\Re_m$ and *mass transfer conductance* K_m of the wall, defined in a manner analogous to similar quantities in fluid flow and heat flow:

$$W_A = K_m[c_{Aw}(R_i) - c_{Aw}(R_o)] = \frac{c_{Aw}(R_i) - c_{Aw}(R_o)}{\Re_m}, \quad (12.85)$$

where $c_{Aw}(r)$ is the concentration of species A in the vessel wall at radial position r.

Solution. *Initial considerations:* We will assume that the vessel wall is nonporous and homogeneous, so the diffusion coefficient is constant. In addition, we will assume that the concentration of species A is small relative to the total molar concentration of the vessel wall.

System definition and environmental interactions: The system of interest in this problem is the vessel wall with boundaries at $r = R_i$ and $r = R_o$. Species A in the vessel wall is assumed to be in local equilibrium with plasma and interstitial fluid at the inside and outside boundaries.

Apprising the problem to identify governing relationships: Since species A is assumed to be dilute, the net flux will be primarily by diffusion. Therefore, Fick's law is the appropriate governing relationship.

Analysis: The concentrations in the vessel wall at the inside and outside radii can be found by assuming phase equilibrium at each solid–liquid interface:

$$c_{Aw}(R_i) = \Phi_{Awp} C_p,$$

$$c_{Aw}(R_o) = \Phi_{Awi} C_i.$$

12.4 Species Transport Within a Single Phase

Fick's Law for diffusion of species A through the vessel wall in the radial direction is:

$$N_A = -D_{Aw} \frac{dc_{Aw}}{dr}.$$

For steady-state diffusion through the cylindrical wall of length L, the flow (not the flux) of species A through the wall will be constant:

$$W_A = N_A(r)S(r) = \left(-D_{Aw} \frac{dc_{Aw}}{dr}\right)(2\pi r L). \tag{12.86}$$

Separating variables and integrating from $r = R_i$ to $r = R_o$, we find:

$$W_A = 2\pi D_{Aw} L \frac{[c_{Aw}(R_i) - c_{Aw}(R_o)]}{\ln(R_o/R_i)}. \tag{12.87}$$

From this, we compute the conductance K_m and resistance $\Re_m$ to mass transfer in the wall:

$$K_m = \frac{1}{\Re_m} = \frac{2\pi D_{Aw} L}{\ln(R_o/R_i)}. \tag{12.88}$$

Examining and interpreting the results: Note that the conductance is not independent of the tube geometry. It increases as the length or radius of the vessel increases and decreases with increasing thickness of the vessel wall. The conductance is directly proportional to the diffusion coefficient of species A in the vessel wall.

Additional comments: The concentration profile $c_{Aw}(r)$ can be found by integrating (12.86) from $r = R_i$ to an arbitrary radial position r, with W_A replaced by (12.87):

$$\frac{[c_{Aw}(r) - c_{Aw}(R_i)]}{[c_{Aw}(R_o) - c_{Aw}(R_i)]} = \frac{\ln(r/R_i)}{\ln(R_o/R_i)}. \tag{12.89}$$

Note that this expression for the concentration profile is independent of the length of the vessel and the diffusion coefficient of species A in the vessel wall.

12.4.5.1 Stokes–Einstein Relation

Einstein (1906) proposed a hydrodynamic theory of diffusion of a dilute solute in an unbounded liquid. He assumed the liquid to be a continuum, but the solute approximated as a solid sphere moving through the liquid. For one-dimensional transport, the force F_z responsible for moving the solute in the positive z-direction is proportional to the negative gradient in chemical potential:

$$F_z = -k_B T \frac{d(\ln c_s)}{dz}, \tag{12.90}$$

where k_B is the Boltzmann constant and T is absolute temperature. If the liquid is stationary and the solid solute moves forward with velocity v_{sz}, then it will experience a drag force in the negative z-direction. Applying (5.109) for a sphere:

$$F_k = -6\pi \mu R_s v_{sz}. \tag{12.91}$$

For steady-state movement of the solute, the solute velocity will be constant, and the sum of forces on the solute must equal zero:

$$F_z + F_k = 0 = -k_B T \frac{d(\ln c_s)}{dz} - 6\pi \mu R_s v_{sz}. \tag{12.92}$$

Solving for the solute velocity:

$$v_{sz} = -\frac{k_B T}{6\pi \mu R_s} \frac{d(\ln c_s)}{dz}. \tag{12.93}$$

Dividing (12.93) by (12.90), we obtain an expression for the mobility of the solute in the liquid, μ_s:

$$\mu_s = \frac{v_{sz}}{F_z} = \frac{1}{6\pi \mu R_s}. \tag{12.94}$$

The solute flux in a stationary liquid is:

$$N_{sz} = c_s v_{sz} = J_{sz} = -D_{s\infty} \frac{dc_s}{dz}, \tag{12.95}$$

where $D_{s\infty}$ is the diffusion coefficient of solute s in the unbounded liquid. This is also known as the *free diffusion coefficient*. Rearranging this expression to solve for the solute velocity:

$$v_{sz} = -D_{s\infty} \frac{1}{c_s} \frac{dc_s}{dz} = -D_{s\infty} \frac{d \ln(c_s)}{dz}. \tag{12.96}$$

Dividing (12.96) by (12.90), we obtain the *Nernst–Einstein equation* that relates the diffusion coefficient to the solute mobility:

$$\mu_s = \frac{D_{s\infty}}{k_B T}. \tag{12.97}$$

12.4 Species Transport Within a Single Phase

Comparing this with (12.94), we find the following relationship between the free diffusion coefficient, solute size, temperature and fluid viscosity:

$$D_{s\infty} = \frac{k_B T}{6\pi \mu R_s}. \tag{12.98}$$

This is known as the *Stokes–Einstein relation*. It indicates that the diffusion coefficient of a solute in a liquid is inversely proportional to the solute size and directly proportional to the ratio T/μ. Recall from Fig. 4.10 that the viscosity of liquids decreases with increasing temperature. Consequently, one would expect the diffusion coefficient of liquids to increase with increasing temperature.

Although this expression was derived using a steady-state analysis, it appears to be applicable in transient applications as well. In addition, it is accurate, even when the solute radius is only 2–3 times larger than the radius of the solvent molecules. The Stokes–Einstein relation is most often used to estimate the radius of solutes from their free diffusion coefficients in liquids, particularly macromolecules.

12.4.6 Mass Transfer Coefficients

In Sect. 12.3, we treated transport of species A across an interface by introducing an overall mass transfer coefficient, P_A. In many cases the primary resistance to species mass transfer between fluids in contact with solid or liquid surfaces occurs across a very thin fluid film in contact with the surface. Consider transport of a species A across a liquid–gas interface into a flowing gas. According to Fick's Law, a suitable driving force for species transport *in a single homogeneous phase* is the negative gradient in mole fraction of the species. Let us assume that the gradient in mole fraction across the film is relatively constant, the film thickness is δ and the gas within the film is nearly stagnant. Integration of Fick's Law from the interface, where the mole fraction is y_{AS}, to the edge of the film, where the mole fraction is $y_{A\infty}$, provides the following relationship for the flux of species A through the film:

$$J_A = D_{AB} c \left(\frac{y_{AS} - y_{A\infty}}{\delta} \right) = \frac{D_{AB} c}{P \delta} [P_{AS} - P_{A\infty}] = \frac{D_{AB}}{RT\delta}[P_{AS} - P_{A\infty}]. \tag{12.99}$$

We have used the ideal gas law to replace P with cRT. Thus, the flux would be expected to be proportional to the partial pressure difference between gas in contact with the liquid interface and the bulk gas far from the interface. The film thickness δ will depend on various factors, including the velocity of the gas.

A similar expression can be obtained in the liquid by integrating Fick's law across the relatively stationary liquid film of thickness δ_l from the inside edge of the film where $x_A = x_{Ai}$ to the liquid–gas interface, where $x_A = x_{AS}$:

$$J_A = D_{Al} c \left(\frac{x_{Ai} - x_{AS}}{\delta_l} \right) = \frac{D_{Al}}{\delta_l}[c_{Ai} - c_{AS}]. \tag{12.100}$$

In reality, neither the gas film nor the liquid film is completely stagnant and the concentration gradients are not linear. In addition, we are generally more interested in estimating the total flux N_A rather than just the diffusive flux J_A. Although (12.99) and (12.100) are not accurate when convection is present, the forms of these equations suggest empirical relationships that can be used to describe interfacial mass transport. For instance, an empirical relation used to describe mass transport in the gas phase across the film is:

$$N_A = k_{AG}[P_{AS} - P_{A\infty}]. \tag{12.101}$$

Equation (12.101) serves as a definition of the *mass transfer coefficient for species A in the gas phase* (k_{AG}). This definition is based on the difference in partial pressure for species A between the gas adjacent to the gas–liquid interface and the gas at edge of the film. Other driving forces for mass transfer are also appropriate, such as the difference in mole fraction of species A, mass fraction of species A, mass concentration of species A or molar concentration of species A. The mass transfer coefficients governing transport for these various driving forces would all be related to k_{GA}. Similarly, in the liquid phase, empirical relations often used are:

$$N_A = k_{Ax}[x_{AS} - x_{A\infty}] = k_{AL}[c_{AS} - c_{A\infty}]. \tag{12.102}$$

Equation (12.102) defines the *mass transfer coefficient for species A in the liquid phase* based on mole fractions (k_{Ax}) or concentrations ($k_{AL} = k_{Ax}/c$). If the concentration difference is selected as the "driving force," then a general empirical expression for gases, liquids or solids can be written as:

$$N_A = k_A[c_{AS} - c_{A\infty}]. \tag{12.103}$$

As written, mass flux is considered positive if directed away from the interface. Comparing these equations with (12.99) and (12.100), we might expect the mass transfer coefficients to be proportional to the diffusion coefficient of A in those materials when convection is small. We might also expect the mass transfer coefficients to increase with higher bulk flow rates, since the film thickness is likely to decrease when velocity increases. Equation (12.99) also suggests that the coefficient k_{GA} might be expected to decrease with increasing temperature. However, since D_{AB} increases with temperature, these effects tend to cancel each other. We will discuss experimental methods for estimating mass transfer coefficients in the next section.

12.4.7 *Experimental Approach to Determining Mass Transfer Coefficients*

In many instances, the relationship between the mass transfer coefficient and the relevant physical parameters and geometry for a particular situation cannot be accurately predicted from theory. This is particularly true in forced convection problems,

12.4 Species Transport Within a Single Phase

where flow regimes can be either laminar or turbulent. The Buckingham Pi method introduced in Chap. 3 can be used to determine the maximum number of independent dimensionless groups that may be relevant for a particular problem. Once these have been identified, experiments can be performed to determine functional relationships between a dimensionless mass transfer coefficient and other relevant dimensionless parameters. Some important relationships are described in the following sections.

12.4.7.1 External Mass Transfer Coefficients

Mass transfer between a solid or an immiscible liquid object and its surrounding fluid is enhanced when the fluid is in motion. Fluid motion reduces the thickness of the relatively stagnant fluid layer surrounding the object, thus increasing the mass transfer coefficient. Convective mass transfer between an object and flowing fluid is generally classified as either forced convection or natural (free) convection. *Forced convection* refers to situations in which the object is forced through a fluid, or fluid is forced, under pressure or gravity, around an object. *Free convection* or *natural convection* refers to fluid motion that results from density differences that occur in the fluid resulting from mass transfer and/or heat transfer between the object and the surrounding fluid. In a gravity field these density differences induce fluid velocities that may further enhance the exchange of a species between the object and fluid.

Forced Convection, Sphere

Consider the case where fluid with approach velocity v_∞ and concentration of species A $[c_{Af}]_\infty$ flows past a sphere with radius R and concentration of $[c_{Af}]_S$ at the fluid surface. Applying the definition of the mass transfer coefficient, the flux is considered positive if directed from the sphere to the fluid, and is related to the difference in concentrations, as follows (12.281):

$$N_A = k_{Af}([c_{Af}]_S - [c_{Af}]_\infty) \text{ (external)}, \quad (12.104)$$

or, since $n_A = M_A N_A$:

$$n_A = k_{Af}[\rho_{AS} - \rho_{A\infty}]. \quad (12.105)$$

The mass transfer coefficient for fluid f is k_{Af}, with fundamental dimensions of L/T. This is expected to depend on the following factors:

Geometry: sphere Diameter, D
Flow property: fluid velocity far from sphere, v_∞
Fluid properties: fluid density ρ, fluid viscosity, μ
Species A properties: species density, ρ_A, species diffusivity in fluid f, D_{Af}

Using the methods of Chap. 3, the important parameters and their fundamental dimensions are:

$$k_{Af}(M^0L^1T^{-1}),$$
$$D(M^0L^1T^0),$$
$$v_\infty(M^0L^1T^{-1}),$$
$$\rho(M^1L^{-3}T^0),$$
$$\mu(M^1L^{-1}T^{-1}),$$
$$\rho_A(M^1L^{-3}T^0),$$
$$D_{Af}(M^0L^2T^{-1}). \tag{12.106}$$

We have seven parameters and three fundamental dimensions, so we might expect four dimensionless groups of parameters. We are interested in excluding k_{Af} from the core, so we can find a dimensionless mass transfer coefficient. Selecting ρ, D_{Af}, and D as the core variables, we find the following four dimensionless groups to be of potential importance for mass flow from a sphere:

$$\begin{aligned}\Pi_{k_{Af}} &= \frac{k_{Af}D}{D_{Af}} = (Nu_m)_D = Sh_D, \\ \Pi_\mu &= \frac{\mu}{\rho D_{Af}} = \frac{\nu}{D_{Af}} = Sc, \\ \Pi_{v_\infty} &= \frac{v_\infty D}{D_{Af}} = Pe_D, \\ \Pi_{\rho_A} &= \frac{\rho_A}{\rho}.\end{aligned} \tag{12.107}$$

The first dimensionless group is the dimensionless mass transfer coefficient based on the sphere diameter. This is called the Nusselt number for mass transfer $(Nu_m)_D$ or alternately the Sherwood number $(Sh)_D$. It is analogous to the Nusselt number for heat transfer $(Nu_D = hD/k)$. The second dimensionless group is the ratio of kinematic viscosity to diffusion coefficient. This is a property of the fluid for a particular species A, and is known as the Schmidt number (Sc). The third dimensionless group is known as the Peclet number and represents the ratio of convective to diffusive mass transport. The Peclet number is the product of the Reynolds number (based on sphere diameter) and Schmidt number.

$$Pe_D = \left(\frac{v_\infty D}{\nu}\right)\left(\frac{\nu}{D_{Af}}\right) = Re_D Sc. \tag{12.108}$$

According to the guidelines outlined in Chap. 3, we can modify any of the dimensionless groups by multiplying or dividing one dimensionless group by any of the other dimensionless groups raised to any power. Dividing the Peclet number by the Schmidt number gives an alternate third dimensionless group, the Reynolds number, Re_D. The last dimensionless group in (12.107) represents a buoyancy factor, which is generally unimportant in forced convection problems, but may be important for free (or natural) convection problems. In general, the dimensionless mass transfer coefficient is a function of three dimensionless groups:

12.4 Species Transport Within a Single Phase

$$Sh_D = \frac{k_{Af}D}{D_{Af}} = f\left(Re_D, Sc, \frac{\rho_A}{\rho}\right) \quad (12.109)$$

For forced convection around a sphere, experiments from many studies were analyzed by Ranz and Marshall (1952). They found the following empirical relationship to fit the data well:

$$Sh_D = 2 + 0.6 Re_D^{\frac{1}{2}} Sc^{\frac{1}{3}} = 2 + 0.6 \left[Re_D Sc^{\frac{2}{3}} \right]^{\frac{1}{2}}. \quad (12.110)$$

A comparison between (12.110) and measured values of Sh_D is shown in Fig. 12.5. It should be understood that this and similar empirical relationships are often only accurate to within 10–20%. In addition, the relationships should not be extrapolated far beyond the experimental range of dimensionless groups that were used to formulate the empirical relationship. In this case, the empirical relationship (12.110) is valid for $Re_D Sc^{\frac{2}{3}} < 40,000$. The data in Fig. 12.5 represent both heat and mass transfer measurements and can also be used to estimate Nu_D, given Pr and Re_D.

The procedure for finding mass flow of species A from the sphere to the flowing fluid is as follows:

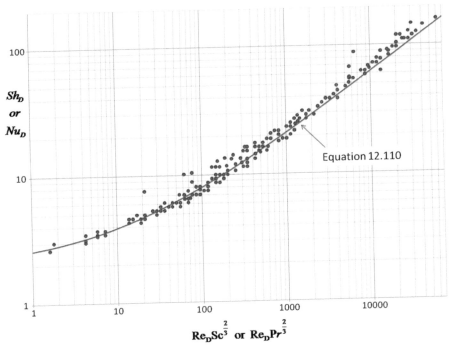

Fig. 12.5 Dimensionless mass transfer coefficient (Sh_D) vs. $Re_D Sc^{2/3}$ and heat transfer coefficient (Nu_D) vs. $Re_D Pr^{2/3}$ for flow around a sphere (redrawn from data in Ranz and Marshall 1952)

1. Compute the Reynolds number and the Schmidt number.
2. Compute the Sherwood number from the appropriate empirical relation, such as (12.110) or Fig. 12.5 in this case. Be sure the empirical relationship is valid at the Re_D and Sc for the problem at hand.
3. Use (12.109) to compute the mass transfer coefficient, k_{Af}.
4. Use (12.103) or (12.104) to compute the molar flux or mass flux of species A.
5. Multiply the flux by the sphere surface area (πD^2) to compute the molar or mass flow rate of A away from the surface.

Example 12.4.6 Dissolution Rate of a Soluble Microsphere.
A stationary microsphere composed of material A has a diameter of 10 μm. The microsphere is suspended in blood plasma having a velocity of 100 μm/s. The density of material A is 0.96 g/ml, the diffusion coefficient of A in plasma is 10^{-7} cm²/s and the partition coefficient for A in plasma is 0.001. What is the rate at which Material A is removed from the microsphere to the flowing plasma? Estimate the time required for the microsphere to completely dissolve.

Solution. *Initial considerations:* The time for the microsphere to dissolve can be estimated by dividing the original mass of the microsphere by the rate that mass dissolves from the surface of the microsphere. We will assume that the rate of dissolution remains constant until all the material has dissolved. We can use empirical data for mass transfer from a sphere to estimate an appropriate mass transfer coefficient for mass transfer of material A from the microsphere surface to the flowing stream. The concentration of material A in the stream far from the microsphere will be assumed to be zero.

System definition and environmental interactions: The system we will analyze to determine the mass rate from the sphere surface is the fluid film surrounding the sphere. Once the rate is known, we will adopt the microsphere mass as our system so we can estimate the time required for the material to completely dissolve.

Apprising the problem to identify governing relationships: To find the mass transfer coefficient, we will use the empirical relationship for mass transfer from a sphere (12.110). We must check to make sure that the computed value is consistent with the flow conditions imposed in this case.

Analysis: Before we can compute the Sherwood number, we must compute the Reynolds number and Schmidt number for this case:

$$Re_D = \frac{\langle v \rangle D}{v} = \frac{\left(100 \frac{\mu m}{s}\right)(10 \mu m)}{10^{-2} \frac{cm^2}{s}} \left(10^{-4} \frac{cm}{\mu m}\right)^2 = 10^{-3},$$

$$Sc = \frac{v}{D_{A,plasma}} = \frac{10^{-2} \frac{cm^2}{s}}{10^{-7} \frac{cm^2}{s}} = 10^5,$$

12.4 Species Transport Within a Single Phase

$$Re_D Sc^{\frac{2}{3}} = 10^{-3}(10^5)^{\frac{2}{3}} = 2.15.$$

Since $Re_D Sc^{\frac{2}{3}} < 40,000$, (12.110) is valid, so the Sherwood number is:

$$Sh_D = 2 + 0.6\left[Re_D Sc^{\frac{2}{3}}\right]^{\frac{1}{2}} = \frac{k_{A,plasma} D}{D_{A,plasma}} = 2 + .6(2.15)^{\frac{1}{2}} = 2.88.$$

Note that the first term (i.e., 2.0) represents the contribution from diffusion and the second term (0.88) results from convection. Most of the contributions to the mass transfer coefficient in this case represent transport by diffusion. The mass transfer coefficient is:

$$k_{A,plasma} = 2.88\left(\frac{D_{A,plasma}}{D}\right) = 2.88 \frac{\left(10^{-7} \frac{cm^2}{s}\right)}{(10^{-3}\ cm)} = 2.88 \times 10^{-4} \frac{cm}{s} = 2.88 \frac{\mu m}{s}.$$

The mass flux of material A from the microsphere surface to plasma is:

$$n_A = k_{A,plasma}(\rho_{A,s} - \rho_{A,\infty}),$$

$$n_A = k_{A,plasma}(\Phi \rho_{microsphere} - 0),$$

$$n_A = 2.88\ \mu m/s(0.001)(0.96\ g/cm^3)(1\ cm/10^4\ \mu m),$$

$$n_A = 2.77 \times 10^{-7}\ g\ cm^{-2}\ s^{-1}.$$

The rate at which mass is lost from the microsphere can be found by multiplying the flux by the surface area of the microsphere:

$$w_A = n_A(\pi D^2),$$

$$w_A = (2.77 \times 10^{-7}\ g\ cm^{-2}\ s^{-1})(\pi)(10^{-3}\ cm)^2,$$

$$w_A = 8.69 \times 10^{-13}\ g/s = 0.869\ pg/s.$$

The initial mass of the microsphere is:

$$m_{microsphere} = \rho_{microsphere}\left(\frac{\pi D^3}{6}\right),$$

$$m_{microsphere} = (0.96\ g\ cm^{-3})\left(\frac{\pi}{6}\right)(10^{-3}\ cm)^3,$$

$$m_{microsphere} = 5.03 \times 10^{-10}\ g.$$

If the mass transfer rates were the same at all times, the sphere would completely dissolve in 579 s or 9.65 min.

Examining and interpreting the results: This estimate is based on a constant rate of dissolution from the surface. However, the mass transfer coefficient increases as the microsphere diameter decreases, so the flux increases with time. On the other hand, the surface area decreases at a rate proportional to the diameter squared. We will account for these factors in Example 13.7.1, where a more accurate estimate of the dissolution time will be provided.

Free or Natural Mass Transfer from a Sphere

In natural convective mass transfer around a sphere, the velocity is induced by the presence of density variations within a gravitational field. Rather than using the ratio ρ_A/ρ, which varies throughout the fluid, a different dimensionless buoyancy factor known as the Grashof number for mass transfer (Gr_{Af}) is most often used:

$$Gr_{Af} \equiv \frac{\zeta_A(\rho_s - \rho_\infty)gD^3}{\rho_\infty \nu^2}, \qquad (12.111)$$

where ρ_s and ρ_∞ are the densities in the fluid at the sphere surface and far upstream from the sphere, g is the acceleration due to gravity, d is the sphere diameter, ν is the fluid kinematic viscosity and ζ_A is the coefficient of compositional expansion, defined as follows:

$$\zeta_A \equiv \frac{1}{\rho}\frac{\partial \rho}{\partial x_A}. \qquad (12.112)$$

Ranz and Marshall (1952) provide the following relationship for free convective transport in the neighborhood of a sphere for $Gr_{Af}Sc^{\frac{4}{3}} < 1.6 \times 10^9$:

$$Sh_D = 2.0 + 0.6Gr_{Af}^{\frac{1}{4}}Sc^{\frac{1}{3}}. \qquad (12.113)$$

The Sherwood number approaches 2.0 for very low values of Gr_{Af}, which as we will see in Chap. 14, is consistent with diffusion from a spherical surface to a completely stagnant fluid.

Forced Convective Mass Transfer from a Cylinder

For flow perpendicular to a cylinder with diameter D, in the range $1 < Re_D < 10{,}000$, the following empirical relationship is valid:

$$Sh_D = 0.57 Re_D^{\frac{1}{2}} Sc^{\frac{1}{3}}. \qquad (12.114)$$

12.4 Species Transport Within a Single Phase

Natural Convection from a Cylinder

For natural convection from the surface of a cylinder with diameter D:

$$Sh_D = a(Gr_{Af}Sc)^m. \qquad (12.115)$$

where the coefficients a and m depend on $Gr_{Af}Sc$ as shown in the table:

$Gr_{Af}Sc$	a	m
$<10^{-5}$	0.49	0
10^{-5}–10^{-3}	0.71	0.04
10^{-3}–1.0	1.09	0.1
1–10^4	1.09	0.2
10^4–10^9	0.53	0.25
$>10^9$	0.13	0.333

Forced Mass Transfer from a Flat Plate

Consider mass transfer from a flat plate with constant surface concentration C_{As} of species A and bulk concentration outside the film of $C_{A\infty}$. The direction of flow is parallel with the plate. The local Sherwood number for $Sc > 0.6$, based on the distance x from the leading edge of the plate is:

$$Sh_{x,loc} = \frac{k_{Af,loc}x}{D_{Af}} = 0.332 Re_x^{\frac{1}{2}} Sc^{\frac{1}{3}}. \qquad (12.116)$$

Thus, the local mass transfer coefficient and the local flux decrease as x increases. Total mass flux over a length L of the plate is found by integration:

$$N_A = \frac{1}{L}\int_{x=0}^{x=L} k_{Af,loc}(C_{As} - C_{A\infty})dx = \frac{1}{L}(C_{As}-C_{A\infty})\int_{x=0}^{x=L} 0.332 D_{Af} x^{-1} Re_x^{\frac{1}{2}} Sc^{\frac{1}{3}}dx. \qquad (12.117)$$

Carrying out the integration:

$$N_A = 0.664 Sc^{\frac{1}{3}} Re_L^{\frac{1}{2}} \left(\frac{D_{Af}}{L}\right)(C_{As}-C_{A\infty}). \qquad (12.118)$$

Note that this is only valid for a constant wall concentration. Defining a mass transfer coefficient in the fluid k_{Af}, then:

$$N_A = k_{Af}(C_{As}-C_{A\infty}). \qquad (12.119)$$

We find from (12.118):

$$k_{Af} = 0.664 Sc^{\frac{1}{3}} Re_L^{\frac{1}{2}} \frac{D_{Af}}{L}, \qquad (12.120)$$

or the overall Sherwood number is:

$$Sh_L = \frac{k_{Af}L}{D_{Af}} = 0.664 Sc^{\frac{1}{3}} Re_L^{\frac{1}{2}}. \qquad (12.121)$$

This is twice the value of the local Sherwood number evaluated at $x = L$.

Many other relationships for mass transfer coefficients in external flow situations can be found in the literature, including natural convection from vertical plates, natural convection from horizontal plates, and natural and forced convection from different shaped objects. In addition, the effects of non-Newtonian fluids, periodic flow and transient flow on mass transfer coefficients have also been studied. Summarizing these studies is beyond the scope of this introductory text.

12.4.7.2 Internal Mass Transfer Coefficients

The mass transfer coefficient for flow through conduits and reactors is defined differently than for external flow situations. As always, the mass transfer coefficient is defined as the ratio of the flux to the driving force, but in this case the driving force is the difference between the wall concentration and the mean or bulk fluid concentration in the fluid, and this will generally depend on axial location from the tube inlet. This relationship was presented in Chap. 2 for flow through a conduit:

$$N_A = k_{Af}([c_{Af}]_w - [c_{Af}]_m) \qquad (12.122)$$

$[c_{Af}]_w$ is the concentration of A in the fluid at the conduit surface, and may vary with axial position. Flux in this case is positive if directed in the positive radial direction from the center toward the wall of the conduit. The *mean* or *bulk fluid concentration* is defined as:

$$[c_{Af}]_m = C_{Ab} = \frac{1}{Q_V} \int_{A_c} v c_{Af} dA_c, \qquad (12.123)$$

where v is the local velocity in the fluid, c_{Af} is the local concentration of species A in the fluid, Q_V is the volumetric flow rate through the conduit and A_c is the cross-sectional area of the conduit. The bulk fluid concentration will also vary with axial location. This is often called the *flow-averaged concentration* or the *mixed mean concentration* or the *mixing cup concentration* because it would equal the concentration measured if the conduit were cut at that axial location and the contents were collected and mixed in a cup.

12.4 Species Transport Within a Single Phase

Forced Convective Mass Transfer in Conduits

Mass transfer in conduits of various shapes is of great importance in physiology and in many biomedical devices. Dimensional analysis can be used to identify the most important dimensionless groups governing mass transfer in a cylinder of diameter D and length L:

$$Sh_D = \frac{k_{Af}D}{D_{Af}} = f(Re_D, Sc, Gr_{Af}, L/D). \tag{12.124}$$

Sh_D depends on the length of the conduit for short conduits, but is independent of L/D for long conduits. For the case of uniform concentration of substance A at the inside surface of a long circular cylinder with $L/D > 0.05 Re_D Sc$, the Sherwood number is constant:

$$Sh_D = 3.658 \text{ (constant surface concentration)} \tag{12.125}$$

If the boundary condition is changed from constant concentration at the surface to constant mass flux at the surface:

$$Sh_D = 4.364 \text{ (constant mass flux at surface)} \tag{12.126}$$

A derivation of this relationship is shown in Sect. 15.5.1.3. Sherwood numbers based on hydraulic diameter of the duct for fully developed ducts with various cross-sections are shown in Table 12.1 for the cases of constant surface concentration and constant surface mass flux.

Table 12.1 Sherwood numbers based on hydraulic diameter for fully developed flow in ducts with various cross-sections (from Kays and Crawford (1993))

Cross-section	Sherwood number (constant flux)	Sherwood number (constant wall concentration)
Circular	4.364	3.66
Equilateral triangle	3.0	2.35
Square	3.61	2.98
Parallel plates, transfer through both plates	8.235	7.54
Parallel plates, transfer through one plate only	5.385	4.86
Rectangular, width/height = 2	4.12	3.39
Rectangular, width/height = 3	4.79	3.96
Rectangular, width/height = 4	5.33	4.44
Rectangular, width/height = 8	6.49	5.60

Forced Convective Mass Transfer in a Packed Column

A common method for removing specific chemical species from a biological fluid is to pass the fluid through a column packed with adsorptive particles. Two new parameters are introduced into the dimensional analysis: the diameter of the particles D_p and the fractional void volume of the column, ε, where

$$\varepsilon \equiv \frac{(\text{Column volume} - \text{Total particle volume})}{\text{Column volume}}. \tag{12.127}$$

In addition, rather than an attempt to compute the average velocity in the neighborhood of the beads, it is customary to use the superficial velocity v_0, defined as:

$$v_0 \equiv \frac{\text{Flow rate}}{\text{Empty column cross - sectional area}}. \tag{12.128}$$

By "empty column cross-sectional area", we refer to the cross-sectional area of the column before the particles are added. Dimensional analysis will show:

$$Sh_{D_p} \equiv \frac{k_{Af} D_p}{D_{Af}} = f\left(Re_{D_p}, Sc, \varepsilon, \frac{D}{D_p}, \frac{L}{D_p}\right), \tag{12.129}$$

where

$$Re_{D_p} \equiv \frac{v_0 D_p}{\nu}.$$

Experiments show that the Sherwood number is independent of D/D_p and L/D_p when these ratios are large. In this case, Wilson and Geankoplis (1966) found:

$$Sh_{D_p} = \frac{1.09}{\varepsilon} \left(Re_{D_p} Sc\right)^{\frac{1}{3}}; \quad Re_{D_p} < 20; \quad Re_{D_p} Sc \gg 1. \tag{12.130}$$

Another popular expression for liquid flow in a packed bed (Cussler, 1997) is:

$$Sh_{D_p} = 25\, Re_{D_p}^{.45} Sc^{.5} \tag{12.131}$$

12.5 Relation Between Individual and Overall Mass Transfer Coefficients

Let us now relate the individual mass transfer coefficients to the overall mass transfer coefficient or permeability P_A discussed in Sect. 12.3. Molar flow of species A across the interface can be found by dividing the molar flow of species A in (12.43) by the interfacial surface area S:

$$N_A = P_A \left[(c_A)_{gas} - \Phi_{A,gas,liquid} (c_A)_{liquid} \right], \tag{12.132}$$

where the gas–liquid partition coefficient can be written in terms of the Henry's Law coefficient ϕ_A:

12.5 Relation Between Individual and Overall Mass Transfer Coefficients

$$\Phi_{A,gas,liquid} = \left[\frac{(c_A)_{gas}}{(c_A)_{liquid}}\right]_{equilibrium} = \frac{c_{gas}}{c_{liquid}}\left[\frac{y_A}{x_A}\right]_{equilibrium} = \left[\frac{c_{gas}}{c_{liquid}}\right]\phi_A. \quad (12.133)$$

Writing (12.85) and (12.86) in terms of concentrations:

$$N_A = \frac{k_{AG}P}{c_{gas}}\left[(c_A)_{gas} - (c_{AS})_{gas}\right]. \quad (12.134)$$

$$N_A = k_{AL}\left[(c_{AS})_{liquid} - (c_A)_{liquid}\right]. \quad (12.135)$$

If all of the resistance to mass flow occurs in the fluid films, then we can assume local equilibrium between the gas and fluid at the interface:

$$(c_{AS})_{gas} = \Phi_{A,gas,liquid}(c_{AS})_{liquid}. \quad (12.136)$$

Substituting this into (12.134) and rearranging:

$$\frac{c_{gas}N_A}{k_{AG}P} = \left[(c_A)_{gas} - \Phi_{A,gas,liquid}(c_{AS})_{liquid}\right]. \quad (12.137)$$

Multiplying (12.135) by the partition coefficient and dividing by k_{LA}:

$$\frac{\Phi_{A,gas,liquid}N_A}{k_{AL}} = \left[\Phi_{A,gas,liquid}(c_{AS})_{liquid} - \Phi_{A,gas,liquid}(c_A)_{liquid}\right]. \quad (12.138)$$

Adding (12.137) and (12.138) eliminates the liquid concentration at the interface. Solving for N_A:

$$N_A = \left[\frac{1}{\frac{\Phi_{A,gas,liquid}}{k_{AL}} + \frac{c_{gas}}{k_{AG}P}}\right]\left[(c_A)_{gas} - \Phi_{A,gas,liquid}(c_A)_{liquid}\right]. \quad (12.139)$$

This is the same as (12.132) if the overall mass transfer coefficient P_A is related to the mass transfer coefficients in the gas and liquid phases as follows:

$$\frac{1}{P_A} = \frac{1}{k_{AL}}\left[\Phi_{A,gas,liquid}\right] + \frac{1}{k_{AG}}\left[\frac{c_{gas}}{P}\right] = \frac{1}{k_{AL}}\left[\frac{m_A c_{gas}}{c_{liquid}}\right] + \frac{1}{k_{AG}}\left[\frac{c_{gas}}{P}\right]. \quad (12.140)$$

If we interpret P_A, k_{GA} and k_{LA} each as representing a conductance, then their inverse represents a resistance. Thus (12.140) states that the overall resistance to mass transfer is the sum of resistances in the liquid and gas phases.

According to (12.45), we can also write the flux of species A in terms of a driving force from the liquid to the gas phase:

$$N_A = -\mathsf{P}_A^*\left[(c_A)_{\text{liquid}} - \Phi_{A,\text{liquid,gas}}(c_A)_{\text{gas}}\right], \tag{12.141}$$

where

$$\frac{1}{\mathsf{P}_A^*} = \frac{1}{k_{\text{AL}}} + \frac{1}{k_{\text{AG}}}\left[\frac{c_{\text{gas}}\Phi_{A,\text{liquid,gas}}}{P}\right] = \frac{1}{k_{\text{AL}}} + \frac{1}{k_{\text{AG}}}\left[\frac{c_{\text{liquid}}}{m_A P}\right]. \tag{12.142}$$

12.6 Permeability of Nonporous Materials

In many biological and physiological situations, we are interested in the flow of solutes across barriers of various types, including skin, various blood–tissue barriers, blood–gas barriers, cell membranes, artificial membranes, hollow fiber walls, etc. In such cases, the overall mass transfer coefficient is often referred to as the *permeability* of the barrier to solute A, P_A. The flux of solute A across the barrier is given by:

$$N_A = \mathsf{P}_A[C_{A1} - \Phi_{A12}C_{A2}]. \tag{12.143}$$

C_{A1} is the bulk concentration of A on side 1 of the barrier, C_{A2} is the bulk concentration of species A on side 2 of the barrier and Φ_{A12} is the partition coefficient for species A on side 1 relative to side 2. The permeability coefficient has dimensions of LT^{-1}, and is often given in units of cm/s. Permeability coefficients reported in the literature generally include contributions from stagnant films on each side of the barrier, in addition to the resistance of the barrier itself.

Equation (12.143) is based on diffusion transport across the barrier, and should not be used if other mechanisms contribute significantly to the overall transport of solute A across the barrier. If the material is porous, then convective transport may be important. Transport across porous barriers is considered in Sect. 13.8. If transport is facilitated or restricted by chemical reactions at the surface or interior of the barrier, (12.143) will not accurately reflect the flux of solute A across the barrier. Some of these mechanisms are considered in Sect. 12.9. Solute flux is assumed to be by diffusion alone in the next two sections.

12.6.1 Membrane Permeability

Figure 12.6 shows a membrane m that separates two liquids, 1 and 2. This can be a cell membrane, an artificial membrane or any solid barrier to the transport of species A that is relatively flat. Our goal is to determine how the membrane permeability is related to the mass transfer coefficients in the fluid films, the

12.6 Permeability of Nonporous Materials

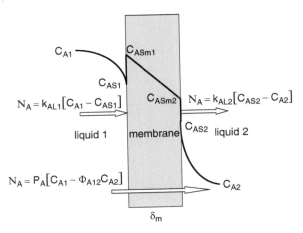

Fig. 12.6 Solute flux across a membrane

partition coefficients for solute A between each fluid and the membrane, the diffusion coefficient in the membrane and the thickness of the membrane.

In liquid 1, flux to the membrane is:

$$N_A = k_{AL1}[C_{A1} - C_{AS1}]. \qquad (12.144)$$

In liquid 2, flux from the membrane is:

$$N_A = k_{AL2}[C_{AS2} - C_{A2}]. \qquad (12.145)$$

Integrating Fick's Law across the membrane, the flux of species A can be written as:

$$N_A = \frac{D_{Am}}{\delta_m}[C_{ASm1} - C_{ASm2}], \qquad (12.146)$$

where D_{Am} is the diffusivity of A in the membrane and δ_m is the membrane thickness. At the interface between fluid 1 and the membrane, we assume a local equilibrium:

$$\Phi_{A1m} = \frac{C_{AS1}}{C_{ASm1}}. \qquad (12.147)$$

Similarly, at the interface between the membrane and fluid 2, we also assume a local equilibrium:

$$\Phi_{A2m} = \frac{C_{AS2}}{C_{ASm2}}. \qquad (12.148)$$

Substituting (12.148) and (12.149) into (12.144) and (12.145):

$$\frac{N_A}{k_{AL1}\Phi_{A1m}} = \left[\frac{C_{A1}}{\Phi_{A1m}} - C_{ASm1}\right], \tag{12.149}$$

$$\frac{N_A}{k_{AL2}\Phi_{A2m}} = \left[C_{ASm2} - \frac{C_{A2}}{\Phi_{A2m}}\right], \tag{12.150}$$

$$\frac{\delta_m N_A}{D_{Am}} = [C_{ASm1} - C_{ASm2}]. \tag{12.151}$$

Adding these three equations:

$$N_A\left[\frac{1}{k_{AL1}\Phi_{A1m}} + \frac{1}{k_{AL2}\Phi_{A2m}} + \frac{\delta_m}{D_{Am}}\right] = \left[\frac{C_{A1}}{\Phi_{A1m}} - \frac{C_{A2}}{\Phi_{A2m}}\right]. \tag{12.152}$$

Note that under equilibrium conditions $C_{A1} = C_{AS1}$, $C_{A2} = C_{AS2}$, $C_{ASm1} = C_{ASm2}$, and:

$$\frac{\Phi_{A1m}}{\Phi_{A2m}} = \frac{C_{AS1}}{C_{ASm1}}\frac{C_{ASm2}}{C_{AS2}} = \left(\frac{C_{A1}}{C_{A2}}\right)_{equilibrium} = \Phi_{A12}. \tag{12.153}$$

That is, the ratio of partition coefficients between the two fluids and the membrane is equal to the partition coefficient between the two fluids. Therefore, (12.152) can be rearranged to be in the form of (12.143):

$$N_A = \frac{[C_{A1} - \Phi_{A12}C_{A2}]}{\left[\dfrac{1}{k_{AL1}} + \dfrac{\Phi_{A12}}{k_{AL2}} + \dfrac{\Phi_{A1m}\delta_m}{D_{Am}}\right]}. \tag{12.154}$$

Consequently, we can write the following expression relating permeability to the membrane parameters and the mass transfer coefficients in the two fluids:

$$\frac{1}{\mathsf{P}_A} = \frac{1}{k_{AL1}} + \frac{\Phi_{A12}}{k_{AL2}} + \frac{\Phi_{A1m}\delta_m}{D_{Am}}. \tag{12.155}$$

If the mass transfer coefficients in the liquid films are high, then the first two terms in (12.155) are negligible. In that case, the permeability is directly related to the diffusion coefficient and inversely related to the membrane thickness:

$$\mathsf{P}_A = \frac{D_{Am}}{\Phi_{A1m}\delta_m} = \frac{D_{Am}\Phi_{Am1}}{\delta_m} \quad \text{(high } k_{AL1}, k_{AL2}\text{)}. \tag{12.156}$$

Equation (12.156) is often used to define membrane permeability. However, if significant resistance is offered by either of the fluid films in contact with the membrane, then (12.156) will provide an overestimate of membrane permeability. In such cases, (12.155) should be used.

12.6.2 Vessel or Hollow Fiber Permeability

Consider the steady-state flow of solute A through the nonporous wall of a cylindrical blood vessel or hollow fiber, such as was treated in Example 12.4.5. Total mass flow through the vessel wall is constant, but mass flux decreases as one progresses from the inside wall to the outside wall of the vessel. The definition of permeability in (12.143) is based on flux. The bulk concentrations and partition coefficient are independent of radial position, but since the flux is different at the inside surface than it is at the outside surface of the vessel, then the permeability must be different at the two surfaces. This can cause confusion when using permeability coefficients reported in the literature. Were they based on the inside radius, the outside radius or somewhere in between? Furthermore, was the driving force for mass transfer considered to be the bulk concentration inside the fiber minus its equilibrium concentration on the outside of the vessel, or was it the outside bulk concentration minus its equilibrium concentration on the inside of the vessel? Let us treat these questions one at a time.

Consider flow in the positive radial direction to be positive (i.e., from inside the vessel toward the outside surface of the vessel). Solute flow through the vessel wall is the product of solute flux and surface area:

$$W_A = N_A S = P_A S[C_{A1} - \Phi_{A12} C_{A2}]. \quad (12.157)$$

If the value of the bracketed term is positive, then solute will flow out of the vessel. Since W_A and the bracketed term are constant for steady-state solute flow, the product of the permeability and the surface area must also be constant:

$$P_A S = P_{Ai} S_i = P_{Ao} S_o. \quad (12.158)$$

Here, the subscripts i and o indicate values at the inside and outside radius of the fiber, respectively. P_{Ai} is known as the permeability based on the inside radius and P_{Ao} is the permeability based on the outside radius.

Some investigators may prefer to use an alternate driving force to describe solute flow, such as $C_{A2} - \Phi_{A21} C_{A1}$. Rearranging (12.157):

$$W_A = -P'_A S[C_{A2} - \Phi_{A21} C_{A1}] \quad (12.159)$$

The permeability P'_A is different than P_A because the driving force is defined differently. If the value of the bracketed term is negative, solute A will flow in the positive direction, from inside to outside. By analogy with (12.158), the product of permeability P'_A and surface area will be constant:

$$\mathsf{P}'_A S = \mathsf{P}'_{Ai} S_i = \mathsf{P}'_{Ao} S_o. \qquad (12.160)$$

The relationship between P_A and P'_A can be found by equating (12.157) and (12.159):

$$\mathsf{P}'_A S[C_{A2} - \Phi_{A21} C_{A1}] = -\mathsf{P}_A S[C_{A1} - \Phi_{A12} C_{A2}]$$
$$= -\mathsf{P}_A S \Phi_{A12} \left[\frac{C_{A1}}{\Phi_{A12}} - C_{A2} \right]. \qquad (12.161)$$

But, since $\Phi_{A21} = 1/\Phi_{A12}$, (12.161) becomes:

$$\mathsf{P}'_A S[C_{A2} - \Phi_{A21} C_{A1}] = \mathsf{P}_A S \Phi_{A12} [C_{A2} - \Phi_{A21} C_{A1}]. \qquad (12.162)$$

After canceling terms, we arrive at the desired relationship:

$$\mathsf{P}'_A = \mathsf{P}_A \Phi_{A12}. \qquad (12.163)$$

Combining (12.163) with (12.158) or (12.160) allows us to relate the permeability P'_A based on the outside surface as defined in (12.159) to the permeability P_A based on the inside surface as defined in (12.159):

$$\mathsf{P}'_{Ao} = \mathsf{P}_{Ai} \frac{S_i}{S_o} \Phi_{A12}. \qquad (12.164)$$

There will be very little difference between these permeability values if the vessel wall is thin and the partition coefficient between the fluids on each side is nearly unity. However, for vessels with thicker walls, it may be important to determine which equation was used to report permeability.

Example 12.6.2.1. Pemeability of a Hollow Fiber.
Hollow fibers are used in hemodialyzers, blood oxygenators and other biological devices. Resistance to mass transfer of species A through a hollow fiber is offered not only by the resistance of the wall, but also by the resistance of the relatively stagnant film layers on each side of the fiber wall. Show how the permeability of a fiber wall, based on the outside radius of the vessel, is related to the properties of the vessel wall and properties of the stagnant films on each side of the vessel wall.

Solution. *Initial considerations:* We follow a procedure similar to that provided in Sect. 12.6.1. Figure 12.7 shows how the concentration of species A varies from the interior of the fiber to the fluid far from the exterior wall. In the central region of the fiber lumen, the concentration in the internal fluid is assumed uniform at the bulk

12.6 Permeability of Nonporous Materials

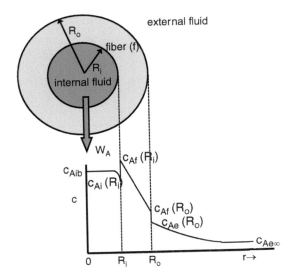

Fig. 12.7 Concentration profile of species A across hollow fiber

concentration, c_{Aib}. Far from the external surface, the concentration of A in the external fluid is $c_{Ae\infty}$. The solubility of A in the internal fluid is assumed to be different than the solubility of A in either the fiber or the external fluid. Consequently, the concentration of A in the fiber at $r = R_i$, $c_{Af}(R_i)$, will be different than the concentration of A in the internal fluid at $r = R_i$, $c_{Ai}(R_i)$. Similarly, the concentration of A in the fiber at $r = R_o$, $c_{Af}(R_o)$, will be different than the concentration of A in the internal fluid at $r = R_o$, $c_{Ao}(R_o)$.

System definition and environmental interactions: There are three systems to be analyzed: the fluid film on the inside lumen of the conduit, the fluid film on the outside surface and the fiber wall. In the steady-state, the flow of species A through all of these systems is the same and the concentrations at the conduit–film interfaces are known, as discussed above.

Apprising the problem to identify governing relationships: The macroscopic relationship for species mass transport through a film, (12.144), will be applied to the internal and external films on the conduit and Fick's law will be applied across the conduit wall. Fluxes will be matched at the wall–fluid interfaces, and local equilibrium will be assumed at each interface.

Analysis: The overall permeability due to fiber and film conductances can be written using (12.157):

$$W_A = P_{Ao}S_o\left[c_{Aib} - \Phi_{A,ie}c_{Ae\infty}\right].$$

Flow of species A in the internal fluid is:

$$W_A = k_{Ai}S_i[c_{Aib} - c_{Ai}(R_i)].$$

Flow of A through the cylindrical fiber wall is

$$W_A = S_o k_{Af}[c_{Af}(R_i) - c_{Af}(R_o)],$$

where

$$k_{Af} = \frac{D_{A,f}}{R_o \ln(R_o/R_i)}, \quad S_o = 2\pi R_o L.$$

Flow of A through the external fluid film is:

$$W_A = k_{Ae} S_o [c_{Ae}(R_o) - c_{Ae\infty}].$$

Assuming local equilibrium at the fluid–fiber interfaces,

$$c_{Af}(R_i) = \Phi_{A,fi} c_{Ai}(R_i),$$
$$c_{Af}(R_o) = \Phi_{A,fe} c_{Ae}(R_o).$$

The flow of A through the fiber can be written in terms of the fluid concentrations at the inner and outer surfaces of the fiber:

$$W_A = S_o k_{Af} [\Phi_{A,fi} c_{Ai}(R_i) - \Phi_{A,fe} c_{Ae}(R_o)] = S_o k_{Af} \Phi_{A,fi} [c_{Ai}(R_i) - \Phi_{A,ie} c_{Ae}(R_o)].$$

Multiplying the flow of A through the external fluid by $\Phi_{A,ie}$:

$$W_A \Phi_{A,ie} = k_{Ae} S_o [\Phi_{A,ie} c_{Ae}(R_o) - \Phi_{A,ie} c_{Ae\infty}].$$

Rearranging the equations for internal fluid, fiber and external fluid, respectively:

$$\frac{W_A}{k_{Ai} S_i} = [c_{Aib} - c_{Ai}(R_i)],$$

$$\frac{W_A}{S_o k_{Af} \Phi_{A,fi}} = [c_{Ai}(R_i) - \Phi_{A,ie} c_{Ae}(R_o)],$$

$$W_A \frac{\Phi_{A,ie}}{k_{Ae} S_o} = [\Phi_{A,ie} c_{Ae}(R_o) - \Phi_{A,ie} c_{Ae\infty}].$$

Adding these three equations:

$$W_A \left[\frac{1}{k_{Ai} S_i} + \frac{1}{k_{Af} S_o \Phi_{A,if}} + \frac{\Phi_{A,ie}}{k_{Ae} S_o} \right] = [c_{Aib} - \Phi_{A,ie} c_{Ae\infty}].$$

12.6 Permeability of Nonporous Materials

Therefore,

$$W_A = \left\{ \cfrac{1}{\left[\cfrac{R_o}{k_{Ai}R_i} + \cfrac{1}{k_{Af}\Phi_{A,if}} + \cfrac{\Phi_{A,ie}}{k_{Ae}}\right]} \right\} S_o \left[c_{Aib} - \Phi_{A,ie} c_{Ae\infty} \right].$$

The quantity in brackets is the overall permeability, which accounts for film resistance and fiber resistance. Multiplying numerator and denominator by $\Phi_{A,fi}$ and substituting for k_{Af}, we find the permeability is:

$$P_o = \cfrac{\Phi_{A,fi}}{\Phi_{A,fi} \cfrac{R_o}{R_i} \cfrac{1}{k_{Ai}} + \cfrac{R_o \ln(R_o/R_i)}{D_{A,f}} + \cfrac{\Phi_{A,fe}}{k_{Ae}}}.$$

Examining and interpreting the results: The overall permeability of species A in the fiber is influenced by the partition coefficients between the fiber and internal fluid and the fiber and external fluid, the inside and outside radii of the fiber, the diffusion coefficient of A in the fiber and the mass transfer coefficients of A in the internal and external fluids. If the internal and external fluids are well-mixed, then k_{Ai} and k_{Ae} will be very large and the permeability will be:

$$P_o = \cfrac{D_{A,f} \Phi_{A,fi}}{R_o \ln(R_o/R_i)} \quad \text{(well mixed)}.$$

The permeability in this case is independent of the partition coefficient between the fiber and external fluid. This same relationship can be derived by substituting $c_{Aib} = \Phi_{A,if} c_{Af}(R_i)$ and $c_{Ae\infty} = \Phi_{A,ef} c_{Af}(R_o)$ for a well-mixed fluid into the expression for W_A through the fiber wall. The permeability is also independent of the partition coefficient between fiber and external fluid if the external fluid is the only fluid that is well-mixed.

12.6.3 Comparison of Internal and External Resistances to Mass Transfer

In some instances, the resistance to mass transfer to a particular species A within a system may be much lower than the resistance to mass transfer at the system boundaries. In such cases, the concentration of A within the system is relatively constant and a well-mixed or "lumped" analysis can be used. We will devote Chap. 13 to the analysis of such cases. Our goal here is to determine the circumstances under which such an approach is valid. In Sect. 12.6.1, we derived an expression for the permeability of a solute that moved from one liquid to another

Fig. 12.8 Mass transfer to an external system

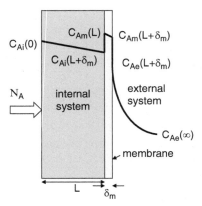

across a membrane. Generalizing this to apply to the movement of a solute from one media, i (gas, liquid or solid) to another, e, across a barrier, as shown in Fig. 12.8:

$$N_A = \mathsf{P}_A\left[c_{Ai}(0) - \Phi_{A,ie}c_{Ae}(\infty)\right]. \tag{12.165}$$

The total resistance to mass transfer between the internal and external media $\Re_{A,tot}$ is equal to the inverse of the permeability P_A. According to (12.155):

$$\Re_{A,tot} = \frac{1}{\mathsf{P}_A} = \Re_{Ai} + \Re_{Am} + \Re_{Ae}, \tag{12.166}$$

where the resistances offered by the internal media, membrane and external media are given as:

$$\Re_{Ai} = \frac{1}{k_{Ai}} = \frac{L}{D_{Ai}}, \tag{12.167}$$

$$\Re_{Am} = \frac{\Phi_{Aim}}{k_{Am}} = \frac{\Phi_{Aim}\delta_m}{D_{Am}}, \tag{12.168}$$

$$\Re_{Ae} = \frac{\Phi_{Aie}}{k_{Ae}}. \tag{12.169}$$

Transport is assumed to be purely by diffusion in the internal media and in the membrane, and a convective mass transfer coefficient k_{Ae} governs mass transfer in the external media. Let us consider first the case where the membrane offers very little resistance to mass transfer. In that case, the ratio of the internal resistance to the external resistance is known as the *Biot number for mass transfer*:

$$Bi_A = \frac{\Re_{Ai}}{\Re_{Ae}} = \frac{k_{Ae}L}{D_{Ai}\Phi_{Aie}}. \tag{12.170}$$

If the Biot number is small, then the resistance to mass transfer is primarily confined to the external media. Consequently, the concentration difference in the internal media will be very small. Under these circumstances, the "lumped" analysis, which neglects concentration variations in the internal media, is justified. Generally, the lumped analysis is presumed valid when $Bi_A < 0.1$.

In nonrectangular systems, the length L in (12.170) represents a characteristic length of the internal system. For a spherical cell, the radius of the cell would be an appropriate choice for the characteristic length. For odd-shaped systems, the ratio of the system volume to surface area can be selected as a characteristic length.

If the membrane resistance is not negligible, then a comparison of internal resistance to the sum of membrane and external resistances is:

$$\frac{\mathfrak{R}_{Ai}}{\mathfrak{R}_{Ae} + \mathfrak{R}_{Am}} = \frac{\frac{k_{Ae}L}{D_{Ai}\Phi_{Aie}}}{1 + \frac{k_{Ae}\delta_m}{D_{Am}\Phi_{Aem}}} = \frac{Bi_A}{1 + \frac{k_{Ae}\delta_m}{D_{Am}\Phi_{Aem}}} \leq Bi_A. \qquad (12.171)$$

Therefore, the Biot number represents an upper limit to the actual resistance ratio. Consequently, a lumped analysis would still be appropriate for $Bi_A < 0.1$, and may in fact be valid for larger values of Bi_A.

12.7 Transport of Electrically Charged Species

Fick's law, which describes how diffusive flux depends on concentration gradient, applies to charged species as well as neutral species. However, if charged species are placed in an electric field, an additional flux will occur. The electrical flux $\vec{J}_A^e$ is found to be proportional to the product of the negative electrical potential gradient $-\vec{\nabla}E$, valence of the species z_A, the species concentration c_A and the species diffusion coefficient D_{AB}, and is inversely proportional to the absolute temperature:

$$\vec{J}_A^e = -\frac{D_{AB}z_A c_A F}{RT}\vec{\nabla}E, \qquad (12.172)$$

where R is the universal gas constant and F is Faraday's constant (96,500 C/g-equivalent). At 37°C, $RT/F = 26.7$ mV. The total passive flux of a charged species is equal to the convective and diffusive fluxes plus the electrical flux:

$$\vec{N}_A = c_A \vec{v} - D_{AB}\vec{\nabla}c_A - D_{AB}z_A c_A \left(\frac{F}{RT}\right)\vec{\nabla}E. \qquad (12.173)$$

The electrical current density $\vec{i}$ caused by the movement of all species is:

$$\vec{i} = F\sum_j z_j \vec{N}_j. \qquad (12.174)$$

Table 12.2 Intracellular and extracellular electrolyte concentrations

Ion	$C_{\text{intracellular}}$ (mM)	$C_{\text{extracellular}}$ (mM)
Chloride	3.5	120
Potassium	140	2.5
Sodium	9.2	120

Example 12.7.1 The Nernst Potential.
The Nernst potential for a particular species A is the potential difference that results across a membrane when all species are at equilibrium. Normal concentrations of sodium, potassium and chloride ions on the inside and outside of a typical muscle cell are shown in Table 12.2. The partition coefficients between intracellular and extracellular fluids can be assumed to be unity for all ions. Compare the Nernst potential for each species with the actual measured potential ($\Delta E = E_{\text{intracellular}} - E_{\text{extracellular}}$) of -90 mV. Discuss any differences.

Solution. *Initial considerations:* If intracellular fluid is in equilibrium with extracellular fluid, the net flux of each species will be zero and the species concentration differences will be consistent with the measured potential difference. We will assume that there is no net convection through the membrane.

System definition and environmental interactions: The system to be analyzed is the cell membrane.

Apprising the problem to identify governing relationships: Since the species all have an electrical charge, we will apply (12.173), with the flux equal to zero, to all species.

Analysis: We begin by applying (12.173) to the flux of an ion species across the cell membrane. Since the membrane is very thin, we can assume one-dimensional flow perpendicular to the surface of the membrane. In addition, we set the net flux equal to zero and assume there is no convection through the membrane. This leads to the following simplification of (12.173):

$$0 = \frac{dC_{Am}}{dx} + \frac{z_A F C_{Am}}{RT}\frac{dE}{dx}.$$

Rearranging:

$$\frac{dC_{Am}}{C_{Am}} = -\frac{z_A F}{RT}dE.$$

Integrating from the inside to outside surface of the cell membrane with $C_{Am,in} = \Phi_{A,m,\text{intracellular}} C_{A,\text{intracellular}}$ and $C_{Am,out} = \Phi_{A,m,\text{extracellular}} C_{A,\text{extracellular}}$ yields:

$$\ln\left(\frac{\Phi_{A,m,\text{extracellular}} C_{A,\text{extracelluar}}}{\Phi_{A,m,\text{intracellular}} C_{A,\text{intracelluar}}}\right) = -\frac{z_A F}{RT}(E_{\text{extracellular}} - E_{\text{intracellular}}).$$

12.7 Transport of Electrically Charged Species

Rearranging to solve for the potential difference and recognizing that $RT/F = 26.7$ mV at $37°C$, and the ratio of partition coefficients is unity, we have:

$$\Delta E = \frac{26.7 \text{ mV}}{z_A} \left[\ln\left(\frac{C_{A,\text{intracelluar}}}{C_{A,\text{extracelluar}}}\right)\right].$$

Examining and interpreting the results: This is known as the Nernst equation. Substituting $z_{Cl} = -1$ and the values for chloride concentrations from Table 12.2 into this expression provides a Nernst potential of -94.4 mV, which is within the experimental error of the measured potential difference. The Nernst potential for potassium ($z_K = +1$) is computed to be -107 mV, which is close to the measured potential difference, but the difference is greater than would be expected from experimental variability. Finally, computation of the Nernst potential for sodium ($z_{Na} = +1$) is $+68.6$ mV, in complete disagreement with the measured value of -90 mV. If the net fluxes of sodium and potassium ions are known to be zero, why are the Nernst potentials for sodium and potassium not equal to the measured potential? The reason is that all ions in the cell are *NOT* in equilibrium. If they were, the cell would be dead. Instead, the cell is in a steady-state. Equation (12.173) describes the *passive* flux of an ion through the membrane, not the total movement of the ion. In the case of chloride ion, there is no active flux, and the ion is in equilibrium. Hence, the Nernst potential for chloride ion is the same as the actual membrane potential. However, for sodium and potassium the passive flux is not zero, but instead just balances the active flux produced by the Na–K–ATP pump.

Additional comments: To account for active transport, (12.173) must be modified as follows:

$$\vec{N}_{A,\text{total}} = \vec{N}_{A,\text{active}} + c_A \vec{v} - D_{AB} \vec{\nabla} C_A - D_{AB} z_A C_A \left(\frac{F}{RT}\right) \vec{\nabla} E. \quad (12.175)$$

If the total flux and the convective flux of sodium are both zero,

$$\frac{N_{Na^+,\text{active}}}{D_{Na^+,m}} = \frac{dC_{Na^+,m}}{dx} + \alpha_{Na^+} C_{Na^+,m} \frac{dE}{dx}, \quad (12.176)$$

where

$$\alpha_{Na^+} \equiv \frac{z_{Na^+} F}{RT}. \quad (12.177)$$

Multiplying both sides of (12.176) by an integrating factor, $e^{\alpha_{Na^+} E}$:

$$\frac{N_{Na^+,\text{active}}}{D_{Na^+,m}} e^{\alpha_{Na^+} E} = e^{\alpha_{Na^+} E} \left[\frac{dC_{Na^+,m}}{dx} + \alpha_{Na^+} C_{Na^+,m} \frac{dE}{dx}\right] = \frac{d}{dx}\left(e^{\alpha_{Na^+} E} C_{Na^+,m}\right). \quad (12.178)$$

Integration of this equation requires knowledge of how the electrical potential varies from one side of the membrane to the other. A common practice is to assume that the potential varies linearly from the inside surface to the outside surface of the cell membrane, with the outside potential being held at zero.

$$E(x) = \Delta E \left(\frac{x}{\delta}\right), \tag{12.179}$$

where δ is the thickness of the cell membrane. Substitution of (12.179) into (12.178) and integrating yields the following expression for the active flux:

$$N_{Na^+,active} = \left(\frac{D_{Na^+,m}}{\delta_m \Phi_{Na^+,extracellular,m}}\right)(\alpha_{Na^+}\Delta E)$$

$$\times \frac{\left[C_{Na^+,extracellular} - \Phi_{Na^+,extracellular,intracellular}C_{Na^+,intracellular}e^{\alpha_{Na^+}\Delta E}\right]}{1 - e^{\alpha_{Na^+}\Delta E}}.$$

The first term in brackets is the passive membrane permeability to sodium (12.156) and the partition coefficient is assumed to be unity. With these definitions, the active transport of sodium ion is:

$$N_{Na^+,active} = P_{Na^+}\alpha_{Na^+}\Delta E \frac{\left[C_{Na^+,extracellular} - C_{Na^+,intracellular}e^{\alpha_{Na^+}\Delta E}\right]}{1 - e^{\alpha_{Na^+}\Delta E}}.$$

This is known as the *Goldman equation*. If the permeability of sodium in the membrane is known, then information in Table 12.2 along with the measured potential difference can be used to compute the active sodium flux. A similar expression can be derived for potassium:

$$N_{K^+,active} = P_{K^+}\alpha_{K^+}\Delta E \frac{\left[C_{K^+,extracellular} - C_{K^+,intracellular}e^{\alpha_{K^+}\Delta E}\right]}{1 - e^{\alpha_{K^+}\Delta E}}.$$

Taking the ratio of the active sodium to potassium fluxes, we obtain the *Goldman–Hodgkin–Katz equation*:

$$\frac{N_{Na^+,active}}{N_{K^+,active}} = \frac{P_{Na^+}}{P_{K^+}} \frac{\left[C_{Na^+,extracellular} - C_{Na^+,intracellular}e^{\left(\frac{\Delta E}{26.7\,mV}\right)}\right]}{\left[C_{K^+,extracellular} - C_{K^+,intracellular}e^{\left(\frac{\Delta E}{26.7\,mV}\right)}\right]}.$$

The ratio of sodium to potassium permeability is about 0.03. Using the concentrations in Table 12.2 and the measured potential difference of -90 mV:

$$\frac{N_{Na^+,active}}{N_{K^+,active}} = 0.03 \frac{\left[120\,mM - 9.2\,mM\left(e^{\left(\frac{-90\,mV}{26.7\,mV}\right)}\right)\right]}{\left[2.5\,mM - 140\,mM\left(e^{\left(\frac{-90\,mV}{26.7\,mV}\right)}\right)\right]} = -1.55.$$

This is consistent with the sodium–potassium–ATP pump which pumps three sodium ions out of the cell for every two potassium ions pumped into the cell.

12.8 Chemical Reactions

Chemical reactions are vital for the survival of all living organisms. Sequences of reactions, most promoted by enzymes, provide metabolic energy for organisms to sustain life. Key reactions which occur on cell surfaces protect the cell from ingestion of toxic substances. Reactions between specific proteins in blood can promote clotting and prevent blood loss. Reactions which take place in erythrocytes can cause considerably higher amounts of oxygen and carbon dioxide to be transported in blood than could physically dissolve in plasma. Any number of such examples can be cited.

Chemical reactions can be classified as being *homogeneous* or *heterogeneous*. Homogeneous chemical reactions take place between species that are distributed in the same phase, while heterogeneous reactions occur at the interface between two phases. A species mass balance performed on a small material volume within a single phase would necessarily include species production by a homogeneous chemical reaction, but it would not include production by a heterogeneous reaction. Material produced by a heterogeneous reaction results in a flux, which does not appear as a source term in a mass balance, but rather as a boundary condition. This will be considered in Sect. 13.5.2.

In this section, we will be concerned with estimating the rate at which individual species are produced or depleted in homogeneous chemical reactions. For example, two reactants A and B combine to form a product C:

$$A + B \rightarrow C. \quad (12.180)$$

If many molecules of A and B are present in solution, the rate of reaction will be proportional to the concentrations of both species A and species B. The rate of formation of C per unit volume R_{cf} (mol L^{-1} s^{-1}) can be written as:

$$R_{Cf} = k_f c_A c_B, \quad (12.181)$$

c_A and c_B are the concentrations of species A and B, and k_f is a proportionality factor known as the forward reaction rate constant. The units for k_f will depend on the number of reactant molecules which combine to form the product(s). In this case, if concentrations are expressed in mol/L, k_f will have units of L^1 mol^{-1} s^{-1}. If the reaction is reversible, then C might decompose to form the original species:

$$C \rightarrow A + B. \quad (12.182)$$

The rate of decomposition might be quite different than the rate of formation of C. The rate of decomposition of C per unit volume R_{Cd} will depend on the concentration of C:

$$R_{Cd} = k_r c_C. \tag{12.183}$$

The factor k_r is known as the reverse reaction rate, and the units of k_r depend on the number of product molecules involved in the decomposition and is generally different than the units for k_f. In this case, a single molecule of C decomposes into its original components, so k_r will have units of 1/s. If k_r is zero or is negligibly small, then the reaction is said to be an irreversible reaction. The forward and reverse reactions are often written in a combined relation, as follows:

$$A + B \underset{k_r}{\overset{k_f}{\rightleftarrows}} C. \tag{12.184}$$

The net rate of production per unit volume of product C is given by the difference between the rate of formation and the rate of decomposition:

$$R_C = R_{Cf} - R_{Cd} = k_f c_A c_B - k_r c_C. \tag{12.185}$$

Following the same procedure to determine the rate of production of species A and B per unit volume:

$$R_A = R_B = -R_C = k_r c_C - k_f c_A c_B. \tag{12.186}$$

Now, let us take a slightly more complicated reaction:

$$4A + B \underset{k_r}{\overset{k_f}{\rightleftarrows}} C + 2D. \tag{12.187}$$

This can also be written as:

$$A + A + A + A + B \underset{k_r}{\overset{k_f}{\rightleftarrows}} C + D + D. \tag{12.188}$$

Following the procedure above, we can find the rate of production of C per unit volume:

$$R_C = k_f c_A c_A c_A c_A c_B - k_r c_C c_D c_D, \tag{12.189}$$

or

$$R_C = k_f c_A^4 c_B - k_r c_C c_D^2. \tag{12.190}$$

12.8 Chemical Reactions

The exponents determine the *order of the reaction* relative to each species. The reaction is first order relative to species C and B, second order relative to D and fourth order relative to species A. From the stoichiometry of the reaction, the rate of production of 1 mole of C is accompanied by the production of 2 moles of D and the disappearance of 4 moles of A and 1 mole of B. Thus, the rates must be related as follows:

$$R_C = \frac{R_D}{2} = -R_B = -\frac{R_A}{4}. \quad (12.191)$$

The general reaction between n reactants R and m products P can be represented as:

$$r_1 R_1 + r_2 R_2 + \cdots + r_n R_n \underset{k_r}{\overset{k_f}{\rightleftarrows}} p_1 P_1 + p_2 P_2 + \cdots + p_m P_m, \quad (12.192)$$

where r_i and p_j are the stoichiometric coefficients. The rates of production of reactants and products are related as follows:

$$\frac{R_{P_1}}{p_1} = \frac{R_{P_2}}{p_2} = \frac{R_{P_m}}{p_m} = -\frac{R_{R_1}}{r_1} = -\frac{R_{R_2}}{r_2} = -\frac{R_{R_n}}{r_n}$$
$$= k_f c_{R_1}^{r_1} c_{R_2}^{r_2} \cdots c_{R_n}^{r_n} - k_r c_{P_1}^{p_1} c_{P_2}^{p_2} \cdots c_{P_m}^{p_m}. \quad (12.193)$$

Chemical equilibrium occurs when the forward and reverse rates for each reacting species are equal. If chemical equilibrium is achieved, the total rate of production of each species is zero:

$$0 = k_f c_{R_1}^{r_1} c_{R_2}^{r_2} \cdots c_{R_n}^{r_n} - k_r c_{P_1}^{p_1} c_{P_2}^{p_2} \cdots c_{P_m}^{p_m}. \quad (12.194)$$

An equilibrium constant K_{eq} can be defined as the ratio of the forward to reverse rate constants:

$$K_{eq} = \frac{k_f}{k_r} = \frac{c_{P_1}^{p_1} c_{P_2}^{p_2} \cdots c_{P_m}^{p_m}}{c_{R_1}^{r_1} c_{R_2}^{r_2} \cdots c_{R_n}^{r_n}}. \quad (12.195)$$

In many biological and physiological applications, the chemical species can be assumed to be in a pseudo-equilibrium. By this, we mean that the forward and reverse rate constants are sufficiently fast so that when the concentration of a reactant or product is changed, the reaction quickly readjusts the concentrations toward new equilibrium values. Important applications involving blood gases, enzyme kinetics and receptor-mediated reactions are discussed in the following sections. Many other examples are equally important.

12.8.1 Hemoglobin and Blood Oxygen Transport

Hemoglobin is a 64,458 molecular weight protein which is contained at high concentration within erythrocytes. Hemoglobin is responsible for one third of the mass of a typical red cell (34 g/dl). Each molecule of hemoglobin contains four heme groups, and each heme group is capable of combining with one molecule of oxygen. The principle function of hemoglobin is to store oxygen as blood passes through the lungs and release needed oxygen as blood passes through tissue capillaries. If hemoglobin were completely saturated with oxygen, so that all of the heme groups were occupied, then the number of moles of oxygen bound to hemoglobin in the form of oxyhemoglobin per liter of erythrocytes would be:

$$\frac{1\,\text{mole HbO}_2}{64,458\,\text{g HbO}_2}\left(\frac{4\,\text{moles O}_2}{\text{mole HbO}_2}\right)\left(\frac{340\,\text{g HbO}_2}{\text{liter RBC}}\right) = 0.021\,\frac{\text{moles O}_2}{\text{liter RBC}}. \quad (12.196)$$

This is the oxygen-binding capacity of hemoglobin in the form of oxyhemoglobin within red blood cells. The oxygen-carrying capacity of blood can be found by multiplying this by the hematocrit value of blood and adding the physically dissolved oxygen in plasma and erythrocytes. The latter value can be computed by applying Henry's law. The Henry's law constant for O_2 in blood is about 0.74 Torr/μM of O_2 and the P_{O_2} of arterial blood is normally about 100 Torr. The amount of O_2 that is physically dissolved in arterial blood is about 135 μM of O_2. For a hematocrit value of 45%, the oxygen capacity of red cells is 9,000 μM of O_2. Clearly, hemoglobin plays a vital role in the storage and transport of oxygen in blood.

The equilibrium relationship between dissolved oxygen and bound oxygen is nonlinear, exhibiting a characteristic sigmoid or "s" shape. A plot of saturation vs. partial pressure of oxygen is known as the oxyhemoglobin dissociation curve. Oxyhemoglobin exists as four species: $Hb(O_2)_1$, $Hb(O_2)_2$, $Hb(O_2)_3$, and $Hb(O_2)_4$. Adair (1925) proposed a four-step model to describe oxyhemoglobin dissociation:

$$\begin{aligned}
Hb(O_2)_4 &\underset{k_{r4}}{\overset{k_{f4}}{\rightleftarrows}} O_2 + Hb(O_2)_3, \\
Hb(O_2)_3 &\underset{k_{r3}}{\overset{k_{f3}}{\rightleftarrows}} O_2 + Hb(O_2)_2, \\
Hb(O_2)_2 &\underset{k_{r2}}{\overset{k_{f2}}{\rightleftarrows}} O_2 + Hb(O_2)_1, \\
Hb(O_2)_1 &\underset{k_{r1}}{\overset{k_{f1}}{\rightleftarrows}} O_2 + Hb.
\end{aligned} \quad (12.197)$$

If oxygen is added to blood (lungs) or removed (tissue), the reactions above bring the species to new near-equilibrium values within a few milliseconds. The normal transit time through capillaries is on the order of 1 s, so it is generally

12.8 Chemical Reactions

assumed that there is a local equilibrium between oxygen and each of the hemoglobin species:

$$\frac{k_{r4}}{k_{f4}} = K_4 = \frac{C_{Hb(O_2)_4}}{C_{O_2} C_{Hb(O_2)_3}},$$

$$\frac{k_{r3}}{k_{f3}} = K_3 = \frac{C_{Hb(O_2)_3}}{C_{O_2} C_{Hb(O_2)_2}},$$

$$\frac{k_{r2}}{k_{f2}} = K_2 = \frac{C_{Hb(O_2)_2}}{C_{O_2} C_{Hb(O_2)_1}},$$

$$\frac{k_{r1}}{k_{f1}} = K_1 = \frac{C_{Hb(O_2)_1}}{C_{O_2} C_{Hb}}. \tag{12.198}$$

Each equilibrium constant has units of M^{-1}. The oxyhemoglobin saturation S_{HbO_2} is equal to the number of heme groups occupied by oxygen relative to the total number of heme groups available, or the number of moles of oxygen bound to hemoglobin relative to the maximum number of moles of oxygen that can bind with hemoglobin:

$$S_{HbO_2} = \frac{C_{Hb(O_2)_1} + 2C_{Hb(O_2)_2} + 3C_{Hb(O_2)_3} + 4C_{Hb(O_2)_4}}{4 C_{Hb,tot}}, \tag{12.199}$$

where $C_{Hb,tot}$ is the total concentration of all hemoglobin species present:

$$C_{Hb,tot} = C_{Hb} + C_{Hb(O_2)_1} + C_{Hb(O_2)_2} + C_{Hb(O_2)_3} + C_{Hb(O_2)_4}. \tag{12.200}$$

Each of the oxyhemoglobin species can be written in terms of the concentration of dissolved oxygen, and Henry's law can be used to write the oxygen concentrations in terms of the partial pressure of oxygen and the solubility of oxygen in blood:

$$C_{O_2} = \alpha_{O_2} P_{O_2}. \tag{12.201}$$

Substituting (12.198), (12.200), and (12.201) into (12.199), we arrive at the Adair equation relating oxyhemoglobin saturation to oxygen partial pressure:

$$S_{HbO_2} = \frac{\left(\frac{a_1}{4}\right) P_{O_2} + \left(\frac{a_2}{2}\right) P_{O_2}^2 + \left(\frac{3a_3}{4}\right) P_{O_2}^3 + a_4 P_{O_2}^4}{1 + a_1 P_{O_2} + a_2 P_{O_2}^2 + a_3 P_{O_2}^3 + a_4 P_{O_2}^4}, \tag{12.202}$$

where the coefficients a_1, a_2, a_3, and a_4 are defined in terms of the four equilibrium constants and the solubility of oxygen in blood:

$$a_1 = K_1 \alpha_{O_2},$$
$$a_2 = K_1 K_2 \alpha_{O_2}^2,$$
$$a_3 = K_1 K_2 K_3 \alpha_{O_2}^3,$$
$$a_4 = K_1 K_2 K_3 K_4 \alpha_{O_2}^4. \qquad (12.203)$$

DeHaven and Deland (1962) found the best-fit values at 37°C and pH 7.4 to be:

$$K_1 \alpha_{O_2} = 0.04 \, \text{Torr}^{-1},$$
$$K_2 \alpha_{O_2} = 0.034 \, \text{Torr}^{-1},$$
$$K_3 \alpha_{O_2} = 0.003 \, \text{Torr}^{-1},$$
$$K_4 \alpha_{O_2} = 0.8 \, \text{Torr}^{-1}. \qquad (12.204)$$

A comparison of experimental data with the prediction based on (12.202) using equilibrium values from (12.204) is shown in Fig. 12.9. This provides an excellent fit over the entire range of partial pressures. Winslow et al. (1983) provide empirical methods for estimating the coefficients a_1, a_2, a_3, and a_4 for different values of 2,3-DPG, pH and P_{CO_2} within the physiological range.

A simpler model of the oxyhemoglobin saturation curve was proposed by Hill (1910). This approach lumps the various oxyhemoglobin species into a single species $(Hb(O_2)_n)$:

$$n(O_2) + Hb \underset{k_r}{\overset{k_f}{\rightleftarrows}} Hb(O_2)_n. \qquad (12.205)$$

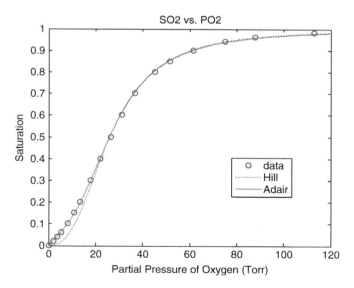

Fig. 12.9 Oxyhemoglobin saturation vs. partial pressure of oxygen

12.8 Chemical Reactions

The rate of production per unit volume of $Hb(O_2)_n$ is:

$$R_{Hb(O_2)_n} = k_f C_{O_2}^n C_{Hb} - k_r C_{Hb(O_2)_n}. \quad (12.206)$$

At equilibrium, the rate of production vanishes and we can write:

$$\frac{k_r}{k_f} = K_{eq} = \frac{C_{O_2}^n C_{Hb}}{C_{Hb(O_2)_n}}. \quad (12.207)$$

The total hemoglobin concentration $C_{Hb,tot}$ is the sum of the unbound hemoglobin concentration and the oxyhemoglobin concentration:

$$C_{Hb,tot} = C_{Hb} + C_{Hb(O_2)_n}. \quad (12.208)$$

Oxyhemoglobin saturation S is:

$$S_{HbO_2} = \frac{C_{Hb(O_2)_n}}{C_{Hb,tot}} = \frac{C_{Hb(O_2)_n}}{C_{Hb} + C_{Hb(O_2)_n}}. \quad (12.209)$$

Substituting (12.201) and (12.207) into (12.209) provides the following relationship between oxyhemoglobin saturation and the partial pressure of oxygen:

$$S_{HbO_2} = \frac{P_{O_2}^n}{P_{O_2}^n + \dfrac{K_{eq}}{\alpha_{O_2}^n}}. \quad (12.210)$$

The $P_{O_2,50}$ is defined as the partial pressure of oxygen when 50% of the hemoglobin is saturated ($S_{HbO_2} = 0.5$). Substituting this into the expression for saturation:

$$S_{HbO_2} = \frac{P_{O_2}^n}{P_{O_2}^n + P_{O_2,50}^n} = \frac{\left(\dfrac{P_{O_2}}{P_{O_2,50}}\right)^n}{1 + \left(\dfrac{P_{O_2}}{P_{O_2,50}}\right)^n}, \quad (12.211)$$

where

$$P_{O_2,50} = \frac{(K_{eq})^{\frac{1}{n}}}{\alpha_{O_2}}. \quad (12.212)$$

Equation (12.211) is known as the Hill equation. Since 4 moles of oxygen can combine with 1 mole of hemoglobin, we might expect the coefficient n to equal four. However, since multiple oxyhemoglobin species are actually present, the best-fit value for n is approximately 2.6 and the best-fit value for $P_{O_2,50}$ is about

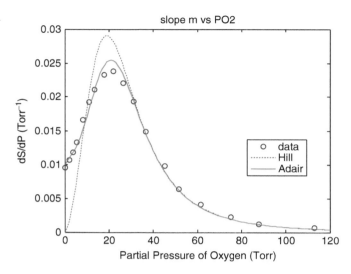

Fig. 12.10 Slope of oxyhemoglobin saturation curve vs. oxygen partial pressure

26 Torr. A comparison of experimental data for S vs. P_{O_2} and the Hill model prediction is also shown in Fig. 12.9. The model provides a good fit to experimental data for partial pressures above 20 Torr but is inaccurate at the low end of the dissociation curve.

As we will see in later chapters, the slope of the oxyhemoglobin dissociation curve is an important factor in predicting blood-tissue oxygen exchange. The slopes of the Hill model and Adair models are compared with the slope from the experimental data in Fig. 12.10. The Adair model represents the data well over the entire range of partial pressures. The Hill model provides a good estimate of the slope for P_{O_2} above about 35 Torr. However, the Hill model predicts a zero slope as P_{O_2} approaches zero, while the Adair model predicts a more realistic slope of $K_1 \alpha_{O_2}/4$.

Improvements to the Hill model have been made by Dash and Bassingthwaighte (2010) which improves the fit at higher saturations and provides for variations in CO_2, pH, temperature and 2,3-DPG over the entire physiological range. Predictions are given in Fig. 12.11, which are in good agreement with the experimental results.

If oxygen gas at standard temperature (273 K) and pressure (760 Torr) were brought into contact with 100 ml blood containing deoxygenated hemoglobin with mass concentration $\rho_{Hb,blood}$, a volume of oxygen would move from the gas phase into the blood. Some of the oxygen would dissolve in the blood and some would combine with hemoglobin. The binding capacity of hemoglobin in blood can be found in a manner similar to that found for the binding capacity of hemoglobin in red cells (12.196):

12.8 Chemical Reactions

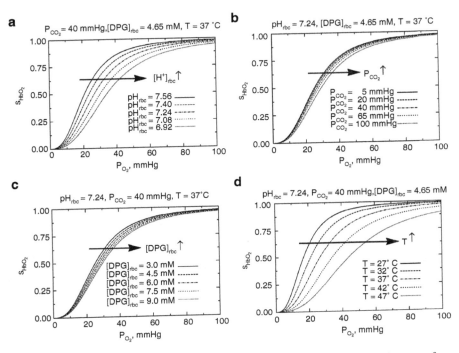

Fig. 12.11 Effect of pH, P_{CO_2}, DPG and temperature on the oxyhemoglobin saturation curve from Dash and Bassingthwaite 2010 with permission

$$\text{Theoretical } O_2 \text{ Binding Capacity of Hb} \left(\frac{\text{ml } O_2(\text{STP})}{\text{dl blood}}\right)$$

$$= 22,394 \frac{\text{ml } O_2}{\text{mol } O_2} \times 4 \frac{\text{mol } O_2}{\text{mol } HbO_2} \times \frac{1 \text{mol } HbO_2}{64,458 \text{ gHbO}_2}$$

$$\times \rho_{Hb,blood} \left(\frac{\text{g } HbO_2}{\text{dl blood}}\right)$$

$$= 1.39 \frac{\text{ml } O_2(\text{STP})}{\text{g } Hb_{blood}} \rho_{Hb,blood}. \tag{12.213}$$

The factor 1.39 ml O_2/g Hb_{blood} represents the maximum volume of O_2 at STP that can combine with 1 g of hemoglobin in blood. The actual binding capacity in blood is usually smaller than this theoretical maximum because some hemoglobin, such as methemoglobin or carboxyhemoglobin, has lost its ability to combine with oxygen. If f_{HbO_2} represents the fraction of total hemoglobin capable of combining with oxygen, then the actual binding capacity will be:

$$\text{Actual } O_2 \text{ Binding Capacity of Hb} \left(\frac{\text{ml } O_2(\text{STP})}{\text{dl blood}}\right)$$

$$= 1.39 \left(\frac{\text{ml } O_2}{\text{g } HbO_2}\right) f_{HbO_2} \rho_{Hb,blood} \left(\frac{\text{g } HbO_2}{\text{dl blood}}\right), \tag{12.214}$$

ordinarily, about 96.4% of the heme groups in blood are able to combine reversibly with oxygen. Therefore, 1 g of hemoglobin can combine with 1.34 ml of oxygen when fully saturated. The total concentration of oxygen in blood at a given P_{O_2} will equal the sum of dissolved oxygen and the oxygen bound to hemoglobin. The bound oxygen is the product of the oxyhemoglobin saturation and the oxygen capacity:

$$C_{O_2}^* = \alpha_{O_2}^* P_{O_2} + 1.34 \rho_{Hb,blood} S_{HbO_2}(P_{O_2}), \qquad (12.215)$$

where $C_{O_2}^*$ is normally expressed in units of (ml O_2(STP))/(dl blood), the mass concentration of hemoglobin is in units of (g Hb)/(dl blood) and the Bunsen solubility coefficient $\alpha_{O_2}^*$ for oxygen in blood is 0.003 (ml O_2(STP))(dl blood)$^{-1}$(Torr)$^{-1}$. The oxyhemoglobin saturation S_{HbO_2} is the fraction of functional heme groups that are bound to oxygen.

12.8.2 Blood CO_2 Transport and pH

Carbon dioxide is transported by three principal mechanisms in blood: (1) physically dissolved in plasma and erythrocytes, (2) in the form of bicarbonate ions and (3) bound to proteins, principally in the form of carbaminohemoglobin. The solubility of CO_2 in blood is more than 20 times greater than the solubility of O_2 in blood. However, depending on the P_{CO_2}, less than 10% of the total CO_2 in blood is present as the physically dissolved species. Approximately 30% of the CO_2 in blood combines with amino groups on proteins. These are different groups from the heme groups which are responsible for O_2 transport, so hemoglobin can transport both O_2 and CO_2. Letting RNH_2 represent a protein with an amino group, the reversible reaction between protein and CO_2 is:

$$RNH_2 + CO_2 \rightleftarrows RNCOOH^- + H^+. \qquad (12.216)$$

In tissue capillaries, where CO_2 from metabolism is high, this reaction is driven to the right. Consequently, dissolved CO_2 is kept low and blood pH is reduced. In the lung, CO_2 diffuses across the alveolar barrier, driving the reaction in the opposite direction. The extent to which this reaction proceeds is about three times greater for hemoglobin than for oxyhemoglobin. Thus, oxygenation of blood containing carbaminohemoglobin in lung capillaries causes a greater release of CO_2, and the loss of O_2 from blood in tissue capillaries promotes the production of carbaminohemoglobin.

The physically dissolved CO_2 is slowly hydrated in plasma to form carbonic acid H_2CO_3, but the extent to which this reaction proceeds is quite small. Erythrocytes, however, contain an enzyme carbonic anhydrase which promotes the rapid hydration of CO_2 and dehydration of H_2CO_3. Thus, when CO_2 concentration is high, carbonic acid is formed and when CO_2 is low carbonic acid is converted to CO_2 and

12.8 Chemical Reactions

water. Carbonic acid is a weak acid which naturally dissociates into to a hydrogen ion and a bicarbonate ion HCO_3^-:

$$CO_2 + H_2O \rightleftarrows H_2CO_3 \rightleftarrows H^+ + HCO_3^-. \quad (12.217)$$
$$\uparrow$$
$$\text{carbonic anhydrase}$$

The bicarbonate ion is more soluble in solution than carbon dioxide, and it rapidly diffuses between plasma and red cells down its concentration gradient. In tissue, bicarbonate ions diffuse from red cells to plasma, promoting further conversion of carbon dioxide to carbonic acid. The process is reversed in lung capillaries, where the loss of CO_2 lowers intracellular bicarbonate ion concentration, causing more bicarbonate to diffuse into the red cells for conversion to CO_2. Approximately 60% of the CO_2 present in blood is present in the form of the bicarbonate ion. The hydrogen ion formed after the conversion of CO_2 to carbonic acid tends to bind with hemoglobin inside red cells. In addition, carbonic acid can combine with other plasma proteins.

The equilibrium relationship between total carbon dioxide concentration and partial pressure of carbon dioxide is known as the carbon dioxide dissociation curve. This is shown in Fig. 12.12.

The relationship is generally nonlinear, but over the range of normal CO_2 partial pressures shown in Fig. 12.12, the relationship between total bulk concentration of CO_2 and P_{CO_2} is nearly linear:

$$C_{CO_2} = A + B P_{CO_2}. \quad (12.218)$$

Data along the dotted line from arterial to venous conditions from Fig. 12.12 for normal human blood yield an intercept A of 23.3 mlCO$_2$/dl blood and slope B of 0.667 mlCO$_2$ dl blood^{-1} mmHg^{-1}. More extensive relationships which account for

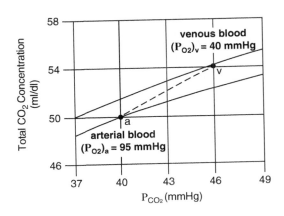

Fig. 12.12 Total CO_2 vs. P_{CO2}

pH, P_{O_2}, bicarbonate ion concentration can also be found in the literature (Dash and Bassingthwaighte 2010).

Blood pH is largely determined by the dissociation of carbonic acid. Under equilibrium conditions, the dissociation constant K is

$$K = \frac{C_{H^+} C_{HCO_3^-}}{C_{H_2CO_3}}. \tag{12.219}$$

Taking the logarithm to the base 10 of both sides and noting that pH $= -\log_{10}(C_{H+})$:

$$\log_{10}(K) = -pK = -6.1 = -pH + \log_{10}\left(\frac{C_{HCO_3^-}}{C_{H_2CO_3}}\right). \tag{12.220}$$

Similarly, the equilibrium reaction between CO_2 and H_2CO_3, catalyzed by carbonic anhydrase, can be written as:

$$K^* = \frac{C_{H_2O} C_{CO_2}}{C_{H_2CO_3}}. \tag{12.221}$$

But since the reaction is virtually complete in the presence of carbonic anhydrase, the ratio K^*/C_{H_2O} is nearly unity. Applying Henry's law, the concentration of carbonic acid is directly proportional to the partial pressure of CO_2:

$$C_{H_2CO_3} = \frac{C_{H_2O} C_{CO_2}}{K^*} = \left(\frac{C_{H_2O}}{K^*}\right) \alpha_{CO_2,blood} P_{CO_2} \approx \alpha_{CO_2,blood} P_{CO_2}. \tag{12.222}$$

Substituting into the expression for pH, we obtain the *Henderson–Hasselbalch equation*:

$$pH = 6.1 + \log_{10}\left(\frac{C_{HCO_3^-}}{C_{CO_2}}\right) = 6.1 + \log_{10}\left(\frac{C_{HCO_3^-}}{\alpha_{CO_2,blood} P_{CO_2}}\right). \tag{12.223}$$

The numerator in the logarithmic expression is largely dictated by regulation of bicarbonate ions in the kidneys while the denominator is regulated by the lungs via respiration. Under normal circumstances the ratio of bicarbonate ion concentration to carbon dioxide concentration in blood is 20:1, so $\log_{10}(20) = 1.3$ and the normal pH is 7.4.

12.8.3 Enzyme Kinetics

Carbonic anhydrase is just one of many different enzymes in the human body. Enzymes are biochemical catalysts that increase the rate of conversion of a

12.8 Chemical Reactions

substrate to a product by reducing the activation energy needed to initiate the reaction. The enzyme may be specific for a single substrate, or may catalyze several different substrates. The enzyme E is a protein that is unchanged by the reaction and can be used to catalyze the reaction over and over. The general reaction for conversion of a single substrate S to a product P is as follows:

$$S + E \underset{k_{rs}}{\overset{k_{fs}}{\rightleftarrows}} ES \underset{k_{rp}}{\overset{k_{fp}}{\rightleftarrows}} E + P. \tag{12.224}$$

ES is known as the enzyme–substrate complex. In many cases, the reverse reaction constant k_{rp} is negligibly small, and the formation of the product is the rate-limiting step. This forms the basis for what is known as Michaelis–Menten kinetics. If the reaction takes place at a steady rate, the amount of substrate bound to the enzyme will remain constant. In that case, with $k_{rp} = 0$, the rate of formation of product P, R_P, will be the same as the rate of removal of substrate, $-R_S$:

$$R_P = k_{fp} c_{ES} = -R_S = k_{fs} c_S c_E - k_{rs} c_{ES}. \tag{12.225}$$

Rearranging:

$$\frac{c_S c_E}{c_{ES}} = \frac{k_{rs} + k_{fp}}{k_{fs}} = K_m. \tag{12.226}$$

The equilibrium coefficient K_m is known as the *Michaelis constant*. The total concentration of enzyme c_{E0} is assumed to remain constant during the reaction and is equal to the sum of unbound and bound enzyme:

$$c_{E0} = c_E + c_{ES} \tag{12.227}$$

Substituting (12.227) into (12.226):

$$K_m = \frac{c_S(c_{E0} - c_{ES})}{c_{ES}} \quad \text{or} \quad c_{ES} = \frac{c_S c_{E0}}{K_m + c_S}. \tag{12.228}$$

Finally, substitution of (12.228) into (12.225) yields the following expression for the rate of production of product:

$$R_P = -R_S = k_{fp} c_{E0} \left(\frac{c_S}{K_m + c_S} \right). \tag{12.229}$$

Thus, the net rate of formation of product is directly proportional to the total concentration of enzyme present and to the forward reaction constant k_{fp} that regulates the conversion of ES to product plus enzyme. Product production rate increases as substrate concentration increases, but levels off at high substrate concentrations because of the limited amount of enzyme present. The maximum rate of conversion V_{max} will occur when all the enzymes are bound to the substrate ($C_{ES} = C_{E0}$), which occurs at high substrate concentrations ($C_S \gg K_m$):

$$V_{max} = R_{P,max} = k_{pf} C_{E0}. \tag{12.230}$$

Substituting the definition of V_{max} into (12.229), and using the bulk substrate concentration C_S, we obtain what is known as the *Michaelis–Menten* equation:

$$R_P = -R_S = V_{max}\left(\frac{C_S}{K_m + C_S}\right). \tag{12.231}$$

The coefficients V_{max} and K_m can be found from the experimental data by measuring the rate of formation of product as a function of substrate concentration. The maximum rate at high concentrations will be V_{max}, and the Michaelis constant will be equal to the substrate concentration that yields a product rate equal to $V_{max}/2$. Although (12.231) is a nonlinear relationship between R_P and C_S, it can be rearranged to provide a linear relationship between $1/R_P$ and $1/C_S$:

$$\frac{1}{R_P} = \frac{1}{V_{max}} + \left(\frac{K_m}{V_{max}}\right)\left(\frac{1}{C_S}\right). \tag{12.232}$$

This is known as the *Lineweaver–Burk equation*. A straight line should result when $1/R_P$ is plotted as a function of $1/C_S$. The intercept will equal $1/V_{max}$ and the slope will equal K_m/V_{max}.

Enzyme reactions can be inhibited by the presence of other substrates that can combine with the enzyme. Competitive substrates combine with the enzyme at the same binding sites as the substrate, thus reducing the *ES* concentration. Noncompetitive inhibitors combine with the enzyme at different sites than the substrate-binding site, but their presence interferes with the binding of the substrate. Binding of a competitive inhibitor is described by:

$$I + E \underset{k_{rI}}{\overset{k_{fI}}{\rightleftarrows}} EI. \tag{12.233}$$

In addition to (12.224) and (12.233), a noncompetitive inhibitor can combine with the *ES* complex, but the inhibitor is assumed to prevent the formation of a product:

$$I + ES \underset{k_{rI}}{\overset{k_{fI}}{\rightleftarrows}} ESI. \tag{12.234}$$

The enzyme–inhibitor complex is assumed to interfere with substrate binding, so there is no need to add a separate equation for the formation of *ESI* from *S* and *EI*. The total enzyme concentration given by (12.227) must be modified to account for enzyme bound with inhibitor. For competitive inhibition:

12.8 Chemical Reactions

$$C_{E0} = C_E + C_{ES} + C_{EI}, \qquad (12.235)$$

and for noncompetitive inhibition:

$$C_{E0} = C_E + C_{ES} + C_{EI} + C_{ESI}. \qquad (12.236)$$

The Michaelis constant for the inhibitor is:

$$K_I = \frac{k_{rI}}{k_{fI}} = \frac{C_I C_E}{C_{EI}}. \qquad (12.237)$$

Combining (12.235), (12.226) and (12.237), we obtain the Lineweaver–Burk equation for *competitive inhibition*:

$$\frac{1}{R_P} = \frac{1}{V_{max}} + \frac{K_m}{V_{max}}\left[1 + \frac{C_I}{K_I}\right]\left(\frac{1}{C_S}\right) \quad \text{(competitive)}. \qquad (12.238)$$

The intercept is the same as the intercept without competitive inhibition, but the slope depends on the inhibitor concentration. The maximum conversion rate at high substrate concentrations is unaffected by the presence of the inhibitor. The equivalent expression for *noncompetitive inhibition* can be found by combining (12.227), (12.234), (12.236), and (12.237):

$$\frac{1}{R_P} = \left[1 + \frac{C_I}{K_I}\right]\left[\frac{1}{V_{max}} + \frac{K_m}{V_{max}}\left(\frac{1}{C_S}\right)\right] \quad \text{(noncompetitive)}. \qquad (12.239)$$

Both the slope and the intercept are influenced by inhibitor concentration in noncompetitive inhibition, so the maximum conversion rate cannot be restored to its noninhibited value by simply adding excess substrate. A comparison of Lineweaver–Burk graphs for no inhibition, competitive inhibition and noncompetitive inhibition are shown in Fig. 12.13.

Many other models of enzyme kinetics have been proposed to account for pH-sensitive enzyme activation, excess substrate inhibition, the influence of multiple binding sites on an enzyme, reactions of multiple substrates, etc. Once the chemical reactions have been determined, the influence of these factors on the rate of conversion can be determined using the methods outlined above.

Example 12.8.3.1 Enzyme Kinetics.

The table below gives values for the rate of conversion of substrate to product in the presence of an enzyme, with and without an inhibitor present. The inhibitor concentration is 5 mM. Find K_m, K_I and V_{max}. Is the inhibitor competitive or noncompetitive?

C_S (M)	R (M/min) (no inhibitor)	R (M/min) (with inhibitor)
10^{-2}	7.5×10^{-5}	5.71×10^{-5}
10^{-3}	7.49×10^{-5}	5.57×10^{-5}
10^{-4}	6×10^{-5}	4.52×10^{-5}
7.5×10^{-5}	5.63×10^{-5}	1.56×10^{-5}

Solution. *Initial considerations:* We must ultimately construct Lineweaver–Burk plots for the two sets of experiments. Therefore, our first task is to compute $1/C_S$ and $1/R$ for all data points.

System definition and environmental interactions: The system is a well-stirred fluid in which the reaction is allowed to proceed. The substrate concentration is measured in the fluid as a function of time and no substrate, product or enzyme is allowed to escape through the boundary (walls) of the system.

Apprising the problem to identify governing relationships: We need to plot the data from both experiments in the form of Lineweaver–Burk plots and compare the results with the expectations for competitive and noncompetitive inhibition, as shown in Fig. 12.13. We can then determine the parameters by analyzing the slopes and intercepts of the plots.

Analysis: Lineweaver–Burk plots for the two experiments are plotted in Fig. 12.14. Both graphs intersect the $1/C_S$-axis at the same point, $-37,500$ M^{-1}, so the inhibitor is noncompetitive. K_m can be estimated as $-1/C_S$ at the $1/R$ zero intercept:

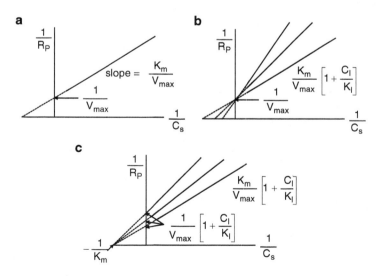

Fig. 12.13 Lineweaver–Burk plots for (**a**) no inhibitor present, (**b**) competitive inhibition and (**c**) noncompetitive inhibition. Graphs are for different inhibitor concentrations

12.8 Chemical Reactions

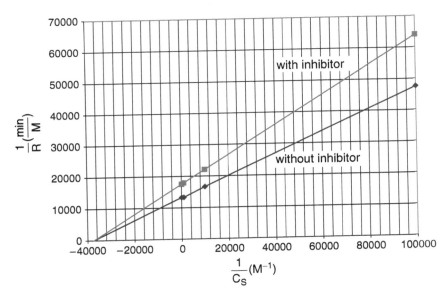

Fig. 12.14 Lineweaver–Burk plots for Example 12.8.3.1

$$K_m = -C_S \left(\frac{1}{R} \to 0\right) = -\frac{1}{(-37,500 \, M^{-1})} = 2.67 \times 10^{-5} \, M$$

The $1/R$ intercept when $1/C_S = 0$ is about 13,100 min/M without inhibitor and 17,500 min/M with the inhibitor. V_{max} can be computed from the intercept with no inhibitor:

$$V_{max} = R\left(\frac{1}{C_S} \to 0\right)\bigg|_{no\ inhibitor} = \frac{1}{13,100\ min/M} = 7.63 \times 10^{-5}\ M/min.$$

K_I can be computed from the intercept with inhibitor:

$$K_I = \frac{C_I}{V_{max}\left(\frac{1}{R}\left(\frac{1}{C_S} \to 0\right)\right)\big|_{inhibitor} - 1} = \frac{5 \times 10^{-3}\ M}{7.63 \times 10^{-5} \frac{M}{min}\left(17,500 \frac{min}{M}\right) - 1}$$
$$= 1.5 \times 10^{-2}\ M.$$

Examining and interpreting the results: The inhibitor is noncompetitive with a K_I that is 5,600 times greater than the Michaelis constant and K_I is three times greater than C_I. Since the slope of the noncompetitive Lineweaver–Burk plot is proportional to $1 + C_I/K_I$, it is essential that C_I not be significantly less than K_I. Otherwise, the two lines would be nearly indistinguishable, and we would not be able to estimate K_I using this method.

12.8.4 Ligand–Receptor Binding Kinetics

Many different intracellular processes are initiated by the binding of extracellular ligands to protein receptors which are present on the extracellular surface of the cell membrane. Ligand–receptor binding is generally a reversible process:

$$L + R \underset{k_r}{\overset{k_f}{\rightleftharpoons}} RL, \tag{12.240}$$

where L represents the extracellular ligand, R the membrane receptor and RL the receptor–ligand complex that is responsible for promoting the intracellular event. The forward reaction rate is k_f and the reverse reaction rate is k_r. Applying the law of mass action to each species:

$$\frac{dc_L}{dt} = \frac{dc_R}{dt} = -\frac{dc_{RL}}{dt} = k_r c_{RL} - k_f c_R c_L. \tag{12.241}$$

Consequently,

$$\frac{d}{dt}(c_R + c_{RL}) = 0. \tag{12.242}$$

Therefore, if the receptor–ligand complex is not internalized, then the total receptor concentration c_{RT} will equal the sum of the bound and unbound receptor concentrations at the membrane surface:

$$c_{RT} = c_R + c_{RL}. \tag{12.243}$$

Another consequence of (12.241) is the following relationship between ligand and ligand-bound receptor concentrations:

$$\frac{d}{dt}(c_L + c_{RL}) = 0. \tag{12.244}$$

Consider the case when all cell receptors are initially devoid of ligand and the cells are exposed to an extracellular ligand concentration of C_{L0} at $t = 0$. Integrating (12.244) from $t = 0$ to an arbitrary time t:

$$c_L + c_{RL} = C_{L0}. \tag{12.245}$$

When dealing with receptor kinetics, molecular biologists often prefer to express the concentration of species A at a cell surface and inside cells as the number of molecules of species A per cell, $\mathbb{N}_A$, rather than use molar concentration, C_A. Let the number of cells in 1 liter be n_c. The number of molecules per mole is given by Avogadro's number, N_{Av}. Therefore, c_A and $\mathbb{N}_A$ are related as follows:

12.8 Chemical Reactions

$$c_A = \frac{n_c}{N_{Av}} \mathbb{N}_A. \qquad (12.246)$$

Rewriting receptor–ligand concentrations in terms of $\mathbb{N}_{RL}$ and substituting (12.245) and (12.243) into (12.241), we find:

$$\frac{d\mathbb{N}_{RL}}{dt} = k_f(\mathbb{N}_t - \mathbb{N}_{RL})\left(C_{L0} - \frac{n_c}{N_{Av}}\mathbb{N}_{RL}\right) - k_r\mathbb{N}_{RL}, \qquad (12.247)$$

where $\mathbb{N}_t$ is the total number of receptors per cell. It is possible to separate variables and integrate (12.247) to obtain an analytic solution for $\mathbb{N}_{RL}$. However, the molar concentration of the ligand–receptor complex is often very small in relation to the initial concentration of ligand in the extracellular space. In that case, (12.247) can be simplified as follows:

$$\frac{d\mathbb{N}_{RL}}{dt} = k_f C_{L0} \mathbb{N}_t - (k_r + k_f C_{L0})\mathbb{N}_{RL}. \qquad (12.248)$$

The solution to (12.248) is:

$$\mathbb{N}_{RL} = \frac{C_{L0}\mathbb{N}_t}{(K_D + C_{L0})}(1 - \exp(-k_f(K_D + C_{L0})t)), \qquad (12.249)$$

where K_D is the dissociation constant:

$$K_D = \frac{k_r}{k_f}. \qquad (12.250)$$

The solution as t approaches infinity is:

$$\mathbb{N}_{RL,\infty} = \frac{\mathbb{N}_t}{\left(1 + \frac{K_D}{C_{L0}}\right)}. \qquad (12.251)$$

How can we identify the unknown quantities k_f, k_r, and $\mathbb{N}_t$ from experimental data? One approach is to perform a series of experiments in which the initial ligandconcentration C_{L0} is varied and the number of ligand–receptor complexes per cell is measured as a function of time. Inserting (12.251) into (12.249) and rearranging, we have:

$$\ln\left[1 - \frac{\mathbb{N}_{RL}}{\mathbb{N}_{RL,\infty}}\right] = -(k_f C_{L0} + k_r)t = -\phi t. \qquad (12.252)$$

Plots of $\ln(1 - \mathbb{N}_{RL}/\mathbb{N}_{RL,\infty})$ vs. t for different values of C_{L0} will be straight lines with slopes ϕ that depend on C_{L0}. Plotting the slopes of these lines vs. C_{L0} will also produce a straight line with intercept k_r and slope k_f. Once k_r and k_f are known, we can compute K_D from (12.250) and can then estimate the total number of receptors from (12.251). An alternate approach for finding K_D and $\mathbb{N}_t$ is to make use of just the steady-state data. The steady-state solution (12.251) can be rewritten in the following form:

$$\frac{\mathbb{N}_{RL,\infty}}{C_{L0}} = \frac{\mathbb{N}_t}{K_D} - \frac{\mathbb{N}_{RL,\infty}}{K_D}. \qquad (12.253)$$

A plot of experimental measurements of $\mathbb{N}_{RL,\infty}/C_{L0}$ as a function of $\mathbb{N}_{RL,\infty}$ is known as a *Scatchard plot*. K_D and $\mathbb{N}_t$ can be found from the slope $(-1/K_D)$ and the intercept (n_t/K_D) of a Scatchard plot. These methods are illustrated in Example 12.8.4.1.

Example 12.8.4.1 Ligand–Receptor Binding Kinetics.
Vascular endothelial cells contain vascular endothelial growth factor receptors (VEGFRs). Blockage of a particular receptor, VEGFR2, promotes rapid regression of blood vessels. Binding kinetics of a monoclonal antibody to VEGFR2 receptors on human microvascular endothelial cells is studied by adding the antibody in four different doses (C_{L0}) to a monolayer of cultured endothelial cells. The number of ligand–receptor complexes per cell is measured as a function of time. Data collected at 5 min intervals are shown in Table 12.3. An additional sample is collected at 90 min, and this is assumed to be the steady-state concentration. Analyze the data to estimate the forward and reverse rate constants, k_f and k_r, the dissociation constant, K_D, and the total number of receptors per cell, $\mathbb{N}_t$.

Solution. *Initial considerations and governing relationships:* We will apply graphical methods to estimate the unknown parameters in (12.252) and (12.253).

Table 12.3 Number of ligand-receptor complexes per cell vs. time for four different initial ligand concentrations

Time (min)	$\mathbb{N}_{RL}$ (C_{L0} = 20 nM)	$\mathbb{N}_{RL}$ (C_{L0} = 40 nM)	$\mathbb{N}_{RL}$ (C_{L0} = 60 nM)	$\mathbb{N}_{RL}$ (C_{L0} = 80 nM)
0	0	0	0	0
5	41,148	70,214	90,754	105,278
10	69,251	104,014	121,546	130,454
15	88,445	120,284	131,994	136,475
20	101,554	128,116	135,539	137,914
25	110,507	131,887	136,741	138,259
30	116,621	133,702	137,149	138,341
35	120,797	134,575	137,288	138,361
40	123,649	134,996	137,335	138,365
45	125,597	135,198	137,351	138,367
50	126,928	135,296	137,356	138,367
55	127,836	135,343	137,358	138,367
60	128,457	135,365	137,359	138,367
90	129,658	135,386	137,359	138,367

12.8 Chemical Reactions

Analysis: The first step is to apply (12.252) to the data in Table 12.3 and plot the results. The plots are shown in Fig. 12.15.

From these graphs, the slopes are measured and plotted in Fig. 12.10 as a function of ligand concentration.

Examining and interpreting the results: The coefficient k_f is equal to the intercept of the line in Fig. 12.16, which is approximately 0.007 min^{-1} or

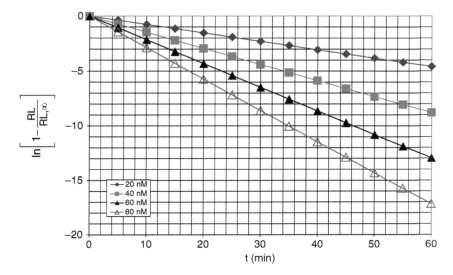

Fig. 12.15 Semilog plots of experimental data

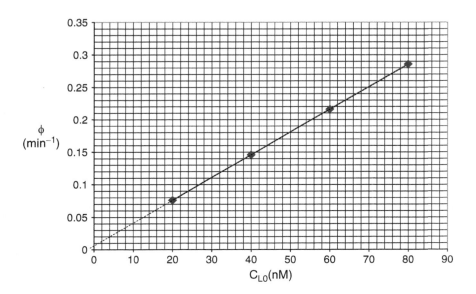

Fig. 12.16 Slopes of graphs in Fig. 12.15 vs. ligand concentration

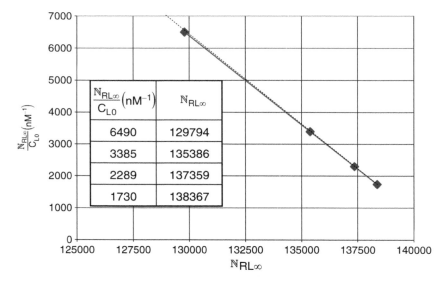

Fig. 12.17 Scatchard plot

1.1×10^{-4} s^{-1}. The rate constant k_r is equal to the slope of the line in Fig. 12.16, which is 5.83×10^4 M^{-1} s^{-1}. The dissociation constant $K_D = k_f/k_r = 1.8 \times 10^{-9}$ M^{-1}. We can also estimate K_D and $\mathbb{N}_t$ from a Scatchard plot, shown in Fig. 12.17. If we use the solid line, based on the slope of the first and last points in Fig. 12.17, we find $K_D = 1.8 \times 10^{-9}$ M^{-1} and $\mathbb{N}_t = 132{,}909$. However, this value of $\mathbb{N}_t$ is smaller than several of the $\mathbb{N}_{RL}$ measurements. If the last two points are used to construct the straight line in the Scatchard plot (dotted line in Fig. 12.17), we compute the same value for K_D, but $\mathbb{N}_t = 140{,}470$. This estimate of the total number of receptors is expected to be more accurate because the data collected at 90 min for the $C_{L0} = 20$ nm and $C_{L0} = 40$ nM cases probably do not reflect true steady-state values, while the data collected at 40 nM and 80 nM have reached a steady state in 90 min. An accurate estimate of $\mathbb{N}_t$ from a Scatchard plot requires true steady-state data.

12.9 Cellular Transport Mechanisms

There are several mechanisms whereby solutes move in and out of cells. Lipid soluble substances and dissolved gases such as oxygen or carbon dioxide can easily diffuse through the lipid bilayer that composes much of the cell membrane. This passive mechanism is covered in some detail in Sects. 12.4–12.6. Some hydrophilic substances are transported by both diffusion and convection through pores created by integral proteins that span the membrane (Sect. 13.8).

12.9 Cellular Transport Mechanisms

Pinocytosis is another transmembrane transport mechanism in which a portion of the membrane forms an invagination, traps extracellular fluid in its interior as it breaks off of the membrane and delivers its contents to specific intracellular locations. Other substances, such as acetylcholine, are delivered from the cell interior to the cell surface via small vesicles. Little is known about the details of vesicle transport, and even less has been done to try to model vesicular transport.

12.9.1 Carrier-Mediated Transport

Many aqueous species essential for cell survival are virtually insoluble in cell membranes. Transport of species such as glucose is mediated by specific proteins that bind to the species and facilitate its movement through the membrane. The protein acts as a mediator in a manner similar to the way that an enzyme facilitates the conversion of product to substrate. Like an enzyme, the protein mediator is unchanged when the process is completed. However, unlike an enzymatic reaction, there is no difference between the substrate and the product. The function of the protein mediator is simply to transport the species from one side of the membrane to the other. Carrier-mediated transport can be classified as being active or passive. Active transport mechanisms require energy to move the species against its electrochemical potential gradient. Passive transport mechanisms do not require an energy source, and consequently cannot be used to move a species against its electrochemical potential gradient. In this section, we consider passive carrier-mediated transport, also known as facilitated diffusion. Our goal is to develop an expression for predicting flux of the transported species in terms of the species concentrations on the two sides of the membrane.

A carrier molecule is usually an integral protein that spans the cell membrane from the extracellular to intracellular sides. Other carrier molecules have been identified which are confined to the membrane, but can ferry substances back and forth by simple diffusion. The carrier combines with a species on one side of the cell membrane to form a carrier-species complex. The complex then undergoes a configuration change which deposits the species on the other side of the membrane, or it diffuses down its concentration gradient toward the other side of the membrane, where the species may dissociate from the carrier. The carrier molecule then returns to its original configuration or diffuses back to the other side so the cycle can be repeated. If the concentration of species were to increase on the other side of the membrane, then the carrier could move the species in the opposite direction. A schematic of the model is shown in Fig. 12.18, where X represents the carrier protein and S the species to be transported. Letting "e" represent conditions on the exterior surface of the cell membrane and "i" conditions on the membrane interior surface, the chemical reactions at the inside and outside surfaces are:

$$S_e + X_e \rightleftarrows (SX)_e, \qquad (12.254)$$

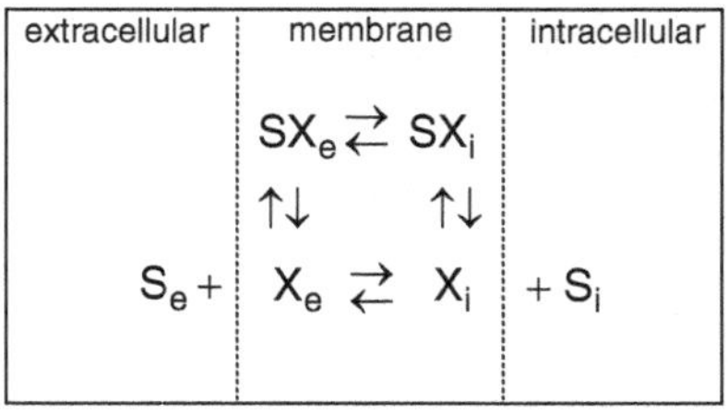

Fig. 12.18 Carrier-mediated transport

extracellular	membrane	intracellular
	$SX_e \rightleftarrows SX_i$	
	$\uparrow\downarrow \qquad \uparrow\downarrow$	
$S_e +$	$X_e \rightleftarrows X_i$	$+ S_i$

$$S_i + X_i \rightleftarrows (SX)_i. \tag{12.255}$$

The reactions which occur at the two surfaces are identical and therefore they will be governed by the same rate constants. Assuming chemical equilibrium at both the interior and exterior surfaces:

$$K = \frac{C_{S_e} C_{X_e}}{C_{SX_e}}, \tag{12.256}$$

$$K = \frac{C_{S_i} C_{X_i}}{C_{SX_i}}, \tag{12.257}$$

where K is the dissociation constant for the carrier-species reactions. The carrier travels across the membrane with velocity v_X. Carrier velocity is assumed to be independent of the direction of transport and is unchanged when the solute is bound to the protein. The net protein flux from the external surface to the internal surface is:

$$N_X = v_X [C_{X_e} - C_{X_i}]. \tag{12.258}$$

If species S is unable to pass through the membrane by other mechanisms, then the flux of S is equal to the flux of the species-carrier complex:

$$N_S = N_{SX} = v_X [C_{SX_e} - C_{SX_i}]. \tag{12.259}$$

In the steady-state, the net flux of carrier and carrier-species complex must be zero. Setting the difference between (12.258) and (12.259) equal to zero, we have:

$$C_{SX_e} + C_{X_e} = C_{SX_i} + C_{X_i}. \tag{12.260}$$

Let the total concentration of carrier in the membrane be C_{XT}. Then conservation of carrier requires:

$$C_{XT} = C_{SX_e} + C_{X_e} + C_{SX_i} + C_{X_i}. \tag{12.261}$$

12.9 Cellular Transport Mechanisms

Substituting (12.260) into (12.261):

$$C_{SX_e} + C_{X_e} = C_{SX_i} + C_{X_i} = \frac{C_{XT}}{2}. \tag{12.262}$$

Substituting (12.262) into (12.256) and (12.257) yields:

$$C_{SX_e} = \frac{C_{XT}}{2}\left[\frac{C_{S_e}}{K + C_{S_e}}\right], \tag{12.263}$$

$$C_{SX_i} = \frac{C_{XT}}{2}\left[\frac{C_{S_i}}{K + C_{S_i}}\right]. \tag{12.264}$$

Finally, we can substitute (12.263) and (12.264) into (12.259) to obtain an expression for the carrier-mediated flux of S:

$$N_S = \frac{v_X C_{XT}}{2}\left[\frac{C_{S_e}}{K + C_{S_e}} - \frac{C_{S_i}}{K + C_{S_i}}\right]. \tag{12.265}$$

The species flux is proportional to the total concentration of carrier molecules in the membrane. However, the dependence of flux on species concentration at either end of the membrane is nonlinear. The flux is independent of extracellular species concentration when $C_{S_e} \gg K$ and is independent of intracellular species concentration when $C_{S_i} \gg K$. This reflects carrier saturation at either the external or internal surfaces, so increasing the concentration will not increase the flux. When the flux is zero, the intracellular and extracellular concentrations will be equal. The maximum inward flux occurs when the concentration at the inside surface is very small and the concentration at the external surface is large:

$$N_{S,\max} = \frac{v_X C_{XT}}{2}. \tag{12.266}$$

The maximum outward flux will equal $-N_{S,\max}$. If both the intracellular and extracellular species concentrations are much lower than K, (12.264) simplifies to:

$$N_S = \frac{v_X C_{XT}}{2K}[C_{S_e} - C_{S_i}]. \tag{12.267}$$

Under these circumstances, the flux resulting from carrier-mediated transport would be indistinguishable from simple diffusion.

Plots of $N_S/N_{S,\max}$ vs. C_{Se}/K are shown in Fig. 12.19 for various values of C_{Si}/K. Solute flux is positive if directed into the cell. The top trace represents the maximum inward flux for a given external solute concentration. This occurs when the solute concentration at the inside surface of the membrane is kept at zero. In practice, this could be caused by rapid utilization of the solute within the cell. The absolute

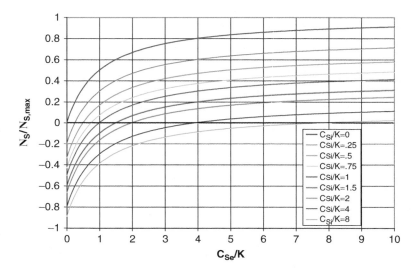

Fig. 12.19 Carrier-mediated solute flux relative to the maximum flux (12.9.1.12) vs. solute concentration on the external surface of the cell membrane

maximum carrier-mediated flux $N_{S,\text{max}}$ is reached when the carrier system is saturated. This does not occur until the solute concentration on the external side of the cell membrane is greater than ten times the equilibrium constant K. This nonlinear carrier saturation behavior is quite different than the linear flux–concentration relation that would result for simple diffusion of the solute through the membrane. The maximum flux attained at high external solute concentrations is reduced as the internal solute concentration increases. For a nonzero internal solute concentration, the flux will be negative (i.e., from inside to outside) at low values of C_{Se}. The maximum negative flux at high internal solute concentrations is $-N_{S,\text{max}}$.

Like an enzyme, the carrier protein can bind with other solutes, either at the same binding site or at other sites that can promote or interfere with transport of the solute of interest. This can lead to competitive or noncompetitive inhibition, similar to that described for enzyme kinetics. In addition, the carrier-mediated model must be modified if the forward and reverse transport rates are different or if the carrier-species complex moves at a different velocity than the carrier alone.

12.9.2 Active Transport

The transport mechanisms discussed thus far are known as *passive transport mechanisms* because they occur spontaneously. *Active transport mechanisms* require energy, normally in the form of ATP, to provide the work necessary to move species, usually ions, across the cell membrane against their electrochemical gradient. A substantial fraction of metabolic energy is utilized by active transport mechanisms.

12.10 Mass Transfer Boundary Conditions

A prime example is the *sodium–potassium–ATP pump*. This is known as *primary active transport* because the energy necessary to move two potassium ions into the cell while simultaneously expelling three sodium ions from the cell interior is supplied directly by the conversion of ATP to ADP. The pump is known as an *electrogenic* pump because of the imbalance of ions moving in each direction. The pump is responsible for maintaining the resting potential and is necessary to prevent *cell lysis*. If the pump is poisoned, the higher osmotic pressure in the cell interior tends to draw extracellular fluid in through the pores in the cell membrane. Ions and other solutes are swept in by convection. Cell volume increases, but cell surface area remains relatively constant. Ultimately, the cell assumes a spherical shape and as more water attempts to enter, the cell loses its ability to regulate its volume, the membrane bursts, and the cell dies.

The gradient in ions caused by the action of the Na^+–K^+–ATP pump can provide the energy necessary to move other species across the cell membrane. This is known as *secondary active transport*. The lipid bi-layer of cell membranes is relatively impermeable to glucose. However, glucose is a primary fuel responsible for a substantial portion of cell metabolic energy. Glucose enters the cell via a co-transporter mechanism in which a carrier protein combines with a molecule of glucose and a sodium ion at the outside surface of the cell, and both the species are transported to the inside surface. This is made possible by the steep Na^+ gradient produced by the pump. Once in the cell, the glucose molecule is converted to glucose-6-phosphate, which is unable to cross the cell membrane.

There are many other important active transport mechanisms present in various cell types. We will not attempt to provide quantitative models of the mechanisms responsible for active transport in this text. Instead, when dealing with active transport, we will assume that sufficient amounts of ATP are present so that the pumps are kept saturated and ion concentration gradients are maintained. Thus active transport will be assumed to induce constant fluxes of the transported species.

12.10 Mass Transfer Boundary Conditions

There are several boundary conditions that are consistently encountered in mass transfer problems. The most common boundary conditions are described below:

12.10.1 Mass or Molar Concentration Specified at a Boundary

In many situations, the concentration C_{As} of species A in the system s is maintained constant at the system boundary, C_0. If the system boundary is located at $y = y_0$, then this boundary condition would be written as:

$$C_{As}(y_0) = C_0. \tag{12.268}$$

If the system is a solid or liquid, and the system boundary located at $y = y_0$ is in contact with a fluid having concentration C_{Af}, then the concentration in the system at the boundary is assumed to be equal to the concentration that would be in equilibrium with the fluid concentration:

$$C_{As}(y_0) = C_0 = \Phi_{Asf} C_{Af}(y_0), \tag{12.269}$$

where Φ_{Asf} is the partition coefficient for A between the system and the fluid.

12.10.2 Mass or Molar Flux Specified at a Boundary

In some cases, the flux of species A is known at a surface, N_{A0}. Assuming the surface is located at y_0 and the flux to be in the y-direction, perpendicular to the surface, the boundary condition would be:

$$N_{A0} = -D_{As} \frac{\partial C_{As}}{\partial y}\bigg|_{y=y_0} + (C_{As} v_y)\big|_{y=y_0}. \tag{12.270}$$

The terms on the right-hand side represent the diffusion and convection of A *inside* the system at the surface. In most cases of interest, the bulk velocity at the surface v_y is zero. In such cases, the appropriate boundary condition is:

$$N_{A0} = -D_{As} \frac{\partial C_{As}}{\partial y}\bigg|_{y=y_0}. \tag{12.271}$$

In other situations, the flux at the surface may not be known explicitly, but is defined in terms of a mass transfer coefficient. For the case without bulk velocity at the surface:

$$k_{Af}\left(C_{Af,\infty} - C_{Af}\big|_{y=y_0}\right) = -D_{As} \frac{\partial C_{As}}{\partial y}\bigg|_{y=y_0}. \tag{12.272}$$

Assuming equilibrium between fluid and system at the surface allows us to write the concentration in the fluid at the surface in terms of the concentration of the system at the surface:

$$k_{Af}\left(C_{Af,\infty} - \frac{C_{As}}{\Phi_{A,sf}}\bigg|_{y=y_0}\right) = -D_{As} \frac{\partial C_{As}}{\partial y}\bigg|_{y=y_0}. \tag{12.273}$$

12.10 Mass Transfer Boundary Conditions

12.10.3 No-Flux Boundary Condition

An impermeable surface is one in which species A cannot penetrate. The net flux of A is zero, and as the diffusion coefficient is not zero, the concentration gradient at the surface must be zero:

$$\left.\frac{\partial C_{As}}{\partial y}\right|_{y=y_0} = 0. \tag{12.274}$$

12.10.4 Concentration and Flux at an Interface

At the interface $y = y_0$ between two systems 1 and 2, two conditions must be specified. First, the concentrations at the interface are assumed to be in equilibrium:

$$C_{A1}(y_0) = \Phi_{A12} C_{A2}(y_0). \tag{12.275}$$

In addition, in the absence of a chemical reaction, the flux of A into the surface from system 1 must be equal to the flux out of the surface into system 2. Assuming that convective fluxes are negligible, the second boundary condition at an interface is:

$$D_{As_1}\left.\frac{\partial C_{A1}}{\partial y}\right|_{y=y_0} = D_{As_2}\left.\frac{\partial C_{A2}}{\partial y}\right|_{y=y_0}. \tag{12.276}$$

12.10.5 Heterogeneous Reaction at a Surface

As mentioned in Sect. 12.8, heterogeneous chemical reactions are reactions that occur at the interface between two phases. The flux of species A to the interfacial surface is related to the concentration of species A at the reacting surface and on the order of the heterogeneous reaction, n:

$$N_{A0} = k\left(C_A|_{y=y_0}\right)^n. \tag{12.277}$$

The proportionality factor k is known as the heterogeneous reaction coefficient. The units of k will depend on the order of the reaction. If flux is in units of mol cm^{-2} s^{-1} and C_A is in units of mol cm^{-3}, then k will have units of mol$^{(1-n)}$ cm$^{(3n-2)}$ s^{-1}.

12.11 Summary of Key Concepts

Important definitions. We introduce the fundamentals of mass transfer of individual species in this chapter. The presence of species A in another medium is generally reported in terms of its concentration. This can be expressed as a molecular concentration, a volume fraction C_A^*, a molar concentration c_A or a mass concentration ρ_A. Several different ways are used to express molar concentrations in biotransport applications. The local molar concentration $c_A(x,y,z,t)$ refers to the concentration at a single point in space (x,y,z) and a particular time t. The average concentration in a system with volume V is $C_A(t) = \int_V c_A dV$. The average concentration over the cross-section A_c of a conduit at axial location z is $\bar{c}_A(z,t) = \int_{A_c} c_A dA_c$. The flow-averaged concentration at a conduit cross-section A_c located at axial position z, also known as the mean concentration, the mixing cup concentration or the bulk concentration, is defined as $C_{Ab}(z,t) = \int_{A_c} c_A v dA_c / \int_{A_c} v dA_c$. Averaged mass or molecular concentrations can also be computed using analogous expressions. The total molar concentration c is the sum of the molar concentrations of all species in a system and mass concentration (i.e., density) ρ is the sum of mass concentrations of all species in a system. The mass fraction is $\omega_A = \rho_A/\rho$ and the mole fraction is $x_A = c_A/c$. If the material is an ideal gas, the equation of state relates the total concentration to the pressure and temperature, $c = P/RT$, and Dalton's law relates the mole fraction of species A to the partial pressure of species A, $x_A = P_A/P$.

Equilibrium. When two systems are in equilibrium, there are no net fluxes of any species or energy between the two systems. If one of the systems is a gas, the equilibrium concentration of species A in the system in contact with the gas $c_{A,s}$ is proportional to the partial pressure of A in the gas. This is known as Henry's law: $c_{A,s} = \alpha_{A,s} P_{A,gas}$, where $\alpha_{A,s}$ is the solubility of gas A in system s. A different form of Henry's law is often used in physiological gas transfer applications, where $c_{A,s}$ is replaced by the volume fraction of gas A in the liquid, and the solubility coefficient is replaced by the Bunsen solubility coefficient (12.32). Although the temperatures of two systems in equilibrium must be the same, the concentrations of a species can be different in two systems that are in equilibrium. This is because the solubility of the species in each system can be different. The ratio of equilibrium concentration of species A in system 1 relative to the equilibrium concentration in system 2 is known as the partition coefficient Φ_{A12}. If system 2 is a gas, the partition coefficient is directly proportional to the solubility coefficient, $\Phi_{A1,gas} = RT\alpha_{A1}$.

Mass transport between systems. A net movement of solute A across a system boundary can only occur if the concentration of A is different than the equilibrium concentration on either side of the boundary between the systems. Molar flow of species A across the boundary from system 1 to system 2 is given by:

$$W_A = \mathsf{P}_A S[C_{A1} - \Phi_{A12}C_{A2}], \tag{12.278}$$

12.11 Summary of Key Concepts

where S is the boundary surface area and P_A is an overall mass transfer coefficient for species A, referred to in the biological literature as the permeability of species A. The permeability depends on the properties of the barrier separating the two systems and on the mass transfer boundary layers that may exist near the interface, as discussed in Sect. 12.6.1 and in Example 12.6.2.1.

Mass transfer in a single phase. The molar flux of species A is defined as $N_A = c_A v_A$, where v_A is the velocity of species A. We can think of the total molar flux as consisting of a bulk flux $c_A v^*$, where $v^* = \sum N_i/c$ is the molar average velocity of all species, and a diffusive flux relative to the molar average velocity $J_A = c_A(v_A - v^*)$. Experiment has shown that for a binary gas mixture of A in B the diffusive flux, or molecular flux, of A relative to the molar average velocity is given by Fick's law:

$$\vec{J}_A^* = -D_{AB} c \vec{\nabla} x_A, \qquad (12.279)$$

where D_{AB} is the diffusion coefficient of species A in gas B. Therefore, the general expression for the flux of A in gas B is:

$$\vec{N}_A = x_A \sum_{i=1}^{N_{\text{species}}} \vec{N}_i - D_{AB} c \vec{\nabla} x_A. \qquad (12.280)$$

This expression is also valid for the flux of A in any medium, as long as species A is dilute. This is further simplified in the case where the total concentration can be considered constant (12.78) or the material is a solid (12.82). For steady-state situations without chemical reaction in the media of interest, we can often separate variables in (12.76) and integrate to find the distribution of mole fraction or concentration of A within the media.

The Stokes–Einstein relation (12.98) can be used to estimate the diffusion coefficient of macromolecules in a liquid from the size of the molecule and viscosity of the liquid. In many instances involving flowing fluids near an exchanging solid surface, we do not have enough information to integrate (12.76), so an empirical approach, first introduced in Chap. 2, is used to determine the flux:

$$N_A = k_{Af}\left([c_{Af}]_s - [c_{Af}]_\infty\right) \text{ (external)}, \qquad (12.281)$$

$$N_A = k_{Af}\left([c_{Af}]_w - [c_{Af}]_m\right) \text{ (internal)}. \qquad (12.282)$$

where k_{Af} is a mass transfer coefficient in the fluid phase, $c_{A\infty}$ is the concentration of species A far from an external surface, c_{AS} is the concentration in the fluid at the external solid surface, $[c_{Af}]_w$ is the concentration of A in fluid at an internal surface and $[c_{Af}]_m$ is the flow-averaged concentration in a conduit. The Sherwood number is a dimensionless mass transfer coefficient ($Sh = k_{Af}L/D_{Af}$, where L is a characteristic length). The Sherwood number is related to Reynolds number and Schmidt number (v_f/D_{Af}) in situations involving forced convection mass transfer. Some useful relationships for Sherwood number are given in Sect. 12.4.7.1 for external flows and Sect. 12.4.7.2 for internal flows.

Solid–liquid mass transfer: internal and external resistance. Transient mass transfer from a solid to a liquid involves the transport through two barriers: the solid and the liquid film surrounding the solid. It is possible to neglect the internal (i) resistance in the solid relative to external resistance of the film (e) if the Biot number for mass transfer $Bi_A = (k_{Ae}L)/(D_{Ai}\Phi_{Aie})$ is small. The macroscopic approach (or "lumped" analysis) can be used when internal resistance is small. However, a microscopic approach will be necessary when Bi_A is large, as significant concentration gradients will be present within the solid.

Chemical reactions. Heterogeneous reactions are chemical reactions that occur on the boundary of a system and homogeneous reactions occur within the system. The net rate of production per unit volume of a species is given by the difference between the rate of formation and the rate of decomposition of that species. Reactions which produce species A in a manner proportional to $(c_A)^n$ is said to be nth order with respect to species A.

Oxygen is transported primarily by hemoglobin inside the red blood cells. Four molecules of oxygen can combine with hemoglobin to form oxyhemoglobin. The saturation of hemoglobin is the percentage of available hemoglobin-binding sites that contain oxygen. The saturation depends on the partial pressure of oxygen in a nonlinear manner, as shown in Fig. 12.9, as well as pH, temperature and P_{CO_2}, as shown in Fig. 12.11.

Some carbon dioxide is physically dissolved in plasma and erythrocytes, but most is transported in the form of bicarbonate ions or is bound to proteins. Total blood concentration is linearly related to the partial pressure of CO_2, as shown by (12.218).

Most biological reactions are promoted by the presence of specific enzymes. The product is proportional to the substrate concentration at low concentrations and saturates at high concentrations (12.231). The maximum rate of conversion and the Michaelis constant can be determined from a Lineweaver–Burk plot (Fig. 12.13a). Binding sites on enzymes can also combine with chemical species other than the normal substrate (competitive inhibition), or the enzyme can combine with inhibitors at other binding sites, interfering with normal substrate–enzyme binding (noncompetitive inhibition). The presence of competitive and noncompetitive inhibition can be quantified with Lineweaver–Burk plots. Ligand–receptor-binding kinetics at cell surfaces can be analyzed in a similar graphical method, as shown in Sect. 12.8.4.

Mass transfer boundary conditions. We close the chapter by listing the common boundary conditions encountered in mass transfer problems. These will be most useful in solving microscopic problems in Chaps. 14 and 15.

12.12 Questions

12.12.1 Distinguish between mass concentration, molar concentration, total molar concentration, density, mole fraction, and mass fraction.

12.12.2 How is molecular concentration related to molar concentration and mass concentration?

12.12.3 What is the difference between average concentration and local concentration?

12.12.4 What is another name for total mass concentration?

12.12.5. Define molar-averaged molecular weight.

12.12.6 How is the total molar concentration related to pressure in an ideal gas?

12.12.7 How is the mole fraction of species A related to the partial pressure of species A in an ideal gas?

12.12.8 If two materials are in equilibrium, are the concentrations of all species the same in both the materials? Why or why not?

12.12.9 How can you compute equilibrium solute concentrations between phases (gas–solid, gas–fluid, and solid/fluid) or between materials in equilibrium? What is Henry's Law? What is a solubility coefficient? What is a partition coefficient? Why are they important?

12.12.10 Show how the partition coefficient for gases between two materials is related to the solubility of the gas in the materials.

12.12.11 What is meant by a volume fraction or volume concentration when referring to dissolved blood gases in plasma?

12.12.12 What is the Bunsen solubility coefficient?

12.12.13 Does the solubility of a gas in a liquid increase or decrease with increasing temperature?

12.12.14 How is the partition coefficient of species A in material B relative to material C related to the partition coefficient of A in C relative to B?

12.12.15 What factors contribute to the flux of a species across an interface?

12.12.16 Is the concentration difference of species A across an interface an appropriate driving force for the flux of A through the interface? Explain.

12.12.17 How is the overall mass transfer coefficient, or permeability, of species A in a barrier P_A defined?

12.12.18 What is the relationship between the molar flux of species A and the velocity of species A?

12.12.19 What is the mass average velocity? What is the molar average velocity?

12.12.20 What is the diffusion velocity of species A? What is the diffusion flux of species A?

12.12.21 What is the meaning of the following fluxes: $N_A, n_A, n, N, j_A, J_A, J_A^*, j_A^*$?

12.12.22 What is the general form of Fick's law for a binary system? How does this simplify for constant molar concentration for a one-dimensional flux?

12.12.23 What form of Fick's law would you use for a gas for a dilute solid?

12.12.24 How would you solve a steady-state diffusion problem for a binary system if the flux of one species is known in terms of the flux of the second species?

12.12.25 The Stokes–Einstein relationship relates the diffusion coefficient of a spherical macromolecule to its radius, the temperature and to what fluid property?

12.12.26 Distinguish between a local mass transfer coefficient and an overall mass transfer coefficient. How do the driving forces differ for local and overall applications?

12.12.27 Is there a difference between the Sherwood Number and the Nusselt Number for mass transfer? What dimensionless numbers does the Sherwood depend on for forced convection mass transfer applications?

12.12.28 Distinguish between the local and overall Sherwood numbers.

12.12.29 Define the bulk or mean fluid concentration, also known as the mixing cup or flow-averaged concentration.

12.12.30 What factors influence the membrane permeability? Overall permeability of a hollow fiber?

12.12.31 How is the permeability of a barrier based on an inside to outside driving force $C_{Ai} - \Phi_{Aio}C_{Ao}$ related to the permeability of the same barrier based on an outside to inside driving force $C_{Ao} - \Phi_{Aoi}C_{Ai}$?

12.12.32 How is the overall mass transfer coefficient based on the outside surface of a hollow fiber related to an overall mass transfer coefficient based on the inside surface of the fiber?

12.12.33 How would you estimate internal and external resistance to species mass flow from a solid to a fluid? If the Biot number for mass transfer is small (say less than 10%), is it necessary to use a microscopic approach, or can a macroscopic approach be used?

12.12.34 What is the meaning of the Nernst equation? How can you use it to determine if an ion is actively transported?

12.12.35 What is the difference between a homogeneous and heterogeneous chemical reaction? How do you express these mathematically? Which of these will appear in a species-conservation equation and which will appear as a boundary condition?

12.12.36 What determines the "order" of a chemical reaction?

12.12.37 Relative to dissolved oxygen, how important is oxygen carried by hemoglobin in helping meet the oxygen demands of consuming tissue? Would plasma be sufficient to transport oxygen and carbon dioxide to/from tissues at normal blood flows?

12.12.38 What is meant by blood oxygen-carrying capacity? Is this increased or decreased in blood doping? Explain the difference between oxygen capacity, oxygen delivery, and oxygen consumption.

12.12.39 Given the partial pressure of oxygen, how would you use the oxyhemoglobin saturation curve to determine blood concentration of oxygen?

12.12.40 What is the meaning of the Michaelis constant K_m and V_{max} for an enzyme-mediated reaction?

12.12.41 What is a Lineweaver–Burk plot and why is it useful?

12.12.42 What is a ligand? A receptor? Carrier-mediated transport?

12.12.43 What mass transfer boundary conditions apply at a solid–liquid interface? What is meant by local equilibrium?

12.13 Problems

12.13.1 Local and Average Concentration

The cumulative number of moles of glucose in tissue between position $z = 0$ and z is found to vary with position as follows: $N_g = Bz^2$, where $B = 2$ μmol/mm². The cross-sectional area of the tissue also varies with position: $A_c = a - bz$, where $a = 10$ mm² and $b = 0.3$ mm. The tissue is 10 mm long. (a) What is the average concentration of glucose in the tissue? (b) What is the local concentration of glucose at $z = 2$ mm and $z = 8$ mm?

12.13.2 Concentrations in an Ideal Gas

An ideal gas at 37°C and 760 mmHg consists by weight of 0.01% helium, 15% oxygen and 5% carbon dioxide, and the balance is nitrogen. Find the following quantities: (a) mass fraction for each species, (b) mass concentration of each species, (c) the partial pressures of each species, (d) mole fraction of each species, (e) molar concentration of each species, (f) mean molecular weight, (g) total mass concentration, and (h) total molar concentration.

12.13.3 Water–Gas Equilibrium

The ideal gas in problem 12.13.2 is brought in contact with a small volume of pure water at 37°C and allowed to equilibrate. The vapor pressure of water at 37°C is 47 mmHg. Find the partial pressure of each species in the gas once equilibrium is established. Neglect loss of gases to the water.

12.13.4 Liquid Equilibrium

For the equilibrium reached in problem 12.13.3, find (a) the concentrations of each of the dissolved gases in the liquid phase expressed as milliliter gas per milliliter liquid and (b) the mole fraction of each species in the liquid phase. Bunsen solubility coefficients (milliliter gas per milliliter liquid per atmosphere partial pressure) for these gases in water at 37°C are He: 0.008, O_2: 0.024, N_2: 0.012 and CO_2: 0.57.

12.13.5 Partition Coefficient, Solubility, Boundary Conditions

Two tissues, A and B, are placed in a gas chamber and equilibrated with carbon monoxide gas at various partial pressures of CO. Equilibrium concentrations at five partial pressures are plotted for each tissue.

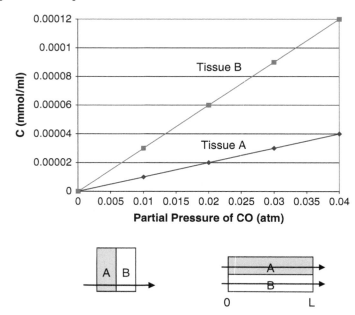

(a) What is the solubility of CO in tissue B?
(b) What is the partition coefficient of CO for tissue A relative to tissue B?
(c) Steady-state diffusion of CO occurs through a tissue consisting of the same two tissues (A and B) in series. There is no resistance to CO transport at the interface. If we were to solve the diffusion equations in each tissue, what conditions must we specify at the interface?
(d) The same two tissues with CO diffusion coefficients $D_B = 3D_A$ are placed in parallel. The ends at $x = 0$ are kept at a P_{CO} of 0.02 atm and the ends at $x = L$ are kept at a P_{CO} of 0.0 atm. What is the steady-state ratio of CO flux in tissue A to the flux in tissue B?

12.13.6 Membrane Permeability

An artificial membrane of thickness 0.1 mm separates two well-mixed fluids containing sucrose. The concentration of sucrose on side A is 10 mM and the bulk concentration on side B is 2 mM. The partition coefficient for sucrose in the

12.13 Problems

membrane relative to fluid A is 0.3 and the partition coefficient for sucrose in the membrane relative to fluid B is 0.2. The diffusion coefficient of sucrose in the membrane is 3.6×10^{-10} m^2/s. Find the permeability of the membrane for transport from side A to side B. Would the magnitude of sucrose flux through the membrane be the same if the fluid concentrations were reversed?

12.13.7 Partition Coefficient

A 1 g sample of tissue is placed in 10 ml of isotonic saline containing radiolabled glucose and allowed to equilibrate. Radioactivity from the equilibrated tissue was measured to be 1,000 counts per minute (CPM), and radioactivity from a 1 ml sample of equilibrated saline was measured to be 1,200 CPM. The density of saline is 1 g/ml and the density of tissue is 1.1 g/ml. What is the partition coefficient for glucose in tissue relative to saline?

12.13.8 Maximum Depth of a Gopher Tunnel

It is thought that the oxygen utilized by gophers while in their tunnels diffuses directly through the soil. If the gopher consumes oxygen at the rate 0.1 mmol O$_2$/min, how far underground (L) can the top of the tunnel be if the gopher is to survive? The tunnel has dimensions of 10 cm high by 20 cm wide by 5 m long. Assume that all oxygen enters through the top of the tunnel, that no gradients in O$_2$ concentration exist in the tunnel or air above the tunnel (well-mixed), that the outside air is 20% O$_2$, that air is an ideal gas at atmospheric pressure (22.4 L/mol) and that the diffusivity of O$_2$ in soil is 0.0002 cm^2/s.

12.13.9 Binary Diffusion Coefficient

Show that $D_{AB} = D_{BA}$ for a binary system consisting of a mixture of gases A and B.

12.13.10 Heterogeneous Reaction

The reaction 3A $\rightarrow$ A$_3$ occurs instantaneously at a catalytic surface ($y = 0$). The concentrations of gases A and A$_3$ are maintained constant at $y = \delta$: $c_A(\delta) = C_{A\delta}$ and $c_{A3}(\delta) = 0$. The diffusion coefficients for the two gases in the film are the same.
(a) Find the steady-state fluxes of A and A$_3$ through the film, assuming other gases

in the film are stationary. (b) Find the concentrations of A and A_3 as functions of position in the film.

12.13.11 Membrane Design

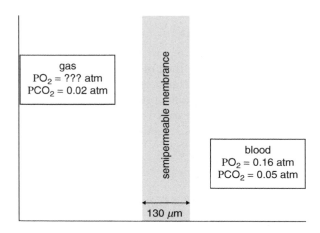

Your company has been contracted to specify the design of a gas exchange device in which oxygen and carbon dioxide will be simultaneously transported, but in opposite directions. The proposed design is a simple rectangular membrane with thickness 130 μm in which blood is contacted on one surface and a gas with controlled composition is contacted on the other surface (see figure). The partial pressures of O_2 and CO_2 on the blood side of the membrane are, respectively, 0.16 atm and 0.05 atm. The material for the membrane has been selected and the membrane CO_2 diffusivity is fivefold greater than the O_2 diffusivity at the operating temperature of 298 K.

(a) If the partial pressure of CO_2 in the gas is 0.02 atm, what must be the O_2 partial pressure in the gas (atm) to ensure equimolar counterdiffusion across the membrane?
(b) What assumptions did you make about the transport mechanism(s) to find the result in part (a)?
(c) If the diffusivity of O_2 through the membrane is 5.1×10^{-6} cm^2/s, what membrane area is required to achieve an O_2 transport rate of 2.8×10^{-3} mol/h?

12.13.12 Empirical Relationships for Mass Transfer

Use the Buckingham Pi theorem to show that the dimensional variables in (12.106) can be reduced to the dimensionless variables given by (12.107). What fundamental

12.13 Problems

dimensions are contained in the parameter set and which variables were used as the core variables?

12.13.13 Mass Transfer from a Sphere

A thin film of material containing species A at a concentration of 10 μM is placed on the surface of a solid sphere with radius 2 mm. Water at 25°C flows past the sphere with an average velocity of 1.0 cm/s. The diffusion coefficient of species A in water is 1.0×10^{-10} m²/s. The partition coefficient for species A in the solid relative to the fluid is 2.0. Find the mass transfer coefficient for species A in water and find the flux of A from the sphere surface if the concentration of A far from the surface is zero.

12.13.14 Internal Mass Transfer Coefficients

Estimate the mass transfer coefficient for the removal of urea from plasma as it passes through a rectangular duct with width = 8 mm and height = 1 mm. The wall concentration is maintained constant at a value near zero. The diffusion coefficient for urea in plasma is 1.8×10^{-9} m²/s.

12.13.15 Membrane Transport of Charged Species

A divalent cation with diffusion coefficient D moves through a cell membrane with thickness L under the action of a concentration gradient and a potential gradient. The potential difference is zero for the first 50% of the membrane length ($0 < x < L/2$), then varies linearly over the rest of the membrane ($L/2 < x < L$). The concentrations measured at $x = 0$ and $x = L/2$ are C_0 and C_1, respectively.

(a) What is the steady-state flux of the cation? (b) Find the concentration profile of the cation in the membrane. (c) What is the concentration of the cation at $x = L$?

12.13.16 Nernst Equation and Donnan Equilibrium

A membrane permeable to Na^+ and Cl^-, but impermeable to Ca^{++}, separates two ideal solutions as shown. Assume that the ions are completely dissociated on both sides of the membrane and that the initial osmotic pressure difference is zero. What is the initial molar concentration of $CaCl_2$ if the initial molar concentration of NaCl is 0.01 mol/L? What will be the final equilibrium molar concentration on each side of the membrane? What will be the potential difference across the membrane? What will be the osmotic pressure difference across the membrane?

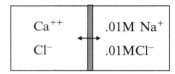

12.13.17 Goldman Equation

The Goldman equation in Example 12.7.1 was derived assuming that the electrical potential varies linearly from one end of the membrane to the other. Derive analogous expressions for the cases where the potential drops linearly over the first 10% or last 10% of membrane thickness, remaining constant for the remaining 90% of the membrane. Compare Cl^- and Na^+ flux computed with these expressions to Goldman flux for potential differences between -100 and $+100$ mV.

12.13.18 Goldman–Hodgkin–Katz Equation

Use the Goldman–Hodgkin–Katz equation derived in Example 12.7.1 to plot membrane potential as a function of external potassium concentration for internal concentration = 150 mM, internal sodium concentration = 15 mM and the sum of external sodium and potassium concentrations = 154 mM. For what range of P_{Na}/P_K and external potassium concentrations might one compute the membrane potential to be the same as the Nernst potential of -90 mV?

12.13.19 Bound and Dissolved Oxygen in Blood

(a) Find the dissolved and bound O_2 in blood at $P_{O_2} = 100$ mmHg, pH = 7.2 and $\rho_{Hb} = 15$ g/dl. Calculate the ratio of dissolved O_2/total O_2. (b) Find the dissolved and bound O_2 in blood at $P_{O_2} = 40$ mmHg, pH = 7.2 and $\rho_{Hb} = 15$ g/dl. Calculate the ratio of dissolved O_2/total O_2.

12.13.20 Blood Hemoglobin

Blood hemoglobin concentration is 15 g/dl and blood pH is 7.3. Arterial P_{O_2} is 100 mmHg and venous P_{O_2} is 10 mmHg. (a) Find the ratio of bound O_2 to dissolved

O_2 in arterial blood. (b) Find the ratio of bound O_2 to dissolved O_2 in venous blood. (c) Find the ratio of dissolved O_2 to total O_2 in arterial blood. (d) Find the ratio of dissolved O_2 to total O_2 in venous blood.

12.13.21 Blood-Doping Ethics

An athlete who lives at high elevation has a 25% greater blood oxygen-carrying capacity than his competitor who lives at sea level. The competitor wishes to level the playing field and is considering the following ways to raise his/her O_2 capacity to the same level: (a) train at high altitude, (b) live in a hypoxic room, (c) stimulate erythropoietin production with a drug, (d) homologous transfusion, and (e) autologous transfusion. Discuss potential dangers and ethical considerations for each. Would you recommend any be banned from competition? Why?

12.14 Challenges

12.14.1 Enzyme Kinetics

Background: Ethylene glycol is used as an antifreeze in automobiles. This is a sweet compound and spilled ethylene glycol is unknowingly ingested by dogs and other animals. By itself, this compound is not harmful to the body and is readily excreted in the urine. However, once ingested, ethylene glycol is converted by the enzyme alcohol dehydrogenase to an aldehyde, which in turn is converted into toxins. *Challenge:* Use your knowledge of enzyme kinetics to search for an inhibitor of alcohol dehydrogenase and to find an appropriate dose that can minimize the conversion of ethylene glycol to aldehydes. *Generate ideas:* Would you prefer to use a competitive or a noncompetitive inhibitor? Perform a literature search to identify potential candidates. Design a set of experiments that could be used to provide an estimate for an appropriate inhibitor dose.

12.14.2 Blood Doping

Background: You have been hired by an international athletic competition oversight committee. Their mission is to establish a set of unambiguous rules that will regulate an athlete's ability to enhance the oxygen-carrying capacity of blood (blood doping). Your ultimate role as a biomedical engineer is to assist the committee by developing a model that predicts the effects of increasing hematocrit value on tissue oxygen delivery and on important cardiovascular variables.

Challenge: Your immediate challenge is to understand blood doping and its physiological effects. *Generate Ideas:* What is meant by the "oxygen-carrying capacity" of blood? What is meant by "blood doping?" What different types of mechanisms are used to elevate hematocrit values? What are the advantages and disadvantages of blood doping? What important cardiovascular variables might be altered when blood hematocrit is changed? Use your findings to plot the following physiological variables as a function of hematocrit: vascular resistance, cardiac output, arterial pressure, arterial total oxygen content and tissue delivery rate of oxygen. Do the high hematocrit values reduce flow enough to decrease oxygen delivery to the tissue? Does increasing the hematocrit to 60% cause blood pressure to rise significantly? Does this rise in hematocrit increase the rate at which muscles can work? Consult the physiology literature to try and answer these questions. The oversight committee needs to decide if athletes should be allowed to alter their hematocrit by various means. They have adopted two primary principles to guide them: Is the intervention fair to all competitors? Is the intervention safe? Based on your findings, comment on these criteria for various interventions, such as training at high altitude, training at sea level in a hypoxic facility, homologous transfusion, autologous transfusion or injections of erythropoietin.

References

Adair GS (1925) The hemoglobin system VI. The oxygen dissociation curve of hemoglobin. J Biol Chem 63:529–545

Cussler EL (1997) Diffusion: mass transfer in fluid systems, 2nd edn. Cambridge University Press, London

Dash RK, Bassingthwaighte JB (2010) Erratum to: Blood HbO_2 and $HbCO_2$ dissociation curves at varied O_2, CO_2, pH, 2, 3-DPG and temperature levels. Ann Biomed Eng 38:1683–1701

DeHaven JC, EC DeLand (1962) Reactions of hemoglobin and steady states in the human respiratory system: an investigation using mathematical models and an electronic computer. The RAND Corporation, RM-3212-PR

Einstein A (1906) Eine neue Bestimmung der Moleküldimensionen [A new determination of molecular size]. Ann Phys 19:289–306

Fick A (1855) On liquid diffusion. Philos Mag 10:30–39

Hill AV (1910) The possible effects of the aggregation of the molecules of haemoglobin on its dissociation curves. J Physiol 40:iv–vii

Kays WM, Crawford W (1993) Convective heat and mass transfer, 3rd edn. McGraw-Hill, New York

Ranz WE, Marshall WR Jr (1952) Evaporation from drops. Parts I & II. Chem Eng Prog 48:141–6; 173–80

Wilson EJ, Geankoplis CJ (1966) Liquid mass transfer at very low Reynolds numbers in packed beds. Ind Eng Chem Fund 5(1):9–14

Winslow RM, Samaja M, Winslow NJ, Rossi-Bernardi L, Shrager RI (1983) Simulation of continuous blood O2 equilibrium curve over physiological pH, DPG, and P_{CO2} range. J Appl Physiol 54:524–529

Chapter 13
Macroscopic Approach to Biomass Transport

13.1 Introduction

Many biological and physiological mass transfer problems can be solved using a macroscopic approach. This is particularly true for fluid systems that can be considered to be reasonably well mixed. Systems with significant spatial gradients in concentration cannot be accurately analyzed with the macroscopic approach. In such cases, the microscopic approach to mass transfer discussed in Chaps. 14 and 15 must be applied. The macroscopic approach is appropriate when we are not interested in the spatial variations within the system, but instead are interested in average transient values or output values. Common objectives of the macroscopic approach are to find the rate of accumulation of a particular species in the system, the mass flow into or out of the system, or the rate at which a species is produced or utilized by chemical reaction within a system. Common applications include the pharmacokinetics of drugs in the body, diagnostic tracer studies, species synthesis in bioreactors, cellular transport, and transvascular transport.

13.2 Species Conservation

Our goal is to derive a general expression that can be used to predict how a single solute, which we shall name "species A," changes with time inside a well-mixed system. We begin by applying the principle of conservation of mass to species A inside a system:

$$\left\{ \begin{array}{c} \text{Rate of} \\ \text{accumulation} \\ \text{of species A} \\ \text{within a system} \end{array} \right\} = \left\{ \begin{array}{c} \text{Net rate species A} \\ \text{enters through} \\ \text{system boundaries} \end{array} \right\} + \left\{ \begin{array}{c} \text{Rate species A} \\ \text{is produced} \\ \text{within the system} \end{array} \right\}. \quad (13.1)$$

Consider the system shown in Fig. 13.1. The rate at which species A (expressed in terms of mass) changes with respect to time is given by:

Fig. 13.1 System with species A entering and leaving through conduits and a leaky wall, and with species A produced by chemical reaction within the system

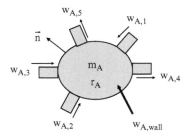

$$\left\{\begin{array}{c} \text{Rate of} \\ \text{accumulation} \\ \text{of species A} \\ \text{within system} \end{array}\right\} = \frac{dm_A}{dt} = \frac{d}{dt}\int_V \rho_A dV. \quad (13.2)$$

Species A can cross system boundaries via two major pathways: through the walls of the system S_{wall} and through K conduits each with its own cross-sectional area S_k. If $\vec{n}$ is an outward-directed unit normal to the surface of the system, then

$$\left\{\begin{array}{c} \text{Net rate species A} \\ \text{enters through} \\ \text{system boundaries} \end{array}\right\} = -\sum_{k=1}^{K}\int_{S_k} (\vec{n}_{Ak}) \cdot (\vec{n}dS) - \int_{S_{\text{wall}}} (\vec{n}_{A,\text{wall}}) \cdot (\vec{n}dS), \quad (13.3)$$

where $\vec{n}_{Ak}$ is the mass flux of species A through a conduit with cross-sectional area S_k and $\vec{n}_{A,\text{wall}}$ is the mass flux of species A through the walls of the system. The negative sign arises from the fact that the unit vector is directed away from the system. The integral of the flux of species A over a surface is simply the mass flow across that surface, w_A. Therefore, (13.3) can be written for N_{inlets} inlets and N_{outlets} outlets as:

$$\left\{\begin{array}{c} \text{Net rate species A} \\ \text{enters through} \\ \text{system boundaries} \end{array}\right\} = \sum_{i=1}^{N_{\text{inlets}}} w_{A,i} - \sum_{j=1}^{N_{\text{outlets}}} w_{A,j} + w_{A,\text{wall}}. \quad (13.4)$$

The net rate of production of species A by chemical reaction is equal to the mass rate of production per unit volume r_A integrated over the volume of the system:

$$\left\{\begin{array}{c} \text{Rate species A} \\ \text{is produced} \\ \text{within the system} \end{array}\right\} = r_{A,\text{tot}} = \int_V r_A dV. \quad (13.5)$$

13.2 Species Conservation

Combining these terms into the conservation relationship:

$$\frac{dm_A}{dt} = w_{A,\text{wall}} + \sum_{i=1}^{N_{\text{inlets}}} w_{A,i} - \sum_{j=1}^{N_{\text{outlets}}} w_{A,j} + r_{A,\text{tot}}. \tag{13.6}$$

One can write expressions similar to (13.6) for each species in the system. Adding the resulting equations would result in the overall conservation of mass statement given by (5.16). Dividing (13.6) by the molecular weight of species A provides a statement of species molar conservation:

$$\frac{dN_A}{dt} = W_{A,\text{wall}} + \sum_{i=1}^{N_{\text{inlets}}} W_{A,i} - \sum_{j=1}^{N_{\text{outlets}}} W_{A,j} + R_{A,\text{tot}}, \tag{13.7}$$

where N_A is the number of moles of A present in the system, $W_{A,\text{wall}}$ is the molar flow of A into the system through the walls, $W_{A,i}$ is the molar flow of A into the system through inlet channel i, $W_{A,j}$ is the molar flow of A out of the system through channel j, and $R_{A,\text{tot}}$ is the molar rate of production of A within the system ($r_{A,\text{tot}}/M_A$). In general, the mass flow of species A through any of the conduits could consist of convective and diffusive components:

$$w_{Ak} = -\int_{S_k} (\vec{n}_{Ak}) \cdot (\vec{n} dS) = -\int_{S_k} (\rho_{Ak}\vec{v}_k - D_{Ak}\vec{\nabla}\rho_{Ak}) \cdot (\vec{n} dS). \tag{13.8}$$

However, in most cases, the diffusive fluxes at the system inlets and outlets will be small in comparison with the convective fluxes. If we can neglect diffusive fluxes at conduit boundaries, then,

$$w_{Ak} \approx -\int_{S_k} (\rho_{Ak}\vec{v}_k) \cdot (\vec{n} dS) = -\rho_{Abk} Q_{Vk}(\vec{e}_k \cdot \vec{n}), \tag{13.9}$$

where $\vec{e}_k$ is a unit vector in the direction of the velocity vector at the conduit-system boundary, $\vec{n}$ is a unit vector directed away from the system surface at the conduit-system boundary, ρ_{Abk} is the flow-averaged or bulk mass concentration in conduit k, and Q_{Vk} is the volumetric flow rate through conduit k. $Q_{Vk}(\vec{e}_k \cdot \vec{n})$ is positive for streams leaving the system and is negative for streams entering the system. Equation (13.9) serves as a definition of the mean mass concentration in the conduit. Introducing this expression into the species conservation equation we find:

$$\frac{dm_A}{dt} = w_{A,\text{wall}} + \sum_{i=1}^{N_{\text{inlets}}} \rho_{Abi} Q_{Vi} - \sum_{j=1}^{N_{\text{outlets}}} \rho_{Abj} Q_{Vj} + r_{A,\text{tot}}. \tag{13.10}$$

If species A is well-mixed inside the system and the system volume is constant, then $m_A = \rho_A V$ and $r_{A,tot} = r_A V$, and the concentrations at all outlets will be the same as the system concentration. The species conservation expression for a well-mixed system with constant volume is:

$$V\frac{d\rho_A}{dt} = w_{A,\text{wall}} + \sum_{i=1}^{N_{\text{inlets}}} \rho_{Abi} Q_{Vi} - \rho_A \sum_{j=1}^{N_{\text{outlets}}} Q_{Vj} + r_A V \quad \text{(well-mixed)}. \quad (13.11)$$

The flow through the system wall can often be written in terms of the permeability-surface area product:

$$w_{A,\text{wall}} = -\mathsf{P}_A S(\rho_A - \Phi_{A,\text{system},\infty} \rho_{A\infty}). \quad (13.12)$$

Here $\mathsf{P}_A S$ is the permeability-surface area product of the system wall for species A, $\Phi_{A,\text{system},\infty}$ is the partition coefficient for species A between the fluid inside the system and the fluid surrounding the system, and $\rho_{A\infty}$ is the mass concentration in the fluid surrounding the system. The negative sign ensures that mass flow is into the system when the driving force is negative. If convective mass exchange occurs between the fluid and a portion of the system boundary, $w_{A,\text{wall}}$ must also include an expression for internal (12.122) or external (12.104) mass transfer of species A. Inserting (13.12) into (13.11), we obtain our final simplified expression for the conservation of species A:

$$V\frac{d\rho_A}{dt} = -\mathsf{P}_A S(\rho_A - \Phi_{A,\text{system},\infty} \rho_{A\infty}) + \sum_{i=1}^{N_{\text{inlets}}} \rho_{Abi} Q_{Vi} - \rho_A \sum_{j=1}^{N_{\text{outlets}}} Q_{Vj} + r_A V. \quad (13.13)$$

This can be expressed in terms of molar bulk concentration of species A by dividing (13.13) by the molecular weight of the species:

$$V\frac{dC_A}{dt} = -\mathsf{P}_A S(C_A - \Phi_{A,\text{system},\infty} C_{A\infty}) + \sum_{i=1}^{N_{\text{inlets}}} C_{Abi} Q_{Vi} - C_A \sum_{j=1}^{N_{\text{outlets}}} Q_{Vj} + R_A V. \quad (13.14)$$

R_A is the molar rate of production of species A per unit volume (i.e., r_A/M_A). C_{Abi} is the bulk (i.e., flow averaged) molar concentration in inlet conduit i.

Equations (13.6) and (13.7) represent the most general forms of the macroscopic species conservation equations, and (13.13) and (13.14) are the species continuity relations if the system volume is constant, the system is well-mixed, transport across the wall is by diffusion, and the concentrations of the inlet streams are uniform over their cross-sections. These equations form the starting point for all of the macroscopic problems considered in this chapter.

13.3 Compartmental Analysis

Compartmental analysis refers to the application of the macroscopic species mass balance to one or more well-mixed systems or compartments. The compartments may be connected in parallel or in series, or in some combination of parallel and series connections. The objective is to find the concentration of species in each compartment as a function of time, given the initial concentrations.

13.3.1 Single Compartment

In this section, we will consider problems that involve species mass flow in and out of a single well-mixed compartment. We will consider only nonreacting species in this section.

13.3.1.1 Single Compartment, Constant Volume, Single Inlet and Outlet, Constant Flow, Constant Rate of Infusion

Consider a well-mixed compartment of constant volume V, with a single inlet and a single outlet. Fluid flows into the compartment at a constant volumetric flow rate Q_V. The concentration of species A in the fluid is initially c_{A0} in the compartment. The walls of the compartment are impermeable to species A. At $t = 0$, the inlet bulk concentration is suddenly changed to C_{A1}. We are interested in finding the concentration of A in the outflow as a function of time.

Equation (13.14) is appropriate for this case, since the compartment volume is constant. There is a single inlet ($N_{inlet} = 1$), a single outlet ($N_{outlet} = 1$), species A is not involved in a chemical reaction ($R_A = 0$) and the wall is impermeable ($PS_{A,wall} = 0$). Under these circumstances, (13.14) reduces to:

$$V \frac{dC_A}{dt} = Q_V (C_{A1} - C_A). \tag{13.15}$$

Since C_{A1} is independent of time, then $dC_A/dt = d(C_A - C_{A1})/dt$, and (13.15) can be rewritten:

$$\frac{d(C_A - C_{A1})}{C_A - C_{A1}} = -\frac{Q_V}{V} dt. \tag{13.16}$$

Integrating this with the initial condition that $C_A(t = 0) = C_{A0}$, we find the solution to be:

$$\frac{C_A - C_{A1}}{C_{A0} - C_{A1}} = e^{-\frac{Q_V}{V} t}. \tag{13.17}$$

Because the system is well-mixed, the concentration at any point in the compartment is given by (13.17), as is the concentration in the outlet conduit. The difference between the compartment and inlet concentration decreases exponentially with time.

13.3.1.2 Single Compartment, Constant Volume, Single Inlet and Outlet, Constant Flow, Bolus Injection

Consider the same single compartment system as above, except that the inlet concentration of species A is zero and a mass m_{A0} of species A is injected into the single compartment at time $t = 0$. The solution would be the same as the solution given in (13.17), except that $C_{A1} = 0$ and the initial concentration would be computed as the ratio of the number of moles (m_{A0}/M_A) injected divided by the volume of the compartment (V):

$$C_{A0} = \frac{m_{A0}}{M_A V}. \qquad (13.18)$$

The solution at the compartment outlet for this case would be:

$$C_A = \frac{m_{A0}}{M_A V} e^{-\frac{Q_v}{V}t} \quad \text{or} \quad \rho_A = \frac{m_{A0}}{V} e^{-\frac{Q_v}{V}t}. \qquad (13.19)$$

13.3.1.3 Single Compartment, Variable Volume, Single Inlet and Outlet, Constant Inlet Flow

In many cases, the compartment volume may not remain constant. For instance, the system shown in Fig. 13.2 is used to perfuse an isolated organ. The inlet tank is the compartment of interest. The resistance to flow out of the tank $\mathfrak{R}_f$ caused by the organ and tubing is assumed to be constant and the fluid ultimately discharges to atmospheric pressure. The initial volume in the tank is V_0 and the cross-sectional area of the tank is A_c. Fluid flow into the collecting tank is initially Q_{V0}, and the mass concentration of metabolite A flowing out of the organ into the tank is ρ_{A0}. At time $t = 0$, the flow is doubled, and the concentration of the metabolite at the inflow to the organ is also doubled. Our goals are to find the concentration of metabolite in the well-mixed tank, the volume of fluid in the tank and the flow rate of fluid out of the tank as a function of time.

The fluid volume and metabolite concentration will both be changing in the collecting tank. We begin by simplifying the general macroscopic species conservation equation, (13.10), for a compartment with a single inlet and single outlet, with no flow through the tank walls, and no homogeneous chemical reaction.

13.3 Compartmental Analysis

Fig. 13.2 Organ perfusion system

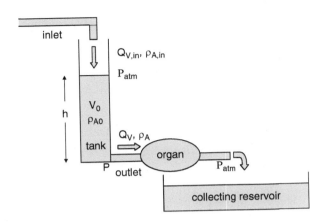

The inlet flow for $t > 0$ is $Q_{\text{v,in}} = 2Q_{\text{v0}}$ and the inlet mean mass concentration is $\rho_{\text{Ab,in}} = 2\rho_{\text{A0}}$:

$$\frac{dm_A}{dt} = \frac{d}{dt}(\rho_A V) = V\frac{d\rho_A}{dt} + \rho_A \frac{dV}{dt} = \rho_{\text{Ab,in}} Q_{\text{V,in}} - \rho_A Q_{\text{V}}. \tag{13.20}$$

This provides one equation for the three unknown quantities: V, Q_V, and ρ_A. A second relationship that is needed is a total mass balance on the tank, provided by (5.11) with no flow through the tank walls:

$$\frac{dm}{dt} = \frac{d}{dt}(\rho_{\text{out}} V) = \rho_{\text{in}} Q_{\text{V,in}} - \rho_{\text{out}} Q_{\text{V}}, \tag{13.21}$$

where ρ_{out} is the total mass density of the fluid flowing out of the tank and ρ_{in} is the density of fluid flowing into the tank. Since the metabolite density is a very small component of the total density, we can assume that $\rho_{\text{out}} = \rho_{\text{in}} = \rho$, so:

$$\frac{dV}{dt} = Q_{\text{V,in}} - Q_{\text{V}}. \tag{13.22}$$

Substituting (13.22) into (13.20), yields:

$$V\frac{d\rho_A}{dt} + Q_{\text{V,in}}(\rho_A - \rho_{\text{Ab,in}}) = 0. \tag{13.23}$$

Separating variables and integrating:

$$\int_{\rho_{\text{A0}} - \rho_{\text{A,in}}}^{\rho_A - \rho_{\text{A,in}}} \frac{d(\rho_A - \rho_{\text{Ab,in}})}{(\rho_A - \rho_{\text{Ab,in}})} = \ln\left(\frac{\rho_A - \rho_{\text{Ab,in}}}{\rho_{\text{A0}} - \rho_{\text{Ab,in}}}\right) = -Q_{\text{V,in}} \int_0^t \frac{dt}{V} \tag{13.24}$$

Before we can compute the mass concentration at the tank outlet, we must determine how the volume varies with time. Flow out of the tank is governed by the difference between tank outlet pressure and atmospheric pressure:

$$Q_V = \frac{P_{\text{out}} - P_{\text{atm}}}{\Re_f}. \tag{13.25}$$

The pressure at the bottom of the tank is primarily due to the weight of the fluid column:

$$P_{\text{out}} - P_{\text{atm}} = \rho g h, \tag{13.26}$$

where h is the height of the fluid column in the tank. Finally, the fluid volume V is equal to the product of the tank cross sectional area A_c and the height of fluid in the tank:

$$V = h A_c. \tag{13.27}$$

Consequently, outlet flow is related to fluid volume as follows:

$$\frac{Q_V}{V} = \frac{Q_{V0}}{V_0} = \frac{\rho g}{\Re_f A_c} \equiv \alpha. \tag{13.28}$$

This indicates that Q_{V0} and V_0 cannot be selected independently. Substituting (13.28) into (13.22), with $Q_{V,\text{in}} = 2 Q_{V0}$:

$$\frac{dV}{dt} + \alpha V = 2\alpha V_0 \quad \text{or} \quad \frac{d}{dt}(V - 2V_0) + \alpha(V - 2V_0) = 0. \tag{13.29}$$

The solution for fluid volume with $V(0) = V_0$ is:

$$V = V_0 [2 - e^{-\alpha t}]. \tag{13.30}$$

The final step is to solve for the mass concentration of metabolite in the tank by substituting (13.30) for V back into (13.24) and integrating. Referring to a table of integrals, we find:

$$\ln\left(\frac{\rho_A - \rho_{Ab,\text{in}}}{\rho_{A0} - \rho_{Ab,\text{in}}}\right) = -\frac{Q_{V,\text{in}}}{V_0} \int_0^t \frac{dt}{[2 - e^{-\alpha t}]} = -\frac{Q_{V,\text{in}}}{V_0}\left[\frac{t}{2} + \frac{\ln(2 - e^{-\alpha t})}{2\alpha}\right]. \tag{13.31}$$

The final solution for $Q_{V,\text{in}} = 2Q_{V0}$ and $\rho_{Ab,\text{in}} = 2\rho_{A0}$ is:

$$\frac{\rho_A}{\rho_{A0}} = 2 - \left(\frac{e^{-\alpha t}}{2 - e^{-\alpha t}}\right). \tag{13.32}$$

13.3 Compartmental Analysis

Solving for the fluid flow using (13.28):

$$\frac{Q_V}{Q_{V0}} = \frac{V}{V_0} = 2 - e^{-\alpha t}. \tag{13.33}$$

The solutions for ρ_A/ρ_{A0} and for $Q_V/Q_{V0} = V/V_0$ are shown in Fig. 13.3 as a function of αt, where α is defined by (13.28).

Notice that all of these quantities will be within 99% of their final values for $\alpha t = 5$. However, the metabolite concentration increases at a faster rate than the tank volume or tank outflow. This can be seen more clearly by comparing the initial rate of change of metabolite relative to the initial amount of metabolite in the tank with the initial rate of change of tank volume relative to the initial tank volume. From (13.20):

$$\frac{1}{m_{A0}}\frac{dm_A}{dt}\bigg|_{t=0} = \frac{(\rho_{Ab,in}Q_{V,in} - \rho_A Q_V)|_{t=0}}{\rho_{A0} V_0} = \frac{3 Q_{V0}}{V_0}, \tag{13.34}$$

and from (13.22):

$$\frac{1}{V_0}\frac{dV}{dt}\bigg|_{t=0} = \frac{(Q_{V,in} - Q_V)|_{t=0}}{V_0} = \frac{Q_{V0}}{V_0}. \tag{13.35}$$

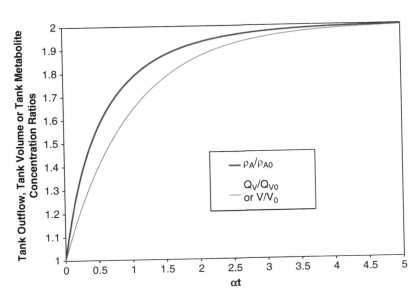

Fig. 13.3 Comparison of the change of tank metabolite concentration with the change in tank fluid volume or fluid outflow relative to the initial values as a function of dimensionless time, αt

Therefore, the relative rate at which metabolite increases in the tank is initially three times the relative rate at which the fluid volume increases.

13.3.1.4 Single Compartment, Constant Volume, Two Inlets and Single Outlet, Constant Inlet Flow

Consider a well-mixed system with a constant volume, V, two inlets, and a single outlet as shown in Fig. 13.4.

The initial concentration of species A in the system is C_{A0}. At $t = 0$, two streams flow into the system: stream 1 with volumetric flow rate $Q_{V,1}$ and mean concentration $C_{A,1}$ and stream 2 with volumetric flow rate $Q_{V,2}$ and mean concentration $C_{A,2}$. Our goal is to find the outlet flow Q_V and the outlet concentration C_A.

We begin by applying an overall mass balance on the system, assuming that the concentrations of species A are dilute so that the fluid density does not change appreciably. If density is constant and there is no fluid or mass transport through the walls of the system, conservation of volume (5.17) yields the following expression for a system with constant volume:

$$\frac{dV}{dt} = 0 = Q_{V,1} + Q_{V,2} - Q_V. \tag{13.36}$$

Consequently, the outlet volumetric flow rate is constant and equal to the sum of the inlet flow rates:

$$Q_V = Q_{V,1} + Q_{V,2}. \tag{13.37}$$

The macroscopic conservation equation for species A in the impermeable system with constant volume is (13.7):

$$V\frac{dC_A}{dt} = C_{A,1}Q_{V,1} + C_{A,2}Q_{V,2} - C_A Q_V, \tag{13.38}$$

where Q_V is given by (13.37). The solution to this ordinary differential equation with constant coefficients and initial condition $C_A(0) = C_{A0}$ is:

$$C_A = C_{A0}e^{-\frac{Q_V}{V}t} + \left(C_{A,1}\frac{Q_{V,1}}{Q_V} + C_{A,2}\frac{Q_{V,2}}{Q_V}\right)\left(1 - e^{-\frac{Q_V}{V}t}\right). \tag{13.39}$$

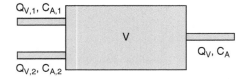

Fig. 13.4 System with constant volume V, two inlets and a single outlet

13.3 Compartmental Analysis

This solution is consistent with our initial condition and after a very long time the new steady-state solution is:

$$C_A(t \to \infty) = \frac{C_{A,1} Q_{V,1} + C_{A,2} Q_{V,2}}{Q_{V,1} + Q_{V,2}}. \tag{13.40}$$

Note that the steady-state solution for species A is independent of the system volume and the initial concentration of species A in the system.

13.3.1.5 Single Compartment, Constant Volume, Single Inlet and Outlet, Oscillating Inlet Flow, Constant Inlet Concentration

Let us return now to the problem analyzed in Sect. 13.3.1.1, but allow the inlet flow to have a sinusoidal component with angular frequency ω, in addition to a steady component:

$$Q_{V,\text{in}} = Q_{V0} + Q_{V1} \sin(\omega t). \tag{13.41}$$

Equation (13.14) reduces to the following equation under these conditions:

$$V \frac{dC_A}{dt} = (Q_{V0} + Q_{V1} \sin(\omega t))(C_{A1} - C_A), \tag{13.42}$$

where C_{A1} is the mean concentration in the inlet stream for $t \geq 0$. Separating variables:

$$\frac{d(C_A - C_{A1})}{(C_A - C_{A1})} = -\frac{(Q_{V0} + Q_{V1} \sin(\omega t))}{V} dt. \tag{13.43}$$

Using the initial condition, the solution can be written:

$$\frac{C_A - C_{A1}}{C_{A0} - C_{A1}} = \exp\left\{ -\frac{Q_{V0}(\omega t)}{\omega V} \left[1 + \frac{Q_{V1}}{Q_{V0}} \left(\frac{1 - \cos(\omega t)}{\omega t} \right) \right] \right\}. \tag{13.44}$$

The solution is written in this manner to show its dependence on the dimensionless time ωt and two dimensionless parameters, $Q_{V0}/(\omega V)$ and Q_{V1}/Q_{V0}. Results are plotted in Fig. 13.5 for various values of $Q_{V0}/\omega V$ and for $Q_{V1}/Q_{V0} = 0$ (no oscillations, light lines) and for $Q_{V1}/Q_{V0} = 1$ (maximal oscillations without reverse flow, heavy lines). The solution approaches the constant flow solution (13.17) when either the oscillating to steady amplitude ratio Q_{V1}/Q_{V0} is small or $Q_{V0}/(\omega V)$ is small. In addition, whenever ωt is a multiple of 2π, the oscillating solution is the same as that of the steady solution. At other values of ωt, the outlet concentration is closer to the inlet concentration for the oscillating

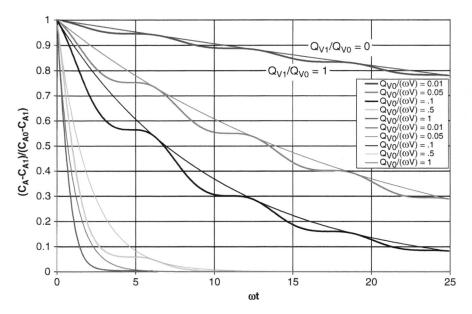

Fig. 13.5 Dimensionless concentration at system outlet for sinusoidal flow rate with $Q_{V1}/Q_{V0} = 1$ vs. constant flow rate at various ratios $Q_{V0}/\omega V$

flow solution than for the constant flow solution, indicating a higher rate of mass transfer. The maximum differences will occur when ωt is an odd multiple of π. This is not to be expected in all oscillating flows, but is directly related to the form of the oscillating flow relationship, (13.41). If the flow were to initially decrease starting at $t = 0$ (e.g., $Q_V = Q_{V0} - Q_{V1}\sin \omega t$), then the initial mass exchange rate will be lower than in the steady flow case, and the constant flow dimensionless concentration will be a lower bound to the oscillating component. If the initial rate of change of oscillations is zero (e.g., $Q_V = Q_{V0} + Q_{V1}\cos \omega t$), the periodic solution will oscillate about the constant flow solution.

13.3.1.6 Single Compartment, Constant Volume, Mass Flow Through a Permeable Wall

Let us analyze the situation when a cell with an initial cytoplasm concentration C_{A0} of species A is placed in an isotonic fluid containing a different concentration of species A C_{A1}. The cell is well-mixed with volume V. The cell membrane has surface area S and is permeable to species A with permeability coefficient P_A. The partition coefficient between intracellular and extracellular fluids is $\Phi_{A,\text{int},\text{ext}}$ for species A. We are interested in finding how the concentration of species A varies with time in the intracellular fluid.

13.3 Compartmental Analysis

Since convection is not present in this case and cell volume remains constant, the species conservation equation becomes:

$$V\frac{dC_A}{dt} = -P_A S (C_A - \Phi_{A,int,ext} C_{A1}). \tag{13.45}$$

Since C_{A1} and $\Phi_{A,int,ext}$ are constant, (13.45) can be rearranged as follows:

$$\frac{d(C_A - \Phi_{A,int,ext} C_{A1})}{(C_A - \Phi_{A,int,ext} C_{A1})} = -\frac{P_A S}{V} dt. \tag{13.46}$$

The solution can be obtained by applying the initial condition $C_A(t=0) = C_{A0}$:

$$\frac{(C_A - \Phi_{A,int,ext} C_{A1})}{(C_{A0} - \Phi_{A,int,ext} C_{A1})} = e^{-\left(\frac{P_A S}{V}\right)t}. \tag{13.47}$$

Hence, the larger the surface to volume ratio or the permeability, the greater will be the rate of exchange of species A across the membrane. In many cases, the partition coefficient between intracellular and extracellular fluids is assumed to equal unity.

13.3.1.7 Single Compartment, Constant Volume, Single Inlet and Outlet, Constant Flow and Inlet Concentration, Permeable Wall

Our final variation on mass transfer in a single compartment is directed at the analysis of mass transfer through the system wall when convection is present. Consider an organ with a single inlet and outlet. The organ is well-mixed with volume V. The wall of the organ has surface area S and is permeable to species A with permeability coefficient P_A. The outside surface of the organ is bathed with interstitial fluid that is maintained at a constant concentration of species A, $C_{A,isf}$. The partition coefficient between blood and interstitial fluid for species A is $\Phi_{A,blood,isf}$. Species A distributes readily in plasma and red cells, so the concentration is assumed uniform and is initially C_{A0}. At $t=0$ blood with mean concentration C_{A1} begins to flow through the organ at a constant volumetric flow rate Q_V. Our goal is to find the concentration of species A at the outlet of the organ.

The macroscopic species conservation equation for this situation is:

$$V\frac{dC_A}{dt} = Q_V(C_{A1} - C_A) - P_A S(C_A - \Phi_{A,blood,isf} C_{A,isf}). \tag{13.48}$$

This can be rewritten in the form

$$\frac{dC_A}{dt} + \beta C_A = \alpha, \tag{13.49}$$

where

$$\alpha \equiv \frac{1}{V}\left[Q_\text{V} C_{\text{A}1} + \text{P}_\text{A} S \Phi_{\text{A,blood,isf}} C_{\text{A,isf}}\right] \quad (13.50)$$

and

$$\beta \equiv \frac{Q_\text{V} + \text{P}_\text{A} S}{V}. \quad (13.51)$$

The solution that satisfies the initial condition is:

$$\frac{C_\text{A} - \dfrac{\alpha}{\beta}}{C_{\text{A}0} - \dfrac{\alpha}{\beta}} = e^{-\beta t}. \quad (13.52)$$

This is similar in form to the solution for exchange across the cell wall in the previous section, and is indeed the same when Q_V is zero. The concentration of species A emerging from the organ after a long period of time is α/β, or:

$$C_\text{A}(t \to \infty) = \left(\frac{Q_\text{V} C_{\text{A}1} + \text{P}_\text{A} S \Phi_{\text{A,blood,isf}} C_{\text{A,isf}}}{Q_\text{V} + \text{P}_\text{A} S}\right). \quad (13.53)$$

Again, as in previous examples, the ultimate steady-state concentration is independent of the organ volume and the initial concentration of species A in the organ.

13.3.2 Two Compartments

In the previous section, we focused on mass exchange without chemical reaction in a single compartment, with each example emphasizing the contribution of different terms or initial conditions in the macroscopic species conservation equation. Problems involving two compartments are analyzed in the following sections.

13.3.2.1 Two Compartments in Series

The chambers of the heart are often modeled as well-mixed compartments (Fig. 13.6). If we inject a bolus containing a mass of species A in the left atrium, what will the concentration of tracer be in the ascending aorta as a function of time after the injection?

To simplify the analysis, we will assume that the heart is isolated and perfused at a constant rate Q_V, so the volume of the left atrium V_LA and the volume of the left ventricle V_LV are both constant. In addition, species A is assumed not to be involved in

13.3 Compartmental Analysis

Fig. 13.6 Simplified model of mixing in the heart

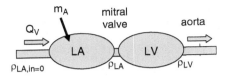

chemical reactions in either chamber and both chambers are assumed to be well-mixed and impermeable to species A. A macroscopic balance must be written for species A in each compartment:

Left atrium:

$$V_{LA}\frac{d\rho_{LA}}{dt} = Q_V(\rho_{LA,in} - \rho_{LA}). \tag{13.54}$$

Left ventricle:

$$V_{LV}\frac{d\rho_{LV}}{dt} = Q_V(\rho_{LV,in} - \rho_{LV}). \tag{13.55}$$

We will assume that species A does not enter the left atrium via blood flow, and the mass concentration of species A leaving the left atrium through the mitral valve must equal the concentration entering the left ventricle:

$$\rho_{LA,in} = 0, \tag{13.56}$$
$$\rho_{LV,in} = \rho_{LA}.$$

The solution to (13.54) for a bolus injection, where $\rho_{LA}(0) = m_A/V_{LA}$ is given by (13.19):

$$\rho_{LA} = \frac{m_A}{V_{LA}} e^{-\left(\frac{Q_V}{V_{LA}}\right)t}. \tag{13.57}$$

Substituting this into (13.55) for $\rho_{LV,in}$:

$$\frac{d\rho_{LV}}{dt} + \frac{Q_V}{V_{LV}}\rho_{LV} = \frac{Q_V}{V_{LV}}\left(\frac{m_A}{V_{LA}}\right) e^{-\left(\frac{Q_V}{V_{LA}}\right)t}. \tag{13.58}$$

The solution to the homogeneous portion of (13.58) is

$$\rho_{LV,h} = Ae^{-\left(\frac{Q_V}{V_{LV}}\right)t}, \tag{13.59}$$

where A is a constant. A particular solution to (13.55) will have the form

$$\rho_{LV,p} = Be^{-\left(\frac{Q_V}{V_{LA}}\right)t}. \tag{13.60}$$

Substitution of (13.60) into (13.58), allows us to find the unknown constant B:

$$B = \frac{m_A}{V_{LA} - V_{LV}}. \tag{13.61}$$

The total solution can be found by adding the homogeneous and particular solutions:

$$\rho_{LV} = \rho_{LV,h} + \rho_{LV,p} = A e^{-\left(\frac{Q_V}{V_{LV}}\right)t} + \left(\frac{m_A}{V_{LA} - V_{LV}}\right) e^{-\left(\frac{Q_V}{V_{LA}}\right)t}. \tag{13.62}$$

Applying the initial condition in the left ventricle, $\rho_{LV}(0) = 0$ allows us to solve for the constant A. The final solution in the ventricle for $V_{LA} \neq V_{LV}$ is:

$$\rho_{LV} = \left(\frac{m_A}{V_{LV} - V_{LA}}\right) \left(e^{-\left(\frac{Q_V}{V_{LV}}\right)t} - e^{-\left(\frac{Q_V}{V_{LA}}\right)t}\right). \tag{13.63}$$

Mass concentrations in the left atrium and left ventricle relative to the initial concentration in the left atrium are shown in Fig. 13.7 for $V_{LA} = 37$ ml, $V_{LV} = 100$ ml, and $Q_V = 60$ ml/s.

If species A is infused at a constant rate w_{Ai} into the left atrium, rather than injected as a bolus, the macroscopic mass balance on the left ventricle (13.55) remains unchanged, but the mass balance on the left atrium for no tracer entering the left atrium via the blood stream ($\rho_{LA,in} = 0$) is altered as follows:

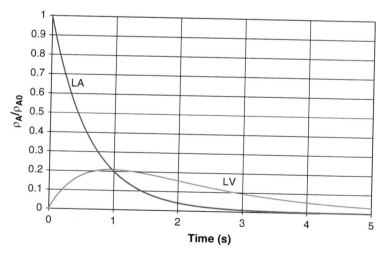

Fig. 13.7 Washout of species A from left atrium and left ventricle following a bolus injection in the left atrium

13.3 Compartmental Analysis

$$V_{LA}\frac{d\rho_{LA}}{dt} = w_{Ai} - \rho_{LA}Q_V. \tag{13.64}$$

The solution in the left atrium for $\rho_{LA}(t=0) = 0$ is:

$$\rho_{LA} = \frac{w_{Ai}}{Q_V}\left[1 - e^{-\frac{Q_V}{V_{LA}}t}\right]. \tag{13.65}$$

Substituting this into (13.55) for $\rho_{LV,in}$:

$$\frac{d\rho_{LV}}{dt} + \frac{Q_V}{V_{LV}}\rho_{LV} = \frac{w_{Ai}}{V_{LV}}\left[1 - e^{-\frac{Q_V}{V_{LA}}t}\right]. \tag{13.66}$$

The homogeneous solution is given by (13.59). A particular solution in this case will have the form:

$$\rho_{LV,p} = A^* + Be^{-\frac{Q_V}{V_{LA}}t}. \tag{13.67}$$

Substituting (13.67) into (13.66) and solving for the constants A^* and B:

$$A^* = \frac{w_{Ai}}{Q_V}, \quad B = \frac{w_{Ai}}{Q_V}\frac{V_{LA}}{(V_{LV} - V_{LA})}. \tag{13.68}$$

Finally, adding the homogeneous and particular solutions and imposing the initial condition that $\rho_{LV}(0) = 0$, we find the solution for ρ_{LV}:

$$\rho_{LV} = \frac{w_{Ai}}{Q_V}\left\{1 - e^{-\left(\frac{Q_V}{V_{LV}}\right)t} - \left(\frac{V_{LA}}{V_{LV} - V_{LA}}\right)\left(e^{-\left(\frac{Q_V}{V_{LV}}\right)t} - e^{-\left(\frac{Q_V}{V_{LA}}\right)t}\right)\right\}. \tag{13.69}$$

The solutions for concentrations in the left atrium and left ventricle relative to the infusion rate divided by the blood flow rate is shown in Fig. 13.8.

The concentration in the left ventricle lags that in the left atrium, as expected, but both mass concentrations eventually level off at a value equal to w_{Ai}/Q_V.

13.3.2.2 Two Compartments in Parallel

The outflow streams from two well-mixed compartments merge at station 3 as shown in Fig. 13.9. The volumetric flow rates to each compartment are constant (Q_{V1} and Q_{V2}) and the initial concentrations of species A in compartments 1 and 2 are C_{A01} and C_{A02}, respectively. The compartment volumes remain constant at V_1 and V_2. At time $t = 0$, the inlet mean concentration in stream 1 is changed from

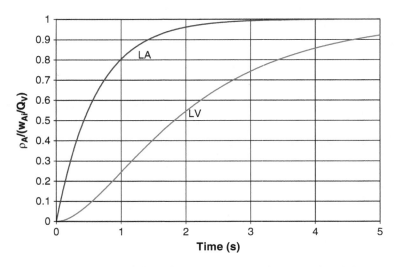

Fig. 13.8 Concentration in left atrium and left ventricle relative to w_{Ai}/Q_V during a constant infusion of species A at rate w_{Ai}

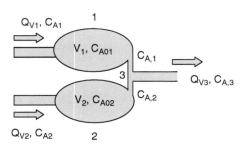

Fig. 13.9 Parallel flow compartments

C_{A01} to C_{A1} and in stream 2 from C_{A02} to C_{A2}. We wish to find the concentration in the mixed outflow at station 3.

The analysis presented in Sect. 13.3.1.1 can be applied to compartments 1 and 2. If the concentrations in compartments 1 and 2 are designated as $C_{A,1}(t)$ and $C_{A,2}(t)$, the solutions for the two compartments from (13.17) are:

$$\frac{C_{A,1} - C_{A1}}{C_{A01} - C_{A1}} = e^{-\frac{Q_{V1}}{V_1}t}, \tag{13.70}$$

$$\frac{C_{A,2} - C_{A2}}{C_{A02} - C_{A2}} = e^{-\frac{Q_{V2}}{V_2}t}. \tag{13.71}$$

Conservation of volume at node 3 will require

$$Q_{V3} = Q_{V1} + Q_{V2}. \tag{13.72}$$

13.3 Compartmental Analysis

A species mass balance at node three gives:

$$C_{A,3}Q_{V3} = C_{A,1}Q_{V1} + C_{A,2}Q_{V2}. \qquad (13.73)$$

Substituting (13.70)–(13.72) into (13.73), and solving for $C_{A,3}$:

$$C_{A,3} = \frac{C_{A1}Q_{V1} + C_{A2}Q_{V2}}{Q_{V1} + Q_{V2}} + \frac{Q_{V1}(C_{A01} - C_{A1})}{Q_{V1} + Q_{V2}}e^{-\frac{Q_{V1}}{V_1}t} \\ + \frac{Q_{V2}(C_{A02} - C_{A2})}{Q_{V1} + Q_{V2}}e^{-\frac{Q_{V2}}{V_2}t}. \qquad (13.74)$$

Examination of (13.74) shows that the concentration has the expected values for $t = 0$ and $t \to \infty$.

13.3.2.3 Exchange Between Two Well Mixed Compartments Across a Membrane

Two stationary compartments, each with constant volume, are separated by a membrane that is permeable to species A (Fig. 13.10). Initially, species A is not present in either compartment. At $t = 0$ species A is introduced as a bolus into compartment 1, with an initial concentration C_{A10}. We wish to find the concentration of species A in each compartment as a function of time.

Species conservation equations for each compartment are:

Compartment 1:

$$V_1 \frac{dC_{A1}}{dt} = -P_A S(C_{A1} - \Phi_{A12}C_{A2}). \qquad (13.75)$$

Compartment 2:

$$V_2 \frac{dC_{A2}}{dt} = P_A S(C_{A1} - \Phi_{A12}C_{A2}), \qquad (13.76)$$

where P_A is the permeability of the membrane to species A and S is the membrane surface area. We thus have two coupled ordinary differential equations subject to the initial conditions:

$$C_{A1}(t = 0) = C_{A10}, \\ C_{A2}(t = 0) = 0. \qquad (13.77)$$

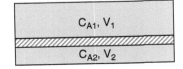

Fig. 13.10 Compartments separated by a membrane

If we were to add (13.75) and (13.76), we would obtain a confirmation that the total number of moles contained in both compartments remains constant:

$$\frac{d}{dt}(C_{A1}V_1 + C_{A2}V_2) = 0. \tag{13.78}$$

Potential solution methods include the use of Laplace transforms, substituting one equation into the other, and the use of ODE solvers in computer packages. In this case, we choose to use substitution. Solving (13.76) for C_{A1}:

$$C_{A1} = \frac{V_2}{P_A S}\frac{dC_{A2}}{dt} + \Phi_{A12}C_{A2}. \tag{13.79}$$

Substituting this into (13.75), we obtain the following homogeneous second-order ODE:

$$\frac{d^2 C_{A2}}{dt^2} + P_A S\left(\frac{1}{V_1} + \frac{\Phi_{A12}}{V_2}\right)\left(\frac{dC_{A2}}{dt}\right) = 0. \tag{13.80}$$

A general solution to (13.80) is $C_{A2} = Ae^{\lambda t}$. Substituting this into (13.80) provides the following quadratic equation for λ:

$$\lambda^2 + P_A S\left(\frac{1}{V_1} + \frac{\Phi_{A12}}{V_2}\right)\lambda = 0, \tag{13.81}$$

which has the two solutions:

$$\lambda_1 = 0 \quad \text{and} \quad \lambda_2 = -P_A S\left(\frac{1}{V_1} + \frac{\Phi_{A12}}{V_2}\right). \tag{13.82}$$

So, C_{A2} has the form:

$$C_{A2} = A + Be^{-P_A S\left(\frac{1}{V_1} + \frac{\Phi_{A12}}{V_2}\right)t}, \tag{13.83}$$

where A and B are constants to be determined by applying the initial conditions. Two initial conditions are required. Substitution of (13.83) into (13.79) when $t = 0$, and using (13.77), provides the two coefficients

$$A = -B = \frac{C_{A10}}{\frac{V_2}{V_1} + \Phi_{A12}}. \tag{13.84}$$

And the final solution for compartment 2 is:

$$C_{A2} = \frac{C_{A10}}{\left(\frac{V_2}{V_1} + \Phi_{A12}\right)}\left[1 - e^{-\frac{P_A S}{V_2}\left(\frac{V_2}{V_1} + \Phi_{A12}\right)t}\right]. \tag{13.85}$$

Substituting this back into (13.79), provides the solution for the concentration in the first compartment:

13.3 Compartmental Analysis

$$C_{A1} = \frac{C_{A10}}{\left(\frac{V_2}{V_1} + \Phi_{A12}\right)} \left[\Phi_{A12} + \frac{V_2}{V_1} e^{-\frac{P_A S}{V_2}\left(\frac{V_2}{V_1} + \Phi_{A12}\right)t}\right]. \quad (13.86)$$

Both expressions satisfy the initial conditions. In addition, as $t \to \infty$, the exponential terms vanish and $C_{A1} = \Phi_{A12} C_{A2}$, as it should for equilibrium. What is the expected concentration in compartment 2 for equilibrium? If $N_{A1\infty}$ and $N_{A2\infty}$ represent the number of moles in compartments 1 and 2, respectively, after a long period of time, the number of moles in each compartment will be:

$$N_{A1\infty} = C_{A1\infty} V_1 = \Phi_{A12} C_{A2\infty} V_1, \quad (13.87)$$

$$N_{A2\infty} = C_{A2\infty} V_2. \quad (13.88)$$

The total number of moles must equal the number introduced at $t = 0$:

$$N_{A1\infty} + N_{A2\infty} = C_{A10} V_1 = \Phi_{A12} C_{A2\infty} V_1 + C_{A2\infty} V_2. \quad (13.89)$$

Solving for $C_{A2\infty}$:

$$C_{A2\infty} = \frac{C_{A10}}{\left(\frac{V_2}{V_1} + \Phi_{A12}\right)}, \quad (13.90)$$

which is consistent with (13.85) for long times. The transient solutions for different values of the partition coefficient are shown in Fig. 13.11 for $V_2/V_1 = 0.5$ and $P_A S/V_2 = 0.004$ s^{-1}. Decreasing the partition coefficient not only changes the equilibrium value but also increases the time required to approach equilibrium.

13.3.2.4 Exchange Between Blood and Tissue Compartments

Of great importance to biomedical engineers is the analysis of the exchange of species A between blood and tissue compartments across a permeable membrane (Fig. 13.12).

We will simplify the analysis by assuming that species A is well-mixed in both the blood (B) and tissue (T) compartments. In addition, the volumes V_B and V_T are assumed constant, as is the volumetric blood flow, Q_B. A mass m_{A0} of species A is injected as a bolus into the blood compartment, and we are interested in following the mass concentrations in each compartment as a function of time. Blood flowing into compartment B does not contain species A. Analysis is similar to that applied in the previous section. Species conservation equations for each compartment are:
Blood:

$$V_B \frac{d\rho_{AB}}{dt} = -\rho_{AB} Q_B - P_A S(\rho_{AB} - \Phi_{ABT} \rho_{AT}). \quad (13.91)$$

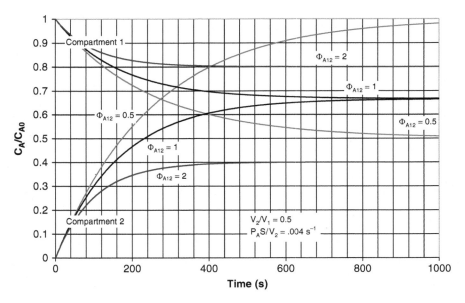

Fig. 13.11 Influence of partition coefficient on mass exchange across a membrane

Fig. 13.12 Two compartment system for blood–tissue exchange

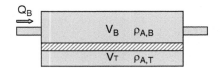

Tissue:

$$V_T \frac{d\rho_{AT}}{dt} = P_A S(\rho_{AB} - \Phi_{ABT}\rho_{AT}). \quad (13.92)$$

These equations are similar to those presented in the previous section, except for the convection term in (13.92). P_A is the permeability of the membrane to species A, Φ_{ABT} is the partition coefficient for species A between blood and tissue, and S is the membrane surface area. Initial conditions are:

$$\rho_{AB}(t=0) = \frac{m_{A0}}{V_B},$$
$$\rho_{AT}(t=0) = 0. \quad (13.93)$$

The solution procedure is identical to that provided in Sect. 13.3.2.3. Rearranging (13.92) to solve for ρ_{AB} in terms of ρ_{AT}, and substituting back into (13.89) yields a second-order homogeneous ODE:

$$\frac{d^2\rho_{AT}}{dt^2} + \alpha \frac{d\rho_{AT}}{dt} + \beta \rho_{AT} = 0 \quad (13.94)$$

13.3 Compartmental Analysis

with coefficients α and β defined as:

$$\alpha = \frac{Q_B}{V_B} + \frac{P_A S}{V_T}\left[\Phi_{ABT} + \frac{V_T}{V_B}\right], \tag{13.95}$$

$$\beta = \left(\frac{P_A S}{V_T}\right)\left(\frac{Q_B}{V_B}\right)(\Phi_{ABT}). \tag{13.96}$$

The general solution to (13.94) is:

$$\rho_{AT} = A_1 e^{\lambda_1 t} + A_2 e^{\lambda_2 t}, \tag{13.97}$$

where,

$$\lambda_1 = -\frac{\alpha}{2} + \frac{\sqrt{\alpha^2 - 4\beta}}{2},$$
$$\lambda_2 = -\frac{\alpha}{2} - \frac{\sqrt{\alpha^2 - 4\beta}}{2} \tag{13.98}$$

and A_1 and A_2 can be found by applying the initial conditions:

$$A_1 = -A_2 = \left(\frac{m_{A0}}{V_B}\right)\left(\frac{P_A S}{V_T}\right)\frac{1}{(\lambda_1 - \lambda_2)}, \tag{13.99}$$

if the membrane permeability is greater than zero, then $\beta > 0$, and from (13.98), λ_1 and λ_2 must both be negative and $(\lambda_1 - \lambda_2) > 0$. The final solution for the intravascular and extravascular concentrations in terms of λ_1 and λ_2 are:

$$\rho_{AT} = \frac{m_{A0}}{V_B}\left(\frac{\frac{P_A S}{V_T}(e^{\lambda_1 t} - e^{\lambda_2 t})}{(\lambda_1 - \lambda_2)}\right), \tag{13.100}$$

$$\rho_{AB} = \frac{m_{A0}}{V_B}\left[\frac{\left(\lambda_1 + \frac{P_A S}{V_T}\Phi_{ABT}\right)e^{\lambda_1 t} - \left(\lambda_2 + \frac{P_A S}{V_T}\Phi_{ABT}\right)e^{\lambda_2 t}}{(\lambda_1 - \lambda_2)}\right]. \tag{13.101}$$

If the tracer remains intravascular ($P_A S = 0$), then $\lambda_1 = 0$, $\lambda_2 = -\alpha = -Q_B/V_B$, (13.100) reduces to $\rho_{AT} = 0$, and the intravascular solution for ρ_{AB} (13.101) reduces to the single compartment solution given by (13.19). If the membrane were highly permeable, then we would expect the tissue concentration to be nearly in equilibrium with the intravascular concentration $\rho_{AT} = \rho_{AB}/\Phi_{ABT}$. The intravascular and extravascular solutions for various values of $P_A S/Q_B$ ranging from 0 to ∞ are shown in Fig. 13.13 for $Q_B = 10$ ml/s, $V_B = V_T = 100$ ml, $\Phi_{ABT} = 1.0$.

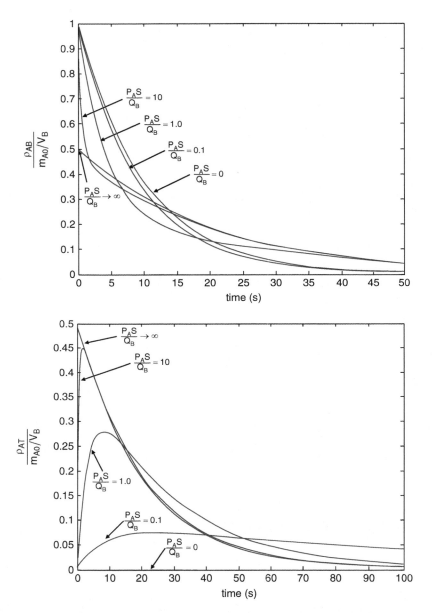

Fig. 13.13 Response of intravascular (*top*) and extravascular (*bottom*) compartments for the system in Fig. 13.11 to a bolus injection at various values of $P_A S/Q_B$. $Q_B = 10$ ml/s, $V_B = V_T = 100$ ml, $\Phi_{ABT} = 1.0$. Note that the ordinate is cut in half and the abscissa doubled for the extravascular plot

13.3 Compartmental Analysis

Notice that the intravascular curves for $P_AS/Q_B = 0$ and $P_AS/Q_B \to \infty$ are both exponential. The higher P_AS, the more rapid is the exchange between the compartments. The influence of changing the partition coefficient from 1.0 to 0.5 is shown in Fig. 13.14. For high P_AS, $\rho_{AT} = \rho_{AB}/\Phi_{ABT}$, as expected.

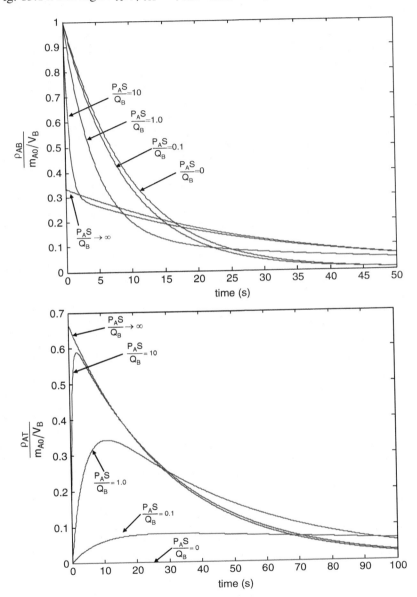

Fig. 13.14 Response of intravascular (*top*) and extravascular (*bottom*) compartments for the system in Fig. 13.12 to a bolus injection at various values of P_AS/Q_B. $Q_B = 10$ ml/s, $V_B = V_T = 100$ ml, $\Phi_{ABT} = 0.5$. Note that the time interval is doubled for the extravascular plot

13.3.3 Multiple Compartments

Many systems, including the human circulatory system, can be modeled to a first approximation as a number of compartments connected in various ways. Blood that enters the right heart passes through the right atrium, the right ventricle, the lung, the left atrium, and the left ventricle before being distributed to the systemic circulation. Patients with an atrial septal defect have an opening in the septum that connects the right and left atria (Fig. 13.15). Since the mean pressure in the left atrium is higher than in the right atrium, some oxygenated blood is shunted back into the right atrium, and thus back through the lungs. Consequently, outflow of the left heart Q_{LH} is less than from the right heart Q_{RH}. If venous return Q_{VR} and compartment volumes are known, it is possible to estimate the shunt flow Q_S by injecting a bolus of an intravascular tracer in the right atrium and measuring its rate of disappearance. The tracer is assumed to remain within the intravascular space during the measurement.

The macroscopic species continuity equations for each of the compartments following the bolus are:

Right atrium (RA):

$$V_{RA}\frac{dC_{RA}}{dt} = Q_S C_{LA} - (Q_S + Q_{VR})C_{RA}. \tag{13.102}$$

Right ventricle (RV):

$$V_{RV}\frac{dC_{RV}}{dt} = (Q_S + Q_{VR})(C_{RA} - C_{RV}). \tag{13.103}$$

Lungs (L):

$$V_L\frac{dC_L}{dt} = (Q_S + Q_{VR})(C_{RV} - C_L). \tag{13.104}$$

Left atrium (LA):

$$V_{LA}\frac{dC_{LA}}{dt} = (Q_S + Q_{VR})C_L - Q_S C_{LA} - Q_{VR} C_{LA}. \tag{13.105}$$

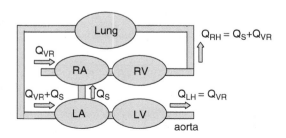

Fig. 13.15 Compartments and flows for atrial-septal defect

13.4 Indicator Dilution Methods

Left ventricle (LV):

$$V_{LV} \frac{dC_{LV}}{dt} = Q_{VR}(C_{LA} - C_{LV}). \quad (13.106)$$

The initial conditions required to solve this set of equations are:

$$C_{RA}(0) = C_{RA0}, \quad C_{RV}(0) = C_L(0) = C_{LA}(0) = C_{LV}(0) = 0. \quad (13.107)$$

Although an exact solution can be found by taking Laplace transforms, the easiest way to solve a set of coupled first-order differential equations such as this is to use an ODE solver such as ODE45 in Matlab. We will provide examples in Sect. 13.6. However, we can often obtain useful information without seeking complete analytic solutions to a system of equations. For the purposes of estimating the shunt flow, we need to only consider the equation governing tracer transport in the right atrium. For small times, tracer has not made its way around the circuit to the left atrium. In this case, we can set $C_{LA} = 0$ in (13.102). The resulting equation is easily solved with the initial condition given in (13.107):

$$\ln\left(\frac{C_{RA}}{C_{RA0}}\right) = -\left(\frac{Q_S + Q_{VR}}{V_{RA}}\right)t. \quad (13.108)$$

If we measure the concentration in the right atrium and plot the left-hand side of (13.107) as a function of time (for small times), the initial slope will be:

$$\text{slope}|_{t=0} = -\left(\frac{Q_S + Q_{VR}}{V_{RA}}\right). \quad (13.109)$$

The shunt flow Q_S can be computed from the measured slope and known values for V_{RA} and Q_{VR}. This is an example of an indicator dilution experiment. Other dilution methods can be used to find system volumes and flows as discussed in the next section.

13.4 Indicator Dilution Methods

Indicator dilution methods can be used to measure system flow and properties such as system volume or the product of tracer permeability and system surface area, $P_A S$. Generally, a tracer or mixture of tracers with known compositions are injected as a bolus at the inlet to the system and the concentration of each tracer is measured at the outlet of the system as a function of time. Tracers are often labeled with radioisotopes or fluorescent markers for easy detection.

13.4.1 Stewart–Hamilton Relation for Measuring Flow Through a System

The steady flow rate through a system such as an organ shown in Fig. 13.16 can be measured by injecting a tracer which does not permanently leave the system at the inlet of the system and measuring its concentration at the system outlet. The tracer must uniformly tag the inlet and outlet flow streams. The general macroscopic mass balance for the tracer is given by (13.10):

$$\frac{dm_A}{dt} = w_{A,\text{wall}} + \sum_{i=1}^{N_{\text{inlets}}} \rho_{Ab,i} Q_{Vi} - \sum_{j=1}^{N_{\text{outlets}}} \rho_{Ab,j} Q_{Vj} + r_{A,\text{tot}}. \tag{13.110}$$

For the problem at hand, we are dealing with a system with a single inlet and outlet, the system walls are impermeable to the tracer, and the tracer is nonreactive. In addition, no tracer enters the system after the introduction of the bolus, so (13.110) simplifies to:

$$\frac{dm_A}{dt} = -\rho_{Ab,\text{out}} Q_V. \tag{13.111}$$

Integrating (13.111) from $t = 0$ to $t \to \infty$, and recognizing that all of the tracer will flow out of the system after a long time:

$$\int_{m_{A0}}^{0} dm_A = -m_{A0} = -Q_V \int_{0}^{\infty} \rho_{Ab,\text{out}} dt. \tag{13.112}$$

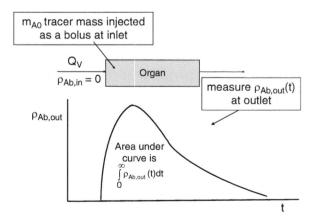

Fig. 13.16 Indicator dilution method for finding blood flow to an organ

13.4 Indicator Dilution Methods

Rearranging this expression to solve for flow rate gives the Stewart–Hamilton equation for measuring flow with an indicator:

$$Q_V = \frac{m_{A0}}{\int_0^\infty \rho_{Ab,out}\,dt}. \tag{13.113}$$

Thus the flow can be estimated by taking the ratio of the tracer mass injected to the area under the tracer mass concentration–time curve. We might test the validity of (13.113) by applying it to any of the examples for bolus injection considered in Sect. 13.3. For instance, inserting the solution for a bolus injection in a single well-mixed compartment, (13.19), into (13.113):

$$Q_V = \frac{m_{A0}}{\int_0^\infty \left(\frac{m_{A0}}{V}e^{-\frac{Q_V}{V}t}\right)dt} = \frac{m_{A0}}{m_{A0}/Q_V}. \tag{13.114}$$

The Stewart–Hamilton equation can also be used for tracers that can diffuse across the blood–tissue barrier, as long as the tracer is not sequestered in the extravascular space. The Stewart–Hamilton equation and its thermal analog (e.g., Example 9.4.2.2) have been used extensively to measure cardiac output in catheterized patients. The tracer is injected in the right atrium and detected either in the pulmonary artery or in the ascending aorta. There are two practical limitations that tend to reduce accuracy of the technique. First, the circulatory system, by design, circulates blood from the venous system back to the arterial system. Tracer that flows out through the venous system will eventually return to the organ inlet and this recirculated tracer will cause the integral in the denominator to grow indefinitely. Corrections to eliminate recirculated tracer are made by plotting the outflow concentration vs. time on a semi-log plot (Fig. 13.17). Recirculation appears as a second (and sometimes third) upward bump on the plot. The slope ($-k$) of the semi-log plot is usually very constant from shortly after the peak of the curve to just before the appearance of the recirculated tracer, t_R. The remainder of the curve is assumed to decline exponentially at the same rate, $-k$.

Hence the area under the curve without recirculation can be estimated from:

$$\int_0^\infty \rho_{Ab,out}\,dt = \int_0^{t_R} \rho_{Ab,out}\,dt + \rho_{Ab,out}(t_R)\int_{t_R}^\infty e^{-kt}\,dt = \int_0^{t_R} \rho_{Ab,out}\,dt + \frac{\rho_{Ab,out}(t_R)e^{-kt_R}}{k}. \tag{13.115}$$

The second source of error in using the Stewart–Hamilton equation to compute flow in physiological situations is that blood flow is not constant, but is pulsatile in nature. Therefore, pulling Q_V out of the integral in (13.113) is not strictly valid. The error can be considerable at low frequencies, but is generally minor at cardiac frequencies (Bassingthwaighte et al. 1970).

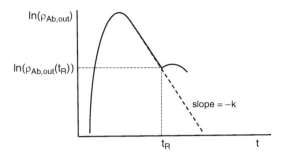

Fig. 13.17 Correcting for recirculation

If the system of interest is well-mixed with multiple inlets and outlets, then the Stewart–Hamilton equation can still be used with the area based on concentrations measured at any of the outlet streams. However, the computed flow will be the total flow through the system, not the flow in the stream where concentration is measured.

13.4.2 Volume Measurements

The concentration–time curve collected at the outlet of a system following a bolus injection at the inlet (e.g., Fig. 13.16) represents a residence time distribution of tracer within the system. According to the mean value theorem, the mean residence time can be found using the following expression:

$$\bar{t} = \frac{\int_0^\infty (\rho_{Ab,out} t) dt}{\int_0^\infty \rho_{Ab,out} dt}. \tag{13.116}$$

If tracer particles are swept through the system just like intravascular fluid particles, then the mean transit time for an intravascular tracer $\bar{t}_v$ should equal the mean residence time for fluid in the system. This is simply the ratio of the intravascular volume V to the fluid flow rate Q_V:

$$\frac{V}{Q_V} = \bar{t}_v = \frac{\int_0^\infty (\rho_{vb,out} t) dt}{\int_0^\infty \rho_{vb,out} dt}. \tag{13.117}$$

The subscript v indicates a tracer that remains intravascular. Since the flow and mean transit time can be computed from the downstream tracer concentration data, (13.117) can be used to estimate the intravascular volume of the system.

13.4 Indicator Dilution Methods

To illustrate this, let us again take the solution for a bolus injection in a single compartment. Substituting (13.19) for $\rho_{vm,out}$ into (13.117):

$$\bar{t}_v = \frac{\int_0^\infty \left(\left(\frac{m_{A0}}{V}e^{-\frac{Q_v}{V}t}\right)t\right)dt}{\int_0^\infty \left(\frac{m_{A0}}{V}e^{-\frac{Q_v}{V}t}\right)dt} = \frac{\left(\frac{m_{A0}}{Q_v}\right)}{\left(\frac{m_{A0}V}{Q_v^2}\right)} = \frac{V}{Q_v}. \quad (13.118)$$

If the tracer is able to diffuse out of the intravascular space, we will use a subscript e to designate an extravascular tracer. The mean transit time of an extravascular tracer will be computed in the same manner as in (13.117):

$$\bar{t}_e = \frac{\int_0^\infty (\rho_{eb,out}t)dt}{\int_0^\infty \rho_{eb,out}dt}. \quad (13.119)$$

The mean transit time for an extravascular tracer will be greater than the mean transit time of an intravascular tracer because it is able to distribute in a volume that is larger than the intravascular volume. The additional transit time will be directly proportional to the extravascular volume and inversely proportional to the partition coefficient of the extravascular tracer A between the vascular and extravascular space, Φ_{Ave}:

$$\bar{t}_e - \bar{t}_v = \frac{V_e}{Q_v \Phi_{Ave}}. \quad (13.120)$$

The extravascular volume of an organ can be estimated using (13.120) if an intravascular and a diffusible tracer are injected simultaneously in a bolus at the inlet of the organ and their concentrations are measured at the outlet of the organ. Mean transit times are computed using (13.116) for each tracer and the flow can be estimated using the Stewart–Hamilton equation for either tracer. The partition coefficient is often assumed to equal unity. Tracer recirculation will strongly influence the mean transit time calculations. The same exponential extrapolation method used to correct the Stewart–Hamilton calculation is also used to correct the mean transit time calculation for each tracer.

13.4.3 Permeability-Surface Area Measurements

Indicator dilution methods can be used to estimate the product of permeability P_D and surface area S for a diffusible tracer D. The simplest method is to inject a tracer that diffuses from blood to tissue, but is unable to diffuse back from tissue to the intravascular space. A tracer that reacts rapidly with another species in the tissue is an ideal choice. In that case, the concentration of the diffusing tracer in the tissue is very small and the model presented in Sect. 13.3.1.7 is appropriate. For a bolus injection with no recirculation of tracer:

$$V\frac{d\rho_D}{dt} = -(P_D S + Q_V)\rho_D. \tag{13.121}$$

Solving this with the initial condition $\rho_{D0} = m_{D0}/V$:

$$\rho_D = \rho_{D0} e^{-\frac{(P_D S + Q_V)}{V}t}. \tag{13.122}$$

This can be compared with the solution for an intravascular or reference tracer R with an initial tracer concentration $\rho_{R0} = m_{R0}/V$:

$$\rho_R = \rho_{R0} e^{-\frac{Q_V}{V}t}. \tag{13.123}$$

Taking the logarithm of the ratio:

$$\ln\left(\frac{\rho_D/\rho_{D0}}{\rho_R/\rho_{R0}}\right) = -\frac{(P_D S)}{V}t. \tag{13.124}$$

The ratio of initial concentrations ρ_{D0}/ρ_{R0} will be equal to the known ratio of tracer masses in the injected bolus, m_{D0}/m_{R0}. To compute $P_D S$ for the diffusible tracer, we would begin by plotting the left-hand side of (13.124) as a function of time. This graph should produce a straight line with a slope equal to $P_D S/V$. The intravascular volume can be found from the mean transit time of the intravascular tracer (13.117). Unless the surface area is known or can be measured by another method, it is generally not possible to compute the diffusible tracer permeability by itself. The flow rate Q_V can be computed from the Stewart–Hamilton equation applied to the intravascular tracer. An error in computed flow would occur if the If the Stewart–Hamilton were applied to the diffusible tracer because some of the tracer is permanently extracted by the tissue.

A measure of the loss of the diffusible tracer relative to the intravascular tracer is known as the *extraction E*. If concentrations of the intravascular and diffusible tracers in the injected bolus are ρ_{Ri} and ρ_{Di}, respectively, extraction is defined as:

$$E = \frac{\rho_R/\rho_{Ri} - \rho_D/\rho_{Di}}{\rho_R/\rho_{Ri}}. \tag{13.125}$$

Consider simultaneous infusion of an intravascular reference tracer and a diffusible tracer into the inlet of an organ. Both tracers are infused at constant rates and the diffusible tracer does not return to the vascular system. This problem was solved

13.4 Indicator Dilution Methods

in Sect. 13.3.1.7 for a well-mixed intravascular compartment. If the tissue concentration is zero and partition coefficient is unity, the solution at the organ outlet after a long period of time (13.53) is:

$$\frac{\rho_D}{\rho_{Dbi}} = \left(\frac{Q_V}{Q_V + P_D S}\right), \quad (13.126)$$

where ρ_{Dbi} is the inlet mass bulk concentration at the organ inlet. An intravascular reference tracer at the organ outlet would have the same concentration as at the inlet, since $P_R S = 0$:

$$\frac{\rho_R}{\rho_{Rbi}} = 1. \quad (13.127)$$

Substituting (13.126) and (13.127) into (13.125) and rearranging, we can compute the permeability-surface area in terms of measured extraction and flow rate:

$$\frac{P_D S}{Q_V} = \frac{E}{1-E}. \quad (13.128)$$

We will present additional indicator dilution methods for measuring capillary permeability in Chap. 15.

Example 13.3.1 Tracer Measurement of Left Ventricular Volumes.
Two grams of Tracer "A" are rapidly injected into the left ventricle during diastole. The concentration of tracer "A" is measured in the aorta after each successive beat of the heart. The data are shown in the table below, where N is the number of cardiac cycles following the first ejection. Estimate end diastolic volume and stroke volume from the data provided.

N	ρ_A (mg/ml)
0	9.96
1	6.08
2	3.63
3	2.06
4	1.23
5	0.79
6	0.48

Solution. *Initial considerations:* Our goal is to use the tracer data collected after each contraction of the left ventricle to estimate end diastolic volume (V_{ED}) and stroke volume (V_s). End diastolic volume is the volume of blood contained in the left ventricle at the end of diastole, just before the ventricle contracts. Stroke volume is the volume of blood ejected from the ventricle during a single contraction.

System definition and environmental interactions: The system to be analyzed in this case is the blood within the ventricle. It will be useful for us to consider the system under two separate phases of the cardiac cycle: the filling phase of diastole and the ejection phase of systole. During the filling phase of diastole, the aortic valve is closed and no blood flows out of the ventricle. Blood enters the left ventricle from the left atrium through the mitral valve, which is open during diastole. In the ejection phase of systole, the mitral valve is closed, blocking inflow from the left atrium, and the aortic valve is open, allowing blood to be ejected into the aorta.

Apprising the problem to identify governing relationships: The appropriate governing relationship is the conservation of tracer A, which is to be applied to tracer in the left ventricle. We will make the following assumptions:

1. Blood is an incompressible fluid (constant density, ρ).
2. The tracer in the left ventricle is well mixed.
3. Blood does not pass through the walls of the left ventricle.
4. Blood does not flow out of the ventricle during diastole.
5. Blood does not flow into the ventricle during systole.
6. The tracer is injected as a bolus during diastole ($t = 0$).
7. No tracer is introduced into the ventricle after the initial bolus (no recirculation of tracer).
8. The tracer does not leave the ventricle through the walls of the ventricle.
9. The tracer is inert. It does not participate in chemical reactions in the ventricle.
10. Stroke volume and end diastolic volume do not change from beat to beat of the heart over the seven beats in which data were collected.

Analysis: Because the ventricle is assumed to be well mixed, we are justified in using a macroscopic approach. A good starting point is the general macroscopic expression for the conservation of mass, (5.2):

$$\frac{dm}{dt} = w_{\text{wall}} + w_{\text{in}} - w_{\text{out}}.$$

The mass of blood in the ventricle is equal to the product of the ventricular volume V and the blood density, ρ. From Assumption 3, the term w_{wall} is zero. The inlet and outlet mass rates equal the products of the blood density with the inlet and outlet volumetric flow rates, $Q_{V,\text{in}}$ and $Q_{V,\text{out}}$. Since blood density is constant, the conservation of mass statement becomes a conservation of blood volume within the left ventricle:

$$\frac{dV}{dt} = Q_{V,\text{in}} - Q_{V,\text{out}}.$$

If the tracer is denoted by the symbol "A," conservation of tracer in the ventricle is given by (13.10) for a system with a single inlet and a single outlet:

$$\frac{dm_A}{dt} = w_{A,\text{wall}} + \rho_{Ab,\text{in}} Q_{V,\text{in}} - \rho_{Ab,\text{out}} Q_{V,\text{out}} + r_{A,\text{tot}}.$$

13.4 Indicator Dilution Methods

Applying assumptions 2 ($\rho_{Ab,out} = \rho_A$), 7 ($\rho_{Ab,in} = 0$), 8 ($w_{A,wall} = 0$), and 9 ($r_{A,tot} = 0$), this simplifies to:

$$\frac{dm_A}{dt} = \frac{d}{dt}(\rho_A V) = -\rho_A Q_{V,out}.$$

Diastole: Let us now apply the conservation of tracer expression during diastolic filling. According to assumption 4, $Q_{V,out} = 0$ during diastole, so:

$$\frac{dm_A}{dt} = 0.$$

Consequently, the mass of tracer in the ventricle remains constant during diastole. If the initial mass of tracer injected as a bolus is m_{A0}, then at the end of the initial filling period, the product of tracer concentration and end diastolic volume must equal the mass of tracer injected:

$$\rho_{A0} V_{ED} = m_{A0}.$$

Systole: During systole, the inlet flow is zero (Assumption 5). Conservation of mass and conservation of tracer during the ejection phase of systole are:

$$\frac{dV}{dt} + Q_{V,out} = 0,$$

$$V \frac{d\rho_A}{dt} + \rho_A \frac{dV}{dt} = -\rho_A Q_{V,out}.$$

Applying the chain rule of differentiation to the tracer equation and inserting conservation of mass yields:

$$\frac{d\rho_A}{dt} = -\frac{\rho_A}{V}\left(\frac{dV}{dt} + Q_{V,out}\right) = 0.$$

Therefore, the tracer concentration will remain constant throughout the duration of each systolic contraction. The initial ejection will have the same tracer concentration as the initial end diastolic tracer concentration:

$$\rho_{A0} = \frac{m_{A0}}{V_{ED}}.$$

Since the mass of tracer injected was 2,000 mg and the initial tracer concentration is 9.96 mg/ml, the end diastolic volume can be estimated from this single measurement as 2,000 mg/9.96 mg/ml = 201 ml.

The mass of tracer ejected in the initial systolic contraction will equal the tracer concentration times the stroke volume:

$$m_{A,\text{ejected}} = \rho_{A0} V_s.$$

The mass of tracer remaining in the ventricle after the initial ejection will be:

$$m_{A1} = m_{A0} - m_{A,\text{ejected}}.$$

Replacing the masses with the products of concentration and volume:

$$\rho_{A1} V_{ED} = \rho_{A0} V_{ED} - \rho_{A0} V_s.$$

Solving for the concentration at the end of the next ejection:

$$\rho_{A1} = \rho_{A0} \frac{(V_{ED} - V_s)}{V_{ED}}.$$

After rearranging this equation, the stroke volume can be estimated from the first two tracer concentration measurements provided in the table:

$$V_s = V_{ED} \left(1 - \frac{\rho_{A1}}{\rho_{A0}}\right) = (201 \text{ ml}) \left(1 - \frac{6.08 \text{ mg/ml}}{9.96 \text{ mg/ml}}\right) = 78.2 \text{ ml}.$$

The mass ejected during the next contraction is:

$$\rho_{A2} V_{ED} = \rho_{A1} (V_{ED} - V_s) = \rho_{A0} \left(\frac{(V_{ED} - V_s)}{V_{ED}}\right)(V_{ED} - V_s).$$

Solving for ρ_{A2}:

$$\rho_{A2} = \rho_{A0} \left(\frac{(V_{ED} - V_s)}{V_{ED}}\right)^2.$$

Indeed, the nth measurement of concentration ρ_{An} will be related to the initial concentration using the following recursion formula:

$$\rho_{An} = \rho_{A0} \left(\frac{(V_{ED} - V_s)}{V_{ED}}\right)^n.$$

Taking the natural logarithm of both sides yields the following relation:

$$\ln(\rho_{An}) = \ln(\rho_{A0}) + n \ln\left(\frac{(V_{ED} - V_s)}{V_{ED}}\right).$$

13.4 Indicator Dilution Methods

Therefore, a plot of $\ln(\rho_{An})$ vs. n should produce a straight line with an intercept equal to $\ln(\rho_{A0})$ and a slope equal to $\ln(1 - V_s/V_{ED})$. By finding the best least-squares fit of the data, we can estimate V_s and V_{ED} from the slope and intercept of the best-fit straight line as follows:

$$V_{ED} = \frac{m_{A0}}{e^{\text{intercept}}},$$

$$V_s = V_{ED}(1 - e^{\text{slope}}) = m_{A0}\frac{(1 - e^{\text{slope}})}{e^{\text{intercept}}}.$$

Applying this method to the data in the table:

N	ρ_A (mg/ml)	$\ln(\rho_A)$
0	9.96	2.299
1	6.08	1.805
2	3.63	1.289
3	2.06	0.723
4	1.23	0.207
5	0.79	−0.236
6	0.48	−0.734

We can use a spreadsheet to find the best fit values:

Intercept = 2.293
slope = −0.509

Solving for the ventricular volumes:

$$V_{ED} = \frac{m_{A0}}{e^{\text{intercept}}} = \frac{2{,}000 \text{ mg}}{e^{2.293} \frac{\text{mg}}{\text{ml}}} = 202 \text{ ml},$$

$$V_s = V_{ED}(1 - e^{\text{slope}}) = 202 \text{ ml}(1 - e^{-509}) = 80.6 \text{ ml}.$$

Examining and interpreting the results: The advantage of estimating V_{ED} and V_s using the curve fit method is that all of the data are used in finding the slope and intercept. In contrast, only the first two measured points were used in the previous estimates. The latter method would be expected to be more accurate.

Additional comments: End diastolic volume is estimated to be 202 ml for this patient. Normal values computed from data reported by Braunstein et al. (1967) for 11 human subjects range from 82.6 ml for a small female to 183.3 ml for a large male. Our measurement of end diastolic volume is higher than the average of 138.4 ml. Unless the data are from a large individual or from an athlete, the patient may have an enlarged heart. The stroke volume of 80.6 ml is within the normal range (49–89 ml), and is only slightly larger than the average of 75.3 ml. The ejection fraction for our patient is $V_s/V_{ED} = 0.4$, which is lower than the normal range of 0.48–0.7. Injection fractions below 0.5 have been reported to be a marker of cardiac disease and a predictor of mortality (Cooper et al. 1991).

13.5 Chemical Reactions and Bioreactors

The term accounting for the rate of production by homogeneous chemical reaction in the species continuity equation has been set to zero in all of the previous examples. We shall now consider compartmental analysis in which chemical reactions occur. When dealing with chemical reactions, molar concentrations are more commonly used than mass concentrations.

13.5.1 Homogeneous Chemical Reactions

Homogeneous reactions occur within the compartment of interest rather than on the surface of the compartment.

13.5.1.1 Zeroth Order Homogeneous Reaction

Let us begin by considering a cell that suddenly begins to produce substance A at a constant rate per unit volume R_A (zeroth order kinetics). The cell interior is assumed to be well mixed with no substance A initially present in the cell. The concentration of substance A in the fluid surrounding the cell is maintained at zero. There is no bulk flow through the cell membrane, the cell volume V is constant, and the cell membrane has surface area S and permeability P_A. Conservation of substance A can be found by applying (13.14):

$$V\frac{dC_A}{dt} = -\mathsf{P}_A S C_A + R_A V. \tag{13.129}$$

The solution is:

$$C_A = \frac{R_A V}{\mathsf{P}_A S}\left[1 - e^{-\left(\frac{\mathsf{P}_A S}{V}\right)t}\right]. \tag{13.130}$$

The final steady-state concentration is directly proportional to the rate of production of substance A by the cell. It is also proportional to the ratio of the volume to surface area of the cell and inversely proportional to the membrane permeability. The time constant for exchange of substance A is $V/\mathsf{P}_A S$. If experimental data for C_A vs. t are available, we can estimate the rate of production of A as follows. First, we find the steady-state concentration $C_{A\infty}$. We can then plot $\ln[(C_{A\infty} - C_A)/C_{A\infty}]$ vs. t. The slope of this line is $-(\mathsf{P}_A S/V)$. Multiplying $c_{A\infty}$ by the negative of the slope will provide an estimate of R_A.

13.5 Chemical Reactions and Bioreactors

13.5.1.2 First-Order Irreversible Homogeneous Reaction

Consider the following irreversible first-order chemical reaction, which occurs between substance A and a plasma protein P:

$$A + P \rightarrow AP. \tag{13.131}$$

The rate of production of A per unit volume is:

$$R_A = -k_f C_P C_A, \tag{13.132}$$

where k_f is the forward rate constant and C_P is the bulk protein concentration. The negative sign indicates that substance A is lost from solution because of the reaction. We will analyze the case where substance A is injected as a bolus into a well-mixed vascular compartment with volume V and flow rate through the compartment Q_V. The concentration of the reacting plasma protein is assumed to be constant ($C_P = C_{P0}$). This reduces the order of the reaction from second-order to first-order, since now $R_A = $ constant $\times C_A$. The initial concentration of substance A following the bolus injection is C_{A0}. The species continuity equation for this case (with no species A entering in the inlet stream) is:

$$V \frac{dC_A}{dt} = -C_A Q_V + R_A V = -C_A Q_V - (k_f C_{P0} C_A) V. \tag{13.133}$$

This is a first-order homogeneous equation with solution:

$$C_A = c_{A0} e^{-\left(\frac{Q_V}{V} + k_f C_{P0}\right)t}. \tag{13.134}$$

Thus, the rate of removal of A from the compartment is more rapid than would be caused by simple convective washout. If the flow were zero, virtually all of substance A would combine with the plasma protein in a time equal to five time constants or $5/(k_f C_{P0})$.

13.5.1.3 Second-Order Reversible Homogeneous Reaction

If the reaction in the previous section were reversible:

$$A + P \rightleftarrows AP, \tag{13.135}$$

then the rate of production of each species per unit volume is:

$$R_A = R_P = -R_{AP} = k_r C_{PA} - k_f C_P C_A, \tag{13.136}$$

where k_r is the reverse rate constant. The protein concentration is not constant in this case, so the forward reaction is second order. Let us consider the case where there is no flow into a well-mixed compartment containing plasma proteins. At time $t = 0$ a bolus of substance A, which reacts reversibly with the plasma proteins, is introduced into the system and the concentration of A is measured with time. The species conservation equations for the three species A, P, and PA are:

$$V\frac{dC_A}{dt} = (k_r C_{PA} - k_f C_P C_A)V, \quad (13.137)$$

$$V\frac{dC_{PA}}{dt} = (k_f C_P C_A - k_r C_{PA})V, \quad (13.138)$$

$$V\frac{dC_P}{dt} = (k_r C_{PA} - k_f C_P C_A)V. \quad (13.139)$$

The system volume can be eliminated from each of these equations, so the concentrations are independent of V. Furthermore, adding (13.138) and (13.139) yields:

$$\frac{d}{dt}(C_P + C_{PA}) = 0. \quad (13.140)$$

Consequently, the sum of the bound and unbound plasma species must be constant. Since $C_{PA}(0) = 0$ and $C_P(0) = C_{P0}$, integration of (13.140) gives:

$$C_{PA} = C_{P0} - C_P. \quad (13.141)$$

Subtracting (13.137) from (13.139) and applying the initial conditions $c_P(0) = c_{P0}$ and $c_A(0) = c_{A0}$, we find:

$$C_P = C_A + C_{P0} - C_{A0}. \quad (13.142)$$

After a very long time, the concentrations will attain their equilibrium values. We can find the equilibrium value of C_A (i.e., $C_{A\infty}$) by setting the time derivative equal to zero in (13.137), and using (13.140) and (13.141). This results in the following quadratic equation:

$$k_f C_{A\infty}^2 + [k_r + k_f(C_{P0} - C_{A0})]C_{A\infty} - k_r C_{A0} = 0. \quad (13.143)$$

The solution for the equilibrium concentration $C_{A\infty}$ is:

$$C_{A\infty} = \frac{-[k_r + k_f(C_{P0} - C_{A0})] + \phi}{2k_f}, \quad (13.144)$$

13.5 Chemical Reactions and Bioreactors

where ϕ is defined as

$$\phi = \sqrt{[k_r + k_f(C_{P0} - C_{A0})]^2 + 4k_f k_r C_{A0}}, \qquad (13.145)$$

turning now to the transient solution, we can substitute (13.141) and (13.142) into (13.137) to obtain an expression involving only a single dependent variable C_A:

$$\frac{dC_A}{dt} = k_r(C_{A0} - C_A) - k_f(C_A + C_{P0} - C_{A0})C_A. \qquad (13.146)$$

Separating variables and integrating:

$$\int_{C_{A0}}^{C_A} \frac{dC_A}{k_r C_{A0} - [k_f(C_{P0} - C_{A0}) + k_r]C_A - k_f C_A^2} = \int_0^t dt. \qquad (13.147)$$

With the aid of a table of integrals and (13.144), we find the transient solution to be:

$$C_A = \frac{(C_{A\infty} - C_{A0})(C_{A\infty} - 2\phi/k_f)e^{-\phi t} - C_{A\infty}(C_{A\infty} - C_{A0} - 2\phi/k_f)}{(C_{A\infty} - C_{A0})e^{-\phi t} - (C_{A\infty} - C_{A0} - 2\phi/k_f)}. \qquad (13.148)$$

This reduces to $C_A = C_{A0}$ at $t = 0$ and $C_A = C_{A\infty}$ for large times. Solutions for various dimensionless ratios of the reverse to forward reaction rate coefficients are shown in Fig. 13.18 for the case where $C_{A0}/C_{P0} = 0.001$ and $k_f C_{P0} = 0.001$ s^{-1}.

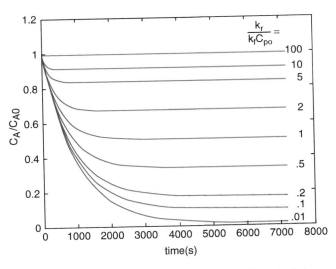

Fig. 13.18 Concentration relative to initial concentration of substance A in a system where substance A reacts reversibly with a plasma protein for various values of the ratio of the reverse to forward reaction rates

13.5.1.4 Second-Order Reversible Homogeneous Reaction with Convection

Now consider the same reversible reaction, but let us add the effects of convection in the compartment by allowing plasma to flow through the compartment at a constant flow rate Q_V. The inlet mean plasma protein concentration is constant (C_{P0}) and a bolus of substance A is injected at time $t = 0$, such that the initial concentration is C_{A0}. The inlet concentrations for substance A and the complex are both zero. Species conservation equations become:

$$V\frac{dC_A}{dt} = -C_A Q_V + (k_r C_{PA} - k_f C_P C_A)V, \tag{13.149}$$

$$V\frac{dC_{PA}}{dt} = -C_{PA} Q_V + (k_f C_P C_A - k_r C_{PA})V, \tag{13.150}$$

$$V\frac{dC_P}{dt} = Q_V(C_{P0} - C_P) + (k_r C_{PA} - k_f C_P C_A)V. \tag{13.151}$$

Proceeding as earlier, where convection was absent, we can add (13.150) and (13.151) to obtain

$$\frac{d}{dt}(C_{PA} + C_P - C_{P0}) + \frac{Q_V}{V}(C_{PA} + C_P - C_{P0}) = 0. \tag{13.152}$$

The general solution to (13.152) is:

$$C_{PA} + C_P - C_{P0} = A e^{-\frac{Q_V}{V}t}, \tag{13.153}$$

where the constant A is to be determined by applying the initial conditions. At $t = 0$ the left side of (13.153) is zero, so the constant $A = 0$. Consequently, we once again arrive at (13.141), which is simply a conservation statement for bound and unbound proteins in the system.

In a similar manner, we can subtract (13.149) from (13.151), solve the resulting ODE, and apply the initial conditions to derive the following relationship between reacting protein concentration and species A:

$$C_P = C_{P0} + C_A - C_{A0} e^{-\frac{Q_V}{V}t}. \tag{13.154}$$

Substituting (13.141) and (13.154) for C_{PA} and C_P, respectively, into the species conservation equation for substance A (13.149), we arrive at the following ODE that involves a single dependent variable, C_A:

13.5 Chemical Reactions and Bioreactors

$$\frac{dC_A}{dt} = -\frac{Q_V}{V}C_A + k_r\left(C_{A0}e^{-\frac{Q_V}{V}t} - C_A\right) - k_f\left(C_A + C_{P0} - C_{A0}e^{-\frac{Q_V}{V}t}\right)C_A.$$
(13.155)

Comparing this to (13.146) (no convection), we see that the terms involving the initial concentration of A now decrease exponentially in time with a time constant of V/Q_V and an additional term involving the same time constant appears in (13.155). Thus the concentration of substance A is changing not only because of the chemical reaction, but also because substance A and the AP complex are being swept out of the system over time.

Rather than attempt to find an analytic solution to (13.155), it is much easier to solve it numerically using an ODE solver package. To avoid solving the problem repeatedly each time that a parameter is changed, it is useful to make the equation dimensionless by dividing the dependent and independent variables by characteristic values. In this problem, we might select C_{A0} as a typical value for the dependent variable, C_A, and the forward reaction time constant $1/k_f C_{P0}$ as a typical time:

$$C_A^* = \frac{C_A}{C_{A0}},$$
(13.156)
$$\tau = k_f C_{P0} t.$$

Substituting these dimensionless variables into (13.155) yields the following dimensionless equation:

$$\frac{dC_A^*}{d\tau} = -\alpha C_A^* + \beta\left(e^{-\alpha\tau} - C_A^*\right) - \left(1 + \gamma\left[C_A^* - e^{-\alpha\tau}\right]\right)C_A^*,$$
(13.157)

where the dimensionless group α compares the rate of elimination with the forward reaction rate, the group β compares the reverse rate with the forward rate, and the third group γ is the ratio of initial concentrations of substance A and plasma protein:

$$\alpha = \frac{Q_V/V}{k_f C_{p0}},$$
$$\beta = \frac{k_r}{k_f C_{p0}},$$
$$\gamma = \frac{C_{A0}}{C_{P0}}.$$
(13.158)

The Matlab code used to solve (13.157) is very straightforward. The code is shown below. The function ConvectionPlusReaction represents the right side of the equation. The solution is given in Fig. 13.19 for $C_A^*(0) = 1$, $\beta = 1$, $\gamma = 0.001$, and various values of α ranging from 0 to 10. As expected, the concentration of substance A will decrease more rapidly as the parameter α is increased.

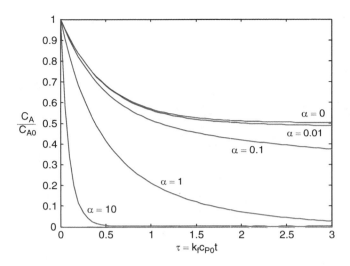

Fig. 13.19 Concentration relative to initial concentration of substance A in a flowing system where substance A reacts reversibly with a plasma protein. See text for explanation

```
% Matlab solution to (13.157)
% Input parameters:
  alpha = [0, 0.01, 0.1, 1, 10];
  beta = 1;
  gamma = 0.001;
  cA0_star = 1;
  tau = 0: 0.05: 3;
  options = [];
  options = odeset (options, "AbsTol," 1e-8, "RelTol," 1e-5);
% Solve Model Equations & Plot
for i = 1: length (alpha)
[t, cA] = ode45 ("ConvectionPlusReaction", tau, cA0_star, options,
   alpha(i), beta, gamma);
  plot (t, cA);
  hold on
end
function ydot = ConvectionPlusReaction (t, y, options, alpha, beta,
   gamma)
ydot = −alpha*y + beta* (exp(−alpha*t)−y) − y* (1 + gamma* (y − exp
   (−alpha*t)));
```

13.5.1.5 Oxygen-Hemoglobin Reactions

Oxygenation and Deoxygenation of Red Cells

Red blood cells are responsible for carrying oxygen from the lungs to the tissues. If the partial pressure of oxygen surrounding a red cell is suddenly changed, how long does it take the cell interior to come within 99% of the newly imposed partial pressure?

13.5 Chemical Reactions and Bioreactors

As a first approximation, let us assume that the cell interior is well mixed and that the initial cell oxyhemoglobin concentration is $c_{HbO_2 0}$. Oxygen solubility in the cytoplasm is assumed to be the same as in the extracellular fluid, so the partition coefficient is unity. The cell is placed in an extracellular medium in which the oxygen concentration is held constant at $C_{O_2,e}$. Intracellular oxyhemoglobin is assumed to be initially in equilibrium with cell oxygen. This will depend on the initial partial pressure of oxygen according to the Hill or Adair equations presented in Sect. 12.8.1 The cell is impermeable to oxyhemoglobin, but has a permeability coefficient for oxygen of P_{O_2}. The species conservation equations for oxygen and oxyhemoglobin in a cell with volume V_c and surface area S_c are:

$$V_c \frac{dC_{O_2}}{dt} = P_{O_2} S_c \left(C_{O_2,e} - C_{O_2} \right) + R_{O_2}, \tag{13.159}$$

$$V \frac{dC_{Hb(O_2)_4}}{dt} = R_{Hb(O_2)_4}. \tag{13.160}$$

We assume here that oxyhemoglobin exists as a single species, rather than as the four distinct species as discussed in Sect. 12.8.1. For every mole of oxyhemoglobin produced, 4 moles of oxygen must be lost from the cell interior:

$$R_{O_2} = -4 R_{Hb(O_2)_4}. \tag{13.161}$$

Multiplying (13.160) by 4 and adding it to (13.159):

$$\frac{d}{dt} \left(C_{O_2} + 4 C_{Hb(O_2)_4} \right) = \frac{P_{O_2} S_c}{V_c} \left(C_{O_2,e} - C_{O_2} \right). \tag{13.162}$$

If we assume that the rate at which oxygen combines with hemoglobin is much faster than the rate at which oxygen passes across the cell membrane, then a local equilibrium will exist between oxygen and oxyhemoglobin, or:

$$C_{Hb(O_2)_4} = S_{HbO_2}(P_{O_2}) C_{Hb,tot}, \tag{13.163}$$

where S_{HbO_2} is the oxyhemoglobin saturation, which is a function of the partial pressure of oxygen, and $C_{Hb,tot}$ is the total concentration of hemoglobin (saturated plus unsaturated) in the cell. Substituting this into (13.162):

$$\frac{dC_{O_2}}{dt} + 4 C_{Hb,tot} \frac{dS_{HbO_2}}{dt} = \frac{P_{O_2} S_c}{V_c} \left(C_{O_2,e} - C_{O_2} \right). \tag{13.164}$$

But the rate of change of oxyhemoglobin saturation is directly related to the rate of change of oxygen concentration:

$$\frac{dS_{HbO_2}}{dt} = \left(\frac{\partial S_{HbO_2}}{\partial C_{O_2}}\right)\frac{dC_{O_2}}{dt}. \tag{13.165}$$

Let us define $m(P_{O_2})$ as the local slope of the oxyhemoglobin saturation curve:

$$m = \frac{\partial S_{HbO_2}}{\partial P_{O_2}}. \tag{13.166}$$

Then, with the help of Henry's law (12.201), we can write (13.165) as:

$$\frac{dS_{HbO_2}}{dt} = \left(\frac{m}{\alpha_{O_2}}\right)\frac{dC_{O_2}}{dt}, \tag{13.167}$$

where α_{O_2} is the solubility coefficient for oxygen in the cytoplasm. Substituting this back into (13.164) provides the following ordinary differential equation for the concentration of oxygen in the cell:

$$\frac{dP_{O_2}}{dt} = \frac{P_{O_2}S_c}{V_c}\left(\frac{\alpha_{O_2}}{\alpha_{O_2} + 4mC_{Hb,tot}}\right)(P_{O_2,e} - P_{O_2}). \tag{13.168}$$

If the slope of the oxyhemoglobin concentration curve can be considered constant in the partial pressure range between the initial intracellular partial pressure, $P_{O_2,0}$, and the extracellular partial pressure $P_{O_2,e}$, then (13.168) can be easily solved:

$$\frac{P_{O_2} - P_{O_2,e}}{P_{O_2,0} - P_{O_2,e}} = \exp\left\{-\frac{P_{O_2}S_c}{V_c}\left(\frac{\alpha_{O_2}}{\alpha_{O_2} + 4mC_{Hb,tot}}\right)t\right\} \quad (\text{constant } m). \tag{13.169}$$

An estimate of the time it takes for the partial pressure difference to be 1% of the initial difference is:

$$t_{1\%} = -\frac{\ln(0.01)}{\dfrac{P_{O_2}S_c}{V_c}\left(\dfrac{\alpha_{O_2}}{\alpha_{O_2} + 4mC_{Hb,tot}}\right)}. \tag{13.170}$$

The permeability of cell membranes to oxygen is in the range of 22–125 cm/s (Subczynski et al. 1989), so we will select 50 cm/s as a representative value. Assuming the cell to be a circular disc with radius of 4 μm and height of 2 μm, the surface to volume ratio will be 1.5×10^4 cm^{-1} and $P_{O_2}S_c/V_c = 75 \times 10^4$ s^{-1}. The intracellular solubility of oxygen is assumed to be approximately that of water; $\alpha_{O_2} = 1.71 \times 10^{-9}$ mol ml^{-1} Torr^{-1} and the intracellular hemoglobin concentration of red cells is 5×10^{-6} mol/ml. The slope of the oxyhemoglobin saturation curve from Fig. 12.10 in the neighborhood of $P_{O_2} = 40$ Torr is 0.01 Torr^{-1}. Substituting these values into (13.170) yields an oxygen exchange time of about 0.73 ms. This time is quite short relative to normal capillary transit times of 500–1,000 ms, so an assumption of local plasma-red cell oxygen–oxyhemoglobin equilibrium as cells progress along the capillary should be appropriate.

13.5 Chemical Reactions and Bioreactors

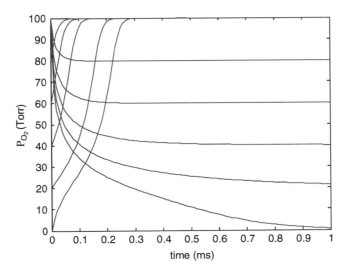

Fig. 13.20 Numerical solution to (13.168) for oxygenation of cell with initial intracellular P_{O_2} of 0, 20, 40, 60, or 80 Torr and extracellular P_{O_2} of 100 Torr, or deoxygenation of a cell with initial intracellular P_{O_2} of 100 Torr and external P_{O_2} of 80, 60, 40, 20 or 0 Torr

The slope of the oxyhemoglobin saturation curve is not constant, as is evident from Fig. 12.10. To examine the effects of changes in slope as the partial pressure is altered, it is necessary to solve (13.168) numerically. We compare numerical solutions in Fig. 13.20 with solutions assuming constant slope in Fig. 13.21. Two sets of graphs are shown in each figure. In one set, the cell is oxygenated from an initial P_{O_2} of 0, 20, 40, 60 or 80 Torr to an external P_{O_2} of 100 Torr. In the other set, the cell is deoxygenated from an initial cellular P_{O_2} of 100 Torr and external P_{O_2} values of 0, 20, 40, 60, or 80 Torr. The slope m used for each of the curves in Fig. 13.21 is the average of the slopes based on the initial intracellular and extracellular values of P_{O_2}.

During oxygenation, the more accurate numerical solutions all plateau within 0.3 ms, faster than the solutions based on the constant slope assumption. However, when the cell is deoxygenated, the opposite behavior is observed. The numerical solution indicates that it takes longer to deoxygenate the cell over the same range of P_{O_2} than to oxygenate it. For example, oxygenation of the cell from a P_{O_2} of 0 to 100 Torr takes less than 0.3 ms, while deoxygenation from 100 to 0 Torr takes slightly more than 1 ms. The constant slope solution predicts the opposite behavior. The constant slope model also predicts that a red cell oxygenates more quickly when the P_{O_2} changes from 0 to 100 Torr than it does if it changes from 20 to 100 Torr, which is an artifact of averaging initial and final slopes, rather than taking a true average over the appropriate range of partial pressures. For those situations, where the slope is relatively constant, for example, from 80 to 100 Torr, the constant slope and numerical solutions are in reasonable agreement.

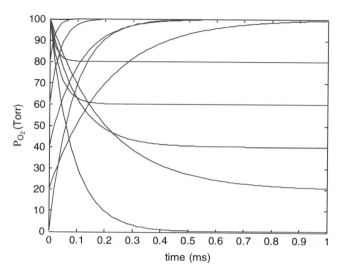

Fig. 13.21 Constant slope solution (13.169) for oxygenation of cell with initial intracellular P_{O_2} of 0, 20, 40, 60, or 80 Torr and extracellular P_{O_2} of 100 Torr, or deoxygenation of a cell with initial intracellular P_{O_2} of 100 Torr and external P_{O_2} of 80, 60, 40, 20, or 0 Torr

In any event, these models predict that red cell oxygenation and deoxygenation should be a relatively rapid processes, generally taking a millisecond or less to complete. However, both models are based on the assumption that the cell interior is well mixed. In reality, oxygen must diffuse through the cell cytoplasm in addition to traversing the cell membrane. This additional intracellular resistance will increase the time needed to approach equilibrium. The relative influence of internal vs. external resistance will be discussed in Chap. 15. Oxygen must diffuse through additional barriers in tissues, including blood plasma, the capillary membrane, and the interstitial space

Pulmonary Shunt Fraction

Under normal circumstances, blood that emerges from the lungs is nearly in equilibrium with alveolar gas. Normal alveolar P_{O_2} is about 100 Torr, and this is equal to the normal P_{O_2} in arterial blood. However, in some circumstances, areas of the lung are not well ventilated. The P_{O_2} in these alveoli eventually become equilibrated with the incoming venous blood, as shown in Fig. 13.22. Therefore, these areas form what is known as a physiological shunt. The fraction of cardiac output that passes through the physiological shunt is known as the shunt fraction. The shunt fraction can be estimated by measuring the P_{O_2} of alveolar gas, arterial blood, and venous blood. Assuming steady-state, a macroscopic mass balance at the downstream node for total oxygen content is:

13.5 Chemical Reactions and Bioreactors

Fig. 13.22 Physiological shunt

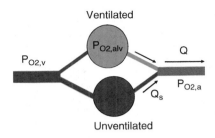

$$C_{O_2,tot,v} Q_s + C_{O_2,tot,alv} (Q - Q_s) - C_{O_2,tot,a} Q = 0. \quad (13.171)$$

The first term represents the molar flow of oxygen into the node from the unventilated alveoli, the second term represents the molar flow into the node from the ventilated alveoli, and the last term represents the molar flow out of the node as mixed arterial blood. The total molar concentration includes both the dissolved and bound oxygen (mol/L):

$$C_{O_2,tot} = \alpha_{O_2} P_{O_2} + 4 H S_{HbO_2} C_{Hb,RBC}. \quad (13.172)$$

α_{O_2} is the oxygen solubility coefficient in blood, H is the fractional hematocrit, S_{HbO_2} is the fractional oxyhemoglobin saturation, $C_{Hb,RBC}$ is the erythrocyte intracellular hemoglobin concentration, and the factor of four indicates the maximum number of moles of oxygen that can be carried per mole of hemoglobin. The shunt fraction can be found by substituting (13.172) into (13.171):

$$\frac{Q_s}{Q} = \frac{C_{O_2,tot,alv} - C_{O_2,tot,a}}{C_{O_2,tot,alv} - C_{O_2,tot,v}}$$

$$= \frac{\alpha_{O_2} (P_{O_2,alv} - P_{O_2,a}) + 4 H C_{Hb,RBC} (S_{HbO_2,alv} - S_{HbO_2,a})}{\alpha_{O_2} (P_{O_2,alv} - P_{O_2,v}) + 4 H C_{Hb,RBC} (S_{HbO_2,alv} - S_{HbO_2,v})}. \quad (13.173)$$

However, as was shown in Sect. 12.8.1, the dissolved amount is very small in comparison to the amount bound to hemoglobin. Therefore, to a good approximation, we can estimate the shunt fraction with the following simple relationship:

$$\frac{Q_s}{Q} \approx \frac{S_{HbO_2,alv} - S_{HbO_2,a}}{S_{HbO_2,alv} - S_{HbO_2,v}}. \quad (13.174)$$

Thus, if we can measure the P_{O_2} of alveolar gas, arterial blood, and venous blood, then we can calculate the saturation values for the two inlet streams and the outlet stream and use (13.174) to compute the shunt fraction.

Cardiac Output Using the Fick Principle

Adolph Fick applied the conservation relationship to oxygen transport across the lung in 1870 to estimate cardiac output. The method, known as the *Fick Principle*, assumes that oxygen is delivered from lung alveoli to pulmonary blood at a constant rate and that the rate of utilization can be measured by measuring the ventilation rate and partial pressure of oxygen in inspired and expired gas. In addition, the partial pressure of oxygen can be measured in venous and arterial blood. Equation (13.7) for this case becomes:

$$C_{O_2,\text{tot},v}Q_{CO} + W_{O_2,\text{alv}} - C_{O_2,\text{tot},a}Q_{CO} = 0, \tag{13.175}$$

where $C_{O_2,\text{tot}}$ is total molar oxygen concentration in blood, the subscripts v and a refer to venous and arterial blood, respectively, $W_{O_2,\text{alv}}$ is the molar flow rate of oxygen from alveolar gas to pulmonary blood, and Q_{CO} is cardiac output. This can be rearranged to compute cardiac output:

$$Q_{CO} = \frac{W_{O_2,\text{alv}}}{C_{O_2,\text{tot},a} - C_{O_2,\text{tot},v}}. \tag{13.176}$$

Physiologists normally express total oxygen concentration in terms of the volume of gaseous oxygen stored per 100 ml of blood, $C^*_{O_2,\text{tot}}$. In addition, the consumption rate of oxygen is generally expressed in (ml O_2)/min, rather than in (moles O_2)/s. Rewriting (13.176) to solve for cardiac output in units of L/min:

$$Q_{CO}(\text{L/min}) = \left(\frac{1\,\text{L}}{10\,\text{dl}}\right)\left[\frac{\dot{V}_{O_2,\text{alv}}(\text{ml }O_2/\text{min})}{C^*_{O_2,\text{tot},a}(\text{ml }O_2/\text{dl}) - C^*_{O_2,\text{tot},v}(\text{ml }O_2/\text{dl})}\right]. \tag{13.177}$$

Finally, writing $C^*_{O_2,\text{tot}}$ in terms of the hemoglobin concentration and oxyhemoglobin saturation using (12.215):

$$Q_{CO}\left(\frac{L}{\min}\right) = 0.1\left[\frac{\dot{V}_{O_2,\text{alv}}}{\alpha^*_{O_2}(P_{O_2,a} - P_{O_2,v}) + 1.34\rho_{\text{Hb,blood}}\left[S_{HbO_2}(P_{O_2,a}) - S_{HbO_2}(P_{O_2,v})\right]}\right]. \tag{13.178}$$

Therefore, to estimate cardiac output we need to measure the oxygen consumption rate (assumed to equal the net rate of delivery of oxygen to the lungs), and take mixed venous and mixed arterial blood samples. Hemoglobin concentration and the partial pressures of oxygen can be measured from the blood samples, and saturations determined from the oxyhemoglobin saturation curve. For example, the normal oxygen consumption rate is about 250 ml O_2/min, the normal blood hemoglobin concentration is about 15 g/dl, and the normal arterial and venous P_{O_2}

13.5 Chemical Reactions and Bioreactors

values are 100 Torr and 40 Torr, respectively. At a pH of 7.4, the arterial and venous fractional saturation levels are 0.97 and 0.74, respectively. Substituting these values into (13.178) gives a value of 5.21 L/min for cardiac output. If the dissolved O_2 is neglected, the computed cardiac output would be 5.41 L/min, an overestimate of only 3.9%.

13.5.1.6 Enzyme Kinetics

Consider the case where an enzyme is added to a solution containing substrate S at time $t = 0$. The solution is well mixed and the enzyme converts the substrate to product according to Michaelis–Menten kinetics:

$$S + E \underset{k_{rs}}{\overset{k_{fs}}{\rightleftarrows}} ES \overset{k_{fp}}{\rightarrow} E + P. \tag{13.179}$$

The species conservation equations for substrate S, enzyme E, enzyme–substrate complex ES, and product P, after dividing by the system volume V are:

Substrate:

$$\frac{dC_S}{dt} = R_S = k_{rs} C_{ES} - k_{fs} C_S C_E. \tag{13.180}$$

Enzyme:

$$\frac{dC_E}{dt} = \left(k_{rs} + k_{fp}\right) C_{ES} - k_{fs} C_S C_E. \tag{13.181}$$

Enzyme–substrate complex:

$$\frac{dC_{ES}}{dt} = k_{fs} C_S C_E - \left(k_{rs} + k_{fp}\right) C_{ES}. \tag{13.182}$$

Product:

$$\frac{dC_P}{dt} = k_{fp} C_{ES}. \tag{13.183}$$

Adding (13.182) and (13.181), we find that the sum of the enzyme and enzyme-complex concentrations must be constant. If the enzyme–substrate complex is initially zero and the initial enzyme concentration is C_{E0}, then:

$$C_E + C_{ES} = C_{E0}. \tag{13.184}$$

Introducing a dimensionless time and dimensionless concentrations:

$$\tau = k_{\text{fS}} C_{\text{S0}}, \quad C_S^* = \frac{C_S}{C_{\text{S0}}}, \quad C_E^* = \frac{C_E}{C_{\text{E0}}}, \quad C_P^* = \frac{C_P}{C_{\text{S0}}}. \quad (13.185)$$

Equations (13.180), (13.181), and (13.183), with the aid of (13.184) and (13.185), can be written in dimensionless form:

$$\frac{dC_S^*}{d\tau} = \alpha_0 \left[\alpha_S \left(1 - C_E^* \right) - C_S^* C_E^* \right], \quad (13.186)$$

$$\frac{dC_E^*}{d\tau} = (\alpha_S + \alpha_P)\left(1 - C_E^*\right) - C_S^* C_E^*, \quad (13.187)$$

$$\frac{dC_P^*}{d\tau} = \alpha_0 \alpha_P \left(1 - C_E^* \right), \quad (13.188)$$

where the dimensionless parameters α_S, α_P, and α_0 are defined as follows:

$$\alpha_S = \frac{k_{\text{rS}}}{k_{\text{fS}} C_{\text{S0}}}, \quad \alpha_P = \frac{k_{\text{fP}}}{k_{\text{fS}} C_{\text{S0}}}, \quad \alpha_0 = \frac{C_{\text{E0}}}{C_{\text{S0}}}. \quad (13.189)$$

The nonlinear nature of (13.186) and (13.187) leads us to seek a numerical solution. The effects of varying α_0 and α_P for $\alpha_S = 1$ are shown in Fig. 13.23. The effect of varying α_S is similar to varying α_P when α_0 is small. In most applications, the enzyme concentration will be small relative to the substrate concentration, so α_0 will be small. In such cases, the enzyme concentration remains relatively constant after a short transient. The enzyme is said to be in a *quasi steady-state* in which the rate of change of the enzyme concentration or the enzyme–substrate complex concentration is very nearly zero.

Thus, for low values of α_0 we can set the left side of (13.182) or (13.183) to zero. This leads to the following expression:

$$\frac{C_S C_E}{C_{\text{ES}}} = \left(\frac{k_{\text{rs}} + k_{\text{fp}}}{k_{\text{fs}}} \right) = K_m, \quad (13.190)$$

where K_m is the Michaelis constant defined by (12.226). The rate at which substrate is produced can be found by combining (13.190), (13.184), and the (12.225):

$$R_S = -\frac{k_{\text{fP}} C_{\text{E0}} C_S}{K_m + C_S}. \quad (13.191)$$

This is the same rate as that given in (12.229) for a steady-state utilization of substrate. The maximum rate of utilization of substrate will occur at high substrate concentrations where $C_S \gg K_m$:

13.5 Chemical Reactions and Bioreactors

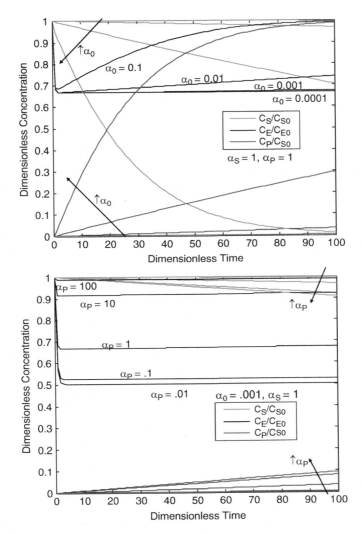

Fig. 13.23 Enzyme kinetics: *Top*: effect of changing α_0 from 0.00001 to 0.1 on the transient solution for C_S/C_{S0} (*red*), C_P/C_{S0} (*blue*), and C_E/C_{E0} (*black*) with $\alpha_S = 1$ and $\alpha_P = 1$. *Bottom*: effect of changing α_P from 0.01 to 100 with $\alpha_S = 1$ and $\alpha_0 = 0.001$

$$R_{S,\max} = -k_{fP} C_{E0} = -V_{\max}. \tag{13.192}$$

$V_{\max}$ is the maximum rate of production of product defined by (12.230) or (13.192). Substituting (13.192) into (13.191) gives the classic Michaelis–Menten equation, which characterizes quasi-steady enzyme reactions when $dC_E/dt = dC_{ES}/dt \approx 0$:

$$R_S = -R_P = -\frac{V_{\max} C_S}{K_m + C_S}.\quad(13.193)$$

At high substrate concentrations ($C_S \gg K_m$), the rate of conversion of substrate is constant (zeroth order), but at low substrate concentrations ($C_S \ll K_m$), the rate of conversion is first order. Substituting this approximate expression for R_S back into the species conservation equation, (13.180):

$$\frac{dC_S}{dt} = -\frac{V_{\max} C_S}{K_m + C_S}.\quad(13.194)$$

Separating variables and integrating provides an implicit solution that relates the substrate concentration to time:

$$\ln\left(\frac{C_S}{C_{S0}}\right) - \frac{(C_{S0} - C_S)}{K_m} = -\left(\frac{V_{\max}}{K_m}\right)t.\quad(13.195)$$

In terms of the dimensionless parameters defined in (13.185) and (13.189):

$$\ln(C_S^*) - \frac{(1 - C_S^*)}{\alpha_S + \alpha_P} = -\left(\frac{\alpha_0 \alpha_P}{\alpha_S + \alpha_P}\right)\tau.\quad(13.196)$$

This quasi-steady solution would not be expected to be valid at high values of α_0, but even for $\alpha_0 = 0.1$, the agreement between the quasi-steady solution and the numerical solution is good, as shown in Fig. 13.24. For values of α_0 of 0.01 and

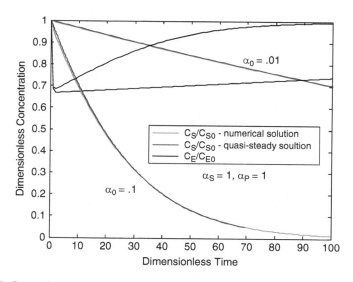

Fig. 13.24 Comparison of quasi-steady solution (13.195) with the numerical solution of (13.186)

13.5 Chemical Reactions and Bioreactors

below, the quasi-steady solution is indistinguishable from the numerical solution. The agreement is good, even though the rate of change of enzyme concentration is not zero.

13.5.2 Heterogeneous Reactions

All of the chemical reactions considered in the previous sections dealt with homogeneous chemical reactions, where the reactions took place within the system boundaries. We will now consider reactions that occur on the boundaries of the system itself. The major difference is that homogeneous reactions appear as a production term in the species conservation equation but heterogeneous reactions do not. Instead, heterogeneous reactions generally appear as a boundary condition. However, in the well-mixed systems that we consider in this chapter, time is the only independent variable. Position within the system is immaterial, and consequently boundary conditions must be accounted for by the term $W_{A,\text{wall}}$, which accounts for the molar flow of A into the system through the walls. Assuming that species A is converted to another species along the entire system boundary S and the reaction rate is nth order, then

$$W_{A,\text{wall}} = -kN_{A,\text{wall}}S = -kSC_A^n. \qquad (13.197)$$

The species conservation equation 13.14 can be generalized to include heterogeneous reactions:

$$\frac{d(C_A V)}{dt} = -P_A S(C_A - \Phi_{A,\text{system},\infty} C_{A,\infty}) - kSC_A^n + \sum_{i=1}^{N_{\text{inlets}}} C_{A,i} Q_{Vi}$$

$$- C_A \sum_{j=1}^{N_{\text{outlets}}} Q_{Vj} + R_A V. \qquad (13.198)$$

13.5.2.1 Heterogeneous Reactions at an Endothelial Surface

Consider a toxic chemical species A that is rapidly converted at sites on the endothelial surface of an organ to a nontoxic species. The reaction is first order ($n = 1$) and the organ walls are impermeable to species A ($P_A = 0$). The organ is well-mixed, has a single inlet and outlet, is perfused at a constant flow rate Q_V, and has an intravascular volume V with surface area S. Species A does not react with other intravascular substances. The inlet mean concentration of the toxic material is $C_{Am,\text{in}}$. With these assumptions, (13.198) can be simplified as follows:

$$V\frac{dC_A}{dt} = -kSC_A + Q_V(C_{Am,in} - C_A). \tag{13.199}$$

The solution is:

$$\frac{C_A}{C_{Am,in}} = \frac{Q_V}{(Q_V + kS)}\left[1 - \exp\left(-\left(\frac{Q_V + kS}{V}\right)t\right)\right]. \tag{13.200}$$

We are interested in finding the maximum flow rate to the organ if 95% of the toxic material is to be removed under steady-state conditions. Setting the left side of (13.198) equal to zero or letting $t \rightarrow \infty$ in (13.200) provides the following steady-state solution for the toxic material at the organ outlet:

$$\frac{C_A}{C_{Am,in}} = \frac{Q_V}{(Q_V + kS)}. \tag{13.201}$$

For the case we are examining, the left side of (13.201) equals 0.05, so the maximum flow rate will equal $(0.05/0.95)kS = 0.053(kS)$. In many cases, we may be more interested in measuring kS from experimental data. If we can measure the inlet and outlet concentrations and know the flow rate, then we can use (13.201) to compute kS. As was the case with the permeability-surface area product, it is difficult to design experiments that can be used to separate k from S.

13.6 Pharmacokinetics

Pharmacokinetics is concerned with the appropriate delivery of drugs to target tissues. All drugs require a minimum local concentration before it can have a therapeutic effect. This is called the *minimum effective concentration* or MEC. In addition, most drugs have a toxic effect if the concentration rises above what is known as the *maximum safe concentration* or MSC. Drugs are normally administered orally, intramuscularly, subcutaneously, intravenously, via inhalation, or through the skin via a patch. Pharmacokinetic analysis is used to determine how much of the drug must be administered to provide a therapeutic effect and how often the drug must be taken to keep the concentration between the MEC and the MSC. A regimen with ineffective periods, as shown in the top of Fig. 13.25, is to be avoided, in favor of a regimen like that shown at the bottom.

Important factors that influence the distribution of the drug include the amount of drug per administration, the frequency of administration, the degree to which the drug reacts with plasma proteins, drug solubility in plasma, erythrocytes and tissue, blood perfusion rate to the target organ, capillary permeability of the target organ, drug metabolism, and renal excretion.

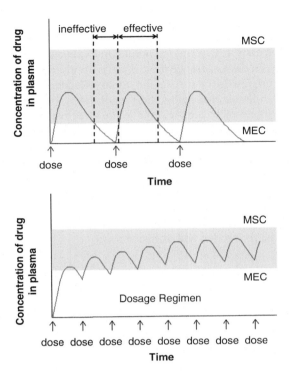

Fig. 13.25 Less effective (*top*) and more effective (*bottom*) dosage regimens

13.6.1 Renal Excretion

One of the primary mechanisms by which drugs are removed from the body is by renal excretion. A certain fraction of plasma that enters the kidneys leaves the bloodstream by passing across the glomerular membrane into the proximal tubules of nephrons. The rate at which the fluid passes through the barrier is known as the glomerular filtration rate or GFR. The driving force for fluid flow across the glomerular membrane consists of a hydrostatic component and an osmotic component. According to Starlings law:

$$Q_G = K_f[P_P - P_T - \sigma_d(\Pi_P - \Pi_T)], \qquad (13.202)$$

where Q_G is the glomerular filtration rate, K_f is the filtration coefficient, P_P and P_T are the hydrostatic pressures on the plasma and tubule sides of the membrane, and Π_P and Π_T are the protein osmotic pressures on the plasma and tubule sides of the membrane. The osmotic reflection coefficient σ_d is ordinarily set equal to unity since proteins do not ordinarily cross the glomerular barrier. Consequently, $\Pi_T = 0$. The filtration coefficient is approximately 15 ml min^{-1} Torr^{-1}, glomerular capillary pressure is normally 45 Torr, proximal tubule pressure is approximately 10 Torr, and plasma protein osmotic pressure in glomerular capillaries is 27 Torr. Thus, the net filtration pressure is about 8 Torr and the resulting GFR amounts to approximately 120 ml/min in normal individuals. Most of the fluid is reabsorbed by blood as it passes through more distal portions of the nephron, so that only about 1% of the original filtrate enters the urinary bladder as urine.

Solute flux across the glomerular barrier can be restricted because of the size of solute molecules relative to the effective pore sizes of the barrier and also by the electrical charge of the solute relative to that of the glomerular barrier. The degree of restriction is characterized by a reflection coefficient σ_A, which is zero for small unrestricted solutes and one for large macromolecules. A sieving coefficient θ, defined as $1 - \sigma_A$ is often used in place of the reflection coefficient. A plot of the sieving coefficient vs. the Stokes–Einstein radius of neutral Ficoll molecules in the rat kidney barrier is shown in Fig. 13.26.

Solute transport across the porous glomerular membrane W_{AG} is a combination of convection and diffusion:

$$W_{AG} = (1 - \sigma_A)C_{AP}Q_G + \mathsf{P}_{AG}S_G(C_{AP} - \Phi_{APT}C_{AT}). \tag{13.203}$$

C_{AP} and C_{AT} are the concentrations of solute A in the glomerular capillary and proximal tubule, respectively, P_{AG} is the permeability of the glomerular membrane to solute A, and S_G is the total surface area of all of the glomerular membranes. For many solutes, the contribution of diffusion is small relative to the convective transport, so

$$W_{AG} \approx (1 - \sigma_A)C_{AP}Q_G. \tag{13.204}$$

The concentration in the proximal tubule can be estimated by dividing the solute flow by the fluid flow:

$$C_{AT} = \frac{W_{AG}}{Q_G} = (1 - \sigma_A)C_{AP}. \tag{13.205}$$

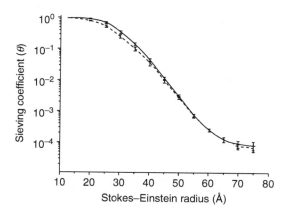

Fig. 13.26 Sieving coefficient $\theta = 1 - \sigma_A$ vs. molecular size for neutral Ficoll molecules in the rat kidney for GFR = 1.2 ml/min (solid) and 2.0 ml/min (dotted) (Data from Rippe et al. 2006)

13.6 Pharmacokinetics

Additional exchange can occur in the nephron between the fluid inside the tubules and the surrounding blood. As fluid is reabsorbed from the tubules, the concentration of all solutes becomes higher, and depending on the permeability of the walls of the tubules and blood vessels, some solute may passively diffuse back into the bloodstream at a rate:

$$W_{AN} = P_{AN}S_N(C_{AN} - \Phi_{ANP}C_{AP}), \qquad (13.206)$$

where the subscript "N" refers to portions of the nephron tubule-blood other than the glomerular membrane. Since the initial concentration of solute in the proximal tubule is proportional to the plasma concentration, then the downstream concentration should also be proportional to the glomerular plasma concentration:

$$C_{AN} = KC_{AP}. \qquad (13.207)$$

The factor K may be large. If the nonglomerular portion of the tubule were impermeable to the solute, then C_{AN} would approach the concentration in urine, which would be about 100 times greater than the plasma concentration. Consequently, the flow out of the non-glomerular portion of the membrane is proportional to the plasma solute concentration:

$$W_{AN} = P_{AN}S_N(K - \Phi_{ANP})C_{AP}. \qquad (13.208)$$

Some materials, particularly toxic materials, can be actively secreted into the tubules, $W_{A,active}$. Thus, the net rate of exchange of solute A between the blood and kidney will equal the rate at which A is excreted into the urine, $W_{A,u}$:

$$W_{A,u} = (1 - \sigma_A)C_{AP}Q_G - P_{AN}S_N(K - \Phi_{ANP})C_{AP} + W_{A,active}. \qquad (13.209)$$

Let us define a kidney excretion coefficient k'_e with dimensions of volume per unit time:

$$k'_e = (1 - \sigma_A)Q_G - P_{AN}S_N(K - \Phi_{ANP}). \qquad (13.210)$$

Then the rate at which the kidney removes solute A simplifies to:

$$W_{A,u} = k'_e C_{AP} + W_{A,active}. \qquad (13.211)$$

Consider a substance such as inulin that is not actively secreted ($W_{inulin,active} = 0$), can easily pass through the glomerular membrane ($\sigma_{inulin} = 0$), but is not reabsorbed ($P_{inulin,N} = 0$). In that case (13.210) can be used to show that the kidney excretion coefficient k'_e is equal to the glomerular filtration rate Q_G. In addition, urine inulin flow is equal to the product of urine volumetric flow Q_u and urine inulin concentration $C_{inulin,u}$:

$$W_{inulin,u} = k'_e C_{inulin,P} = Q_G C_{inulin,P} = Q_u C_{inulin,u}. \qquad (13.212)$$

If an experiment is designed such that inulin plasma concentration is maintained constant, and urine flow and urine inulin concentration are measured, then one can use (13.212) to compute glomerular filtration rate. The sieving coefficient for substances that are not reabsorbed or actively secreted, such as the Ficoll molecules illustrated in Fig. 13.26, can be measured by comparing their plasma and urine concentrations to inulin concentrations:

$$\theta = (1 - \sigma_A) = \left(\frac{C_{A,u}}{C_{A,P}}\right)\left(\frac{C_{\text{inulin},P}}{C_{\text{inulin},u}}\right). \tag{13.213}$$

It is difficult to measure the reabsorption term in (13.210) for solutes with nonzero tubular permeabilities. It is much easier to measure k'_e directly for a particular solute by analyzing a bolus experiment. Let us assume for the moment that kidney excretion is the only mechanism by which solute A is removed from plasma following a bolus injection into the bloodstream. Assuming that solute A distributes only in plasma (not in red cells) and is not actively secreted into the kidney tubules, conservation of solute A in the plasma can be written:

$$V_P \frac{dC_{AP}}{dt} = k'_e C_{AP}. \tag{13.214}$$

The solution is:

$$c_{AP} = \left(\frac{N_{AP,0}}{V_P}\right) e^{-\frac{k'_e}{V_P}t} = C_{AP0} e^{-k_e t}, \tag{13.215}$$

where $N_{AP,0}$ is the number of moles of solute A injected as a bolus at time $t = 0$, C_{AP0} is the initial concentration in plasma following the bolus, and k_e is the excretion rate constant expressed in dimensions of reciprocal time:

$$k_e = \frac{k'_e}{V_P}. \tag{13.216}$$

If the solute concentration is measured in plasma as a function of time, then we can estimate the two unknown quantities by plotting $\ln(c_{AP})$ vs. time. Extrapolation of the experimental data back to zero time will give the initial concentration. Dividing $N_{AP,0}$ by the initial concentration provides an estimate of the plasma volume, V_P. The excretion coefficient for solute A can be estimated by measuring the slope of the semilog plot: $k_e = -\text{slope}$.

This procedure for estimating k_e would also be appropriate if the solute distributes in both plasma and red cells, but the volume so computed would be the blood volume rather than the plasma volume, and this volume should be used in the definition of k_e in (13.216). If the solute is eliminated from the bloodstream by other first order mechanisms, such as removal by cells in the liver, the vascular concentration will still have a single exponent, as in (13.215), but k_e would reflect both

13.6 Pharmacokinetics

kidney and liver elimination. In some cases, the vascular concentration will exhibit multiple time constants, and the method of feathering or *peeling off exponentials* must be used to identify the most important ones (see Sect. 9.5.1). Excretion by the kidney is usually associated with the fastest time constant ($\tau = V/k_e$, where V is the vascular volume of distribution).

13.6.2 Drug Delivery to Tissue, Two Compartment Model

In this section, we will model the delivery of a drug to a tissue in order to establish an optimal dosage regimen. We will consider oral and intravenous administration of the antibiotic ampicillin. Ampicillin belongs to the penicillin group and is used to fight various types of infections, including urinary infections, middle ear infections, and pneumonia. Ampicillin is capable of entering gram-negative bacteria and preventing them from growing. Early studies by Jusko and Lewis (1973) indicate that ampicillin distributes into two compartments in the human body: a central compartment and a peripheral compartment. Elimination of ampicillin from the peripheral compartment is neglected. A schematic is shown in Fig. 13.27.

13.6.2.1 Bolus Injection

Conservation of ampicillin following a bolus injection in the central compartment is shown below for the central (c) and peripheral (p) compartments.

Central compartment:

$$\frac{d\rho_{Ac}}{dt} = -\frac{P_A S}{V_c}\left(\rho_{Ac} - \Phi_{Acp}\rho_{Ap}\right) - k_{el}\rho_{Ac}. \tag{13.217}$$

Peripheral Compartment:

$$\frac{d\rho_{Ap}}{dt} = +\frac{P_A S}{V_p}\left(\rho_{Ac} - \Phi_{Acp}\rho_{Ap}\right) \tag{13.218}$$

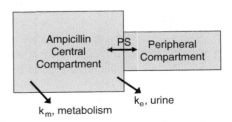

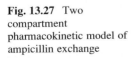

Fig. 13.27 Two compartment pharmacokinetic model of ampicillin exchange

Fig. 13.28 Classical depiction of 2 compartment pharmacokinetic model

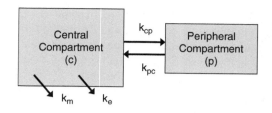

and the total excretion coefficient is:

$$k_{el} = k_e + k_m. \tag{13.219}$$

Equations (13.217) and (13.218) are variations on the more classical presentation of the two compartment model that can be found in the pharmacokinetics literature depicted in Fig. 13.28.

The coefficients k_{cp} and k_{pc} represent rate constants for first-order transport from compartments c to p and p to c, respectively, with dimensions of reciprocal time. The differential equations used to describe transport in the two compartments are described in terms of drug mass rather than concentration:

Central compartment:

$$\frac{dm_{Ac}}{dt} = -(k_{cp} + k_{el})m_{Ac} + k_{pc}m_{Ap}. \tag{13.220}$$

Peripheral compartment:

$$\frac{dm_{Ap}}{dt} = k_{cp}m_{Ac} - k_{pc}m_{Ap}. \tag{13.221}$$

Comparison of (13.220) and (13.221) with (13.217) and (13.218) with $m_{Ac} = V_c \rho_{Ac}$ and $m_{Ap} = V_p \rho_{Ac}$ provides a physical interpretation for the coefficients k_{cp} and k_{pc}:

$$k_{cp} = \frac{P_A S}{V_c},$$

$$k_{pc} = \Phi_{Acp} \frac{P_A S}{V_p} = \left(\Phi_{Acp} \frac{V_c}{V_p}\right) k_{cp}. \tag{13.222}$$

The solution for a single iv bolus injection can be found using the same method presented in Sect. 13.3.2.4. Equations (13.216) and (13.217) and (13.221) and (13.222) are completely analogous to (13.91) and (13.92) using the substitutions in Table 13.1.

Thus, the mass concentrations of ampicillin in the central and peripheral compartments following an iv bolus injection are biexponential in nature and the

13.6 Pharmacokinetics

Table 13.1 Analogous variables and parameters for three 2 compartment models

Equations (13.91) and (13.92)	Equations (13.217) and (13.218)	Equations (13.220) and (13.221)
ρ_{AB}, ρ_{AT}	ρ_{Ac}, ρ_{Ap}	m_{Ac}/V_c, m_{Ap}/V_p
V_B, V_T	V_c, V_p	V_c, V_p
$P_A S$	$P_A S$	$V_c k_{cp}$
Φ_{ABT}	Φ_{Acp}	$(V_p k_{pc})/(k_{cp} V_c)$
Q_B/V_B	k_{el}	k_{el}

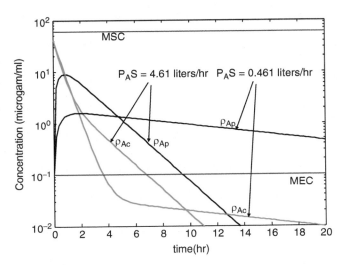

Fig. 13.29 Logarithmic plot of transient central (*red*) and peripheral (*black*) concentrations following a bolus injection of 500 mg of ampicillin for $P_A S = 4.61$ L/h and 0.461 L/h

solutions are given by (13.100) and (13.101). The graphical solutions given in Figs. 13.13 and 13.14 are also applicable to this problem if Q_B is replaced with $k_{el} V_c$.

Jusko and Lewis (1973) fit (13.220) to experimental data and found $k_{cp} = 0.384$ h^{-1}, $k_{pc} = 0.733$ h^{-1}, $V_c = 12$ L, and the total distribution volume for ampicillin $(V_c + V_p) = 17.9$ L. The central compartment is larger than the average human vascular volume of about 5 L, so it probably includes interstitial fluid in organs with highly permeable microvessels. The total elimination rate constant from the central compartment was found to be $k_{el} = 1.73$ h^{-1}. Ninety percent of the injected ampicillin was recovered in the urine, indicating that $k_e = 0.9 \times (1.73$ h$^{-1}) = 1.55$ h^{-1}. The rate constant for elimination by metabolism, biliary excretion, and other mechanisms is $k_m = 0.1 \times (1.73$ h$^{-1}) = 0.173$ h^{-1}. Using Table 13.1 to convert these to parameters used in the species conservation equations: $V_c = 12$ L, $V_p = 5.9$ L, $P_A S = 4.61$ L/h, $\Phi_{Acp} = 0.939$, and $k_{el} = 1.73$ h^{-1}.

In Fig. 13.29 we examine the effect of changing the permeability of the central-peripheral barrier for a bolus injection of 500 mg of ampicillin. The minimum effective concentration of ampicillin depends on the bacteria strain. For the purpose of illustration, we choose a hypothetical strain with MEC = 0.1 μg/ml and MSC = 60 μg/ml. The predicted concentration in the central compartment for $\mathsf{P_A}S$ = 4.61 L/h agree well with the plasma measurements of Jusko and Lewis. The advantage of using the model is that it can also predict the concentration of ampicillin in the peripheral compartment, which includes the site of infection. This is not easily measured in humans and is often different than the plasma concentration. Note that the predicted concentration in the peripheral compartment rises above the plasma concentration after about an hour for the normal permeability case and after about 2 h for the low permeability case. The plasma concentration drops below the minimum effective concentration after about 7 h, and the peripheral concentration falls below the MEC after another 2.5 h.

Contrast the normal permeability simulation with the simulation for a low permeability drug with the same MEC and MSC, as shown in Fig. 13.29. Because of the significantly lower permeability, one might expect the same dose to be much less effective. However, once some of the drug has passed into the tissue, the low permeability prevents rapid loss from the peripheral compartment and provides a significant biexponential character to the concentration in the central compartment. After less than 6 h, the drug concentration in the peripheral compartment is higher than it is for the drug with normal permeability. Although the plasma concentration drops below the MEC after only 4 h, the tissue concentration, where the infection is to be treated, does not fall below the MEC for well over 20 h. Thus, in some cases, it might be beneficial to chemically reduce the permeability of a drug, as long as the MEC is low and the potency of the modified drug is unaltered. An optimal drug delivery scheme might be one in which the permeability is high when the drug moves from plasma to tissue, but once in the tissue the drug permeability is lowered, perhaps by reaction with tissue components or a second drug infused shortly after the first.

Measurements made by Dalla Costa et al. (1998) in the interstitial fluid of rats using injections of piperacillin and tazobactam have confirmed that it is possible to use a two compartment model to predict tissue concentrations based on parameters derived from plasma concentration-time data. In addition, they showed that for both of these substances the peripheral concentration is higher than the plasma concentration within 10 min of a bolus injection.

13.6.2.2 Constant Infusion

If a patient has an intravenous line inserted into a peripheral vein, a relatively stable plasma concentration can be attained by infusing the drug at a constant rate. If ampicillin is infused at a rate $w_A(t)$, the central compartment species conservation equation (13.217) must be modified to include the infusion rate:

$$\frac{d\rho_{Ac}}{dt} = \frac{w_A(t)}{V_c} - \frac{\mathsf{P_A}S}{V_c}\left(\rho_{Ac} - \Phi_{Acp}\rho_{Ap}\right) - k_{el}\rho_{Ac}. \tag{13.223}$$

13.6 Pharmacokinetics

Let us consider the case where w_A is constant. A long time after beginning the infusion, the concentrations in the central and peripheral compartments will be constant. Setting the left sides of (13.218) and (13.223) equal to zero, we can compute the ultimate concentrations in the central and peripheral compartments:

$$\rho_{Ap}(t \to \infty) = \frac{\rho_{Ac}(t \to \infty)}{\Phi_{Acp}}, \tag{13.224}$$

$$\rho_{Ac}(t \to \infty) = \frac{w_A}{k_{el}V_c}. \tag{13.225}$$

To be effective, the steady-state tissue concentration must be above the MEC and below the MSC. Consequently, the infusion rate must be constrained as follows:

$$\text{MEC} < \frac{w_A}{\Phi_{Acp}k_{el}V_c} < \text{MSC}. \tag{13.226}$$

An analytic solution to the coupled equations (13.218) and (13.223) with constant infusion rate can be found using the method outlined in Sect. 13.3.2.4. The solution for concentration in the central compartment is:

$$\rho_{Ac} = \frac{w_A}{k_{el}V_c} + Ae^{\lambda_1 t} + Be^{\lambda_2 t}, \tag{13.227}$$

where the coefficients λ_1 and λ_2 are both negative and can be computed from (13.98) using the values of α and β in (13.95) and (13.96) with Q_B/V_B replaced by k_{el}. The constants A and B can be found by applying the initial conditions $\rho_{Ac}(0) = \rho_{Ap}(0) = 0$. The final solution is:

$$\rho_{Ac} = \frac{w_A}{k_{el}V_c}\left\{1 + \frac{(k_{el} + \lambda_2)}{(\lambda_1 - \lambda_2)}e^{\lambda_1 t} - \frac{(k_{el} + \lambda_1)}{(\lambda_1 - \lambda_2)}e^{\lambda_2 t}\right\}. \tag{13.228}$$

The transient solution for concentration in the peripheral compartment can be found by substituting (13.228) into (13.223). The time required to closely approximate a steady-state is independent of w_A. Plots of the transient central and peripheral concentrations for the case of a constant infusion of ampicillin (500 mg/h) are shown in the left panel of Fig. 13.30. The flow of ampicillin from the central compartment to the peripheral compartment (i.e., $P_A S(\rho_{Ac} - \Phi_{Acp}\rho_{Ap})$) is shown in the right panel of Fig. 13.30. A steady-state is predicted to occur within 7–8 h.

13.6.2.3 Loading Dose Followed by Constant Infusion

The time required to approach a steady-state can be reduced by introducing a loading dose of ampicillin as a bolus at the same time that the infusion is started. The mass of ampicillin in the bolus that is necessary to bring the initial concentration up to the steady-state concentration is:

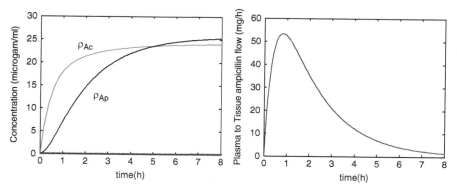

Fig. 13.30 Response to a constant infusion of ampicillin at the rate of 500 mg/h. *Left*: central and peripheral compartment mass concentrations. *Right*: ampicillin flow from central to peripheral compartment

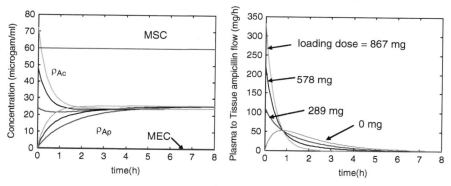

Fig. 13.31 Response to various loading doses followed by a constant infusion of ampicillin at the rate of 500 mg/h. *Left*: central and peripheral compartment mass concentrations. *Right*: ampicillin flow from central to peripheral compartment

$$\rho_A(t=0) = \frac{m_{A0}}{V_c} = \frac{w_A}{k_{el}V_c}. \tag{13.229}$$

Therefore, if the plasma concentration is to be initially raised to the steady-state value, the mass of ampicillin in the loading dose should be:

$$m_{A0} = \frac{w_A}{k_{el}}. \tag{13.230}$$

The time to reach a steady-state is reduced by administering a loading dose. The left panel of Fig. 13.31 shows the effect of administering loading doses that are a multiple of the value computed with (13.230) (289 mg), followed by a constant infusion of 500 mg/h. The right panel shows the flow of drug from central to peripheral compartments. The higher the loading dose, the shorter will be the time necessary to reach a steady state. A loading dose three times higher than is necessary to bring the initial plasma concentration to its steady-state value reduces the time required to reach a steady state to 2 h. However, the initial concentration in the central compartment exceeds the maximum safe concentration. This can be avoided by making sure that the loading dose is constrained so the initial concentration in the central compartment falls below the MSC:

$$\frac{m_{A0}}{V_c} < \text{MSC}. \tag{13.231}$$

We can obtain an estimate of the time required to reach a steady-state when a loading dose similar to the one computed from (13.230) is administered. If the central compartment concentration was maintained constant, then we can solve the species conservation equation for ampicillin in the peripheral compartment directly:

$$\rho_{Ap} = \frac{\rho_{Ac}}{\Phi_{Acp}} \left(1 - \exp\left(-\frac{P_A S \Phi_{Acp}}{V_p}\right)\right). \tag{13.232}$$

The time constant for the transient in (13.232) is proportional to the volume of the peripheral compartment and inversely proportional to the product of $P_A S$ and the partition coefficient. Using the values from the examples above, the time constant is $(5.9\ \text{L})/(4.61\ \text{L/h} \times 0.939) = 1.36$ h. The transient period is nearly complete within 5 time constants or 6.82 h, which is similar to the time required following a loading dose of 289 mg as shown in Fig. 13.31.

13.6.2.4 Oral Administration

Oral administration of a drug is usually in the form of either a liquid or a capsule. In both cases, the drug must pass through the walls of the gastrointestinal system before entering the blood stream, where it is ultimately delivered to the target tissue. Drugs that are delivered nasally, via inhalation or via a skin patch, must also pass through physiological barriers before entering the blood stream. The analysis will be similar for each of these methods of administration.

Consider the simple three compartment model of oral administration as shown in Fig. 13.32. We assume that drug absorption is by simple diffusion. The GI system is

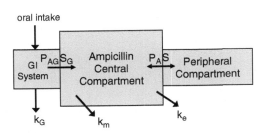

Fig. 13.32 Model for oral administration of ampicillin

modeled as a simple well-mixed compartment with volume V_G, permeability P_{AG}, and surface area for drug absorption S_G. Some of the drug is not absorbed and passes out of the system at a rate $k_G \rho_{AG}$ in to the feces. Neglecting any back diffusion of drug from blood to gut, conservation of ampicillin in the gut for a bolus infusion leads to:

$$V_G \frac{d\rho_{AG}}{dt} = -P_{AG} S_G \rho_{AG} - k_G \rho_{AG}. \quad (13.233)$$

Since back diffusion was neglected, this equation is not coupled to the central or peripheral concentrations and can be solved directly to give:

$$\frac{\rho_{AG}}{\rho_{AG}(0)} = \exp\left[-\left(\frac{P_{AG} S_G + k_G}{V_G}\right)t\right] = e^{-k_a t}. \quad (13.234)$$

The unknown parameters $P_{AG} S_G$, V_G, and k_G can be estimated from a bolus experiment, if the concentration of drug is measured in the GI system and the unabsorbed mass of drug in the feces is measured. The GI compartment volume can be computed by dividing the dose of ampicillin swallowed m_{A0} by the initial concentration:

$$V_G = \frac{m_{A0}}{\rho_{AG}(0)}. \quad (13.235)$$

The absorption coefficient k_a can be estimated from the slope of a plot of ln $(\rho_{AG}/\rho_{AG}(0))$ vs. time. Therefore, the unknown coefficients k_G and $P_{AG} S_G$ can be written in terms of the measured quantities k_a and V_G:

$$P_{AG} S_G + k_G = k_a V_G. \quad (13.236)$$

The mass flow rate of drug from the GI system to the central compartment is

$$w_A = P_{AG} S_G \rho_{AG}(0) e^{-k_a t}. \quad (13.237)$$

The fraction of the dose swallowed that passes into the bloodstream f is:

$$f = \frac{\int_0^\infty w_A dt}{m_{A0}} = \frac{P_{AG} S_G}{V_G} \int_0^\infty e^{-k_a t} dt = \frac{P_{AG} S_G}{P_{AG} S_G + k_G}. \quad (13.238)$$

The mass of drug in the feces divided by the initial dose is equal to $1 - f$, so the fraction of the initial dose that is absorbed can be measured. Thus the measured

13.6 Pharmacokinetics

values of f, k_a, and V_G can be used to compute $P_{AG}S_G$ and k_G using (13.236) and (13.238):

$$P_{AG}S_G = fk_a V_G, \qquad (13.239)$$

$$k_G = k_a V_G (1 - f). \qquad (13.240)$$

With these values known, we are now in a position to substitute them into (13.237) to compute the drug mass flow from the GI system to the central compartment in terms of measured quantities:

$$w_A = k_a f m_{A0} e^{-k_a t}. \qquad (13.241)$$

We can now return to the two compartment model and substitute w_A from (13.241) as a source term into the central compartment conservation equation, (13.223):

$$\frac{d\rho_{Ac}}{dt} = \frac{k_a f m_{A0}}{V_c} e^{-k_a t} - \frac{P_A S}{V_c}\left(\rho_{Ac} - \Phi_{Acp}\rho_{Ap}\right) - k_{el}\rho_{Ac}. \qquad (13.242)$$

This equation, along with the peripheral compartment conservation relationship, (13.218), must be solved simultaneously. Jusko and Lewis found the average fraction of ampicillin absorbed following oral administration to be 0.32. They were unable to accurately measure k_a, so we will use a value of 1.02 h^{-1} reported by Arancibia et al. (1980) for a similar drug, amoxicillin. The numerical solution for a single oral dose of 500 mg is shown in Fig. 13.33. The peak concentrations

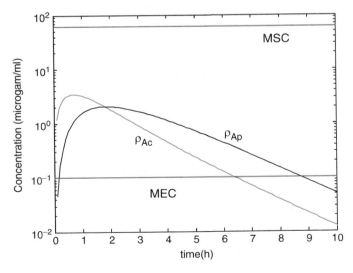

Fig. 13.33 Response to a single oral dose of 500 mg of ampicillin

in both compartments are lower than for a bolus iv injection shown in Fig. 13.29, since only 32% of the ampicillin is absorbed. The concentration in the peripheral compartment becomes higher than in the central compartment after about 2 h, and the tissue concentration falls below the minimum effective concentration after about 9 h.

To maintain the central compartment concentration between the MEC and MSC, 500 mg of ampicillin should be taken every 6 h or four times a day. This should ensure that the peripheral compartment concentration is kept above the MEC. Alternately, a larger dose could be taken less frequently. A simulation where a patient skips a 500 mg dose at 18 h is shown in Fig. 13.34. Based on the plasma concentration, the missed dose resulted in about 5 h of ineffective drug use, but the tissue was below the MEC for only about 3 h.

These results indicate that although the plasma ampicillin concentration is sometimes lower and sometimes higher than the tissue concentration, the maximum tissue concentration will be below the maximum plasma concentration and the minimum tissue concentration will be above the minimum plasma concentration. Therefore, if blood levels are kept between the MEC and MSC with an oral regimen, tissue levels will also oscillate between acceptable levels. Unfortunately, this cannot be generalized to all drugs. Drugs with high permeabilities and high tissue solubility will have peaks that are actually higher in the peripheral compartment than in the central compartment. The opposite will occur if the solubilities are reversed.

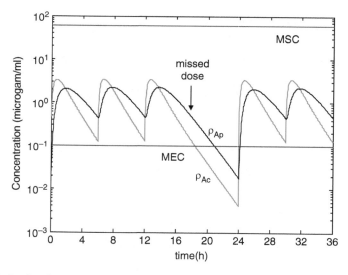

Fig. 13.34 Predicted central and peripheral compartment concentrations of ampicillin for multiple oral doses of 500 mg. One dose is missed at 18 h

13.6.3 More Complex Pharmacokinetics Models

Two compartment models of drug delivery with elimination from the central compartment are used to describe many different drugs. However, some drugs get distributed in more than two compartments and others are partially removed from a peripheral compartment. A three compartment model in which the peripheral compartment is divided into an interstitial region and a cellular region is shown in Fig. 13.35. The drug diffuses across the cell membrane and undergoes an irreversible first-order chemical reaction with an intracellular component. Our primary interest is in properly setting up the conservation equations rather than seeking analytic solutions for each concentration. The species conservation relations applicable for a bolus injection of drug into the central compartment are:

Central compartment (c):

$$V_c \frac{d\rho_{Ac}}{dt} = -P_{AcI}S_{cI}(\rho_{Ac} - \Phi_{AcI}\rho_{AI}) - k_e V_c \rho_{Ac}. \quad (13.243)$$

Interstitial fluid compartment (I):

$$V_I \frac{d\rho_{AI}}{dt} = P_{AcI}S_{cI}(\rho_{Ac} - \Phi_{AcI}\rho_{AI}) - P_{AIT}S_{IT}(\rho_{AI} - \Phi_{AIT}\rho_{AT}). \quad (13.244)$$

Target cells (T):

$$V_T \frac{d\rho_{AT}}{dt} = P_{AIT}S_{IT}(\rho_{AI} - \Phi_{AIT}\rho_{AT}) - k_m V_T \rho_{AT}. \quad (13.245)$$

Coupled first-order differential equations such as these are easily solved using numerical methods similar to that presented in Sect. 13.5.1.4. The solution in each compartment is shown in Fig. 13.36 for a 500 mg bolus of a drug with the following properties similar to ampicillin: $V_c = 12$ L, $V_I = 3.2$ L, $V_T = 2.7$ L, $P_{AcI}S_{cI} = 4$ L/h, $P_{AIT}S_{IT} = 2$ L/h, $\Phi_{AcI} = 0.939$, $\Phi_{AIT} = 1.1$, $k_e = 1.55$ h^{-1}, and $k_m = 0.173$ h^{-1}. After about 3 h, the concentration of drug in the target tissue remains about ten times

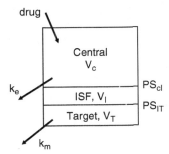

Fig. 13.35 Three compartment model

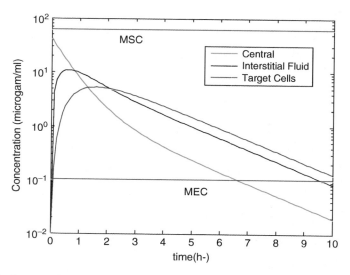

Fig. 13.36 Concentrations in central, interstitial, and cellular compartments following a 500 mg bolus injection. See text for parameters used

higher than the concentration in plasma, and the concentration in the target cells remains above the MEC for about 3.5 h longer than in the plasma. So, once again, plasma concentration should not be confused with the concentration in the target tissue.

13.7 Mass Transfer Coefficient Applications

The mass transfer coefficients presented in Sect. 12.4.7 can be combined with the macroscopic approach to solve unsteady-state mass transfer problems. This approach is particularly useful if we want to estimate the amount of time it takes for some fraction of material to dissolve or be transported from a solid surface to a flowing fluid.

Consider, for instance, a solid object that contains a layer of material deposited on its surface containing species A. The solid is immersed in a fluid, where the transport of A in the fluid is governed by a mass transfer coefficient k_A. The rate at which species A is removed from the solid is equal to the rate at which it is transported away from the surface:

$$\frac{dm_A}{dt} = -k_A S[\rho_{A0} - \rho_{A\infty}] = -k_A S\left[\Phi_{A,\text{liqid,solid}}\rho_{A,\text{solid}} - \rho_{A\infty}\right]. \quad (13.246)$$

S is the surface area of the solid, ρ_{A0} is the mass concentration of A in the fluid next to the surface, and $\rho_{A\infty}$ is the mass concentration of A far from the surface.

13.7 Mass Transfer Coefficient Applications

Local equilibrium of A between solid and fluid is assumed at the surface. If species A is uniformly distributed in the surface layer, then $m_A = \rho_{A,\text{solid}} V$, where V is the volume of the surface layer. Substituting this into (13.246), we have an expression for the rate of change of material volume:

$$\frac{dV}{dt} = -k_A S \left[\Phi_{A,\text{liqid,solid}} - \frac{\rho_{A\infty}}{\rho_{A,\text{solid}}} \right]. \tag{13.247}$$

Equation (13.247) can be solved to determine what volume of material containing species A remains in the solid as a function of time. This is relatively straightforward if k_A and S remain relatively constant. In many instances, however, V, S and k_A all change with time, as illustrated in the following example.

Example 13.7.1: Time for a Soluble Microsphere to Dissolve.

In Example 12.4.6, we estimated that a soluble microsphere with initial diameter of 10 μm would dissolve in plasma after about 579 s when plasma with a velocity of 100 μm/s moves past the sphere. This estimate was based on the assumption that the rate of removal of the material from the sphere remained constant. Use conservation of mass for the soluble material to obtain a better estimate of the time required for the microsphere to dissolve completely.

Solution. *Initial considerations.* The original estimate for the dissolution time was based on the assumptions that the surface area and mass transfer coefficients were constant. In reality, the diameter of the sphere will change with time. Consequently, neither k_A nor S is constant.

System definition and environmental interactions. There are two systems of interest: the fluid film surrounding the microsphere and the microsphere itself. The rate at which material is gained by mass transfer to the fluid is equal to the rate at which material is lost from the surface of the microsphere.

Apprising the problem to identify governing relationships. The mass transfer coefficient, determined from empirical data, governs the rate at which material is transferred from the microsphere surface to the flowing fluid. Conservation of mass applied to the microsphere can be used to predict the time required for the microsphere to completely dissolve.

Analysis. For a sphere of diameter D, $S = \pi D^2$ and $V = \pi D^3/6$. Substituting these into (13.247) with $\rho_{A\infty} = 0$:

$$\frac{d}{dt}\left[\frac{\pi D^3}{6}\right] = -k_{A,\text{plasma}} \Phi_{A,\text{plasma,sphere}} \left[\pi D^2\right].$$

Since $dD^3/dt = 3D^2 dD/dt$, this can be written:

$$\frac{dD}{dt} = -2k_{A,\text{plasma}} \Phi_{A,\text{plasma,sphere}}.$$

If $k_{A,\text{plasma}}$ is constant, then $D(t)$ would decrease linearly with time:

$$D(t) - D(0) = -2k_{A,\text{plasma}} \Phi_{A,\text{plasma,sphere}} t.$$

This is shown as the top graph in Fig. 13.37 using values from Example 12.4.6. The time t^* required for D to go to zero with constant $k_{A,\text{plasma}}$ is:

$$t^* = \frac{D(0)}{2k_{A,\text{plasma}} \Phi_{A,\text{plasma,sphere}}} = \frac{10^{-3}\,\text{cm}}{2(2.88 \times 10^{-4}\,\text{cm/s})(0.001)} = 1,736\,\text{s}.$$

This is longer than the estimate of 579 s in Example 12.4.6. That estimate was based on constant mass rate from the surface. Since the surface area is actually decreasing with time, the mass rate is reduced. However, a more accurate estimate of the dissolution time would also account for the dependence of the mass transfer coefficient on the diameter of the sphere. According to (12.110):

$$k_{A,\text{plasma}} = \frac{D_{A,\text{plasma}}}{D}\left[2 + 0.6\left(\frac{\langle v \rangle D}{\nu}\right)^{\frac{1}{2}} Sc^{\frac{1}{3}}\right].$$

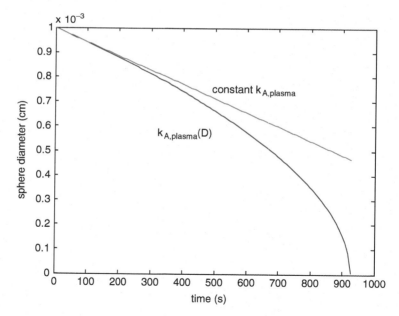

Fig. 13.37 Sphere diameter vs. time computed on the basis of constant $k_{A,\text{plasma}}$ (*top*) and $k_{A,\text{plasma}}$ computed as a function of diameter (*bottom*)

13.8 Solute Flow Through Pores in Capillary Walls

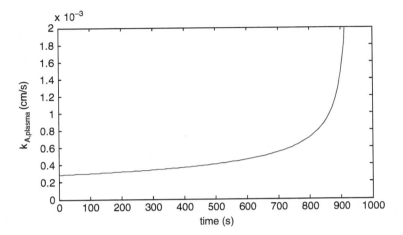

Fig. 13.38 Dependence of $k_{A,plasma}$ on time

Substituting this into the expression for dD/dt:

$$\frac{dD}{dt} = -2D_{A,plasma}\Phi_{A,plasma,sphere}\left[2\left(\frac{1}{D}\right) + 0.6\left(\frac{\langle v \rangle}{\nu}\right)^{\frac{1}{2}}Sc^{\frac{1}{3}}\left(\frac{1}{D^{\frac{1}{2}}}\right)\right].$$

Examining and interpreting the results. The solution to this ordinary differential equation can be found using the ode45 function in Matlab. This is shown as the lower graph in Fig. 13.37. The diameter reaches zero at about 928 s. This is shorter than the time predicted for constant $k_{A,plasma}$ because $k_{A,plasma}$ increases with time, particularly as the sphere diameter becomes very small, as shown in Fig. 13.38.

13.8 Solute Flow Through Pores in Capillary Walls

In Sects. 5.17 and 6.3.5, we examined the flow of fluid through pores that span the capillary blood–tissue barrier. Fluid flow through a pore is governed by the Starling equation (5.154):

$$Q_{Vp} = K_f\{\Delta P - \sigma_d \Delta \Pi\},$$

where ΔP is the pressure difference across the pore, $\Delta \Pi$ is the osmotic pressure difference across the pore, and σ_d is the reflection coefficient. For a Newtonian fluid with viscosity μ flowing through a circular pore with length Δz and radius R_p, the filtration coefficient K_f is equal to the product of the hydraulic conductivity of the pore L (6.106) and the pore surface area, πR_p^2:

$$K_f = \frac{\pi R_p^4}{8\mu \Delta z}.$$

Many solutes are also transported passively from plasma to tissue through these same capillary pores. Their passage is by a combination of convection and diffusion. In the following sections, we will develop expressions that can be used to describe solute flow through equivalent pores.

13.8.1 Small Solute Transport

If the solutes are small relative to the pore size, the flux in the z-direction from the plasma side to interstitial side of the pore is given by:

$$N_{sz} = c_s v_z - D_{s\infty} \frac{dc_s}{dz}, \qquad (13.248)$$

where $D_{s\infty}$ is the free diffusion coefficient of solute s in plasma and v_z is the mass average velocity of fluid in the pore. For steady-state transport, N_{sz} is constant and (13.248) can be integrated from the plasma side to the interstitial side. If the length of the pore is Δz and solute bulk concentration is maintained at C_{sP} on the plasma side and C_{sI} on the interstitial side, the flux is:

$$N_{sz} = v_z \left[\frac{C_{sP} - C_{sI} e^{-\frac{v_z \Delta z}{D_{s\infty}}}}{1 - e^{-\frac{v_z \Delta z}{D_{s\infty}}}} \right]. \qquad (13.249)$$

The molar rate at which solute s flows through the pore, W_{sz}, can be found by integrating the solute flux over the cross-section of the pore:

$$W_{sz} = \int_A N_{sz} dA = \left(\int_A v_z dA \right) \left[\frac{C_{sP} - C_{sI} e^{-\frac{v_z \Delta z}{D_{s\infty}}}}{1 - e^{-\frac{v_z \Delta z}{D_{s\infty}}}} \right] = Q_v \left[\frac{C_{sP} - C_{sI} e^{-\frac{v_z \Delta z}{D_{s\infty}}}}{1 - e^{-\frac{v_z \Delta z}{D_{s\infty}}}} \right]. \qquad (13.250)$$

The dimensionless group $v_z \Delta z / D_{s\infty}$ is the Peclet number based on the pore length, $Pe_{\Delta z}$. If convection dominates solute transport from plasma to tissue, the solute flux will equal $C_{sP} v_z$. Comparing the actual flux to the convective flux, we have:

$$\frac{N_{sz}}{C_{sP} v_z} = \frac{1 - (C_{sI}/C_{sP}) e^{-Pe_{\Delta z}}}{1 - e^{-Pe_{\Delta z}}}. \qquad (13.251)$$

13.8 Solute Flow Through Pores in Capillary Walls

If the Peclet number is greater than about 3, the solute flux will be nearly equal to the convective flux through the pore. Therefore, if $v_z \gg 3D_{sl}/\Delta z$, the diffusive flux can be ignored.

> **Example 13.8.1.1: Urea Transport in a Small Pore.**
> Examine the transport of urea ($D_{\text{urea,water}} = 1.8 \times 10^{-5}$ cm^2/s) through a water-filled pore with length of 100 nm and diameter of 40 nm under the conditions specified in Example 6.3.5.2.

Solution. *Initial considerations.* In Example 6.3.5.2, we estimated a net effective pressure drop, $\Delta P - \sigma_d \Delta \Pi$, of 18.7 mmHg across the pore and a flow rate through the pore of 1.56×10^{-12} ml/s.

System definition and environmental interactions. The system of interest is the fluid within a pore with length of 100 nm and diameter of 40 nm. The walls of the pore are impermeable and urea concentrations are maintained constant at each end of the pore.

Apprising the problem to identify governing relationships. The Starling equation was used in Example 6.3.5.2 to find the flow through the pore. The average velocity can be found by dividing the flow by the pore cross-sectional area. Since the solute (urea) is small relative to the pore size, we can use (13.251) to find the flux of urea through the pore.

Analysis. We estimate the average fluid velocity in the pore to be about 0.124 cm/s. Consequently, the Peclet number would be about 0.069.

Examining and interpreting the results. On the one hand, since the Peclet number is small, diffusion is the dominant mechanism for urea transport in the pore. On the other hand, if the solute is replaced by a protein the size of fibrinogen ($D_{\text{fibrinogen,water}} = 2 \times 10^{-7}$ cm^2/s) moving through the same pore, the Peclet number would be estimated to be approximately 6.2, so convection would dominate. In reality, the hydrodynamic radius of fibrinogen (10.6–12.7 nm) is of the same order of magnitude as the radius of the pore (20 nm). Consequently, we need to take a different approach for finding the flow of large solutes in pores, as illustrated in the next section.

13.8.2 *Large Solute Transport Through Pores*

When the size of the solute is of the same order of magnitude as the narrowest dimension of the channel through which it moves, we must abandon the continuum approach upon which Fick's Law is based. Instead, we can adopt the hydrodynamic approach, as in Sect. 12.4.5.1, where the solute is treated as a solid body moving through a liquid that is modeled as a continuum. The drag force on the solid can be estimated using the methods from Sect. 5.10. If the solute is spherical with radius

R_s, and the pore has a cylindrical shape, the drag force F_k is related to the fluid velocity, fluid viscosity, and the sphere velocity as given in (5.110):

$$F_k = -6\pi\mu R_s (\mathsf{K}_1 v_{sz} - \mathsf{K}_2 v_z). \tag{13.252}$$

The coefficients K_1 and K_2 are drag coefficients that depend on the radius of the particle R_s relative to the radius of the pore R_p, and on the position of the sphere center relative to the axis of the tube. This drag force is assumed to be balanced by a thermodynamic driving force arising from a gradient in chemical potential in the axial direction:

$$F_z = -k_B T \frac{d \ln(c_s)}{dz}, \tag{13.253}$$

where k_B is the Boltzmann constant and T is the absolute temperature. The sum of forces on the solute is zero in the steady state. Adding (13.252) and (13.253), we obtain:

$$\mathsf{K}_1 v_{sz} - \mathsf{K}_2 v_z + \frac{k_B T}{6\pi\mu R_s} \frac{d \ln(c_s)}{dz} = 0. \tag{13.254}$$

Solving for the solute velocity and using the Stokes–Einstein relationship (12.98), we find:

$$v_{sz} = \frac{\mathsf{K}_2}{\mathsf{K}_1} v_z - \frac{D_{s\infty}}{\mathsf{K}_1} \frac{d \ln(c_s)}{dz}, \tag{13.255}$$

where $D_{s\infty}$ is the free diffusion coefficient for the solute in the liquid. The flux of solute is:

$$N_{sz} = c_s v_{sz} = \left(\frac{\mathsf{K}_2}{\mathsf{K}_1} v_z\right) c_s - \frac{D_{s\infty}}{\mathsf{K}_1} \frac{dc_s}{dz}. \tag{13.256}$$

Comparing this expression with (13.248) for the flux of a small solute, we find that the convective flux of a large solute is modified by the factor $\mathsf{K}_2/\mathsf{K}_1$ and the diffusive flux is modified by a factor $1/\mathsf{K}_1$. The flux in (13.256) applies to the solute flux in the central portion of the pore ($0 < r < R_p - R_s$). Since the solute cannot come any closer to the pore wall than its radius, the region between $r = R_p - R_s$ and $r = R_p$ will be modeled to have zero solute concentration, and therefore zero flux. Assuming the solute concentration is uniform in the core region of the pore, and equal to $C_s(z)$, the average solute flow through the pore can be found by integrating over the available cross sectional area for solute in the pore, A_{core}:

$$W_{sz} = \int_{A_{\text{core}}} N_{sz} dA_{\text{core}} = C_s \int_{A_{\text{core}}} \frac{\mathsf{K}_2}{\mathsf{K}_1} v_z dA_{\text{core}} - \frac{dC_s}{dz} \int_{A_{\text{core}}} \frac{D_{s\infty}}{\mathsf{K}_1} dA_{\text{core}}. \tag{13.257}$$

13.8 Solute Flow Through Pores in Capillary Walls

Let us now define the following quantities:

$$1 - \sigma_s \equiv \frac{1}{Q_V} \int_{A_{core}} \left(\frac{K_2}{K_1} v_z\right) dA_{core} = \frac{\int_0^{R_p - R_s} \left(\frac{K_2}{K_1} v_z\right) r\, dr}{\int_0^{R_p} v_z r\, dr}, \qquad (13.258)$$

$$D_{se} = P_s \Delta z \equiv \frac{1}{A_p} \int_{A_{core}} \frac{D_{s\infty}}{K_1} dA_{core} = \frac{2 D_{s\infty}}{R_p^2} \int_0^{R_p - R_s} \left(\frac{1}{K_1}\right) r\, dr, \qquad (13.259)$$

where A_p is the pore cross-sectional area. The integrations extend from $r = 0$ to $r = R_p - R_s$ since solute concentration is assumed zero for $r > R_p - R_s$. The coefficient σ_s is known as the solute reflection coefficient. D_{se} is the *effective diffusivity* of the solute in the pore. D_{se} divided by the pore length Δz is also defined as the permeability of the pore P_s. Substituting these definitions into the solute flow equation, (13.257):

$$W_{sz} = C_s(1 - \sigma_s)Q_V - P_s A_p \Delta z \frac{dC_s}{dz}. \qquad (13.260)$$

The first term on the right side represents convection and the second term is solute diffusion. Integrating (13.260) from the plasma side of the pore where $C_s(0) = C_{sP}$ to the interstitial side of the pore, where $C_s(\Delta z) = C_{sI}$, we obtain the analog to (13.249) for the flux of a large solute through a pore:

$$W_{sz} = (1 - \sigma_s)Q_V \left[\frac{C_{sP} - C_{sI} e^{-\beta_s}}{1 - e^{-\beta_s}}\right], \qquad (13.261)$$

where the modified Peclet number β_s is defined as:

$$\beta_s = (1 - \sigma_s)\left(\frac{Q_V}{A_p}\right)\left(\frac{\Delta z}{D_{se}}\right) = \frac{(1 - \sigma_s)Q_V}{P_s A_p} = \frac{(1 - \sigma_s)\langle v_z \rangle}{P_s}. \qquad (13.262)$$

Under equilibrium conditions, $Q_V = 0$, $W_{sz} = 0$, $\beta_s = 0$, and $C_{sP} = C_{sI}$ for all solutes with radii smaller than the pore radius. However, steady-state conditions, not equilibrium conditions, normally exist in the microcirculation, so the interstitial concentration is lower than plasma concentration for positive transvascular flow. The dimensionless concentration in the pore can be found as a function of axial position by solving (13.260) and using (13.261):

$$\frac{C_{sP} - C_s(z)}{C_{sP} - C_{sI}} = \frac{e^{\beta_s \left(\frac{z}{\Delta z}\right)} - 1}{e^{\beta_s} - 1}. \qquad (13.263)$$

Graphs of dimensionless concentration are shown in Fig. 13.39 as a function of $z/\Delta z$ for various values of the modified Peclet number. It is tempting to ask how much of the solute transport is by convection and how much is by diffusion. Note

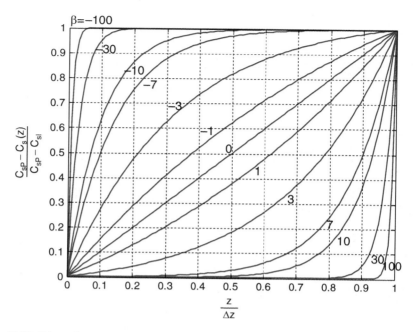

Fig. 13.39 Dimensionless concentration inside a pore as a function of dimensionless position and modified Peclet number, β_s

that for positive values of β_s (i.e., flow from plasma to interstitial side of the pore), the concentration gradient is steeper at the interstitial end of the pore than it is at the plasma end of the pore. Consequently, the fractions of solute transport by diffusion and convection are not constant, but depend on axial position in the pore. The diffusion fraction increases as solute moves through the pore from plasma to interstitial fluid. At high Peclet numbers, transport is predominantly by convection at the plasma end and, for low ratios of C_{sI}/C_{sP}, predominantly by diffusion at the interstitial fluid end of the pore.

Since solutes are confined to the core region near the center of the pore, one approach is to approximate the drag coefficients K_1 and K_2 by the centerline values for spheres in a cylinder, K_{10} and K_{20}, which are tabulated as a function of R_s/R_p (Paine and Scherr 1975). If in addition the velocity profile is assumed parabolic, (i.e., $v_z(r) = 2 <v_z> (1 - r^2/R_p^2)$), we can integrate (13.258) to obtain:

$$1 - \sigma_s = \frac{K_{20}}{K_{10}}\left\{1 - (1 - \Phi_s)^2\right\} = \frac{K_{20}}{K_{10}}[1 - \sigma_d]. \qquad (13.264)$$

Integration of (13.259) with the assumption of constant K_{10} leads to:

$$\frac{D_{se}}{D_{s\infty}} = \frac{P_s \Delta z}{D_{s\infty}} = \left(\frac{\Phi_s}{K_{10}}\right). \qquad (13.265)$$

13.8 Solute Flow Through Pores in Capillary Walls

The steric partition coefficient Φ_s for solute between intrapore fluid and bulk fluid is given by (6.104):

$$\Phi_s = \left(1 - \frac{R_s}{R_p}\right)^2. \tag{13.266}$$

Since K_{20}, K_{10}, and Φ_s are each functions of R_s/R_p, the reflection coefficient is just a function of R_s/R_p. Note that this simplified model predicts that the solute reflection coefficient σ_s given in (13.264) is generally not equal to the osmotic reflection coefficient σ_d for the same solute. A more complex theory that accounts for the actual shape of the solute, correct velocity profile, off-center drag coefficients, etc. would presumably agree with the non-equilibrium thermodynamic prediction that $\sigma_s = \sigma_d$ in a binary system. The steric partition coefficient, diffusivity relative to the free diffusion coefficient, the solute reflection coefficient, and the osmotic reflection coefficient are plotted in Fig. 13.40 as a function of the ratio of solute radius to pore radius. Note that the effective diffusion coefficient drops quickly as R_s/R_p increases. The effective diffusion curve in Fig. 13.40 is fit well by the following polynomial (Renkin 1954):

$$\frac{D_{se}}{D_{s\infty}} = \Phi_s\left[1 - 2.1\left(\frac{R_s}{R_p}\right) + 2.09\left(\frac{R_s}{R_p}\right)^3 - 0.95\left(\frac{R_s}{R_p}\right)^5\right]. \tag{13.267}$$

The ratio of drag coefficients has been approximated by Verniory et al. (1973):

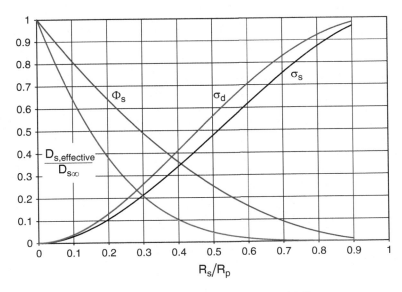

Fig. 13.40 Dependence of $D_{se}/D_{s\infty}$, σ_s, σ_d, and Φ_s as functions of R_s/R_p

$$\frac{K_{20}}{K_{10}} = \frac{1 - \frac{2}{3}\left(\frac{R_s}{R_p}\right)^2 - 0.20217\left(\frac{R_s}{R_p}\right)^5}{1 - 0.75857\left(\frac{R_s}{R_p}\right)^5}. \qquad (13.268)$$

Example 13.8.2.1: Albumin and Globulin Flow Through a Pore.
Estimate albumin and globulin flow rates through an equivalent pore that spans a junction between endothelial cells in a capillary. Compare convective and diffusive transport at each end of the pore. The equivalent pore radius is 20 nm and its length Δz is 100 nm. The hydrostatic pressure on the plasma side of the pore is 20 mmHg higher than the pressure on the interstitial side. The concentrations of albumin and globulin on the plasma (p) and interstitial (i) sides of the pore are: $C_{alb,p} = 0.682$ mM, $C_{g,p} = 0.192$ mM, $C_{alb,i} = 0.435$ mM, and $C_{g,i} = 0.083$ mM. The radii of albumin and globulin molecules are approximately 3.6 nm and 4.5 nm, respectively. The free diffusion coefficients for albumin and globulin in plasma are 9.08×10^{-7} cm^2/s and 7.26×10^{-7} cm^2/s, respectively.

Solution. *Initial considerations, system definition and environmental interactions.* These were discussed in solution of the original example problem. What is new in this problem is the determination of protein flux through the pore.

Apprising the problem to identify governing relationships. The flow rate through the pore is determined using the Starling relationship and protein flow through a restrictive pore can be found using (13.261).

Analysis. In Example 6.3.5.2, we found the fluid flow rate through the same pore under these conditions to be 1.54×10^{-12} ml/s. The average velocity will be:

$$\langle v_z \rangle = \frac{Q_V}{\pi R_p^2} = \frac{1.54 \times 10^{-12} \text{ ml/s}}{\pi (20 \times 10^{-7} \text{ cm})^2} = 0.123 \frac{\text{cm}}{\text{s}}.$$

We also found the partition coefficients between the pore and bulk fluid to be 0.672 for albumin and 0.601 for globulin. The ratio $R_{alb}/R_p = 3.6$ nm/20 nm $= 0.18$. From Fig. 13.40, we find $\sigma_{alb} = 0.09$ and $D_{alb,effective}/D_{alb\infty} = 0.4$. From (13.265) we estimate the permeability of the pore to albumin:

$$P_{alb} = \frac{D_{alb\infty}}{\Delta z}\left(\frac{D_{alb,effective}}{D_{alb\infty}}\right) = \frac{9.08 \times 10^{-7} \text{ cm}^2/\text{s}}{10^{-5} \text{ cm}}(0.4) = 3.63 \times 10^{-2} \text{ cm/s}.$$

The modified Peclet number for albumin transport from (13.262) is:

$$\beta_{alb} = \frac{(1 - \sigma_{alb})\langle v_z \rangle}{P_{alb}} = \frac{(0.91)(0.123 \text{ cm/s})}{3.63 \times 10^{-2} \text{ cm/s}} = 3.07$$

and the molar flow of albumin from (13.261) is:

13.8 Solute Flow Through Pores in Capillary Walls

$$W_{alb} = (1 - \sigma_{alb})Q_V \left[\frac{c_{alb,P} - c_{alb,I}e^{-\beta_{alb}}}{1 - e^{-\beta_{alb}}} \right]$$

$$= (0.91)\left(\frac{1.54 \times 10^{-12} \text{ ml/s}}{1,000 \text{ ml/L}} \right)$$

$$\times \left[\frac{0.683 \text{ mM} - 0.435 \text{ mM}(e^{-3.07})}{1 - e^{-3.07}} \right],$$

$$W_{alb} = 9.74 \times 10^{-16} \text{ mmol/s}$$

The transport of albumin by diffusion relative to total transport is:

$$\frac{W_{s,\text{diffusion}}}{W_s} = \frac{\left(1 - \frac{C_{sI}}{C_{sP}}\right) e^{\beta_s \left(\frac{z}{\Delta z}\right)}}{e^{\beta_s} - \frac{C_{sI}}{C_{sP}}}, \quad (13.269)$$

at the plasma end of the pore, $z = 0$, so:

$$\left(\frac{W_{alb,\text{diffusion}}}{W_{alb}} \right)_{z=0} = \frac{\left(1 - \frac{0.435}{0.682}\right)}{e^{3.07} - \frac{0.435}{0.682}} = 0.017.$$

At the interstitial end of the pore, $z = \Delta z$:

$$\left(\frac{W_{alb,\text{diffusion}}}{W_{alb}} \right)_{z=\Delta z} = \frac{\left(1 - \frac{0.435}{0.682}\right) e^{3.07}}{e^{3.07} - \frac{0.435}{0.682}} = 0.397.$$

Repeating the procedure for globulin, the ratio $R_g/R_p = 4.5 \text{ nm}/20 \text{ nm} = 0.225$. From Fig. 13.39, we find $\sigma_g = 0.13$ and from either Fig. 13.40 or (13.267), $D_{alb,\text{effective}}/D_{alb\infty} = 0.34$. From (13.265), we estimate the permeability of the pore to globulin:

$$P_g = \frac{7.26 \times 10^{-7} \text{ cm}^2/\text{s}}{10^{-5} \text{ cm}} (0.34) = 2.47 \times 10^{-2} \text{ cm/s}.$$

The modified Peclet number for albumin transport from (13.262) is:

$$\beta_g = \frac{(0.87)(0.123 \text{ cm/s})}{2.47 \times 10^{-2} \text{ cm/s}} = 4.33$$

and the molar flow of globulin from (13.261) is:

$$W_g = (0.87)\left(\frac{1.54 \times 10^{-12} \text{ ml/s}}{1,000 \text{ ml/L}} \right) \left[\frac{0.192 \text{ mM} - 0.083 \text{ mM}(e^{-4.33})}{1 - e^{-4.33}} \right]$$

$$= 2.6 \times 10^{-16} \text{ mmol/s}.$$

Examining and interpreting the results. Therefore, the flow of albumin by diffusion through the pore is less than 2% of the total albumin flow at the plasma end and is nearly 40% at the interstitial end. Consequently, convection is dominant within the pore, but much more so at the inlet than at the outlet. The fraction of globulin transport by diffusion at each end of the pore can be found from (13.269) to be 0.7% at the plasma end and 57.5% at the interstitial end.

13.9 Summary of Key Concepts

Species Conservation. The general macroscopic conservation equation for species A produced at a rate $r_{A,tot}$ in a system with multiple inlets and outlets is:

$$\frac{dm_A}{dt} = w_{A,\text{wall}} + \sum_{i=1}^{N_{\text{inlets}}} w_{A,i} - \sum_{j=1}^{N_{\text{outlets}}} w_{A,j} + r_{A,\text{tot}}. \qquad (13.270)$$

This can be further simplified for incompressible systems, systems with constant volume, systems with impermeable walls, and systems without a homogeneous chemical reaction. Mass flow through the system wall can often be written in terms of the permeability-surface area product, or mass transfer from a system wall can be written in terms of a mass transfer coefficient. For external flow around an object with mass transfer at the surface or internal flow through a conduit with mass exchange at the wall, the methods introduced in Sect. 13.7 can be used to solve the species continuity equation. For compartmental analysis, where fluid within a system of constant volume is assumed to be well-mixed, the general species conservation relationship is:

$$V\frac{d\rho_A}{dt} = -\mathbf{P}_A S\left(\rho_A - \Phi_{A,\text{system},\infty}\rho_{A\infty}\right) + \sum_{i=1}^{N_{\text{inlets}}} \rho_{Abi}Q_{Vi} - \rho_A \sum_{j=1}^{N_{\text{outlets}}} Q_{Vj} + r_A V. \qquad (13.271)$$

Compartmental analysis is used to model drug and tracer exchange between blood and tissue and other idealized systems in the body. Many examples of the application of this expression without chemical reactions are provided in Sect. 13.3 for single and multiple compartments. Examples that include chemical reaction are presented in Sect. 13.5, including a method to measure pulmonary shunt fraction (13.174), cardiac output using the Fick principle (13.178), and the analysis of reactions catalyzed by enzymes (Sect. 13.5.1.6).

Indicator Dilution. A tracer experiment in which a tracer is introduced at the inlet to a system and measured at the outlet can be used to determine flow through a system with the Stewart–Hamilton equation (13.113), which is based on conservation of a nonreacting tracer in the system. The volume of distribution of a tracer can also be estimated from the mean transit time (13.116), so comparison of the mean transit times of a tracer that can cross the blood–tissue barrier with the one that cannot provides an estimate of the extravascular volume of distribution (13.120). The permeability-surface area product of the barrier to a tracer can also be determined by measuring the loss of the tracer relative to that of a tracer confined to the vascular system (Sect. 13.4.3).

Pharmacokinetics. The field of pharmacokinetics is based on the compartmental analysis of drugs introduced into the body via oral, nasal, intramuscular or subdural injections, intravenous administration, or application to the skin. Most drugs are ineffective below a minimum effective concentration (MEC) and may be toxic above a maximum safe concentration (MSC). The goal of pharmacokinetic analysis is to provide a safe and effective regimen of drug delivery. The analysis in Sect. 13.6 shows that concentration measurements made in blood may not be representative of concentrations in the target tissue, so model predictions can be beneficial.

Convective and diffusive flux through small pores. Solute transport across the microvascular barrier is often modeled as flow through equivalent pores in the barrier. Because of the additional drag caused by the pore walls, large solutes move more slowly than small solutes through these small diameter pores. Solute flow through equivalent pores is governed by (13.261) and depends on an effective Peclet number defined by (13.262).

13.10 Questions

13.10.1. How would you simplify the general macroscopic species continuity equation (13.6) for a nonreacting species A in a well-mixed system with constant volume V? The system has two inlets with volumetric flow rates Q_{Vi1} and Q_{Vi2} and two outlets with flows Q_{Vo1} and Q_{Vo2}. The species mass concentration at the first inlet is $\rho_{A1}(t)$ and there is no species A in the fluid entering the system via the second inlet.

13.10.2. Answer Question 13.10.1 if the system volume changes with time.

13.10.3. Does the concentration need to be uniform across the inlet and outlet streams in order to use (13.10)?

13.10.4. Given the velocity and the concentration as functions of radial position in a conduit with circular cross section, how would you compute the mean or mixing cup concentration?

13.10.5. Fluids from five different inlet streams converge at the same location to form a single outlet vessel. What is the outflow and mean concentration in the vessel in terms of the flow and mean concentration of each of the converging inlet vessels?

13.10.6. How would you compute tracer concentration vs. time downstream of a single well-mixed compartment following a bolus injection or constant infusion into the compartment?

13.10.7. Bolus injection of tracer into a single well-mixed compartment leads to a first-order differential equation, while addition of a second well-mixed compartment in series leads to a second-order differential equation. Can you show that addition of a third compartment in series leads to a third-order differential equation. What auxiliary conditions are needed in order to solve this equation?

13.10.8. How might you estimate flow through an organ by injecting a tracer at the inlet of the organ and measuring the tracer concentration at the outflow? What conservation principle is this method based on? What are its limitations?

13.10.9. What is meant by "recirculation" and how can its effects be removed?

13.10.10. Explain how you might use a tracer to estimate blood volume in an organ.

13.10.11. What experiment might you use to estimate extravascular volume in an organ?

13.10.12. What is meant by tissue extraction of a tracer, and how is extraction related to tissue permeability?

13.10.13. When oxygen dissociates from oxyhemoglobin, what is the relationship between the molar rate of production of oxygen and the molar rate of production of oxyhemoglobin?

13.10.14. Is the Fick method for computing cardiac output valid in the presence of nonzero pulmonary shunt fraction?

13.10.15. What is meant by a quasi-steady state in relation to enzyme kinetics?

13.10.16. For Michaelis–Menten enzyme kinetics, what is the relationship between the rate of production of product, the rate of production of substrate and the maximum rate of production of product V_{max}.

13.10.17. How is a heterogeneous chemical reaction accounted for in a macroscopic species conservation relationship?

13.10.18. What is meant by minimum effective concentration and maximum safe concentration? What is the therapeutic range?

13.10.19. What is meant by the sieving coefficient of the glomerular membrane?

13.10.20. What is the glomerular filtration rate and how can it be measured with inulin?

13.10.21. A drug is administered orally as a liquid, and is rapidly transported into the circulation. It must pass from the bloodstream to a target site in a specific tissue. Discuss the various factors that must be considered in selecting the appropriate dose and frequency of administration.

13.10.22. Explain how a pharmacokinetic model of drug delivery can be developed from a description of how the drug is eliminated from and is transported across various components of the system. What conservation principle is generally applied?

13.10.23. At $t = 0$ a mass of drug m is injected as a bolus into a well-mixed fluid volume V initially devoid of the drug. We are interested in predicting the mass concentration of drug in the compartment as a function of time. What initial condition would you use for the mass concentration of the drug?

13.10.24. At $t = 0$ tracer is introduced at a constant rate, $\rho_{in}Q_V$, into a well-mixed fluid volume V initially devoid of tracer. We are interested in predicting the mass concentration of tracer in the compartment as a function of time. What initial condition would you use for the mass concentration of the tracer?

13.10.25. What is the method of "peeling off exponentials?" How can this be used to estimate compartment flows and volumes?

13.10.26. Would you expect tissue concentrations of a drug to be nearly the same as blood concentrations following a bolus injection in the blood? Explain.

13.10.27. What is a loading dose? What are the potential benefits and problems that a loading dose can introduce over a simple constant infusion?

13.10.28. Under what circumstances does solute concentration in a pore depend linearly on axial position in the pore?

13.10.29. What is the solute reflection coefficient? Steric partition coefficient? Effective diffusivity of a solute in a pore? When are these important?

13.11 Problems

13.11.1 Unsteady-State Mass Transfer

A polymer cylinder is saturated with a drug solution having uniform initial concentration of 200 mol/m^3. The cylinder diameter is 2 mm and the length is 4 cm. At $t = 0$, the surface of the cylinder is washed with pure water. Neglect mass transfer resistance in the polymer. The mass transfer coefficient at the polymer-fluid surface is 2.22×10^{-7} m/s. Find the amount of drug delivered from the polymer cylinder (mol) after 1 h.

13.11.2 Formaldehyde and Eye Irritation

Thirty people smoke cigarettes at the rate of two per hour within an enclosed space with dimensions $12 \times 12 \times 4$ m. One of the gases liberated during cigarette smoking is formaldehyde. Presume that each smoked cigarette produces 1.35 mg of formaldehyde. Starting at time $t = 0$, fresh air without formaldehyde is delivered into the enclosed space at a rate of 800 m^3/h. Assume that the incoming air is instantly mixed with the air in the enclosed space. Air with formaldehyde is vented from the enclosed space at the same volumetric flow rate as the incoming air.

(a) What is the steady-state concentration of formaldehyde (mg/m^3) in the enclosed space? (b) If the initial formaldehyde concentration is 1 mg/m^3 at $t = 0$, how long does it take for the formaldehyde concentration to drop to 0.3 mg/m^3?

(c) If the threshold for eye irritation due to formaldehyde exposure is 0.05 mg/m^3, what is the minimum flow rate of fresh air (m^3/h) necessary to prevent eye irritation due to formaldehyde exposure in the enclosed space?

13.11.3 Hematocrit Value

Blood with hematocrit H_1 flows from tank 1 to tank 2, when the clamp is released. Tank 2 initially contains blood plasma. Starting from conservation of species, using words, derive an expression for the hematocrit value in Tank 2 in terms of time t after the clamp is removed and in terms of the following parameters: the hematocrit value in tank 1 H_1, the initial flow rate $Q_{V0} = Q_V(0)$ from tank 1 to tank 2, the initial height of tank 2 $h_2(0)$, the cross-sectional areas of each tank A_1 and A_2, and the parameter α defined as:

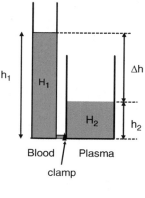

$$\alpha = \frac{\rho g}{\Re}\left[\frac{1}{A_1} + \frac{1}{A_2}\right],$$

where ρ is the density of blood (assumed to be the same as plasma), and $\Re$ is the flow resistance of the tubing segment between the two tanks. Assume $H_2(0) = 0$.

13.11.4 Hematocrit Value

For the mixing problem in problem 13.11.3, plot H_2 as a function of time for $0 \le t \le 600$ s using the following parameters:

Fluid densities: $\rho_1 = \rho_2 = \rho = 1{,}000$ kg/m^3
Tubing resistance: $\Re = 30$ (cm H$_2$O)/(ml/s)
Cross-sectional areas: $A_1 = 5$ cm^2, $A_2 = 10$ cm^2
Initial fluid levels: $h_1(t = 0) = 20$ cm, $h_2(t = 0) = 5$ cm
Hematocrit: $H_1(t = 0) = 40\%$, $H_2(t = 0) = 0\%$

13.11.5 Hematocrit Value

We wish to use the apparatus in problem 13.11.3 to mix blood in tank 1 at $H = 40\%$ with plasma in tank 2 so that the final composition in tank 2 has a hematocrit value of 18%. Find (a) the volume of blood that must pass from tank 1 to tank 2, (b) the time it takes for tank 2 to reach a hematocrit of 18%, and (c) Repeat (a) and (b) for desired hematocrit values of 10% and 21%. Discuss.

13.11.6 Hematocrit Value

Derive an expression for the hematocrit value in tank 2 in problem 13.11.3, if we account for different densities of blood and plasma. Will the final height of the two tanks be the same in this case? Explain.

13.11.7 Macroscopic Mass Transfer

A spherical cell with radius $R = 10$ μm contains a toxic substance "A." In an effort to remove the toxin, a biologist places the cell in a large beaker of a toxin-free isotonic solution.

The flux of toxin out of the cell is governed by the equation:

$$N_A = k_m \left(C_{A,\text{intracellular}}(R) - \Phi C_{A,\text{extracellular},\infty} \right), \quad k_m = 20 \ \mu\text{m/s}, \quad \Phi = 0.8,$$

where k_m is a mass transfer coefficient (permeability) for substance "A." Assume that concentration gradients inside and outside the cell are negligible. How long will it take for the cell to lose half of the toxin?

13.11.8 Steady-State Removal of a Toxin

Endothelial cells contain a surface enzyme that converts a toxin A into a harmless species at a rate that is proportional to the perfused surface area and the toxin concentration squared (conversion rate = $k'' S c_A^2$), where $k'' = 9 \times 10^5$ cm^4 mol^{-1} min^{-1} and $S = 60$ cm^2. Blood flow to the organ segment is 6 ml/min. Find the toxin concentration at the outlet of the organ segment if the inlet concentration is 0.01 mol/L. Assume the toxin distributes equally in plasma and red cells, and that the blood volume in the organ segment is well-mixed.

13.11.9 Toxic Waste

A toxic waste product A is to be removed from a fluid in a well-stirred holding tank with volume V by adding a quantity of material B to the tank. Material B reacts with A and the resulting compound C is nontoxic. The rate at which B reacts with A is proportional to the concentrations of both A and B. The constant of proportionality is k. Find the time required for the concentration of A to drop to one hundredth of its original concentration if the initial amount of A is 1 mole and B is continuously added to the tank to maintain 2 moles of B at all times. How might this time be reduced? $k = 1.1 \times 10^5 \text{ L}^2 \text{ mol}^{-1} \text{ min}^{-1}$, $V = 5{,}000$ L.

13.11.10 Toxic Waste

Repeat problem 13.11.9, with 2 moles of B added at $t = 0$ and no additional B is added after that.

13.11.11 Unsteady-State Mass Transfer from a Cell

A spherical cell is equilibrated with isotonic saline containing a 4 μM concentration of toxin A. After equilibration, the cell concentration is found to be 3 μM. At time $t = 0$, the same cell is immersed in a large volume of isotonic saline that does not contain toxin. The flux of toxin away from the outside surface of the cell is $N_A = kC_{As}$, where k is a mass transfer coefficient equal to 10^{-4} cm/s and C_{As} is the concentration of A in the saline at the cell surface. The cell radius is $R = 10$ μm and the diffusion coefficient of toxin in the cytoplasm is $D = 10^{-6}$ cm^2/s. The cell membrane is very thin, and its resistance to the flow of toxin A is negligible.

(a) What is the partition coefficient of toxin in the cell cytoplasm relative to saline?
(b) Estimate the resistance to the flow of toxin through the cytoplasm if the concentration gradient inside the cell is linear. Compare this to the resistance to toxin flow in the saline. Based on this comparison, is there a macroscopic approach for estimating toxin concentration in the cell justified? Explain why or why not.
(c) Assuming the macroscopic approach to be valid, how long will it take for half of the toxin to flow out of the cell?

13.11 Problems

13.11.12 Blood Flow and Tissue Volumes of White and Gray Matter in the Brain

^{85}Kr is used as a tracer to estimate blood flow and tissue volumes in the brain. The tracer is assumed to equilibrate instantly between blood and tissue compartments. Data collected downstream of the brain following a bolus injection of ^{85}Kr are shown below. Develop a compartmental model of tracer exchange and use the method of "peeling off exponentials" to estimate blood flow and tissue volumes of white matter (slow compartment) and gray matter (fast compartment). The following quantities are known from other experiments:

Total blood flow = 250 ml/min, Blood/Tissue partition coefficient = 0.8 for tracer in white matter relative to gray matter (i.e., at equilibrium $C_B = 0.8\, C_T$). Counts per second vs. time (min) are provided below:

Time (min)	CPS
0	1,000
0.25	716.8
0.5	521.6
0.75	386.5
1	292.4
1.25	226.2
1.5	179.3
1.75	145.6
2	121
2.25	102.6
2.5	88.7
2.75	77.9
3	69.2
3.25	62.2
3.5	56.3
3.75	51.3
4	46.9
4.25	43.1
4.5	39.7
4.75	36.6
5	33.8
5.25	31.3
5.5	29
5.75	26.8
6	24.9
6.25	23.1
6.5	21.4
6.75	19.8
7	18.4
7.25	17.1
7.5	15.8
7.75	14.7
8	13.6

13.11.13 Pulmonary Circulation

The right ventricle (V_1) is connected in series to the lung (V_2). The vascular compartment for each organ is assumed to be well-mixed and the flow F is constant. Ten percent of the flow is shunted directly to the left side of the heart without passing through the lung. Starting at $t = 0$, tracer is infused at a constant rate w (mmol/s) into the right ventricle and the concentration of the tracer is measured downstream of the lung-shunt junction. The tracer cannot leave the cardiopulmonary system and there is no recirculation of the tracer.

(a) Derive a differential equation for the concentration of tracer as a function of time $C(t)$, measured downstream of the lung-shunt junction.
(b) Show that the solution to part a is $C(t) = w/F + A \exp(-b_1 t) + B \exp(-b_2 t)$, where A and B are constants and $b_1 = 0.9\, F/V_1$ and $b_2 = 0.9\, F/V_2$.
(c) How might you make use of experimentally measured tracer concentration to estimate flow and compartment volumes? Be specific.

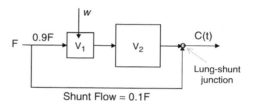

13.11.14 Compartmental Analysis

The right ventricle is connected in series with the lung. The vascular compartment of each organ is assumed to be well-mixed and the flow Q is constant. A mass m of tracer is injected as a bolus into the heart at $t = 0$ and the concentration of the tracer is measured downstream of the lung. If the tracer remains in the blood stream:

(a) Find the concentration of tracer measured downstream of the lung as a function of time.
(b) Find the area under the $C(t)$ curve.
(c) How might you make use of this tracer information to estimate flow and compartment volumes?

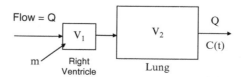

13.11.15 Cocaine Exchange

Develop a model of cocaine exchange across the placenta. Assume that the maternal and fetal circulatory systems are well-mixed compartments with the placenta serving as a barrier to cocaine transport. Use your model to predict fetal cocaine concentration in response to an iv bolus of 1 mg per kg of body weight in the maternal circulation. Make reasonable estimates of the circulatory volumes of mother and fetus, and assume biological half-lives of 1 h for cocaine in the maternal circulation and 1.35 h for cocaine in the fetal circulation. Consult the paper by Zhou et al. (2001) for experimental data in Rhesus monkeys.

13.11.16 Exchange in a Well-Mixed Hemodialyzer

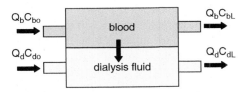

The blood and dialysis compartments of the hemodialyzer above are each modeled as well-mixed compartments. The barrier between them has a permeability-surface area product for urea of P_uS. When run in the cocurrent mode, as shown above, the concentration of urea at the blood compartment outlet is given as:

$$\frac{C_{bL} - \Phi C_{d0}}{C_{b0} - \Phi C_{d0}} = \frac{\alpha + \beta}{1 + \alpha + \beta}; \quad \text{where} \quad \beta = \frac{Q_b}{PS} \quad \text{and} \quad \alpha = \frac{Q_d}{\Phi Q_b}$$

and Φ is the partition coefficient for urea between blood and dialysis fluid.

(a) Use a macroscopic balance to show that the expression for blood concentration at the outlet of the hemodialyzer is the same when run in the counter-current mode, but C_{d0} is replaced by C_{dL}.
(b) Derive an expression for C_{d0} in terms of α, β, Φ, and the known concentrations C_{b0} and C_{dL}.
(c) Derive an expression for the exchange rate of urea across the membrane.

13.11.17 Compartmental Modeling

While light is applied to tissue in compartment 1, the tissue is stimulated to produce species A at a constant rate per unit volume R_A. If light is not applied, the production of species A ceases. If light is applied during the interval between

$t = 0$ and $t = t_1$, derive differential equations that can be used to describe concentrations in each compartment for $t < t_1$ and for $t > t_1$ (do not solve).

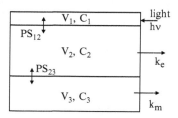

13.11.18 Flow Measurement

A method for measuring cardiac output is proposed. A known quantity of inert gas is dissolved in 100 ml of saline and at time $t = 0$ the mixture begins to flow under gravity (laminar flow) through an infusion tube into the vena cava. The tube is initially filled with saline. The blood and infused saline mix in the right heart, and enter the lung via the pulmonary artery. During passage through the lung, some of the inert gas diffuses through the alveolar membrane and is lost from the blood stream. There is no recirculation of the inert gas back into the vena cava. Blood in the lung, right heart, and left heart can be assumed to be well-mixed compartments, as can gas in the respiratory system.

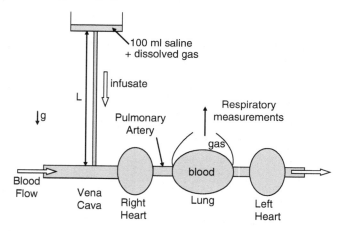

We can measure blood concentration of the inert gas in the right heart and in the left heart. In addition, we can measure respiratory flow and the inert gas as it is expired from the respiratory system.

(a) How would you use the measurements to estimate total blood flow (cardiac output)?

(b) How would you use the experimental data to estimate an overall lung mass transfer coefficient K_m for the inert gas that is defined in terms of the difference in partial pressures across the alveolar membrane: gas mass flow = $K_m(P_{gas,blood} - P_{gas,alvelolar})$?

(c) One potential matter of concern in making the above measurements is that all of the fluid in the saline bag might flow out of the bag before steady-state concentrations can be reached. As long as the fluid remains in the bag, the infusion rate Q_i can be assumed to be a constant. How would you go about estimating the time it should take for a steady-state to be established? What physiological variables might be needed to make this estimate?

(d) How would you use the time estimated in part c to restrict the diameter of the infusion tubing?

13.11.19 Flow Measurement

Tracer concentration measurements are made in a vein, which drains two well-mixed compartments. At $t = 0$ a mass of indicator m_i is injected into one of the compartments with known volume V_1. If tracer concentration is zero in the streams entering each compartment, show how the downstream concentration measurement $C(t)$ can be used to find flow through each compartment.

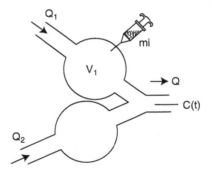

13.11.20 Oxygen Transport to a Bioartificial Organ

Half of a bioartificial organ is shown in the figure below. Cells are contained in a rectangular chamber of height $2\delta = 400$ μm and bounded by a membrane in contact with blood at $x = 0$ and $x = +2\delta$. The arterial and venous oxygen partial pressures in the blood are 95 mmHg and 40 mmHg, respectively, and the blood pH is 7.4. The cells are islets of Langerhans and can be presumed to be spherical with a diameter of 150 μm and an oxygen consumption rate of 25.9 μM/s. The device contains 750,000 islets and the fraction of the volume occupied by cells is 0.15.

The oxygen permeability of the membrane is 4×10^{-3} cm/s. Consider operation of this device at steady state and answer the following questions.

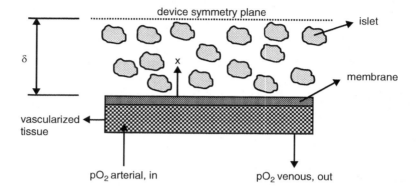

(a) What is the oxygen consumption rate μmols/s by all islets in this device?
(b) What fraction of the inlet oxygen is bound to hemoglobin?
(c) What is the fractional saturation of venous hemoglobin?
(d) What is the minimum total blood flow rate (cm³/min) required to provide adequate oxygen to the islets?
(e) What is the total volume of the device?

13.11.21 Physiological Shunt, Neglecting Dissolved Oxygen

Estimate physiological shunt in the lung by neglecting the contribution of dissolved oxygen. Measured partial pressures of oxygen (37°C) are:

- Systemic venous blood: 40 mmHg, pH = 7.4
- Systemic arterial blood: 90 mmHg, pH = 7.4
- Alveolar gas: 105 mmHg

13.11.22 Physiological Shunt, Including Dissolved Oxygen

Repeat problem 13.11.21 above when dissolved oxygen is not neglected and $\rho_{Hb} = 15$ g/100 ml blood (solubility of O_2 in blood = 0.003 mlO_2/100 ml blood). How much error is introduced by using the approximation in problem 13.11.21?

13.11 Problems

13.11.23 Oxygen Delivery to an Isolated Perfused Organ

An isolated organ at 37°C is perfused at a rate of 20 ml/s with an extracorporeal system containing blood with a pH of 7.4, a hematocrit value of 45%, and an arterial P_{O_2} of 95 mmHg. Red cell hemoglobin concentration is 34 g/dl. (a) What is the total delivery rate of oxygen to the organ? (b) What is the oxygen consumption rate if venous P_{O_2} is 60 mmHg and pH is 7.2?

13.11.24 Heart Muscle O_2 Consumption

Estimate the minimum hemoglobin concentration (g/dl) in blood at 37°C required to supply heart muscle with O_2 at pH = 7.4. Blood flow rate through the heart muscle is 750 ml/min and the O_2 consumption rate in this perfused tissue is 60 ml O_2/min. The $P_{O_2,in}$ and $P_{O_2,out}$ are 100 mmHg and 10 mmHg, respectively.

13.11.25 Oxygen Exchange and Organ Resistance for an Isolated Perfused Organ

The hind limb of a rat consists primarily of muscle with a volume of 2 ml. The muscle consumes oxygen at a rate of 0.2 ml O_2 min^{-1} ml^{-1}. The hind limb is perfused at 37°C with an extracorporeal system containing blood with a pH of 7.2, a hematocrit value of 45%, and an arterial P_{O_2} of 100 mmHg. Red cell hemoglobin concentration is 34 g/dl. What blood flow rate is necessary to maintain the venous P_{O_2} at 20 mmHg? Neglect dissolved oxygen.

13.11.26 Pharmacokinetics

N moles of drug A are injected as a bolus in the blood stream having volume V_b. The drug is removed from the blood stream by the following mechanisms:

1. Some drug binds very slowly and irreversibly with vascular endothelial cells at a constant rate Q_A(mol/s).
2. Some drug is eliminated by the kidney at a rate proportional to drug concentration, $k_e C_{Ab}$. Units of k_e are cm^3 s^{-1}.
3. Some drug passes across the endothelial barrier with permeability-surface area PS (ml/s) and is rapidly utilized by tissue, so tissue concentration can be assumed to be zero.

The minimum effective concentration of the drug is $0.1 \times N/V_b$. For how long after drug injection is the blood concentration at or above the MEC?

13.11.27 Absorption of Aspirin from the Gut

A 100 mg aspirin pill is taken orally at time zero. Assume that the pill is delivered immediately to the gut where it dissolves completely in a volume of 100 ml. The aspirin moves passively across the gut microvascular barrier ($PS = 0.5$ ml/min, as measured from gut to plasma) into the plasma (Volume $= 2,000$ ml) where it is eventually eliminated at a rate of (50 ml/min) × (plasma concentration). The plasma/chyme partition coefficient for aspirin is 0.7 (i.e., plasma concentration $= 0.7$ × chyme concentration under equilibrium conditions). How quickly after taking the aspirin can the person expect for a headache relief if the aspirin is effective only at plasma concentrations greater than 0.0031 mg/ml? How long will it take before the aspirin level drops back down below the effective level?

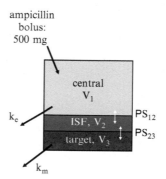

13.11.28 Pharmacokinetics of Ampicillin

The body is idealized as three fluid compartments separated by permeable membranes as shown in the figure. Five hundred milligrams of ampicillin are injected as a bolus into the central compartment. Ampicillin is eliminated by first-order kinetics from both the central compartment (k_e) and the target tissue compartment (k_m). The equations needed to solve for the distribution of ampicillin in the central compartment, interstitial fluid compartment (ISF), and the target tissue compartment are given by (13.243)–(13.245).

(a) Use Matlab or another ordinary differential equation solver to find and plot the concentrations as a function of time in each compartment using the following parameters:
$V_1 = 12$ L, $V_2 = 3.2$ L, $V_3 = 2.7$ L, $PS_{12} = 4$ L/h, $PS_{23} = 2$ L/h, $k_e V_1 = 2$ L/h, $k_m V_3 = 18$ L/h. Equilibrium: $\rho_{A2} = 0.7\, \rho_{A1}$, $\rho_{A3} = 1.1\, \rho_{A2}$. Compare your results with Fig. 13.29 and explain any differences.

(b) Use your model to examine the effects of systematically changing values for PS_{12}, PS_{23}, and k_e. Discuss your results.

13.11.29 Oral Drug Administration

A drug contained in chewing gum is released into the gut at a constant rate R mg/h, while it is chewed. Once the drug enters the gut (well mixed), some of it passes across the gut/blood barrier and the rest is swept out of the gut along with the outlet flow Q (ml/h). Drug that enters the blood can (1) pass across a second barrier and enter the tissue compartment, (2) be removed from the blood by kidney and liver at a rate of k_b times mass concentration of drug in the blood, or (3) pass back into the gut. Drug that enters the tissue can be irreversibly bound at a rate equal to k_t times the mass concentration of drug in tissue, or it can pass back into the blood stream.

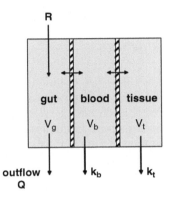

(a) Write differential equations that describe the rate of change of drug concentration in the gut, blood, and tissue.
(b) Assign reasonable values to each of the variables and use Matlab or another initial value ode solver to solve for concentrations in each compartment as a function of time.

13.11.30 Pharmacokinetics

A time-release skin patch releases drug "A" at a constant rate W_A (moles/s) into a small volume of skin, V_s. The drug is assumed to be well mixed in this volume. The drug passes from the skin into the blood stream through a microvascular barrier with permeability-surface area $(P_A S)_s$. The drug is removed from the blood stream via excretion at a rate proportional to blood concentration, $k_e C_b$. Some drug diffuses across the blood-tissue barrier with permeability-surface area $(P_A S)_T$ into the tissue with volume V_T, where it reacts with tissue components at a rate proportional to tissue concentration, $k_T C_T$. The partition coefficients for the drug between each compartment are 1.0.

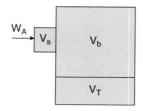

(a) Use species mass balances to derive ordinary differential equations for C_s, C_b, and C_T. (b) Provide expressions that can be used to compute these quantities a long time after the patch is in place.

13.11.31 *Pharmacokinetics*

Blood flows into an organ at a flow rate Q (ml/min). The organ is composed of a vascular space with volume V_1 and an extravascular space with volume V_2. At time $t = 0$, a drug D is introduced at a constant rate R (mg/min) at the inlet of the blood compartment. The drug is metabolized in the tissue compartment at a rate that is proportional to the concentration of drug in the tissue ($k_m C_2$). The barrier between blood and tissue compartments has a permeability-surface area product PS for the drug. The drug solubility is the same in tissue and blood.

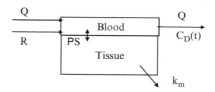

(a) Provide equations that can be used to describe the concentration of the drug in the two compartments.
(b) What initial conditions are necessary to solve for the concentrations?
(c) Derive an expression for the concentration of the drug emerging from the vascular space as a function of time.

13.11.32 *Pharmacokinetics*

A mass m_A of drug A ($m_A = M_A C_{A0} V_0$) is introduced as a *bolus* into the blood stream at time zero, where M_A is the molecular weight of A, C_{A0} is the concentration of A in the injectate, and V_0 is the volume of material injected. The drug does

13.11 Problems

not enter red cells, the plasma is assumed well-mixed, and the total plasma volume is V_P. The drug leaves the plasma by two mechanisms: (1) elimination by the kidney, with the rate of elimination being proportional to plasma concentration of the drug (kC_A); and (2) transport into the tissue through the microvascular barrier, which has permeability P and surface area S. Once the drug passes across the microvascular barrier, it is instantly utilized by tissue cells, so the tissue concentration of drug A can be assumed to be zero for all time.

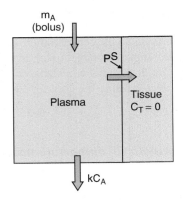

(a) Formulate a model that describes drug concentration in the plasma as a function of time.
(b) The drug concentration in the plasma must remain above a concentration C_{MEC} to be effective. For what time interval will the drug remain effective after its initial injection?

13.11.33 Pharmacokinetics

Drug A is continuously administered intravenously into compartment B with constant volume at a constant rate R. The drug is eliminated from compartment B to compartment C by a single mechanism with first-order kinetic rate constant k_2.

(a) Write the differential equation that describes the rate of change in the amount of drug in compartment B.
(b) Integrate the differential equation from part (a) to obtain an algebraic equation for the amount of drug in compartment B as a function of time. Presume that there is no drug in compartment B at $t = 0$.
(c) A patient receives a constant infusion of aminophylline at a rate of 45 mg/h. After 48 h of infusion, the patient's drug concentration in compartment B is 12 mg/L. The volume of compartment B is 5 L. What is the apparent first-order kinetic rate constant (h^{-1}) for drug elimination from compartment B?

13.11.34 *Pharmacokinetics*

Starting at time $t = 0$, drug A is infused into the plasma at a constant rate $w_{A,\text{inf}}$. The drug is not initially present in the body and does not react with plasma proteins. The drug is eliminated from the plasma by the following mechanisms: (1) it is eliminated by the kidney at a rate proportional to plasma concentration, (2) some diffuses into red cells, where it is rapidly converted into an inactive drug, (3) some diffuses into the target tissue, which metabolizes the drug at a rate proportional to the concentration in the tissue, and (4) some diffuses into nontarget tissue, where it is metabolized at a rate proportional to the concentration. (a) Construct a model of the pharmacokinetics of the drug in plasma, target tissue and nontarget tissue. Define all terms. (b) Solve the resulting differential equations. (c) Find the steady-state concentrations in each region after infusing the drug for a long period of time.

13.11.35 *Pharmacokinetics/Compartmental Analysis*

A bolus containing a soluble drug is injected into a feeding tube where it mixes with a volume of fluid in the stomach. The mixture moves at a constant flow rate through the digestive tract and some of the drug is passively absorbed into the bloodstream from the duodenum (just downstream of the stomach). Immediately after introducing the drug, water is infused into the feeding tube at the same rate that fluid flows out of the stomach.

The drug is eliminated from blood by the following two mechanisms: (1) clearance via the kidney and liver (1st order) and (2) passive diffusion from bloodstream to extravascular space of the whole body. Most cells in the body (including red cells and white cells) are impermeable to the drug. However, cells in the target organ have surface receptors that combine with the drug and transport the receptor–drug complex to the inside surface of the cell at a rate proportional to drug–receptor complex concentration. When the drug–receptor complex reaches the inside surface, it becomes immobile. The drug and receptors dissociate at the inside cell surface, with drug entering the cytoplasm and receptors returning to the outside cell surface at the same rate as they entered. (Note, the receptor–drug complex at the inside surface cannot return to the outside cell surface and receptors without drug cannot move to the inside surface. Also, surface reactions are rapid, so receptor and drug at each surface are in equilibrium.)

The internalized drug binds irreversibly with an intracellular species that is the source of the patient's symptoms. Enough of this toxic species must be neutralized to reduce its intracellular concentration to C_{safe}. Our objective is to estimate how much drug must be injected into the feeding tube in the initial bolus.

Develop a mathematical model of this biotransport system, being careful to define all your dependent variables and parameters. Associate equation(s) and initial conditions for each process described earlier. *Do not attempt to solve.*

13.11 Problems

As part of this model, be sure to include the following:

1. Derive an expression for the concentration of drug entering the duodenum as a function of time.
2. Derive an expression for the flux of drug into the cell, assuming carrier-mediated transport.

13.11.36 *Complex Chemotherapy Model*

Our goal is to construct a model that allows us to predict cell death in both tumor cells and normal cells after administration of doxorubicin. Use the information below to constrain your model. The number of unknown dependent variables should be equal to the number of equations you propose for your model. Explicitly list the unknown variables and number the equations that you would use to form the model. Also, list the parameters that need to be specified. *Do not attempt to solve the system of equations you develop.*

- The body is characterized as consisting of a central blood region with volume V_c, normal tissue, and tumor. The tissue and tumor regions each consist of three compartments: a vascular region, an interstitial region, and an extravascular cellular region. The mass of the tumor relative to the mass of normal tissue is known and equals f. The vascular, interstitial, and extravascular cellular volumes of the tissue region are V_v, V_I, and V_e, respectively. Tumor compartmental volumes are fV_v, fV_I, and fV_e, respectively. The surface area for normal tissue blood–interstitial fluid exchange is S and tumor surface area is fS.
- Fractional blood hematocrit is H in all vascular compartments.
- Doxorubicin (DOX) is not initially present in the body.
- At time $t = 0$ a known amount of DOX is injected as a bolus in the central blood compartment.
- DOX does not enter red blood cells.
- DOX combines with a plasma protein. Free DOX and protein-bound DOX are assumed to be in equilibrium with an equilibrium constant K_1.
- Proteins and protein-bound DOX cannot leave the blood stream.
- Free DOX in plasma is removed by liver and kidneys at a rate per unit volume equal to the product of k_e and free DOX plasma concentration.
- Free DOX can diffuse across the vascular-interstitial barriers in both the normal tissue and tumor regions. Normal tissue blood–interstitial fluid permeability to DOX is P_1 and tumor blood–interstitial permeability is nP_1, where n is much greater than one.
- DOX in the interstitial compartment does *not* combine with interstitial proteins. It can move in and out of the interstitial compartment by only two mechanisms: diffusion across the blood–interstitial barrier or carrier-mediated transport across cell membranes. The number of carriers per cell is different for tumor and normal cells.

- Intracellular DOX reacts irreversibly with an intracellular component X to form DOX-X with rate k_f. Intracellular concentration of X is constant.
- Cell death is proportional to the intracellular concentration of DOX-X.

13.11.37 Dissolution of a Sucrose Rod

A solid sucrose rod with diameter of 0.5 cm is immersed in water at 25°C, flowing with a velocity of 5 cm/s perpendicular to the axis of the rod. Estimate the time it takes for the rod to completely dissolve. The solubility of sucrose in water is 2.0 g/ml and $\rho = 1.587$ g/ml.

13.11.38 Two Pore System

Consider a 100-nm thick membrane that contains two pore populations. One population consists of 10^5 pores, each with a 20 nm radius. The second population of pores consists of 10^{13} pores with a radius of 2 nm. Globulin and albumin concentrations on each side of the membrane are maintained at the values specified in Examples 6.3.5.2 and 13.8.2.1. Find the total fluid flow and solute flows through the membrane for hydrostatic pressure differences of 20, 15, and 10 mmHg.

13.11.39 Steric Partition Coefficient

Consider the flow of a spherical solute with radius R_s through the space between two endothelial cells. The space can be considered a slit with height $2H$, length L, and width w. The width is much greater than H, so this is equivalent to the flow between two parallel plates. The solute center can come no closer to the walls of the slit than its radius. Derive an expression for the steric partition coefficient Φ_s between solute in the slit and solute in the bulk fluid at the entrance to the slit.

13.11.40 Solute Flow Through a Narrow Slit

Consider solute flow through the slit in problem 13.11.39. The concentrations of solute in plasma and interstitial fluid are C_{sp} and C_{si}. Hydrostatic and osmotic pressures in the bulk fluids at either end of the slit are known. Use methods described in Sect. 13.8 to show that that the same expression can be used to describe solute flow through a slit as was derived for solute flow through a pore, but the definitions of σ_s, the solute drag coefficient, and P_s, the permeability, must be modified. As with the case of flow in a pore, assume the velocity profile in the slit is parabolic. Assume the centerline drag coefficients K'_{10} and K'_{20} for a sphere flowing between parallel plates are valid throughout the core region.

13.11.41 Dual Tracer Study

Equal masses of two tracers are mixed together and injected into the blood stream of an organ as a bolus. The tracers are assumed to be well mixed in the intravascular region. Tracer A remains in the intravascular region, but tracer B can diffuse across the blood–tissue barrier and enters the extravascular space. Assume the permeability of the barrier to tracer B is very high, so the intravascular and extravascular concentrations of tracer B are in equilibrium at all times. Tracers A and B are labeled with different radioactive materials. The concentration of each tracer can be computed from radioactivity measured in blood at the outlet of the organ. In addition, radioactivity from a blood-free extravascular region is measured, so the extravascular concentration of tracer B can be computed as a function of time. Show how you would use the concentration measurements to estimate: (a) blood flow, (b) vascular volume, (c) extravascular volume, and (d) partition coefficient between blood and extravascular fluid.

13.12 Challenges

13.12.1 Maternal–Fetal Exchange Across the Placenta

Background: The circulatory systems of the mother and fetus are separate, but blood from both mother and fetus pass through the placenta, where exchange of blood gases, nutrients, waste products, and other chemicals occur. In some cases, drugs are purposely introduced into the maternal circulatory system, which is meant to benefit the fetus in some way. In other cases, drugs such as cocaine are injected into the maternal circulation and are inadvertently passed on to the fetus through the placenta. *Challenge:* Develop a compartmental model of maternal–fetal exchange of a substance like cocaine that can be used to predict fetal blood concentration of the drug after it is injected into the maternal blood stream. *Generate ideas:* What is the nature of the barrier between maternal and fetal blood? What are the important considerations in developing a compartmental model of maternal–fetal exchange? Perform a literature search to find important parameters that characterize exchange across the placenta.

13.12.2 *Pharmacokinetics of Aspirin*

Background: Aspirin, also known as acetylsalicylic acid (ASA), is taken orally to get relief from headache, particularly migraine headaches and other minor pain. *Challenge:* Develop a model of the pharmacokinetics of aspirin that can be used to estimate an appropriate oral dose for relief of a migraine headache for individuals with various body weights. *Generate ideas:* Where is ASA most likely to be absorbed in the GI system? How is the solubility of ASA in the stomach and small intestine influenced by pH? Does ASA combine with plasma proteins, and if so, is the ASA–protein compound effective, or is only the unbound ASA active? Is ASA carried by erythrocytes, or is it only present in plasma? How is ASA eliminated from the body? Does it appear to follow first-order kinetics? Consult the physiological and pharmacological literature to find minimum effective concentrations, maximum safe concentrations, permeability-surface area, physiological half life, and other information needed in the operation of your model.

13.12.3 *Chemotherapy*

Background: Successful chemotherapy selectively poisons cancer cells while inducing minimal damage to normal cells. Once inside cells, chemotoxins are often just as likely to poison a normal cell as to kill a cancer cell. Therefore, it is desirable to select a toxin that is taken up much more rapidly by cancer cells than by normal cells. Folate receptors are known to be over-expressed in cancer cells. Chemotoxins that can be linked to folic acid, such as doxorubicin, can combine with cell folate receptors and be internalized via endosomes. However, normal cells also have a small number of folate receptors on their cell surface, so high drug concentrations can kill normal cells. *Challenge:* Develop a model that will allow us to control the delivery of doxorubicin to the body such that the uptake by cancer cells is lethal, but the uptake by normal cells is minimal. *Generate ideas:* What compartments need to be considered in this model? What conservation equations need to be applied? What transport mechanisms are involved? Perform a literature search to learn more about how doxorubicin is internalized and how cell death is related to internal concentration of doxorubicin. Based on your literature findings and model results, what would be the minimum effective concentration and maximum safe concentration? What time course for intravenous delivery would you suggest?

References

Arancibia A, Guttmann J, Gonzalez G et al (1980) Absorption and disposition kinetics of amoxicillin in normal human-subjects. Antimicrob Agents Chemother 17:199–202

Bassingthwaighte JB, Knopp TJ, Anderson DU (1970) Flow estimation by indicator dilution (bolus injection) – reduction of errors due to time-averaged sampling during unsteady flow. Circ Res 27:277–291

References

Braunstein N, Braunstein M, Levinson GE, Frank MJ (1967) Studies of cardiopulmonary blood volume: measurement of left ventricular volume by dye dilution. Circulation 35:1038–1048

Cooper R, Ghali J, Simmons BE, Castaner A (1991) Elevated pulmonary artery pressure. An independent predictor of mortality. Chest 99:112–120

Dalla Costa T, Nolting A, Kovar A, Derendorf H (1998) Determination of free interstitial concentrations of piperacillin–tazobactam combinations by microdialysis. J Antimicrob Chemother 42:769–778

Jusko WJ, Lewis GP (1973) Comparison of ampicillin and hetacillin pharmacokinetics in man. J Pharm Sci 62:69–76

Paine PL, Scherr P (1975) Drag coefficients for the movement of rigid spheres through liquid-filled cylindrical pores. Biophys J 15:1087–1091

Renkin EM (1954) Filtration, diffusion, and molecular sieving through porous cellulose membranes. J Gen Physiol 38:225–243

Rippe C, Asgeirsson D, Venturoli D, Rippe A, Rippe B (2006) Effects of glomerular filtration rate on Ficoll sieving coefficients (θ) in rats. Kidney Int 69:1326–1332

Subczynski WK, Hyde JS, Kusumi A (1989) Oxygen permeability of phosphatidylcholine–cholesterol membranes. Proc Natl Acad Sci USA 86:4474–4478

Verniory A, Dubois R, Decoodt P, Gassee JP, Lambert PP (1973) Measurement of permeability of biological-membranes. J Gen Physiol 62:489–507

Zhou M, Song Z, Lidow MS (2001) Pharmacokinetics of cocaine in maternal and fetal rhesus monkeys at mid-gestation. J Pharmacol Exp Ther 297:556–562

Chapter 14
Shell Balance Approach for One-Dimensional Biomass Transport

14.1 Introduction

Applications of the macroscopic species conservation equation discussed in Chap. 13 are used extensively in biotransport. However, the macroscopic approach has important practical restrictions. It is limited to predicting concentrations, fluxes, or flows that are spatially averaged. If concentrations or fluxes have significant spatial variations, a different approach must be applied. Rather than apply the species conservation principle to the entire system, a microscopic portion of the system is analyzed. The resulting expression will be a differential equation that is valid at any position within the boundaries of the system. Boundary conditions that are specific to the problem at hand must be applied to find a solution for a particular system. Applications include axial variations of oxygen and carbon dioxide in capillaries, axial variations in salt concentration in the Loop of Henle, radial concentration variations of urea in tissue or hemodialyzers, solute concentration variations in porous microcapsules, etc.

In this chapter, we will deal primarily with steady-state mass transfer in situations when a concentration gradient develops in a single spatial dimension. We will treat a few important unsteady-state problems at the end of the chapter. A more general approach for 2D and 3D problems, steady and unsteady-state, will be considered in Chap. 15.

14.2 Microscopic Species Conservation

To find steady-state 1D spatial variations in concentration, we can apply a general procedure similar to that described for momentum transport in Sect. 6.2:

1. Define a microscopic volume or "shell" that lies within the system of interest. If species A flows in the z direction, then the shell is defined as the system volume between positions z and $z + \Delta z$. If transport is in the radial direction, then the shell will consist of material bounded by planes at r and $r + \Delta r$.
2. Apply the species conservation equation to species A within the shell.

3. Divide the resulting equation by the volume of the shell and let the volume approach zero. This results in a differential equation that applies to the flux of species A at any position within the system. In many cases, the resulting differential equation can be solved to provide an algebraic expression for the flux of species A in the material.
4. If convective flux is negligible, then $N_A = J_A$ and we can find concentration of species A by applying Fick's law to the flux solution found in step 3. Alternatively, we can substitute Fick's law for N_A into the differential equation derived in step 3 before integrating. If diffusion is negligible, then the flux is the product of the bulk concentration and average velocity, $N_A = C_{Ab}\langle v \rangle$. If convection and diffusion are both important, we must use the appropriate expression for N_A provided for liquids and gases in Sect. 12.4.5.
5. Apply appropriate boundary conditions to solve for unknown constants of integration. At this point, the concentration and flux will have been determined as a function of position within the system.
6. Total mass flow of species A into or out of the system can be found by integrating the known flux over the entire surface of the system.

We will apply this procedure to each problem considered in the remainder of the chapter, beginning with simple diffusion problems, and progressing to more complicated problems that include chemical reaction and convection.

14.3 One-Dimensional Steady-State Diffusion Through a Membrane

Let us begin with the analysis of diffusion of a waste product A through the wall of a spherical cell. The concentration at the inside surface of the cell membrane is C_{Ami} and the concentration on the outside surface of the cell membrane is C_{Amo}. Species A moves radially through the membrane by diffusion alone. The inside surface of the membrane is at $r = R_i$ and the outside surface is at $r = R_o$.

Our first step is to select a shell with surfaces perpendicular to the direction of flow. The shell must represent a small volume bounded by a surface at r and a surface at $r + \Delta r$, where $R_i \leq r \leq R_o$. That is, the shell must be within the membrane, as shown in Fig. 14.1.

Applying the species conservation equation to the shell:

$$\left\{ \begin{array}{c} \text{Rate of} \\ \text{accumulation} \\ \text{of species A} \\ \text{within shell} \end{array} \right\} = \left\{ \begin{array}{c} \text{Net rate species A} \\ \text{enters through} \\ \text{shell boundaries} \end{array} \right\} + \left\{ \begin{array}{c} \text{Rate species A} \\ \text{is produced} \\ \text{within the shell} \end{array} \right\}. \quad (14.1)$$

In the steady state, nothing changes with time; so the accumulation term is zero. In addition, species A is not involved in a chemical reaction within the system; so

14.3 One-Dimensional Steady-State Diffusion Through a Membrane

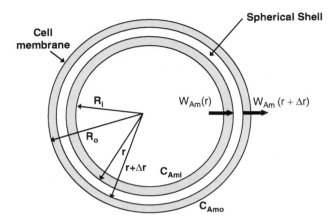

Fig. 14.1 Shell for transport through a cell membrane

the production term is zero. Consequently, molar flow of species A through the shell boundary at r must be balanced by molar flow out through the boundary at $r + \Delta r$:

$$\left\{ \begin{array}{c} \text{Net rate species A} \\ \text{enters through} \\ \text{shell boundaries} \end{array} \right\} = 0 = W_{Am}|_r - W_{Am}|_{r+\Delta r}. \qquad (14.2)$$

The symbol W_{Am} refers to the molar flow of species A through the membrane. Dividing by the shell volume ($4\pi r^2 \Delta r$) and taking the limit as the shell thickness Δr approaches zero, we obtain an equation that is valid at any single point r between $r = R_i$ and $r = R_o$:

$$0 = \lim_{\Delta r \to 0} \left\{ \frac{W_{Am}|_r - W_{Am}|_{r+\Delta r}}{4\pi r^2 \Delta r} \right\} = -\frac{1}{4\pi r^2} \frac{dW_{Am}}{dr}. \qquad (14.3)$$

Since r cannot be infinitely large, the only way that (14.3) can be zero is if:

$$\frac{dW_{Am}}{dr} = 0. \qquad (14.4)$$

Consequently, the molar flow of species A through the membrane is independent of radial position. Since molar flow is equal to molar flux multiplied by surface area, we can now introduce Fick's Law:

$$W_{Am} = 4\pi r^2 N_{Am} = 4\pi r^2 \left(-D_{Am} \frac{dc_{Am}}{dr} \right). \qquad (14.5)$$

Rearranging this expression for a constant diffusion coefficient, D_{Am}:

$$\frac{dc_{Am}}{dr} = -\frac{W_{Am}}{4\pi D_{Am} r^2}. \tag{14.6}$$

Integrating, we find the concentration as a function of radial position in terms of the unknown flow W_{Am} and a constant of integration C:

$$c_{Am} = \frac{W_{Am}}{4\pi D_{Am} r} + C. \tag{14.7}$$

Applying the boundary conditions at $r = R_i$ and $R = R_o$, and eliminating the constant of integration C, we find the flow of species through the membrane to be directly proportional to the concentration difference across the membrane:

$$W_{Am} = \frac{4\pi R_o R_i D_{Am}}{(R_o - R_i)}(C_{Ami} - C_{Amo}). \tag{14.8}$$

The concentration as a function of radial position in the membrane is:

$$\frac{C_{Am}(r) - C_{Ami}}{C_{Amo} - C_{Ami}} = \frac{R_o}{r}\frac{(r - R_i)}{(R_o - R_i)}. \tag{14.9}$$

However, since the cell membrane is very thin, R_o/r for all values of r in the membrane is nearly equal to one, and $R_o - R_i$ is the membrane thickness δ_m. Therefore, the concentration profile is very close to being linear from the inside to outside surface of the membrane:

$$\frac{C_{Am}(r) - C_{Ami}}{C_{Amo} - C_{Ami}} \approx \frac{r - R_i}{\delta_m}. \tag{14.10}$$

This same linear relationship will be valid for cells with nonspherical geometry, as long as the membrane thickness is small relative to the cell size. The flow of species A through the membrane from (14.8) for $R_i \approx R_o$ will be:

$$W_{Am} = \frac{4\pi R_i^2 D_{Am}}{\delta_m}(C_{Ami} - C_{Amo}) = \left(\frac{D_{Am}}{\delta_m}\right)S(C_{Ami} - C_{Amo}), \tag{14.11}$$

where S is the surface area of the cell membrane. The ratio of the diffusion coefficient for species A in the membrane to membrane thickness is the membrane permeability, P_{Am}. In addition, the membrane is assumed to be in equilibrium with the cytoplasm and extracellular fluid at the inside and outside surfaces, respectively. Written in terms of the fluid concentrations, rather than membrane concentrations, (14.11) becomes:

$$W_{Am} = P_{Am}S(\Phi_{Aim}C_{Ai} - \Phi_{Aom}C_{Ao}). \tag{14.12}$$

14.3 One-Dimensional Steady-State Diffusion Through a Membrane

Finally, from (12.153), $\Phi_{Aoi} = \Phi_{Aom}/\Phi_{Aim}$, so this can be written as:

$$W_{Am} = P_{A,cell} S(C_{Ai} - \Phi_{Aoi} C_{Ao}), \tag{14.13}$$

where the cell permeability to species A, $P_{A,cell}$, is the product of the membrane permeability and the cytoplasm-membrane partition coefficient, $P_{Am}\Phi_{Aim}$. The transport of species A across a membrane is more often computed from (14.13) than (14.11) because the concentrations within the membrane at the inside and outside surface are not generally known. Experimental measurements of cellular permeability will also include effects of any unstirred layers near the inside and outside surfaces of the membrane.

> **Example 14.3.1 Diffusion Through a Heterogeneous Barrier.**
> Consider diffusion of species A from one well-mixed tank (tank 1) to another (tank 2) through a wall composed of two different materials B and C. We wish to determine how the molar concentration of species A varies with position from one side of the wall to the other. In addition, we would like to determine whether the resistance to mass transfer is different when the two materials are placed in series than when they are placed in parallel. The barrier thickness L and the amount of each material used are the same in each case. The two cases are illustrated in Fig. 14.2.

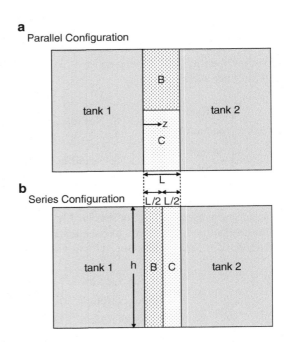

Fig. 14.2 (a) Parallel and (b) series barriers

Solution. *Initial considerations*: We will assume that species A is dilute, flux is in the z-direction only, and convective flux of species A through either material B or material C is negligible. In addition, species A is not produced or removed in materials B or C. Finally, we will assume that the flux of A through each material is steady.

System definition and environmental interactions: The composite wall is the ultimate system to be analyzed. We will need to divide it into two systems since the wall consists of two different materials. The system interactions will be different for the parallel and series arrangements. If the materials are in parallel, each material will be exposed to the same fluid concentrations at the two fluid–solid interfaces. We will assume local equilibrium between the fluid and solid at each interface. In the series configuration, only one end of each material is in contact with the fluid. The other interface is the contact surface between materials B and C. At that surface we will assume local equilibrium. The other condition that must be applied at the solid–solid interface is that the flux of species A out of material B must be the same as the flux of species A into material C. Although our ultimate goal is to determine the flux of species A through a macroscopic system, the composite wall, the systems selected for the initial analysis must be microscopic systems consisting of very small shells of thickness Δz within each material. This will allow us to determine how the concentration of species A varies with position in each material.

Apprising the problem to identify governing relationships: We will apply conservation of species A to a microscopic shell located within material B and to another shell located within material C. Since convection is negligible, we can use Fick's law to relate the flux in each material to the concentration gradient. The resulting conservation relationships can be integrated and the boundary conditions applied to determine constants of integration.

Analysis: Consider first, the parallel flow case. We begin by selecting a shell within material B between z and $z + \Delta z$. We then apply conservation of species A within the shell. In words:

$$\left\{\begin{array}{c} \text{Rate of} \\ \text{accumulation} \\ \text{of species A} \\ \text{within shell} \end{array}\right\} = \left\{\begin{array}{c} \text{Net rate species A} \\ \text{enters through} \\ \text{shell boundaries} \end{array}\right\} + \left\{\begin{array}{c} \text{Rate species A} \\ \text{is produced} \\ \text{within the shell} \end{array}\right\}.$$

In the steady state, nothing changes with time; so the accumulation term is zero. In addition, species A is not involved in a chemical reaction within the barriers; so the production term is zero. Consequently, molar flow of species A through the shell boundary at z must be balanced by molar flow out through the boundary at $z + \Delta z$:

$$\left\{\begin{array}{c} \text{Net rate species A} \\ \text{enters through} \\ \text{shell boundaries} \end{array}\right\} = 0 = W_{AB}|_z - W_{AB}|_{z+\Delta z}.$$

14.3 One-Dimensional Steady-State Diffusion Through a Membrane

The symbol W_{AB} refers to the molar flow of species A through material B. Dividing by the shell volume $(w/2)h\Delta z$ and letting the shell volume shrink to zero:

$$0 = \lim_{\Delta z \to 0} \left[\frac{W_{AB}|_z - W_{AB}|_{z+\Delta z}}{(w/2)h\Delta z} \right] = -\frac{1}{(w/2)h} \frac{dW_{AB}}{dz} = -\frac{dN_{AB}}{dz}.$$

N_{AB} is the molar flux of species A in material B. This expression tells us that N_{AB} is independent of z. Since molar flow is steady and one dimensional, N_{AB} must be constant. Fick's Law for 1D diffusion through a solid material (12.82) is:

$$J_{AB} = N_{AB} = -D_{AB} \frac{dc_{AB}}{dz}.$$

Integrating this from $z = 0$ to $z = L$:

$$N_{AB} = \frac{D_{AB}}{L}(c_{AB}(0) - c_{AB}(L)).$$

If fluid in each tank is well mixed, the concentrations in material B at $z = 0$ and $z = L$ are related through partition coefficients to the bulk concentrations in tanks 1 and 2 as follows:

$$c_{AB}(0) = \Phi_{AB1} C_{A1},$$
$$c_{AB}(L) = \Phi_{AB2} C_{A2}.$$

Consequently, the flux of species A through material B is related to known quantities as follows:

$$N_{AB} = \frac{D_{AB}}{L}(\Phi_{AB1} C_{A1} - \Phi_{AB2} C_{A2}) = \frac{D_{AB}}{L \Phi_{A1B}}(C_{A1} - \Phi_{A12} C_{A2}).$$

To find the concentration distribution in material B, we integrate Fick's law between $z = 0$ and $z = z$, and use the expression above for N_{AB}.

$$\frac{C_{A1} - \Phi_{A1B} c_{AB}(z)}{C_{A1} - \Phi_{A12} C_{A2}} = \frac{z}{L}.$$

Therefore, the concentration varies linearly from one end of material B to the other. By analogy, the flux and concentration distribution for species A in material C are:

$$N_{AC} = \frac{D_{AC}}{L \Phi_{A1C}}(C_{A1} - \Phi_{A12} C_{A2})$$

and

$$\frac{C_{A1} - \Phi_{A1C}\, c_{AC}(z)}{C_{A1} - \Phi_{A12}\, C_{A2}} = \frac{z}{L}.$$

The total molar flow of species A through the composite barrier composed of materials B and C configured in parallel is:

$$W_A = \frac{hw}{2}[N_{AB} + N_{AC}] = \left(\frac{hw}{2}\right)\left(\frac{D_{AB}\Phi_{AB1} + D_{AC}\Phi_{AC1}}{L}\right)(C_{A1} - \Phi_{A12}C_{A2})$$

(parallel).

Note that $hw/2$ is the surface area of either barrier; so the second term on the right represents the overall mass transfer coefficient or permeability of the composite membrane in the parallel configuration.

Turning now to the series configuration (Fig. 14.2b), integration of Fick's law for each material yields:

$$N_{AB} = \frac{2D_{AB}}{L}\left(c_{AB}(0) - c_{AB}\left(\frac{L}{2}\right)\right) = \frac{2D_{AB}}{L}\left(\Phi_{AB1}C_{A1} - c_{AB}\left(\frac{L}{2}\right)\right),$$

$$N_{AC} = \frac{2D_{AC}}{L}\left(c_{AC}\left(\frac{L}{2}\right) - c_{AC}(L)\right) = \frac{2D_{AC}}{L}\left(c_{AC}\left(\frac{L}{2}\right) - \Phi_{AC2}C_{A2}\right).$$

In the steady state, two conditions must be satisfied at the interface, $z = L/2$ (Sect. 12.10.4):

$$N_{AB} = N_{AC},$$

$$c_{AB}\left(\frac{L}{2}\right) = \Phi_{ABC}\, c_{AC}\left(\frac{L}{2}\right).$$

Substituting these boundary conditions into the expressions above, we find the interfacial concentration and total flux to be:

$$c_{AC}\left(\frac{L}{2}\right) = \frac{D_{AC}\Phi_{AC2}C_{A2} + D_{AB}\Phi_{AB1}C_{A1}}{D_{AC} + D_{AB}\Phi_{ABC}},$$

$$N_A = N_{AC} = N_{AB} = \frac{2D_{AB}D_{AC}\Phi_{AB1}}{L(D_{AC} + D_{AB}\Phi_{ABC})}\left(C_{A1} - \frac{\Phi_{ABC}\Phi_{AC2}}{\Phi_{AB1}}C_{A2}\right).$$

Using the definition of Φ_{Aij}, the flux expression can be further reduced to:

$$N_A = \frac{2D_{AB}D_{AC}\Phi_{AB1}}{L(D_{AC} + D_{AB}\Phi_{ABC})}(C_{A1} - \Phi_{A12}C_{A2}).$$

14.3 One-Dimensional Steady-State Diffusion Through a Membrane

The total molar flow of species A for the series configuration is:

$$W_A = (wh)\left(\frac{2D_{AB}D_{AC}\Phi_{AB1}}{L(D_{AC} + D_{AB}\Phi_{ABC})}\right)(C_{A1} - \Phi_{A12}C_{A2}) \quad \text{(series)}.$$

The second term represents the permeability or overall mass transfer coefficient for the parallel configuration. Comparing the flow of species A through the series barrier to flow through the parallel barrier:

$$\frac{W_{A,\text{series}}}{W_{A,\text{parallel}}} = \frac{4}{\left(1 + \dfrac{D_{AB}\Phi_{ABC}}{D_{AC}}\right)\left(1 + \dfrac{D_{AC}}{D_{AB}\Phi_{ABC}}\right)}.$$

Examining and interpreting the results: Note that the flow ratio is independent of C_{A1}, C_{A2}, and Φ_{A12}. This ratio depends only on a single dimensionless parameter, $\dfrac{D_{AC}}{\Phi_{ABC}D_{AB}}$. This is plotted on a log scale in Fig. 14.3.

Results of our analysis indicate that the parallel arrangement of barriers B and C allows the greater mass transfer. Consider the case where the diffusion coefficient for species A is almost zero for material C, but is non-zero for material B. In the parallel arrangement, there would be essentially no mass flow through material C, but significant mass flow through material B. However, in the series case, material C would significantly resist mass flow through the entire barrier. The series to parallel mass flow ratio would be nearly zero, as reflected in Fig. 14.3 for very small

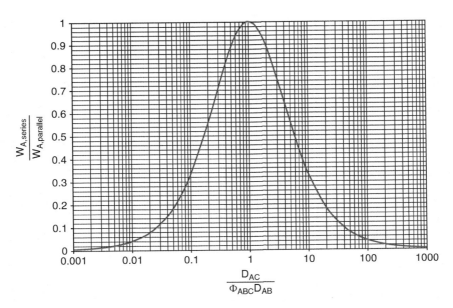

Fig. 14.3 Ratio of flow through barriers in series relative to barriers in parallel

values of $\dfrac{D_{AC}}{\Phi_{ABC} D_{AB}}$. If we reversed the diffusion coefficients of the two materials, material B would limit flow through the series barrier, but species A would still move through material C in the parallel arrangement. Thus the series to parallel mass flow ratio for a high value of $\dfrac{D_{AC}}{\Phi_{ABC} D_{AB}}$ would also approach zero, as shown in Fig. 14.3. Similarly, if the diffusion coefficients D_{AB} and D_{AC} are nearly equal, and the partition coefficient Φ_{ABC} is very large or very small, flow will be significantly reduced through the series barrier, but will only be cut in half for barriers in parallel. If the barriers have the same partition coefficients and the same diffusion coefficients, then the two configurations are equivalent, and the mass flow ratio will be unity.

14.4 1D Diffusion with Homogeneous Chemical Reaction

The application of species conservation using the shell balance approach is very useful for solving steady-state, one-dimensional problems. In this section, we treat problems that involve simultaneous diffusion and homogeneous chemical reaction for several different geometries.

14.4.1 Zeroth Order Reaction

Let us begin our analysis by analyzing the consumption of oxygen by a cell. We will model this as one-dimensional transport through the cell. In addition, we will assume that the cell consumes oxygen at a constant rate per unit volume, Q_{O_2}. The rate of production of oxygen per unit volume has the same magnitude as the rate of consumption of oxygen, but the opposite sign:

$$R_{O_2} = -Q_{O_2}. \tag{14.14}$$

We will consider three different cell shapes: rectangular, cylindrical, and spherical.

14.4.1.1 Rectangular Shaped Cell

Consider a columnar epithelial cell having a rectangular shape with length $2L$, height h, and width w. The cell consumes oxygen at a constant rate per unit volume Q_{O_2}. The concentrations at the two ends of the cell are maintained constant at C_L. If we take a coordinate system with an origin at the center of the cell, then the oxygen concentration will be symmetrical about $x = 0$, since the flux of oxygen will be zero at the center of the cell. We therefore only need to solve for the oxygen

14.4 1D Diffusion with Homogeneous Chemical Reaction

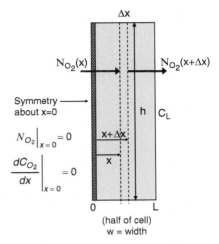

Fig. 14.4 Shell for analysis of a consuming cell of length 2 L

concentration between $x = 0$ and $x = +L$, as shown in Fig. 14.4. Applying conservation of oxygen to the shell between x and $x + \Delta x$:

$$\left\{\begin{array}{c}\text{Rate of}\\ \text{accumulation}\\ \text{of } O_2\end{array}\right\} = \left\{\begin{array}{c}\text{Rate at}\\ \text{which } O_2\\ \text{enters shell}\end{array}\right\} - \left\{\begin{array}{c}\text{Rate at}\\ \text{which } O_2\\ \text{leaves shell}\end{array}\right\} + \left\{\begin{array}{c}\text{Rate of}\\ O_2 \text{ production}\\ \text{within shell}\end{array}\right\}. \quad (14.15)$$

We are interested in the steady-state movement of oxygen into the consuming cell; so the rate of accumulation will be zero. The molar rate at which oxygen enters the shell is $W_{O_2}|_x$ and the rate at which oxygen leaves the shell $W_{O_2}|_{x+\Delta x}$. The molar conservation equation is:

$$0 = W_{O_2}|_x - W_{O_2}|_{x+\Delta x} - Q_{O_2}(wh\Delta x). \quad (14.16)$$

where $wh\Delta x$ is the volume of the shell and the product of $-Q_{O_2}$, and the shell volume is the rate of production of oxygen in the shell. The molar rate W_{O_2} is equal to the molar flux N_{O_2} multiplied by the area perpendicular to the direction of movement, hw. Consequently, the species conservation equation for oxygen becomes:

$$0 = wh\left(N_{O_2}|_x - N_{O_2}|_{x+\Delta x}\right) - Q_{O_2}(wh\Delta x). \quad (14.17)$$

Dividing (14.17) by the shell volume and letting the volume approach zero, we derive an ordinary differential equation that describes the conservation of oxygen at any point in the rectangular cell:

$$0 = \lim_{\Delta x \to 0} \frac{\left(N_{O_2}|_x - N_{O_2}|_{x+\Delta x}\right)}{\Delta x} - Q_{O_2} = -\frac{dN_{O_2}}{dx} - Q_{O_2}. \quad (14.18)$$

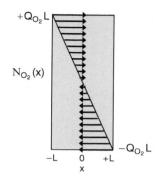

Fig. 14.5 Oxygen flux in rectangular shaped cell

Integrating this equation and applying the boundary condition at $x = 0$ that $N_{O_2} = 0$:

$$N_{O_2} = -Q_{O_2} x. \tag{14.19}$$

Therefore, the flux becomes progressively more negative as x becomes more positive and the flux becomes progressively more positive as x becomes more negative. At $x = -L$, the flux is $+Q_{O_2}L$ and at $x = +L$, the flux is $-Q_{O_2}L$. This indicates that the flux at each boundary is directed inward toward the cell interior, as shown in Fig. 14.5.

The total inward flow is equal to the flux at each surface multiplied by the surface area:

$$W_{O_2,\text{tot}} = 2hLwQ_{O_2}. \tag{14.20}$$

Thus, the total inflow of oxygen is exactly equal to the rate at which oxygen is consumed by the cell. If the flux into the cell is by diffusion alone, then we can introduce Fick's law for N_{O_2} in (14.19):

$$-D_{O_2,\text{cell}} \frac{dC_{O_2}}{dx} = -Q_{O_2} x. \tag{14.21}$$

The general solution to (14.21) is:

$$C_{O_2} = \frac{Q_{O_2}}{D_{O_2,\text{cell}}} \frac{x^2}{2} + C_1. \tag{14.22}$$

The constant C_1 is to be determined by applying the boundary condition that at $x = L$, $C_{O_2} = C_L$:

$$C_1 = C_L - \frac{Q_{O_2}}{D_{O_2,\text{cell}}} \frac{L^2}{2}. \tag{14.23}$$

14.4 1D Diffusion with Homogeneous Chemical Reaction

Thus, the concentration in the cell varies with position as follows:

$$C_{O_2} = C_L - \frac{Q_{O_2}L^2}{2D_{O_2,\text{cell}}}\left(1 - \frac{x^2}{L^2}\right). \tag{14.24}$$

The concentration profile has a parabolic shape. We can find the location of the minimum concentration by setting the derivative of C_{O_2} with respect to x equal to zero. This will show that the concentration is a minimum at $x = 0$. Let us examine the solution as we raise the consumption rate. Figure 14.6 shows that as the consumption rate is raised, the concentration of oxygen at the center of the cell drops.

Note that at a consumption rate of 0.2×10^{-8} moles O_2 ml^{-1} s^{-1}, the oxygen concentration near the center of the cell is predicted to go below zero. Is this possible? No. Negative concentrations are physically unrealistic. So what did we do wrong?

When a model predicts a physically unrealistic result, we need to go back and check to make sure our unit conversions are correct, that we integrated properly, that we used realistic values for our parameters, that we did not make a math error, and that our original assumptions were correct. In this case, we must conclude that one of our model assumptions is inappropriate when the oxygen concentration falls to zero. At that point there is not any oxygen available for consumption and therefore the consumption rate cannot remain constant. Consequently, two different

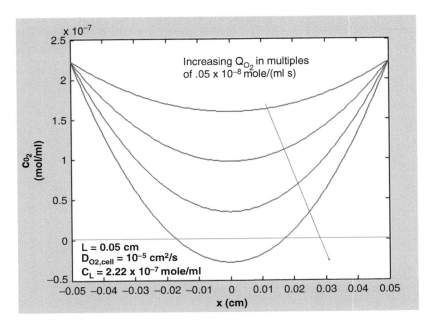

Fig. 14.6 Concentration of oxygen in a consuming cell. Effect of increased consumption rate

models of oxygen consumption in the cell must be applied at high consumption rates. The model we have developed above is appropriate as long as the oxygen concentration is above zero. However, once the concentration drops to zero, the consumption rate must also drop to zero. So a second model of oxygen transport must be developed for the central region of the cell. The procedure would be exactly the same as that followed above, except we would set $Q_{O_2} = 0$. According to (14.19), the flux of oxygen throughout the central region would be $N_{O_2} = 0$, and integration of (14.21) with the boundary condition that the concentration is zero at the edge of the central region leads to the solution that $C_{O_2} = 0$ everywhere in the central region. This is the expected result. Now, how do we find out when we need to break the problem into two regions, and if we must, where is the boundary between the two regions located?

If the oxygen consumption rate is constant and oxygen concentration is greater than zero everywhere in the cell, then our original model is valid. However, if the consumption rate is held constant and the half length of the cell L is increased, a maximum cell half length $L_{\max}$ will occur when the oxygen concentration at the center of the cell just drops to zero. This can be found by setting the concentration at $x = 0$ equal to zero in (14.24):

$$L_{\max} = \sqrt{\frac{2D_{O_2,\text{cell}} C_L}{Q_{O_2}}}. \qquad (14.25)$$

If the actual cell thickness is greater than $2L_{\max}$, then the cell must be split into three regions as shown in Fig. 14.7: a core region, which is devoid of oxygen between $-L_{\text{crit}} < x < L_{\text{crit}}$ and two peripheral regions, one in which the concentration drops from C_L at $x = -L$ to zero at $x = -L_{\text{crit}}$ and a second where the

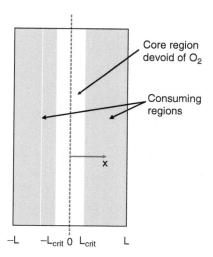

Fig. 14.7 Cell with a central core region devoid of oxygen and two peripheral regions with a constant consumption rate per unit volume

14.4 1D Diffusion with Homogeneous Chemical Reaction

concentration drops from C_L at $x = L$ to zero at $x = L_{crit}$. How can we find $\pm L_{crit}$, the locations in the cell interior where the oxygen concentration just falls to zero? We can find L_{crit} by applying a shell balance to the region $L_{crit} \leq x \leq L$. This leads to (14.18). Integrating (14.18) between x and L_{crit}, and applying the boundary condition that $N_{O_2} = 0$ at $x = L_{crit}$, we find the following expression for the flux of oxygen:

$$N_{O_2}(x) = -Q_{O_2}(x - L_{crit}) = -D_{O_2,cell} \frac{dC}{dx}. \tag{14.26}$$

Integrating (14.26) between $x = L_{crit}$, where $C_{O_2} = 0$, and $x = L$, where $C_{O_2} = C_L$, yields the following expression:

$$C_L = \frac{Q_{O_2}}{2D_{O_2,cell}} (L - L_{crit})^2. \tag{14.27}$$

Solving for L_{crit} and taking the root that ensures that $L > L_{crit}$, we find:

$$L_{crit} = L - \sqrt{\frac{2D_{O_2,cell} C_L}{Q_{O_2}}} = L - L_{max}. \tag{14.28}$$

Because of the symmetry in this problem, if $L > L_{max}$, then the cell will be devoid of oxygen in the region between $-L_{crit} < L < L_{crit}$:

$$C_{O_2}(x) = 0 \text{ if } L \geq \sqrt{\frac{2D_{O_2,cell} C_L}{Q_{O_2}}} \text{ and } -L + \sqrt{\frac{2D_{O_2,cell} C_L}{Q_{O_2}}} \leq x \leq L - \sqrt{\frac{2D_{O_2,cell} C_L}{Q_{O_2}}}. \tag{14.29}$$

Since L_{crit} is now known, we can find the concentration in the peripheral region $L_{crit} \leq x \leq L$ by integrating (14.26) and applying the boundary condition that $C_{O_2} = 0$ at $x = L_{crit}$:

$$L_{crit} \leq x \leq L : C_{O_2} = \left(\frac{Q_{O_2} L_{crit}^2}{2D_{O_2,cell}}\right) \left(\frac{x^2}{L_{crit}^2} - 2\frac{x}{L_{crit}} + 1\right). \tag{14.30}$$

Applying a shell balance to the region between $-L$ and $-L_{crit}$, we find:

$$N_{O_2}(x) = -Q_{O_2}(x + L_{crit}) = -D_{O_2,cell} \frac{dC}{dx}. \tag{14.31}$$

The concentration in the peripheral region $-L \leq x \leq -L_{crit}$ can be found by integrating (14.31) and applying the boundary condition that the concentration of oxygen is zero at $x = -L_{crit}$:

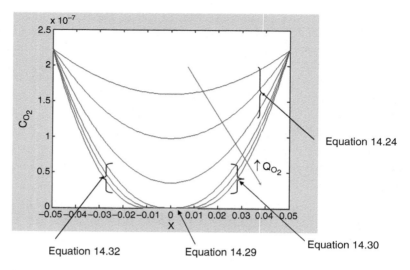

Fig. 14.8 Predicted oxygen concentration in a rectangular shaped cell

$$-L \leq x \leq -L_{\text{crit}} : C_{O_2} = \left(\frac{Q_{O_2} L_{\text{crit}}^2}{2D_{O_2,\text{cell}}}\right) \left(\frac{x^2}{L_{\text{crit}}^2} + 2\frac{x}{L_{\text{crit}}} + 1\right). \quad (14.32)$$

The solution as Q_{O_2} increases is shown in Fig. 14.8. When $L_{\max} > L$, a transition occurs between the single region solution (14.24), and the three region solution given by (14.29), (14.30), and (14.32). When this occurs, note that the oxygen concentration and the oxygen flux are both zero at the edges of the core region. If $L > L_{\max}$, the total flow of oxygen into the cell is found by multiplying the fluxes at $x = L$ from (14.26) and $x = -L$ from (14.31) by their respective surface areas (wh) and adding them together: $W_{O_2} = 2Q_{O_2}(L - L_{\text{crit}})wh$. As expected, this is exactly equal to the consumption rate per unit volume Q_{O_2} multiplied by the volume of consuming tissue, $2(L - L_{\text{crit}})wh$.

14.4.1.2 Cylindrical Cell

Next, let us analyze oxygen transfer to a cylindrically shaped muscle cell. The cell has a radius R, Length L, concentration C_R at $r = R$, and constant consumption rate per unit volume, Q_{O_2}. A cross section of the cell is shown in Fig. 14.9.

The flux of oxygen will be in the negative radial direction. We therefore select a shell with length L that lies within the cell with surfaces that are perpendicular to the radial direction and are a very small distance Δr apart. Our motivation for selecting such a shell is that we wish to apply conservation of oxygen to the shell, then let Δr shrink to zero so the resulting equation applies at any location

14.4 1D Diffusion with Homogeneous Chemical Reaction

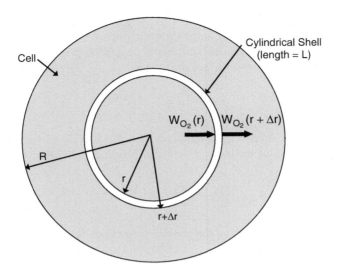

Fig. 14.9 Cylindrical cell and shell selection

r within the cell. Applying the species conservation equation (14.15) to the shell in Fig. 14.9:

$$0 = W_{O_2}|_r - W_{O_2}|_{r+\Delta r} - Q_{O_2} V_{\text{shell}}, \tag{14.33}$$

where the shell volume V_{shell} is:

$$V_{\text{shell}} = \pi(r + \Delta r)^2 L - \pi r^2 L = 2\pi r L \Delta r + \pi(\Delta r)^2 L \approx 2\pi r L \Delta r. \tag{14.34}$$

As the shell thickness approaches zero, the term involving $(\Delta r)^2$ becomes negligible. Dividing (14.33) by the shell volume in (14.34), and taking the limit as Δr approaches zero yields:

$$\frac{1}{2\pi r L} \lim_{\Delta r \to 0} \left\{ \frac{W_{O_2}|_{r+\Delta r} - W_{O_2}|_r}{\Delta r} \right\} = -Q_{O_2} = \frac{1}{2\pi r L} \frac{dW_{O_2}}{dr}. \tag{14.35}$$

We need a boundary condition before we can solve (14.35). We know the concentration at $r = R$, but this will not help solve for the oxygen flow. However, since this problem is symmetrical about $r = 0$, the flux of oxygen must be zero at the center of the cell. Therefore, W_{O_2} is zero at $r = 0$. Applying this boundary condition, we can integrate (14.35) to obtain:

$$W_{O_2}(r) = -\pi r^2 L Q_{O_2}. \tag{14.36}$$

Thus the oxygen flow vs. radial position has a parabolic shape. The flow of oxygen at $r = R$ is found to be $-Q_{O_2}(\pi R^2 L)$, which is simply the product of the consumption rate per unit volume and the volume of the cell. The minus sign indicates that oxygen is flowing in the direction opposite to the positive r direction or toward the cell center. The oxygen flux at any radial position can be found by dividing (14.36) by the local surface area, $2\pi rL$:

$$N_{O_2}(r) = \frac{W_{O_2}}{2\pi rL} = -\left(\frac{Q_{O_2}}{2}\right)r. \tag{14.37}$$

So the flux varies linearly with radial position. Finally, we can apply Fick's law for the case where all of the transport is by diffusion:

$$-D_{O_2}\frac{dC_{O_2}}{dr} = -\left(\frac{Q_{O_2}}{2}\right)r. \tag{14.38}$$

Integrating (14.38) and using the boundary condition that $C_{O_2}(R) = C_R$:

$$C_{O_2} = C_R - \left(\frac{Q_{O_2}R^2}{4D_{O_2}}\right)\left(1 - \frac{r^2}{R^2}\right). \tag{14.39}$$

As with transport in the rectangular epithelial cell, the shape of the concentration profile in the cylindrical muscle cell is parabolic, similar to that shown in Fig. 14.6. The maximum radius $R_{\max}$ that the cell can have and still be able to supply oxygen to all portions of the cell is:

$$R_{\max} = \sqrt{\frac{4D_{O_2}C_R}{Q_{O_2}}}. \tag{14.40}$$

Compare this with the maximum half length of a rectangular shaped cell with the same concentration at the outside surface of the cell, the same diffusivity, and the same oxygen consumption rate:

$$R_{\max} = \sqrt{2}L_{\max}. \tag{14.41}$$

Therefore, a cylindrically shaped cell can have its surface farther away from its center than a rectangular cell without depriving portions of the cell from oxygen. If the cell radius is larger than $R_{\max}$, we must divide the cell into a core region which is devoid of oxygen, and a peripheral region which contains oxygen, as was done for the epithelial cell.

Let us take an alternative approach to solving this problem. Rather than sequentially solving for W_{O_2}, N_{O_2}, and C_{O_2}, we begin by substituting (14.37), which relates oxygen flow to oxygen flux, into the shell balance, (14.33):

$$\lim_{\Delta r \to 0}\left\{\frac{(2\pi rLN_{O_2})|_{r+\Delta r} - (2\pi rLN_{O_2})|_r}{2\pi rL\Delta r}\right\} = -Q_{O_2}. \tag{14.42}$$

14.4 1D Diffusion with Homogeneous Chemical Reaction

Great care must be exercised in taking the limiting process. Many students want to cancel the $2\pi rL$ terms from numerator and denominator. This is fine for $2\pi L$, which is a true constant, but the radial position in the first term of the numerator, $r + \Delta r$, is different than the radial term in the denominator r. To see this more clearly, let us expand the two terms in the numerator:

$$\lim_{\Delta r \to 0} \left\{ \frac{(2\pi rLN_{O_2}(r+\Delta r) + 2\pi \Delta rLN_{O_2}(r+\Delta r)) - (2\pi rLN_{O_2}(r))}{2\pi rL\Delta r} \right\} = -Q_{O_2}, \quad (14.43)$$

where $N_{O_2}(r + \Delta r)$ is interpreted as the oxygen flux evaluated at $r = r+\Delta r$ and $N_{O_2}(r)$ is the oxygen flux evaluated at r. We can now cancel terms from the numerator and the denominator to find:

$$\lim_{\Delta r \to 0} \left\{ \frac{(N_{O_2}(r+\Delta r)) - (N_{O_2}(r))}{\Delta r} + \frac{N_{O_2}(r+\Delta r)}{r} \right\} = -Q_{O_2}. \quad (14.44)$$

Now we can carry out the limiting process to obtain:

$$\frac{dN_{O_2}}{dr} + \frac{N_{O_2}}{r} = \frac{1}{r}\frac{d}{dr}[rN_{O_2}] = -Q_{O_2}. \quad (14.45)$$

Introducing Fick's law for the flux of oxygen, we obtain a second-order ordinary differential equation for oxygen concentration:

$$\frac{1}{r}\frac{d}{dr}\left[r\frac{dC_{O_2}}{dr}\right] = \frac{Q_{O_2}}{D_{O_2,\text{cell}}}. \quad (14.46)$$

This can be integrated twice to obtain:

$$C_{O_2} = \frac{r^2 Q_{O_2}}{4D_{O_2,\text{cell}}} + A\ln(r) + B, \quad (14.47)$$

where the constants of integration A and B are to be determined using the two boundary conditions of (1) known concentration C_R at $r = R$ and (2) zero flux at $r = 0$ ($-D_{O_2,\text{cell}}\, dC_{O_2}/dr = 0$). Consequently, we find:

$$A = 0, B = C_R - \frac{R^2 Q_{O_2}}{4D_{O_2,\text{cell}}}. \quad (14.48)$$

Substituting these back into (14.47) provides the same expression for $C_{O_2}(r)$ as was given by (14.39).

14.4.1.3 Spherical Cell

Many cells in the human body, such as leukocytes, have a nearly spherical shape. If we consider oxygen diffusion in a spherical cell with a constant rate of consumption per unit volume, the shell will consist of planes at r and $r + \Delta r$, similar to those for a cylindrical shell. The steady-state species conservation equation (14.16) becomes:

$$0 = \left.(4\pi r^2 N_{O_2})\right|_r - \left.(4\pi r^2 N_{O_2})\right|_{r+\Delta r} - Q_{O_2}\left[\frac{4}{3}\pi(r+\Delta r)^3 - \frac{4}{3}\pi(r)^3\right]. \quad (14.49)$$

The first term on the right represents the flow of oxygen into the shell at radial position r, the second term represents the flow of oxygen out through the surface at $r + \Delta r$, and the third term is the rate of consumption per unit volume multiplied by the volume of the shell. The oxygen flows have been written in terms of oxygen fluxes multiplied by the surface area of the shell. The shell volume is equal to the volume of a sphere at $r + \Delta r$ minus the volume of a sphere at r. After expanding the terms in square brackets and neglecting terms containing $(\Delta r)^2$ and $(\Delta r)^3$, we find the volume of the shell can be approximated as the product of the surface area and the thickness of the shell, $V_{\text{shell}} = 4\pi r^2 \Delta r$. Substituting this for the volume difference in (14.49) and rearranging:

$$0 = \lim_{\Delta r \to 0}\left\{\frac{\left.(r^2 N_{O_2})\right|_r - \left.(r^2 N_{O_2})\right|_{r+\Delta r}}{r^2 \Delta r}\right\} - Q_{O_2}. \quad (14.50)$$

For reasons outlined in the previous section, the r^2 terms in the numerator and denominator do not cancel. Taking the limit as Δr approaches zero, but with r remaining constant:

$$\frac{1}{r^2}\frac{d}{dr}\left(r^2 N_{O_2}\right) = -Q_{O_2}. \quad (14.51)$$

Substituting Fick's law for the flux:

$$\frac{1}{r^2}\frac{d}{dr}\left(r^2 \frac{dC_{O_2}}{dr}\right) = \frac{Q_{O_2}}{D_{O_2,\text{cell}}}. \quad (14.52)$$

Integrating twice:

$$C_{O_2} = \frac{Q_{O_2}}{6 D_{O_2,\text{cell}}}r^2 - \frac{A}{r} + B. \quad (14.53)$$

Applying the boundary conditions of (1) known concentration C_R at $r = R$ and (2) zero flux at $r = 0$, we find:

$$A = 0, B = C_R - \frac{R^2 Q_{O_2}}{6 D_{O_2,\text{cell}}}. \quad (14.54)$$

14.4 1D Diffusion with Homogeneous Chemical Reaction

The final concentration profile in the spherical cell is:

$$C_{O_2} = C_R - \left(\frac{Q_{O_2} R^2}{6 D_{O_2}}\right)\left(1 - \frac{r^2}{R^2}\right). \tag{14.55}$$

Note the similarities between the solutions for the rectangular shaped cell (14.24), the cylindrically shaped cell (14.39), and the spherically shaped cell (14.55). All have parabolic concentration profiles. The only difference is the value of the constant in the denominator.

The largest radius the cell can have before a zero concentration core region must be included can be found by setting the concentration at the center equal to zero:

$$R_{\max} = \sqrt{\frac{6 D_{O_2} C_R}{Q_{O_2}}}. \tag{14.56}$$

This is larger than the maximum half length of a rectangular shaped cell (14.25) and is larger than the maximum radius of a cylindrical cell (14.40). Thus, if oxygen delivery to the cell center is the primary factor limiting cell survival and oxygen consumption rate per unit volume is a constant, we should be able to estimate the maximum cell size for various shaped cells. Taking the following parameter estimates for oxygen delivery to cells:

- $D_{O_2} = 10^{-5}$ cm^2 s^{-1}
- $Q_{O_2} = 1.7 \times 10^{-8}$ mol ml^{-1} s^{-1}
- $C_R = C_L = 9.4 \times 10^{-7}$ mol ml^{-1}

Substitution of these values into (14.25), (14.40), and (14.56) provides the following estimates of maximum cell size:

Rectangular cell: $L_{\max} = 332$ μm
Cylindrical cell: $R_{\max} = 470$ μm
Spherical cell: $R_{\max} = 576$ μm

The ovum is the largest cell in the human body with a diameter of about 1 mm or a radius of 500 μm. This agrees fairly well with the predicted maximum size for a spherical cell. The predicted value was based on the assumption that the oxygen consumption rate was constant. In the next section, we will examine the case where the consumption rate is first order.

14.4.2 First-Order Reaction

Let us return to the treatment of oxygen delivery to a rectangular cell, but consider the case where the consumption rate per unit volume is first order instead of zeroth order:

$$Q_{O_2} = -R_{O_2} = k C_{O_2}. \tag{14.57}$$

Using the shell shown in Fig. 14.4, the conservation of oxygen is given by:

$$0 = wh\left[N_{O_2}|_x - N_{O_2}|_{x+\Delta x}\right] - kC_{O_2}(wh\Delta x). \tag{14.58}$$

Dividing by the shell volume and taking the limit as the volume approaches zero, we obtain:

$$0 = -\frac{dN_{O_2}}{dx} - kC_{O_2}. \tag{14.59}$$

Inserting Fick's law for the flux, we arrive at an ordinary differential equation for the oxygen concentration:

$$\frac{d^2C_{O_2}}{dx^2} = \frac{k}{D_{O_2,\text{cell}}}C_{O_2}. \tag{14.60}$$

This equation cannot be solved by separating variables followed by integration. Let us try a solution of the form:

$$C_{O_2} = Ae^{\lambda x}. \tag{14.61}$$

Substituting (14.61) into (14.60) leads to the following quadratic equation:

$$\lambda^2 = \frac{k}{D_{O_2,\text{cell}}}. \tag{14.62}$$

Therefore, there are two solutions for λ:

$$\lambda_1 = \sqrt{\frac{k}{D_{O_2,\text{cell}}}}, \quad \lambda_2 = -\sqrt{\frac{k}{D_{O_2,\text{cell}}}}. \tag{14.63}$$

The general solution for the concentration is:

$$C_{O_2} = A_1 e^{\lambda_1 x} + A_2 e^{\lambda_2 x}, \tag{14.64}$$

where A_1 and A_2 are constants to be determined from the boundary conditions. Since the flux is zero at the center of the cell and the diffusion coefficient is not zero, then:

$$\left.\frac{dC_{O_2}}{dx}\right|_{x=0} = 0 = \lambda_1 A_1 + \lambda_2 A_2 = \lambda_1(A_1 - A_2). \tag{14.65}$$

14.4 1D Diffusion with Homogeneous Chemical Reaction

Therefore, the coefficients A_1 and A_2 are equal. Substituting this back into (14.64):

$$C_{O_2} = A_1 \left(e^{\lambda_1 x} + e^{\lambda_2 x} \right) = A_1 \left(e^{\lambda_1 x} + e^{-\lambda_1 x} \right) = 2A_1 \cosh\left(x \sqrt{\frac{k}{D_{O_2,\text{cell}}}} \right), \quad (14.66)$$

where $\cosh(z)$ is the hyperbolic cosine of z. Applying the boundary condition at $x = L$ or $x = -L$ provides a relationship for the unknown coefficient A_1:

$$C_L = 2A_1 \cosh\left(L \sqrt{\frac{k}{D_{O_2,\text{cell}}}} \right). \quad (14.67)$$

Dividing (14.66) by (14.67) gives the final solution for first-order kinetics:

$$\frac{C_{O_2}}{C_L} = \frac{\cosh\left(x \sqrt{\frac{k}{D_{O_2,\text{cell}}}} \right)}{\cosh\left(L \sqrt{\frac{k}{D_{O_2,\text{cell}}}} \right)}. \quad (14.68)$$

Since the reaction rate decreases as the oxygen concentration decreases, the concentration at the center of the cell never actually reaches zero:

$$C_{O_2}(x=0) = \frac{C_L}{\cosh\left(L \sqrt{\frac{k}{D_{O_2,\text{cell}}}} \right)}. \quad (14.69)$$

The concentration profiles for a thin cell and a thick cell, each with the same first-order consumption rate, are shown in Fig. 14.10.

For a cylindrical cell with first-order kinetics, the species conservation shell balance becomes:

$$0 = (2\pi r L N_{O_2})|_r - (2\pi r L N_{O_2})|_{r+\Delta r} - k C_{O_2} 2\pi r \Delta r L. \quad (14.70)$$

Taking the limit as the shell volume goes to zero:

$$0 = -\frac{1}{r} \frac{d}{dr}(r N_{O_2}) - k C_{O_2}. \quad (14.71)$$

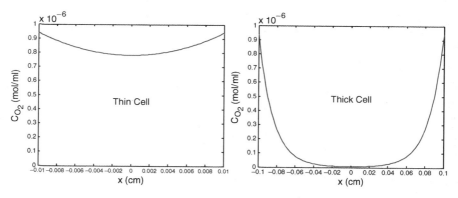

Fig. 14.10 Comparison of concentration profiles in a thin cell (0.02 cm) and a thick cell (0.2 cm) for the same values of $k = 0.04$ s^{-1}, $D_{O_2,\text{cell}} = 10^{-5}$ cm^2/s, and $C_L = 9.4 \times 10^{-7}$ mol/ml

Substituting Fick's law:

$$0 = \frac{1}{r}\frac{d}{dr}\left(r\frac{dC_{O_2}}{dr}\right) - \frac{k}{D_{O_2,\text{cell}}}C_{O_2}. \tag{14.72}$$

If we introduce a dimensionless radial position r^*:

$$r^* = r\sqrt{\frac{k}{D_{O_2,\text{cell}}}}, \tag{14.73}$$

then (14.72) becomes:

$$\frac{d^2 C_{O_2}}{dr^{*2}} + \frac{1}{r^*}\frac{dC_{O_2}}{dr^*} - C_{O_2} = 0. \tag{14.74}$$

This equation is known as the modified Bessel equation and is commonly encountered in problems dealing with cylindrical coordinates. The general solution is:

$$C_{O_2} = AI_0(r^*) + BK_0(r^*), \tag{14.75}$$

where A and B are constants and I_0 and K_0 are functions of r^* known as modified zeroth order bessel functions of the first kind and second kind, respectively. Both functions oscillate in a manner similar to sine or cosine functions, but neither the amplitude nor the frequency of oscillations are constant for bessel

14.4 1D Diffusion with Homogeneous Chemical Reaction

functions. The bessel function K_0 becomes infinitely large as the argument approaches zero. Since the concentration at $r = 0$ must be finite, the coefficient B must be zero. The coefficient A can be found by applying the boundary condition at the outside radial position. Substituting (14.73) back for r^*, the final solution is:

$$\frac{C_{O_2}}{C_L} = \frac{I_0\left(r\sqrt{\frac{k}{D_{O_2,\text{cell}}}}\right)}{I_0\left(R\sqrt{\frac{k}{D_{O_2,\text{cell}}}}\right)}. \tag{14.76}$$

The bessel function $I_0(x)$ is available as besseli $(0, x)$ in Matlab. Plots of (14.76) for cells with radii the same as the half widths in Fig. 14.10 are shown in Fig. 14.11.

If the substance is being generated within the cell according to the first-order kinetics, rather than being consumed, then the sign preceding the last term in (14.74) would be positive. The resulting equation is known as the Bessel equation of zeroth order. The solution when a species such as CO_2 is produced by a first-order chemical reaction would have the same form as (14.76), but with the ordinary bessel function J_0 replacing the modified bessel function I_0.

How does the concentration profile for a cylindrical cell compare with that of a spherical cell with the same radius? The appropriate species continuity equation can be obtained by replacing Q_{O_2} in (14.52) with kC_{O_2}:

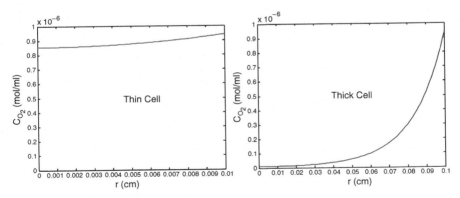

Fig. 14.11 Comparison of concentration profiles in a thin cylindrical cell ($R = 0.01$ cm) and a thick cell ($R = 0.1$ cm) for the same values of $k = 0.04$ s^{-1}, $D_{O_2,\text{cell}} = 10^{-5}$ cm^2/s, and $C_L = 9.4 \times 10^{-7}$ mol/ml

$$\frac{1}{r^2}\frac{d}{dr}\left(r^2\frac{dC_{O_2}}{dr}\right) = \frac{k}{D_{O_2,\text{cell}}}C_{O_2}. \tag{14.77}$$

After some manipulation, this equation can be rewritten as:

$$\frac{d^2}{dr^2}(rC_{O_2}) = \frac{k}{D_{O_2,\text{cell}}}(rC_{O_2}). \tag{14.78}$$

The student can show that the two equations are equivalent. If we let $f = rC_{O_2}$, (14.78) becomes:

$$\frac{d^2 f}{dr^2} = \frac{k}{D_{O_2,\text{cell}}}f. \tag{14.79}$$

This equation has the same form as (14.60) which was derived for the rectangular cell. The general solution is:

$$f = rC_{O_2} = A_1 \exp\left(r\sqrt{\frac{k}{D_{O_2,\text{cell}}}}\right) + A_2 \exp\left(-r\sqrt{\frac{k}{D_{O_2,\text{cell}}}}\right). \tag{14.80}$$

The condition of zero flux at the center of the cell requires that the coefficient A_2 be equal to $-A_1$. Therefore, we can write:

$$C_{O_2} = \frac{A_1}{r}\left\{\exp\left(r\sqrt{\frac{k}{D_{O_2,\text{cell}}}}\right) - \exp\left(-r\sqrt{\frac{k}{D_{O_2,\text{cell}}}}\right)\right\} = \frac{2A_1}{r}\sinh\left(r\sqrt{\frac{k}{D_{O_2,\text{cell}}}}\right). \tag{14.81}$$

The boundary condition that $C_{O_2} = C_R$ at $r = R$ can be used to find A_1. The final solution for the spherical cell with the first-order kinetics is:

$$\frac{C_{O_2}}{C_R} = \left(\frac{R}{r}\right)\left\{\frac{\sinh\left(r\sqrt{\frac{k}{D_{O_2,\text{cell}}}}\right)}{\sinh\left(R\sqrt{\frac{k}{D_{O_2,\text{cell}}}}\right)}\right\}. \tag{14.82}$$

The solution is shown in Fig. 14.12 for the same cell radii as shown for the cylindrical cells in Fig. 14.11. Concentrations in the spherical cell are slightly higher than in the cylindrical cell with the same radius.

14.4 1D Diffusion with Homogeneous Chemical Reaction

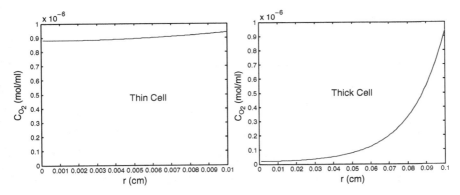

Fig. 14.12 Comparison of concentration profiles in a thin spherical cell ($R = 0.01$ cm) and a thick cell ($R = 0.1$ cm) for the same values of $k = 0.04$ s^{-1}, $D_{O_2,\text{cell}} = 10^{-5}$ cm^2/s, and $C_L = 9.4 \times 10^{-7}$ mol/ml

14.4.3 Michaelis–Menten Kinetics

Most cellular reactions are regulated by enzymes and the resulting rate of conversion follows Michaelis–Menten kinetics. The rate of utilization of the substrate is equal in magnitude and opposite in sign to the rate of production of product. According to (13.193):

$$R_P = -R_S = V_{\max}\left(\frac{C_S}{K_m + C_S}\right). \tag{14.83}$$

Substituting this for the substrate oxygen in a shell balance for a rectangular cell, the species conservation equation (14.15) becomes:

$$0 = wh\left(N_{O_2}|_x - N_{O_2}|_{x+\Delta x}\right) - V_{\max}\left(\frac{C_{O_2}}{K_m + C_{O_2}}\right)(wh\Delta x). \tag{14.84}$$

Taking the limit as the shell volume approaches zero:

$$\frac{dN_{O_2}}{dx} = -V_{\max}\left(\frac{C_{O_2}}{K_m + C_{O_2}}\right). \tag{14.85}$$

This is coupled to Fick's law, which can be rearranged as follows:

$$\frac{dC_{O_2}}{dx} = -\frac{N_{O_2}}{D_{O_2,\text{cell}}}. \tag{14.86}$$

The solution to the coupled set of first-order equations (14.85) and 14.86 must be determined numerically using a boundary value solver such as bvp4c in Matlab. The Matlab code is given below.

```
% parameters
L=.04;              % cell thickness, cm
Vmax = 1.7e-8;      % O2 consumption rate mole/(ml s)
D = 1e-5;           % O2 diffusivity (cm^2/s)
Km=4e-7;            % Michaelis constant (moles/ml)
options = [];
% Boundary Conditions
CL=9.4e-7;          % concentration at x=L
J0=0;               % flux at x=0
%Initial guess:
N_guess = -.0002e-6;
C_guess =6e-8;
% Initialize
  solinit = bvpinit(linspace(0,L,10), [N_guess C_guess]);
% Solve ode (BVP)
  sol = bvp4c(@M_M_ode,@M_M_bc,solinit,options,L,Vmax,D,Km,J0,CL);
% Plot flux
  x = linspace(0,L);
  y = deval(sol,x);
  figure(1);
  plot(x,y(1,:));
% Plot concentration
  figure(2);
  plot(x,y(2,:));

function dydx = M_M_ode(x,y,L,Vmax,D,Km,J0,CL)
% y(1) = NO2 = oxygen flux
% y(2) = CO2 = oxygen concentration
% odes:
% dNO2/dx = -[Vmax*CO2/(Km+CO2)] (conservation of mass)
% dCO2/dx = -NO2/D (Fick's law)
dydx = [-Vmax*y(2)/(Km+y(2))
        -y(1)/D];

function res = M_M_bc(ya,yb,L,Vmax,D,Km,J0,CL)
res = [ ya(1)       % BC1: at x = 0: J =0 (ya(1) = 0)
        yb(2)-C0]   % BC2: at x = L:C = C0 (yb(2) - C0 = 0
```

The effect of changing K_m and $Q_{max} = V_{max}$ are shown in Fig. 14.13. Increasing Q_{max} causes an increase in consumption rate and a decrease in concentration, while increasing K_m decreases consumption rate and increases oxygen concentration.

14.4.4 Diffusion and Reaction in a Porous Particle Containing Immobilized Enzymes

Some bioreactors function by immobilizing specific enzymes responsible for converting substrate(s) to product(s) within porous particles. Substrates are converted to product by enzymes which are not themselves altered by the reaction. Specific enzymes can be selected for converting toxic target substances in blood into harmless products. It is advantageous to immobilize the enzymes so that they

14.4 1D Diffusion with Homogeneous Chemical Reaction

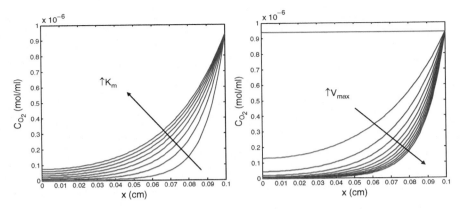

Fig. 14.13 Michaelis–Menten kinetics in a rectangular cell with half length 0.1 cm. *Left*: effect of increasing K_m in increments of 2×10^{-7} mol/ml while holding Q_{max} at 1.7×10^{-8} mol ml^{-1} s^{-1}. *Right*: effect of increasing Q_{max} in increments of 0.3×10^{-8} mol ml^{-1} s^{-1} while holding $K_m = 4 \times 10^{-7}$ mol/ml

remain inside the bioreactor. This can introduce significant savings, since the isolation of cells and the purification of enzymes are expensive processes.

In Sect. 14.6.3, we will analyze the conversion of a toxic material in a continuous feed bioreactor. One of the key components of the analysis involves the diffusion and reaction of the toxic substance within the pores of the particle. An exact analysis would require detailed knowledge of the particle geometry, the nature of the porous pathways, and the distribution of enzymes within the particles. Let us assume that the particles are spherical in shape, that pores are uniformly distributed in the sphere, and that the enzymes are homogenously distributed along the pore walls. Transport of substrate is assumed to be by diffusion alone. Because the pores are tortuous and the substrate interacts with the pore walls, the effective diffusivity of substrate within the pores will be less than the diffusion coefficient in free solution. The effective diffusivity for solute s, D_{se}, is proportional to free diffusion coefficient and the porosity of the particle (pore volume relative to particle volume), and inversely proportional to the tortuosity of the pores. Radial flux through the particle pores can be expressed in terms of a modification of Fick's Law, which accounts for restricted diffusion in the porous particle:

$$N_{sp} = -D_{se} \frac{dC_{sp}}{dr}. \qquad (14.87)$$

Let us perform a shell balance for steady-state transport of species s within the pores of one of the spherical particles, as shown in Fig. 14.14:

$$0 = N_{sp}S\big|_r - N_{sp}S\big|_{r+\Delta r} + R_{sp}V. \qquad (14.88)$$

Fig. 14.14 Porous particle with immobilized enzymes

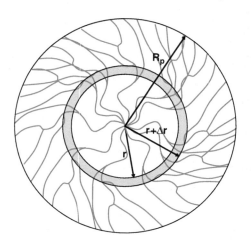

S_r and $S_{r+\Delta r}$ are the surface area of pores on the particle at radial positions r and $r + \Delta r$, respectively, V is the volume of pores within the particle between r and $r + \Delta r$, respectively, and R_{sp} is the rate of production of species s per unit pore volume. Let the fraction of the particle occupied by pores be ϕ_p and let this porosity be uniform at all radial positions. Then, at any radial position, r, the pore surface area and volume will be:

$$S = 4\pi\phi_p r^2, \quad V = 4\pi\phi_p r^2 \Delta r. \tag{14.89}$$

Substituting these expressions into the species shell balance, and letting $\Delta r \to 0$, yields:

$$-\frac{d}{dr}\left(r^2 N_{sp}\right) + r^2 R_{sp} = 0. \tag{14.90}$$

Note that this expression is independent of the particle porosity. Immobilized enzyme located along the surface of the pores within the particle irreversibly converts the toxic substance s to a nontoxic product as follows:

$$S + E \underset{k_{rs}}{\overset{k_{fs}}{\rightleftarrows}} ES \overset{k_p}{\to} E + P. \tag{14.91}$$

If the reaction follows Michaelis–Menten kinetics, as discussed in Sect. 12.8.3, the rate of production of species s per unit volume is:

$$R_{sp} = -\frac{V_{\max} C_{sp}}{K_m + C_{sp}}. \tag{14.92}$$

14.4 1D Diffusion with Homogeneous Chemical Reaction

C_{sp} is the species concentration within the pores. R_{sp} is negative in this case because species s is lost by chemical reaction. If species s were produced by chemical reaction, the sign in (14.92) would be positive. The Michaelis constant K_m (12.226) is

$$K_m = \frac{C_{sp}C_{Ep}}{C_{Esp}} = \frac{k_{rs} + k_p}{k_{fs}}. \tag{14.93}$$

C_{Ep} is the number of moles of enzyme per unit pore volume. Enzymes are assumed to be uniformly distributed along the surface of the pores such that the initial number of moles per unit pore volume is C_{Ep0}. The maximum rate of conversion of species s to product is V_{max}, and this rate will occur at high C_{sp} when the enzyme is saturated with species s:

$$V_{max} = k_p C_{Ep0}. \tag{14.94}$$

Substituting (14.87) and (14.92) into (14.90), we arrive at the following ordinary differential equation for the transport of species s within the pores of the particle:

$$D_{se} \frac{d}{dr}\left(r^2 \frac{dC_{sp}}{dr}\right) = r^2 \left(\frac{V_{max} C_{sp}}{K_m + C_{sp}}\right). \tag{14.95}$$

This must be solved subject to the following boundary conditions at the particle center, $r = 0$, and the particle outside boundary, $r = R_p$:

$$r = 0, \quad \frac{dC_{sp}}{dr} = 0,$$
$$r = R_p, \quad C_{sp} = C_{sp}(R_p). \tag{14.96}$$

Let us introduce the following dimensionless independent and dependent variables which should range from 0 to 1:

$$r^* = \frac{r}{R_p},$$
$$C_{sp}^* = \frac{C_{sp}}{C_{sp}(R_p)} = \frac{C_{sp}}{\Phi_{spf} C_{sf}}. \tag{14.97}$$

$C_{sp}(R_p)$ is the concentration of species s just inside the pores at $r = R_p$. This is equal to the concentration of species s in the fluid film C_{sf} just outside the pores multiplied by the partition coefficient Φ_{spf} for species s between the pores and the film. Now let us define two dimensionless parameters:

$$\beta = \frac{C_{sp}(R_p)}{K_m}, \tag{14.98}$$

$$\phi_T = \frac{R_p}{3}\sqrt{\frac{V_{max}}{D_{se}K_m}}. \qquad (14.99)$$

The dimensionless parameter β is the ratio of species s concentration at $r = R_p$ to the Michaelis constant K_m. The dimensionless group ϕ_T is known as the *Thiele modulus*, which is related to the ratio of reaction rate to diffusion rate. The factor $R_p/3$ is the ratio of the volume to the surface area of the sphere. In this analysis, we will assume that the concentration inside the pore at $r = R_p$ is known. In an actual bioreactor, this concentration will depend on the axial position in the device as we will see in Sect. 14.6.3. Substituting the dimensionless groups into (14.95) and (14.96):

$$\frac{1}{r^{*2}}\frac{d}{dr^*}\left(r^{*2}\frac{dC_{sp}^*}{dr^*}\right) = 9\phi_T^2\left(\frac{C_{sp}^*}{1+\beta C_{sp}^*}\right), \qquad (14.100)$$

$$r^* = 0, \quad \frac{dC_{sp}^*}{dr} = 0,$$
$$r^* = 1, \quad C_{sp}^* = 1. \qquad (14.101)$$

Equation (14.100) is a nonlinear second-order ordinary differential equation that does not have a known analytic solution. However, it can be solved numerically using a Runge–Kutta method, similar to that used in Sect. 14.4.3, which breaks (14.100) into two coupled first-order ordinary differential equations:

$$\Psi^* = r^{*2}\frac{dC_{sp}^*}{dr^*}, \qquad (14.102)$$

$$\frac{1}{r^{*2}}\frac{d\Psi^*}{dr^*} = 9\phi_T^2\left(\frac{C_{sp}^*}{1+\beta C_{sp}^*}\right), \qquad (14.103)$$

$$r^* = 0, \quad \Psi^* = 0,$$
$$r^* = 1, \quad C_{sp}^* = 1. \qquad (14.104)$$

The concentration of species s in the particle pores will depend on radial position and on the dimensionless parameters ϕ_T and β. Concentration profiles are shown in Fig. 14.15 for nine different combinations of β and ϕ_T. As the Thiele modulus becomes small, the concentration at all radial positions becomes closer to the concentration at $r = R_p$. This condition represents cases where D_e/R_p^2 is large relative to V_{max}/K_m; so species s can diffuse rapidly to the sites of all the immobilized enzymes. The higher the ratio of $C_{sp}(R_p)/K_m$ (i.e., β), the more efficient is the conversion process. For high values of the Thiele modulus, the reaction rate is high relative to the rate at which species s can be delivered toward the particle center. Consequently, enzymes in the central portions of the particle do not

14.4 1D Diffusion with Homogeneous Chemical Reaction

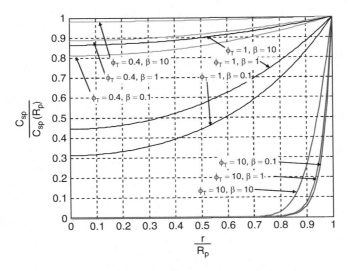

Fig. 14.15 Concentration profiles for Michaelis–Menten kinetics in a spherical particle

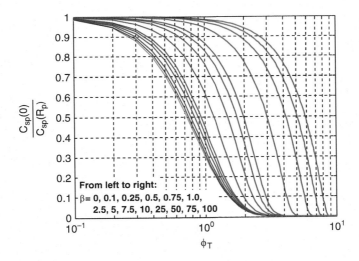

Fig. 14.16 Centerline concentration relative to concentration at $r = R_p$ vs. Theile modulus for various values of β

participate in the conversion process. The region near the particle surface where chemical conversion occurs becomes smaller as β becomes smaller. The concentration at the center of the particle relative to the edge is shown in Fig. 14.16 as a function of Thiele modulus for several values of β.

14.4.4.1 Simplification for High and Low Values of $\beta = C_{sp}(R_p)/K_m$

Let us consider two limiting cases: (1) $C_{sp} \gg K_m$ (i.e., $\beta \gg 1$) and (2) $C_{sp} \ll K_m$ (i.e., $\beta \ll 1$). If the species concentration inside the pore is everywhere large relative to the Michaelis constant, then (14.100) reduces to:

$$\frac{1}{r^{*2}} \frac{d}{dr^*}\left(r^{*2} \frac{dC_{sp}^*}{dr^*}\right) = \frac{9\phi_T^2}{\beta}. \tag{14.105}$$

This has the same form as the expression solved in Sect. 14.4.1.3 for O_2 consumption in a spherical cell. The solution is:

$$C_{sp}^* = 1 - \frac{3}{2}\left(\frac{\phi_T^2}{\beta}\right)(1 - r^{*2}). \tag{14.106}$$

When written in terms of the original variables, the concentration profile has a parabolic shape and is independent of the Michaelis constant K_m:

$$C_{sp}(r) = C_{sp}(R_p) - \left(\frac{V_{max}R_p^2}{6D_e}\right)\left(1 - \frac{r^2}{R_p^2}\right) \quad (C_{sp} \gg K_m). \tag{14.107}$$

Let us now consider the case where $K_m \gg C_{sp}$. The reaction term is first order in that case, and (14.100) becomes:

$$\frac{1}{r^{*2}} \frac{d}{dr^*}\left(r^{*2} \frac{dC_{sp}^*}{dr^*}\right) = 9\phi_T^2 C_{sp}^*. \tag{14.108}$$

Using the procedure outlined in Sect. 14.4.2, the solution to this differential equation with boundary conditions found in (14.101) is:

$$C_{sp}^*(r) = \left(\frac{1}{r^*}\right)\left\{\frac{\sinh(3\phi_T r^*)}{\sinh(3\phi_T)}\right\}. \tag{14.109}$$

In terms of the original variables:

$$C_{sp}(r) = C_{sp}(R_p)\left(\frac{R_p}{r}\right)\left\{\frac{\sinh\left(r\sqrt{\frac{V_{max}}{D_e K_m}}\right)}{\sinh\left(R_p\sqrt{\frac{V_{max}}{D_e K_m}}\right)}\right\} \quad (C_{sp} \ll K_m). \tag{14.110}$$

14.4 1D Diffusion with Homogeneous Chemical Reaction

The concentration at the center of the particle can be found by applying L'Hospital's rule at $r^* = 0$:

$$C_{sp}^*(0) = \frac{3\phi_T}{\sinh(3\phi_T)}. \tag{14.111}$$

This is plotted as the red line ($\beta = 0$) in Fig. 14.16.

14.4.4.2 Effectiveness Factor, η

For a given substrate concentration at the particle surface, the maximum possible rate of conversion of substrate to product would occur if all enzymes in the particle were exposed to the surface concentration. In the steady state, the production rate of substrate is equal to the rate that species s flows in the positive r-direction through the particle surface. The maximum possible molar flow of species s to the particle under those circumstances would be:

$$W_{s,max} = \phi_p V_p R_{sp,max} = -\frac{\phi_p V_p V_{max} C_{sp}(R_p)}{K_m + C_{sp}(R_p)}. \tag{14.112}$$

where $\phi_p V_p$ is the pore volume in a single spherical particle and $R_{sp,max}$ is the maximum rate of production of s per unit pore volume. In this case the production rate and the substrate flow rate are both negative. $W_{s,max}$ would be the same as the molar flow rate out of a single particle if the substrate were well mixed throughout the pore volume. The actual rate of conversion will be lower because the flow of substrate within the particle is limited by diffusion:

$$W_s = \phi_p S_p N_{sp} = \phi_p S_p \left(-D_{se} \frac{dC_{sp}}{dr}\bigg|_{r=R_p}\right). \tag{14.113}$$

The *effectiveness factor* η is defined as the actual rate of conversion relative to the maximum rate of conversion. In the steady state, this will equal $W_s/W_{s,max}$:

$$\eta = \frac{W_s}{W_{s,max}} = \frac{S_p N_{sp}}{V_p R_{sp,max}} = \frac{3 D_{se}}{R_p} \left(\frac{K_m + C_{sp}(R_p)}{V_{max} C_{sp}(R_p)}\right) \left(\frac{dC_{sp}}{dr}\bigg|_{r=R_p}\right). \tag{14.114}$$

In terms of the dimensionless groups in (14.102)–(14.104), this can be written as:

$$\eta = \left(\frac{1+\beta}{3\phi_T^2}\right) \Psi^*\big|_{r^*=1}. \tag{14.115}$$

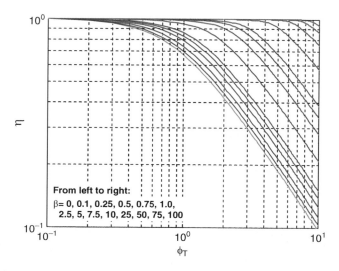

Fig. 14.17 Effectiveness factor vs. Theile modulus for various values of β

Numerical solutions for η vs. ϕ_T for various values of β are shown in Fig. 14.17. The particles are more efficient in converting substrate to product as β is increased or ϕ_T is decreased. From a practical standpoint, we are limited in how the Thiele modulus defined by (14.99) can be decreased. Reducing the radius of the particles would decrease ϕ_T proportionately. Increasing the effective diffusion coefficient in the pores by altering the particle porosity would reduce ϕ_T, but in a nonlinear manner. Reducing $V_{\max}$ by reducing the number of enzymes in each particle would decrease ϕ_T and increase the effectiveness. However, according to (14.114), reducing $V_{\max}$ would also reduce the maximum conversion rate. Therefore, we must be careful in not confusing the effectiveness factor, which reflects efficiency of conversion, with the actual conversion rate. Increasing the number of enzymes in a particle will increase the conversion of substrate to product, but may decrease the efficiency, since enzymes near the particle core will be exposed to lower substrate concentrations.

Two limiting cases occur at high and low values of β. If $C_s(R_p) \gg K_m$, we can substitute (14.107) into (14.114) to show that when β is large:

$$\eta = 1, \quad C_s(R_p) \gg K_m. \tag{14.116}$$

Consequently, the effectiveness factor is unity when species s concentration is large relative to the Michaelis constant. For the opposite case, where $C_s(R_p) \ll K_m$, we can substitute (14.110) into (14.114) to obtain the solution for $\beta \to 0$:

$$\eta = \frac{1}{\phi_T}\left(\coth(3\phi_T) - \frac{1}{3\phi_T}\right) \quad C_s(R_p) \ll K_m. \tag{14.117}$$

14.5 Convection and Diffusion

This is shown as the red line in Fig. 14.17, representing a lower limit on the effectiveness for any value of the Thiele modulus.

14.5 Convection and Diffusion

Continuous-flow or continuous-feed mass exchangers are devices that are used to selectively add or remove substances from fluid as the fluid flows through the device. This exchange process generally involves both convection and diffusion, and sometimes chemical reaction. Mass exchangers that involve chemical reactions will be treated in Sect. 14.6. In this section, we will focus on mass exchange between the flowing fluid and the internal surface of the exchanger. For a given concentration gradient at the exchanger boundary, the amount of mass exchange will be directly proportional to the total surface area of the device. Therefore, mass exchangers are generally composed of many parallel pathways which serve to increase fluid residence time and increase the overall surface to volume ratio of the device. These pathways may consist of the regions between parallel sheets or the interior of conduits with various cross-sectional shapes. Analysis of a device with parallel pathways can be reduced to analyzing the exchange in a single pathway, then multiplying by the number of total pathways.

Consider steady-state flow Q_b through a single pathway with constant cross-sectional area in a continuous-feed mass exchanger. Species A is dissolved in the fluid and it is neither produced nor removed by chemical reaction within the fluid. Let us define a shell composed of fluid in the pathway between axial positions x and $x + \Delta x$, as shown in Fig. 14.18.

The accumulation of substance A is zero in the steady state. We will assume that diffusion in the x-direction is negligible in comparison with axial convection. The fraction of the total conduit surface in contact with the shell is $S_i \Delta x/L$, where S_i is the total internal surface area of the conduit and L is the conduit length. The rate at which substance A leaves the shell across this surface is $N_A S_i \Delta x/L$. The mean or bulk concentration at any position x is $C_{Ab}(x)$. The rate at which species A enters the

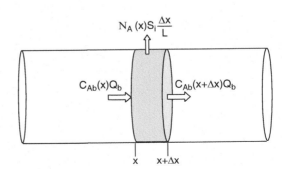

Fig. 14.18 Shell for species conservation equation in conduit with constant cross-sectional area

shell at x is $C_{Ab}(x)Q_b$, and the rate at which it leaves by convection at $x + \Delta x$ is $C_{Ab}(x + \Delta x) Q_b$. Species conservation for substance A in the shell is:

$$0 = C_{Ab}(x)Q_b - C_{Ab}(x+\Delta x)Q_b - N_A(x)S_i\left(\frac{\Delta x}{L}\right). \tag{14.118}$$

Dividing by the volume of the shell $V_{shell} = A_c\Delta x$, where A_c is the cross-sectional area of blood in the conduit, we find:

$$0 = \frac{Q_b}{A_c} \lim_{\Delta x \to 0} \left\{\frac{C_{Ab}(x) - C_{Ab}(x+\Delta x)}{\Delta x}\right\} - \frac{N_A(x)S_i}{A_c L}. \tag{14.119}$$

After taking the limit and rearranging, we obtain the following first-order ordinary differential equation:

$$\frac{dC_{Ab}}{dx} = -\frac{N_A(x)S_i}{LQ_b}. \tag{14.120}$$

The flux out of the shell $N_A(x)$ depends on the nature of the conduit wall. In the following sections, we will consider two fundamental types of exchangers: those that maintain species A at a constant concentration along the wall of the conduit, and those with permeable walls in which exchange takes place with another fluid across the conduit wall.

14.5.1 Conduits with Constant Wall Concentration

Species mass exchange in some devices can be regulated at the conduit surface. For instance, enzymes can be imbedded in the wall of a bioreactor, causing the concentration of substrate at the wall to be essentially zero along the entire length of the device. Such a device would be useful in removing a toxin from blood. The wall of a different type of device may contain a slightly soluble substance that is to be released slowly as fluid flows through the device. Assuming equilibrium at the solid–fluid interface, the concentration in the fluid at the wall will be constant along the length of the device. Such a device could be used to provide time release of a slightly soluble drug into blood.

In cases where the wall concentration is maintained constant, the flux of substance A from the bulk fluid to the wall is governed by a local mass transfer coefficient k_A:

$$N_A(x) = k_A(x)[C_{Ab}(x) - C_{Aw}]. \tag{14.121}$$

Note that C_{Aw} is the concentration of A in the fluid at the wall, not the concentration in the wall material itself. Substituting the flux from (14.121) into (14.120) gives:

14.5 Convection and Diffusion

$$\frac{dC_{Ab}}{dx} = -\frac{k_A S_i}{L Q_b}[C_{Ab} - C_{Aw}]. \qquad (14.122)$$

Let us define a dimensionless concentration C^*_{Ab} based on C_{Aw} and the inlet concentration of species A, C_{A0}:

$$C^*_{Ab} = \frac{C_{Ab} - C_{Aw}}{C_{A0} - C_{Aw}}. \qquad (14.123)$$

Thus C^*_{Ab} ranges from 1 at the inlet to 0 at the conduit wall. Substituting (14.123) into (14.122), separating variables, and integrating, we obtain:

$$\int_0^1 \frac{dC^*_{Ab}}{C^*_{Ab}} = -\frac{S_i}{L Q_b} \int_0^L k_A dx \qquad (14.124)$$

or

$$C^*_{Ab} = \exp\left\{-\frac{S_i}{Q_b}\left[\frac{1}{L}\int_0^L k_A dx\right]\right\}. \qquad (14.125)$$

The quantity in square brackets is defined as the *mean mass transfer coefficient* k_{Am}:

$$k_{Am} \equiv \frac{1}{L}\int_0^L k_A dx. \qquad (14.126)$$

It depends on the length and cross section of the conduit, the viscosity of the fluid, and the diffusion coefficient of species A in the flowing fluid. A more rigorous treatment for estimating k_{Am} will be presented in Sect. 15.5. For sufficiently long conduits, the mass transfer coefficient is constant and can be found from Sherwood numbers tabulated in Table 12.1. These constant values for the Sherwood number are valid if:

$$\left(\frac{1}{Re_{D_h} Sc}\right)\frac{L}{D_h} > 0.05. \qquad (14.127)$$

Once k_{Am} is known, the bulk concentration at the conduit outlet can be evaluated from:

$$\frac{C_{Ab}(L) - C_{Aw}}{C_{A0} - C_{Aw}} = \exp\left\{-\frac{k_{Am} S_i}{Q_b}\right\} = \exp\left\{-\frac{k_{Am}}{\langle v \rangle}\frac{S_i}{A_c}\right\}. \qquad (14.128)$$

The exponent in (14.128) is seen to be the product of two dimensionless parameters. The first is the ratio of the mean mass transfer coefficient to the mean fluid velocity. This parameter depends on fluid properties and geometry. The second parameter is the ratio of conduit internal surface area to cross-sectional area, which depends only on geometry. Equation (14.128) can be expressed in terms of other standard mass transfer dimensionless groups:

$$\frac{C_{Ab}(L) - C_{Aw}}{C_{A0} - C_{Aw}} = \exp\left\{-\frac{4Sh_m}{Re_{D_h} Sc} \frac{L}{D_h}\right\}. \quad (14.129)$$

Both the mean Sherwood number and the Reynolds number in (14.129) are based on the hydraulic diameter, D_h.

Example 14.5.1 Drug Delivery to Blood in a Rectangular Duct.
A mass exchanger is composed of many ducts with rectangular cross sections of 0.1 mm × 0.2 mm in parallel. The concentration of a soluble drug is maintained constant at the conduit walls. Drug-free blood enters each conduit at a rate of 1 ml/h. Estimate the conduit length necessary to bring the exit concentration of drug to 95% of the wall concentration. The diffusion coefficient of the drug in blood is 9×10^{-6} cm²/s.

Solution. *Initial considerations*: We will simplify this problem by assuming that we can neglect variations of the concentration of the drug perpendicular to the flow direction. This is not exactly true because there must be a gradient at the wall if drug is to move into the fluid from the wall. We will examine the effects of including concentration gradients perpendicular to the flow direction in Chap. 15. For now, we will assume that we can treat this problem using a one-dimensional approach to analyze the flow-averaged drug concentration as a function of axial position. We will also treat this as a steady-state problem.

System definition and environmental interactions: Our system consists of the fluid flowing through the rectangular conduit.

Apprising the problem to identify governing relationships: The analysis in Sect. 14.5, based on a 1D microscopic shell balance, applies to solute flow through the conduit. Equation (14.128) governs exchange in this case.

Analysis: The surface area of the conduit is $2(0.01 \text{ cm} + 0.02 \text{ cm})L$, where L is the length of the duct. The flow rate is $\langle v \rangle A_c$, which is 1/3600 cm³/s = 2.78×10^{-4} cm³/s. The Sherwood number, based on the hydraulic diameter for a long duct with an aspect ratio of two, is estimated to equal 3.39 from Table 12.1. The hydraulic diameter is four times the cross-sectional area divided by the wetted perimeter of the duct: $D_h = 4A_c/P_w = 4(0.0002 \text{ cm}^2)/(0.06 \text{ cm}) = 0.0133$ cm. Assuming that the mass transfer coefficient k_{Am} can be approximated as the mass transfer coefficient for a long duct, $k_{A\infty}$, we find:

$$k_{Am} \approx k_{A\infty} = \frac{Sh_{D_h} D_{Ab}}{D_h} = 3.39\left(\frac{9 \times 10^{-6} \text{cm}^2/\text{s}}{0.0133 \text{ cm}}\right) = 2.29 \times 10^{-3} \text{cm/s}.$$

14.5 Convection and Diffusion

Rearranging (14.128) to solve for the length of the conduit, where $S_i = P_w L$:

$$L = -\frac{Q_b}{P_w k_{Am}} \ln\left[\frac{C_{Ab}(L) - C_{Aw}}{C_{A0} - C_{Aw}}\right]$$

$$= -\frac{2.78 \times 10^{-4} \text{cm}^3/\text{s}}{(0.06\,\text{cm})(2.29 \times 10^{-3}\,\text{cm/s})} \ln\left[\frac{0.95\,C_{Aw} - C_{Aw}}{0 - C_{Aw}}\right] = 6.06\,\text{cm}.$$

Examining and interpreting the results: We can now check our assumption that k_{Am} can be approximated as $k_{A\infty}$. Applying the criterion in (14.127), we find:

$$\left(\frac{1}{Re_{D_h} Sc}\right)\frac{L}{D_h} = \frac{D_{Ab} A_c L}{Q_b D_h^2} = \frac{(9 \times 10^{-6}\,\text{cm}^2\,\text{s}^{-1})(0.01\,\text{cm})(0.02\,\text{cm})(6.06\,\text{cm})}{(2.78 \times 10^{-4}\,\text{cm}^3\,\text{s}^{-1})(0.0133\,\text{cm})^2}$$

$$= 0.222 > 0.05.$$

Therefore, the criterion is met and the fiber length computation should be reasonably accurate. If the criterion were not met, a more appropriate choice for k_{Am} would need to be made using methods similar to those outlined in Sect. 15.5.

14.5.2 Hollow Fiber Devices

Hollow fibers with various compositions are used to promote mass exchange in many different devices including dialyzers and blood oxygenators. The concept is straightforward. Blood enters the lumen of fiber with a particular concentration of solute. If the substance is to be removed from the bloodstream, such as urea in a dialyzer, the concentration of the substance is made very low on the outside of the fiber; so the substance diffuses out of the blood and into the surrounding fluid. If the substance is to be added to the blood, such as oxygen in a blood oxygenator, the concentration surrounding the fiber is maintained higher than on the inside, promoting an inward movement of solute across the fiber wall.

14.5.2.1 Solute Exchange with a Well-Mixed External Compartment

We begin by examining a single fiber in a blood dialysis unit. Blood enters with a bulk concentration C_{Ab} of solute A (e.g., urea) which is to be removed as it passes through the fiber. The bulk concentration of solute A in the fluid surrounding the fiber is maintained at a constant value, C_{Ad}, preferably near zero.

In Example 12.6.2.1, we derived an expression for solute flow through a cylindrical hollow fiber, based on the inside surface area:

$$W_A = \mathsf{P}_A S_i [C_{Ab} - \Phi_{Abd} C_{Ad}]. \tag{14.130}$$

The permeability is based on the inside surface area $S_i = 2\pi R_i L$:

$$P_A = \frac{1}{\dfrac{R_i \ln(R_o/R_i)}{D_{Aw}\Phi_{Awb}} + \dfrac{1}{k_{Ab}} + \dfrac{R_i \Phi_{Abd}}{R_o k_{Ad}}}. \tag{14.131}$$

The quantities k_{Ab} and k_{Ad} represent the individual mass transfer coefficients in the blood and dialysis fluid, respectively, D_{Aw} is the diffusion coefficient of A in the fiber wall, Φ_{Abd} is the partition coefficient of A between blood and dialysis fluid, Φ_{Awb} is the partition coefficient of A between fiber wall and blood, R_i is the inside radius of the fiber, and R_o is the outside radius of the fiber. If x is the axial coordinate, the local flux of solute at the inside surface from blood to dialysis fluid at a given value of x is:

$$N_A(x) = P_A[C_{Ab}(x) - \Phi_{Abd}C_{Ad}(x)]. \tag{14.132}$$

We are interested in finding the length of fiber necessary to remove a substantial portion of substance A from the blood. The shell balance approach described in Sect. 14.5 also applies in this case. Substituting (14.132) for the flux through the fiber wall into the species continuity equation, (14.120), we find:

$$\frac{dC_{Ab}}{dx} = -\frac{P_A S_i}{L Q_b}[C_{Ab} - \Phi_{Abd}C_{Ad}]. \tag{14.133}$$

Since Φ_{Abd} and C_{Ad} are both constant in this problem, (14.133) can be rewritten as:

$$\frac{d}{dx}[C_{Ab} - \Phi_{Abd}C_{Ad}] = -\frac{P_A S_i}{L Q_b}[C_{Ab} - \Phi_{Abd}C_{Ad}]. \tag{14.134}$$

This is easily integrated with the boundary condition $C_{Ab}(x=0) = C_{AB0}$ to find an expression for the variation of concentration in the blood with axial position:

$$\frac{C_{Ab}(x) - \Phi_{Abd}C_{Ad}}{C_{Ab0} - \Phi_{Abd}C_{Ad}} = \exp\left[-\left(\frac{P_A S_i}{Q_b}\right)\left(\frac{x}{L}\right)\right]. \tag{14.135}$$

The concentration at the fiber outlet, $x = L$, can be computed from:

$$\frac{C_{Ab}(L) - \Phi_{Abd}C_{Ad}}{C_{Ab0} - \Phi_{Abd}C_{Ad}} = e^{-\left(\frac{P_A S_i}{Q_b}\right)}. \tag{14.136}$$

Note the similarity of (14.136) with the expression for constant wall concentration, (14.128). The two expressions are identical when all of the resistance to mass

14.5 Convection and Diffusion

exchange occurs in the blood. In that case, (14.131) reduces to $P_A = k_{Ab}$ and equilibrium will exist between fluid at the inside wall and the dialysis fluid, so $C_{Aw} = \Phi_{Abd} C_{Ad}$.

The permeability of the fiber to substance A can be estimated by measuring the inlet and outlet concentrations for a given flow rate. The fiber length needed for $C_{Ab}(L)$ to drop to some fraction f of the inlet concentration C_{Ab0}, will be:

$$L = \frac{S_i}{2\pi R_i} = -\frac{Q_b}{2\pi R_i P_A} \ln\left(\frac{fC_{Ab0} - \Phi_{Abd}C_{Ad}}{C_{Ab0} - \Phi_{Abd}C_{Ad}}\right). \tag{14.137}$$

The length of a given fiber can be minimized by keeping the dialysis fluid concentration as low as possible, reducing blood flow per fiber and maximizing fiber permeability.

It is instructive to compare the solution above to the steady-state solution when the blood in the fiber is assumed to be well mixed. In that case the concentration everywhere in the fiber, including the fiber exit, would be given by (13.53). Using the notation for the current problem:

$$C_{Ab}(L) = \left(\frac{Q_b C_{Ab0} + P_A S_i \Phi_{Abd} C_{Ad}}{Q_b + P_A S_i}\right). \tag{14.138}$$

Putting this in the dimensionless form of (14.136):

$$\frac{C_{Ab}(L) - \Phi_{Abd}C_{Ad}}{C_{Ab0} - \Phi_{Abd}C_{Ad}} = \frac{Q_b}{P_A S_i + Q_b}. \tag{14.139}$$

Let us define a dimensionless variable β as the ratio of blood flow to permeability surface area product:

$$\beta \equiv \frac{Q_b}{P_A S_i}. \tag{14.140}$$

Then the concentration at the fiber exit for the well-mixed and distributed cases can be written as follows.

Distributed (14.136):

$$\frac{C_{Ab}(L) - \Phi_{Abd}C_{Ad}}{C_{Ab0} - \Phi_{Abd}C_{Ad}} = e^{-\left(\frac{1}{\beta}\right)}. \tag{14.141}$$

Well mixed (14.139):

$$\frac{C_{Ab}(L) - \Phi_{Abd}C_{Ad}}{C_{Ab0} - \Phi_{Abd}C_{Ad}} = \frac{\beta}{1+\beta}. \tag{14.142}$$

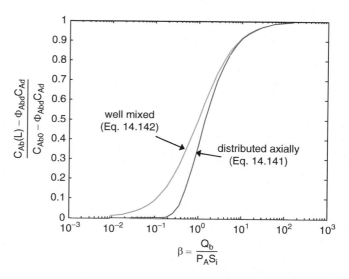

Fig. 14.19 Dimensionless concentrations at the exit of a hollow fiber for well-mixed and axially distributed models

Predictions of the two models are shown in Fig. 14.19 as a function of β. Although one might expect the well-mixed model to provide a more efficient exchange of solute A, this is not the case. For all values of β, the outlet concentration of the distributed model falls below that of the well-mixed model, indicating a more efficient exchange. The models are in good agreement for $\beta > 10$.

The reason for the difference in efficiency is illustrated in Fig. 14.20. This shows how the dimensionless bulk concentration varies with axial position in the two models for $\beta = 1$. The well-mixed model requires that entering solute be instantly diluted by mixing with the entire blood volume within the fiber. The concentration difference between blood and dialysis fluid is the same at all values of x. However, for the distributed model, the concentration difference is very high near the fiber entrance, causing a large efflux of substance A. The concentration decreases with axial position, but does not drop below the well-mixed concentration until it nears the fiber exit. Consequently the net flux is greater for the distributed case.

Another way to look at this is to compute the total mass flow out of the fiber for the two cases. The theoretical maximum mass flow across the fiber for a fixed dialysis fluid concentration C_{Ad0} would occur if the blood concentration was held equal to the inlet blood concentration at all axial positions.

$$W_{A,\max} = P_A S_i [C_{Ab0} - \Phi_{Abd} C_{Ad0}]. \quad (14.143)$$

The actual predicted mass flow across the membrane can be found by subtracting the outflow of species A $Q_b C_{Ab}(L)$ from the inflow $Q_b C_{Ab0}$. The ratio of the predicted

14.5 Convection and Diffusion

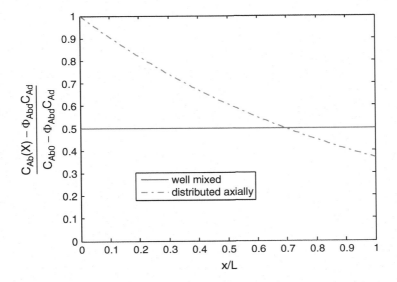

Fig. 14.20 Comparison of the dimensionless concentration vs. axial position in a hollow fiber for well-mixed and axially distributed models

mass flow to theoretical maximum for the well-mixed and distributed models are as follows:

Well-mixed model:

$$\frac{W_A}{W_{A,max}} = \frac{\beta}{1+\beta}. \qquad (14.144)$$

Distributed model:

$$\frac{W_A}{W_{A,max}} = \beta\left[1 - \exp\left(-\frac{1}{\beta}\right)\right]. \qquad (14.145)$$

Taking the ratio:

$$\frac{W_{A,\,wellmixed}}{W_{A,\,axially distributed}} = \frac{1}{[1+\beta]\left[1 - \exp\left(-\frac{1}{\beta}\right)\right]}. \qquad (14.146)$$

As $\beta \to 0$ the ratio approaches 1, and a Taylor expansion can be used to show that as $\beta \to \infty$ the molar flow ratio also approaches 1. Mass flow ratios from (14.144)–(14.146) are plotted in Fig. 14.21. The axially distributed model is more efficient than the well-mixed model of exchange across the walls of the hollow

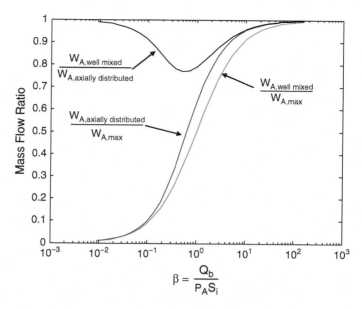

Fig. 14.21 Solute exchange rate across the walls of a hollow fiber relative to the maximum exchange rate for a well-mixed model and an axially distributed model of exchange with constant dialysis fluid solute concentration. The ratio of well mixed to axially distributed exchange rate is also plotted

fiber, particularly for values of β between 0.01 and 100. However, the worst agreement between the two models occurs when β is approximately 0.56, where the well-mixed model predicts that the total exchange of solute is about 77% of the exchange predicted with the axially distributed model. Therefore, the simpler well-mixed model can be used to obtain a reasonable estimate of the solute exchange rate, particularly at high or low values of β.

14.5.2.2 Cocurrent Mass Exchanger

It is often difficult to maintain a constant solute concentration on the outside of the hollow fiber. A more realistic situation is shown in Fig. 14.22, where both blood and dialysis fluid enter a hemodialyzer in a cocurrent fashion at $x = 0$. For simplicity, we will treat solute exchange across a single fiber, but an actual hemodialyzer contains many such fibers in parallel. Initially, the blood solute concentration is high and dialysis fluid solute concentration is low (or zero). As the blood flows in the axial direction, solute flows from blood to dialysis fluid. The concentration of solute in the blood will decrease and solute concentration in the dialysis fluid will increase as the fluids move through the hemodialyzer. Therefore, the concentration gradient across the fiber is smaller than the gradient achieved when the dialysis fluid solute concentration is maintained at the inlet value.

14.5 Convection and Diffusion

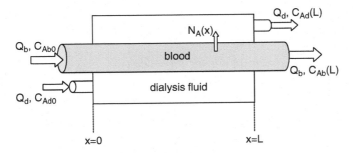

Fig. 14.22 Hemodialyzer or mass exchanger in cocurrent configuration

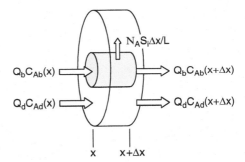

Fig. 14.23 Shells for cocurrent analysis

We now perform steady-state species mass balances on solute A in the blood and dialysis fluid. The fluid shells are confined between planes at x and $x + \Delta x$ as shown in Fig. 14.23. Mathematical statements of species conservation are as follows:

Blood:

$$0 = Q_b C_{Ab}(x) - Q_b C_{Ab}(x + \Delta x) - N_A S_i \frac{\Delta x}{L}. \quad (14.147)$$

Dialysis fluid:

$$0 = Q_d C_{Ad}(x) - Q_d C_{Ad}(x + \Delta x) + N_A S_i \frac{\Delta x}{L}. \quad (14.148)$$

Dividing each equation by the shell volume, letting the shell volume approach zero, and substituting (14.132) for the flux of solute between blood and dialysis fluid, we obtain the following coupled ordinary differential equations:

$$\frac{dC_{Ab}}{dx} + \frac{P_A S_i}{Q_b L}[C_{Ab} - \Phi_{Abd} C_{Ad}] = 0, \quad (14.149)$$

$$\frac{dC_{Ad}}{dx} - \frac{P_A S_i}{Q_d L}[C_{Ab} - \Phi_{Abd} C_{Ad}] = 0. \quad (14.150)$$

These can be reduced to a single expression by multiplying (14.150) by Φ_{Abd} and subtracting the result from (14.149):

$$\frac{d}{dx}[C_{Ab} - \Phi_{Abd}C_{Ad}] = \frac{P_A S_i}{Q_b L}\left(1 + \frac{\Phi_{Abd}Q_b}{Q_d}\right)[C_{Ab} - \Phi_{Abd}C_{Ad}]. \tag{14.151}$$

Defining α as:

$$\alpha = \frac{Q_d}{\Phi_{Abd}Q_b}, \tag{14.152}$$

and using the definition of β from (14.140), the solution to (14.151) with inlet concentrations C_{Ab0} and C_{Ad0} is:

$$C_{Ab} - \Phi_{Abd}C_{Ad} = [C_{Ab0} - \Phi_{Abd}C_{Ad0}]\exp\left(-\left(\frac{1+\alpha}{\alpha\beta}\right)\frac{x}{L}\right). \tag{14.153}$$

Substituting this back into (14.149):

$$\frac{dC_{Ab}}{dx} = \frac{d}{dx}[C_{Ab} - \Phi_{Abd}C_{Ad0}]$$

$$= -\frac{1}{\beta L}[C_{Ab0} - \Phi_{Abd}C_{Ad0}]\exp\left(-\left(\frac{1+\alpha}{\alpha\beta}\right)\frac{x}{L}\right). \tag{14.154}$$

Separating variables and integrating between $x = 0$ and x, we obtain the final result for the concentration of solute in the blood as a function of axial position:

$$C_{Ab}(x) = \Phi_{Abd}C_{Ad0} + \left(\frac{C_{Ab0} - \Phi_{Abd}C_{Ad0}}{1+\alpha}\right)$$
$$\times \left[1 + \alpha\exp\left(-\left(\frac{1+\alpha}{\alpha\beta}\right)\frac{x}{L}\right)\right]. \tag{14.155}$$

Substituting (14.155) into (14.150), we can obtain the expression for the concentration in the dialysis fluid as a function of position:

$$C_{Ad}(x) = C_{Ad0} + \left(\frac{C_{Ab0} - \Phi_{Abd}C_{Ad0}}{\Phi_{Abd}(1+\alpha)}\right)\left[1 - \exp\left(-\left(\frac{1+\alpha}{\alpha\beta}\right)\frac{x}{L}\right)\right]. \tag{14.156}$$

Plots of $C_{Ab}(x)/C_{Ab0}$ and $C_{Ad}(x)/C_{Ab0}$ vs. x/L are shown in Fig. 14.24 for $C_{d0} = 0$, $\Phi_{Abd} = 1$, $\beta = 1$ and for three values of α. The higher the dialysis fluid

14.5 Convection and Diffusion

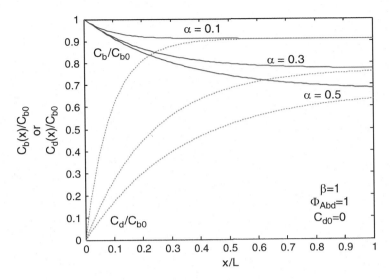

Fig. 14.24 Blood (solid) and dialysis fluid (dotted) concentrations vs. position for three ratios of dialysis flow to blood flow

flow relative to blood flow (i.e., α), the lower is the blood solute concentration at the fiber outlet and the greater is the solute exchange. If dialysis fluid flow is very low, the dialysis fluid solute concentration rises rapidly, preventing solute flux to occur over a significant portion of the fiber.

The dimensionless solute concentration in the blood at the outlet of the fiber ($x = L$) is:

$$\frac{C_{Ab}(L) - \Phi_{Abd}C_{Ad0}}{C_{Ab0} - \Phi_{Abd}C_{Ad0}} = \left(\frac{1}{1+\alpha}\right)\left[1 + \alpha \exp\left(-\left(\frac{1+\alpha}{\alpha\beta}\right)\right)\right]. \quad (14.157)$$

A macroscopic mass balance taken on solute in the blood shows:

$$W_A = Q_b[C_{Ab}(L) - C_{Ab0}]. \quad (14.158)$$

Substituting (14.157) for $C_{Ab}(L)$ into (14.158) and dividing by $W_{A,max}$ from (14.143):

$$\frac{W_{A,\text{co-current}}}{W_{A,max}} = \left(\frac{\alpha\beta}{1+\alpha}\right)\left[1 - \exp\left(-\left(\frac{1+\alpha}{\alpha\beta}\right)\right)\right]. \quad (14.159)$$

We will compare cocurrent solute flow with solute flow when the dialysis fluid concentration is fixed at C_{d0} and with solute flow for the counter-current case in the next section.

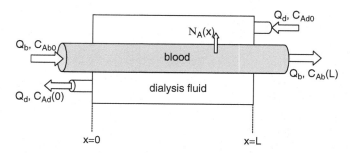

Fig. 14.25 Hemodialyzer or mass exchanger in counter-current configuration

14.5.2.3 Counter-Current Mass Exchanger

Now, let us reverse the flow of dialysis fluid so it enters with concentration C_{d0} at $x = L$ rather than at $x = 0$. The counter-current configuration is shown in Fig. 14.25.

The species conservation equation for solute in the blood is unchanged. Reversing the direction of dialysis fluid flow in Fig. 14.23 leads to the following species conservation equation for the dialysis fluid:

$$0 = Q_d C_{Ad}(x + \Delta x) - Q_d C_{Ad}(x) + N_A S_i \frac{\Delta x}{L}. \tag{14.160}$$

Taking the limit as the shell volume approaches zero:

$$\frac{dC_{Ad}}{dx} + \frac{P_A S_i}{Q_d L}[C_{Ab} - \Phi_{Abd} C_{Ad}] = 0. \tag{14.161}$$

This differs from the cocurrent case (14.150) by the sign preceding the second term. Substituting for $\Phi_{Abd} C_{Ad}$ from (14.149) into (14.161) yields a second-order ordinary differential equation for C_{Ab}:

$$\frac{d^2 C_{Ab}}{dx^2} + \frac{P_A S_i}{Q_b L}\left[1 - \frac{\Phi_{Abd} Q_b}{Q_d}\right] \frac{dC_{Ab}}{dx} = 0. \tag{14.162}$$

The solution to this homogeneous ODE for $\alpha \neq 1$ has the form:

$$C_{Ab}(x) = A + B \exp\left[-\frac{1}{L}\left(\frac{\alpha - 1}{\alpha \beta}\right) x\right], \tag{14.163}$$

where α and β have the same definitions as for the cocurrent case [i.e., (14.152) and (14.140)]. The constants A and B can be found by applying the two boundary conditions: (1) $C_{Ab}(0) = C_{Ab0}$ and (2) $C_{Ad}(L) = C_{Ad0}$. The final solution for $\alpha \neq 1$ is:

14.5 Convection and Diffusion

$$\frac{C_{Ab}(x) - \Phi_{Abd}C_{Ad0}}{C_{Ab0} - \Phi_{Abd}C_{Ad0}} = \frac{\alpha \exp\left[\left(\frac{\alpha-1}{\alpha\beta}\right)\left(1-\frac{x}{L}\right)\right] - 1}{\alpha \exp\left(\frac{\alpha-1}{\alpha\beta}\right) - 1} \quad (\alpha \neq 1). \quad (14.164)$$

The solution for $\alpha = 1$ can be found in problem 14.10.15. The concentration at the outlet of the fiber is:

$$C_{Ab}(L) = \Phi_{Abd}C_{Ad0} + \left[\frac{(\alpha-1)}{\alpha \exp\left(\frac{\alpha-1}{\alpha\beta}\right) - 1}\right][C_{Ab0} - \Phi_{Abd}C_{Ad0}]. \quad (14.165)$$

Solute flow from the tube to the shell, relative to the maximum possible solute flow, can be found by substituting this into (14.158) and dividing by $W_{A,\max}$:

$$\frac{W_{A,\text{counter-current}}}{W_{A,\max}} = \frac{\beta\alpha\left[\exp\left(\frac{\alpha-1}{\alpha\beta}\right) - 1\right]}{\alpha \exp\left(\frac{\alpha-1}{\alpha\beta}\right) - 1}. \quad (14.166)$$

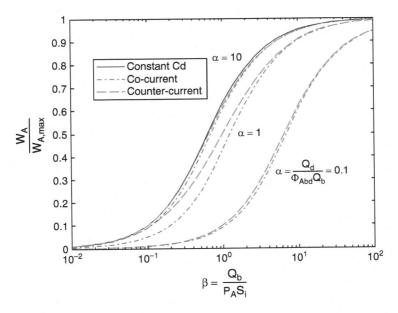

Fig. 14.26 Solute flow from blood to dialysis fluid vs. β for three models of exchange: constant dialysis fluid concentration (14.145), cocurrent (14.159), and counter current (14.166) exchangers. Comparisons are made at three values of α

A comparison of solute exchange between blood and dialysis fluid for a cocurrent exchanger, counter-current exchanger, and an exchanger with a constant dialysis fluid concentration is shown in Fig. 14.26. The exchanger with constant dialysis fluid solute concentration is the most efficient, and its exchange rate is independent of α. The counter-current exchanger is more efficient than the cocurrent exchanger, but both are nearly as efficient as the constant dialysis solute concentration model at high values of α. Neither cocurrent nor counter-current exchangers are very efficient at low values of α and low values of β. The most effective exchange rates occur at high values of α and high values of β.

14.5.2.4 Effect of Axial Diffusion on the Rate of Solute Exchange

In the previous examples, we have neglected the effect of axial diffusion. In this section, we will examine the effect that axial diffusion of solute in the blood has on overall mass exchange. We will assume that the dialysis fluid concentration is maintained constant and that the concentration in the blood does not depend on radial position. In addition to the convective flux introduced in Sect. 14.5.2.1, we must also consider the diffusive flux in the axial direction, as shown in Fig. 14.27:

Conservation of solute within the shell is expressed as:

$$0 = C_{Ab}(x)Q_b + J_{Ax}(x)A_c - C_{Ab}(x+\Delta x)Q_b - J_{Ax}(x+\Delta x)A_c - N_A(x)S_i\left(\frac{\Delta x}{L}\right), \quad (14.167)$$

where J_{Ax} is the diffusive flux in the x-direction and A_c is the cross-sectional area of the hollow fiber. Dividing by the shell volume $A_c\Delta x$:

$$0 = \frac{Q_b}{A_c}\frac{[C_{Ab}(x) - C_{Ab}(x+\Delta x)]}{\Delta x} + \frac{[J_{Ax}(x) - J_{Ax}(x+\Delta x)]}{\Delta x} - \left(\frac{N_A(x)S_i}{A_cL}\right). \quad (14.168)$$

Fig. 14.27 Solute fluxes into and out of a shell between x and $x + \Delta x$

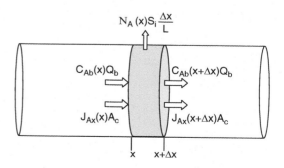

14.5 Convection and Diffusion

In the limit as Δx approaches zero:

$$0 = -\frac{Q_b}{A_c}\frac{dC_{Ab}}{dx} - \frac{dJ_{Ax}}{dx} - \left(\frac{N_A(x)S_i}{A_c L}\right). \tag{14.169}$$

Substituting Fick's Law for J_{Ax} and (14.132) for $N_A(x)$:

$$0 = -\frac{Q_b}{A_c}\frac{dC_{Ab}}{dx} - \frac{d}{dx}\left(-D_{Ab}\frac{dC_{Ab}}{dx}\right) - \left(\frac{P_A S_i}{A_c L}\right)[C_{Ab}(x) - \Phi_{Abd} C_{Ad0}]. \tag{14.170}$$

This can be conveniently written in dimensionless terms:

$$\frac{1}{Pe}\left(\frac{d^2 C^*}{dx^{*2}}\right) - \frac{dC^*}{dx^*} - \frac{1}{\beta}C^* = 0, \tag{14.171}$$

where the dimensionless variables are defined as follows:

$$C^* = \frac{C_{Ab}(x) - \Phi_{Abd} C_{Ad0}}{C_{Ab0} - \Phi_{Abd} C_{Ad0}}, \quad x^* = \frac{x}{L}, \quad \beta = \frac{Q_b}{P_A S_i}, \quad Pe = \frac{Q_b}{A_c}\frac{L}{D_{Ab}} = \frac{\langle v \rangle L}{D_{Ab}}. \tag{14.172}$$

The general solution to (14.171) is:

$$C^* = A e^{\lambda_1 x^*} + B e^{\lambda_2 x^*}, \tag{14.173}$$

where A and B are constants and λ_1 and λ_2 are:

$$\lambda_1 = \frac{Pe}{2}\left[1 + \sqrt{1 + \frac{4}{\beta Pe}}\right], \quad \lambda_2 = \frac{Pe}{2}\left[1 - \sqrt{1 + \frac{4}{\beta Pe}}\right]. \tag{14.174}$$

It is clear from (14.174) that $\lambda_1 > 0$ and $\lambda_2 < 0$. The problem must be bounded for large Pe or small β, and therefore A must equal zero. Applying the boundary condition at $x^* = 0$ that $C^* = 1$, then $B = 1$. The final solution in terms of the dimensionless variables is:

$$\frac{C_{Ab}(x) - \Phi_{Abd} C_{Ad0}}{C_{Ab0} - \Phi_{Abd} C_{Ad0}} = \exp\left\{\frac{Pe}{2}\left[1 - \sqrt{1 + \frac{4}{\beta Pe}}\right]\frac{x}{L}\right\}. \tag{14.175}$$

The dimensionless concentration at the outlet of the fiber will be:

$$\frac{C_{Ab}(L) - \Phi_{Abd} C_{Ad0}}{C_{Ab0} - \Phi_{Abd} C_{Ad0}} = \exp\left\{\frac{Pe}{2}\left[1 - \sqrt{1 + \frac{4}{\beta Pe}}\right]\right\}. \tag{14.176}$$

The total solute flow through the walls of the hollow fiber can be found by integrating the flux over the entire surface area of the fiber:

$$W_{A,\text{with diffusion}} = \frac{P_A S_i}{L} \int_0^L (C_{Ab}(x) - \Phi_{Abd} C_{Ad0}) dx,$$

$$W_{A,\text{with diffusion}} = P_A S_i (C_{Ab0} - \Phi_{Abd} C_{Ad0}) \int_0^1 C^* dx^*.$$

(14.177)

The factor multiplying the integral term is simply $W_{A,\text{max}}$. After performing the integration:

$$\frac{W_{A,\text{with diffusion}}}{W_{A,\text{max}}} = \frac{e^{\lambda_2} - 1}{\lambda_2}.$$

(14.178)

Comparing this to the solute flow in the distributed model without axial diffusion (14.145):

$$\frac{W_{A,\text{with diffusion}}}{W_{A,\text{no diffusion}}} = \frac{e^{\lambda_2} - 1}{\lambda_2 \beta \left[1 - \exp\left(-\frac{1}{\beta}\right)\right]}.$$

(14.179)

The influence of axial diffusion on solute flow through the fiber is shown in Fig. 14.28. Axial diffusion enhances solute flow through the fiber wall, but the

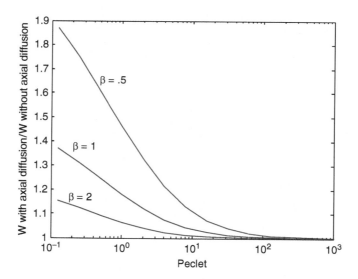

Fig. 14.28 Comparison of solute flow with axial diffusion to solute flow without axial diffusion as a function of the Peclet number

14.5 Convection and Diffusion

effect becomes smaller at higher values of Pe and β. Does axial diffusion significantly enhance mass transfer? Consider a fiber with length 10 cm, average velocity of 1 cm/s, $\beta = 1$, and solute diffusivity of 1.4×10^{-5} cm^2/s. The Peclet number for this case is 7.1×10^5. Figure 14.28 shows that diffusion provides negligible enhancement in this case.

14.5.3 Capillary Exchange of Non-Reacting Solutes

There are many different mechanisms responsible for the movement of various substances across the microvascular barrier. The barrier consists of an endothelial cell layer, the junctions between cells, and the basement membrane on the ablumenal side of the capillary. Some substances combine with integral proteins on the lumenal side of the capillary barrier and enter the cells via carrier-mediated transport. Others might enter via receptor-mediated endocytosis. Still others might diffuse across the membrane. Many substances that cross the endothelial cell surface, by whatever mechanism, are modified by intracellular enzymes and do not leave the cell, at least not as the same molecular species. In this section, we will analyze only those substances that diffuse passively through the lumenal and ablumenal cell membrane and through the cell cytoplasm without reacting with any other species. Generally, this is limited to small lipid-soluble solutes and dissolved gases.

14.5.3.1 Small Solute and Inert Gas Exchange in Lung Capillaries

Some gases and liquids are introduced into the respiratory airways to serve as drugs or tracers. Still others are introduced in the airways as anesthesia gases during surgery. A quantitative understanding of gas and small solute transport across the blood–gas barrier is essential for designing therapies that can improve gas and aerosol drug delivery.

Consider the steady-state exchange of a small non-reacting solute A which passes across the barrier separating alveolar gas from pulmonary blood. The concentration in entering pulmonary blood is C_{Ab0} and the concentration in the alveolar space is $C_{A,alv}$. We are interested in finding the concentration of A at the downstream end of the capillary. The solution procedure follows that presented in Sect. 14.5.2.1 and the solution is the same as given by (14.135) for exchange across a hollow fiber dialyzer with the subscript "d" replaced with "alv":

$$\frac{C_{Ab}(x) - \Phi_{A,b,alv}C_{A,alv}}{C_{Ab0} - \Phi_{A,b,alv}C_{A,alv}} = \exp\left[-\left(\frac{P_A S_i}{Q_b}\right)\left(\frac{x}{L}\right)\right]. \tag{14.180}$$

If the solute is an inert gas, this can be rewritten in terms of the partial pressures of the gas A in the alveolar space and in the blood. From (12.35), we can rewrite the partition coefficient between a liquid and an ideal gas in terms of the solubility coefficient for the gas in the liquid:

$$\Phi_{A,b,alv} = RT\alpha_{A,b}. \tag{14.181}$$

The concentration in the blood can be written in terms of the partial pressure of A in the blood $P_{A,b}$ using Henry's Law:

$$C_{Ab} = \alpha_{A,b} P_{A,b}. \tag{14.182}$$

The concentration of gas A in an ideal gas is related to the partial pressure of A in the gas:

$$C_{A,alv} = \frac{P_{A,alv}}{RT}. \tag{14.183}$$

Substituting (14.181)–(14.183) into (14.135) completely eliminates the dependence on the solubility coefficient or partition coefficient:

$$\frac{P_{A,b}(x) - P_{A,alv}}{P_{A,b0} - P_{A,alv}} = \exp\left[-\left(\frac{P_A S_i}{Q_b}\right)\left(\frac{x}{L}\right)\right]. \tag{14.184}$$

The dimensionless partial pressure at the outlet of the pulmonary capillary depends on a single parameter, $P_A S_i / Q_b$:

$$\frac{P_{A,b}(L) - P_{A,alv}}{P_{A,b0} - P_{A,alv}} = \exp\left[-\left(\frac{P_A S_i}{Q_b}\right)\right]. \tag{14.185}$$

If $P_A S_i / Q_b$ is known, (14.185) can be used to predict the partial pressure of gas A at the outlet of a pulmonary capillary. Alternatively, (14.185) can be used to estimate the product of permeability and surface area from measured partial pressures in the blood and alveoli at a known blood flow.

14.5.3.2 Solute Removal by Tissue Capillaries

Consider a capillary that removes a nonreacting waste product, solute A, from the tissue that surrounds it. Solute A is produced at a constant rate per unit volume in the tissue, and it is removed only by diffusion into the capillary. Let us apply a mass balance on a small tissue shell, as shown in Fig. 14.29.

14.5 Convection and Diffusion

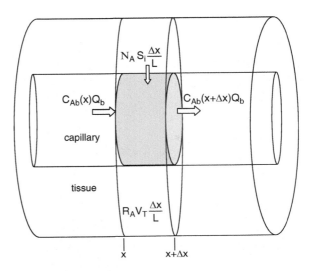

Fig. 14.29 Tissue capillary exchange of a solute produced in the tissue

A species mass balance in the tissue confined between x and $x + \Delta x$ is:

$$\left\{ \begin{array}{c} \text{Rate of} \\ \text{accumulation} \\ \text{of species A} \\ \text{within shell} \end{array} \right\} = \left\{ \begin{array}{c} \text{Net rate species A} \\ \text{enters through} \\ \text{shell boundaries} \end{array} \right\} + \left\{ \begin{array}{c} \text{Rate species A} \\ \text{is produced} \\ \text{within the shell} \end{array} \right\}. \quad (14.186)$$

For steady-state transport, solute A, produced in the tissue, leaves across the capillary barrier:

$$0 = -N_A S_i \frac{\Delta x}{L} + R_A V_T \frac{\Delta x}{L}. \quad (14.187)$$

Dividing by the shell volume and letting the volume approach zero, we can solve for the flux:

$$N_A = \frac{R_A V_T}{S_i}. \quad (14.188)$$

The right side of (14.188) is constant, and therefore the flux from tissue to capillary is constant if species A is produced at a constant rate per unit volume of tissue. Applying (14.186) to the small capillary shell in Fig. 14.29:

$$0 = Q_b C_{Ab}(x) - Q_b C_{Ab}(x + \Delta x) + N_A S_i \frac{\Delta x}{L}. \quad (14.189)$$

Dividing by the shell volume and letting the shell volume approach zero:

$$\frac{dC_{Ab}}{dx} = \frac{N_A S_i}{Q_b L}. \tag{14.190}$$

The solution with boundary condition that $C_b(0) = C_{b0}$ is:

$$C_{Ab}(x) = C_{Ab0} + \frac{N_A S_i}{Q_b}\left(\frac{x}{L}\right). \tag{14.191}$$

Therefore, the solute concentration increases linearly with x. Using (14.188), the concentration at the outlet is:

$$C_{Ab}(L) = C_{Ab0} + \frac{R_A V_T}{Q_b}. \tag{14.192}$$

If the flow rate and concentrations at the inlet and outlet can be measured, then the production rate of solute A, $R_A V_T$ can be computed from (14.192).

14.6 Convection, Diffusion, and Chemical Reaction

There are many transport situations that involve simultaneous convection, diffusion, and chemical reaction. Oxygen and carbon dioxide exchange in lung and tissue capillaries involve convection by blood to and from the capillary bed, chemical reaction with other species within the capillary bed, and diffusion across the capillary barrier. Similar processes occur in blood oxygenators and various artificial organs. Bioreactors are another common bioengineering application in which a solute is brought by convection into contact with a stationary phase, diffuses to a reaction site, is converted to another species, diffuses back into the mobile phase, and is transported by convection out of the device.

14.6.1 Transcapillary Exchange of O_2 and CO_2

The exchange of the respiratory gases oxygen and carbon dioxide between tissues and the atmosphere is essential for the sustenance of human life. A quantitative understanding of the transport across the blood–gas barrier and the blood–tissue barrier is essential for designing therapies that can assist patients with respiratory problems. The same principles apply to the exchange of these gases between the gas and liquid phases in blood oxygenators and other devices.

14.6 Convection, Diffusion, and Chemical Reaction

14.6.1.1 Oxygen Exchange in Lung Capillaries

As was discussed in Sect. 12.8.1, oxygen is transported by two primary mechanisms in blood: physically dissolved oxygen and oxygen transported in the form of oxyhemoglobin. At normal hematocrit values, the greatest proportion of oxygen transport is via oxyhemoglobin. Most of the oxygen that diffuses across the pulmonary capillary barrier combines with hemoglobin. We must therefore account for both the dissolved and bound oxygen species. Applying steady-state species mass balances for oxygen and oxyhemoglobin over a section of a pulmonary capillary represented by Fig. 14.18:

Oxygen:

$$0 = C_{O_2,b}(x)Q_b - C_{O_2,b}(x+\Delta x)Q_b + N_{O_2}(x)S_{cap}\left(\frac{\Delta x}{L}\right) + R_{O_2}V_{cap}\left(\frac{\Delta x}{L}\right). \tag{14.193}$$

Oxyhemoglobin:

$$0 = C_{HbO_2,b}(x)Q_b - C_{HbO_2,b}(x+\Delta x)Q_b + R_{HbO_2}V_{cap}\left(\frac{\Delta x}{L}\right). \tag{14.194}$$

Hemoglobin is confined to the blood stream; so there is no flux of hemoglobin across the pulmonary capillary barrier. V_{cap} is the capillary volume, S_{cap} is the capillary internal surface area, R_{O_2} is the rate of production of oxygen per unit volume inside the capillary, and R_{HbO_2} is the rate of production of oxyhemoglobin per unit volume within the capillary. Since the reaction between oxygen and oxyhemoglobin is rapid, we can assume a local equilibrium between oxygen and oxyhemoglobin. Since each mole of oxyhemoglobin can combine with four moles of oxygen:

$$R_{O_2} = -4R_{HbO_2}. \tag{14.195}$$

Multiplying (14.194) by 4 and adding it to (14.193), then taking the limit as the shell volume approaches zero yields the following expression:

$$\frac{dC_{O_2,b}}{dx} + 4\frac{dC_{HbO_2,b}}{dx} = \frac{N_{O_2}(x)S_{cap}}{Q_bL}. \tag{14.196}$$

Instead of dealing with the oxyhemoglobin concentration, it is more common to use the oxyhemoglobin saturation $S_{HbO_2} = C_{HbO_2}/C_{Hb,tot}$, where $C_{Hb,tot}$ is the total hemoglobin concentration (unbound plus bound). Rewriting (14.196):

$$\frac{dC_{O_2,b}}{dx} + 4C_{Hb,tot}\frac{dS_{HbO_2}}{dx} = \frac{N_{O_2}S_{cap}}{Q_bL}. \tag{14.197}$$

The oxyhemoglobin saturation is a strong function of the partial pressure of oxygen in blood, as shown in Fig. 12.9. If the dependency on other factors, such as pH, pCO$_2$, etc., can be neglected as we travel down a single capillary, then as the axial position changes, the resulting change in P_{O_2} will alter the oxyhemoglobin saturation as follows:

$$\frac{dS_{HbO_2}}{dx} = \left(\frac{\partial S_{HbO_2}}{\partial P_{O_2}}\right)\frac{dP_{O_2}}{dx} = m(P_{O_2}(x))\frac{dP_{O_2}}{dx}, \quad (14.198)$$

where $m(P_{O_2})$ is the slope of the oxyhemoglobin saturation curve shown in Fig. 12.10. Henry's law can be used to relate the partial pressure of oxygen to the concentration of dissolved oxygen:

$$C_{O_2,b} = \alpha_{O_2,b} P_{O_2,b}. \quad (14.199)$$

Substituting (14.199) and (14.198) into (14.197) provides a total oxygen conservation statement in terms of the partial pressure of oxygen in the capillary as a function of axial position:

$$\frac{dP_{O_2,b}}{dx} = \frac{\left(\frac{S_{cap}}{Q_b L}\right) N_{O_2}}{\alpha_{O_2,b} + 4 C_{Hb,tot} m(P_{O_2,b})}. \quad (14.200)$$

This expression is valid for both lung capillaries and tissue capillaries. The difference lies in the expression used to describe the flux across the capillary barrier, N_{O_2}. If oxygen is consumed at a constant rate per unit volume in tissue, then $-N_{O_2}$ will be a constant given by (14.188). If we are dealing with a pulmonary capillary, where the oxygen concentration is maintained constant in the alveoli by ventilation, then the flux is expressed by (14.132):

$$N_{O_2}(x) = -\mathsf{P}_{O_2,cap}\left[C_{O_2,b}(x) - \Phi_{Abd} C_{O_2,alv}(x)\right]. \quad (14.201)$$

The flux can be written in terms of partial pressures in the alveoli and the blood stream:

$$N_{O_2}(x) = \mathsf{P}_{O_2,cap}\alpha_{O_2,b}\left[P_{O_2,alv}(x) - P_{O_2,b}(x)\right]. \quad (14.202)$$

Substituting this into (14.200):

$$\frac{dP_{O_2,b}(x)}{dx} = \left\{\frac{\left(\frac{\mathsf{P}_{O_2,cap} S_{cap}}{Q_b L}\right)}{\left(1 + \frac{4 C_{Hb,tot} m(P_{O_2,b})}{\alpha_{O_2,b}}\right)}\right\}\left[P_{O_2,alv} - P_{O_2,b}(x)\right]. \quad (14.203)$$

14.6 Convection, Diffusion, and Chemical Reaction

In many applications, the slope m of the oxyhemoglobin concentration vs. oxygen partial pressure curve is assumed to be a constant. The solution for constant m and initial blood partial pressure equal to $P_{O_2,b0}$ is:

$$\frac{P_{O_2,\text{alv}} - P_{O_2,b}(x)}{P_{O_2,\text{alv}} - P_{O_2,b0}} = \exp\left\{-\frac{\left(\dfrac{P_{O_2,\text{cap}} S_{\text{cap}}}{Q_b}\right)}{\left(1 + \dfrac{4 C_{\text{Hb,tot}} m}{\alpha_{O_2,b}}\right)} \left(\frac{x}{L}\right)\right\} \quad \text{(constant } m\text{)}. \quad (14.204)$$

The slope m is generally taken as an average of the slopes $m\,(P_{O_2,\text{alv}})$ and $m\,(P_{O_2,b0})$. Equation (14.204) reduces to (14.184) for a nonreacting solute ($m = 0$).

A more accurate approach would be to solve (14.203) numerically using data for the slope m from Fig. 12.10 or by taking the slopes of the Hill or Adair equations. A comparison between solutions using the constant slope approach vs. the actual slope is shown in Fig. 14.30. The partial pressure of oxygen at the pulmonary capillary inlet is 40 Torr and the alveolar partial pressure is 90 Torr for each graph. $P_{O_2,\text{cap}} S_{\text{cap}}/Q_b = 120$, $C_{\text{Hb,tot}} = 2.2$ mM, and the solubility coefficient $\alpha_{O_2,b}$ for oxygen in blood is 1.29×10^{-6} M/Torr. Curves are shown for the constant slope model using values of m computed at P_{O_2} values of 40 Torr (inlet P_{O_2}), 90 Torr (alveolar P_{O_2}), and 65 Torr (average of inlet and alveolar P_{O_2}). None of the constant slope curves approximate the numerical solution very well. Using the slope at $P_{O_2} = 40$ Torr (0.0128 Torr^{-1}) results in a prediction of oxygen exchange that is much too slow, while using the slope at a P_{O_2} of 90 Torr (0.0011 Torr^{-1}) results in an exchange rate that is too rapid. Use of a value for m computed at the average

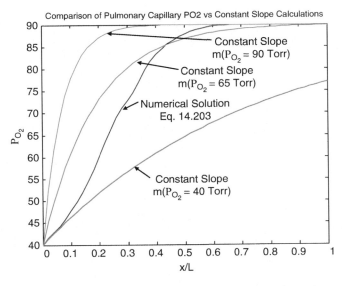

Fig. 14.30 Comparison of the numerical solution for oxygen exchange in a pulmonary capillary (14.203) with the prediction based on the constant slope model (14.204) for three different slopes

value at $P_{O_2} = 65$ Torr (0.0031Torr^{-1}) results in a prediction that is initially too rapid, then ultimately too slow. The average value could be used to provide an estimate of the minimum length of a hollow fiber needed to oxygenate blood to a particular P_{O_2}.

The total amount of oxygen exchanged in the lung is given by:

$$W_{O_2} = Q_b \left[\alpha_{O_2,b} \left(P_{O_2,a} - P_{O_2,v} \right) + 1.34 \rho_{Hb} \left(S_{O_2,a} - S_{O_2,v} \right) \right]. \quad (14.205)$$

For arterial and venous partial pressures of oxygen of 100 mmHg and 40 mmHg, respectively, $\rho_{Hb} = 15.3$ g Hb/dl blood and a cardiac output of 5 L/min, $W_{O_2}^* = 244$ mlO$_2$/min, which agrees well with the normally cited value of 250 mlO$_2$/min.

14.6.1.2 Oxygen Exchange in Tissue Capillaries

Consider now the exchange of oxygen from a capillary supplying a consuming tissue. If the consumption of oxygen is constant, then from (14.188), the flux of oxygen into the capillary is given by:

$$N_{O_2} = -\frac{R_{O_2} V_T}{S_{cap}}. \quad (14.206)$$

Substituting this into (14.197):

$$\frac{d}{dx} \left[C_{O_2,b} + 4 C_{Hb,tot} S_{HbO_2} \right] = \frac{-R_{O_2} V_T}{Q_b L}. \quad (14.207)$$

Integrating (14.207) from $x = 0$ to an arbitrary value of x:

$$\alpha_{O_2,b} \left[P_{O_2,b}(x) - P_{O_2,b0} \right] + 4 C_{Hb,tot} \left[S_{HbO_2}(x) - S_{HbO_2,0} \right] = -\frac{R_{O_2} V_T}{Q_b} \left(\frac{x}{L} \right). \quad (14.208)$$

Equation (14.208) can be solved implicitly by selecting a value of x, computing the right-hand side of the equation, making a guess for $P_{O_2,b}(x)$, which determines the left-hand side of the equation. Modifications to the guess for $P_{O_2}(x)$ are made until the left-hand side of the equation agrees with the right-hand side of the equation. This can be accomplished using the Matlab function fzero. An alternate approach would be to integrate (14.200) after substituting (14.206) for N_{O_2}:

$$\int_{P_{O_2,b0}}^{P_{O_2,b}} \left[\alpha_{O_2,b} + 4 C_{Hb,tot} m(P_{O_2,b}) \right] dP_{O_2,b} = -\left(\frac{R_{O_2} V_T}{Q_b} \right) \frac{x}{L}. \quad (14.209)$$

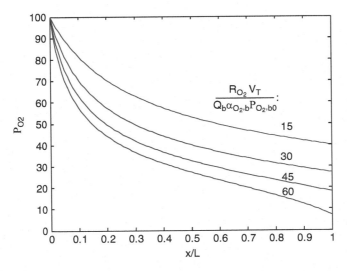

Fig. 14.31 Partial pressure in a tissue capillary vs. axial position for four different values of the dimensionless consumption rate. Inlet P_{O_2} is 100 Torr

The solution to (14.208) for four different values of the dimensionless consumption rate is shown in Fig. 14.31 when the inlet P_{O_2} is 100 Torr. When the dimensionless consumption rate rises above 65.8, the model predicts negative partial pressures at the capillary outlet. This impossibility is a consequence of the assumption that the tissue consumption rate per unit volume is constant. In reality, the consumption rate will approach first-order kinetics at low P_{O_2}, and a first-order model should be applied.

Internal Resistance to Oxygen Exchange in Capillaries

For many years, the radial distribution of oxygen was considered to be uniform in blood at any given axial position in a capillary. The analysis in the previous section, for instance, treats hemoglobin as though it were uniformly distributed throughout the capillary, both radially and axially. In reality, nearly all of the oxygen supplied to tissues dissociates from oxyhemoglobin, which is confined to red cells. Erythrocytes, in turn, are confined to the central or core region of the capillary. Therefore, an additional resistance to mass transfer is present in the peripheral region, caused by the resistance of the erythrocyte membrane and plasma gap. A more exact analysis, which includes discrete cell effects, was conducted by Hellums et al. (1996). They found that transport through this layer is characterized by a simple mass transfer coefficient k_c that depends primarily on tube hematocrit, microvessel size, and oxyhemoglobin saturation. The flux of oxygen from the core region with concentration $C_{O_2}(0)$ to the vessel wall at $r = R_c$, where the concentration is $C_{O_2}(R_c)$ is:

$$N_{O_2} = k_c(C_{O_2}(0) - C_{O_2}(R_c)) = k\alpha_{O_2}(P_{O_2}(0) - P_{O_2}(R_c)). \quad (14.210)$$

Henry's law is used in (14.210) to convert to partial pressures in the plasma gap region, where hemoglobin is not present. The mass transfer coefficient is usually presented in dimensionless form as the Sherwood number:

$$Sh = \frac{k_c(2R_c)}{D_{O_2}}. \quad (14.211)$$

The dependency of the Sherwood number for mass transfer on HbO_2 saturation levels is weak for small vessels like capillaries (Fig. 14.32a). The dependency on tube hematocrit for small capillaries is shown in Fig. 14.32b.

For a capillary with a diameter of 10 μm and tube hematocrit of 25%, the Sherwood number is about 2. In this case, the mass transfer coefficient for flux in the plasma gap is approximately equal to the ratio of the diffusion coefficient for oxygen in blood to the capillary radius:

$$k_c = \frac{ShD_{O_2}}{2R_c} \approx \frac{D_{O_2}}{R_c}. \quad (14.212)$$

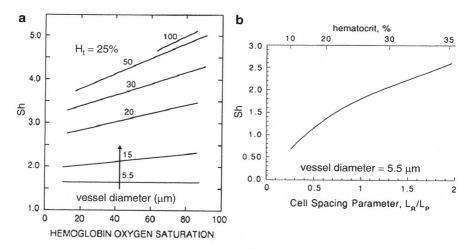

Fig. 14.32 Sherwood number for oxygen exchange between core and wall regions in microvessels, from Hellums et al. 1996 with permission. (**a**) Dependency on vessel size and oxyhemoglobin saturation levels for a tube hematocrit of 25%. (**b**) Dependency on tube hematocrit for capillaries with diameters of 5.5 μm

14.6 Convection, Diffusion, and Chemical Reaction

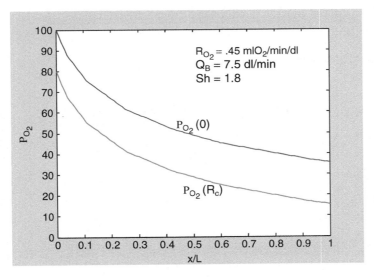

Fig. 14.33 Comparison of the partial pressure of oxygen in the core region and region near the capillary wall

If the rate of oxygen consumption in tissue is constant, then the flux of oxygen from blood to tissue can be found by combining (14.206) and (14.210):

$$N_{O_2} = \frac{R_{O_2} V_T}{S_{cap}} = k_c \alpha_{O_2} (P_{O_2}(0) - P_{O_2}(R_c)). \quad (14.213)$$

Therefore, the partial pressure drop across the plasma gap is predicted to be constant and equal to:

$$P_{O_2}(0) - P_{O_2}(R_c) = \frac{R_{O_2} V_T}{k_c \alpha_{O_2} S_{cap}}. \quad (14.214)$$

Figure 14.33 uses (14.208) to predict the axial distribution of $P_{O_2}(0)$ and (14.214) to compute $P_{O_2}(R_c)$ for $Sh = 1.8$. Note that there is a significant partial pressure drop of 20 mmHg caused by internal resistance within the capillary.

14.6.1.3 Carbon Dioxide Exchange in Lung Capillaries

The amount of carbon dioxide that can dissolve in plasma is considerably greater than the amount of oxygen that dissolves in plasma. Nevertheless, CO_2 is transported in the blood stream in the form of several different species, as discussed in Sect. 12.8.2. The relationship between total CO_2 in blood and the partial pressure of CO_2 is linear over the normal physiological range of carbon dioxide partial pressures, as shown in Fig. 12.12.

Excess carbon dioxide is eliminated from the blood as it passes through the lungs. A species balance on CO_2 over a section of lung capillary of length Δx and width w is:

$$(C_{CO_2,\text{total}} Q_b)_x - (C_{CO_2,\text{total}} Q_b)_{x+\Delta x} = N_{CO_2} w \Delta x. \quad (14.215)$$

The flux N_{CO_2} of carbon dioxide through the blood–gas barrier is:

$$N_{CO_2}(x) = \mathcal{P}_{CO_2,\text{alv}} \alpha_{CO_2,\text{alv}} \left[P_{CO_2,\text{blood}}(x) - P_{CO_2,\text{alveolus}} \right]. \quad (14.216)$$

$\mathcal{P}_{CO_2,alv}$ is the permeability of the blood–gas barrier to CO_2 and $\alpha_{CO_2,alv}$ is the solubility of CO_2 in the blood–gas barrier. The relation between total blood concentration of CO_2 and the partial pressure in blood is (12.218):

$$C_{CO_2,\text{total}} = A + B P_{CO_2}, \quad (14.217)$$

where, from Fig. 12.12, A is approximately 23.3 (ml CO_2)/(dl blood) and $B = 0.667$ ml CO_2 (dl blood)$^{-1}$ mmHg^{-1}. Substituting (14.216) and (14.217) into (14.215), and setting the surface area of the capillary S_{alv} equal to the capillary length L multiplied by the capillary width w, provides the following equation:

$$B Q_b \left[(P_{CO_2,\text{blood}})_x - (P_{CO_2,\text{blood}})_{x+\Delta x} \right]$$
$$= \frac{\mathcal{P}_{CO_2,\text{alv}} S_{\text{alv}} \alpha_{CO_2,\text{alv}}}{L} \left[P_{CO_2,\text{blood}}(x) - P_{CO_2,\text{alveolus}} \right] \Delta x. \quad (14.218)$$

Dividing by Δx and taking the limit as Δx approaches zero and applying the boundary condition that the partial pressure of carbon dioxide at the capillary entrance is that of venous blood, $P_{CO_2}(x=0) = P_{CO_2,v}$, the solution is:

$$\frac{P_{CO_2,\text{blood}}(x) - P_{CO_2,\text{alveolus}}}{P_{CO_2,v} - P_{CO_2,\text{alveolus}}} = \exp\left\{ -\left(\frac{\mathcal{P}_{CO_2,\text{alv}} S_{\text{alv}}}{Q_b} \right) \left(\frac{\alpha_{CO_2,\text{alv}}}{B} \right) \left(\frac{x}{L} \right) \right\}. \quad (14.219)$$

At the capillary outlet, $x = L$, $P_{CO_2,\text{blood}} = P_{CO_2,a}$ the partial pressure of arterial blood:

$$\frac{P_{CO_2,a} - P_{CO_2,\text{alveolus}}}{P_{CO_2,v} - P_{CO_2,\text{alveolus}}} = \exp\left\{ -\left(\frac{\mathcal{P}_{CO_2,\text{alv}} S_{\text{alv}}}{Q_b} \right) \left(\frac{\alpha_{CO_2,\text{alv}}}{B} \right) \right\}. \quad (14.220)$$

From a macroscopic balance, the total flow of CO_2 across the blood–gas barrier for a single capillary is:

$$W_{CO_2} = Q_b B \left[P_{CO_2,v} - P_{CO_2,a} \right]. \quad (14.221)$$

14.6 Convection, Diffusion, and Chemical Reaction

Substituting (14.220) into (14.221):

$$W_{CO_2} = Q_b B \left[P_{CO_2,v} - P_{CO_2,a} \right] \left[1 - \exp\left\{ -\left(\frac{P_{CO_2,alv} S_{alv}}{Q_b} \right) \left(\frac{\alpha_{CO_2,alv}}{B} \right) \right\} \right]. \quad (14.222)$$

Under normal circumstances, the exponential term is very small so that $P_{CO_2,a}$ is approximately $P_{CO_2,alveolus}$.

$$W_{CO_2} \approx Q_b B \left[P_{CO_2,v} - P_{CO_2,alveolus} \right]. \quad (14.223)$$

The rate of exchange across the entire lung for a cardiac output of 5 L/min is:

$$\begin{aligned} W_{CO_2} &\approx (5,000 \text{ ml/min})(0.00667 \text{ mlO}_2 \text{ ml}^{-1} \text{ mmHg}^{-1})[6 \text{ mmHg}] \\ &= 200 \text{ ml/min}. \end{aligned} \quad (14.224)$$

The volume of CO_2 expired per minute is 80% of the volume of O_2 inspired per minute (250 ml/min). The *respiratory quotient*, defined as the ratio of carbon dioxide expired to oxygen inspired, is normally about 0.8.

14.6.1.4 Carbon Dioxide Exchange in Tissue Capillaries

If the rate of production of carbon dioxide per unit volume R_{CO_2} is constant in tissue, then the one-dimensional shell balance method described in Sect. 14.5.3.2 leads to the following relationship for the concentration of carbon dioxide in the capillary:

$$C_{CO_2,b}(x) = C_{CO_2,b}(0) + \frac{R_{CO_2} V_T}{Q_b} \left(\frac{x}{L} \right). \quad (14.225)$$

The total concentration of CO_2 is related to the partial pressure of CO_2 using (14.217). This allows us to find the partial pressure of carbon dioxide as a function of axial position:

$$P_{CO_2,b}(x) = P_{CO_2,a} + \frac{R_{CO_2} V_T}{B Q_b} \left(\frac{x}{L} \right). \quad (14.226)$$

Therefore, the partial pressure of carbon dioxide is predicted to increase linearly with distance along the capillary. At the capillary outlet:

$$P_{CO_2,v} - P_{CO_2,a} = \frac{R_{CO_2} V_T}{B Q_b}. \quad (14.227)$$

This expression can be rearranged to provide an estimate for the rate of production per unit volume of CO_2 in the tissue from measured values:

$$R_{CO_2} = \frac{Q_b}{V_T} B \left(P_{CO_2,v} - P_{CO_2,a} \right). \quad (14.228)$$

For example, if the blood flow per unit volume is 0.6 min^{-1}, $B = 0.00667$ mlCO$_2$ ml^{-1} mmHg^{-1}, and the partial pressure difference is 6 mmHg, $R_{CO_2} = 0.024$ mlCO$_2$min^{-1}ml^{-1}.

14.6.2 *Tissue Solute Exchange, Krogh Cylinder*

The capillary bed in tissue, particularly muscle, is often modeled as though each capillary has the same size and flow rate. In addition, each capillary is assumed to be equidistant from its neighboring capillaries as shown in the top panel of Fig. 14.34. A tissue cylinder can be defined with a radius R_T equal to half the distance between adjacent capillaries. The tissue surrounding a particular capillary is assumed to be exclusively supplied with nutrients by that capillary, and waste products are assumed to be exclusively removed from the tissue cylinder by the same capillary. Consequently, from the standpoint of mass exchange in the tissue, each tissue cylinder can be treated as though it is a functional unit without any interactions with adjacent units. In mass transfer terms, a boundary condition that applies at the edge of the cylinder ($r = R_T$) is that the flux of any solute is zero.

This ideal tissue unit, shown in the bottom panel of Fig. 14.34, is known as a Krogh Cylinder. It is named after August Krogh (1919) who first proposed it for muscle. The Krogh cylinder is, of course, an idealization of true mass exchange tissue. It excludes some tissue contained in the regions outside the boundaries of

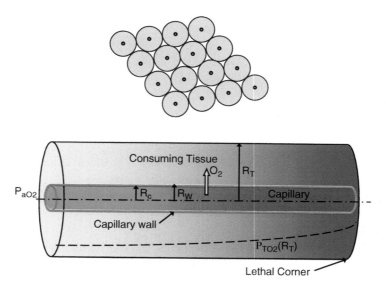

Fig. 14.34 Krogh cylinder. *Top*: tissue cross-section with evenly distributed capillaries. *Bottom*: longitudinal section of a single capillary and its surrounding tissue

14.6 Convection, Diffusion, and Chemical Reaction

adjacent cylinders (top panel of Fig. 14.34). However, it is an extremely informative model that allows us to learn much about solute exchange in tissue that would otherwise be virtually intractable. Although this is a two-dimensional problem, the axial dependence is dictated by the intracapillary solution; so the one-dimensional approach treated in this chapter is appropriate.

14.6.2.1 Oxygen Exchange in a Krogh Cylinder

Our goal in this section is to predict the partial pressure of oxygen in tissue as a function of position in the tissue. We will make the following assumptions:

- Tissue consumption of oxygen follows zeroth order kinetics
- Axial diffusion of oxygen is negligible
- There is no oxygen flux at the outer tissue surface ($r = R_T$)
- Radial symmetry
- $P_{O_2}(R_T) > 0$
- Allow for capillary wall resistance but neglect oxygen consumption in capillary wall

We begin our analysis with conservation of oxygen in a segment of tissue with length Δx, bounded in the radial direction between $r = R_W$ and $r = R_T$, as shown in Fig. 14.35. Since axial diffusion is assumed negligible, oxygen enters the tissue segment at a rate ΔW_{O_2} through the capillary wall. All the oxygen that enters the tissue segment is consumed at a rate per unit volume R_{O_2}, since it cannot leave through the surface at $r = R_T$ and does not leave by diffusion through surfaces at x

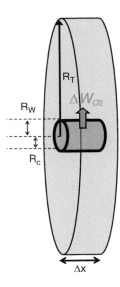

Fig. 14.35 Capillary segment

or $x + \Delta x$. A steady-state molar balance on oxygen exchange in the tissue segment yields:

$$\Delta W_{O_2,T} = R_{O_2} \frac{V_T}{L} \Delta x = R_{O_2} \left(\pi R_T^2 - \pi R_W^2 \right) \Delta x. \quad (14.229)$$

Dividing by Δx:

$$\frac{\Delta W_{O_2,T}}{\Delta x} \equiv W'_{O_2} = R_{O_2} \frac{V_T}{L} = R_{O_2} \left(\pi R_T^2 - \pi R_W^2 \right), \quad (14.230)$$

where W'_{O_2} is a constant that can be interpreted as the oxygen flow from capillary to tissue per unit capillary length or the tissue consumption rate of oxygen per unit capillary length. This is constant by virtue of the assumption of a constant consumption rate of oxygen per unit volume.

The partial pressure in the capillary as a function of position has already been computed in Sects. 14.6.1.1 and 14.6.1.2. Substituting (14.230) into (14.208), the partial pressure of oxygen in the core region is given by:

$$P_{O_2,b}(x,0) - P_{O_2,b0} = \frac{4C_{\text{Hb,tot}}}{\alpha_{O_2,b}} \left[S_{\text{HbO}_2,0} - S_{\text{HbO}_2}(x) \right] - \frac{W'_{O_2} L}{Q_b \alpha_{O_2,b}} \left(\frac{x}{L} \right). \quad (14.231)$$

The partial pressure drop across the plasma gap is (14.214):

$$P_{O_2,b}(x,0) - P_{O_2,b}(x,R_c) = \frac{W'_{O_2}}{2\pi R_c k_c \alpha_{O_2,b}}. \quad (14.232)$$

The partial pressure in the capillary at $r = R_c$ will be used as a boundary condition at the inside surface of the capillary wall. The radial distribution of oxygen in tissue can be found by applying conservation of oxygen to a shell bounded by x, $x + \Delta x$, r, and $r + \Delta r$, as shown in Fig. 14.36:

$$W_{O_2}|_r - W_{O_2}|_{r+\Delta r} - R_{O_2}(2\pi r \Delta x \Delta r) = 0. \quad (14.233)$$

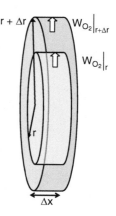

Fig. 14.36 Capillary tissue segment

14.6 Convection, Diffusion, and Chemical Reaction

Writing the oxygen flow in terms of the oxygen flux:

$$2\pi \Delta x \left[(rN_{O_2,T})_r - (rN_{O_2,T})|_{r+\Delta r} \right] - R_{O_2}(2\pi r \Delta x \Delta r) = 0. \tag{14.234}$$

Dividing by the shell volume and letting the volume approach zero, we obtain the following partial differential equation:

$$\frac{1}{r}\frac{\partial}{\partial r}(rN_{O_2,T}) = -R_{O_2}. \tag{14.235}$$

Integrating:

$$N_{O_2,T} = -R_{O_2}\frac{r}{2} + \frac{f(x)}{r}, \tag{14.236}$$

where $f(x)$ is an unknown function of x. Applying the no flux boundary condition at $r = R_T$ for all values of x shows that the function f is $R_{O_2}R_T^2/2$, which is actually independent of x. This is consistent with the assumption that axial diffusion in the tissue is negligible. The flux of oxygen in the tissue depends only on the radial position:

$$N_{O_2,T}(r) = \frac{R_{O_2}}{2}\left[\frac{R_T^2}{r} - r\right]. \tag{14.237}$$

We can follow the same procedure for oxygen exchange in the capillary wall. The flux must only depend on radial position and we neglect oxygen consumption in the capillary wall:

$$N_{O_2,W}(r) = \frac{C_1}{r}. \tag{14.238}$$

C_1 is a constant, which can be determined by applying the boundary condition at the inside radius of the capillary wall. At $r = R_c$, the flux into the capillary wall must equal the flux out of the capillary blood at $r = R_c$. The flux out of the capillary is:

$$N_{O_2,c}(R_c) = \frac{W_{O_2,\text{total}}}{2\pi R_c L} = \frac{R_{O_2}\pi L(R_T^2 - R_W^2)}{2\pi R_c L} = \frac{W'_{O_2}}{2\pi}. \tag{14.239}$$

Using the boundary condition at $r = R_c$, we find the radial dependence of oxygen flux in the capillary wall to be:

$$N_{O_2,W}(r) = \frac{W'_{O_2}}{2\pi r}. \tag{14.240}$$

Applying Fick's law and Henry's law:

$$-D_{O_2,W}\frac{\partial C_{O_2}}{\partial r} = -D_{O_2,W}\alpha_{O_2,W}\frac{\partial P_{O_2}}{\partial r} = \frac{W'_{O_2}}{2\pi r}. \quad (14.241)$$

Integrating, we obtain an expression for the partial pressure of oxygen in the capillary wall:

$$P_{O_2,W}(x,r) = -\frac{W'_{O_2}}{2\pi\alpha_{O_2,W}D_{O_2,W}}\ln(r) + g(x). \quad (14.242)$$

The function $g(x)$ can be found using the boundary condition at $r = R_c$, where the partial pressure of oxygen in the capillary wall must equal the partial pressure of oxygen in the capillary blood at the wall, $P_{O_2,\text{blood}}(x)$:

$$g(x) = P_{O_2,\text{blood}}(x,R_c) + \frac{W'_{O_2}}{2\pi\alpha_{O_2,W}D_{O_2,W}}\ln(R_c). \quad (14.243)$$

The final expression for $P_{O_2,W}$ is:

$$P_{O_2,W}(x,r) = P_{O_2,\text{blood}}(x,R_c) - \frac{W'_{O_2}}{2\pi\alpha_{O_2,W}D_{O_2,W}}\ln\left(\frac{r}{R_c}\right). \quad (14.244)$$

Applying Fick's law and Henry's law to the expression for flux in the tissue, (14.237):

$$N_{O_2,T}(r) = -D_{O_2,T}\frac{\partial C_{O_2,T}}{\partial r} = -D_{O_2,T}\alpha_{O_2,T}\frac{\partial P_{O_2,T}}{\partial r} = \frac{R_{O_2}}{2}\left[\frac{R_T^2}{r} - r\right]. \quad (14.245)$$

Integrating and applying the boundary condition at $r = R_W$, we obtain the final expression for the partial pressure of oxygen in the tissue as a function of axial position and radial position:

$$P_{O_2,T}(x,r) - P_{O_2,W}(x,R_W) = \frac{R_{O_2}R_T^2}{4\alpha_{O_2,T}D_{O_2,T}}\left[\frac{r^2}{R_T^2} - \frac{R_W^2}{R_T^2} - 2\ln\left(\frac{r}{R_W}\right)\right]. \quad (14.246)$$

A schematic of the radial variation of partial pressure of oxygen in the Krogh cylinder at a single axial location is shown in Fig. 14.37. $P_{O_2,T}(x,r)$ is given by (14.246), $P_{O_2,W}(x,r)$ is given by (14.244), $P_{O_2,b}(x,R_c)$ can be found from (14.232), and $P_{O_2,b}(x,0)$ can be computed from (14.231).

The distribution of P_{O_2} in a Krogh cylinder as a function of x and r is shown in Fig. 14.38 for the following conditions: $R_c = 2.5$ μm, $R_W = 2.8$ μm, $R_T = 18$ μm, $L = 0.05$ cm, $P_{O_2}(x=0) = 100$ mmHg, $R_{O_2} = 0.44$ mlO$_2$ min^{-1} ml^{-1},

14.6 Convection, Diffusion, and Chemical Reaction

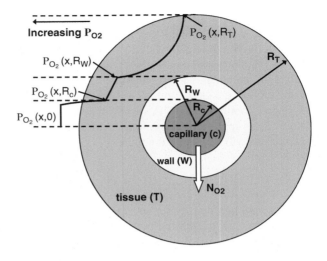

Fig. 14.37 Distribution of oxygen partial pressure in blood, capillary wall, and tissue

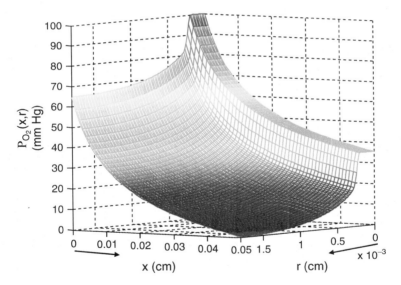

Fig. 14.38 Partial pressure of oxygen as a function of axial and radial position in a Krogh cylinder

$Q_b = 3.83 \times 10^{-3}$ μl/min, and $Sh = 2.5$. The P_{O_2} in the capillary is assumed to be constant in the core, and this is assumed to comprise 80% of the capillary radius. The P_{O_2} drops over the last 20% to the value at the interior capillary wall. If a Sherwood number of 2.5 is accurate, the drop in partial pressure between the core and capillary radius is about 20 mmHg, which is slightly more than the drop between the capillary wall and the edge of the tissue cylinder. The P_{O_2} at the edge of the cylinder at the venous end is where the tissue P_{O_2} is lowest. This is known as the "lethal corner." If the P_{O_2} is near zero in this region, tissue cells may not receive enough oxygen to remain viable.

The capillary parameters for a single Krogh cylinder (R_W, R_T, Q_b, W'_{O_2}) can be related to parameters more commonly associated with an entire muscle. Capillary radius R_c is considered to be a known quantity, generally in the neighborhood of 5–8 μm, capillary length is approximately 500 μm, and capillary wall thickness $R_W - R_c$ is approximately 0.3 μm. A tissue that contains a number N_c of parallel capillaries will have a total volume V_{total}:

$$V_{\text{total}} = N_c \pi R_T^2 L. \tag{14.247}$$

The capillary density for several tissues has been measured. This is defined as the number of capillaries per unit cross-sectional area of a tissue, generally expressed as the number of capillaries per square millimeter. For a tissue with a cross-sectional area of V_{total}/L:

$$\text{Capillary Density} = \frac{N_c}{V_{\text{total}}/L} = \pi R_T^2. \tag{14.248}$$

Therefore, the tissue radius in micrometers can be estimated from the capillary density:

$$R_T(\mu m) = \sqrt{\frac{\text{Capillary Density (capillaries mm}^{-2})}{\pi}}. \tag{14.249}$$

For example, a tissue with a capillary density of 982 capillaries/mm² would have a tissue radius of about 18 μm. The number of capillaries per ml of tissue with a capillary density of 982 capillaries/mm² would be 1.96×10^6 per ml. The volume of tissue that is consuming oxygen $V_{T,\text{total}}$ relative to the total volume (including capillary volume and capillary wall volume) is:

$$\frac{V_{T,\text{total}}}{V_{\text{total}}} = 1 - \frac{R_W^2}{R_T^2}. \tag{14.250}$$

If $R_W = 2.8$ μm and $R_T = 18$ μm, the consuming tissue would comprise 97.5% of the total tissue volume. This excludes the volume occupied by large blood vessels. Total blood flow, $Q_{b,\text{total}}$, and total oxygen exchange in tissue $W_{O_2,\text{total}}$ are:

$$Q_{b,\text{total}} = N_c Q_b, \tag{14.251}$$

$$W_{O_2,\text{total}} = N_c W'_{O_2} L = R_{O_2} V_{T,\text{total}}. \tag{14.252}$$

14.6 Convection, Diffusion, and Chemical Reaction

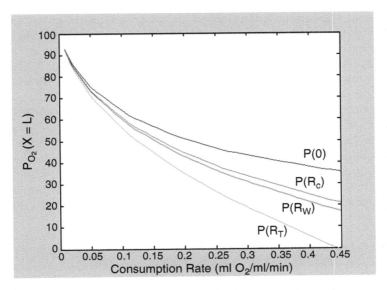

Fig. 14.39 Effect of tissue consumption rate of oxygen on the partial pressures at the venous end of the Krogh cylinder

The capillary parameters in (14.230) can be eliminated and replaced with tissue-specific terms by combining it with these two equations:

$$\frac{W'_{O_2} L}{Q_b} = \frac{W_{O_2,\text{total}}}{Q_{b,\text{total}}} = \frac{R_{O_2} V_{T,\text{total}}}{Q_{b,\text{total}}} = \frac{R_{O_2} V_{\text{total}}}{Q_{b,\text{total}}} \left[1 - \frac{R_W^2}{R_T^2} \right]. \quad (14.253)$$

The Krogh cylinder solution shown in Fig. 14.38 corresponds to a tissue with a volume of 100 ml, blood flow of 750 ml/min, and R_{O_2} of 0.44 mlO$_2$ min^{-1} ml^{-1}. The effect of changing the consumption rate on the partial pressure of oxygen at the venous end of the capillary is shown in Fig. 14.39. As the consumption rate approaches 0.44 mlO$_2$ min^{-1} ml^{-1}, the partial pressure at the lethal corner approaches zero. The model is not valid for consumption rates above this value because the assumption that consumption kinetics is zeroth order will lead to predictions of negative partial pressures for oxygen in the tissue. First order or Michaelis–Menten kinetics will provide more accurate results at high consumption rates. However, in those cases, the solution in the capillary described in Sect. 14.6.1.2 can no longer be uncoupled from the solution in the tissue.

14.6.3 Bioreactors

The term "bioreactor" is used to describe at least two separate types of devices. One device is used to control the conditions under which cells are cultured, generally for use in tissue engineering. Some of the variables controlled in these devices include the cell type, reactor materials, culture media composition, oxygen concentration (for aerobic cells), stir rate (batch processes), flow rate (continuous feed processes), pH, temperature, and removal of byproducts. One of the most difficult tasks engineers face with this type of bioreactor is scaling the device from what works well in the laboratory to a larger commercial bioreactor.

A second type of bioreactor is a device in which cells or substances derived from cells are used to promote specific chemical reactions. Common examples include the anaerobic fermentation of sugars in malt, grapes, or corn. Yeast cells are mixed with these substances and enzymes within the cells convert the sugars to ethanol found in beer, wine, or fuel additives. Drugs, such as insulin, can now be produced by genetically engineered bacteria using a similar process. These are known as "batch" processes, in which the cells and raw materials are allowed to react in large vessels for some time before undergoing additional processing.

An alternative type of bioreactor is known as a "continuous feed" bioreactor. Raw materials are fed at a constant rate at the inlet of these reactors and the desired product flows from the outlet. Cells are usually immobilized in a bioreactor by encapsulating the cells inside a polymer framework. The cells are constrained from moving, but substrate and product can easily diffuse between the mobile and stationary phases of the bioreactor. Antibiotics such as penicillin and ampicillin have been produced in immobilized cell bioreactors. The encapsulation concept is also used in the design of implantable artificial organs. Cells encapsulated in artificial organs must also be protected from the immune response of the host. Consequently, the barrier that separates encapsulated cells from the mobile phase must be porous enough to allow easy passage of the substrates and products, but must prevent the movement of large molecules such as antibodies and complement.

Other bioreactors function by immobilizing the enzymes responsible for converting the substrate(s) to product(s), rather than encapsulation of living cells. Immobilized urease and sulfide oxidase can be immobilized in a bioreactor to simultaneously remove urea and harmful sulfides from wastewater. Immobilized enzymes such as heparinase have been used for blood detoxification. A recently introduced therapeutic approach is to load red blood cells or liposomes with exogenous enzymes; so these carriers behave as intravascular bioreactors with virtually no immune response.

14.6.3.1 Analysis of an Imbedded Enzyme Bioreactor

Consider a bioreactor in which a toxic material s is converted to a nontoxic material P with the assistance of an enzyme E. The bioreactor is a continuous feed device consisting of a stationary phase containing the immobilized enzymes and a mobile phase that is devoid of enzyme. The toxic substance, dissolved in the mobile phase, flows into the bioreactor, where it comes in contact with particles which comprise the stationary phase. The particles are porous, with enzymes located in the walls of the tortuous pores, similar to those analyzed in Sect. 14.4.4. Some of the toxic

14.6 Convection, Diffusion, and Chemical Reaction

substrate diffuses from the mobile phase into one of these pores, eventually coming in contact with an enzyme, where it is converted to the nontoxic product. The product then diffuses away from the enzyme surface through the same pore traversed by the substrate, eventually diffusing back into the mobile phase, where it is swept out of the bioreactor.

How can we model this process? Diffusion in the particle pores has been analyzed in Sect. 14.4.4. This is fundamentally different than convection in the mobile phase. In addition, transport of substrate and product between the mobile and stationary phases takes place across a nearly stationary fluid film. Consequently, we should account for all three of these regions in a realistic model. We shall use a subscript "b" to designate the bulk fluid, a subscript "p" to designate fluid within particle pores, and a subscript "f" to designate fluid in the static film between the bulk fluid and the pore fluid.

14.6.3.2 Bioreactor: Analysis of the Mobile Phase

Let us begin by applying a steady-state species mass balance on the mobile phase. The shell shown in Fig. 14.40 contains all the bulk fluid between planes at x and $x + \Delta x$. It excludes fluid contained within particle pores and the surrounding static film.

$$\left\{ \begin{array}{c} \text{Rate substrate} \\ \text{flows into shell} \end{array} \right\} - \left\{ \begin{array}{c} \text{Rate substrate} \\ \text{flows out of shell} \end{array} \right\} + \left\{ \begin{array}{c} \text{Rate substrate} \\ \text{produced in shell} \end{array} \right\} = 0. \tag{14.254}$$

Substrate enters the shell by convection and leaves by two mechanisms. Some leaves by convection at $x + \Delta x$ and some is transported into the film layer, where it diffuses into the pore system of the particles. Species s does not react with any other species in the bulk fluid, so the last term in (14.254) is zero. The steady-state species shell balance is:

$$(Q_V C_{sb})|_x - (Q_V C_{sb})|_{x+\Delta x} - N_{sf} \frac{S_p}{L} \Delta x = 0. \tag{14.255}$$

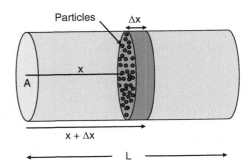

Fig. 14.40 Mobile phase shell

C_{sb} is the substrate concentration in the bulk fluid, away from the particle surfaces, Q_V is the volumetric flow rate in the mobile phase, N_{sf} is the flux of species s from the bulk fluid to the film, and S_p/L is the ratio of the surface area of all particles to the length of the particle bed. This assumes that the film thickness is very small; so the contact surface area between bulk fluid and film fluid is $S_p \Delta x/L$. Dividing by Δx and letting the shell volume approach zero:

$$\frac{dC_{sb}}{dx} = -\frac{S_p}{Q_V L} N_{sf}(x). \quad (14.256)$$

The flow of species s through the film must be the same as the flow into the pores:

$$W_s = N_{sf} S_p = -N_{sp} \phi_p S_p. \quad (14.257)$$

The negative sign arises from the convention that a positive flux inside the particle is a flux in the positive r-direction. Therefore, $-N_{sp}$ is the flux into a particle. Replacing N_{sf} with $-\phi_p N_{sp}$ in (14.256) yields:

$$\frac{dC_{sb}}{dx} = \frac{S_p \phi_p}{Q_V L} N_{sp}(x). \quad (14.258)$$

The outward flux at the particle surface can be found in terms of the dimensionless radius and concentration from Sect. 14.4.4, (14.97):

$$N_{sp} = -D_{se} \frac{dC_{sp}}{dr}\bigg|_{r=R_p} = -\left(\frac{D_{se} C_{sp}(R_p)}{R_p}\right) \frac{dC_{sp}^*}{dr^*}\bigg|_{r^*=1} = -\left(\frac{D_{se} C_{sp}(R_p)}{R_p}\right) \Psi^*\big|_{r^*=1}. \quad (14.259)$$

The function Ψ^* is defined by (14.102). It will depend on radial position in the particle, the Thiele modulus, and $\beta = C_{sp}(R_p)/K_m$. For a given reactor, the Thiele modulus is a constant. Therefore, Ψ^* at $r^* = 1$ will depend only on β at any given axial location. If species s is lost by chemical reaction in the particles, $\Psi^*|_{r^*=1}$ will be positive, and if species s is produced by chemical reaction $\Psi^*|_{r^*=1}$ will be negative. Since β can vary with x, $\Psi^*|_{r^*=1}$ will vary with axial position in the reactor. Substituting (14.259) into (14.258):

$$\frac{dC_{sb}}{dx} = \frac{S_p \phi_p}{Q_V L}\left(-\left(\frac{D_{se} C_{sp}(R_p)}{R_p}\right) \Psi^*\big|_{r^*=1}\right). \quad (14.260)$$

At any given axial location, the flux of substrate across the film separating the bulk fluid with concentration C_{sb} from the fluid in the film at a particle surface with concentration C_{sf} can be expressed as:

14.6 Convection, Diffusion, and Chemical Reaction

$$N_{sf} = k(C_{sb} - C_{sf}), \qquad (14.261)$$

where k is a mass transfer coefficient in the film. The Sherwood number for mass transfer in a packed bead bed is given by (12.131) as a function of the diffusion coefficient of the substance in the film D_{sf}, the particle radius R_p, the kinematic viscosity of the fluid v, and the superficial velocity in the reactor ($v_0 = Q_v/A_{\text{reactor}}$):

$$Sh = \frac{k(2R_p)}{D_{sf}} = 25 \left(\frac{2R_p v_0}{v} \right)^{0.45} \left(\frac{v}{D_{sf}} \right)^{0.5}. \qquad (14.262)$$

Fluid in the film is in equilibrium with fluid in the particle pore at the surface:

$$C_{sf} = \Phi_{sfp} C_{sp}(R_p) = \frac{C_{sp}(R_p)}{\Phi_{spf}}. \qquad (14.263)$$

Therefore, since the flux in the film and the pore are related as $N_{sf} = -\phi_p N_{sp}$, then:

$$k\left(C_{sb} - \frac{C_{sp}(R_p)}{\Phi_{spf}}\right) = -\phi_p \left(\frac{-D_{se} C_{sp}(R_p)}{R_p} \right) \Psi^*|_{r^*=1}. \qquad (14.264)$$

Solving for $C_{sp}(R_p)$ in terms of C_{sb}:

$$C_{sp}(R_p) = \frac{\Phi_{spf} C_{sb}}{\left(\frac{\Psi^*|_{r^*=1}}{Bi_s} + 1 \right)}, \qquad (14.265)$$

where the Biot number, Bi_s, represents the ratio of transport resistance inside the particle to resistance in the film outside the particle:

$$Bi_s = \frac{kR_p}{\phi_p D_{se} \Phi_{spf}}. \qquad (14.266)$$

Substituting (14.265) back into (14.260) yields a differential equation in terms of the bulk solute concentration:

$$\frac{dC_{sb}}{dx} = -\left(\frac{\Psi^*|_{r^*=1}}{(\Psi^*|_{r^*=1} + Bi_s)} \right) \frac{S_p k}{Q_V} \frac{C_{sb}}{L}. \qquad (14.267)$$

In general, (14.267) must be solved numerically, since $\Psi^*|_{r^*=1}$ is a function of β, which in turn is a function of axial position. One method of solving this is provided in Sect. 14.6.3.5. If $Bi_s \gg \Psi^*|_{r^*=1}$, the right-hand side of (14.267) will be independent of the mass transfer coefficient in the film, k. Before treating the general case, let us look at the limiting cases where β is large (zeroth-order reaction) and where β is small (first-order reaction).

14.6.3.3 Zeroth-Order Reaction in the Stationary Phase ($\beta \gg 1$)

For the case where the concentration of species s is high relative to the Michaelis constant, the flux of species s out of the particle can be computed by substituting (14.107) into (14.87):

$$N_{sp} = -D_{se} \frac{dC_{sp}}{dr}\bigg|_{r=R_p} = -\left(\frac{R_p}{3}\right)V_{max}. \tag{14.268}$$

Consequently, the flux of species s from bulk fluid into the particles is independent of axial position as long as the concentration is high relative to the Michaelis constant in the particles. Thus, for the case of zero-order reaction, we can substitute (14.268) into (14.258) to obtain:

$$\frac{dC_{sb}}{dx} = -\frac{S_p \phi_p}{Q_v L}\left(\frac{R_p}{3}\right)V_{max}. \tag{14.269}$$

Since the right-hand side is constant, species s concentration decreases linearly with position in the device:

$$C_{sb}(x) = C_{sb}(0) - \left[\frac{S_p V_{max} \phi_p}{Q_v}\left(\frac{R_p}{3}\right)\right]\frac{x}{L}. \tag{14.270}$$

The product of the surface area of all of the spherical particles and $(R_p/3)$ is equal to the volume of all of the spherical particles in the device, V_p. The volume of the particles can be expressed in terms of the volume of the device, V_d (mobile phase plus stationary phase), and the void volume of the device, ε:

$$S_p\left(\frac{R_p}{3}\right) = V_p = V_d(1 - \varepsilon). \tag{14.271}$$

In addition, if the device volume per unit length is the cross-sectional area of the device, and the flow divided by the cross-sectional area is the superficial velocity in the device, v_0. Consequently:

$$\frac{V_d}{Q_v L} = \frac{1}{v_0}. \tag{14.272}$$

14.6 Convection, Diffusion, and Chemical Reaction

Substituting (14.271) and (14.272) into (14.269) yields:

$$C_{sb}(x) = C_{sb}(0) - \left[\frac{(1-\varepsilon)\phi_p}{v_0}V_{max}\right]x. \quad (14.273)$$

Applying (14.272), we find the concentration of toxic substance at the outlet of the device to be:

$$C_{sb}(L) = C_{sb}(0) - \frac{(1-\varepsilon)V_d\phi_p V_{max}}{Q_V}. \quad (14.274)$$

The solution is independent of the mass transfer coefficient k in the film, the diffusion coefficient of the species in the film D_{sf}, the effective diffusion coefficient of the species in the particle pores D_e, the partition coefficient Φ_{spf}, and the Michaelis constant K_m. Note that (14.274) predicts negative concentrations for inlet concentrations less than $(1-\varepsilon)V_d\phi V_{max}/Q_V$. Of course, negative concentrations are not possible. Equation (14.274) is only valid when the concentration of toxic material within the particles is much greater than the Michaelis constant. As we move toward the exit of the bioreactor, the concentration in the bulk fluid drops. Consequently, the original assumption that $C_{sp}(R_p) \gg K_m$ may not be valid beyond some axial position in the reactor. The concentration of toxic material outside the particles will always be higher than inside the particle pores. If the Michaelis constant is high, then relatively high outlet concentrations of species s in the bulk fluid might be required for (14.274) to be valid. This is not compatible with the purpose of the device, which is to remove as much toxic material as is practical. However, if K_m is much lower than the desired exit concentration, the zero-order analysis may be appropriate along the entire length of the device.

14.6.3.4 First-Order Reaction in the Stationary Phase ($\beta \ll 1$)

The flux of species s out of a particle in which $C_{sp}(R_p) \ll K_m$ can be computed from Fick's law and (14.110):

$$N_{sp} = -\gamma \frac{D_e}{R_p} C_{sp}(R_p), \quad (14.275)$$

where γ is defined as:

$$\gamma = 3\phi_T \coth(3\phi_T) - 1. \quad (14.276)$$

Comparing the flux in (14.275) for the first-order case with the general flux in (14.259), we find that:

$$\Psi^*|_{r^*=1} = \gamma \quad (\beta \ll 1). \tag{14.277}$$

Substituting this into the general shell balance relationship, (14.267):

$$\frac{dC_{sb}}{dx} = -\left(\frac{\gamma}{Bi_s + \gamma}\right)\frac{kS_p}{Q_V}\frac{C_{sb}}{L}. \tag{14.278}$$

The solution for the first-order case is:

$$C_{sb}(x) = C_{sb}(0)\exp\left\{-\left(\frac{\gamma}{\gamma + Bi_s}\right)\frac{kS_p}{Q_V}\frac{x}{L}\right\}. \tag{14.279}$$

The concentration of the toxic species drops exponentially with position. The lower the value of k, the longer must be the bioreactor to remove the same amount of toxic material. S_p is the surface area of all particles, $S_p = 4\pi n_p R_p^2$, where n_p is the number of spherical particles in the bioreactor. The concentration of toxic substance at the outlet of the device can be found by letting $x/L = 1$ in (14.279).

Note that the mass transfer coefficient k appears explicitly in this expression. If the film was truly stationary, the mass transfer coefficient would equal the diffusion coefficient of species s in the film, D_{sf}, divided by the film thickness, δ. The Biot number in that case would be:

$$Bi_s = \left(\frac{1}{\phi_p}\right)\left(\frac{1}{\Phi_{spf}}\right)\left(\frac{D_{sf}}{D_{se}}\right)\left(\frac{R_p}{\delta}\right) \quad \text{(stationary film)}. \tag{14.280}$$

Each term in parentheses on the right-hand side of (14.280) is greater than unity; so it is likely that the Biot number is relatively high. If $Bi_s \gg \gamma$, then the pore concentration at the surface will be in equilibrium with the bulk concentration:

$$C_{sp}(R_p) = \Phi_{spf}C_{sb}, \quad (Bi_s \gg \gamma). \tag{14.281}$$

In that case, the bulk concentration will be independent of k:

$$C_{sb}(x) = C_{sb}(0)\exp\left\{-\left(\left(\frac{D_{se}S_p}{R_p Q_V}\right)\phi_p \gamma \Phi_{spf}\right)\frac{x}{L}\right\}, \quad (Bi_s \gg \gamma). \tag{14.282}$$

14.6.3.5 Michaelis–Menten Kinetics in the Stationary Phase

For the general case where the concentration in the pore near the surface of the particles is of the same order of magnitude as K_m, we must apply a numerical solution to (14.267). One method for solving this would be:

1. Make an initial guess at $C_p(R_p)$ at the inlet of the bioreactor. A good starting point would be to assume equilibrium between the bulk fluid at the inlet and the concentration just inside the particle pores:

$$C_{sp}(R_p)\big|_{x=0} = \Phi_{spf} C_{sb}(0) \quad \text{(initial guess)}. \qquad (14.283)$$

2. Compute β at the inlet:

$$\beta(x=0) = \frac{C_{sp}(R_p)\big|_{x=0}}{K_m} \quad \text{(initial guess)}. \qquad (14.284)$$

3. Knowing β and ϕ_T, compute $\Psi^*\big|_{r^*=1}$ at $x=0$ by solving (14.102)–(14.104) or by finding the effectiveness factor from Fig. 14.15 and using (14.115) to compute $\Psi^*\big|_{r^*=1}$.
4. Compute the concentration just inside the particle pores at $x=0$ using (14.265):

$$C_{sp}(R_p)\big|_{x=0} = \frac{\Phi_{spf} C_{sb}(0)}{\left(1 + \dfrac{\Psi^*\big|_{r^*=1}}{Bi_s}\right)}. \qquad (14.285)$$

5. Compare the computed concentration from (14.285) with the initial guess in (14.283). If the two agree within 1%, move on to step six. Otherwise make a new guess by splitting the difference, and repeat steps 2–4 until the computed value agrees with the initial guess to within 1%.
6. Divide the length of the reactor into an equal number of segments, n. The length of each segment will be $\Delta x = L/n$. We can now use a finite difference form of (14.267) to compute the concentration in the bulk fluid at $x = \Delta x$:

$$C_{sb}(x + \Delta x) = \left\{1 - \left(\frac{\Psi^*\big|_{r^*=1}}{\Psi^*\big|_{r^*=1} + Bi_s}\right) \frac{S_p k}{Q_V} \frac{\Delta x}{L}\right\} C_{sb}(x). \qquad (14.286)$$

As long as Δx is small, we can use the value of $\Psi^*\big|_{r^*=1}$ computed at x to estimate a new value of C_{sb} at $x + \Delta x$.

7. Use the new value of C_{sb} to compute $C_{sp}(R_p)$ from (14.265), compute β at $x = x + \Delta x$ and $\Psi^*|_{r^*=1}$ as in steps 2 and 3 above. Substitute these into (14.286) to compute the next value of $C(x + \Delta x)$ and repeat until the end of the reactor is reached.

This procedure is illustrated in the Matlab code provided in Example 14.6.1.

Example 14.6.1 Removal of Urea.
Chen and Chiu (1999) designed a prototype device for removing urea from blood. This is a cylindrical bioreactor with diameter of 1 cm and length of 15 cm. The bioreactor is filled with 700 μm diameter chitosan beads containing immobilized urease. The beads have a porosity of 0.85 and they occupy 60% of the bioreactor volume. The average pore diameter in the beads is 0.15 μm. Assume the effective diffusivity of urea in the bead pores to be 20% of the free diffusion coefficient. Urease is covalently bound (25 μg/ml of beads) to the walls of the pores. Urea is converted to carbon dioxide and ammonia by urease:

$$\text{Urea} + \text{H}_2\text{O} + \text{Urease} \underset{}{\overset{K_m}{\rightleftharpoons}} \text{Complex} \overset{k_{cat}}{\rightarrow} \text{Urease} + \text{CO}_2 + 2\text{NH}_3,$$

where $K_m = 12$ mM and $k_{cat} = 0.49$ mmol NH_3 min^{-1} mg^{-1} of enzyme. The device is to be tested at 20°C with water at various flow rates ranging between 0.5 and 10 ml/min. The inlet concentration of urea is 0.2 mg/ml. Provide an estimate of the outlet urea concentration at each flow rate. What changes might you suggest for scaling this up as a blood dialysis device?

Solution. *Initial considerations:* Our goal is to predict the performance of this bioreactor under the conditions specified and to use those results to scale the device up for use in dialyzing blood of humans. We will assume flow to be steady and a 1D analysis to be appropriate.

System definition and environmental interactions: The system under investigation is the fluid within the mobile phase of the reactor. To analyze this system, we must also analyze the fluid within the pores of the chitosan beads and must match the flux into the beads with the flux out of the mobile phase.

Apprising the problem to identify governing equations: The general equations describing simultaneous diffusion and reaction within the pores of spherical beads are (14.102)–(14.104). The primary result from this analysis is $\Psi^*|_{r^*=1}$, a dimensionless parameter proportional to the flux of urea at the surface of the beads. This must be re-evaluated at different axial locations within the reactor. The bulk concentration in the mobile phase of the reactor can be determined as a function of axial position using (14.267), following the procedure outlined in Sect. 14.6.3.4. A simpler analytical analysis would be possible if the ratio of urea concentration to Michaelis constant is either very high [(14.282), $\beta \ll 1$] or very low [(14.273),

14.6 Convection, Diffusion, and Chemical Reaction

$\beta \gg 1$] throughout the length of the reactor. The molecular weight of urea is 60, so the inlet concentration in the bulk fluid C_{sb0} is (0.2 mg/ml)/(60 g/mol) = 3.33 mM. The ratio C_{sb0}/K_m = 3.33/12 = 0.278, so β is relatively small at the inlet and will get progressively smaller as the fluid approaches the outlet. The first-order approximation might be reasonable in this case. To be safe, we will compute the outlet concentration using the general procedure and compare it with the first-order approximation.

Analysis: In addition to β (which changes with axial position), we need to compute the other relevant dimensionless parameters for this problem: the Thiele modulus, ϕ_T, the Biot number, Bi, and the ratio kS_p/Q_v. Let us start by computing kS_p/Q_v. The total surface area of the beads can be computed from the bead volume, which constitutes 60% of the bioreactor volume. The number of beads n_b is:

$$n_b = (1-\varepsilon)\left(\frac{\pi L d^2}{4}\right) \Big/ \frac{4\pi R_p^3}{3} = (1-0.4)\left(\frac{3}{16}\right)\left(\frac{(15\,\text{cm})(1\,\text{cm})^2}{(3.5\times 10^{-2}\text{cm})^3}\right)$$

$$= 3.94 \times 10^4.$$

The total bead surface area is:

$$S_p = n_b\left(4\pi R_p^2\right) = (3.94\times 10^4)\left(4\pi(3.5\times 10^{-2}\text{cm})^2\right) = 607\,\text{cm}^2.$$

The mass transfer coefficient in the film fluid (water) can be found from (14.262). The superficial velocity for the low flow case (0.5 cm/min) is:

$$v_0 = \frac{4Q_v}{\pi d^2} = \frac{4}{\pi}\frac{\left(0.5\,\frac{\text{cm}^3}{\text{min}}\right)}{(1\,\text{cm})^2} = 0.636\,\frac{\text{cm}}{\text{min}}.$$

The free diffusion coefficient for urea in water at 20°C is 1.08×10^{-3} cm²/min and the kinematic viscosity is 0.6 cm²/min. The mass transfer coefficient:

$$k = 25\frac{D_{sf}}{2R_p}\left(\frac{2R_p v_0}{\nu}\right)^{0.45}\left(\frac{\nu}{D_{sf}}\right)^{0.5},$$

$$k = 25\frac{(1.08\times 10^{-3}\text{cm}^2\,\text{min}^{-1})}{(7\times 10^{-2}\text{cm})}\left[\frac{(7\times 10^{-2}\text{cm})(0.636\,\text{cm}\,\text{min}^{-1})}{0.6\,\text{cm}^2\,\text{min}^{-1}}\right]^{0.45}$$

$$\times \left[\frac{0.6\,\text{cm}^2\,\text{min}^{-1}}{1.08\times 10^{-3}\,\text{cm}^2\,\text{min}^{-1}}\right]^{0.5} = 2.82\,\text{cm}\,\text{min}^{-1}.$$

Therefore,

$$\frac{kS_p}{Q_V} = \frac{(2.82 \text{ cm min}^{-1})(607 \text{ cm}^2)}{0.5 \text{ cm}^3 \text{ min}^{-1}} = 3423.$$

Next, let us work on computing the Biot number from (14.266). Since the pore diameter (120 µm) is much larger than the diameter of urea, we will assume the partition coefficient between the pore and film Φ_{spf} is equal to 1.0. The effective diffusion coefficient is assumed to be 20% of the free diffusion coefficient, $D_e = 2.16 \times 10^{-4}$ cm²/min. Consequently, the Biot number for the low flow case is:

$$Bi = \frac{kR_p}{\phi_p D_e \Phi_{spf}} = \frac{(2.82 \text{ cm min}^{-1})(3.5 \times 10^{-2} \text{cm})}{(0.85)(2.16 \times 10^{-4} \text{ cm}^2 \text{ min}^{-1})(1.0)} = 538.$$

The final dimensionless parameter, the Thiele modulus, can be computed from (14.99). The parameter V_{max}, the maximum rate of conversion of urea to $CO_2 + 2$ NH_3, must be determined before we can compute ϕ_T.

The total enzyme concentration in the pores E_0 is equal to the total enzyme mass per unit pore volume. The enzyme mass per unit bead volume is 25 mg/ml. Therefore:

$$E_0 = \left(\frac{\text{Enzyme mass}}{\text{ml bead volume}}\right)\left(\frac{\text{Bead volume}}{\text{Pore volume}}\right) = (25 \text{ µg ml}^{-1})\left(\frac{1}{0.85}\right)$$
$$= 29.4 \text{ µg ml}^{-1}.$$

The maximum rate of production of NH_3 is:

$$(V_{max})|_{NH_3} = k_{cat}E_0 = (0.49 \text{ µmol µg}^{-1} \text{ min}^{-1})(29.4 \text{ µg ml}^{-1})$$
$$= 14.4 \text{ mM min}^{-1}.$$

Since two molecules of NH_3 are produced for every one molecule of urea hydrolyzed, the maximum reaction rate for urea is half that of the maximum rate of NH_3:

$$(V_{max})|_{urea} = \frac{1}{2}(V_{max})|_{NH_3} = \frac{14.4 \text{ mM min}^{-1}}{2} = 7.2 \text{ mM min}^{-1}.$$

The Thiele modulus is:

$$\phi_T = \frac{R_p}{3}\sqrt{\frac{V_{max}}{D_e K_m}} = \frac{3.5 \times 10^{-2} \text{cm}}{3}\sqrt{\frac{7.2 \text{ mM min}^{-1}}{(2.16 \times 10^{-4} \text{ cm}^2 \text{ min}^{-1})(12 \text{ mM})}} = 0.615.$$

14.6 Convection, Diffusion, and Chemical Reaction

Using the first-order analysis, we find γ from (14.276) to be:

$$\gamma = 3\phi_T \coth(3\phi_T) - 1 = 3(0.615)\coth((3)(0.615)) - 1 = 0.94.$$

Finally, we can estimate the concentration at the exit if the reaction were first order from (14.279) at $x/L = 1$:

$$C_{sb}(L) = C_{sb}(0)\exp\left\{-\left(\frac{\gamma}{\gamma + Bi}\right)\frac{kS_p}{Q_v}\right\}$$

$$= 3.33\,\text{mM}\left[\exp\left\{-\left(\frac{0.94}{0.94 + 538}\right)3423\right\}\right] = 8.5\,\mu\text{M}.$$

Clearly, the Biot number is much larger than γ in this case, so (14.282) could have been used to compute the concentration at the exit. Treating the bioreactor as though it follows first-order kinetics predicts that 99.7% of the urea will be converted to ammonia and carbon dioxide when flow through the bioreactor is 0.5 ml/min. The procedure above should then be repeated for several bioreactor flows between 0.5 and 10 ml/min. Exit concentrations are plotted (square symbols) as a function of the bioreactor flow rate in Fig. 14.41. The full procedure outlined in Sect. 14.6.3.5, not assuming first-order kinetics, was also followed, with the Matlab code provided below Fig. 14.41. The results for the Michaelis–Menten (MM) case are also shown in Fig. 14.41 (diamond symbols).

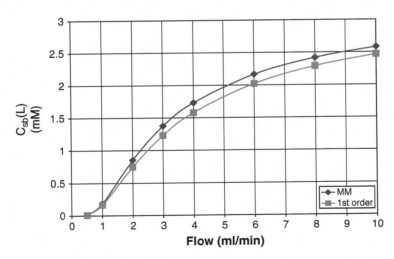

Fig. 14.41 Outlet urea concentration vs. bioreactor flow rate for inlet concentration of 3.33 mM

Matlab code for Example 14.6.1

```
function bioreactor_section_14_6_1
%Input parameters ---------------
Csb0=3.3; % urea concentration at reactor inlet (x=0)(mM)
Rp=3.5e-2; %particle radius (cm)
phi_spf=1.;% pore-film partition coefficient for urea
Dsf=1.8*60.e-5; %urea free diffusion coefficient (cm^2/min)
De=.2*Dsf; %Effective diffusion coefficient for urea in pores, cm^2/min
E_over_V=25; %mg enzyme/volume wet bead (ml)
kcat=.490; %reaction rate (millimoles min^-1 mg^-1 enzyme)
Km=12;%Michaelis constant for urea-urease (mM)
porosity = .85; %pore volume/wet bead volume
nb=3.94e4; %number of beads
L= 15; %length of reactor (cm)
d=1; % diameter of reactor (cm)
Qv=0.5; %flow through reactor (ml/min)
nu=0.6; %kinematic viscosity of mobile phase (cm^2/min)[water]
% calculations
E0=E_over_V/porosity % Enzyme concentration in pores (micrograms/ml)
Vmax=kcat*E0/2 % urea maximum rate (mM/min)
Area=pi*d^2/4% cross-sectional area of bead bed (cm^2)
v0=Qv/Area %superficial velocity (cm/min)
Re=2*Rp*v0/nu %Reynolds number based on bead diameter
k=(Dsf/(2*Rp))*25*(Re^.45)*(nu/Dsf)^.5%mass transfer coefficient in film
(cm/min)
Sp=nb*4*pi*Rp^2
%Compute dimensionless parameters:
Thiele=(Rp/3)*sqrt(Vmax/(Km*De)) %Thiele modulus
Bi=(k*Rp)/(porosity*De*phi_spf) %Biot number
kSp_Qv=k*Sp/Qv

repeat=1;%1 percent tolerance on Csp(0)
%-------------------------------
Np = 21;
Nx=51;
dx=L/(Nx-1);
x=linspace(0,1,Nx);
rinitial=linspace(0,1,Np);
solinit = bvpinit(rinitial,[rinitial(:,1)  rinitial(:,1)]);
options=bvpset('AbsTol',1e-7,'RelTol',1e-6,'Stats','on');
Csp0=phi_spf*Csb0;
rint = linspace(0,1);
%----if first order kinetics ----
gamma=3*Thiele*coth(3*Thiele)-1;
coeff=-(gamma/(gamma+Bi))*kSp_Qv;
Csb_first=exp(coeff.*x);%Csb/Csb0 for first order kinetics
while repeat>=1; %Recompute if repeat > 1%
    beta(1) = Csp0/Km;
    sol = bvp4c(@rjrode,@rjrbc,solinit,options,Thiele,beta(1));
    yint = deval(sol,rint);
    Psi=yint(2,length(rint)); %Psi(r=Rp)
    Csp1=(phi_spf*Csb0)/(1+Psi/Bi);
    repeat=100*abs(Csp1-Csp0)/Csp0;
    Csp0=(Csp0+Csp1)/2;
end
```

14.6 Convection, Diffusion, and Chemical Reaction

```
Csb(1)=Csb0;
Csp(1)=Csp1;
beta(1)=Csp1/Km;
Csb1(1)=Csb0;
for i=2:Nx
   Csb(i)=Csb(i-1)*(1-((kSp_Qv/Bi)*(dx/L))*(Bi*Psi/(Bi+Psi)));
   Csp(i)=(phi_spf*Csb(i))/(1+Psi/Bi);
   beta(i)=Csp(i)/Km;
   sol = bvp4c(@rjrode,@rjrbc,solinit,options,Thiele,beta(i));
   yint = deval(sol,rint);
   Psi=yint(2,length(rint)); %Psi(r*=1)
end
Csb(length(Csb))
Csb0*Csb_first(length(Csb_first))
% -----------------------------------------------
function dydr = rjrode(r,y,T,B)
% dy1/dr = y2/r^2
% dy2/dr = 9*T^2*r^2*y1/(1+B*y1))
if r==0
   dydr= [0
       9*(T^2)*(r^2)*y(1)/(1+B*y(1))];
else
   dydr = [ y(2)/r^2
       9*(T^2)*(r^2)*y(1)/(1+B*y(1))];
end
% -----------------------------------------------
function res = rjrbc(ya,yb,T,B)
% ya: r=0 -> y(2) = 0
% yb: r=Rp -> y(1) = 1
res = [ ya(2)
       yb(1)-1 ];
%-----------------------------------------------
```

Examining and interpreting the results: Results in Fig. 14.41 indicate that the first-order kinetics model (14.279) provides a reasonable approximation to the more complete model under the conditions specified in the problem. The first-order model overestimates the removal of urea by about 12% at a flow of 2 ml/min and 4% at a flow of 10 ml/min. We can make a quick estimate of how to scale this device up so it can be used for blood dialysis. Patients who need dialysis might have a plasma concentration of urea as high as 30 mM or more. A target post-dialysis urea level for such patients might be about 8 mM. The excess amount of urea dissolved in 5 L of blood would be 110 mmol. The rate of removal of urea from the device is:

$$\text{Rate of removal of urea} = Q_V(C_{sb0} - C_{sb}(L)).$$

Using the data in Fig. 14.41, the highest rate of urea removal is 7.47×10^{-3} mmol/min, which occurs at the highest flow rate tested (10 ml/min). We estimate the time required to remove 110 mmol of urea with this device at this rate to be 245 h! We might conclude from this that if the actual dialysis is to be completed in 4 h, the number of beads in the device would need to be increased by a factor of 61.4. However, there are a couple of flaws in this reasoning, as discussed below.

Additional comments: We cannot simply construct a scale factor by dividing the desired removal rate in the dialysis unit by the measured removal rate in the prototype device. Altering the inlet concentration of urea will affect the rate of exchange, even in the exact same device. There are two good reasons for this. First, changing the bulk concentration will change the transport rate to the bead surface and diffusion gradients in the pores of the beads. Second, altering the bulk concentration will change the value of β at all axial positions, thus altering the rate of conversion.

For example, let us change the inlet concentration of urea to the same device analyzed above from 3.33 to 33.3 mM. The concentration at the outlet and the removal rates are shown in Figs. 14.42 and 14.43, respectively. Again we compare the solution from the first-order kinetic model with the solution for the general Michaelis–Menten model. In contrast to the 3.33 mM case, the outlet concentrations for the 33.3 mM case are quite different for the two models. This is because β at the inlet is ten times larger for the 33.3 mM case than the 3.33 mM case. The first-order model significantly overpredicts the rate of conversion in the 33.3 mM case. Although the MM model prediction is much lower than the first-order model prediction, the MM model still predicts the rate of conversion to be about four times higher than the rate of conversion for the lower inlet concentration case. In addition, increasing the flow rate above about 2 ml/min has very little effect on the overall removal rate. Finally, in actual operation, the inlet concentration of urea will drop during the treatment; so the removal rate will also drop.

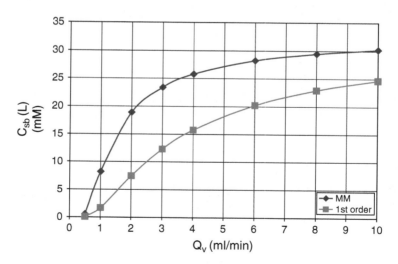

Fig. 14.42 Outlet urea concentration vs. bioreactor flow rate for inlet concentration of 33.3 mM

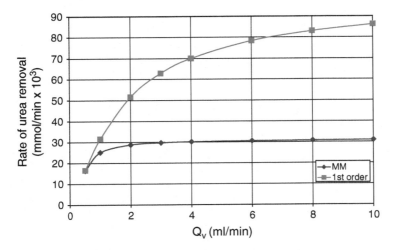

Fig. 14.43 Rate of removal of urea vs. bioreactor flow rate for inlet concentration of 33.3 mM

14.7 One-Dimensional Unsteady-State Shell Balance Applications

In Sect. 12.6.3, we showed that a macroscopic approach can be used to track concentration changes in a system as long as the Biot number for mass transfer is small. In that case, the spatial concentration gradient in the system is small and the concentration can be considered to depend on time alone. This formed the basis of the macroscopic approach for analyzing mass transfer presented in Chap. 13. However, if the resistance to mass transfer in the system is not small relative to the resistance to mass transfer through the system boundary, then both spatial and temporal variations in concentration will exist simultaneously in the system. There are many biological mass transfer situations where we must account for both spatial and temporal variations in concentration. We will consider cases in this section that involve unsteady-state diffusion through tissue where the mass flux is one dimensional.

14.7.1 Diffusion to Tissue

When a tissue, such as skin, is suddenly exposed to a substance, such as a drug or a pollutant, the substance will begin to diffuse from the fluid–tissue interface toward the interior of the tissue. We are generally interested in predicting the flux of the substance into the tissue and the concentration of the substance in the tissue as a function of position and time, $C(x, t)$. These factors will depend on the geometry of the tissue, the properties of the tissue, mass transfer resistance at the boundaries of the tissue, and the initial distribution of substance within the tissue. In the cases examined below, we will assume that the substance is initially uniformly distributed within the tissue, so $C_A(x, 0) = C_{A0}$, a constant. In addition, we will assume the substance is neither produced nor consumed within the tissue.

14.7.1.1 Diffusion in a Semi-Infinite Slab

Let us first consider the case where a thick slab of tissue is exposed to a constant concentration C_{AS} of species A at one surface. Initially, the concentration of species A is C_{A0} which is assumed uniform throughout the slab. Far from the surface, the concentration in the slab is assumed to remain constant at C_{A0}. We begin our analysis by conducting a species shell balance on a small segment of tissue contained between x and $x + \Delta x$:

$$\left\{ \begin{array}{c} \text{Rate of} \\ \text{accumulation} \\ \text{of species A} \\ \text{within shell} \end{array} \right\} = \left\{ \begin{array}{c} \text{Net rate species A} \\ \text{enters through} \\ \text{shell boundaries} \end{array} \right\} + \left\{ \begin{array}{c} \text{Rate species A} \\ \text{is produced} \\ \text{within the shell} \end{array} \right\}. \quad (14.287)$$

Species A is not produced in the tissue; so the last term will be zero. The rate of accumulation of species A in the shell that is Δx long having a cross-sectional area A is:

$$\left\{ \begin{array}{c} \text{Rate of} \\ \text{accumulation} \\ \text{of species A} \\ \text{within shell} \end{array} \right\} = \frac{\partial}{\partial t}[c_A(x,t)A\Delta x]. \quad (14.288)$$

Since convective flux is zero, the net rate that species A enters through the shell boundaries is by diffusion:

$$\left\{ \begin{array}{c} \text{Net rate species A} \\ \text{enters through} \\ \text{shell boundaries} \end{array} \right\} = J_A(x,t)A - J_A(x+\Delta x, t)A. \quad (14.289)$$

Substituting (14.288) and (14.289) into (14.287), dividing by the shell volume, $A\Delta x$, and letting the volume approach zero, we obtain the following partial differential equation:

$$\frac{\partial c_A}{\partial t} = -\frac{\partial J_A}{\partial x}. \quad (14.290)$$

14.7 One-Dimensional Unsteady-State Shell Balance Applications

Finally, introducing Fick's law for the diffusive flux, and assuming the diffusion coefficient is independent of position, we find:

$$\frac{\partial c_A}{\partial t} = D_A \frac{\partial^2 c_A}{\partial x^2}. \tag{14.291}$$

This is known as *Fick's second law of diffusion*. The boundary and initial conditions are:

$$c_A(0,t) = C_{As},$$
$$c_A(\infty,t) = C_{A0},$$
$$c_A(x,0) = C_{A0}. \tag{14.292}$$

If we define a dimensionless concentration C_A^*:

$$C_A^* = \frac{c_A - C_{A0}}{C_{As} - C_{A0}}. \tag{14.293}$$

Then the problem becomes:

$$\frac{\partial C_A^*}{\partial t} = D_A \frac{\partial^2 C_A^*}{\partial x^2},$$
$$C_A^*(0,t) = 1,$$
$$C_A^*(\infty,t) = 0,$$
$$C_A^*(x,0) = 0. \tag{14.294}$$

Note that this problem is identical to the fluids problem dealing with a suddenly moved wall in Sect. 6.4, with C_A^* replacing v, x replacing y, D_A replacing ν, and $V = 1$ in (6.115) and (6.116). Therefore, the solution, by analogy with (6.123), is:

$$\frac{c_A - C_{A0}}{C_{As} - C_{A0}} = \text{erfc}\left(\frac{x}{2\sqrt{D_A t}}\right). \tag{14.295}$$

Since $\text{erfc}(z) = 1 - \text{erf}(z)$, this can be written as:

$$\frac{c_A - C_{As}}{C_{A0} - C_{As}} = \text{erf}\left(\frac{x}{2\sqrt{D_A t}}\right). \tag{14.296}$$

The surface concentration in the slab is assumed to be in equilibrium with the contacting medium at $x = 0$. Therefore, $C_{As} = \Phi_{Asf} C_{Af}$, where C_{Af} is the concentration in the region $x < 0$, and Φ_{Asf} is the partition coefficient between the slab material and the contacting medium.

Let us define the penetration distance $x = \delta$ to be the layer near the tissue surface over which the actual concentration difference $c_A(\delta, t) - C_{As}$ is within 99% of the maximum concentration difference, $C_{A0} - C_{As}$. Since erf $(1.82) = 0.99$, then the penetration distance is given by:

$$\delta(t) = 3.64\sqrt{D_A t}. \tag{14.297}$$

This is analogous to the penetration distance given for molecular momentum transfer (6.124), where ν replaces D_A. A similar relationship for heat transfer via conduction can be found by replacing D_A with the thermal diffusivity, α.

The flux of solute A at any given position and time in the semi-infinite slab is:

$$J_{Ax} = -D_A \frac{\partial c_A}{\partial x} = -\frac{D_A(C_{A0} - C_{As})}{\sqrt{\pi D_A t}} e^{-\frac{x^2}{4 D_A t}}. \tag{14.298}$$

The flux into the tissue at the surface ($x = 0$) is:

$$J_{Ax}(0, t) = \sqrt{\frac{D_A}{\pi t}}(C_{As} - C_{A0}) \tag{14.299}$$

The total number of moles of A that enters the surface in time t can be found by multiplying (14.299) by the tissue surface area A_s and integrating:

$$N_A(t) = 2 A_s \sqrt{\frac{D_A t}{\pi}}(C_{As} - C_{A0}). \tag{14.300}$$

The number of moles transported into the tissue grows at a rate proportional to $(D_A t)^{1/2}$, similar to the penetration thickness.

> **Example 14.7.1 Carbon Dioxide Transport in Plasma.**
> Plasma, initially devoid of CO_2, is suddenly exposed to a gas at atmospheric pressure containing CO_2 with a mole fraction of 0.5. Assume the gas is well mixed, so the concentration at the plasma surface remains constant. The diffusion coefficient for CO_2 in plasma is 1.46×10^{-5} cm^2/s, the Henry law constant H for CO_2 in plasma is 2,100 atm, and the surface area of the gas–plasma interface is 100 cm^2. Assume the molar concentration of plasma is 0.05517 mol/cm^3. Find the penetration thickness and the number of moles of CO_2 in plasma after 1 s, 1 min, and 1 h.

Solution. *Initial considerations*: This is an unsteady-state one-dimensional diffusion problem for CO_2 in plasma.

System definition and environmental interactions: The plasma, bounded at the top by the gas–liquid interface and at the bottom by a solid wall, is the system of interest. The plasma can be considered as a semi-infinite medium until the advancing front of CO_2 reaches the bottom of the container.

14.7 One-Dimensional Unsteady-State Shell Balance Applications

Apprising the problem to identify governing relationships: The analysis for unsteady-state diffusion in a semi-infinite medium is appropriate for this problem.

Analysis: The penetration distance for carbon dioxide in plasma from (14.297) is 0.14 mm after 1 s, 1.1 mm after 1 min, and 8.4 mm after 1 h. Carbon dioxide is assumed to be in equilibrium between gas and liquid at the interface. From Henry's law, we find the concentration of CO_2 in plasma at the gas surface to be:

$$c_{CO_2,s} = \frac{c_{plasma} P_{CO_2}}{H_{CO_2,plasma}} = \frac{(0.05517 \text{ mol cm}^{-3})(0.5 \text{ atm})}{2100 \text{ atm}} = 1.31 \times 10^{-5} \text{ mol cm}^{-3}.$$

The number of moles of CO_2 that move from gas to plasma as a function of time can be determined using (14.300) with $C_{A0} = 0$:

$$N_A(t) = \left(\frac{2}{\sqrt{\pi}} A_s C_{As} \sqrt{D_A}\right) \sqrt{t}$$

$$= \frac{2}{\sqrt{\pi}}(100 \text{ cm}^2)(1.3 \times 10^{-5} \text{ mol cm}^{-3})(1.46 \times 10^{-5} \text{ cm}^2 \text{ s}^{-1})^{\frac{1}{2}} \sqrt{t},$$

$$N_A [\text{moles}] = 5.6 \times 10^{-6} \sqrt{t[s]}.$$

Therefore, 5.6 mmol of CO_2 enter the plasma in 1 s, 43.5 mmol enter in 1 min, and 336 mmol enter in 1 h.

Examining and interpreting the results: The penetration thickness and the rate of accumulation of CO_2 in the plasma both slow down considerably with time. If the plasma layer is thicker than about 1 cm, the above analysis could be used to predict mass exchange of CO_2 for at least an hour. Therefore, the assumption of infinite thickness is not necessarily a limiting factor for many solutes, particularly solutes with very low diffusion coefficients.

14.7.1.2 Diffusion Between Two Semi-Infinite Slabs

When a "thick" slab of tissue is brought into contact with another slab of material containing solute A at a different concentration, a flux of A will generally occur. The direction and magnitude of the flux will depend on several factors, including the partition coefficient of A between the two materials, the initial concentration difference, the diffusion coefficients of A in the two materials, and time.

An analysis of mass exchange between the two materials begins with a shell balance performed in each material, similar to the one shown in the previous section. We take the origin ($x = 0$) at the interface. The material on the left side of the interface is designated as material "2" and material on the right side of the interface is material "1," as shown in Fig. 14.44. Shell balances performed on each material yield:

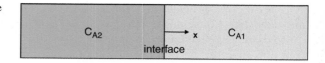

Fig. 14.44 Two semi-infinite slabs of material with different initial concentrations of species A

$$\frac{\partial c_{A1}}{\partial t} = D_{A1}\frac{\partial^2 c_{A1}}{\partial x^2}, x \geq 0, \tag{14.301}$$

$$\frac{\partial c_{A2}}{\partial t} = D_{A2}\frac{\partial^2 c_{A2}}{\partial x^2}, x \leq 0. \tag{14.302}$$

The appropriate initial and boundary conditions are:

$$c_{A2}(x=0,t) = \Phi_{A21}c_{A1}(x=0,t),$$

$$J_{A1}(x=0,t) = J_{A2}(x=0,t) \text{ or } D_{A1}\frac{\partial c_{A1}}{\partial x}\bigg|_{x=0} = D_{A2}\frac{\partial c_{A2}}{\partial x}\bigg|_{x=0},$$

$$c_{A1}(x=\infty,t) = C_{A10},$$

$$c_{A1}(x,t=0) = C_{A10},$$

$$c_{A2}(x=-\infty,t) = C_{A20},$$

$$c_{A2}(x,t=0) = C_{A20}. \tag{14.303}$$

By analogy with (14.296), the general solutions to (14.301) and (14.302) are:

$$c_{A1} = A_1 + B_1\mathrm{erf}\left(\frac{x}{2\sqrt{D_{A1}t}}\right), x \geq 0,$$

$$c_{A2} = A_2 + B_2\mathrm{erf}\left(\frac{x}{2\sqrt{D_{A2}t}}\right), x \leq 0. \tag{14.304}$$

Since erf $(0) = 0$, erf $(\infty) = 1$ and erf $(-\infty) = -1$, the auxiliary conditions can be used to find the unknown coefficients A_1, B_1, A_2, and B_2. Since the initial concentration and the concentration far from the interface are the same in each region, the six conditions in (14.303) collapse to four when applied to (14.304):

$$\begin{aligned}
x=0: \quad & c_{A2} = \Phi_{A21}c_{A1} &&\rightarrow&& A_2 = \Phi_{A21}A_1 \\
x=0: \quad & J_{A2} = J_{A1} \\
x\rightarrow\infty \text{ or } t=0: \quad & c_{A1} = C_{A10} &&\rightarrow&& C_{A10} = A_1 + B_1 \\
x\rightarrow-\infty \text{ or } t=0: \quad & c_{A2} = C_{A20} &&\rightarrow&& C_{A20} = A_2 - B_2.
\end{aligned} \tag{14.305}$$

14.7 One-Dimensional Unsteady-State Shell Balance Applications

Solving for the unknown coefficients and substituting back into (14.304) yield the final expressions for the concentrations in the two regions:

$$c_{A1} = \frac{\left(C_{A10} + C_{A20}\sqrt{\frac{D_{A2}}{D_{A1}}}\right) + \sqrt{\frac{D_{A2}}{D_{A1}}}(C_{A10}\Phi_{A21} - C_{A20})\mathrm{erf}\left(\frac{x}{2\sqrt{D_{A1}t}}\right)}{1 + \Phi_{A21}\sqrt{\frac{D_{A2}}{D_{A1}}}}, x \geq 0,$$

$$c_{A2} = \frac{\Phi_{A21}\left(C_{A10} + C_{A20}\sqrt{\frac{D_{A2}}{D_{A1}}}\right) + (C_{A10}\Phi_{A21} - C_{A20})\mathrm{erf}\left(\frac{x}{2\sqrt{D_{A2}t}}\right)}{1 + \Phi_{A21}\sqrt{\frac{D_{A2}}{D_{A1}}}}, x \leq 0.$$

(14.306)

The flux at the interface ($x = 0$) is:

$$J_{Ax}(x=0) = \frac{D_{A2}}{\sqrt{D_{A1}\pi t}}\left(\frac{C_{A20} - \Phi_{A21}C_{A10}}{1 + \Phi_{A21}\sqrt{\frac{D_{A2}}{D_{A1}}}}\right). \tag{14.307}$$

Note that the concentration in each material at the interface ($x = 0$) remains constant with time:

$$c_{A1}(0,t) = \frac{\left(C_{A10} + C_{A20}\sqrt{\frac{D_{A2}}{D_{A1}}}\right)}{1 + \Phi_{A21}\sqrt{\frac{D_{A2}}{D_{A1}}}},$$

$$c_{A2}(0,t) = \Phi_{A21}c_{A1}(0,t). \tag{14.308}$$

14.7.1.3 Diffusion to a Semi-Infinite Slab with Finite External Resistance to Mass Transfer

In Sect. 14.7.1.1, we analyzed the situation where a slab of material is exposed to a sudden change in concentration at the surface $x = 0$, and the surface concentration is maintained at that value for all time. This implies negligible resistance to mass transfer in the contacting material. In reality, there is always some finite resistance, and a better boundary condition at the contacting surface might be that the mass flux from the surface is governed by a mass transfer coefficient. If the slab occupies the space $x \geq 0$ and a fluid f occupies the space $x \leq 0$, then the flux of A from the fluid into the slab surface will be:

$$N_{Af}(0,t) = k_{Af}(c_{Af\infty} - c_{Afs}), \qquad (14.309)$$

where k_{Af} is a mass transfer coefficient for species A and the subscript f indicates values in the fluid phase. The flux into the slab by diffusion is:

$$N_{A1}(0,t) = -D_{A1}\frac{\partial c_{A1}}{\partial x}\bigg|_{x=0}, \qquad (14.310)$$

where the subscript 1 indicates the semi-infinite slab material. At the interface, we assume equilibrium between the two phases, so the fluid concentration at the interface will be:

$$c_{Afs} = \frac{c_{A1}(0,t)}{\Phi_{A1f}}. \qquad (14.311)$$

Setting the flux into the interface from the fluid equal to the flux out of the interface into the slab and applying the equilibrium condition at the interface, we obtain the following boundary condition at the interface:

$$\frac{\partial c_{A1}}{\partial x}\bigg|_{x=0} = \frac{k_{Af}}{D_{A1}\Phi_{A1f}}(c_{A1}(0,t) - \Phi_{A1f}c_{Af\infty}). \qquad (14.312)$$

This replaces the constant concentration boundary condition in (14.292), and the solution to Fick's second law under these circumstances is:

$$\frac{c_{A1} - \Phi_{A1f}c_{Af\infty}}{c_{A0} - \Phi_{A1f}c_{Af\infty}} = \mathrm{erf}\left(\frac{x}{2\sqrt{D_{A1}t}}\right) + \exp\left(\frac{k_{Af}}{D_{A1}\Phi_{A1f}}\left(x + \left(\frac{k_{Af}}{\Phi_{A1f}}\right)t\right)\right)$$
$$\times \left[\mathrm{erfc}\left(\frac{1}{2\sqrt{D_{A1}t}}\left(x + \left(\frac{2k_{Af}}{\Phi_{A1f}}\right)t\right)\right)\right]. \qquad (14.313)$$

Note that in contrast to the case of diffusion between two slabs (14.306), the interfacial concentration is not constant, but changes with time:

$$c_{A1}(0,t) = \Phi_{A1f}c_{Af\infty} + (c_{A0} - \Phi_{A1f}c_{Af\infty})\exp(\tau)\left[\mathrm{erfc}\left(\sqrt{\tau}\right)\right], \qquad (14.314)$$

where τ is a dimensionless time defined as:

$$\tau \equiv \left(\frac{k_{Af}}{\Phi_{A1f}}\right)^2 \frac{t}{D_{A1}}. \qquad (14.315)$$

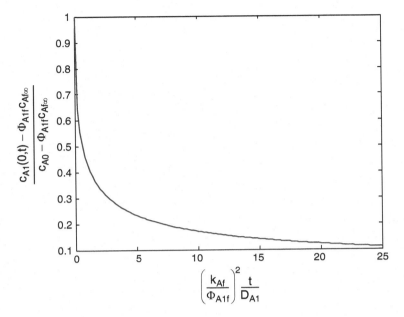

Fig. 14.45 Dimensionless concentration at the interface versus dimensionless time after the surface is exposed to a fluid with mass transfer coefficient k_f

This relationship is shown in Fig. 14.45. Initially, the interfacial concentration is C_{A0}, but as τ becomes very large, the concentration in the slab at the boundary approaches its equilibrium value with the fluid, $\Phi_{A1f}C_{Af\infty}$. However, the dimensionless time τ must be on the order of 1,000 before the interfacial concentration is nearly in equilibrium with the fluid concentration far from the interface. The higher the mass transfer coefficient, k_f, the more quickly will the interfacial concentration approach its final equilibrium concentration

14.7.1.4 Unsteady-State Diffusion to a Slab with Finite Thickness and with Non-zero Surface Resistance

We turn our attention now to one-dimensional unsteady-state diffusion of solute "A" to a slab of material with thickness $2\,L$. We will assume that the initial concentration in the slab is uniform, and that the surfaces at $x = -L$ and $x = +L$ are suddenly exposed to a fluid containing solute A at a concentration $C_{Af\infty}$ far from the solid material. Mass transfer in the fluid is governed by a mass transfer coefficient k_{Af}. Fick's second law, (14.291), is to be solved subject to the following boundary conditions and initial condition:

$$-D_{A1}\frac{\partial c_{A1}}{\partial x}\bigg|_{x=-L} = \frac{k_{Af}}{\Phi_{A1f}}(\Phi_{A1f}C_{Af\infty} - c_{A1}(-L,t)), \qquad (14.316)$$

$$-D_{A1} \left. \frac{\partial c_{A1}}{\partial x} \right|_{x=+L} = \frac{k_{Af}}{\Phi_{A1f}} (c_{A1}(+L,t) - \Phi_{A1f} c_{Af\infty}), \tag{14.317}$$

$$c_{A1}(x,0) = c_{A0}. \tag{14.318}$$

The signs are reversed on the right sides of (14.316) and (14.317) to ensure that a positive flux is in the positive x-direction. An alternate boundary condition to (14.316) would be that the concentration is symmetrical about the center of the slab, or there is no flux at $x = 0$:

$$\left. \frac{\partial c_{A1}}{\partial x} \right|_{x=0} = 0. \tag{14.319}$$

Two assumptions are implicit in the boundary conditions. First, the fluid in the region $x < -L$ is the same as the fluid in the region $x > +L$; so both fluids have the same partition coefficient and mass transfer coefficient for solute A. Second, the fluid concentrations of solute far from each surface are also identical and equal to $c_{Af\infty}$. Let us introduce the following dimensionless quantities:

$$Y = \frac{c_{A1} - \Phi_{A1f} c_{Af\infty}}{c_{A0} - \Phi_{A1f} c_{Af\infty}}, \tag{14.320}$$

$$n = \frac{x}{L}, \tag{14.321}$$

$$X = \frac{D_{A1} t}{L^2} \quad \text{(Fourier Number)}, \tag{14.322}$$

$$m = \frac{D_{A1} \Phi_{A1f}}{k_{Af} L} \quad \text{(1/Biot Number)}. \tag{14.323}$$

The quantities X and $1/m$ are the Fourier and Biot numbers, respectively, for mass transfer of species A. Substituting these values into Fick's second Law and into the applicable boundary and initial conditions yields:

$$\frac{\partial Y(n,X)}{\partial X} = \frac{\partial^2 Y(n,X)}{\partial n^2}, \tag{14.324}$$

$$\left. \frac{\partial Y}{\partial n} \right|_{n=0} = 0$$

$$\left. m \frac{\partial Y}{\partial n} \right|_{n=+1} = -Y(+1, X) \tag{14.325}$$

$$Y(n, t=0) = 1.$$

14.7 One-Dimensional Unsteady-State Shell Balance Applications

This is the same set of equations that arose for heat transfer from a slab to fluid at a different temperature (Sect. 10.6.1). The solution is an infinite series that involves the dependent variable Y, the two independent variables X and n, and the parameter m:

$$Y = \sum_{k=1}^{\infty} A_k \cos(\lambda_k n) e^{-\lambda_k^2 X}, \qquad (14.326)$$

where

$$A_k = \frac{2 \sin(\lambda_k)}{\lambda_k + \sin(\lambda_k) \cos(\lambda_k)}, \qquad (14.327)$$

and λ_k is the kth solution of the transcendental equation:

$$\lambda_k \tan(\lambda_k) = \frac{1}{m}. \qquad (14.328)$$

The solution can be displayed in graphical form by plotting Y as a function of X for a particular location n and specified values of m. The resulting graphs, known as the Gurney–Lurie or Heisler charts, were first published by Gurney and Lurie (1923) and expanded by Heisler (1947). Appropriate graphs that can be used either for heat transfer or mass transfer from a slab can be found in Appendix D.

The value of Y at the center of the tissue as a function of X, with m as a parameter is shown in Fig. D.1. The error in using just the first term in the series to approximate the centerline concentration is shown in Fig. 14.46.

For $m > 3$ and $X > 0.01$, the error in using just the first term in the series will always be less than 5%. Even for small values of m, if $X > 0.12$, the error will be less than 5%. Consequently, a good approximation for $X > 0.12$ is:

$$Y(n, X) \approx A_1 \cos(\lambda_1 n) e^{-\lambda_1^2 X}, \; X > 0.12. \qquad (14.329)$$

Therefore, the dimensionless concentration at an arbitrary radial position relative to the dimensionless concentration at the center of the slab is:

$$\frac{Y(n, X)}{Y(0, X)} \approx \cos(\lambda_1 n), \; X > 0.12. \qquad (14.330)$$

Note that this ratio is independent of dimensionless time, but does depend on m since λ_1 is a function of m. A graph of $Y(n, X)/Y(0, X)$ vs. n is provided in Fig. D.2 for several values of m. To find the concentration at any position n and time $X > 0.12$ in the slab, first find $Y(0, X)$ from the centerline chart, Fig. D.1, then find $Y(n, X)/Y(0, X)$ from the axial profile chart and multiply the two together to find $Y(n, X)$.

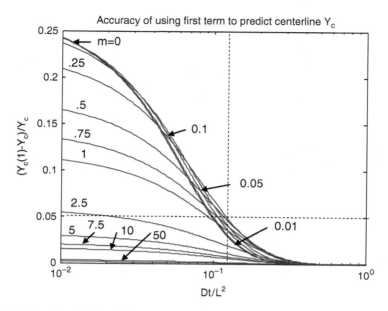

Fig. 14.46 Error in prediction of dimensionless concentration at the center of a slab using only the first term of the infinite series solution (14.329) with $n = 0$

For low values of m and X, the dimensionless concentration in the media relative to the value at $x = 0$ is higher than that predicted using the method above. This is illustrated in Fig. 14.47 for $m = 0.05$ (i.e., $Bi_A = 20$). The blue lines represent the actual concentrations vs. location in the material at the dimensionless times specified on the chart. The red line is the single term approximation given by (14.330). The difference between the blue lines and the red line represents the error introduced using the single term approximation. The error is small for $X > 0.12$, but can be rather large for small values of X. Additional plots for $m = 0.02, 0.1, 1.0$, and 10 are shown in Fig. D.3.

In Sect. 12.6.3, we stated that internal resistance could be neglected when $Bi_A < 0.1$. This is equivalent to stating that concentration profiles in the slab are relatively flat when $m > 10$. Indeed, the concentration profiles shown in Figs. D.2 and D.3 support this assertion.

The molar flux of substance A at the interface is:

$$N_{AS} = k_{Af}[c_{AfS} - c_{Af\infty}] = \frac{k_{Af}}{\Phi_{Asf}}[c_{AS} - \Phi_{Asf}c_{Af\infty}]. \tag{14.331}$$

The maximum molar flux occurs at $t = 0$ when c_{AS} is c_{A0}:

$$N_{A\max} = \frac{k_{Af}}{\Phi_{Asf}}[c_{A0} - \Phi_{Asf}c_{Af\infty}]. \tag{14.332}$$

14.7 One-Dimensional Unsteady-State Shell Balance Applications

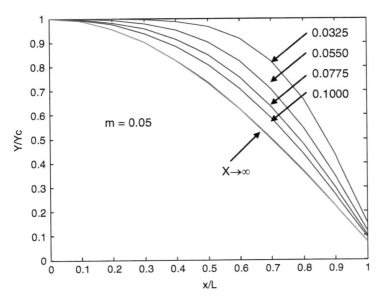

Fig. 14.47 Error in predicting concentration profile for small m and X using the single term approximation (14.330)

Let us define Ψ as the molar flux at the surface relative to the maximum molar flux at time zero. Dividing (14.331) by (14.332):

$$\Psi \equiv \frac{N_{AS}}{N_{A\max}} = \left[\frac{c_{AS} - \Phi_{Asf}c_{Af\infty}}{c_{A0} - \Phi_{Asf}c_{Af\infty}}\right] = Y(n=1, X). \qquad (14.333)$$

Therefore (14.333) can be used to find both the flux at the surface and the surface concentration. If the numerator and denominator are both multiplied by the molecular weight of substance A, then Ψ is also the ratio of mass flux at the surface to the maximum mass flux at $t = 0$. The surface flux relative to the maximum flux is plotted in Fig. D.4 as a function of dimensionless time (Fourier number, X) for various values of m.

After a long period of time, the concentration in the solid will eventually come into equilibrium with the concentration in the fluid far from the solid. The maximum number of moles that can exchange between solid and fluid is:

$$N_{A\infty} = V[c_{A0} - \Phi_{Asf}c_{Af\infty}]. \qquad (14.334)$$

V is the volume of the material. Let us define Σ as the number of moles that cross the surface S in time t, relative to the maximum:

$$\Sigma \equiv \frac{N_A(t)}{N_{A\infty}} = \frac{m_A(t)}{m_{A\infty}} = \frac{k_{Af}}{\Phi_{Asf}}\left(\frac{S}{V}\right)\int_0^t \left[\frac{c_{AS} - \Phi_{Asf}c_{Af\infty}}{c_{A0} - \Phi_{Asf}c_{Af\infty}}\right]dt. \qquad (14.335)$$

As indicated in (14.335), the molar ratio is also equal to the mass m_A of solute A that crosses the boundaries after time t relative to the mass of solute that cross the boundaries after a long period of time, $m_{A\infty}$. The surface areas at $x = L$ and $x = -L$ each equal wh, where w is the width of the slab and h is the height of the slab. Therefore, the total surface to volume ratio for the slab is $1/L$. Using the definitions of X and Y, Σ can be rewritten as follows:

$$\Sigma = \frac{1}{m} \int_0^X Y(n=1, X) \, dX = \frac{1}{m} \sum_{k=1}^\infty \frac{A_k \cos(\lambda_k)}{\lambda_k^2} \left[1 - e^{-\lambda_k^2 X}\right]. \tag{14.336}$$

The cumulative mass ratio Σ is plotted in Fig. D.5 as a function of dimensionless time X for various values of m.

Example 14.7.2 CO_2 Diffusion from a Thin Slab of Tissue.
A tissue segment at 1 atm is equilibrated with CO_2 at a partial pressure of 60 mmHg. The dimensions of the tissue are 100 μm thick, 0.5 cm wide, and 0.5 cm high. The partial pressure of CO_2 in the fluid surrounding the tissue is suddenly changed to 40 mmHg. The solubility of CO_2 in tissue is 3.35×10^{-8} mole CO_2 ml^{-1} mmHg^{-1}. The diffusivity of CO_2 in tissue is 1.46×10^{-5} cm^2/s, the mass transfer coefficient for CO_2 in the film is 0.973×10^{-3} cm/s, and the partition coefficient for CO_2 between tissue and fluid is 1.0. Find the following quantities:

1. How long will it take for P_{CO_2} at the center of tissue to drop to 45 mmHg?
2. When will the surface flux drop to 10% of the initial value?
3. When will the slab lose 95% of the excess CO_2?
4. When does surface P_{CO_2} reach 50 mmHg?
5. What is P_{CO_2} at $x = 25$ μm and $t = 5$ s?
6. How many moles of CO_2 cross the surface in 1 s?

Solution. *Initial considerations*: Since the tissue is much thinner than it is wide or long, we can treat this problem as a 1D unsteady-state diffusion problem in a slab. Each of the above questions can be answered using the charts in Appendix D.2.

System definition and environmental interactions: The slab of tissue is the system of interest. It is initially at a P_{CO_2} of 60 mmHg and at time zero the P_{CO_2} in the fluid is reduced to 40 mmHg.

Apprising the problem to identify governing relationships: CO_2 concentration in the tissue is initially uniform, and mass transfer at the tissue surface is limited by the value of the mass transfer coefficient. Therefore, the analysis presented in Sect. 14.7.1.4 (14.324)–(14.325) applies to this problem, and consequently the charts in Appendix D.2 can be used.

14.7 One-Dimensional Unsteady-State Shell Balance Applications

Analysis: First, we compute the value m for this case:

$$m = \frac{D_{CO_2,T}\Phi_{CO_2,Tf}}{k_f L} = \frac{(1.46 \times 10^{-5}\,\text{cm}^2\,\text{s}^{-1})(1)}{(0.973 \times 10^{-3}\,\text{cm s}^{-1})(50 \times 10^{-4}\,\text{cm})} = 3.0.$$

Next, we need to express Y in terms of partial pressures, rather than concentrations. Since the reflection coefficient is just the ratio of solubilities:

$$Y = \frac{c_{CO_2,T} - \Phi_{CO_2,Tf}c_{CO_2f\infty}}{c_{CO_2,T0} - \Phi_{CO_2,Tf}c_{CO_2f\infty}} = \frac{(\alpha_{CO_2,T}P_{CO_2,T}) - \left(\dfrac{\alpha_{CO_2,T}}{\alpha_{CO_2,f}}\right)(\alpha_{CO_2,f}P_{CO_2,f\infty})}{(\alpha_{CO_2,T}P_{CO_2,T0}) - \left(\dfrac{\alpha_{CO_2,T}}{\alpha_{CO_2,f}}\right)(\alpha_{CO_2,f}P_{CO_2,f\infty})}$$

$$= \frac{P_{CO_2,T} - P_{CO_2,f\infty}}{P_{CO_2,T0} - P_{CO_2,f\infty}}.$$

Here $P_{CO_2,T}$ is the partial pressure of CO_2 in the tissue, $P_{CO_2,T0}$, is the initial partial pressure of CO_2 in the tissue (60 mmHg), and $P_{CO_2,f\infty}$ is the partial pressure of CO_2 in the fluid far from the tissue surface. To find how long it takes for the partial pressure of CO_2 at the center of the tissue to reach 45 mmHg, we compute $Y(0, X)$:

$$Y(0,X) = \frac{45\,\text{mmHg} - 40\,\text{mmHg}}{60\,\text{mmHg} - 40\,\text{mmHg}} = 0.25.$$

We now use Fig. D.1 to find the intersection of a line drawn at $Y(0, X) = 0.25$ and the $m = 3.0$ graph. The intersection occurs at $X = 4.8$. Converting this to time:

$$t = X\left(\frac{L^2}{D_{CO_2,T}}\right) = 4.8\left(\frac{(50 \times 10^{-4}\,\text{cm})^2}{1.46 \times 10^{-5}\,\text{cm}^2\,\text{s}^{-1}}\right) = 4.8\,(1.71\,\text{s}) = 8.22\,\text{s}.$$

The time required for the flux to drop to 10% of its initial value can be found from the intersection of $\Psi = 0.1$ and $m = 3.0$ in Fig. D.4. The intersection occurs at $X = 7.3$. The time when the flux drops to 10% of the initial flux is $7.3 \times 1.71\,\text{s} = 12.5\,\text{s}$.

Ninety-five percent of the excess CO_2 is lost when $\Sigma = 0.95$. The intersection of $\Sigma = 0.95$ and $m = 3.0$ in Fig. D.5 occurs at $X = 10$. The time for 95% of the excess CO_2 to leave the tissue is $10 \times 1.71\,\text{s} = 17.1\,\text{s}$.

Since $Y(1, X) = \Psi$, the time for the partial pressure of carbon dioxide to reach 50 mmHg can be found by first computing $Y(1, X)$:

$$Y(1,X) = \frac{50\,\text{mmHg} - 40\,\text{mmHg}}{60\,\text{mmHg} - 40\,\text{mmHg}} = 0.5.$$

The intersection of $\Psi = 0.5$ and $m = 3.0$ in Fig. D.4 is about 2.05; so the time required for the surface to reach 50 mmHg is 2.05×1.71 s = 3.51 s.

To find the P_{CO_2} at $x = 25$ μm and $t = 5$ s, we must find the centerline partial pressure of CO_2 at $t = 5$ s, then use either Fig. D.2 or D.3 to estimate the value at 25 μm. The first step is to find X at 5 s:

$$X = \frac{D_{CO_2,T} t}{L^2} = \frac{(1.46 \times 10^{-5} \text{cm}^2 \text{ s}^{-1})(5 \text{ s})}{(50 \times 10^{-4} \text{cm})^2} = 8.56.$$

The dimensionless partial pressure at the center is found from the intersection of $X = 8.56$ and $m = 3.0$ in Fig. D.1 to be $Y(0, 8.56) = 0.08$. Since $X \gg 0.12$, we can use Fig. D.2 to estimate the partial pressure of CO_2 at any radial position. The intersection of $n = 0.5$ with the graph for $m = 3.0$ occurs at $Y(0.5)/Y(0) = 0.97$. The dimensionless partial pressure at $x = 25$ μm is:

$$Y(0.5, 8.56) = \frac{Y(0.5)}{Y(0)} Y(1, 8.56) = (0.97)(0.08) = 0.0776.$$

Converting this to partial pressure:

$$Y(n, X) = \frac{P_{CO_2,T}(x,t) - P_{CO_2,f\infty}}{P_{CO_2,T0} - P_{CO_2,f\infty}} = \frac{P_{CO_2,T}(x,t) - 40 \text{ mmHg}}{20 \text{ mmHg}} = 0.0776.$$

Solving for P_{CO_2}, we find the partial pressure of CO_2 at $x = 25$ μm and $t = 5$ s to be 41.6 mmHg.

The number of moles of CO_2 that cross the surface after 1 s can be found from Fig. D.5. The value of X at $t = 1$ s is 0.585. The intersection of $X = 0.585$ and $m = 3.0$ in Fig. D.5 is $\Sigma = 0.17$. Therefore, 17% of the total excess CO_2 leaves the tissue in 1 s. The number of moles of CO_2 that would cross the tissue surfaces after a long time would be:

$$N_{A\infty} = V \alpha_{CO_2,T} [P_{CO_2,T0} - P_{CO_2,f\infty}],$$

$$N_{A\infty} = [(100 \times 10^{-4} \text{cm})(0.5 \text{ cm})(0.5 \text{ cm})] \left[3.35 \times 10^{-8} \frac{\text{molCO}_2}{\text{ml mmHg}} \right] [20 \text{ mmHg}]$$

$$= 1.68 \text{ nmol CO}_2.$$

The number of moles which cross in the first 1 s is 17% of this, or 0.285 nanomoles of CO_2.

Examining and interpreting the results: The exchange of CO_2 between the thin slab of tissue and surroundings is rapid, with 95% completion within 17.1 s. We have neglected the loss of CO_2 from the other four sides of the tissue; so this will be an overestimate of the actual time for removal of 95% of the excess CO_2.

14.7 One-Dimensional Unsteady-State Shell Balance Applications

Additional comments: Students often ask why we go to the trouble of setting up problems in nondimensional form. This problem clearly demonstrates the benefits. A problem posed in dimensionless form only needs to be solved once, and the results can often be displayed graphically. The solution for any particular set of parameters can be obtained from the dimensionless graphical solution. Physiologists, physicians, and others who may know nothing about solving partial differential equations, and practicing bioengineers who may have forgotten how to solve them, can still solve unsteady-state diffusion problems using the graphical methods illustrated here and in Appendix D.

14.7.1.5 Unsteady-State Diffusion in a Long Cylinder

We consider now unsteady-state diffusion through a material having a cylindrical shape with radius R and length L. The length is assumed to be much larger than the radius ($L \gg R$), so mass transfer is in the radial direction only. The concentration of A in the cylinder is initially c_{A0}, and at time $t = 0$ the surface of the cylinder is exposed to a fluid with a concentration $c_{Af\infty}$. Mass transfer of species A from the cylinder surface is governed by a mass transfer coefficient k_f.

We begin by applying species conservation to the shell in Fig. 14.9:

$$\left\{ \begin{array}{c} \text{Rate of} \\ \text{accumulation} \\ \text{of species A} \\ \text{within shell} \end{array} \right\} = \left\{ \begin{array}{c} \text{Net rate species A} \\ \text{enters through} \\ \text{shell boundaries} \end{array} \right\} + \left\{ \begin{array}{c} \text{Rate species A} \\ \text{is produced} \\ \text{within the shell} \end{array} \right\}. \quad (14.337)$$

Species A is not produced in the tissue, so the last term will be zero. The rate of accumulation of species A in the shell with volume $2\pi r L \Delta r$ is:

$$\left\{ \begin{array}{c} \text{Rate of} \\ \text{accumulation} \\ \text{of species A} \\ \text{within shell} \end{array} \right\} = \frac{\partial}{\partial t}[c_{Ac}(x,t) 2\pi r L \Delta r], \quad (14.338)$$

where C_{Ac} is the concentration of A in the cylinder. Since convective flux of species A in the cylinder is zero, the net rate that species A enters through the shell boundaries is by diffusion:

$$\left\{ \begin{array}{c} \text{Net rate species A} \\ \text{enters through} \\ \text{shell boundaries} \end{array} \right\} = 2\pi r L J_{Ac}(r,t) - 2\pi(r + \Delta r) L J_{Ac}(r + \Delta r, t)$$

$$= 2\pi r L \left[((rJ_{Ac})|_r - (rJ_{Ac})|_{r+\Delta r}) \right]. \quad (14.339)$$

Substituting (14.338) and (14.339) into (14.337), dividing by the shell volume, and letting the volume approach zero, we obtain the following partial differential equation:

$$\frac{\partial c_{Ac}}{\partial t} = -\frac{1}{r}\frac{\partial(rJ_{Ac})}{\partial r}. \tag{14.340}$$

Finally, introducing Fick's law for the diffusive flux, and assuming the diffusion coefficient is independent of position, we find:

$$\frac{\partial c_{Ac}}{\partial t} = \frac{D_{Ac}}{r}\frac{\partial}{\partial r}\left(r\frac{\partial c_{Ac}}{\partial r}\right). \tag{14.341}$$

This is to be solved subject to the following boundary conditions and initial condition:

$$\left.\frac{\partial c_{Ac}}{\partial r}\right|_{r=0} = 0, \tag{14.342}$$

$$-D_{Ac}\left.\frac{\partial c_{Ac}}{\partial r}\right|_{r=R} = \frac{k_{Af}}{\Phi_{Acf}}(c_{Ac}(R,t) - \Phi_{Acf}c_{Af\infty}), \tag{14.343}$$

$$c_{Ac}(r,0) = c_{A0}. \tag{14.344}$$

Introducing the following dimensionless variables:

$$Y = \frac{c_{Ac} - \Phi_{Acf}c_{Af\infty}}{c_{A0} - \Phi_{Acf}c_{Af\infty}}, \tag{14.345}$$

$$n = \frac{r}{R}, \tag{14.346}$$

$$X = \frac{D_{Ac}t}{R^2}, \tag{14.347}$$

$$m = \frac{D_{Ac}\Phi_{Acf}}{k_{Af}R}. \tag{14.348}$$

After converting the differential equation (14.341) and auxiliary conditions (14.342)–(14.344) to dimensionless equations, we find the solution to be an infinite series (Crank 1956):

$$Y = \sum_{k=1}^{\infty} A_k J_0(\lambda_k n)e^{-\lambda_k^2 X}. \tag{14.349}$$

14.7 One-Dimensional Unsteady-State Shell Balance Applications

J_0 is a Bessel function of the first kind and zero order (not to be confused with the symbol used for diffusive flux). The coefficients A_k are given by:

$$A_k = \frac{2J_1(\lambda_k)}{\lambda_k\left[J_0^2(\lambda_k) + J_1^2(\lambda_k)\right]}. \tag{14.350}$$

J_1 is a Bessel function of the first kind and first order. The value λ_k represents the kth solution of the transcendental equation:

$$\lambda_k \frac{J_1(\lambda_k)}{J_0(\lambda_k)} = \frac{1}{m}. \tag{14.351}$$

The solution for the dimensionless concentration at the center of the cylinder as a function of dimensionless time X for various values of m is given in Fig. D.6. In addition, radial concentration profiles can be estimated for dimensionless times $X > 0.15$, where the solution is well represented by just the first term of the series:

$$\frac{Y(n, X)}{Y(0, X)} \approx J_0(\lambda_1 n), \; X > 0.15. \tag{14.352}$$

This is plotted in Fig. D.7 and more accurate plots for $X < 0.15$ are found in Fig. D.8. The graphs in Appendix D.3 can be used to estimate either heat transfer or mass transfer from a cylinder.

The flux from the surface ($n = 1$) relative to the maximum flux is given by (14.333). For the cylinder:

$$\Psi = Y(n = 1, X) = \sum_{k=1}^{\infty} A_k J_0(\lambda_k) e^{-\lambda_k^2 X}. \tag{14.353}$$

Equation (14.353) can be used to find both the flux at the surface and the surface concentration. Ψ is plotted in Fig. D.9 as a function of dimensionless time (Fourier number, X) for various values of m.

The number of moles of A exchanged between cylinder and fluid relative to the number of moles exchanged after a long time is given by (14.335) with $S/V = 2/R$ for a cylinder. Using dimensionless variables Y and X:

$$\Sigma = \frac{N_{Ac}}{N_{A\infty}} = \frac{m_{Ac}}{m_{A\infty}} = \frac{2}{m}\int_0^X Y(n=1, X)\, dX = \frac{2}{m}\sum_{k=1}^{\infty}\frac{A_k J_0(\lambda_k)}{\lambda_k^2}\left[1 - e^{-\lambda_k^2 X}\right]. \tag{14.354}$$

This ratio is plotted in Fig. D.10 as a function of dimensionless time X for various values of m. The solution procedure for problems in which unsteady-state mass transfer occurs between a flowing fluid and a long stationary cylinder is similar to the procedure illustrated in Example 14.7.2 for exchange from a finite slab.

14.7.1.6 Unsteady-State Diffusion in a Sphere

Finally, we consider unsteady-state diffusion through a spherically shaped material having with radius R. The concentration of A in the sphere is initially c_{A0}, and at time $t = 0$ the surface of the sphere is exposed to a fluid with a concentration $c_{Af\infty}$. Mass transfer of species A from the sphere surface is governed by a mass transfer coefficient k_f. Mass transfer is assumed to be in the radial direction only.

We begin by applying species conservation to a shell confined between planes of r and $r + \Delta r$:

$$\left\{\begin{array}{c} \text{Rate of} \\ \text{accumulation} \\ \text{of species A} \\ \text{within shell} \end{array}\right\} = \left\{\begin{array}{c} \text{Net rate species A} \\ \text{enters through} \\ \text{shell boundaries} \end{array}\right\} + \left\{\begin{array}{c} \text{Rate species A} \\ \text{is produced} \\ \text{within the shell} \end{array}\right\}. \quad (14.355)$$

Since species A is not produced in the tissue, the last term will be zero. The rate of accumulation of species A in the shell with volume $4\pi r^2 \Delta r$ is:

$$\left\{\begin{array}{c} \text{Rate of} \\ \text{accumulation} \\ \text{of species A} \\ \text{within shell} \end{array}\right\} = \frac{\partial}{\partial t}\left[c_{As}(x,t) 4\pi r^2 \Delta r\right], \quad (14.356)$$

where c_{As} is the concentration of A in the sphere. Since convective flux of species A in the sphere is zero, the net rate that species A enters through the shell boundaries is by diffusion:

$$\left\{\begin{array}{c} \text{Net rate species A} \\ \text{enters through} \\ \text{shell boundaries} \end{array}\right\} = 4\pi\left[\left((r^2 J_{As})\big|_r - (r^2 J_{As})\big|_{r+\Delta r}\right)\right]. \quad (14.357)$$

Substituting (14.356) and (14.357) into (14.355), dividing by the shell volume, and letting the volume approach zero, we obtain the following partial differential equation:

$$\frac{\partial c_{As}}{\partial t} = -\frac{1}{r^2}\frac{\partial (r^2 J_{As})}{\partial r}. \quad (14.358)$$

Finally, introducing Fick's law for the diffusive flux, and assuming the diffusion coefficient is independent of position, we find:

14.7 One-Dimensional Unsteady-State Shell Balance Applications

$$\frac{\partial c_{As}}{\partial t} = \frac{D_{As}}{r^2} \frac{\partial}{\partial r}\left(r^2 \frac{\partial c_{As}}{\partial r}\right). \tag{14.359}$$

This is to be solved subject to the same boundary conditions specified for the cylinder (14.342)–(14.344) with the subscript s (for sphere) replacing the subscript c (for cylinder). We can also introduce the dimensionless variables that are identical to those introduced for the cylinder (14.345)–(14.348), again with s replacing c. The solution to the dimensionless form of (14.359) in this case is (Crank 1956):

$$Y = \sum_{k=1}^{\infty} A_k \left(\frac{\sin \lambda(\beta_k n)}{n}\right) e^{-\lambda_k^2 X}. \tag{14.360}$$

The coefficients A_k are given by:

$$A_k = \frac{2(m-1)}{\lambda_k \cos(\lambda_k)\left[m^2 \lambda_k^2 + 1 - m\right]}, \tag{14.361}$$

and λ_k represents the kth solution of the transcendental equation:

$$\lambda_k \cot(\lambda_k) = 1 - \frac{1}{m}. \tag{14.362}$$

Solutions for the dimensionless concentration at the center of the sphere as a function of dimensionless time X for various values of m are given in Fig. D.11. In addition, radial concentration profiles are shown in Fig. D.12 for dimensionless times when $X > 0.15$. This is based on the assumption that the solution is adequately represented by just the first term in the infinite series:

$$\frac{Y(n, X)}{Y(0, X)} \approx \frac{\sin(\lambda_1 n)}{\lambda_1 n \cos(\lambda_1)}, \quad X > 0.15. \tag{14.363}$$

Radial concentration profiles are plotted in Fig. D.13 for four values of m and for $X < 0.15$.

The flux of solute from the sphere surface relative to the maximum flux is given by (14.333):

$$\Psi = \frac{N_{As}}{N_{A\max}} = Y(n=1, X) = \sum_{k=1}^{\infty} A_k \sin(\lambda_k) e^{-\lambda_k^2 X}. \tag{14.364}$$

This is plotted in Fig. D.14 as a function of dimensionless time X for various values of m. The maximum number of moles that can cross the surface of the sphere after an infinitely long time, relative to the total number that eventually cross is given by (14.335) with $S/V = 3/R$ for a sphere:

$$\Sigma = \frac{N_{Ac}}{N_{A\infty}} = \frac{m_{Ac}}{m_{A\infty}} = \frac{3}{m}\int_0^X Y(n=1,X)\,dX = \frac{3}{m}\sum_{k=1}^{\infty}\frac{A_k \sin(\lambda_k)}{\lambda_k^2}\left[1 - e^{-\lambda_k^2 X}\right].$$

(14.365)

This ratio is plotted in Fig. D.15 as a function of dimensionless time X for various values of m.

14.7.2 Unsteady-State 1D Convection and Diffusion

Steady-state convective mass transfer analysis is appropriate when the input concentration and flow to the system under investigation remain constant. If either of these changes significantly with time, then an unsteady-state analysis is necessary. The transient response to changes in input concentration can often be used to identify system or solute properties which may be difficult to measure by other means. In this section, we will analyze transient situations in which solute concentration varies with time and position, while flow through the system remains constant. Two important examples are the characterization of vascular properties using multiple indicator dilution experiments and the separation of macromolecules using size-exclusion chromatography.

14.7.2.1 Indicator Dilution Applications

In Sect. 13.4, we introduced indicator dilution methods for determining flow, volume, and permeability in well-mixed organ systems. In this section, we will analyze the exchange of nonreacting tracers as they move through a single capillary. The tracer is introduced as a bolus at the capillary inlet, $C_{Ac}(0, t) = f(t)$. We will assume that the flow is steady and the velocity profile is flat. We neglect axial diffusion relative to axial convection in the capillary.

We can apply the unsteady-state expression for conservation of tracer in the capillary:

$$\left\{\begin{array}{c}\text{Rate of}\\\text{accumulation}\\\text{of species A}\\\text{within shell}\end{array}\right\} = \left\{\begin{array}{c}\text{Net rate species A}\\\text{enters through}\\\text{shell boundaries}\end{array}\right\} + \left\{\begin{array}{c}\text{Rate species A}\\\text{is produced}\\\text{within the shell}\end{array}\right\}. \quad (14.366)$$

Referring to Fig. 14.48, this becomes:

$$\frac{\partial}{\partial t}[c_{Ab}(x,t)A_c\Delta x] = Q_b c_{Ab}(x,t) - Q_b c_{Ab}(x,t) - N_A S_i \frac{\Delta x}{L}. \quad (14.367)$$

14.7 One-Dimensional Unsteady-State Shell Balance Applications

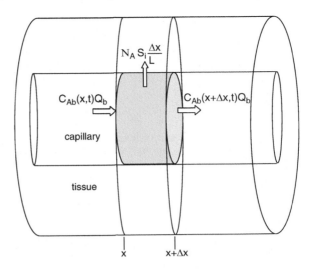

Fig. 14.48 Capillary-tissue exchange

S_i is the internal surface area of the capillary, Q_b is blood flow through the capillary, and A_c is the cross-sectional area of the capillary. Dividing by the shell volume $A_c \Delta x$ and letting the volume shrink to zero, we find:

$$\frac{\partial c_{Ab}}{\partial t} = -\langle v \rangle \frac{\partial c_{Ab}}{\partial x} - \frac{N_A S_i}{A_c L}. \qquad (14.368)$$

We have replaced Q_b/A_c with the average velocity $\langle v \rangle$. Replacing the flux with the standard expression for transcapillary transport:

$$\frac{\partial c_{Ab}}{\partial t} + \langle v \rangle \frac{\partial c_{Ab}}{\partial x} = -\frac{P_A S_i}{V_c} [c_{Ab} - \Phi_{AbT} c_{AT}]. \qquad (14.369)$$

c_{AT} is the tracer concentration in the tissue, Φ_{AbT} is the blood–tissue tracer partition coefficient, and $V_c = A_c L$ is the intracapillary volume. Applying (14.366) to the stationary tissue, we find:

$$\frac{\partial c_{AT}}{\partial t} = +\frac{P_A S_i}{V_T} [c_{Ab} - \Phi_{AbT} c_{AT}]. \qquad (14.370)$$

V_T is the tissue volume surrounding the capillary. Substituting (14.370) into (14.369) yields:

$$\frac{\partial c_{Ab}}{\partial t} + \langle v \rangle \frac{\partial c_{Ab}}{\partial x} + \frac{V_T}{V_c} \frac{\partial c_{AT}}{\partial t} = 0. \qquad (14.371)$$

Let us now consider three special tracers: (1) a tracer that cannot move through the capillary wall; (2) a tracer that does not diffuse back into the intravascular space after entering the tissue; and (3) a tracer that rapidly equilibrates between the capillary and tissue space.

Intravascular Tracer, c_R

For an intravascular tracer or a reference tracer, we let species A in the capillary be represented by the symbol "R". Letting capillary permeability be zero in (14.369), we find:

$$\frac{\partial c_R}{\partial t} + \langle v \rangle \frac{\partial c_R}{\partial x} = 0. \tag{14.372}$$

Taking the Laplace transform of (14.372), where $\bar{C}_R$ is the Laplace transform of c_R:

$$s\bar{C}_R - c_R(0,x) + \langle v \rangle \frac{d\bar{C}_R}{dx} = 0. \tag{14.373}$$

Assuming there is no tracer initially present in the capillary and the tracer at the capillary inlet has an input function $C_R(t, 0) = C_{R0}(t)$, then the solution to (14.386) is:

$$\bar{C}_R(x) = \bar{C}_{R0} e^{-\frac{sx}{\langle v \rangle}}. \tag{14.374}$$

Inverting this back into the time domain, we find:

$$c_R(x,t) = 0, t < \frac{x}{\langle v \rangle},$$

$$c_R(x,t) = c_{R0}\left(t - \frac{x}{\langle v \rangle}\right), t \geq \frac{x}{\langle v \rangle}. \tag{14.375}$$

The solution is shown in Fig. 14.49. In words, the solution tells us that the input function $c_R(t)$ will travel through the capillary undistorted at a velocity $\langle v \rangle$ and will appear at any location x at a time $x/\langle v \rangle$. The input function will begin to appear at the outlet ($x = L$) after a time equal to $t_R = L/\langle v \rangle$, the mean transit time through the capillary. The capillary volume can be estimated from the mean transit time of the reference curve:

$$V_c = Q_b t_R. \tag{14.376}$$

Non-Returning Diffusible Tracer, c_D

Let us turn our attention now to a tracer that can pass from the capillary space into the tissue space, but back-diffusion is prevented. Such behavior is referred to as

14.7 One-Dimensional Unsteady-State Shell Balance Applications

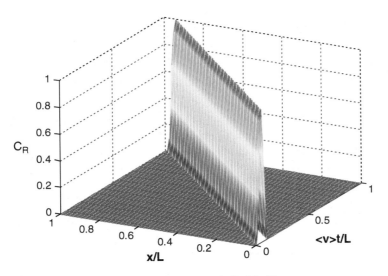

Fig. 14.49 Intravascular reference tracer solution $C_R(x/L, \langle v \rangle t/L)$

diffusion to an infinite sink. The tracer might react with another species in the tissue, be internalized by cells, be rapidly shunted to the lymphatic system, or simply be greatly diluted in the tissue. We will call this the diffusible tracer and denote its concentration in the capillary as $c_D(x, t)$ and the capillary permeability as P_D. Setting the tissue concentration equal to zero in (14.369):

$$\frac{\partial c_D}{\partial t} + \langle v \rangle \frac{\partial c_D}{\partial x} = -\left(\frac{P_D S_i}{V_c}\right) c_D. \tag{14.377}$$

Taking the Laplace Transform of (14.377) with $c_D(x, 0) = 0$:

$$s\bar{c}_D + \langle v \rangle \frac{d\bar{c}_D}{dx} + \frac{P_D S_i}{V_c} \bar{c}_D = 0. \tag{14.378}$$

The solution to (14.378) for an input function $c_D(t, 0) = c_{D0}(t)$ is:

$$\bar{C}_D(x) = \bar{C}_{D0} e^{-\left(\frac{P_D S_i}{V_c} + s\right)\left(\frac{x}{\langle v \rangle}\right)}. \tag{14.379}$$

Inverting this back into the time domain:

$$c_D(x, t) = 0, t < \frac{x}{\langle v \rangle},$$
$$c_D(x, t) = c_{D0}\left(t - \frac{x}{\langle v \rangle}\right) e^{-\left(\frac{P_D S_i}{Q_b}\right)\left(\frac{x}{L}\right)}, t \geq \frac{x}{\langle v \rangle}. \tag{14.380}$$

At the capillary exit, where $x = L$, the emerging tracer concentration for $t > L/\langle v \rangle$ will be:

$$c_D(L,t) = c_{D0}\left(t - \frac{L}{\langle v \rangle}\right) e^{-\left(\frac{P_D S_i}{Q_b}\right)}. \qquad (14.381)$$

If the diffusible tracer and reference tracers are injected simultaneously, the input functions will have the same shape, but may have different magnitudes, depending on the concentrations of reference tracer and diffusible tracer in the injected bolus, c_{Ri} and c_{Di}:

$$\frac{c_{D0}(t)}{c_{R0}(t)} = \frac{c_{Di}}{c_{Ri}}. \qquad (14.382)$$

Comparing the diffusible and reference tracer concentrations at the exit of the capillary, we find:

$$\frac{c_D(L,t)/c_{Di}}{c_R(L,t)/c_{Ri}} = e^{-\left(\frac{P_D S_i}{Q_b}\right)}. \qquad (14.383)$$

Therefore, if we compare the normalized diffusion tracer concentration at the capillary exit with the normalized reference curve at the tracer outlet, every point will be attenuated by a factor equal to $\exp(-P_D S_i/Q_b)$. The permeability surface area product can then be estimated from (14.383) by comparing the magnitudes of diffusible and reference indicator curves at the capillary exit:

$$P_D S_i = -Q_b \ln\left[\frac{c_D(L,t)/c_{Di}}{c_R(L,t)/c_{Ri}}\right] = -Q_b \ln[1 - E]. \qquad (14.384)$$

The extraction E is defined by (13.125).

Most diffusible tracers will eventually diffuse back into the vascular system after the bolus has passed. The solution to the general problem (14.370) and (14.371) can be easily found in the Laplace domain, but inversion to the time domain, even for an impulse input function, is complex (Sangren and Sheppard 1953). The normalized reference and diffusible tracer curves will cross when back-diffusion is present, as shown in Fig. 14.50. However, if the tracer is a diffusion-limited or barrier-limited tracer (i.e., low capillary permeability) (14.383) contains the dominant term in the solution. Therefore, the above approach for estimating $P_D S_i$ is valid during the initial exchange period when the tissue tracer concentration is nearly zero. A common approach is to assume that back-diffusion is negligible until after the peak of the reference curve has emerged from the capillary ($t = t_p$). In that case, the extraction E in (14.384) is replaced by the *integral extraction*, E_I:

14.7 One-Dimensional Unsteady-State Shell Balance Applications

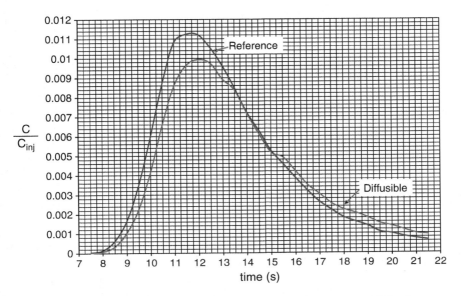

Fig. 14.50 Comparison of reference and diffusible normalized tracer curves following tracer injection in the pulmonary artery and collection from the ascending aorta. The reference curve is a composite constructed from ^{51}Cr-erythrocytes and ^{125}I-albumin. The diffusible tracer is ^{14}C-urea

$$E_{\mathrm{I}} = \frac{\int_0^{t_p} \left[\dfrac{c_{\mathrm{R}}(L,t)}{c_{\mathrm{R}0}(L,t)} - \dfrac{c_{\mathrm{D}}(L,t)}{c_{\mathrm{D}0}(L,t)} \right] \mathrm{d}t}{\int_0^{t_p} \dfrac{c_{\mathrm{R}}(L,t)}{c_{\mathrm{R}0}(L,t)} \mathrm{d}t}. \tag{14.385}$$

Example 14.7.3 Lung Urea Permeability Surface Area.
The tracer data in Fig. 14.50 were collected at the outlet of an isolated canine lung preparation. Estimate the flow through the lungs and the permeability surface area product for urea exchange in the lung from the normalized data in the figure. The injected bolus containing both tracers had a volume of 1 ml.

Solution. *Initial considerations*: Tracer data can be used to estimate both the flow rate and the product of lung permeability to urea and the exchanging surface area. Urea and intravascular reference data must first be made dimensionless by normalizing the data by the appropriate tracer concentration in the injected bolus.
System definition and environmental interactions: The system consists of the blood in lung capillaries. Urea can pass through the walls of lung capillaries, while

the intravascular tracer cannot. Since red cells move faster through capillaries than plasma, a composite intravascular curve is constructed from radiolabeled red cells and radiolabeled albumin.

Apprising the problem to identify governing relationships: Conservation of species for either tracer in the lung circulation leads to the Stewart–Hamilton equation that can be used to estimate flow rate through the lung. This was derived using the macroscopic approach in Sect. 13.4.3. The permeability surface area for urea can be estimated by measuring the extraction and using (14.384).

Analysis: The Stewart–Hamilto equation for the reference curve is:

$$Q_b = \frac{m_{Ri}}{\int_0^\infty \rho_R(t)dt} = \frac{V_{inj}}{\int_0^\infty \left(\frac{\rho_R(t)}{\rho_{R,inj}}\right)dt} = \frac{V_{inj}}{\int_0^\infty \left(\frac{c_R(t)}{c_{R,inj}}\right)dt}.$$

V_{inj} is the volume of the injected bolus of tracer and $c_{R,inj}$ is the concentration of reference tracer in the injected bolus. Integration can be performed numerically. However, since the concentration does not return to zero when data collection was terminated, we need to first estimate the remaining area under the curve using the extrapolation method shown in Fig. 13.17. A plot of $\ln(c_R/c_{R,inj})$ vs. time is linear for $t > 14.5$ s with a slope $= -k = -0.3426$ s^{-1}. We can then divide the area calculation into two parts as indicated in (13.115): numerical integration between 0 and 14.5 s and the area under an exponentially falling curve with slope $-k$ for $t > 14.5$ s. A simple rectangular integration scheme can be used with intervals of $\Delta t = 0.5$ s. The area under the reference curve is:

$$\int_0^\infty \left(\frac{c_R(t)}{c_{R,inj}}\right)dt = \int_0^{14.5 s} \left(\frac{c_R(t)}{c_{R,inj}}\right)dt + \int_{14.5 s}^\infty \left(\frac{c_R(t)}{c_{R,inj}}\right)dt$$

$$= \sum_{t=0}^{t=14.5 s} \left(\frac{c_R(t)}{c_{R,inj}}\right)\Delta t + \frac{1}{k}\left(\frac{c_R(t=14.5)}{c_{R,inj}}\right)$$

$$= 0.0479 \text{ s} + 0.0183 \text{ s} = 0.0662 \text{ s}.$$

The flow rate, computed using the Stewart–Hamilton equation, is (1 ml)/(0.0662 s) $=$ 15.1 ml/s. The integral extraction of urea can be computed using (14.385). We use numerical integration of the diffusible and reference curves from $t = 0$ to $t = 11.75$ s, the peak of the reference curve. The area under the normalized reference curve to the peak is 0.0241 s, and the area under the diffusible curve from $t = 0$ to $t = 11.75$ s is 0.01874 s. The integral extraction for urea is $(0.0241 - 0.01874)/(0.0241) = 0.222$. The permeability surface area product for urea in the isolated, perfused dog lung, computed using (14.384) is:

$$P_{urea}S_i = -Q_b \ln[1 - E] = -(15.1 \text{ ml/s}) \ln(1 - 0.222) = 3.79 \text{ ml/s}.$$

14.7 One-Dimensional Unsteady-State Shell Balance Applications

Examining and interpreting the results: Since $P_{urea}S_i/Q_b$ is relatively small, urea can be considered a barrier-limited tracer. A tracer with a much lower permeability than urea would be very difficult to distinguish experimentally from a reference tracer.

Flow-Limited Diffusible Tracer, c_D

If the capillary permeability is very high for a particular tracer, the tracer is said to be flow-limited. In that case a local equilibrium will exist between tracer in the capillary and tracer at the same axial location in the tissue:

$$c_{Ab} = \Phi_{AbT} c_{AT}. \tag{14.386}$$

Substituting this into (14.371) and letting C_{Ab} be designated as the intravascular concentration of diffusible tracer, C_D:

$$\left[1 + \left(\frac{1}{\Phi_{AbT}} \frac{V_T}{V_c}\right)\right] \frac{\partial c_D}{\partial t} + \langle v \rangle \frac{\partial c_D}{\partial x} = 0. \tag{14.387}$$

Let us define γ as a dimensionless tissue volume:

$$\gamma \equiv \frac{1}{\Phi_{AbT}} \frac{V_T}{V_c}. \tag{14.388}$$

Then (14.387) can be written as:

$$\frac{\partial c_D}{\partial t} + \frac{\langle v \rangle}{1+\gamma} \frac{\partial c_D}{\partial x} = 0. \tag{14.389}$$

The solution to (14.389) can be found by analogy with (14.372) and (14.375). The flow-limited tracer concentration at any axial position within the capillary will be:

$$c_D(x,t) = 0, t < \frac{(1+\gamma)x}{\langle v \rangle},$$

$$c_D(x,t) = c_{D0}\left(t - \frac{(1+\gamma)x}{\langle v \rangle}\right), t \geq \frac{(1+\gamma)x}{\langle v \rangle}. \tag{14.390}$$

Therefore, a flow-limited tracer measured at the capillary exit $x = L$ will have a normalized concentration–time curve that is shifted to the right relative to a reference tracer by a factor of $1 + \gamma$.

Since the capillary volume can be found from the mean transit time of the reference tracer, we can use the mean transit time of the flow-limited diffusible tracer t_D to estimate the tissue volume:

$$V_T = Q_b \Phi_{AbT}(t_D - t_R). \tag{14.391}$$

This is the same as (13.120), which was derived in Sect. 13.4.2.

14.7.2.2 Chromatography

Size-exclusion chromatography is used to separate macromolecules according to their size. A bolus containing a mixture of macromolecules is introduced at the inlet to the column and macromolecules emerge from the column at different times, dependent upon their size. The column is packed with beads containing materials that are permeable to a select range of macromolecules. The columns can be either preparative, in which case the different macromolecules are collected at the column outlet for future use, or analytic, the purpose which is to accurately detect the amount of each protein emerging from the column.

We can separate the material in the column into a mobile phase with concentration of macromolecule A, c_{Am}, and a stationary phase with concentration c_{As}. Unsteady-state mass balances performed on macromolecule A in the mobile and stationary portions of a thin segment of the column (Fig. 14.40) are identical to the equations derived for a diffusible tracer flowing through a capillary:

$$\frac{\partial c_{Am}}{\partial t} + \langle v \rangle \frac{\partial c_{Am}}{\partial x} = -\frac{P_A S_i}{V_m}[c_{Am} - \Phi_{Ams} c_{As}], \tag{14.392}$$

$$\frac{\partial c_{As}}{\partial t} = +\frac{P_A S_i}{V_s}[c_{Am} - \Phi_{Ams} c_{As}]. \tag{14.393}$$

The void volume of the column ε is defined as:

$$\varepsilon = \frac{V_m}{V_m + V_s}. \tag{14.394}$$

The superficial velocity v_0 is related to the average velocity in the mobile phase $\langle v \rangle$ as follows:

$$\langle v \rangle = \frac{v_0}{\varepsilon}. \tag{14.395}$$

Substituting (14.393) to (14.395) into (14.392):

14.7 One-Dimensional Unsteady-State Shell Balance Applications

$$\frac{\partial c_{Am}}{\partial t} + \frac{v_0}{\varepsilon}\frac{\partial c_{Am}}{\partial x} + \left(\frac{1-\varepsilon}{\varepsilon}\right)\frac{\partial c_{As}}{\partial t} = 0. \tag{14.396}$$

Finally, let us assume that flow through the column is very slow, so the mobile and stationary phases are in local equilibrium; so $c_{As} = c_{Am}/\Phi_{Ams}$. Equation (14.396) then becomes:

$$\frac{\partial c_{Am}}{\partial t} + \frac{v_0 \Phi_{Ams}}{1 - \varepsilon(1 - \Phi_{Ams})}\frac{\partial c_{Am}}{\partial x} = 0. \tag{14.397}$$

Note that this is the same as the expression for a flow-limited tracer in a capillary (14.389), if we redefine γ as:

$$\gamma = \frac{1-\varepsilon}{\varepsilon \Phi_{Ams}}. \tag{14.398}$$

Therefore, the solution at the column outlet can be found from (14.390):

$$c_m(L,t) = c_{m0}\left(t - \frac{(1+\gamma)L}{\langle v \rangle}\right), t \geq \frac{(1+\gamma)L}{\langle v \rangle}. \tag{14.399}$$

Each protein can be identified by its appearance time at the column exit: t_a.

$$t_a = \frac{L}{v_0}\left(\frac{1 + \varepsilon(\Phi_{Ams} - 1)}{\Phi_{Ams}}\right). \tag{14.400}$$

The values of ε, L, and v_0 are the same for all tracers; so the difference between appearance times for different macromolecules is determined by the partition coefficient between the mobile and stationary phases. The partition coefficient, Φ_{Ams}, generally increases with increasing molecular size. In preparative work, the beads constituting the stationary phase are composed of aqueous gels like Sephadex or Agarose. The gel fibers are randomly distributed throughout the stationary volume. By their very presence, gel fibers exclude macromolecules from distributing into all portions of the gel. For instance, consider a single cylindrical gel fiber of length L_f and radius R_f in the stationary phase. The center of a solute molecule with radius R_A can come no closer to the center of the fiber than a distance equal to the sum of the fiber radius and the solute radius. Consequently, the center of the solute is excluded from a volume equal to $V_e = n_f \pi (R_f + R_A)^2 L_f$, where n_f is the number of fibers in the gel. The partition coefficient between the mobile and stationary phases, based on simple steric exclusion is:

$$\Phi_{Asm} = \frac{1}{\Phi_{Ams}} = \frac{V_s - V_e}{V_s}. \tag{14.401}$$

If the gel fibers are randomly distributed in the gel, the closeness of overlapping fibers alters the steric exclusion of solute and a more commonly used expression is:

$$\Phi_{Asm} = \frac{1}{\Phi_{Ams}} = \exp\left\{-\frac{n_f \pi L_f (R_A + R_f)^2}{V_s}\right\} = \exp\left\{-\frac{V_f}{V_s}\left(1 + \frac{R_A}{R_f}\right)^2\right\}. \tag{14.402}$$

V_f/V_s is the fraction of the stationary phase occupied by gel fibers, which is small. Note that for very small solutes Φ_{Asm} approaches unity, while for large solutes Φ_{Asm} becomes very small. Consequently, large macromolecules will be unable to penetrate into the stationary phase at all, while small solutes can distribute into the entire volume within the stationary phase.

In chromatography applications, the volume of fluid which passes through the column, before the solute of interest emerges, is known as the elution volume, $V_{A,el}$. This is equal to the transit time of the solute multiplied by the flow rate through the column:

$$V_{A,el} = Q_b t_A = V_m \left(1 + \Phi_{Asm}\left(\frac{1-\varepsilon}{\varepsilon}\right)\right) = V_m + \Phi_{Asm} V_s. \tag{14.403}$$

Therefore, the largest molecules with Φ_{Asm} equal to zero will be the first to elute from the column and will emerge in an elution volume equal to the volume of the mobile phase. A *distribution coefficient* K_{Ad} is defined in the literature as:

$$K_{Ad} = \frac{V_{A,el} - V_0}{V_t - V_0} = \frac{V_{A,el} - V_m}{V_s}, \tag{14.404}$$

where the void volume of the column is V_0, which is the same as the mobile phase volume, and the total column volume is V_t, which is identical to $V_m + V_s$. Very small solutes, with Φ_{Asm} nearly equal to unity will elute in a volume equal to $V_m + V_s$. A column is calibrated with molecules of known molecular weight. The useful range of a column is the range of molecular weights that can be accurately differentiated. A calibration curve for a high performance liquid chromatography column used for separating proteins is shown in Fig. 14.51. The breakthrough point occurs when the elution volume is equal to the void volume, or the mobile phase volume, of the column. The molecular weight reached at breakthrough is 316,000. All proteins larger than this will elute together in the void volume. The exclusion limit is reached when the elution volume is equal to the total volume of the column, $V_s + V_m$. All proteins with molecular weights less than about 5,000 will elute together when the elution volume equals the column volume. This column is capable of distinguishing between proteins with molecular weights between 5,000 and 316,000. This is called the useful range of the column. If proteins outside this range are to be detected, a different column would need to be used.

14.7 One-Dimensional Unsteady-State Shell Balance Applications

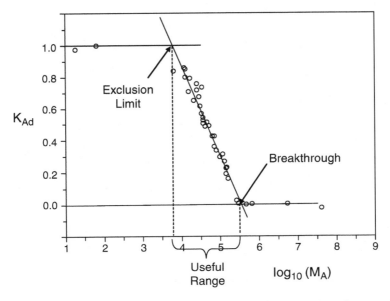

Fig. 14.51 Calibration curve for a size exclusion TSK G3000SW high performance liquid chromatography column

Example 14.7.4 Chromatography Column Application.
The column calibrated in Fig. 14.51 is used to separate a mixture of proteins. The void volume of the column is 8 ml and the total volume of the column is 25 ml. Three protein peaks appear at the following elution volumes: 8.1, 15, and 20 ml. Estimate the molecular weights of the three proteins.

Solution. *Initial considerations*: We will assume that the column is uniformly perfused and the packing is uniformly distributed. In addition, we assume that local equilibrium exists between proteins in the mobile and stationary phases.

System definition and environmental interactions: The system of interest in this problem is the mobile fluid in the column.

Apprising the problem to identify governing relationships: The chromatography relationships derived in Sect. 14.7.2.2 are assumed to be valid.

Analysis: From the data given, the void volume is 8 ml, the total volume is 25 ml, and the stationary phase volume is 17 ml. The distribution coefficient for the first protein is (8.1 ml − 8 ml)/17 ml = 0.006. It is impossible to distinguish this protein from any of the proteins that elute with the void fraction; so it has a molecular weight at or above 316,000. The second protein has a distribution coefficient of (15 ml − 8 ml)/17 ml = 0.41. The calibration curve indicates that $\log_{10}(M_A) = 4.9$, or $M_A = 10^{4.9} = 79{,}400$. Finally, the third protein has a distribution coefficient of

(20 ml − 8 ml)/17 ml = 0.71. The calibration curve indicates that $\log_{10}(M_A) = 4.3$, or $M_A = 10^{4.3} = 20{,}000$.

Examining and interpreting the results: The last two estimates are well within the useful range of the column. However, a different column is needed if a more accurate molecular weight is desired for the high molecular weight protein.

14.8 Summary of Key Concepts

Mass transfer by diffusion is relatively slow, producing significant spatial variations in concentration. Consequently, a microscopic approach is necessary to analyze problems where molecular diffusion is an important transport mechanism. The shell balance approach for one-dimensional mass transfer problems is based on the application of conservation of species to a small portion of the system of interest, as described in Sect. 14.2. For steady-state problems, this procedure results in an ordinary differential equation. Appropriate conditions must be specified at the system boundaries if a meaningful solution is to be found. This procedure is illustrated in Sect. 14.3 for mass flow through a cell membrane.

Oxygen diffusion in consuming cells. The effect of oxygen consumption in cells is examined in Sect. 14.4 for various cell shapes and for different relationships between the rate of oxygen consumption and oxygen concentration, including zeroth order, first order, and Michaelis–Menten kinetics. Cell size is ultimately limited by the ability of oxygen to diffuse to all portions of the cell.

Bioreactors. Bioreactors are often used to convert a common substrate into a drug or other commercial product. Many bioreactors are composed of porous pellets packed in a reactor bed. The pores are lined with an enzyme that catalyzes the conversion of substrate to product. The reactor can be run in batch mode or as a flow-through device. Analysis requires shell balance analysis of diffusion and reaction in the pores of the beads (Sect. 14.4.4), a shell balance applied to the bulk fluid in the device, and mass transfer through the relatively stationary film between the beads and the bulk fluid (Sect. 14.6.3). Performance of the device depends on the Thiele modulus (14.99), which is a measure of the reaction rate relative to the diffusion rate, and on the local product concentration relative to the Michaelis constant. Solution for substrate concentration in a single bead is shown in Fig. 14.15. The actual rate of conversion in the bead pores relative to the maximum possible rate of conversion, known as the effectiveness factor, is analyzed in Sect. 14.4.4.2.

1D convection. The microscopic approach is also appropriate for analyzing the exchange of species A between fluid flowing through a device and materials confined within the boundaries of the system (Sect. 14.5). The exchange of waste products like urea in a hollow fiber mass exchanger is a prime example. We examine the effects of mixing, cocurrent and counter-current flow, and axial diffusion on mass exchange through the fiber walls in Sect. 14.5.2. Counter-current operation is found to be more efficient than cocurrent operation, and axial diffusion, in the case of small solutes like urea, is predicted to be negligible. A 1D shell balance analysis can be applied to the exchange of anesthesia gases between lung capillaries and alveolar gas (Sect. 14.5.3) or the exchange of respiratory gases and solutes between tissue capillaries and the surrounding tissue (Krogh cylinder, Sect. 14.6.2). Inclusion of intracapillary resistance to oxygen exchange is shown to affect the delivery of O_2 to tissue (Fig. 14.33).

1D Unsteady-state diffusion. The one-dimensional species continuity equation for a nonreacting species A is:

$$\frac{\partial c_A}{\partial t} = D_A \frac{\partial^2 c_A}{\partial x^2}.$$

The solution to this partial differential equation depends on the applied boundary conditions. We show solution procedures for several cases of biological relevance in Sect. 14.7. Since molecular diffusion is slow, the analysis for diffusion to a semi-infinite slab is often relevant to diffusion in materials that are not very thick (Sects. 14.7.1.1 and 14.7.1.3). We analyze the flux of species A when two materials with different concentrations are brought together in Sect. 14.7.1.2. Unsteady-state diffusion of species A into a slab of finite thickness, a cylinder, and a sphere are analyzed in Sects. 14.7.1.4–14.7.1.6. When put in dimensionless terms, these are shown to have the same form as the corresponding problems for heat transfer. Detailed graphical solutions are provided in Appendix D.

1D Unsteady-state convection with radial transport. We close this chapter with the analysis of unsteady-state transport of species A across vessel walls or into the stationary phase of a device as fluid flows at a steady rate through the vessel or device. A multiple indicator dilution experiment, where different tracers are injected as a bolus at the inlet to a capillary and their concentrations measured at the outlet of the capillary, can be used to analyze permeability characteristics of the capillary wall. If one of the tracers remains intravascular (zero permeability), the permeability surface area product for the other tracer(s) can be computed with (14.384). A flow-limited tracer is one with a very high permeability, so transport to the tissue is limited only by its flow rate through the capillary. Comparison of the outlet curve for a flow-limited tracer and an intravascular tracer can be used to estimate the tissue volume surrounding the capillary (14.391). Finally, the same principles can be used to analyze the elution of proteins from a size-exclusion chromatography column. Once a column is calibrated with proteins of known molecular weight, the molecular weight of unknown proteins in a sample can be determined by the time they elute from the column.

14.9 Questions

14.9.1. When is it necessary to use a microscopic approach to solve a mass transfer problem?
14.9.2. What general procedure is used to set up a 1D steady-state species transport problem if convection is negligible? Why do we initially analyze a shell within the system, rather than the entire system?

14.9.3 How would you use Fick's Law and Henry's Law to find the concentration profile and flux of a dissolved gas in a biological membrane if the dissolved gas concentrations are known in the fluids that are separated by the membrane? What boundary conditions are needed?

14.9.4 How would you extend the procedure above to the case where diffusion occurs through several materials that are placed in series? What equation applies for each material and what boundary conditions apply at the interface between each material?

14.9.5. What factors influence the permeability of a membrane to a dissolved gas?

14.9.6. How would you use Fick's law to compute the diffusive flow of a non-reacting gas through the solid wall of a blood vessel?

14.9.7. State the species conservation equation in words for the unsteady-state transport of oxygen through a portion of a spherical cell between r and $r + \Delta r$.

14.9.8. What major factors limit cell size?

14.9.9. If oxygen consumption in a cell were truly zeroth order, explain what would happen to oxygen concentration as cell thickness increases. Is this also true for first-order oxygen consumption?

14.9.10. If oxygen consumption in a cell obeys Michaelis–Menten kinetics, under what circumstances is it valid to assume that the reaction is zeroth order? When can oxygen consumption be approximated as being a first-order reaction?

14.9.11. What key assumptions were made in analyzing product formation in a porous bioreactor containing immobilized enzymes in Sect. 14.4.4?

14.9.12. What is the physical interpretation of the Thiele modulus ϕ_T?

14.9.13. What is the meaning of the effectiveness factor η for a bioreactor?

14.9.14. How is the mean mass transfer coefficient for internal mass transfer related to the local mass transfer coefficient?

14.9.15. What factors contribute to the permeability of species A through the wall of a hollow fiber?

14.9.16. Is exchange through a hollow fiber more efficient if the fluid inside the fiber is mixed so there is no axial concentration gradient? Explain.

14.9.17. Which is more efficient: a cocurrent or counter-current mass exchanger?

14.9.18. Is neglecting axial diffusion in a hollow fiber mass exchanger likely to cause significant error? Explain.

14.9.19. What factors influence the exchange of a nonreacting gaseous species between alveolar gas and pulmonary blood in the lung? What additional factors affect the exchange of oxygen and carbon dioxide?

14.9.20. Can you design an experiment to measure the rate of production of species A in an organ?

14.9.21. Would plasma be sufficient to transport oxygen to tissues? Explain.

14.9.22. What boundary conditions are used for oxygen exchange in pulmonary capillaries and how do they differ from boundary conditions in tissue capillaries?

14.9.23 What is a Krogh cylinder? What are the major assumptions made in adopting the Krogh cylinder as a model for blood–tissue exchange? Is this model appropriate for species other than oxygen? What boundary conditions are used at each interface in the Krogh cylinder? What is the lethal corner? What is its significance?

14.9.24. What is meant by intracapillary resistance to mass exchange? How can one account for intracapillary resistance?

14.9.25. How would you distinguish between a batch bioreactor and a continuous feed bioreactor?

14.9.26. What key assumptions were made in Sect. 14.6.3 for the analysis of a continuous feed bioreactor?

14.9.27. How thin can a material be if we are to apply the solution for a semi-infinite media, (14.296), at a time t^*?

14.9.28. How can the graphical solutions in Appendix D for unsteady exchange between a solid and fluid with non-zero surface resistance be applied to the case where the surrounding fluid concentration is well mixed (constant concentration at the solid boundary)?

14.9.29. How can the graphs in Appendix D be used to find concentration as a function of time at the center or surface of a slab, cylinder, or sphere? How can they be used to find surface flux or the accumulation of mass in the solid?

14.9.30. What key assumptions are made in deriving (14.384) for estimating the permeability surface area product of a tracer from a multiple indicator dilution experiment?

14.9.31. What is the exclusion limit and breakthrough volume for a size-exclusion chromatography column?

14.10 Problems

14.10.1 Steady-State Diffusion of an Inert Gas Through the Wall of a Tube

The concentration of a dissolved inert gas A is kept constant inside a blood vessel by maintaining a high rate of flow through the vessel. Under equilibrium conditions, the concentration of A in the vessel wall is found to be three times greater than the concentration of A in the blood, and the concentration of A in the tissue surrounding the vessel is half that in the vessel wall.

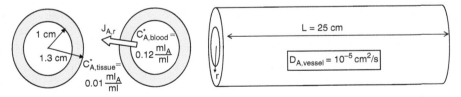

(a) What are the concentrations of A at the inside and outside boundaries of the vessel wall if the concentrations in the blood and tissue are maintained at the values shown above?

(b) Show that a species shell balance in the wall leads to $d(rJ_{A,r})/dr = 0$.
(c) Apply Fick's Law to find the concentration of A as a function of radial position in the vessel wall.
(d) Find the total flow of A (ml$_A$/s) through the vessel wall?

14.10.2 Bioreactor

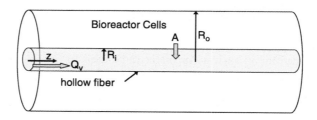

A bioreactor is constructed of a mass of specialized cells that produces a drug A at a constant rate per unit volume R_A (mol s^{-1} cm^{-3}). The cell mass is permeated by a number of straight horizontal fibers with walls permeable to the drug.

Assume the bioreactor can be subdivided into a number of Krogh cylinders with inner radius R_i, outer radius R_o, and length L. Each fiber is perfused at a rate Q_V. No drug is present in the perfusate at the fiber inlet and the drug does not combine chemically with other species in the perfusate. Neglect the effects of intra-fiber radial variations in drug concentration, axial diffusion, and mass transfer resistance at the fiber wall. Find: (a) an expression for the flux of drug from tissue to perfusate and (b) an expression for the concentration of drug in the fluid within each fiber as a function of axial location z.

14.10.3 Steady-State Removal of a Toxin

Endothelial cells are grown on the inside surface of a hollow fiber with radius 100 μm and length 15 cm. The cells contain a surface enzyme that converts a toxin A into a harmless species at a rate that is proportional to the perfused surface area and the toxin concentration squared:

Conversion rate per unit surface area $= k''c_A^2$, where $k'' = 8 \times 10^4$ cm^4 mol^{-1} min^{-1} and blood flow to the organ segment is 0.75 ml/min. Use a 1D shell balance to find an expression for the concentration of toxin as a function of position. Find the toxin concentration at the outlet of the organ segment if the inlet concentration is 0.01 mol/L. Assume the toxin distributes equally in plasma and red cells.

14.10.4 CO_2 Diffusion in Cell Culture Media

A spherical cell with radius R is placed in a large volume of cell culture media. The surface area of the cell is $4\pi R^2$ and the cell volume is $4\pi R^3/3$. Carbon dioxide is produced at a constant rate per unit volume R_{CO_2} in the cell. The concentration of CO_2 in the media far from the cell ($r \to \infty$) is zero. Carbon dioxide is not produced or consumed in the cell media.

(a) What is the flux of CO_2 from the cell surface into the culture media?
(b) Use a shell balance to find the steady-state concentration of CO_2 in the culture media ($r \geq R$) as a function of radial position.

14.10.5 Anesthetic Gas Exchange in the Lung

An anesthesiologist wishes to maintain blood systemic arterial partial pressure of gas A at 0.01 atm by breathing a gas mixture containing gas A. Systemic venous partial pressure of gas A is maintained at zero. What alveolar partial pressure would provide this arterial level? Ignore capillary curvature and assume each capillary to be 1 mm long. Alveolar membrane thickness δ_m is 0.5 μm, total capillary surface area is 100 m^2, and total blood flow to the lungs is 100 ml/s. Assume alveolar membrane diffusivity, $D_{Am} = 2.5 \times 10^{-6}$ cm^2/s, total blood flow, $Q = 100$ ml/s, $\alpha^*_{A,blood} = 0.47$ ml/(ml blood* atm), and $\alpha^*_{A,m} = 2.3$ ml/(ml blood*atm).

14.10.6 Inert Gas Exchange in Lung Capillaries

The partial pressure at the midpoint of pulmonary capillaries is found to be 70% of the partial pressure of gas A in alveolar gas. What is the permeability surface area product of the blood–lung barrier to inert gas A if lung blood flow is 100 ml/s, inlet partial pressure of gas A is zero, and capillary length is 1 mm.

14.10.7 Exchange of Inert Gas in Lungs

Helium gas is maintained at a partial pressure in alveolar gas of $P_{He,alv} = 10$ mmHg. Blood flow through the lungs is 5 L/min, the permeability surface area of the alveolar blood–gas barrier is $P_{He,M}S = 100$ ml/s, and Bunsen solubility coefficients for He in blood and barrier are $\alpha^*_{He,blood} = 0.008$ ml O_2/(ml blood*atm) and $\alpha^*_{He,M} = 0.08$ ml O_2/(ml blood*atm), respectively. Helium is not present in the inlet blood.
(a) Find the He flow across the microvascular barrier of the lung. (b) Compare this

with the maximum helium exchange for very high flow rates (diffusion limited). (c) Compare the result in part (a) with the He flow when the permeability is very high (flow-limited).

14.10.8 O_2 Consumption by Cells

A layer of cells with thickness of δ_c is placed at the bottom of a glass beaker. The glass is impermeable to O_2. The cells consume oxygen at a constant rate per unit volume Q_{O_2}. Cell culture medium (no oxygen consumption) is placed on top of the cell layer to a thickness of δ_f. The Henry law constants for oxygen in cell culture medium and in cells are α_f and α_c, respectively. The partial pressure of oxygen in air is $P_{O_2,\text{air}}$. The diffusion coefficients for O_2 in medium and in cells are D_f and D_c.

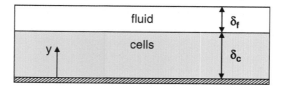

(a) Provide an expression for the flux of oxygen from air to the top surface of the culture medium in terms of known quantities.
(b) What is the flux of oxygen from the culture medium to the cells?
(c) What is the flux of oxygen at the bottom of the cell layer?
(d) Derive an expression for the distribution of the partial pressure of O_2 in the medium and in the cell layer as a function of position, y.
(e) Sketch P_{O_2} vs. y from $y = 0$ to $y = \delta_c + \delta_f$.
(f) What fluid layer thickness is necessary to lower the partial pressure of oxygen at the bottom of the cell layer to zero? Is your solution valid for thicker fluid layers? Explain.

14.10.9 Steady-State Capillary Filtration (1D)

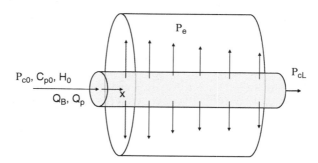

As blood flows through a water-permeable capillary, some water flows through the capillary wall, concentrating protein and red cells in the axial direction. This leads to an axial gradient in osmotic pressure, tending to slow down the flow of water through the capillary wall as x increases. Assume that hydrostatic pressure in the fluid surrounding the capillary is constant (P_e), and this fluid is identical in composition to plasma, except that it contains no protein. Assume further that van't Hoff's law can be used to relate osmotic pressure to protein concentration in the plasma ($\Pi = RTC_p$, where R = gas constant, T = absolute temperature, and C_p = protein concentration). The capillary filtration coefficient is K_f (dimensions of flow per unit pressure), capillary length is L, and total blood flow Q_B is proportional to the capillary pressure gradient:

$$Q_B(x) = \frac{Q_p(x)}{(1-H(x))} = -\frac{L}{\Re(x)}\left(\frac{dP_c(x)}{dx}\right),$$

where H is the hematocrit value, expressed as a fraction. Assume the local fluid resistance $\Re$ is a function of the local hematocrit value. Plasma flow (Q_p), protein concentration, and hematocrit value at the capillary inlet are Q_{p0}, C_{p0}, and H_0, respectively. Outlet pressure is P_{cL}. Red cells retain constant volume as they move downstream.

Perform 1D fluid, protein, and red cell balances on a small section of the capillary between x and $x + \Delta x$ to derive a coupled set of ODEs for P_c, H, and C_p. Give all information necessary to solve these equations and describe how you would solve them. What boundary conditions are needed?

14.10.10 Distributed Consumption Rate

A spherical cell with radius R contains a vacuole in the center with radius R_1. The glucose concentration at $r = R$ is held constant at ρ_0. Glucose is not able to penetrate through the vacuole boundary. Mitochondria are distributed in the cell such that the consumption of glucose per unit volume Q_g is proportional to the radial position:

$$Q_g = Q_{max}\frac{r}{R}, r > R_1,$$

where Q_{max} is a constant equal to the maximum consumption rate per unit volume. Use a shell balance to find an ode for ρ_g vs. radial position. Find the following expressions:

(a) Mass concentration of glucose vs. radial position
(b) Mass flux of glucose vs. radial position
(c) Mass flow of glucose at $r = R$
(d) Mass concentration of glucose at $r = R_1$

(e) What is the maximum value that Q_{max} can have if your solution is to make physical sense?

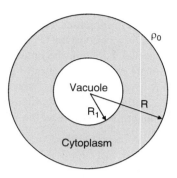

14.10.11 Diffusion of Drug in a Tumor

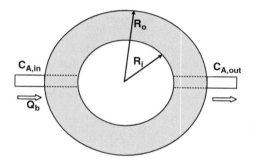

A solid tumor has the shape of a spherical shell with inside radius R_i and outside radius R_o. The central portion of the tumor is highly vascularized, and can be assumed to be filled with blood that is continually supplied at the arterial end and leaves at the same rate at the venous end. A drug "A" is maintained at concentration $C_{A,in}$ in arterial blood. The tumor tissue reacts with the drug at a constant rate per unit volume, K. Drug cannot pass through the outside radius at $r = R_o$.

(a) Use a macroscopic balance on the tumor tissue to find the steady-state flow of drug from blood to tissue in terms of the geometry and reaction rate per unit volume.

(b) If blood flow to the tissue is Q_b, what is the minimum concentration of drug in arterial blood necessary to ensure that all portions of the tumor reacts with the drug?

(c) Use a shell balance to find an expression for the steady-state flow of drug in the tissue as a function of radial position, $W_A(r)$. Show that the flow of drug into the tissue at $r = R_i$ is the same as you found in part (a).

14.10.12 Radial Variation in Consumption Rate

Mitochondria are distributed in a spherical cell such that the rate of consumption of oxygen depends linearly on radial position:

$$\dot{Q}_{O_2} = \dot{Q}_{O_2,\max}\left(\frac{r}{R}\right).$$

$\dot{Q}_{O_2,\max}$ is a constant consumption rate per unit volume, r is radial position, and R is the cell radius. The oxygen concentration at the cell surface is kept at a constant value, C_0. Assume $D_{O_2} = 10^{-5}$ cm²/s, $\dot{Q}_{O_2,\max} = 1.7 \times 10^{-8}$ mol/(ml s), $C_0 = 2.22 \times 10^{-7}$ mol/ml.

(a) Derive an expression for the concentration of oxygen in the cell as a function of radial position.
(b) Derive an expression for the flux of oxygen in the cell as a function of radial position.
(c) What is the flux at $r = R$?
(d) Find the radius of the cell if the oxygen concentration at the center is zero.

14.10.13 Steady-State Removal of a Waste Product from Tissue

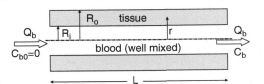

A metabolic waste product, substance "A", is produced in a cylindrical tissue (inside radius $= R_i$, outside radius $= R_o$, length $= L$) at a rate per unit volume R_A (mol/s) that is inversely proportional to its radial position in tissue:

$$R_A = k\left(\frac{R_0}{r}\right).$$

Well-mixed blood with concentration C_b bathes the tissue surface at $r = R_i$, and the surface at $r = R_o$ is impermeable to substance A. The blood–tissue partition coefficient is 1.0.

(a) Derive expressions for the steady-state flux of substance A and the concentration of substance A in the tissue as a function of radial position.
(b) What is the total flow of substance A from tissue to blood?
(c) If blood flow to the tissue is Q_b, and substance A is not present in the inlet blood, what is the outlet concentration C_b of substance A in venous blood?

14.10.14 Disk-Shaped Particles with Immobilized Enzymes

Consider porous particles that are disk-shaped with radius R and length L. The edge along the circumference is impermeable, but both faces of the disk are permeable to substrate. A toxic substrate is to be removed by enzymes, which are covalently bound to pores in the disk. Use the methods of Sect. 14.4.4 to derive the governing differential equation for the concentration of substrate in the pores of the disk. Solve this for the special case of first-order kinetics. Find the flux at the surfaces of the disk and the effectiveness factor (use volume/surface area as the characteristic length in the definition of Thiele modulus).

14.10.15 Counter-Current Mass Exchanger

Show that when $\alpha = Q_d/\Phi_{Abd}Q_b = 1$ in a counter-current mass exchanger, the solute concentration in the tube varies with axial position as shown below:

$$\frac{C_{Ab}(x) - \Phi_{Abd}C_{Ad0}}{C_{Ab0} - \Phi_{Abd}C_{Ad0}} = \frac{1 + \beta - \dfrac{x}{L}}{1 + \beta}.$$

14.10.16 Dialysis Fluid Concentration in a Counter-Current Mass Exchanger

Derive an expression for dialysis fluid concentration vs. position in a counter-current mass exchanger. Plot blood and dialysis fluid concentration, relative to initial blood concentration, as a function of axial position in the exchanger. Use the same conditions as shown in Fig. 14.24 for a cocurrent exchanger.

14.10.17 Kidney Dialysis

Perform a literature search to characterize hollow fibers or membranes used for dialysis and to find blood concentrations of metabolic waste products for dialysis patients before and after dialysis. Model a device constructed of these fibers in either the cocurrent or counter-current configuration. Fiber length is 15 cm. Select an appropriate fiber radius, wall thickness, dialysis fluid flow, and blood flow per fiber so that 95% of the selected substance (e.g., urea) is removed from the incoming blood. Use an unsteady-state macroscopic balance on blood in the body to determine the number of fibers and total flow rates needed to remove 85% of the selected substance from blood in 4 h. List values you use for initial concentration, blood volume, and any other data needed to make your calculation.

14.10.18 Oxygen Exchange in an Organ

Blood flows through an organ at a rate of 0.5 L/min. The blood has a hemoglobin concentration of 10 g/dl, pH $= 7.2$, temperature of 37°C, and a Bunsen solubility coefficient of 0.003 ml O_2 (dl blood)$^{-1}$ (mmHg)$^{-1}$. Arterial P_{O_2} is 70 mmHg and venous P_{O_2} is 20 mmHg. Find:

(a) The ratio of dissolved to bound oxygen in venous blood
(b) The consumption rate of the tissue

14.10.19 Hollow Fiber Reactor

A bioreactor is used to remove a potentially toxic substance ("substance A") from the blood stream. The device contains a number of cylindrical hollow fibers that span the width of the device filled with cells that can react with substance A and remove it from the blood. Blood flows on the outside of the fibers, perpendicular to the longitudinal axis of the fibers. Flux of the toxin from the blood to the cell surface is governed by a mass transfer coefficient, k_A: $N_A = k_A(C_{blood} - C_s)$. The cells remove substance A at a constant rate per unit volume, Γ.

(a) *Fiber Radius.* We wish to minimize cell volume by selecting the radius of each hollow fiber such that the concentration of substance A approaches zero at the fiber center as the concentration of substance A approaches C_{min} in the blood. Derive an expression for the fiber radius, R, in terms of known quantities.
(b) *Device Length.* Derive an expression that can be used to compute the length of the reactor if Reactor height is H, reactor width is W, blood flow rate is Q, inlet concentration is C_{A0}, outlet concentration is C_{min}, and the number of fibers per unit volume is N.

14.10.20 Blood Doping

We wish to quantify the advantages and disadvantages of blood doping for tissue supplied by a single capillary. When corrected for the glycocalyx, the capillary has an effective diameter of 10 μm. Assume the in vitro Fahraeus–Lindqvist relationship can be applied to the corrected vessel size. The hematocrit value is increased from 45% before doping to 60% after doping and flow in the capillary is observed to remain the same during maximal exercise. Arterial P_{O_2} is 90 Torr in both cases, but blood pH changes from 7.4 before doping to 7.2 after doping. Plasma viscosity is the same before and after doping. Find the percent change in oxygen delivery rate (e.g., mlO_2/s) to the capillary and the percent change in the capillary pressure drop. Neglect dissolved oxygen. Discuss the effects of doping on hemodynamics and oxygen delivery to tissue.

14.10.21 CO_2 Exchange in Tissue Capillaries

CO_2 is produced at a constant rate per unit volume R_{CO_2} (ml CO_2/ml tissue/min) in the tissue surrounding a cylindrical muscle capillary. Capillary flow is Q_c and muscle tissue volume of the Krogh cylinder is V_T. Derive an expression for the P_{CO_2} in blood at the capillary exit. Use (12.218) to account for total CO_2.

14.10.22 Carbon Dioxide Exchange in the Lung

Plot blood partial pressure of CO_2 versus dimensionless axial position in a capillary, x/L, for blood flows of 5, 10, 20, and 30 L/min. Assume $P_{CO_2,\text{lung}} S_{\text{lung}} = 2 \times 10^5$ ml/s, inlet CO_2 partial pressure = 45 mmHg, tissue solubility = 0.0057 mlCO_2/ml*mmHg, blood solubility = 0.00057mlCO_2/ml*mmHg, and alveolar partial pressure of CO_2 = 40 mmHg.

14.10.23 Facilitated Diffusion with Consumption

Oxygen is transported across a thin layer of smooth muscle by simple diffusion and by facilitated diffusion of myoglobin. The muscle consumes O_2 at a constant rate per unit volume, Q_{O_2}. The partial pressures of O_2 at each end of the layer are constant. Myoglobin combines linearly with O_2 over this range of partial pressures (bound $O_2 = kP_{O_2}$). Derive an expression for total steady-state O_2 content (bound + dissolved) as a function of position in the layer. Myoglobin is confined to the muscle.

14.10.24 Facilitated Transport of Oxygen in a Hemoglobin Solution

Find the steady-state O_2 flux by direct diffusion and by HbO_2 carrier-mediated transport through a Hb solution which is bound by two thin membranes a distance L apart. Make this calculation for two sets of boundary conditions: (a) $P_{O_2}(x = 0) = 20\,\text{mmHg}, P_{O_2}(x = L) = 40\,\text{mmHg}$ and (b) $P_{O_2}(x = 0) = 80\,\text{mmHg}$,

14.10 Problems

$P_{O_2}(x = L) = 100$ mmHg. Compare the flux of bound O_2 to the flux of dissolved O_2 in each case. Make the following assumptions:

(a) The membranes are highly permeable to oxygen and present no resistance to O_2 flux
(b) The HbO$_2$ saturation curve for 37°C and pH = 7.4 is appropriate
(c) $D_{O_2} = 1.8 \times 10^{-5}$ cm^2/s; $D_{HbO_2} = 0.8 \times 10^{-5}$ cm^2/s; $L = 1$ mm
(d) Total hemoglobin concentration = 0.1 g/ml
(e) Henry's law constant for dissolved $O_2 = 3 \times 10^{-5}$ mlO$_2$/(ml mmHg)
(f) Binding capacity of HbO$_2$ = 1.34 mlO$_2$/g.

14.10.25 Urea Production: Krogh Cylinder

Urea is produced in liver tissue at a constant rate per unit volume, $R_{urea} = 0.2$ µmol/(ml hr). Assume that the capillary and tissue are consistent with a Krogh cylinder with $Q_b/V_T = 0.6$ min^{-1}, $R_c = 2.5$ µm, $R_W = 2.8$ µm, $R_T = 18$ µm, $L = 500$ µm. No urea enters the cylinder at the capillary inlet.

(a) Use a species shell balance to derive an expression for urea concentration $C(z)$ in the capillary.
(b) Use a shell balance to derive an expression for urea concentration $C(r, z)$ in the capillary wall.
(c) Derive an expression for urea concentration $C(r, z)$ in the tissue.
(d) Using results from (a) to (c) above, estimate the highest urea concentration in the capillary and in the tissue. Where do these occur? Assume urea partition coefficients between blood, capillary wall, and tissue to be 1.0.

14.10.26 Production of Species in a Bioreactor

Find an example in the literature where an immobilized enzyme reactor is used to produce a specific substance (i.e., it is not designed to remove a substance like urea). Find the Michaelis constant K_m for this enzyme reaction and the production rate constant (k_p in (14.94) or k_{cat} in Example 14.6.1).

(a) How is (14.257), which was derived to predict removal of a substrate s, modified for production of a product p?
(b) Design an experimental procedure, using an apparatus similar to the bioreactor in Example 14.6.1, in which your reactor can be run under conditions in which the reaction is essentially zeroth order over the entire length of the reactor. Be sure to list all assumptions or sources for variables used in your analysis (such as E_0, etc.).
(c) Compute the outlet concentration and the product yield rate.

14.10.27 Batch Reactor

Example 14.6.1 treats the case of a continuous feed reactor. Consider, instead, the case of a batch reactor in which the same beads in the example are added to 200 ml of water with an initial urea concentration of 1 mM. Assume the reaction is first order and the reactor is well mixed. Use a macroscopic mass balance to derive an expression for urea concentration vs. time in the reactor. How long will it take to convert half of the urea to NH_3 and CO_2?

14.10.28 Immobilized Enzyme Bioreactor

A toxic substance S is converted to a nontoxic product P in a bioreactor packed with disk-shaped porous pellets containing an immobilized enzyme. The rate of conversion follows the Michaelis–Menten equation:

$$R_P = -R_S = V_{max} C_S / (K_m + C_S).$$

Our ultimate goal is to find the length L of the reactor such that 90% of the toxic substance is converted to P as it passes through the reactor. Mass flux of S between bulk fluid and fluid next to a pellet surface is governed by a mass transfer coefficient, k. Reactor flow is Q, inlet concentration is C_{bs0}, void volume of the reactor is ε_R, the partition coefficient between pellets and bulk fluid is Φ, effective diffusion coefficient in the pellet is D, and the cross-sectional area of the reactor is A. Use the following procedure to find the concentration of S in the pellet:

(a) Write the general species conservation equation for S in the solid pellet using a cylindrical coordinate system with *origin at the center of the disk*. Assume that all transport of S within the pellet is by diffusion and that there are no changes of concentration with respect to θ.
(b) If the disk thickness $2\Delta t$ is much smaller than the disk radius, use scaling to simplify the equation found in part (a). Flux of S will be negligible in which direction compared to the other?
(c) Solve the equation found in part (b) for the case where the concentration of S is low everywhere in the pellet, relative to K_m. Explicitly state the boundary conditions used in your solution.
(d) Find a relation for the effectiveness factor for this case.

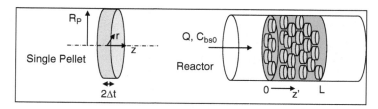

14.10 Problems   1143

14.10.29 Mobile Phase of Reactor

For Problem 14.10.28 above, we wish to find the bulk concentration vs. position in the reactor. Use the procedure below to provide a relationship that can be used to find the bulk concentration far away from the pellet surface C_{sb} in terms of the bulk fluid concentration right next to the pellet surface, $C_{sb,surface}$:

(a) Use a shell balance to derive an ODE for the bulk concentration of S as a function of axial position z in the reactor. Assume no variations in the r or θ directions, neglect diffusion, use the effectiveness factor derived in part (d) above, and assume K_m is much greater than the bulk concentration at any position z'.
(b) Solve the ODE in part (f) to provide an expression for the length of reactor required to remove 90% of the toxic substance.

14.10.30 Krogh Cylinder

Consider a tissue in which the oxygen concentration is accurately modeled by a Krogh cylinder representation. For an increase in each of the parameters that follow, identify if the outlet (venous) blood P_{O_2} would increase, decrease, remain the same, or if the relationship between the parameter and the outlet (venous) blood P_{O_2} cannot be determined. Presume that the blood enters the tissue cylinder with a P_{O_2} of 85 mmHg. The parameters of interest are (a) blood flow, (b) capillary length, (c) oxygen consumption rate, (d) tissue radius, and (e) hematocrit value. Justify your answers.

14.10.31 Oxygen Exchange from a HbSS Solution

We wish to design a device that can deoxygenate a solution of sickle hemoglobin (HbSS). HbSS will polymerize when the P_{O_2} drops to 40 mmHg or below. The device consists of two parallel plates (0.01 mm thick, 10 cm wide). The hemoglobin solution flows between the plates at a rate of 50 ml/s. Oxygen can diffuse through the plates ($D_{O_2} = 10^{-5}$ cm^2/s). The partial pressure of oxygen in the incoming HbSS solution is 70 mmHg, and the P_{O_2} in the gas phase on the outside surface of the plates is maintained at 20 mmHg. The Bunsen solubility of O_2 in the plate material is the same as the Bunsen solubility of O_2 in the hemoglobin solution [$\alpha^* = 3 \times 10^{-5}$ mlO$_2$/(ml mmHg)]. The relationship between total oxygen concentration and P_{O_2} for the HbSS solution is assumed to be linear ($C_{O_2} = mP_{O_2}$, see graph below).

(a) What is the oxygen permeability of the plates?
(b) Use Fick's Law to derive an expression for the steady-state flux of O_2 through the plates $J_{O_2}(x)$, in terms of P_{O_2} in the HbSS solution and P_{O_2} in the surrounding gas. Note: the flux depends on x, but not on y.
(c) Derive an expression for $P_{O_2}(x)$ in the HbSS solution between the plates. Neglect variations in the y-direction. How long should the plates be if HbSS just begins to polymerize as the solution exits from the device? (i.e., $L = ?$). Don't be surprised if you find the plates need to be very long.

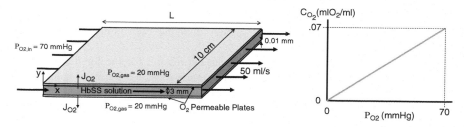

14.10.32 *Carbon Dioxide Transport in a Bioreactor*

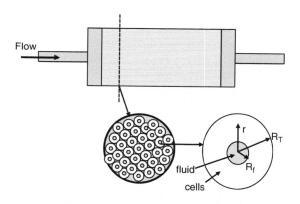

Consider the steady-state transport of carbon dioxide from cells imbedded in the stationary phase of a bioreactor to perfusion fluid in the mobile phase. *No chemical reactions occur in the perfusion fluid.* The mobile phase consists of perfusion fluid flowing through parallel hollow fibers at a rate per fiber of Q_f. The space surrounding each fiber can be treated as a Krogh cylinder with radius R_T. Each cell produces CO_2 at a constant rate, R_{CO_2}. Cells are distributed radially in the stationary phase, and *the number of cells per unit volume is directly proportional to the radial position: $N = br$, where b is constant*. Assume that the fiber wall is very thin with permeability P_{CO_2}, and the mobile/stationary phase partition coefficient is 1.

(a) Find an expression for the steady-state flux of CO_2 from stationary (cells) to mobile (fluid) phase at the fiber radius $r = R_f$.

(b) Find an expression for the concentration of CO_2 versus axial position in the fiber if the inlet concentration is zero.
(c) Find an expression for the concentration of CO_2 in the stationary phase as a function of radial and axial positions.

14.10.33 Mass Transfer from a Muscle Fiber

An isolated muscle with cylindrical shape is stimulated repeatedly until the concentration of lactate in the muscle reaches $C_0 = 1$ mol/m^3. At that point, stimulation ceases and the muscle is allowed to cool in lactate-free saline.

Muscle radius = 0.5 cm, muscle length = 30 cm, mass transfer coefficient for lactate in saline $k_A = 2 \times 10^{-7}$ m/s, lactate muscle-saline partition coefficient = $\Phi_{\text{lactate,muscle,saline}} = 0.5$, lactate diffusion coefficient = 2×10^{-8} m^2/s.

(a) How many seconds will it take for the lactate concentration at the center of the muscle to drop to one-tenth the initial value?
(b) Find the initial rate of lactate loss from the muscle (in mol/s)

14.10.34 Mass Transfer from a Finite Slab

A 5-cm long slab of tissue with thickness 2 mm is initially equilibrated with pure N_2 at one atmosphere. At $t = 0$, the tissue is immersed in a stream of water that is completely devoid of N_2. Velocity of the stream is parallel to the tissue length with a magnitude of 0.1 cm/s. Use the charts for a slab to find the partial pressure of N_2 at the center of the tissue for a few times after immersion and plot the partial pressure as a function of time. Assume the Sherwood number can be computed from:

$$Sh_L = 0.664 \, Re_L^{\frac{1}{2}} Sc^{\frac{1}{3}}.$$

The diffusion coefficient for N_2 in water or tissue is 2.6×10^{-5} cm^2/s. The kinematic viscosity of water is 0.01 cm^2/s. The partition coefficient for N_2 between tissue and water is unity.

14.10.35 Transient Inert Gas Exchange from Blood to Gas in the Lung

A bolus of plasma containing an inert gas g is injected at the inlet to a pulmonary capillary bed. If the inert gas in the alveolus is maintained at zero concentration, estimate what percentage of the capillary length the bolus would travel before 99% of the gas passes across the membrane. $P_g S = 300$ ml/s, $Q_V = 50$ ml/s.

14.10.36 Indicator Dilution and Chromatography

Species A is injected at the inlet of a chromatographic column with constant flow and with concentration specified as a function of time $f(t)$. It binds slowly and irreversibly with the column material at a rate proportional to species concentration and bead surface area. How would you analyze the outlet data to estimate the binding rate?

14.10.37 Multiple Indicator Dilution Experiment

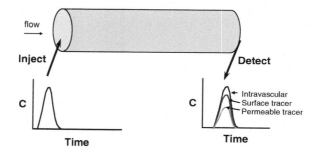

A volume V_i of a uniform mixture of three indicators, each with the same initial molar concentration C_i, is injected upstream of a capillary, and the concentration of each is monitored in the blood at the capillary outlet. The concentration of each indicator in the capillary is initially zero. Concentration of each indicator in the capillary is independent of radial position, but depends on time and axial position for $t > 0$. The capillary has cylindrical geometry, with unknown length L. Indicator 1 is a reference tracer that remains in the bloodstream. Indicator 2 is a permeable tracer that can pass through the capillary wall and is immediately sequestered by cells in the tissue, so it does not return to the bloodstream. Indicator 3 reacts irreversibly with receptors on the endothelial cell surface, such that the flux to the surface is proportional to concentration in the capillary: $N_3 = kC_3$. The proportionality factor k is a known quantity.

(a) Show how you would estimate capillary flow and volume from measured quantities
(b) Compute capillary surface area and capillary length
(c) Show how you would compute capillary permeability of indicator 2 by measuring $C_1(t, L)$, $C_2(t, L)$ and $C_3(t, L)$.

14.11 Challenges

14.11.1 Cell Size

Background: Living cells are generally very small. The largest aerobic cells in the human body have a diameter of about 1 mm. *Challenge:* Why are living cells so small? What factors limit the size of aerobic cells? *Generate ideas*: To answer this question, develop a model of essential nutrient transport to the cell.

- What system is relevant for this problem?
- What molecular species are involved in cell energy production?
- How does the system interact with the environment?
- What transport mechanism is dominant?
- What are the governing principles?
- What are appropriate constitutive relations?
- How can you use this information to formulate a model?
- What additional information do you need?
- What is the influence of cell shape?
- What is the influence of reaction rate kinetics?
- Where should you search for the needed information?

14.11.2 Membrane Resistance to Oxygen Transport

Background: Oxygen must diffuse through the cell membrane before entering the cell cytoplasm and being used in energy production. *Challenge*: Does the membrane offer significant resistance to oxygen exchange compared to the resistance to transport in the cytoplasm? *Generate ideas*: How would you model the steady-state transport of oxygen in the membrane and in the cell interior? What boundary conditions are appropriate? What factors determine membrane resistance and internal cell resistance to oxygen transport? Perform a literature search to find appropriate properties and geometry to estimate the ratio of membrane resistance to internal cell resistance.

14.11.3 Modeling Blood Doping

Background: You have been hired by an International Athletic Competition Oversight Committee. Their mission is to establish a set of unambiguous rules that will regulate an athlete's ability to enhance the oxygen carrying capacity of blood (blood doping). *Challenge*: Your role as a biomedical engineer is to assist the committee by developing a model that predicts the effects of increasing hematocrit value on tissue oxygen delivery and on important cardiovascular variables. *Generate Ideas*: What is a Krogh cylinder and what are the inherent assumptions made when adopting a Krogh cylinder model? Is it appropriate for modeling oxygen exchange in tissue? Use biotransport principles and experimental data to develop a model that can be used to study the effects of blood doping on tissue oxygen delivery. What equations are appropriate for oxygen transport in the blood, capillary wall, and muscle tissue? Does oxygen concentration vary both in the radial and axial directions in the tissue? How would you account for intracapillary radial resistance to oxygen transport? What factors contribute to the intracapillary mass transfer coefficient? What parameters would you select for your model? What predictions will you make with your model? How will you account for the effects of hematocrit on pressure and flow through your model? What key points will you make in your report to the oversight committee?

References

Chen JP, Chiu SH (1999) Preparation and characterization of urease immobilized onto porous chitosan beads for urea hydrolysis. Bioprocess Eng 21:323–330

Crank J (1956) The mathematics of diffusion. Clarendon Press, Oxford

Gurney HP, Lurie J (1923) Charts for estimating temperature distributions in heating or cooling solid shapes. Ind Eng Chem 15:1170–1172

Heisler MP (1947) Temperature charts for induction and constant temperature heating. Trans ASME 69:227–36

Hellums JD, Nair PK, Huang NS, Ohshima N (1996) Simulation of intraluminal gas transport processes in the microcirculation. Ann Biomed Eng 24:1–24

Sangren WC, Sheppard CW (1953) A mathematical derivation of the exchange of a labeled substance between a liquid flowing in a vessel and an external compartment. Bull Math Biophys 15:387–394

Chapter 15
General Microscopic Approach for Biomass Transport

15.1 Introduction

The macroscopic approach introduced in Chap. 13 is useful for solving mass transfer problems with temporal concentration variations, but without spatial variations. The shell balance method introduced in Chap. 14 can be used to analyze problems with one-dimensional spatial variations. However, many biomass transfer problems are multidimensional or involve concentration changes in both position and time. In such cases, it is useful to develop a general equation that accounts for spatial and temporal concentration variations. Once developed, this general equation should be applicable to any mass transfer problem, including those treated in previous chapters. The approach for solving the general equations would involve applying conditions that are appropriate to a specific problem, such as symmetry, initial spatial concentration distribution, and appropriate conditions at the system boundaries.

15.2 3-D, Unsteady-State Species Conservation

Consider the system shown in Fig. 15.1. We shall analyze a microscopic shell within the system surrounding a point (x,y,z) with sides Δx, Δy, and Δz as shown.
Conservation of species A can be written as follows:

$$\left\{ \begin{array}{c} \text{rate of} \\ \text{accumulation} \\ \text{of A in shell} \end{array} \right\} = \left\{ \begin{array}{c} \text{rate at} \\ \text{which A} \\ \text{enters shell} \end{array} \right\} - \left\{ \begin{array}{c} \text{rate at} \\ \text{which A} \\ \text{leaves shell} \end{array} \right\} + \left\{ \begin{array}{c} \text{rate of} \\ \text{production of A} \\ \text{within shell} \end{array} \right\}.$$

(15.1)

The rate of accumulation of moles of A in the shell between time t and time $t + \Delta t$ is the number of moles in the shell at time $t + \Delta t$ minus the number of moles in the shell at time t divided by the time interval as that interval becomes very short:

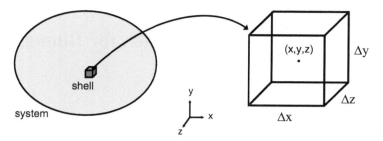

Fig. 15.1 Microscopic subsytem

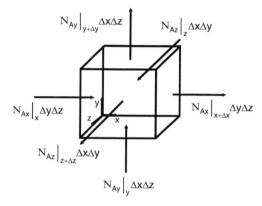

Fig. 15.2 Species mass balance on shell

$$\left\{\begin{array}{c} \text{rate of} \\ \text{accumulation} \\ \text{of A in shell} \end{array}\right\} = \lim_{\Delta t \to 0} \left\{\frac{c_A(x,y,z,t+\Delta t)\Delta x \Delta y \Delta z - c_A(x,y,z,t+\Delta t)\Delta x \Delta y \Delta z}{\Delta t}\right\}$$

$$= \left(\frac{\partial c_A}{\partial t}\right)\Delta x \Delta y \Delta z.$$

(15.2)

The number of moles entering and leaving through the boundaries of the shell per unit time are shown in Fig. 15.2. The direction of each flux is assumed to be in the positive coordinate direction. The rates at which moles of species A enter and leave the shell per unit time are:

$$\left\{\begin{array}{c} \text{rate at} \\ \text{which A} \\ \text{enters shell} \end{array}\right\} = N_{Ax}|_x \Delta y \Delta z + N_{Ay}|_y \Delta x \Delta z + N_{Az}|_z \Delta x \Delta y, \quad (15.3)$$

15.2 3-D, Unsteady-State Species Conservation

$$\left\{\begin{array}{c}\text{rate at}\\ \text{which A}\\ \text{leaves shell}\end{array}\right\} = N_{Ax}|_{x+\Delta x}\Delta y \Delta z + N_{Ay}|_{y+\Delta y}\Delta x \Delta z + N_{Az}|_{z+\Delta z}\Delta x \Delta y. \quad (15.4)$$

Finally, the molar rate of production of A within the shell is:

$$\left\{\begin{array}{c}\text{rate of}\\ \text{production of A}\\ \text{within shell}\end{array}\right\} = R_A \Delta x \Delta y \Delta z. \quad (15.5)$$

Here R_A is the molar rate of production of A per unit volume. Substituting (15.2)–(15.5) into (15.1) and dividing by the volume of the shell, we have:

$$\left(\frac{\partial c_A}{\partial t}\right) = \frac{N_{Ax}|_x - N_{Ax}|_{x+\Delta x}}{\Delta x} + \frac{N_{Ay}|_y - N_{Ay}|_{y+\Delta y}}{\Delta y} + \frac{N_{Az}|_z - N_{Az}|_{z+\Delta z}}{\Delta z} + R_A. \quad (15.6)$$

Finally, in the limit as the volume of the shell approaches zero (i.e., $\Delta x \to 0$, $\Delta y \to 0$, $\Delta z \to 0$):

$$\left(\frac{\partial c_A}{\partial t}\right) = -\frac{\partial N_{Ax}}{\partial x} - \frac{\partial N_{Ay}}{\partial y} - \frac{\partial N_{Az}}{\partial z} + R_A. \quad (15.7)$$

But the molar flux can be written as a vector:

$$\vec{N}_A = N_{Ax}\vec{i} + N_{Ay}\vec{j} + N_{Az}\vec{k} \quad (15.8)$$

and

$$\vec{\nabla} \cdot \vec{N}_A = \frac{\partial N_{Ax}}{\partial x} + \frac{\partial N_{Ay}}{\partial y} + \frac{\partial N_{Az}}{\partial z}. \quad (15.9)$$

Substituting (15.9) into (15.8) provides a more compact form for the 3-D species conservation equation:

$$\left(\frac{\partial c_A}{\partial t}\right) = -\vec{\nabla} \cdot \vec{N}_A + R_A. \quad (15.10)$$

This can be rewritten in terms of mass concentration by multiplying by the molecular weight of A, M_A:

$$\left(\frac{\partial \rho_A}{\partial t}\right) = -\vec{\nabla} \cdot \vec{n}_A + r_A. \quad (15.11)$$

Adding the species continuity equations for all species (N_species):

$$\left(\frac{\partial}{\partial t}\sum_{i=1}^{N_\text{species}} \rho_i\right) = -\vec{\nabla} \cdot \left(\sum_{i=1}^{N_\text{species}} \vec{n}_i\right) + \sum_{i=1}^{N_\text{species}} r_i. \quad (15.12)$$

The last term in (15.12) is zero since the net mass of products and reactants is unchanged in chemical reactions. The summation in the first term is simply the total mass density ρ and the summation in the second term is the net mass flux $\rho\vec{v}$, which is based on the definition of the mass-average velocity (12.51). Consequently, (15.12) reduces to the continuity equation, (7.4):

$$\left(\frac{\partial \rho}{\partial t}\right) = -\vec{\nabla} \cdot (\rho\vec{v}). \quad (15.13)$$

Returning to (15.10) and splitting the molar flux into its convective and diffusive components (12.60):

$$\left(\frac{\partial c_A}{\partial t}\right) = -\vec{\nabla} \cdot \left(c_A \vec{v} + \vec{J}_A\right) + R_A = -c_A \vec{\nabla} \cdot \vec{v} - \vec{v} \cdot \vec{\nabla} c_A - \vec{\nabla} \cdot \vec{J}_A + R_A. \quad (15.14)$$

If the material is incompressible, then (15.14) can be written:

$$\frac{\partial c_A}{\partial t} = -\vec{v} \cdot \vec{\nabla} c_A - \vec{\nabla} \cdot \vec{J}_A + R_A. \quad (15.15)$$

Equation (15.15) states that species concentration can change by convection (first term), by diffusion (second term) and/or by chemical reaction (last term). Introducing the definition of the substantial derivative (7.39) and Fick's Law (12.80) for an incompressible material into (15.15):

$$\frac{Dc_A}{Dt} = \vec{\nabla} \cdot \left(D_{AB} \vec{\nabla} c_A\right) + R_A. \quad (15.16)$$

This is the most general expression of the species conservation equation for an incompressible material. If the diffusion coefficient is independent of position, then (15.16) becomes:

$$\frac{Dc_A}{Dt} = D_{AB} \nabla^2 c_A + R_A. \quad (15.17)$$

Expanding this in the rectangular coordinate system:

$$\frac{\partial c_A}{\partial t} + v_x \frac{\partial c_A}{\partial x} + v_y \frac{\partial c_A}{\partial y} + v_z \frac{\partial c_A}{\partial z} = D_{AB} \left[\frac{\partial^2 c_A}{\partial x^2} + \frac{\partial^2 c_A}{\partial y^2} + \frac{\partial^2 c_A}{\partial z^2}\right] + R_A. \quad (15.18)$$

15.2 3-D, Unsteady-State Species Conservation

In cylindrical coordinates, the species continuity equation for constant ρ and D_{AB} is:

$$\frac{\partial c_A}{\partial t} + v_r \frac{\partial c_A}{\partial r} + \frac{v_\theta}{r}\frac{\partial c_A}{\partial \theta} + v_z \frac{\partial c_A}{\partial z} = D_{AB}\left\{\frac{1}{r}\frac{\partial}{\partial r}\left(r\frac{\partial c_A}{\partial r}\right) + \frac{1}{r^2}\frac{\partial^2 c_A}{\partial \theta^2} + \frac{\partial^2 c_A}{\partial z^2}\right\} + R_A.$$
(15.19)

In spherical coordinates, the species continuity equation is:

$$\frac{\partial c_A}{\partial t} + v_r \frac{\partial c_A}{\partial r} + \frac{v_\theta}{r}\frac{\partial c_A}{\partial \theta} + \frac{v_\phi}{r\sin\theta}\frac{\partial c_A}{\partial \phi}$$
$$= D_{AB}\left\{\frac{1}{r^2}\frac{\partial}{\partial r}\left(r^2\frac{\partial c_A}{\partial r}\right) + \frac{1}{r^2 \sin\theta}\frac{\partial}{\partial \theta}\left(\sin\theta\frac{\partial c_A}{\partial \theta}\right) + \frac{1}{r^2\sin^2\theta}\frac{\partial^2 c_A}{\partial \phi^2}\right\} + R_A.$$
(15.20)

These are the general conservation equations for species A in an incompressible material with constant diffusion coefficient.

The complete general equations are rarely solved in practice. They can be further simplified for solid materials ($\vec{v} = 0$), for steady flow ($\partial c_A/\partial t = 0$), for negligible diffusion ($D_{AB} \approx 0$), and for no chemical reaction ($R_A = 0$). In addition, reasonable assumptions are often made such as symmetry about a particular axis, negligible flux in certain directions, etc.

Example 15.2.1 Simplification of the General Species Continuity Equation.
The vasoactive drug serotonin combines with receptors on endothelial cells and is rapidly internalized and metabolized. Serotonin (species A) is infused into the inlet of a blood vessel at a constant rate. Assume the drug is uniformly distributed at the inlet, with concentration c_{A0}. The flux of serotonin to the endothelial cell surface is proportional to the concentration of serotonin at the surface. Simplify the species continuity equation for serotonin in the blood and provide appropriate auxiliary conditions necessary to solve for the serotonin concentration.

Solution. *Initial considerations*: Let us apply the following assumptions:

1. Flow is steady
2. Blood flow is laminar and fully developed in the vessel
3. The blood vessel is cylindrical with radius R and length L
4. Serotonin concentration at the vessel inlet is constant
5. Concentration is symmetrical about the vessel center
6. At the endothelial wall, $J_{Ar}(R,z) = kc_A(R,z)$

System definition and environmental interactions: The system is the blood in the blood vessel and the flux of serotonin out of the blood is proportional to the concentration at the wall.

Apprising the problem to identify governing relationships: The general microscopic species conservation equation is the appropriate expression to apply in this case.

Analysis: Applying the above assumptions to the species continuity equation in cylindrical coordinates, (15.19):

$$\overset{1}{\cancel{\frac{\partial c_A}{\partial t}}} + \overset{2}{\cancel{v_r \frac{\partial c_A}{\partial r}}} + \overset{2,5}{\cancel{\frac{v_\theta}{r}\frac{\partial c_A}{\partial \theta}}} + v_z \frac{\partial c_A}{\partial z} = D_{AB} \left\{ \frac{1}{r}\frac{\partial}{\partial r}\left(r\frac{\partial c_A}{\partial r}\right) + \overset{5}{\cancel{\frac{1}{r^2}\frac{\partial^2 c_A}{\partial \theta^2}}} + \frac{\partial^2 c_A}{\partial z^2} \right\} + \overset{6}{\cancel{R_A}}$$

Examining and interpreting the results: Note that the reaction occurs at the endothelial surface, not in the blood. Consequently, R_A, the rate of production of serotonin per unit volume in blood, is zero. The resulting partial differential equation is second order in each of the independent variables r and z. Therefore, two boundary conditions are needed in r and two in z. Assumption 6 provides a condition at $r = R$. Assumption 5 relating to radial symmetry requires that there should be no radial flux in the center of the vessel. The concentration at the inlet of the vessel is constant according to assumption 4. This leaves one more condition in z that must be specified. If the vessel were very long, the concentration would approach zero at the exit. Although this condition cannot be used for a short vessel, we know that the concentration must decrease with axial position, so the rate of change of concentration at the vessel exit is less than or equal to zero. This is usually sufficient to eliminate potential solutions that predict concentration increases with increasing axial position (e.g., see Sect. 14.5.1.1). Consequently, for a parabolic velocity profile in the vessel, this problem reduces to the following:

$$2\langle v \rangle \left[1 - \left(\frac{r}{R}\right)^2\right] \frac{\partial c_A}{\partial z} = D_{AB}\left\{\frac{1}{r}\frac{\partial}{\partial r}\left(r\frac{\partial c_A}{\partial r}\right) + \frac{\partial^2 c_A}{\partial z^2}\right\}$$

$$r = 0: \frac{\partial c_A}{\partial r} = 0$$

$$r = R: -D_{AB}\frac{\partial c_A}{\partial r} = kc_A$$

$$z = 0: c_A = c_{A0}$$

$$z = L: \frac{\partial c_A}{\partial z} \leq 0.$$

Additional comments: For nearly all applications, scaling will indicate that the axial diffusion term is very small in comparison to the axial convection term or the

15.2.1 Comparison of the General Species Continuity Equation and the One-Dimensional Shell Balance Approach

For steady, one-dimensional transport in the z-direction, the species continuity equation in rectangular coordinates reduces to:

$$v_z \frac{dc_A}{dz} = D_{AB} \frac{d^2 c_A}{dz^2} + R_A. \tag{15.21}$$

Many of the steady-state relationships derived using 1-D rectangular shell balances in Chap. 14 can be obtained by simplifying (15.21) for the case at hand. However, there are exceptions to this. Shells in Chap. 14 were allowed to extend to system boundaries in directions that are perpendicular to the direction of mass flow. Mass exchange at those system boundaries was included in the overall conservation statements. Exchange at system boundaries are not included in the general species conservation relationships. Consider, for example the problem of mass exchange of species A to fluid flowing through a hollow fiber, treated in Sect. 14.5.2. Neglecting axial diffusion and production of species A via chemical reaction, we found:

$$\frac{dC_{Ab}}{dz} = -\frac{N_A(z) S_i}{L Q_b}. \tag{15.22}$$

Here z is the axial direction, C_{Ab} is the bulk concentration of species A, Q_b is the fluid flow rate, $N_A(z)$ is the flux of species A at the fluid–fiber interface, L is the fiber length, and S_i is the surface area of the fiber. Simplification of the one-dimensional general species conservation expression, (15.21), for this case yields:

$$v_z \frac{dc_A}{dz} = 0. \tag{15.23}$$

The only solution to (15.23) is a constant concentration, which is clearly incorrect and is different from (15.22). The reason for this discrepancy is that this is not really a one-dimensional problem. Since radial flux occurs at the interface between the fiber wall and the flowing fluid, the concentration in the fluid must depend on both radial and axial positions. For steady, fully developed flow ($v_r = v_\theta = 0$) with radial symmetry ($\partial/\partial \theta = 0$), no chemical reaction, and negligible axial diffusion ($\partial^2 c_A/\partial z^2 = 0$), (15.19) reduces to:

$$v_z \frac{\partial c_A}{\partial z} = D_{AB} \frac{1}{r} \frac{\partial}{\partial r} \left(r \frac{\partial c_A}{\partial r} \right). \tag{15.24}$$

Since v_z is independent of z and since $J_{Ar}(z,r) = -D_{AB}\partial c_A/\partial r$, this can be written:

$$\frac{\partial}{\partial z}(v_z c_A) = -\frac{1}{r}\frac{\partial}{\partial r}(r J_{Ar}(r,z)). \tag{15.25}$$

Multiplying by an element of area and integrating (15.25) from the center of the tube to the fiber wall ($r = R$):

$$\frac{d}{dz}\int_0^R (v_z c_A)(2\pi r dr) = -\int_0^R \frac{1}{r}\frac{\partial}{\partial r}(r J_{Ar}(r,z))(2\pi r dr). \tag{15.26}$$

Using the definition of the bulk concentration:

$$C_{Ab} = \frac{\int_A (v_z c_A) dA}{\int_A v_z dA}. \tag{15.27}$$

Integration of terms in (15.26) yields:

$$<v>(\pi R^2)\frac{dC_{Ab}}{dz} = -2\pi[R J_{Ar}(R,z)]. \tag{15.28}$$

Letting $N_A(z) = -J_{Ar}(R,z)$, $S_i = 2\pi R L$, and $Q_b = \pi R^2 <v>$, we arrive at the same expression obtained with the shell balance approach. Either approach can be used, but care must be exercised when reducing the general species continuity equation. The primary difference between the two approaches is that the shell used in the 1-D shell balance extends all the way to the system boundary, while the 3-D shell excludes the system boundary. Flux at the boundary is included in the 1-D conservation equation, while flux at the boundary is a boundary condition for the general species conservation equation. Radial variations in concentration are ignored in the one-dimensional shell balance approach, so the solution is found in terms of the bulk concentration. Solution of the general species conservation equation is in terms of the local concentration, which will depend on both radial and axial position. We will examine the solution to (15.24) for constant wall flux in Sect. 15.5.1.3.

Example 15.2.2 Comparison Between Shell Balance and General Species Continuity Equations.
Simplify the species continuity equation for the case of steady O_2 consumption in a cylindrical cell. Assume O_2 consumption is first order, so $R_{O_2} = -k c_{O_2}$. Compare your result with the result obtained using the 1-D shell balance approach, (14.72).

15.2 3-D, Unsteady-State Species Conservation

Solution. *Initial considerations*: We begin by listing the assumptions:

1. Steady-state
2. Axisymmetric
3. No convection
4. Radial diffusion only (neglect CO_2 transport to cell ends)
5. $R_{O_2} = -kc_{O_2}$

System definition, environmental interactions, and governing relationships: Since the system is a cylindrical cell, we use the species continuity equation in cylindrical coordinates, (15.19).

Analysis: Eliminating terms that are zero or negligible according to our list of assumptions:

$$\overset{1}{\cancel{\frac{\partial c_A}{\partial t}}} + \overset{3}{\cancel{v_r\frac{\partial c_A}{\partial r}}} + \overset{2,3}{\cancel{\frac{v_\theta}{r}\frac{\partial c_A}{\partial \theta}}} + \overset{3}{\cancel{v_z\frac{\partial c_A}{\partial z}}} = D_{AB}\left\{\frac{1}{r}\frac{\partial}{\partial r}\left(r\frac{\partial c_A}{\partial r}\right) + \overset{2}{\cancel{\frac{1}{r^2}\frac{\partial^2 c_A}{\partial \theta^2}}} + \overset{4}{\cancel{\frac{\partial^2 c_A}{\partial z^2}}}\right\} + R_A$$

Finally, using assumption 5, identifying species A as O_2, and recognizing that c_{O_2} depends only on radial position, we obtain:

$$0 = D_{O_2,\text{cell}}\frac{1}{r}\frac{d}{dr}\left(r\frac{dc_{O_2}}{dr}\right) - kc_{O_2}.$$

Examining and interpreting the results: This is the same expression derived using a one-dimensional shell balance (14.72). Agreement between the 1-D shell balance and simplification of the general species continuity equation will occur when there is no exchange of material through system boundaries that are perpendicular to the transport direction. In this case, the shell used to derive (14.72) consisted of boundaries at $r = r$, $r = r + \Delta r$, $z = 0$ and $z = L$, where L is the length of the cell. The two surfaces at $r = r$ and $r = r + \Delta r$ do not include the system boundary, but the surfaces at $z = 0$ and $z = L$ are system boundaries. However, since we assume no mass exchange at these surfaces, the two expressions agree.

Additional comments: If, however, we account for the inward fluxes of O_2 through the two ends of the cell, $N_{z0}(r)$ and $N_{zL}(r)$, a 1-D shell balance would produce the following expression:

$$0 = D_{O_2,\text{cell}}\frac{1}{r}\frac{d}{dr}\left(r\frac{dc_{O_2}}{dr}\right) - kc_{O_2} + \frac{(N_{z0} + N_{zL})}{L}.$$

The general species continuity equation for this case (15.15) can be written:

$$0 = D_{O_2,\text{cell}}\frac{1}{r}\frac{\partial}{\partial r}\left(r\frac{\partial c_{O_2}}{\partial r}\right) - \frac{\partial J_{O_2 z}}{\partial z} - kc_A.$$

Integrating this from one end of the cell to the other with boundary conditions $J_{O_2 z}(0,r) = N_{z0}$ and $J_{O_2 z}(L,r) = -N_{zL}$ yields:

$$0 = D_{O_2,\text{cell}} \frac{1}{r} \frac{d}{dr}\left(r \frac{d\bar{c}_{O_2}}{dr}\right) - k\bar{c}_{O_2} + \frac{(N_{z0} + N_{zL})}{L},$$

where

$$\bar{c}_{O_2} = \frac{1}{L} \int_0^L c_{O_2} dz.$$

Therefore, the two approaches agree if we interpret the concentration in the 1-D shell balance as the average concentration over the length of the cell at any given radial position. Of course, if the fluxes at the ends are substantial, then the 1-D approach is not a good approximation, and the general species continuity equation must be solved to find $c_{O_2}(r,z)$.

15.3 Diffusion

Let us consider an important classification of problems in which the media is stationary ($\vec{v} = 0$) and there is no chemical reaction occurring in the media ($R_A = 0$). Under these circumstances, the species conservation equation becomes:

$$\frac{\partial c_A}{\partial t} = D_{AB} \nabla^2 c_A. \tag{15.29}$$

This is known as the diffusion equation. Because (15.29) is a partial differential equation, analytic solutions are not always possible, particularly for complicated geometry or boundary conditions. Even for relatively simple geometries and boundary conditions, solution procedures can be elaborate. Entire textbooks are devoted to providing analytic and numerical solutions to the diffusion equation for various geometries and boundary conditions (e.g., Crank 1956, Carslaw and Jaeger 1959). A productive approach to finding analytic solutions has been to use the separation of variables method. This method was used to produce the solutions presented in Sects. 14.7.1.4–14.7.1.6. We will walk through the separation of variables procedure in Sect. 15.3.1, but thereafter will focus on the proper formulation of problems, provide the solutions, and refer the student to the extensive diffusion literature for solution details. In the following sections, we analyze diffusion problems of particular interest to biomedical engineers.

15.3.1 Steady-State, Multidimensional Diffusion

Let us begin our discussion of multidimensional diffusion with a problem that involves steady-state diffusion in two dimensions. A rectangular shaped tissue sample

15.3 Diffusion

Fig. 15.3 Steady-state diffusion through tissue

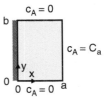

is placed between two glass coverslips, as shown in Fig. 15.3. The glass is impermeable to species A, so there is no mass transport in the z-direction. The surface at $x = 0$ is also impermeable to species A, the surface at $x = a$ is maintained at constant concentration C_a, and the surfaces at $y = 0$ and $y = b$ are maintained at zero concentration of species A. Convective transport and the rate of production of species A in the tissue are both zero. We are interested in the steady-state concentration and flux of species A within the tissue. The diffusion equation and the appropriate boundary conditions for this problem are:

$$\frac{\partial^2 c_A}{\partial x^2} + \frac{\partial^2 c_A}{\partial y^2} = 0, \tag{15.30}$$

$$\left. \frac{\partial c_A}{\partial x} \right|_{x=0} = 0, \tag{15.31a}$$

$$c_A(a, y) = C_a, \tag{15.31b}$$

$$c_A(x, 0) = 0, \tag{15.31c}$$

$$c_A(x, b) = 0. \tag{15.31d}$$

Let us attempt a *separation of variables* approach by assuming a product solution of the form:

$$c_A(x, y) = X(x)Y(y). \tag{15.32}$$

Substituting this into (15.30) and dividing by XY gives:

$$\frac{1}{X}\frac{d^2 X}{dx^2} = -\frac{1}{Y}\frac{d^2 Y}{dy^2}. \tag{15.33}$$

The left-hand side of (15.33) depends only on x and the right-hand side depends only on y. This is possible only if each side is a constant, say λ. Thus (15.33) can be expressed as two ordinary differential equations:

$$\frac{d^2 X}{dx^2} - \lambda X = 0, \tag{15.34}$$

$$\frac{d^2Y}{dy^2} + \lambda Y = 0. \tag{15.35}$$

The general solutions to these equations are:

$$X = A_X \sinh\left(\sqrt{\lambda}x\right) + B_X \cosh\left(\sqrt{\lambda}x\right), \tag{15.36}$$

$$Y = A_Y \sin\left(\sqrt{\lambda}y\right) + B_Y \cos\left(\sqrt{\lambda}y\right). \tag{15.37}$$

The constants A_X, B_X, A_Y, and B_Y are to be determined using the boundary conditions, (15.31a)–(15.31d). Applying boundary condition (15.31a):

$$\left.\frac{\partial c_A}{\partial x}\right|_{x=0} = Y(y)\left.\frac{dX}{dx}\right|_{x=0} = Y(y)\left[A_X\sqrt{\lambda}\right] = 0. \tag{15.38}$$

In general, $Y(y)$ and λ are not zero, so (15.38) requires that $A_X = 0$. Applying boundary condition (15.31c) at $y = 0$:

$$c_A(x,0) = X(x)[B_Y] = 0. \tag{15.39}$$

Since $X(x)$ is not zero, then $B_Y = 0$. Letting the product of the two unknown coefficients $A_Y B_X = B$, (15.32) can be written:

$$c_A(x,y) = B \cosh\left(\sqrt{\lambda}x\right) \sin\left(\sqrt{\lambda}y\right). \tag{15.40}$$

We can now use the remaining two boundary conditions to evaluate the unknowns B and λ. Applying the boundary condition at $y = b$, (15.31d) becomes:

$$c_A(x,b) = 0 = B \cosh\left(\sqrt{\lambda}x\right) \sin\left(\sqrt{\lambda}b\right). \tag{15.41}$$

The coefficient B cannot be zero, otherwise the only solution would be $c_A(x,y) = 0$, which is incorrect. The hyperbolic cosine term is not zero, since it depends on x. Therefore, the sine term must be zero. This will be zero for an infinite number of different values of λ, with each value given by the following expression for n equal to integer values starting with 0:

$$\lambda_n = \left(\frac{n\pi}{b}\right)^2. \tag{15.42}$$

Therefore, there are an infinite number of solutions that satisfy the three boundary conditions that we have applied thus far. The quantity λ_n is known as an *eigenvalue*, and (15.40), with λ replaced by λ_n, is an example of an *eigenfunction*.

15.3 Diffusion

Since the diffusion equation is linear, the sum of all solutions is also a solution, so we can write:

$$c_A(x, y) = \sum_{n=0}^{\infty} B_n \cosh\left(n\pi \frac{x}{b}\right) \sin\left(n\pi \frac{y}{b}\right). \tag{15.43}$$

Our final task is to determine values for each coefficient, B_n, such that the fourth boundary condition, given by (15.31b), is satisfied. This appears to be an impossible task, since we have an infinite number of coefficients and only one unsatisfied boundary condition. However, there is only one set of coefficients B_n, which produce a solution to the original partial differential equation and all of the boundary conditions. Each coefficient can be found using the following procedure:

1. Apply the boundary condition at $x = a$ (15.31b)
2. Multiply both sides of (15.31b) by $\sin(m\pi y/b)$
3. Integrate both sides with respect to y over the interval from $y = 0$ to $y = b$
4. Recognize that the functions $\sin(m\pi y/b)$ and $\sin(n\pi y/b)$ are orthogonal:

$$\int_0^b \sin\left(n\pi \frac{y}{b}\right) \sin\left(m\pi \frac{y}{b}\right) dy = 0, \; n \neq m, \tag{15.44}$$

$$\int_0^b \sin\left(n\pi \frac{y}{b}\right) \sin\left(m\pi \frac{y}{b}\right) dy = \frac{b}{2}, \; n = m. \tag{15.45}$$

Applying this procedure, we find:

$$C_A \int_0^b \sin\left(m\pi \frac{y}{b}\right) dy = \sum_{n=0}^{\infty} B_n \cosh\left(n\pi \frac{a}{b}\right) \int_0^b \sin\left(n\pi \frac{y}{b}\right) \sin\left(m\pi \frac{y}{b}\right) dy. \tag{15.46}$$

Application of (15.44) shows that all terms on the right-hand side of (15.46) are zero, except for the case when $n = m$. This allows us to solve directly for the coefficients B_m:

$$B_m = \left(\frac{2C_A}{m\pi}\right) \left(\frac{1 - \cos(m\pi)}{\cosh\left(m\pi \frac{a}{b}\right)}\right). \tag{15.47}$$

The factor $1-\cos(m\pi)$ in the numerator is zero for even values of m and equals 2 for odd values. Replacing m with $2k + 1$, we retain only the non-zero values of B_m. The final solution can be written as:

$$c_A(x, y) = \frac{4C_A}{\pi} \sum_{k=0}^{\infty} \left(\frac{1}{2k+1}\right) \frac{\cosh\left((2k+1)\pi \frac{x}{b}\right)}{\cosh\left((2k+1)\pi \frac{a}{b}\right)} \sin\left((2k+1)\pi \frac{y}{b}\right). \tag{15.48}$$

Fick's law can be used to find the components of the flux of species A in the tissue as a function of position:

$$J_{Ax}(x,y) = -D_{AB}\frac{\partial c_A}{\partial x}$$

$$= -\frac{4D_{AB}C_A}{b}\sum_{k=0}^{\infty}\frac{\sinh\left((2k+1)\pi\frac{x}{b}\right)}{\cosh\left((2k+1)\pi\frac{a}{b}\right)}\sin\left((2k+1)\pi\frac{y}{b}\right), \quad (15.49)$$

$$J_{Ay}(x,y) = -D_{AB}\frac{\partial c_A}{\partial y}$$

$$= -\frac{4D_{AB}C_A}{b}\sum_{k=0}^{\infty}\frac{\cosh\left((2k+1)\pi\frac{x}{b}\right)}{\cosh\left((2k+1)\pi\frac{a}{b}\right)}\cos\left((2k+1)\pi\frac{y}{b}\right). \quad (15.50)$$

The flux of species A is a vector quantity that depends on both x and y:

$$\vec{J}_A(x,y) = J_{Ax}(x,y)\vec{i} + J_{Ay}(x,y)\vec{j}. \quad (15.51)$$

The expressions above are complex functions, and it is not obvious by simply looking at the analytic solution just how the concentration or the flux varies with position. A contour plot of concentration is shown in the top panel of Fig. 15.4. Each contour line represents a concentration that is 0.5 µM different than its neighboring contour lines. The boundary at $x = 1$ cm is maintained at 10 µM and the concentrations at $y = 0$ and $y = 2$ cm are 0 µM. As expected, the concentration is symmetrical about the center at $y = 1$ cm. The concentration in the center of the tissue at $x = 0$ cm is about 4.5 µM. Concentration gradients near the walls at $y = 0$ and $y = 2$ cm are steep, particularly in the corners near $x = 1$ cm.

In the bottom two panels of Fig. 15.4, flux vectors are superimposed on the concentration contour plot near the central portion of the tissue at $x = 0$ and in the corner region near $x = 1$ cm, $y = 2$ cm. The lengths of the vectors are proportional to the magnitudes of the flux, but the magnitudes of the vectors in the left panel have been multiplied by a factor of 5 relative to those in the right panel. Note how the flux vectors are perpendicular to the lines of constant concentration. The steep gradient in concentration at the two corners $(a,0)$ and (a,b) induces high fluxes. The flux in the x-direction at $x = 0$ is zero for all values of y (boundary condition), but there is an outward flux in the y-direction ay $x = 0$.

15.3.2 Steady-State Diffusion and Superposition

The diffusion equation is a linear partial differential equation. By this we mean that if c_1 and c_2 are both solutions to the diffusion equation, then any linear combination $c_3 = Ac_1 + Bc_2$, where A and B are constants, is also a solution to the diffusion equation. This makes it possible for us to solve new problems by superimposing

15.3 Diffusion

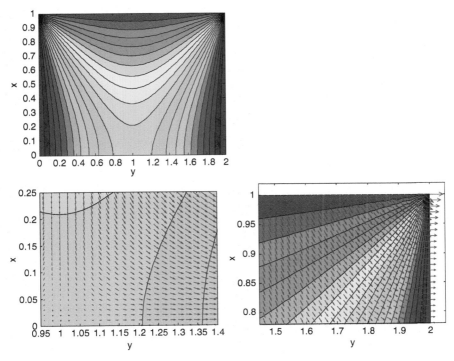

Fig. 15.4 Graphical solution to 2-D diffusion problem. *Top panel*: Contour plot of concentration vs. position (in cm). *Red line* at $x = 1$ cm represents a concentration of 10 µM, blue lines at $y = 0$ and $y = 2$ cm represent 0 µM. Each line differs from its neighbor by 0.5 µM. *Bottom Left Panel*: This shows a portion of the tissue near the *bottom* with flux vectors superimposed on the contour plot. *Bottom Right Panel*: This shows a portion of the tissue at the *top right corner* with flux vectors superimposed on the contour plot

solutions to problems that have already been solved. Of course it is not sufficient that the superimposed solutions satisfy the diffusion equation. The boundary conditions to the previously solved problems, when superimposed, must be identical to the boundary conditions for the new problem. We illustrate this with an example.

> **Example 15.3.1 Superposition for 2-D Diffusion.**
> Show that if c_{A1} and c_{A2} are solutions to the two steady-state diffusion problems on the right side of Fig. 15.5, $c_A = c_{A1} + c_{A2}$ would be the solution to the problem on the left side of the figure.

Solution. Substituting $c_A = c_{A1} + c_{A2}$, the problem on the left side of Fig. 15.5 is:

$$\frac{\partial^2 c_A}{\partial x^2} + \frac{\partial^2 c_A}{\partial y^2} = \left(\frac{\partial^2 c_{A1}}{\partial x^2} + \frac{\partial^2 c_{A1}}{\partial y^2}\right) + \left(\frac{\partial^2 c_{A2}}{\partial x^2} + \frac{\partial^2 c_{A2}}{\partial y^2}\right) = 0 + 0 = 0,$$

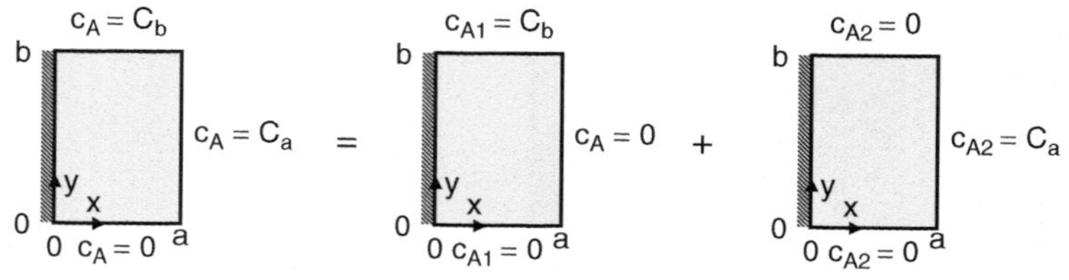

Fig. 15.5 Superposition of diffusion problems

$$\left.\frac{\partial c_A}{\partial x}\right|_{x=0} = \left.\frac{\partial c_{A1}}{\partial x}\right|_{x=0} + \left.\frac{\partial c_{A2}}{\partial x}\right|_{x=0} = 0 + 0 = 0,$$

$$c_A(a, y) = c_{A1}(a, y) + c_{A2}(a, y) = 0 + C_a = C_a,$$

$$c_A(x, 0) = c_{A1}(x, 0) + c_{A2}(x, 0) = 0 + 0 = 0,$$

$$c_A(x, b) = c_{A1}(x, b) + c_{A2}(x, b) = C_b + 0 = C_b.$$

Since the sum of the differential equations and the sum of the boundary conditions for the two problems at the right are identical to the differential equation and boundary conditions on the left, the solution to the problem on the left is equal to the sum of the solutions to the two problems on the right.

15.3.3 Unsteady-State, Multidimensional Diffusion

In Sect. 14.7, we analyzed the unsteady-state diffusion of a solute into a slab of material with finite thickness. However, the slab was assumed to be very wide and very high, so the mass flux was in a single direction. Let us now consider the case where the material is a parallelepiped of finite dimensions, with length $2L$, height $2h$, and width $2w$, as shown in Fig. 15.6. The origin of the coordinate system is at the center of the slab.

The concentration of solute A in material B is initially C_{A0}, and at time $t = 0$, the material is immersed in a fluid with uniform concentration C_{Af} far from the material. A mass transfer coefficient k_{Af} governs the flow of solute A between the solid and the fluid. We are interested in how the concentration of solute A changes with position in the material and time after immersion. Since there is no convection or chemical reaction occurring within the material, the species continuity equation is the unsteady-state diffusion equation:

$$\frac{\partial c_A}{\partial t} = D_{AB}\left[\frac{\partial^2 c_A}{\partial x^2} + \frac{\partial^2 c_A}{\partial y^2} + \frac{\partial^2 c_A}{\partial z^2}\right]. \tag{15.52}$$

15.3 Diffusion

Fig. 15.6 Diffusion in slab of finite dimensions

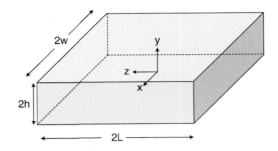

Defining dimensionless dependent and independent variables:

$$c_A^* = \frac{c_A - \Phi_{Asf} c_{Af}}{c_{A0} - \Phi_{Asf} c_{Af}}, \tag{15.53}$$

$$z^* = \frac{z}{L}, \quad y^* = \frac{y}{h}, \quad x^* = \frac{x}{w}. \tag{15.54}$$

With these definitions, the problem and appropriate auxiliary conditions become:

$$\frac{\partial c_A^*}{\partial t} = \frac{D_{AB}}{w^2} \frac{\partial^2 c_A^*}{\partial x^{*2}} + \frac{D_{AB}}{h^2} \frac{\partial^2 c_A^*}{\partial y^{*2}} + \frac{D_{AB}}{L^2} \frac{\partial^2 c_A^*}{\partial z^{*2}}, \tag{15.55}$$

$$c_A^* \big|_{t=0} = 1, \tag{15.56a}$$

$$\frac{\partial c_A^*}{\partial z^*} \bigg|_{z^*=1} = -\frac{1}{m_z} c_A^* \big|_{z^*=1}, \tag{15.56b}$$

$$\frac{\partial c_A^*}{\partial z^*} \bigg|_{z^*=0} = 0, \tag{15.56c}$$

$$\frac{\partial c_A^*}{\partial y^*} \bigg|_{y^*=1} = -\frac{1}{m_y} c_A^* \big|_{y^*=1}, \tag{15.56d}$$

$$\frac{\partial c_A^*}{\partial y^*} \bigg|_{y^*=0} = 0, \tag{15.56e}$$

$$\frac{\partial c_A^*}{\partial x^*} \bigg|_{x^*=1} = -\frac{1}{m_x} c_A^* \big|_{x^*=1}, \tag{15.56f}$$

$$\frac{\partial c_A^*}{\partial x^*} \bigg|_{x^*=0} = 0, \tag{15.56g}$$

where m_x, m_y, and m_z are inverse Biot numbers, defined as follows:

$$m_x = \frac{D_{AB}\Phi_{Asf}}{k_{Af}w}, \quad m_y = \frac{D_{AB}\Phi_{Asf}}{k_{Af}h}, \quad m_z = \frac{D_{AB}\Phi_{Asf}}{k_{Af}L}. \quad (15.57\text{a,b,c})$$

Let us anticipate a product solution of the form:

$$c_A^*(x^*, y^*, z^*, t) = Y_x(x^*, t)Y_y(y^*, t)Y_z(z^*, t). \quad (15.58)$$

Substituting this into the original partial differential equation, dividing by $Y_x Y_y Y_z$, and grouping terms, the diffusion equation becomes:

$$\frac{1}{Y_z}\left[\frac{\partial Y_z}{\partial t} - \frac{D_{AB}}{L^2}\frac{\partial^2 Y_z}{\partial z^{*2}}\right] + \frac{1}{Y_y}\left[\frac{\partial Y_y}{\partial t} - \frac{D_{AB}}{h^2}\frac{\partial^2 Y_y}{\partial y^{*2}}\right] + \frac{1}{Y_x}\left[\frac{\partial Y_x}{\partial t} - \frac{D_{AB}}{w^2}\frac{\partial^2 Y_x}{\partial x^{*2}}\right] = 0. \quad (15.59)$$

Let us define three dimensionless times:

$$t_x^* = \frac{D_{AB}t}{w^2}, \quad t_y^* = \frac{D_{AB}t}{h^2}, \quad t_z^* = \frac{D_{AB}t}{L^2}. \quad (15.60\text{a,b,c})$$

Equation (15.59) will be satisfied if each expression in square brackets is equal to zero. Using the dimensionless times above, these expressions become:

$$\frac{\partial Y_x}{\partial t_x^*} - \frac{\partial^2 Y_x}{\partial x^{*2}} = 0, \quad \frac{\partial Y_y}{\partial t_y^*} - \frac{\partial^2 Y_y}{\partial y^{*2}} = 0, \quad \frac{\partial Y_z}{\partial t_z^*} - \frac{\partial^2 Y_z}{\partial z^{*2}} = 0. \quad (15.61\text{a,b,c})$$

We need to determine how the boundary conditions for the problem expressed in terms of the dimensionless concentration c_A^* translate to boundary conditions for the functions Y_x, Y_y, and Y_z. Let us look first at the boundary condition at $x^* = 0$:

$$\left.\frac{\partial c_A^*}{\partial x^*}\right|_{x^*=0} = Y_y Y_z \left.\frac{\partial Y_x}{\partial x^*}\right|_{x^*=0} = 0. \quad (15.62)$$

If Y_y or Y_z were zero, then the only solution to the problem would be $c_A^* = 0$, which is not correct. Therefore, the third term in the product must be zero:

$$\left.\frac{\partial Y_x}{\partial x^*}\right|_{x^*=0} = 0. \quad (15.63)$$

The boundary condition at $x^* = 1$ in terms of Y_x, Y_y, and Y_z is:

$$Y_y Y_z \left.\frac{\partial Y_x}{\partial x^*}\right|_{x^*=1} = -\frac{1}{m_x}\left(Y_y Y_z Y_x|_{x^*=1}\right). \quad (15.64)$$

Dividing by Y_y and Y_z:

$$\left.\frac{\partial Y_x}{\partial x^*}\right|_{x^*=1} = -\left(\frac{1}{m_x}\right) Y_x|_{x^*=1}. \quad (15.65)$$

15.3 Diffusion

By analogy, the boundary conditions at $y^* = 0$, $y^* = 1$, $z^* = 0$, and $z^* = 1$ become:

$$\left.\frac{\partial Y_y}{\partial y^*}\right|_{y^*=0} = 0, \tag{15.66}$$

$$\left.\frac{\partial Y_y}{\partial y^*}\right|_{y^*=1} = -\frac{1}{m_y} Y_y\big|_{y^*=1}, \tag{15.67}$$

$$\left.\frac{\partial Y_z}{\partial z^*}\right|_{z^*=0} = 0, \tag{15.68}$$

$$\left.\frac{\partial Y_z}{\partial z^*}\right|_{z^*=1} = -\frac{1}{m_z} Y_z\big|_{z^*=1}. \tag{15.69}$$

The initial condition is satisfied if we let:

$$c_A^*(x^*, y^*, z^*, 0) = Y_x(x^*, 0) = Y_y(y^*, 0) = Y_z(z^*, 0) = 1. \tag{15.70}$$

Consequently, the problem for c_A^* can be broken into three separate well-defined problems, each having the form:

$$\frac{\partial Y}{\partial X} = \frac{\partial^2 Y}{\partial n^2}, \tag{15.71}$$

$$\left.\frac{\partial Y}{\partial n}\right|_{n=0} = 0, \tag{15.72}$$

$$\left.\frac{\partial Y}{\partial n}\right|_{n=1} = -\left(\frac{1}{m}\right) Y\big|_{n=1}, \tag{15.73}$$

$$Y(n, 0) = 1. \tag{15.74}$$

Y_x is the solution to this problem when $n = x^*$, $X = t_x^*$ and $m = m_x$. Similarly, Y_y is the solution when $n = y^*$, $X = t_y^*$, and $m = m_y$, and Y_z is the solution when $n = z^*$, $X = t_z^*$, and $m = m_z$. But the Solution for Y given by (15.71)–(15.74) was found previously in Sect. 14.7.1.4 to be (14.326):

$$Y = \sum_{k=1}^{\infty} A_k \cos(\lambda_k n) e^{-\lambda_k^2 X}, \tag{15.75}$$

where A_k and λ_k are given by (14.327) and (14.328). The final solution is found by multiplying the solutions for Y_x, Y_y, and Y_z:

$$\frac{c_A - \Phi_{Asf} c_{Af}}{c_{A0} - \Phi_{Asf} c_{Af}} = \sum_{i=1}^{\infty} \sum_{j=1}^{\infty} \sum_{k=1}^{\infty} A_i A_j A_k \cos\left(\lambda_i \frac{x}{w}\right) \cos\left(\lambda_j \frac{y}{h}\right) \cos\left(\lambda_k \frac{z}{L}\right)$$
$$\times \exp\left(-D_{AB} t \left[\frac{\lambda_i^2}{w^2} + \frac{\lambda_j^2}{h^2} + \frac{\lambda_k^2}{L^2}\right]\right). \quad (15.76)$$

The coefficients are found using the following relationships:

$$\lambda_i \tan(\lambda_i) = \frac{1}{m_x} = \frac{k_{Af} w}{D_{AB} \Phi_{Asf}}, \quad \lambda_j \tan(\lambda_j) = \frac{1}{m_y} = \frac{k_{Af} h}{D_{AB} \Phi_{Asf}}, \quad \lambda_k \tan(\lambda_k) = \frac{1}{m_z} = \frac{k_{Af} L}{D_{AB} \Phi_{Asf}},$$
$$(15.77\text{a,b,c})$$

$$A_i = \frac{2 \sin(\lambda_i)}{\lambda_i + \sin(\lambda_i) \cos(\lambda_i)}, \quad A_j = \frac{2 \sin(\lambda_j)}{\lambda_j + \sin(\lambda_j) \cos(\lambda_j)}, \quad A_k = \frac{2 \sin(\lambda_k)}{\lambda_k + \sin(\lambda_k) \cos(\lambda_k)}.$$
$$(15.78\text{a,b,c})$$

The analytic solution is useful if we have a computer available and select enough terms in each series to approximate the final solution. However, an important result from the analysis is that the final solution can be found by multiplying the solutions for three slabs of finite thickness. Therefore, the graphical solutions generated in Appendix D can also be used to find the concentration at a particular point in space and time within a material with a parallelepiped shape. This is illustrated in the following example.

Example 15.3.2 CO_2 Diffusion in a Small Volume of Tissue.
A tissue segment with sides 100 μm × 200 μm × 300 μm is initially equilibrated with carbon dioxide at a partial pressure of 60 mmHg. At $t = 0$ the tissue is placed in a fluid with P_{CO_2} of 40 mmHg. Find the partial pressure at the center of the tissue segment 5 s after immersion. Compare this to the partial pressure of CO_2 after 5 s for a tissue segment with the same 100 μm thickness, but with much larger cross-section. Use the same material properties as in Example 14.7.2.

Solution. *Initial considerations*: We solved this problem using graphical methods in Example 14.7.2 for the case where the cross-section of the tissue was much greater than its thickness. In this case we will use the same graphs, but allow for CO_2 loss from all six surfaces. We will ignore any carbon dioxide production in the tissue during the time of the measurement.

System definition and environmental interactions: The system is a tissue with dimensions shown in Fig. 15.7.

15.3 Diffusion

Fig. 15.7 Tissue segment

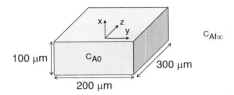

Apprising the problem to identify governing relationships: We will use the graphical approach described for a slab in Sect. 14.7.1 for 1-D transport in each of the three mutually perpendicular directions, then multiply the solutions together as suggested by (15.76).

Analysis: This problem can be solved by applying Fig. D.1 three times to find $Y_x(0,5\text{ s})$, $Y_y(0,5\text{ s})$, and $Y_z(0,5\text{ s})$, then multiplying the three together to find c_A^* (0,0,0,5 s). To find $Y_x(0,5\text{ s})$ we first compute t_x^* and m_x:

$$X = t_x^* = \frac{D_{AB}t}{w^2} = \frac{(1.46 \times 10^{-5} \frac{\text{cm}^2}{\text{s}})(5\text{ s})}{(50 \times 10^{-4}\text{cm})^2} = 2.92,$$

$$m = m_x = \frac{D_{AB}\Phi_{Asf}}{k_{Af}w} = \frac{\left(1.46 \times 10^{-5} \frac{\text{cm}^2}{\text{s}}\right)(1)}{\left(0.973 \times 10^{-3} \frac{\text{cm}}{\text{s}}\right)(50 \times 10^{-4}\text{cm})} = 3.0.$$

The intersection of $X = 2.92$ with the $m = 3$ line in Fig. D.1 gives $Y_x(0,2.92) = 0.43$. Repeating this for the Y_y problem, we find:

$$X = t_y^* = \frac{D_{AB}t}{h^2} = \frac{(1.46 \times 10^{-5} \frac{\text{cm}^2}{\text{s}})(5\text{ s})}{(100 \times 10^{-4}\text{cm})^2} = 0.73,$$

$$m = m_y = \frac{D_{AB}\Phi_{Asf}}{k_{Af}h} = \frac{\left(1.46 \times 10^{-5} \frac{\text{cm}^2}{\text{s}}\right)(1)}{\left(0.973 \times 10^{-3} \frac{\text{cm}}{\text{s}}\right)(100 \times 10^{-4}\text{cm})} = 1.5.$$

The intersection of $X = 0.73$ with the $m = 1.5$ line in Fig. D.1 gives Y_y (0,0.73) = 0.73. Repeating this procedure for the Y_z problem, we find:

$$X = t_z^* = \frac{D_{AB}t}{L^2} = \frac{(1.46 \times 10^{-5} \frac{\text{cm}^2}{\text{s}})(5\text{ s})}{(150 \times 10^{-4}\text{cm})^2} = 0.324,$$

$$m = m_z = \frac{D_{AB}\Phi_{Asf}}{k_{Af}L} = \frac{\left(1.46 \times 10^{-5} \frac{\text{cm}^2}{\text{s}}\right)(1)}{\left(0.973 \times 10^{-3} \frac{\text{cm}}{\text{s}}\right)(150 \times 10^{-4}\text{cm})} = 1.0.$$

The intersection of $X = 0.324$ with the $m = 1.0$ line in Fig. D.1 gives Y_z (0,0.324) = 0.9. Finally, we multiply the values found from the graphs to compute $c^*_{CO_2}$:

$$c^*_{CO_2}(0,0,0,5 \text{ s}) = Y_x(0,5 \text{ s})Y_y(0,5 \text{ s})Y_z(0,5 \text{ s}) = (0.43)(0.73)(0.9) = 0.28.$$

In terms of partial pressure:

$$\frac{P_{CO_2}(0,0,0,5 \text{ s}) - 40 \text{ mmHg}}{60 \text{ mmHg} - 40 \text{ mmHg}} = 0.28.$$

Examining and interpreting the results: Solving for the partial pressure at the center of the tissue segment, we find $P_{CO2}(0,0,0,5 \text{ s}) = 45.6$ mmHg. If the tissue segment was very wide and very high, the slab solution would be $Y_x = 0.43$. The partial pressure of carbon dioxide at the center of the tissue in that case would be 48.6 mmHg. Therefore, loss of CO_2 through the surfaces at $y = \pm h$ and $z = \pm L$ helps to bring the P_{CO2} in the center down faster than loss through surfaces only at $x = \pm w$.

15.4 Diffusion and Chemical Reaction

If a chemical reaction occurs in a stationary media, the species continuity equation becomes:

$$\frac{\partial c_A}{\partial t} = D_{AB} \nabla^2 c_A + R_A. \tag{15.79}$$

The initial concentration of species A in the material will be a function of position, $c_{A0}(x,y,z)$. After a long period of time, the concentration will reach a new steady-state value $c_{A\infty}(x,y,z)$. Let us define a function $u(x,y,z,t)$ as the difference between c_A and $c_{A\infty}$. Then:

$$c_A(x,y,z,t) = c_{A\infty}(x,y,z) + u(x,y,z,t). \tag{15.80}$$

Substituting (15.80) into (15.79), we find that (15.79) will be satisfied if:

$$D_{AB}\nabla^2 c_{A\infty} + R_A = 0 \tag{15.81}$$

and

$$\frac{\partial u}{\partial t} = D_{AB}\nabla^2 u. \tag{15.82}$$

The separation of the species continuity equation into these two relationships is independent of the order of the rate of production of species A per unit volume, R_A. If R_A is second order or higher, or R_A is characterized by Michaelis–Menten kinetics, the nonlinearity is confined to the steady-state problem for $c_{A\infty}$, while

15.4 Diffusion and Chemical Reaction

the transient problem in u is linear. Assuming we can find a solution to (15.81), the initial condition for u is:

$$u(x,y,z,0) = c_{A0}(x,y,z) - c_{A\infty}(x,y,z). \tag{15.83}$$

Let us consider problems where the boundary conditions for the original problem in c_A are linear. For instance, a common boundary condition at a surface $x = L$ will have the form:

$$\alpha \frac{\partial c_A}{\partial x} + \beta c_A = \gamma. \tag{15.84}$$

Substituting (15.80) into (15.84), we see that this boundary condition is satisfied if:

$$\alpha \frac{\partial c_{A\infty}}{\partial x} + \beta c_{A\infty} = \gamma \tag{15.85}$$

and

$$\alpha \frac{\partial u}{\partial x} + \beta u = 0. \tag{15.86}$$

If the other boundary conditions are linear in c_A, then all boundary conditions for u will be homogeneous, similar to the one in (15.86). Therefore, the solution to the unsteady-state diffusion-reaction problem, (15.79), can be found by adding the steady-state diffusion-reaction solution to the solution of the diffusion equation, (15.82) with initial condition given by (15.83) and boundary conditions of the form given in (15.86). We examine such a case in the following example.

Example 15.4.1 Oxygen Consumption and Diffusion in the Cornea.
Before oxygen-permeable contact lenses were developed, glass contact lenses were in common use. Many who wore glass contact lenses found it painful to wear them for extended periods of time, presumably because an inadequate amount of oxygen was supplied to cells in the cornea. Let us analyze the transport of oxygen to the cornea after an oxygen-impermeable contact lens is placed on the cornea. Under normal circumstances, oxygen is supplied to the cornea from two sources. The vitreous humor, in contact with the posterior surface, has a partial pressure of oxygen around 55 mmHg. The anterior surface is in contact with a tear layer that has a P_{O_2} of about 155 mmHg. Approximate metabolic consumption of oxygen in the cornea as being constant, with $R_{O_2} = -8.25 \times 10^{-6}$ mlO$_2$ ml^{-1} s^{-1}. The cornea is assumed to be flat with a thickness $L = 0.6$ mm. The diffusion coefficient for O$_2$ in the cornea is 1×10^{-5} cm^2/s, and the Bunsen solubility coefficient for oxygen in the cornea is 3.16×10^{-5} ml O$_2$ ml^{-1} mmHg^{-1} (values estimated from Weissman, Fatt, and Rasson, 1981).

Solution. *Initial considerations*: This is an unsteady-state problem with oxygen consumption occurring in the cornea. We can assume that oxygen diffusion is in one direction and can neglect the curvature of the cornea as a first approximation.

System definition and environmental interactions: The system of interest is the cornea, modeled as a thin slab of tissue.

Apprising the problem to identify governing relationships: The appropriate governing equation is the one-dimensional unsteady-state species conservation equation with no flux at the surface in contact with the lens and constant concentration at the surface in contact with the vitreous humor.

Analysis: Using Henry's law and assuming one-dimensional diffusion in the cornea, the species continuity equation can be written in terms of the partial pressure of oxygen in the cornea, P_{O_2}:

$$\frac{\partial P_{O_2}}{\partial t} = D_{O_2,cornea} \frac{\partial^2 P_{O_2}}{\partial x^2} + \frac{R_{O_2}}{\alpha^*_{O_2,cornea}}.$$

Let $x = 0$ be the cornea-air interface and $x = L$ be the location of the vitreous humor-cornea interface. When the contact lens is present, the initial and boundary conditions are:

$$P_{O_2}(x,0) = P_{O_2,0}(x),$$

$$\frac{\partial P_{O_2}(0,t)}{\partial x} = 0,$$

$$P_{O_2}(L,t) = P_L.$$

By analogy with (15.80), we split the solution into steady-state and transient components:

$$P_{O_2}(x,t) = P_{O_2,\infty}(x) + u(x,t).$$

Introducing this into the original diffusion reaction equation and auxiliary conditions allows us to separate the transient problem for P_{O_2} into two separate problems. The first problem for $P_{O_2,\infty}$ is:

$$0 = D_{O_2,cornea} \frac{d^2 P_{O_2,\infty}}{dx^2} + \frac{R_{O_2}}{\alpha^*_{O_2,cornea}},$$

$$\frac{dP_{O_2}(0)}{dx} = 0,$$

$$P_{O_2,\infty}(L) = P_L.$$

15.4 Diffusion and Chemical Reaction

The solution to the steady-state problem a long time after the lens is placed on the eye is:

$$P_{O_2,\infty}(x) = P_L + \frac{R_{O_2} L^2}{2 D_{O_2,\text{cornea}} \alpha^*_{O_2,\text{cornea}}} \left(1 - \left(\frac{x}{L}\right)^2\right).$$

The second problem to be solved is the diffusion equation for the function $u(x,t)$. The no flux boundary condition at $x = 0$ is a special case of (15.86) with $\alpha = 1$ and $\beta = 0$. The boundary condition at the interface with the vitreous humor is also a special case of (15.86) with $\alpha = 0$ and $\beta = 1$. The problem statement for $u(x,t)$ is:

$$\frac{\partial u}{\partial t} = D_{O_2,\text{cornea}} \frac{\partial^2 u}{\partial x^2},$$

$$\frac{\partial u(0,t)}{\partial x} = 0,$$

$$u(L,t) = 0,$$

$$u(x,0) = P_{O_2,0}(x) - P_{O_2,\infty}(x).$$

$P_{O_2,0}(x)$ is the partial pressure of oxygen in the tissue before the contact lens is placed on the eye. Assuming the cornea is exposed to air for a long time before the contact lens is introduced, $P_{O_2,0}(x)$ can be found by solving the steady-state diffusion-reaction equation with boundary conditions $P_{O_2,0}(L) = P_L = 55$ mmHg and $P_{O_2,0}(0) = P_0 = 155$ mmHg. The solution is:

$$P_{O_2,0}(x) = P_0 + \frac{x}{L} \left\{ (P_L - P_0) + \left(\frac{R_{O_2} L^2}{2 D_{O_2,\text{cornea}} \alpha^*_{O_2,\text{cornea}}}\right) \left(1 - \frac{x}{L}\right) \right\}.$$

The initial condition $u(x, 0)$ is found by subtracting $P_{O_2,\infty}(x)$ from $P_{O_2,0}(x)$:

$$u(x,0) = \left[P_0 - P_L - \frac{R_{O_2} L^2}{2 D_{O_2,\text{cornea}} \alpha^*_{O_2,\text{cornea}}}\right] \left(1 - \frac{x}{L}\right).$$

Thus, the initial distribution is linear. The product solution method described in Sect. 15.3.1 can be used to find $u(x,t)$:

$$u(x,t) = 2\left[P_0 - P_L - \frac{R_{O_2} L^2}{2 D_{O_2,\text{cornea}} \alpha^*_{O_2,\text{cornea}}}\right] \sum_{n=0}^{\infty} \frac{\cos\left(\frac{\phi_n x}{L}\right)}{\phi_n^2} \exp\left(-\phi_n^2 \left(\frac{D_{O_2,\text{cornea}} t}{L^2}\right)\right),$$

where,

$$\phi_n = \left(\frac{2n+1}{2}\right)\pi.$$

Examining and interpreting the results: Adding the functions $u(x,t)$ and $P_{O_2,\infty}(x)$, and inserting the values for P_0, P_L, R_{O_2}, $D_{O_2,\text{cornea}}$, $\alpha^*_{O_2,\text{cornea}}$, and L, we obtain the solution displayed in Fig. 15.8. The green line at $t = 0$ represents the initial distribution of $P_{O_2,0}(x)$ and the red line for $t \rightarrow \infty$ is the function $P_{O_2,\infty}(x)$. The model predicts that the P_{O_2} at the surface of the eye drops about 9 mmHg in the first second after the contact lens is placed on the eye. However, for $x/L > 0.5$, the partial pressure is virtually unchanged for the first 10 s. After about 2 min, the partial pressure of O_2 is relatively uniform across the entire thickness of the cornea. Within 10 min, the P_{O_2} approaches the final steady-state distribution. The slopes of all curves when $t > 0$ are zero at $x/L = 0$, since the contact lens is impermeable to O_2. Cells near the surface are supplied with oxygen from the vitreous humor side of the cornea, but the steady-state P_{O_2} is only 8 mmHg. If the stroma were to swell just a small amount after the contact lens is placed on the eye, the P_{O_2} near the anterior surface could quickly approach zero, causing damage to epithelial cells and pain to the subject.

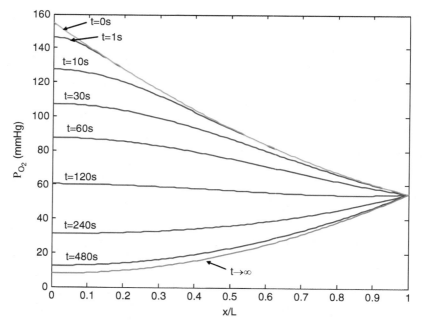

Fig. 15.8 Partial pressure of oxygen in the cornea as a function of position for several times after an oxygen-impermeable contact lens is placed on the surface of the eye ($x/L = 0$)

15.5 Convection and Diffusion

Intracellular transport of respiratory gases, metabolic substances, and waste products is dominated by diffusion. However, tissues cannot rely on diffusion alone to deliver an adequate supply of nutrients. As nutrients diffuse into the tissue, some will be utilized in the metabolic process. As needed nutrients penetrate further into tissue, the concentration decreases, until at some point, all nutrients have been converted to waste products. Only very thin tissues or multicellular organisms, on the order of a thousand microns or so, would be able to survive if diffusion were the only transport mechanism available.

Organisms larger than a few hundred microns exist because nutrients are delivered to tissues via a convective transport mechanism. Oxygen and nutrients are delivered to tissue via blood flowing in arteries, and CO_2 and waste products are removed by blood flowing through veins. Combined convection and diffusion transport mechanisms occur in tissue capillaries and in other specialized capillaries. For instance, nutrient uptake from the gut, gas exchange in the lungs, and waste product removal in the liver and kidneys all involve convection and diffusion in capillaries. Convection in the respiratory system delivers O_2 and removes CO_2 from the body.

To illustrate the importance of convection relative to diffusion, let us perform two thought experiments. In each we carefully bring a tube filled with pure water into contact with a reservoir containing an albumin solution at a concentration C_0. In the first experiment, there is no pressure gradient across the tube and in the second there is a slight pressure gradient, enough to induce an average velocity of 1 mm/min. We monitor the average concentration of albumin at a location 1 cm downstream from the end of the tube. Our goal is to estimate how long it takes before the concentration at the sampling site is 50% of the concentration in the reservoir. In the first experiment, the only transport mechanism is diffusion and in the second experiment, we neglect diffusion and assume that the only transport mechanism is convection.

We solved the diffusion problem in Sect. 14.7.1.1. Using (14.295) with the diffusion coefficient for albumin in water equal to 9×10^{-7} cm^2/s and since erfc $(0.48) = 0.5$, we find the diffusion time to be 1,157,000 s, or 13.4 days! For the second experiment we assume laminar flow with a parabolic velocity profile:

$$v_z(r,t) = \frac{\partial (z(r,t))}{\partial t} = 2<v> \left(1 - \left(\frac{r}{R}\right)^2\right). \tag{15.87}$$

The variable $z(r,t)$ is the distance that a fluid element located at position r travels in time t. At any given time t^*, $z(r,t^*)$ represents the location of the interface between the water and albumin solution. Integrating (15.87), noting that the interface is $z = 0$ for all fluid particles at $t = 0$, we find:

$$z(r,t) = 2<v>t \left(1 - \left(\frac{r}{R}\right)^2\right). \tag{15.88}$$

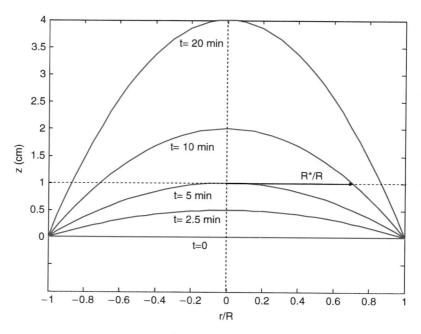

Fig. 15.9 Location of the interface between an albumin solution and water for laminar flow in a tube. The region below a line representing a particular time contains albumin solution and the region above that line contains water. The position $r/R = R^*/R$ is the location of the front when the average concentration at $z = 1$ cm is 50% of the inlet concentration

The location of the interface separating the albumin solution from water is shown in Fig. 15.9 for various times t.

We are interested in the particular time t^* when the average concentration of albumin at $z = 1$ cm is $C_0/2$. The average concentration of species A over the cross section is given by:

$$\bar{c}_A = \int_0^R c_A r\,dr \Big/ \int_0^R r\,dr. \tag{15.89}$$

The interface between the two fluids at the center of the tube reaches any axial position z in time $z/[2<v>]$. After the albumin front at $r = 0$ passes a particular axial position, two regions will exist: a central region ($r \leq R^*$) containing albumin solution with $c_A = C_0$, and a peripheral region ($R^* < r \leq R$) composed of water, i.e., $c_A = 0$. Substituting this information into (15.89):

$$\frac{C_0}{2} = \int_0^{R^*} C_0 r\,dr \Big/ \int_0^R r\,dr. \tag{15.90}$$

15.5 Convection and Diffusion

From this we find that if the average concentration is $C_0/2$ at a position z, the radius of the front R^* at that position is:

$$R^* = R\big/\sqrt{2}. \tag{15.91}$$

Finally, we can substitute R^* back into (15.88) to estimate the time t^* required for the average concentration to reach $C_0/2$:

$$t^* = \frac{z(R^*,t^*)}{2<v>\left(1-\left(\frac{R^*}{R}\right)^2\right)} = \frac{z(R^*,t^*)}{<v>}. \tag{15.92}$$

Since $z(R^*,t^*) = 1$ cm and $<v>$ is 1 mm/min, we find $t^* = 10$ min. Compare this with the diffusion time of 19,280 min. Convection is faster in this case by a factor of nearly 2,000! Keep in mind that the average velocity in this example is very low, so convection is an even more important transport mechanism than diffusion when the flows are higher. Consequently, axial diffusion can often be neglected relative to axial convection. However, diffusion perpendicular to the direction of flow can ultimately cause significant variations in concentration from what would be expected from convection alone.

Consider, for example the case where a mass m of tracer A is instantaneously and uniformly deposited over the cross-section of a tube at an axial position $z = 0$. Flow in the tube is laminar. The axial distribution of tracer by convection alone at various times would be given by the lines in Fig. 15.9. In the absence of diffusion, tracer-free fluid would exist on each side of the lines. In reality, even if we neglect axial diffusion, the steep gradients will cause tracer to diffuse radially in both the positive and negative directions. Tracer that diffuses toward the central portion of the tube will diffuse into a fluid layer moving at a faster velocity and tracer that diffuses toward the wall will diffuse into a slower layer of fluid. This combination of axial convection and radial diffusion leads to a distortion of the concentration profile, commonly known as dispersion. G.I. Taylor (1953) analyzed this phenomenon and found that for a sufficient distance downstream, the average mass concentration over the cross-section of a tube with radius R is given by:

$$\bar{\rho}_A = \frac{m}{2\pi^{3/2}R^2\sqrt{D^*t}} \exp\left[-\frac{(z-<v>t)^2}{4D^*t}\right], \tag{15.93}$$

where D^* is known as the *Taylor dispersion coefficient*:

$$D^* = D_{AB} + \frac{<v>^2 R^2}{48 D_{AB}}. \tag{15.94}$$

An observer moving with the average velocity would find the average concentration decreasing with time as $1/\sqrt{t}$. This is caused by tracer loss via radial diffusion to both the faster and slower moving fluids relative to the average. Note that the second term is inversely proportional to the diffusion coefficient, so that the lower the diffusion coefficient, the greater the dispersion. This is illustrated in Fig. 15.10.

Equation (15.93) is only valid at locations far downstream from where the tracer was originally deposited. If L is the distance between the injection and detection sites, Taylor found the relationship to be valid if:

$$\frac{LD_{AB}}{R^2 <v>} \gg 0.1. \tag{15.95}$$

For the dispersion of glucose in blood flowing through an artery with $R = 0.5$ cm, $<v> = 50$ cm/s (assumed steady) and $D_{AB} = 6.9 \times 10^{-6}$ cm^2/s, we find that the artery must be longer than 1,800 m before Taylor dispersion can be applied! This might be fine for long pipelines, but clearly, Taylor dispersion should not be applied to tracer distribution in arteries or veins. However, for glucose flowing in a capillary ($R = 5 \times 10^{-4}$ cm, $<v> = 0.1$ cm/s) with an assumed parabolic velocity profile, we find $L \gg 3.6 \times 10^{-4}$ cm. Therefore, if glucose was introduced as a bolus at the entrance of a capillary, Taylor dispersion would be observed for distances greater than a few diameters downstream of the capillary entrance and the Taylor diffusion coefficient would be about twice the diffusivity of glucose.

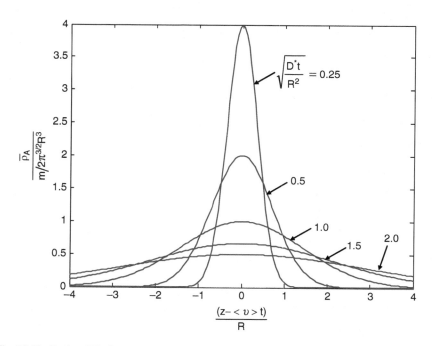

Fig. 15.10 Taylor diffusion: Tracer distribution far from tube inlet following a bolus injection of tracer of mass m

15.5.1 Steady-State, Multidimensional Convection and Diffusion

There are several situations in biomedical engineering where we are interested in analyzing steady-state mass transfer involving both convection and diffusion. In the following sections, we consider common applications involving external flow past a surface and internal flow in conduits.

15.5.1.1 Mass Transfer with Flow Past a Flat Surface

Let us consider the case where fluid flows past a solid surface containing species A. We will assume that the surface is flat with the coordinate direction x being in the direction of flow and the coordinate direction y perpendicular to the surface. The approaching fluid has a constant velocity v_0 and contains species A at a concentration c_{A0}. The concentration of species A in the solid at the surface is assumed to be constant and equal to c_{AS}. Our goal is to find the steady-state concentration of species A in the flowing fluid as a function of x and y.

For steady, two-dimensional flow in the fluid with no production of species A and negligible axial diffusion, (15.18) becomes:

$$v_x \frac{\partial c_A}{\partial x} + v_y \frac{\partial c_A}{\partial y} = D_{Af} \frac{\partial^2 c_A}{\partial y^2}, \tag{15.96}$$

with boundary conditions:

$$\begin{aligned} x &= 0 : c_A = c_{A0}, \\ y &= 0 : c_A = \Phi_{Afs} c_{As}, \\ y &\to \infty : c_A = c_{A0}. \end{aligned} \tag{15.97}$$

In Example 7.13.3, we used scaling to analyze the velocity components in the fluid and found that the components were of the same order of magnitude in the vicinity of the surface. Consequently, there will be convection of species A in both the x and y directions and diffusion of A in the y-direction. Introducing a dimensionless concentration $C^*_A = (c_A - \Phi_{Afs} c_{As})/(c_{A0} - \Phi_{Afs} c_{As})$, substituting (7.103) for v_x and (7.104) for v_y, and assuming a combination of variables η given by (7.102), the problem reduces to solution of the following ordinary differential equation:

$$\frac{d^2 C^*_A}{d\eta^2} + \frac{1}{2}\left(\frac{\nu}{D_{Af}}\right) f(\eta) \frac{d C^*_A}{d\eta} = 0. \tag{15.98}$$

The ratio ν/D_{Af} is the Schmidt number Sc, and the function $f(\eta)$ was defined in Example 7.13.3 and is plotted in Fig. 7.25. The dimensionless boundary conditions are

$$\begin{aligned} C^*_A(0) &= 0, \\ C^*_A(\infty) &= 1. \end{aligned} \tag{15.99}$$

Integrating (15.98) twice:

$$\frac{c_A - \Phi_{Afs} c_{As}}{c_{A0} - \Phi_{Afs} c_{As}} = C_1 \int_0^\eta \exp\left(-\frac{Sc}{2} \int_0^\eta f(\eta) d\eta\right) d\eta + C_2. \tag{15.100}$$

The boundary conditions can be applied to evaluate the constants C_1 and C_2. The final solution, first derived by Pohlhausen (1921), is:

$$\frac{c_A - \Phi_{Afs} c_{As}}{c_{A0} - \Phi_{Afs} c_{As}} = \frac{\int_0^\eta \exp\left(-\frac{Sc}{2} \int_0^\eta f(\eta) d\eta\right) d\eta}{\int_0^\infty \exp\left(-\frac{Sc}{2} \int_0^\eta f(\eta) d\eta\right) d\eta}. \tag{15.101}$$

Since $f(\eta)$ is the solution to (7.101), it is a known function of η. Therefore, the right side of (15.101) can be computed as a function of Sc and η. An alternative approach is to integrate the ordinary differential equations for C^*_A and f, using Matlab or another differential equation solver. The code is given below and the relationship between C^*_A and η is shown in Fig. 15.11 for several values of the Schmidt number.

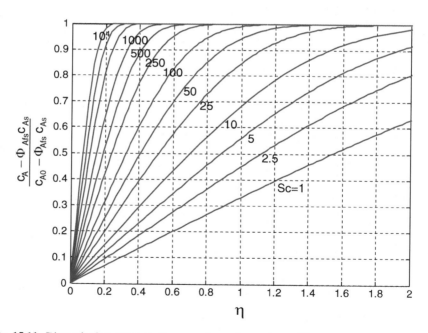

Fig. 15.11 Dimensionless concentration vs. η for various values of Sc

15.5 Convection and Diffusion

Matlab code for Blasius and Pohlhausen solutions:

Matlab code for Blasius and Pohlhausen solutions:
```
% Pohlhausen solution, flow past a flat plate, no pressure gradient
% Intermediate variables: V=vx/v0, F=d²f/dη², G=dC*/dη
% Set of 5 first order equations
% y(1)=f:    df/dη = V
% y(2)=V:    dV/dη = F
% y(3)=F:    dF/dη = -f*F/2
% y(4)=C*:   dC*/dη = -G
% y(5)=G:    dG/dη = -(Sc*f/2)*G
clear
%Boundary condition at infinity applied at η=10
Inf = 10;
%Boundary conditions
f0=0;
V0=0;
Vinf=1;
C0=0;
Cinf =1;
Sc=[1, 2.5, 5, 10, 25,50,100,250,500,1000,2500,5000,10000];
x=0:.05:Inf;
for i=1:length(Sc)
    solinit = bvpinit(x, [0 0 0 0 0]);
    options=bvpset('RelTol',1e-4);
    sol = bvp4c(@Pohlhausen_ode,@Pohlhausen_bc,solinit,...
              options,f0,V0,Vinf,C0,Cinf,Sc(i));
    X=sol.x;
    C=sol.y(4,:);
    plot (X,C)
    grid on
    hold on
end
%------------------------------------
function dydx = Pohlhausen_ode(x,y,f0,V0,Vinf,C0,Cinf,Sc)
dydx = [y(2)
        y(3)
        -y(1)*y(3)/2
        -y(5)
        -Sc*y(1)*y(5)/2];
%------------------------------------
function res = Pohlhausen_bc(ya,yb,f0,V0,Vinf,C0,Cinf,Sc)
res = [ ya(1)-f0
        ya(2)-V0
        yb(2)-Vinf
        ya(4)-C0
        yb(4)-Cinf];
```

The flux of species A at the surface, $y = 0$, is:

$$N_A|_{y=0} = -D_{Af} \frac{\partial c_A}{\partial y}\bigg|_{y=0} = -D_{Af}(c_{A0} - \Phi_{Afs} c_{As})\sqrt{\frac{v_0}{vx}} \frac{dC_A^*}{d\eta}\bigg|_{\eta=0} \qquad (15.102)$$

or

$$N_A|_{y=0} = -\frac{1}{Sc}\sqrt{\frac{v_0 v}{x}} \frac{(c_{A0} - \Phi_{Afs}c_{As})}{\int_0^\infty \exp\left(-\frac{Sc}{2}\int_0^\eta f(\eta)d\eta\right)d\eta}. \quad (15.103)$$

We can use the Matlab solution given above to find $dC^*_A/d\eta$ at $\eta = 0$. However, in most applications, the Schmidt number is much greater than one. In that case the hydrodynamic boundary layer is much thicker than the boundary layer thickness for species A. If we assume the velocity gradient is constant in the region over which the concentration varies (Kays and Crawford 1980), then from Fig. 7.25:

$$\frac{d^2 f}{d\eta^2} = 0.332. \quad (15.104)$$

Integrating this twice and noting that $v_x(0) = v_0(df/d\eta) = 0$ and $f(0) = 0$, we find:

$$f = \frac{0.332}{2}\eta^2 (Sc \gg 1). \quad (15.105)$$

Therefore, for $Sc \gg 1$:

$$\int_0^\eta f(\eta)d\eta = 0.166 \int_0^\eta \eta^2 d\eta = 0.0553\eta^3 \quad (15.106)$$

Substituting this into the integral expression in (15.103):

$$\int_0^\infty \exp\left(-\frac{Sc}{2}\int_0^\eta f(\eta)d\eta\right)d\eta = \int_0^\infty \exp\left[-0.0277(Sc)(\eta^3)\right]d\eta. \quad (15.107)$$

From a table of integrals we find:

$$\int_0^\infty \exp(-a\eta^3)d\eta = \frac{a^{-1/3}}{3}\Gamma\left(\frac{1}{3}\right), \quad (15.108)$$

where $\Gamma(x)$ is the Gamma function and $\Gamma(1/3) = 2.6789$. The factor a is equal to $0.0277Sc$. Substituting this back into (15.103), we find:

$$N_A|_{y=0} = -0.339 Sc^{-2/3}\sqrt{\frac{v_0 v}{x}}(c_{A0} - \Phi_{Afs}c_{As}). \quad (15.109)$$

This is very close to a more accurate solution that is obtained using numerical integration:

$$N_A|_{y=0} = -0.332 Sc^{-2/3}\sqrt{\frac{v_0 v}{x}}(c_{A0} - \Phi_{Afs}c_{As}). \quad (15.110)$$

15.5 Convection and Diffusion

We can also write the flux in terms of a mass transfer coefficient, k_{Af}:

$$N_A|_{y=0} = -k_{Af}(c_{A0} - \Phi_{Afs}c_{As}). \tag{15.111}$$

The local Sherwood number can be obtained from (15.110) and (15.111):

$$\text{Sh}_x = \frac{k_{Af}x}{D_{Af}} = 0.332 Sc^{\frac{1}{3}} \text{Re}_x^{\frac{1}{2}}. \tag{15.112}$$

The mean Sherwood number Sh_L over a distance L along the plate is:

$$\text{Sh}_L = \frac{1}{L}\int_0^L \frac{k_{Af}x}{D_{Af}}\,dx = 0.664 Sc^{\frac{1}{3}} \text{Re}_L^{\frac{1}{2}}. \tag{15.113}$$

Therefore, the mean Sherwood number up to an axial position $x = L$ is twice the local Sherwood number at the location $x = L$.

15.5.1.2 Analogies Between Momentum Transport and Convective Heat and Mass Transport

Let us now compare momentum transport in fluid moving past a flat surface (Example 7.13.3), heat transport to fluid flowing past a flat surface (Sect. 8.3.3.2), and mass transport to fluid flowing past a flat surface (Sect. 15.5.1.1). For laminar flow past a smooth flat plate, we found the friction factor to be related to the square root of the local Reynolds number:

$$\frac{f}{2} = 0.332 \text{Re}_x^{-\frac{1}{2}}. \tag{15.114}$$

For laminar flow past a plate with constant temperature, the local Nusselt number is:

$$\text{Nu}_x = \frac{hx}{k} = 0.332 \text{Pr}^{\frac{1}{3}} \text{Re}_x^{\frac{1}{2}}. \tag{15.115}$$

Finally, for laminar flow past a flat plate with constant wall concentration, the local Sherwood number is:

$$\text{Sh}_x = \frac{k_{Af}x}{D_{Af}} = 0.332 Sc^{\frac{1}{3}} \text{Re}_x^{\frac{1}{2}}. \tag{15.116}$$

Dividing (15.115) by $\text{Re}_x \text{Pr}^{1/3}$, we find:

$$\frac{\text{Nu}_x}{\text{Pr}^{\frac{1}{3}}\text{Re}_x} = 0.332 \text{Re}_x^{-\frac{1}{2}}. \tag{15.117}$$

The left hand side of (15.117) is independent of x and is defined as the *j factor for heat transfer*, j_H:

$$j_H \equiv \frac{h}{\rho c_p v_0} Pr^{\frac{2}{3}}. \tag{15.118}$$

Dividing the Sherwood number in (15.112) by $Re_x Sc^{1/3}$, we find:

$$\frac{Sh_x}{Sc^{\frac{1}{3}} Re_x} = 0.332 Re_x^{-\frac{1}{2}}. \tag{15.119}$$

The left hand side of (15.119) is independent of x and is defined as the *j factor for mass transfer*, j_D:

$$j_D \equiv \frac{k_{Af}}{v_0} Sc^{\frac{2}{3}}. \tag{15.120}$$

Finally, recognizing that the right hand sides of (15.114), (15.117), and (15.119) are all the same, we find:

$$j_H = j_D = \frac{f}{2}. \tag{15.121}$$

This is known as the Chilton–Colburn analogy (Chilton and Colburn 1934). It relates the momentum, heat, and mass transfer coefficients f, h, and k_{Af}. If any one of these coefficients is known, the other two can be computed from (15.121). Although the derivation applies strictly to the case of flow past a flat surface, where the entire drag force is caused by viscous effects near the surface, empirical studies have shown the relationship is valid wherever form drag is small relative to skin friction. Even in cases where form drag is significant, such as flow past a cylinder or a sphere, the analogy for heat and mass transfer ($j_H = j_D \neq f/2$) is found to be valid for $0.6 \leq Sc \leq 2{,}500$ and $0.6 \leq Pr \leq 100$.

> **Example 15.5.1 Application of the Chilton–Colburn analogy.**
> Experimental data are available for convective heat transfer from a red blood cell oriented perpendicular to the flow field:
>
> $$Nu_d = \frac{hd}{k_f} = B Pr^{\frac{1}{3}} Re_d^n,$$
>
> where B and n are constants, k_f is the thermal conductivity of the fluid, h is the heat transfer coefficient, and d is the cell diameter. Use the Chilton–Colburn analogy to estimate the dimensionless mass transfer coefficient that governs exchange of an inert substance A from the cell surface.

15.5 Convection and Diffusion

Solution. *Initial considerations, system and governing relationships*: The Chilton–Colburn analogy is appropriate for converting a heat transfer relationship to a species mass transfer relationship for a system with the same or similar geometry. In this case, the analogy is appropriate, since we are applying the analogy to the same system, a human red blood cell.
Analysis: From (15.118) and (15.119):

$$\frac{k_{Af}}{v_0} Sc^{\frac{2}{3}} = \frac{h}{\rho c_p v_0} Pr^{\frac{2}{3}}. \tag{15.122}$$

Solving for k_{Af}, and using the empirical relationship for h:

$$k_{Af} = \frac{1}{\rho c_p} \left(\frac{k_f}{d} B Pr^{\frac{1}{3}} Re_d^n \right) \frac{Pr^{\frac{2}{3}}}{Sc^{\frac{2}{3}}}.$$

Inserting $\rho c_p v / k_f$ for the Prandtl number, and recognizing that $Sc = v/D_{Af}$, we obtain:

$$Sh_d = \frac{k_{Af} d}{D_{Af}} = B Sc^{\frac{1}{3}} Re_d^n.$$

Examining and interpreting the results: In general, multiplying both sides of (15.122) by a characteristic length, L, and rearranging, we find the Chilton–Colburn analogy between heat and mass transfer can be written:

$$Sh_L = Nu_L \left(\frac{Sc}{Pr} \right)^{\frac{1}{3}}. \tag{15.123}$$

15.5.1.3 Constant Solute Flux to Fluid Flowing in a Tube

In Sect. 14.5.3.2 we used a one-dimensional shell balance to analyze steady-state waste product removal from tissue by blood as it flows through a capillary. The solute A was assumed to be produced at a constant rate per unit volume in the tissue, and therefore the steady-state flux of species A into the capillary will be constant. If the capillary is directed in the positive z-direction, the solution was found to be (14.191):

$$C_{Ab}(z) = C_{Ab0} + \frac{N_A S_i}{Q_b} \left(\frac{z}{L} \right), \tag{15.124}$$

where C_{Ab} is the bulk concentration of solute A in the blood, C_{Ab0} is the bulk concentration of A at the entrance to the capillary, N_A is the flux of A from tissue to blood, S_i is the capillary inside surface area, Q_b is the blood flow through the capillary, and L is the length of the capillary. In some instances we are interested in the concentration of species A at specific radial and axial positions, rather than the bulk concentration. The multidimensional species continuity equation must be used to find local concentrations. Let us make the following assumptions:

1. Transport of species A is steady-state.
2. Species A does not undergo chemical reaction in the blood.
3. Blood flow is fully developed in the capillary ($v_r = v_\theta = 0$).
4. Concentration is symmetrical about the center of the tube.

The appropriate starting point for a cylindrical capillary would be (15.19). Applying the assumptions above:

$$\cancel{\frac{\partial c_A}{\partial t}}^1 + \cancel{v_r \frac{\partial(c_A)}{\partial r}}^3 + \cancel{\frac{v_\theta}{r}\frac{\partial(c_A)}{\partial \theta}}^3 + v_z \frac{\partial(c_A)}{\partial z} = D_{AB}\left\{\frac{1}{r}\frac{\partial}{\partial r}\left(r\frac{\partial c_A}{\partial r}\right) + \cancel{\frac{1}{r^2}\frac{\partial^2 c_A}{\partial \theta^2}}^4 + \frac{\partial^2 c_A}{\partial z^2}\right\} + \cancel{R_A}^2 \qquad (15.125)$$

where the velocity is:

$$v_z = 2\langle v\rangle\left[1 - \left(\frac{r}{R_c}\right)^2\right]. \qquad (15.126)$$

R_c is the capillary radius and $\langle v\rangle$ is the average velocity in the capillary. The diffusion coefficient D_{AB} is the diffusion coefficient of species A in blood, D_{Ab}. Defining dimensionless radial and axial dimensions, $r^* = r/R_c$ and $z^* = z/L$, (15.125) becomes:

$$\frac{2\langle v\rangle R_c^2}{D_{Ab}L}[1 - r^{*2}]\frac{\partial c_A}{\partial z^*} = \frac{1}{r^*}\frac{\partial}{\partial r^*}\left(r^*\frac{\partial c_A}{\partial r^*}\right) + \frac{R_c^2}{L^2}\frac{\partial^2 c_A}{\partial z^{*2}}. \qquad (15.127)$$

Since $R_c \ll L$, the last term is very small and therefore $\partial^2 c_A/\partial z^2$ can be neglected, except near the inlet of the capillary where z is of the same order of magnitude as R_c. Neglecting axial diffusion, (15.127) becomes:

$$2\langle v\rangle\left[1 - \left(\frac{r}{R_c}\right)^2\right]\frac{\partial c_A}{\partial z} = D_{Ab}\frac{1}{r}\frac{\partial}{\partial r}\left(r\frac{\partial c_A}{\partial r}\right). \qquad (15.128)$$

If axial diffusion is actually zero, then $\partial^2 c_A/\partial z^2 = 0$, and $\partial c_A/\partial z$ would just be a function of radial position, say $f_1(r)$.

$$\frac{\partial c_A}{\partial z} = f_1(r). \qquad (15.129)$$

Integrating (15.129):

$$c_A(r,z) = f_1(r)z + f_2(r). \qquad (15.130)$$

15.5 Convection and Diffusion

Substituting this back into (15.128):

$$\frac{2\langle v \rangle}{D_{AB}}\left[1 - \left(\frac{r}{R_c}\right)^2\right]f_1(r) = z\frac{1}{r}\frac{d}{dr}\left(r\frac{df_1}{dr}\right) + \frac{1}{r}\frac{d}{dr}\left(r\frac{df_2}{dr}\right). \qquad (15.131)$$

Note that the left-hand side and the last term on the right-hand side depend on radial position alone, while the first term on the right hand side also depends on axial position, z. Therefore, this term must be zero. Since z is not zero, then:

$$\frac{d}{dr}\left(r\frac{df_1}{dr}\right) = 0. \qquad (15.132)$$

Integrating:

$$f_1(r) = C_1 \ln(r) + C_2, \qquad (15.133)$$

where C_1 and C_2 are constants. However, since the concentration in the center of the tube is bounded, or since the concentration is symmetrical about the center of the capillary, C_1 must be zero and f_1 is just a constant, C_2. Substituting this back into (15.131):

$$\frac{2\langle v \rangle C_2}{D_{Ab}}\left[1 - \left(\frac{r}{R_c}\right)^2\right] = \frac{1}{r}\frac{d}{dr}\left(r\frac{df_2}{dr}\right). \qquad (15.134)$$

Integrating this expression twice yields:

$$f_2(r) = \frac{\langle v \rangle C_2}{2D_{Ab}}\left[r^2 - \frac{r^4}{4R_c^2}\right] + C_3 \ln(r) + C_4, \qquad (15.135)$$

where C_3 and C_4 are constants of integration. Once again, since c_A is bounded at $r = 0$, C_3 must be zero. Substituting this back into (15.130), we find:

$$c_A(r,z) = C_2 z + \frac{\langle v \rangle C_2}{2D_{Ab}}\left[r^2 - \frac{r^4}{4R_c^2}\right] + C_4. \qquad (15.136)$$

Applying the constant flux boundary condition at $r = R_c$ for an inward flux N_A:

$$-N_A = -D_{Ab}\frac{\partial c_A}{\partial r}\bigg|_{r=R_c}. \qquad (15.137)$$

We find:

$$C_2 = \frac{2N_A}{\langle v \rangle R_c}. \qquad (15.138)$$

Since our solution is not valid near $z = 0$, we cannot use the fact that the concentration is known at the inlet as a boundary condition. However, we can use a macroscopic mass balance between the inlet and axial position z to determine C_4:

$$\int_A c_A v_z dA \bigg|_{z=0} - \int_A c_A v_z dA \bigg|_{z=z} + 2\pi R_c z N_A = 0. \tag{15.139}$$

If the bulk concentration at the inlet is C_{Ab0}, and the capillary flow rate is Q_b, then substitution of (15.136) into (15.139) yields:

$$C_{Ab0} Q_b - \int_0^{R_c} \left[C_2 z + \frac{\langle v \rangle C_2}{2 D_{Ab}} \left(r^2 - \frac{r^4}{4 R_c^2} \right) + C_4 \right] \left[2\langle v \rangle \left(1 - \frac{r^2}{R_c^2} \right) \right] [2\pi r dr]$$
$$+ 2\pi R_c z N_A = 0. \tag{15.140}$$

Using (15.77a,b,c) for C_2, we find C_4 to be:

$$C_4 = C_{Ab0} - \frac{7}{24} \frac{R_c N_A}{D_{Ab}}. \tag{15.141}$$

Substituting C_2 and C_4 back into (15.136), the final solution for concentration as a function of radial and axial position is:

$$c_A(r,z) = C_{Ab0} + \frac{2 N_A}{\langle v \rangle R_c} z + \frac{N_A}{D_{Ab} R_c} \left[r^2 - \frac{r^4}{4 R_c^2} - \frac{7}{24} R_c^2 \right]. \tag{15.142}$$

Note that this solution is not valid near the inlet of the capillary. Setting $r = 0$ and $z = 0$ yields a centerline concentration at the inlet that is different from the bulk concentration. This is certainly true downstream from the inlet, but the concentration is actually uniform at the inlet. The bulk concentration at any axial position is:

$$C_{Ab}(z) = \frac{\int_0^{R_c} v_z c_A (2\pi r dr)}{\int_0^{R_c} v_z (2\pi r dr)} = C_{Ab0} + \frac{2 N_A z}{\langle v \rangle R_c}. \tag{15.143}$$

Note that this is exactly the expression that one would obtain by performing a 1-D shell balance on a portion of the tube with flow rate $Q_b = \pi R_c^2 \langle v \rangle$ and surface area $2\pi R_c L$ (15.124). The average concentration over the cross-section at any axial position is:

$$\bar{c}_A = \frac{\int_0^{R_c} c_A (2\pi r dr)}{\int_0^{R_c} (2\pi r dr)} = C_{Ab0} + \frac{2 N_A z}{\langle v \rangle R_c} + \frac{N_A R_c}{8 D_{Ab}}. \tag{15.144}$$

The relationship between average concentration and bulk concentration for the case of constant wall flux to fluid flowing in a tube is:

$$\bar{c}_A = C_{Ab} + \frac{N_A R_c}{8 D_{Ab}}. \tag{15.145}$$

15.5 Convection and Diffusion

This relation is independent of axial position and independent of flow. The concentration at the capillary wall at any axial position is:

$$c_{Ab}(z, R_c) = C_{Ab0} + \frac{2N_A z}{<v> R_c} + \frac{11}{24}\left(\frac{N_A R_c}{D_{Ab}}\right). \tag{15.146}$$

Therefore, the concentration difference between the wall and the bulk fluid is a constant:

$$c_{Ab}(z, R_c) - C_{Ab} = \frac{11}{24}\left(\frac{N_A R_c}{D_{Ab}}\right). \tag{15.147}$$

The difference between the concentration at the wall and the concentration at the center of the capillary is also constant and equal to $3N_A R_c/4D_{Ab}$. Recall that the mass transfer coefficient k_{Ab} governing the flow of species A from the tube wall to the bulk fluid (blood) is defined by the following relation:

$$N_A = k_{Ab}(c_{Ab}(z, R_c) - C_{Ab}). \tag{15.148}$$

Therefore, for the case of constant mass influx at the tube wall:

$$k_{Ab} = \frac{24}{11}\frac{D_{Ab}}{R_c}. \tag{15.149}$$

The Sherwood number for constant mass flux at the wall, based on the capillary diameter d_c, is:

$$\text{Sh}_{d_c} = \frac{2R_c k_{Ab}}{D_{Ab}} = \frac{48}{11} = 4.364. \tag{15.150}$$

This derivation of the Sherwood number from the application of the general species conservation equation illustrates how the dimensionless mass transfer coefficients given in Table 12.1 are derived for conduits with different cross-sections and different boundary conditions.

> **Example 15.5.2 Constant Flux of Urea from Tissue to Capillary.**
> Urea is produced when excess amino acids in the blood are broken down by cells in the liver. Urea diffuses across capillary walls in the liver, is removed from the circulation by the kidneys, and removed from the body via the urine. Consider the situation when the flux per liver capillary is 0.071 µmol min^{-1} cm^{-2}. The average velocity in each capillary is 0.1 cm/s, the capillary radius is 4×10^{-4} cm, and capillary length is 0.1 cm. The diffusion coefficient for urea in blood is 1.8×10^{-5} cm^2/s. What is the difference in concentration between the wall and center of the capillary? How does this compare with the difference in urea concentration between the outlet and inlet of the capillary?

Solution. *Initial considerations*: We will assume that a liver capillary can be modeled as a circular tube and that the flux of urea into the blood at the capillary wall is constant along the entire length of the capillary.

System definition and environmental interactions: The system of interest is the blood inside a single capillary. We will assume that urea is not produced within the blood in the capillary.

Apprising the problem to identify governing relationships: The species conservation equation for urea is the appropriate governing relationship. Since the wall flux is constant, we can use the solution developed in Sect. 15.5.1.3. The concentration at any axial and radial location in the capillary (far from the inlet) is given by (15.142).

Analysis: From (15.142), the difference between the concentration at the wall and the concentration at the center of the capillary is equal to $3N_A R_c/4D_{Ab}$:

$$c_{Ab}(z, R_c) - c_{Ab}(z, 0) = \frac{3N_A R_c}{4D_A} = \frac{3\left(0.071 \frac{\mu mol}{min\, cm^2}\right)(4 \times 10^{-4} cm)}{4\left(1.8 \times 10^{-5} \frac{cm^2}{s}\right)\left(60 \frac{s}{min}\right)}$$

$$= 0.0197 \frac{\mu mol}{ml} = .0197 \text{ mM}.$$

The difference in bulk concentration between outlet and inlet from (15.124) or (15.143) is:

$$c_{Ab}(L) - c_{Ab0} = \frac{N_A S_i}{Q_b} = N_A \frac{2\pi R_c L}{<v>\pi R_c^2} = \frac{2N_A L}{<v> R_c}.$$

Substituting the values:

$$c_{Ab}(L) - c_{Ab0} = \frac{2\left(0.071 \frac{\mu mol}{min\, cm^2}\right)(0.1 \text{ cm})}{\left(0.1 \frac{cm}{s}\right)\left(60 \frac{s}{min}\right)(4 \times 10^{-4} cm)} = 5.92 \frac{\mu mol}{ml} = 5.92 \text{ mM}.$$

Examining and interpreting the results: The urea concentration difference between outlet and inlet is 300 times greater than the radial variation in urea concentration between the wall and center of the capillary. From (15.147), the difference in concentration between urea at the wall and bulk urea concentration at any axial position is 0.012 mM. If the inlet urea concentration is zero, the difference between wall and bulk concentrations will be less than 10% of the bulk concentration for all axial locations beyond 2% of the capillary length. In other words, the difference is negligible over 98% of the capillary length, so the 1-D approach should provide reasonably accurate results.

15.5.1.4 Flow Between Parallel Plates with Constant Wall Concentration

Blood flow through alveolar capillaries has been likened to flow between parallel plates. Let us consider steady-state exchange of an anesthetic gas in the lung as blood passes through alveolar capillaries. The concentration of gas at the capillary inlet is c_{A0} and the concentration of gas in blood at the capillary wall is c_{Aw}, which is assumed constant. The capillary walls have a length L and the distance between walls is $2H$. Selecting a rectangular coordinate system with origin between the plates at the inlet, we can define the following dimensionless dependent and independent variables:

$$c_A^* = \frac{c_A - c_{Aw}}{c_{A0} - c_{Aw}}, \quad y^* = \frac{y}{H}, \quad z^* = \frac{z}{L}. \tag{15.151}$$

Assuming fully developed laminar flow, no production of gas in blood, and neglecting axial diffusion, the dimensionless species continuity equation becomes:

$$\frac{3}{2} \frac{\langle v \rangle}{L} (1 - y*^2) \frac{\partial c_A^*}{\partial z^*} = \frac{D_{Ab}}{H^2} \frac{\partial^2 c_A^*}{\partial y^{*2}}. \tag{15.152}$$

The appropriate dimensionless boundary conditions are:

$$z^* = 0 : c_A^* = 1,$$
$$y^* = 0 : \frac{\partial c_A^*}{\partial y} = 0, \tag{15.153}$$
$$y^* = 1 : c_A^* = 0.$$

Nusselt (1923) solved this problem using a separation of variables method, assuming a product solution of the form $c^*_A = Z(z^*) Y(y^*)$. The solution is:

$$\frac{c_A - c_{Aw}}{c_{A0} - c_{Aw}} = \sum_{m=1}^{\infty} B_m Z_m Y_m = \sum_{m=1}^{\infty} B_m \left\{ \exp\left(-\lambda_m^2 \left[\frac{2 D_{Ab} z}{3 \langle v \rangle H^2}\right]\right) \left[\sum_{n=0}^{\infty} \left[A_{m,n} \left(\frac{y}{H}\right)^n \right]\right] \right\}, \tag{15.154}$$

where the coefficients B_m and $A_{n,m}$ are given by:

$$B_m = -\frac{2}{\lambda_m \left(\frac{\partial Y_m}{\partial \lambda_m}\right)_{y^*=1}}, \tag{15.155}$$

$$A_{n,m} = 0 \text{ for odd } n; \quad A_{2,m} = -\frac{\lambda_m^2}{2}; \quad A_{n,m} = \frac{A_{n-2,m} - A_{n-4,m}}{n(n-1)} \text{ for } n \geq 4. \tag{15.156}$$

The first 10 values of λ_m and $(\partial Y/\partial y)_{y=1}$ have been computed to ten decimal places and tabulated by Brown (1960). The solution to this problem can be displayed on a single dimensionless plot of concentration versus position if we define a new dimensionless axial position based on the hydraulic diameter, as first suggested by Graetz (1885):

$$Gz(z) \equiv \left(\frac{1}{\text{Re}_{D_h}\text{Sc}}\right)\frac{z}{D_h} = \frac{D_{Ab}z}{16\langle v\rangle H^2}. \quad (15.157)$$

The hydraulic diameter is equal to $4H$ for flow between parallel plates. The dimensionless axial position Gz is known as the *Graetz number*. Since it is based on the hydraulic diameter, the Graetz number can also be used as a dimensionless axial parameter for ducts of different cross-sections. Plots of c^*_A vs. Gz for several y^* are shown in Fig. 15.12 for the case of flow between parallel plates. Note that the maximum value of $Gz = Gz(L)$ will occur when the end of the duct is reached. Only the portion of the plot between $Gz = 0$ and $Gz = L/[D_h\text{Re}_{D_h}\text{Sc}]$ will apply to a particular problem. The dimensionless bulk or mixing-cup concentration is also shown in Fig. 15.12. Since the velocity is greater at low values of y/H, the bulk concentration weights concentrations near the center more than concentrations near the walls.

In the inlet region between parallel plates, where Gz is smaller than approximately 3×10^{-2}, there can be significant variations in concentration between the center and the walls. Figure 15.12 indicates that the concentration everywhere in

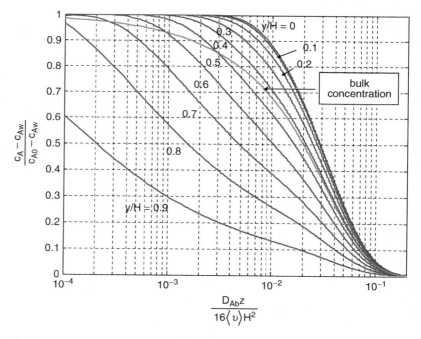

Fig. 15.12 Dimensionless concentration vs. dimensionless axial position for flow between parallel plates with constant wall concentration c_{AW} and constant inlet concentration c_{A0}. Each curve is for a different position in the fluid relative to the center ($y/H = 0$). Dimensionless bulk concentration is also shown

15.5 Convection and Diffusion

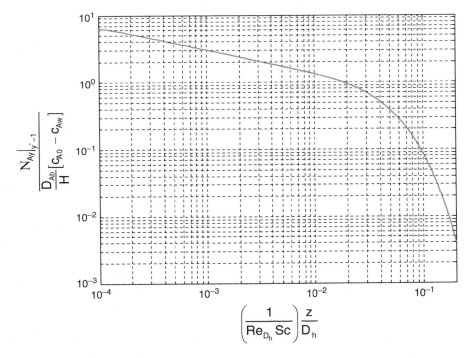

Fig. 15.13 Dimensionless wall flux of species A for flow through horizontal plates with constant wall concentration

the fluid has come within 1% of the wall concentration for $Gz > 0.2$. If this is the desired condition, then the length of the plates must exceed $0.2\, D_h Re_{D_h} Sc$. We can use this to estimate the minimum length that lung capillaries must have for pulmonary blood to become nearly equilibrated with a nonreacting anesthetic gas. For a gas with $D_{Ab} = 10^{-5}$ cm^2/s, blood average velocity $<v>$ of 0.5 mm/s and distance between walls of the capillary sheet of $2H = 10$ μm, we find $L > 40$ μm. This is much shorter than the average length of a pulmonary capillary (approximately 1,000 μm). Therefore, the anesthetic gas should be nearly equilibrated by the time it travels 4% of the length of the capillary.

The flux of species A at either wall can be computed from (15.154). The dimensionless flux as a function of dimensionless axial position is shown in Fig. 15.13. The flux decreases as the difference between bulk concentration and wall concentration decreases. Initially, the flux decreases slowly with axial position, but it drops off quickly for $Gz > 3 \times 10^{-2}$.

These results can be used to compute the dimensionless local mass transfer coefficient Sh_z and the dimensionless mean mass transfer coefficient Sh_m for plates of length L. Using the hydraulic diameter $4H$ as the characteristic length, these are defined as follows:

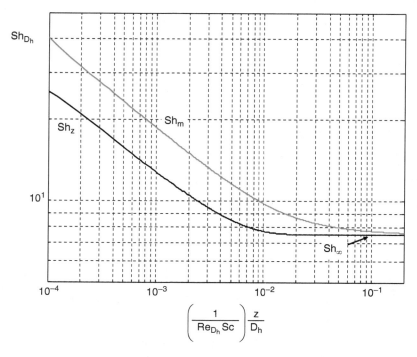

Fig. 15.14 Local and mean Sherwood numbers vs. Graetz number, flow between parallel plates with constant inside wall concentration

$$Sh_z = \frac{4Hk_{Ab}}{D_{Ab}} = \frac{4H}{D_{Ab}}\left(\frac{N_{Ay}|_{y^*=1}}{C_{Ab} - c_{Aw}}\right), \quad (15.158)$$

$$Sh_m = \frac{1}{L}\int_0^L Sh_z dz. \quad (15.159)$$

These are plotted in Fig. 15.14. The mean Sherwood number is higher than the local Sherwood number because the mass transfer coefficient decreases with position. However, for very long plates, both converge to the same constant asymptotic value, $Sh_\infty = 7.54$, which is the value listed in Table 12.1 for parallel plates.

In Sect. 14.5.1, we used a shell balance approach to predict the bulk concentration of species A in fluid flowing through a conduit with constant wall concentration. We neglected concentration variations perpendicular to the direction of flow and neglected the velocity profile. We found that the dimensionless bulk concentration at the exit of the conduit is given by:

$$\frac{C_{Ab}(L) - c_{Aw}}{c_{A0} - c_{Aw}} = \exp\left\{-\frac{S_i}{Q_b}\left[\frac{1}{L}\int_0^L k_{Ab}dx\right]\right\} = \exp\left\{-\frac{1}{\langle v \rangle H}\int_0^L k_{Ab}dz\right\}. \quad (15.160)$$

Which Sherwood number should be used to compute the mass transfer coefficient, k_{Ab}: Sh_z, Sh_m or Sh_∞? Predictions of dimensionless bulk concentration using

15.5 Convection and Diffusion

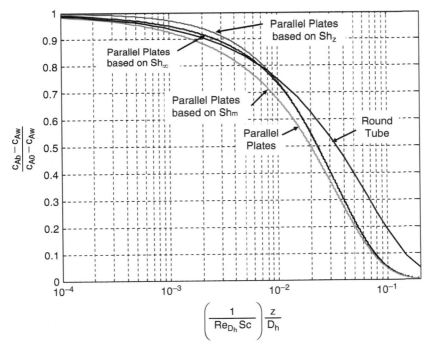

Fig. 15.15 Dimensionless bulk concentration vs. Graetz number for flow between parallel plates and flow in a round tube with constant wall concentration. Also shown are predictions for the parallel plates case based on three different values for Sherwood number

each definition of the Sherwood number are shown in Fig. 15.15 as a function of dimensionless position from the inlet. Note that the concentration predictions based on Sh_z and Sh_∞ are higher (i.e., predict less exchange) for $Gz < 0.2$ than the value predicted with the more extensive theory presented in this section. When Sh_m is used to compute k_A in (15.160), the simpler theory is shown to correspond exactly with the more extensive theory. Proof for this can be found by inserting (15.159) into (15.160) and using the definition of the Graetz parameter, (15.157):

$$\frac{c_{Ab}(L) - c_{Aw}}{c_{A0} - c_{Aw}} = \exp\{-4 Sh_m Gz(L)\}. \quad (15.161)$$

This is identical to (14.129).

Mean Sherwood numbers for conditions other than constant wall concentration and for different geometries can be found in the literature (see Shah and London, 1978 for a summary). The dimensionless bulk concentration in a round tube with constant wall concentration is also shown as a function of Graetz number in Fig. 15.15. Note that the parallel plate and round tube predictions are different, and that the dimensionless concentration for a tube is higher than the dimensionless concentration for parallel plates at all values of Gz. This does not mean that exchange is less efficient in the round tube than for parallel plates, as is shown in Example 15.5.3.

Example 15.5.3 Comparison Between Mass Exchange in a Tube and Mass Transfer Between Plates with Constant Wall Concentration.
A toxic substance in blood is to be removed by passing blood at a rate of 500 ml/min through a bioreactor. Surfaces of the conduit walls in the reactor contain an enzyme that rapidly converts the toxic material to harmless byproducts. Thus, the wall concentration of the toxin can be assumed to be zero. As a young biomedical engineer your company asks you to predict the performance of two different devices. One is to be constructed of parallel pathways through hollow fibers and the other with parallel pathways between sheets of the enzyme-embedded material. The length of each device is 20 cm and the exchange area of each device is 4,000 cm². The fiber inside diameter (D) is 100 μm and the distance between parallel plates ($2H$) is also 100 μm. The width of each plate is 5 cm, the kinematic viscosity of blood is 0.04 cm²/s and the diffusion coefficient for the toxin in blood is 5×10^{-6} cm²/s.

Solution. *Initial considerations*: Since the wall concentration is maintained constant, we can use the analysis developed in Sect. 15.5.1.4.

System definition and environmental interactions: We are analyzing two different systems. The first consists of the blood inside a single hollow fiber and the second is the blood contained between two parallel plates. The inlet concentration is known and the concentration at the wall is maintained at zero by the enzyme.

Apprising the problem to identify governing relationships: We can find the bulk concentration at the exit of the device by computing the Graetz number at the device exit and then using Fig. 15.15 to determine the outlet concentration of the toxic material.

Analysis: The average velocity in each pathway must be known before we can compute Gz. Our first task is to compute the number of parallel pathways for each device. The number of pathways equals the total surface area divided by the surface area of a single pathway:

$$n_{\text{tubes}} = \frac{S_{\text{tot}}}{\pi D L} = \frac{4{,}000 \text{cm}^2}{\pi (0.01 \text{ cm})(20 \text{ cm})} = 6{,}366,$$

$$n_{\text{channels}} = \frac{S_{\text{tot}}}{2wL} = \frac{4{,}000 \text{cm}^2}{2(5 \text{ cm})(20 \text{ cm})} = 20.$$

Velocities in each tube and channel are:

$$\langle v \rangle_{\text{tube}} = \frac{(Q_{\text{total}}/n_{\text{tubes}})}{(\pi D^2/4)} = \frac{(500 \text{ ml/min}/6{,}366)}{\left(\pi (0.01 \text{ cm})^2/4\right)} = 1{,}000 \text{ cm/min},$$

15.5 Convection and Diffusion

$$\langle v \rangle_{channel} = \frac{(Q_{total}/n_{channels})}{(2Hw)} = \frac{(500 \text{ ml/min}/20)}{(0.01 \text{ cm})(5 \text{ cm})} = 500 \text{ cm/min}.$$

The hydraulic diameters are:

$$(D_h)_{tube} = \frac{4A_c}{P_w} = \frac{4\left(\frac{\pi}{4}D^2\right)}{\pi D} = D = 0.01 \text{ cm},$$

$$(D_h)_{channel} = \frac{4A_c}{P_w} = \frac{4(2Hw)}{2w} = 4H = 0.02 \text{ cm}.$$

Reynolds numbers based on the hydraulic diameter are:

$$(Re_{D_h})_{tube} = \frac{\langle v \rangle_{tube} D}{\nu} = \frac{(1{,}000/60 \text{cm/s})(0.01 \text{cm})}{0.04 \text{ cm}^2/\text{s}} = 4.167,$$

$$(Re_{D_h})_{channel} = \frac{\langle v \rangle_{channel}(4H)}{\nu} = \frac{(500/60 \text{cm/s})(0.02 \text{cm})}{0.04 \text{ cm}^2/\text{s}} = 4.167.$$

The Schmidt number is:

$$Sc = \frac{\nu}{D_{toxin,blood}} = \frac{0.04 \text{cm}^2/\text{s}}{5 \times 10^{-6} \text{ cm}^2/\text{s}} = 8{,}000.$$

Finally, Graetz numbers at the outlets of the reactors are:

$$Gz_{tube} \equiv \left(\frac{1}{(Re_{D_h})_{tube} Sc}\right) \frac{L}{(D_h)_{tube}} = \frac{1}{(4.167)(8{,}000)} \left(\frac{20 \text{cm}}{0.01 \text{ cm}}\right) = 0.06,$$

$$Gz_{channel} \equiv \left(\frac{1}{(Re_{D_h})_{channel} Sc}\right) \frac{L}{(D_h)_{channel}} = \frac{1}{(4.167)(8{,}000)} \left(\frac{20 \text{cm}}{0.02 \text{ cm}}\right) = 0.03.$$

Examining and interpreting the results: Since the wall concentration is zero in this case, the dimensionless bulk concentration in Fig. 15.15 is simply the ratio of the outlet to inlet bulk concentrations of the toxin. For toxin emerging from the round tube, Fig. 15.15 predicts an outlet concentration of 33% relative to the inlet concentration. For toxin emerging from the parallel plate bioreactor, Fig. 15.15 predicts an outlet concentration of about 37% of the inlet concentration. The parallel plate reactor is slightly less efficient than the tube reactor.

Additional comments: To increase the efficiency of either reactor, we must increase the Graetz number at the reactor outlet. Other than adjusting the temperature of the blood, we have little control over the Schmidt number. However, we can decrease the concentration of toxin leaving the device by decreasing the hydraulic diameter or decreasing the velocity per pathway. This can be achieved by decreasing the overall blood flow rate through the device or by adding tubes or channels.

15.6 Convection, Diffusion, and Chemical Reaction

Many biological and physiological mass transfer applications involve simultaneous diffusion, convection, and chemical reaction. Analytic solutions are rare and in most cases numerical solutions are required. These are generally published in graphical form and the usefulness of the solutions dictates that the problem be expressed in relevant dimensionless variables and parameters.

15.6.1 Blood Oxygenation in a Hollow Fiber

The transfer of oxygen to blood flowing through a hollow fiber in a blood oxygenator is a classic example. Oxygen surrounding the fiber must first dissolve in the fiber material, then diffuse through the fiber wall before entering the flowing blood. Once in the blood, oxygen diffuses through plasma and then through the erythrocyte membrane, before chemically combining with hemoglobin in the red cell cytoplasm. A detailed analysis of oxygen exchange in this case would require writing mass balances for oxygen in the fiber, plasma, and red cell membrane; and oxyhemoglobin and oxygen balances in the red cell cytoplasm. Since there are millions of red cells, all with different boundaries that are moving, this approach would not be successful. A reasonable set of simplifying assumptions must be introduced if the analysis is to be fruitful. As a starting point we will make the following assumptions:

1. Blood can be treated as a homogeneous fluid rather than a suspension of erythrocytes in plasma. In effect, we treat blood as a hemoglobin solution with a hemoglobin concentration $C_{Hb,blood} = H\, C_{Hb,rbc}$, where H is the hematocrit value. We use an effective diffusion coefficient for oxygen in blood, which accounts for the additional resistance of the red cell membrane and neglect any facilitated diffusion caused by the diffusion of oxyhemoglobin.
2. Oxygen exchange in the hollow fiber is symmetrical about the fiber axis (i.e., is independent of θ).
3. Axial diffusion is negligible, both in the blood and the fiber.
4. Analysis is performed under steady-state conditions.
5. Flow in the fiber is fully developed, so v_r and v_θ are zero and v_z depends only on radial position.
6. Oxygen and oxyhemoglobin are in local equilibrium everywhere within the blood.

15.6 Convection, Diffusion, and Chemical Reaction

7. Oxygen concentration at the outside radius of the fiber R_{fo} is maintained constant at C_o.

With these assumptions, the species continuity equations for the fiber and the blood are:

Oxygen in fiber:
$$0 = D_{O_2,f}\left\{\frac{1}{r}\frac{\partial}{\partial r}\left(r\frac{\partial c_{O_2,f}}{\partial r}\right)\right\}. \tag{15.162}$$

Oxygen in blood:
$$v_z\frac{\partial c_{O_2,b}}{\partial z} = D_{O_2,b}\left\{\frac{1}{r}\frac{\partial}{\partial r}\left(r\frac{\partial c_{O_2,b}}{\partial r}\right)\right\} + R_{O_2}. \tag{15.163}$$

Oxyhemoglobin in blood:
$$v_z\frac{\partial c_{HbO_2,b}}{\partial z} = R_{HbO_2}. \tag{15.164}$$

Let us begin by converting oxygen concentrations to partial pressures in the hollow fiber using Henry's law. Equation (15.162) becomes:

$$\frac{\partial}{\partial r}\left(r\frac{\partial P_{O_2,f}}{\partial r}\right) = 0. \tag{15.165}$$

Integrating this twice, we find:

$$P_{O_2,f}(r,z) = f_1(z)\ln(r) + f_2(z), \tag{15.166}$$

where f_1 and f_2 are arbitrary functions of z to be determined by application of appropriate boundary conditions. Applying the boundary conditions at the outside radius, $P_{O_2,f} = P_o$, and at the inside radius, $P_{O_2,f} = P_{O_2,b}(R_{fi},z)$, we find:

$$\frac{P_{O_2,f}(r,z) - P_o}{P_{O_2,b}(R_{fi},z) - P_o} = \frac{\ln(r/R_{fo})}{\ln(R_{fi}/R_{fo})}. \tag{15.167}$$

Thus, if we can find the partial pressure of O_2 in the blood at the inside surface of the fiber, $P_{O_2,b}(R_{fi},z)$, we can find the partial pressure anywhere in the fiber.

Returning to analysis in the blood, we know that for every mole of HbO_2 produced, 4 moles of O_2 are lost:

$$R_{O_2} = -4R_{HbO_2} = -4\left(v_z\frac{\partial c_{HbO_2,b}}{\partial z}\right). \tag{15.168}$$

Substituting this into (15.163)

$$v_z\left(\frac{\partial c_{O_2,b}}{\partial z} + 4\frac{\partial c_{HbO_2,b}}{\partial z}\right) = D_{O_2,b}\left\{\frac{1}{r}\frac{\partial}{\partial r}\left(r\frac{\partial c_{O_2,b}}{\partial r}\right)\right\}. \tag{15.169}$$

But the oxyhemoglobin concentration gradient can be written in terms of a gradient in oxyhemoglobin saturation, which in turn is related to a gradient in partial pressure of oxygen in blood:

$$\frac{\partial c_{HbO_2,b}}{\partial z} = c_{Hb,tot} \frac{\partial}{\partial z}\left(\frac{c_{HbO_2,b}}{c_{Hb,tot}}\right) = c_{Hb,tot} \frac{\partial S_{HbO_2}}{\partial z} = c_{Hb,tot} \frac{\partial S_{HbO_2}}{\partial P_{O_2,b}} \frac{\partial P_{O_2,b}}{\partial z}. \quad (15.170)$$

$c_{Hb,tot}$ is the total hemoglobin concentration in blood (with and without bound oxygen). Substituting (15.170) into (15.169) and using Henry's law $c_{O_2,b} = \alpha_{O_2,b} P_{O_2,b}$ to convert oxygen concentrations to partial pressures:

$$v_z \left(1 + 4 \frac{c_{Hb,tot}}{\alpha_{O_2,b}} \frac{\partial S_{HbO_2}}{\partial P_{O_2}}\right) \frac{\partial P_{O_2,b}}{\partial z} = D_{O_2,b}\left\{\frac{1}{r}\frac{\partial}{\partial r}\left(r \frac{\partial P_{O_2,b}}{\partial r}\right)\right\}. \quad (15.171)$$

This is more often written in terms of the mass concentration of total hemoglobin $\rho_{Hb,tot}$ (g/dl) and the Bunsen solubility coefficient, $\alpha^*_{O_2,b}$ (mlO$_2$ dl^{-1} mmHg^{-1}), as discussed in Sect. 12.2.1. Using this formulation, and assuming laminar flow with a parabolic velocity profile in the fiber:

$$2\langle v \rangle\left(1 - \left(\frac{r}{R}\right)^2\right)\left(1 + 1.34 \frac{\rho_{Hb,tot}}{\alpha^*_{O_2,b}} \frac{\partial S_{HbO_2}}{\partial P_{O_2}}\right)\frac{\partial P_{O_2,b}}{\partial z} = D_{O_2,b}\left\{\frac{1}{r}\frac{\partial}{\partial r}\left(r \frac{\partial P_{O_2,b}}{\partial r}\right)\right\}. \quad (15.172)$$

Let us introduce a dimensionless radial position relative to the inside radius of the fiber $r^* = r/R_{fi}$ and a dimensionless axial position X, which is related to the Graetz number in (15.157):

$$X = \frac{D_{O_2,b} z}{2\langle v \rangle R_{fi}^2} = \frac{\pi D_{O_2,b} z}{2 Q_b} = \left(\frac{2}{Re_{D_h} Sc}\right)\frac{z}{D_h} = 2Gz. \quad (15.173)$$

A dimensionless partial pressure of oxygen in the blood can be written in terms of the minimum P_{O_2}, which is present at the fiber inlet and the maximum P_{O_2} at the fiber outside radius:

$$P^*_{O_2} = \frac{P_{O_2}(r^*, X) - P_o}{P_{O_2,in} - P_o}. \quad (15.174)$$

With these definitions, the dimensionless form of (15.172) is:

$$(1 + m^*)(1 - r^{*2})\frac{\partial P^*_{O_2,b}}{\partial X} = \left\{\frac{1}{r^*}\frac{\partial}{\partial r^*}\left(r^* \frac{\partial P^*_{O_2,b}}{\partial r^*}\right)\right\}, \quad (15.175)$$

15.6 Convection, Diffusion, and Chemical Reaction

where m^* is the dimensionless slope of the oxyhemoglobin saturation curve:

$$m^* = \frac{1.34 \rho_{Hb,tot}}{\alpha^*_{O_2,b}} \frac{\partial S_{HbO_2}}{\partial P_{O_2,b}} = \frac{1.34 \rho_{Hb,tot} m}{\alpha^*_{O_2,b}}. \tag{15.176}$$

The slope m of the oxyhemoglobin saturation curve is plotted in Fig. 12.10. To find the partial pressure of oxygen in the blood, we must solve (15.175) subject to the following boundary conditions:

1. At $z = 0$: $P_{O_2} = P_{O_2,in}$
2. At $r = 0$: $\partial P_{O_2,b}/\partial r = 0$ \hfill (15.177)
3. At $r = R_{fi}$: $-D_{O_2,b}\alpha_{O_2,b}\partial P_{O_2,b}/\partial r = -D_{O_2,f}\alpha_{O_2,f}\partial P_{O_2,f}/\partial r$

Substituting $P_{O_2,f}$ from (15.167), the third boundary condition can be rewritten in terms of the partial pressure in the blood. The dimensionless forms of the boundary conditions are:

1. At $z^* = 0$: $P^*_{O_2} = 1$
2. At $r^* = 0$: $\partial P^*_{O_2}/\partial r^* = 0$ \hfill (15.178)
3. At $r^* = 1$: $\dfrac{\partial P^*_{O_2,b}}{\partial r^*} = -\dfrac{1}{\gamma} P^*_{O_2,b}$

where γ is a dimensionless fiber wall resistance to the transport of oxygen:

$$\gamma = \left(\frac{D_{O_2,b}\alpha_{O_2,b}}{D_{O_2,f}\alpha_{O_2,f}}\right) \ln\left(\frac{R_{fo}}{R_{fi}}\right). \tag{15.179}$$

The third boundary condition is analogous to the boundary condition for convective mass transfer at a surface [e.g., (14.325)]. An early approach to solving this problem was to use an average value of m in (15.175). In the case where m is constant we can define a new axial variable $X^* = X/(1 + m^*)$, and (15.175) becomes:

$$(1 - r^{*2}) \frac{\partial P^*_{O_2,b}}{\partial X^*} = \left\{\frac{1}{r^*} \frac{\partial}{\partial r^*}\left(r^* \frac{\partial P^*_{O_2,b}}{\partial r^*}\right)\right\}. \tag{15.180}$$

The solution to (15.180) for the boundary conditions in (15.178) was provided by Buckles et al. (1968) and is given in Fig. 15.16 for four different values of γ.

The relationship between $1 + m^*$ and $P_{O_2,b}$ is shown in Fig. 15.17 for blood with a hemoglobin content of 15 g/dl. The Adair relationship, (12.202), was used to compute the slope of the oxyhemoglobin saturation curve (see Fig. 12.10). Note that the values of $1 + m^*$ are generally much greater than one, particularly in the neighborhood of 20 mmHg. Even at a P_{O_2} of 120 mmHg, the value of $1 + m^*$ is

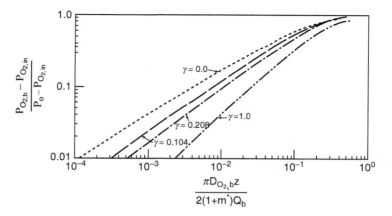

Fig. 15.16 Dimensionless P_{O_2} vs. dimensionless axial position for flow in tube with wall resistance, assuming constant m, from Buckles et al. 1968 with permission

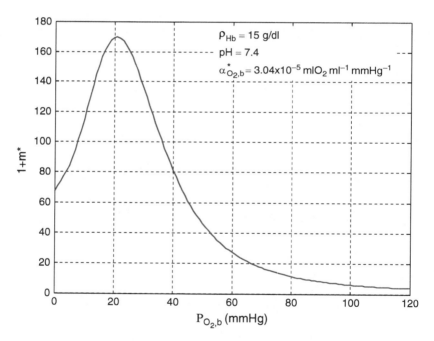

Fig. 15.17 Expected value of $1 + m^*$ as a function of P_{O_2} in normal blood

nearly four times the value if all hemoglobin is fully saturated (i.e., when $m^* = 0$). In addition, the relationship is highly nonlinear, so the notion that a single average value of m^* can be used in (15.175) is not likely to reflect reality for oxygen

15.6 Convection, Diffusion, and Chemical Reaction

transport to blood in a hollow fiber. This approach is more appropriate for CO_2 exchange, since the relationship between total CO_2 concentration and P_{CO_2} is linear (see Problem 15.9.7). However, such an approach can provide a rough quantitative estimate of oxygen exchange. A better approach might be to break the vessel axially into a number of segments, where Fig. 15.16 would be used to find the P_{O2} at the outlet of each segment and Fig. 15.17 used to estimate the appropriate value of $1 + m^*$ to use for each segment. This would necessarily require an iterative scheme where the P_{O_2} at the outlet of each segment would be consistent with the value of m^* used for that segment.

Because of the nonlinear nature of the oxyhemoglobin saturation curve, (15.175) and (15.178) must be solved numerically. Buckles et al. (1968) solved (15.175) for the case where the inlet blood was completely deoxygenated (P_{O_2},in $= 0$), the surrounding gas was pure oxygen ($P_{O_2} = 760$ mmHg), blood hemoglobin concentration was 16 g/dl and $\gamma = 0.2$. The solution is shown in Fig. 15.18, which shows P_{O_2} as a function of dimensionless axial position.

Let us estimate the fiber length required to bring the P_{O_2} from zero at the inlet to 100 mmHg at the outlet if flow through the fiber is 0.4 ml/min. Note that from (15.173), the dimensionless axial position can be computed without knowing the fiber diameter. We will look at three cases: (1) blood with nonfunctional hemoglobin flows through the fiber, (2) blood flows through the fiber under the conditions

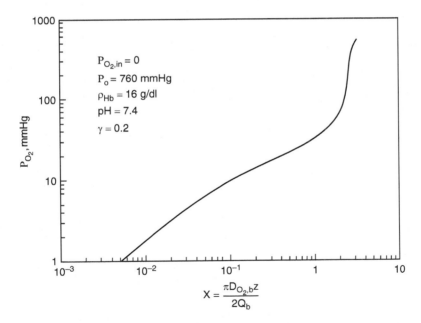

Fig. 15.18 Mixing-cup partial pressure of oxygen vs. axial position in blood flowing through a round tube, from Buckles et al. 1968 with permission

that apply in Fig. 15.18, and (3) blood with an assumed constant value of m flows through the same fiber. In each case we will use the diffusion coefficient for O_2 in blood reported by Weissman and Mockros (1969) of 0.89×10^{-5} cm^2/s. The dimensionless partial pressure at the outlet in Fig. 15.16 for blood with nonfunctioning hemoglobin ($m = 0$) is 100 mmHg/760 mmHg = 0.132. The dimensionless axial position corresponding to this dimensionless P_{O_2} and $\gamma = 0.2$ is about 0.02, corresponding to a fiber length of 9.52 cm. For blood, Fig. 15.18 shows that a partial pressure of 100 mmHg is reached when $X = 2.2$, which corresponds to a fiber length of 1,049 cm. Since the amount of oxygen bound to hemoglobin at 100 mmHg is approximately 100 times greater than the amount that is dissolved in plasma, it is not surprising that the fiber would need to be about 100 times longer to bring blood from 0 to 100 mmHg. If we used Fig. 15.16 based on an average slope from 0 to 100 mmHg ($1 + m^* \approx 70$), our estimated fiber length would be 70 times longer than the nonfunctioning hemoglobin case (70×9.52 cm = 666 cm). To predict the correct length, we would need to use $1 + m^* = 110$, which is well above the average value. Thus, the constant slope assumption will generally underestimate the fiber length necessary to bring the outlet P_{O_2} to an established level. The opposite would be true if the fiber were used to deoxygenate blood.

Numerical solutions of blood oxygenation in fibers show very steep radial gradients in hemoglobin saturation, as shown in Fig. 15.19. At low values of X, a thin annulus of blood near the wall is fully oxygenated, while blood in the core region is still at the inlet saturation value. The fully oxygenated region grows radially as the blood moves downstream. Since the velocity in the central portion

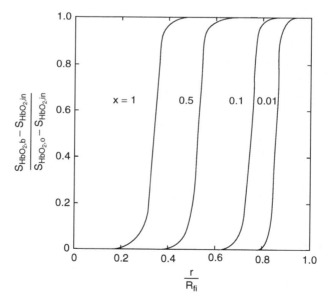

Fig. 15.19 Radial profiles for dimensionless oxyhemoglobin saturation, adapted from Colton and Drake, 1971

of the fiber is faster than near the wall, the mixing-cup saturation value will be heavily weighted by the inlet saturation. This behavior has led Lightfoot (1968) to propose an "Advancing Front" theory, which assumes a step function for the oxyhemoglobin saturation curve to predict the growth of the saturated region with position. This theory will generally overestimate the fiber length necessary to bring the outlet P_{O_2} to an established level during oxygenation. When the numerical solution is not available, a comparison of the two methods will provide upper and lower bounds on the necessary fiber length.

The conditions represented in Fig. 15.18 represent an extreme case. In reality, the inlet P_{O_2} is closer to 40 mmHg, so hemoglobin in blood is nearly 75% saturated when it enters the fiber. Weissman and Mockros (1969) provide the numerical solution for the case where the inlet saturation is 75% and the surrounding P_{O_2} is 735 mmHg. They neglected tube wall effects ($\gamma = 0$), since they found no significant experimental differences in exchange in tubes with outside to inside wall radius ratios that varied from 1.06 to 1.56. A comparison between theoretical and experimental mixing-cup saturation vs. dimensionless axial position is shown in Fig. 15.20.

Returning to our original example, we can use Fig. 15.20 to estimate the fiber length required to bring blood, flowing at a rate of 0.4 ml/min, from 75% to 97% (i.e., $P_{O_2} = 100$ mmHg). The change in saturation is 22%, which corresponds to $X \approx 0.25$. Using $D_{O_2,b} = 0.89 \times 10^{-5}$ cm^2/s, we compute a fiber length of 119 cm, which is considerably shorter than the length of 1,049 cm required to

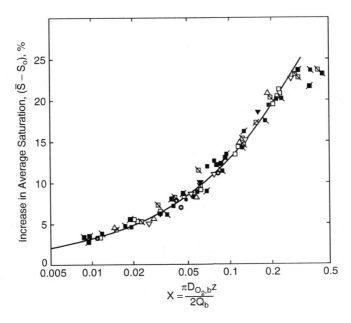

Fig. 15.20 Change in mixing-cup blood oxyhemoglobin saturation in a round hollow fiber. Inlet saturation = 75%, $\rho_{Hb} = 16$ g/dl, $\gamma = 0$, and the surrounding P_{O_2} is 735 mmHg, from Weissman and Mockros 1969 with permission

bring the oxyhemoglobin saturation up from 0% to 97%. Let us compare this with the estimate of fiber length based on the constant slope method. The dimensionless partial pressure for the ordinate in Fig. 15.16 is (100–40)/(735–40) = 0.086. The value of the abscissa is 0.003 for $\gamma = 0$. From Fig. 15.17, the average value of $1 + m^*$ between partial pressures of 40 mmHg and 100 mmHg is about 30. Solving for the length of the fiber, we find 42.9 cm, which is less than half of the necessary length. Therefore, one should be cautious when using the constant slope method for oxygen transport. However, it provides a reasonable order of magnitude estimate when the numerical solution is not available for a particular set of input conditions.

15.7 Summary of Key Concepts

The general unsteady-state, 3-D microscopic species conservation equation is derived in Sect. 15.2. For an incompressible material with constant diffusion coefficient:

$$\frac{Dc_A}{Dt} = D_{AB}\nabla^2 c_A + R_A. \tag{15.17}$$

The general species continuity equation is expanded in rectangular coordinates (15.18), in cylindrical coordinates (15.19), and in spherical coordinates (15.20). The species continuity equation must be coupled with the momentum equation and continuity equation for cases where convection is an important transport mechanism. Imposing important information about the problem at hand will allow the species continuity equation to be simplified for transport in solid materials ($\vec{v} = 0$), for steady flow ($\partial c_A/\partial t = 0$), for negligible diffusion ($D_{AB} \approx 0$), and for no chemical reaction ($R_A = 0$). In addition, reasonable assumptions can be made which further simplify the equation, such as symmetry about an axis or negligible flux in specific directions. Once the equation has been simplified, it cannot be solved it until an initial condition (for unsteady-state transport) and an appropriate set of boundary conditions (Sect. 12.10) are applied. The primary goal of this chapter is to assist the student in identifying and simplifying the appropriate partial differential equation(s) and auxiliary conditions. Analytic solutions are often difficult to obtain, so numerical methods may be required. We present some of the solution methods most commonly used in solving multidimensional problems of interest in bioengineering. We assume that the student has not attempted to solve partial differential equations in the past, so provide relatively detailed explanations for each new procedure.

Diffusion. An example of the use of the separation of variables method and the application of superposition of solutions for two-dimensional diffusion problems are introduced in Sect. 15.3. We also show that the unsteady-state diffusion in tissue of finite dimensions can be found by multiplying solutions from one dimensional unsteady-state diffusion in a slab or cylinder. Such problems can be solved by repeated use of the graphical solutions developed in Appendix D. Problems involving diffusion and chemical reaction can also be approached using the principle of superposition, as illustrated in Sect. 15.4.

Convection, Diffusion, and Reaction. Mass transport in extracorporeal devices and in the major physiological systems, such as the respiratory system, circulatory system, renal system, or the gastrointestinal tract, involves both convection and diffusion, and for many species, chemical reactions as well. Mass transfer from a flat surface with constant concentration to a flowing fluid can be solved by using a combination of variables approach, since a characteristic length does not appear in the problem formulation. The solution for various values of the Schmidt number is shown in Fig. 15.11. Momentum, heat, and mass transport are analogous, and for laminar flow situations the Chilton–Colburn analogy (15.121) can be used to estimate mass transfer coefficients from heat transfer coefficients and vice versa. Mass transfer between fluid flowing in conduits with various shapes and the wall of the conduit are very common in bioengineering applications. Solutions for constant wall flux and constant wall temperature are shown in Sects. 15.5.1.3 and 15.5.1.4, respectively. Graphical solutions for the constant wall concentration case are given in Fig. 15.15 for parallel plates and round tubes. Analysis of blood–tissue oxygen exchange is complex because it must include axial convection, radial diffusion, wall resistance, and hemoglobin oxygen exchange. Methods for estimating the partial pressure of oxygen at the outlet of a blood vessel or hollow fiber are discussed in Sect. 15.6. Numerical solutions are generally required since oxygen exchange depends on blood hemoglobin content and inlet saturation levels.

15.8 Questions

15.7.1. What assumptions were made in the derivation of the general species continuity equation, (15.17)?

15.7.2. Is the general species continuity equation, (15.7), appropriate for non-Newtonian and Newtonian fluids?

15.7.3. Can you distinguish between the derivatives $\partial c_A/\partial t$, dc_A/dt, and Dc_A/Dt, and give examples of each?

15.7.4. Simplify the general species continuity equation for conversion of species A into a nontoxic species inside a spherical cell. The reaction occurs in the cytoplasm and is first order. Species A is introduced into the media surrounding the cell at time $t = 0$. The reaction takes place uniformly in the cytoplasm of a spherical cell. What assumptions are appropriate? What initial and boundary conditions might you suggest?

15.7.5. Simplify the general species continuity equation for the case of 1-D steady-state convection and diffusion of species A in fluid flowing through a cylindrical tube.

15.7.6. Simplify the general species continuity equation for the case when a solid bar with square cross-section, initially devoid of species A, is placed in a solution containing species A. Transport in the bar is by diffusion only and the sides of the bar are much smaller than the length of the bar. What initial condition and boundary conditions are appropriate?

15.7.7. What is the difference between the shell used in the derivation of (15.22) and the shell used to derive (15.24)?

15.7.8. What is the principle of superposition? Can it generally be applied to nonlinear equations?

15.7.9. How would you apply the graphical methods in Appendix D to solve the problem described in Question 15.7.6? What additional information is needed before a solution can be found?

15.7.10. Under what circumstances is it valid to neglect axial diffusion relative to axial convection?

15.7.11. Under what physiological circumstances is Taylor dispersion likely to be important?

15.7.12. What is the Chilton–Colburn analogy and under what circumstances does the analogy break down?

15.7.13. In what region of a tube is the constant wall flux solution presented in Sect. 15.5.1.3 not valid? Why?

15.7.14. Why is the average concentration in a tube with constant wall flux different from the bulk or mixing-cup concentration at the same axial position?

15.7.15. Explain why the bulk concentration curve in Fig. 15.12 (constant wall concentration, flow between parallel plates) is closer to the curves for high y/H at small values of z and closer to the curves for lower values of y/H for higher values of z.

15.7.16. What key assumptions were made in the development of (15.172) for oxygen exchange in blood vessels?

15.7.17. Under what circumstances would the solution to (15.180) (Fig. 15.16) provide a good estimate of the partial pressure of oxygen at the outlet of a blood vessel?

15.9 Problems

15.9.1 Tissue CO_2 Exchange

Simplify the general microscopic equations for steady-state CO_2 exchange in the capillary and tissue regions of a Krogh cylinder. Assume CO_2 is produced at a constant rate per unit volume R_{CO2} (ml CO_2/(ml tissue/min)) in the tissue. Do not neglect axial diffusion. What boundary conditions are needed to solve the problem? Neglect resistance caused by the presence of the capillary wall.

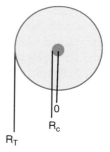

15.9 Problems

15.9.2 Diffusion of Carbon Dioxide in a Tapered Lung Capillary Channel

The steady-state flow of blood between the walls of adjacent alveoli can be modeled as flow between two membranes that converge as you move downstream. The width of the channel, W, is much greater than the height of the channel, $h(x) = b - ax$, so there is no velocity in the z-direction. Assume blood is Newtonian and all properties are constant. CO_2 concentration is zero at the walls and C_0 at the inlet.

(a) Simplify the general microscopic continuity equation, the Navier–Stokes equations for flow in the channel, and the species continuity equation or CO_2 transport in the channel. Neglect diffusion in the x and z directions. Explain why each term is either retained or neglected.
(b) Provide boundary conditions needed to solve the Navier–Stokes equations and species continuity equation. (Do not attempt to integrate and solve these equations)

15.9.3 Tracer Diffusion Through a Vessel Wall

The transient diffusion of an inert tracer through a blood vessel is studied by applying a step change in tracer concentration to the inside surface of the vessel. The vessel wall has thickness δ, which is thin compared with its radius, so curvature can be neglected. Define a new variable $y = r - R_i$, where R_i is the inside radius of the vessel. The concentration of tracer in the fluid bathing the inside surface ($y = 0$) is C_{i1} and the concentration of tracer in the fluid bathing the outside surface ($y = \delta$) is C_0. A steady-state concentration is established in the vessel wall $C_0(y)$. At time $t = 0$, the tracer concentration in the inner fluid is suddenly changed to C_{i2}. Equilibrium partition coefficients ($\Phi = C_{vessel}/C_{fluid}$) are Φ_i at the inner surface and Φ_o at the outer surface.

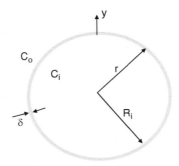

(a) Find the initial distribution of tracer in the vessel wall, $C(y,0)$.
(b) Find the final distribution of tracer in the vessel wall, $C(y,\infty)$.
(c) Transform this problem so that separation of variables can be used to solve for $C(y,t)$. Provide the initial and boundary conditions needed for the transformed variables, and outline how you would use separation of variables to solve the problem (*do not solve*).

15.9.4 Mass Transfer from a Finite Cylinder: Product Solution

A finite solid cylinder with radius R and length $2L$, initially at a concentration C_{As0}, is dropped into a fluid with concentration $C_{Af\infty}$. The mass transfer coefficient in the fluid is k_f, the diffusion coefficient for A in the solid is D_{As}, and the solid is in equilibrium with the fluid at the cylinder surface ($C_{As} = \Phi_{Asf}C_{Af}$). Show that the solution for dimensionless concentration in the finite cylinder can be expressed as the product of the solution for an infinitely long cylinder with radius R and a slab with thickness $2L$. Show that the dimensionless diffusion equation, initial condition, and all boundary conditions are satisfied by the product solution. (Do not solve the pde).

15.9.5 Mass Transfer from a Finite Cylinder

Use the information in problem 15.9.4 to find the concentration of N_2 at the center of a cylindrically shaped tissue with $R = 1$ mm and $L = 4$ mm. The tissue is initially equilibrated with pure N_2 at one atmosphere. At $t = 0$ the tissue is immersed in a stream of water that is completely devoid of N_2. Velocity of the stream is 0.1 cm/s. Use the charts for a slab and a cylinder to find the partial pressure of N_2 in the center of the finite cylinder for a few times after immersion and plot the partial pressure of N_2 vs. time at the center of the tissue. Compare this with P_{N_2} in the center of an infinitely long cylinder with the same radius. Assume the Sherwood number can be computed from:

$$Sh_D = 0.57 Re_D^{\frac{1}{2}} Sc^{\frac{1}{3}}. \tag{15.181}$$

The diffusion coefficient for N_2 in water or tissue is 2.6×10^{-5} cm^2/s. The kinematic viscosity of water is 0.01 cm^2/s. The partition coefficient for N_2 between tissue and water is unity.

15.9.6 Constant Mass Flux to Fluid Flowing Between Parallel Plates

A bioreactor consists of thin parallel membranes, which separate cells from a fluid that flows between the membranes. The cells produce CO_2 at a constant rate per unit

15.9 Problems

volume, so there is a constant flux of CO_2 into the fluid as it flows through the device. If the height of a channel is $2h$, follow the procedure in Sect. 15.5.1.3 to find the concentration profile for CO_2 in the channel and the Sherwood number based on $2h$. Compare this with the value listed in Table 12.1 for parallel plates, where the flux is through both plates.

15.9.7 Carbon Dioxide Transport for Blood Flowing in a Hollow Fiber

Show that (15.175) and Fig. 15.16 can be used to predict blood-CO_2 exchange in a hollow fiber with constant wall P_{CO_2} if the subscripts for oxygen transport are replaced with subscripts for CO_2 transport and $1 + m^* = B/\alpha_{CO_2,b}$, where B is the slope of total CO_2 concentration vs. P_{CO_2} (12.218). Start with species conservation equation for dissolved CO_2 in terms of mass concentration and a second equation for CO_2 carried by all other species. Neglect radial diffusion of all species other than dissolved CO_2 and assume the mass rate of formation of dissolved CO_2 is equal to the rate CO_2 is lost from the carrier species. Add the two species equations and use (12.218) to convert from total molar concentration to partial pressure.

15.9.8 Mass Transfer from a Finite Cylinder: Product Solution

A finite solid cylinder with radius R and length $2L$, initially at a concentration C_{As0}, is dropped into a fluid with concentration $C_{Af\infty}$. The mass transfer coefficient in the fluid is k_f, the diffusion coefficient for A in the solid is D_{As}, and the solid is in equilibrium with the fluid at the cylinder surface ($C_{As} = \Phi_{Asf} C_{Af}$). Show that the solution for dimensionless concentration in the finite cylinder can be expressed as the product of the solution for an infinitely long cylinder with radius R and a slab with thickness $2L$. Show that the dimensionless diffusion equation, initial condition, and all boundary conditions are satisfied by the product solution. (Do not solve the pde).

15.9.9 Design of a Membrane Dialysis Device

You are asked to design a dialysis device that can filter blood as it flows through the device at 20 ml/s. The design criterion is that the device must remove 90% of waste products from the blood. The device will be composed of many functional units in parallel. Each functional unit consists of two parallel membranes that are 10 cm long (L) by 10 cm wide (W), and are separated by 0.05 cm ($2d$). Blood flows between the membranes and dialysis fluid is pumped rapidly along the outside surface of each membrane. The concentration of waste products can be assumed to be zero along the outside surface.

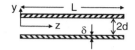

(a) Simplify the species continuity equations for steady-state concentration of waste product within the blood and membrane phases of a single functional unit. Give all boundary conditions necessary to solve for these concentrations, but do not attempt to obtain an analytic solution.

(b) The first term of the infinite series solution is considered to be a good approximation to the total solution near the outlet of the dialyzer, particularly if the membrane resistance to diffusion is small. Find the maximum flowrate per functional unit required to meet the design criterion if the mean concentration at the outlet is given by:

$$C/C_0 = \exp(-1.726(D_A L/(<v>d^2))),$$

where $<v>$ is average blood velocity and D_A is diffusivity of waste product in blood (10^{-6} cm^2/s)

(c) How many functional units are need to meet the design blood flow, and what total blood volume is needed to prime the dialysis unit? What problems do you see with this design and how might you suggest improving it?

15.9.10 Diffusion of CO_2 from Lung Capillaries to Alveolar Gas

Consider steady-state exchange of CO_2 from blood to alveolar gas. As CO_2 diffuses across the alveolar membrane, more CO_2 is produced by homogeneous chemical reactions within the blood, so that:

$$R_{CO_2} = kP_{b,CO_2},$$

where P_{b,CO_2} is the partial pressure of CO_2 in the blood and k is a constant. The Bunsen solubility coefficient (mol ml^{-1} atm^{-1}) for CO_2 in blood is α_b and the solubility coefficient for CO_2 in the alveolar wall is α_w. Lung capillaries can be modeled as channels with constant height h, width W, and length L. No chemical reaction occurs in alveolar walls, which are of constant thickness δ. Assume $W \gg L \gg h \gg \delta$. The P_{CO_2} in alveolar gas is maintained constant at $P_{CO_2,A}$. The P_{CO_2} of blood at the capillary inlet is $P_{CO_2,v}$. Assume blood to be a Newtonian fluid and the pressures at the inlet and outlet of the capillary are kept constant at P_{in} and P_{out}, respectively.

$$R_{CO2} = kP_{b,CO2}$$

```
                              alveolar gas
                              P_CO2 = P_CO2,A
         y      ↓δ      ┌─────────────────────┐
inlet blood    ─→ x  ↑h       blood
P_CO2 = P_CO2,v         
FLOW = Q        δ↑     └─────────────────────┘
                              alveolar gas
                              P_CO2 = P_CO2,A
              ←─────────── L ──────────→
```

(a) Write the species conservation equations for blood (b) and for the alveolar wall (w) in terms of the partial pressure of carbon dioxide. Simplify these in light of the assumptions above (*do not solve*).
(b) Provide appropriate boundary conditions sufficient to solve for the velocity, pressure, and P_{CO2} in the capillary and the P_{CO2} in the alveolar wall.

15.9.11 Hollow Fiber Design for O_2 Transport

Estimate the number of parallel hollow fibers with length 20 cm and inside diameter of 100 μm that are required to bring blood with $\rho_{Hb} = 16$ g/dl from 75% saturation to 95% saturation. Total flow through the device is 1 L/min and the P_{O2} surrounding the fiber is 735 Torr. Neglect wall resistance. How many fibers are needed if the fiber diameter is 200 μm? Explain. What is the average velocity through the device and the blood volume in the device for each fiber diameter?

15.9.12 Hollow Fiber Design for CO_2 Transport

Is it possible to use the same hollow fiber device in the previous problem to reduce the P_{CO_2} from 46 Torr at the inlet to 40 Torr at the outlet of the device? Assume wall resistance is negligible. If it is possible, what partial pressure of CO_2 should be applied around the fibers?

15.10 Challenges

15.10.1 Kidney Dialysis Device

Background. You are asked to design a parallel plate dialysis device that uses cellulose membranes to remove waste product from blood in patients with kidney failure. *Challenge*: Design a dialysis device that can be used three times a week for about 4 h to keep toxic waste products at safe levels. *Generate ideas*: Which solutes should be eliminated and which solutes should be retained? What should be the composition of the dialysis fluid? What quantity of each waste product needs to be removed with each visit? What size membranes will you select? Perform a literature search to estimate the permeability of the membrane to various solutes. How many membranes in parallel will be necessary? What is the total volume of your device? With what fluid will you prime your device before connecting it to the patient? What flow rates for blood and dialysis fluid are appropriate?

15.10.2 Extracorporeal Membrane Oxygenation (ECMO)

Background: The lungs of some premature infants are not developed sufficiently to provide adequate oxygenation. *Challenge*: Design a hollow fiber oxygenator that can be used continuously to keep premature infants alive for several weeks until their lungs become sufficiently developed. *Generate ideas*: What exchange rate is needed for oxygen and carbon dioxide? Where would you connect the inlet and outlet of your device to the infant's circulatory system? What material will you select for the hollow fibers? What is their inside and outside diameter and length? What is the normal blood volume of a premature infant? What is the total volume of your device? What fluid will you fill your device with before it is connected to the infant? What is the permeability of the fibers to oxygen and carbon dioxide? How many fibers will be needed in parallel? What flow rate is needed for air and blood? What gas composition will you use?

References

Brown GM (1960) Heat or mass transfer in a fluid in laminar flow in a circular or flat conduit. AIChE J 6:179–183
Buckles RG, Merrill EW, Gilliland ER (1968) An analysis of oxygen absorption in a tubular membrane oxygenator. AIChE J 25:703–708
Carslaw HS, Jaeger JC (1959) Conduction of heat in solids, 2nd edn. Oxford University Press, London, 1577
Chilton TH, Colburn AP (1934) Mass transfer (absorption) coefficients: prediction from data on heat transfer and fluid friction. Ind Eng Chem 26:1183–1187
Colton CK, Drake RF (1971) Effect of boundary conditions on oxygen transport to blood flowing in a tube. Chem Eng Prog Symp Ser 67(114):88–95
Crank J (1956) The mathematics of diffusion. Clarendon Press, Oxford
Graetz L (1885) Uber die Warmeleitungs fahigkeit van Flussigkeiten. Ann Phys Chem 25:337–357
Kays WM, Crawford W (1980) Convective heat and mass transfer, 2nd edn. McGraw-Hill, New York

References

Lightfoot EN (1968) Low-order approximations for membrane blood oxygenators. AIChE J 14: 669–670

Nusselt W (1923) Der Warmeaustausch am Berieselungskuhler. Z Vereines Deut Ing 67:206–210

Shah RK, London AL (1978) Laminar flow forced convection in ducts. In: Advances in heat transfer. Academic Press, New York

Taylor GI (1953) Dispersion of soluble matter in solvent flowing slowly through a tube. Proc R Soc A 219:186–203

Weissman MH, Mockros LF (1969) Oxygen and carbon dioxide transfer in membrane oxygenators. Med Biol Eng 7:169–184

Appendix

Appendix A Nomenclature

Symbol	Meaning	Fundamental dimensions						SI units	First appearance (bold: definition)
		M	L	T	N	Θ	I		
a	Radius of compliant vessel	0	1	0	0	0	0	m	(7.91)
a_0	Reference radius of compliant vessel	0	1	0	0	0	0	m	(7.91)
a_w	Activity of water in bulk fluid	0	0	0	0	0	0	None	(5.149)·
a'_w	Activity of water in channel	0	0	0	0	0	0	None	(5.149)
A	Area perpendicular to direction of flux	0	2	0	0	0	0	m^2	(2.5)
A	Intercept C_{CO_2} vs. P_{CO_2}	0	−3	0	1	0	0	$mol\ m^{-3}$	**(12.218)**
A	Thermal injury frequency factor	0	0	−1	0	0	0	s^{-1}	**(11.54)**
A	Virial coefficient	0	3	0	−1	0	0	$m^3\ mol^{-1}$	**(5.161)**
A^*	Area available for flow in a bead bed	0	2	0	0	0	0	m^2	(5.90)
A_c	Cross-sectional area	0	2	0	0	0	0	m^2	(2.20)
A_f	Projected frontal area	0	2	0	0	0	0	m^2	(2.38)
A_r	Cylindrical surface area at position r ($2\pi rL$)	0	2	0	0	0	0	m^2	(8.5)
B	Characteristic dimension	0	1	0	0	0	0	m	Sect. 5.6, Sect. 5.8
B	Virial coefficient	0	6	0	−2	0	0	$m^6\ mol^{-2}$	**(5.161)**
B	Slope C_{CO_2} vs. P_{CO_2}	−1	−2	2	1	0	0	$mol\ s^2\ m^{-2}\ kg^{-1}$	**(12.218)**
Bi	Biot number for heat transfer	0	0	0	0	0	0	None	**Table 8.2**
Bi_A	Biot number for mass transfer	0	0	0	0	0	0	None	**(12.170)**

(continued)

Symbol	Meaning	Fundamental dimensions						SI units	First appearance (bold: definition)
		M	L	T	N	Θ	I		
Bi_s	Biot number for mass transfer, porous particle	0	0	0	0	0	0	None	(**14.266**)
c	Total molar concentration	0	−3	0	1	0	0	mol m^{-3}	(5.160), (**12.7**)
c_A	Molar concentration of species A	0	−3	0	1	0	0	mol m^{-3}	(2.7), (**12.4**)
$\bar{c}_A$	Volume averaged molar concentration of species A	0	−3	0	1	0	0	mol m^{-3}	(**12.2**)
c_{Af}	Molar concentration of species A in fluid f	0	−3	0	1	0	0	mol m^{-3}	(2.45)
$[c_{Af}]_m$	Molar bulk concentration of species A in fluid	0	−3	0	1	0	0	mol m^{-3}	(2.43)
$[c_{Af}]_s$	Molar concentration of species A in fluid at solid interface	0	−3	0	1	0	0	mol m^{-3}	(2.42)
$[c_{Af}]_w$	Molar concentration of species A in fluid at conduit wall	0	−3	0	1	0	0	mol m^{-3}	(2.43)
$[c_{Af}]_\infty$	Molar concentration of species A in fluid far from solid surface	0	−3	0	1	0	0	mol m^{-3}	(2.42)
c_{As}	Molar concentration of species A in solid s	0	−3	0	1	0	0	mol m^{-3}	(2.45)
$[c_{As}]_s$	Molar concentration of species A in solid at fluid interface	0	−3	0	1	0	0	mol m^{-3}	(2.48)
$[c_{As}]_o$	Molar concentration of species A in conduit wall at outside surface	0	−3	0	1	0	0	mol m^{-3}	(2.49)
$[c_{As}]_w$	Molar concentration of species A in conduit wall at inside fluid interface	0	−3	0	1	0	0	mol m^{-3}	(2.47)
c_P	Heat capacity per unit mass at constant pressure (specific heat)	0	2	−2	0	−1	0	J kg^{-1} K^{-1}	(2.15)
c_0	Speed of light	0	1	−1	0	0	0	m s^{-1}	(8.74) **Appendix B.1**
C	Compliance	−1	4	2	0	0	0	m^3 Pa^{-1}	(5.124)
C	Heat capacity rate = wc_p	1	2	−3	0	−1	0	J s^{-1} K^{-1}	(10.76)
C_{AB}	Bulk concentration of species A in material B	0	−3	0	1	0	0	mol m^{-3}	(2.52), (**12.123**)
$C_{Hb,tot}$	Total hemoglobin concentration	0	−3	0	1	0	0	mol m^{-3}	(**12.200**)
C_r	Heat capacity ratio	0	0	0	0	0	0	None	(10.79)

(*continued*)

Appendix A Nomenclature

Symbol	Meaning	Fundamental dimensions						SI units	First appearance (bold: definition)
		M	L	T	N	Θ	I		
C_T	Thermal capacitance	1	2	−2	0	−1	0	J K^{-1}	**(9.24)**
C_T	Total concentration of solutes other than water in bulk fluid	0	−3	0	1	0	0	mol m^{-3}	**(5.161)**
CEM 43	Cumulative equivalent minutes of exposure at 43°C	0	0	1	0	0	0	s	(11.57)
C^*	Dimensionless concentration	0	0	0	0	0	0	None	(3.5)
C_A^*	Volumetric fraction of species A	0	0	0	0	0	0	None	**(12.31)**
d	Characteristic dimension	0	1	0	0	0	0	m	Sect. 5.6
d_1, d_2	Major and minor axes of heart valve	0	1	0	0	0	0	m	(3.13)
D, d	Tube diameter	0	1	0	0	0	0	m	(4.43)
D_{AB}	Diffusion coefficient for species A in material B	0	2	−1	0	0	0	m^2 s^{-1}	**(2.7)**
D_h	Hydraulic diameter	0	1	0	0	0	0	m	**(5.62)**
D_p	Particle diameter	0	1	0	0	0	0	m	(5.92)
D_s	Sphere diameter	0	1	0	0	0	0	m	(5.105)
D_{se}	Effective diffusion coefficient for solute s	0	2	−1	0	0	0	m^2 s^{-1}	**(13.259)**
$D_{s\infty}$	Free diffusion coefficient for solute s in an infinite medium	0	2	−1	0	0	0	m^2 s^{-1}	(12.95)
D_{xy}	Rate of deformation in the x-y plane	0	0	−1	0	0	0	s^{-1}	(7.22)
D_0	Time for cell population to drop to $1/e$ of original number	0	0	1	0	0	0	s	(11.58)
D^*	Taylor diffusion coefficient	0	2	−1	0	0	0	m^2 s^{-1}	**(15.94)**
$\vec{e}_i$	Unit vector in the direction of flow for conduit i	0	0	0	0	0	0	None	(5.20)
$\vec{e}_x, \vec{e}_y, \vec{e}_z$ $\vec{e}_r, \vec{e}_\theta, \vec{e}_z$ $\vec{e}_r, \vec{e}_\theta, \vec{e}_\phi$	Unit vectors in rectangular, cylindrical and spherical coordinate systems	0	0	0	0	0	0	None	Sect. 7.11
E	Total energy	1	2	−2	0	0	0	J	**(5.34)**
E	Electrical potential	1	2	−3	0	0	−1	kg m^2 A^{-1} s^{-3}	(2.11)
E	Extraction	0	0	0	0	0	0	None	**(13.125)**
E_b	Black body emissive power	1	0	−3	0	0	0	W m^{-2}	**(8.73)**
E_I	Integral extraction	0	0	0	0	0	0	None	**(14.385)**
E_V	Rate at which energy is lost by friction	1	2	−3	0	0	0	W	(5.47)

(*continued*)

Symbol	Meaning	Fundamental dimensions						SI units	First appearance (bold: definition)
		M	L	T	N	Θ	I		
E_λ	Emissive power at wavelength λ	1	0	−3	0	0	0	W m^{-2}	(8.77)
$E_{\lambda,b}$	Black body monochromatic emissive power at wavelength λ	1	0	−3	0	0	0	W m^{-2}	(**8.74**)
$\dot{E}$	Rate of change of energy in system	1	2	−3	0	0	0	W	(**11.44**)
$\hat{E}$	Total energy per unit mass	0	2	−2	0	0	0	J kg^{-1}	(**5.34**)
$\hat{E}_V$	Energy lost by friction per unit mass	0	2	−2	0	0	0	J kg^{-1}	(5.48), (**5.49**), (5.73)
f	Friction factor	0	0	0	0	0	0	None	(2.37), (**5.55**)
$f(\eta)$	Blasius function	0	0	0	0	0	0	None	(7.101)
f_{HbO_2}	Fraction of hemoglobin capable of combining with O_2	0	0	0	0	0	0	None	(12.214)
$\vec{f}_S$	Force per unit surface area	1	−1	−2	0	0	0	Pa	(5.37)
F	Applied force	1	1	−2	0	0	0	N	(4.18)
F	Faraday's constant	0	0	1	−1	0	1	A s mol^{-1}	(**12.172**), **Appendix B.1**
$F(\eta)$	Blasius function = $df/d\eta$	0	0	0	0	0	0	None	(7.99)
F_D	Drag force	1	1	−2	0	0	0	N	(2.38)
F_f	Friction force	1	1	−2	0	0	0	N	Ex. 7.12.2
F_k	Kinetic force	1	1	−2	0	0	0	N	(5.54)
$F_{m \to n}$	Shape factor	0	0	0	0	0	0	None	**Sect. 8.4.4**
F_P	Pressure force	1	1	−2	0	0	0	N	Ex. 7.12.2
F_s	Force of solid on fluid	1	1	−2	0	0	0	N	(2.37)
F_z	Force on solute	1	1	−2	0	0	0	N	(12.90)
Fo	Fourier number	0	0	0	0	0	0	None	**Table 8.2**
g	Gravitational acceleration	0	1	−2	0	0	0	m s^{-2}	**Sect. 5.3**, **Appendix B.1**
$\vec{g}$	Gravitational vector	0	1	−2	0	0	0	m s^{-2}	(5.26)
G	Shear modulus	1	−1	−2	0	0	0	Pa	(4.2)
G	Irradiation	1	0	−3	0	0	0	W m^{-2}	**Sect. 8.4.3**
G_λ	Irradiation at wavelength λ	1	0	−3	0	0	0	W m^{-2}	**Sect. 8.4.3**
Gr	Grashof number for heat transfer	0	0	0	0	0	0	None	(**8.22**)
Gr_{Af}	Grashof number for mass transfer	0	0	0	0	0	0	None	(**12.111**)
Gz	Graetz number	0	0	0	0	0	0	None	**Table 8.2**, (15.157)
h	Planck constant	1	2	−1	0	0	0	J s	(8.74), **Appendix B.1**

(*continued*)

Appendix A Nomenclature

Symbol	Meaning	Fundamental dimensions						SI units	First appearance (bold: definition)
		M	L	T	N	Θ	I		
h	Convective heat transfer coefficient	1	0	−3	0	−1	0	W m^{-2} K^{-1}	(2.40)
h	Height above datum	0	1	0	0	0	0	m	Sect. 5.4
h	Film thickness	0	1	0	0	0	0	m	Sect. 6.2
h_{ij}	Position of boundary between lung zones i and j	0	1	0	0	0	0	m	(5.144), (5.145)
$\bar{h}$	Average heat transfer coefficient	1	0	−3	0	−1	0	W m^{-2} K^{-1}	Ex 3.2.6.2, (**8.14**)
H	Hematocrit value	0	0	0	0	0	0	None	Sect. 4.9.2
H, h	Distance between parallel plates	0	1	0	0	0	0	m	(3.2), (4.22)
H_A	Henry law constant for species A	1	−1	−2	0	0	0	Pa	(**12.27**)
H_D	Discharge hematocrit	0	0	0	0	0	0	None	(4.44), (**4.47**)
H_F	Feed hematocrit	0	0	0	0	0	0	None	Sect. 4.9.2
H_T	Tube hematocrit	0	0	0	0	0	0	None	(**4.46**)
H_0	Hematocrit in core region	0	0	0	0	0	0	None	(4.48)
$\hat{H}$	Enthalpy per unit mass	0	2	−2	0	0	0	J kg^{-1}	(5.45), (**5.46**)
$\vec{i}$	Current density (vector)	0	−2	0	0	0	1	A m^{-2}	(2.12), (**12.174**)
$\vec{i}, \vec{j}, \vec{k}$	Unit vectors in x, y, z directions, also $\vec{e}_x, \vec{e}_y, \vec{e}_z$	0	0	0	0	0	0	None	(2.8)
$<i>$	Average current density	0	−2	0	0	0	1	A m^{-2}	(**2.29**)
i_s	Number of dissociated ions	0	0	0	0	0	0	None	(5.164)
i_x	Current density in x-direction	0	−2	0	0	0	1	A m^{-2}	(**2.11**)
I	Fundamental dimension of current	0	0	0	0	0	1	A	**Table 3.2**
I or I_x	Electrical current in x-direction	0	0	0	0	0	1	A	(**2.11**)
I	Inertance	1	−4	0	0	0	0	kg m^{-4}	(5.136), (**5.137**)
I	Laser beam intensity	1	0	−3	0	0	0	W m^{-2}	(11.63)
$\bar{I}_z$	Centroidal moment of inertia	1	2	0	0	0	0	kg m^2	(7.14)
$\vec{j}_A$	Mass flux of species A relative to the mass average velocity (diffusive flux)	1	−2	−1	0	0	0	kg m^{-2} s^{-1}	(**12.55**)
$\vec{j}_A^*$	Mass flux of species A relative to the molar average velocity (diffusive flux)	1	−2	−1	0	0	0	kg m^{-2} s^{-1}	(**12.58**)
j_D	j factor for mass transfer	0	0	0	0	0	0	None	(**15.120**)
j_H	j factor for heat transfer	0	0	0	0	0	0	None	(**15.118**)

(*continued*)

Symbol	Meaning	Fundamental dimensions						SI units	First appearance (bold: definition)
		M	L	T	N	Θ	I		
J	Radiosity	1	0	−3	0	0	0	W m^{-2}	(**8.91**)
$\vec{J}_A$	Molar flux of species A relative to mass average velocity (diffusive flux)	0	−2	−1	1	0	0	mol m^{-2} s^{-1}	(2.8), (**12.57**)
$\vec{J}_A^e$	Electrical charge flux contributed by species A	0	−2	0	0	0	1	A m^{-2}	(**12.172**)
$\vec{J}_A^*$	Molar flux of species A relative to molar average velocity (diffusive flux)	0	−2	−1	1	0	0	mol m^{-2} s^{-1}	(**12.56**)
J_{Ax}	Diffusive flux of species A in the x-direction	0	−2	−1	1	0	0	mol m^{-2} s^{-1}	(2.7)
k, k_f	Thermal conductivity, material f	1	1	−3	0	−1	0	W m^{-1} K^{-1}	(2.9)
k_A, k_{Af}	Convective mass transfer coefficient for species A in fluid f (subscript f sometimes dropped)	0	1	−1	0	0	0	m s^{-1}	(2.42), (**12.103**)
$k_{AB,loc}$	Local convective mass transfer coefficient for species A in material B (subscript B sometimes dropped)	0	1	−1	0	0	0	m s^{-1}	(12.116)
k_{AG}	Mass transfer coefficient for species A in gas phase	−1	−1	1	1	0	0	mol s^{-1} m^{-2} Pa^{-1}	(**12.101**)
k_{AL}	Mass transfer coefficient for species A in liquid phase	0	1	−1	0	0	0	m s^{-1}	(**12.102**)
k_{Am}	Mean mass transfer coefficient, species A	0	1	−1	0	0	0	m s^{-1}	(**14.126**)
k_{As}	Convective mass transfer coefficient for species A in solid s	0	1	−1	0	0	0	m s^{-1}	(2.49)
k_{Ax}	Mass transfer coefficient for species A based on mole fraction	0	−2	−1	1	0	0	mol s^{-1} m^{-2}	(**12.102**)
k_B	Boltzman constant	1	2	−2	0	−1	0	kJ K^{-1}	(**8.74**), Appendix B.1
k_e	Electrical conductivity	−1	−3	3	0	0	2	A^2 s^3 kg^{-1} m^{-3}	(2.11)
k_{eff}	Effective thermal conductivity	1	1	−3	0	−1	0	W m^{-1} K^{-1}	(**8.61**), (**8.64**)
k_f, k	Forward reaction rate constant (nth order)	0	3n−3	−1	1−n	0	0	mol^{1-n} s^{-1} m^{3n-3}	(12.181)
k_r	Reverse reaction rate constant, (mth order)	0	3m−3	−1	1−m	0	0	mol^{1-m} s^{-1} m^{3m-3}	(12.183)

(*continued*)

Appendix A Nomenclature

Symbol	Meaning	Fundamental dimensions						SI units	First appearance (bold: definition)
		M	L	T	N	Θ	I		
k_{T_m}	Thermal conductivity evaluated at the mean fluid temperature	1	1	−3	0	−1	0	$W\,m^{-1}\,K^{-1}$	(8.34)
K_1, K_2	Drag coefficients	0	0	0	0	0	0	None	(5.110)
K	Kinetic energy per unit volume	1	−1	−2	0	0	0	$J\,m^{-3}$	(5.54)
K	Flow consistency index, power law fluid and Herschel–Bulkley fluid	1	−1	n−2	0	0	0	$kg\,m^{-1}\,s^{n-2}$	**(4.26)**, **(4.34)**
K_{eq}	Equilibrium coefficient (nth order forward, mth order reverse)	0	3(n−m)	0	m−n	0	0	$mol^{m-n}\,m^{-3(m-n)}$	**(12.195)**
K_f	Filtration coefficient	−1	4	1	0	0	0	$s^{-1}\,Pa^{-1}\,m^3$	(5.154)
K_m	Mass transfer conductance	0	3	−1	0	0	0	$m^3\,s^{-1}$	(12.85)
K_m	Michaelis constant	0	−3	0	1	0	0	$mol\,m^{-3}$	(12.226)
K_w	Friction loss factor	0	0	0	0	0	0	None	(5.72)
$\hat{K}$	Kinetic energy per unit mass	0	2	−2	0	0	0	$J\,kg^{-1}$	(5.34)
L	Fundamental dimension of length	0	1	0	0	0	0	m	Table 3.2
L	Length of control volume	0	1	0	0	0	0	m	(2.29)
L	Electrical inductance	1	2	−2	0	0	−2	$kg\,m^2\,A^{-2}\,s^{-2}$	(5.138)
L_{crit}	Distance from cell center to position of zero oxygen	0	1	0	0	0	0	m	(14.26), (14.28)
L_e	Hydrodynamic entry length	0	1	0	0	0	0	m	(5.70)
L_{et}	Thermal entry length	0	1	0	0	0	0	m	(8.28)
L_i	Hydraulic conductance of channel i	−1	2	1	0	0	0	$kg^{-1}\,m^2\,s$	(5.148)
L_{max}	Cell half thickness for no oxygen at center	0	1	0	0	0	0	m	(14.25)
L'	Fluid conductance per unit length	−1	1	1	0	0	0	$kg^{-1}\,m\,s$	(5.146)
m	System mass	1	0	0	0	0	0	kg	Sect. 2.2
m	Slope of oxyhemoglobin saturation curve	−1	1	2	0	0	0	Pa^{-1}	(13.166)
m	Inverse Biot number	0	0	0	0	0	0	None	**Appendix D**, (14.323)
m, n, p	Finite element indices for x, y, t	0	0	0	0	0	0	None	(11.29), (11.30)
m_A	Mass of species A	1	0	0	0	0	0	kg	(12.1)
m^*	Dimensionless slope of oxyhemoglobin saturation curve	0	0	0	0	0	0	None	(15.176)

(continued)

Symbol	Meaning	Fundamental dimensions						SI units	First appearance (bold: definition)
		M	L	T	N	Θ	I		
M	Fundamental dimension of mass	1	0	0	0	0	0	kg	**Table 3.2**
M	Local average molecular weight	1	0	0	−1	0	0	kg mol^{-1}	**(12.9)**
M_A	Molecular weight of species A	1	0	0	−1	0	0	kg mol^{-1}	(12.1)
M_0	Geometric coefficient for flow in conduits	0	0	0	0	0	0	None	**(5.83)**
M_z	Moment of forces about the z-axis	1	2	−2	0	0	0	kg m^2 s^{-2}	(7.14)
MEC	Minimum effective concentration	0	−3	0	1	0	0	mol m^{-3}	**Sect. 13.6**
MSC	Maximum safe concentration	0	−3	0	1	0	0	mol m^{-3}	**Sect. 13.6**
n	Power law index	0	0	0	0	0	0	None	**(4.26)**, **(4.34)**
n	Dimensionless position	0	0	0	0	0	0	None	**Appendix D**, **(14.321)**
$\vec{n}$	Unit outward normal to system boundary	0	0	0	0	0	0	None	(5.4)
$\vec{n}$	Total mass flux	1	−2	−1	0	0	0	kg m^{-2} s^{-1}	**(12.63)**
$n_A, \vec{n}_A$	Total mass flux of species A	1	−2	−1	0	0	0	kg m^{-2} s^{-1}	**(12.50)**
n_c	Number of cells per liter	0	−3	0	0	0	0	m^{-3}	**(12.246)**
N	Fundamental dimension of molar quantity	0	0	0	1	0	0	mol	**Table 3.2**
N_A	Number of moles of species A	0	0	0	1	0	0	mol	Sect. 2.5.1.3
$\vec{N}$	Total molar flux	0	−2	−1	1	0	0	mol m^{-2} s^{-1}	**(12.62)**
$\vec{N}_A$	Total molar flux of species A	0	−2	−1	1	0	0	mol m^{-2} s^{-1}	**(12.49)**
$\langle N_A \rangle$	Average molar flux of species A	0	−2	−1	1	0	0	mol m^{-2} s^{-1}	(2.36)
$\vec{N}_{A,\text{active}}$	Active flux of species A	0	−2	−1	1	0	0	mol m^{-2} s^{-1}	(12.175)
N_{AV}	Avogadro's number	0	0	0	−1	0	0	Molecules mol^{-1}	(12.2), **Appendix B.1**
N_{Ax}	Molar flux of species A in the x-direction	0	−2	−1	1	0	0	mol m^{-2} s^{-1}	(2.26), **(12.48)**
N_D	Number of fundamental dimensions in a problem	0	0	0	0	0	0	None	(3.17)
N_{inlets}	Number of system conduit inlets	0	0	0	0	0	0	None	(5.16)
N_{outlets}	Number of system conduit outlets	0	0	0	0	0	0	None	(5.16)
N_{species}	Number of molecular species present in system	0	0	0	0	0	0	None	(12.8)

(*continued*)

Appendix A Nomenclature

Symbol	Meaning	Fundamental dimensions						SI units	First appearance (bold: definition)
		M	L	T	N	Θ	I		
N_v	Number of variables in a problem	0	0	0	0	0	0	None	(3.17)
N_Π	Number of independent dimensionless groups in a problem	0	0	0	0	0	0	None	(3.17)
NTU	Number of transfer units	0	0	0	0	0	0	None	(10.78)
Nu_L	Nusselt number for heat transfer based on dimension L	0	0	0	0	0	0	None	**(8.19)**
$(Nu_L)_m$	Nusselt number for mass transfer based on dimension L (Sherwood number)	0	0	0	0	0	0	None	**(12.107)**
Nu_x	Local Nusselt number	0	0	0	0	0	0	None	**(8.43)**
$\vec{p}$	System momentum	1	1	−1	0	0	0	kg m s^{-1}	**(5.19)**
p_i	Stoichiometric coefficient for product i	0	0	0	0	0	0	None	**(12.192)**
$p_{x|n}$	Flux of x-momentum in the n-direction	1	−1	−2	0	0	0	Pa	(4.10)
pH	$\log_{10}(C_{H^+}(M))$	0	0	0	0	0	0	None	(12.223)
P_A, P_A'	Permeability coefficient of species A	0	1	−1	0	0	0	m s^{-1}	(2.50), **(12.159)**
P_A^*	Permeability based on inside to outside driving force	0	1	−1	0	0	0	m s^{-1}	(12.46)
P	Pressure	1	−1	−2	0	0	0	Pa	(2.30)
P_a	Arterial pressure	1	−1	−2	0	0	0	Pa	(5.141)
P_A	Partial pressure of species A	1	−1	−2	0	0	0	Pa	(12.22)
P_A	Alveolar pressure	1	−1	−2	0	0	0	Pa	(5.143)
P_e	External or surrounding pressure	1	−1	−2	0	0	0	Pa	(5.123)
P_I	Interstitial pressure	1	−1	−2	0	0	0	Pa	(6.99)
P_L	Outlet pressure in tube of length L	1	−1	−2	0	0	0	Pa	Sect. 6.3
P_{LA}	Left atrial pressure	1	−1	−2	0	0	0	Pa	(5.142)
P_p	Plasma pressure	1	−1	−2	0	0	0	Pa	(6.99)
$P_{O_2,50}$	Partial pressure of O_2 when 50% of hemoglobin is saturated	1	−1	−2	0	0	0	Pa	(12.212)
P_{PA}	Pulmonary artery pressure	1	−1	−2	0	0	0	Pa	(5.141)
P_R	Pressure at arbitrary reference state	1	−1	−2	0	0	0	Pa	(5.46)
P_{tm}	Transmural pressure	1	−1	−2	0	0	0	Pa	(5.123)
P_v	Venous pressure	1	−1	−2	0	0	0	Pa	(5.142)
$P_{vap,A}$		1	−1	−2	0	0	0	Pa	(12.24)

(*continued*)

Symbol	Meaning	Fundamental dimensions						SI units	First appearance (bold: definition)
		M	L	T	N	Θ	I		
	Vapor pressure of species A								
P_W	Wetted perimeter	0	1	0	0	0	0	m	(5.55)
P_0	Pressure at tube inlet	1	-1	-2	0	0	0	Pa	Sect. 6.3
P^*	Dimensionless pressure	0	0	0	0	0	0	None	**(7.90)**
P'_i	Pressure inside channel at fluid interface	1	-1	-2	0	0	0	Pa	(5.148)
Pe_L	Peclet number based on a characteristic length L	0	0	0	0	0	0	None	**(3.12)**
Pr	Prandtl number	0	0	0	0	0	0	None	**(3.10)**
$\vec{q}$	Heat flux (vector)	1	0	-3	0	0	0	W m^{-2}	(2.10)
$<q>$	Average heat flux	1	0	-3	0	0	0	W m^{-2}	**(2.34)**
q_s	Surface heat flux	1	0	-3	0	0	0	W m^{-2}	(10.31)
q_x	Heat flux in x-direction	1	0	-3	0	0	0	W m^{-2}	**(2.9)**
$\dot{q}_e$	Rate of heat production per unit volume (electric)	1	-1	-3	0	0	0	W m^{-3}	Ex. 10.3.1
$\dot{q}_{met}$	Rate of heat production per unit volume (metabolic)	1	-1	-3	0	0	0	W m^{-3}	(10.3)
Q	Electrical charge	0	0	1	0	0	1	A s	(2.11)
Q_B	Blood flow	0	3	-1	0	0	0	m^3 s^{-1}	(5.146)
Q_G	Glomerular filtration rate	0	3	-1	0	0	0	m^3 s^{-1}	(13.202)
Q_{O_2}	Rate of consumption of oxygen per unit volume	0	-3	-1	1	0	0	mol s^{-1} m^{-3}	(14.14)
Q_V	Volumetric flow rate	0	3	-1	0	0	0	m^3 s^{-1}	(2.30), **(5.7)**, (6.84) (tube)
Q_{wall}	Volumetric inward flow through system wall	0	3	-1	0	0	0	m^3 s^{-1}	(5.12)
Q'_V	Flow per unit width	0	2	-1	0	0	0	m^2 s^{-1}	(7.56)
$\dot{Q}_{conv}$	Convective heat flow	1	2	-3	0	0	0	W	(10.39)
$\dot{Q}_{gen}$	Rate heat is generated within system	1	2	-3	0	0	0	W	(5.36)
$\dot{Q}_{max}$	Maximum possible heat exchange	1	2	-3	0	0	0	W	(10.66)
$\dot{Q}_S$	Heat flow through surface S	1	2	-3	0	0	0	W	(2.40)
$\dot{Q}_x$	Heat flow in x-direction	1	2	-3	0	0	0	W	(2.9)
$\dot{Q}_{1 \to 2}$	Radiation heat exchange between surfaces 1 and 2	1	2	-3	0	0	0	W	(8.94)
r	Radial position	0	1	0	0	0	0	m	(4.46)
r_A		1	-3	-1	0	0	0	kg s^{-1} m^{-3}	**(13.5)**

(*continued*)

Appendix A Nomenclature

Symbol	Meaning	Fundamental dimensions						SI units	First appearance (bold: definition)
		M	L	T	N	Θ	I		
	Mass rate of production of species A per unit volume								
$r_{A,tot}$	Mass rate of production of species A	1	0	−1	0	0	0	kg s^{-1}	**(13.5)**
r_i	Stoichiometric coefficient for reactant i	0	0	0	0	0	0	None	**(12.192)**
R	Universal gas constant	1	2	−2	−1	−1	0	kg m^2 s^{-2} mol^{-1} K^{-1}	(5.149), **Appendix B.1**
R	Tube radius	0	1	0	0	0	0	m	(4.16)
$\vec{R}$	Force by fluid on system walls	1	1	−2	0	0	0	N	(5.26)
R_A	Net molar rate of production per unit volume of species A	0	−3	−1	1	0	0	mol s^{-1} m^{-2}	**(12.185)**
R_{Ad}	Molar rate of decomposition of species A per unit volume	0	−3	−1	1	0	0	mol s^{-1} m^{-2}	(12.183)
R_{Af}	Molar rate of formation of species A per unit volume	0	−3	−1	1	0	0	mol s^{-1} m^{-2}	(12.181)
$R_{A,tot}$	Molar rate of production of species A	0	0	−1	1	0	0	mol s^{-1}	(13.7)
R_c	Red cell radius	0	1	0	0	0	0	m	(4.48)
R_c	Capillary radius	0	1	0	0	0	0	m	Sect. 14.6.1
R_i	Inside radius	0	1	0	0	0	0	m	(6.56)
R_{max}	Cell radius for no oxygen at center	0	1	0	0	0	0	m	(14.40), (14.56)
R_o	Outside radius	0	1	0	0	0	0	m	(6.56)
R_p	Pore radius	0	1	0	0	0	0	m	Sect. 6.3.5
R_s	Sphere radius	0	1	0	0	0	0	m	(5.108)
R_s	Solute radius	0	1	0	0	0	0	m	Sect. 6.3.5
R_T	Krogh cylinder tissue radius	0	1	0	0	0	0	m	Sect. 14.6.1
R_y	Yield radius	0	1	0	0	0	0	m	**(6.87)**
R_W	Outside radius of capillary wall	0	1	0	0	0	0	m	Sect. 14.6.1
Ra_L	Rayleigh number	0	0	0	0	0	0	None	**Table 8.2**
Re_L	Reynolds number based on a characteristic length L	0	0	0	0	0	0	None	**(3.7)**
s	Laplace variable	0	0	−1	0	0	0	s^{-1}	(6.118)
S	Surface area	0	2	0	0	0	0	m^2	(2.39)
S	Casson fluid parameter	1/2	−1/2	−1/2	0	0	0	Pa$^{0.5}$ s$^{0.5}$	**(4.31)**

(*continued*)

Symbol	Meaning	Fundamental dimensions						SI units	First appearance (bold: definition)
		M	L	T	N	Θ	I		
S_{HbO_2}	Oxyhemoglobin saturation	0	0	0	0	0	0	None	**(12.199)**
S_p	Surface area of all particles in a packed bed	0	2	0	0	0	0	m^2	(14.255)
S_w	Wetted surface area	0	2	0	0	0	0	m^2	(2.39)
Sc	Schmidt number	0	0	0	0	0	0	None	Table 8.2, **(12.107)**
Sh_L	Sherwood number based on length L	0	0	0	0	0	0	None	Table 8.2, **(12.107)**
$Sh_{x,loc}$	Local Sherwood number based on position x	0	0	0	0	0	0	None	**(12.116)**
Ste	Stephan number	0	0	0	0	0	0	None	Ex. 10.4.1
t	Time	0	0	1	0	0	0	s	(2.3)
t_a	Appearance time	0	0	1	0	0	0	s	(14.400)
$\bar{t}_A$	Mean transit time for tracer A	0	0	1	0	0	0	s	**(13.116)**
T	Fundamental dimension of time	0	0	1	0	0	0	s	Table 3.2
T	Temperature	0	0	0	0	1	0	K	(2.9)
$\bar{T}$	Average temperature	0	0	0	0	1	0	K	(11.19)
T_b, T_m	Bulk or mixing cup fluid temperature	0	0	0	0	1	0	K	(2.41), **(8.31)**
T_f	Film temperature	0	0	0	0	1	0	K	Sect. 8.3.3.2
T_i	Initial temperature	0	0	0	0	1	0	K	Sect. 9.4.1
T_0	Slab surface temperature ($x = 0$)	0	0	0	0	1	0	K	(8.2)
T_L	Slab surface temperature ($x = L$)	0	0	0	0	1	0	K	(8.2)
T_R	Reference temperature	0	0	0	0	1	0	K	Sect. 2.5.1.3, (2.23)
T_S	Surface temperature	0	0	0	0	1	0	K	(2.40)
T_W	Wall temperature	0	0	0	0	1	0	K	(2.41)
T_∞	Fluid temperature far from a solid surface	0	0	0	0	1	0	K	(2.40)
T^*	Dimensionless temperature	0	0	0	0	0	0	None	(3.4)
T^2	One half the sum of squares of stress components	2	−2	−4	0	0	0	Pa^2	**(7.119)**
U	Internal energy	1	2	−2	0	0	0	J	Sect. 2.5.1.3
$\bar{U}$	Overall heat transfer coefficient	1	0	−3	0	−1	0	$W\ m^{-2}\ K^{-1}$	(10.42)
$\hat{U}$	Internal energy per unit mass	0	2	−2	0	0	0	$J\ kg^{-1}$	**(5.34)**
$\tilde{U} = \rho\hat{U}$	Internal energy per unit volume	1	−1	−2	0	0	0	$J\ m^{-3}$	Sect. 2.5.1.3
U_i	Overall heat transfer coefficient based on inside surface area	1	0	−3	0	0	0	$W\ m^{-2}\ K^{-1}$	(10.51)

(*continued*)

Appendix A Nomenclature

Symbol	Meaning	Fundamental dimensions						SI units	First appearance (bold: definition)
		M	L	T	N	Θ	I		
U_o	Overall heat transfer coefficient based on outside surface area	1	0	−3	0	0	0	W m^{-2} K^{-1}	(10.52)
$\vec{v}$	Mass average velocity vector	0	1	−1	0	0	0	m s^{-1}	(5.4), (**12.51**)
$<v>$	Average velocity	0	1	−1	0	0	0	m s^{-1}	(2.32), (6.83) (tube)
$<v^k>$	Average of v^k over the cross section of a conduit	0	k	−k	0	0	0	m^k s^{-k}	(**5.23**)
$v_A, \vec{v}_A$	Local velocity of species A	0	1	−1	0	0	0	m s^{-1}	(12.48), (12.49)
v_{max}	Maximum velocity	0	1	−1	0	0	0	m s^{-1}	(6.82) (tube)
v_s	Solid velocity sphere velocity	0	1	−1	0	0	0	m s^{-1}	(2.37), (5.109)
v_x, v_y, v_z	Velocity in the x-, y-, and z-directions	0	1	−1	0	0	0	m s^{-1}	(2.13), (7.2)
$\bar{v}_x$	Laplace transform of v_x	0	1	0	0	0	0	m	(6.118)
v_z'	Velocity in core region	0	1	−1	0	0	0	m s^{-1}	(6.97)
v_z''	Velocity in peripheral region	0	1	−1	0	0	0	m s^{-1}	(6.98)
v_0	Superficial velocity	0	1	−1	0	0	0	m s^{-1}	(5.94), (**12.128**)
v_0	Terminal velocity	0	1	−1	0	0	0	m s^{-1}	Ex. 5.10.1, (**12.111**)
v_∞	Fluid velocity far from a solid surface	0	1	−1	0	0	0	m s^{-1}	(2.38)
$\vec{v}^*$	Molar average velocity vector	0	1	−1	0	0	0	m s^{-1}	(**12.53**)
v^*	Dimensionless velocity	0	0	0	0	0	0	None	(3.3)
V	System volume	0	3	0	0	0	0	m^3	(5.3)
V_{max}	Maximum conversion rate per unit volume	0	−3	−1	1	0	0	mol s^{-1} m^{-3}	(12.230)
V_w	Partial molar volume of water	0	3	0	−1	0	0	m^3 mol^{-1}	(5.149)
V^*	Fluid volume in bead bed	0	3	0	0	0	0	m^3	(5.90)
w	Mass flow	1	0	−1	0	0	0	kg s^{-1}	(2.31)
w_A	Mass flow of species A	1	0	−1	0	0	0	kg s^{-1}	(12.44)
w_{in}	Total mass flow into system via conduits	1	0	−1	0	0	0	kg s^{-1}	(5.2), (**5.5**)
w_{out}	Total mass flow out of system via conduits	1	0	−1	0	0	0	kg s^{-1}	(5.2), (**5.6**)
w_{wall}	Total mass flow into system through system walls	1	0	−1	0	0	0	kg s^{-1}	(5.2), (**5.4**)
W	Weight	1	1	−2	0	0	0	N	Ex. 5.10.1
W	Shell width	0	1	0	0	0	0	m	(6.2)

(continued)

Symbol	Meaning	Fundamental dimensions						SI units	First appearance (bold: definition)
		M	L	T	N	Θ	I		
W_A, W_{Ax}	Molar flow of species A in x-direction	0	0	-1	1	0	0	mol s^{-1}	(2.35)
W'_{O_2}	Oxygen consumption per unit capillary length	0	-1	-1	1	0	0	mol s^{-1} m^{-1}	(**14.230**)
$\dot{W}$	Rate work is done by the system on the surroundings	1	2	-3	0	0	0	W	(5.36), (**5.37**)
$\dot{W}_f$	Rate work is done by friction forces applied to the surroundings	1	2	-3	0	0	0	W	(5.39)
$\dot{W}_s$	Rate shaft work is done by the system on the surroundings	1	2	-3	0	0	0	W	(5.38)
We	Weber number	0	0	0	0	0	0	None	**Table 8.2**
x, y, z	Rectangular coordinate directions	0	1	0	0	0	0	m	Sect. 2.3.1
x_A	Mole fraction of species A	0	0	0	0	0	0	None	(5.157), (**12.12**)
x^*, y^*, z^*	Dimensionless rectangular coordinate directions	0	0	0	0	0	0	None	Sect. 3.2.5
x_0, y_0, z_0	Coordinates of a point in space	0	1	0	0	0	0	m	Sect. 2.3.1
X	Extensive property	–	–	–	–	–	–	–	(2.2)
X	Fourier number	0	0	0	0	0	0	None	**Appendix D**, (**14.322**)
$\tilde{X}$	Extensive property per unit volume (an intensive property)	–	–	–	–	–	–	–	(2.16)
y_A	Mole fraction of species A in a gas	0	0	0	0	0	0	None	(**12.20**)
Y	Dimensionless concentration or temperature	0	0	0	0	0	0	None	**Appendix D**, (**14.320**)
z_A	Valence of species A	0	0	0	0	0	0	None	(12.172)
Greek symbols									
α	Thermal diffusivity	0	2	-1	0	0	0	m^2 s^{-1}	(**2.15**)
α	Womersley number	0	0	0	0	0	0	None	(**5.140**)
α	Compliance factor	-1	2	2	0	0	0	m Pa^{-1}	(7.91)
α	Coefficient of absorption	0	0	0	0	0	0	None	(8.79)
α	Rate constant	0	0	-1	0	0	0	s^{-1}	(**13.28**)
α	Relative viscosity exponent	0	0	0	0	0	0	None	(4.44), (**4.45**)
α	Cone angle	0	0	0	0	0	0	None	Ex. 7.16.3
α	Dimensionless dialysis fluid flow rate	0	0	0	0	0	0	None	(**14.152**)
α, β	Inclination angles	0	0	0	0	0	0	None	Sect. 6.2, Sect. 6.3.3

(*continued*)

Appendix A Nomenclature

Symbol	Meaning	Fundamental dimensions						SI units	First appearance (bold: definition)
		M	L	T	N	Θ	I		
α_i	Electrical charge factor	−1	−2	3	0	0	1	$kg^{-1} m^{-2} A s^3$	**(12.177)**
$\alpha_{A,B}$	Solubility of species A in material B	−1	−2	2	1	0	0	$mol\ m^{-3} Pa^{-1}$	**(12.28)**
α_z	Angular acceleration about the z-axis	0	0	−2	0	0	0	s^{-2}	(7.14)
$\alpha^*_{A,B}$	Bunsen solubility coefficient for species A in material B	−1	1	2	0	0	0	Pa^{-1}	**(12.32)**
β	Coefficient of thermal expansion	0	0	0	0	−1	0	K^{-1}	(8.16)
β	Concentration of solute inside pore relative to Michaelis constant	0	0	0	0	0	0	None	**(14.98)**
β	Blood flow relative to permeability-surface area	0	0	0	0	0	0	None	**(14.140)**
β	Absorption coefficient	0	−1	0	0	0	0	m^{-1}	**(11.63)**
β_s	Modified Peclet number for solute s in a pore	0	0	0	0	0	0	None	**(13.262)**
δ	Boundary layer or film thickness	0	1	0	0	0	0	m	Ex. 6.2.6.1, (14.6.1)
δ_m	Membrane thickness	0	1	0	0	0	0	m	(12.146)
δV	Critical continuum volume	0	3	0	0	0	0	m^3	Sect. 2.3.1
δN_i	Number of molecules of species i in volume δV	0	0	0	0	0	0	None	(12.5)
Δc_A	Concentration difference of species A (inlet minus outlet)	0	−3	0	1	0	0	0	(2.35)
ΔE	Electrical potential difference between two points (inlet minus outlet)	1	2	−3	0	0	−1	$kg\ m^2 A^{-1} s^{-3}$	2.28
ΔE	Activation energy for injury	1	2	−2	−1	0	0	$J\ mol^{-1}$	(11.54)
ΔL	Length of fluid element that enters control volume in time Δt	0	1	0	0	0	0	m	(2.20)
Δm	Change in mass	1	0	0	0	0	0	kg	Sect. 2.3.1
ΔP	Pressure difference (inlet minus outlet)	1	−1	−2	0	0	0	Pa	(2.30)
$\Delta P'$	Pressure difference in core region	1	−1	−2	0	0	0	Pa	(6.97)
$\Delta P''$	Pressure difference in peripheral region	1	−1	−2	0	0	0	Pa	(6.96)
Δr	Cylindrical shell thickness	0	1	0	0	0	0	m	(6.33)
Δt	Time increment	0	0	1	0	0	0	s	Sect. 2.4
ΔT	Temperature difference (inlet minus outlet)	0	0	0	0	1	0	K	(2.33)

(*continued*)

Symbol	Meaning	Fundamental dimensions						SI units	First appearance (bold: definition)
		M	L	T	N	Θ	I		
ΔT_{lm}	Log mean temperature difference	0	0	0	0	1	0	K	(**10.63**)
ΔU	Change in internal energy	1	2	−2	0	0	0	J	(2.23)
ΔV	Volume difference	0	3	0	0	0	0	m^3	Sect. 2.3.1
$\Delta x, \Delta y, \Delta z$	Rectangular shell dimensions	0	1	0	0	0	0	m	(6.2), (7.1)
Δz	Pore length	0	1	0	0	0	0	m	(13.249)
$\Delta \Pi_s$	Partial osmotic pressure difference, solute s	1	−1	−2	0	0	0	Pa	(6.108)
ε	Void fraction	0	0	0	0	0	0	None	(5.90), (**12.127**)
ε	Emissivity	0	0	0	0	0	0	None	(**8.78**)
ε	Effectiveness of heat exchanger	0	0	0	0	0	0	None	**Sect. 10.4.3.2**
ε_{fin}	Effectiveness of heat exchange from a fin	0	0	0	0	0	0	None	(**10.100**)
ε_x	Strain in the x-direction	0	0	0	0	0	0	None	(7.25)
ε_λ	Emissivity at wavelength λ	0	0	0	0	0	0	None	(**8.77**)
$\dot{\varepsilon}_x$	Strain rate in the x-direction $= D_{xx}$	0	0	−1	0	0	0	s^{-1}	(**7.26**)
ϕ	Angular coordinate, spherical coordinates	0	0	0	0	0	0	None	Sect. 7.3, **Sect. 7.11**
ϕ	Dimensionless function	0	0	0	0	0	0	None	(3.14)
ϕ_A	Henry's law constant	0	0	0	0	0	0	None	(**12.26**)
ϕ_p	Fraction of pellet volume occupied by pores	0	0	0	0	0	0	None	(14.89)
ϕ_T	Thiele modulus	0	0	0	0	0	0	None	(**14.99**)
Φ	Fluence rate	1	0	−3	0	0	0	W m^{-2}	(11.69)
Φ_{A12}	Equilibrium partition coefficient of species A in material 1 relative to material 2	0	0	0	0	0	0	None	(**2.44**)
Φ_s	Steric partition coefficient	0	0	0	0	0	0	None	(**6.104**)
$\hat{\Phi}$	Potential energy per unit mass	0	2	−2	0	0	0	J kg^{-1}	(**5.34**)
γ	Shear strain	0	0	0	0	0	0	None	(4.1)
γ	Combined friction loss factor	0	0	0	0	0	0	None	(**5.79**)
γ	Dimensionless parameter, first-order reaction	0	0	0	0	0	0	None	(**14.288**)
γ	Dimensionless extravascular volume	0	0	0	0	0	0	None	(**14.388**)
γ	Dimensionless wall resistance to O$_2$ transfer	0	0	0	0	0	0	None	(**15.179**)
γ_X	Transport coefficient for flux of X	0	2	−1	0	0	0	m^2 s^{-1}	(2.16)

(*continued*)

Appendix A Nomenclature

Symbol	Meaning	Fundamental dimensions						SI units	First appearance (bold: definition)
		M	L	T	N	Θ	I		
$\dot{\gamma}, \dot{\gamma}_{nx}$	Shear rate on a plane of constant n in the x-direction	0	0	−1	0	0	0	s^{-1}	(2.13), **(4.4)**
$\dot{\gamma}_0$	Constant $= 1 \text{ s}^{-1}$	0	0	−1	0	0	0	s^{-1}	Ex. 4.8.3.1
$\dot{\gamma}^2$	Twice the sum of squares of rate of deformation components (total shear rate squared)	0	0	−2	0	0	0	s^{-2}	**(7.118)**
η	Apparent viscosity or effective viscosity	1	−1	−1	0	0	0	Pa s	(4.12)
η	Combination of variables	0	0	0	0	0	0	None	**(7.100)**, (7.102)
η	Effectiveness factor	0	0	0	0	0	0	None	**(14.114)**
η_r	Relative viscosity	0	0	0	0	0	0	None	**(4.40)**
κ	Ratio of outside to inside radius, annulus	0	0	0	0	0	0	None	(6.57)
λ	Wavelength	0	1	0	0	0	0	m	(8.74)
λ_{max}	Wavelength at which maximum emission occurs	0	1	0	0	0	0	m	(8.75)
λ_k	Eigenvalue	0	0	0	0	0	0	None	(10.121)
K_{ki}	Integration factor, velocity profile	0	0	0	0	0	0	None	(5.22), **(5.23)**
Λ	Latent heat of fusion for water	0	2	−2	0	0	0	$J\text{ kg}^{-1}$	Ex. 10.4.1
μ	Fluid viscosity	1	−1	−1	0	0	0	Pa s	(2.13)
μ_a	Absorption coefficient	0	−1	0	0	0	0	m^{-1}	(11.69)
μ_p	Plasma viscosity	1	−1	−1	0	0	0	Pa s	(4.40)
μ_s	Viscosity evaluated at surface temperature	1	−1	−1	0	0	0	Pa s	(8.38)
μ_s	Mobility of solute s	−1	0	1	0	0	0	$s\text{ kg}^{-1}$	**(12.94)**
μ_0	Viscosity-like parameter in Bingham model	1	−1	−1	0	0	0	Pa s	**(4.28)**
μ_{T_m}	Viscosity evaluated at mean fluid temperature	1	−1	−1	0	0	0	Pa s	(8.38)
ν	Kinematic viscosity	0	2	−1	0	0	0	$m^2\text{ s}^{-1}$	(2.14)
π	Ratio of circle circumference to diameter $= 3.14159...$	0	0	0	0	0	0	None	—
Π	Osmotic pressure	1	−1	−2	0	0	0	Pa	**(5.150)**
Π_I	Interstitial osmotic pressure	1	−1	−2	0	0	0	Pa	(6.99)
Π_p	Plasma osmotic pressure	1	−1	−2	0	0	0	Pa	(6.99)
Π_v	Dimensionless group containing excluded variable v	0	0	0	0	0	0	None	(3.19)

(*continued*)

Symbol	Meaning	Fundamental dimensions						SI units	First appearance (bold: definition)
		M	L	T	N	Θ	I		
Π'	Osmotic pressure inside channel	1	−1	−2	0	0	0	Pa	(**5.151**)
θ	Angular coordinate, cylindrical coordinates	0	0	0	0	0	0	None	Sect. 6.3, **Sect. 7.11**
θ	Angular coordinate, spherical coordinates	0	0	0	0	0	0	None	Sect 7.3, **Sect. 7.11**
θ	Temperature difference, $T-T_\infty$	0	0	0	0	1	0	K	(9.22)
θ	Sieving coefficient	0	0	0	0	0	0	None	(13.213), Fig. 13.26
θ_i	Temperature difference, T_i-T_∞	0	0	0	0	1	0	K	(9.22)
θ_{in}	Temperature difference, $T_s-T_{m,in}$	0	0	0	0	1	0	K	(10.36)
θ_{lm}	Log mean temperature difference	0	0	0	0	1	0	K	(10.41)
θ_{out}	Temperature difference, $T_s-T_{m,out}$	0	0	0	0	1	0	K	(10.36)
Θ	Fundamental dimension of temperature	0	0	0	0	1	0	K	**Table 3.2**
ρ	Mass density or total mass concentration	1	−3	0	0	0	0	kg m^{-3}	(2.1), (**12.8**)
ρ	Coefficient of reflection	0	0	0	0	0	0	None	(8.79)
ρ_A	Mass concentration of species A	1	−3	0	0	0	0	kg m^{-3}	(4.42), (**12.6**)
$\bar{\rho}_A$	Volume averaged mass concentration of species A	1	−3	0	0	0	0	kg m^{-3}	**12.3**
σ	Surface tension	1	0	−2	0	0	0	kg s^{-2}	Table 8.2
σ	Stefan–Boltzmann constant	1	0	−3	0	−4	0	W m^{-2} K^{-4}	(8.73), **Appendix B.1**
σ_d	Overall osmotic reflection coefficient	0	0	0	0	0	0	None	(5.156)
σ_{di}	Overall osmotic reflection coefficient for channel i	0	0	0	0	0	0	None	(5.152), (**5.153**)
σ_{ds}	Osmotic reflection coefficient for solute s	0	0	0	0	0	0	None	(6.108)
σ_s	Solute reflection coefficient	0	0	0	0	0	0	None	(13.258)
Σ	Dimensionless quantity of heat or mass that crosses a solid–fluid boundary	0	0	0	0	0	0	None	**Appendix D**, (**14.335**)
τ	Coefficient of transmission	0	0	0	0	0	0	None	(8.79)
τ	Thermal injury exposure time	0	0	1	0	0	0	s	(11.55)
τ_T	Thermal time constant	0	0	1	0	0	0	s	(9.26)

(*continued*)

Appendix A Nomenclature

Symbol	Meaning	Fundamental dimensions						SI units	First appearance (bold: definition)
		M	L	T	N	Θ	I		
τ_w	Wall shear stress	1	−1	−2	0	0	0	Pa	(4.16)
τ_y	Yield stress	1	−1	−2	0	0	0	Pa	**(4.28)**
τ_{yx}	Flux of x-momentum in the y-direction, or shear stress in the x-direction on a plane of constant y	1	−1	−2	0	0	0	Pa	(2.13)
ω	Angular frequency	0	0	−1	0	0	0	s^{-1}	Ex. 5.3.1
ω_A	Mass fraction of species A	0	0	0	0	0	0	None	**(12.13)**
Ω	Cone angular velocity	0	0	−1	0	0	0	s^{-1}	Ex. 7.16.3
Ω	Arrhenius thermal injury function	0	0	0	0	0	0	(None)	(11.54)
Ψ	Intensive property	–	–	–	–	–	–	–	(2.6)
Ψ	Stream function	0	2	−1	0	0	0	$m^2\,s^{-1}$	Sect. 7.9
Ψ	Dimensionless flux of heat or mass from solid to fluid	0	0	0	0	0	0	(None)	**Appendix D, (14.333)**
Ψ^*	Dimensionless concentration in pore	0	0	0	0	0	0	(None)	**(14.102)**
ζ_A	Coefficient of compositional expansion	0	0	0	0	0	0	None	**(12.112)**
Special symbols									
$\vec{\nabla}$	Nabla or del operator	0	−1	0	0	0	0	m^{-1}	(2.8), **(7.38)**
∇^2	Laplacian operator	0	−2	0	0	0	0	m^{-2}	**(7.46)**
ℓ	Tube length	0	1	0	0	0	0	m	(5.137)
$\mathbb{N}_A$	Number of molecules of species A	0	0	0	0	0	0	None	(12.1)
$\wp$	Modified pressure	1	−1	−2	0	0	0	Pa	**(5.50)**
$\Re, \Re_f$	Fluid resistance	1	−4	−1	0	0	0	$kg\,m^{-4}\,s^{-1}$	**(2.30)**, (5.84) (tube)
$\Re'$	Resistance per unit length	1	−5	−1	0	0	0	$kg\,m^{-5}\,s^{-1}$	**(5.132)**
$\Re_{cap}$	Capillary resistance	1	−4	−1	0	0	0	$kg\,m^{-4}\,s^{-1}$	(5.143)
$\Re_{AB}$	Resistance to transport of species A in material B (subscript B sometimes dropped)	0	−3	1	0	0	0	$s\,m^{-3}$	**(2.35)**, Sect. 2.6
$\Re_e'$	Electrical resistivity	1	3	−3	0	0	−2	$kg\,m^3\,A^{-2}\,s^{-3}$	(2.11)
$\Re_e$	Electrical resistance	1	2	−3	0	0	−2	$kg\,m^2\,A^{-2}\,s^{-3}$	(2.28)
$\Re_T$	Thermal resistance	−1	−2	3	0	1	0	$K\,W^{-1}$	(2.33)
$\Re_{T,cond}$	Thermal resistance via conduction	−1	−2	3	0	1	0	$K\,W^{-1}$	(8.8)
$\Re_{T,conv}$	Thermal resistance via convection	−1	−2	3	0	1	0	$K\,W^{-1}$	(8.69)
$\mathfrak{J}$	Torque	1	2	−2	0	0	0	$kg\,m^2\,s^{-2}$	Ex. 7.12.1

Appendix B.1 Physical Constants

Symbol	Description	Value
c_0	Speed of light	2.998×10^8 m s^{-1}
e	Electron charge	1.602×10^{-19} C
F	Faraday's constant	96,485.34 C mol^{-1}
g	Gravitational acceleration	9.807 m s^{-1}
h	Planck constant	6.636×10^{-34} J s
k_B	Boltzman constant	1.381×10^{-23} J K^{-1}
R	Gas constant	8.315 J mol^{-1} K^{-1}
		82.057 ml atm mol^{-1} K^{-1}
		62,364 ml mmHg mol^{-1} K^{-1}
N_{AV}	Avogadro's number	6.022×10^{23} molecules mol^{-1}
σ	Stefan–Boltzmann constant	5.67×10^{-8} W m^{-2} K^{-4}

Appendix B.2 Prefixes and Multipliers for SI Units

Prefix	Multiplier	Scale	Symbol
Yocto	10^{-24}	Septillionth	y
Zepto	10^{-21}	Sextillionth	z
Atto	10^{-18}	Quintillionth	a
Femto	10^{-15}	Quadrillionth	f
Pico	10^{-12}	Trillionth	p
Nano	10^{-9}	Billionth	n
Micro	10^{-6}	Millionth	μ
Milli	10^{-3}	Thousandth	m
Centi	10^{-2}	Hundredth	c
Deci	10^{-1}	Tenth	d
	10^{0}	One	
Deca	10^{1}	Ten	da
Hecto	10^{2}	Hundred	h
Kilo	10^{3}	Thousand	k
Mega	10^{6}	Million	M
Giga	10^{9}	Billion	G
Tera	10^{12}	Trillion	T
Peta	10^{15}	Quadrillion	P
Exa	10^{18}	Quintillion	E
Zetta	10^{21}	Sextillion	Z
Yotta	10^{24}	Septillion	Y

Appendix B.3 Conversion Factors

Common units used in biotransport with conversion to SI base units. (Fundamental dimensions and SI units are in bold).

Physical quantity	Common unit	Symbol	In terms of basic SI units
Area	Square meter	$1\ m^2$	$1\ m^2$
	Square foot	$1\ ft^2$	$0.0929\ m^2$
	Square inch	$1\ in^2$	$6.45 \times 10^{-4}\ m^2$
	Square centimeter	$1\ cm^2$	$1 \times 10^{-4}\ m^2$
Concentration (molar)	Moles per cubic meter	$1\ mol\ m^{-3}$	$1\ mol\ m^{-3}$
	Kilomoles per cubic meter	$1\ kmol\ m^{-3}$	$1 \times 10^{3}\ mol\ m^{-3}$
	Moles per milliliter	$1\ mol\ ml^{-1}$ or $1\ mol\ cm^{-3}$	$1 \times 10^{6}\ mol\ m^{-3}$
	Moles per liter	$1\ mol\ L^{-1} = 1\ M$	$1 \times 10^{3}\ mol\ m^{-3}$
	Pound mole per cubic foot	$1\ lb\text{-}mol\ ft^{-3}$	$1.602 \times 10^{4}\ mol\ m^{-3}$
Density or mass concentration	Kilograms per cubic meter	$1\ kg\ m^{-3}$	$1\ kg\ m^{-3}$
	Grams per milliliter	$1\ g\ ml^{-1}$ or $1\ g\ cm^{-3}$	$1{,}000\ kg\ m^{-3}$
	Pound mass per cubic foot	$1\ lb_m\ ft^{-3}$	$16.02\ kg\ m^{-3}$
Diffusion coefficient, kinematic viscosity, thermal diffusivity	–	$1\ m^2\ s^{-1}$	$1\ m^2\ s^{-1}$
		$1\ cm^2\ s^{-1}$	$1 \times 10^{-4}\ m^2\ s^{-1}$
		$1\ ft^2\ s^{-1}$	$0.0929\ m^2\ s^{-1}$
Electrical capacitance	Farad	$1\ F$	$1\ A^2\ s^4\ kg^{-1}\ m^{-2}$
Electrical charge	Coulomb	$1\ C$	$1\ A\ s$
Electrical charge density	–	$1\ C\ m^{-3}$	$1\ A\ s\ m^{-3}$
Electrical current	**Ampere**	$1\ A$	$1\ A$
Electrical current density	–	$1\ A\ m^{-2}$	$1\ A\ m^{-2}$
Electrical inductance	Henry	$1\ H$	$kg\ m^2\ A^{-2}\ s^{-2}$
Electrical potential	Volt	$1\ V$	$1\ kg\ m^2\ A^{-1}\ s^{-3}$
Electrical resistance	Ohm	$1\ \Omega$	$1\ kg\ m^2\ A^{-2}\ s^{-3}$
Energy or heat (see also work)	Joule	$1\ J$	$1\ kg\ m^2\ s^{-2}$
	Erg	$1\ erg$	$1 \times 10^{-7}\ kg\ m^2\ s^{-2}$
	Kilowatt hour	$1\ kW\ h$	$3.6 \times 10^{6}\ kg\ m^2\ s^{-2}$
	British thermal units	$1\ Btu$	$1{,}054\ kg\ m^2\ s^{-2}$
	Calorie	$1\ cal$	$4.184\ kg\ m^2\ s^{-2}$
	Kilocalorie	$1\ kcal$	$4{,}184\ kg\ m^2\ s^{-2}$
Force	Newton	$1\ N$	$1\ kg\ m\ s^{-2}$
	Dyne	$1\ dyn$	$1 \times 10^{-5}\ kg\ m\ s^{-2}$
	Pound force	$1\ lb_f$	$4.448\ kg\ m\ s^{-2}$
	Poundal	$1\ lb_m\ ft\ s^{-2}$	$0.13826\ kg\ m\ s^{-2}$
	–		$1\ m^2\ s^{-2}\ K^{-1}$

(continued)

Physical quantity	Common unit	Symbol	In terms of basic SI units
Heat capacity per unit mass (Specific heat)		$1\ \text{J kg}^{-1}\ \text{K}^{-1}$ or $1\ \text{W s kg}^{-1}\ \text{K}^{-1}$	
		$1\ \text{kJ kg}^{-1}\ \text{K}^{-1}$	$1{,}000\ \text{m}^2\ \text{s}^{-2}\ \text{K}^{-1}$
		$1\ \text{Btu lb}_m^{-1}\ °\text{F}^{-1}$	$4{,}184\ \text{m}^2\ \text{s}^{-2}\ \text{K}^{-1}$
		$1\ \text{cal g}^{-1}\ \text{K}^{-1}$	$4{,}184\ \text{m}^2\ \text{s}^{-2}\ \text{K}^{-1}$
Heat transfer coefficient	–	$1\ \text{J s}^{-1}\ \text{m}^{-2}\ \text{K}^{-1}$ or $1\ \text{W m}^{-2}\ \text{K}^{-1}$	$1\ \text{kg s}^{-3}\ \text{K}^{-1}$
		$1\ \text{cal cm}^{-2}\ \text{s}^{-1}\ \text{K}^{-1}$	$4.184 \times 10^4\ \text{kg s}^{-3}\ \text{K}^{-1}$
		$1\ \text{lb}_m\ \text{s}^{-3}\ °\text{F}^{-1}$	$0.816\ \text{kg s}^{-3}\ \text{K}^{-1}$
		$1\ \text{Btu ft}^{-2}\ \text{h}^{-1}\ °\text{F}^{-1}$	$5.68\ \text{kg s}^{-3}\ \text{K}^{-1}$
Length	Meter	$1\ \textbf{m}$	$1\ \text{m}$
	Angstrom	$1\ \text{A}$	$1 \times 10^{-10}\ \text{m}$
	Foot	$1\ \text{ft}$	$0.3048\ \text{m}$
	Inch	$1\ \text{in}$	$0.0254\ \text{m}$
Mass	**Kilogram**	$1\ \textbf{kg}$	$1\ \text{kg}$
	Pound mass	$1\ \text{lb}_m$	$0.45359\ \text{kg}$
	Slug	$1\ \text{slug} = 32.17\ \text{lb}_m$	$14.59\ \text{kg}$
Mass transfer coefficient, permeability, velocity	–	$1\ \text{m s}^{-1}$	$1\ \text{m s}^{-1}$
		$1\ \text{cm s}^{-1}$	$0.01\ \text{m s}^{-1}$
		$1\ \text{cm h}^{-1}$	$2.778 \times 10^{-6}\ \text{m s}^{-1}$
		$1\ \text{ft s}^{-1}$	$0.3048\ \text{m s}^{-1}$
		$1\ \text{ft h}^{-1}$	$1{,}097\ \text{m s}^{-1}$
Molar quantity	Mole	$1\ \textbf{mol}$	$1\ \text{mol}$
	Kilo mole	$1\ \text{kmol}$	$1 \times 10^3\ \text{mol}$
	Pound mole	$1\ \text{lb-mol}$	$453.59\ \text{mol}$
Momentum	–	$1\ \text{kg m s}^{-1}$	$1\ \text{kg m s}^{-1}$
		$1\ \text{g cm s}^{-1}$	$1 \times 10^{-5}\ \text{kg m s}^{-1}$
		$1\ \text{lb}_m\ \text{ft s}^{-1}$	$0.13825\ \text{kg m s}^{-1}$
Overall mass transfer coefficient, permeability-surface area (mass units)	–	$1\ \text{kg s}^{-1}\ \text{m}^{-2}$	$1\ \text{kg s}^{-1}\ \text{m}^{-2}$
		$1\ \text{g cm}^{-2}\ \text{s}^{-1}$	$10\ \text{kg s}^{-1}\ \text{m}^{-2}$
		$1\ \text{lb}_m\ \text{ft}^{-2}\ \text{s}^{-1}$	$4.8824\ \text{kg s}^{-1}\ \text{m}^{-2}$
		$1\ \text{lb}_m\ \text{ft}^{-2}\ \text{h}^{-1}$	$1.3562 \times 10^{-3}\ \text{kg s}^{-1}\ \text{m}^{-2}$
Overall mass transfer coefficient, permeability-surface area (molar units)	–	$1\ \text{mol s}^{-1}\ \text{m}^{-2}$	$1\ \text{mol s}^{-1}\ \text{m}^{-2}$
		$1\ \text{mol cm}^{-2}\ \text{s}^{-1}$	$1 \times 10^4\ \text{mol s}^{-1}\ \text{m}^{-2}$
		$1\ \text{lb-mol ft}^{-2}\ \text{s}^{-1}$	$4882.4\ \text{mol s}^{-1}\ \text{m}^{-2}$
		$1\ \text{lb-mol ft}^{-2}\ \text{h}^{-1}$	$1.3562\ \text{mol s}^{-1}\ \text{m}^{-2}$
Power	Watt	$1\ \text{W} = 1\ \text{Js}^{-1}$	$1\ \text{kg m}^2\ \text{s}^{-3}$
	–	$1\ \text{cal s}^{-1}$	$4.184\ \text{kg m}^2\ \text{s}^{-3}$
	–	$1\ \text{kcal s}^{-1}$	$4{,}184\ \text{kg m}^2\ \text{s}^{-3}$
	–	$1\ \text{Btu s}^{-1}$	$1{,}054\ \text{kg m}^2\ \text{s}^{-3}$
	–	$1\ \text{ft lb}_f\ \text{s}^{-1}$	$1.356\ \text{kg m}^2\ \text{s}^{-3}$
	Horsepower	$1\ \text{hp}$	$745.7\ \text{kg m}^2\ \text{s}^{-3}$
Pressure or stress or momentum flux	Pascal	$1\ \text{Pa} = 1\ \text{Nm}^{-2}$	$1\ \text{kg m}^{-1}\ \text{s}^{-2}$
	Atmosphere	$1\ \text{atm}$	$1.0133 \times 10^5\ \text{kg m}^{-1}\ \text{s}^{-2}$
	Pounds per Square inch	$1\ \text{psi}$	$6{,}895\ \text{kg m}^{-1}\ \text{s}^{-2}$
		$1\ \text{cmH}_2\text{O}$	$98.06\ \text{kg m}^{-1}\ \text{s}^{-2}$

(*continued*)

Appendix B.3 Conversion Factors

Physical quantity	Common unit	Symbol	In terms of basic SI units
	Centimeter of water		
	Millimeter of mercury, Torr	1 mmHg or 1 Torr	133.32 kg m^{-1} s^{-2}
	Dynes per square centimeter	1 dyn cm^{-2}	0.1 kg m^{-1} s^{-2}
Temperature (absolute)	**Kelvin**	**1 K**	1 K
	Degee Rankine	1°R	1.8 K
Temperature (relative)	Degree Celsius	1°C	T(K) − 273.15
	Degree Fahrenheit	1°F	1.8 T(K) − 459.67
Thermal conductivity	—	1 W m^{-1} K^{-1} or 1 N °C^{-1} s^{-1}	1 kg m s^{-3} K^{-1}
		1 kW m^{-1} K^{-1}	1,000 kg m s^{-3} K^{-1}
		1 cal s^{-1} cm^{-1} K^{-1}	418.4 kg m s^{-3} K^{-1}
		1 kcal h^{-1} m^{-1} °C^{-1}	1.163 kg m s^{-3} K^{-1}
		1 erg s^{-1} cm^{-1} K^{-1}	1 × 10^{-5} kg m s^{-3} K^{-1}
		1 Btu h^{-1} ft^{-1} K^{-1}	1.731 kg m s^{-3} K^{-1}
Time	**Second**	1 s	1 s
	Minute	1 m	60 s
	Hour	1 h	3,600 s
Viscosity	Pascal second	1 Pa s	1 kg m^{-1} s^{-1}
	Centipoise	1 cP	1 × 10^{-3} kg m^{-1} s^{-1}
	Poise	1 P	0.1 kg m^{-1} s^{-1}
	—	1 lb$_m$ ft^{-1} s^{-1}	1.49 kg m^{-1} s^{-1}
Volume	Cubic meter	1 m^3	1 m^3
	Liter	1 L	1 × 10^{-3} m^3
	Milliliter or cubic centimeter	1 ml or 1 cm^3	1 × 10^{-6} m^3
	Cubic inch	1 in^3	1.639 × 10^{-5} m^3
	Cubic foot	1 ft^3	0.0283 m^3
	Gallon	1 gal	3.785 × 10^{-3} m^3
Work	Newton meter	1 N m or 1 J	1 kg m^2 s^{-2}
	Foot pounds	1 ft lb$_f$	1.3558 kg m^2 s^{-2}
	Dyne centimeter	1 dyn cm or 1 erg	1 × 10^{-7} kg m^2 s^{-2}

Appendix C Transport Properties

Properties listed in these appendices have been compiled from measurements from many different sources. We report representative values where more than one value is reported in the literature.

Fluid Properties

Flow Properties of Selected Fluids

Fluid	$T(°C)$	$\rho(kg/m^3)$	$\mu \times 10^6$ (Pa s)	$\nu \times 10^6$ (m^2/s)
Gases				
Air	0	1.292	17.36	13.44
	10	1.247	17.87	14.33
	20	1.204	18.36	15.25
	25	1.184	18.62	15.73
	30	1.164	18.86	16.20
	35	1.146	19.10	16.67
	37	1.140	19.20	16.84
	40	1.127	19.34	17.16
CH_4	25	0.657	11.19	16.98
	37	0.632	11.57	18.34
CO	25	1.145	17.65	15.42
	37	1.101	18.18	16.51
CO_2	25	1.809	14.93	8.26
	37	1.738	15.51	8.93
H_2	25	0.082	8.915	108.7
	37	0.079	9.159	115.9
He	25	0.164	19.85	121.0
	37	0.157	20.39	129.9
N_2	25	1.145	17.81	15.56
	37	1.101	18.36	16.68
N_2O	25	1.840	14.79	8.038
	37	1.739	15.35	8.827
O_2	25	1.309	20.46	15.63
	37	1.258	21.12	16.79
Liquids				
Water	20	998.2	1,002	1.004
	25	997.0	890	0.893
	30	995.7	798	0.801
	35	994.0	719	0.723
	37	993.3	692	0.697
	40	992.2	653	0.658
Normal saline (0.155 M NaCl)	25	1,003.8	904	0.901
	37	1,000.0	777	0.777
Glycerine	25	1,258	1.183×10^6	940.4
Ethanol	25	789	1,074	1.361
Methanol	25	786	544	0.692

(*continued*)

Appendix C Transport Properties

Fluid	$T(°C)$	$\rho(kg/m^3)$	$\mu \times 10^6$ (Pa s)	$\nu \times 10^6$ (m^2/s)
Biological fluids				
Blood plasma	25	1,027	1,403	1.366
Blood plasma	37	1,027	1,390	1.353
Blood, $H = 45$, $\dot{\gamma} = 125 s^{-1}$	22	1,060	4,270	4.028
Blood, $H = 44$, $\dot{\gamma} = 100 s^{-1}$	37	1,060	3,390	3.198
Cerebral spinal fluid	37	1,007	716	0.711

Example: Find the kinematic viscosity of air at 25°C
$\nu(m^2/s) \times 10^6 = 15.73$, or $\nu = 15.73 \times 10^{-6} m^2/s = 0.1573 cm^2/s$

Normal Blood Perfusion Rates in Human Tissue

Organ	Weight (g)	Blood flow (ml/min)	Blood perfusion rate[a] (ml min^{-1} g^{-1})
Adrenal glands	14	25	1.79
Brain	1,500	750	0.50
Cardiac muscle	330	265	0.8
Fat	15,000	400	0.027
Intestines	1,000	900	0.90
Kidneys	300	1,200	4.0
Liver	1,500	1,500	1.0
Lungs (pulmonary)	1,000	5,000	5.0
Skeletal muscles	28,000	750	0.027
Skeletal system	11,200	250	0.022
Skin	4,100	200	0.049
Spleen	180	220	1.2
Stomach and esophagus	190	75	0.4
Thyroid gland	12	50	4.16

[a]Blood flow relative to total tissue mass, including blood mass

Thermal Properties

Thermal Properties of Selected Materials

Material	T (°C)	ρ (kg/m^3)	$k \times 10^2$ (W m^{-1} K^{-1})	$c_p \times 10^{-3}$ (J kg^{-1} K^{-1})	$\alpha \times 10^6$ (m^2/s)	Pr
Gases						
Air	0	1.292	2.410	1.005	18.56	0.724
	10	1.247	2.488	1.005	19.85	0.722
	20	1.204	2.566	1.005	21.21	0.719
	25	1.184	2.604	1.005	21.88	0.719
	30	1.164	2.643	1.005	22.59	0.717
	35	1.146	2.681	1.005	23.28	0.716
	37	1.140	2.696	1.005	23.53	0.716

(*continued*)

Material	T (°C)	ρ (kg/m³)	$k \times 10^2$ (W m⁻¹ K⁻¹)	$c_p \times 10^{-3}$ (J kg⁻¹ K⁻¹)	$\alpha \times 10^6$ (m²/s)	Pr
	40	1.127	2.719	1.005	24.01	0.715
CH_4	25	0.657	3.430	2.232	23.39	0.726
	37	0.632	3.598	2.260	25.19	0.728
CO	25	1.145	2.648	1.042	22.19	0.695
	37	1.101	2.729	1.042	23.79	0.694
CO_2	25	1.809	1.664	0.851	10.82	0.763
	37	1.738	1.762	0.862	11.76	0.759
H_2	25	0.082	18.49	14.31	158	0.688
	37	0.079	19.06	14.35	168	0.690
He	25	0.164	15.53	5.193	190	0.693
	37	0.157	15.96	5.193	195	0.666
N_2	25	1.145	2.574	1.041	21.60	0.720
	37	1.101	2.653	1.041	23.15	0.721
N_2O	25	1.840	1.723	0.883	10.61	0.758
	37	1.739	1.738	0.894	11.18	0.790
O_2	25	1.309	2.651	0.920	22.01	0.710
	37	1.258	2.747	0.921	23.73	0.708
Liquids						
Water	20	998.2	59.84	4.184	0.143	7.02
	25	997.0	60.72	4.181	0.146	6.12
	30	995.7	61.55	4.180	0.148	5.41
	35	994.0	62.33	4.179	0.150	4.82
	37	993.3	62.63	4.179	0.151	4.62
	40	992.2	63.06	4.179	0.152	4.33
Normal saline	25	1,003.8	60.72	4.082	0.148	6.09
	37	1,000.0	62.63	4.081	0.153	5.08
Glycerine	25	1,258	23	2.41	0.076	12,374
Ethanol	25	789	17	2.44	0.088	15.5
Methanol	25	786	21	2.535	0.105	6.59
Biological fluids						
Blood plasma	25	1,027	57.8	4.147	0.136	10.04
	37	1,027	47.7	3.932	0.118	11.47
Blood, $H = 45$	22	1,060	50.7	3.559	0.134	30.06
Blood, $H = 44$	37	1,060	48.8	3.74	0.123	26.0
Solids						
Non-Metals						
Acrylic plastic	25	1,190	18.6	1.46	0.107	
Carbon	25	1,950	170	0.711	1.23	
Cellulose acetate	25	1,280	26	1.5	0.135	
Cork	25	140	4.3	1.80	0.171	
Felt insulation	25	90	25	0.71	3.91	
Fiberglass	25	32	3.8	0.835	1.42	
Glass	25	2,700	105	0.84	0.463	
Ice	0	920	218	2.04	1.16	
Polyethylene terephthalate	25	1,350	25	1.17	0.158	
Polypropylene	25	855	22	1.920	0.134	
Polystyrene	25	1,120	3.3	1.340	0.022	
Polysulfone	25	1,240	26	1.300	0.161	
Polyvinyl chloride	25	1,350	19	0.90	0.156	

(*continued*)

Appendix C Transport Properties

Material	T (°C)	ρ (kg/m³)	$k \times 10^2$ (W m⁻¹ K⁻¹)	$c_p \times 10^{-3}$ (J kg⁻¹ K⁻¹)	$\alpha \times 10^6$ (m²/s)	Pr
Rubber	25	1,100	16	2.01	0.072	
Wood (Oak)	25	545	17	2.385	0.131	
Metals						
Aluminum	25	2,700	25,000	0.903	102.5	
Brass	25	8,520	5,400	0.355	17.85	
Copper	25	8,940	40,100	0.385	116.5	
Gold	25	19,300	31,000	0.129	124.5	
Lead	25	11,342	3,500	0.129	23.92	
Platinum	25	21,450	7,000	0.133	24.54	
Silver	25	10,500	42,900	0.235	173.9	
Stainless steel	25	7,900	1,490	0.477	3.95	
Tin	25	7,300	6,700	0.222	41.34	
Zinc	25	7,140	11,600	0.388	41.87	
Firefighter protective clothing						
Aralite (thermal liner)	20	70	3.4	0.7	0.694	
	55		4.4		0.898	
Kevlar (outer shell)	20	324	5.7	1.637	0.107	
	55		7.6		0.143	
Nomex (moisture barrier)	20	300	4.6	0.26	0.59	
	55		6.0		0.769	
Fabrics						
Cotton	25	80	6	1.3	0.577	
Nylon	25	1,140	25.0	1.67	0.131	
Polyester	25	1,390	4.5	1.3	0.0789	
Wool	25	380	7.0	1.26	0.146	
Tissues						
Adrenal gland	37	1,025	42.2	(3.7)	(0.111)	
Aorta	35	1,089	47.6	3.47	0.126	
Bone, cortical	37	1,850	56	1.3	0.233	
Brain	20	1,050	52.7	3.77	0.133	
Fat tissue	37	916	23	2.30	0.109	
Kidney (cortex)	37	1,040	49.9	3.64	0.132	
Kidney (medulla)	37	1,040	49.9	3.79	0.127	
Liver	25	1,051	56.4	3.41	0.157	
Muscle (cardiac)	37	1,060	53.7	3.71	0.137	
Muscle (skeletal)	37	1,041	46	3.81	0.116	
Skin (dermis)	37	1,200	29.3	3.22	0.076	
Skin (epidermis)	37	1,200	20.9	3.60	0.048	
Spleen	37	1,060	54.3	3.69	0.139	
Thyroid gland	37	1,050	53	(3.7)	(0.136)	
Tooth (enamel)	37	2,970	82	0.75	0.368	
Tooth (dentine)	37	2,140	59	1.17	0.236	

Mass Transfer Properties

Diffusion Coefficients in Gases at Atmospheric Pressure

$T(°C)$	Solute A	Medium B	$D_{AB} \times 10^4$ (m^2 s^{-1})
9	Argon	Air	0.177
25	Benzene	Air	0.096
0	Carbon disulfide	Air	0.088
9	CH$_4$	Air	0.196
0	Chlorine	Air	0.124
9	CO	Air	0.196
0	CO$_2$	Air	0.139
10	CO$_2$	Air	0.149
20	CO$_2$	Air	0.160
25	CO$_2$	Air	0.165
30	CO$_2$	Air	0.170
35	CO$_2$	Air	0.176
40	CO$_2$	Air	0.181
25	Ethanol	Air	0.132
0	Ethyl acetate	Air	0.071
25	H$_2$	Air	0.770
37	H$_2$	Air	0.825
0	H$_2$O vapor	Air	0.209
10	H$_2$O vapor	Air	0.225
20	H$_2$O vapor	Air	0.242
25	H$_2$O vapor	Air	0.251
30	H$_2$O vapor	Air	0.260
35	H$_2$O vapor	Air	0.268
40	H$_2$O vapor	Air	0.277
9	He	Air	0.658
25	Methanol	Air	0.162
0	NH$_3$	Air	0.198
0	O$_2$	Air	0.179
10	O$_2$	Air	0.191
20	O$_2$	Air	0.203
25	O$_2$	Air	0.209
30	O$_2$	Air	0.215
35	O$_2$	Air	0.221
40	O$_2$	Air	0.227
27	SF$_6$	Air	0.098
0	SO$_2$	Air	0.122
25	Toluene	Air	0.0844
15	CO	N$_2$	0.192
43	CO	N$_2$	0.240
25	CO$_2$	N$_2$	0.165
20	O$_2$	CH$_4$	0.215
43	O$_2$	H$_2$	0.891
20	O$_2$	CO$_2$	0.153
60	O$_2$	CO$_2$	0.193
20	O$_2$	H$_2$O (vapor)	0.240
60	O$_2$	H$_2$O (vapor)	0.339

(*continued*)

Appendix C Transport Properties

$T(°C)$	Solute A	Medium B	$D_{AB} \times 10^4$ (m² s⁻¹)
0	O_2	N_2	0.181
20	O_2	N_2	0.219
60	O_2	N_2	0.274

Example: At 35°C: $D_{O_2,air} \times 10^4 = 0.221$ m²/s or $D_{O_2,air} = 2.21 \times 10^{-5}$ m²/s

Diffusion Coefficients and Bunsen Solubility Coefficients for Dissolved Gases in Various Media at Atmospheric Pressure

$T(°C)$	Gas A	Medium B	$D_{AB} \times 10^9$ (m² s⁻¹)	α^*_{AB} (ml$_A$ml$_B^{-1}$ atm⁻¹)
37	Ar	Blood ($H = 44\%$)	–	0.0305
37	Ar	Brain tissue	–	0.0327
37	Ar	Muscle	–	0.0229
37	Ar	Normal saline	–	0.0296
37	Ar	Olive oil	–	0.16
37	Ar	Water	3.4	0.0298
25	CH_4	Water	1.49	0.03395
25	Cl_2	Water	1.25	2.236
37	CO	Cell membrane[a]	–	0.103
20	CO	Hb solution (18.8 g/dl)	0.51	–
20	CO	Hb solution (32.2 g/dl)	0.37	–
37	CO	Olive oil	–	0.0858
25	CO	Water	2.03	0.02334
37	CO_2	Aorta wall	0.65	–
38	CO_2	Blood ($H = 44\%$)	–	0.488
22	CO_2	Brain tissue	1.0	0.97
37	CO_2	Cell membrane[a]	–	0.99
38	CO_2	Hb solution (16 g/dl)	1.6	–
38	CO_2	Hb solution (33 g/dl)	1.14	0.44
22	CO_2	Muscle	1.07	0.78
25	CO_2	Normal saline	–	0.742
37	CO_2	Normal saline	–	0.550
37	CO_2	Olive oil	–	1.25
25	CO_2	Plasma	–	0.681
37	CO_2	Plasma	–	0.515
22	CO_2	Skin	0.95	0.73
20	CO_2	Water	1.76	0.878
25	CO_2	Water	1.94	0.759
30	CO_2	Water	2.20	0.665
35	CO_2	Water	2.93	0.592
37	CO_2	Water	2.96	0.567
37	H_2	Cell membrane[a]	–	0.026
37	H_2	Muscle	2.57	0.0218
37	H_2	Olive oil	3.7	0.0484
20	H_2	Water	3.63	0.01941
25	H_2	Water	4.05	0.01913
30	H_2	Water	4.48	0.01895
37	H_2	Water	5.07	0.01887

(*continued*)

$T(°C)$	Gas A	Medium B	$D_{AB} \times 10^9$ (m² s⁻¹)	α^*_{AB} (ml$_A$ml$_B^{-1}$ atm⁻¹)
20	H$_2$S	Water	1.74	2.792
25	H$_2$S	water	1.95	2.510
30	H$_2$S	water	2.22	2.085
35	H$_2$S	water	2.75	1.791
38	He	Blood ($H = 44\%$)	–	0.0088
37	He	Muscle	3.94	0.012
37	He	Olive oil	–	0.0159
25	He	Water	6.28	0.015
30	Kr	Water	2.17	0.060
38	N$_2$	Blood ($H = 44\%$)	–	0.013
37	N$_2$	Cell membrane*	–	0.106
20	N$_2$	Hb solution (18.8 g/dl)	0.52	–
20	N$_2$	Hb solution (32.2 g/dl)	0.36	–
25	N$_2$	Normal saline	–	0.0141
38	N$_2$	Normal saline	–	0.0122
37	N$_2$	Olive oil	–	0.067
38	N$_2$	Plasma	–	0.0117
10	N$_2$	Water	1.29	0.0186
25	N$_2$	Water	2.01	0.0143
37	N$_2$	Water	2.20	0.0123
37	N$_2$O	Blood ($H = 44\%$)	0.465	0.412
37	N$_2$O	Brain tissue	–	0.497
19.3	N$_2$O	Hb solution (35 g/dl)	0.231	–
37	N$_2$O	Muscle	1.27	0.512
37	N$_2$O	Normal saline	–	0.452
37	N$_2$O	Olive oil	–	1.40
37	N$_2$O	Plasma	–	0.454
20	N$_2$O	Water	1.84	0.6788
25	N$_2$O	Water	1.88	0.5937
30	N$_2$O	Water	1.93	0.5241
37	N$_2$O	Water	2.60	0.467
37	Ne	Blood	–	0.0093
37	Ne	Brain	–	0.0115
37	Ne	Cell membrane[a]	–	0.013
37	Ne	Normal saline	–	0.0111
37	Ne	Olive oil	–	0.027
30	Ne	Water	3.81	0.0158
25	NH$_3$	Water	1.64	312.7
20	NO	Water	2.07	0.05046
25	NO	Water	2.21	0.04708
30	NO	Water	3.96	0.04430
37	O$_2$	Aortic wall	0.90	–
37	O$_2$	Blood ($H = 44\%$)	1.33	0.0223
19	O$_2$	Cell membrane	0.727	–
37	O$_2$	Cell membrane[a]	–	0.124
25	O$_2$	Hb solution (33 g/dl)	0.838	0.033
25	O$_2$	Hb solution (8 g/dl)	1.87	0.025
37	O$_2$	Muscle	1.5	0.0235
25	O$_2$	Normal saline	–	0.0272
37	O$_2$	Normal saline	–	0.0227
37	O$_2$	Olive oil	–	0.130

(continued)

$T(°C)$	Gas A	Medium B	$D_{AB} \times 10^9$ (m² s⁻¹)	α^*_{AB} (ml$_A$ml$_B^{-1}$ atm⁻¹)
20	O_2	Perflurocarbon (FC-40)	8.29	–
25	O_2	Plasma	1.21	0.0257
37	O_2	Plasma	1.63	0.0214
10	O_2	Water	1.54	0.0383
25	O_2	Water	2.20	0.0283
37	O_2	Water	2.89	0.0239
40	O_2	Water	3.33	0.0231
37	SF_6	Blood	–	0.0075
37	SF_6	Brain tissue	–	0.0165
37	SF_6	Cell membrane[a]	–	0.166
37	SF_6	Muscle	–	0.012
37	SF_6	Olive oil	–	0.275
37	SF_6	Plasma	–	0.0056
37	SF_6	Water	–	0.0044
30	Xe	Water	1.02	0.090

[a]Cell membrane solubilities are based on data from red cell ghosts with an assumed membrane density of 1,100 kg/m³
Example: At 25°C: $D_{\text{albumin,water}} \times 10^9 = 0.069$ m²/s or $D_{\text{albumin,water}} = 6.9 \times 10^{-11}$ m²/s

Diffusion Coefficients and Solubility Coefficients for Non-Gaseous Solutes in Various Media at Atmospheric Pressure

$T(°C)$	Solute A	Medium B	$D_{AB} \times 10^9$ (m² s⁻¹)	Solubility (mg$_A$/ml$_B$)
25	Acetic acid	Water	1.19	Miscible
25	Acetone	Water	1.28	Miscible
25	L-alanine	Water	0.91	167.2
25	Albumin	Water	0.069	40
25	Caffeine	Dichloromethane	–	140
20	Caffeine	Ethanol	–	15.2
25	Caffeine	Water	0.63	21.7
20	Catalase	Water	0.041	–
20	Carbonic anhydrase	Water	0.09	–
25	Ethyl alcohol	Water	1.24	miscible
37	Fibrinogen	Water	0.020	–
37	Fibrinogen	Plasma	–	23
20	γ-globulin	Water	0.0384	–
25	Glucose	Water	0.69	910
25	Glycerol	Water	0.93	miscible
25	Glycine	Water	1.06	249.9
25	Hb	Water	0.069	–
20	Hb	Hb solution (18.8 g/dl)	0.0295	–
20	Hb	Hb solution (32.2 g/dl)	0.0177	–
25	Histidine	Water	0.73	41.9
20	Insulin	Water	0.0745	–
25	Lactose	Water	0.49	189
25	L-leucine	Water	0.73	24.26

(*continued*)

T(°C)	Solute A	Medium B	$D_{AB} \times 10^9$ (m² s⁻¹)	Solubility (mg$_A$/ml$_B$)
15	Methyl alcohol	Water	1.26	miscible
20	Myoglobin	Water	0.113	–
20	Myosin	Water	0.0087	–
20	Pepsin	Water	0.090	–
25	L-phenylalanine	Water	0.705	29.65
25	L-proline	Water	0.88	1,623
25	DL-serine	Water	0.88	50.3
25	Sucrose	Water	0.524	2,114
25	L-tryptophan	Water	0.660	11.36
25	L-tyrosine	Water	0.0453	0.453
12	Urea	Ethanol	0.54	200
25	Urea	Water	1.38	1,000
20	Urease	Water	0.0346	–
25	L-valine	Water	0.77	88.5
25	Vitamin B12	Cellulose	0.008	–
25	Water (self)	Water	2.54	–

Example: At 25°C: $D_{\text{albumin,water}} \times 10^9 = 0.069$ m²/s, or $D_{\text{albumin,water}} = 6.9 \times 10^{-11}$ m²/s

Partition Coefficients for Solute A in Material B Relative to Material C at 37°C ($\Phi_{ABC} = (c_{AB})_{eq}/(c_{AC})_{eq} = 1/\Phi_{ACB}$)

Solute A	Material B	Material C	Φ_{ABC}
Ar	Oil	Water	5.3
Chloroform	Oil	Water	110
	Liver tissue	Blood	0.9
	Brain tissue	Blood	1.1
Cyclopropane	Oil	Water	35
	Blood	Water	2.24
	Muscle tissue	Blood	0.92
	Liver tissue	Blood	1.34
Ethyl ether	Oil	Water	3.2
	Blood	Water	0.961
	Brain tissue	Blood	1.14
Ethylene	Oil	Water	14.4
	Blood	Water	1.57
	Brain tissue	Blood	1.2
	Heart tissue	Blood	1.0
Halothane	Oil	Water	315
	Blood	Water	3.24
	Brain tissue	Blood	2.6
	Liver tissue	Blood	2.6
	Kidney	Blood	1.6
	Muscle	Blood	3.5
	Fat	Blood	60.
H$_2$	Oil	Water	3.1
He	Oil	Water	1.7
	Blood	Water	1.01
Kr	Oil	Water	9.6

(*continued*)

Solute A	Material B	Material C	Φ_{ABC}
N_2	Blood	Water	1.021
	Brain tissue	Blood	1.1
	Liver tissue	Blood	1.1
	Fat	Blood	5.2
	Oil	Water	5.2
N_2O	Oil	Water	3.2
	Brain tissue	Blood	1.0
	Heart tissue	Blood	1.0
	Lung tissue	Blood	1.0
	Blood	Water	1.014
Profofol	Blood	Water	35
	Brain tissue	Blood	3.23
	Heart tissue	Blood	5.94
Xe	Olive oil	Water	21.55
	Plasma	Water	1.1
	Red blood cells	Plasma	2.894
	Adipose tissue	Plasma	18.82
	Skeletal muscle	Plasma	1.213
	Heart tissue	Plasma	1.413
	Normal saline	Water	0.940

References

Altman PL, Dittmer DS (eds) (1971) Biological handbooks: respiration and circulation. Federation of American Societies for Experimental Biology, Bethesda

Battino R, Clever HL (1966) The solubility of gases in liquids. Chem Rev 66:395–463

Blake AST, Petley GW, Deakin CD (2000) Effects of changes in packed cell volume on the specific heat capacity of blood: implications for studies measuring heat exchange in extracorporeal circuits. Br J Anaesth 84:28–32

Brown AM, Stubbs DW (eds) (1983) Medical physiology. Wiley, New York

Chen RY, Fan FC, Kim S, Jan KM, Usami S, Chien S (1980) Tissue-blood partition coefficient for xenon: temperature and hematocrit dependence. J Appl Physiol 49:178–183

Cinar Y, Senyol AM, Duman K (2001) Blood viscosity and blood pressure: role of temperature and hyperglycemia. Am J Hypertens 14:433–438

Cohen ML (1977) Measurement of the thermal properties of human skin. A review. J Invest Dermatol 69:333–338

Crane Company (1988) Flow of fluids through valves, fittings, and pipe. Technical Paper No. 410 (TP 410)

Cussler EL (1997) Diffusion: mass transfer in fluid systems, 2nd edn. Cambridge University Press, London

Diller KR, Valvano JW, Pearce JA (2005) Bioheat transfer. In: Kreith F, Goswami Y (eds) The CRC handbook of mechanical engineering, 2nd edn. CRC Press, Boca Raton

Duck FA (1990) Physical properties of tissue: a comprehensive reference book. Academic, London

Eckmann DM, Bowers S, Stecker M, Cheung AT (2000) Hematocrit, volume expander, temperature, and shear rate effects on blood viscosity. Anesth Analg 91:539–545

Engineering Toolbox (2010) Tools and basic information for design, engineering and construction of technical applications. http://www.engineeringtoolbox.com/

ThermExcel (2003) Physical characteristics of water (at the atmospheric pressure). http://www.thermexcel.com/english/tables/eau_atm.htm

Engineers Edge Solutions by Design (2010) Fluid characteristics chart. http://www.engineersedge.com/fluid_flow/fluid_data.htm

Ferrell RT, Himmelblau DM (1967) Diffusion coefficients of nitrogen and oxygen in water. J Chem Eng Data 12:111–115

Frydrych I, Dziworska G, Bilska J (2002) Comparative analysis of the thermal insulation properties of fabrics made of natural and man-made cellulose fibres. Fibres Text East Eur Oct/Dec:40–44

González-Alonso J, Bjørn Quistorff PK, Bangsbo J, Saltin B (2000) Heat production in human skeletal muscle at the onset of intense dynamic exercise. J Physiol 524:603–615

Guyton AC (1968) Textbook of medical physiology, 3rd edn. W.B Saunders, Philadelphia

Haduk W, Laudie H (1974) Prediction of diffusion coefficients for nonelectrolytes in dilute aqueous solutions. AIChE J 20:611–615

Henriques FC, Moritz AR (1947) Studies of thermal injury. I. The conduction of heat to and through skin and the temperatures attained therein. A theoretical and an experimental investigation. Am J Pathol 23:531–549

Johnson AT (1999) Biological process engineering: an analog approach to fluid flow, heat transfer, and mass transfer applied to biological systems. Wiley, New York

Lango T, Morland T, Brubakk AO (1996) Diffusion coefficients and solubility coefficients for gases in biological fluids and tissues: a review. Undersea Hyperb Med 23:247–272

Larson CP, Eger EI, Severinghous JW (1962) The solubility of halothane in blood and tissue homogenates. Anesthesiology 23:349–355

Lawrence JS (1950) The plasma viscosity. J Clin Pathol 3:332–334

Lemmon EW, McLinden MO, Friend DG (2010) Thermophysical properties of fluid systems. In: Linstrom PJ, Mallard WG (eds) NIST chemistry WebBook, NIST Standard Reference Database Number 69. National Institute of Standards and Technology, Gaithersburg, p 20899. http://webbook.nist.gov

Liu M, Nicholson JK, Parkinson JA, Lindon JC (1997) Measurement of biomolecular diffusion coefficients in blood plasma using two dimensional 1H-1H diffusion-edited total-correlation NMR spectroscopy. Anal Chem 69:1504–1509

Longmuir IS, Roughton FJW (1952) The diffusion coefficients of carbon monoxide and nitrogen in haemoglobin solutions. J Physiol 118:264–275

Marrero TR, Mason EA (1972) Gaseous diffusion coefficients. J Phys Chem Ref Data 1:1–118

Naka S, Kamata Y (1977) Determining the thermal conductivity of fabrics by non-steady state method. Trans J Textile Mach Soc Jpn 30:T30–T44

Power GG, Stegall H (1970) Solubility of gases in human red cell ghosts. J Appl Physiol 29:145–149

Rosenson RS, McCormick A, Uretz EF (1996) Distribution of blood viscosity values and biochemical correlates in healthy adults. Clin Chem 42:1189–1195

Shargel L, Yu ABC (1985) Applied biopharmaceutics and pharmacokinetics, 2nd edn. Appleton-Century-Croft, Norwalk

Turner MJ, MacLeod IM, Rothberg AD (1989) Effects of temperature and composition on the viscosity of respiratory gases. J Appl Physiol 67:472–477

Vettori RL (2005) Estimates of thermal conductivity for unconditioned and conditioned materials used in fire fighters' protective clothing. NISTIR 7279; p 33

Von Antroff A (1910) The solubility of xenon, krypton, argon, neon, and helium in water. Proc R Soc Lond A83:474–482

Weaver BMQ, Staddon GE, Mapleson WW (2001) Tissue/blood and tissue/water partition coefficients for propofol in sheep. Br J Anaesth 86:693–703

Weast RC (ed) (1970) Handbook of chemistry and physics, 51st edn. CRC, Cleveland

West ES, Todd WR, Mason HS, Van Bruggen JT (1968) Textbook of biochemistry, 4th edn. Macmillan, New York

Wilhelm E, Battino R, Wilcock RJ (1977) Low-pressure solubility of gases in liquid water. Chem Rev 77:219–262

Wise DL, Houghton G (1968) Diffusion coefficients of neon, krypton, carbon monoxide and nitric oxide in water at 10–60°C. Chem Eng Sci 23:1211–1216

Appendix D Charts for Unsteady Conduction and Diffusion

D.1 Introduction

Charts are presented in this appendix that can be used to estimate transport of heat or mass from a solid material to a fluid. The solid initially has a uniform temperature T_0 or concentration C_{A0} and is immersed in a fluid with temperature $T_{f\infty}$ or $C_{Af\infty}$. Transport at the interface is governed by a heat transfer coefficient h or a mass transfer coefficient k_{Af}. The solid and fluid are assumed to be at equilibrium at the fluid-solid interface, so $T_S = T_{fS}$ and $C_{AS} = \Phi_{Asf} C_{AfS}$, where Φ_{Asf} is the partition coefficient for substance A between the solid and the fluid.

The problems of unsteady-state conduction and diffusion in a slab, cylinder or sphere can be analyzed using the same procedure if we use dimensionless independent variables Y, Ψ and Σ as defined in Table D.1, dimensionless dependent variables X and n, and a dimensionless parameter m as defined in Table D.2. All terms are defined in Table D.3.

Table D.1 Dimensionless dependent variables

Description	Symbol	Heat	Molar	Mass
Change in transport variable	Y	$\dfrac{T - T_{f\infty}}{T_0 - T_{f\infty}}$	$\dfrac{c_A - \Phi_{Asf} C_{Af\infty}}{C_{A0} - \Phi_{Asf} C_{Af\infty}}$	$\dfrac{\rho_A - \Phi_{Asf}\rho_{Af\infty}}{\rho_{A0} - \Phi_{Asf}\rho_{Af\infty}}$
Flux relative to maximum flux	Ψ	$\dfrac{q_S}{h(T_0 - T_{f\infty})}$	$\dfrac{J_{AS}}{\dfrac{k_{Af}}{\Phi_{Asf}}(C_{A0} - \Phi_{Asf}C_{Af\infty})}$	$\dfrac{j_{AS}}{\dfrac{k_{Af}}{\Phi_{Asf}}(\rho_{A0} - \Phi_{Asf}\rho_{Af\infty})}$
Accumulated heat or mass relative to maximum	Σ	$\dfrac{Q}{\rho c_P V(T_0 - T_{f\infty})}$	$\dfrac{N_A}{V[C_{A0} - \Phi_{Asf}C_{Af\infty}]}$	$\dfrac{m_A}{V\left[\rho_{A0} - \Phi_{Asf}\rho_{Af\infty}\right]}$

Table D.2 Dimensionless independent variables and parameters

Description	Symbol	Object	Heat	Molar or mass
Dimensionless time (Fourier number)	X	Slab	$\dfrac{\alpha t}{L^2}$	$\dfrac{D_{As} t}{L^2}$
		Cylinder or sphere	$\dfrac{\alpha t}{R^2}$	$\dfrac{D_{As} t}{R^2}$
Dimensionless position	n	Slab	$\dfrac{x}{L}$	
		Cylinder or sphere	$\dfrac{r}{R}$	
Relative resistance (inverse Biot number)	m	Slab	$\dfrac{k}{hL}$	$\dfrac{D_{As}\Phi_{Asf}}{k_{Af}L}$
		Cylinder or sphere	$\dfrac{k}{hR}$	$\dfrac{D_{As}\Phi_{Asf}}{k_{Af}R}$

Table D.3 Variables and parameters for unsteady state transport

t	Time
x	Position in slab relative to center of slab
r	Radial position in cylinder or sphere
L	Slab half width
L_c	Cylinder length
R	Radius of cylinder or sphere
V	Volume of solid material: $2Lwh$ (slab), $\pi R^2 L$ (cylinder), $4\pi R^3/3$ (sphere)

Conduction		Diffusion	
T	Temperature in solid material	C_A	Molar concentration of A in solid
		ρ_A	Mass concentration of A in solid
T_0	Initial temperature in solid	C_{A0}	Initial concentration of A in solid
		ρ_{A0}	
$T_{f\infty}$	Fluid temperature far from solid	$C_{Af\infty}$	Fluid concentration far from solid
		$\rho_{Af\infty}$	
h	Heat transfer coefficient	k_{Af}	Mass transfer coefficient
q_S	Heat flux at solid surface	N_{AS}	Molar flux at solid surface
		n_{AS}	Mass flux at solid surface
Q	Heat in solid relative to initial heat in solid	N_A	Moles or mass in solid relative to initial value
		m_A	
α	Thermal diffusivity of solid	D_{As}	Diffusion coefficient, A in solid
k	Thermal conductivity of solid	Φ_{Asf}	Solid/fluid partition coefficient
ρ	Density of solid		
c_p	Specific heat of solid		

The charts in Sects. D.2, D.3 and D.4 can be used to find the following quantities:

1. The concentration or temperature at the center of the slab, cylinder or sphere as a function of time.
2. The surface temperature or concentration of the solid at any time.
3. The temperature or concentration at any radial position (within 5%) for dimensionless times greater than $X = 0.15$. Additional charts are provided for finding radial profiles when dimensionless times less are than 0.15, for specific values of m.
4. The flux of heat or mass across the solid-fluid interface at any time.
5. The amount of heat or mass which has accumulated in the solid after a given time.

The procedure for finding each of these quantities is given below.

D.1.1 Finding the Concentration or Temperature at the Center of the Material at a Given Time

Use the appropriate expressions in Table D.2 to compute m and X. The variables composing these dimensionless numbers are different for heat transfer and mass transfer problems, and the length scale is different for a slab than for a cylinder or sphere. Once m and X are computed, read the value of $Y(0,X;m)$ from the

Appendix D Charts for Unsteady Conduction and Diffusion 1253

appropriate chart. The appropriate chart for a slab is Fig. D.1, for a cylinder is Fig. D.6, and for a sphere is Fig. D.11. The centerline temperature can be computed from:

$$T(0,t) = T_{f\infty} + (T_0 - T_{f\infty})Y(0,X;m)$$

The centerline concentration can be computed from:

$$c_A(0,t) = \Phi_{Asf}C_{Af\infty} + (C_{A0} - \Phi_{Asf}C_{Af\infty})Y(0,X;m)$$

D.1.2 Finding the Surface Concentration or Temperature of the Solid at a Particular Time

Use the appropriate expressions in Table D.2 to compute m and X. Then go to the appropriate chart that provides the flux at the surface relative to the maximum flux, and read the value of $\Psi(X;m)$. The appropriate graph for a slab is Fig. D.4, for a cylinder is Fig. D.9, and for a sphere is Fig. D.14. Since $\Psi(X;m) = Y(1,X;m)$, then the surface temperature can be found from:

$$T_S(t) = T_{f\infty} + (T_0 - T_{f\infty})\Psi(X;m)$$

or the surface concentration can be computed from:

$$c_{AS}(t) = \Phi_{Asf}C_{Af\infty} + (C_{A0} - \Phi_{Asf}C_{Af\infty})\Psi(X;m)$$

D.1.3. Finding the Temperature or Concentration at a Position in the Material Other than at the Center or the Surface

If $X \geq 0.15$, the first term in the infinite series provides an estimate of the solution for Y that is within 95% of the actual solution for all values of m. If, in addition, m is greater than three, the first term is accurate for all values of X. The ratio of the first term $Y_1(n,X;m)$ at any position within the material relative to the first term at the center of the material $Y_1(0,X;m)$ is independent of X. This ratio is plotted for a slab, a cylinder and a sphere in Fig. D.2, Fig. D.7 and Fig. D.12 respectively. To estimate the temperature at a dimensionless position n, Find $Y(n,X;m)/Y(0,X;m)$ from the appropriate figure and follow the first procedure above to find the centerline value, $Y(0,X;m)$. Compute the temperature $T(r,t)$ or $T(x,t)$ using the following expression:

$$T(r,t) = T_{f\infty} + (T_0 - T_{f\infty})\left(\frac{Y(n;m)}{Y(0;m)}\right)Y(0,X;m)$$

The analogous expression for concentration at any position in the material is:

$$c_{\text{AS}}(t) = \Phi_{\text{Asf}} C_{\text{Af}\infty} + (C_{\text{A0}} - \Phi_{\text{Asf}} C_{\text{Af}\infty}) \left(\frac{Y(n;m)}{Y(0;m)} \right) Y(0, X; m)$$

This method will not be accurate to within 5% if $m < 3$ and $X < 0.15$. In that case, the ratio will depend on dimensionless time. Comparisons between actual profiles, which include 30 terms of the series solution, and profiles predicted using the first term (red lines) are shown in Fig. D.3 (slab), Fig. D.8 (cylinder) and Fig. D.13 (sphere) for several small values of X and four values of m (0.01, 0.1, 1.0, 10.0). The actual profiles are flatter than the prediction based on the first term. These figures, along with Fig. D.2, Fig. D.7 and Fig. D.12 can be used to estimate temperature or concentration at various positions within the material for $X < 0.15$ and $m < 3$.

D.1.4 Finding the Flux of Heat or Mass Across the Solid-Fluid Interface at Any Time

The flux at the surface relative to the maximum flux $\Psi(X;m)$ is plotted for a slab, cylinder and sphere in Fig. D.4, Fig. D.9 and Fig. D.14, respectively. The actual heat flux by conduction at a particular time can computed from

$$q_{\text{S}}(t) = h(T_0 - T_{\text{f}\infty}) \Psi(X; m)$$

The analogous expression for molar flux by diffusion at time t is:

$$N_{\text{A}}(t) = \frac{k_{\text{f}}}{\Phi_{\text{Asf}}} (C_{\text{A0}} - \Phi_{\text{Asf}} C_{\text{Af}\infty}) \Psi(X; m)$$

D.1.5 Finding the Amount of Heat or Mass Which Has Accumulated in the Solid After a Given Time

The quantity of heat or mass that crosses the solid-fluid boundary relative to the amount which would cross after a long period of time $\Sigma(X;n)$ is plotted for a slab, cylinder and sphere in Fig. D.5, Fig. D.10 and Fig. D.15 respectively. The actual amount of heat that has crossed the surface in time t can be found from:

$$Q(t) = \rho c_{\text{P}} V (T_0 - T_{\text{f}\infty}) \Sigma(X; m)$$

D.2 Charts for a Slab

and the analogous expressions for molar or mass transfer are:

$$N_A = V[C_{A0} - \Phi_{Asf}C_{Af\infty}]\Sigma(X;m)$$

$$m_A = V[\rho_{A0} - \Phi_{Asf}\rho_{Af\infty}]\Sigma(X;m)$$

where V is the volume of the solid material:

$$V = 2whL \text{ (slab)}$$

$$V = \pi R^2 L \text{ (cylinder)}$$

$$V = \frac{4}{3}\pi R^3 \text{ (sphere)}$$

D.2 Charts for a Slab

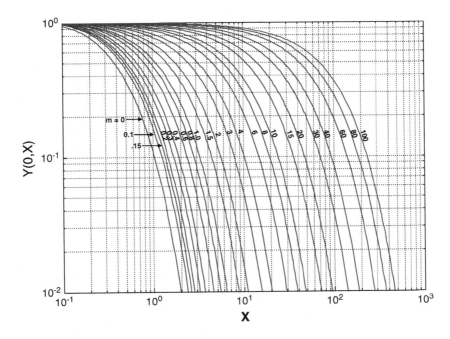

Fig. D.1 Slab: centerline value

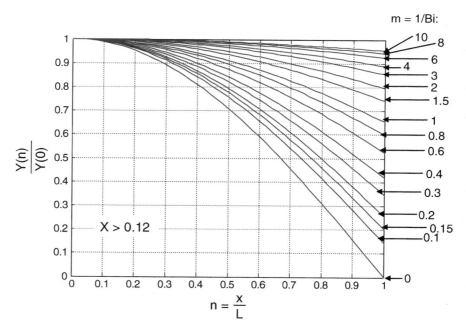

Fig. D.2 Slab: value relative to centerline value vs. position n ($X > 0.12$)

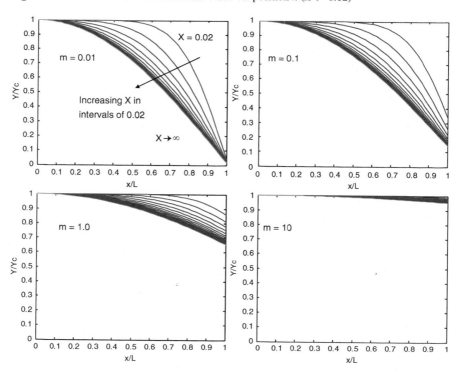

Fig. D.3 Slab: value relative to centerline value vs. position n (small X) for 4 values of m

D.2 Charts for a Slab

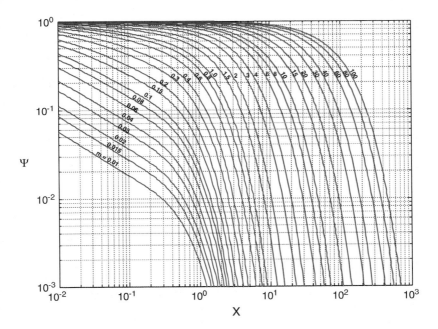

Fig. D.4 Slab: flux relative to maximum flux

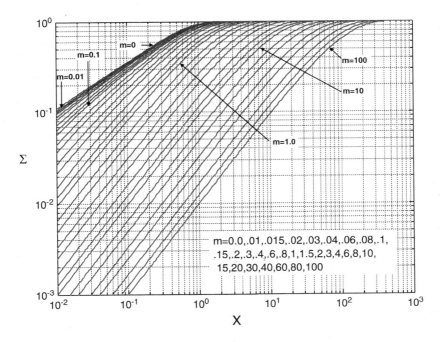

Fig. D.5 Slab: accumulation relative to maximum accumulation

D.3 Charts for a Cylinder

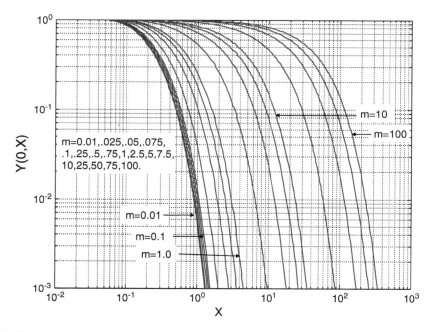

Fig. D.6 Cylinder: centerline value

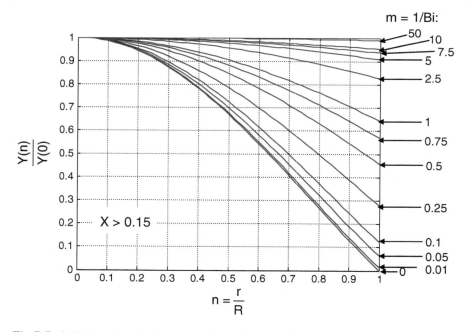

Fig. D.7 Cylinder: value relative to centerline value vs. position n ($X > 0.15$)

D.3 Charts for a Cylinder

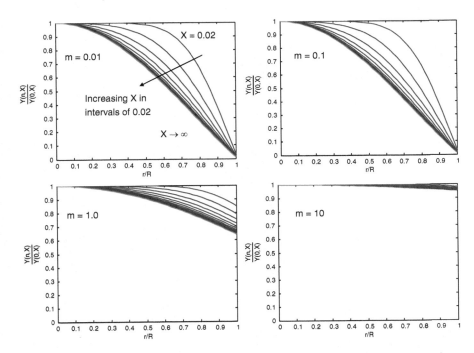

Fig. D.8 Cylinder: value relative to centerline value vs. position n (small X) for 4 values of m

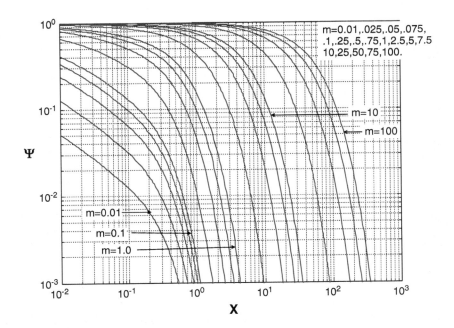

Fig. D.9 Cylinder: flux relative to maximum flux

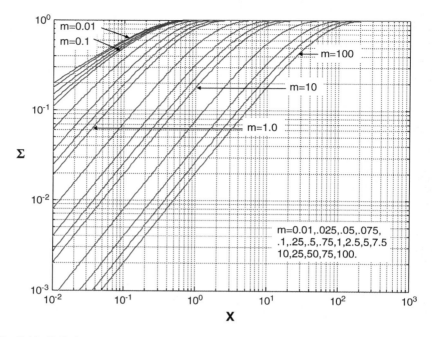

Fig. D.10 Cylinder: accumulation relative to maximum accumulation

D.4 Charts for a Sphere

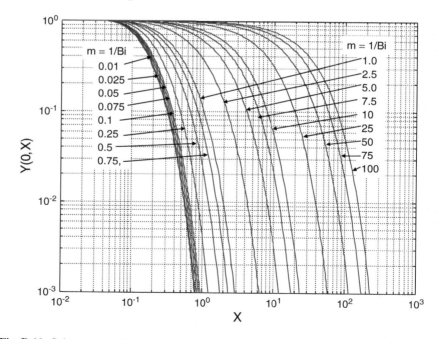

Fig. D.11 Sphere: centerline value

D.4 Charts for a Sphere

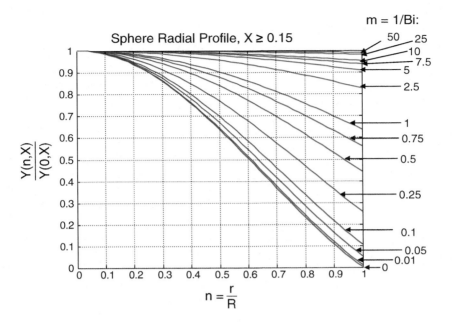

Fig. D.12 Sphere: value relative to centerline value vs. position n ($X > 0.15$)

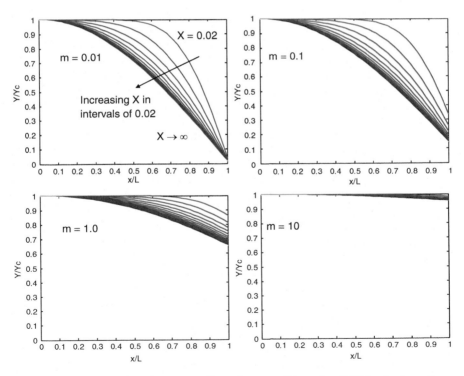

Fig. D.13 Sphere: value relative to centerline value vs. position n (small X) for 4 values of m.

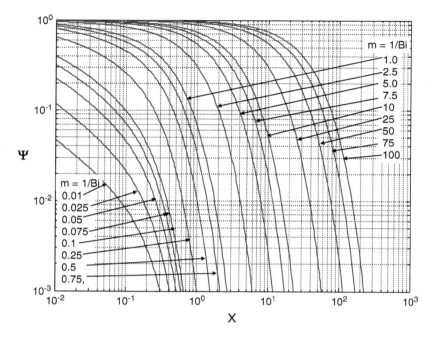

Fig. D.14 Sphere: flux relative to maximum flux

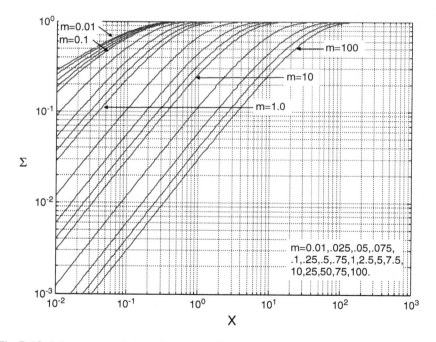

Fig. D.15 Sphere: accumulation relative to maximum accumulation

Index

A
Absorptivity, 532
Active transport, 880–881
Adaptive expertise, HPL methodology
   definition, 4
   knowledge and innovation, 4
   learning, 5
Advection, 49
Albumin and globulin flow through pore, 978–976
Alveolar fluxes, 813–817
Alveolar gas composition, 793–795
Amniotic fluid, 141
Analogies: momentum, heat and mass transfer, 1183–1185
Aneurysm, 298
Annulus flow, 477
Apparent viscosity, 113
Arrhenius thermal injury model, 752
Automotive radiator, 648
Average overall heat transfer coefficient, 643
Average value, 50

B
Backward finite difference method, 747
Behavior index, 459
Bends, diving, 800–801
Bernoulli equation, 194–195
Bingham fluid, 461–462
Bingham fluid model
   constitutive relation, 129
   velocity profile, 129–130
Biofluid transport
   macroscopic approach (*see* Macroscopic approach, biofluid transport)
   microscopic approach (*see* Microscopic approach, biofluid transport)
   shell balance approach (*see* Shell balance approach, one-dimensional biofluid transport)
Bioheat equation, 679
Bioheat transport
   macroscopic approach (*see* Macroscopic approach, bioheat transport)
   microscopic approach (*see* Microscopic approach, bioheat transport)
   shell balance approach (*see* Shell balance approach, one-dimensional bioheat transport)
Biomass transport
   macroscopic approach (*see* Macroscopic approach, biomass transport)
   microscopic approach (*see* Microscopic approach, biomass transport)
   shell balance approach (*see* Shell balance approach, one-dimensional biomass transport)
Bioreactor
   1D biomass transport
      first-order reaction, stationary phase, 1085–1086
      imbedded enzyme bioreactor analysis, 1080–1081
      Michaelis–Menten kinetics, stationary phase, 1087–1091
      mobile phase analysis, 1081–1084
      types, 1080
      urea removal, 1088–1091
      zeroth-order reaction, stationary phase, 1084–1085
   first order kinetics, 1085–1086
   Michaelis–Menten kinetics, 1087–1091
   mobile phase, 1081–1084
   zeroth order kinetics, 1084–1085

Biorheology and disease
   cancer, 156
   cystic fibrosis, 157–158
   polycythemia, 155
   sickle cell anemia, 156–157
Biot number, 523–524, 591, 684, 736
Biot number for mass transfer, 850
Biot number for mass transfer, bioreactor, 1083
Biotransport
   biological systems, 56–58
   challenges, 32–33
   conservation principles, 39–40
   convective transport mechanism
      fluid flow, control volume, 48
      total fluxes, 49
   interphase transport
      cell membrane/capillary wall, 55
      external and internal flow, 53
      momentum transport, 52
      partition coefficient, 54
      proportionality factor, 53
      relationships with, 56
      thermal boundary layer, 53
   macroscopic transport coefficients
      charge flow, electrical wire, 49
      flow of thermal energy/heat flow, 51
      relationships, 52
      volumetric flow rate, 50
   molecular transport mechanism
      constitutive equations, 43–45
      1D analogies, 46–48
      1D flux, negative gradient, 42–43
      diffusion, 41
      energy and momentum, 42
      flux and gradient, 41
      flux, n-direction, 41–42
      properties, 45–46
   system and its environment
      boundary identification, 34
      closed system, 34
      definition, 34
      extensive properties, 35
      intensive properties, 34
      open system, 34–35
   system's equilibrium, 40
   transport, biological systems, 24–25
   transport scales, time and space
      continuum concepts, 37–39
      feedback control systems, 37
      length scales, 36
      macroscopic approach, 36
      microscopic approach, 35
      molecular interaction, 35
      nonlinear properties, 37
Biotransport problems
   dimensional analysis
      cylindrical cell, 100–101
      cylindrical tube, 100
      elliptical blood vessel, 101
      lung alveoli, 99
      parallel plates, 99–100
   empirical approach
      advantages and disadvantages, 68
      Buckingham Pi theorem (*see*
         Buckingham Pi theorem)
      flow between reservoirs, 97
      flow measurement, 101
      forced convection, circular cylinder, 101–102
      heat loss, environment, 103
      hematocrit, reservoir, 63
      natural convection, circular cylinder, 102
      pressure drop, 98–99
      tapered tube, 98
      temperature, reservoir, 36
   theoretical approach
      advantages and disadvantages, 68
      geometry, 69–70
      GIM, 68–69
      governing equations, 70
      graphical presentations, results, 72
      mass flow and heat transfer, tapered bronchiole, 78–83
      scaling, 72–75
      solution procedures, 70–72
      species conservation, bioreactor, 75–78
   transport principles, 98
   transport problem, 96–97
Black body emissive power, 526–529
Blake–Kozeny equation, 229
Blood CO_2 transport and pH
   carbonic acid dissociation, 866
   dissociation curve, 865
   Henderson–Hasselbalch equation, 866
   principal mechanisms, 864
Blood doping, 166
Blood flow
   hollow fiber device, 238–239
   microvessels, 235
   organs, 270–272
   small hollow fiber, 235–237
Bolus injection, 957–960
Boundary conditions
   heat transfer, 547–549
   mass transfer, 881–883
Boundary layer, 86–87

Boundary layer thickness, 376, 453–455
Brain freeze, 554–555
Buckingham Pi theorem
   application, 84–92
   fundamental dimensions, 84
   permeability, porous membrane, 92–94
Bulk concentration, 1155
Bunsen solubility coefficient, 797
Burke–Plummer equation, 230

C

Cancer, 156
Capillary viscometer, 116
Casson fluid, 462–463
Casson fluid model
   constitutive relation, 130, 131
   velocity profile, 131–132
Casson fluid parameters, 162–163
Cellular transport mechanisms
   acetylcholine, 873
   active transport, 880–881
   carrier mediated transport, 877–880
   passive mechanism, 876
   pinocytosis, 877
CEM43, cumulative equivalent minutes at 43°C, 761
Cerebrospinal fluid (CSF), 139
Chemical reactions, mass transfer
   blood CO_2 transport and pH
      carbonic acid dissociation, 866
      dissociation curve, 865
      Henderson–Hasselbalch equation, 866
      principal mechanisms, 864
   enzyme kinetics
      biochemical catalysts, 866
      inhibitor concentrations, 869–870
      Lineweaver–Burk equation, 868
      Michaelis constant, 867
      rate of, substrate-product conversion, 869–871
   equilibrium constant, 857
   hemoglobin and blood oxygen transport
      actual binding capacity, 863
      bound oxygen, 864
      dissociation curve, 858
      effect of pH, P_{CO2}, DPG and temperature, 862–863
      Hill equation, 861
      Hill model *vs*.Adair models, 862
      principle function, 858
      saturation *vs*. partial pressure of oxygen, 860
   heterogeneous, 855
   homogeneous, 855
   ligand–receptor binding kinetics, 872–876
   order of the reaction, 857
   reverse reaction rate, 856
Chilton–Colburn analogy, 1184–1185
Chromatography, 1124–1127
Chromatography, distribution coefficient, 1127–1128
Cocurrent mass exchangers, 1050–1054
Coefficient of compositional expansion, 836
Collapsible tube flow, 252–256
Compartmental analysis, 598
Compliance and resistance, flexible conduits
   transmural pressure, 243
   volume and diameter, 243–244
Concentration, bulk or mean, 838
Concentration, flow-averaged or mixing cup, 838
Concentration, mixing cup or bulk, 54
Concentration, volume, 803
Conduction heat transfer
   constant temperature gradient, finite sized system, 490–491
   definition, 489
   Fourier's law, 490
   steady state heat flow, hollow cylinder, 491–492
   thermal conductivity, 490
   thermal resistance, 492–493
Cone and plate viscometer, 116, 117, 471–474
Conservation of energy, 527, 559, 630–631, 734
Conservation of energy, biofluid transport
   accumulation, 189
   definition, 187–189
   friction, 192
   inlet and outlet conduits, 191
   measurement, 189
   net energy production, 190
   pressure, 192–193
   steady-state energy balance, 191–192
   work rate, 190–191
Conservation of energy, bioheat transport
   combined conduction and convection
      energy accumulation rate, 724
      energy production rate, 725
      initial and boundary conditions, 726
   energy equation
      cooling, cylinder, 727–728
      steady-state conduction, tissue, 727
      steady-state flow, tube, 728–729
Conservation of mass, biofluid transport
   definition, 169
   density, 171
   mass flux, 170

Conservation of mass, biofluid transport (cont.)
  mechanisms, 169
  steady-state flow, 171–172
  volumetric flow, 170–171
Conservation of momentum, biofluid transport
  external forces, 183–184
  force estimation, 180
  K_{ki}, value of, 181–182
  rate of change, 181
Constitutive property, 42
Constitutive relationship
  Newtonian fluid
    rate of deformation, 396
    shear strain, 395
    viscous stress, 396–398
  non-Newtonian fluid
    apparent viscosity, 457
    Bingham fluid, 461–462
    Casson fluid, 462–463
    deformation components, 457, 458
    Herschel–Bulkley fluid, 463
    power law fluid, 459–461
    strain rate, 458, 459
    stress-strain rate relationships, 458, 460
Constitutive relationship, definition, 41
Continuity equation, 216, 391
Convection and diffusion
  chemical reaction, 1D biomass transport
    bioreactors, 1079–1095
    O_2 and CO_2, transcapillary exchange, 1062–1072
    tissue solute exchange, Krogh cylinder, 1072–1079
  Chilton–Colburn analogy, 1184–1185
  constant solute flux-fluid flow, 1185–1189
  mass transfer, flat surface, 1179–1183
  momentum vs. convective heat and mass transport, 1183–1184
  parallel plates, constant wall concentration, 1191–1195
  shell balance approach, 1D biomass transport
    constant wall concentration, 1042–1044
    continuous-flow mass exchangers, 1041
    drug delivery, rectangular duct, 1044–1045
    hollow fiber devices, 1045–1059
    non-reacting solutes, capillary exchange, 1059–1062
    single pathway exchange, 1041–1042
  tube vs. plates, 1196–1198
  unsteady-state shell balance applications
    chromatography, 1124–1127
    column application, chromatography, 1127–1128
    indicator dilution applications, 1116–1118
    lung urea permeability surface area, 1121–1123
    mass transfer analysis, 1116
    urea constant flux, 1189–1190
Convection coefficient, range of values, 496
Convection heat transfer
  Biot number, 523–524
  dimensionless parameters
    biotransport applications, 500–501
    Grashof number, 502
    Nusselt number, 500
    Reynolds number, 502
  forced analysis
    external flow geometries, 510–513
    internal flow geometries, 503–510
  free processes
    closed cavities, 516–522
    cold exposure weather advisory, 518–521
    external flow, 521–522
    growth, vertical cooled plate, 514–515
    horizontal cylinder, 515–516
    horizontal plate, 515
    sphere, 516
    vertical plate, 514–515
  Newton's law of cooling, 495–497
  principle characteristics, 494–495
  temperature and velocity boundary layers, 497–500
  thermal resistance, 522–523
Convection principles, 494–495
Convective and diffusive transport, 811–812
Convective heat flux, 49
Convective heat transfer coefficient, 53, 592–594
Convective mass flux, 48
Convective mass transfer coefficient, 54
Convective mass transfer, constant flux, 1185–1190
Convective mass transfer, constant wall concentration, 1191–1198
Convective mass transfer, flat plate, 1179–1183
Convective molar flux, 48
Convective momentum flux, 48
Convective transport, 48–49
Convective transport mechanism
  fluid flow, control volume, 48
  total fluxes, 49

Core variables, 87
Couette flow
  cytoplasm, 468–471
  Newtonian fluid, 420–423
Couette viscometer, 116–117
Counter-current mass exchangers, 1054–1056
Creeping flow, 423–429
Cylindrical vessel wall diffusion, 826–829
Cystic fibrosis, 157–158
Cytoplasm, 139

D

Dalton's law, 794
Decaffeination process, 805–806
Density, 37
Density, local, 38
Density, total mass concentration, 791
Diffusion
  and chemical reaction, 1170–1174
  CO_2 diffusion, tissue, 1168–1170
  convection
    Chilton–Colburn analogy, 1184–1185
    constant solute flux-fluid flow, 1185–1189
    mass transfer, flat surface, 1179–1183
    momentum vs. convective heat and mass transport, 1183–1184
    parallel plates, constant wall concentration, 1191–1195
    tube vs. plates, 1196–1198
    urea constant flux, 1189–1190
  equation, 1158
  fluxes and velocities, 810–811
  multidimensional, 1158–1162
  oxygen consumption, cornea, 1171–1174
  steady-state, multidimensional study
    coefficient, 1161
    2D problem, 1162, 1163
    eigenvalue and eigenfunction, 1160
    rectangular shaped tissue sample, 1158–1159
    separation of variables approach, 1159
    superposition, 1162–1164
  unsteady-state, multidimensional study
    boundary conditions, 1167
    coefficients, 1168
    dimensionless dependent and independent variables, 1165
    inverse Biot numbers, 1166
    mass transfer coefficient, 1164–1165
    slab, finite dimensions, 1164
  velocity, 810–811
Diffusivity or diffusion coefficient, 818

Dilatant fluid, 124
Dimensional analysis, biotransport problems
  cylindrical cell, 100–101
  cylindrical tube, 100
  elliptical blood vessel, 101
  lung alveoli, 99
  parallel plates, 99–100
Dimensionless parameters, 499
Dissolved oxygen, 802
Drag and lift, external flow
  definition, 230
  friction factor, 231–232
  Stokes' law, 232
  types, 230–231
Drag force, sphere, 232
Drug delivery, tissue
  ampicillin exchange, pharmacokinetic model, 957
  bolus injection, 957–960
  constant infusion, 960–961
  loading dose, 961–963
  oral administration, 963–966
Dufour effect, 44

E

Effective diffusivity, 977, 1033
Effectiveness factor, 1039–1041
Effectiveness, of fin, 674
Effectiveness, of heat exchanger, 657
Effective viscosity, 114
Electrically charged species transport, 851–855
Electrical resistance analogy, 563
Electrical resistance analogy, radiation, 539–547
Electrical resistance heating, 635
Emissivity, 529
Entrance length, 503–504
Enzyme inhibition
  competitive, 868–871
  non-competitive, 868–871
Enzyme kinetics
  biochemical catalysts, 866
  effects, dimensionless parameters, 944–945
  inhibitor concentrations, 869–870
  Lineweaver–Burk equation, 868
  Michaelis constant, 867
  numerical solution vs. quasi-steady solution, 946
  quasi steady-state, 944
  rate of, substrate-product conversion, 869–871
  species conservation equations, 943
Equation of state, 792

Equations of motion, 393, 793
Equilibrium, 40
Equilibrium constant, 857
Ergun equation, 230
Erythrocytes
　hemoglobin solution, 143, 144
　viscosity vs. protein concentration, 143
Excluded variables, 87
Extended surfaces, 669–674
External flow, forced convection, 510
External forced convection, mass transfer coefficients
　cylinder, 836
　dissolution rate, soluble microsphere, 834–836
　flat plate, 837–838
　free/natural mass transfer, sphere, 836
　natural convection, cylinder, 837
　sphere, 836
Extracorporeal blood cooler and warmer
　conservation of energy, 664
　cooling process, 662
　efficiency, 666
　heat transfer coefficient, 663
　number of passes, 669
　operating characteristics, 667
　overall exchanger thermal performance, 663
　residence time, 668
　Reynolds number, 665
　shell and tube heat exchanger arrangement, 661
Extraction, 928
Extravascular body fluids, 139–142

F

Fahraeus–Lindquist effect, 149–150
Falling ball viscometer, 115–116
Falling film, 477
Fanning friction factor (f), 199
Fick principle, cardiac output, 946–947
Fick's law, 818
Fick's law of diffusion, 43
Fick's second law of diffusion, 1097
Filtration coefficient, 277
Finger, heat transfer, 670
Finite difference method
　advantage, 730
　backward, 747
　forward
　　vs. backward, 747–751
　　Biot number, 734
　　conservation of energy, 732, 734
　　cooling, hot plate, 737–740
　　Fourier number, 733
　　grid geometries, 734, 735
　　skin temperature, burn, 740–746
　　second order difference, 732
　　temperature gradient, 732
　　two dimensional physical system, 730–731
Fins, 669
Fire fighter burnover shelter
　area, shelter, 609
　characteristic radial length, 611
　conservation of energy equation, 607–609
　convective heat transfer, 606, 611
　effective thermal conductivity, 611
　emissive power, 609
　first iteration, 605
　four shield design, 612
　inhalation burn, 607
　personal shelter, 604
　radiation shield, 605
　Rayleigh number, 611
First order irreversible homogeneous reactions, 935
Flow consistency index, 125
Flow in a tapered vessel, 439–446
Flow in networks, 235–242
Flow rate
　mass, 170
　volumetric, 170
Flow through a leaky vessel, 351–356
Flow through pores, capillary walls
　large solute transport, 973–978
　small solute transport, 972–973
Fluid inertia, 261–262
Fluid mass balance, 322–323
Flux, definition, 41
Forced convection analysis
　circular cylinder, 101–102
　external flow geometries
　　boundary layer, 510
　　impingent flow, planar perpendicular surface, 513
　　laminar flow, flat plate, 511
　　perpendicular cylinder, 511–512
　　sphere, 512–513
　　turbulent flow, flat plate, 511
　external mass transfer coefficients
　　cylinder, 836
　　dissolution rate, soluble microsphere, 834–836
　　flat plate, 837–838
　　free/natural mass transfer, sphere, 836
　　natural convection, cylinder, 837

Index 1269

sphere, 831–834
internal flow geometries
  correlation equations, 507
  mean temperature, 506
  temperature boundary layer regions, 505
  turbulent, 509–510
  velocity boundary layer regions, 504
internal mass transfer coefficients
  conduits, 838
  mean/bulk fluid concentration, 838
  packed column, 839–840
Forward finite difference method
  vs. backward, 747–751
  Biot number, 736
  conservation of energy, 734, 736
  cooling, hot plate, 737–740
  Fourier number, 733
  grid geometries, 735
  skin temperature, burn, 740–746
Fourier law, 550
Fourier number, 490, 684, 700, 733, 749
Fourier's law of conduction, 43
Free convection processes
  closed cavities
    cold exposure weather advisory, 518–521
    concentric spheres, 517
    enclosed straight sided spaces, 518
    external flow heat transfer, 521–522
    horizontal concentric cylinders, 516–518
  growth, vertical cooled plate, 514
  horizontal cylinder, 514
  horizontal plate, 515
  sphere, 516
  vertical plate, 514–515
Free diffusion coefficient, 828
Free/natural convection, 831
Friction factor
  interphase transport, 52
  packed bed, 229
  sphere, 232
Friction loss, conduits
  Bernoulli equation, 200–201
  Buckingham Pi theorem, 199
  fanning friction factor (f), 199–200
  hydraulic diameter (D_h), 201
  hydrodynamic entry length, 202
  kinetic force, 200
  Moody diagram, 203, 204
  Reynolds number, 203–204
  straight conduits, 199

Friction loss factors
  flow through fittings, 213–215
  sudden expansion, 216–219
Frontal area, 53

G
Gas diffusion, heterogeneous chemical reaction, 823
Gas–solid equilibrium, 803–806
General species continuity equation, 1155–1158
Generate ideas model (GIM)
  analysis, 10–11
  definition, 9
  initial considerations, 10
  methodology steps, 0
  osmotic lysis, cell, 17
  solution development, 11–12
  system analysis defining, 13–15
  theoretical approach, 68–69
  what to do next step, 11–12
Glomerular filtration rate (GFR), 953
Goldman equation, 854, 890
Goldman–Hodgkin–Katz equation, 854, 894
Gradient, definition, 42
Graetz number, 1192
Grashof number
  for free convection, 502
  for mass transfer, 836

H
Hagen–Poiseuille equation, 235, 350
Heat capacity rate, 659
Heat capacity ratio, 659
Heat exchangers
  automotive radiator, 649
  co-current and counter-current, 650–656
  design, 649
  NTU analysis method
    co-current concentric tube heat exchanger, 659, 660
    counter-current concentric tube heat exchanger, 659, 660
    effectiveness, definition, 657
    effectiveness function, 658
    extracorporeal blood cooler and warmer, 660–669
    heat capacity rate ratio, 659
    log mean temperature difference analysis, 656
  overall heat transfer, 649
Heat flow, 51
Heat transfer fundamentals

Heat transfer fundamentals (*cont.*)
  boundary conditions, 546–547
  conduction
    constant temperature gradient, finite sized system, 488–489
    definition, 487
    Fourier's law, 488
    steady state heat flow, hollow cylinder, 489–490
    thermal conductivity, 488
    thermal resistance, 490–491
  convection
    Biot number, 521–522
    dimensionless parameters, 498–501
    forced analysis, 501–512
    free processes, 512–520
    Newton's law of cooling, 493–495
    principle characteristics, 492–493
    temperature and velocity boundary layers, 495–498
    thermal resistance, 520–521
  thermal radiation
    electrical resistance model, 537–546
    electromagnetic wave propagation, 522
    geometric sizes, shapes, separation and orientation, role of, 530–536
    surface properties, role of, 527–530
    surface temperature, role of, 522–527
Heisler charts
  cylinder, 702, 704
  internal spatial distribution, 702
  long aluminum rod cooling, 703–708
  short aluminum rod cooling, 708–709
  slab, 702–703
  sphere, 702, 705
  temperature determination, 85–90
  transient behavior, 702
Hematocrit value, 142, 146
Hemoglobin and blood oxygen transport
  binding capacity, 863
  bound oxygen, 864
  dissociation curve, 858
  effect of pH, P_{CO_2}, DPG and temperature, 862–863
  Hill equation, 861
  Hill model *vs.* Adair models, 862
  principle function, 858
  saturation *vs.* partial pressure of oxygen, 862
Henderson-Hasselbalch equation, 866
Henry's law, 796
Herschel–Bulkley fluid model, 132–133, 463
Heterogeneous chemical reactions, 951–952
Hill equation, 861
Hollow fiber devices
  axial diffusion effect, solute exchange rate, 1056–1059
  blood flow, 238–239
  cocurrent mass exchanger, 1050–1054
  counter-current mass exchanger, 1054–1056
  well-mixed external compartment, solute exchange, 1045–1050
Hollow fiber permeability, 845–849
Homogeneous chemical reaction diffusion, 1D biomass transport
  first-order reaction, oxygen delivery
    cylindrical cell, 1027–1030
    rectangular cell, 1025–1027
    spherical cell, 1030–1031
  immobilized enzymes
    effective diffusivity, 1033
    effectiveness factor (η), 1039–1041
    Michaelis–Menten kinetics, 1033–1034
    process, 1032–1033
    species concentration, 1036
    steady-state transport, of species, 1033–1034
    Thiele modulus, 1036
    toxic material conversion analysis, 1033
  Michaelis–Menten kinetics, 1031–1032
  zeroth order reaction, oxygen consumption
    cylindrical shaped cell, 1020–1023
    rectangular shaped cell, 1014–1020
    spherical cell, 1020–1025
Homogeneous chemical reactions
  convection, 938–940
  enzyme kinetics, 943–947
  first order irreversible, 935
  oxygen-hemoglobin
    cardiac output, fick principle, 946–947
    pulmonary shunt fraction, 944–945
    red cells oxygenation and deoxygenation, 940–944
  second order reversible, 935–936
  zeroth order, 934
Hot wire anemometer, 569
How people learn (HPL) methodology
  adaptive expertise
    definition, 4
    knowledge and innovation, 4
    learning, 5
  challenge-based instruction
    lecture format, 6–8
    structured learning environment, 7
  effective learning, principles of, 5–6

Index 1271

innovation development
  generate ideas model (GIM), 9–25
    usage, 26–27
  routine expertise, 5
  STAR.Legacy (SL) Cycle, 8–9
  understanding concept
    biotransport, 28
    curriculum and course design, 28
    enduring, 27
    Wiggins and McTighe's approach, 27
Human skin structure, 752
Human thermoregulation
  blood flow distribution, 615
  conservation of energy, 616–617
  direct thermal conduction, 615
  interactive garments, 617
  operational mechanisms, 615
  transient blood perfusion, 615
Hydraulic conductance/permeability, 275
Hydraulic diameter (D_h), 201
Hydrodynamic boundary layers, 449–455
Hydrodynamic entry length, 202
Hydrostatics, 404–405
Hypertonic solution, 279
Hypotonic solution, 279

I
Immiscible fluids, 477
Impingement jet heat transfer. *See* Spray cooling
Indicator dilution
  capillary permeability, 1120–1124
  capillary volume, 1117–1120
  tissue volume, 1124–1125
Indicator dilution methods
  permeability-surface area measurements, 927–929
  Stewart–Hamilton relation
    blood flow through organ, 924
    recirculation corrections, 925–926
    tracer measurement, left ventricular volumes, 929–933
    volume measurements, 926–927
Innovation development, HPL methodology
  generate ideas model (GIM)
    analysis, 10–11
    definition, 9
    initial considerations, 10
    methodology steps, 10
    osmotic lysis, cell, 17
    solution development, 11–12
    system analysis defining, 12–16

    what to do next step, 11–12
    usage, 26–27
Insulation, radiation, 585–586
Internal flow convection
  constant heat flux boundary condition, 646–648
  constant temperature boundary condition
    average overall heat transfer coefficient, 643
    cooling solution, lab, 644–646
    log mean temperature difference, 643
  correlation equations
    fully developed laminar flow, 507–508
    laminar flow entrance length, 508–509
    laminar flow patterns, 507
    turbulent, 509–510
  process, 644–646
Internal flow, forced convection, 503
Internal forced convection, mass transfer coefficients
  conduits, 839
  mean/bulk fluid concentration, 838
  packed column, 839–840
Internal thermal gradients, transient diffusion processes
  accumulation rate, 681
  graphical methods, 698–709
  semi-infinite geometry, 692–698
  symmetric geometries, 682–692
Interphase transport
  cell membrane/capillary wall, 55
  external and internal flow, 53
  momentum transport, 52
  partition coefficient, 54
  proportionality factor, 53
  relationships with, 56
  thermal boundary layer, 53
Interstitial fluid (ISF), 139
Isotonic solution, 279

J
Jet impingement convection, 513
j-factor
  heat transfer, 1184
  mass transfer, 1184

K
Kinematic viscosity, 46, 122
Krogh cylinder
  definition, 1072–1073
  oxygen exchange, 1073–1079

L

Laminar flow and flow resistance, noncircular conduits
  dimensionless coefficient, 224
  friction factor, 223
  nonhorizontal tubes, 224
  Reynolds number, 225–226
Laser fluence rate, 767
Laser irradiation, tissue
  distributed energy absorption
    absorption coefficient, 765
    conservation of energy terms, 765, 766
  surface cooling
    diffusion, 772
    freezing, 773
    geometry, 769, 771
    Reynolds number, 771
    spray cooling, 771
    Stefan number, 771
    transient heat transfer process, 771–772
    Weber number, 770
  time constant analysis
    diffusion equation, 766
    Fourier number, 767
Latent heat, 588, 771
Leukocytes, 144–145
Ligand–receptor binding kinetics, 872–874
Lineweaver–Burk equation, 868
Liquid–gas equilibrium
  Bunsen solubility coefficient, 797
  diving and bends, 800–801
  Henry's law, 796
  plasma-CO_2 solubility and partition coefficient, 798–800
  trout survival, warm waters, 801–803
  vapor pressure, 796
Liquid–liquid equilibrium
  decaffeination process, 805–806
  distribution coefficient, 803
  plasma-artificial membrane, 804–805
Liquid–solid equilibrium, 803–805
Local equilibrium, 61
Log mean temperature difference, 643
Lumped analysis, 849–851
Lumped parameter analysis
  Biot number, 592
  classic exponential decay process, 590
  convective heat transfer coeffcient, 592–594
  quenching, 589
  time constant, 591
Lung capillary exchange
  carbon dioxide, 1069–1071
  oxygen, 1063–1066
  small inert solutes, 1059–1060
Lung volumes
  tidal volume, 63–64
  expiratory reserve capacity, 63–64
  functional residual capacity, 63–64
  total lung capacity, 63–64
  inspiratory capacity, 63–64
  residual volume, 63–64
  vital capacity, 63–64

M

Macroscopic approach, biofluid transport
  Bernoulli equation, 194–195
  blood flow
    hollow fiber device, 238–239
    microvessels, 235
    organs, 270–272
    small hollow fiber, 235–237
  catheter flushing, 172–178
  cell membrane, 280–282
  cell velocity, in centrifuge, 233–235
  collapsible tube flow, 252–256
  compliant artery, periodic flow, 266–270
  conduit size estimation
    flow rate, 211–213
    velocity, 209–211
  conservation of energy
    accumulation, 189
    definition, 188–189
    friction, 192
    inlet and outlet conduits, 191
    measurement, 189
    net energy production, 190
    pressure, 192–193
    steady-state energy balance, 191–192
    work rate, 190
  conservation of mass
    definition, 169
    density, 171
    mass flux, 170
    mechanisms, 169
    steady-state flow, 171–172
    volumetric flow, 170–171
  conservation of momentum
    external forces, 183–184
    force estimation, 180
    K_{ki}, value of, 181–182
    rate of change, 181
  elliptical cross section, flow through vein, 226–227
  external flow, drag and lift
    definition, 231

Index 1273

friction factor, 231–232
Stokes' law, 232
types, 230–231
flexible conduits, compliance and resistance
transmural pressure, 243–244
volume and diameter, 243–244
fluid inertia, 261–262
forced expiration, respiratory system, 256–258
friction, closed system, 195–196
friction loss, conduits
Bernoulli equation, 200–201
Buckingham Pi theorem, 199
fanning friction factor (f), 199–200
hydraulic diameter (D_h), 201
hydrodynamic entry length, 202–203
kinetic force, 200
Moody diagram, 203, 204
Reynolds number, 203–204
straight conduits, 199
friction loss factor
flow through fittings, 213–215
sudden expansion, 216–219
glomerular filtration, 284–285
height change, tank, 179–180
hollow fiber device, blood flow, 238–239
laminar flow and flow resistance, noncircular conduits
dimensionless coefficient, 224
friction factor, 223
nonhorizontal tubes, 224
Reynolds number, 225
left ventricle work, 196–197
left ventricular force, 184–188
microtubules, resistance, 239–242
microvessels, blood flow, 235
mitral valve, pressure difference, 219–221
oncotic pressure, protein mixtures, 285–289
osmotic pressure and flow
albumin, 279
aquaporin, 277
cell lysis, 279
definition, 276
Gibbs–Donnan effect, 279
oncotic pressure difference, 279
osmotic reflection coefficient, 276
Starling equation, 277
van't Hoff's law, 278
volumetric flow, 275
water chemical potential, 275–276
packed bed flow
Blake–Kozeny equation, 229

Burke–Plummer equation, 230
Ergun equation, 230
friction factor, 228, 229
superficial velocity, 229
void fraction, 228
void volume, 228
passive cell water loss, 282–284
pressure contribution, enthalpy, 193–194
pulmonary resistance, 273–274
pump power rating, 197–198
respiratory flow, exercise, 258–261
respiratory flow model, 244–249
respiratory system, pressure drop, 204–206
rigid artery, periodic flow, 263–266
rigid conduits, steady flow, 237–238
small hollow fiber, blood flow, 235–237
square duct, pressure drop, 206–207
startup flow, rigid tube, 262–263
steady-state mixing, 178–179
syringe-needle system, 221–223
terminal velocity, of cell, 233
velocity and flow, pressure drop, 207–209
Windkessel model, arterial pressure, 249–252
Macroscopic approach, bioheat transport
blood warmer, 622–623
conservation of energy, 620
energy relation, 559–561
heart–lung machine perfusion, 625–626
heat exchange, man and environment, 621–622
heat exchanger, blood coagulation, 782
insulating properties, clothing, 621
multiple system interactions
conduction and convection, 601–602
convection, 598–601
fire fighter burnover shelter (see Fire fighter burnover shelter)
human thermoregulation (see Human thermoregulation)
radiation, flame burn injury, 602–604
postmortem interval, 624–625
scalding water, bath tub, 620–621
steady conduction, multilayered skin, 623
steady state energy balance applications
convective heat transport, 574
energy exchange, blood vessel, 574–575
heat flux, skin, 563–565
heat loss, heart lung machine, 565–568
heat transfer coefficients, 568–569
hot wire anemometer, 569–570
phase change, 587–588

stream mixing, 575
thermal radiation (*see* Thermal radiation)
thermal resistances, 561–563
WCF, 570–574
steady-state temperature distribution, 719
thermal mixing, blood, 623
transient processes, 618–619
unsteady-state heat transfer applications
  lumped parameter analysis (*see* Lumped parameter analysis)
  thermal compartmental analysis, 595–598
Macroscopic approach, biomass transport
  chemical reactions and bioreactors
    heterogeneous, 951–952
    homogeneous, 934–951
  flow through pores, capillary walls
    large solute transport, 973–980
    small solute transport, 972–973
  indicator dilution methods
    permeability-surface area measurements, 927–929
    Stewart–Hamilton relation, 924–926
    tracer measurement, left ventricular volumes, 929–933
    volume measurements, 926–927
  mass transfer coefficient applications, 968–971
  multiple compartmental analysis
    atrial-septal defect, 922
    macroscopic species continuity equations, 922
    ODE45, 923
  pharmacokinetics
    complex models, 967–968
    drug delivery, tissue, 957–966
    effective dosage regimens, 952, 953
    renal excretion, 953–957
  single compartmental analysis, 901–910
  species conservation, 897–900
  two compartmental analysis
    blood-tissue exchange, 917–921
    parallel, 913–915
    series, 910–913
    two well mixed membrane exchange, 915–917
Macroscopic transport coefficients
  charge flow, electrical wire, 49
  flow of thermal energy/heat flow, 51
  relationships, 52
  volumetric flow rate, 50
Marginal zone theory
  blood and plasma layers, 332, 333
  no-slip boundary condition, 334
  velocity profile, 335
  viscosity, 333
Mass average velocity, 809
Mass concentration, average, 789
Mass concentration, flow averaged, 899
Mass concentration, local, 791
Mass exchange, constant wall concentration, 1042–1045
Mass flow, 50
Mass flux
  convective, 811
  diffusive, 811
  relative to mass average velocity, 810
  relative to molar average velocity, 811
  species transport, single phase, 809
  total, 812
Mass fraction, 791
Mass transfer coefficients, 829–830
Mass transfer coefficients, gas phase, 830
Mass transfer coefficients, liquid phase, 830
Mass transfer fundamentals
  average, local mass, molar concentrations
    alveolar gas composition, 793–795
    equation of state, 792
    local average molecular weight, 791
    molecular species, closed volume, 789–790
  boundary conditions
    concentration and flux, interface, 883
    mass/molar concentration, 881–882
    mass/molar flux, 882
    no-flux, 883
    surface heterogeneous reaction, 883
  cellular transport mechanisms
    acetylcholine, 877
    active transport, 880–881
    carrier mediated transport, 877–880
    passive mechanism, 876
    pinocytosis, 877
  chemical reactions
    blood CO_2 transport and pH, 864–966
    enzyme kinetics, 866–871
    equilibrium constant, 857
    hemoglobin and blood oxygen transport, 858–864
    homogeneous, 855
    ligand–receptor binding kinetics, 872–876
    order of the reaction, 857
    reverse reaction rate, 856
  coefficients

Index

external forced convection, 831–838
independent dimensionless groups, 831
internal forced convection, 838–840
electrically charged species transport, 851–855
individual-overall relation, 840–842
non-porous material permeability
  internal vs. external resistances, 849–851
  membrane, 842–845
  vessel/hollow fiber, 845–849
phase equilibrium
  liquid–gas equilibrium, 795–803
  liquid–liquid, gas–solid, liquid–solid, solid–solid equilibrium, 803–806
species transport
  between phases, 806–808
  single phase, 808–840
Mass transfer, Nusselt number, 832
Materials interface, 549
Maximum safe concentration (MSC), 952
Mean mass transfer coefficient, 1043
Mean transit time, 926
Membrane permeability, 842–845
Membrane steady-state diffusion, 1D biomass transport
  boundary conditions, 1008–1009
  heterogeneous barrier diffusion, 1009–1014
  molar flow, 1007–1008
  shell species conservation equation, 1006–1007
  waste product diffusion analysis, 1006
Metacognition, 11
Michaelis–Menten equation, 868
Michaelis–Menten kinetics, 867, 1031–1032
Michaelis–Menten kinetics, stationary phase, 1087–1088
Microscopic approach, biofluid transport
  Bingham fluid, Couette viscometer, 483
  blood flow
    alveolar lung wall, 480
    convergent channel, 479–480
    lung microcirculation, 481
  conservation of linear momentum
    applied forces, 392
    bulk fluid motion, x-momentum, 391, 392
    equations of motion, 393
    momentum equations, 393
    viscous molecular transport, x-momentum, 392
  conservation of mass
    bulk fluid motion, 389, 390

continuity equation, 391
three dimensional fluid shell, 389, 390
constitutive relationship
  Newtonian fluid (see Newtonian fluid)
  non-Newtonian fluid (see Non-Newtonian fluid)
equations of motion, Newtonian fluids, 400–401
modified pressure, 400
moment equations, 394–395
momentum equations, non-Newtonian fluids, 455–457
Navier–Stokes equations
  Couette flow, Newtonian fluid, 420–423
  creeping flow, 423–429
  cylindrical and spherical coordinate systems, 415–420
  elliptical cross section, Newtonian fluid flow, 409–415
  equations of motion, 406
  flow, tapered blood vessel, 439–446
  hydrodynamic boundary layer, 450–455
  hydrostatics, 404
  parallel plates, Newtonian fluid flow, 406–409
  periodic flow, tube, 429–436
  pressure transducer, 404–405
  scaling, 436–438
  Womersley number, 446–450
non-Newtonian problems
  Bingham fluid, 464–468
  cone and plate viscometer, 471–474
  Couette flow, cytoplasm, 468–471
  procedure, 464
parallel plate bioreactor, 478–479
pressure drop, single red cell, 478
stream function and streamlines, 402–403
substantial derivative, 398–400
Microscopic approach, bioheat transport
  conservation of energy
    combined conduction and convection, 723–726
    cooling, cylinder, 727–728
    steady-state conduction, tissue, 727
    steady-state flow, tube, 728–729
  heat transfer, hollow fiber, 781
  hyperthermia therapy, tumors, 781–782
  laser irradiation, tissue
    distributed energy absorption, 764–766
    latent storage, 764
    sensible storage, 764
    surface cooling, 769–776
    time constant analysis, 766–769

Microscopic approach, bioheat transport (*cont.*)
   metabolic heat generation, 780
   neonatal thermoregulatory function, 783
   numerical methods, transient conduction
      backward finite difference method, 747
      finite element method, 730
      forward finite difference method (*see* Forward finite difference method)
      second order difference, 732
      temperature gradient, 732
      two dimensional physical system, 730–731
   temperature gradient, 780–781
   thermal injury mechanisms and analysis
      burn injury, 751–760
      therapeutic applications, hyperthermia, 761–764
   vulcanization process, 779
Microscopic approach, biomass transport
   blood oxygenation, hollow fiber, 1198–1206
   convection and diffusion, 1175–1198
   diffusion
      chemical reaction, 1170–1171
      CO_2 diffusion, tissue, 1168–1170
      convection, 1175–1178
      oxygen consumption, cornea, 1171–1174
      steady-state, multidimensional study, 1158–1162
      superposition, 1162–1164
      unsteady-state, multidimensional study, 1164–1168
   unsteady-state species conservation
      accumulation rate, 1149
      mass balance, shell, 1150
      microscopic subsytem, 1149–1150
      molar flux, 1151
      shell balance *vs.* species continuity equations, 1155–1158
      species continuity equation, 1153–1155
Minimum effective concentration (MEC), 952
Mixed fluid temperature, 640
Mobility, 828
Modified pressure, 195, 361
Molar average velocity, 810
Molar concentration, average, 789
Molar concentration, flow averaged, 900
Molar concentration, local, 790
Molar concentration, total, 790
Molar flow, 51
Molar flux
   convective, 811
   diffusive, 811
   relative to mass average velocity, 811
   relative to molar average velocity, 811
   species transport, single phase, 809
   total, 812
Molecular diffusion and Fick's law of diffusion, 817–829
Molecular transport mechanism
   constitutive equations
      Dufour effect, 44
      Newtonian fluids, 44
      relationships, 44–45
      Soret effect, 44
      thermal diffusion, 43
   1D analogies, 46–48
   1D flux, negative gradient, 42–43
   diffusion, 41
   energy and momentum, 42
   flux and gradient, 41
   flux, n-direction, 41–42
   properties
      inverse solution method, 45
      specific heat, 46
      thermal diffusivity, 46
      viscosity, Newtonian fluids, 45
Molecular weight
   molar averaged, 791
   number averaged, 791
Mole fraction, 791
Momentum equations, 393
Mucus, 140–141
Multidimensional diffusion
   steady-state
      coefficient, 1161
      2D problem, 1162, 1163
      eigenvalue and eigenfunction, 1160
      rectangular shaped tissue sample, 1158–1159
      separation of variables approach, 1159
   unsteady-state
      boundary conditions, 1167
      coefficients, 1168
      dimensionless dependent and independent variables, 1165
      inverse Biot numbers, 1166
      mass transfer coefficient, 1164
      slab, finite dimensions, 1164, 1165
Multiple compartmental analysis
   atrial-septal defect, 922
   macroscopic species continuity equations, 922
   ODE45, 923

Index   1277

Multiple system interactions
  conduction and convection, 601–602
  convection, 598–601
  fire fighter burnover shelter
    area, shelter, 610
    characteristic radial length, 611
    conservation of energy equation,
      609–611
    convective heat transfer, 606–607, 611
    effective thermal conductivity, 611
    emissive power, 609
    first iteration, 605
    four shield design, 612–613
    inhalation burn, 607
    personal shelter, 604
    radiation shield, 605
    Rayleigh number, 611
  human thermoregulation
    blood flow distribution, 615
    conservation of energy, 616–618
    direct thermal conduction, 615
    interactive garments, 617
    operational mechanisms, 615
    transient blood perfusion, 615
  radiation, flame burn injury, 602–604

N
Natural convection, circular cylinder, 102
Navier–Stokes equations
  Couette flow, Newtonian fluid, 420–423
  creeping flow
    continuity equation, 424
    net pressure force, 427
    no-slip boundary conditions, 425
    Stokes law, 428
    stream function, 425
    velocity components, 426
  cylindrical and spherical coordinate
      systems
    Cartesian coordinates, 417
    continuity equation, 416–417
    del operator, substantial derivative and
      Laplacian operator, 415, 416
    Newton's law of viscosity, 418–419
    stream functions, 419–420
  elliptical cross section, 409–415
  equations of motion, 406
  flow, tapered blood vessel
    axial velocity profile, 443–444
    dimensionless radial velocity, 440–441
    flow rate, 442–443
    self-similar profiles, 446
    tapered tube predictions, 444, 445

hydrodynamic boundary layer
  Blasius solution, 453, 454
  boundary conditions, 452
  continuity equation, 451
  friction factor, 455
hydrostatics, 404
incompressible Newtonian fluid, 398
Newtonian fluid, elliptical cross section
  continuity equation, 410
  Navier–Stokes equation, 410–411
  no-slip boundary condition, 412
  resistance, 414–415
  velocity profile, 413
Newtonian fluid flow, parallel plates
  axial velocity profiles $vs.$ x/h, 409
  boundary conditions, 408
  centerline pressure, 409
  continuity equation, 407
parallel plates, 406–409
periodic flow, tube
  boundary conditions, 431
  ODE, 433
  PDE, 431–433
  pressure gradient, 429, 430
  pressure gradient $vs.$ fraction of cardiac
    cycle, 435
  shear stress, 436
  velocity profiles, 434
pressure transducer, 404–405
scaling, 436–438
Womersley number
  velocity profiles, 449, 450
  velocity, tube center, 446, 447
Nernst–Einstein equation, 828
Nernst equation, 853
Nernst potential, 852–855
Newtonian fluid flow
  annulus
    no-slip boundary condition, 357
    relative velocity $vs.$ k, 358
    shear stress, 359
  circular cylinder
    flow resistance, 350–351
    fluid shell, 347
    shear stress, 349
    z-momentum equation, 347–348
  inclined channel
    parabolic velocity profile, 339, 340
    pressure distribution, 338
    x-momentum equation, 337–338
  parallel plates
    axial velocity profiles $vs.$ x/h, 409
    boundary conditions, 408

centerline pressure, 409
continuity equation, 407
Newtonian fluid model
  kinematic viscosity
    vs. mass fraction, 120, 122
    vs. temperature, 120, 122
  Newton's law of viscosity, 120
  viscosity measurement, 121–124
  viscosity *vs.* temperature, 120, 121
Newtonian fluids
  rate of deformation, 396
  shear strain, 395
  viscous stress, 396–398
Newton's law of cooling, 53, 493, 495
Newton's law of viscosity, 44, 47
Non-Newtonian fluid
  apparent viscosity, 457
  Bingham fluid, 461–462
  Casson fluid, 462–463
  deformation components, 457, 458
  Herschel–Bulkley fluid, 463
  model
    Bingham fluid model, 129–130
    Casson fluid model, 130–132
    Herschel–Bulkley fluid model, 132–133
    power law model, 125–128
  power law fluid, 459–461
  strain rate, 458, 459
  stress-strain rate relationships, 458, 460
Non-porous material permeability
  internal *vs.* external resistances, 849–851
  membrane, 842–845
  vessel/hollow fiber, 845–849
Non-reacting solutes, capillary exchange
  lung capillaries, small solute and inert gas exchange, 1059–1060
  solute removal, tissue capillaries, 1060–1062
NTU analysis method
  co-current concentric tube heat exchanger, 660, 661
  counter-current concentric tube heat exchanger, 660, 662
  effectiveness definition, 657
  effectiveness function, 658
  extracorporeal blood cooler and warmer
    conservation of energy, 664
    cooling process, 662
    efficiency, 666
    heat transfer coefficient, 663
    number of passes, 669
    operating characteristics, 667
    overall exchanger thermal performance, 663
    residence time, 668
    Reynolds number, 665
    shell and tube heat exchanger arrangement, 661
  heat capacity rate ratio, 659
  log mean temperature difference analysis, 656
Number of transfer units (NTU), 659
Nusselt number, 500, 502, 592

O

O_2 and CO_2, transcapillary exchange
  lung capillaries, 1063–1066, 1069–1071
  tissue capillaries, 1066–1069, 1071–1072
Ohm's law, 44
Ohm's law, thermal analogy, 492
Oncotic pressure, protein mixtures, 285–289
One-dimensional biofluid transport, shell balance approach
  annulus, Newtonian fluid flow
    no-slip boundary condition, 357
    relative velocity *vs.* k, 358
    shear stress, 359
  application, fluid constitutive relation, 328–329
  appropriate shell selection, 321–322
  blood flow, 320
  boundary conditions, air-mucus interface, 381–382
  Casson fluid flow, circular cylinder
    Casson flow/Newtonian flow, 365
    shear stress distribution, 363
    velocity profile, 364, 365
    yield radius, 363
  circular cylinder, Newtonian fluid flow
    flow resistance, 350–351
    fluid shell, 347
    shear stress, 349
    z-momentum equation, 347–348
  falling film analysis, 343–346
  film thickness, blood, 330–332
  flow, inclined tube
    geometry, 360
    modified pressure, 361
  fluid flow, cylindrical vessel
    dimensionless pressure, 353–354
    leakage flow, 355–356
    steady-state mass balance, 352
  fluid mass balance, 322–323
  fluid momentum balance

Index

convection and viscous molecular motion, 326, 327
potential forces, 323, 324
shear stress distribution, 328
y-momentum equation, 324–325
inclined channel, Newtonian fluid flow
parabolic velocity profile, 339, 340
pressure distribution, 338
x-momentum equation, 337–338
lung microcirculation, 383
marginal zone theory
blood and plasma layers, 332, 333
no-slip boundary condition, 334
velocity profile, 335
viscosity, 333
mucus plug, 382
Newtonian film oxygenator, 384
non-Newtonian film oxygenator, 384
osmotic pressure and flow, cylindrical pore
boundary conditions, 367
capillary pore, 366, 367
hydraulic conductivity, 368
multiple solutes, 370–371
steady flow, equivalent pore., 371–373
velocity profile, restrictive pore, 369–370
parallel flow, 319
power law fluid flow, inclined plane, 340–343
respiratory mucus transport, 381
shear stress and velocity, 329–330
Starling's law, 387
thin film blood oxygenator, 382
unsteady-state 1-D shell balances
boundary layer thickness $vs.$ time, 377
kinematic viscosity, 374
Laplace transform, 375
penetration thickness and velocity profile, 374
velocity profile $vs.$ time, 376
One-dimensional bioheat transport, shell balance approach
definition, 629
1D heat conduction and convection
extended surfaces, 669–678
heat exchange, tissue, 678–680
heat exchangers
automotive radiator, 648
co-current and counter-current, 650–656
design, 649
NTU analysis method, 656–669
overall heat transfer, 649
internal flow convection
constant heat flux boundary condition, 646–648
constant temperature boundary condition, 642–646
process, 640–641
steady-state conduction-heat generation
cylinder, 633–639
slab, 630–633
sphere, 639–640
transient diffusion processes, internal thermal gradients
accumulation rate, 681
graphical methods, 698–709
semi-infinite geometry, 692–698
symmetric geometries, 682–692
One-dimensional biomass transport, shell balance approach
chemical reaction, convection, and diffusion
bioreactors, 1079–1095
O_2 and CO_2, transcapillary exchange, 1062–1072
tissue solute exchange, Krogh cylinder, 1072–1079
convection and diffusion
constant wall concentration, 1042–1044
drug delivery, rectangular duct, 1044–1045
hollow fiber devices, 1045–1059
mass exchangers, 1041
non-reacting solutes, capillary exchange, 1059–1062
single pathway exchange, 1041–1042
homogeneous chemical reaction diffusion
first-order reaction, 1025–1031
immobilized enzymes, 1032–1041
Michaelis–Menten kinetics, 1031–1032
zeroth order reaction, 1014–1025
membrane steady-state diffusion
boundary conditions, 1008–1009
heterogeneous barrier diffusion, 1009–1014
molar flow, 1007–1008
shell species conservation equation, 1006–1007
waste product diffusion analysis, 1006
microscopic species conservation, 1005–1006
one-dimensional unsteady-state shell balance applications
convection and diffusion, 1116–1128
tissue, diffusion to, 1095–1116

One-dimensional heat conduction and convection
  extended surfaces
    effectiveness, 674
    factors, 669
    finned surface geometries, 670–671
    geometric and thermal symmetry, 673
    pin fin, short length, 670, 672
    thermal behavior, finger, 674–678
  heat exchange, tissue
    bioheat transfer process, 678
    blood/tissue thermal interaction, 679
    effects of metabolism, 679
    physiological systems, 680
One-dimensional unsteady-state shell balance applications
  convection and diffusion
    chromatography, 1124–1127
    column application, chromatography, 1128
    indicator dilution applications, 1117–1122
    lung urea permeability surface area, 1122–1124
    mass transfer analysis, 1116
  macroscopic approach, 1095–1096
  tissue, diffusion to
    carbon dioxide transport, plasma, 1099–1100
    finite external resistance to mass transfer, 1102–1104
    long cylinder, unsteady-state diffusion, 1111–1114
    semi-infinite slab, 1096–1098
    sphere, unsteady-state diffusion, 1114–1116
    thin slab, CO_2 diffusion, 1108–1111
    two semi-infinite slabs, 1100–1102
    unsteady-state diffusion, 1104–1108
Osmotic lysis, cell
  conservation of momentum equation, 20
  inverse problem, 24
  mass fluxes, 20
  mechanical properties, membrane, 25
  permeability, 17
  permeability and water concentration, 22
  potential energy storage, 21
  semipermeable barrier, 17
  strain energy, 23
  system and boundary, 18–19
Osmotic pressure and flow
  albumin, 281
  aquaporin, 278
  cell lysis, 280
  cell volume *vs.* time, 276
  definition, 277
  Gibbs–Donnan effect, 280
  oncotic pressure difference, 280–281
  osmotic reflection coefficient, 277–278
  Starling equation, 278–279
  van't Hoff's law, 279–280
  water chemical potential, 277
O_2 transport, driving force, 807–808
Overall heat transfer coefficient, 650
Overall mass transfer coefficient, 807
Oxygen consumption, 1014–1032
Oxygen-hemoglobin reactions, 940–944
Oxygen transfer in blood, 1198–1206
Oxyhemoglobin saturation, 859

P

Packed bead bed flow
  Blake–Kozeny equation, 229
  Burke–Plummer equation, 230
  Ergun equation, 230
  friction factor, 228, 229
  superficial velocity, 229
  void fraction, 228
  void volume, 228
Packed bed
  flow, 228–230
  mass transfer, 839–840
Parallel plate bioreactor, 478
Partial pressure, 793
Partition coefficient, 54, 798
Partition coefficient, steric, 977
Peclet number, 75, 832
Peeling off exponentials, 600
Pennes equation, 679
Periodic flow
  compliant artery, 266–270
  rigid artery, 263–266
  rigid tube, 429–436
Permeability, 56, 803, 807, 842
  hollow fiber, 846–849
  porous membrane, 92–94
Pharmacokinetics
  complex models, 967–968
  drug delivery, tissue
    ampicillin exchange, pharmacokinetic model, 957
    bolus injection, 957–960
    constant infusion, 960–961
    loading dose, 961–963
    oral administration, 963–966
    effective dosage regimens, 952–953

Index

renal excretion
  Ficoll molecules, 956
  glomerular filtration rate (GFR), 953
  kidney excretion coefficient, 955
  peeling off exponentials, 957
  sieving coefficient *vs.* Stokes–Einstein radius, 954
Phase change, 587–588
Phase equilibrium, 795
Pin fin, 670
Pinocytosis, 877
Planck distribution, 525
Plasma-artificial membrane equilibrium, 804–8050
Plasma-CO_2 solubility and partition coefficient, 798–800
Pneumotachograph, 64
Polycythemia, 155
Power law fluid, 459–461
Power law model
  behavior index, 125
  coordinate system, 127–128
  velocity distribution, 125–127
Prandtl number, 74, 502
Pressure drop, red blood cell, 478
Pressure transducer, 309
Pressure transducer, calibration, 404–405
Properties, extensive or extrinsic, 34, 56, 192
Properties, intensive or intrinsic, 34, 278
Pseudoplastic fluid, 124
Pulmonary resistance, 273–274
Pulmonary shunt fraction, 944–945

Q
Quasi steady-state, 948

R
Radiation, thermal, 524
Radiosity, 540
Raoult's law, 796
Red cells oxygenation and deoxygenation, 940–944
Reflectivity, 530
Relative viscosity, 146
Respiratory flow model, 244–249
Respiratory mucus transport, 381
Respiratory quotient, 1071
Restricted diffusion, 1033
Reynolds number, 74, 502
  external flow, 512
  internal flow, 504
Rheology, biological fluids
  apparent viscosities, 162

biorheology and disease
  cancer, 156
  cystic fibrosis, 157–158
  polycythemia, 155
  sickle cell anemia, 156–157
blood plasma, 142, 143
boundary conditions, 110–111
Casson fluid parameters, 162–163
constitutive model equations
  characterization, 134–136
  rheological coeffcients, 136–138
cytoplasm viscosity, 161
erythrocytes
  hemoglobin solution, 143, 144
  viscosity *vs.* protein concentration, 143
extravascular body fluids, 139–141
laminar flow, 110
leukocytes, 144–145
Newtonian fluid model
  kinematic viscosity *vs.* mass fraction, 120, 122
  kinematic viscosity *vs.* temperature, 120, 122
  Newton's law of viscosity, 120
  viscosity measurement, 121–124
  viscosity *vs.* temperature, 120, 121
non-Newtonian fluid model
  Bingham fluid model, 129–130
  Casson fluid model, 130–132
  Herschel–Bulkley fluid model, 132–133
  power law model, 125–128
solids and fluids
  deformation, 107, 108
  deformation *vs.* time, 108, 109
  instantaneous shear rate, 109
  shear modulus, 108
  viscoelastic materials, 108
turbulent flow, 110
viscometers
  Reynolds number, 119
  types, 115–117
  wall shear stress, 118
viscous momentum flux and shear stress
  constitutive relationship, 115
  effective viscosity, 113–114
  sign conventions, 114–115
viscous properties, fluids
  block movement, 112
  velocity profiles, 112, 113
  viscous force, 112
whole blood
  Casson plot, 147, 148

discharge hematocrit *vs.* tube diameter, 153
Fahraeus–Lindquist effect, 149
fibrinogen concentration *vs.* yield stress, 149
hematocrit value, 146
microvascular hematocrit, 154
relative blood viscosity *vs.* tube diameter, 149–151
relative viscosity *vs.* shear rate, 147, 148
shear rate dependence, 146–147
in vivo *vs.* in vitro resistance, 154, 155
yield stress, 161

S

Safe touch temperature, 695
Saliva, 140
Scatchard plot, 874
Schmidt number, 832
Schmidt plot
　conservation of energy, 699
　discrete approximation method, 701
　finite difference analysis, 702
　homogeneous semi-infinite material, 699–700
Second order reversible homogeneous reactions, 935–937
Sensible heat, 588
Separation of variables, 1158–1162
Shape factor, 532, 535–539
Shape factor laws, 533
Shear modulus, 108
Shear rate, 109
Shear strain, 107
Shell balance approach, one-dimensional biofluid transport
　annulus, Newtonian fluid flow
　　no-slip boundary condition, 357
　　relative velocity *vs.* κ, 358
　　shear stress, 359
　application, fluid constitutive relation, 328–329
　appropriate shell selection, 321–322
　blood flow, 320
　boundary conditions, air-mucus interface, 381–382
　Casson fluid flow, circular cylinder
　　Casson flow/Newtonian flow, 365
　　shear stress distribution, 363
　　velocity profile, 364, 365
　　yield radius, 363
　circular cylinder, Newtonian fluid flow
　　flow resistance, 351

fluid shell, 347
shear stress, 349
z-momentum equation, 347–348
falling film analysis, 343–346
film thickness, blood, 330–332
flow, inclined tube
　geometry, 360
　modified pressure, 361
fluid flow, cylindrical vessel
　dimensionless pressure, 354–355
　leakage flow, 355–356
　steady-state mass balance, 352
fluid mass balance, 322–323
fluid momentum balance
　convection and viscous molecular motion, 326, 327
　potential forces, 323, 324
　shear stress distribution, 328
　y-momentum equation, 324–325
inclined channel, Newtonian fluid flow
　parabolic velocity profile, 339, 340
　pressure distribution, 338
　x-momentum equation, 337
lung microcirculation, 383
marginal zone theory
　blood and plasma layers, 332, 333
　homogeneous blood layer, 335
　no-slip boundary condition, 334
　viscosity, 333
mucus plug, 382
Newtonian film oxygenator, 384
Non-Newtonian film oxygenator, 384
osmotic pressure and flow, cylindrical pore
　boundary conditions, 367
　capillary pore, 366, 367
　hydraulic conductivity, 368
　multiple solutes, 370–371
　steady flow, equivalent pore., 371–373
　velocity profile, restrictive pore, 369–370
parallel flow, 319
power law fluid flow, inclined plane, 340–343
respiratory mucus transport, 381
shear stress and velocity, 329–330
Starling's law, 387
thin film blood oxygenator, 382
unsteady-state 1-D shell balances
　boundary layer thickness *vs.* time, 376–377
　kinematic viscosity, 374
　Laplace transform, 375

Index

penetration thickness and velocity profile, 374
velocity profile *vs.* time, 376
Shell balance approach, one-dimensional bioheat transport
definition, 629
1D heat conduction and convection
extended surfaces, 669–678
heat exchange, tissue, 678–680
heat exchangers
automotive radiator, 648
co-current and counter-current, 650–656
design, 649
NTU analysis method, 656–669
overall heat transfer, 649
internal flow convection
constant heat flux boundary condition, 646–648
constant temperature boundary condition, 642–646
process, 640–641
steady-state conduction-heat generation
cylinder, 633–639
slab, 630–633
sphere, 639–640
transient diffusion processes, internal thermal gradients
accumulation rate, 681
graphical methods, 698–709
semi-infinite geometry, 692–698
symmetric geometries, 682–692
Shell balance approach, one-dimensional biomass transport
chemical reaction, convection, and diffusion
bioreactors, 1079–1095
O_2 and CO_2, transcapillary exchange, 1062–1072
tissue solute exchange, Krogh cylinder, 1072–1079
convection and diffusion
constant wall concentration, 1042–1044
drug delivery, rectangular duct, 1044–1045
hollow fiber devices, 1045–1059
mass exchangers, 1041
non-reacting solutes, capillary exchange, 1059–1060
single pathway exchange, 1040
homogeneous chemical reaction diffusion
first-order reaction, 1025–1031
immobilized enzymes, 1032–1041

Michaelis–Menten kinetics, 1031–1032
zeroth order reaction, 1014–1025
membrane steady-state diffusion
boundary conditions, 1008
heterogeneous barrier diffusion, 1009–1014
molar flow, 1007
shell species conservation equation, 1006–1007
waste product diffusion analysis, 1006
microscopic species conservation, 1005–1006
one-dimensional unsteady-state shell balance applications
convection and diffusion, 1116–1128
tissue, diffusion to, 1095–1116
Sherwood number, 832
Sickle cell anemia, 156–157
Single compartmental analysis
constant inlet flow
constant volume, two inlets and single outlet, 906–907
variable volume, single inlet and outlet, 902–906
constant volume, single inlet and outlet
bolus injection, 902
constant flow and inlet concentration, permeable wall, 909–910
constant rate of infusion, 901–902
oscillating inlet flow, constant inlet concentration, 907–908
mass flow through a permeable wall, constant volume, 908–909
Sliding plate viscometer, 116, 117
Solid–solid equilibrium, 803–806
Solubility coefficient, 797
Solute diffusion through membrane, 825–826
Solute reflection coefficient, 975
Solving non-Newtonian problems, 464–474
Soret effect, 44
Space blankets, 585
Species conservation equation, 1149–1153
Species transport, single phase
alveolar fluxes, 813–817
convective and diffusive transport, 811–812
cylindrical vessel wall diffusion, 826–829
diffusion fluxes and velocities, 810–811
gas diffusion, heterogeneous chemical reaction, 823–824
mass transfer coefficients, 829–830
molecular diffusion and Fick's law of diffusion, 817–829

Species transport, single phase (*cont.*)
  solute diffusion through membrane, 825–826
  species fluxes and velocities, 809–810
  total mass and molar fluxes, 812–813
  water evaporation, 820–823
Specific absorption rate, 762
Spray cooling, 769
STAR.Legacy (SL) Cycle, 8–9
Starling resistor, 253
Steady-state conduction-heat generation
  cylinder
    conservation of energy, 634
    1D cylindrical system, 633–634
    electrical current treatment, 635–639
    electrical resistivity, 637
    finite difference technique, 639
    thermal conductivity, 638
    thermal injury, 636
  slab
    1D symmetric Cartesian system, 630–631
    parabolic distribution, 633
    second-order differential equation, 632
  sphere, 639–640
Steady-state, multidimensional diffusion
  coefficient, 1161
  2D problem, 1162, 1163
  eigenvalue and eigenfunction, 1160
  rectangular shaped tissue sample, 1158–1159
  separation of variables approach, 1159
Stefan–Boltzmann law, 524
Stefan number, 646, 771
Stewart–Hamilton equation, 924–926
Stewart–Hamilton relation
  blood flow through organ, 924
  pulsatile blood flow, 925
  recirculation corrections, 925–926
Stokes–Einstein relation
  dilute solute diffusion, 827
  free diffusion coefficient, 828
  solute mobility, 828
  solute radius estimation, 829
Stokes law, 232, 428
Stream function, 402
Streamlines, 402–403
Superficial velocity, 229
Superposition, 395–489, 1162–1164
Surface heterogeneous reaction, 883
Symmetry boundary condition, 111
Synovial fluid, 139–140
System
  closed definition, 34
  open definition, 34
System analysis, GIM
  centrifugal pump, 13
  closed system, rotor, 16
  entire pump, 14–15
  Medtronic bio-pump, 14–15
  open system, pump, 16
  viscous interface, rotor *vs.* blood, 17

T

Taylor dispersion, 1178
Temperature, film, 510
Temperature, fluid, 510
Temperature, mean, 506
Temperature, mixing cup or flow-averaged, 53
Theoretical approach, biotransport problems
  advantages and disadvantages, 68
  geometry, 69–70
  GIM, 68–69
  governing equations, 70
  graphical presentations, results, 72
  mass flow and heat transfer, tapered bronchiole, 78–83
  scaling, 72–75
  solution procedures, 70–72
  species conservation, bioreactor, 75–78
Thermal boundary layer, 53
Thermal compartmental analysis
  thermal dilution technique, 596–598
  well-mixed compartment, 595–596
Thermal conductivity, 490
Thermal diffusivity, 46, 681
Thermal dose, cancer therapy, 761
Thermal inertia, 696
Thermal injury mechanisms and analysis
  burn injury
    activation energy *vs.* natural log of frequency factor, 754, 756
    computation, 759–760
    exposure time and constant surface temperature, 753, 754
    human skin structure, 752
    injury function, 754
    injury rate *vs.* temperature, 756–759
    kinetic coefficients, 754, 755
    threshold levels, 752
  therapeutic applications, hyperthermia
    burn injury function, 762
    cell survival data, 762, 763
    specific absorption rate (SAR), 762
    thermal dose, 761
Thermal radiation

convection and radiation, 581–585
electrical resistance model, 539–547
electromagnetic wave propagation, 524
environmental radiation load, human body, 580–581
geometric sizes, shapes, separation and orientation, role of, 532–539
principles, 524
radiation barrier, 586–587
radiation insulation, 585–586
steady-state radiation exchange, 576–580
surface properties, role of
   absorption, reflection, and transmission phenomena, 530
   emissivity, 529
   isothermal enclosure, small bodies, 531
   Kirchhoff's identity, 532
surface temperature, role of
   characteristics, 526
   emissive power, Sun, 526–529
   spectral blackbody emissive power, 525–526
   Stefan–Boltzmann law, 524
Thermal resistances
   composite systems, 561
   conduction, 492–493
   convection, 523
   heat conduction, skin, 562
Thermoregulation, 614
Thiele modulus, 1036
Time constant, laser irradiation, 766
Tissue capillary exchange
   carbon dioxide, 1071–1072
   oxygen, 1066–1067
   small inert solutes, 1059–1060
Tissue capillary, solute exchange, 1072–1073
Tissue diffusion, unsteady-state shell balance applications
   carbon dioxide transport, plasma, 1098–1099
   finite external resistance to mass transfer, 1101–1103
   long cylinder, unsteady-state diffusion, 1111–1114
   semi-infinite slab, 1096–1098
   sphere, unsteady-state diffusion, 1114–1116
   thin slab, CO_2 diffusion, 1108–1111
   two semi-infinite slabs, 1099–1101
   unsteady-state diffusion, 1103–1108
Transient diffusion processes, internal thermal gradients
   accumulation rate, 681

graphical methods
   Heisler charts, 702–709
   Schmidt plot, 699–702
semi-infinite geometry
   1D coordinate system, 692–693
   definition, 692
   direct contact, media, 695–696
   Gaussian error function, 694
   safe touch temperature analysis, 697–698
   surface vs. internal temperature distributions, 695
   thermal inertia., 696
symmetric geometries
   coefficients, single term approximation, 686–687
   Fourier number, 684
   thermal interactions, 682–683
   tissue storage, vitrified state, 688–692
Transmisivity, 530
Transmural pressure, 243–244
Transport
   analogies, 46–48
   of carbon dioxide, 864–866
   of charged species, 851–855
   mechanisms, 40–49
   mechanisms, molecular, 41–48
   of oxygen, 858–864
   scales, time and space
      continuum concepts, 37–39
      feedback control systems, 37
      length scales, 36
      macroscopic approach, 36
      microscopic approach, 35
      molecular interaction, 35
      nonlinear properties, 37
Trout survival, 801–803
Turbulent flow, 110
Two compartmental analysis
   blood-tissue exchange, 917–921
   parallel, 913–915
   series, 910–913
   two well mixed membrane exchange, 915–917

U
Unsteady-state 1-D shell balances
   boundary layer thickness vs. time, 376–377
   kinematic viscosity, 374
   Laplace transform, 375
   penetration thickness and velocity profile, 374
   rate of accumulation, 373

Unsteady-state 1-D shell balances (*cont.*)
  velocity profile *vs.* time, 376
Unsteady-state mass transfer
  cylinder, 1111–1114
  semi-infinite slab, 1096–1103
  slab, 1103–1111
  sphere, 1114–1116
Unsteady-state, multidimensional diffusion
  boundary conditions, 1167
  coefficients, 1168
  dimensionless dependent and independent variables, 1165
  inverse Biot numbers, 1166
  mass transfer coefficient, 1165
  slab, finite dimensions, 1164
Unsteady-state species conservation
  accumulation rate, 1149
  mass balance, shell, 1150
  microscopic subsytem, 1149–1150
  molar flux, 1151
  shell balance *vs.* species continuity equations, 1155–1158
  species continuity equation, 1153–1155
Urea transport, 973

V
Vapor pressure, 796

Velocity, average, 171
Velocity gradient, 109, 174
Velocity profile, 112
Venous occlusion, 311
Virial coefficients, 278
Viscoelastic materials, 108
Viscometers, 115–119
Viscosity, 121
Viscosity, PEG, 122
Viscous force, 112

W
Wall shear stress, 118
Water evaporation, stagnant gas film, 820–823
WCF. *See* Wind chill factor
Weber number, 770
Wetted perimeter, 201
Wien's law, 525
Wind chill factor (WCF), 570–574
Windkessel model, arterial pressure, 249–252
Womersley number, 265, 433, 446–450

Y
Yield stress, 129

Z
Zeroth order homogeneous reactions, 934